CW00376784

Handbook of Immunohistochemistry and *in situ* Hybridization of Human Carcinomas, Volume 1

Handbook of Immunohistochemistry and *in situ* Hybridization of Human Carcinomas, Volume 1

Molecular Genetics; Lung and Breast Carcinomas

Edited by

M.A. Hayat
Distinguished Professor
Department of Biological Sciences
Kean University
Union, New Jersey

ELSEVIER
ACADEMIC
PRESS

Amsterdam • Boston • Heidelberg • London • New York • Oxford
Paris • San Diego • San Francisco • Singapore • Sydney • Tokyo

Acquisition Editor: *Hilary Rowe*
Project Manager: *Justin Palmeiro*
Editorial Assistant: *Erin LaBonte-McKay*
Marketing Manager: *Kristin Banach*
Cover Design: *Cate Barr*
Full Service Provider: *Graphic World Publishing Services*
Composition: *Cepha Imaging Pvt., Ltd.*

Elsevier Academic Press
200 Wheeler Road, Burlington, MA 01803, USA
525 B Street, Suite 1900, San Diego, California 92101-4495, USA
84 Theobald's Road, London WC1X 8RR, UK

This book is printed on acid-free paper.

Copyright © 2004, Elsevier Inc. All rights reserved.

No part of this publication may be reproduced or transmitted in any form or by any
means, electronic or mechanical, including photocopy, recording, or any information
storage and retrieval system, without permission in writing from the publisher.

Permissions may be sought directly from Elsevier's Science & Technology
Rights Department in Oxford, UK: phone: (+44) 1865 843830, fax: (+44) 1865 853333,
e-mail: permissions@elsevier.com.uk. You may also complete your request on-line
via the Elsevier homepage (http://elsevier.com), by selecting "Customer Support"
and then "Obtaining Permissions."

Library of Congress Cataloging-in-Publication Data
Immunohistochemistry and in situ hybridization of human carcinomas / edited by M.A. Hayat.
   p. ; cm.
  Includes index.
  ISBN 0-12-333941-3
   1. Cancer–Immunodiagnosis. 2. Immunohistochemistry. 3. In situ hybridization. 4.
 Fluorescence in situ hybridization. I. Hayat, M.A.,
  [DNLM: 1. Neoplasms–diagnosis. 2. Genetic Techniques. 3. In Situ
 Hybridization–methods. QZ 241 I333 2004]
 RC270.3.I46I465 2004
 616.994'0756–dc22

                                                                 2004040365

British Library Cataloguing in Publication Data
A catalogue record for this book is available from the British Library

ISBN: 0-12-333941-3

For all information on all Academic Press publications
visit our website at www.academicpressbooks.com

Printed in China
04 05 06 07 08 09   9 8 7 6 5 4 3 2 1

To

Molecular Geneticists/Clinical Pathologists

Contents

I Introduction

*Senior author

II Molecular Pathology

Contents

Authors and Coauthors of Volume 1

Geza Acs (351)
Department of Pathology and Laboratory Medicine,
University of Pennsylvania School of Medicine,
University of Pennsylvania Medical Center,
6 Founders Pavilion, 3400 Spruce Street, Philadelphia,
PA 19104

N.J. Agnantis (425)
Department of Pathology, Medical School, University of
Iannina, 45110 Ioannina, Greece

Debbie Altomare (307)
Human Genetics Program, Fox Chase Cancer Center,
7701 Burholme Avenue, Philadelphia, PA 19111

J.P. Aubert (115)
Service d'Anatomie Pathologique, Hospital Calmette,
CHRU-U Lille, Bd J Leclercq 59037, Lille Cedex,
France

Sarah S. Bacus (307)
Ventana Medical Systems, Inc./QDL, 610 Oakmont
Lane, Westmont, IL 60559

Ana-Maria Bamberger (337)
Department of Gynecopathology, Institute of Pathology,
University Hospital Hamburg-Eppendorf, Matinistrassse
52, 20246 Hamburg, Germany

Christoph M. Bamberger (337)
Department of Medicine I, University Hospital
Hamburg-Eppendorf, Matinistrassse 52, 20246
Hamburg, Germany

Sushanta K. Bannerjee (409)
Cancer Research Unit, Research Division, V.A. Medical
Center, 4801 Linwood Boulevard, Kansas City, MO 64128

Marika Bogdani (371)
Free University of Brussels (VUB), Academic Hospital,
Department of Pathology, Laarrbeeklaan 103, Jette,
Brussels, 1090, Belgium

Cecilia Bozzetti (267)
Department of Oncology, University Hospital, Via
Gransci 14, 43100 Parma, Italy

Elisabeth Brambilla (105, 133)
Laboratoire de Pathologie Cellulaire, CHU Albert
Michallon, BP 217, Grenoble, Cedex, 09, France

Marianne Briffod (523)
Service d'Anatomie et de Cytologie Pathologiques,
Centre Rene Huguenin, 35 rue Dailly, Saint-Cloud,
92211, France

M.P. Buisine (115)
Service d'Anatomie Pathologique, Hospital Calmette,
CHRU-U Lille, Bd J Leclercq 59037, Lille Cedex,
France

Yong-Jig Cho (75)
Vanderbilt-Ingram Cancer Center, Department of
Cancer Biology, School of Medicine, Vanderbilt
University, Nashville, TN

Cynthia Cohen (289)
Immunohistochemistry, Emory University Hospital,
1364 Clifton Road, Atlanta, GA, 30322

M.C. Copin (115)
Service d'Anatomie Pathologique, Hospital Calmette,
CHRU-U Lille, Bd J Leclercq 59037, Lille Cedex,
France

Nadia Dandachi (279)
Department of Internal Medicine, Division of Oncology,
Medical University Hospital Graz, Auenbruggerplatz 15,
A-8036 Graz, Austria

L. Devisme (115)
Service d'Anatomie Pathologique, Hospital Calmette,
CHRU-U Lille, Bd J Leclercq 59037, Lille Cedex,
France

Adhemar Longatto Filho (513)
Universidade do Minho Escola de Ciências de Saude
Campus de Gualtar 4710-057, Braga, Portugal

Sylvie Gazzeri (127, 133)
Groupe de Recherche sur le Cancer de Poumon,
INSERM U578, Institut Albert Bonniot, 38706
La Tronche, France

Antonio Giordano (223)
Temple University, Sbarro Institute, Biolife Science Building, Suite 333, 1900 North 12th Street, Philadelphia, PA 19122

Vassilis G. Gorgoulis (141)
Molecular Carcinogenesis Group, Department of Histology-Embryology, Medical School, National Kapodistrian University of Athens, Athens, Greece

A. Goussia (425)
Department of Pathology, Medical School, University of Iannina, 45110 Ioannina, Greece

William Grizzle (181)
Department of Nutrition Sciences, Division of Nutritional Biochemistry and Molecular Biology, University of Alabama, University Station, Birmingham, AL 35294

Tatyana A. Grushko (385)
Department of Medicine, Section of Hematology/Oncology, University of Chicago, 5841 S. Maryland Ave., MC-2115, E-203, Chicago IL, 60637

Jean Guillot (487)
University of Clermont I, Faculty of Pharmaceutical Sciences, Department of Cellular Biology, EA 2416, 28 Place Henri Dunant, Clermont-Ferrand, 6300, France

Jean-Marc Guinebretiere (523)
Service d'Anatomie et de Cytologie Pathologiques, Centre Rene Huguenin, 35 rue Dailly, Saint-Cloud, 92211, France

Aaron C. Han (343)
The Reading Hospital Regional Cancer Center, 6th Avenue and Spruce Street, West Reading, PA 19612

Cornelia Hauser-Kronberger (279)
University of California at San Diego, Department of Pathology 0612, 9500 Gilman Drive, La Jolla, CA 92093

M.A. Hayat (1, 43, 49, 99, 231)
Department of Biological Sciences, Kean University, 1000 Morris Avenue, Union, NJ, 07083

Qimin He (463)
Department of Oncology Clinical Research Laboratory, Clinical Research Center (KFC), Huddinge University Hospital, Karolinska Institute, 14186, Stockholm Sweden

David G. Hicks (57)
Cleveland Clinic Foundation, Department of Anatomic Pathology, L-25, 9500 Euclid Avenue, Cleveland, OH 44195

Chao-Chi Ho (163)
Department of Pathology, National Taiwan University Hospital, Taipei, Taiwan

Su-Ming Hsu (163)
Department of Pathology, National Taiwan University Hospital, Taipei, Taiwan

Pei-Hsin Huang (163)
Department of Pathology, National Taiwan University Hospital, Taipei, Taiwan

David Huntsman (493)
Hereditary Cancer Program, British Columbia Cancer Agency, Vancouver, BC, V5Z 4H2, Canada

Keiichi Iwaya (299)
Department of Pathology, Tokyo Medical University, Shinjuku 6-1-1, Shinjuku-ku, Tokyo, 160-8402, Japan

Roger S. Jackson II (75)
Vanderbilt-Ingram Cancer Center, Department of Cancer Biology, School of Medicine, Vanderbilt University, Nashville, TN

Antonio Juretic (471)
Clinics for Oncology, Clinical Hospital Center, Zagreb, Croatia

Anna Kádár (321)
Second Department of Pathology, Semmelweis University of Medicine, Ulloi ut 93, 1091 Budapest, Hungary

Rajko Kavalar (471)
Department of Pathology, Maribor General Hospital, Moribor, Slovenia

Oichi Kawanami (189)
Department of Molecular Pathology, Nippon Medical School, Graduate School, Institute of Gerontology, 1-396, Kosugi-cho, Nakahara-ku, Kawasaki 211-8533, Japan

Athanassios Kotsinas (141)
Molecular Carcinogenesis Group, Department of Histology-Embryology, Medical School, National Kapodistrian University of Athens, Athens, Greece

Janina Kulka (321)
Second Department of Pathology, Semmelweis University of Medicine, Ulloi ut 93,1091 Budapest, Hungary

Fumiyauki Kumaki (205)
Pathology Core, National Heart, Lung, and Blood Institute, National Institutes of Health, Building 10/2N240, 10 Center Drive, MSC-1518, Bethesda, MD 20892-1518

Nancy Lane Smith (505)
Brody School of Medicine at East Carolina University, Department of Pathology and Laboratory Medicine, 7S-10 Brody Building, Greenville, NC, 27858

Wen-Ying Lee (361)
Department of Pathology, National Cheng Kung University Hospital, 138 Sheng Li Road, Tainan 70428, Taiwan

Peng Liang (75)
Vanderbilt-Ingram Cancer Center, Department of
Cancer Biology, School of Medicine, Vanderbilt
University, Nashville, TN

Ricardo V. Lloyd (27)
Department of Laboratory Medicine and Pathology,
Mayo Clinic, 200 First Street, SW Rochester, MN
55905

Yongrong Mao (463)
Department of Pathology, Hubei Cancer Hospital,
430069, Wuhan, China

Valeria Masciullo (223)
Jefferson Medical College, Sbarro Institute,
Department of Pathology, Anatomy and Cell Biology,
1020 Locust Street, Room 226, Philadelphia,
PA, 19107

Larry E. Morrison (213)
Vysis, Inc., 3100 Woodcreek Drive, Downers Grove, IL,
60515

P.A. Mote (449)
Westmead Institute for Cancer Research, University of
Sydney, Westmead Hospital, Westmead, NSW, 2145,
Australia

Kiyoshi Mukai (299)
Department of Pathology, Tokyo Medical University,
Shinjuku 6-1-1, Shinjuku-ku, Tokyo, 160-8402,
Japan

Beatrix Müller (199)
Department of Dermatology, University Hospital of
Zurich, Gloriastrasse 31, 8091, Zurich, Switzerland

Rakesh Naidu (395)
International Medical University, Sesama Centre,
Plaza Komenwel, Bukit Jalil, Kuala Lumpur, 5700,
Malaysia

Lydia Nakopoulou (249)
Department of Pathology, The Athens National
University, Medical School, 75 Mikras Asias Str. Goudi,
GR-115 27 Athens, Greece

Toshiro Niki (155)
The University of Tokyo, Department of Human
Pathology, Hongo 7-3-1, Bunkyo-ku, Tokyo, 113-0033,
Japan

Alain Niveleau (181)
Department of Nutrition Sciences, Division of
Nutritional Biochemistry and Molecular Biology,
University of Alabama, University Station, Birmingham,
AL 35294

Olufunmilayo I. Olapade (385)
Section of Hematology/Oncology, Department of
Medicine, Committees on Genetics and Cancer Biology,
University of Chicago, 5841 S. Maryland Avenue,
Chicago IL, 60637-1463

Akishi Ooi (259)
Department of Pathology, University of Yamanishi,
Tamaho, Yamanishi 409-3898, Japan

Chandrika J. Piyathilake (181)
Department of Nutrition Sciences, Division of
Nutritional Biochemistry and Molecular Biology,
University of Alabama, University Station, Birmingham,
AL 35294

N. Porchet (115)
Service d'Anatomie Pathologique, Hospital Calmette,
CHRU-U Lille, Bd J Leclercq 59037, Lille Cedex, France

Xiang Qian (27)
Department of Laboratory Medicine and Pathology,
Mayo Clinic, 200 First Street, SW Rochester, MN
55905

Karin K. Ridderstråle (385)
University of Chicago, Department of Medicine,
Section of Hematology/Oncology, 5841 S. Maryland
Avenue, Chicago, IL, 60637

Bozena Sarcevic (471)
University Hospital for Tumors, Zagreb, Croatia

Philippa T.K. Saunders (439)
MRC Human Reproductive Sciences Unit, 37 Chalmers
Street, Edinburgh, EH3 9ET, UK

Zuo-rong Shi (13)
Zymed Laboratories, 561 Eccles Avenue,
South San Francisco, CA 94080

Marek Skacel (57)
Cleveland Clinic Foundation, Department of Anatomic
Pathology, L-25, 9500 Euclid Avenue, Cleveland,
OH 44195

Bradley L. Smith (307)
Ventana Medical Systems, Inc./QDL, 610 Oakmont
Lane, Westmont, IL, 60559

Irina A. Sokolova (213)
Vysis Inc., 3100 Woodcreek Drive, Downers Grove, IL,
60515

Poul H.B. Sorenson (493)
Children's and Women's Health Center of British
Columbia, Departments of Pathology and Laboratory
Medicine and Pediatrics, 950 West 28th Avenue,
Vancouver, BC, V5Z 4H4, Canada

Giulio C. Spagnoli (471)
Department of Surgery, Division of Research,
University Hospital, Basel, Switzerland

Neil L. Spector (307)
Ventana Medical Systems, Inc./QDL, 610 Oakmont
Lane, Westmont, IL, 60559

Susanne Stein (75)
Vanderbilt-Ingram Cancer Center, Department of
Cancer Biology, School of Medicine, Vanderbilt
University, Nashville, TN

Yuko Sugiyama (67)
Department of Surgery, General Surgery,
University of Texas Medical Branch, 6.200 John Sealy
Annex, 301 University Blvd., Galveston,
TX 77555-0536

Careen K. Tang (415)
Lombardi Cancer Center, Georgetown University
Medical Center, Research Building, Suite E512, 3970
Reservoir Road, NW, Washington,
DC 20007-2197

Luigi Terracciano (471)
Department of Pathology, University of Basel, Basel,
Switzerland

Cristina Tognon (493)
Research Institute for Women and Children's Health,
Department of Pediatrics and Oncology, 950 West
28th Avenue, Rm. 3064, Vancouver, BC, V5Z 4H2,
Canada

Anna-Mária Tōkés (321)
Second Department of Pathology, Semmelweis
University of Medicine, Ulloi ut 93, 1091 Budapest,
Hungary

Raymond R. Tubbs (57)
Cleveland Clinic Foundation, Department of Clinical
Pathology (LII), 9500 Euclid Avenue, Cleveland, OH
44195-5153

Mirjana Urosevic (199)
Department of Dermatology, University Hospital of
Zurich, Gloriastrasse 31, 8091, Zurich,
Switzerland

Jessica Wang-Rodriguez (477)
UC San Diego School of Medicine, Department of
Pathology 0612, 9500 Gilman Drive, La Jolla, CA,
92093

Jianping Wu (463)
Department of Pathology, Hubei Cancer Hospital,
430069, Wuhan, China

Rina Wu (13)
Professor of Medicine, Senior Scientist, Director
Immunohistochemistry, Zymed Laboratories, 561
Eccles Avenue, South San Francisco, CA 94080

Pan-Chyr Yang (163)
Department of Pathology, National Taiwan University
Hospital, Taipei, Taiwan

Yasushi Yatabe (169)
Department of Pathology and Molecular Diagnostics,
Aichi Cancer Center, 1-1 Kanokoden, Chikusa-ku,
Nagoya 464-8681, Japan

Shui Qing Ye (85)
Division of Pulmonary and Critical Care Medicine,
Johns Hopkins University School of Medicine, 5200
Eastern Avenue, Mason Lord Building, Center
Tower-Room 664, Baltimore, MD 21224

Panayotis Zacharatos (141)
Molecular Carcinogenesis Group, Department of
Histology-Embryology, Medical School, National
Kapodistrian University of Athens, Athens, Greece

P. Zagorianakou (425)
Department of Pathology, Medical School, University of
Iannina, 45110 Ioannina, Greece

Foreword

According to mortality data from the National Center for Health Statistics, approximately 1,334,100 new cases of cancer will have been diagnosed, and 556,500 people will have died from cancer in the United States by the end of 2003. Though the number of cancer-related deaths has been on the decline since 1992, the incidence has increased over the same period. This increase is largely due to the implementation of improved screening techniques that have in turn been made possible by advances in immunochemical diagnostic testing. As immunochemical techniques, such as *in situ* hybridization (ISH) and immunohistochemistry (IHC), continue to be refined; their use in improving patient care through research and improved methods of diagnosis is becoming ever more valuable.

In situ hybridization is a well-established approach for identifying the organization and physical position of a specific nucleic acid within the cellular environment, by means of hybridizing a complimentary nucleotide probe to the sequence of interest. The use of deoxyribonucleic acid (DNA) and ribonucleic acid (RNA) as probes to assay biological material has been in use for approximately 30 years. However, recently, advances in ISH have seen a replacement of radioactive detection by more adaptable colorimetric and fluorescent (FISH) methods for the interrogation of nuclei, metaphase chromosomes, DNA fibers, patient tissue, and, most recently, deriving information from patient samples using DNA microarrays. Technological advances, including array comparative genomic hybridization, spectral karyotyping, and multi-color banding, have provided a refinement in the study of genome organization and chromosomal rearrangements. In addition, ISH using RNA has allowed for a determination of the expression pattern and the abundance of specific transcripts on a cell-to-cell basis. Advances in DNA and RNA ISH have migrated from the research setting and becoming routine tests in the clinical setting permitting examination of the steps involved in tumorigenesis, which would not have been possible by the use of classical cytogenetic analysis.

Since the introduction of monoclonal antibodies, immunohistochemistry has developed into a vital tool, which is now extensively used in many research laboratories as well as for clinical diagnosis. Immunohistochemistry is a collective term for a variety of methods, which can be used to identify cellular or tissue components by means of antigen-antibody interactions. Immunostaining techniques date back to the pioneering work by Albert Coons in the early 1940s, using fluorescein-labelled antibodies. Since then, developments in the techniques have permitted visualization of antigen-antibody interactions by conjugation of the antibody to additional fluorophores, enzyme, or radioactive elements. As there are a wide variety of tissue types, antigen availabilities, antigen-antibody affinities, antibody types, and detection methods it is essential to select antibodies almost on a case-to-case basis. The consideration of these factors has lead to the identification of several key antibodies that have great utility in the study and diagnosis of tumors.

The scientific advances in the field of immunochemistry have necessitated rapid developments in microscopy, image capture, and analytical software in order to objectively quantify results. These cutting edge experimental systems have already produced many significant differences between cancers that might not have been distinguished by conventional means.

The focus of these volumes is the use of ISH and IHC to study the molecular events occurring at the DNA, RNA, and protein levels during development and progression of human carcinomas. Continued investment of time and expertise by researchers worldwide has contributed significantly to a greater understanding of the disease processes. As the technical requirements for many immunochemical techniques is quite demanding and as the methodology itself poses many pitfalls, the step-by-step methods provided in these volumes will serve as an excellent guide for both clinical and basic researchers studying human malignancies.

Simon Hughes
Ontario Cancer Institute
Princess Margaret Hospital
Toronto, Canada

Preface to Volume 1

One of the primary objectives of this volume is discussion of procedures of immunohistochemistry (IHC) and *in situ* hybridization (ISH), including fluorescence *in situ* hybridization (FISH), as they are used in the field of pathology, especially cancer diagnosis. The practical importance of the antigen-retrieval protocol in IHC was realized in 1991, and since then it has been used routinely in pathology laboratories. Many chapters in this volume contain the details of this protocol. In this volume, IHC, ISH, and FISH of two major carcinomas (lung and breast) are presented. Other major cancers will be discussed in Volumes 2 and 3. The procedures are explained in a detailed step-by-step fashion so that the reader can use them without additional references. Materials required to carry out the procedures are also included.

Another objective of this volume is the discussion of the role of molecular pathology (molecular genetics, molecular medicine, molecular morphology) to understand and achieve correct diagnosis and therapy in neoplastic disease. Molecular pathology–genetics has the advantage of assessing genes directly. Knowledge of the genetic basis of disease will, in turn, allow more specific targeting of the cause, rather than the symptoms, of the disease. The time is overdue to apply our knowledge of molecular genetics, in conjunction with IHC and histology, to diagnostic, therapeutic, and prognostic decisions.

Genetic information will improve the prognosis used to monitor both the efficacy of treatment and disease recurrence. Molecular markers, largely from tumors, but also from the germline, have great potential for diagnosis, for directing treatment, and as indicators of the outcome. Indeed, methods of molecular testing of tumors are well established and are discussed in this volume. They are of considerable importance to clinical practitioners. The role of mutations in cancer is emphasized, for the characteristics of the tumor depend on the mutations that lead to their emergence. Widespread molecular testing is the future for clinical practice.

Unfortunately, clinical practice has lagged behind the current knowledge of research in molecular genetics. Both technicians and pathologists need to be aware of the importance of molecular pathology testing. Somatic mutations are rarely performed, though some histopathology and cytogenetics laboratories have done limited testing of chromosomal rearrangements in lymphoma. Molecular testing should be regarded as a means of complementing, rather than replacing, established methods such as IHC and FISH.

There are several reasons for the limited use of molecular genetics in clinical practice. One reason is the high cost of establishing facilities for molecular techniques; another is our comparatively meager understanding of the nature of many diseases, including cancer. Although equipment for molecular testing is available, some investment is needed. Another reason is the dearth of clinician–scientist training programs, resulting in limited clinician–scientists. Also, an inequity in pay exists between those working in clinical practice versus a research faculty. Accordingly, the differential in pay may be a disincentive for choosing a full-time career in medical research. The length of time (8 years as an average) to receive the M.D./Ph.D. is probably also a barrier in the development of new clinician–scientists. Also, many clinician–scientist trainees are married or in stable relationships, and personal time for family life and children is increasingly important. Narrowing the gap in income between clinical practitioners and full-time medical researchers would provide a positive incentive for this profession.

Pathologists are well advised to adapt to modern therapeutic shifts (i.e., morphological interpretation needs to be combined with molecular diagnostic modalities). The latter protocols can provide a second level of testing that is particularly useful for the analysis of neoplasms for which histologic and immunophenotypic data are inconclusive. Therapies already are beginning to move more and more toward specific molecular targets. Examples are DNA-microarrays, differential display of gene expression, serial analysis of gene expression, comparative genomic hybridization, rolling circle amplification, reverse transcription-polymerase chain reaction, FISH, Southern Blot hybridization, and specific cloned probes; most of these methods are discussed in this volume. Flow cytometry technology is also presented. We already are down a path that has the potential to alter oncology–clinical practice. My hope, through this volume, is to expedite the translation of molecular genetics into clinical practice.

I am indebted to the authors of the chapters for their promptness and appreciate their dedication and hard work in sharing their expertise with the readers. In most cases the protocols presented were either introduced or refined by the authors and are routinely used in their clinical pathology laboratories. The methods presented here offer much more detailed information than is available in scientific journals. Because of its relatively recent emergence from the research laboratory, many molecular pathology protocols are still found in scientific journals and have not appeared in a book. Each chapter provides unique, individual, practical knowledge based on the expertise of the author. As with all clinical laboratory testing, the results obtained should be interpreted in conjunction with other established and proven laboratory data and clinical findings.

This volume has been developed through the efforts of 45 authors, representing 11 countries. The high quality of each manuscript made my work as the editor an easy one. The authors were gracious and prompt. This volume is intended for use in research and clinical laboratories by medical technicians and pathologists, especially in the field of oncology. This volume will also be of interest and help to medical students. I appreciate the cooperation extended to me by Hilary Rowe, a valued publishing editor. I am grateful to Elizabeth McGovern, Meredith Schmitt, and Kevin King Sang Tse for their help in preparing this volume.

M.A. Hayat

Preface to Volumes 2 and 3

Cancer is ultimately a genetic disease and as such the focus of much cancer research has been directed toward understanding which and how many oncogenes are activated and how tumor suppressor genes become dysfunctional in human malignancies. Recently, the range of methods to examine these abnormalities has enormously widened, and many new and powerful molecular, immunohistochemical, and *in situ* hybridization techniques have become available. These include the detection of mutations using the polymerase chain reaction, reverse transcription-polymerase chain reaction and DNA sequencing, and comparative genomic hybridization on genomic microarrays to detect gene amplifications and detentions on a genome-wide basis. Other relevant techniques include serial analysis of gene expression, suppression subtractive hybridization, and flow cytometry.

Various signal amplification approaches have been introduced to increase the sensitivity, accompanied by reduced nonspecific background staining, of immunohistochemistry. Similarly, the conventional polymerase chain reaction method has been improved through quantitative real-time polymerase chain reaction. Standard *in situ* hybridization has also been improved by its modifications such as fluorescence *in situ* hybridization and chromogenic *in situ* hybridization. The details of the aforementioned and other techniques are presented in Volume 1 of *Immunohistochemistry and In Situ Hybridization of Human Carcinomas* and will also be discussed in Volumes 2 and 3.

One of the goals of Volumes 2 and 3 is to provide, not only step-by-step protocols, but also both advantages and limitations of the methods used for cancer diagnosis. In addition to the molecular genetics techniques mentioned earlier, the use of immunohistochemistry and *in situ* hybridization in the diagnosis of four major cancers (colorectal, prostate, ovarian, and pancreatic) is described in detail. Each chapter is organized to provide an introduction, required materials (including reagents, antibodies, and apparatus), step-by-step details of the protocol, and an interpretation of the results obtained. The results of each method are shown by including a color photomicrograph that contains useful immunohistochemical diagnostic information. Special attention has been paid to the antigen-retrieval methods. A literature review from the early 1990s to the present of the subject matter is also included. Each chapter is comprehensive and stands alone in terms of determination of cancer diagnosis, so that the user does not have to scour multiple places in the book or consult outside sources.

It was challenging to bring some semblance of order to the vast body of information in the field of molecular genetics (molecular pathology), which has become available primarily in scientific journals during the past decade. The contributions of expert authors in each of their respective disciplines have made it possible to accept this challenge.

Contents of Volumes 2 and 3

Prostate Carcinoma

Liver Carcinoma

Introduction to Volumes 2 and 3

The elucidation of the genetic events underlying the initiation and progression of malignancy has been hampered by limitations inherent in both *in vitro* and *in vivo* methods of study. The limitation of in vitro-based system is that genetic information obtained from cell lines may not accurately represent the molecular events occurring in the actual tissue milieu from which they were derived. On the other hand, *in vivo* genetic analysis is limited because of the inability to procure pure populations of cells from complex, heterogeneous tumor tissue.

The development and use of molecular-based therapy for human malignancies will require a detailed molecular genetic analysis of patient tissue, including solving the two previously mentioned limitations. This objective can be achieved, for example, by using cDNA microarray analysis. This technique facilitates high-throughput genetic analysis of cancer and allows monitoring *in vitro* gene expression levels in normal, invasive, and metastatic cell populations. In other words, it provides a powerful approach to elucidate molecular events responsible for the development and progression of malignancy. This technology, with its capacity for simultaneous monitoring of thousands of genes, provides a unique opportunity for high-throughput genetic analysis of cancer. These *in vivo* gene expression profiles can be verified by quantitative real-time polymerase chain reaction and immunohistochemistry. Although most microarray studies are carried out with *in vitro*-derived genetic material, microarray in combination with laser capture microdissection protocol can be performed with *in vivo*-derived genetic material originating from morphologically distinct cellular subpopulations within neoplastic tissue (tumor).

The laser capture microdissection technique allows for the rapid, reliable, and accurate procurement of cells from specific microscopic regions of tissue sections under direct visualization. Thus, molecular genetic analysis of pure populations of malignant cells in their native tissue environment can be performed.

This technical advancement overcomes the limitation associated with conventional in vitro and in vivo approaches.

Another method that will be updated in Volumes 2 and 3 is the polymerase chain reaction (PCR) and its variations. The PCR technique, with both its high accuracy and sensitivity, is widely used in the areas of gene expression analysis, identification of microorganisms, polymorphism studies, and detection of a variety of genetic diseases. It has also been extensively used for detecting and quantifying virus-derived nucleic acids in clinical specimens. In many diagnostic real-time PCR assays, the nucleic acids are tested against reference standard nucleic acids, which are amplified in parallel to the clinical samples. When quantitation is performed, the reference nucleic acids are applied in graded amounts as external quantitation standards.

To further improve the PCR performance, a number of its variations have been introduced. One such variation is TagMan used for quantitative analysis of gene expression. This protocol is a variation of real-time PCR, which utilizes the 5′-3′ exonuclease activity of TAG polymerase for enzymatic digestion of the fluorescent oligonucleotide probe during PCR amplification. One of the advantages of the TagMan approach is that it requires as few as 1000 molecules of RNA, which is 2000-fold less than that typically required for microarray measurements. Thus, the TagMan technique allows quantification of mRNA levels of numerous genes of interest in small tissue samples.

Four major cancers—colorectal, pancreatic, prostate, and ovarian—are discussed in Volume 2, currently in press. Pancreatic cancer is a devastating disease with a very poor prognosis and continues to have one of the highest mortality rates of any malignancy. Each year, an average of 28,000 patients are diagnosed with pancreatic cancer, and nearly all of them will die of their disease. The 5-year survival rate of patients with ductal adenocarcinoma of the pancreas is 4%, which is one of the lowest of any neoplasm. The reason that this disease is devastating is that the vast majority of patients are diagnosed at an advanced stage of disease that is incurable with existing therapy. Currently no tumor markers are known that provide reliable screening for pancreatic cancer at an earlier, potentially curable stage.

Pancreatic cancer is a particularly serious problem for persons with a strong familial history of this disease, who may have up to 57-fold greater risk of developing this cancer in their lifetime. The urgent need for reliable tumor markers of pancreatic cancer is obvious. In other words, the identification of genes differentially expressed in pancreatic cancer is critical to the development of novel, effective therapeutics and new markers to detect disease at an earlier, potentially curable stage.

After identifying differentially expressed genes, for example, in infiltrating pancreatic cancer using RNA-based global gene expression profiling biotechniques, they can be confirmed by immunohistochemistry, ISH, or RT-PCR. Although the reasons for the aggressive growth and metastatic behavior of pancreatic cancer are not yet fully understood, the information presented in Volume 2 is a step forward in elucidating the pathogenesis of this disease, which may lead to the development of screening markers and therapeutic targets.

Prologue

We possess scientific and industrial knowledge in new biotechniques, including human genetic technologies. However, ethical and social implications of these advances must be addressed. Such concerns are especially relevant in some of the applications of genetic engineering, such as pharmacogenetics; gene therapy; predictive diagnostics including prenatal genetic diagnosis; therapeutic cloning; cloning of humans and other animals; human tissue banking; transplanting; and patenting of inventions that involve elements of human origin, including stem cells. Bioethics should be a legitimate part of governmental control or supervision of these technologies. Scientific and industrial progress in this field is contingent on the extent to which it is acceptable to the cultural values of the public. In addition, in medical research on human subjects, considerations related to the well-being of human subjects should take precedence over the interests of science and industry. Any form of discrimination against a person based on genetic heritage is prohibited.

M.A. Hayat

Selected Definitions

Definitions of some of the commonly used terms in this volume and other publications follow.

Alternative Splicing: Genes with new functions often evolve by gene duplication. Alternative splicing is another means of evolutionary innovation in eukaryotes, which allows a single gene to encode functionally diverse proteins (Kondrashov *et al.*, 2001). Alternative splicing can produce variant proteins and expression patterns as different as the products of different genes. Alternative splicing either substitutes one protein sequence segment for another (substitution alternative splicing) or involves insertion or deletion of a part of the protein sequence (length-difference–alternative splicing). Thus, alternative splicing is a major source of functional diversity in animal proteins. Very large types and numbers of proteins are required to perform immensely diverse functions in a eukaryote.

Lack of correlation between the high complexity of an organism and the number of genes can be partially explained if a gene often codes for more than one protein. Individual genes with mutually alternative exons are capable of producing many more protein isoforms than there are genes in the entire genome. A substantial amount of exon duplication events lead to alternative splicing, which is a common phenomenon. Indeed, alternative splicing is widespread in multicellular eukaryotes, with as many as one (or more) in every three human genes producing multiple isoforms (Mironov *et al.*, 1999). In other words, alternative splicing is an ubiquitous mechanism for the generation of multiple protein isoforms from single genes, resulting in the increased diversity in the proteomic world.

Clinical Guidelines: Clinical guidelines are statements aimed to assist clinicians in making decisions regarding treatment for specific conditions. They are systematically developed, evidence-based, and clinically workable statements that aim to provide consistent and high-quality care for patients. From the perspective of litigation, the key question has been whether guidelines can be admitted as evidence of the standard of expected practice, or whether this would be regarded as hearsay. Guidelines may be admissible as evidence in the United States if qualified as authoritative material or a learned treatise, although judges may objectively scrutinize the motivation and rationale behind guidelines before accepting their evidential value (Samanta *et al.*, 2003). The reason for this scrutiny is the inability of guidelines to address all the uncertainities inherent in clinical practice. However, clinical guidelines should form a vital part of clinical governance.

Diagnosis: Diagnosis means the differentiation of malignant from benign disease or of a particular malignant disease from others. A tumor marker that helps in diagnosis may be helpful in identifying the most effective treatment plan.

DNA Methylation: Genetic mutations or deletions often inactivate tumor suppressor genes. Another mechanism for silencing genes involves DNA methylation. In other words, in addition to genetic alterations, epigenetics controls gene expression, which does not involve changes of genomic sequences. DNA methylation is an enzymatic reaction that brings a methyl group to the 5th carbon position of cystine located 5′ to guanosine in a CpG dinucleotide within the gene promoter region. This results in the prevention of transcription. Usually multiple genes are silenced by DNA methylation in a tumor. DNA methylation of genes, on the other hand, is not common in normal tissues. Gene methylation profiles, almost unique for each tumor type, can be detected in cytologic specimens by methylation-specific polymerase chain reaction (Pu *et al.*, 2003).

In the human genome, ~80% of CpG dinucleotides are heavily methylated, but some areas remain unmethylated in GC-rich CpG island (Bird, 2002). In cancer cells, aberrant DNA methylation is frequently observed in normally unmethylated CpG islands, resulting in the silencing of the function of normally expressed genes. If the silencing occurs in genes critical to growth inhibition, the epigenetic alteration could promote tumor progression because of uncontrolled cell growth. However, pharmacological demethylation can restore gene function and promote death of tumor cells (Shi *et al.*, 2003).

Epigenetics can be defined as the study of mitotically and/or meiotically heritable changes in gene function that cannot be explained by changes in DNA sequence. Processes less irrevocable than mutation fall under the umbrella term *epigenetic*. Known molecular mechanisms involved in epigenetic phenomenon include DNA methylation, chromatin remodeling, histone modification, and RNA interference. Patterns of gene expression regulated by chromatin factors can be inherited through the germ line (Cavalli *et al.*, 1999). The evidence that heritable epigenetic variation is common raises questions about the contribution of epigenetic variation to quantitative traits in general (Rutherford *et al.*, 2003).

Genomic Instability: It takes many years to get a cancer. Approximately 20 years may elapse from the time of exposure to a carcinogen to the development of a clinically detectable tumor. During this duration, tumors are characterized by genomic instability, resulting in the progressive accumulation of mutations and phenotypic changes. Some of the mutations bypass the host-regulatory processes that control cell location, division, expression, adaptation, and death. Genetic instability is manifested by extensive heterogeneity of cancer cells within each tumor.

Destabilized DNA repair mechanisms can play an important role in genomic instability. Human cells may use at least seven different repair mechanisms to deal with DNA lesions that represent clear danger to survival and genomic stability. For example, homologous recombination repair, nonhomologous end-joining, and mismatch repair mechanisms normally act to maintain genetic stability, but if they are deregulated, genomic instability and malignant transformation might occur (Pierce *et al.*, 2001). Also, because the human genome contains ~ 500,000 members of the *Alu* family, increased levels of homologous/homologous recombination events between such repeats might lead to increased genomic instability and contribute to malignant progression (Rinehart *et al.*, 1981).

In addition, BCR/ABL oncogenic tyrosine kinase allows cells to proliferate in the absence of growth factors, protects them from apoptosis in the absence of growth factors, protects them from apoptosis in the absence of external survival factors, and promotes invasion and metastasis. The unrepaired and/or aberrantly repaired DNA lesions resulting from spontaneous and/or drug-induced damage can accumulate in BCR/ABL-transformed cells, which may lead to genomic instability and malignant progression of the disease (Skorski, 2002).

Laser-Capture Microdissection: Tissue heterogeneity and the consequent need for precision before specimen analysis present a major problem in the study of disease. Even a tissue biopsy consists of a heterogenous population of cells and extracellular material, and analysis of such material may yield misleading or confusing results. Cell cultures can be homogenous but not necessarily reflect the in vivo condition. Therefore, a strategy is required to facilitate selective purification of relevant homogenous cell types.

The technology of laser-capture microdissection allows extraction of single cells or defined groups of cells from a tissue section. This technique is important for characterizing molecular profiles of cell populations within a heterogeneous tissue. In combination with various downstream applications, this method provides the possibility of cell-type or even cell-specific investigation of DNA, RNA, and proteins (Mikulowska-Mennis *et al.*, 2002).

Loss of Heterozygosity: In the majority of cases in which the gene mutation is recessive, tumor cells often retain only the mutated allele and lose the wild-type one. This loss is known as loss of heterozygosity.

Metastasis: Initially tumor growth is confined to the tissue of origin, but eventually the mass grows sufficiently large to push through the basement membrane and invade other tissues. When some cells loose adhesiveness, they are free to be picked up by lymph and carried to lymph nodes and/or may invade capillaries and enter blood circulation. If the migrant cells can escape host defenses and continue to grow in the new location, a metastasis is established. Approximately more than half of cancers have metastasized by the time of diagnosis. Usually it is the metastasis that kills the person rather than the primary (original) tumor.

Metastasis itself is a multi-step process. The cancer must break through any surrounding covering (capsule) and invade the neighboring (surrounding) tissue. Cancer cells must separate from the main mass and be picked up by the lymphatic or vascular circulation. The circulating cancer cells must lodge in another tissue. Cancer cells traveling through the lymphatic system must lodge in a lymph node. Cancer cells in vascular circulation must adhere to the endothelial cells and pass through the blood vessel wall into the tissue. For cancer cells to grow, they must establish a blood supply to bring oxygen and nutrients; this usually involves angiogenesis factors. All of these events must occur before host defenses can kill the migrating cancer cells.

If host defenses are to be able to attack and kill malignant cells, they must be able to distinguish between cancer and normal cells. In other words, there must be immunogens on cancer cells not found on normal cells. In the case of virally induced cancer circulating cells, viral antigens are often expressed, and such cancer cells can be killed by mechanisms similar to those for virally infected tissues. Some cancers do express antigens specific for those cancers (tumor-specific antigens), and such antigens are not expressed by normal cells.

As stated earlier, metastasis is the principal cause of death in individuals with cancer, yet its molecular basis is poorly understood. To explore the molecular difference between human primary tumors and metastases, Ramaswamy *et al.* (2002) compared the gene-expression profiles of adenocarcinoma metastases of multiple tumor types to unmatched primary adenocarcinomas. They found a gene-expression signature that distinguished primary from metastatic adenocarcinomas. More importantly, they found that a subset of primary tumors resembled metastatic tumors with respect to this gene-expression signature. The results of this study differ from most other earlier studies in that the metastatic potential of human tumors is encoded in the bulk of a primary tumor. In contrast, some earlier studies suggest that most primary tumor cells have low metastatic potential, and cells within large primary tumors rarely acquire metastatic capacity through somatic mutation (Poste *et al.*, 1980). The emerging notion is that the clinical outcome of individuals with cancer can be predicted using the gene profiles of primary tumors at diagnosis.

Monitoring: Monitoring means repeated assessment if there is early relapse or the presence of other signs of disease activity or progression. If early relapse of the disease is identified, a change in patient management will be considered, which may lead to a favorable outcome for the patient.

Prognosis: Prognosis is defined as the prediction of how well or how poorly a patient is likely to fare in terms of response to therapy, relapse, survival time, or other outcome measures.

Screening: Screening is defined as the application of a test to detect disease in a population of individuals who do not show any symptoms of their disease. The objective of screening is to detect disease at an early stage, when curative treatment is more effective.

Serial Analysis of Gene Expression (SAGE) is an approach that allows rapid and detailed analysis of thousands of transcripts. The LongSAGE method (Saha *et al.*, 2002) is similar to the original SAGE protocol (Velculescu *et al.*, 1995), but produces longer transcript tags. The resulting 21 bp consists of a constant 4 bp sequence representing the restriction site at which the transcript has been cleaved, followed by a unique 17 bp sequence derived from an adjacent sequence in each transcript. This improved method was used for characterizing ~ 28,000 transcript tags from the colorectal cancer cell line DLD-1. The SAGE method was also used for identifying and quantifying a total of 303,706 transcripts derived from colorectal and pancreatic cancers (Zhang *et al.*, 1997). Metastatic colorectal cancer showed multiple copies of the PRL-3 gene that was located at chromosome 8q24.3 (Saha *et al.*, 2001). Several genes and pathways have been identified in breast cancer using the SAGE method (Porter *et al.*, 2001). The SAGE method is particularly useful for organisms whose genome is not completely sequenced because it does not require a hybridization probe for each transcript and allows new genes to be discovered. Because SAGE tag numbers directly reflect the abundance of the mRNAs, these data are highly accurate and quantitative. For further details, see Chapter 26 by Dr. Ye in this volume.

Tumor Markers: Tumor markers are molecular entities that distinguish tumor cells from normal cells. They may be unique genes or their products are found only in tumor cells, or they may be genes or gene products that are found in normal cells but are aberrantly expressed in unique locations in the tumor cells, or are present in abnormal amounts, or function abnormally in response to cellular stress or to environmental signals (Schilsky *et al.*, 2002). Tumor markers may be located intracellularly (within the nucleus, in the cytoplasm, or on the membrane), on the cell surface, or secreted into the extracellular space, including into circulation. Tumor markers usually are used for monitoring and detecting early response in asymptomatic patients. For example, tissue-based estrogen receptor and HER-2/*neu* amplification/overexpression markers in breast cancer have been validated to predict response to therapy in breast cancer. Other examples are prostate-specific antigen (PSA) that is a marker for early detection of prostate cancer and carcino-embryonic antigen, which is used for detecting colon cancer.

References

Birch, J.M., Alston, R.D., Kelsey, A.M., Quinn, M.J., Babb, P., and McNally, R.J.Q. 2002. Classification and incidence of cancers in adolescents and young adults in England 1979–1997. *Br. J. Cancer 87*:1267–1274.

Bird, A. 2002. DNA methylation patterns and epigenetic memory. *Genes Dev. 16*:6–21.

Cavalli, G., and Paro, R. 1999. Epigenetic inheritance of active chromatin after removal of the main transactivator. *Science 286*:955–958.

Kondrashov, F.A., and Koonin, E.V. 2001. Origin of alternative splicing by tandem exon duplication. *Hum. Mol. Genet. 10:* 2661–2669.

Mikulowska-Mennis, A., Taylor, T.B., Vishnu, P., Michie, S.A., Raja, R., Horner, N., and Kunitake, S.T. 2002. High quality RNA from cells isolated by laser capture microdissection. *Biotechniques 33*:1–4.

Mironov, A.A., Fickett, J.W., and Gelfand, M.S. 1999. Frequent alternative splicing of human genes. *Genome Res. 9*:1288–1293.

Pierce, A.J., Stark, J.M., Araujo, F.D., Moynahan, M.E., Berwick, M., and Jasin, M. 2001. Double-strand breaks and tumorigenesis. *Trends Cell Biol. 11*:S52–S59.

Porter, D.A., Krop, I.E., Nasser, S., Sgroi, D., Kaelin, C.M., Marks, J.R., Riggins, G., and Polyak, K. 2001. A sage (Serial analysis of gene expression) view of breast tumor progression. *Cancer Res. 61:*5697–5702.

Poste, G., and Fidler, I.J. 1980. The pathogenesis of cancer metastasis. *Nature 283:*139–146.

Pu, R.T., and Clark, D.P. 2003. Detection of DNA methylation. Potential applications to diagnostic cytopathology. *Acta. Cytol. 47:*247–252.

Ramaswamy, S., and Golub, T.R. 2002. DNA microarrays in clinical oncology. *J. Clin. Oncol. 20:*1932–1941.

Rinehart, F.P., Ritch, T.G., Deininger, P., and Schmid, C.W. 1981. Renaturation rate studies of a single family of interspersed repeated sequences in human deoxyribonucleic acid. *Biochemistry 20:*3003–3010.

Rutherford, S.L., and Henikoff, S. 2003. Quantitative epigenetics. *Nat. Genetics 33:*6–8.

Saha, S., Bardelli, A., Buckhaults, P., Velculescu, V.E., Rago, C., Croix, B. St., Romans, K.E., Choti, A., Lengauer, C., Kinzler, K.W., and Vogelstein, B. 2001. A phosphate associated with metastasis of colorectal cancer. *Science 294:* 1343–1345.

Saha, S., Sparks, A.B., Rago, C., Akmaev, V., Wang, C.J., Vogelstein, B., Kinzler, K.W., and Velculescu, V.E., 2002. Using the transcriptome to annotate the genome. *Nat. Biotechnol. 19:*508–512.

Samanta, A., Samanta, J., and Gunn, M. 2003. Legal considerations of clinical guidelines: Will NICE make a difference? *J. R. Soc. Med. 96:*133–138.

Schilsky, R.L., and Taube, S.E. 2002. Introduction: Tumor markers as clinical cancer tests—are we there yet? *Sem. Oncol. 29:*211–212.

Shi, H., Maier, S., Nimmrich, I., Yan, P.S., Caldwell, C.W., Olek, A., and Huang, T.H.-M. 2003. Oligonucleotide-based microarray for DNA methylation analysis: Principles and applications. *J. Cellular Biochem. 88:*138–143.

Skorski, T. 2002. BCR/ABL regulates response to DNA damage: The role in resistance to genotoxic treatment and in genomic instability. *Oncogene 21:*8591–8604.

Velculescu, V.E., Zhang, L., Vogelstein, B., and Kinzler, K.W. 1995. Serial analysis of gene expression. *Science 270:*484–487.

Zhang, L., Zhou, W., Velculescu, V.E., Kern, S.E., Hruban, R.H., Hamilton, S.R., Vogelstein, B., and Kinzler, K.W. 1997. Gene expression profiles in normal and cancer cells. *Science 276:*1268–1272.

Classification Scheme of Human Cancers

Leukemias
Acute Lymphoid Leukemia (ALL)
Acute Myeloid Leukemia (AML)
Chronic Myeloid Leukemia (CML)
Other and Unspecified Leukemia (Other Leuk)
      Other and unspecified lymphoid leukemias
      Other and unspecified myeloid leukemias
      Other specified leukemias, NEC

Lymphomas
Non-Hodgkin's Lymphoma (NHL)
      Non-Hodgkin's lymphoma, specified subtype
      Non-Hodgkin's lymphoma, subtype not specified
Hodgkin's Disease (HD)
      Hodgkin's disease, specified subtype
      Hodgkin's disease, subtype not specified

Central Nervous System and Other Intracranial and Intraspinal Neoplasms (CNS Tumors)
Astrocytoma
      Specified low-grade astrocytoma
      Glioblastoma and anaplastic astrocytoma
      Astrocytoma not otherwise specified
Other Gliomas
Ependymoma
Medulloblastoma and Other Primitive Neuroectodermal Tumors (Medulloblastoma)
Other and Unspecified Malignant Intracranial and Intraspinal Neoplasms (Other CNS)
      Other specified malignant intracranial and intraspinal neoplasms
      Unspecified malignant intracranial and intraspinal neoplasms
Non-Malignant Intracranial and Intraspinal Neoplasms
      Specified non-malignant intracranial or intraspinal neoplasms
      Unspecified intracranial or intraspinal neoplasms

Osseous and Chondromatous Neoplasms, Ewing's Tumor, and Other Neoplasms of Bone (Bone Tumors)
Osteosarcoma
Chondrosarcoma
Ewing's Tumor
Other Specified and Unspecified Bone Tumors (Other Bone Tumors)
      Other specified bone tumors
      Unspecified bone tumors

Soft Tissue Sarcomas (STS)
Fibromatous Neoplasms (Fibrosarcoma)
Rhabdomyosarcoma
Other Soft Tissue Sarcomas
        Other soft tissue sarcomas
        Unspecified soft tissue sarcomas

Germ Cell and Trophoblastic Neoplasms (Germ Cell Tumors)
Gonadal Germ Cell and Trophoblastic Neoplasms
Germ Cell and Trophoblastic Neoplasms of Non-Gonadal Sites
        Intracranial germ cell and trophoblastic tumors
        Other non-gonadal germ cell and trophoblastic tumors

Melanoma and Skin Carcinoma
Melanoma
Skin Carcinoma

Carcinomas (except of skin)
Carcinoma of Thyroid
Other Carcinoma of Head and Neck
        Nasopharyngeal carcinoma
        Carcinoma of other sites in lip, oral cavity, and pharynx
        Carcinoma of nasal cavity, middle ear, sinuses, larynx, and other ill-defined sites in head and neck
Carcinoma of Trachea, Bronchus, Lung, and Pleura
Carcinoma of Breast
Carcinoma of Genitourinary (GU) Tract
        Carcinoma of kidney
        Carcinoma of bladder
        Carcinoma of ovary and testis
        Carcinoma of cervix and uterus
        Carcinoma of other and ill-defined sites in GU tract
Carcinoma of Gastrointestinal (GI) Tract
        Carcinoma of colon and rectum
        Carcinoma of stomach
        Carcinoma of liver and ill-defined sites in GI tract
Carcinomas of Other and Ill-Defined Sites Not Elsewhere Classified (NEC)
        Adrenocortical carcinoma
        Other carcinomas NEC

Miscellaneous Specified Neoplasms NEC
Embryonal Tumors NEC
        Wilm's tumor
        Neuroblastoma
        Other embryonal tumors NEC
Other Rare Miscellaneous Specified Neoplasms
        Paraganglioma and glomus tumors
        Other specified gonadal tumors NEC
        Myeloma, mast cell tumors, and miscellaneous reticuloendothelial

Lung and Breast Carcinomas

Conventional classification of cancer is based either on the type of tissue in which the disease originates (histologic type) or on the primary site or location in the body where the cancer first develops. The following classification is based on the histologic type. On the basis of histology, there are hundreds of different cancers that are grouped into five categories: carcinoma, sarcoma, myeloma, leukemia, and lymphoma.

Carcinomas are malignant neoplasms of epithelial origin, or cancers of the internal or external lining of the body. Malignancies of epithelial tissue account for 80–90% of all cancer cases. Epithelial tissue is found throughout the body and is present in the skin, the covering and lining of organs, and internal passageways, such as the gastrointestinal tract.

Carcinomas are divided into two major subtypes: *adenocarcinoma,* which develops in an organ or gland, and *squamous cell carcinoma,* which originates in the squamous epithelium. Adenocarcinomas generally occur in mucous membranes and are first seen as a thickened plaque-like white mucosa. They often spread easily through the soft tissue where they occur. Squamous cell carcinomas occur in many areas of the body. Most carcinomas affect organs or glands capable of secretion, such as breasts that produce milk, or lungs that secrete mucus, or prostate or bladder.

Examples of Breast Carcinoma[a]

Ductal

Intraductal *(in situ),* Invasive with Predominant Intraductal Component, Invasive, Comedo

Inflammatory, Medullary with Lymphocytic Infiltrate, Mucinous (Colloid), Papillary, Scirrhous, Tubular, and Others

Lobular

***In situ,* Invasive with Predominant *in situ* Component, and Invasive**

Nipple

Paget's disease, Paget's disease with Intraductal Carcinoma, and Paget's disease with Invasive Ductal Carcinoma

Examples of Lung Carcinoma[b]

Small-Cell Lung Cancer
Non–Small-Cell Lung Cancer
Squamous Cell (Epidermoid) Carcinoma
　　Spindle cell variant
Adenocarcinoma
　　Acinar, papillary, bronchoalveolar, and solid
　　　tumor with mucin
Large-Cell Carcinoma
　　Giant cell and clear cell
Adenosquamous Carcinoma
Undifferentiated Carcinoma

　　Sarcomas are cancers that originate in supportive and connective tissues such as bones, tendons, cartilage, muscles, and fat. They generally occur in young adults.

---

[a]For further details, see Breast. In: *American Joint Committee on Cancer: A JCC Cancer Staging Manual*. Philadelphia: Lippincott-Raven Publishers, 1999.

[b]For further details, see International Histologic Classification of Tumors: No. 1. *Histological Typing of Lung Tumors*. Geneva: WHO, 1981.

The most common sarcoma often develops as a painful mass on the bone. Sarcoma tumors usually resemble the tissue in which they grow.

  Neoplasms NEC

  Other specified neoplasms NEC

Unspecified Malignant Neoplasms NEC

Reference

Birch, J.M., Alston, R.D., Kelsey, A.M., Quinn, M.J., Babb, P., and McNally, R.J.Q. 2002. Classification and incidence of cancers in adolescents and young adults in England 1979–1997. *Br. J. Cancer* 87:1267–1274.

I

Introduction

1

Comparison of Immunohistochemistry, *in situ* Hybridization, Fluorescence *in situ* Hybridization, and Chromogenic *in situ* Hybridization

M.A. Hayat

Immunohistochemical Technology

The usefulness and power of a technology are measured by its continued successful use for a long time. Immunohistochemistry (IHC) more than satisfies this criterion, for the basic steps today are more or less the same as those used more than three decades ago. A primary antibody is applied to cells or a tissue section to detect a target antigen by using a detection system and a visualization reagent (enzyme or fluorochrome). The antigen may be a protein, a glycoprotein, a lipoprotein, or even a carbohydrate (Hayat, 2002).

It is important to study antigens rather than genes only, because protein levels and activities can differ significantly from those of ribonucleic acid (RNA). Considerable regulation of gene activity occurs at the protein level, including protein stability, modification, and location. A difference between studying antigens and genes is that many functions of the latter can be studied only using biochemical techniques.

Currently, IHC is being extensively used as a diagnostic tool for determining the presence or absence of particular proteins and certain carbohydrates in routinely fixed and embedded tissue specimens. It is inexpensive, relatively easy to perform, and can help visualize cell types that may harbor abnormalities. This antibody-based detection system for specific antigens is a versatile and powerful tool for molecular and cellular analyses. The power of this system originates from the considerable specificity of monoclonal antibodies for particular antigenic epitopes.

In general, this technique can provide useful information regarding the location of both normal and abnormal gene products (either between different cell types or within cellular compartments), and on the level of gene expression of such products in tumor cells compared with normal cells. This approach connects the molecular biology of any cancer cell under investigation with its histological characteristics and behavior. Immunohistochemistry can be termed

Handbook of Immunohistochemistry and *in situ* Hybridization of Human Carcinomas, Volume 1: Molecular Genetics; Lung and Breast Carcinomas

3

Copyright © 2004 by Elsevier (USA)
All rights reserved.

molecular histological technique. Molecular histology, therefore, is an explanation of the morphological characteristics of a tissue in terms of the molecules present and the functional interactions among them.

In the light of aforementioned advantages of IHC, it has become well established as the most popular methodology in the diagnosis of cancer in intact tissues. Several reasons justify the usefulness of this technique in the field of surgical pathology. It facilitates the identification and localization of specific molecular constituents in cells *in situ*. The technique provides an excellent specificity and sensitivity and can be performed using chemically fixed cells, frozen cells, or archival tissues. The advantage of using chemically fixed tissues is that they show good preservation of cell morphology and anchored antigens, which are important requirements for disease determination. In contrast, unfixed, frozen tissue sections tend to show both antigen displacement and less-than-good preservation of cell morphology.

The ability to retrieve antigens in formalin-fixed, paraffin-embedded tissues provides additional impetus to the use-fulness of this method, although the precise mechanism responsible for antigen recovery is still being debated (Hayat, 2002). Routine IHC is relatively easy to carry out in any laboratory equipped with basic instruments. Also, the technique has been automated. No other method can claim these advantages.

Although IHC is an indispensable methodology in modern pathology, both in basic research and diagnosis, it does not lend itself to standardization, even with attempts to automate it. Because immunostaining results can directly determine therapeutic decisions, reliability and reproducibility of the results are necessary. A large number of factors can influence staining results, causing a high degree of interlaboratory variability (inconsistency) in the results obtained with this technology. It is applied to a wide variety of tissues that are processed in different ways, involving a number of crucial steps. Each of these steps can affect the final results. Different results yielded by identical antigens are due to numerous factors, including the use of different reagents, retrieval methods, and detection systems. The extent of antigen degradation or masking (reversibly or irreversibly) during processing varies from specimen to specimen depending on the time of intraoperative anoxia, time elapsed from resection to fixation with formalin, type and amount of fixative used and its pH, duration and temperature of fixation, type of buffer used, and penetration of the fixative into the whole tissue specimen.

To improve the reliability of IHC, processing conditions must be optimized for each immunohistochemical test. This requirement is especially important for tests involving an antigen that has diagnostic significance only when expressed to its maximum level. This goal is best accomplished by optimizing first the processing conditions by using a specimen of known characteristics. Immunohistochemical results, in the final analysis, are operator dependent.

During the history of IHC technology, constant efforts have been made to improve sensitivity for detection of antigens in the formalin-fixed, paraffin-embedded tissues. A number of strategies, including direct peroxidase conjugates, peroxidase-anti-peroxidase (PAP), avidin-biotin complex (ABC), LSAB, and polymer-based methods, are being used for greater gains in sensitivity. Several newer techniques, such as tyramide amplification and rolling circle amplification (see chapter by Wheeler in Volume 2 of this series) provide even greater gains in sensitivity.

It has been suggested that the biotinyl tyramine-based method improves the detection limit of enzyme immunoassay by greater than 200-fold (Bobrow *et al.*, 1989). This protocol can be applied to IHC to amplify weak signals caused by masked or reduced antigenicity in formalin-fixed, paraffin-embedded tissues. Tyramide signal amplification involves conventional horseradish peroxidase (HRP) IHC, but in place of a chromogen (e.g., DAB), the peroxidase catalyzes the deposition of a tyramide conjugated to a fluorochrome, a biotin, or gold particles (Stanarius *et al.*, 1997; Toda *et al.*, 1999). Further increases in sensitivity can be achieved by incubating the tissue section in ABC preceded by treatment with biotinylated tyramide conjugate.

Although this signal amplification protocol provides significant enhancement in the sensitivity, its application to diagnostic pathology has not been widely accepted. The main reason is its nonspecific background staining that increases in parallel with enhanced detection sensitivity. The biotinylated antibody incubation step and the HRP-conjugated streptavidin step are the critical steps that introduce nonspecific background staining. Such staining is thought to be because of the binding of HRP-conjugated streptavidin to macromolecules through nonspecific ionic or hydrophobic interactions. Complete elimination of nonspecific reaction of the secondary antibody and HRP-conjugated streptavidin is critical in minimizing background staining. Another source of nonspecific background staining is the binding of conjugated streptavidin to endogenous biotin or biotin-like proteins.

The background staining is not induced by the biotinyl tyramine itself, but such staining by the secondary antibody and HRP-conjugated streptavidin is gradually amplified by biotinyl tyramine. Nonfat dry milk (skim milk) and trypton (8%) are the most effective blocking agents for both secondary antibody and HRP-conjugated streptavidin. In addition, aqueous Tween-20 (0.03%) is an effective rinsing solution to avoid background staining (Kim *et al.*, 2003). Biotinyl tyramine can be prepared as described by Kerstens *et al.* (1995).

In situ Hybridization

In situ hybridization (ISH) is one of the basic methods of developmental biology and provides the advantage of visualizing and even quantifying clinically relevant molecules in a morphological context. It is one of the most important techniques to visualize gene expression at the cellular level in tissues. The identification of gene expression patterns in tissues can provide critical spatial and temporal information and is the first step in understanding gene function.

After the initial description of ISH (Gall *et al.*, 1969; John *et al.*, 1969), radioactive labeling became the norm, allowing sensitive semiquantitative detection of nucleotide sequences. If sensitivity is the highest priority, ISH using radioactive-labeled probes remains the preferred choice. If radioactive labeling is necessary, [^{35}S]-uridine triphosphatase (UTP) has proven to be the most sensitive and reliable label for detecting mRNA because it provides the greatest compromise between time and resolution of signal. However, ^{35}S-labeled probes emit electrons of an intermediate energy and a range that permits less-accurate cellular localization of a transcript by the autoradiographic protocol than that obtained with nonradioactive probes. The use of such labels (^{3}H, ^{32}P, ^{35}S, ^{125}I) is hazardous, expensive, and time-consuming. Furthermore, these labels have a limited shelf life.

For routine use of ISH in pathology, nonradioactive probes are considered better than radioactive ones because the use of the latter is tedious. Nonisotopic techniques to amplify fluoroescent or colorimetric detection are of comparable sensitivity to that yielded by radioactive labeling in the detection of relatively abundant genes in a single cell (Breininger and Baskin, 2000). If still greater sensitivity is desired to detect single copies of specific mRNAs, it can be obtained by using super-sensitive ISH in conjunction with streptavidin-nanogold-silver staining (Hacker *et al.*, 1997). In any case, sensitivity depends primarily on the detection system used.

Techniques for ISH of messenger RNA (mRNAs) have become a tremendously powerful tool for analyzing patterns of gene expression. The power of the *in situ* detection of mRNA sequences is that the cells that synthesize and accumulate sequences of interest can be identified and placed in proper morphological context. Thus, changes in gene expression as measured in homogenates can be corrected for a proper tissue base, and potential cellular interactions can be assessed. In other words, the advantage of localizing mRNA with ISH is that mRNA is present in the cell expressing the gene, whereas the site of peptide synthesis is not always in the same cell as the mature protein (antigen) identified with IHC.

Various methods are available to improve the sensitivity of conventional ISH. Recently, Hrabovszky and Petersen (2002) demonstrated that enhanced corrected signal density (total density of signal area minus background density) can be obtained by using concentrations of probe and/or dextran sulfate several-fold higher than those used conventionally. They used concentrations of probe and dextran sulfate $> 4 \times 10^4$ cpm/µl and >10%, respectively. Prolonged hybridization reaction (>16 hr) also augmented the signal density. Nonspecific probe binding was greatly reduced, and corrected signal density was enhanced by including 750–1000 mM dithiothreitol in the hybridization buffer. This study was carried out in rat brain tissue. Its advantages in other tissue types is awaited.

Player *et al.* (2001) have developed a branched deoxyribonucleic acid (bDNA) ISH method for detecting DNA and mRNA in whole cells. This method is a signal amplification system in which target nucleic acid sequences are hybridized to a series of synthetic oligionucleotide probes and visualized through generation of chromogenic or fluorescent signals in an alkaline phosphatase (AP)-catalyzed reaction.

The application of nucleic acid target and signal amplification techniques to ISH can allow detection of as few as one or two copies of specific DNA molecules in cell preparations. The bDNA-ISH method can be used to achieve this goal, as it is highly sensitive and can detect one or two copies of DNA per cell. It is specific and provides precise localization, yielding positive signals that are retained within the subcellular compartments in which the target nucleic acid is located. Damage to cell morphology caused by heat exposure is minimized because this method does not require repeated cycling at elevated temperatures. Background noise caused by endogenous biotin is avoided because the procedure does not use an avidin-biotin reporter system (Kenny *et al.*, 2002). Diffusion of the signal is not a problem because reporter probes are physically linked to the spatially fixed nucleic acid targets. Both cell lines and tissues, including clinical specimens, can be studied with the bDNA-ISH method.

Recently, Weisheit *et al.* (2002) combined conventional ISH of tissue sections with the technique of whole-mount ISH to detect gene activity. An advantage of this method is the strict control of hybridization conditions by using tightly sealed reaction vessels. This approach eliminates liquid evaporation and results in constant salt and formamide concentrations, even at high temperatures. Theoretically, each section may be analyzed with a different probe, facilitating analysis of multiple genes in sections from a single specimen. Miniature glass slides for mounting sections are cut from regular microscope slides and handled for ISH in laboratory-made 2-ml containers.

A new approach to detect small changes in gene expression, as well as to optimize expression profiles of genes of low abundance, was reported by Ky and Shughrue (2002). To increase the sensitivity of isotopic ISH for detection of rare mRNAs, they used cryostat sections and labeled the probe with both [^{35}S]-UTP and [^{35}S]-adenosine triphosphate (ATP). In addition, sections were hybridized with probe for two nights instead of one night, which significantly enhances signal intensity because the probe has a greater chance to hybridize to its complementary mRNA. These two methods of enhancement—independently or in combination—demonstrated increased signal. However, this protocol tends to be accompanied by an increase in background noise.

Holm (2000) evaluated the sensitivity of three detection systems: AP-antialkaline phosphatase (APAAP), streptavidin-fluorescein isothiocyanate (STAV-FITC), and tyramide signal amplification (TSA), using biotin-labeled human papillomavirus 16, and found that the APAAP method provided the best sensitivity as single viral copies were detected in formalin-fixed, paraffin-embedded tissue specimens.

Because of the biological significance of sensitive *in situ* mRNA detection in conjunction with high spatial resolution, Moorman *et al.* (2001) developed a nonradioactive ISH protocol to detect mRNA sequences in tissue sections. The procedure is essentially based on the whole-mount ISH method that is made suitable for ISH on sections. The sensitivity of this procedures is improved by increasing the hybridization temperature to 70°C. Higher temperatures facilitate better tissue penetration by the probe.

Novel detection methods are available to enhance signal detection; one is based on the peroxidase-mediated deposition of haptenized or fluorochromized tyramide, termed catalyzed reporter deposition (CARD), and the other is TSA (Bobrow *et al.*, 1992; Speel *et al.*, 1998). In the TSA system, HRP is covalently linked to the anti-digoxigenin Fab fragments, which allows deposition of biotinylated tyramide that is subsequently detected with streptavidin-conjugated AP. However, this protocol may or may not improve the signal-to-noise ratio, depending on the type of study.

Methodological Considerations

Factors that influence the sensitivity of ISH include the type, quality, and method of detection of the hybridization probe, the effects of tissue fixation on target mRNA preservation and accessibility to probe, the efficiency of hybrid formation, the stability of *in situ*-formed hybrids during post-hybridization treatments, and background noise masking or confusing the hybridization signals. Strategies designed to optimize these variables can improve the sensitivity of ISH.

Reliable detection and localization of mRNAs in histological sections using ISH require the use of well-fixed, but not over-fixed, tissues of minimize mRNA degradation. RNA degradation in fresh tissues is well known. Storage of paraffin sections of formalin-fixed tissues for weeks or longer diminishes the volume of a radioactive signal. The decrease in signal intensity is not dependent on the type of the tissue. Diminished signal intensity is more pronounced in the case of mRNA present in abundant quantity. Such decrease is less consistent for the mRNA present in low to moderate amounts (Lisowski *et al.*, 2001). For these reasons, it is recommended to use freshly prepared sections of formalin-fixed tissues. If storage of slides is unavoidable, they should be stored at a cold temperature. Antigenicity is also diminished in paraffin sections stored for prolonged duration at room temperature (Hayat, 2002).

Fluorescence *in situ* Hybridization

Fluorescence *in situ* hybridization (FISH) is a powerful molecular cytogenetic method. It facilitates the detection of specific DNA sequences in intact cells and chromosomes, and so is extensively applied to the histochemical mapping of specific nucleic acid sequences on chromosomes in metaphase and to the interphase nuclei. Because FISH enables selective staining of various sequences in interphase nuclei, detection, analysis, and quantification of specific numerical and structural chromosomal abnormalities within these nuclei can be accomplished. It can also be used for detecting mRNA and allows simultaneous detection of multiple mRNA species in a single tissue section or cell.

Determination of the gene amplification levels and its pattern can also be facilitated by FISH. Clustered signals and multiple scattered signals shown by FISH correspond respectively to amplicons in homogeneous staining regions and in double-minute chromosomes found in conventional cytogenetics. Using dual-color FISH, differentiation of low-level amplification from increased gene number by polysomy is also possible. Another advantage of FISH is that it can detect gene amplification not only in isolated nuclei and imprinted cells, but also in formalin-fixed, paraffin-embedded tissues. The latter advantage allows detection of gene amplification in archival tumor samples. The sensitivity of FISH can be increased to detect low-abundance mRNA by amplifying the signal through combining TSA and enzyme-labeled fluorescent AP substrate. In conclusion, applications of FISH are diverse, including karyotype analysis, gene mapping, DNA replication and recombination, clinical diagnosis, and monitoring of disease in clinical trials.

Points of light with variable intensities resulting either from hybridization or background fluorescent noise compose FISH images. The identification of specific nucleic acid sequences with FISH reveals sites of mRNA processing, transport, and cytoplasmic localization. Recognition of these sites of hybridization is possible only when sufficient concentrations of the target sequence provide contrast with regions of lesser or no signal.

The FISH method is based on the hybridization of specific DNA sequences to the target genome, involving both histochemical and solid matrix hybridization techniques. A permeabilization step is required to facilitate penetration of the fluorescence-labeled probe, and a denaturation step is necessary for converting double-stranded DNA in tissue sections to single-stranded DNA to allow reannealing of the probe and target genome.

To obtain clearly defined images, FISH relies on automatic focusing. The counting of signals (dots or spots) in these images is used for studying numerical chromosomal aberrations, for example, in hematopoietic neoplasias, various solid tumors, prenatal diagnosis, and disease-related chromosomal translocations. However, counting of dots is required in a large number of cells for an accurate estimation of the distribution of chromosomes, especially in applications involving a relatively low frequency of abnormal cells. Because visual evaluation of large numbers of cells and enumeration of hybridization signals are tedious and time-consuming, FISH analysis with dot counting can be expedited by using an automatic system (Netten et al., 1997). This system exploits three-dimensional information of cells contained in the sample and obtains the sharpest image along the Z-axis. This mechanism has to be activated for each field of view. Automatic acquisition depends on finding the sharpest image, which can fail if the mechanism focuses on a source of noise such as debris or background fluorescence or if the field of view is empty.

As an alternative to the use of an autofocusing mechanism, FISH dot counting can be based on a neural network classifier that discriminates between in- and out-of-focus images taken at different focal planes of the same field of view (Lerner et al., 2001). This signal classification system is an accurate and efficient alternative to the autofocusing mechanism.

Because FISH procedures are based on the air-drying step of chromosome preparations, the results are well-spread metaphases and flattening of the originally spherical interphase nuclei. However, the flattening of the nuclei may lead to questionable results. If this is a problem in a particular study, it can be avoided by performing the entire procedure on a cell suspension by placing it on a polished concave slide as the final step of the procedure. This protocol allows examination of almost three-dimensional structures of chromosomes (Steinhaeuser et al., 2002).

Simultaneous identification of proteins and DNA probes is difficult to accomplish. Recently, Gośalvez et al. (2002) developed a method that allows simultaneous detection of DNA probes and surface antigens (CD3) in peripheral blood samples. The method involves the use of microwave heating-FISH for fluorescent detection of antigens. Hybridization of the whole genome probe is more efficient using microwave energy for denaturation, and embedding of cells in a low–melting-point agarose gel avoids the need of using cytospins to attach cells to the slides. This method facilitates simultaneous tricolor fluorescent visualization and represents a reasonable alternative to other methods that use precipitation of AP to detect proteins (Speel et al., 1994).

Most protocols for FISH analysis of fixed, paraffin-embedded tissues are carried out using histological sections. Although such sections preserve tissue architecture, they impede the enumeration of signals because of the frequent overlap or slicing of nuclei. The use of isolated nuclei prepared from embedded tissues largely avoids this problem, and thus provides more accurate quantitative results. Recently, Schurter et al. (2002) described the application of FISH to isolated interphase nuclei from B5- or formalin-fixed archival tissues that were associated with previously documented cytogenetic abnormalities; thus, facilitating follow-up confirmation of a previously identified chromosomal abnormality. It should be noted that precipitating fixatives such as B5 (mercuric chloride-based) are associated with unsatisfactory results using conventional FISH protocols.

Similarly, FISH is useful in the field of effusion cytology. Body cavity effusion specimens are better suited for FISH than are paraffin-embedded tissue sections, because nuclei remain intact, thereby avoiding the problem of nuclear sectioning inherent in FISH performed on standard surgical specimens. For example, for detecting homozygous CDKN2A deletion in malignant mesothelioma in body cavity effusions. Homozygous deletion of this gene is present in >70% of mesothelioma tumors. The usefulness of this approach becomes clear when one considers that distinction between benign reactive mesothelial cells and malignant mesothelial cells in serous effusions is difficult. There are no generally accepted markers to distinguish between those two types of mesotheliomas. Such studies improve the diagnostic accuracy of effusion cytology in the initial management of patients with suspected mesothelioma.

Many factors affect the detection of fluorescent signals produced by the FISH method. These factors include the type and duration of fixation, artifacts caused by sectioning of tissue blocks, thickness of sections, storage of embedded tissues or sections, and probe penetration. Prolonged fixation (longer than one week) with formalin adversely affects FISH analysis.

Repeated claims have been made that FISH is superior to IHC and *vice versa*. Both methods have comparative advantages and limitations. The FISH method allows visualization of the number of gene copies present in tumor cells and provides a sensitive, accurate, and reproducible measure of gene amplification, even in archival tissue specimens. This method is superior to some other techniques, such as Southern Blotting, for detecting gene amplification for clinical application. As in the FISH method, IHC can also be carried out using formalin-fixed, paraffin-embedded archival tissues. In addition, IHC can be used for identifying the origin of metastatic carcinoma of unknown primary origin (primary cancer). Such metastatic carcinomas are a common problem for the surgical pathologist, accounting for up to 10–15% of all solid tumors at presentation (Srodon and Westra, 2002). The IHC method can be used to rule out unlikely primary sites, especially in patients who have no clinical evidence of a primary neoplasm. Recently, Lewis *et al.* (2002) have used this technique for identifying breast carcinoma as the primary cancer in a patient with brain metastases.

Tissue treatment with microwave heating is useful in both FISH and IHC (Hayat, 2002). However, IHC is more familiar, less expensive, and simple compared with FISH. Furthermore, the former can identify cases in which the gene product (protein) is overexpressed in the absence of detectable gene amplification. On the other hand, genes can be amplified without detectable protein overexpression. However, in general, amplification is associated with overexpression. It should be noted that FISH and IHC do not provide the same information. Therefore, if possible, both techniques should be used in clinical trials to establish the best method for each specific treatment. Gene status is useful not only as a prognostic factor in determining clinical outcome but also as a predictive factor in projecting the response to adjuvant therapies. For additional information on the FISH technology, see Part I, Chapter 2; Part III, Chapter 14; and Part IV, Chapters 3 and 4.

Immunohistochemistry versus *in situ* Hybridization in Tumor Pathology

The IHC is *in situ*–based technology that evaluates the protein level of cells in tissue section. It has been the predominant method for many diagnostic applications in cancer. IHC can be carried out using frozen, fixed, or archival specimens. It is relatively easy to perform, has rapid turnaround time, and is relatively inexpensive. However, IHC assays have demonstrated wide variation in sensitivity and specificity in formalin-fixed, paraffin-embedded tissue sections. The effect of specimen processing, including tissue fixation, has a substantial effect on IHC staining. Antigen retrieval and immunodetection systems currently used have improved sensitivity of antibodies (Shi *et al.*, 1991, 2001). The overall problem with any IHC method is that a negative result may not be reliable because there is no positive internal control to check whether the epitope of interest is not destroyed or masked during processing (Ross *et al.*, 1999). Positive results of IHC may not be reliable due to the lack of a universal scoring system or interobserver difference in interpretation of IHC results (Allred *et al.*, 1998). For example, a recent study of 394 invasive primary breast cancers demonstrated that if there is no standardization of the procedure, HER-2 evaluation by IHC is not a reproducible technique (Gancberg *et al.*, 2002).

The ISH technique allows visualization of cellular DNA or RNA in tissue section, single cell, or chromosome preparations. The ISH method using DNA probes is often applied in tumor pathology to quantify the number of gene copies of cells in a tissue section. Unlike IHC, which is a relatively subjective test for protein expression, ISH objectively demonstrates gene status in tumors. The utility of ISH has been apparent in pathology since the development of nonradioactive procedures, which yield significantly better cellular resolution than that provided by radiolabeled assays. The FISH method, one of the nonradioactive procedures, can detect gene amplification, deletion, chromosome translocation, and chromosome aneuploidy in frozen formalin-fixed tumor tissue sections. Chromogenic ISH (CISH), another non-radioactive procedure, has the same ability as FISH to detect genetic alterations in tumor pathology. It was developed three years ago in the Zymed Laboratories, Inc., (San Francisco, CA) and has gained more attention recently. The difference between CISH and FISH is the detection step. Instead of fluorescence detection and fluorescence microscopy used in FISH, CISH is based on chromogenic detection, usually with HRP-substrate and/or an AP-substrate reaction, using bright-field microscopy.

In breast cancer, overexpression of specific oncogenes is often observed. Protein overexpression may be the result of gene amplification (e.g., overexpression of HER-2 protein generally results from HER-2 gene amplification [Pauletti et al., 1996; Slamon et al., 1989]). Clinically, sensitivity and specificity of HER-2 testing with FISH or CISH has been shown to be superior to those provided by IHC in fixed tissue sections. Patients who are positive by FISH but negative by IHC had a worse survival than those who had HER-2 overexpression but absence of gene amplification (Pauletti et al., 2000). Therefore, HER-2 amplification might provide more meaningful prognostic information than HER-2 overexpression in breast cancer patients, although consensus on this view is lacking.

Protein overexpression may also be related to tumor proliferation rate (e.g., in primary breast cancer, expression of topoisomerase type II alpha [TopoIIα], a key enzyme in DNA replication, is proliferation-dependent and associated with TopoIIα gene amplification as well as chromosome 17 aneuploidy [Depowski et al., 2000]). In TopoIIα testing, clinical studies showed that TopoIIα gene amplification is a more specific predictor than TopoIIα expression determined by IHC for clinical response to TopoIIα inhibitor, a widely used chemotherapy drug in metastatic breast cancer (Coon et al., 2002).

Therefore, based on the reasons previously mentioned, although protein is the final functional molecule in tumorigenesis and/or tumor progression, identifying gene status along with its protein expression may give more insight into tumor pathology.

In situ Hybridization versus Fluorescence in situ Hybridization

The ISH method visualizes gene activity at the cellular level and, indeed, the subcellular segregation of specific RNAs. It can be carried out on tissue sections or on intact tissue pieces. ISH has the advantage of not fading or bleaching and does not require fluorescence optics. It permits direct visualization of gene amplification in a morphological correlative context, using conventional microscopy. Use of stable and permanent archival slides is an additional advantage.

The FISH method can be very helpful because the probe in this method is typically large and spans long (>100 kb) regions of DNA, allowing detection of chromosomal breakpoints that are widely dispersed. This assay using a single probe can detect breakpoints within a chromosome, provides presumptive evidence of translocation, and is helpful diagnostically. However, disruption of the gene does not prove the presence of a partner chromosome. For this purpose, two FISH probes labeled with different fluorescent colors (usually green and red) are required. The topics of ISH and FISH are also discussed elsewhere in this volume.

Chromogenic in situ Hybridization versus Fluorescence in situ Hybridization

Although FISH used for formalin-fixed tissue sections is a valid alternative to IHC in many cases, it is not used frequently in noncytogenetic diagnostic pathology. One reason for this is that evaluation of FISH requires an expensive fluorescence microscope equipped with high-quality 60 or 100 oil immersion objectives and multi-band pass fluorescence filters, which are not universally available to surgical pathologists responsible for histological diagnosis. Another reason is that fluorescence signals can fade within several weeks. In addition, FISH results need to be recorded with an expensive CCD camera, making analysis of FISH data time-consuming.

Because of the limitation using FISH in fixed tissue sections already mentioned, the use of CISH assay enables routine evaluation of gene amplification/deletion, aneuploidy, and chromosomal translocation in archived tumor specimens using light microscopy. This approach is greatly beneficial in surgical pathology. In the past, ISH using digoxigenin (DIG), biotin, FITC-labeled DNA probes, synthetic oligonucleotide probes, and complementary RNA (cRNA) probes followed by chromogenic (enzymatic) detection has been mainly applied for identification of virus infection, detection of immuno-globulin mRNA, and chromosome centromere in the tissue sections. These studies are carried out because of the abundance of targeted sequences to be detected (Takarabe et al., 2001).

Application of CISH in tumor pathology using conventional ISH probe to detect gene amplification/deletion and chromosome translocation was limited due to the low ratio of signal to background staining. To overcome this limitation, DNA probes generated by Subtracted Probe Technology (SPT) was developed (Zymed). With SPT technology, repetitive DNA sequences (e.g., alu, and LINE elements, which may consist of up to more than 40% of template and cause unspecific hybridization) are quantitatively removed from the probe. Therefore, the final probe is very specific and the need for blocking nonspecific hybridization with Cot-1 DNA used in traditional FISH probes is avoided. The tissue preparation and probe hybridization in CISH is similar to the FISH procedure.

They differ in the method of probe detection, in which CISH requires IHC-like chromogen detection, whereas FISH is based on fluorescence detection. According to our studies and those of others, CISH using the SPT probe and an optimized sensitive chromogenic detection system (Zymed) gives much brighter and crisper signals than ordinary FISH probe (Tanner *et al.*, 2000; Zhao *et al.*, 2002).

When comparing CISH with FISH, both have distinctive advantages and disadvantages. One substantial advantage of the FISH approach is that probe detection can be accomplished by direct detection. In contrast, CISH generally requires at least two additional steps beyond probe hybridization, prior to nuclear counterstaining. For example, a probe can be labeled with FITC or rhodamine and then detected using the FISH method. The same probe may be labeled with DIG or biotin, and then detected by sequential incubations with mouse anti-DIG and goat anti-mouse-HRP/diaminobenzidine (DAB) reaction or streptavidin-HRP/DAB, respectively, using the CISH method. Another advantage of FISH is its easy-to-engage multicolor detection, whereas CISH is restricted to single or dual colors due to the limitation of current chromogenic detection technology.

CISH, on the other hand, has several advantages that are particularly relevant for fixed tissue sections. First, interpretation of CISH is performed using a standard light microscope and permits simultaneous evaluation of gene copies and tissue morphology on the same slide. Large regions of the tissue section can be scanned rapidly in CISH using conventional counterstain such as hematoxylin. Morphological details and CISH signals are readily apparent using 10, 20, or 40 objective lenses. Fluorescence signals and counterstainings, on the other hand, are generally only appreciated at substantially higher magnifications (60 or 100 objectives), and tissue morphology with FISH is not optimal for fixed specimens. Therefore, review of hematoxylin-eosin (H&EO) stained sections may be necessary. This can be a particular problem in distinguishing invasive breast cancer and breast carcinoma *in situ*, where HER-2 gene amplification or protein overexpression has different clinical significance. Second, CISH signals are not subject to rapid fading and the slides can therefore be archived, whereas FISH slides must be stored at 4°C or lower and are subject to quenching of the fluorescent signal. Third, with FISH, cellular and extracellular proteins can contribute to a dull, generalized, autofluorescence that often obscures FISH signals in paraffin section. These limitations make the use of FISH for fixed tissue sections cumbersome for routine work.

References

Allred, D.C., Harvey, J.M., Berardo, M., and Clark, G.M. 1998. Prognostic and predictive factors in breast cancer by immunohistochemical analysis. *Mod. Pathol. 11:*155–168.

Bobrow, M.N., Harris, T.D., Shaughnessy, K.J., and Litt, G.J. 1989. Catalyzed reporter deposition, a novel method of signal amplification: Application to immunoassays. *J. Immunol. Methods 125:*279–285.

Bobrow, M.N., Shaughnes, K.J., and Litt, G.J. 1992. The use of catalyzed reporter deposition as a means of signal amplification in a variety of formats. *J. Immunol. Methods 150:* 145–149.

Breininger, J.F., and Baskin, D.G. 2000. Fluorescence *in situ* hybridization of scarce leptin receptor mRNA using the enzyme-labeled fluorescent substrate method and tyramide signal amplification. *J. Histochem. Cytochem. 48:*1593–1599.

Coon, J.S., Marcus, E., Gupta-Burt, S., Seeling, S., Jacobson, K., Chen, S., Renta, V., Fronda, G., and Preisler, H.D. 2002. Amplification and overexpression of topoisomerase II alpha predict response to anthracycline-based therapy in locally advanced breast cancer. *Clin. Cancer Res. 8:*1061–1067.

Depowski, P.L., Rosenthal, S.I., Brien, T.P., Stylos, S., Johnson, R.L., and Ross, J.S. 2000. Topoisomerase II alpha expression in breast cancer: Correlation with outcome variables. *Mod. Pathol. 13:*542–547.

Gall, J.G., and Purdue, M.L. 1969. Formation and detection of RNA-DNA hybrid molecules in cytological preparations. *Proc. Natl. Acad. Sci. USA 63:*378–383.

Gancberg, D., Di Leo, A., Cardoso, F., Rouas, G., Pedrocchi, M., Paesmans, M., Verhest, A., Bernard-Marty, C., Peccart, M.J., and Larsimont, D. 2002. Comparison of HER-2 status between primary breast cancer and corresponding distant metastatic sites. *Ann. Oncol. 13:*1036–1043.

Gósalvez, J., Torre, D.L., Pita, M., Martinez-Ramirez, A., Lopez-Fernandez, C., Goyanes, V., and Fernandez, J.L. 2002. Fishing in the microwave: The easy way to preserve proteins. 1. Colocalization of DNA probes and surface antigens in human leukocytes. *Chromos. Res. 10:*137–143.

Hacker, G.W., Harser-Kronberger, C., Zehbe, I., Su, H., Schiechl, A., and Dietze, O. 1997. *In situ* localization of DNA and RNA sequences: Supersensitive *in situ* hybridization using streptavidin Nanogold-silver staining: Minireview, protocols and possible applications. *Cell Vision 4:*54–65.

Hayat, M.A. 2002. *Microscopy, immunohistochemistry, and antigen retrieval methods.* Kluwer AcademicPlenum Publishers, New York.

Holm, R. 2000. A highly sensitive nonisotopic detection method for *in situ* hybridization. *Appl. Immunohist. Mol. Morphol. 8:*162–165.

Hrabovszky, E., and Petersen, S.L. 2002. Increased concentrations of radioisotopically-labeled complementary ribonucleic acid probe, dextran sulfate, and dithiothreitol in the hybridization buffer can improve results of *in situ* hybridization histochemistry. *J. Histochem. 50:*1389–1400.

John, H.A., Birnstiel, M.L., and Jones, K.W. 1969. RNA-DNA hybrids at the cytological level. *Nature 223:*582–587.

Kenny, D., Shen, L.-P., and Kolberg, J.A. 2002. Detection of viral infection and gene expression in clinical tissue specimens using branched DNA (bDNA) *in situ* hybridization. *J. Histochem. Cytochem. 50:* 1219–1227.

Kerstens, M.J., Poddighe, P.J., and Hanselaar, A.G.J.M. 1995. A novel *in situ* hybridization amplification method based on the deposition of biotinylated tyramine. *J. Histochem. Cytochem.* 43:347–352.

Kim, S.H., Shin, Y.K., Lee, K.M., Lee, J.S., Yun, J.H., and Lee, S.M. 2003. An improved protocol of biotinylated tyramine-based immunohistochemistry minimizing nonspecific background staining. *J. Histochem. Cytochem.* 51:129–132.

Ky, B., and Shugrue, P.J. 2002. Methods to enhance signal using isotopic *in situ* hybridization. *J. Histochem. Cytochem.* 50:1031–1037.

Lerner, B., Clocksin, W.F., Dhanjal, S., Hulten, M.A., and Bishop, C.M. 2001. Automatic signal classification in fluorescence *in situ* hybridization images. *Cytometry* 43:87–93.

Lewis, E.H., Nashid, N., Ludwig, M.E., and Cartun, R.C. 2002. New onset visual disturbances in a 37-year-old female with metastatic carcinoma from unknown primary. *J. Histotechnol.* 25:279–281.

Lisowski, A.R., English, M.L., Opsahl, A.C., Bunch, R.T., and Blomme, E.A.G. 2001. Effect of the storage period of paraffin sections of the detection of mRNAs by *in situ* hybridization. *J. Histochem. Cytochem.* 49:927–928.

Moorman, A.F.M., Houweling, A.C., deBoer, P.A.J., and Christoffels, V.M. 2001. Sensitive nonradioactive detection of mRNA in tissue sections: Novel applications of the whole mount *in situ* hybridization protocol. *J. Histochem. Cytochem.* 49:1–8.

Netten, H., Young, I.T., van Vliet, H.J., Tanke, H.J., Vroljik, H., and Sloos, W.C.R. 1997. FISH and chips: Automation of fluorescence dot counting in interphase cell nuclei. *Cytometry* 28:1–10.

Pauletti, G., Dandekar, S., Rong, H., Ramos, L., Peng, H., Seshadri, R., Slamon, D.J. 2000. Assessment of methods for tissue-based detection of the HER-2/neu alteration in human breast cancer: A direct comparison of fluorescence *in situ* hybridization and immunohistochemistry. *J. Clin. Oncol.* 18:3651–3664.

Pauletti, G., Goddphin, W., Press, M.F., and Slamon, D.J. 1996. Detection and quantitation of HER-2/neu gene amplification in human breast cancer archival material using fluorescence *in situ* hybridization. *Oncogene 13:*63–72.

Player, A.N., Shen, L.P., Kenny, D., Antao, V.P., and Kolberg, J.A. 2001. Single copy gene detection using branched DNA (bDNA) *in situ* hybridization. *J. Histochem. Cytochem.* 49:603–612.

Ross, J.S., Yang, F., Kallakury, B.V., Sheehan, C.E., Ambros, R.A., and Muraca, P.J. 1999. HER-2/neu oncogene amplification by fluorescence *in situ* hybridization in epithelial tumors of the ovary. *Am. J. Clin. Pathol.* 111:311–316.

Schurter, M.J., LeBurn, D.P., and Harrison, K.J. 2002. Improved technique for fluorescence *in situ* hybridization analysis of isolated nuclei from archival B5 or formalin-fixed, paraffin-wax embedded tissue. *J. Clin. Pathol. Mol. Pathol.* 55:121–124.

Shi, S.-R., Cote, R.J., and Taylor, C.R. 2001. Antigen retrieval techniques: Current prospectives. *J. Histochem. Cytochem.* 49:931–937.

Shi, S.-R., Key, M.E., and Kalra, K.L. 1991. Antigen retrieval in formalin-fixed, paraffin-embedded tissue: An enhancement method for immunohistochemical staining based on microwave oven heating of tissue sections. *J. Histochem. Cytochem.* 39:741–748.

Slamon, D.J., Godolphin, W., Jones, L.A., Holt, J.A., Wong, S.G., Keith, D.E., Levin, W.J., Stuart, S.G., Udove, J., Ullrich, A., and Press, M.F. 1989. Studies of the HER-2/neu protooncogene in human breast and ovarian cancer. *Science 244:*707–712.

Speel, E.J.M., Herbergs, J., Remae-kers, F.C., and Hopman, A.H. 1994. Combined immunocytochemistry and fluorescence *in situ* hybridization for simultaneous tricolor detection of cell cycle, genomic, and phenotypic parameters of tumor cells. *J. Histochem. Cytochem.* 7:961–966.

Speel, E.J.M., Saremaslani, P., Roth, J., Hopman, A.H.N., and Komminoth, P. 1998. Improved mRNA *in situ* hybridization of formaldehyde-fixed and paraffin-embedded tissue using signal amplification with different haptinized tyramides. *Histochem. Cell Biol.* 110:571–577.

Srodon, M., and Westra, W.H. 2002. Immunohistochemical staining for thyroid transcription factor-1: A helpful aid in discerning primary site of tumor origin in patients with brain metastases. *Hum. Pathol.* 33:642–645.

Stanarius, A., Topel, I., Schulz, S., Noack, H., and Wolfe, G. 1997. Immunocytochemistry of endothelial nitric oxide synthase in the rat brain: A light and electron microscopical study using the tyramide signal amplification technique. *Acta. Histochem.* 99:411–429.

Steinhaeuser, U., Starke, H., Nietzel, A., Lindenau, J., Ullmann, P., Claussen, U., and Liehr, T. 2002. Suspension (S)-FISH, a new technique for interphase nuclei. *J. Histochem. Cytochem.* 50:1697–1698.

Takarabe, T., Tsuda, H., Okada, S., Fukutomi, T., and Hirohashi, S. 2001. Detection of numerical alterations of chromosome 1 in cytopathological specimens of breast tumors by chromogen *in situ* hybridization. *Pathol. Int.* 51:786–791.

Tanner, M., Jarvinen, T.A.H., and Kauraniemi, P. 2000. Topoisomerase 11α amplification and deletion predict response to chemotherapy in breast cancer. *Proc. Am. Assoc. Cancer Res.* 41:803–abstract.

Toda, M., Kono, K., Abriu, H., Kokuryo, K., Endo, M., Yaegashi, H., and Fukumoto, M. 1999. Application of tyramide signal amplification system to immunohistochemistry: A potent method to localize antigens that are not detectable by ordinary method. *Pathol. Intern.* 49:479–483.

Weisheit, G., Mertz, K.D., Schilling, K., and Viebahn, C. 2002. An efficient *in situ* hybridization protocol for multiple tissue sections and probes on miniaturized slides. *Dev. Genes Evol.* 212:403–406.

Zhao, J., Wu, R., Au, A., Marquez, A., Yu, Y., and Shi, Z. 2002. Determination of HER-2 gene amplification by chromogenic *in situ* hybridization (CISH) in archival breast carcinoma. *Mod. Pathol.* 15: 657–665.

2

Comparison of Chromogenic *in situ* Hybridization, Fluorescence *in situ* Hybridization, and Immunohistochemistry

Rina Wu and Zuo-rong Shi

Introduction

Immunohistochemistry versus *in situ* Hybridization in Tumor Pathology

Immunohistochemistry (IHC) is *in situ*–based technology that evaluates the protein level of cells in tissue section. It has been the predominant method for many diagnostic applications in breast cancer. The IHC technology can be carried out using frozen, fixed, or archival specimens. It is relatively easy to perform, has rapid turnaround time, and is relatively inexpensive. However, IHC assays have demonstrated wide variation in sensitivity and specificity in formalin-fixed, paraffin-embedded tissue sections. The effect of specimen processing, including tissue fixation, has a substantial effect on IHC staining. Antigen retrieval and immunodetection systems currently used have improved sensitivity of IHC staining (Shi *et al.,* 1991, 2001). The overall problem with any IHC method is that a negative result may not be reliable, because there is no positive internal control to check whether the epitope of interest is not destroyed or masked during processing (Ross

et al., 1999). Positive results of IHC may not be reliable due to the lack of a universal scoring system or interobserver difference in interpretation of IHC results (Allred *et al.,* 2000). For example, a recent study of 394 invasive primary breast cancers demonstrated that if there is no standardization of the procedure, HER-2 evaluation by IHC is not a reproducible technique (Gancberg *et al.,* 2002).

In situ hybridization (ISH) technique allows visualization of cellular DNA or RNA in a tissue section, single cell, or chromosome preparations. ISH using DNA probes is often applied in tumor pathology to quantify the number of gene copies of cells in a tissue section. Unlike IHC, which is a relatively subjective test for protein expression, ISH objectively demonstrates gene status in tumors. The utility of ISH is apparent in pathology since the development of nonradioactive procedures, which yield significantly better cellular resolution than that provided by radiolabeled assays. Fluorescent ISH (FISH), one of the nonradioactive procedures, can detect gene amplification, deletion, chromosome translocation, and chromosome aneuploidy in frozen or formalin-fixed tumor tissue sections.

Handbook of Immunohistochemistry and *in situ* Hybridization of Human Carcinomas, Volume 1: Molecular Genetics; Lung and Breast Carcinomas

13

Copyright © 2004 by Elsevier (USA)
All rights reserved.

Chromogenic ISH (CISH), another nonradioactive procedure, has the same ability as FISH to detect genetic alterations in tumor's pathology. The CISH procedure was developed three years ago in the Zymed Laboratories, Inc. (South San Francisco, CA) and has gained more attention recently. The difference between CISH and FISH is the detection step. Instead of fluorescence detection and fluorescence microscopy used in FISH, CISH is based on chromogenic detection, usually with horseradish peroxidase (HRP)-substrate and/or an alkaline phosphatase (AP)-substrate reaction, using bright-field microscopy.

In breast cancer, overexpression of specific oncogenes is often observed. Protein overexpression may be the result of gene amplification (e.g., overexpression of HER-2 protein generally results from HER-2 gene amplification [Pauletti *et al.*, 1996; Slamon *et al.*, 1989]). Clinically, sensitivity and specificity of HER-2 testing with FISH or CISH has been shown to be superior to those provided by IHC in fixed tissue sections. Patients who were positive by FISH but negative by IHC had a worse survival than those who had HER-2 overexpression but absence of gene amplification (Pauletti *et al.*, 2000). Therefore, HER-2 amplification might provide more meaningful prognostic information than HER-2 overexpression in breast cancer patients, although consensus on this view is lacking.

Protein overexpression may also be related to tumor proliferation rate (e.g., in primary breast cancer, expression of topoisomerase type II alpha [TopoIIα], a key enzyme in DNA replication, is proliferation-dependent and associated with TopoIIα gene amplification as well as chromosome 17 aneuploidy [Depowski *et al.*, 2000]). In TopoIIα testing, clinical studies showed that TopoIIα gene amplification is a more specific predictor than TopoIIα expression determined by IHC for clinical response to TopoIIα inhibitor, a widely used chemotherapy drug in metastatic breast cancer (Coon *et al.*, 2002).

Therefore, based on the reasons already mentioned, although protein is the final functional molecule in tumorigenesis and/or tumor progression, identifying gene status along with its protein expression may give more insight into tumor pathology.

General Procedure of Chromogenic *in situ* Hybridization (CISH) in Formalin-Fixed and Paraffin-Embedded (FFPE) Tissue Section or Bone Marrow/Blood Smear

MATERIALS

1. 4–5-μm thick formalin-fixed, paraffin-embedded sections mounted on Superfrost/Plus microscope slides, or fresh bone marrow/or blood smear on Superfrost/Plus microscope slides.

2. Xylene.

3. 100% ethanol.

4. Deionized water (dH_2O).

5. Spot-Light Tissue Pretreatment Kit (Zymed) (not required for CISH in bone marrow/blood smear).

6. 70%, 85%, 95%, and 100% ethanol at room temperature.

7. Spot-Light DNA probe (Zymed).

8. Coverslips or CISH UnderCover Slips (Zymed).

9. [optional] 5-ml syringe, rubber cement, and 18G$\frac{1}{2}$ inch needle (when humidity slide chamber or CISH UnderCover Slips is not used).

10. 0.5 × SSC (saline-sodium citrate buffer, 0.015-M sodium citrate and 0.15-M NaCl).

11. 3% H_2O_2 in absolute methanol.

12. 1 × PBS.

13. 1 × PBS/Tween 20 (0.025%) (PBS/T).

14. A CISH Polymer Detection Kit, or CISH Centromere Detection Kit, or CISH Translocation Detection Kit, or CISH Bone Marrow/Blood Smear Detection Kit (Zymed).

15. 0.45-μm filter and 5-ml syringe (need for CISH Translocation Detection Kit and CISH Bone Marrow/Blood Smear Detection Kit).

16. Mayer's hematoxylin.

17. 60°C oven.

18. Timer.

19. Hood.

20. Hot plate, a 500-ml glass beaker and aluminum foil; or pressure cooker with pressure indicator (Decloaking Chamber, Biocare Medical, Walnut Creek, CA); or microwave with temperature probe.

21. Coplin jar.

22. Slide rack.

23. Slide incubation chamber.

24. Tissue paper.

25. Polymerase chain reaction (PCR) machine with slide block; or heating block with digital temperature display, humidity slide chamber, and 37°C incubator.

26. P20, and P1000 pipettes.

27. Pipette tips.

28. Waterbath.

29. 37°C incubator.

30. Microscope with 10X, 20X, and 40X dry objectives, and 100X oil lens.

PROCEDURES

I. Slide Pretreatment Before Hybridization
  A. For FFPE Tissue Section
    1. Bake slides at 60°C for 2 hrs to overnight in an oven.

2. Equilibrate Tissue Heat Pretreatment Solution and pepsin (Spot-Light) Tissue Pretreatment Kit (Zymed), to room temperature.
3. Deparaffinize the slides in 2 changes of xylene for 5 min each.
4. Clear xylene from the slides in 3 changes of 100% ethanol for 3 min each.
5. Clear ethanol from the slides in 3 changes of deonized water (dH_2O) for 2 min each.

Note: If next step cannot proceed immediately, air-dry slides after 3 changes of ethanol instead of using dH_2O.

6. Heat the slides for 15 min in CISH Tissue Heat Pretreatment Solution at 100°C (at least ≥ 98°C).
   This may be achieved by one of the following methods:
   a. When using hot plate and glass beaker,
      i. Place slides in a slide rack.
      ii. Heat CISH Tissue Heat Pretreatment Solution in a beaker until it is boiled. Cover the beaker with aluminum foil.
      iii. Place slides in the slide rack in the boiling solution and continue boiling for 15 min. Cover the beaker with aluminum foil.
   b. When using the pressure cooker with pressure indicator,
      i. Place slides in a plastic Coplin jar containing CISH Tissue Heat Pretreatment Solution and screw cap loosely.
      ii. Put steam nozzle weight on the pressure cooker.
      iii. Set timer for 10 min according to the instruction.
      iv. Open the pressure cooker when the pressure is reduced to the level that it can be opened safely according to the instructions.
   c. When using microwave with temperature probe,
      i. Set the temperature at 93°C (199°F).
      ii. Place slides in a plastic Coplin jar containing CISH Tissue Heat Pretreatment Solution and screw cap loosely.
      iii. Place temperature probe in another plastic Coplin jar containing tap water (uncapped).
      iv. Set timer for 15 min after the probe temperature in the uncapped jar has reached 93°C (199°F).
      When the uncapped Coplin jar with temperature probe reaches 93°C (199°F), the Coplin jar containing the slides has

reached above 98°C, since it is capped to prevent the buffer evaporation.
7. Transfer slides immediately to dH_2O at room temperature. Wash slides in 3 changes of dH_2O, for 2 min each.
8. Blot excess dH_2O from the slides. Place slides in the slide incubation chamber.
9. Depending on the size of each section, add 2–10 drops (100–500 µl) of pepsin to each section, and incubate for 10 min at room temperature.
10. Transfer slides to dH_2O and wash them in 3 changes of dH_2O for 2 min each.
11. Dehydrate slides through 70%, 85%, 95%, and twice 100% ethanol for 2 min each.
12. Air-dry slides for 20 min.
13. Proceed denaturation and hybridization steps.

B. For Bone Marrow or Blood Smear
1. Air-dry the fresh bone marrow or blood smear slide at room temperature for 1–2 hrs.
2. Prepare 1X fixative solution (1% paraformaldehyde) with dH_2O using 10X fixative provided as reagent G in CISH Bone Marrow/Blood Smear Detection Kit.
3. Place the slides in 1X fixative solution and soak for 15 min at room temperature.
4. Transfer slides to dH_2O. Wash slides in 3 changes of dH_2O for 2 min each.
5. Dehydrate slides through 70%, 85%, 95%, and twice 100% ethanol for 2 min each.
6. Air-dry slides for 20 min.
7. Proceed denaturation and hybridization step.

II. Denaturation and Hybridization
A. Add 13–15 µl of Ready-To-Use Spot-Light probe to the center of a 22 × 22 mm coverslip, or 17–20 µl of probe to the center of 24 × 32 mm coverslip.
B. Place coverslip, probe side down, to the appropriate area of the tissue sample on slide.
C. If PCR machine is used for denaturation and hybridization, seal coverslip to prevent evaporation during incubation, place the slides in the slide block, denature the probe and DNA in tissue section at 95°C for 5 min, and then incubate overnight the slides at 37°C.

The slides can be sealed by one of the following methods:
1. Using a 5-ml syringe containing rubber cement and topped with an 18G½ inch needle, carefully apply a thin layer of rubber cement to the edges of the coverslip, slightly overlapping onto the slide. Allow rubber cement to dry (~10 min).
2. Using UnderCover Slips, peel tape/coverslip off paper backing, add the probe to the center

of the coverslip, and stick to the appropriate area of the slide, covering tissue sample. Do not press the edges of the coverslip, which may cause probe leaking out.

D. If heating block is used, set the heating block at 95°C, place the slides on the heating block for 5 min, then transfer the slides to a humidity chamber prewarmed at 37°C and incubate overnight at 37°C.

III. Stringent Washing

A. Prepare two Coplin jars each containing 40-ml 0.5X SSC (saline-sodium citrate), one at room temperature, the other heated to 75°C in a water-bath according to the number of slides per Coplin jar. For more than two slides, increase temperature 1°C per two slides. For example, if six slides are used, set temperature at 77°C. Do not exceed 80°C.

B. Remove rubber cement and coverlsip. Briefly rinse the slides or place the slides in the jar containing 0.5X SSC at room temperature and soak for 1–5 min.

C. Immerse the slides in the jar containing 0.5X SSC at 75–80°C for 5 min.

D. Transfer the slides immediately to dH$_2$O at room temperature. Wash slides in 3 changes of dH$_2$O for 2 min each.

IV. Chromogenic Detection

Note: All steps are carried out at room temperature.

A. Detection of digoxigenin- (DIG)-labeled Spot-Light DNA Probe

1. Equilibrate CISH Polymer Detection Kit to room temperature.

2. Immerse the slides in 3% H$_2$O$_2$ in absolute methanol for 10 min.

3. Wash slides in 3 changes of 1X phosphate buffer saline (PBS) for 2 min each.

4. Drain excess PBS from slides with tissue paper. Place slides in the slide incubation chamber. Do not let slides dry.

5. Add 2–5 drops (100–250 μl) of CAS-Block (CISH Polymer Detection Kit) to the section, and incubate for 10 min.

6. Tap off CAS-Block from slides with tissue paper. Do not rinse and do not let slides dry.

7. Add 2–5 drops (100–250 μl) of mouse-anti-DIG antibody (CISH Polymer Detection Kit) to the section, and incubate for 30 min.

8. Wash slides in 3 changes of PBS/T for 2 min each.

9. Drain excess PBS/T from slides with tissue paper. Do not let slides dry.

10. Add 2–5 drops (100–250 μl) of polymerized HRP-anti-mouse antibody (CISH Polymer Detection Kit) to the section, and incubate for 30 min.

11. Wash slides in 3 changes of PBS/T for 2 min each.

12. Prepare DAB working solution: add 20X substrate buffer, 20X diaminobenzidine (DAB), and 20X H$_2$O$_2$, each 1 drop (50 μl) in 1 ml dH$_2$O (CISH Polymer Detection Kit). Mix.

13. Drain excess PBS/T from slides with tissue paper. Do not let slides dry.

14. Add 2–5 drops (100–250 μl) of DAB working solution to the section, and incubate for 30 min.

15. Wash the slides with running tap water for 2 min.

16. Proceed to counterstaining and coverslipping step.

B. Detection of Biotin-labeled Spot-Light DNA Centromeric Probe

1. Equilibrate CISH Centromere Detection Kit to room temperature.

2. Immerse the slides in 3% H$_2$O$_2$ in absolute methanol for 10 min.

3. Wash slides in 3 changes of 1X PBS for 2 min each.

4. Drain excess PBS from slides with tissue paper. Place slides in the slide incubation chamber. Do not let slides dry.

5. Add 2–5 drops (100–250 μl) of CAS-Block (CISH Centromere Detection Kit) to the section, and incubate for 10 min.

6. Tap off CAS-Block from slides with tissue paper. Do not rinse and do not let slides dry.

7. Add 2–5 drops (100–250 μl) of HRP-streptavidin (CISH Centromere Detection Kit) to the section, and incubate for 30 min.

8. Wash slides in 3 changes of PBS/T for 2 min each.

9. Prepare DAB working solution: add 20X substrate buffer, 20X DAB, and 20X H$_2$O$_2$, each 1 drop (50 μl) in 1-ml dH$_2$O (CISH Centromere Detection Kit). Mix.

10. Drain excess PBS/T from slides with tissue paper. Do not let slides dry.

11. Add 2–5 drops (100–250 μl) of DAB working solution to the section, and incubate for 30 min.

12. Wash the slides with running tap water for 2 min.

13. Proceed to counterstaining and coverslipping step.

C. Detection of DIG- and Biotin-labeled Spot-Light DNA Translocation Probe Pair or Chromosome X/Y Cocktail

1. Equilibrate CISH Translocation or Bone Marrow/ Blood Smear Detection Kit to room temperature.
2. Immerse the slides in 3% H_2O in absolute methanol for 10 min.
3. Wash slides in 3 changes of 1X PBS for 2 min each.
4. Drain excess PBS from slides with tissue paper. Place slides in the slide incubation chamber. Do not let slides dry.
5. Add 2–5 drops (100–250 μl) of CAS-Block (CISH Translocation or Bone Marrow/Blood Smear Detection Kit) to the section, and incubate for 10 min.
6. Tap off CAS-Block from slides with tissue paper. Do not rinse. Do not let slides dry.
7. Add equal volume of HRP-streptavidin and AP-anti-DIG, total 2–5 drops (100–250 μl), (CISH Translocation or Bone Marrow/Blood Smear Detection Kit) to the section, and incubate for 30 min.
8. Wash slides in 3 changes of PBS/T for 2 min each.
9. Prepare DAB working solution: Add 20X substrate buffer, 20X DAB, and 20X H_2O_2, each 1 drop (50 μl) in 1-ml dH_2O (CISH Translocation or Bone Marrow/Blood Smear Detection Kit) Mix.
10. Drain excess PBS/T from slides with tissue paper. Do not let slides dry.
11. Add 2–5 drops (100–250 μl) of DAB working solution to the section, and incubate for 30 min.
12. Prepare Fast Red working solution: dissolve one tablet of Fast Red in 5 ml Fast Red buffer (CISH Translocation or Bone Marrow/Blood Smear Detection Kit), then filter the Fast Red working solution using a 0.45-μm filter.
13. Wash the slides with running tap water for 2 min.
14. Drain excess water from slides with tissue paper. Do not let slides dry.
15. Add 2–5 drops (100–250 μl) of Fast Red working solution to the section, and incubate for 10 min.
16. Tap off Fast Red from slides with tissue paper. Do not rinse and do not let slides dry.
17. Repeat preceding two steps two more times.
18. Wash the slides with running tap water for 2 min.
19. Proceed to counterstaining and coverslipping step.

V. Counterstaining and Coverslipping
   A. Counterstain tissue section with hematoxylin for 10 sec to 1 min. (Counterstaining time is dependent on tissue used. Dark counterstaining is not recommended, as it may obscure positive staining signals).
   B. Wash with running tap water.
   C. For the slides without Fast Red staining:
      1. Dehydrate the slides in graded ethanol series (70%, 85%, 95%, 100%, and 100%) for 2 min each.
      2. Immerse the slides in xylene twice for 2 min each.
      3. Coverslip, using Histomount Mounting Solution (CISH Kits).
   D. For the slides with Fast Red staining:
      1. Drain excess water from slides with tissue paper.
      2. Apply ClearMount Solution (CISH Translocation or Bone Marrow/Blood Smear Detection Kit) on the section or smear.
      3. Incubate the slides at 37°C for 2 hrs to solidify ClearMount Solution.
      4. Transfer the slides to room temperature. Immerse the slides in xylene briefly and coverslip using Histomount mounting Solution (CISH Kits).

Bright-Field Microscopy

Examine CISH signal and tissue morphology simultaneously, using bright-field microscopy. Use 40X dry objective to evaluate single-color CISH. Use 100X oil lens to evaluate dual-color CISH, except chromosome X/Y cocktail probe, which can be evaluated using 40X dry objective.

Interpretation of CISH Results

Criteria for interpretation of HER-2 CISH results in 4–5 μm FFPE tissue section has been suggested in several studies. Tumors with HER-2 gene amplification appear typically as large peroxidase-positive intranuclear gene copy clusters, or as numerous individual peroxidase-positive small signals, or as a mixture of clusters and individual gene copies (Figure 1A and 1B). Tumors with HER-2 low amplification typically show 6–10 gene copies per nucleus in >50% of tumor cells, which can be confirmed by CISH using chr.17cen probe on an adjacent section. Tumors with normal HER-2 gene typically show 1–2 dots per nucleus in >50% of tumor cells (Figure 1C). Tumors with chromosome 17 polysomy typically show 3–5 HER-2 gene copies per nucleus in >50% of tumor cells.

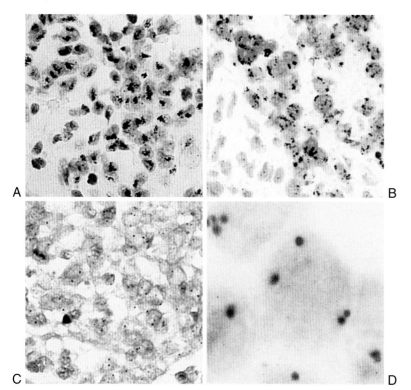

Figure 1 **A–C:** Example of CISH staining in breast cancer using STP HER-2 DNA probe. A typical HER-2 amplification appears either as large peroxidase-positive intranuclear gene copy clusters (**A:** 40X), as numerous individual peroxidase-positive small signals, or as a mixture of clusters and individual gene copies. (**B:** 40X). Tumors with normal HER-2 gene typically showed 1–2 dots per nucleus (**C:** 40X). **D:** CISH using SPT C-Myc translocation probe pair in Raji cell blocks, 100X.

Criteria for interpretation of TopoIIα CISH results in 4–5 μm thick formalin-fixed, paraffin-embedded tissue section is as follows. Tumors with TopoIIα gene amplification typically appear as large peroxidase-positive intranuclear gene copy clusters, or as numerous individual peroxidase-positive small signals, or as a mixture of clusters and individual gene copies. Tumors with low ToppIIα amplification typically show 6–10 gene copies per nucleus in >50% of tumor cells, which can be confirmed by CISH using chromosome 17 centromeric (chr.17cen) probe on an adjacent section. Tumors without TopoIIα amplification or deletion show 1–5 copies of TopoIIα gene copies per nucleus in >50% of tumor cells, and TopoIIα gene copy number is equal to chr.17cen copy number detected in adjacent section. Tumors with TopoIIα deletion show the TopoIIα gene copy number is less than chr.17cen copy number in >50% of tumor cells.

Criteria for interpretation of gene translocation detected by CISH using split-apart probe pair in 4–5-μm thick formalin-fixed, paraffin-embedded tissue section: The cells without the chromosome translocation were displayed when both probe pairs of brown and red dots (CISH) were in juxtaposition. The cells carrying the chromosome translocation display one pair of probes, brown and red dots (CISH) separated (Figure 1D). The tissues with the chromosome

translocation were defined when there were more than 10 cells having one pair of probe, brown and red dots, in juxtaposition and another pair of probes separated in distance greater than or equal to one-half the diameter of that nucleus. Due to cutting artifacts, seven hybridization patterns on the tissue section with the C-Myc translocation could be seen (Figure 2). For bone marrow/blood smear specimens, cutting artifacts do not exist. Cells without the translocation in the region tested by CISH using a split-apart probe pair should show both pair of dots in juxtaposition (Figure 3), while cells with the translocation should display one

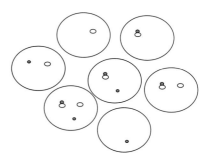

Figure 2 Schematic illustration of C-Myc translocation detected by dual-color CISH using split-apart C-Myc translocation probe pair on 4–5 μm tissue section.

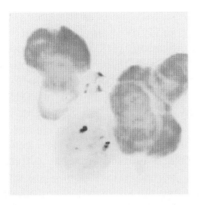

Figure 3 CISH using SPT BCR/ABL translocation probe pairs in a normal bone marrow smear, 100X.

pair of dots in juxtaposition, and the other pair split-apart. In the case when centromeric material of the gene in the translocated chromosome is deleted, like that which occurs in 10% of chronic myeloid leukemia cases, cells should show one pair of dots in juxtaposition, and the red dot (centromeric material of the gene) of slip-apart probe pair is missing.

I. Troubleshooting
  A. The heat pretreatment and pepsin digestion steps are the most critical for successful and optimized CISH performance.
    1. The specimen must be boiled or heated above 98°C for 15 min in CISH Tissue Heat Pretreatment Solution. Boiling the specimen in a baker on a hot plate yields the best results.
    2. The proper incubation time for pepsin digestion is crucial. Depending on the tissue type and fixation method, different incubation times (+/− 5 min at room temperature) may be required. For most standard FFPE breast tissues, 10-min pepsin digestion at room temperature will produce the best CISH results. Evaluation of the pepsin digestion can be performed when the CISH procedure is completed. If nuclei are not counterstained and there is an absence of CISH signal or a weak CISH signal in the nuclei, it is most likely due to a loss of nuclei by excessive digestion. If nuclei are counterstained and there is an absence of CISH signal or a weak CISH signal, it may be due to insufficient pepsin digestion, or the CISH signal may be obscured by overcounterstaining. Although enzyme digestion for 10 min at room temperature is recommended, digestion at 37°C for 3 min may also be used.

  B. Strong counterstaining is not recommended because the hematoxylin can obscure the CISH signal.
  C. When dual-color CISH is performed, DAB must be added before Fast Red. Fast Red is alcohol-soluble. Do not dehydrate tissue section after counterstaining and before coverslipping.
  D. When Fast Red chromogen is used for CISH, Clearmount (provided in the dual-color CISH Detection Kit) should be used for mounting the slide. After slides are counterstained and washed with water, they should be mounted with Clearmount immediately. The tissue section should be wet when Clearmount is added.
  E. When evaluating dual-color CISH, a 100X objective is required since it can be difficult to distinguish red (Fast Red) from brown (DAB) colors at lower objectives. It is important to adjust the condenser focus, condenser aperture diaphragm ring, field diaphragm, and amount of light for 100X oil lens when changing objective from lower power to 100X. Normally, the condenser scale is indicated on the 100X objective. The investigator needs to determine the size of the aperture as well as the amount of light that gives the best resolution for red and brown colors.

Chromogenic *in situ* Hybridization versus Fluorescence *in situ* Hybridization

Although FISH used for formalin-fixed tissue sections is a valid alternative to IHC in many cases, it is not used frequently in noncytogenetic diagnostic pathology. One reason for this is that evaluation of FISH requires an expensive fluorescence microscope equipped with high-quality 60X or 100X oil immersion objectives and multiband pass fluorescene filters, which are not universally available to surgical pathologists responsible for histological diagnosis. Another reason is that fluorescence signals can fade within several weeks. In addition, FISH results need to be recorded with an expensive CCD camera, making analysis of FISH data time-consuming.

Because of the limitation of the use of FISH in fixed-tissue section previously mentioned, the use of CISH assay enables routine evaluation of gene amplification/deletion, aneuploidy, and chromosomal translocation in archived tumor specimens using light microscopy. This approach is greatly beneficial in surgical pathology. In the past, ISH using DIG, biotin, or FITC-labeled DNA probes, synthetic oligonucleotide probes, and cRNA probes followed by chromogenic (enzymatic) detection has been mainly applied for identification of virus infection, detection of immunoglobulin mRNA, and chromosome centromere in the tissue sections. These studies are carried out due to the abundance of targeted sequences to be detected (Takarabe *et al.*, 2001).

Application of CISH in tumor pathology using conventional ISH probe to detect gene amplification/ deletion and chromosome translocation was limited due to the low ratio of signal to background staining. To overcome this limitation, DNA probes generated by Subtracted Probe Technology (SPT) were developed (Zymed). With SPT technology, repetitive DNA sequences (e.g., *Alu* and LINE elements, which may consist of up to more than 40% of template and cause nonspecific hybridization) are quantitatively removed from the probe. Therefore, the final probe is very specific and the need for blocking nonspecific hybridization with Cot-1 DNA used in traditional FISH probes is avoided. The tissue preparation and probe hybridization in CISH are similar to FISH procedures. They differ in the method of probe detection, in which CISH requires IHC-like chromogen detection, whereas FISH is based on fluorescence detection. According to our and other studies, CISH using the SPT probe and an optimized sensitive chromogenic detection system (Zymed) gives much brighter and crispier signals than an ordinary FISH probe (Tanner *et al.*, 2000; Zhao *et al.*, 2002).

When comparing CISH with FISH, both have distinctive advantages and disadvantages. One substantial advantage of the FISH approach is that probe detection can be accomplished by direct detection. In contrast, CISH generally requires at least two additional steps beyond probe hybridization, prior to nuclear counter-staining. For example, a probe can be labeled with FITC or rhodamine and then detected directly using the FISH method. The same probe may be labeled with DIG or biotin and then detected by sequential incubations with mouse-anti-DIG and goat-anti-mouse-HRP/ diaminobenzidine (DAB) reaction or streptavidin-HRP/DAB, respectively. Another advantage of FISH is its easy-to-engage multicolor detection, whereas CISH is restricted to single or dual colors due to the limitation of current chromogenic detection technology.

On the other hand, CISH has several advantages that are particularly relevant in formalin-fixed, paraffin-embedded sections. First, interpretation of CISH is performed using a standard light microscope and permits simultaneous evaluation of gene copies and tissue morphology on the same slide. Large regions of the tissue section can be scanned rapidly in CISH using conventional counterstain such as hematoxylin. Morphological details and CISH signals are readily apparent using 10X, 20X, or 40X objective lens.

Fluorescence signals and counterstaining, on the other hand, are generally only appreciated at substantially higher magnification (60X or 100X objectives), and tissue morphology with FISH is not optimal for formalin-fixed, paraffin-embedded specimens. Therefore, review of hematoxylin-and-eosin (H&E)-stained sections may be necessary. This can be a particular problem in distinguishing invasive breast cancer and breast carcinoma *in situ*, where HER-2 gene amplification or protein overexpression has different clinical significance. Second, CISH signals are not subject to rapid fading and the slides can therefore be archived, whereas FISH slides must be stored at 4°C or lower and are subject to quenching of the fluorescent signal. Third, with FISH, cellular and extracellular proteins can contribute to a dull, generalized, autofluorescence that often obscures FISH signals in paraffin sections. These limitations make the use of FISH for fixed tissue sections cumbersome for routine work (Gupta *et al.*, 2003)

HER-2 Status in Breast Carcinoma Using CISH, FISH, and IHC

Introduction

The HER-2 oncogene is a member of the epidermal growth factor receptor family, and its amplification is known to be one of the most common genetic alterations associated with human breast cancer and some other cancers (Sahin *et al.*, 2000). Identification of HER-2 status is important for determining prognosis of patients who have invasive breast cancer as well as selecting subgroups with HER-2 overexpression in metastasis for therapy with trastuzumab (Herceptin, Genentech, Inc., South San Francisco, CA) (Shak *et al.*, 1999), a humanized monoclonal antibody to the HER-2 protein. Trastuzumab has been found to be effective only in patients whose tumors show HER-2 gene amplification and/or HER-2 protein overexpression. Therefore, accurate, consistent, and straightforward methods for evaluating HER-2 status have become increasingly important.

Currently, IHC and FISH of formalin-fixed, paraffin-embedded tissues are the two methodologies approved by the Food and Drug Administration for use in HER-2 testing. To accurately identify gene status by bright-field microscopy, and overcome FISH limitations, CISH using the HER-2 DNA probe generated by SPT technology and chr.17cen probes was developed in 1999 (Zymed). Using this particular CISH technique, the SPT HER-2 DNA probe (Zymed) was first reported to evaluate against FISH (PathVision,

Vysis, Downers Grove, IL) and a HER-2 antibody CB11 (Novocastra, Newcastle, U.K.) in 157 breast cancer tissue specimens (Tanner *et al.*, 2000). CISH was performed on fixed sections, whereas FISH was used on imprints of whole tumor nuclei prepared from fresh tissue samples. A chr.17cen probe was used in the FISH assay but not in the CISH assay. For interpretation of CISH results, unaltered HER-2 gene copy number was defined as 1–5 signals per nucleus; low-level amplification was defined as 6–10 signals per nucleus in >50% of cancer cells, or when a small gene cluster was found. Amplification of HER-2 was defined when a large cluster in >50% of carcinoma cells or numerous (>10) separated gene copies were seen. Differences in amplification rates (FISH, 23.6%; CISH, 17.2%) were thought to result in part from differences in preparations (e.g., sectioned versus whole nuclei). The conclusion of this study was that CISH is a useful alternative for detection of HER-2 amplification in formalin-fixed tumor samples, especially for confirming the IHC staining results. This study served to define the criteria for amplification of HER-2 in fixed tissue section by the CISH method.

Subsequently, the study by Kumamoto *et al.* (2001) compared the results of CISH (Zymed) with HercepTest (Dako, Carpinteria, CA) for 18 breast carcinomas. Four of 18 tumors had a definitively positive 3+ IHC score, and 4 of 18 had the problematic 2+ score. Four of these nine cases (3+ and 2+ cases) showed high-level amplification by CISH, and three showed low-level amplification. The criteria used in this study for low-level amplification was modified slightly from that used by Tanner *et al.* (2000), and probe for chr.17cen were not used to address the possibility of chromosome 17 polysomy. Kumamoto *et al.* (2001) concluded that CISH could be a useful adjunct to IHC analysis.

More recently, we studied 62 breast carcinomas using SPT HER-2 DNA probe and CISH assay in formalin-fixed, paraffin-embedded tissue sections. In this study, the DIG-labeled DNA probe is detected first by FISH using FITC-conjugated anti-DIG antibody, and then followed by CISH using HRP-conjugated anti-FITC antibody and a simple IHC-like peroxidase reaction on the same slide. Further, a biotin-labeled chr.17cen probe and CISH centromere detection kit (Zymed) were used as a reference probe to assist in distinguishing HER-2 amplification from chromosomal polysomy. In order to know the chromosome 17 polysomy in the patients studied, chr.17cen probe was also applied to the breast cancer tissue with HER-2 high amplification. The revised standard described by Tanner *et al.* (2000) was used to interpret the HER-2 FISH and CISH results, since chr.17cen probe was

added in our study. To further evaluate and assess the clinical usefulness of CISH using SPT HER-2 probe, three commonly used and commercially available HER-2 antibodies: monoclonal TAB250 (Zymed) and CB11 (Novocastra), which recognize the external and internal domains of HER-2 oncoprotein, respectively, and A0485 (Dako), a rabbit polyclonal antibody that recognizes internal domains, were compared in this study with CISH. The scoring proposed by the HercepTest (Dako) was used to interpret the immunoreactivity of these three antibodies.

RESULTS

FISH and CISH: Criteria for successful FISH and CISH analysis included identification of at least one copy of the HER-2 gene per nucleus in most cancer cells and appropriate pepsin digestion, as evidenced by well-preserved cell morphology. FISH and CISH were successful in 85% of cases in the first run. At the end, FISH and CISH for HER-2 were successful in all 62 cases. These protocols were also successful for all cases using chr.17cen (samples with more than diploid number of HER-2 CISH signals, total 34 cases). Pepsin digestion was adjusted to 10 sec for one case and 3 min with 10X pepsin for another case. The FISH signals were seen by 40X objective lens and were easily identified by 100X objective lens. The CISH staining results were clearly seen using a 40X objective lens in tissue sections counterstained with hematoxylin. Tumors with HER-2 gene amplification appeared typically as large peroxidase-positive intranuclear gene copy clusters, as numerous individual peroxidase-positive small signals, or as a mixture of clusters and individual gene copies (Figures 1A and 1B). Tumors with normal HER-2 gene typically showed 1–2 dots per nucleus in >50% of tumor cells (Figure 1C). Tumors with HER-2 low amplification typically showed 6–10 gene copies per nucleus in >50% of tumor cells, which was confirmed by CISH using chr.17cen probe on an adjacent section. Tumors with polysomy showed typically 3–5 HER-2 gene copies per nucleus in >50% of tumor cells, which was also confirmed by CISH using chr.17cen probe on an adjacent section.

Correlation of FISH with CISH: The concordance between FISH and CISH is 100%.

IHC: HER-2 protein high overexpression (3+) was clearly seen (using a 10X or 20X objective lens) in tissue sections counterstained with hematoxylin. HER-2 protein weak overexpression (2+) was clearly distinguishable using a 20X or 40X objective lens. HER-2 protein no overexpression (0, 1+) was distinguished using a 20X and 40X objective lens.

In the 62 breast cancers studied, the prevalence of HER-2 protein overexpression was 19% by TAB250, 23% by CB11, and 36% by A0485. There was a 97% concordance between the results obtained with the TAB250 and CB11 antibodies. The concordance between the results obtained with TAB250 and A0485 or CB11 and A0485 is 86%.

Correlation of CISH with IHC: The results of TAB250 and CISH were discordant in two cases: one case had HER-2 gene amplification but was negative by TAB250; the other case did not have HER-2 gene amplification but was 2+ by TAB250. The results of CB11 and CISH were discordant in four cases: one case had HER-2 gene amplification but was negative for CB11; the other three cases did not have HER-2 gene amplification but were 2+ by CB11. Two of these three cases had chromosome 17 polysomy. The results of A0485 and CISH were discordant in 10 cases: all the cases had overexpression of the protein but were absent of gene amplification. Eight of these 10 cases had chromosome 17 polysomy.

With one exception, 11 of 12 cases with HER-2 gene amplification detected by CISH definitely showed positive staining by all three antibodies. With two exceptions, all the cases with 3+ positive for A0485 had HER-2 amplification. Six of eight (75%) cases with 2+ positive by A0485 were negative by TAB250, as well as by CB11, and were chromosome 17 polysomy and lacked HER-2 amplification. The concordance between the results of TAB250 and CISH, CB11 and CISH, and A0485 and CISH were 97%, 94%, and 84%, respectively.

DISCUSSION

For routinely used HER-2 testing, accuracy, and ease, it is essential to start trastuzumab (Herceptin) therapy for metastatic breast cancer. In our 62 cases, results of FISH and CISH have 100% agreement. The current version of HER-2 FISH is based on single-color detection, which is similar to the INFORM FISH test (Ventana, Tucson, AZ). The HER-2 CISH is a continuation of FISH with IHC-like reaction. The complete agreement of FISH and CISH results demonstrated that HER-2 CISH is as sensitive as HER-2 FISH. The current study, together with a previously published paper (Tanner et al., 2000), demonstrated the utility of CISH, a novel methodology, in the determination of HER-2 amplification in fixed tumor samples.

The current HER-2 CISH is based on single-color detection. To avoid misinterpretation of gene amplification with polysomy, we used the chr.17cen probe on the adjacent section of the tumor tissues, which have 3–10 HER-2 gene copies per nucleus in >50% of tumor cells. The increased HER-2 copy number was evaluated both as absolute numbers of HER-2 or as clusters of dots per nucleus, and as HER-2 copy numbers relative to chr.17cen copy numbers. By using the HER-2 probe and chr.17cen probe on serial sections, simultaneous detection of oncogene and chromosome copy numbers was ascertained. To evaluate HER-2 CISH, the criteria suggested by Tanner et al. (2000) were used. We defined amplification as clusters of dots or more than six gene copies per nucleus in >50% of invasive tumor cells, polysomy as 3–5 gene copies per nucleus in >50% of invasive tumor cells, and normal as HER-2 CISH shows 1–2 gene copies per nucleus in >50% of invasive tumor cells. The chr.17cen probe was applied to the tumors that had 3–10 gene copies per nucleus, as well as those tumors with HER-2 gene high amplification, in order to know the frequency of chromosome 17 polysomy. It was only informative for one case, which had 4–7 HER-2 gene copies per nucleus, since the use of the centromeric probe confirmed that it was due to chromosome 17 polysomy. Although chr.17cen probe confirmed that the tumor with 6–10 HER-2 gene copies per nucleus was due to HER-2 gene low amplification and the tumors (total 12) with 3–5 HER-2 gene copies per nucleus were due to chromosome 17 polysomy, chr.17cen probe did not provide more information for these tumors. As suggested by Tubbs et al. (2001), we suggest that the chr.17cen probe is only necessary when there are 6–10 HER-2 gene copies per nucleus in >50% of cells to confirm HER-2 gene low amplification, which is relatively rare (one out of 62 cases, 1.6%). We do not think that 2–4 HER-2 gene copies per nucleus can even be due to gene amplification. In the invasive breast cancers we studied, chromosome 17 polysomy was frequent (31%), which is similar to the results reported by others (Gancberg et al., 2000) and even more frequent (67%) in the invasive breast cancers with HER-2 amplification.

In the current study, correlations between HER-2 CISH with 2+ and 3+ cases of TAB250, CB11, and 3+ cases of A0485 on invasive breast cancers using fixed tissue were generally good, and are similar to that seen in prior comparison studies of HER-2 gene amplification with HER-2 protein expression (Jacobs et al., 1999). Our study showed that TAB250 had the lowest misclassification rate compared with the CISH result. All tumors except one with HER-2 gene amplification demonstrated overexpression of HER-2 protein with the three antibodies used in this study. This outlying case may present a small, undetermined percentage that amplified HER-2 without protein overexpression. The high rate of concordant results obtained by the two monoclonal antibodies was shown in this study. The monoclonal antibodies detected HER-2 overexpression in the absence of gene amplification in 2% (1 of

50, TAB250) to 6% (3 of 50, CB11) of the cases. These results are in agreement with a published range of other studies and could represent single-copy overexpression at the transcriptional level and/or beyond. Alternatively, it may be due to gene amplification that is below the detection level of FISH and CISH. Compared with the monoclonal antibodies, the high level of overexpression of 36% (22 of 62) was detected by the polyclonal antibody A0485, which is in higher range of 10–34% HER-2 overexpression reported in the literature (Ross et al., 1999). Although the HercepTest kit was not used in the current study, the same polyclonal DAKO antibody provided in the kit with standard heat-induced epitope retrieval was applied. Among the tumors with HER-2 overexpression detected by A0485, 36% (8 of 22) did not have HER-2 gene amplification and were negative by TAB250 and CB11. These discrepancies were mainly represented by tumors detected as 2+ by A0485. This observation is in agreement with the recent findings of Ridolfi et al. (2000) and Lebeau et al. (2001). When closely investigating the tumors stained 2+ by A0485, 6 of 8 (75%) cases had chromosome 17 polysomy but without HER-2 gene amplification. The current study demonstrated again that A0485 has higher sensitivity than other commercially available antibodies, and it might detect lower levels of protein expression as suggested by others. It has also been reported by several investigators in independent laboratory facilities that the HercepTest or A0485 results in significant false-positive cases. As discussed by Pauletti (2000), in a subgroup analysis of the phase III clinical trials that led to approval of trastuzumab (Benz, 1998), patients with 2+ IHC score did not seem to benefit significantly from trastuzumab therapy (Check, 1999), and the beneficial treatment effects were largely limited to patients with the highest levels of HER-2 protein overexpression (3+). Some investigators suggested a revised scoring system by subtracting the level of staining of nonneoplastic epithelium to improve HercepTest specificity (Jacobs et al., 1999). However, many tumor specimens in our series did not contain nonneoplastic epithelium; therefore, we could not apply this scheme in the current study. A major advantage of CISH and IHC as an in situ-based technology is their ability to combine molecular diagnosis with histological examination of the tissue. A combinatorial strategy using IHC and CISH should provide comprehensive and valuable information on both HER-2 protein concentrations and gene amplification to help clinicians make crucial management decisions.

In summary, we have documented a complete agreement between CISH and FISH, and a high level of concordance between CISH and IHC, especially between CISH and monoclonal antibody TAB250 or CB11, in the evaluation of HER-2 status on invasive breast carcinoma. Sixty-seven percent of the tumors that were scored as 2+ by polyclonal antibody A0485 were negative by TAB250 and CB11 and did not show HER-2 gene amplification but showed chromosome 17 polysomy. In agreement with the study by Tanner et al. (2000), the current study confirmed the validity of CISH methodology. If such consistency can be reproduced in other laboratories, CISH could prove to be truly valuable in clinical practice.

A recent study by Dandachi et al. (2002) compared results of CISH, the HercepTest, and PCR in 173 formalin-fixed, paraffin-embedded breast tumor specimens. Discrepant cases (between CISH and IHC) were then assayed using the FISH PathVision assay (Vysis, Downer's Grove, IL); concordance between CISH and FISH was 100% in 38 cases studied. The authors concluded that CISH was a promising alternative to IHC or FISH in the routine diagnostic setting. Finally, the most recent studies by Gupta et al. (2003) compared Zymed CISH assay with PathVision FISH in 31 cases of infiltrating breast carcinoma. To the end, authors support Dandachi et al. (2002) and our findings on CISH (Zhao et al., 2002) and concluded that CISH, by itself, is a practical, economical, and definitive means for routinely assessing HER-2 in breast tumors.

Application of Chromogenic *in situ* Hybridization (CISH) in Tumor Pathology

Three years after SPT HER-2 probe and CISH assay were introduced to the market, they have gained more and more recognition, as was evident in the published papers and studies presented in the United States and Canadian Academy of Pathology Annual Meeting in 2003.

Application of CISH assay has not only shown its accuracy in HER-2 testing, as previously discussed, but also in other testings, such as the detection of TopoIIα gene amplification or deletion in archival breast carcinoma (Park et al., 2003), detection of epidermal growth factor receptor (EGFR) gene amplification in a series of archival human cancers (Quezado et al., 2003), detection of C-Myc gene translocation in archival lymphomas (Chu et al., 2003), detection of PLAG1 translocation lipoblastoma (Gisselsson et al., 2001), and detection of chromosome aneuploidy in archival human tumors (Ronchetti et al., 2003).

TopoIIα is a key enzyme in DNA replication and is a target for TopoIIα inhibitors such as anthracyclines, commonly used chemotherapeutic drugs in breast cancer. TopoIIα gene is located closely to HER-2 gene on the long arm of chromosome 17, and its amplification or deletion was found in breast cancers with HER-2 gene amplification (Järvinen *et al.*, 1999). Recent clinical studies showed that TopoIIα amplification is a more specific predictor than TopoIIα expression by IHC and HER-2 gene amplification for clinical response to TopoIIα inhibitor in breast cancer (Isola *et al.*, 2000). In our study (Wu *et al.*, 2001), using Zymed SPT HER-2 and TopoIIα DNA probes and CISH Polymer Detection KIT, TopoIIα gene amplification and deletion were identified in 20 (44%) and 3 (7%) of 44 breast cancers with HER-2 gene amplification, respectively, TopoIIα amplification or deletion was not found in 60 breast cancers without HER-2 gene amplification. This finding is in agreement with findings by Coon *et al.* (2002). TopoIIα deletion was further confirmed by dual-color CISH using cocktail of TopoIIα and chr.17cen probes (Figure 4). This study, together with another recently published study (Park *et al.*, 2003) further demonstrated that CISH is a reliable, practical, and economical approach to detect gene amplification and deletion in archival tumor tissue.

Alteration of EGFR at the gene and/or protein level has been found in diverse carcinoma types. EGFR gene amplification is a common feature of primary malignant gliomas and is associated with increased tumor grading, worse clinical outcome, and resistance to radio- and chemotherapy. It is still unclear whether EGFR gene amplification or protein overexpression is an independent prognostic marker for cancer patients. Phase III clinical trials are ongoing now for a number of EGFR-targeted therapies, yet, it is still under investigation on which testing method is more specific to select the patients who have the best chance to respond. Therefore, an easy, reliable assay to detect EGFR gene amplification will facilitate future clinical

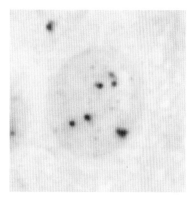

Figure 4 Dual-color CISH evaluation of TopoIIα gene (DAB) deletion with chromosome 17 centromere (Fast Red) in breast cancer known to have HER-2 gene amplification, 100X.

studies. In the study done by Marquez *et al.* (2003), CISH using SPT EGFR DNA probe (Zymed) was successfully performed in 245 of 248 (99%) cases of archival malignant tumor specimens. Glioblastoma multiforme and Astrocytoma Grade III had the highest rate of EGFR gene amplification (33% and 30%, respectively), which was similar to what was reported by Quezado *et al.* (2003) and others. Esophageal carcinoma also had a relatively high rate of EFGR gene amplification (18%). Comparatively, EGFR gene amplification occurred relatively low in lung and head and neck carcinomas (8%), and was rarely detected in colon cancer (1.8%) and breast cancers (0 in 44). The EGFR expression was determined by IHC using monoclonal antibody 31G7 (Zymed). High expression of EGFR was found in head and neck carcinoma (88%), esophageal carinoma (73%), gliomas (58%), lung cancer (37%), colon cancer (16%), and breast cancer (12%). The study showed that tumors with amplification of the EGFR gene were associated with high expression of the EGFR protein, but the high protein expression did not necessarily result from the gene amplification. Similar findings were also reported by others (Ekstrand *et al.*, 1991; Quezado *et al.*, 2003). This is especially important because EGFR high expression is present in certain normal tissues and it is hard to distinguish high expression from normal high expression in these tissues. Therefore, EGFR amplification detected by the reliable assay such as CISH might provide more meaningful predictive value than EGFR high expression in the tumors.

In the study conducted by Chu *et al.* (2003), dual-color CISH developed by Zymed was applied in archival lymphoma tissue sections using SPT C-Myc translocation probe pair (Zymed). This probe pair is located on chromosome 8q24 and has a gap of more than 500 kb on either side of the C-Myc gene; therefore, it can detect all the C-Myc translocations reported so far. Dual-color CISH procedure is similar to single-color CISH. After FFPE tissue sections were heat-treated and enzyme-digested, tissue sections were denatured and hybridized with the C-Myc probe pair labeled with biotin and DIG. After stringent washing, the probe pair was detected with HRP-streptavidin/DAB and alkaline phosphotase-anti-DIG/Fast Red chromogenic reactions. The CISH signal and morphology of the tissues were easily visualized simultaneously with a 100X objective under a bright-field microscope in hematoxylin-counterstained sections. Cells lacking t(8;n)(q24;n) translocation showed a probe pair of brown and red colored dots in juxtaposition, whereas cells carrying t(8;n)(q24;n) translocation displayed a split of the C-Myc probe pair (Figure 1D). C-Myc translocations were screened in 98 lymphomas. Ninety of 98

(92%) cases had successful CISH results. C-Myc translocation was detected in 30 of 33 (91%) cases of Burkitt lymphoma, 3 of 18 cases (16.6%) of diffused large B-cell lymphoma, 0 of 34 other types of lymphomas. Conclusion of this study was that CISH assay provides a feasible and sensitive tool for routine screening of the C-Myc translocations in archival lymphoma tissues.

Acknowledgment

The authors would like to thank Mr. Bill Arnold for preparation of images used in this chapter.

References

Allred, D.C., and Swanson, P.E. 2000. Testing for erbB-2 by immunohistochemistry in breast cancer. *Am. J. Clin. Pathol. 113:*171–175.

Benz, C.C. 1998. The molecular biology of HER-2. Presented at *21st San Antonio Breast Cancer Symposium*, Mini-Symposium I, San Antonio, Texas, December 12–15.

Check, W. 1999. More than one way to look for HER-2. *CAP Today 13:*48–52.

Chu, W., Wu, R., Fisher, S., Wei, M.Q., Zhao, J., Marquez, A., Shi, Z., Aguilera, N., and Abbondanzo, S.L. 2003. Detection of C-Myc oncogene translocations and protein expression in Burkitt and other non-Burkitt Lymphomas. *U.S. and Canadian Academy of Pathology Annual Meeting, Abstr. 1043.*

Coon, J.S., Marcus, E., Cupta-Bart, S., Seelig, S., Jacobson, K., Chen, S., Renta, V., Fronda, G., and Preisler, H. 2002. Amplification and over-expression of topoisomerase $II\alpha$ predict response to anthracycline-based therapy in locally advanced breast cancer. *Clin. Cancer Res. 8:* 1061–1067.

Dandachi, N., Dietze, O., and Hauser-Kronberger, C. 2002. Chromogenic *in situ* hybridization: A novel approach to a practical and sensitive method for the detection of HER-2 oncogene in archival human breast carcinoma. *Lab. Invest. 82:* 1007–1014.

Depowski, P., Rosenthal, S.I., Brien, T.P., Stylos, S., Johnson, R.L., and Ross, J.S. 2000. Topoisomerase $II\alpha$ expression in breast cancer: Correlation with outcome variables. *Mod. Pathol. 13:*542–547.

Ekstrand, A.J., James, C.D., Cavenee, W.K., Seliger, B., Pettersson, R.F., and Collins, V.P. 1991. Genes for epidermal growth factor receptor, transforming growth factor alpha, and epidermal growth factor and their expression in human gliomas in vivo. *Cancer Res. 51:*2164–2172.

Gancberg, D., Järvinen, T., Di Leo, A., Rouas, G., Cardoso, F., Paesmans, M., Verhest, A., Piccart, M.J., Isola, J., and Larsimont, D. 2002. Evaluation of HER-2/neu protein expression in breast cancer by immunohistochemistry: An interlaboratory study assessing the reproducibility of HER-2/neu testing. *Breast Cancer Res. Treat. 74:* 113–120.

Gancberg, D., Lespagnard, L., Rouas, G., Paesmans, M., Piccart, M., Di Leo, A., Nogaret, J.M., Hertens, D., Verhest, A., and Larsimont, D. 2000. Sensitivity of HER-2/neu antibodies in archival tissue samples of invasive breast carcinomas. *Am. J. Clin. Pathol. 113:*675–682.

Gisselsson, D., Hibbard, M.K., Dal Cin, P., Sciot, R., Hsi, B., Kozakewich, H.P., and Fletcher, J., 2001. PLAG1 alterations in liposarcoma. *Am. J. Pathol. 159:*955–962.

Gupta, D., Middleton, L.P., Whitaker, M.J., and Abrams, J. 2003. Comparison of fluorescence and chromogenic *in situ* hybridization for detection of HER-2/neu oncogene in breast cancer. *Am. J. Clin. Pathol. 119:*381–387.

Isola, J.J., Tanner, M., Holli, K., and Joensuu, H. 2000. Amplification of topoisomerase $II\alpha$ is a strong predictor of response to epirubicin-based chemotherapy in HER-2/neu positive metastatic breast cancer. *Breast Cancer Res. Treat. 68:*31.

Jacobs, T.W., Gown, A.M., Yaziji, H., Barnes, M.J., and Schnitt, S.J. 1999. Comparison of fluorescence *in situ* hybridization and immunohistochemistry for the evaluation of HER-2/neu in breast cancer. *J. Clin. Oncol. 17:*1974–1982.

Jacobs, T.W., Gown, A.M., Yaziji, H., Barnes, M.J., and Schnitt, S.J. 1999. Specificity of HercepTest in determining HER-2/neu status of breast cancers using the United States Food and Drug Administration-approved scoring system. *J. Clin. Oncol. 17:*1983–1987.

Järvinen, T.H., Tanner, M., Barlund, M., Borg, A., and Isola, J. 1999. Characterization of topoisomerase $II\alpha$ gene amplification and deletion in breast cancer. *Genes Chromosomes Cancer 26:*142–150.

Kumamoto, H., Sadano, H., Takahiro, T., Suzuki, T., Moriya, T., and Ichinohasama, R. 2001. Chromogenic *in situ* hybridization analysis of HER-2/neu status in breast carcinoma: Application in screening of patients for trastuzumab (Herceptin) therapy. *Pathol. Int. 51:*579–584.

Lebeau, A., Deimling, D., Kalz, C., Sendelhofert, A., Iff, A., Luthardt, B., Untch, M., and Löhrs, U. 2001. HER-2/neu analysis in archival tissue samples of human breast cancer: Comparison of immunohistochemistry and fluorescence *in situ* hybridization. *J. Clin. Oncol. 19:*354–363.

Marquez, A., Wu, R., Zhao, J., Yu, Y., Tao, J., and Shi, Z. 2003. Detection of EGFR gene amplification and protein expression by chromosome *in situ* hybridization (CISH) and immunohistochemistry (IHC) in a series of archival human cancers. *U.S. and Canadian Academy of Pathology Annual Meeting Abstr. 1486.*

Park, K., Kim, J., Lim, S., and Han, S. 2003. Topoisomerase II-alpha (topoII) and HER-2 amplification in breast cancers and response to preoperative doxorubicin chemotherapy. *Eur. J. Cancer. 39:*631–634.

Pauletti, G., Dandekar, S., Rong, H.M., Ramos, L., Peng, H.J., Seshadri, R., and Slamon, D.J. 2000. Assessment of methods for tissue-based detection of the HER-2/neu alteration in human breast cancer: A direct comparison of fluorescence *in situ* hybridization and immunohistochemistry. *J. Clin. Oncol. 18:*3651–3664.

Pauletti, G., Godolphin, W., Press, M.F., and Slamon, D.J. 1996. Detection and quantitation of HER-2/neu gene amplification in human breast cancer archival material using fluorescence *in situ* hybridization. *Oncogene 13:*63–72.

Quezado, M.M., Ronchetti, R.D., Harris, C., Perez, J.L., Tavora, F., Patrocinio, R., and Ghatak, N. 2003. Correlation of EGFR gene amplification and overexpression by chromogenic *in situ* hybridization (CISH), and immunohistochemistry in high-grade gliomas. *U.S. and Canadian Academy of Pathology Annual Meeting. Abstr. 1334.*

Ridolfi, R.L., Jamehdor, M.R., and Arber, J.M. 2000. HER-2/neu testing in breast carcinoma: A combined immunohistochemical and fluorescence *in situ* hybridization approach. *Mod. Pathol. 13:*866–873.

Ronchetti, R.D., Torres-Cabala, C.A., Chian-Garcia, C.A., Eyler, R., Linehan, W.M., and Merino, M.J. 2003. Detection of chromosome 7 and 17 aneuploidy in type 1 papillary renal cell carcinoma by chromogenic *in situ* hybridization (CISH).

A comparison with other renal cell tumors. *U.S. and Canadian Academy of Pathology Annual Meeting. Abstr. 761.*

Ross, J.S., and Fletcher, J.A. 1999. HER-2/neu (c-erb-B2) gene and protein in breast cancer. *Am. J. Clin. Pathol. 112:*S53–S67.

Sahin, A.A. 2000. Biologic and clinical significance of HER-2/neu (c-erbB2) in breast cancer. *Adv. Anat. Pathol. 7:*158–166.

Shak, S. 1999. Overview of the trastuzumab (Herceptin) anti-HER-2 monoclonal antibody clinical program in HER-2-overexpressing metastatic breast cancer. Herceptin multinational investigator study group. *Cancer Res. 6:*71–77.

Shi, S.R., Key, M.E., and Kalra, K.L. 1991. Antigen retrieval in formalin-fixed, paraffin-embedded tissues: An enhancement method for immunohistochemical staining based on microwave oven heating of tissue sections. *J. Histochem. Cytochem. 39:* 741–748.

Shi, S.R., Cote, R.J., and Taylor, C.R. 2001. Antigen retrieval techniques: Current perspectives. *J. Histochem. Cytochem. 49:*931–937.

Slamon, D.J., Godolphin, W., Jones, L.A., Holt, J.A., Wong, S.G., Keith, D.E., Levin, W.J., Stuart, S.G., Udove, J., Ullrich, A., and Press, M.F. 1989. Studies of the HER-2/neu proto-oncogene in human breast and ovarian cancer. *Science 244:*707–712.

Takarabe, T., Tsuda, H., Okada, S., Fukutomi, T., and Hirohashi, S. 2001. Detection of numerical alterations of chromosome 1 in cytopathological specimens of breast tumors by chromogen *in situ* hybridization. *Pathol. Int. 51:*786–791.

Tanner, M., Gancberg, D., Di Leo, A., Larsimont, D., Rouas, G., Piccart, M.J., and Isola, J. 2000. Chromogenic *in situ* hybridization: A practical alternative for fluorescence *in situ* hybridization to detect HER-2/neu oncogene amplification in archival breast cancer samples. *Am. J. Pathol. 157:*1467–1472.

Tubbs, R., Pettay, J., Roche, P., Stoler, M.H., Jenkins, R.B., and Grogan, T.M. 2001. Discrepancies in clinical laboratory testing of eligibility for trastuzumab therapy: Apparent immunohistochemical false-positives do not get the message. *J. Clin. Oncol. 19:* 2714–2721.

Wu, R., Zhao, J., Marquez, A., Au, A., and Shi, Z. 2001. Determination of topoisomerase IIα (TopoIIα) gene amplification and deletion by chromogenic *in situ* hybridization (CISH) in archival breast carcinoma. *J. Mol. Diagn. 3:*207 (S11).

Zhao, J., Wu, R., Au, A., Marquez, A., Yu, Y., and Shi, Z. 2002. Determination of HER-2 gene amplification by chromogenic *in situ* hybridization (CISH) in archival breast carcinoma. *Mod. Pathol. 15*(6):657–665.

3

Target and Signal Amplification to Increase the Sensitivity of *in situ* Hybridization

Xiang Qian and Ricardo V. Lloyd

Introduction

In situ hybridization (ISH) is a very powerful molecular tool used in research and diagnosis. ISH has greatly advanced the study of gene structure and expression at the level of individual cells in complex tissues (Jin *et al.*, 1997). ISH has contributed substantially to the diagnosis and understanding of viral and neoplastic diseases (McNicol *et al.*, 1997) and has provided invaluable insights into hormone regulation, storage, and secretion (DeLellis, 1994). In addition to morphologic identification of gene targets within cells, ISH allows for quantification of observations, for example, with respect to tumor burden or viral load (Lisby, 1999) and identification of infectious organisms, such as bacteria and fungi, in the routine fixed tissues (Hayden *et al.*, 2001a, b; Schonhuber *et al.*, 1997). However, wide applicability is limited by relatively low sensitivity, especially with non-radioactive reporters (Jin *et al.*, 2001; Yang *et al.*, 1999). In most conventional nonisotopic ISH protocol, the threshold level for messenger RNA (mRNA) is ~ 20 copies per cell (Hougaard *et al.*, 1997).

ISH applies the technology of nucleic acid hybridization to the single-cell level, and in combination with immunocytochemistry permits the maintenance of morphology and the identification of cellular markers and allows for the localization of sequences to specific cells within populations, such as tissues (Speel, 1999b) and blood samples (Muratori *et al.*, 1996). However, chromogenic ISH is limited primarily to the detection of nongenomic material (e.g., RNA), reiterated genes, or multiple genomes, since the limits of detection in most situations are several copies of the target nucleic acid per cell. Due to the presence of higher copy number, hybridization for RNA is considerably more sensitive than for DNA detection. In addition, other factors that affect the sensitivity of the technique for RNA targets are the strandedness of the target molecule and lack of a complementary sequence proximal to the target sequences (Mitsuhashi, 1996).

The most promising methods for target sequence amplification before ISH include *in situ* polymerase chain reaction (PCR) (Long, 1998; Nuovo *et al.*, 1991) and primed *in situ* labeling (PRINS) (Coullin *et al.*, 2002; Wilkens *et al.*, 1997). Methods for signal amplification after ISH include the catalyzed reporter deposition (Evans *et al.*, 2002; Hopman *et al.*, 1998). The detection of intracellular mRNAs (Kriegsmann *et al.*, 2001)

Handbook of Immunohistochemistry and *in situ* Hybridization of Human Carcinomas, Volume 1: Molecular Genetics; Lung and Breast Carcinomas

27

Copyright © 2004 by Elsevier (USA)
All rights reserved.

or viral RNAs (Murakami *et al.*, 2001), as well as endogenous and foreign DNAs (Speel *et al.*, 2000), by these methods has many potential applications when the level of gene expression or the number of DNA copies is below that detectable by conventional ISH. Conventional self-sustained sequence replication (3SR) protocols (Mueller *et al.*, 1997), which thus far have not found broad acceptance in diagnostic laboratories, along with 3SR *in situ*, may be replaced by the newer alternative strategies. These strategies are becoming available for the intracellular analysis of nucleic acids. Signal amplification methods by catalyzed reporter deposition using biotinylated tyramine may influence significantly future directions in the development of ISH. The combined efforts to target sequence and signal amplification are the key strategies for the future to make ISH an easy-to-perform, fast, highly sensitive, and accurate method, which would increase its use in diagnostic laboratories (Qian *et al.*, 2003).

In this chapter, we will review the principles, methods, applications, and limitations of *in situ* polymerase chain reaction as examples of target amplification methods and catalyzed reporter deposition (CARD) using biotinylated tyramine as an approach to signal amplification during ISH.

In Situ (RT-) PCR for Target Amplification

MATERIALS

In Situ PCR—Required Materials and Solutions

This is a list of all materials needed for *in situ* PCR and *in situ* reverse transcription PCR (RT-PCR). For all solutions diethyl pyrocarbonate (DEPC) treated water is used.

1. Equipment for making paraffin-embedded sections, cell culture, and cytospin.

2. Thermo Hybaid OmniSlide Thermal Cycler (Fisher Scientific Middlesex, U.K.).

3. Charged micro-slides (Cat. No. 48311-703, VWR International, Inc., West Chester, PA).

4. Cytoseal XYL mounting medium (xylene-based, Richard-Allan Scientific, Kalamazoo, MI).

5. Ethanol (100%, 95%, 75%).

6. Xylene.

7. Hydrophobic pen (PAP pen, Lipshaw Corp., Pittsburgh, PA).

8. Parafilm (American National Can, Chicago, IL).

9. Coplin jars.

10. DEPC RNase-free H_2O made with diethyl pyrocarbonate-treated deionized, distilled water.

11. Phosphate buffer saline (PBS): 10 mM PBS includes 138 mM NaCl, 2.7 mM KCl (pH 7.4) (Sigma, St. Louis, MO). Store at room temperature.

12. 4% Paraformaldehyde (Sigma) in phosphate buffer saline (PBS) (pH 7.4). Prepared freshly by dissolving 12 g of paraformaldehyde in 300 ml of PBS by heating to near boiling point, then cooling rapidly to room temperature. Filter through a Whatman paper prior to use.

13. 5–25 µg of Proteinase K in PBS (Roche Applied Science, Indianapolis, IN).

14. DNase solution: 10X reaction buffer (400 mM Tris-HCl, 100 mM $MgSO_4$, 10 mM $CaCl_2$, pH 8.0), RQ1 RNase-Free DNase (1 U/µl) (Cat. No. M6101, Promega, Madison, WI).

15. 20X standard saline citrate buffer (SSC): Weigh 175.3 g sodium chloride and 88.2 g sodium citrate, add DEPC H_2O to make 1000 ml, and adjust to pH 7.2.

16. Buffer A: Weigh 58.44 g sodium chloride, 15.76 g Tris-HCl, and 407 mg magnesium chloride; add 5-ml 10% Tween-20 and distilled water to make 1000 ml, and adjust to pH 7.5.

17. Buffer C: Weigh 5.84 g sodium chloride, 12.11 g Tris-Base, and 5.1 g magnesium chloride, add distilled water to 1000 ml and adjust to pH 9.5.

18. Blocking buffer: 1% normal swine serum (Vector, Burligame, CA) in buffer A with 0.3% Triton X-100 (Roche). Prepare fresh.

19. 4-Nitroblue tetrazolium chloride (NBT) and 5-bromo-4-chloro-3-indolyl-phosphate (BCIP) were purchased from Roche Applied Science (Cat. No. 1681451). Add 200 µl of NBT/BCIP stock solution and 80 µl Levamisole (Vector SP-5000) to 10 ml buffer C. Prepare prior to use and keep out of direct light.

20. Terminal deoxynucleotidyl transferase (TdT) (15–30 U/µl, store at −20°C; Cat. No. M1871, Promega, Madison, WI).

21. dATP (dilute 100 pM in DEPC H_2O) (Cat. No. 1051440, Roche Applied Science).

22. Digoxigenin 11-dUTP (Cat. No. 1558706, Roche Applied Science).

23. 3.0 M sodium acetate: Weigh 24.6 g sodium acetate, add DEPC H_2O to 100 ml and adjust to pH 6.0 using acetic acid (store at 4°C).

24. Glycogen (20 mg/ml, aliquotted at −20°C; Cat. No. 901393, Roche Applied Science).

25. RNase Inhibitor (RNase Block), (Stratagene).

26. StrataScript reverse transcriptase (Stratagene).

27. Monoclonal anti-digoxigenin-alkaline phosphatase (AP) conjugated antibody (Roche Applied Science).

28. Acetylation solution: add 1.2 ml acetic anhydride to 200 ml of 0.1 M triethanolamine.

29. 10X PCR reaction buffer without $MgCl_2$: 500 mM KCl, 100 mM Tris-HCl (pH 9.0 at 25°C) and 1% TritonX-100, and separate 25 mM $MgCl_2$ solution (Promega).

30. Taq DNA polymerase (5 U/µl, Promega).

31. Reverse-transcription reaction mixture include 50 mM Tris-HCl (pH 8.3), 75 mM KCl, 3 mM $MgCl_2$, 1 mM of each dNTP, 500 ng antisense primer, and 50 U of StrataScript reverse-transcriptase in a 50-µl reaction volume. Prepare fresh.

32. PCR mix containing 10 mM Tris-HCl, 50 mM KCl, 1.5 mM $MgCl_2$, 1 mM DTT, 2 mM of each of the deoxynucleotides dATP, dCTP, dGTP, dTTP. 500 ng sense and antisense primers, 10 U Taq DNA polymerase in a 100-µl reaction volume.

33. Oligonucleotides used for localization of human prolactin (PRL) and epidermal growth factor receptor (EGFR) by *in situ* RT-PCR. PRL: antisense primer 5' TGG GAA TCC CTG CGC AGG CA 3', sense primer 5' CCT GAA GAC AAG GAA CAA GCC 3' and hybridization probe 5' CCT TCG AGA AGC CGC TTG TTT TGT TCC TC 3'. EGFR: antisense primer: 5' AAT ATT CTT GCT GGA TGC GTT TCT GTA 3', sense primer 5' TTT CGA TAC CCA GGA CCA AGG CAC AGC AGG 3' and hybridization probe 5' GTG GGT CTA GAA GCT AAT GCG GGC ATG GCT 3'.

34. The GH_3 and A431 cell lines (human epidermoid carcinoma) were obtained from the American Type Culture Collection (Rockville, MD).

35. αT_3 cell line, a rat pituitary cell line that produces α-subunit but not PRL, was provided by Dr. P. Mellon, University of California, San Diego, CA.

36. All cell lines were grown and maintained separately in Dulbecco's modified Eagle's medium (DMEM) supplemented with 15% horse serum, 2.5% fetal calf serum, 1 µg/ml insulin, and 1% antibiotics (100 U/ml penicillin, 100 µg/ml streptomycin, and 0.25 µg/ml fungizone) (Gibco BRL, Grand island, NY). Cells were then harvested by 0.25% trypsin (Gibco BRL) for cytospin preparation (Jin *et al.*, 1995).

METHODS

in situ PCR and *in situ* Reverse Transcription (RT)-PCR

1. Tissue Section and Cell Cytospin Preparation

(1) Paraffin-embedded tissue: Routinely formalin-fixed paraffin-embedded tissue sections can be quite successful. Cut 4–5 µm paraffin-embedded sections on ⊕ charged slides (Cat. No. 48311-703, VWR International, Inc., West Chester, PA) or silane-coated slides and melt tissue in 60°C oven for 30 min. Deparaffinize sections in xylene for 10 min and dehydrate in alcohol series, 100% for 2 min, two changes and 95% alcohol for 2 min, two changes. Air-dry slides at room temperature.

(2) Cytospin: The culture cells were harvested. Make cells cytospin on ⊕ charged slides and fix in 4% paraformaldehyde (pH 7.4) for 20 min, then wash slides in 2X standard saline citrate (SSC), and dehydrate in alcohol. Air-dry and store at −70°C.

(3) Cryosections: Cut sections as thin as possible (~ 5–8 µm, fix slides in 4% paraformaldehyde (pH 7.4) for 20 min or in ethanol: acetic acid in a 3:1 ratio for 15 min. Rinse slides in 1X PBS for 2 min twice and dehydrate with 100% ethanol. It is possible to use cryosections for *in situ* (RT-) PCR; however, the morphology of the tissue following the amplification process is generally not as good as with paraffin sections. It is very important to use tissues that were frozen in liquid nitrogen or placed on dry ice immediately after they were harvested before autolysis began to take place.

2. Proteinase K Treatment

Proteinase K treatment will break down some of the cross-linking that develops secondary to formalin fixation, thereby enabling primers and polymerase to get to the target sequence. At the same time, the remaining cross-links will retain the PCR product, and if tissue is treated for too long diffusion of the PCR product will occur. The concentration and length of the proteinase K treatment also depends on fixation and type of tissue. Tissues fixed in 4% paraformaldehyde or 10% neutral formalin are highly cross-linked and therefore require longer proteinase K treatment than tissues fixed in alcohol. Make a ring around sections using a Pap pen.

(1) Treat the paraffin-embedded sections with 5–25 µg/ml proteinase K in PBS solution at 37°C for 10–20 min, depending on the tissue and the fixative used.

(2) Treat cytospin cells with 1 µg/ml of proteinase K in PBS solution at 37°C for 15 min. Wash slides in 2X SSC, two changes.

3. Acetylation

Nonspecific binding of probes or primers to positively charged amino groups is prevented by acetylation of these residues with acetic anhydride. In many situations this step has no effect on background and is optional.

(1) Place slides in 200 ml 0.1 M triethanolamine (TEA) with 1.2 ml acetic anhydride for 15 min.

(2) Wash in 2X SSC, two changes.

REMARKS

Acetic anhydride is highly unstable. Add acetic anhydride to each change of TEA-acetic anhydride solution immediately before incubation.

4. DNase Treatment
(Only for *in situ* RT-PCR)

DNase treatment is only necessary for *in situ* RT-PCR, and only if it is impossible to use a cDNA specific set of primers. DNase treatment should be avoided whenever possible.

(1) Apply 40 µl of RQl RNase-Free DNase solution (Promega, M6101) per slide.

10X Reaction buffer	4 µl
RQl RNase-Free DNase (1 U/µl)	5 µl
DEPC H$_2$O	31 µl
Total	40 µl

(2) The slides were covered with Parafilm.
(3) Place slides in a humidified chamber at 37°C for minimum 30 min or extend for overnight.
(4) DNase was inactivated by heating (75°C for 5 min) and washing in DEPC H$_2$O twice for 5 min, then in ethanol (100%) for 2 min.

5. Reverse Transcription
(Only for *in situ* RT-PCR)

Reverse transcription is only necessary if *in situ* RT-PCR is performed. First-strand cDNA (RT) was synthesized with StrataScript reverse transcriptase (Stratagene). RT is done after DNase treatment; it is of great importance that the slides are carefully rinsed before reverse transcription. Residual DNase will destroy the DNA synthesized by reverse transcriptase.

10X first strand buffer	5 µl
RNase block	2 µl
100 mM dNTPs	2 µl
Antisense primer (500 ng/µl in stock)	1 µl
StrataScript reverse transcriptase (50 U/µl)	1 µl
DEPC H$_2$O	39 µl
Total	50 µl

The RT reaction mixture (including 50 mM Tris-HCl (pH 8.3), 75 mM KCl, 3 mM MgCl$_2$, 1 mM of each dNTP and 50 U of StrataScript reverse transcriptase in a 50-µl reaction volume) was applied to each slide and covered with Parafilm. After incubation for 2 hr at 42°C in humidified chamber, wash slides in 2X SSC, 1X SSC, 0.5X SSC and DEPC H$_2$O.

Although oligo (dT) primers can be alternatively used to first convert all mRNA populations into cDNA, specific downstream antisense primers may be a first option for generating the gene-specific cDNA. It is advantageous to reverse-transcribe only relatively small fragments of mRNA (<300 bp). Larger fragments may not be completely reverse-transcribed due to the presence of secondary structures or target fragments in formalin-fixed, paraffin-embedded sections. Furthermore, the reverse transcriptase enzymes such as AMVRT and MMLVRT are not very efficient in transcribing large mRNA fragments.

The following additional points should be considered.

(i) The length for both sense and antisense primers should be 18–22 bp.
(ii) At the 3′ ends, primers should contain a GC-type base pairs (e.g., GG, CC, GC, or CG) to facilitate complementary strand formation.
(iii) The preferred GC content of the primers in from 45–55%.
(iv) Try to design primers so they do not form intra- or interstrand base pairs. Furthermore, the 3′ ends should not be complementary to each other or they will form primer dimers.

REMARKS

All reagents for *in situ* RT-PCR should be prepared with RNase-free H$_2$O (i.e., DEPC-treated H$_2$O). In addition, all glassware should be baked in a 95°C oven for overnight before use.

6. PCR Amplification During Thermal Cycling

The *in situ* PCR step was accomplished with a Thermo Hybaid OmniSlide Thermal Cycler. A total volume of 100 µl of PCR reaction mixture was prepared for each slide containing 10 mM Tris-HCl; 50 mM KCl; 1.5 mM MgCl$_2$; 1 mM DTT; 2 mM of each of the deoxynucleotides dATP, dCTP, dGTP, dTTP; 500-ng each sense and antisense primers; 10 U Taq DNA polymerase.

10X PCR buffer	10 µl
MgCl$_2$ (25 mM in stock)	6 µl
Each dNTP (100 mM in stock)	2 µl × 4 = 8 µl
Sense primer (500 ng/µl in stock)	1 µl
Antisense primer (500 ng/µl in stock)	1 µl
Taq DNA polymerase (5 u/µl)	2 µl
DEPC H$_2$O	72 µl
Total	100 µl

For direct detection: add 1.2 µl of dTTP, and 0.8 µl digoxigenin-11-dUTP replaces 2 µl of dTTP.

(1) Samples were assembled without mineral oil or agarose. Place ~ 50–100 µl prewarmed of PCR mixture onto the slide. Cover slide with a glass coverslip

and seal with nail polish to prevent evaporation and reagent loss. While the *in situ* PCR thermocycler was prewarmed to 70°C, the covered slides were placed in the cycler.

(2) The following cycler conditions were used: initial denaturation (94°C for 5 min), 10–60 cycles of denaturation (94°C for 2 min), annealing at 60°C for 2 min, extension (72°C for 2 min). Finally, a terminal elongation step (72°C for 15 min) followed. Annealing temperature for DNA amplification can be chosen according to the following formula: Tm of the primer = 81.5°C + 16.6 (Log M) + 0.41 (G + C)% − 600/n − 0.65 × % formamide, where n = length of primer and M = molarity of the salt in the buffer. Usually, primer annealing is optimal at 2°C above its Tm. Optimal annealing temperature should be carried out first with solution-based PCR. It is important to know the optimal temperature before attempting to conduct *in situ* PCR.

(3) Remove coverslip with a scalpel and scratch out remaining nail polish with a razor blade. Samples were postfixed for 20 sec in 4% paraformaldehyde and washed extensively in 2X SSC and 0.5X SSC at 40°C for 10 min each and then rinse in Buffer A.

7. Detection

Two different detection methods can be used: (1) direct detection where a marker molecule (usually digoxigenin or biotin) is incorporated into the PCR product during *in situ* PCR and subsequently detected by immunohistochemical methods. This method is discouraged due to lack of specificity (Sallstrom *et al.*, 1993). (2) indirect method in which the location of the PCR product is detected by routine ISH. For both methods, three different reporter systems can be used: fluorescein labeling, gold labeling, and enzyme labeling. Direct detection with alkaline phosphatase as the reporter system is described next. This method results in a deep purple precipitate where the PCR products occur.

A. Direct Detection

(1) Make a new ring around sections using a Pap pen.
(2) Block slides in blocking buffer for 10 min.
(3) Apply 100 μl per slide of monoclonal anti-digoxigenin, alkaline phosphatase-conjugated antibody diluted 1:200 in blocking buffer.
(4) Leave in humid chamber at room temperature for 1 hr.
(5) Drain slides and rinse them twice for 15 min in buffer A and buffer C.
(6) Add 100 μl per slide NBT/BCIP solution. Place slides in humid chamber and cover with aluminum foil for 5–60 min.

(7) Monitor development of the purple color and stop by immersing slides into buffer C and rinse slides for 5 min.
(8) Counterstain slides using 1% nuclear Fast Red for 3 min.

B. Indirect Detection

3′-end labeling of internal oligonucleotide probes:

(1) Prepare mixture using RNAse-free reagents and supplies.

5X TdT buffer (Promega: M1871)	8 μl
DEPC H₂O	20.6 μl
dATP (Dilute 100 pM in DEPC H₂O) (Roche: 1051440)	4 μl
Digoxigenin 11-dUTP (1 nM/μl, Roche: 1570013)	5 μl
Oligonucleotide probe (500 ng/μl)	1 μl
TdT (30 U/μl) (Fisher: M1871)	1.4 μl
	40 μl

(2) Incubate at 37°C for 30 min.
(3) Add 160 μl DEPC H₂O.
(4) Add 30 μl 3.0 M sodium acetate (pH 6.0) (store at 4°C).
(5) Add 1 μl glycogen (20 mg/ml, aliquot at −20°C).
(6) Mix well.
(7) Add 600 μl absolute ethanol.
(8) Lightly mix.
(9) Place in −20°C freezer overnight or several hours (rush: −70°C for 3 hr).
(10) Centrifuge at 12,000 rpm for 1 hr (4°C).
(11) Decant supernatant.
(12) Add 300 μl 70% ethanol (cooled to −20°C).
(13) Centrifuge at 12,000 rpm for 30 min (4°C).
(14) Decant supernatant and air-dry by inverting tube for 5 min until semi-dry.
(15) Resuspend in 62-μl DEPC H₂O. (Final probe concentration is 8 ng/μl.)

Routine ISH after PCR amplification:

(1) Prehybridization is performed for 30 min at room temperature with a probe dilute.
(2) After prehybridization, residual prehybridization buffer is removed thoroughly from around the tissue. A 30-μl of digoxigenin-labeled internal oligonucleotide probe between PCR primers (1 ng/μl in probe dilute) is applied to sections.
(3) Slides are covered with a Sigmacote (Sigma)-coated coverglass, heat treated at 95°C for 5 min, and hybridized in a humid environment for 3 hr at 50°C.
(4) Sections are rinsed twice in 2X SSC for 10 min at room temperature, washed in 0.5X SSC at 37°C for 20 min (to remove excess probe).
(5) Rinse twice in blocking buffer for 10 min at room temperature.

(6) The sections are then incubated in a 1:200 dilution of alkaline phosphatase-conjugated anti-digoxigenin Fab fragment, in blocking buffer A, for 30 min at room temperature.

(7) Rinse with buffer A and buffer C.

(8) Sections are subsequently reacted with nitrobluetetrazolium chloride and 5-bromo-4-chloro-3-indolyphosphate, forming an insoluble blue precipitate at the site of reaction.

(9) Sections are then rinsed in buffer C, counterstained with 1% nuclear Fast Red.

8. Mounting

Slides are dehydrated in graded ethanols, cleared in xylene, and "coverslipped" with a xylene-based synthetic mounting medium. Cytoseal XYL Mounting medium is permanent. However, semipermanent mounting can be done in Crystal/Mountor Mowiol, which is a water-soluble mounting medium. The purple precipitate made by the NBT/BCIP method is soluble in many organic solvents and mounting in permanent mounting media like DePeXand. Euparol will result in recrystallization of the precipitate and should not be used.

9. Appropriate Controls

The use of adequate controls is important for *in situ* (RT)-PCR. Failure to provide adequate controls for artifacts may result in significant errors in interpretation of results (Teo *et al.*, 1995). Controls are required to guarantee the quality of tissues, reagents, and technical performance. Fixative type and fixation time, enzyme concentration, and digestion time should be optimized for each tissue. A tabulation of appropriate controls in designing experiments is included in Table 1. Omission of primers in the PCR mixture is the most-suggested control to detect artifacts in direct *in situ* PCR for both DNA and cDNA detection (*in situ* RT-PCR). However, the controls that include use of irrelevant probes for checking hybridization specificity are considered to be very important for indirect PCR method.

RESULTS AND DISCUSSION

Solution-based PCR has proven a valuable tool for basic researchers and clinical scientists (Fredricks *et al.*, 1999). However, one of the major drawbacks of solution-based PCR is that the procedure does not allow the association of amplified signals of a specific gene segment with the histologic cell type(s). Since 1990, several *in situ* PCR protocols with varying modifications have been published (Bagasra *et al.*, 1994; Man *et al.*, 1996). Initially, fixed cells suspended in the PCR reaction mixture were thermal cycled in micro-Eppendorf tubes using conventional block cycler. After PCR, the cells were cytocentrifuged onto glass slides followed by visualization of intracellular PCR products by ISH or immunohistochemistry. *In situ* PCR using cells or tissue sections on glass slides can be performed by two approaches: direct and indirect (Figure 5). When the copy of nucleic acid inside individual cells (< 10 copies per cell) is below the level of routine ISH detection, the *in situ* PCR is the most logical choice due to its exquisite sensitivity. When *in situ* PCR is performed starting with a mRNA template, a reverse-transcription step has to be performed to generate a

Table 1 Controls Required for *In Situ* (RT-) PCR Techniques

Control	Appropriate Design	Interpretation
General	1. Use of known positive and negative tissues	1. Control specificity and sensitivity of methods used
	2. Omission of primary antibody in immunohistochemical detection	2. Control detection step to avoid false negative or background
	3. Collect solution phase from slides after PCR amplification for gel electrophoresis	3. Validate amplification step to avoid false negative
	4. Detect the DNA or RNA of housekeeping gene	4. Control tissue DNA or RNA quality
Method Specific	5. Omission of DNA polymerase	5. Control nonspecific probe hybridization and antibody affinity
	6. Omission of primers	6. Control artifacts related to DNA repair and endogenous priming for direct *in situ* PCR
	7. Use of irrelevant probes for hybridization step	7. Control hybridization probe specificity for indirect *in situ* PCR
	8. Omission of reverse transcriptase and pretreatment tissues with DNase	8. Control mispriming and amplification of endogenous DNA for *in situ* RT-PCR

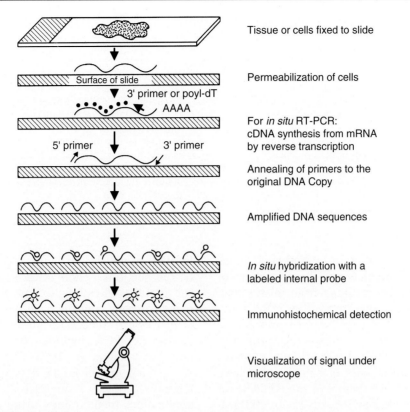

Tissue or cells fixed to slide

Permeabilization of cells

For *in situ* RT-PCR:
cDNA synthesis from mRNA
by reverse transcription

Annealing of primers to the
original DNA Copy

Amplified DNA sequences

In situ hybridization with a
labeled internal probe

Immunohistochemical detection

Visualization of signal under
microscope

Figure 5 Flow diagram shows indirect *in situ* PCR and *in situ* RT-PCR strategies.

cDNA from which all subsequent PCR amplification occurs (Clerico *et al.*, 1999).

In situ PCR methods have been applied to cultured cells (Jin *et al.*, 1995), peripheral blood (Muratori *et al.*, 1996; Nuovo *et al.*, 1994), frozen sections (Zhou *et al.*, 2001) and formalin- and paraffin-embedded section (Martinez *et al.*, 1995; Lau *et al.*, 1994), and to specimens prepared for electron microscopy (Morel *et al.*, 1998). In general, *in situ* PCR requires the critical design of a primer set. The relatively small size (100–500 bp) of the amplification product is frequently chosen for the PCR. Primer pair must be capable of producing an appropriate product in a standard liquid-phase PCR with purified template. Due to the complex physical architecture within the fixed cell, which may impede both the efficiency of the polymerase as well as diffusion and annealing, elongation times are substantially longer than those for liquid conditions. In our hands, the optimum signal is produced with a 15-min elongation time. Reaction components including enzyme, target DNA, and primers must be optimized to varying cofactors (Mg^{2+}, primer-template ratios, etc.) (Nuovo *et al.*, 1993; O'Leary *et al.*, 1996). The importance of the choice of fixation methods is crucial for the detection of sequences generally *in situ* hybridization and are even more so when using

the *in situ* PCR (McAllister *et al.*, 1985; Tournier *et al.*, 1987).

In the direct approach, a labeled nucleotide is incorporated into the PCR products, whereas in the indirect method ISH is performed after *in situ* amplification using labeled oligonucleotide or other probes. However, despite speed and relative simplicity of the direct *in situ* PCR, there are many concerns about its specificity. Nonspecific incorporation of labeled nucleotides and misprimings or DNA repair artifacts may occur, resulting in false-positive results (Teo *et al.*, 1995). It is not possible to interpret *in situ* PCR results without adequate controls. Therefore, many different controls are needed to allow adequate interpretation of the results (Table 1).

In theory, *in situ* PCR techniques should be as reproducible as are conventional PCR techniques. In practice, however, they are associated with many problems (Table 2) such as low amplification efficiency and poor reproducibility. These difficulties are caused by a number of events resulting from PCR amplification *in situ*, such as the diffusion of PCR products (which are bound up with the denaturation steps) from the site of synthesis inside and/or outside the cells, followed by the possible extracellular generation of amplicants. In addition, the final results of direct *in situ* PCR can be

Table 2 Troubleshooting Guide to Improve the *In Situ* (RT)-PCR Technique

Problems	Probable Causes	Recommended Actions
False Positive Signals	• Mispriming • "Diffusion" artifacts • "DNA repair" artifact(s) • Endogenous priming	• Reduce PCR cycles and the concentration Taq enzyme • Hot start technique • Immunoblocking with anti-DNA antibodies • Treat with DNase for *in situ* RT-PCR • Dideoxy blocking
False Negative Signals	• Reduced amplification efficiency • Loss of PCR products	• Increase numbers of cycles • Increase the concentration of Taq enzyme • Check detection protocol • Optimize fixation or permeabilization (Microwave and enzyme digest time, etc.)

influenced by the incorporation of labeled nucleotides into nonspecific PCR products and the generation of nonspecific PCR products resulting from mispriming, fragmented DNA undergoing "repair" by DNA polymerase, "repair" artifacts, or priming of nonspecific DNA or cDNA fragments, "endogenous priming," artifacts (Chou *et al.*, 1992). These artifacts can also be observed in apoptotic cells or samples that have been pretreated with DNAse before *in situ* RT-PCR for mRNA detection (Raatikainen-Ahokas *et al.*, 2001). It is prudent to use many different controls to allow adequate interpretation of *in situ* PCR results and to use indirect *in situ* PCR to increase the specificity of the amplified nucleic acids. *In situ* PCR is considered a rather cumbersome ISH method, in which sample pretreatment consists of fixation and protease digestion in combination with heating (thermal cycling) during nucleic acid amplification by PCR. Moreover, the increase in detection sensitivity compared with conventional ISH is rather limited, even after optimization.

Analyses of PRL and EGFR expression were detected by *in situ* RT-PCR and conventional ISH. There were threefold and sixfold increases in GH3 cells positive for EGFR and prolactin mRNA detected by *in situ* RT-PCR, compared with that obtained with conventional ISH. There was also a significant increase in the intensity of the mRNA signals for PRL and EGFR in frozen tissue sections (Jin *et al.*, 1995). *In situ* PCR has been mainly applied to detect DNA sequences that are not easily detected by conventional ISH (which include human single-copy gene, rearranged cellular genes, and chromosomal translocations) and to map low-copy number of genomic sequences in metaphase chromosomes. The use of *in situ* PCR to detect low-copy number of viral genes, especially HIV (Nuovo *et al.*, 1994) and hepatitis C (Komminoth *et al.*, 1994), has led to significant discoveries about viral infectious diseases. Application of the *in situ* RT-PCR to detect gene expression is still limited primarily to culture cell preparations and frozen sections (Zhou *et al.*, 2001). Relatively few successful applications to paraffin sections with adequate rigorous controls have been reported (Martinez *et al.*, 1995). Successful amplification of mRNA by *in situ* RT-PCR includes hormone, receptor, and oncogenes (Jin *et al.*, 1995; Wesselingh *et al.*, 1997). A one-step *in situ* RT-PCR procedure that compartmentalizes these sequential steps within a single applications methodology using the enzyme rTth has been applied to detect and localize mRNA transcripts for Fas ligand within the immune-privileged placental environment and to provide verification of immunohistochemical localization of gene product (Steele *et al.*, 1998).

Catalyzed Reporter Deposition for Signal Amplification

A. Method for Biotin-Based Amplification System

MATERIALS

1. The CaSki, HeLa 229, and SiHa cell lines (cervical carcinoma) were obtained from the American Type Culture Collection (Rockville, MD). SiHa cells are reported to contain one to two integrated copies of HPV 16; the CaSki cell line, 400–600 integrated copies of HPV 16; and the HeLa 229 cell line, 10–50 integrated copies of HPV 18.

Cell lines were grown and maintained separately in DMEM supplemented with 15% horse serum, 2.5% fetal calf serum, 1 µg/ml insulin, and 1% antibiotics (100 U/ml penicillin, 100 µg/ml streptomycin, and 0.25 µg/ml fungizone) until reaching confluency (Gibco/BRL-Life Technologies, Gaithersburg, MD). Cells were harvested by removing culture medium and then adding 0.25% trypsin (Gibco/BRL) prewarmed to 37°C to the cell culture flask for 5 min to detach the cells. Collected cells were centrifuged for 5 min at 1500 rpm (Beckman model TJ-6, Fullerton, CA) and then fixed for 6 hr at room temperature in 25 ml of 10% neutral buffered formalin (pH 7.2). The cell pellets were carefully removed from the conical tubes, wrapped in lens paper, and placed into plastic tissue-processing cassettes. After tissue processing, three cell pellets were embedded into one paraffin block, sectioned at 5 μm using a rotary microtome, and mounted onto 3-aminopropyltriethoxysilane-coated, positively charged microscope slides. Specimen slides were then dried for 60 min at 75°C and stored at room temperature (Plummer et al., 1998).

2. Biotinylated HPV cDNA probes specific for HPV 6 and 11, HPV 16 and 18, and HPV 31, 33, and 51 were used. Negative control probes were biotinylated pBR322 or pUC18. (Dako).

3. Cocktails of six 30-mer oligonucleotides specific for each HPV 16 and 18 3′-tailed with biotin-11-dUTP (Enzo Diagnostics, Inc.) (Table 3) were used.

4. TBST (Tris-buffered saline/Tween-20). 10X TBST is available from Dako (Cat. No. S3306). Dilute the TBST concentrate 1:10 in deionized water. The diluted solution is stable for one month at 4°C.

5. Tris-buffered saline solution (0.05M Tris-HCl, 0.15M NaCl, pH 7.6). ISH-qualified TBS is available in packets from Dako (Cat. No. S3001).

6. Proteinase K (Roche Applied Science, Indianapolis, IN): Prepare 5–25 μg/ml proteinase K in Tris-buffered saline (TBS) solution (500 μg/ml in stock).

7. Dako GenPoint kit (Cat. No. K0620) provides the major components, which include primary streptavidin/HRP, biotinyl tyramide, secondary streptavidin/HRP, and DAB chromogen concentrate.

8. 10X Target Retrieval Solution (TRS) is available from Dako (Cat. No. S1699). Dilute the TRS concentrate 1:10 in deionized water. The diluted solution is stable for one month at 4°C. Fill a Coplin jar or other suitable container with TRS and heat to 95°C in a water bath (do not boil).

9. Stringent wash solution (Dako, Cat. No. S3500): Dilute the stringent wash concentrate 1:50 in deionized water.

10. DAB is provided Dako GenPoint kit, and was prepared freshly before use by diluting the DAB chromogen concentrate 1:50 in the DAB chromogen diluent.

11. 0.3–3% H_2O_2 in methanol (v/v).

12. Counterstain: Hematoxylin.

Protocol #1

1. Label slides appropriately and place in deparaffinizing rack.

2. Deparaffinize slides in xylene, two changes for 5 min each.

3. Remove xylene with 100% ethanol, two changes for 15 dips each.

4. Hydrate with 95% ethanol, two changes for 15 dips each.

5. Rinse slides in distilled water, three times for 1 min each.

6. Place slides into Coplin jar containing 1X Dako target retrieval solution (preheated in a microwave, 1 min at high power) and microwave for 10 min.

7. Remove slides from microwave (still in Coplin jar) and place at room temperature for 20 min.

8. Rinse as in **step 5**.

9. Treat with 100–300 μl of 25 μg/ml proteinase K for 10 min at room temperature in a Coplin jar on the orbital shaker, prepared fresh by adding 3 ml stock enzyme to 57 ml TBS.

10. Rinse as in **step 5**.

Table 3 Oligonucleotide Probe Cocktail Sequences for HPV 16 and HPV 18

Probe Name	AS Location	Sequence, 5′ TO 3′
HPV16-E6/1	115-86	GTC CTG AAA CAT TGC AGT TCT CTT YYG GTG
HPV16-E6/2	155-126	CTG TGC ATA ACT GTG GTA ACT TTC TGG GTC
HPV16-E6/4	396-367	TCA CAC AAC GGT TTG TTG TAT TGC TGT TCT
HPV16-E6-E7OL	552-523	TGG GTT TCT CTA CGT GTT CTT GAT GAT CTG
HPV16-E7/3	810-781	TAA CAG GTC TTC CAA AGT ACG AAT GTC TAC
HPV16-E7/4	857-828	TAT GGT TTC TGA GAA CAG ATG GGG CAC ACA
HPV18-E6/1	173-144	AGT GTT CAG TTC CGT GCA CAG ATC AGG TAG
HPV18-E6/2	234-205	CCT CTG TAA GTT CCA ATA CTG TCT TGC AAT
HPV18-E6/6	511-482	CCT CTA TAG TGC CCA GCT ATG TTG TGA AAT
HPV18-E6/E7OL	575-546	TTG TGT TTC TCT GCG TCG TTG GAG TCG TTC
HPV18-E7/3	801-772	CTG GCT TCA CAC TTA CAA CAC ATA CAC AAC
HPV18-E7/4	844-815	TGC TCG AAG GTC GTC TGC TGA GCT TTC TAC

11. Quench endogenous peroxidase by incubating in 0.3–3% H_2O_2 for 10 min at room temperature.

12. Rinse as in **step 5**.

13. Additional distilled water rinse for 10 min on a shaker.

14. Air-dry slides.

15. Apply 20–30 μl of each diluted probe to the appropriate slides.

    i. pBR322 negative probe (Dako) is prediluted.

    ii. HPV probes (Enzo) need to be diluted 1/5 with prehybridization buffer. Dilute by adding 20-μl probe to 80-μl prehybridization buffer.

    iii. Or 0.5–2.0 ng/μl oligonucleotide probe cocktails in prehybridization buffer.

16. Coverslip with Sigmacote-coated coverslip and place on a metal slide tray.

17. Denature target DNA and probe together by heating for 5 min in 95°C oven.

18. Place slides in a moisture chamber and hybridize for 2 hr at 37°C.

19. Prepare 1X stringency wash (0.1X SSC + detergent) by adding 2 ml 50X concentrate to 98 ml distilled water. Mix and place in 55°C waterbath for 1 hr prior to stringency washing.

20. After 2 hr hybridization, remove coverglass and place slides in Coplin jar containing 1X TBST.

21. Rinse in TBST twice for 5 min each.

22. 1X stringency wash for 20 min at 55°C (in shaking waterbath).

23. Rinse as in **step 21**.

24. Incubate with primary streptavidin-HRP diluted 1:400 in PBS for 15 min at room temperature. Prepare by adding 2.5 μl primary SA-HRP to 1 ml primary streptavidin diluent (kit) and vortex.

25. Rinse as in **step 21**.

26. Incubate with biotinyl tyramide for 15 min at room temperature. Reagent is prediluted, and use 2–3 drops.

27. Rinse as in **step 21**.

28. Incubate with secondary streptavidin-HRP 15 min at room temperature. Reagent is prediluted, and use 2–3 drops.

29. Rinse as in **step 21**.

30. Apply diluted DAB 100–150 μl per slide and incubate at room temperature for 3–8 min.

31. Rinse as in **step 5**.

32. Counterstain with hematoxylin for 60 sec.

33. Rinse as in **step 5**.

34. Wash slides with running tap water for 1–2 min.

35. Rinse as in **step 5**.

36. Dehydrate with 95% ethanol, two changes for 15 dips each.

37. Dehydrate with 100% ethanol, two changes for 15 dips each.

38. Clear in xylene, two changes for 15 dips each.

39. Coverslip with Cytoseal XYL mounting medium.

B. Method for Biotin-Free Amplification System

The more recently developed dinitrophenyl (DNP) deposition avoids background problems occasionally associated with biotinylated tyramide.

MATERIALS

TSA Plus DNP (HRP or AP) Kit (Cat. No. NEL 746, Perkin Elmer)

1. Dimethyl sulfoxide (DMSO) for molecular biology or HPLC-grade (Sigma, D8418). DMSO may freeze at 4°C. Thaw the stock solution, if necessary, before use.

2. DNP amplification stock solution (provided in kit): DNP (dinitrophenyl) is supplied as a solid (which may have a light yellow color) and needs to be reconstituted. Add 0.3 ml (for NEL746A/NEL747A) or 0.15 ml (for NEL746B/747B) of DMSO for making the DNP stock solution. This stock solution may also exhibit a light yellow color. DNP stock solution, when stored at 4°C, is stable for at least six months.

3. DNP amplification working solution (provided in kit): Before **each** procedure, dilute the DNP stock solution 1:50 using 1X plus amplification diluent to make the DNP working solution. Approximately 100–300 μl of DNP Working Solution is required per slide.

4. TNT wash buffer: 0.1 M Tris-HCl, 0.15 M NaCl (pH 7.5), 0.05% Tween 20.

5. TNB blocking buffer: 0.1 M Tris-HCl, 0.15 M NaCl (pH 7.5), 0.5% blocking reagent (supplied in kit). To dissolve the blocking reagent, heat TNB blocking buffer to 60°C for 1 hr with stirring. TNB blocking buffer may be stored up to 1 month at −20°C.

6. Hydrogen peroxidase: To minimize background, endogenous peroxidase activity can be quenched before the immunostaining protocol. Options for tissue treatment are as follows: 3% H_2O_2 in methanol or PBS for 10 min or 1% H_2O_2 in PBS for 15 min.

7. Anti–fluorescein-HRP (Cat. No. NEF710, Perkin Elmer) diluted 1:250 in TNB buffer.

8. Anti–digoxigenin-HRP (or POD) (Cat. No. 1207733, Roche Applied Science) diluted in 1:100 in TNB buffer.

9. Anti–DNP-alkaline phosphatase-conjugate (provided in kit) diluted 1:100 in TNB buffer.

10. NBT/BCIP solution

11. Nuclear Fast Red is recommended: Mix 0.1-g of nuclear Fast Red with 100 ml of 5% aluminum sulfate solution. Heat the mixture to 75°C to completely dissolve, and filter through Whatman #1 filter paper. Add a grain of thymol for preservation.

Protocol #2:

1. Prepare slides for detection with *TSA* plus DNP system using standard nonradioactive *in situ* hybridization techniques with digoxigenin- or fluorescein-labeled probes up to posthybridization washes as in **Protocol #1** from **step 1 to 22**.

2. Incubate slides with 100 μl TNB blocking buffer in a humid chamber for 30 min at room temperature. The use of a coverslip will reduce evaporation.

3. Incubate slides with 100 μl anti–digoxigenin (1:100) or anti–fluorescein-HRP (1:250) diluted in TNB buffer in a humid chamber for 30 min at room temperature. The use of a coverslip will reduce evaporation.

4. Wash the slides three times for 5 min each in TNT buffer at room temperature with agitation.

5. Pipet 100–300 μl DNP amplification working solution onto each slide. Incubate the slides at room temperature for 30 min. Discard any unused portion of DNP working solution.

6. Wash the slides as in **step 4**.

7. Add ~ 100 μl of diluted and anti-DNP-AP 1:100 diluted in TNB buffer to each slide. Incubate the slides in a humid chamber at room temperature for 60 min.

8. Wash the slides as in **step 4**.

9. Visualize signals with standard AP catalyzed NBT/BCIP substrates for 10–30 min. Monitor slides under microscope every 5–10 min until a strong signal free background (nonspecific signal) is reached.

10. Rinse slides twice in buffer C.

11. Counterstain with nuclear Fast Red.

12. Rinse slides as in **Protocol #1** from **step 35 to 38**.

13. Coverslip with Cytoseal XYL mounting medium.

RESULTS AND DISCUSSION

Catalyzed reporter deposition (CARD) with biotinylated tyramine has been used for ISH amplification in cell preparations and tissue sections (Kerstens *et al.*, 1995; Plummer *et al.*, 1998). This signal amplification technique is based on the deposition of activated biotinylated tyramine onto electron-rich moieties such as tyrosine, phenylalanine, dinitrophenyl (DNP), or tryptophan at or near the site of HRP (Hopman *et al.*, 1998). The binding of tyramide to proteins is believed to be due to the production of free

oxygen radicals by the HRP. The biotin sites on the bound tyramide act as further binding sites (e.g., streptavidin-biotin complexes or enzyme- and fluorochromelabeled streptavidin). In this way, a lot of extra hapten molecules can be introduced at the hybridization site *in situ* (Figure 6). The sensitivity of ISH can be improved to detect: (1) repetitive and single-copy DNA sequences in cell preparations, (2) up to three different DNA sequences (repetitive as well as single-copy) simultaneously in cell preparations, (3) low- and single-copy human papillomaviruses and other microorganisms in cell and tissue preparations (Hayden *et al.*, 2001a, b; Plummer *et al.*, 1998), and (4) rRNA and mRNA ranging from high to low abundance in cell and tissue preparations (Yang *et al.*, 1999).

When the CARD-ISH was performed on control cells using the HPV 16/18 genomic DNA probes and the GenPoint kit, the signals showed 2–3 punctate dots in SiHa cell nuclei, two or more larger dots in ~ 80% of HeLa 229 cell nuclei, and densely positive in greater than 90% of CaSki cell nuclei. Hybridization with biotinylated plasmids (pBR322 or pUC18) was negative. Biotinylated oligonucleotide cocktails specific for HPV 16 or HPV 18 were examined on CaSki, HeLa 229, and SiHa cells. There was positive staining for HPV 16 in SiHa and CaSki cells, as well as HPV 18 in HeLa 229 cells with similar staining patterns (Plummer *et al.*, 1998). Background staining was noted more often with oligonucleotide probes than with the genomic DNA probes, and this was not totally dependent on probe concentration (Qian *et al.*, 2001). An increase in the hybridization temperature to 50°C and the use of an acetylation step with 0.1 M triethanolamine plus 0.6% acetic anhydride further diminished background.

Some cases with negative or weak positive staining with HPV probes for types 6/11, 16/18, and 31/33/51 by conventional ISH became strongly positive, and two cases changed from a weakly to strongly positive reaction (Table 4). The cases that became positive after CARD-ISH exhibited punctate patterns in the nuclei and lacked the heavily stained nuclear pattern, which is usually seen with conventional ISH reactions in strongly positive cases, suggesting that the copy level was very low in these specimens. Cases considered as weakly positive showed a finely punctate pattern with conventional ISH method, but after CARD-ISH exhibited intense staining reactions with an increased number of positively staining cells. The negative controls were consistently negative in these analyses. Digestion of HPV DNA with DNase I prior to hybridization abolished positive hybridization signal. The elimination of the biotinyl tyramide step eliminated positive signals in the SiHa and HeLa 229 cells while significantly decreasing the signal in CaSki cells (Plummer *et al.*, 1998).

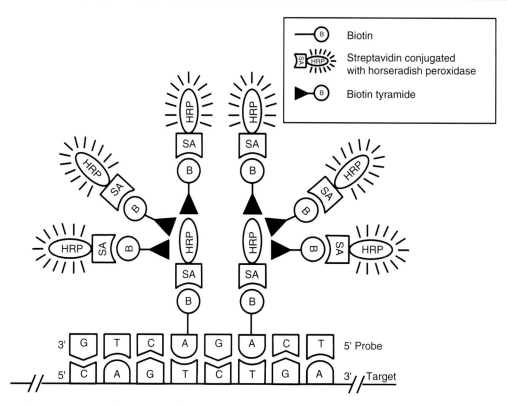

Figure 6 Basic components of the CARD-ISH method. Hybridized biotinylated probes are detected with primary streptavidin conjugated with horseradish peroxidase (SA/HRP) and reacted with hydrogen peroxide and biotinyl tyramide. Newly deposited biotin molecules are then detected with an application of secondary SA/HRP. Peroxidase is finally reacted with hydrogen peroxide and 3,3'-diaminobenzidene to form an insoluble brown precipitate.

The main advantage of the CARD-ISH procedure is that it is performed after probe hybridization and stringent washings, so that the specificity of the probe hybridization is not compromised (Speel *et al.*, 1999a). Therefore, in principle, proper conditions for sample fixation and pretreatment, including endogenous peroxidase inactivation, as well as hybridization *in situ,* do not need to be modified compared with conventional ISH procedures. A number of different combinations of probe detection (one to three detection conjugate layers) and CARD signal amplification systems (using different tyramides, amplification buffers, reaction times, and temperatures) have been applied, and optimization of each detection system has been necessary to obtain a high signal-to-noise ratio.

CARD has been used for diagnostic nonradioactive ISH with digoxigenin or biotin-labeled probes to increase the sensitivity and shorten the overall turnaround time of assays. Biotinylated tyramides are relatively easy and inexpensive to synthesize (Hopman *et al.*, 1998). However, biotin is associated with large numbers of endogenous streptavidin binding sites, such as

Table 4 Comparison of Conventional ISH With CARD-ISH in Detecting HPV in Clinical Specimens

Specimen	Standard ISH Positive			CARD-ISH Positive		
	6/11	16/18	31/33/51	6/11	16/18	31/33/51
Genitourinary (*n* = 17)	0	1	0	0	2	2[b]
Other[c] (*n* = 3)	1[a]	1[a]	0	1	1	0

[a]Result consisted of a weak positive reaction as a blush or fine dotting.

[b]1 of 2 cases represents a double infection with 16/18.

[c]Includes: 1 larynx and 2 eyelid specimens. See reference (Plummer *et al.*, 1998).

in liver or kidney, and high background staining may be encountered. It is therefore desirable to rely on tyramides, labeled with digoxigenin, di- or trinitriphenyl, or fluorochromes (Speel *et al.*, 1998). Moreover, because the tyramide deposition reaction proceeds very quickly, minor differences in amplification reaction time may lead to variations in the final signal intensities. Nevertheless, an amplification factor in the range of five- to tenfold, or possibly higher, together with preservation of distinct localization of ISH signals, seems to be a realistic goal for both DNA and mRNA ISH. Biotin-free chromogenic system can reduce background associated with endogenous biotin and eliminate biotin-quench step.

The practical limitations of CARD are due to a number of factors (Table 5). Because of high sensitivity, CARD may amplify nonspecific background signal, which can result in an unfavorable signal-to-noise ratio. Endogenous peroxidases in human tissue are potent enough to catalyze the CARD reaction. To avoid this unwanted reaction, endogenous peroxidase must be blocked or quenched, and appropriate positive and negative controls should be used. CARD ISH uses biotinylated probes, biotinyl tyramide, and streptavidin-conjugated horseradish peroxidase.

Peptide nucleic acid (PNA) probes combined with CARD detection have been recently introduced (Murakami *et al.*, 2001). PNA consists of a peptide backbone composed of N-(2 aminoethyl) glycine units to which nucleobases are attached by carbonyl methylene linkers, bringing new possibilities for the *in situ* hybridization. PNA is an uncharged molecule, that reduces hybridization time and background significantly. The combination of PNA or oligonucleotide probes and CARD technology greatly simplifies the *in situ* hybridization component of the reaction (Speel *et al.*, 1998; Van de Corput *et al.*, 1998).

Future Prospects

The relatively low sensitivity of ISH is a limiting factor in many applications, especially in diagnostic pathology. Improving the sensitivity of ISH has been a goal of many investigators in this field (Nuovo, 2001; Qian *et al.*, 2003). There are two general approaches to achieve this: amplification of target sequences (Wiedorn *et al.*, 1999) and signal enhancement (Speel *et al.*, 1999a; Wiedorn *et al.*, 2001). Currently, nucleic acid amplification has become both a necessity and a routine procedure in many aspects of molecular biology. Although very versatile and sensitive, *in situ* PCR has limitations, with poor preservation of morphology of some biologic structures and unreliable as a quantitative method (Teo *et al.*, 1995). Clinical application of *in situ* PCR must await the resolution of some of the current limitations. Isothermal methods such as IS-3SR (Mueller *et al.*, 1997) and rolling circle amplification (RCA) (Christian *et al.*, 2001; Zhou *et al.*, 2001) are not appropriate for precise quantification of the target nucleic acid. In contrast, amplifying the signal rather than the target or probes is suitable for quantitative interpretations, as bDNA ISH (Player *et al.*, 2001), which addresses both the detection and quantification of the target, providing valuable information in the molecular diagnosis of pathogens. CARD with a combination of enzyme/substrate is able to generate and deposit large numbers of molecules at the detection site. Therefore, ISH results in considerable improvement in sensitivity. In addition, brighter fluorescent polystyrene microspheres and enzyme-labeled fluorescence signal amplification technology are being developed, which could prove useful in the future for detection (Hakala *et al.*, 1997). CARD signal amplification method appears to be one of the most promising approaches (Wiedorn *et al.*, 1999) because it does not

Table 5 Troubleshooting Guide for CARD Method

Problem	Probable Cause	Recommended Action
1. No or very weak signal	1a. Low copy of target or low hybridization efficiency.	1a. Increase probe concentration and/or hybridization time.
	1b. Suboptimal pretreatment.	1b. Optimize enzyme treatment (e.g., proteinase K concentration and/or incubation time).
	1c. Low efficiency of tyramide deposition.	1c. Increase tyramide concentration and/or lengthen incubation time.
2. High background	2a. Overdevelopment of signal.	2a. Decrease concentration of HRP conjugate and/or incubation time. Decrease substrate development time.
	2b. Endogenous peroxide activity.	2b. Perform peroxide quench.
	2c. Suboptimal probe hybridization.	2c. Titrate probe and/or increase stringency wash.
	2d. Endogenous biotin.	2d. Switch to biotin-free amplification system.
	2e. Others.	2e. Filter buffers and blocking reagents.

require special equipment. Trends in recent developments suggest that it should be possible to use different chromogens for various amplified sequences within the same cell (Speel, 1999b).

Branched DNA (bDNA) ISH method improves nucleic acid detection to increase the signal rather than amplifying the target or probe sequence, and uses a series of nonisotopic oligonucleotide probes to mRNA or DNA (Player et al., 2001). The advantages of this method are its specificity and the avoidance of biotin. The disadvantages of bDNA are less sensitivity compared with target and probe amplification methods. Compared with PCR-based methods, the bDNA signal amplification does not require any DNA or RNA polymerase activity and avoids high-temperature cycling. The diffusion of amplification products away from the target site, which is a concern with PCR-based ISH method, is not a significant concern with bDNA ISH. bDNA ISH provides a rapid, sensitive, and reproducible means for detecting specific DNA and mRNA sequences in various cell types.

RCA is an isothermal nucleic-acid amplification method, but it differs from the polymerase chain reaction and other nucleic-acid amplification schemes in several respects (Christian et al., 2001). The immuno-RCA protocols have been extended to hapten-labeled nucleic acid probes for nuclear DNA sequences or cytoplasmic mRNA molecules in fixed cells and tissues (Gusev et al., 2001; Zhou et al., 2001). The circle DNA, RCA primer, and Φ 29 DNA polymerase can be used for replicating the circle template. The use of additional pairs of circles and probe conjugates can permit the simultaneous detection of multiple targets. RCA in situ is useful for discriminating alleles, determining gene copy number, and quantifying gene expression in single cells (Christian et al., 2001). One advantage of RCA is a significant increase in the signal that is achieved through a single round of enzymatic amplification of a nucleic acid substrate, leading to improvements in relative discrimination as well as absolute amounts of measurable signal (Gusev et al., 2001). The sensitivity, specificity, and speed of RCA may also allow its use for clinical purposes such as prenatal diagnosis and pathologic characterization of tumors.

More sensitive techniques for bright-field detection of hybridization products are being developed. One new approach is the use of molecular beacons (MB) (Antony et al., 2001). MB is a new class of nucleic acid probes that become fluorescent when they bind to a complementary sequence. Because of the sequence specificity of nucleic acid hybridization and the marked sensitivity, MB can be used to detect low-copy number of DNA or RNA molecules. In situ strand displacement amplification has recently been described as an improved technique for the

detecting low-copy nucleic acids. This technique was reported to be as sensitive as in situ PCR (Nuovo et al., 2000). Bright-field assays for assessment of HER-2/neu gene amplification using GoldEnhance gold-based automatography (Nana probes, Inc., Yaphank, NY) (Tubbs et al., 2002), CARD (Zehbe et al., 1997), the use of biotinylated-labeled probes that approached the sensitivity of the FISH technique (Van de Corput et al., 1998; Uchihara et al., 2000) and electron microscopy (Mayer et al., 1999) are also promising recent developments. These new developments indicate that the development of highly sensitive and specific probes or amplification systems for ISH will continue unabated into the near future.

References

Antony, T., and Subramaniam, V. 2001. Molecular beacons: nucleic acid hybridization and emerging applications. *J. Biomol. Struct. Dyn.* 19:497–504.

Bagasra, O., Seshamma, T., Hansen, J., Bobroski, L., Saikumari, P., Pestaner, J.P., and Pomerantz, R.J. 1994. Application of *in situ* PCR methods in molecular biology: I. Details of methodology for general use. *Cell Vis.* 1:324–335.

Chou, Q., Russell, M., Birch, D.E., Raymond, J., and Bloch, W. 1992. Prevention of pre-PCR mispriming and primer dimerization improves low-copy-number amplifications. *Nucleic Acids Res.* 20:1717–1723.

Christian, A.T., Pattee, M.S., Attix, C.M., Reed, B.E., Sorensen, K.J., and Tucker, J.D. 2001. Detection of DNA point mutations and mRNA expression levels by rolling circle amplification in individual cells. *Proc. Natl. Acad. Sci. USA* 98:14238–14243.

Clerico, L., Mancuso, T., Da Prato, L., Marziliano, N., Garagna, S., Pecile, V., Demori, E., Morgutti, M., Citta, A., Parod, S., Amoroso, A., and Crovella, S. 1999. *In situ* RT-PCR allows the detection of ornithine decarboxylase mRNA in paraffin embedded archival human hyperplastic breast tissues. *Eur. J. Histochem.* 43:179–183.

Coullin, P., Roy, L., Pellestor, F., Candelier, J.J., Bed-Hom, B., Guillier-Gencik, Z., and Bernheim, A. 2002. PRINS: The other *in situ* DNA labeling method useful in cellular biology. *Am. J. Med. Genet.* 107:127–135.

DeLellis, R.A. 1994. *In situ* hybridization techniques for the analysis of gene expression: Applications in tumor pathology. *Hum. Pathol.* 25:580–585.

Evans, M.F., Mount, S.L., Beatty, B.G., and Cooper, K. 2002. Biotinyl-tyramide-based *in situ* hybridization signal patterns distinguish human papillomavirus type and grade of cervical intraepithelial neoplasia. *Mod. Pathol.* 15:1339–1347.

Fredricks, D.N., and Relman, D.A. 1999. Application of polymerase chain reaction to the diagnosis of infectious diseases. *Clin. Infect. Dis.* 29:475–486.

Gusev, Y., Sparkowski, J., Raghunathan, A., Ferguson, H. Jr., Montano, J., Bogdan, N., Schweitzer, B., Wiltshire, S., Kingsmore, S.F., Maltzman, W., and Wheeler, V. 2001. Rolling circle amplification: A new approach to increase sensitivity for immunohistochemistry and flow cytometry. *Am. J. Pathol.* 159: 63–69.

Hakala, H., Heinonen, P., Iitia, A., and Lonnberg, H. 1997. Detection of oligonucleotide hybridization on a single microparticle by time-resolved fluorometry: Hybridization assays on polymer particles obtained by direct solid-phase assembly of the oligonucleotide probes. *Bioconjug. Chem.* 8:378–384.

Hayden, R.T., Qian, X., Roberts, G.D., and Lloyd, R.V. 2001a. *In situ* hybridization for the identification of yeastlike organisms in tissue section. *Diagn. Mol. Pathol. 10:*15–23.

Hayden, R.T., Uhl, J.R., Qian, X., Hopkins, M.K., Aubry, M.C., Limper, A.H., Lloyd, R.V., and Cockerill, F.R. 2001b. Direct detection of *Legionella* species from bronchoalveolar lavage and open lung biopsy specimens: Comparison of Light Cycler PCR, *in situ* hybridization, direct fluorescence antigen detection, and culture. *J. Clin. Microbiol.* 39:2618–2626.

Hopman, A.H., Ramaekers, F.C., and Speel, E.J.M. 1998. Rapid synthesis of biotin-, digoxigenin-, trinitrophenyl-, and fluorochrome-labeled tyramides and their application for *in situ* hybridization using CARD amplification. *J. Histochem. Cytochem.* 46:771–777.

Hougaard, D.M., Hansen, H., and Larsson, L.I. 1997. Nonradioactive *in situ* hybridization for mRNA with emphasis on the use of oligodeoxynucleotide probes. *Histochem. Cell. Biol.* 108:335–344.

Jin, L., and Lloyd, R.V. 1997. *In situ* hybridization: Methods and applications. *J. Clin. Lab. Anal.* 11:2–9.

Jin, L., Qian, X., and Lloyd, R.V. 1995. Comparison of mRNA expression detected by *in situ* polymerase chain reaction and *in situ* hybridization in endocrine cells. *Cell. Vis.* 2:314–321.

Jin, L., Qian, X., and Lloyd, R.V. 2001. *In situ* hybridization: detection of DNA and RNA. In Lloyd, R.V. (ed). *Morphology Methods: Cell and Molecular Biology Techniques.* Totowa: Humana Press, 27–47.

Kerstens, H.M., Poddighe, P.J., and Hanselaar, A.G. 1995. A novel *in situ* hybridization signal amplification method based on the deposition of biotinylated tyramine. *J. Histochem. Cytochem.* 43:347–352.

Komminoth, P., Adams, V., Long A.A., Roth, J., Saremaslani, P., Flury, R., Schmid, M., and Heitz, P.U. 1994. Evaluation of methods for hepatitis C virus detection in archival liver biopsies: Comparison of histology, immunohistochemistry, *in situ* hybridization, reverse transcriptase polymerase chain reaction (RT-PCR) and *in situ* RT-PCR. *Pathol. Res. Pract.* 190: 1017–1025.

Kriegsmann, J., Muller-Ladner, U., Sprott, H., Brauer, R., Petrow, P.K., Otto, M., Hansen, T., Gay, R.E., and Gay, S. 2001. Detection of mRNA by nonradioactive direct primed *in situ* reverse transcription. *Histochem. Cell. Biol.* 116:199–204.

Lau, G.K., Fang, J.W., Wu, P.C., Davis, G.L., and Lau, J.Y. 1994. Detection of hepatitis C virus genome in formalin-fixed, paraffin-embedded liver tissue by *in situ* reverse transcription polymerase chain reaction. *J. Med. Virol.* 44:406–409.

Lisby, G. 1999. Application of nucleic acid amplification in clinical microbiology. *Mol. Biotechnol.* 12:75–99.

Long, A.A. 1998. *In situ* polymerase chain reaction: Foundation of the technology and today's options. *Eur. J. Histochem.* 42:101–109.

Man, Y.G., Zhuang, Z.P., Bratthauer, G.L., Bagasra, O., and Tavassoli, F.A. 1996. Detailed RT-IS-PCR protocol for preserving morphology and confining PCR products in routinely processed paraffin sections. *Cell Vis.* 3:389–396.

Martinez, A., Miller, M.J., Quinn, K., Unsworth, E.J., Ebina, M., and Cuttitta, F. 1995. Nonradioactive localization of nucleic acids by direct *in situ* PCR and *in situ* RT-PCR in paraffin-embedded sections. *J. Histochem. Cytochem.* 43:739–747.

Mayer, G., and Bendayan, M. 1999. Immunogold signal amplification: Application of the CARD approach to electron microscopy. *J. Histochem. Cytochem.* 47:421–429.

McAllister, H.A., and Rock, D.L. 1985. Comparative usefulness of tissue fixatives for *in situ* viral nucleic acid hybridization. *J. Histochem. Cytochem.* 33:1026–1032.

McNicol, A.M., and Farquharson, M.A. 1997. *In situ* hybridization and its diagnostic applications in pathology. *J. Pathol. 182:* 250–261.

Mitsuhashi, M. 1996. Technical report: I. Basic requirements for designing optimal oligonucleotide probe sequences. *J. Clin. Lab. Anal.* 10:277–284.

Morel, G., Berger, M., Ronsin, B., Recher, S., Ricard-Blum, S., Mertani, H.C., and Lobie, P.E. 1998. *In situ* reverse transcription-polymerase chain reaction: Applications for light and electron microscopy. *Biol. Cell 90:*137–154.

Mueller, J.D., Putz, B., and Hofler, H. 1997. Self-sustained sequence replication (3SR): An alternative to PCR. *Histochem. Cell Biol.* 108:431–437.

Murakami, T., Hagiwara, T., Yamamoto, K., Hattori, J., Kasami, M., Utsumi, M., and Kaneda, T. 2001. A novel method for detecting HIV-1 by nonradioactive *in situ* hybridization: Application of a peptide nucleic acid probe and catalyzed signal amplification. *J. Pathol. 194:*130–135.

Muratori, L., Gibellini, D., Lenzi, M., Cataleta, M., Muratori, P., Morelli, M.C., and Bianchi, F.B. 1996. Quantification of hepatitis C virus-infected peripheral blood mononuclear cells by *in situ* reverse transcriptase-polymerase chain reaction. *Blood* 88:2768–2774.

Nuovo, G.J. 2000. *In situ* strand displacement amplification: An improved technique for the detection of low-copy nucleic acids. *Diagn. Mol. Pathol.* 9:195–202.

Nuovo, G.J. 2001. Co-labeling using *in situ* PCR: A review. *J. Histochem. Cytochem.* 49:1329–1339.

Nuovo, G.J., Becker, J., Burke, M., Fuhrer, J., and Steigbigel, R. 1994. *In situ* detection of PCR-amplified HIV-1 nucleic acids in lymph nodes and peripheral blood in asymptomatic infection and advanced stage AIDS. *J. Acquired Immun. Def.* 7:916–923.

Nuovo, G.J., Gallery, F., Hom, R., MacConnell, P., and Bloch, W. 1993. Importance of different variables for optimizing *in situ* detection of PCR-amplified DNA. *PCR. Methods. Appl.* 2: 305–312.

Nuovo, G.J., Gallery, F., MacConnell, P., Becker, J., and Bloch, W. 1991. An improved technique for the detection of DNA by *in situ* hybridization after PCR-amplification. *Am. J. Pathol.* 139:1239–1244.

O'Leary, J.J., Chetty, R., Graham, A.K., and McGee, J.O. 1996. *In situ* PCR: Pathologist's dream or nightmare? *J. Pathol. 178:*11–20.

Player, A.N., Shen, L.P., Kenny, D., Antao, V.P., and Kolberg, J.A. 2001. Single-copy gene detection using branched DNA (bDNA) *in situ* hybridization. *J. Histochem. Cytochem.* 49: 603–612.

Plummer, T.B., Sperry, A.C., Xu, H.S., and Lloyd, R.V. 1998. *In situ* hybridization detection of low-copy nucleic acid sequences using catalyzed reporter deposition and its usefulness in clinical human papillomavirus typing. *Diagn. Mol. Pathol.* 7:76–84.

Qian, X., Bauer, R.A., Xu, H.S., and Lloyd, R.V. 2001. *In situ* hybridization detection of calcitonin mRNA in routinely fixed, paraffin-embedded tissue sections: A comparison of different types of probes combined with tyramide signal amplification. *Appl. Immunohistochem. Mol. Morphol.* 9:61–69.

Qian, X., and Lloyd, R.V. 2003. Recent developments in signal amplification methods for *in situ* hybridization. *Diagn. Mol. Pathol.* 12:1–13.

Raatikainen-Ahokas, A.M., Immonen, T.M., Rossi, P.O., Sainio, K.M., and Sariola, H.V. 2001. An artifactual *in situ* hybridization signal associated with apoptosis in rat embryo. *J. Histochem. Cytochem. 48:*955–961.

Sallstrom, J.F., Zehbe, I., Alemi, M., and Wilander, E. 1993. Pitfalls of *in situ* polymerase chain reaction (PCR) using direct incorporation of labeled nucleotides. *Anticancer Res. 13:*1153.

Schonhuber, W., Fuchs, B., Juretschko, S., and Amann, R. 1997. Improved sensitivity of whole-cell hybridization by the combination of horseradish peroxidase-labeled oligonucleotides and tyramide signal amplification. *Appl. Environ. Microb. 63:*3268–3273.

Speel, E.J.M. 1999b. Detection and amplification systems for sensitive, multiple-target DNA and RNA *in situ* hybridization: Looking inside cells with a spectrum of colors. *Histochem. Cell Biol. 112:*89–113.

Speel, E.J.M., Hopman, A.H., and Komminoth, P. 1999a. Amplification methods to increase the sensitivity of *in situ* hybridization: Play CARD(s). *J. Histochem. Cytochem. 47:*281–288.

Speel, E.J.M., Hopman, A.H., and Komminoth, P. 2000. Signal amplification for DNA and mRNA. *Methods Mol. Biol. 123:*195–216.

Speel, E.J.M., Saremaslani, P., Roth, J., Hopman, A.H., and Komminoth, P. 1998. Improved mRNA *in situ* hybridization on formaldehyde-fixed and paraffin-embedded tissue using signal amplification with different haptenized tyramides. *Histochem. Cell Biol. 110:*571–577.

Steele, A., Uckan, D., Steele, P., Chamizo, W., Washington, K., Koutsonikolis, A., and Good, R.A. 1998. RT *in situ* PCR for the detection of mRNA transcripts of Fas-L in the immune-privileged placental environment. *Cell Vis. 5:*13–19.

Teo, I.A., and Shaunak, S. 1995. PCR *in situ*: Aspects which reduce amplification and generate false-positive results. *Histochem. J. 27:* 660–669.

Tournier, I., Bernuau, D., Poliard, A., Schoevaert, D., and Feldmann, G. 1987. Detection of albumin mRNAs in rat liver by *in situ* hybridization, usefulness of paraffin-embedding and comparison of various fixation procedures. *J. Histochem. Cytochem. 35:*453–459.

Tubbs, R., Pettay, J., Skacel, M., Powell, R., Stoler, M., Roche, P., and Hainfeld, J. 2002. Gold-facilitated *in situ* hybridization: A bright-field autometallographic alternative to fluorescence *in situ* hybridization for detection of HER-2/neu gene amplification. *Am. J. Pathol. 160:*1589–1595.

Uchihara, T., Nakamura, A., Nagaoka, U., Yamazaki, M., and Mori, O. 2000. Dual enhancement of double immunofluorescent signals by CARD: Participation of ubiquitin during formation of neurofibrillary tangles. *Histochem. Cell Biol. 114:*447–451.

Van de Corput, M.P., Dirks, R.W., van Gijlswijk, R.P., van Binnendijk, E., Hattinger, C.M., de Paus, R.A., Landegent, J.E., and Raap, A.K. 1998. Sensitive mRNA detection by fluorescence *in situ* hybridization using horseradish peroxidase-labeled oligodeoxynucleotides and tyramide signal amplification. *J. Histochem. Cytochem. 46:*1249–1259.

Wesselingh, S.L., Takahashi, K., Glass, J.D., Griffin, J.W., and Griffin, D.E. 1997. Cellular localization of tumor necrosis factor mRNA in neurological tissue from HIV-1 infected patients by combined reverse transcriptase/polymerase chain reaction *in situ* hybridization and immunohistochemistry. *J. Neuroimmunol. 74:*1–8.

Wiedorn, K.H., Goldmann, T., Henne, C., Kuhl, H., and Vollmer, E. 2001. EnVision+, a new dextran polymer-based signal enhancement technique for *in situ* hybridization (ISH). *J. Histochem. Cytochem. 49:*1067–1071.

Wiedorn, K.H., Kuhl, H., Galle, J., Caselitz, J., and Vollmer, E. 1999. Comparison of *in situ* hybridization, direct and indirect *in situ* PCR as well as tyramide signal amplification for the detection of HPV. *Histochem. Cell Biol.111:*89–95.

Wilkens, L., Tchinda, J., Komminoth, P., and Werner, M. 1997. Single- and double-color oligonucleotide primed *in situ* labeling (PRINS): Applications in pathology. *Histochem. Cell Biol. 108:*439–446.

Yang, H., Wanner, I.B., Roper, S.D., and Chaudhari, N. 1999. An optimized method for *in situ* hybridization with signal amplification that allows the detection of rare mRNAs. *J. Histochem. Cytochem. 47:*431–445.

Zehbe, I., Hacker, G.W., Su, H.C., Hauser-Kronberger, C., Hainfeld, J.F., and Tubbs, R. 1997. Sensitive *in situ* hybridization with catalyzed reporter deposition, streptavidin-nanogold, and silver acetate autometallography-detection of single-copy human papillomavirus. *Am. J. Pathol. 150:*1553–1561.

Zhou, C.J., Kikuyama, S., and Shioda, S. 2001a. Application and modification of *in situ* RT-PCR for detection and cellular localization of PAC1-R splice variant mRNAs in frozen brain sections. *Biotech. Histochem. 76:*75–83.

Zhou, Y., Calciano, M., Hamann, S., Leamon, J.H., Strugnell, T., Christian, M.W., and Lizardi, P.M. 2001b. *In situ* detection of messenger RNA using digoxigenin-labeled oligonucleotides and rolling circle amplification. *Exp. Mol. Pathol. 70:*281–288.

II

Molecular Pathology

1

Polymerase Chain Reaction Technology

M.A. Hayat

Polymerase Chain Reaction Technology

Polymerase chain reaction (PCR) is a powerful technique in molecular biology, for it accomplishes quantitative nucleic acid analysis. With rapid advances in pharmacogenetics and pharmacogenomics, PCR is playing an increasingly important role in medical diagnosis and daily medical care decisions. In fact, PCR is the most frequently used tool for molecular genetic analysis. Sufficient evidence is available indicating that PCR-based testing is a useful adjunct for the diagnosis of malignant tumors. It has significantly advanced the analysis of tumors for the presence of mutated genes. Sensitive techniques of detection are indispensable when the mutation of interest is present only in a small population of cells in tumor samples. With the advent of PCR, *in vitro* amplification of DNA has significantly increased the sensitivity of detecting such mutations.

This method can detect chromosomal translocations in tissue samples. Reciprocal translocations are among the chromosomal abnormalities most strongly correlated with individual tumor types. Such translocations result in the formation of novel chimeric genes whose transcripts encode fusion proteins that combine functional domains not normally juxtaposed in a single protein. This phenomenon leads to dysregulation of gene transcription or alteration of cellular signal transduction pathways.

Thus, the role of PCR protocol in molecular genetic testing in diagnostic surgical pathology is apparent.

Although DNA extracted from formalin-fixed, paraffin-embedded tissues can be used for gene analysis using PCR (Pinto *et al.*, 1998), the nucleic acids extracted from such tissues may not be well preserved and/or may show decline in the efficiency of their amplification (Greer *et al.*, 1994). Organic solvents such as acetone seem to be better than formalin as a fixative for such studies. When archival formalin-fixed, paraffin-embedded tissues are the only samples available, the best approach is to extract DNA by melting the paraffin by microwave heating (Banerjee *et al.*, 1995). This protocol is simple, has a lower contamination risk, and may provide a high yield of long DNA fragments for amplification with PCR. Recently, Sato *et al.* (2001) have successfully extratcted DNA from archival formalin-fixed, paraffin-embedded gastric carcinomas and B-cell lymphomas using microwave heating.

Limitations of PCR

There are many reasons for the difficulty in quantifying the sensitivity and specificity of PCR-based testing. The analytic characteristics of testing are highly dependent on the specimen type (e.g., fresh tissue,

Handbook of Immunohistochemistry and *in situ* Hybridization of Human
Carcinomas, Volume 1: Molecular Genetics; Lung and Breast Carcinomas

45

Copyright © 2004 by Elsevier (USA)
All rights reserved.

fresh cell lines, or formalin-fixed, paraffin-embedded tissue). Also, a subset of a tumor type does not harbor the characteristic chromosomal translocations. In addition, more than one type of tumor can share the same genetic abnormality.

Another problem is the difficulty in discerning if a PCR product is biologically relevant based on its size alone. The reason is that combinatorial generation of fusion proteins, together with the involvement of more than one member of multigenic families, produces a large repertoire of fusion transcripts for many cancers (Hill *et al.*, 2002). Alternate mRNA processing and the occurrence of variant transcripts create an even larger assortment of fusion transcripts, further complicating identification of PCR products by simple gel electrophoresis. Nonspecific amplification of mRNA may reduce specificity. Therefore, the PCR product should be sequenced, facilitating identification of PCR products.

Variations of PCR Technology

As PCR technology finds increased use in various genetic analyses, variations of this technique have emerged to augment the high-resolution genotyping and genetic mapping of various complex genomes. One such variation is real-time PCR or reverse transcription-PCR (RT-PCR) that facilitates PCR amplification using RNA as the starting material. It is similar to DNA-PCR with the modification that PCR amplification is preceded by reverse transcription of RNA into cDNA. The RT-PCR is the most sensitive method for characterizing or confirming gene expression patterns and comparing mRNA levels in different sample populations. In comparison to *in situ* hybridization (ISH), RT-PCR provides a higher level of sensitivity, theoretically allowing the identification of a single gene copy per cell.

The RT-PCR technique is superior to conventional PCR approaches, especially in the setting of a diagnostic laboratory, because post-amplification manipulation of specimens is not necessary, thereby greatly reducing the risk of contamination and the workload. In contrast to all *in situ* methods (e.g., immunohistochemistry and ISH), the reaction conditions and the data evaluation can be easily standardized by well-trained workers in an objective manner for the RT-PCR system. This advantage facilitates comparison of results between different laboratories. Flow cytometry also provides objective data but relies exclusively on freshly collected specimens (see the chapter by Leers in Volume 2 of this series). Several recent applications of the RT-PCR in cancer diagnosis are given next.

The RT-PCR technique was used for detecting circulating tumor cells (Mira *et al.*, 2002). The technique amplifies mRNA sequences unique to malignant cells in the circulation to detect early micrometastasis and minimal residual disease. Vlems *et al.* (2002) have also used RT-PCR for detecting cytokeratin 20 in disseminated tumor cells in healthy blood samples spiked with colon tumor cells. Baba *et al.* (2001) have reported correlation of beta-casein–like protein from mRNA with cervical cancer recurrence using RT-PCR; such results can aid in clinical diagnosis.

Several PCR methods have been developed to detect a broad spectrum of muscosotropic human papilomavirus (HPV) types, using either degenerate or consensus primers. More than 40 types of HPV infect the genital epithelium, and several high-risk types, including HPV 16, 18, 31, 33, and 45, are found in almost all cases of high-grade cervical intraepithelial neoplasia and cervical cancer. Rapid RT-PCR was used for distinguishing between high-risk HPV types 16 and 18 (Cubie *et al.*, 2001). This method can distinguish closely related sequences on the basis of amplification followed by DNA melting temperature analysis. This approach allows the detection of single and mixed samples of HPV 16 and HPV 18 in both cells lines *in vitro* and cervical secretions from patients.

Another variation is quantitative RT-PCR. This approach is a highly sensitive and powerful technique for the quantitation of nucleic acids. It has tremendous potential for the high throughput analysis of gene expression in research and routine diagnosis. Many of the applications of quantitative RT-PCR include measuring mRNA expression levels, DNA copy number, transgene copy number and expression analysis, allelic discrimination, and measuring viral titers. However, successful application of this technique is not trivial. The major hurdle is not necessarily the practical performance of the experiments themselves but rather the efficient evaluation and the mathematical and statistical analyses of the enormous amount of data gained by this technique, as these tasks are not included in the software provided by manufacturers of the detection systems. Muller *et al.* (2002) have explained the mathematical evaluation and analysis of the data generated by this technology, the calculation of the final results, the propagation of experimental variation of the measured values to the final results, and the statistical analysis. There are many other variables that can limit the usefulness of this method. Therefore, careful consideration of the assay design, template preparation, and analytical methods is essential for accurate gene quantification (Ginzinger, 2002).

Because of the high cost of RT-PCR equipment and supplies, especially custom-made probes, quantitative

RT-PCR is a valuable alternative method. Szibor and Morawietz (2002) have developed a protocol using several competitive standards of different sizes to quantify gene expression in a one-tube-amplification procedure. The advantage of this protocol is that efficient quantitation of mRNA expression can be accomplished in microbiopsies (2–5-mg tissue) on a lower cost level and in shorter time.

With the advent of sophisticated PCR machines that incorporate software-driven fluorescence analysis systems, RT-PCR is rapidly becoming a more common method for the analysis and quantification of gene expression. Fluorescence-based, quantitative RT-PCR integrates the amplification and analysis steps of the PCR. Its sensitivity, specificity, and wide dynamic range make it the method of choice for quantitating steady-state mRNA levels (Bustin, 2002). In other words, the method provides the ability to determine precise levels of a target nucleic acid within a given tissue or cell sample as opposed to a mere qualitative demonstration of its presence or absence or even semiquantitative characterization by standard PCR methods. Absolute quantitation of gene expression in single cells can be carried out with this method (Smith et al., 2000, 2003).

To carry out the preceding method, the LightCycler machine is commercially available (Idaho Technology, Inc.), which allows semiquantitative assay. It combines simultaneous PCR amplification with sophisticated computer analysis of the kinetics data generated. This system has the advantage of providing simple, rapid quantification of relative yields of any gene product without using any internal or external standards, provided amplification conditions are specific for the PCR product of interest. Recently, this machine was used for amplification of a serial tenfold dilution series (spanning four orders of magnitude) or a 379-bp cDNA template (Gentle et al., 2001). The PCR product was detected using SYBR Green 1 chemistry. The LightCycler machine was also used for detecting HPV, and HPV-16 and HPV-18 were separated from each other (Cubie et al., 2001).

Another variation is the allele-specific PCR system that allows identification of predetermined point mutations in a minority of mutant DNAs against normal background. It relies on the use of primers homologous to the mutation at its 3′ end, favoring selective amplification of the mutant allele; the mismatch with wild-type alleles is then refractory to amplification (Newton et al., 1989). The design of the primary pair is a critical step in the successful RT-PCR protocol because it is influential in determining the sensitivity and specificity of the assay, and the design should be specific for the mutation that each primer is designed to detect.

The use of certain primer pairs may result in multiple peaks in the melting curve analysis even after several attempts with primer annealing temperatures and hold times. This may indicate nonspecificity of primers or complex transcriptional regulation such as alternate splicing events or detection of novel gene family members (Rajeevan et al., 2001). Therefore, the specificity of the primers should be studied first by solution-phase RT-PCR (Zhou et al., 2001). Highly specific primer pairs should reveal the band of the expected size only for the PCR products, which should be further confirmed by sequence analysis from the total RNA extracted from the sample tissues. PCR products may be recovered from glass capillaries for gel electrophoresis and sequencing for product verification.

The allele-specific PCR method has the potential to be applied to the study of molecular progression of cancer, including diagnosis and detection of residual disease. It can be extended to the in situ detection of aberrant cells. The method has been used in the screening of mutations in extracted DNA from patients with various diseases. Its application in the search for recurrent or minimal residual disease could improve the prognosis of patients by the early detection of tumors with known mutations. Sensitive detection and localization of the origin of mutations detected would further increase our understanding of the tumorigenic process. Recently, Low et al. (2000) have employed this nested procedure for analyzing cells with predetermined p53 mutations.

The allele-specific PCR is also potentially useful in noninvasive clinical diagnostic procedures, such as harvesting cells from saliva specimens, urine sediments, feces, sputum, pancreatic juice, and blood samples, or collected by brush cytology from skin and mucosa for molecular analysis (Low et al., 2000).

Use of RT-PCR for Archival Specimens

Ribonucleic acid (RNA) extracted from formalin-fixed, paraffin-embedded tissues is amenable to RT-PCR amplification of small fragments of genes. The importance of such RT-PCR application becomes apparent when one considers that archival specimens represent a vast collection of well-characterized materials from which diseases can be studied using this or other molecular methods. Use of archival materials allows retrospective study of diseases at the molecular level. Tissues that have undergone antigen retrieval can also be used for RT-PCR. Genes and infectious agents involved in disease process can be identified for both biomedical research and clinical investigation using this technology. The necessity of using archival material is obvious when only such materials are available for investigation.

However, RNA may be difficult to extract and may also show some degradation because of formalin fixation and paraffin embedding. Prolonged storage of fixed specimens may also result in negative effect on RNA for RT-PCR. It is generally true that PCR-based testing of archival tissues is associated with decreased frequency of positive results when compared with those obtained using fresh tissues. Nevertheless, it was recently demonstrated that the RNA extracted from various types of archival histologic and cytologic specimens, including both stained slides and unstained spare sections, is amenable to RT-PCR of short fragments of less than 150 bp (Liu *et al.*, 2002). This protocol is also useful to study minute cell populations microdissected from stained sections. Recently, it was reported that treatment of cryosections with RNALater instead of formalin results in improved preservation of histologic integrity as well as RNA and subsequent extraction of this nucleic acid.

References

Baba, T., Koizumi, M., Suzuki, T., Yamanaka, I., Yamashita, S., and Kudo, R. 2001. Specific detection of circulating tumor cells by reverse transcriptase-polymerase chain reaction of a beta-casein-like protein, preferentially expressed in malignant neoplasms. *Anticancer Res. 21*:2547–2552.

Banerjee, S.K., Makdisi, W.F., Weston, A.P., Mitchell, S.M., and Campbell, C.R. 1995. Microwave-based DNA extraction from paraffin-embedded tissue for PCR amplification. *Biotechniques 18*:768–773.

Bustin, S.A. 2002. Quantification of mRNA using real time RT-PCR: Trends and problems. *J. Mol. Endocrinal. 28*:23–39.

Cubie, H.A., Seagar, A.L., McGoogan, E., Whitehead, J., Brass, A., Arenas, M.J., and Whitley, M.W. 2001. Rapid real time PCR to distinguish between high risk human papilloma virus types 10 and 18. *Mol. Pathol. 54*:24–29.

Gentle, A., Anastasopoulos, F., and McBrien, N.A. 2001. High-resolution semiquantitative real-time PCR without the use of a standard curve. *Biotechniques 31*:502–508.

Ginzinger, D.G. 2002. Gene quantification using real-time quantitative PCR: An emerging technology hits the mainstream. *Exp. Hematol. 30*:503–512.

Greer, C.E., Wheeler, C.M., and Manos, M.M. 1994. Sample preparation and PCR amplification from paraffin-embedded tissues. *PCR Methods Appl. 3*:S113–S112.

Hill, D.A., O'Sullivan, M.J., Zhu, X., Vollmer, R.T., Humphrey, P.A., Dehner, L.P., and Pfeifer, J.D. 2002. Practical application of molecular genetic testing as an aid to the surgical pathologic diagnosis of sarcoma. *Am. J. Surg. Pathol. 26*:965–977.

Liu, H., Huang, X., Zhang, Y., Ye, H., Hamidi, A.E., Kocjan, G., Dagan, A., Isaacson, P.G., and Du, M.-Q. 2002. Archival fixed histologic and cytologic specimens, including stained and unstained materials, are amenable to RT-PCR. *Diag. Mol. Pathol. 11*:222–227.

Low, E.O., Jones, A.M., Gibbins, J.R., and Walker, D.M. 2000. Analysis of the amplification refractory mutation allele-specific polymerase chain reaction system for sensitive and specific detection of p3 mutations in DNA. *J. Pathol. 190*:512–515.

Mira, E., Lacalle, R.A., Gomez-Mouton, C., Leonardo, E., and Manes, S. 2002. Quantitative determination of tumor cell intravasation in a real-time polymerase chain reaction-based assay. *Clin. Exp. Metastasis 19*:313–318.

Muller, P.Y., Janobjak, H., Misrez, A.K., and Dobbie, A. 2002. Processing of gene expression data generated by quantitative real-time RT-PCR. *Biotechniques 32*:1372–1379.

Newton, C.R., Graham, A., Heptinstall, L.E., Powell, S.J., Summers, C., Kasdheker, N., Smith, J.C., and Markham, A.F. 1989. Analysis of any point mutation in DNA. The amplification refractory mutation system (ARMS). *Nucleic Acids Res. 17*: 2503–2516.

Pinto, A.P., and Villa, L.L. 1998. A spin cartridge system for DNA extraction from paraffin wax embedded tissues. *J. Clin. Pathol. Mol. Pathol. 51*:48–49.

Rajeevan, M.S., Ranamukhaarachchii, D.G., Vernon, S.D., and Unger, E.R. 2001. Use of real-time quantitative PCR to validate the results of cDNA array and differential display PCR technologies. *Methods 25*:443–451.

Sato, Y., Sugie, R., Tsuchiya, B., Kameya, T., Natori, M., and Mukai, K. 2001. Comparison of the DNA extraction methods for polymerase chain reaction amplification from formalin-fixed and paraffin-embedded tissues. *Diag. Mol. Pathol. 10*:265–271.

Smith, R.D., Brown, B., Ikonomi, P., and Schechter, A.N. 2003. Exogenous reference RNA for normalization of real-time quantitative PCR. *Biotechniques 34*:1–4.

Smith, R.D., Malley, J.D., and Schechter, A.N. 2000. Quantitative analysis of globin gene induction in single human erythroleukemic cells. *Nucleic Acids Res. 28*:4998–5004.

Szibor, M., and Morawietz, H. 2002. Serial competitive RT-PCR using multiple standards. *Biotechniques 33*:744–748.

Vlems, F., Soong, R., Depstra, H., Punt, C., Wobbes, T., Tabiti, K., and van Muijen, G. 2002. Effect of blood sample handling and reverse transcriptase polymerase chain reaction assay sensitivity on detection of CK20 expression in healthy donor blood. *Diagn. Mol. Pathol. 11*:90–97.

Zhou, C.J., Kikuyama, S., and Shioda, S. 2001. Application and modifications of *in situ* RT-PCR for detection and cellular localization of PAC-R splice variant mRNAs in frozen brain sections. *Biotechnic. Histochem. 76*:75–83.

2

DNA Microarrays Technology

M.A. Hayat

The recent sequencing of the human genome has and will provide us with a wealth of information to advance our understanding of biological systems, especially cancer development. The immediate benefit of having access to large amounts of sequence information is the ability to use microarrays that contain either complete complements of expressed complementary deoxyribonucleic acid (cDNA) or large numbers of known genes. Microarray technology is being used in a high-throughput approach to study gene expression and sequence variation on a genomic scale. In fact, microarray has become invaluable in identifying subsets of genes that appear to have different degrees of expression in various disease stages. More targeted gene expression profiling will become increasingly relevant to clinical diagnosis as correlations between expression patterns and cell states are established. Changes in expression of a defined subset of genes can be indicative of specific disease type or stage. Expression profiling has led to identifying sets of gene markers to detect several types of cancers are summarized in the following. The application of microarray to cancer detection is also discussed by other authors in this volume.

Through the use of the principle of specific DNA base pairing (i.e., A-T and G-C) DNA microarrays facilitate large-scale analysis either of messenger ribonucleic acid (mRNA) abundance (an indicator of gene expression), polymorphism within a population, or detection of new genes as unknown DNA sequences can be analyzed. Probes are of two types. cDNAs are often selected, either from databases or at random, and

allow the use of anonymous sequences. However, these are polymerase chain reaction (PCR) products that can be both costly and time-consuming to produce. Oligonucleotides or peptide nucleic acids are other types. These are smaller than cDNAs, allowing more spots per array, but potentially may give confusing results of sequences occurring repetitively through the genome, and may cross-hybridize.

Oligonucleotides or peptide nucleic acids require the identification and sequencing of regions of interest before they can be synthesized but do not require PCR. Peptide nucleic acids are thought to have higher affinity for the target and so may yield more accurate results. The selected probes are deposited and immobilized on a suitable substrate. Glass, silicon, nylon, nitrocellulose membranes, gels, and beads have all been used. Glass is particularly suited, being durable and nonporous, allowing the covalent attachment of probes and having a low background fluorescence. The deposition of probes is computer-aided, with up to 10,000 cDNA or 250,000 oligonucleotide spots now being routinely produced per cm^2 (Maughan *et al.*, 2001). The probes are either spotted by pins or capillary tubes or blown onto the substrate. Oligonucleotides can be synthesized *in situ* onto the substrate, using a process similar to photolithography.

DNA Microarrays in Medical Practice

Today DNA microarrays are of tremendous importance for various applications regarding nucleic acid

Handbook of Immunohistochemistry and in situ Hybridization of Human Carcinomas, Volume 1: Molecular Genetics; Lung and Breast Carcinomas

49

Copyright © 2004 by Elsevier (USA)
All rights reserved.

analyses, such as the monitoring of mRNA expression, the sequencing of DNA fragments for the genotyping of single-nucleotide polymorphism, and the detection of viruses and other pathogens. The DNA microarray has become a diagnostic assay system of ever-increasing importance to a wide range of biotechnical and biomedical applications. This technology promises to enhance cancer research by allowing improved conservation of tissue resources and experimental reagents, improved internal experimental control, and increased tissue sample throughput.

Technical advances in DNA microarrays have made it possible to miniaturize this DNA probe detection. Instead of detecting and studying one gene at a time, microarrays allow thousands of specific DNA or RNA sequences to be detected simultaneously on a glass or silica slide only 1–2 cm square. Thus, through global analysis of gene expression, the function of genes previously identified only by their DNA sequence is now being discovered almost as a matter of routine. The increased clinical information provided by microarrays should ensure their entry into routine clinical practice within the next 2–3 years. For example, DNA microarrays have recently been applied for classification of human tumors. The high-density tissue microarray technology was used for examining E-cadherin in a wide spectrum of well-characterized prostate cancers ranging from relatively low-grade clinically localized prostate tumors to hormone-refractory metastatic prostate tumors (Rubin *et al.*, 2001). E-cadherin was down-regulated in localized prostate cancer, while it showed strong expression in metastatic cancer, as determined by anti–E-cadherin antibody (HECD-1). In addition, cDNA microarrays are used for the diagnosis of hepatocellular and colorectal carcinomas and of neuroblastoma (Beheshtel *et al.*, 2003).

Applications of DNA Microarrays to Molecular Pathology

The applications of the microarray technology to elucidate disease mechanisms are vast. It has important applications to the diagnosis of disease, especially cancer. These arrays have begun to provide unprecedented amounts of data regarding the genetic alterations found in carcinogenesis. New genes up- or down-regulated in carcinomas themselves can also be investigated. This technology has proven capable of identifying genes involved in tumor development, invasiveness, and metastasis. Thus, possible genes and proteins for drug and gene targeting may be identified, with enormous therapeutic implications.

Microarray technology can provide analysis of the prognostic benefit of a myriad of potential targets on large cohorts of patient samples. This technology can be used in RNA expression studies as a means of identifying signaling pathways regulated by key genes implicated in tumorigenesis. Molecular expression profiles using oligonucleotide or cDNA-based microarrays have been used to derive molecular-based identification and classification of a number of cancer types; a number of recent examples are given in the following.

Approximately 90% of human cancers are epithelial in origin and display marked aneuploidy, multiple gene amplifications and deletions, and genetic instability, making resulting downstream effects difficult to study with traditional methods. This complexity explains the clinical diversity of histologically similar tumors. Therefore, a comprehensive understanding of the genetic alterations present in all tumors is required.

Sequencing of the human genome coupled with technologic advances make it possible to understand the molecular pathology of human cancers in a global fashion. Tools are currently available or are being developed for the identification of molecular changes that occur in cancer at the DNA, RNA, and protein levels. DNA microarrays analysis of mRNA expression (expression profiling) is especially helpful in clinical oncology.

Molecular express profiles using oligonucleotide or cDNA-based microarrays have been used to derive a molecular-based classification of several cancer types, including B-cell lymphomas, malignant melanomas, and breast cancer (Perou *et al.*, 2000). A large number of novel genes with distinct patterns in high-grade and low-grade gliomas have been identified using oligonucleotide microarray analysis (Rickman *et al.*, 2001). This study provides a molecular profile for astrocytic neoplasma, which may be relevant to diagnosis and therapy. DNA microarray analysis has also been reported to be a highly useful system for identifying molecular markers for various stages of chronic myeloidleukemia (Ohmine *et al.*, 2001).

Giordano *et al.* (2002) have used this technique for generating transcriptional profiles of benign and malignant adrenocortical tumors, as well as normal adrenal cortex and macronodular hyperplasia. Hendriks *et al.* (2002) have shown that immunohistochemistry (IHC) and tissue microarrays can be reliably used to simultaneously screen a large number of tumors from suspected hereditary nonpolyposis colorectal cancer patients. Microarray studies by Alcorta *et al.* (2002) have shown significant gene expression changes in circulating leukocytes in patients with renal diseases such as systemic lupus erythematosus and IgA nephropathy.

Delpuech *et al.* (2001) have used cDNA microarray analysis for identifying the genes related to the progression of hepatocellular carcinoma. This carcinoma is one of the most common malignant tumors worldwide. This study compared the gene expression profiles of well, moderately, or poorly differentiated hepatocellular carcinomas. Most cases of this cancer develop in a setting of chronic inflammation, with chronic active hepatitis or cirrhosis; however, a subset of this cancer occurs in the absence of cirrhosis. Numerous etiological factors, including hepatitis B virus and hepatitis C virus infections, alcoholism, chemical carcinogens, and iron overload, have been associated with this disease. Infections with these viruses are the major risk factors for the development of this cancer, and a number of genetic alterations caused by these viruses have been reported (Tornillo et al., 2000).

cDNA microarray comparison studies by Al Moustafa et al. (2002) demonstrated significant changes in the expression of a large number of genes in the head and neck carcinogenesis. This cancer is the sixth most common malignancy in both males and females. In this study of 12,530 human genes, 213 genes showed marked changes; 91 genes were up-regulated, while 122 genes were down-regulated. More specifically, this study provided the first evidence that claudin-7 and connexin 31.1 were down-regulated in these cancerous cells, compared with normal cells. These genes include those associated with signal transduction, apoptosis, and cell-cell adhesion. This comprehensive, systematic approach toward marker identification should be useful in finding new targets to prevent and/or treat this cancer.

Waldmüller et al. (2002) have developed a DNA chip capable of detecting a set of 12 mutations known to cause hypertrophic cardiomyopathy. An important advantage of this chip hybridization protocol over other mutation-scanning techniques is the use of long target-DNA molecules for study. Mutations can even be detected even in amplicons as long as 800 bp. This methodology increases the detection speed of disease-related mutations.

Sanchez-Carbayo et al. (2002) used cDNA technology for validating new targets involved in bladder tumor progression. According to this study, keratin 10 and caveolin-1 define squamous differentiation and might become useful markers to further stratify bladder tumors. E-cadherin, moesin, and zyxin are also associated with tumor progression, revealing the relevance of deregulation of cell adhesin in bladder cancer progression.

Moesin expression is thought to be a significant prognostic factor associated with patient survival. The inactivation of p53, RB, and INK4A pathways has been shown to be required for the transformation and immortalization of uroepithelial cells, and their alterations are common and of predictive nature in clinical

studies of bladder cancer (Cordon-Cardo et al., 1997). Bladder cancer is one of the most common malignancies in developed countries, ranking as the sixth most frequent neoplasm.

Collagen-induced arthritis as a model of rheumatoid arthritis has been studied (Durie et al., 1994). Such studies have been used for understanding the underlying mechanisms of autoimmune arthritis because of their clinical, histologic, immunologic, and genetics similarity to rheumatoid arthritis. Global gene expression, using DNA microarrays, was analyzed in early and late collagen-induced arthritis (Thornton et al., 2002). This study revealed genes that are novel to arthritis and are involved in cell proliferation, inflammation, apoptosis, and angiogenesis.

DNA microarrays have also been used for analyzing gene expression of human histiocytic lymphoma cell line U937 at 6 hr after 1-MHz ultrasound treatment (Tabuchi et al., 2002). This study of acoustic cavitation in biological systems is of interest because of the widespread application of ultrasound in medical diagnosis and therapeutic application.

Another application of cDNA microarrays is to analyze the effects of stimuli on gene expression in the cardiovascular system (McCormick et al., 2003). It is important to know how cells and tissues respond to biochemical and mechanical stimuli. A large number of genes are differentially expressed in shear-stressed vascular endothelial cells, and these genes play a role in response to a specific stimulus. Different types of mechanical stimuli use different signaling pathways, leading to a better understanding of the mechanisms of atherosclerosis and of vascular biology.

Ultraviolet radiation (UV) is the most important physical carcinogen in the environment, and the skin is its main target. Effects of UV on the skin depends on its wavelength; 2,900,320-nm wavelength (UVB) is the most important, and it is considered the causative agent of many effects, including skin cancer. Oligonucleotide microarrays containing 6000 genes were used for quantitatively assessing changes in gene expression in primary keratinocytes after UVB irradiation (Sesto et al., 2002). In this study, 539 regulated transcripts were found, and classification of these genes revealed that several biological processes are globally affected by UVB. Significant increases were seen in the expression of genes involved in basal transcription, splicing, and translation, as well as in the proteosome-mediated degradation category. In contrast, the transcripts belonging to the metabolism and adhesion categories were strongly down-regulated.

The oligonucleotide microarrays method was used for analyzing lung adenocarcinoma in smokers and nonsmokers (Powell et al., 2002). This study indicated that the expression of GPC3 gene was lower in the healthy lung tissue of smokers than in nonsmokers, and was lower in tumors than in healthy tissue. These data suggest that GPC3 is a lung tumor suppressor gene, the expression of which may be regulated by exposure to tobacco smoke. This gene encodes glypcin 3, a glycosylphosphatidyl inositol-linked heparan sulfate proteoglycan.

Brain tumors are the third most frequent cause of cancer-related death in middle-aged males and the most frequent cause of cancer death in children. Glioma is the most common brain tumor. In spite of increasing biological and molecular information regarding such tumors, the success of their treatment remains limited.

Rickman et al. (2001) have used DNA microarrays for identifying gene expression differences between high-grade and low-grade glial tumors. They compared the transcriptional profile of 45 astrocytic tumors, including 21 glioblastoma multiformes (GBM) and 19 pilocytic astrocytomas using oligonucleotide-based microarrays. Of the 6800 genes that were analyzed, a set of 360 genes provided a molecular signature that distinguished between these two types of gliomas. A large number of novel genes were identified with distinct expression patterns in high-grade and low-grade gliomas. Overexpression of laminin α 4 chain in GBM and astrocytoma grade II by gene microarray analysis has been confirmed by semiquantitative reverse transcription-PCR and immunohistochemistry (Ljubimova et al., 2001). Up-regulation of α 4 chain-containing laminins (laminin-8) could be important for the development of glioma-induced neovascularization and glial tumor progression. Overexpression of laminin-8 may be predictive of glioma recurrence. Immunohistochemical analysis has localized the protein products of specific genes of interest to the neoplastic cells of high-grade astrocytomas. Glioblastoma multiformes, which is synonymous with grade IV astrocytomas and which is the most common type of brain tumor, is considered to be one of the most malignant human neoplasms.

Ohmine et al. (2001), using microarray analysis with the purified Blast Bank samples, have identified molecular markers for various states of chronic myeloid leukemia, as well as insight into the molecular mechanism of transformation. Hoos et al. (2001) demonstrated that fat tissue microarray-based profiling allows identification of molecular markers in Hürthle cell tumors. High Ki-67 proliferative index was associated with adverse outcome in this neoplasm. Together with down-regulation of bcl-2, high Ki-67 index may be useful for diagnosing widely invasive Hürthle cell carcinoma. Molecular alterations in the p53 gene pathway plays a role in this tumorigenesis,

but other unidentified molecular changes seem to be required to induce the malignant phenotype. Hürthle cell tumors are rare thyroid neoplasms with variable biological behavior. Recently, the microarrays method was also used for identifying downstream transcriptional targets of the BRCA1 tumor-suppressor gene as a means of defining its function (Mullan et al., 2001). This gene is implicated in the predisposition to early onset of breast and ovarian cancers.

Protein Microarrays

A variation of the DNA microarray technology is protein microarrays. This variation is expected to play a key role in bridging the gap between genomics and proteomics by providing the functional information about gene expression. The importance of protein microarrays becomes clear when one considers that in contrast to nucleic acid analysis, no biochemical target amplification scheme (e.g., PCR) is available.

It is known that mRNA levels and protein expression do not necessarily correlate. This is due to post-translation control of protein translation, a number of post-translational modifications of protein, as well as protein degradation by proteolysis. Protein functionality is often dependent on post-translational processing of the precursor protein, and regulation of cellular pathways frequently occurs by specific interaction between proteins and/or by reversible covalent modification such as phosphorylation.

While post-transcriptional control probably affects only a small number of proteins, recent estimations suggest more than 200 types of protein modifications (Meri and Baumann, 2001). Some mammalian genes code for proteins that modify other proteins (Miklos and Maleszka, 2001). This complexity in the human genome proteome may range from 100,000 to several million different protein molecules, in contrast to a considerably lower number of genes. The situation is additionally complicated by the fact that no function is known for more than 75% of the predicted proteins in multicellular organisms, and that the dynamic range of protein expression can be as large as 10^7 (Edwards et al., 2000: Kettman et al., 2002).

To obtain detailed information about a complex biological system, information on the state of many proteins is required. High-throughput protein analysis methods allow a fast, direct, and quantitative detection. Efforts are underway to expand microarray technology beyond DNA chips and establish array-based approaches to characterize proteomes (Templin et al., 2002). Such protein microarrays can be applied to understand enzyme-substrate (MacBeath and Schreiber, 2002),

DNA-protein (Bulky *et al.*, 1999), and different types of protein-protein (Zhu *et al.*, 2001) interactions.

The high-performance microarray-based analysis system, for example, was used for achieving the highest sensitivity and precision for simultaneous quantification of different cytokines (Pawlak *et al.*, 2002). Protein microarrays have also been used for highly parallel detection and quantitation of specific proteins and antibodies in complex solutions such as serum (Haab *et al.*, 2001). Advantages and problems of antibody microarrays have been reviewed by Kusnezow and Hoheisel (2002).

Limitations of Microarrays Technology

As previously discussed, conventional microarrays are well-suited for global profiling studies. They require relatively large amounts of starting material and are limited to analysis of two specimen sources, typically one control and the other experimental population. A technical challenge in the technology is that the affected tissue area or region may be too small to provide enough RNA for microarray analysis, or that extra manipulation may compromise the quality of harvested RNA. If the RNA is diluted during extraction from the assayed tissue samples, it is difficult to attribute specific gene expression alterations to the pathological change. To address this problem, Hamedeh *et al.* (2002) have evaluated the sensitivity of cDNA microarrays in detecting diluted known gene expression alterations, thus simulating relatively minor changes in the context of total tissue.

Other limitations of microarray are with respect to sensitivity and multiplexing capabilities. These limitations can be avoided by using electronic microarrays (Weidenhammer *et al.*, 2002). The electronic microarray method provides specific target detection and quantification with advantages over currently available techniques for targeted gene expression profiling and combinatorial genomics testing.

The electronic microarrays protocol comprises independent microelectrode test sites that can be electronically biased positive or negative, or left neutral, to move and concentrate charge molecules such as DNA and RNA to one or more test sites. Multiplexed gene expression profiling of mRNA targets that use electronic field-facilitated hybridization on electronic microarrays can be accomplished in as little as 5 hr. It should be noted that although real-time PCR offers tremendous sensitivity and is ideal for quantitative analysis, it is accomplished only in a few targets, and is not easily multiplexed for panels of candidate genes.

Another problem that may arise is the saturation of the hybridization signal of high-abundant transcripts. This problem arises from the truncation of the laser fluorescence signal. When the hybridization signal on the microarray is very strong, this truncation can result in serious consequences that may not be readily apparent to the user. To rectify this potential problem, Hsiao *et al.* (2002) have developed a filtering procedure that identifies a subset of genes least affected by the signal saturation. This procedure facilitates proper comparison and clustering of the gene expression data. The procedure can be obtained at www.hugeindex.org.

At least three possible solutions for improving the detection limit of microarray hybridization are available (Cho and Tiedje, 2002): (1) Increase the amount of DNA (probe) immobilized on the microarray substratum. This can be accomplished by improving microarray fabrication or developing new substrata to bind larger amounts of probes on the microarray; (2) Develop a microarray signal detection system with higher sensitivity; (3) Develop a protocol that can selectively enrich genomes or genes of interest.

In addition to the problems already discussed, variability of results obtained using microarray analysis is another limitation (Ramasaway and Golub, 2002). Sources that may cause variability in results include genetic heterogeneity within a tumor, difference in the cellular composition among tumors, difference in specimen processing, nonspecific cross-hybridization of probes, and differences among individual microarrays. Increasing the specimen number can help in understanding the range of biologic variation in an experiment. Variations in results due to technical factors can be addressed by replicating specimen preparation or array hybridization. The absence of standards for the design and implementation of expression databases is also a problem. Consequently, it is difficult to compare databases generated in different laboratories. Databases are required for efficient storage and retrieval of the information.

Statistical analysis of microarrays is still in its infancy, and methods are not yet available to facilitate automated high-throughput analysis without human intervention. As a result, difficulties arise primarily from many potential sources of random and systematic measurement error in the microarray process, and from the small number of specimens relative to the large number of variables (probes). Nadon and Shoemaker (2002) have presented an overview of statistical methods, tailored to microarrays. Standardization of this technology is required to ensure the high quality of results and discard misleading ones. Powerful computers will be needed to analyze the plethora of generated data.

References

Alcorta, D., Preston, G., Munger, W., Sullivan, P., Yang, J.J., Wage, I., Jennette, J.C., and Falk, R. 2002. Microarray studies of gene expression in circulating leukocytes in kidney diseases. *Exp. Nephrol. 10:*139–149.

Al Moustafa, A-E., Alaoui-Jamali, M.A., Batist, G., Hernandez-Perez, M., Serruta, C., Alpert, L., Black, M.J., Sladek, R., and Foulkes, W.D. 2002. Identification of genes associated with head and neck carcinogenesis by cDNA microarray comparison between matched primary normal epithelial and squamous carcinoma cells. *Oncogene 21:*2634–2640.

Beheshtel, B., Braude, L., Marrano, P., Thorner, P., Zielenska, M., and Squire, J.A. 2003. Chromosomal localization of DNA amplifications in neuroblastoma tumors using cDNA microarray comparative genomic hybridization. *Neoplasia 5:*53–62.

Bulky, M.L., Gentalen, E., Lackhart, D.J., and Church, G.M. 1999. Quantifying DNA-protein interactions by double-stranded DNA arrays. *Nat. Biotechnol. 17:*573–577.

Cho, J.-C., and Tiedje, J.M. 2002. Quantitative detection of microbial genes by using DNA microarrays. *Appl. Environ. Microbiol. 68:*1425–1430.

Cordon-Cardo, C., Zhang, Z.F., Drobnjak, M., Charytonowicz, E., Hu, S.X., Xu, H.J. 1997. Cooperative effects of p53 and pRB alterations in primary superficial bladder tumors. *Cancer Res. 57:*1217–1221.

Delpeuch, O., Trabut, J.-B., Carnot, F., Feuillard, J., Brechot, C., and Kremsdorf, D. 2001. Identification, using cDNA microarray analysis, of distinct gene expression profiles associated with pathological and virological features of hepatocellular carcinoma. *Oncogene 21:*2926–2937.

Durie, F.H., Fava, R.A., and Noelle, R.J. 1994. Collagen-induced arthritis as a model of rheumatoid arthritis. *Clin. Immunol. Immunopathol. 73:*11–18.

Edwards, A.M., Arrowsmith, C.H., and des Palliers, B. 2000. Proteomics: New tools for new era. *Mod. Drug Discovery 5:*35–44.

Giordano, T.J., Thomas, D.G., Kuick, R., Lizyness, M., Misel, D.E., Smith, A.L., Sanders, D., Aljundi, R.T., Gauger, P.G., Thompson, N.W., Taylor, J.M.G., and Hanash, S.M. 2002. Distinct transcriptional profiles of adrenocortical tumors uncovered by DNA microarray analysis. *Am. J. Pathol. 162:*521 531.

Haab, B.B., Dunham, M.J., and Brown, P.O. 2001. Protein microarrays for highly parallel detection and quantitation of specific proteins and antibodies in complex solutions. *Genome Biol. 2:*41.

Hamedeh, H.L., Bushel, P., Tucker, C.J., Martin, K., Paules, R., and Afshari, C.A. 2002. Detection of diluted gene expression alterations using cDNA microarrays. *Biotechniques 32:*1–6.

Hendriks, Y., Franken, P., Kierssen, J.W., de Leeuw, W., Wijnen, J., Dreef, E., Tops, C., Breuning, M., Brocker-Vriends, A., Vasen, H., Fodde, R., and Morreau, H. 2002. Conventional and tissue microarray immunohistochemical expression analysis of mismatch repair in hereditary colorectal tumors. *Am. J. Pathol. 162:*469–477.

Hoos, A., Urist, M.J., Stojadinovic, A., Mastorides, S., Dudas, M.E., Leung, D.H.Y., Kuo, D., Brennan, M.F., Lewis, J.L., and Cordon-Cardo, C. 2001. Validation of tissue microarrays for immunohistochemical profiling of cancer specimens using the example of human fibroblastic tumors. *Am. J. Pathol. 158:*1245–1251.

Hsiao, L.L., Jensen, R.V., Yoshida, T., Clark, K.E., Blumenstock, J.E., and Gullans, S.R. 2002. Correcting for signal saturation errors in the analysis of microarray data. *Biotechniques 32:*1–5.

Kettman, J.R., Coleclough, C., Frey, J.R., and Lefkovits, I. 2002. Clonal proteomics: One gene-family of proteins. *Proteomics 2:*624–631.

Kusnezow, W., and Hoheisel, J.D. 2002. Antibody microarrays: Promises and problems. *Biotechniques 33:*14–23.

Ljubimova, J.Y., Lakhter, A.J., Loksh, A., Yong, W.H., Riedinger, M.S., Miner, J.H., Sorokin, L.M., Ljubimova, A.V., and Black, K.L. 2001. Overexpression of α4 chain-containing laminins in human glial tumors identified by gene microarray analysis. *Cancer Res. 61:*5601–5610.

MacBeath, G., and Schreiber, S.L. 2000. Printing proteins as microarrays for high-throughput function determination. *Science 289:*1760–1763.

Maughan, N.J., Lewis, F.A., and Smith, V. 2001. An introduction to arrays. *J. Pathol. 195:*3–6.

McCormick, S.M., Frye, S.R., Eskin, S.G., Teng, C.L., Lu, C.-M., Russell, C.-G., Chittur, K.K., and McIntire, L.V. 2003. Microarray analysis of shear-stressed endothelial cells. *Biorheology 40:*5–11.

Meri, S., and Baumann, M. 2001. Proteomics: posttranslational modifications, immune responses, and current analytical tools. *Biomol. Eng. 18:*213–220.

Miklos, G.L., and Maleszka, R. 2001. Protein functions and biological contexts. *Proteomics 1:*169–178.

Mullan, P.B., McWilliams, S., Quinn, J., Andrews, H., Gilmore, P., and McCabe, N. 2001. Uncovering BRCA1-regulated signaling pathways by microarray based expression profiling. *Biochem. Soc. Trans. 29:*678–683.

Nadon, R., and Shoemaker, J. 2002. Statistical issues with microarrays: Processing and analysis. *Trends Genet. 18:*265–271.

Ohmine, K., Ota, J., Ueda, M., Ueno S.-I., Yoshida, K., Yamashita, Y., Kirito, K., Imagawa, S., Nakamura, Y., Saito, K., Akutsu, M., Mitani, K., Kano, Y., Komatsu, N., Ozawa, K., and Mano, H. 2001. Characterization of stage progression in chronic myeloid leukemia by DNA microarray with purified hematopoietic stem cells. *Oncogene 20:*8249–8257.

Pawlak, M., Schick, E., Bopp, M.A., Schneider, M.J., Oroszlan, P., and Ehrat, M. 2002. Zeposens' protein microarrays: A novel high-performance microarray platform for low abundance protein analysis. *Proteomics 2:*383–393.

Perou, C.M., Sorlie, T., Johnsen, H., Akslen, L.A., Fleige, O., Pergomenschikov, A., Williams, C., Zhu, S.X., Lonning, P.E., Borrensen-Dale, A.L., Brown, P.O., and Botstein, D. 2000. Molecular portraits of human breast tumors. *Nature 406:* 747–752.

Powell, C.A., Xu, G., Filmus, J., Busch, S., Brody, J.S., and Rothman, P.B. 2002. Oligonucleotide microarray analysis of lung adenocarcinoma in smokers and nonsmokers identifies GPC3 as a potential lung tumor suppressor. *Chest 121:*74–75.

Ramsaway, S., Ross, K.N., Lander, E.S., and Golub, T.R. 2003. A molecular signature of metastasis in primary solid tumors. *Nat. Genetics 33:*49–54.

Rickman, D.S., Bobek, M.P., Misek, D.E., Kuick, R., Blairos, M., Kurnit, D.M., Taylor, J., and Hanash, S.M. 2001. Distinctive molecular profiles of high-grade and low-grade gliomas based on oligonucleotide microarray analysis. *Cancer Res. 61:*6885–6891.

Rubin, M.A., Mucci, N.R., Figurski, J., Fecko, A., Pienta, K.J., and Day, M.L. 2001. E-cadherin expression in prostate cancer: A broad survey using high-density tissue microarray technology. *Hum. Pathol. 32:*690–697.

Sanchez-Carbayo, M., Socci, N.D., Charytonowicz, E., Lu, M., Prystowsky, M., Childs, G., and Cordon-Cardo, C. 2002. Molecular profiling of bladder cancer using cDNA microarrays: Defining histogenesis and biological phenotypes. *Cancer Res. 62:*6973–6980.

Sesto, A., Navarro, M., Burslem, F., and Jorcano, J.L. 2002. Analysis of the ultraviolet B response in primary human keratinocytes using oligonucleotide microarrays. *Proc. Nat. Acad. Sci. USA* 99:2965–2970.

Tabuchi, Y., Kondo, T., Ogawa, R., and Mori, H. 2002. DNA microarray analyses of genes elicited by ultrasound in human U937 cells. *Biochem. Biophys. Res. Commun. 290:*498–503.

Templin, M.F., Stoll, D., Schrenk, M., Traub, P.C., Vöhringer, C.F., and Joos, T.O. 2002. Protein microarray technology. *Trends Biotechnol. 20:*160–166.

Thornton, S., Sowders, D., Aronow, B., Witte, D.P., Brunner, H.I., Giannini, E.H., and Hirsch, R. 2002. DNA microarray analysis reveals novel gene expression profiles in collagen-induced arthritis. *Clin. Immunol. 105:*155–168.

Tornillo, L., Carafa, V., Richter, J., Sauter, G., Moch, H., Bianchi, L., Vecchione, R., and Terracciano, L.M. 2000. Marked genetic similarities between hepatitis B virus-positive and hepatitis C virus-positive hepatocellular carcinomas. *J. Pathol. 192:* 307–312.

Waldmüller, S., Freund, P., Mauch, S., Toder, R., and Vosberg, H-P. 2002. Low-density DNA microarray are versatile tools to screen for known mutations in hypertrophic cardiomyopathy. *Hum. Mutat. 19:*560–569.

Weidenhammer, E.M., Kahl, B.F., Wang, L., Duhon, M., Jackson, J.A., Slater, M., and Xu, X. 2002. Multiplexed, targeted gene expression profiling and genetic analysis on electronic microarrays. *Clin. Chem. 48:*1873–1882.

Zhu, H., Bilgin, M., Bangham, R., Hall, D., Casamayor, A., Bertone, P., Lane, N., Jansen, R., Bidlingmaier, S., Houfek, T., Mitchell, T., Miller, P., Dean, R.A., Gerstein, M., and Snyder, M. 2001. Global analysis of protein activities using proteome chips. *Science 293:*2101–2105.

3

Tissue Microarrays and Their Modifications in High-Throughput Analysis of Clinical Specimens

Marek Skacel, David G. Hicks, and Raymond R. Tubbs

Introduction

The recent availability of new high-throughput molecular genetic analyses offers new opportunities to decipher the complex nature of human diseases at the molecular level. The (near) complete knowledge of the human genome sequence has started to generate countless exciting discoveries in molecular biology, but also introduced numerous new challenges for both basic science and translational research. The great expectations in this "postgenome" era include a better understanding of the diseases and the identification of new diagnostic and prognostic markers and therapeutic targets. There is currently much interest in implementing newly discovered genomic platforms in numerous clinical applications, including global profiling of gene and protein expression, studying protein-protein interactions, identifying biomarkers, and conducting a comparative analysis of normal and disease-derived tissues and cell populations, as well as in drug target discovery. Because an important step in the validation of a newly discovered target/marker is its confirmation and introduction as a routine test in a hospital setting, it is critical to study various

molecular genetic events *in situ* (i.e., at the level of cells or tissues obtained from clinical or experimental samples).

Until recently, pathologists examined sections of archival diseased tissues from patients with known follow-up information slide by slide. However, the capability to study high numbers of clinical specimens in a rapid fashion is an important prerequisite for demonstrating the clinical benefits for patients and timely translation of the new discoveries from basic science to clinical practice. The conventional approach of subjecting hundreds of separate tissue sections to immunohistochemical staining or *in situ* hybridization (ISH) is time-consuming, laborious, and expensive. The tissue microarray (TMA) technology allows simultaneous molecular profiling of thousands of samples arrayed onto glass slides at the DNA, RNA, and protein level. This technique enables researchers to perform large-scale studies and use immunohistochemistry (IHC), fluorescence *in situ* hybridization (FISH), or RNA *in situ* hybridization.

Using TMAs, the analysis of the correlation between molecular alterations, morphologic, and clinical features of the studied entities can be performed

Handbook of Immunohistochemistry and in situ Hybridization of Human Carcinomas, Volume 1: Molecular Genetics; Lung and Breast Carcinomas

57

Copyright © 2004 by Elsevier (USA)
All rights reserved.

substantially faster and at dramatically lower costs compared with conventional approaches. Importantly, this approach allows performing all analyses under laboratory conditions identical for all samples in the array, resulting in improved standardization. Furthermore, this technology is less exhausting for the finite original donor material, and as such allows performing higher number of assays per sample. This technique represents an important tool for rapid and cost-effective correlation between the global changes in gene/protein expression patterns and the *in situ* changes occurring at the cellular/tissue level. Given the invaluable preservation of the topomorphologic correlates in the experimental or clinical samples used to assemble the tissue arrays, the changes resulting from genomic alterations, transcriptional, post-transcriptional or post-translational modifications, as well as shifts of proteins occurring between different cellular compartments, can be assessed in a highly precise fashion.

This chapter provides a review of the tissue microarray technology, discusses its modifications and suitability for various applications, and emphasizes several technical aspects of the TMA design and construction. The validity of the results generated by using tissue microarrays in clinical research is also briefly discussed, and recent publications on this subject are provided for reference.

Historical Background and Evolution of Tissue Microarrays

The basic concept of the tissue microarray is a novel modification of the original method of a multitumor tissue block proposed and published by Hector Battifora in 1986 and known colloquially as the "sausage block" (Battifora, 1986). This early predecessor of today's tissue microarrays was limited to embedding relatively large specimens without the use of specialized supplies. Although improved a few years later as a "checkerboard tissue block" the construction of such a block was still time-consuming and labor-intensive, requiring melting and re-embedding of the studied tissue samples (Battifora and Mehta, 1990). In addition, only larger specimens were suitable for such processing, and special equipment such as a multiblade knife, an embedding mold, and a cubal cassette were needed to prepare the necessary "tissue rods." This laborious rod preparation was replaced by a direct punching of tissue cores from a paraffin block using a cannula (Wan *et al.*, 1987), an approach also adopted in the contemporary tissue array concept.

The Contemporary Tissue Microarrays

The tissue microarray workstation significantly advanced the precision and sophistication of the concept of using a large number of small tissue segments in a single paraffin block. Using a custom-built instrument equipped with stainless steel thin-wall needles (Beecham Instruments, Hackensack, NJ; Figure 7A), core tissue biopsies are taken from carefully selected morphologically representative areas of the original paraffin blocks ("donor" blocks) and arrayed into a new "recipient" paraffin block (Figure 7B). A set of digital micrometers allows precise placement and spacing of the tissue cores in the recipient block. The resulting tissue microarray can contain several hundred to more than a thousand tissue biopsies, depending on the core diameter used for construction. Such a block provides approximately 100 consecutive 5-µm sections of the identical set of tissue samples, depending on the thickness of the tissue in the original paraffin blocks.

Technical Aspects of Tissue Microarray Construction and Processing

The construction of a tissue microarray is a relatively uncomplicated procedure, although it requires a certain amount of practice to generate good-quality arrays. With sufficient experience, a construction of an average medium-density microarray can be accomplished within a few hours to several days (average punch speed: 30–70 cores/hr). There are several important issues an investigator should be familiar with in order to perform a successful TMA-based study.

Selection of an Adequate Tissue Core Diameter

The diameter of the sampling needle should be carefully selected to reflect the particular objectives of the study, for which the array will be used. The needle diameter does not only affect the number of cores that can be hosted by one microarray block but, more importantly, it determines the amount of tissue available for analysis (Figures 7C and 7D). The core diameter of the needles supplied with the Beecher arrayer ranges from 0.6 mm to 2.0 mm. Although the key concept of TMAs is to increase the number of samples in a single paraffin block to a maximum using the smallest sample core possible, this extreme may not be necessary for many studies involving a relatively modest number of specimens (up to a few hundred). For large studies and centralized multicenter analyses of thousands of cases, the high-density arrays certainly offer the best utilization

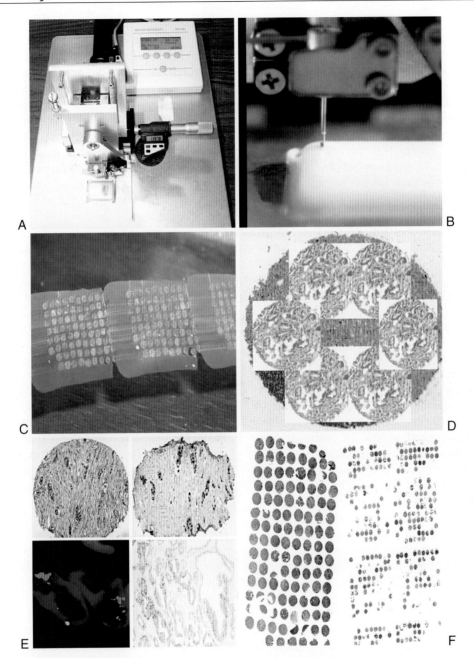

Figure 7 A: The tissue microarray workstation (Beecham Instruments, Hackensack, NJ): recipient paraffin block (bottom center), two stainless steel needle cores in a mobile arm, and digital precision device (in this case equipped with a motorized arm; right). **B:** Inserting a tissue core into an array using a 0.6-mm-diameter sampling core. **C:** Cutting a ribbon from a tissue microarray block using a microtome and a routine waterbath technique. An excellent preservation of virtually all of the tissue discs can be noted. **D:** An illustrative comparison of the tissue content of a 1.5-mm tissue disc (background) and that of a 0.6-mm disc (foreground). **E:** Techniques applicable to tissue microarrays: A breast cancer disc stained with hematoxylin-eosin (upper left; 100X), immunostained for estrogen receptor (upper right; 100X), and hybridized with probes to the HER-2/neu gene (red) and chromosome 17 centromere (green) using FISH (lower left). A prostate cancer tissue disc submitted to mRNA ISH with a prostate-specific antigen (PSA) riboprobe (lower right). **F:** Immunohistochemically stained tissue microarray sections containing tissue cores of small and large diameter (left—1.5-mm, right—0.6-mm).

of the TMA approach. On the other hand, one paraffin block can easily accommodate more than 200 larger-diameter cores, making the analysis of larger tissue discs easier. If a study includes only a moderate number of samples, this "lower-density" approach may be preferable. Even if several TMAs need to be constructed, this still represents a tremendous savings in time and labor compared with conventional sectioning.

It is important to realize that the construction and sectioning of larger-core arrays is substantially easier, and the larger tissue discs are subject to fewer artifacts, which can occur during processing the TMA. This is particularly important because the tissue disks may become partially folded or fragmented during sectioning. In such situations, the small discs may preclude reliable analysis, whereas the larger discs usually contain sufficient intact tissue for evaluation. In addition, IHC may yield indeterminate results when the tissue discs become disrupted or the lesion is situated at the edge of the tissue disc, which is subject to edge artifact. Using larger-core diameter samples also provides better margin of error because small areas of interest can be missed more easily when using a small needle core. This may be less of a problem in diffusely growing neoplasms (e.g., lymphomas) than in the cases where specific areas of the neoplasm ought to be analyzed (e.g., infiltrating component of carcinomas). On the other hand, using the small-diameter sampling core enables selective "microdissection" of different components of the studied lesion (e.g., specific Gleason patterns of prostatic adenocarcinoma) and, thus, allows differential analysis (Skacel *et al.*, 2001). Another advantage of using the small core is that the original specimen remains relatively intact and available for further diagnostic or research uses in the archives. This is true for the use of the larger-diameter core only when the original lesions are of sufficient size, such as those from most resection specimens. Thus, when selecting the core diameter, both the nature of the tumor to be studied, the size of the tissue in the donor blocks, as well as the specific aims of the study, should be thoroughly considered.

Placement of Tissue Cores into the Recipient Block

An extraordinary attention to a precise placement of each tissue core in an appropriate depth in the recipient block is the key factor affecting the availability of the tissue discs in the sections obtained from the array. Because the length of the tissue cores retrieved from archival tissues will inherently vary as a result of the nature of the original tissues, as well as the number of sections previously cut from each of the donor paraffin blocks, the cores from a significant proportion of cases

may not span the entire depth of the recipient paraffin block. It is imperative to ensure that the insertion of each tissue core into the array takes place under a close visual control in order to avoid placing the cores too deep into the block. There is no effective means of correcting such mistake without violating the integrity of the array. Therefore, it is preferable to leave the cores protruding minimally above the paraffin surface, rather than to undergo the risk that the tissue will not be accessible until a significant portion of the array has been cut away.

When the construction is complete, a very gentle pressure can be applied to the surface of the array using either a glass slide (pushing) or a thick pencil (rolling over the array surface) to push the cores deeper into the paraffin block. This pressure cannot be too excessive and the extent of the protrusion of the cores above the paraffin surface cannot be more than a fraction of a millimeter to avoid any disruption of the paraffin mass, which could violate the firm support of the cores in the array. The latter may result in "traveling" of the tissue discs across the array sections, which can make the orientation within the array extremely difficult.

Sectioning of the Tissue Microarray

One of the most challenging steps technically in the tissue microarray processing is sectioning of the recipient blocks. The original method refers to using a paraffin sectioning aid system that includes using adhesive-coated slides, adhesive tape, and a UV lamp (Instrumedics, Hackensack, NJ) (Kononen *et al.*, 1998). Although this may be the first method a novel user of tissue microarrays might try, using a routine cutting method employing a water bath and electrostatically charged slides such as those routinely used for IHC provides superior results. Several measures proved to be helpful when using the latter approach, including short-term soaking of the blocks in a low-concentration detergent solution followed by cooling on ice prior to cutting. Using this method, generally between 90–95% of the tissue cores remain in the array sections when processing the small diameter core microarrays, and close to 100% of tissue disks can be routinely transferred when the large diameter cores are used. In respect to both IHC and FISH, avoiding the use of the adhesive slides provides much cleaner results (Skacel *et al.*, 2002).

Recent Advances in TMA Construction Automation

Recently, two modifications of the original manual workstation have been offered by Beecher Instruments (www.beecherinstruments.com). The first is a semiautomated version equipped with motorized arms, which allows automatic spacing between the cores in the array at a predefined interval (such motorized arms can also be attached to the existing manual arrayer). Although significantly increasing the costs of the arrayer compared with the basic manual device, this improved version obviates the need for tiresome manipulation with the micrometers and thus eliminates one of the most time-consuming components of the arraying process, which can be a source of potential construction errors.

The second recently available alternative is a fully automated system consisting of a computer and a construction unit, which accepts a full range of punch sizes from 0.6 mm to 3 mm, and can produce up to 26 replicate array blocks in a single run. An optional autoloader can increase the system's capacity to 500 donor blocks and 30 recipients. The construction is a video-monitored robotized operation, for which the investigator can choose any combination of recipient and donor blocks to fit a particular study design. The system comes equipped with a high-performance PC preloaded with custom software for mapping donor cores to a single or multiple recipient cassettes.

Advanced options allow the user full control over many of the process variables that optimize the successful creation of recipients given a large diversity of donor tissue characteristics. Areas of interest are selected from the donor block using a corresponding hematoxylin-eosin–stained slide viewed through a stereomicroscope and visualized on the computer screen. As the user moves the stage of the microscope, the donor block moves under its needle as well. After the entire platen of blocks is mapped, the machine will begin to punch cores from the donor blocks, transfer them to the recipient block, and replace the holes in the donor blocks by a core of blank paraffin from the recipient block. This process, termed "back filling," increases the stability of the tissue in the donor block and prevents it from collapsing when many adjacent cores are punched out during the array construction. This highly sophisticated instrument can transfer between 120 and 180 cores per hour, and it has been shown to reduce the time required for construction of three replicas of an approximately 500 core arrays from 6–7 weeks to about 4 days (Schumacher and Rubin, 2003).

As might be expected, the fully automated arrayer also allows far more precise placement of cores in both x-y and z-planes (i.e., the depth of the cores is much more uniform compared with the manual instrument). When used to create replicate arrays, this system allows for maximum utilization of the donor blocks. One reported potential problem resulting from arraying unusually hard tissues (bone marrow or bone) is the potential for bending the sampling needles, which may result in permanent damage to the recipient block (Schumacher and Rubin, 2003). Although the fully automated arrayer allows researchers to markedly shorten the tissue array construction time and lower the technician-time costs, the initial investment on this system is probably beyond the reach of most low-volume TMA users.

Modifications and Variations of the Tissue Microarray Concept

MaxArray

A technique somewhat different from the Beecher Instruments system is used by a commercial service called MaxArray (Zymed Laboratories Inc., San Francisco, CA). Although the mode of obtaining the cores from the donor paraffin block is similar to the one described for the TMA workstation, punching the areas of interest is performed by the investigator using a hand-held device (provided as part of this service). Properly labeled tissue cores are then shipped to the company in microfuge tubes. The subsequent embedding of the cores takes place at the commercial facility and is performed according to a map prepared by the investigator. The MaxArray is constructed by melting all tissue cores together into a new homogenous paraffin block, a major difference from the "cold paraffin" approach, which forms the basis of the Beecher system. Although using this commercially made TMA may represent a smaller investment compared with the acquisition of the Beecher system, the relatively high cost of MaxArrays could be cost-prohibitive in the long-term. Further information about this service is available at www.zymed.com.

Only limited data exist comparing the "cold" and "warm" approach in terms of TMA characteristics. Some authors have suggested that the MaxArray system provides lower rates of tissue disc loss during processing (Rudiger et al., 2002). However, because the quality of the resulting TMA constructed on the Beecher manual arrayer is largely user-dependent, a direct comparison of both systems is difficult.

Homemade Tissue Microarrays

A group of investigators in Europe recently reported their experience with an inexpensive alternative to the commercial TMA construction systems. Using a routinely used steel embedding mold ($37 \times 24 \times 5\,\text{mm}$), a plastic embedding cassette, and a computer numerical control-drilling machine with a drill stand, template holes can be drilled into the recipient paraffin block. Following the removal of the steel mold, tissue cores from the donor paraffin blocks can be manually transferred into the recipient block using a hypodermic needle or disposable tissue puncher. According to the authors, this approach allows construction of TMAs containing up to 700 cores, with a total investment of less than $300 (Vogel *et al.*, 2003).

Midiarrays

Tissue midiarrays (TMDA) are another inexpensive alternative that have been shown to provide a practical solution mainly for prospective lower-throughput applications, which benefit from using redundant surgical pathology material. It can also be used in studies of archival material involving lesions of sufficient homogeneity. This system consists of metal cassette holder, which accommodates recipient blocks of 2-cm depth. A novel manual coring device of 3-mm in diameter is used to obtain tissue cores from tissue blocks prepared from redundant tissue. An important prerequisite of this approach is to identify areas sufficiently homogeneous, to ensure the presence of representative tissue throughout the full 2-cm depth of the TMDA block. Such sampling can be achieved by coring the paraffin blocks from their side instead of from the top surface. These cores are then embedded into a recipient block in a specific fashion using hot paraffin. An excellent stability of the tissue cores within the array, an essentially complete elimination of loss of tissue discs during cutting, and the ability to extend the life of such array blocks to prepare an excess of 300 consecutive sections are all important features of this approach. A preliminary validation study performed by IHC and FISH on a series of breast carcinomas suggested that TMDAs are useful for numerous practical applications, including preparation of comprehensive control blocks for IHC, reagent validations, and interlaboratory quality assurance surveys (Hicks *et al.*, 2003).

Frozen Tissue Microarrays

One difficulty with paraffin-embedded tissue relates to the antigenic changes in proteins and mRNA degradation induced by the fixation and embedding process. The construction of a TMA from frozen tissue would obviate these problems; however, it is much more challenging compared with that made from paraffin-embedded blocks (Fejzo and Slamon, 2001). The technical challenges associated with such procedures include a build-up of ice on the working surfaces due to a necessity to maintain a low temperature of the entire construction unit, the variable homogeneity of frozen material, as well as the fact that the supplies of frozen tissues are much less abundant than those of paraffin-embedded specimens. A limited selection of premade frozen TMAs is available from Diomeda Life Sciences, Inc. (www.diomeda.com), and a frozen tissue microarray workstation is currently being developed at Beecher Instruments.

Techniques Applicable to Tissue Microarrays

Virtually any technique applicable to conventional histologic sections can be applied to the TMA sections. Besides staining with hematoxylin-eosin, detection of DNA, RNA, or protein targets in each of the hundreds of specimens in the array using IHC, FISH, or RNA ISH is possible (Figures 7A and 7B). This allows researchers to perform parallel analysis of an excess of 100 targets of interest in each microarray block. In addition since all of the tissue samples in the array are pretreated and stained/hybridized under identical conditions, such analysis can be performed under highly standardized conditions, minimizing run-to-run variability in staining.

Data Management

One of the most powerful capacities of TMAs is their employment in studies correlating the morphologic, genotypic, and phenotypic characteristics of the studied entities in respect of their clinical features and disease outcome. Therefore, every well-designed study will include creating a comprehensive database amenable to comfortably importing data from each particular tissue array experiment. The exact identification of the tissue cores from each patient is imperative in this process. In a manual format, the utilization of Excel or Access tracking sheets represents the most widely used method for recording the results for a subsequent statistical analysis. The results from a TMA can be read out quickly and entered simultaneously into a computer database containing all remaining data on the study subjects. This allows a rapid transition of the results into standard statistical software packages. Each antibody or probe used to analyze a given array not only adds to that array value, but also increases the data complexity.

Because it is possible to perform hundreds of different stains per case using TMAs, there will be a cumulative accrual of data linked to each of the thousands of individual cases, leading to an urgent need for tools to handle this data efficiently and effectively. Therefore, more sophisticated approaches have been recently developed, which allow researchers to export the data into Web-based database structures, create formats suitable for hierarchial clustering analysis, and facilitate interinstitutional collaborations by providing investigators online access to the complete raw data (Liu et al., 2002; Manley et al., 2001). Such systems include links to high-resolution images of staining results, and thus allow reliable evaluating and scoring of the staining patterns in every particular core of the array, even by collaborators from remote locations. An online demonstration of such a system is available at www.genome.stanford. edu/TMA/explore.shtml.

Validity of the Tissue Microarray Results and Tissue Heterogeneity Issues

Since the seminal description of the tissue microarray by Kononen et al. (1998), multiple studies have been published that detailed the utility of this approach to the cost-effective and reliable evaluation of a large number of tissue samples (Camp et al., 2000; Hoos et al., 2001; Manley et al., 2001; Moch et al., 1999; Richter et al., 2000, Sallinen et al., 2000; Schraml et al., 1999; Skacel et al., 2002). Initial concerns about technical difficulties in creating and staining the TMAs were rapidly put to rest. It is the experience of most investigators that TMA paraffin blocks can be handled like regular blocks for cutting and staining (see the section on technical aspects of TMAs).

A more significant logical concern was the demonstration of the validity of the tissue microarray approach compared with the analysis of the conventional tissue. Due to the known heterogeneity of tumors, it was feared that the limited sampling represented by the tissue microarray may not fully represent the original tumors. Although it is possible that some alterations may not be detected when restricting the samples to the 0.6-mm core diameter, most tissue microarray studies published in the literature to date have confirmed the data obtained previously from conventional studies. To name at least a few, in a multitumor microarray study using FISH, Schraml et al. (1999) demonstrated that 92% of the known gene amplifications can be detected using this approach when at least 25 cases per tumor type are arrayed. Evaluating replica arrays and comparing their results with those obtained from conventional sections, the work by Moch et al. (1999) suggests that tumor heterogeneity may not significantly influence the identification of prognostic markers in breast and bladder cancers. Sallinen et al. (2000) reported little variation in p53 overexpression by IHC when comparing standard sections and arrayed tissues from gliomas. Camp et al. (2000) validated the results of arrayed breast carcinomas with conventional sections for estrogen (ER) and progesteron (PR)

receptors and HER-2/neu oncogene expression by IHC, and found that using one or two 0.6-mm discs is comparable to analysis of a whole tissue section in more than 95% of cases. Proverbs-Singh et al. (2001) in a Ki-67 IHC analysis of prostate cancer established three 0.6-mm cores as optimal sampling to reflect the full range of expression. Similarly, Nocito et al. (2001) studied over 2000 bladder cancers and showed that results from 4 replica TMAs containing 0.6-mm cores from different areas of each of the individual tumors yielded highly concordant information on tumor grade and proliferative activity, when compared with the whole sections of the tumors studied. Importantly, Tohorst et al. (2001) found that in a series of 553 breast cancers, even a single core of tumor was sufficient to demonstrate the prognostic significance of estrogen and progesteron receptor expression, as well as p53 immunostaining, and that the immunostaining of a single core was equivalent to staining of whole sections.

Based on the evidence in the literature, obtaining three 0.6-mm cores from each case can be generally recommended to ensure adequate amount of material (minimum of two well-preserved cores) in at least 95% of cases. The relative abundance of tissue in the three cores compensates for an occasional loss of tissue disc that may occur during sectioning or can result from improper sampling. When using 1.5-mm cores, obtaining two cores per case suffices to generate representative data from virtually 100% of cases. The loss of a core during cutting or as a result of inadequate sampling occurs less commonly with the large-diameter cores than when using the small 0.6-mm cores. Even if a single core remains available for study, as a result of its large diameter, it almost always provides adequate amount of representative tissue for evaluation (i.e., twice as much as all three 0.6-mm cores together). In a well-constructed array, the loss of both large-diameter cores per case almost never occurs (Skacel et al., 2002).

Tissue Microarray Applications

There are a vast number of potential applications of TMAs. The main areas in which TMAs have already been successfully employed or where the use of this technique can be anticipated include the following:

"Clinical Outcome" Tissue Arrays, Validation of Prognostically Important Targets Identified by Genome-Wide Microarray Studies

Use of TMAs containing tissues from patients with a known treatment information and clinical outcome provides a logical follow-up platform for rapid screening of large number of potential targets identified by using cDNA or oligoncleotide microarrays. Given the potential differences between mRNA and protein expression levels, both IHC and mRNA ISH can be applied to TMAs to establish the correlation between the global expression data, topographic localization of the expression of particular proteins within the tissues, and their ultimate prognostic relevance. In addition, the utilization of custom-made riboprobes for mRNA ISH will allow for the assessment of gene expression levels in situations where either antibodies do not currently exist or existing antibodies do not currently work well in formalin-fixed, paraffin-embedded tissue.

A similar complementary approach is represented by performing FISH on TMAs for verification of DNA targets identified by metaphase- or microarray-based comparative genomic hybridization studies.

Multitumor and Multitissue Arrays: Study of Expression Patterns, Identification of New Diagnostic Markers, and Therapeutic Targets

The use of TMAs provide a unique opportunity to study the expression pattern of new markers by RNA or DNA probes or by IHC simultaneously in a spectrum of human malignancies and normal tissues. Also suitable for this purpose are cell lines, which can be embedded in paraffin and incorporated into a TMA (Hoos and Cordon-Cardo, 2001). It can be anticipated that the phenomenon of newly described diagnostic markers being highly specific in the first 6 months after they are described, with a progressive decline in specificity as more data are accumulated in the literature, will soon end with the accessibility of high-density multitumor TMAs. Similar to their utility in the evaluation of diagnostic markers, TMAs can be used for effective screening for drug targets in the pharmacogenomics industry.

Quality Assurance in Pathology

The TMA technique is a cost-effective tool for standardizing numerous applications in diagnostic and research IHC or ISH. The use of TMAs containing negative and positive controls, as well as tissues fixed under different conditions, allows rapid work-up of new antibodies, analysis of day-to-day variations in staining patterns, comparison between different lots of antibodies, and interinstitutional staining quality assessment.

Future Perspectives

The strategy of combining expression arrays (cDNA or oligonucleotide microarrays) or comparative genomic

hybridization arrays with TMAs is a powerful approach to facilitate the translation of new molecular biology findings to clinical specimens on large scale. The development of large-scale arraying resources collaborating with a large number of investigators will eventually generate a new paradigm for clinical cancer research (Moch *et al.*, 2001; Skacel *et al.*, 2002). In the future, the retrospective or prospective construction of TMA repositories from routinely processed tumors or other non-neoplastic disease entities could provide powerful tools for pathoepidemiological studies. A comprehensive acquisition of all corresponding clinical data in conjuction with the utilization of highly flexible Web-based data management systems, including digital image collection, will maximize the important resource that TMAs represent. In addition, it is not beyond the realm of possibility that someday pathologists will add tissue cores of each new tumor diagnosed at their institution to comprehensive TMAs, which in a sense could represent a tissue tumor registry for a particular institution. These archival tumor TMAs could be used to screen new diagnostic, prognostic, and treatment response markers as they become available, in a comprehensive fashion, helping to establish their utility for a given institution.

Further development of this technology includes the utilization of digital imaging and automation, preliminary experience with which was reported (Bubendorf *et al.*, 1999). Automated microscope scanners or high-resolution phosphor-imagers will enable quantitative analysis of the mRNA and protein expression or DNA content in human tissues, enable precise localization of proteins in subcellular compartments, and allow researchers to assess the shifts of proteins between these compartments in response to the disease (Camp and Rimm, 2002). This technology will enable pathologists to maintain their integral role in the application of the basic science data to the clinical specimens at the tissue level and it will further advance our ability to effectively evaluate the clinical relevance of molecular changes in human neoplasia.

References

Battifora, H. 1986. The multitumor (sausage) tissue block: Novel method for immunohistochemical antibody testing. *Lab Invest.* 55:244–248.

Battifora, H., and Mehta, P. 1990. The checkerboard tissue block: An improved multitissue control block. *Lab. Invest. 63:* 722–724.

Bubendorf, L., Kononen, J., Barlund, M., Kallioniemi, A., Grigirian, A., Sauter, G., Dougherty, E.R., and Kallioniemi, O.P. 1999. Tissue microarray FISH and digital imaging: Toward automated analysis of thousands of tumors with thousands of probes. *The American Society of Human Genetics, 49th Annual Meeting*, Abstract.

Camp, R.L., Charette, L.A., and Rimm, D.L. 2000. Validation of tissue microarray technology in breast carcinoma. *Lab. Invest.* *80:*1943–1949.

Camp, R.L., and Rimm, D.L. 2002. Automated subcellular localization and quantification of protein expression in tissue microarrays. *Nat. Med. 11:*1323–1327.

Fejzo, M.S., and Slamon, D.J. 2001. Frozen tumor tissue microarray technology for analysis of tumor RNA, DNA, and proteins. *Am. J. Pathol. 159:*1645–1650.

Hicks, D.G., McDonald, L., Kotschi, H., Skacel, M., and Tubbs, R.R. 2003. Design, construction and preliminary validation of tissue midi-arrays. *Program of the 29th Annual Symposium of the National Society for Histotechnology*, Abstract.

Hoos, A., and Cordon-Cardo, C. 2001. Tissue microarray profiling of cancer specimens and cell lines: Opportunities and limitations. *Lab. Invest. 81:*1331–1338.

Hoos, A., Urist, M.J., Stojadinovic, A., Mastorides, S., Dudas, M.E., Leung, D.H., Kuo D., Brennan, M.F., Lewis, J.L., and Cordon-Cardo, C. 2001. Validation of tissue microarrays for immunohistochemical profiling of cancer specimens using the example of human fibroblastic tumors. *Am. J. Pathol. 158:* 1245–1251.

Kononen, J., Bubendorf, L., Kallioniemi, A., Barlund, M., Schraml, P., Leight, S., Tohorst, J., Mihatsch, M.J., Sauter, G., and Kallioniemi, O.P. 1998. Tissue microarrays for high-throughput molecular profiling of tumor specimens. *Nat. Med. 4:*844–847.

Liu, C.L., Prapong, W., Natkunam, Y., Alizadeh, A., Montgomery, K., Gilks, C.B., and van de Rijn, M. 2002. Software tools for high-throughput analysis and archiving of immunohistochemistry staining data obtained with tissue microarrys. *Am. J. Pathol. 161:*1557–1565.

Manley, S., Mucci, N.R., De Marzo, A.M., and Rubin, M.A. 2001. Relational database structure to manage high-density tissue microarray data and images for pathology studies focusing on clinical outcome: The prostate specialized program of research excellence model. *Am. J. Pathol. 159:*837–843.

Moch, H., Kononen, J., Kallioniemi, A., and Sauter, G. 2001. Tissue microarrays: What will they bring to molecular and anatomic pathology? *Adv. Mol. Pathol. 8:*14–20.

Moch, H., Schraml, P., Bubendorf, L., Mirlacher, M., Kononen, J., Gasser, T., Mihatch, M.J., Kallioniemi, O.P., and Sauter, G. 1999. High throughput tissue microarray analysis to evaluate genes uncovered by cDNA microarray screening in renal cell carcinoma. *Am. J. Pathol. 154:*981-986.

Nocito, A., Bubendorf, L., Tinner, M.E., Suess, K., Wagner, U., Forster, T., Kononen, J., Fijan, A., Bruderer, J., Schmid, U., Ackermann, D., Maurer, R., Alund, G., Knonagel, H., Rist, M., Anabitarte, M., Hering, F., Hardmeier, T., Schonenberger, A., Flury, R., Jager, P., Fehr, L.J., Schraml, P., Moch, H., Mihatch, M.J., Gasser, T., and Sauter, G. 2001. Microarrays of bladder cancer tissue are highly representative of proliferative index and histological grade. *J. Pathol. 194:*349–357.

Proverbs-Singh, T., Mucci, N.R., Strawdermann, M., and Rubin, M.A. 2001. Prostate carcinoma biomarker analysis using tissue microarrays: Optimization of a tissue sampling strategy for proliferation labelling index. *Mod. Pathol. 14:* 117A (Abstract).

Richter, J., Wagner, U., Kononen, J., Fijan, A., Bruderer, J., Schmid, U., Ackermann, D., Maurer, R., Alund, G., Knonagel, H., Rist, M., Wilber, K., Anabitarte, M., Hering, F., Hardmeier, T., Schonenberger, A., Flury, R., Jager, P., Fehr, J., Sehraml, P., Moch, H., Mihatch, M.J., Gasser, T., Kallioniemi, O.P., and Sauter, G. 2000. High-throughput tissue microarray analysis of cyclin E gene amplification and overexpression in bladder cancer. *Am. J. Pathol. 157:*787–794.

Rudiger, T., Hofler, H., Kreipe, H.H., Nizze, H., Pfeifer, U., Stein, H., Dallenbac, F.E., Fischer, H.P., Mengel, M., von Waielewski, R., and Muller-Hermelink, H.K. 2002. Quality assurance in immunohistochemistry: Results of an interlaboratory trial involving 172 pathologists. *Am. J. Surg. Pathol. 26:* 873–882.

Sallinen, S.L., Sallinen, P.K., Haapasalo, H.K., Helin, H.J., Helen, P.T., Schraml., P., Kallioniemi, O.P., and Kononen, J. 2000. Identification of differently expressed genes in human gliomas by DNA microarray and tissue chip techniques. *Cancer Res. 60:*6617–6622.

Schraml, P., Kononen, J., Bubendorf, L., Moch, H., Bissig, H., Nocito, A., Mihatch, M.J., Kallioniemi, O.P., and Sauter, G. 1999. Tissue microarrays for gene amplification surveys in many different tumor types. *Clin. Cancer Res. 5:*1966–1975.

Schumacher, L., and Rubin, M.A. 2003. Utility of the automated tissue microarray machine. *Mod. Pathol. 16:*327A (Abstract).

Skacel, M., Ormsby, A.H., Pettay, J.D., Tsiftsakis, E.K., Lious, L.S., Klein, E.A., Levin, H.S., Zippe, C.D., and Tubbs, R.R. 2001. Aneusomy of chromosomes 7, 8, and 17, HER-2/neu and EGFR amplification in Gleason score 7 prostate carcinoma: A differential fluorescent *in situ* hybridization study of Gleason pattern 3 and 4 using tissue microarray. *Hum. Pathol. 32:* 1392–1397.

Skacel, M., Skilton, B., Pettay, J., and Tubbs, R.R. 2002. Tissue microarrays: A powerful tool in anatomic pathology—a review of the method and validation data. *Appl. Immun. Mol. Morph. 10:*1–6.

Tohorst, J., Bucher, C., Kononen, J., Haas P., Zuber, M., Kochli, O.R., Mross, F., Dietrich, H., Moch, H., Mihatch, M.J., Kallioniemi, O.P., and Sauter, G. 2001. Tissue microarrays for rapid linking of molecular changes to clinical endpoints. *Am. J. Pathol. 159:*2249–2256.

Vogel, U., and Bueltmann, B. 2003. A simple and cheap method to produce homemade high-density multitissue arrays. *Mod. Pathol. 16:*328A (Abstract).

Wan, W.H., Fortuna, M.B., and Furmanski, P. 1987. A rapid and efficient method for testing immunohistochemical reactivity of monoclonal antibodies against multiple tissue samples simultaneously. *J. Immuno. Methods 103:*121–129.

4

Gene Expression Profiling Using Microdissection in Cancer Tissues

Yuko Sugiyama

Introduction

Endometrial carcinomas are a common malignancy of the female genital tract. Approximately 5% of such carcinomas are classified as an endometrial papillary serous carcinoma (EPSC), which is characterized by rapid progression and a poor prognosis (Carcangiu et al., 1992; Kato et al., 1995). Identification of the genes responsible for such a feature may provide some clues about how to control this aggressive disease. There are a number of reports describing alterations of tumor-related genes in various endometrial carcinomas, including EPSC (Baker et al., 1996; Berchuck et al., 1995; Burton et al., 1998). The recent advent of the gene expression array method has made it possible to analyze thousands of genes simultaneously using in vitro or in vivo specimens (Alizadeh et al., 2000; Perou et al., 1999; Schena et al., 1995). This method may be of great value for high-throughput analysis of gene expression in clinical specimens. However, clinically resected cancer tissues contain not only cancer cells but also diverse kinds of stromal cells (e.g., fibroblasts, lymphocytes, and vascular endothelial cells). Such cells may affect the gene expression patterns and prevent an accurate analysis of gene expression in cancer cells per se. To avoid the contamination of stromal cells, microdissection is essential for gene expression profiling of clinical specimens, as described previously (Sugiyama et al., 2002). This study describes the gene expression profile of an EPSC case using the gene expression array method in combination with the laser capture microdissection (LCM) method. No amplification procedures, such as polymerase chain reaction (PCR) or T7 polymerase reaction (Luo et al., 1999; Sgroi et al., 1999), were employed in an effort to obtain proportionally quantitative results in gene expression.

MATERIALS AND METHODS

I. Tissue Preparation

    A. Immediately after hysterectomy, a few pieces of fresh tissue measuring $3 \times 3 \times 3$ mm were excised, with informed consent, from the endometrial tumor and the adjacent normal endometrium.

    B. The pieces of tissue were frozen in liquid nitrogen as rapidly as possible and were stored at $-80°C$ until the performance of LCM.

    C. Part of the tissue was fixed in formalin, embedded in paraffin, sectioned, and used for immunohistochemical studies or hematoxylin and eosin staining for histologic diagnosis.

Handbook of Immunohistochemistry and in situ Hybridization of Human Carcinomas, Volume 1: Molecular Genetics; Lung and Breast Carcinomas

67

Copyright © 2004 by Elsevier (USA)
All rights reserved.

II. Laser Capture Microdissection
   A. A piece of frozen tissue was cut into 10–20 serial sections with a thickness of 10 µm, which were mounted on uncoated glass slides.
   B. One section was stained with hematoxylin and eosin for histologic diagnosis and for use as a guide during LCM.
   C. Other sections were fixed in 70% ethanol for 1 min, followed by stepwise dehydration with 95% ethanol twice and 100% ethanol twice for 1 min each, and incubation in xylene for 1 min.
   D. After air-drying the sections, target tumor cell nests were delineated with markers on the monitor and selectively microdissected from the section according to the standard laser capture procedure (Bonner *et al.*, 1997; Emmert-Buck *et al.*, 1996) using a PixCell II LCM system (Arcturus Engineering, Mountain View, CA).
   E. Normal endometrial cells were similarly dissected from the sections of normal endometrium.
III. RNA Extraction
   A. Laser-captured cell nests were resuspended with RLT lysis buffer (Qiagen, Valencia, CA).
   B. Total ribonucleic acid (RNA) was extracted using the RNeasy Kit (Qiagen), according to the manufacturer's protocol. During this step, DNase I (Roche, Basel, Switzerland) was added to remove any contaminating genomic deoxyribonucleic acid (DNA).
IV. Expression Array Method
   A. Synthesis of complementary DNA (cDNA) probe and its hybridization to an expression array membrane were performed using the Atlas Human Cancer 1.2 Array system (Clontech, Palo Alto, CA), basically according to the manufacturer's protocol.
   B. For the synthesis of the cDNA probe, all of the total RNA obtained from the tumor cells or normal endometrial cells was reverse-transcribed by Superscript II (Gibco/BRL-Life Technologies, Gaithersburg, MD) in the presence of 35 mCi of phosphorus 32 alpha-labeled deoxyadenosine triphosphate ([alpha-^{32}P]dATP [Amersham Life Science, Arlington Heights, IL]) and CDS primer mixture (Clontech) provided with the system.
   C. The synthesized probes were purified with a Chroma Spin-200 DEPC-H_2O column (Clontech), denatured, mixed with Cot-1 DNA, and added to ExpressHyb (Clontech).
   D. An expression array membrane, prespotted with 1176 cancer-related genes, was hybridized with the probes overnight at 68°C.

   E. The membrane was washed four times in 2× saline-sodium citrate (SSC) containing 1% sodium dodecyl sulfate (SDS) at 68°C for 30 min each, in 0.1× SSC containing 0.5% SDS at 68°C for 30 min, and in 2× SSC at room temperature for 5 min.
V. Analysis and Comparison of Expression Array
   A. An imaging plate (Fuji Imaging Plate; Fuji Film, Tokyo, Japan) was exposed to the hybridized membrane for a time that was dependent on the signal intensity obtained with a test exposure for 24 hr.
   B. The plate was scanned using the Fuji BAS 2500 system (Fuji Film). Detection and quantification of signals were performed with Image Gauge Software (Fuji Film).
   C. The signal intensity of each gene was represented as the actually measured signal intensity minus the background signal intensity.
   D. The values were further normalized by the mean signal intensity of five *stably* expressed housekeeping genes. For comparison of gene expression between two membranes, the signal ratio of each gene corresponding between the two membranes was calculated by the application of Excel 98 software (Microsoft, Redmond, WA). When the ratio of the signal intensity of a gene to the background intensity (S/N ratio) was less than 1.5 on both membranes, it was excluded from analysis.
VI. Immunohistochemistry
   A. 4-µm–thick sections from the formalin-fixed, paraffin-embedded tissue sample were used.
   B. Sections were deparaffinized and rehydrated through a graded ethanol series (100% ethanol twice and 95% ethanol twice for 1 min each), incubated in 10 mM citrate buffer (pH 6), and heated in a microwave oven for 5 min at 500 W.
   C. After cooling in the buffer at room temperature, sections were incubated in methanol containing 0.3% H_2O_2 to inhibit endogenous peroxidase.
   D. As a primary antibody, the monoclonal antibody was used against P-cadherin (clone: 56, BD Transduction Laboratories, Lexington, KY), E-cadherin (clone: 36, BD Transduction Laboratories) and MMP-7 (clone: 141-7B2, Daiichi Fine Chemical, Toyama, Japan), and polyclonal antibody against VEGF-C (Santa Cruz Biotechnology, Santa Cruz, CA), with a dilution of 1:200, 1:500, 1:200, and 1:200, respectively.
   E. Sections were incubated at 4°C for 12 hr with the primary antibody.

F. Antibody-antigen complexes were detected by the streptavidin-biotin technique (Nichirei SAB-PO Kit, Tokyo, Japan). Color was developed with 0.03% 3,3′-diaminobenzidine tetrahydrochloride (Merck KGaA, Darmstadt, Germany) in 50 mM Tris-HCl buffer (pH 7.6) containing 0.006% H_2O_2.

G. Counterstaining was performed with hematoxylin. Negative controls were obtained by normal serum in the same solution as the diluted primary antibody.

VII. Evaluation of Immunohistochemical Results

A. The percentage of positive cells was evaluated on 10 consecutive high-magnification power fields (40×) by two observers. Mean values were obtained by averaging 10 counts per tissue section.

B. To evaluate the protein expression, the results were graded as follows: positive when more than 30% of the epithelial cells were stained, heterogeneously positive when up to 30% of the epithelial cells were stained, and negative when completely negative.

RESULTS

Performance of the LCM and the Expression Array

Fresh tumor tissue and normal endometrial tissue were excised after patients underwent hysterectomy. According to the histologic criteria (Kurman *et al.,* 1994), the tumor was diagnosed as endometrial papillary serous carcinoma (EPSC) (Figure 8). The primary tumor was rather small (10 × 8 mm) on the surface of the endometrium, and there was little invasion into the myometrium. However, there was severe metastasis to the para-aortic and pelvic lymph nodes. By dissecting target cell nests from frozen tissue sections using the LCM method, I obtained 5000–10,000 cells per section and a total of 1.0×10^5 and 1.2×10^5 cells from the tumor and the adjacent normal endometrium, respectively. I prepared 1.2 μg and 0.8 μg of total RNA using all the cells obtained from respective tissues. The synthesized cDNA probes were then labeled with [α–^{32}P] dATP. The synthesized cDNA probes were hybridized to a membrane on which cDNA of 1176 cancer-related genes was prespotted. To compare the pattern of gene expression between the tumor cells and the normal endometrial cells, the signal ratio of the former to the latter (T/N ratio) was calculated for each gene. When the ratio of the signal intensity of a gene relative to that of the background was less than 1.5 on both membranes, the sample was excluded from the analysis. Consequently, I could analyze 855 (72%) of the 1176 spotted genes, as follows.

Genes Up-Regulated in the Tumor Cells

Fifty-nine genes were up-regulated in the tumor cells, with a T/N ratio of more than four. Among them, seven genes were up-regulated to more than 30-fold. Of these strongly up-regulated genes, CDC28 protein kinase 1 is an essential component of the cyclin-dependent protein kinases that regulates mitosis, and transcription factor AP-4 activates both viral and cellular genes. Five genes related to tumor invasion and metastasis were also up-regulated, although the T/N ratios were not so high, including placental (P-) cadherin, integrin β6, integrin α3, vascular endothelial growth factor C precursor (VEGF-C), and matrix metalloproteinase (MMP)-7. Up-regulation of such genes may account for the aggressive biological feature of EPSC. Moreover, it was notable that six genes related to the metabolic pathways were up-regulated to more than 10-fold, possibly indicating a hypermetabolic state of the cancer cells.

Genes Down-Regulated in the Tumor Cells

There were 35 genes in the tumor cells, with a T/N ratio of less than 0.25. Among them, six genes were markedly down-regulated less than one-tenth, including insulin-like growth factor binding protein 2 (IGFBP2), which might prevent insulin-like growth factor (IGF) from transferring mitogenic signals. Interestingly, insulin-like growth factor binding protein 4 (IGFBP4), which is another inhibitor of IGF, was also down-regulated (T/N = 0.16). Concerning cell adhesion, epithelial (E-) cadherin, which plays an important role in the suppression of tumor invasion by promoting cell-cell adhesion, was down-regulated in the tumor cells (T/N = 0.22). Furthermore, tissue inhibitor of metalloproteinase 3 (TIMP-3) and two matrix metalloproteinases (MMP-2 and MMP-14) were also down-regulated. Several genes that are related to apoptosis or the inhibition of cell proliferation in various cell lines were also down-regulated, including neuroendocrine-DLG, interferon-inducible protein (AIM2), MDM2-like p53-binding protein (MDMX), and growth inhibitory factor.

Immunohistochemical Staining of P-Cadherin, VEGF-C, MMP-7, and E-Cadherin in the Tumor and the Normal Endometrium

The signal intensity on the expression array basically reflects the transcription of each gene. To confirm at protein level the differences of expression seen on the arrays, I carried out immunohistochemical staining, focusing on the molecules related to cell adhesion and invasion. In the tumor, P-cadherin, VEGF-C, and MMP-7, which were observed up-regulated in the tumor by the expression array, were stained positive in the cytoplasm of the tumor cells (Figure 8C, E, G). On the other hand,

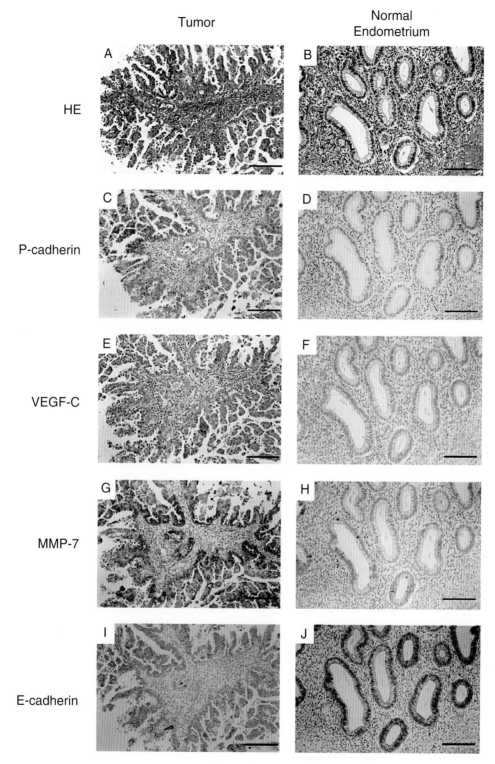

Figure 8 Immunohistochemical staining of the tumor (**A, C, E, G, I**) and the endometrium (**B, D, F, H, J**). Hematoxylin-eosin staining (**A, B**), P-cadherin (**C, D**), VEGF-C (**E, F**), MMP-7 (**G, H**), E-cadherin (**I, J**). Scale bar = 100 μm.

in the normal endometrium, P-cadherin was stained negative, VEGF-C and MMP-7 were stained heterogeneously positive (Figure 8D, F, H). E-cadherin, which was down-regulated in the expression array of the tumor cell, was stained heterogeneously positive in the cytoplasm of the scattered tumor cells (Figure 8I), whereas it was stained positive along the lateral membranes of the normal endometrium (Figure 8J). Thus, the immunohistochemical study supported the data obtained by the expression array method with the LCM.

DISCUSSION

Gene expression profile with the availability of a gene expression array method is a powerful technique for simultaneous analysis of the expression of thousands of genes. This high-throughput method is reliable when it is used to examine homogenous and abundant samples such as cultured cell lines. It is also expected to be applicable for the genetic examination of clinical specimens such as cancer tissues. However, clinically resected cancer tissue contains not only cancer cells but also various stromal cells, which may prevent accurate analysis of gene expression profile of the cancer cell *per se*. LCM is a sophisticated method for microscopical dissection of target cells from heterogeneous tissue samples (Bonner *et al.*, 1997; Emmert-Buck *et al.*, 1996), and a combination of the LCM method with the gene expression array method can avoid the problem of the complexity of the clinical specimens (Sugiyama *et al.*, 2002). In this study, the quantity of RNA obtained by the LCM was rather small (~ 1 μg/sample) despite the time-consuming procedure. However, I did not employ any enzymatic amplification procedures to enrich cDNA probes, such as the PCR or the T7 polymerase reaction (Luo *et al.*, 1999; Sgroi *et al.*, 1999), because I wished to quantitate gene expression as linearly as possible and make the data proportional to that obtained with Northern hybridization or dot blot hybridization.

Cell adhesion molecules are considered to play an important role in tumor invasion and metastasis. In the current study, I observed up-regulation of P-cadherin, integrin β6, and integrin α3, as well as down-regulation of E-cadherin.

Immunohistochemically, P-cadherin was stained positive in the tumor cells, whereas it was stained negative in the normal endometrium. It is expressed physiologically in basal cells and is often detected in squamous cell carcinomas and cervical adenocarcinomas (Aaron *et al.*, 2000; Wakita *et al.*, 1998). Furthermore, this protein is expressed preferentially in invasive rather than *in situ* lesions (Aaron *et al.*, 2000). E-cadherin is expressed normally in the adherent junction of the epithelial cells and is recognized as a powerful suppressor of invasion.

Reduction of E-cadherin expression was observed in the tumor cells at both the mRNA and protein level. Reduction of the E-cadherin expression already has been reported in uterine endometrial carcinoma (Fujimoto *et al.*, 1998).

Integrins are usually expressed in the basal surface of cells, where they form heterodimers that mediate adhesion to the extracellular matrix and thus promote cell migration (Fujimoto *et al.*, 1998). In addition, these molecules are involved in signal transduction that leads to cell transformation (Breuss *et al.*, 1995). Although I did not confirm the integrins expression using immunohistochemistry (IHC), these molecules are speculated to be involved in the invasiveness of the EPSC.

The MMPs belong to a gene family of zinc-dependent endopeptidases and are considered to be involved in the degradation of the extracellular matrix substrate in association with tumor cell invasion. Enhanced production and activation of MMP-7 was also reported in human endometrial carcinomas (Ueno *et al.*, 1999). Also in the current case, messenger RNA (mRNA) expression of MMP-7 was up-regulated and stained positively in the tumor cells immunohistochemically, indicating the invasive nature of EPSC. Moreover, TIMP-3, which binds to and inhibits MMPs, was down-regulated in the tumor cells supporting a previous study (Smid-Koopman *et al.*, 2000). However, MMP-2 and MMP-14 were down-regulated in the tumor cells in the current case, contrary to the published data (Irwin *et al.*, 1996; Maatta *et al.*, 2000). In addition, I also observed other several contradictory findings and speculate that such discrepancies might be caused by whether stromal cells were excluded by the usage of LCM or not.

The VEGFs are well known as stimulators of vascular endothelial cell proliferation, migration, and permeability (Alon *et al.*, 1995; Dvorak *et al.*, 1995). VEGF-C binds to VEGF receptor (VEGFR)-2 and -3 and induces proliferations of endothelial cells of blood vessels and lymphatic endothelial cells via a respective receptor (Jeltsch *et al.*, 1997). In gastric cancer, VEGF-C is closely related to lymphatic invasion and lymph node metastasis (Yonemura *et al.*, 1999). In this case, VEGF-C was up-regulated in the tumor cells at both the mRNA and protein levels, suggesting that VEGF-C may be involved in the lymph node metastasis in EPSC.

IGFBP-2 and -4 are inhibitors of IGF, which stimulate the proliferation and differentiation of various types of cells, including endometrial cells. They were down-regulated in this study, as demonstrated in other types of endometrial carcinoma (Roy *et al.*, 1999). Transcription factor p53 plays a pivotal role in cell growth arrest and apoptosis, and its mutations are frequently observed in various cancers, including EPSC

(King *et al.,* 1996; Tashiro *et al.,* 1997). In this study, however, I found no alterations in mRNA expression of p53 or of downstream genes such as p21$^{WAF1/Cip1}$ and Bax. There are numerous kinds of genes, which are regulated post-transcriptionally or post-translationally, such as adaptor proteins and transcription factors. As for such genes, the expression array method does not appear to be useful, especially for functional evaluation. Both MDMX and MDM2 interact directly with p53 but modulate the p53 function differently: namely, degradation of p53 is facilitated by MDM2, whereas it is inhibited by MDMX (Shvarts *et al.,* 1996). Thus, it is speculated that the down-regulation of MDMX observed in the tumor cells of this case indicates the post-translational modification of p53.

Taken together, alterations of the expression in multiple genes related to cell adhesion, invasion, cell proliferation, transcription, and general metabolism are attributed to the aggressive nature of EPSC. To confirm the current finding and reveal the carcinogenesis of EPSC, more data accumulation obtained by the expression array method with the LCM will be needed. An improvement of the techniques for both the microdissection method and the gene expression array method will be required for their clinical applications.

References

Aaron, C.H., Mitchell, I.E., Alejandro, P.S., Karen, A.K., Beatriz, L.M., Beanard, C., Norman, G.R., and Salazar, H. 2000. Cadherin expression in glandular tumors of the cervix. *Cancer* 15:2053–2058.

Alizadeh, A.A., Eisen, M.B., Davis, R.E., Ma, C., Lossos, I.S., Rosenwald, A., Boldrick, J.C., Sabet, H., Tran, T., Yu, X., Powell, J.I., Yang, L., Marti, G.E., Moore, T., Hudson, Jr. J., Lu, L., Lewis, D.B., Tibshirani, R., Sherlock, G., Chan, W.C., Greiner, T.C., Weisenburger, D.D., Armitage, J.O., Warnke, R., Levy, R., Wilson, W., Grever, M.R., Byrd, J.C., Botstein, D., Brown, P.O., and Staudt, L.M. 2000. Distinct types of diffuse larger B-cell lymphoma identified by gene expression profiling. *Nature* 403:503–511.

Alon, T., Hemo, I., Itin, A., Pe'er, J., Stone, J., and Keshet, E. 1995. Vascular endothelial growth factor acts as a survival factor for newly formed retinal vessels and has implications for retinopathy of prematurity. *Nat. Med.* 10:1024–1028.

Baker, V.V. 1996. The molecular biology of endometrial adenocarcinoma. *Clin. Obstet. Gynecol.* 39:707–715.

Berchuck, A., and Boyd, J. 1995. Molecular basis of endometrial cancer. *Cancer* 76:2034–2040.

Bonner, R.F., Emmert-Buck, M., Cole, K., Pohida, T., Chuaqui, R., Goldstein, S., and Liotta, L.A. 1997. Laser capture microdissection: Molecular analysis of tissue. *Science* 278:1481–1483.

Breuss, J.M., Gallo, J., Delisser, H.M., Klimanskaya, I.V., Folkesson, H.G., Pittet, J.F., Nishimura, S.L., Aldape, K., Landers, D.V., Carpenter, W., Gillett, N., Sheppart, D., Matthay, M.A., Albelda, S.M., Kramer, R.H., and Pytela, R. 1997. Expression of the β6 integrin subunit in development, neoplasia, and tissue repair suggests a role in epithelial remodeling. *J. Cell Sci.* 108:2241–2251.

Burton, J.L., and Wells, M. 1998. Recent advances in the histopathology and molecular pathology of carcinoma of the endometrium. *Histopathology* 33:297–303.

Carcangiu, M.L., and Chambers, J.T. 1992. Uterine papillary serous carcinoma: A study on 108 cases with emphasis on the prognostic significance of associated endometrioid carcinoma, absence of invasion, and concomitant ovarian carcinoma. *Gynecol. Oncol.* 47:298–305.

Dvorak, H.F., Brown, L.F., Detmar, M., and Dvorak, A.M. 1995. Vascular permeability factor/vascular endothelial growth factor, microvascular hyperpermeability, and angiogenesis. *Am. J. Pathol.* 146:1029–1039.

Emmert-Buck, M.R., Bonner, R.F., Smith, P.D., Chuaqui, R.F., Zhuang, Z., Goldstein, S.R., Weiss, R.A., and Liotta, L.A. 1996. Laser capture microdissection. *Science* 274:998–1001.

Fujimoto, J., Ichigo, S., Hori, M., and Tamaya, T. 1998. Expression of E-cadherin and α- and β-catenin mRNAs in uterine endometrial cancers. *Eur. J. Gynaecol. Oncol.* 19:78–81.

Irwin, J.C., Kirk, D., Gwatkin, R.B.L., Navre, M., Cannon, P., and Giudice, L.C. 1996. Human endometrial matrix metalloproteinase-2, a putative menstrual proteinase. *J. Clin. Invest.* 97:438–447.

Jeltsch, M., Kaipainen, A., Joukov, V., Meng, X., Lakso, M., Rauvala, H., Swartz, M., Fukumura, D., Jain, R.K., and Alitalo, K. 1997. Hyperplasia of lymphatic vessels in VEGF-C transgenic mice. *Science* 276:1423–1425.

Kato, D.T., Ferry, J.A., Goodman, A., Sullinger, J., Scully, R.E., Goff, B.A., Fuller, A.F., and Rice, L.W. 1995. Uterine papillary serous carcinoma (UPSC): A clinicopathologic study of 30 cases. *Gynecol. Oncol.* 59:384–389.

King, S.A., Adas, A.A., LiVolsi, V.A., Takahashi, H., Behbakht, K., McGovern, P., Benjamin, I., Rubin, S.C., and Boyd, J. 1995. Expression and mutation analysis of the p53 gene in uterine papillary serous carcinoma. *Cancer* 75:2700–2705.

Kurman, R.J., Zaino, R.J., and Norris, H.J. 1994. Endometrial carcinoma. In Kurman R.J. (ed.) *Blaustein's Pathology of the Female Genital Tract.* 4th ed. New York: Springer-Verlag, 466–471.

Luo, L., Salunga, R.C., Guo, H., Bittner, A., Joy, K.C., Galindo, J.E., Xiao, H., Rogers, K.E., Wan, J.S., Jackson, M.R., and Erlander, M.G. 1999. Gene expression profiles of laser-captured adjacent neuronal subtypes. *Nat. Med.* 1:117–122.

Maatta, M., Soini, Y., Liakka, A., and Autio-Harmainen, H. 2000. Localization of MT1-MMP, TIMP-1, TIMP-2, and TIMP-3 messenger RNA in normal, hyperplastic, and neoplastic endometrium. *Am. J. Clin. Pathol.* 114:402–411.

Perou, C.M., Jeffrey, S.S., Rijn, M., Rees, C.A., Eisen, M.B., Ross, D.T., Pergamenschikov, A., Williams, C., Zhu, S.X., Lee, J.C.F., Lashkari, D., Shalon, D., Brown, P.O., and Botstein, D. 1999. Distinctive gene expression patterns in human mammary epithelial cells and breast cancers. *Proc. Natl. Acad. Sci. USA* 96:9212–9217.

Roy, R.N., Gerulath, A.H., Cecutti, A., and Bhavnani, B.R. 1999. Discordant expression of insulin-like growth factors and their receptor messenger ribonucleic acids in endometrial carcinomas relative to normal endometrium *Mol. Cell Endocrinol.* 153:19–27.

Schena, M., Shalon, D., Davis, R.W., and Brown, P.O. 1995. Quantitative monitoring of gene expression patterns with a complementary DNA microarray. *Science* 270:467–470.

Sgroi, D.C., Teng, S., Robinson, G., LeVangie, R., Hudson Jr., J.R., and Elkahloun, A.G. 1999. *In vivo* gene expression profile analysis of human breast cancer progression. *Cancer Res.* 59:5656–5661.

Shvarts, A., Steegenga, W.T., Riteco, N., van Laar, T., Dekker, P., Bazuine, M., van Ham, R.C.A., van der Houven van Oordt, W.,

Hateboer, G., van der Eb, A.J., and Jochemsen, A.G. 1996. MDMX: A novel p53-binding protein with some functional properties of MDM2. *EMBO J. 15:*5349–5357.

Smid-Koopman, E., Blok, L.J., Chadha-Ajwani, S., Helmerhorhorst, T.J.M., Brinkmann, A.O., and Huikeshoven, F.J. 2000. Gene expression profiles of human endometrial cancer samples using a cDNA-expression array technique: Assessment of an analysis method. *Brit. J. Cancer 83:*246–251.

Sugiyama, Y., Sugiyama, K., Hirai, Y., Akiyama, F., and Hasumi, K. 2002. Microdissection is essential for gene expression profiling of clinically resected cancer tissues. *Am. J. Clin. Pathol. 117:*109–116.

Tashiro, H., Isacson, C., Levine, R., Kurman, R.J., Cho, K.R., and Hedrick, L. 1997. p53 gene mutations are common in uterine serous carcinoma and occur early in their pathogenesis. *Am. J. Pathol. 150:*177–185.

Ueno, H., Yamashita, K., Azumano, I., Inoue, M., and Okada, Y. 1999. Enhanced production and activation of matrix metalloproteinase-7 (matrilysin) in human endometrial carcinomas. *Int. J. Cancer 84:*470–477.

Wakita, H., Shirahama, S., and Furukawa, F. 1998. Distinct P-cadherin expression in cultured normal human keratinocytes and squamous cell carcinoma cell lines. *Microsc. Res. Tech. 43:*218–223.

Yonemura, Y., Endo, Y., Fujita, H., Fushida, S., Ninomiya, I., Bandou, E., Taniguchi, K., Miwa, K., Ohoyama, S., Sugiyama, K., and Sasaki, T. 1999. Role of vascular endothelial growth factor C expression in the development of lymph node metastasis in gastric cancer. *Clin. Cancer Res. 5:*1823–1829.

5

Differential Display of Gene Expression in Human Carcinomas

Roger S. Jackson II, Susanne Stein, Yong-Jig Cho, and Peng Liang

Introduction

Cancer is a disease state caused by multiple genetic alterations that progressively lead to cellular transformation and unregulated cell proliferation (Hanahan and Weinberg, 2000). Disruption of normal cellular homeostasis through activating mutations of proto-oncogenes and inactivating mutations of tumor suppressor genes generally results in altered signal transduction. This can lead to sustained growth, disruption of genomic damage repair systems, evasion of programmed cell death (apoptosis), and an increased propensity at the stage of tumor development for angiogenesis, invasion, and metastasis (Gray *et al.*, 2000; Hanahan *et al.*, 2000).

Although much has been learned to date about carcinogenesis, the molecular pathology of cancer remains to be elucidated fully. Oncologists and pathologists are continuously in search of novel cancer biomarkers—molecular signatures of a cellular phenotype that will aid in early cancer detection, pathological grading and staging, and risk assessment—in order to enable use and development of molecularly targeted therapies tailored to the patient, and to increase the rate of survival due either to earlier detection or improved therapies. Because most available biomarkers

of carcinogenesis are either applicable only to late-stage cancers or have lower predictive capability than desired, it is necessary to continue the search for new markers (Negm *et al.*, 2002).

Gene expression directs all aspects of cell biology, including apoptosis, cell cycle, development, differentiation, growth, homeostasis, and responses to the environment. Pathologic changes, such as those observed in human cancers, are driven most often by changes in gene expression resulting from epigenetic (methylation) and genetic mechanisms, such as acquisition of gene mutation(s), chromosomal alterations, and genomic instability (Hanahan *et al.*, 2000). Immunohistochemistry (IHC) and *in situ* hybridization (ISH) techniques are often employed to assess gene expression at the protein and mRNA levels, respectively, in combination with cellular localization (indicated by extracellular, cell surface, cytoplasmic, or nuclear staining patterns). Although effective, both techniques are laborious, time-consuming, and expensive processes used to screen for differential gene expression patterns and/or identify novel biomarkers of disease, such as cancer. With the exception of post-transcriptionally or post-translationally regulated genes, alterations in gene transcription generally correlate with alterations in gene translation (i.e., induction or

Handbook of Immunohistochemistry and *in situ* Hybridization of Human Carcinomas, Volume 1: Molecular Genetics; Lung and Breast Carcinomas

75

Copyright © 2004 by Elsevier (USA)
All rights reserved.

repression of an mRNA usually results in the corresponding induction or repression of the protein for which it encodes). Moreover, it is well known that altered signal transduction and/or gene expression often precedes morphological and histological changes that can be detected by IHC or ISH (Negm *et al.*, 2002). The differential display method measures the expression of a gene transcript (steady-state mRNA levels), but does not assess protein expression, and thus may only imply a correlative change in the protein level based on gene induction or repression. Therefore, mRNA differential display can be employed as a quick and reliable primary screening strategy that will facilitate, focus, and complement further validation and characterization studies better done by techniques such as IHC or ISH.

In this chapter, the procedure of differential display (DD) is presented, along with a discussion of some of the critical factors affecting the method's accuracy.

Since its introduction in 1992, DD methodology traditionally has been based on ^{33}P isotopic labeling of cDNA bands (Liang *et al.*, 1992a). This is the most commonly used DD technology because of its sensitivity, simplicity, versatility, and reproducibility. Since its inception, numerous differentially expressed genes have been identified successfully across diverse biological systems and fields of study (reviewed by Liang and Pardee, 1995). More recently, the introduction of fluorescent primer labeling and robotic technology has enabled the establishment of an accurate, high-throughput fluorescent differential display (FDD) method (Figure 9) with similar sensitivity and reproducibility to the isotopic method, yet with greater accuracy (Cho *et al.*, 2001).

Either total or poly-A RNA can be used for DD. Although RNA isolated from cell lines is the easiest to obtain with uniformly high quality, methods have been

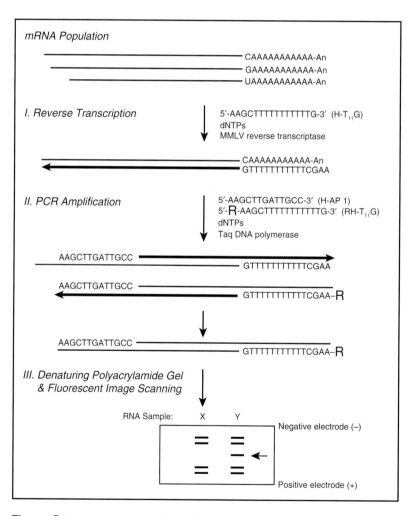

Figure 9 Schematic representation of fluorescent differential display. (Illustration courtesy of GenHunter Corporation, Nashville, TN.)

devised to isolate RNA from fresh, frozen, and paraffin-embedded tissues, such as tumors (referenced by Liang et al., 1995). Because it is essential that only high-quality RNA be used, all chromosomal DNA contamination must be removed from the RNA samples with DNase-I (with the addition of RNase inhibitor of RNase-free dH_2O) before carrying out FDD. Other strategies have been developed to reduce the amount or RNA source material necessary to do a complete FDD screen, such as using a more proficient reverse transcriptase (Bosch et al., 2000), through pre-amplification of the cDNA pool from a limited number of cells (Jing et al., 2000; Zhao et al., 1998), or through the combination of a noncompetitive carrier molecule with the source RNA (Melichar et al., 2000). Ultimately, use of such techniques may result in the further reduction of the amount of mRNA required for a complete FDD screen to the lowest possible limit. This will likely enable increased use, in both clinical and molecular studies, of valuable clinical tissue specimens that are often limited by both accessibility and quantity. Moreover, the combination of material acquisition techniques (e.g., laser capture microdissection) with gene expression methods (e.g., FDD) may allow the long-awaited region-specific analysis of tissues and/or tumors. This could lend important insight into studies of local tumor interactions with the surrounding normal tissue and either remove or reduce the influence of tissue/tumor heterogeneity on studies conducted.

The principle of FDD is to detect differential gene expression patterns by reverse transcription–polymerase chain reaction (RT-PCR). Using one of three one-base–anchored oligo-dT primers that anneal to the beginning of a subpopulation of the RNA poly-A tails, mRNA is reverse-transcribed into cDNA. This anchored oligo-dT ($H-T_{11}V$) primer consists of 11 Ts (T_{11}) with a $5'$ HindIII (AAGCTT) site, plus one additional $3'$ base V (where V may be dG, dA, or dC) that provides specificity. For FDD, fluorescent (rhodamine, red) labeled anchored $R-H-T_{11}V$ primers are combined with various arbitrary primers (H-AP primers; 13mer containing a $5'$ HindIII site) in PCR steps. Amplified PCR products up to 700 bp can be separated on a denaturing polyacrylamide sequencing gel. The FDD image can be obtained using a fluorescent laser scanner. Side-by-side comparisons of cDNA patterns between or among relevant RNA samples would reveal differences in gene expression. The cDNA fragments of interest can be retrieved from the gel, purified, and reamplified with the same set of primers (lacking fluorescent labeling of $H-T_{11}V$ anchor primer) under the same PCR conditions as in the initial FDD-PCR reactions. For further molecular characterization, the obtained reamplified PCR fragments can be cloned and sequenced.

DNA sequence analysis of these cDNA fragments by GenBank Blast search (www.ncbi.nlm.nih.gov/BLAST) may provide information as to whether a gene identified by FDD is a known, homologous to known, or novel gene. The final step of the FDD procedure is to confirm the differential expression of the obtained partial cDNAs by Northern Blot. The result of this analysis provides not only an independent confirmation of differential gene expression, but also information regarding the transcript size of the gene of interest. After confirmation, the cloned cDNA probe can be used to screen a cDNA library for a full-length clone, or as a starting point for rapid amplification of cDNA ends PCR (RACE-PCR), which is helpful for the functional characterization of the gene.

MATERIALS

1. RNApure Reagent (GenHunter, Nashville, TN).
2. MessageClean Kit (GenHunter), 10X reaction buffer: 10 mM Tris-Cl, pH 8.3, 50 mM potassium chloride (KCl), 1.5 mM magnesium chloride ($MgCl_2$).
3. Diethyl pyrocarbonate (DEPC)-treated dH_2O (GenHunter).
4. RNA loading dye (GenHunter).
5. 10X 3-(N-Morpholino)-propanesulfonic acid (MOPS) buffer: 0.4-M MOPS, pH 7.0, 0.1 M sodium acetate, 0.01 M disodium ethylenediaminetetraacetic acid (EDTA), store at room temperature in the dark.
6. 5X reverse transcription buffer: 125 mM Tris-Cl, pH 8.3, 188 mM KCl, 7.5 mM $MgCl_2$, 25 mM dithiothreitol (DTT) (GenHunter).
7. Murine Moloney Leukemia Virus (MMLV) reverse transcriptase (100 units/µl) (GenHunter).
8. Deoxyribonucleotide triphosphate (dNTP) mix (2.5 mM) (GenHunter).
9. $H-T_{11}V$ anchor primer (V = A, C, G) (2 µM) (GenHunter).
10. $R-H-T_{11}V$ anchor primer (V = A, C, G) (8 µM) (GenHunter). (Note: Rhodamine-labeled primers are light sensitive.)
11. H-AP 13mer primers (1 to 160) with 50–70% GC content (2 µM) (GenHunter).
12. 10X PCR buffer: 100 mM Tris-HCl, pH 8.4, 500 mM KCl, 15 mM $MgCl_2$, 0.01% gelatin (GenHunter).
13. FDD loading dye: 99% formamide, 1 mM EDTA, pH 8.0, 0.009% xylene cyanol FF, 0.009% bromophenol blue (GenHunter).
14. Rhodamine locator dye (GenHunter).
15. 10X Tris-Boric acid-EDTA (TBE) buffer (for 1L): 108 g Trizma base, 55 g boric acid, 3.7 g EDTA.
16. Autoclaved millipure deionized H_2O (dH_2O).
17. Taq DNA polymerase (5 units/µl) (Qiagen).

18. 20X Saline-sodium citrate (SSC) buffer: 3 M sodium chloride, 0.3 M sodium citrate, pH 7.0.

19. 20X Saline-sodium phosphate-EDTA (SSPE) (1L): 3.6 M sodium chloride, 0.2 M sodium dihydrophosphate, 0.02 M EDTA, pH 7.4.

20. 50X Denhardt's solution (for 500 mL; store at −20°C): 5 g Ficoll 400, 5 g polyvinylpyrrolidone (molecular weight 360000), 5 g bovine serum albumin (BSA) fraction V.

21. Hybridization buffer: 5X SSPE, 50% formamide, 5X Denhardt's solution, 0.1% sodium dodecyl sulfate (SDS) (store at −20°C), to which heat-denatured sheared nonhomologous salmon sperm DNA is added to 100 μg/ml just before use.

22. HotPrime DNA labeling Kit (GenHunter).

METHOD

RNA Isolation from Cell Cultures

Total RNA can be isolated with a one-step acid-phenol extraction method using the RNApure Reagent.

1. For example, remove the cell culture medium from p150 cell culture plates (Sarstedt), wash with 10–20 ml cold 1X phosphate buffer saline (PBS), and set the plates on ice. Lyse the cells by adding 2 ml RNApure reagent to each plate, spread the solution across the plates, and incubate on ice for 10 min. Scrape the cells off the plate using a cell scraper, and pipette the cell lysate into sterile 1.5-ml Eppendorf tubes. Add 150 μl chloroform per milliliter of cell lysate, and mix well by vortexing for about 10 sec. Freeze the tubes at −80°C or proceed to **step 2**.

2. Spin the tubes at maximum speed (14,000g) in an Eppendorf centrifuge at 4°C for 10 min.

3. Carefully remove the upper (aqueous, RNA-containing) phase and save into a new sterile Eppendorf tube.

4. Precipitate the RNA by adding an equal volume of 100% isopropanol to the aqueous phase, mix well by vortexing, and incubate on ice for 10 min. Spin the RNA down at maximum speed for 10 min at 4°C. Rinse the RNA pellet with 0.5–1 ml cold 70% ethanol (made with DEPC-dH$_2$O). Spin down at maximum speed again for 10 min at 4°C. Remove the ethanol and resuspend the RNA pellet in 20–50 μl DEPC-dH$_2$O. Make RNA aliquots and store at −80°C.

5. Before treatment with DNase-I, measure the RNA concentration at OD$_{260}$ with a spectrophotometer, and check the integrity (18S and 28S rRNA bands) of the RNA samples by running 2 μg of each RNA sample on a 1% agarose 7% formaldehyde gel in 1X MOPS buffer.

DNase I Treatment of Total RNA

Removal of all contaminating chromosomal DNA from the RNA sample is absolutely essential for successful DD. The MessageClean Kit is specifically designed for the complete digestion of single and double-stranded DNA.

1. Incubate 50 μl (10–50 μg) of total cellular RNA (use DEPC-dH$_2$O when diluting RNA) with 10 units (1 μl) of DNase-I (RNase free) in 5.7 μl 10X reaction buffer for 30 min at 37°C.

2. Inactivate DNase-I by adding an equal volume of phenol:chloroform (3:1) to the sample. Mix by vortexing and leave the sample on ice for 10 min. Centrifuge the sample at maximum speed in an Eppendorf centrifuge for 5 min at 4°C.

3. Save the supernatant and ethanol precipitate the RNA by adding 3 volumes of ethanol and 0.1 volumes 3 M sodium acetate pH 5.2.

4. After incubation at −80°C for 1 hr (overnight to a few days at −80°C is recommended), pellet the RNA by centrifuging at maximum speed at 4°C for 10 min. Rinse the RNA pellet with 0.5 ml of 70% ethanol (made with DEPC-dH$_2$O) and dissolve the RNA in 20 μl of DEPC-dH$_2$O.

5. Measure the RNA concentration at OD$_{260}$ with a spectrophotometer. Check the integrity of the RNA samples before and after cleaning with DNase-I by running 2-μg samples of each RNA on a 1% agarose 7% formaldehyde gel in 1X MOPS buffer. It is recommended to store the RNA samples as 1–2 μg aliquots at −80°C before using for DD to minimize freeze-thaw cycles and preserve RNA stability.

Reverse Transcription of mRNA

The success of the differential display technique is dependent on the integrity of the RNA and that it is free of chromosomal DNA contamination. Upon completion of DNase-I treatment of the RNA, the mRNA then can be reverse transcribed into cDNA.

1. Set up three reverse transcription reactions for each RNA sample in three PCR tubes (0.2–0.5 ml size, thin-walled). Each should contain one of the three different one-base-anchored H-T$_{11}$V primers (where V may be A, C, or G). For a final volume of 20 μl, combine: 9.4-μl dH$_2$O, 4-μl 5X reverse transcription buffer, 1.6-μl dNTP mix (2.5 mM), 2-μl total RNA (0.1 μg/μl freshly diluted in dH$_2$O), and 2-μl H-T$_{11}$V primer (2 μM). To minimize pipetting errors, it is recommended to use a core mix without an RNA template for each anchored oligo-dT primer, especially if two or more RNA samples are to be compared.

2. Program the thermocycler to 65°C for 5 min, 37°C for 60 min, 75°C for 5 min, and 4°C for 5 min.

3. Add 1 μl MMLV reverse transcriptase to each tube 10 min after incubation begins at 37°C in order to initiate the RT reaction. At the end of the reaction, spin the tubes briefly to collect condensation. Set tubes on ice for FDD-PCR or store at −20°C for later use.

Fluorescent Differential Display

The RNAspectra Red Kit can be used for this step as well as the previous reverse transcription reactions (also see Notes 1 to 3 at end of chapter). This step can be automated with a robotic liquid handling workstation such as the BioMek 2000 (Beckman), which can significantly increase the throughput and accuracy.

1. Set up on ice (in dim light) a 20 μl PCR reaction in thin-walled reaction tubes. For each primer set combination, use the following formula: 4.2 μl dH$_2$O, 2 μl 10X PCR buffer, 1.6-μl dNTP mix (2.5 mM), 8 μl H-AP-primer (2 μM), 2 μl R-H-T$_{11}$V (2 μM) (it has to contain the same H-T$_{11}$V primer used for RT-PCR), 2 μl of a completed RT-PCR reaction mix (template), and 0.2 μl Taq DNA polymerase. Use core mixes as much as possible to avoid pipetting errors.

2. Mix well by pipetting up and down. PCR conditions are as follows: 94°C for 20 sec, 40°C for 2 min, 72°C for 1 min. After 40 cycles, follow with 72°C for 5 min and 4°C for 5 min. After completion of PCR, store the samples in the dark and either on ice or at −20°C until ready to run the gel.

3. Prepare a 6% denaturing polyacrylamide sequencing gel in 1X TBE buffer. Let the gel polymerize for about 2 hr before use. It is recommended that one glass plate be treated with Sigmacote (Sigma, St. Louis, MO) to facilitate the separation of the plates after running. Pre-run the gel for 30 min to prepare it for sample loading. The wells must be flushed completely just before sample loading as removal of urea from the wells after the pre-run is critical.

4. Mix each PCR reaction with 8 μl FDD loading dye and incubate at 80°C for 2–3 min immediately before loading onto the gel.

5. Electrophorese for 2 hr at 60 W constant power until the xylene dye (the slower-moving dye) reaches the bottom.

6. After the gel is run, clean the exterior of the plates well with water and 70% ethanol.

7. Scan the gel on a fluorescent laser scanner using a 585-nm filter (for Rhodamine).

8. Cut out the bands of interest after careful separation of the glass plates. For orientation of the lanes,

it is very helpful to use the Rhodamine locator dye during scanning.

Purification and Reamplification of cDNA Bands from FDD

1. Soak the FDD gel slice (containing the cDNA band of interest) in 1 ml dH$_2$O for 30 min, mixing gently by finger-tipping.

2. Remove the water without taking the gel slice, and add 50 μl dH$_2$O. Boil the tube with its cap closed (and parafilm sealed) for 15 min to elute the DNA from the gel slice. Spin the tubes after cooling for 2 min at maximum speed to collect condensation and pellet the gel. Transfer the supernatant to a new tube and keep for the reamplification reaction. The tube with the gel slice can also be saved for the reamplification reaction.

3. Reamplification PCR should be carried out in a total volume of 40 μl using the same primer combination and concentration (4 μl of each 2-μM primer), but with an anchor primer (H-T$_{11}$V) lacking the fluorescent label. The PCR conditions also should be kept the same, except for the dNTP concentration, which is changed, so use 1 μl, 250 μM dNTP mix instead. The following can be used as DNA templates: a) 4–5 μl of supernatant (**step 2**) and/or b) the gel slice (**step 2**), which still contains small traces of the removed DNA.

4. Check 30 μl of each PCR sample on a 1.5% agarose gel stained with ethidium bromide. Save the remaining PCR samples at −20°C for future experiments (e.g., cloning, Northern blot). Compare the size of the reamplified PCR products with that originally found on the FDD gel. Extract the positive reamplified cDNA fragments from the agarose gel using the Qiaex II Gel Extraction Kit (Qiagen).

Sequencing and Cloning of PCR Products

One crucial advantage of FDD is the rapid identification of the cDNA sequence by direct sequencing of the PCR products without subcloning these fragments. Alternatively, after gel purification, reamplified cDNA probes can be ligated into various cloning vector systems and then subjected to DNA sequence analysis.

Confirmation of Differential Gene Expression by Northern Blot

Confirmation of differential gene expression patterns identified by FDD can be obtained through the use of independent methods such as Northern blot. Northern blots are extremely sensitive and provide additional details such as the transcript size of the gene of interest through the use of the cDNA fragments isolated from the FDD screen as probes. Northern blots can be performed following the standard procedure (Asubel *et al.*, 1993) with the use of the HotPrime DNA labeling Kit.

Although DD can be used to detect amplification, reduction, or loss of gene expression, one should be reminded that this method is unlikely to detect mutations at the DNA level directly. For diseases caused by single gene mutations that have a clear genetic component, chromosome mapping of the mutation locus should be a method of choice. It should be emphasized that the method is only a simple screening tool. Upon confirmation by Northern blot, however, the differentially expressed cDNA probe(s) might release a series of molecular studies leading to a better understanding of complex pathways.

Applications of Fluorescent Differential Display

Although FDD has the power and ease to assess differential gene expression between multiple related cell lines or tissues, the achievement of physiologically relevant results is highly dependent on the quality of the experimental system implemented with all necessary controls. For example, gene expression studies by DD or FDD can be done using RNA isolated from normal and aberrant tissues of patients with the same disease; from different developmental stages (such as tumor progression from normal epithelium to benign adenoma, carcinoma *in situ*, and metastatic carcinoma); along a temporal course of treatment or between different related treatments in a cell line of interest; and so on (Liang *et al.*, 1995). The FDD technique is most successful when the RNA samples to be compared are from closely matched, homologous populations of primary cells or cell lines (Martin *et al.*, 1999). Proper experimental design will allow quicker identification, isolation, and functional studies of genes that may be important for the process of interest, while simultaneously eliminating false positives that may randomly vary between experimental groups assessed (Liang *et al.*, 1995). Side-by-side comparison of RNA from many different yet related treatment groups may shed light on specific processes or pathways involved, rather than just simply being informative about differential gene expression.

RESULTS AND DISCUSSION

Differential display is a globally used method for identifying differentially expressed genes in eukaryotic cells. Ease of use and rapid results makes fluorescent differential display an ideal screening tool to use in order to focus and complement clinical and diagnostic studies that traditionally use IHC and/or ISH techniques. Unlike IHC and ISH, which are sporadic screening methods based on the availability of antibodies or riboprobes, DD is a systematic screening of all expressed genes in a cell through the use of multiple arbitrary and anchor primer combinations. The DD technique also allows identification of new genes and potential biomarkers for which no antibodies or probes are available (i.e., novel and rare messages can be identified). Indeed, DD has proven useful particularly for finding diagnostic markers for pathological processes in which altered gene expression has been shown to play a role. Examples of such biomarkers identified by DD include cyclin G, Mob-1, and macrophage lectin (Liang *et al.*, 1997). The DD procedure sets the stage for IHC and/or ISH studies because it allows detection of the known, homologous to known, or novel genes of interest for the process or system being studied that then can be followed up by IHC or ISH for further characterization.

The DD method has been used successfully by many groups in the study of human carcinomas, including those of the breast, colon, and prostate (reviewed by Gray *et al.*, 2000; Guan *et al.*, 2000; Martin *et al.*, 1998). More specifically, the DD method has been used to identify tumor promotor (oncogenes) or tumor suppressor genes, candidate biomarkers, and cell cycle regulated genes (reviewed by Gray *et al.*, 2000; Martin *et al.*, 1998). In addition, our lab and others have used differential display to identify and clone genes regulated by Ras (e.g., Wang *et al.*, 2002; Zhang *et al.*, 1997; Zhang *et al.*, 1998) and p53 (reviewed in Stein *et al.*, 2002; Stein *et al.*, 2003), both of which are mutated frequently in carcinomas.

Our comprehensive FDD screening using hundreds of FDD primer combinations yielded more than two dozen inducible or repressible p53-regulated genes (to be published elsewhere), in addition to the detection of known target genes, including p21[wafl] (Figure 10), HDM2, and PIG3. This provides strong validation of our nonbiased and exhaustive screening strategy for p53 target genes by FDD. Furthermore, FDD also allows the analysis of digital gene expression profiling and precise quantification of gene expression differences, unlike IHC or ISH (Cho *et al.*, 2001; Stein *et al.*, 2003). Interestingly, more than 50% of these genes

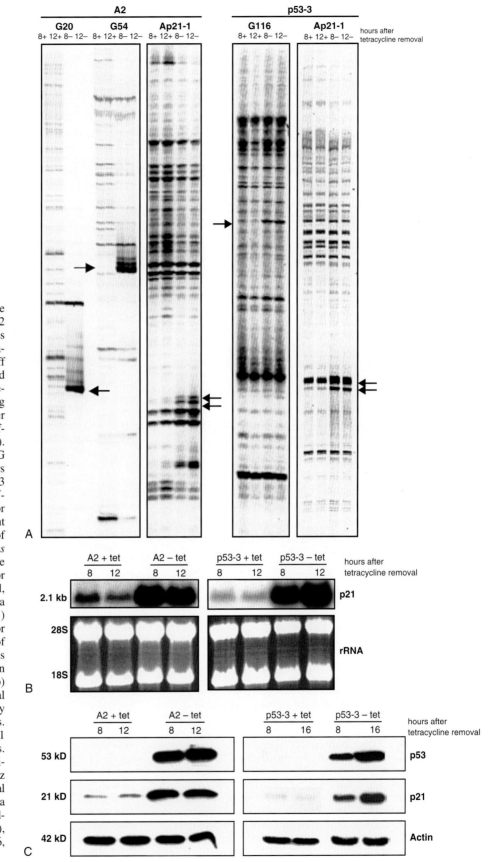

Figure 10 Differential gene expression of p53 and p21^{waf1} in A2 and p53-3 cell lines. **A:** Examples of fluorescent displays from comprehensive screening of A2 (tet-off p53 DLD-1 colon carcinoma) and p53-3 (tet-off p53 H1299 lung adenocarcinoma) cell lines showing induction of p53 and p21 after removal of tetracycline (tet) at different time points (8 hr and 12 hr). Displays of p53 used the HT$_{11}$G anchor primer with H-AP primers 20 and 54 for A2 and 116 for p53-3 cells (since the vector-imposed 3′-tail of p53 in the constructs used for these two cell lines are different from each other and from that of endogenous wild type p53). *Arrows* are used to indicate the inducible p53 band or p21 bands (doublet) for each display. As a positive control, p21 induction was displayed with a p21-specific 13mer primer (p21-1) in combination with HT$_{11}$A anchor primer. **B:** Northern blots results of p21 induction in A2 and p53-3 cells done through the use of a human p21-specific cDNA probe (474 bp) with equivalent loading (10-μg total RNA per lane) demonstrated by 28S and 18S rRNA band intensities. **C:** Western analysis of p53 and p21 protein levels in A2 and p53-3 cells. Antibodies used include the anti-p21 antibody (C-19, Santa Cruz Biotechnology, CA) and a polyclonal pAB 1801 anti-p53 antibody. As a control for equivalent protein loading (50-μg total protein per lane), antiactin antibody was used (A2066, Sigma-Aldrich, St. Louis, MO).

represent novel and/or previously uncharacterized genes (to be published elsewhere). This is in contrast to methodology of DNA microarrays, which can recognize only gene sequences present on the chip and may be limited because of improper cross-hybridization. Other advantages of DD over DNA microarrays are the requirement of much less RNA, the ability to compare more than two different RNA samples simultaneously, and also the ability to detect rare mRNAs. In summary, our results and those from other colleagues provide evidence that DD is an elegant methodology to identify and quantify changes in gene expression, elucidate candidate biomarkers, and discover novel genes involved in important biological and pathological pathways.

Notes

1. All materials or products for DD and FDD technology are commercially available from GenHunter Corporation (Nashville, TN). The company also offers automated FDD services (from reverse transcription reactions to FDD results).

2. The DD method is widely used to identify and isolate differentially expressed genes, and now the automated FDD method can increase the throughput of screening for many more differentially expressed genes. At the moment, some problems may appear because of flaws in the technical equipment. For example, check the glass plates for evenness as this is a critical requirement for a correct scan. Also, the comb must fit well between the two glass plates to avoid lane leakage. The major drawback to FDD is the expense of a fluorescent laser scanner.

3. Information to minimize extrinsic and intrinsic factors can be found in some recently published papers or reviews (Cho *et al.*, 2002; Liang *et al.*, 1995; Liang, 1998).

Acknowledgments

This work was supported in part by NIH grants CA76969 and CA74067 to P.L., by a grant from the Deutsche Akademie der Naturforscher-Leopoldina (Halle, Germany) to S.S., and by a NIH Cellular, Biochemical, and Molecular Sciences Training Grant #GM08554-7 to R.J. The authors greatly appreciate the help of E. Thomas in the proofreading of the manuscript. We thank GenHunter Corporation (Nashville, TN) for permission to adapt the protocol from its RNASpectra Fluorescent Differential Display kit and for use of the Hitachi FMBIO II fluorescent laser scanner. We also thank Dr. C. Prives for the tet-off regulated p53 H1299 (p53-3) cell line, Dr. J. Pietenpol for the p53 pAB1801 antibody, and Dr. B. Vogelstein for the tet-off regulated p53 DLD-1 (A2) cell line.

References

Asubel, F., Brent, R., Kingston, R.E., Moore, D.D., Seidman, J.G., Smith, J.A., and Struhl, K. 1993. *Current Protocols in Molecular Biology. Vol 1.* New York: Greene & Wiley, 4.9.1–4.9.8.

Bosch, L., Melichar, H., and Pardee, A.B. 2000. Identification of differentially expressed genes from limited amounts of RNA. *Nucl. Acids Res. 28*:e27 I–IV.

Cho, Y.J., Meade, J.D., Walden, J.C., Chen, X., Guo, Z., and Liang, P. 2001. Multicolor fluorescent differential display. *BioTechniques 30*:562–572.

Cho, Y.J., Prezioso, V.R., and Liang, P. 2002. Systematic analysis of intrinsic factors affecting differential display. *BioTechniques 32*:1–4.

Gray, J.W., and Collins, C. 2000. Genome changes and gene expression in human solid tumors. *Carcinogenesis 21:* 443–452.

Guan, R.J., Ford, H.L., Fu, Y., Li, Y., Shaw, L.M., and Pardee, A.B. 2000. Drg-1 as a differentiation-related, putative metastatic suppressor gene in human colon cancer. *Can. Res. 60:* 749–755.

Hanahan, D., and Weinberg, R.A. 2000. The hallmarks of cancer. *Cell 100*:57–70.

Jing, C., Rudland, P.S., Foster, C.S., and Ke, Y. 2000. Microquantity differential display: A strategy for a systematic analysis of differential gene expression with a small quantity of starting RNA. *Anal. Biochem. 287*:334–337.

Liang, P. 1998. Factors ensuring successful use of differential display. *METHODS: A Companion to Methods in Enzymology 16*:361–364.

Liang, P., and Pardee, A.B. 1992a. Differential display of eukaryotic messenger RNA by means of the polymerase chain reaction. *Science 257*:967–971.

Liang, P., and Pardee, A.B. 1995. Recent advances in differential display. *Curr. Opin. Immunol. 7*:274–280.

Liang, P., Wang, F., Zhu, W., O'Connell, R.P., and Averboukh, L. 1997. Identification of novel diagnostic markers by differential display. In Reischl, U. (ed.). *Methods in Molecular Medicine Vol 13: Molecular Diagnosis of Infectious Disease*. Totowa, NJ: Humana Press, pg. 3–13.

Martin, K.J., Kwan, C.P., Nagasaki, K., Zhang, X., O'Hare, M.J, Kaelin, C.M., Burgeson, R.E., Pardee, A.B., and Sager, R. 1998. Down-regulation of laminin-5 in breast carcinoma cells. *Mol. Med. 4*:602–613.

Martin, K.J., and Pardee, A.B. 1999. Principles of differential display. *Methods Enzymol. 303*:234–258.

Melichar, M., Bosch, I., Molnar, G.M., Huang, L., and Pardee, A.B. 2000. Detection of eukaryotic cDNA in differential display is enhanced by the addition of *E. coli* RNA. *BioTechniques 28*:76–82.

Negm, R.S., Verma, M., and Srivastava, S. 2002. The promise of biomarkers in cancer screening and detection. *Trends Mol. Med. 8*:288–293.

Stein, S., and Liang, P. 2002. Differential display analysis of gene expression in mammals: A p53 story. *Cell. Mol. Life Sci. 59*:1274–1279.

Stein, S., Cho, Y., Jackson II, R.S., and Liang, P. 2003. Identification of p53 target genes by fluorescent differential display. In Deb, S., and Deb, S.P., (eds.). *Methods in Molecular Biology Vol 234:* p53 Protocols. Totowa, NJ: Humana Press.

Wang, M., Tan, Z., Zhang, R., Kotenko, S.V., and Liang, P. 2002. Interleukin 24 (MDA-7/MOB-5) signals through two heterodimeric receptors, IL-22R1/IL-20R2 and IL-20R1/IL-20R2. *J. Biol. Chem. 277:*7341–7347.

Zhang, R., Averboukh, L., Zhu, W., Zhang, H., Jo, H., Dempsey, P.J., Coffey, R.J., Pardee, A.B., and Liang, P. 1998. Identification of rCop-1, a new member of the CCN protein family, as a negative regulator for cell transformation. *Mol. Cell Biol. 18:*6131–6141.

Zhang, R., Zhang, H., Zhu, W., Pardee, A.B., Coffey, R.J., and Liang, P. 1997. Mob-1, a Ras target gene, is overexpressed in colorectal cancer. *Oncogene 14:*1607–1610.

Zhao, S., Molnar, G., Zhang, J., Zheng, L., Averboukh, L., and Pardee, A.B. 1998. 3′-End cDNA pool suitable for differential display from a small number of cells. *BioTechniques 24:*842–852.

6

Serial Analysis of Gene Expression in Human Diseases

Shui Qing Ye

Introduction

Serial analysis of gene expression (SAGE) (Velculescu *et al.*, 1995) and DNA microarray technology (Schena *et al.*, 1995) have become popular high-throughput platforms for the genomewide analysis of gene expressions and functions in normal and diseased conditions as the large-scale human genomic sequencing effort comes to fruition. Previously, our understanding of the molecular basis of complex syndromes such as cancer, coronary artery disease, and diabetes relied heavily on the study of individual genes and how changes in specific genes affected phenotype. However, it is now generally recognized that phenotypic changes in a tissue are the result of changes in the spatial and temporal expression of dozens or even hundreds of genes. The molecular basis for disease is thus difficult to discern based simply on the study of individual genes. The research paradigm is shifting from the traditional search for a single disease-specific gene to the current understanding of the biochemical and molecular functioning of a variety of genes and how complicated networks of interaction can lead to the pathogenesis of various human diseases. In an attempt to look more globally at gene expression changes, methods such as cDNA subtraction, mRNA differential display, and expression sequence tag (EST) have been used. These, however, can analyze only a limited number of genes and are not quantitative. Two high-throughput technologies, SAGE (Velculescu *et al.*, 1995) and oligonucleotide (Lockhart *et al.*, 1996) or cDNA microarray (Schena *et al.*, 1995), allow researchers to determine the expression pattern of thousands of genes simultaneously in normal or diseased tissues. With the development of SAGE or DNA microarray-based research, more accurate diagnosis and treatment of various human diseases based on an individual patient's gene expression profile are burgeoning. Two attractive features of SAGE compared with DNA microarray are its ability to quantify gene expression without prior sequence information and its provision of gene expression result in a digital format. The SAGE technique has been intensively applied to the gene expression profiling of a number of human diseases. In this chapter, the first part will draw up a short synopsis of the SAGE technique and its contrast to DNA microarray; the second part will expound the protocol of the SAGE technique; the third part will discuss new biological insights derived from the application of SAGE in human diseases; and perspectives will be offered in the last part.

Handbook of Immunohistochemistry and in situ Hybridization of Human Carcinomas, Volume 1: Molecular Genetics; Lung and Breast Carcinomas

85

Copyright © 2004 by Elsevier (USA)
All rights reserved.

SAGE Technique Overview and Its Contrast to DNA Microarray

The SAGE technique was originally developed by Velculescu *et al.* (1995). Two basic principles underlie the SAGE methodology: 1) a short sequence tag (10 bp) in a defined position in the cDNA that contains sufficient information to uniquely identify a transcript and 2) the concatenation of tags, which allows for efficient sequence-based analysis of transcription. The detailed introduction of the SAGE technique can be found in two comprehensive reviews (Bertelsen *et al.*, 1998;

Madden *et al.*, 2000). The schematic of the SAGE procedure is depicted in Figure 11A. Briefly, poly(A)$^+$ RNA is isolated by oligo-dT column chromatography. Then cDNA is synthesized from poly(A)$^+$ RNA using a primer of biotin-5'-T$_{18}$-3'. The cDNA is cleaved with an anchoring enzyme (e.g., NlaIII), and the 3'-terminal cDNA fragments are bound to streptavidin-coated beads. An oligonucleotide linker containing recognition sites for a tagging enzyme (e.g., BsmFI) is linked to the bound cDNA. The tagging enzyme is a class II restriction endonuclease that cleaves the DNA at a constant number of bases 3' to the recognition site. This results

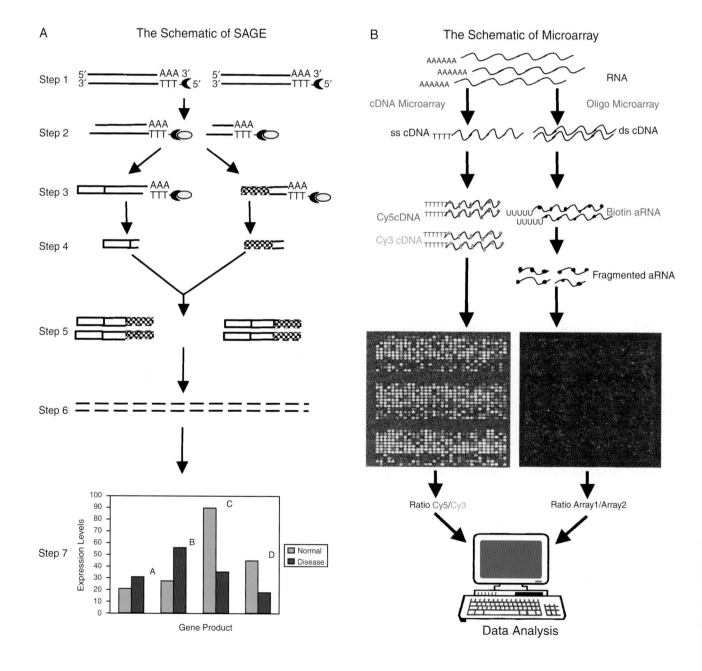

in the release of a short tag plus the linker from the beads after digestion with BsmFI. The 3′-ends of the released tags plus linkers are then blunted and ligated to one another to form 102-bp linked ditags. After polymerase chain reaction (PCR) amplification of the 102-bp ditags, the linkers and tags are release by digestion with the anchoring enzyme. The 26–28 mer tags are then gel purified, concatenated, and cloned into a sequence vector. Sequencing the concatemers enables individual tags to be identified and the abundance of the transcripts for a given cell line or tissue to be determined (Velculescu *et al.*, 1995).

Although SAGE has become an extremely powerful technique for global analysis of gene expression, its requirement for a large amount of input mRNA (2.5–5.0 μg, which is equivalent to 250–500 μg total RNA) limits its utility. Several laboratories have attempted SAGE gene profiling using smaller amounts of RNA, but these attempts have all involved either PCR amplification of starting cDNA materials, such as SAGE-Lite and PCR-SAGE (Neilson *et al.*, 2000; Peters *et al.*, 1999), or PCR reamplification of SAGE ditags generated by a first round of PCR amplification, such as microSAGE and SAGE adaptation for down-sized extracts (SADE) (Datson *et al.*, 1999; Virlon *et al.*, 1999). These additional PCR amplifications potentially introduce bias and compromise the quantitative aspects of the SAGE method. A recently modified method, miniSAGE, was successfully applied to profile gene expression of human fibroblasts from 1-μg total RNA without extra PCR amplification (Ye *et al.*, 2000).

Three key modifications contributed to the establishment of the miniSAGE: 1) The application of Phase Lock Gel (Eppendorf) to purify DNA after phenol extraction to increase significantly the recovery of DNA material and get purer DNA; 2) the addition of 25-fold less linkers (10 ng/per reaction) in the ligation reaction to cDNA, thus reducing its interference with the SAGE ditag amplification and increasing the SAGE ditag yield; 3) the employment of the mRNA Capture Kit (Boehringer Manneheim) to carry out the initial steps depicted in Figure 11A, which include mRNA isolation, cDNA synthesis, enzyme cleavage of cDNA, and the binding of the cleaved biotin-cDNA to the streptavidin-magnetic beads, ligating linkers to the bound cDNA, and the release of cDNA tags in one tube to reduce significantly the loss of the material between the successive steps. Recently, Invitrogen (www. invitrogen.com) developed a convenient I-SAGE kit, which packages all necessary quality-controlled reagents together, enough for five SAGE library syntheses. These modifications have contributed to the broad application of SAGE. Three major Web sites for SAGE technology and data information are listed in Table 6.

The SAGE technique is a rival to DNA microarray technology. The latter was derived from an initial report in the mid-1970s (Burgess, 2001) that gene expression could be monitored by nucleic acid molecules attached to a solid support. A typical cDNA microarray operational scheme is illustrated in Figure 11B, right. Microarrays are created using a precise xyz robot that is programmed to spot cDNA samples onto a solid

Figure 11 Schematics of the serial analysis of gene expression (SAGE, A) and microarray (B). **A:** The schematic of SAGE: **Step 1**, cDNA synthesis using biotin (●)-oligo dT: poly(A)⁺ RNA is isolated by an oligo-dT column and cDNA is synthesized from poly(A)⁺ RNA using the biotin-oligo-dT. **Step 2**, Anchoring enzyme digestion and streptavidin beads (○) binding: the cDNA is cleaved by the anchoring enzyme (e.g., NlaIII) and 3′-portion of the cDNA is captured by the streptavidin-coated magnetic beads. **Step 3**, Add linker 1 (▭) and linker 2 (▩): two different linkers containing a type II restriction enzyme (tagging enzyme such as BsmFI) are ligated to two aliquots of the captured cDNA, respectively. **Step 4**, Tagging enzyme digestion: short tags plus linkers are released from the cDNA by tagging enzyme digestion. **Step 5**, Blunting ends, ligation and PCR amplification: after the ends of the released short tags are blunted by the treatment of the DNA polymerase I, large (Klenow) fragment, the linker-tag molecules are ligated tail to tail to form ditags. Ditags are amplified by PCR. **Step 6**, Anchoring enzyme digestion, concatenation by ligase, cloning and DNA sequencing: ditags are released from linkers by the anchoring enzyme digestion and concatenated by ligase. The concatemers are then cloned into a sequencing vector and are subjected to DNA sequencing. **Step 7**, Data analysis: SAGE tag information is analyzed by SAGE software to eventually lead to identifying the genes of interest in specific physiological or identifying disease states. **B:** The schematic of microarray: Depicted on the left side is the cDNA microarray schema. RNA from two different tissues or cell populations is used to synthesize single-stranded cDNA in the presence of nucleotides labeled with two different fluorescent dyes (e.g., Cy3 and Cy5). Both samples are mixed in a small volume of hybridization buffer and hybridized to the array surface, usually by stationary hybridization under a coverslip, resulting in competitive binding of differentially labeled cDNAs to the corresponding array elements. High-resolution confocal fluorescence scanning of the array with two different wavelengths corresponding to the dyes used provides relative signal intensities and ratios of mRNA abundance for the genes represented on the array. Depicted on the right side is the oligonucleotide microarray schema. The RNA from different tissues or cell populations is used to generate double-stranded cDNA carrying a transcriptional start site for T7 DNA polymerase. During *in virto* transcription, biotin-labeled nucleotides are incorporated into the synthesized cRNA molecules. Each target sample is hybridized to a separate probe array and target binding is detected by staining with a fluorescent dye coupled to streptavidin. Signal intensities of probe array element sets on different arrays are used to calculate relative mRNA abundance for the genes represented on the array. This figure was adopted from Ye *et al.* (2002). Courtesy of J. Biomed. Sci. (www.karger.com/jbs).

Table 6 SAGE Technology and Data Information Internet

Web Site	URL	Feature
SAGEnet	www.sagenet.org	SAGE introduction, software, protocols, references, and link to other SAGE resources
SAGEmap	www.ncbi.nlm.nih.gov/SAGE	Online SAGE data analysis, tag to gene mapping, download CGAP SAGE data, submitting SAGE data to GEO
Genzyme Molecular Oncolocy	www.genzyme.com/sage	SAGE technology and applications information for commercial users

substrate, usually a glass microscope slide, in a high-density pattern. Similarly, cDNA arrays are also produced on nylon membranes but with spots of larger size in a lower-density pattern. Patrick Brown's laboratory at Stanford University created the first xyz arrayer, and instructions on how to build an arrayer can be found on their Web site (www.cmgm.stanford.edu/pbrown). Many companies now produce cDNA microarrayers commercially. These machines differ primarily in the way the spot is placed on the substrate. Spotted arrays allow a greater degree of flexibility in the choice of arrayed elements, particularly for the preparation of smaller, customized samples. Accordingly, cDNA gridded arrays have been the technique most frequently used. With prices for oligonucleotide synthesis becoming more reasonable for large-scale studies, spotted long-oligonucleotide arrays could be a viable alternative to full-length cDNA array.

Oligonucleotide microarrays, another type of microarray (Schulze *et al.*, 2001), use photolithography or ink-jet technology to synthesize oligonucelotides *in situ*, on silicon wafers. A typical oligonucleotide microarray operational schema is illustrated in Figure 11B right. High-density-oligonucleotide microarrays from Affymetrix (www.affymetrix.com) are examples of the use of photolithography technology. A similar procedure developed by Rosetta Inpharmatics (www.rii.com) and licensed to Agilent Technologies represents an example of the use of ink-jet technology. Alternatively, presynthesized oligonucleotides can be printed onto glass slides. Methods based on synthetic oligonucleotides offer the advantage that sequence information alone is sufficient to generate the DNA to be arrayed; cDNAs do not have to be produced. Affymetrix (Santa Clara, CA) has pioneered the use of this form of array production with the development of the GeneChip. Its newest GeneChip, Human Genome U133 Set (HG-U133A and HG-U133B, released in January 2002), is composed of two microarrays containing more than 1,000,000 unique oligonucleotide features covering more than 39,000 transcript variants, which in turn represent greater than 33,000 of the best characterized human genes. Affymetrix released (March 2003) its latest GeneChip Mouse Expression Set 430 (MOE430 A and B), which can analyze the expression level of more than 39,000 mouse transcripts and variants, including more than 34,000 well-substantiated mouse genes, and Rat Expression Set 230 (RAE230A and B), which can analyze the expression level of more than 30,200 rat transcripts and variants, including more than 28,000 well-substantiated rat genes. These newly empowered GeneChips will facilitate the gene expression profiling of human diseases in animal models.

Ishii *et al.* (2000) compared the quantitative accuracy of oligonucleotide arrays and SAGE using identical RNA specimens prepared from human blood monocytes and macrophages stimulated with granulocyte-macrophage colony-stimulating factor (GM-CSF). Results for the unstimulated monocytes and stimulated macrophages were similar. The correlation was better for genes that were more highly expressed or more differentially expressed. Nacht *et al.* (1999) demonstrated the strength of combining the two approaches to help elucidate pathways in breast cancer progression by comparing primary breast cancers, metastatic breast cancers, and normal mammary epithelial cells. They reported that combining SAGE and custom array technology allowed for the rapid identification and validation of the clinical relevance of genes potentially involved in breast cancer progression.

However, each technology has its pros and cons. Table 7 lists several major differences between SAGE and Microarray Technology. Because the preparation of microarrays requires prior knowledge of the sequence of the gene transcripts to be analyzed, an advantage of SAGE is that it can identify novel genes and be used to analyze gene expression in organisms whose genomes are largely uncharacterized. This is a serious limitation for microarrays, even for organisms with completely sequenced genomes such as humans,

Table 7 Differences Between SAGE and Microarray Technology

Parameters	SAGE	Microarray
New gene identification	Yes	No
Data format	Digital	Analog
Sensitivity	Higher	Lower
Setup cost	Low	High
Operation procedure	Relatively cumbersome	Simpler
Application to multiple samples	Higher sequencing cost involved	Economical

because genome annotation and gene prediction remain technically challenging. There are additional advantages to SAGE. The data SAGE provides are in the digital format, whereas microarray data are analog. Although SAGE data are directly comparable, the differences in microarray formats and normalization methodologies make direct comparison of data sets between microarray platforms somewhat difficult. The SAGE technology not only can accurately determine the absolute abundance of mRNAs but also can detect even slight differences in expression levels between samples. Microarray is only reliable in detecting genes whose expression differences are relatively large. Setup cost for SAGE is low. It does not require expensive instruments other than a DNA sequencer, which is available at most institutions. Microarrays need expensive robotic arrayers and scanners, currently available only in core facilities of major institutions. Microarrays, on the other hand, are relatively easy to use and more suitable for high-throughput applications. Also, SAGE requires several enzymatic manipulation steps, which are relatively taxing, especially for the novice, although Invitrogen has recently marketed a kit, the I-SAGE kit (www.invitrogen.com), which helps new SAGE users. Finally, although the cost of DNA sequencing keeps falling, the relative high cost for sequencing a SAGE library is still a concern for potential SAGE users. Despite the great power of SAGE using a 10-bp tag as a transcript identifier, a small percentage of tags are ambiguous. A single tag may identify multiple genes or multiple tags may identify a single gene (Neilson et al., 2000). This remains a challenge for identifying genes. Chen et al. (2002) tried to solve this problem by extending the SAGE tags into 3′ cDNAs. They have improved their original generation of longer cDNA fragments from SAGE tags for gene identification (GLGI) technique into a high-throughput procedure for simultaneous conversion of a large number of SAGE tags into corresponding 3′ DNAs. This improved GLGI converting novel SAGE tags into 3′ cDNAs was demonstrated to be efficient in identifying the correct gene for SAGE tags with multiple matches and identifying novel genes in large numbers. Applying this high-throughput procedure should accelerate the rate of gene identification significantly in human and other eukaryotic genomes. Because microarrays use well-characterized immobilized sequences (cDNAs or oligonucleotides), identification of expressed genes is less of a problem.

Saha et al. (2002) developed a long SAGE method that generates 21-bp tags, instead of 14-bp, from the 3′ ends of transcripts. This method is similar to the original SAGE approach (Velculescu et al., 1995) but uses a different type IIS restriction endonuclease (MmeI) and incorporates other modifications to produce longer transcript tags. More than 75% of 21-bp tags, but not 14-bp tags, can be uniquely assigned to the human genome based on actual sequence information from ~16,000 known genes, although 14-bp tags do allow such assignment to ESTs and previously characterized mRNAs (Caron et al., 2001; Lal et al., 1999; Velculescu et al., 1995).

The choice of gene expression technique is determined by the question being asked. Expression profiling of hundreds of disease samples is certainly more efficient using microarrays. However, SAGE seems to be a better choice for identifying new genes. In addition, SAGE is useful in analyzing previously uncharacterized organisms. Also, SAGE provides for more sensitive quantification of gene expression. Regular 14-bp tag SAGE is still useful for the quantification of mRNA level, while 21-bp tag SAGE (long SAGE) is more suitable for identifying new genes.

The SAGE Protocol

The standard SAGE protocol presented as follows is mainly based on Velculescu et al. (1995, 2000), and Saha et al. (2002). Some modifications and updates improving the performance of the standard SAGE protocol are incorporated. Because of space limitations, micro- or mini-SAGE protocol will not be described here.

All necessary reagents needed for the SAGE protocol is listed one by one in the first part of this section. Catalog number and company name for all reagents are provided. However, equivalent quality reagents from other resources can be substituted. Investigators may use the I-SAGE Kit (Invitrogen, Cat. No. T500001), which contains enough reagents to generate five libraries, each consisting of >100,000 tags/library starting from 5 µg total RNA. One advantage to using the I-SAGE Kit rather than assembling the reagents on your own is to use optimized and quality-controlled enzymes, adapters, primers, and buffers to save time.

MATERIALS

1. *Commercially available kits:* Total RNA Isolation Kit (Promega, Cat. No. Z5110); MessageMaker mRNA Kit (Invitrogen, Cat. No. 10298-016); Superscript Choice System cDNA Synthesis Kit (Invitrogen, Cat. No. 18090-019); Dynabeads M-280 Streptavidin Slurry (Dynal, Cat. No. 112.05); Magnet (Dynal, Cat. No. 120.04).

2. *Enzymes:* BsmFI (NEB, Cat. No. 572S); NlaIII (NEB, Cat. No. 125S), which should be shipped on dry ice and stored at −80°C; SphI (NEB, Cat. No. 182S); Klenow fragment of DNA polymerase I (USB Corp., Cat. No. 27-0929-01); T4 Ligase (Invitrogen, Cat. No. 15224-041, 5 U/µl); T4 Ligase (Invitrogen, Cat. No. 15224-017, 1 U/µl); PLATINUM Taq DNA polymerase (Invitrogen, Cat. No. 10966-034); T4 polynucleotide kinase (NEB, Cat. No. 182S); MmeI, which is used for the long SAGE and is not commercially available. The following is the contact information for obtaining MmeI enzyme: Prof. Anna J. Podhajska, Dept. Microbiology, University of Gdansk, ul. Kladki 24, 80-822 Gdansk, Poland, Phone: + 48-58 301-28-07, Fax: + 48-58 301-28-07, e-mail: podhajsk@biotech.univ. gda.pl.

3. *Chemicals:* Glycogen (Roche, Cat. No. 901393); 10 mM dNTP mix (Invitrogen, Cat. No. 18427-013); 100 mM ATP solution (Amersham Biosci., Cat. No. 27-2056-01); DMSO (Sigma, Cat. No. D2650); 7.5 M ammonium acetate (Sigma; Cat. No. A2706); ethanol (Sigma, Cat. No. E702-3, molecular biology grade); TEMED (Sigma, T8133); 40% polyacrylamide (37.5:1 acrylamide:bis, Bio-Rad, Cat No. 161-0148); 40% polyacrylamide (37.5:1 acrylamide:bis, Bio-Rad, Cat. No. 161-0148); Streptavidin (Sigma, Cat. No. S-4762); ethidium bromide (Sigma, Cat. No. E-1510, 10 µg/µl); ammonium persulfate (Sigma, Cat. No. A9164); 50× tris acetate buffer (Quality Biological, Cat. No. 330-008-161); Phase Lock Gel (PLG, light in 2-ml microfuge tube, Eppendof, Cat. No. 0032005.105).

4. *Cloning vector:* pZERO-1 plasmid (Invitrogen, Cat. No. K2500-01).

5. *2× Binding and washing (2× B + W) buffer:* 10 mM Tris-HCl, pH 7.5, 1 mM EDTA, 2.0 M Nacl, stored at room temperature.

6. *Low concentration of Tris-EDTA solution (LoTE):* 3 mM Tris-HCl, pH 7.5, 0.2 mM EDTA in dH$_2$O (pH 7.5), stored at 4°C.

7. *Phenol: Chloroform reagent (PC8):* Prewarm 480-ml phenol for 0.5 hr in a 65°C water bath then add 320-ml 0.5 M Tris-HCl, pH 8.0 and 640 ml chloroform in sequence. Shake to mix the mixture gently and store at 4°C. Shake it again in 2–3 hr. After another 2–3 hr aspirate aqueous layer to be discarded, aliquot the organic phase into 1.5-ml microfuge tubes, and store them at −20°C.

8. *10× PCR buffer:* 166 mM (NH$_4$)$_2$SO$_4$, 670 mM Tris, pH 8.8, 67 mM MgCl$_2$, 100 mM β-mercaptoethanol, aliquot into 0.5-ml microfuge tubes and store them at −20°C.

9. *10% Ammonium persulfate (APS) solution:* Dissolve 0.1 g ammonium persulfate powder in 1 ml H$_2$O. It should be prepared immediately before use.

10. *12% PAGE gel for isolating PCR products and ditags:* Add 40% polyacrylamide (19:1 acrylamide:bis, Biop-Rad, Cat. No. 161-0144) 10.5 ml, dH$_2$O 23.5-ml, 50× tris acetate buffer (Quality Biological, Cat. No. 330-008-161) 700 µl, 10% APS 350-µl, and TEMED 30 µl into a beaker, mix, and add to a vertical gel apparatus (Owl Scientific, Model No. P9DS) with 1.5-mm spacers and comb and let gel sit at least 30 min to polymerize.

11. *8% PAGE gel for separating concatemers:* Add 40% polyacrylamide (37.5:1 acrylamide:bis, Bio-Rad, Cat. No. 161-0148) 10.5 ml, dH$_2$O 27 ml, 50× Tris acetate buffer (Quality Biological, Cat. No. 330-008-161) 700 µl, 10% APS 350 µl, and TEMED 30 µl into a beaker, mix up and add to a vertical gel apparatus (Owl Scientific, Model No. P9DS) with 1.5-mm spacers and comb and let gel sit at least 30 min to polymerize.

12. *20% Novex gel (Novex, Cat. No. EC6315).*

13. *Linkers and primers:*

For 14-bp tag SAGE:
▲ Linker 1A: 5′ TTT GGA TTT GCT GGT GCA GTA CAA CTA GGC TTA ATA GGG ACA TG 3′
▲ Linker 1B: 5′ TCC CTA TTA AGC CTA GTT GTA CTG CAC CAG CAA ATC C[amino mod. C7] 3′
▲ Linker 2A: 5′ TTT CTG CTC GAA TTC AAG CTT CTA ACG ATG TAC GGG GAC ATG 3′
▲ Linker 2B: 5′ TCC CCG TAC ATC GTT AGA AGC TTG AAT TCG AGC AG[amino mod. C7] 3′

▲ Linker 1B and 2B should also be modified with 5′ phosporylation. All linkers should be purified by PAGE after synthesis

▲ Primer 1: 5′ GGA TTT GCT GGT GCA GTA CA 3′

▲ Primer 2: 5′ CTG CTC GAA TTC AAG CTT CT 3′

▲ Primer 1 and primer 2 should be labeled with two sequential biotins on the 5′ end and purified by PAGE after synthesis.

▲ Biotinylated oligo dT, 5′ [biotin]T18, should be purified after synthesis.

▲ M13 Forward: 5′ GTA AAA CGA CGG CCA GT 3′

▲ M13 Reverse: 5′ GGA AAC AGC TAT GAC CAT G 3′

For 21-bp tag SAGE:

▲ Linker 3A: 5′-TTTGGATTTGCTGGT-GCAGTACAACTAGGCTTAATATCCGA-CATG-3′

▲ Linker 3B: 5′-TCGGATATTAAGCCTAGTTG-TACTGCACCAGCAAATCC Amino-modified C7-3′

▲ Linker 3A: 5′-TTTCTGCTC-GAATTCAAGCTTCTAACGATGTACGTCC-GACATG-3′

▲ Linker 3B: 5′-TCGGACGTACATCGTTA-GAAGCTTGAATTCGAGCAG Amino-modified C-7-3′

▲ As with 14-bp tag SAGE, linker 3B and 4B should also be modified with 5′ phosphorylation. All linkers should be purified by PAGE after synthesis.

▲ Primer 3: 5′-GTG CTC GTG GGA TTT GCT GGT GCA GTA CA-3′

▲ Primer 4: 5′-GAG CTC GTG CTG CTC GAA TTC AAG CTT CT-3′

▲ As with 14-bp tag SAGE, primer 3 and primer 4 should be labeled with two sequential biotins on the 5′ end and purified by PAGE after synthesis.

14. *Instruments:* Agarose and polyacrylamide gel electrophoresis apparatus, power supply, microfuge, water bath, PCR thermal cycler, oligonucleotide synthesizer and DNA sequencer, which may be available in the institutional DNA synthesis and analysis core facility.

15. *Software for SAGE data analysis:* SAGE 2000 software version 4.12 can be used to extract SAGE tags, tabulate tag counts, and remove duplicate ditags, linker sequences and 1-bp variation sequences. It is freely available to academic users (refer to www.sagenet.org for details). SAGE commercial inquires should be directed to sage@genzyme.com. Several other SAGE software programs are also available, such as Expression profile viewer (ExProView, Larsson *et al.*, 2000), eSAGE (Margulies *et al.*, 2000), USAGE (van Kampen *et al.*, 2000), POWER_SAGE (Man *et al.*, 2000), SAGE Genie (Boon *et al.*, 2002). Each software program has its feature. Because of the space limitation, no detailed descriptions of SAGE data analysis and the software programs are presented.

METHODS

I. PC8 extraction to purify DNA:
   A. Add equal volume PC8 to sample in a 1.5-ml microfuge tube and vortex for 10 sec to mix.
   B. Transfer the mixture into a 2-ml microfuge tube with PLG (Eppendorf, Cat. No. 0032005.105), which has been prespun for 30 sec at $13,000 \times g$ on a microfuge.
   C. Centrifuge at $13,000 \times g$ for 5 min to separate the phases. The PLG will form a barrier between the aqueous (top) and organic phases (bottom).
   D. Carefully pipet off nucleic-acid–containing aqueous top phase to a fresh tube, DNA will be concentrated either by ethanol precipitation or by spin column on heated DNA Speedvac (Savant, Model DNA 110) for downstream usage.

II. Ethidium bromide dot quantitation of DNA in low concentration:
   A. Use any solution of pure DNA to prepare the following standards: 0 ng/μl, 1 ng/μl, 2.5 ng/μl, 5 ng/μl, 7.5 ng/μl, 10 ng/μl, 20 ng/μl.
   B. Use 1 μl of sample DNA to make 1/5, 1/25, and 1/125 dilutions in LoTE.
   C. Add 4 μl of each standard or 4 μl of each diluted sample to 4 μl 1 μg/ml ethidium bromide and mix it well.
   D. Place a sheet of plastic wrap on a UV transilluminator, spot each 8-μl mix on plastic wrap, photograph under UV light, and estimate DNA concentration by comparing the intensity of the sample with the standards.

III. Testing biotinylation of biotin-oligo dT and primers:
   A. Add several hundred nanograms of biotin-oligo dT or primers to 1 μg streptavidin (Sigma Cat. No. S-4762) in a 0.5-ml microfuge tube and in a control tube without 1 μg streptavidin.
   B. Incubate 5 min at room temperature.
   C. Both the oligo bound to streptavidin and the oligo alone are run on a 20% Novex gel (Cat. No. EC6315).

If the oligo is well-biotinylated, the entire amount of oligo in the streptavidin-containing tube should be shifted to higher molecular weight.

Alternatively, increasing amounts of oligo (from several hundred nanograms to several micrograms) can be incubated with or without separate aliquots of 100 µl of Dynabeads (Dynal). After 15 min, the beads are separated from the supernatant using a magnet, the supernatant is removed, and DNA can be quantified by reading optical density at the wavelength of 260 nm. At low amounts of oligo, when bead-binding capacity is not saturated, the ratio of unbound oligo to the total oligo will indicate the percent of oligo that is not biotinylated.

SAGE METHODS

Genzyme Molecular Oncology (www.genzymemolecularoncology.com/sage) has licensed the SAGE technology from The Johns Hopkins University for commercial purposes. So several steps of the SAGE method, if not published before, are not described in detail here. However, the technology is freely available to academia for research purposes and the detailed protocol can be obtained on the Internet via www.sagenet.org. Key steps in SAGE protocol are outlined below.

1. mRNA Isolation and cDNA Synthesis.

mRNA isolation and cDNA synthesis are not unique to the SAGE procedure. Various reagents and kits can be used for these purpose at the investigator's choice. For example, total RNA from tissues or cells of interest can be isolated using the Total RNA Isolation Kit (Promega, Cat. No. Z5110) followed by the MessageMaker mRNA Kit (Invitrogen, Cat. No. 10298–016) to isolate mRNA per supplier's instruction. The mRNA can also be isolated directly from tissues or cells of interest. Superscript choice System cDNA Synthesis Kit (Invitrogen, Cat. No. 18090-019) or other equivalent reagents can be used for cDNA synthesis. The standard SAGE protocol requires 5 µg mRNA as a starting material.

2. Generation of the Linked 14-bp SAGE Tags

Cleavage of the biotinylated cDNA with the anchoring enzyme NlaIII. Synthesized cDNA is digested with the anchoring enzyme NlaIII (NEB, Cat. No. 125S). Other anchoring enzymes can be used. For LongSAGE (21-bp tag), cDNA should be digested using the enzyme MmeI. Digested cDNA can be purified using PC8 extraction, as previously described. Biotin-labeled cDNA of the 3′ end can be captured by the streptavidin-coated magnetic

beads (Dynabeads M-280 Streptavidin Slurry, Dynal, Cat. No. 112.05). Divide the captured cDNA equally into the two tubes, tube 1 and tube 2.

Ligating linkers to bound cDNA. The linker 1A with the complementary strand linker 1B and the linker 2A with the complementary strand linker 2B needs to be annealed for the 14-bp tag SAGE protocol, while the linker 3 and linker 4 need to be prepared for the 21-bp tag SAGE protocol. The annealed linkers 1 and 2 (or linker 3 and linker 4) are added to tube 1 and tube 2, respectively, to be ligated to the bound cDNA using the high concentration of T4 ligase.

Release of the cDNA tags using the tagging enzyme BsmFI. After the ligation, linked cDNA tags are released by the digestion with the tagging enzyme BsmFI (NEB, Cat. No. 572S).

3. Formation of 26–28mer SAGE Ditags

Blunt the ends of the released linked SAGE tags. After the BsmFI digestion, the released linked cDNA (SAGE) tags are purified using PC8 extraction, as previously described. The nucleic acids are precipitated in ethanol. The DNA ends are blunted using the Klenow fragment of DNA polymerase I (USB Corp., Cat. No. 27-0929-01). The DNAs are again purified using PC8 extraction and precipitated in ethanol.

Ligation of the linked SAGE tages to form the linked ditags. After the blunting reaction, sample 1 in tube 1 and sample 2 in tube 2 are ligated in the same tube to form the linked ditags. The negative control without ligase in the reaction was simultaneously carried out.

The relatively detailed descriptions of the remaining steps are adopted from Ye *et al.* (2000).

PCR amplification of the linked ditags. After the ligation, the volume is raised to 20 µl by additon of 14 µl LoTE. A series of dilutions, 0-, 10-, 20, 100-, and 200-fold, are prepared. 1 µl of each diluted ligation mixture is used as an input in a 100-µl PCR reaction containing 5-U PLATINUM Taq polymerase (Cat. No. 10996-026, GIBCOBRL), 1.5 mM $MgCl_2$, 0.2 mM dNTPs, and 0.8 µM biotinylated SAGE primers 1 and 2 (or primers 3 and 4 for the LongSAGE) and amplified for 28 cycles of 30 sec at 95°C, 1 min at 55°C, and 1 min at 70°C with an initial heat activation of the enzyme for 4 min at 95°C and a final extension of 5 min at 70°C on a 96-well PCR Thermal Cycler. 288 PCR reactions are needed for each SAGE library.

Isolation of the linked ditags. 28.8-ml PCR products combined from 288 tubes are split equally among six 15-ml tubes containing Light Phase Lock Gel (LPLG, Cat. No. 0032005.209, Eppendorf Sci., Inc., Westbury, NY), which have been spun for 1 min at $1500 \times g$ immediately before use. 4.8 ml PC8 is then added into each tube and thoroughly mixed up by inverting tubes 5–7 times. The aqueous and organic phases are separated by centrifugation at $1500 \times g$, 22°C for 5 min. The linked ditags contained in the aqueous upper phase are pipeted into a fresh tube. The nucleic acids from each tube are precipitated in ethanol (4.8-ml sample, 90 µl glycogen (1 µg/µl), 2.4 ml, 7.5 M ammonium acetate and 15 ml ethanol) in a 30-ml tube (Cat. No. 3114-0030, Nalge Co., Rochester, NY) by centrifugation at $17,500 \times g$ for 30 min at 4°C on the rotor F0630, GS-15 R centrifuge (Beckman Coulter, Inc., Fullteron, CA). The pellets from each tube are washed twice with 70% ethanol, air-dried at room temperature for 10 min, and resuspended in 35 µl LoTE. Total volume of 210 µl sample is obtained.

54-µl $1 \times$ Gel loading solution (GLS, Cat. No. G2526, Sigma, St. Louis, MO) is added to the sample. 12 µl sample mixture is loaded on each lane for total 22 lanes on two 12% polyacrylamide gels with a 10-bp ladder (GIBCOBRL) as a marker on each gel. After electrophoresis, gels are stained with SYBR Green I at 1:10,000 dilution in 1× running buffer. The gel regions with the same relative mobility of the 100-bp standard from all 22 sample lanes are excised and loaded into six 0.5-ml tubes, pierced with a 21-gauge needle, inserted in a 1.5-ml tube. The gel pieces are fragmented by spinning for 5 min at $14,000 \times g$, room temperature. The DNA is eluted from the gel fragments by adding 300 µl LoTE and incubating for 15 min at 65°C, followed by the removal of the polyacrylamide gel on Spin X columns (Costar) by spinning in a microfuge at $13,000 \times g$, 22°C for 5 min.

Purification of the SAGE ditags. The eluate from Spin X column is ethanol-precipitated, as previously described, and the pellet is resuspended in 70 µl LoTE. The linked ditags are digested with the anchoring enzyme NlaIII (NEB, Cat. No. 125S or MmeI for the long SAGE) for 1 hr at 37°C in 100-µl reaction volume containing 70 µl sample, 10 µl NlaIII (10 U/µl), 10 µl BSA (1 µg/µl), 10 µl $10 \times$ NEbuffer 4. After the digestion, PC8 extraction, PLG separation, and ethanol precipitation (with extra step of dry ice/ethanol bath for 10 min before spinning) are carried out as previously described, the pellet is resuspended in 32 µl LoTE.

The pellets, after mixing with 8 µl loading dye, are loaded on four lanes, each lane with 10 µl, on a 12% polyacrylamide gel with a 10-bp DNA standard. After the electrophoresis, the ditag band running at 22–26 bp is excised and eluted, as described, except that the incubation is performed at 37°C instead of 65°C. The pellet is resuspended in 100 µl LoTE.

Removal of any contaminating linkers using Dynal streptavidin-coated magnetic beads (Dynabeads M-280 streptavidin, Dynal A.S., Oslo, Norway) can be carried

out exactly according to the description of Powell (1998). The purified SAGE ditags are subject to PC8 extraction, PLG separation, and ethanol precipitation (with extra step of dry ice/ethanol bath for 10 min before spinning) as previously described; the pellet is resuspended in 7 µl LoTE. Alternatively, Damgaard *et al.* (2003) introduced the purification of 26-bp ditags by reverse-phase high-performance liquid chromatography (HPLC) using polystyrene/divinylbenzene columns and tetraethylammonium acetate buffer with acetonitrile as mobile phase. Ditags purified by HPLC readily ligate to high-molecular-weight concatemers leading to their efficient cloning.

4. Cloning, Sequencing, and Analyzing of SAGE Tags

Ligation of SAGE ditags to form concatemers. Purified ditags are ligated to concatemers for 3 hr at 16°C in a 10-µl reaction with 1 µl T4 DNA ligase (5 U/µl, Cat. No. 15224-041, GIBCOBRL), 2-µl $5 \times$ ligase buffer, 7 µl SAGE ditag samples. After the ligation, the sample is heated for 5 min at 65°C (Kenzelmann *et al.*, 1999), mixed with 2.5 µl $5 \times$ GLS and run in a single lane on an 8% polyacrylamide gel with a 100-bp ladder as a marker. After staining with SYBR Green I, the gel regions between 600 and 1200 bp are excised and the concatemers are purified as previously described. After ethanol precipitation steps, the pellet is resuspended in 6 µl LoTE.

Cloning and sequencing of concatemers (SAGE tags). pZErO-1 vector (Cat. No. K2500-01, Invitrogen, Carlsbad, CA) is digested for 15 min in a 20-µl reaction volume containing 1 µl (1 µg) vector, 1 µl SphI (10 U/µl, New England Biolabs), 2 µl BSA (1 µg/µl), 2 µl $10 \times$ NEbuffer 2 and 14 µl LoTE. After the digestion, PC8 extraction, PLG separation, and ethanol precipitation are carried out as described, the pellet is resuspended in 90 µl LoTE. The final concentration of the digested vector is about 10 ng/µl.

Ligation of concatemers into the SphI-digested vectors is performed at 16°C for 30 min in a 10 µl reaction with 1 µl T4 DNA ligase (5 U/µl, Cat. No. 15224-041, GIBCOBRL), 2 µl $5 \times$ ligase buffer, 1 µl digested vector, and 6 µl concatemer sample. After the ligation, the samples are purified by PC8 extraction, PLG separation, and ethanol precipitation steps, the pellet is resuspended in 6 µl LoTE.

3 µl Resuspended DNA is transfected into ElectroMax DH10Bs (Cat. No. 18290-015, Invitrogen) using Invitrogen Cell-Porator with the setting of capacitance: 330 µF, impedance: low Ω; charge rate: fast, voltage: 400. The voltage is delivered up to 2.5 kV with Invitrogen Voltage Booster which is set to 4 kΩ

resistance. After the transfection, 1/10 (about 100 μl) of the transfected bacteria is plated onto each 10-cm Zeocine-containing agar plate and grown for 12–16 hr at 37°C. Positive bacterial colonies are identified by PCR using vector-specific primers. Only PCR products >650 bp, which should contain at least 30 tags, are selected for sequence analysis. Direct sequencing of PCR products can be performed in the individual lab using the automated DNA sequencer, or preferably in The DNA Analysis Core Facility.

Sequence files can be analyzed using any SAGE software listed in the Methods section by following the software developer's instruction.

Application of SAGE to Human Diseases

The SAGE technique has been used for analyzing various human diseases with the aim of deciphering pathways involved in the pathogenesis and identifying novel diagnostic tools, prognostic markers, and potential therapeutic targets. Two recent reviews (Polyak et al., 2001; Riggins, 2001) covered in detail the application of SAGE to identify tumor markers and antigens. Here we illustratively highlight some discoveries and the new biological insights that have emerged from SAGE's applications to human diseases.

Insights from Transcriptome Map Analyses

Because SAGE can evaluate the expression pattern of thousands of genes in a quantitative manner without prior sequence information, it has been readily applied to characterize transcriptomes from a number of cells and tissues with different conditions. In addition, SAGE is one of the techniques used in the National Cancer Institute–funded Cancer Genome Anatomy Project (CGAP). This project is an interdisciplinary program to develop technology and gather information needed to understand the molecular basis of cancer. As part of CGAP's goal of creating a Tumor Gene Index, SAGE was added as a strategic analytical technique. A database with archived SAGE tag counts and online query tools was created and is now the largest source of public SAGE data. To date, millions of tags from 162 different libraries have been deposited on the National Center for Biotechnology Education/CGAP SAGEmap Web site (www.ncbi.nlm.nih.gov/SAGE/, Feb.13, 2003). This public depository has made the transcriptome analysis convenient. Fujii et al. (2002) constructed a genome-wide transcriptome map of non–small cell lung carcinomas based on gene-expression profiles generated by SAGE using primary tumors and bronchial epithelial cells of the lung. Using the human genome working draft and the public databases, 25,135 nonredundant UniGene clusters were mapped onto

unambiguous chromosomal positions. Of the 23,056 SAGE tags that appeared more than once among the nine SAGE libraries, 11,156 tags representing 7097 UniGene clusters were positioned onto chromosomes. A total of 43 and 55 clusters of differentially expressed genes were observed in squamous cell carcinoma and adenocarcinoma, respectively. Examination of clusters identified in squamous cell lung cancer suggested that 9 of 15 clusters with overexpressed genes and 13 of 28 clusters with underexpressed genes were concordant with previously reported cytogenetic, comparative genomic hybridization or loss of heterozygosity studies. At least a portion of the gene clusters identified via the transcriptome map most likely represented the transcriptional or gentic alterations that occurred in the tumors. Integrating chromosomal mapping information with gene expression profiles may help reveal novel molecular changes associated with human lung cancer.

Caron et al. (2001) integrated gene chromosome mapping data with genome-wide messenger RNA expression profiles of 2.45 million SAGE transcript tags derived from 12 tissue types. They developed algorithms to assign these tags to UniGene clusters and their chromosomal position. The resulting Human Transcriptome Map generates gene expression profiles for any chromosomal region in 12 normal and pathologic tissue types. The map reveals a clustering of highly expressed genes to specific chromosomal regions. One interesting observation is that chromosomes 4, 13, 18, and 21 seem to be gene-poor and transcribed at low levels, whereas others such as chromosomes 19 and 22 are relatively gene-rich and highly transcribed. Although this unequal distribution may disappear when all of the expressed genes have been identified and mapped, it is worth noting that the vast majority of constitutional trisomies arise from chromosomes 13, 18, and 21. Their low gene density may explain why people with these syndromes survive birth.

A human transcriptome based on the analysis of 3.5 million transcripts from 19 normal and diseased tissues was published (Velculescu et al., 1999). It identified the expression of about 84,000 transcripts and revealed that more than 43,000 genes are expressed in a single cell type. In addition, these analyses identified specific genes uniquely expressed in individual cell types, genes that are expressed in all cell types, and a small number of genes that are uniformly elevated in many cancers compared with their normal counterparts. The analyses produced a heretofore-unavailable global picture of gene expression patterns and a wealth of information of gene functions and clinical pathogenesis. Scientists have begun to characterize transcriptomes of various organs such as kidney, liver, and thyroid. Comprehensively characterizing and contrasting gene expression patterns in normal and diseased organs will provide an alternative strategy to identify organ-specific candidate pathways, which regulate disease susceptibility and progression, and novel targets for therapeutic intervention.

Revelation of Molecular Pathophysiology from the "Signature" Gene Expression Profile

The SAGE technique has been employed to reveal previously unrecognized molecular pathophysiology important in cancer, neurodegeneration, and cardiovascular disease. Schwering *et al.* (2003) observed the loss of the B-lineage–specific "signature" gene expression program in Hodgkin and Reed-Sternberg cells of Hodgkin lymphoma. Hodgkin and Reed-Sternberg (HRS) cells represent the malignant cells in classical Hodgkin lymphoma. Because their immunophenotype cannot be attributed to any normal cell of the hematopoietic lineage, the origin of HRS cells has been controversially discussed, but molecular studies established their derivation from germinal center B cells. Schwering *et al.* (2003) compared gene expression profiles generated by SAGE and DNA chip microarrays from HL cell lines with those of normal B-cell subsets and found that mRNA levels for nearly all established B-lineage–specific genes were decreased in (HRS) cells. For nine of these genes, lack of protein expression was histochemically confirmed. Down-regulation of genes affected multiple components of signaling pathways active in B cells, including B-cell receptor (BCR) signaling. This suggests that the lost B-lineage identity in HRS cells may explain their survival without BCR expression and reflect a fundamental defect in maintaining the B-cell differentiation state in HRS cells.

The "signature" gene expression profiles provided molecular details of the pathophysiology of neurodegeneration in the GM2 gangliosidoses. Tay-Sachs and Sandhoff diseases are lysosomal storage disorders characterized by the absence of beta-hexosaminidase activity and the accumulation of GM2 ganglioside in neurons. In each disorder, a virtually identical course of neurodegeneration begins in infancy and leads to demise generally by 4–6 years of age. Through SAGE, Myerowitz *et al.* (2002) determined gene expression profiles in cerebral cortex from a Tay-Sachs patient, a Sandhoff disease patient, and a pediatric control. Examination of genes that showed altered expression in both patients revealed molecular pathophysiology of the disorders relating to neuronal dysfunction and loss. A large fraction of the elevated genes in the patients could be attributed to activated macrophages/microglia and astrocytes. They included class II histocompatability antigens, the proinflamatory cytokine osteopontin, complement components, proteinases and inhibitors, galectins, osteonectin/SPARC, and prostaglandin D2

synthase. The results are consistent with a model of neurodegeneration that includes inflammation as a factor leading to the precipitous loss of neurons in individuals with these disorders.

The application of SAGE to the cardiovascular disease field is still in its infancy. Activation of human arterial endothelial cells (ECs) is an early event in the pathogenesis of atherosclerosis. To identify the repertoire of genes that are differentially expressed after activation, de Waard *et al.* (1999) used SAGE to compare the mRNA spectrum of quiescent ECs with that of ECs activated for 6 hr with a strong atherogenic stimulus, oxidized LDL. The authors analyzed a moderate number of tags, about 12,000, but about 5% of the tags were derived from differentially expressed genes (at least fivefold up- or down-regulated). These transcript tags mapped to 56 genes, close to 1% of the total number of analyzed genes. Among these 56 differentially expressed genes are 42 known genes, including the hallmark endothelial cell activation markers interleukin 8 (IL-8), monocyte chemoattractant protein 1 (MCP-1), vascular cell adhesion molecule 1 (VCAM-1), plasminogen activator inhibitor 1 (PAI-1), Gro-α, Gro-β, and E-selectin. Differential transcription of a selection of the up-regulated genes was confirmed by Northern Blot analysis. A novel observation is the up-regulation of activin β_A mRNA, a member of the transforming growth factor beta family. The authors demonstrated that a moderate number of transcript tags are sufficient to reveal the significant alterations of EC transcription and the activation of a particular pathway resulting from a strong atherogenic stimulus.

Discovery of Candidate Disease Markers

One of the goals of the SAGE studies has been to identify disease markers.

Saha *et al.* (2001) reported in *Science* that the PRL-3 protein tyrosine phosphatase, revealed by SAGE analysis, was associated with metastasis of colorectal cancer. To gain insights into the molecular basis for metastasis, they compared the global gene expression profile of metastatic colorectal cancer with that of primary cancers, benign colorectal tumors, and normal colorectal epithelium. Among the genes identified, the PRL-3 protein tyrosine phosphatase gene was of particular interest. It was expressed at high levels in each of 18 cancer metastases studied but at lower levels in nonmetastatic tumors and normal colorectal epithelium. In 3 of 12 metastases examined, multiple copies of the PRL-3 gene were found within a small amplicon located at chromosome 8q24.3. These data suggest that the PRL-3 gene is important for colorectal cancer metastasis and provide a new therapeutic target for these intractable lesions.

Fibrillary astrocytoma, the most common primary central nervous system neoplasm, is infiltrating, rapidly proliferating, and almost invariably fatal. This contrasts with the biologically distinct pilocytic astrocytoma, which is circumscribed, often cystic, slowly proliferating, and associated with a favorable long-term outcome. Diagnostic markers for distinguishing pilocytic astrocytomas from infiltrating anaplastic astrocytomas are currently not available. The SAGE analysis using the NCBI database showed a higher level of expression of apoD RNA in pilocytic astrocytoma than in any of the other 94 neoplastic and non-neoplastic tissues in the database. cDNA microarray analysis also showed that apolipoprotein D (apoD) expressed 8.5-fold higher in pilocytic astrocytomas over infiltrating anaplastic astrocytomas. By immunohistochemistry, 10 of 13 pilocytic astrocytomas stained positively for apoD, while none of 21 infiltrating astrocytomas showed similar staining. Additionally, ApoD immunostaining was seen in 9 of 14 of gangliogliomas, 4 of 5 subependymal giant cell astrocytomas (SEGAs), and a single pleomorphic xanthoastrocytomas (PXAs). By *in situ* hybridization, pilocytic astrocytomas, in contrast with infiltrating astrocytomas, showed widespread increased apoD expression. The appearance ApoD is associated with decreased proliferation in some cell lines, and apoD is the protein found in highest concentration in cyst fluid from benign cystic disease of the breast. In addition ApoD might play a role in either decreased proliferation or cyst formation in pilocytic astrocytomas, gangliogliomas, *SEGAs*, and PXAs. Thus, apoD is emerged as a potential marker for pilocytic tumors (Hunter *et al.*, 2002).

Approximately 6% of human beings harbor an unruptured intracranial aneurysm. Each year in the United States, more than 30,000 people suffer a ruptured intracranial aneurysm, resulting in subarachnoid hemorrhage. Despite the high incidence and catastrophic consequences of a ruptured intracranial aneurysm and the fact that there is considerable evidence that predisposition to intracranial aneurysm has a strong genetic component, very little is known with regard to the pathology and pathogenesis of this disease. To begin characterizing the molecular pathology of intracranial aneurysm, Peters *et al.* (2001) used a global gene expression analysis approach (SAGE-Lite) in combination with a novel data-mining approach to perform a high-resolution transcript analysis of a single intracranial aneurysm, obtained from a 3-year-old girl. SAGE-Lite provided a detailed molecular snapshot of a single intracranial aneurysm. The study indicated that the aneurysmal dilation resulted in a highly dynamic cellular environment in which extensive wound healing and tissue/extracellular matrix remodeling were taking place. Specifically, the authors observed significant overexpression of genes encoding extracellular matrix components (e.g., COL3A1, COL1A1, COL1A2, COL6A1, COL6A2, elastin) and genes involved in extracellular matrix turnover (TIMP-3, OSF-2), cell adhesion and antiadhesion (SPARC, hevin), cytokinesis (PNUTL2), and cell migration (tetraspanin-5). Although these were preliminary data, representing analysis of only one individual, the study presented a unique first insight into the molecular basis of aneurysmal disease and defined numerous candidate markers for future biochemical, physiological, and genetic studies of intracranial aneurysm.

Future Perspectives

A "completed" human genome has been achieved, and although DNA microarray is more suitable for large-scale gene expression profiling of human disease, SAGE remains useful for the quantification of low-abundance transcripts and the analysis of gene expression patterns in organisms in which gene or cDNA sequence information is lacking or incomplete. The advantage of SAGE over DNA microarray is that it can identify genes without knowing the sequence of the gene. This utility will remain attractive to medical professionals in identifying diagnostic and prognostic markers and therapeutic targets of human diseases, including the identification or fingerprinting of unknown infectious agents. The long SAGE could be used to facilitate the annotation of genomes of other microorganisms whose genome sequences have been determined but whose transcript databases are less extensive than those for humans. It is anticipated that the combined application of gene profiling by SAGE or DNA microarray and genetic strategies will emerge as one of the more valuable approaches to facilitating the intelligent and productive use of gene expression data in the near future. By interpreting genome-wide expression profiles in a genetic context, it may be possible to extract more information than that gained from conventional studies of a more descriptive or correlational nature. It is hoped that the combination of genetic strategies with genome-wide expression profiling will be helpful in unraveling the pathogenesis of complex human disorders such as cancer, coronary artery disease, and diabetes.

Acknowledgments

Parts of the studies cited in this chapter were in part supported by Johns Hopkins Institutional Research Grant (Ye, SQ), a pilot project (Ye, SQ) in The Hopkins DK center for the Analysis of

Gene Expression (R24DK58757-01, NIDDK), and the Dorothy Wallis Wagner Charitable Trust (Ye, SQ).

References

Bertelsen, A.H., and Velculescu, V.E. 1998. High-throughput gene expression analysis using SAGE. *DDT 3:*152–159.

Boon, K., Osorio, E.C., Greenhut, S.F., Schaefer, C.F., Shoemaker, J., Polyak, K., Morin, P.J., Buetow, K.H., Strausberg, R.L., De Souza, S.J., and Riggins, G.J. 2002. An anatomy of normal and malignant gene expression. *Proc. Natl. Acad. Sci. USA. 99(17):* 11287–11292.

Burgess, J.K. 2001. Gene expression studies using microarrays. *Clin. Exp. Pharmacol. Physiol. 28:*321–328.

Caron, H., van Schaik, B., van der Mee, M., Baas, F., Riggins, G., van Sluis, P., Hermus, M.C., van Asperen, R., Boon, K., Voute, P.A., Heisterkamp, S., van Kampen, A., and Versteeg, R. 2001. The human transcriptome map: Clustering of highly expressed genes in chromosomal domains. *Science 291:*1289–1292.

Chen, J., Lee, S., Zhou, G., and Wang, S.M. 2002. High-throughput GLGI procedure for converting a large number of serial analysis of gene expression tag sequences into 3' complementary DNAs. *Genes Chromosomes Cancer 33:*252–261.

Damgaard Nielsen, M., Millichip M., and Josefsen K. 2003. High-performance liquid chromatography purification of 26-bp serial analysis of gene expression ditags results in higher yields, longer concatemers, and substantial time savings. *Anal. Biochem. 313(1):*128–132.

Datson, N.A., Jong, J., van der Perk-de van den Berg, M.P., de Kloet, E.R., and Vreugdenhil, E. 1999. MicroSAGE: A modified procedure for serial analysis of gene expression in limited amounts of tissue. *Nucl. Acids Res. 27:*1300–1307.

de Waard, V., van den Berg, B.M., Veken, J., Schultz-Heienbrok, R., Pannekoetk, H., and van Zonneveld, A.J. 1999. Serial analysis of gene expression to assess the endothelial cell response to an atherogenic stimulus. *Gene. 226:*1–8.

Fujii, T., Dracheva, T., Player, A., Chacko, S., Clifford, R., Strausberg, R.L., Buetow, K., Azumi, N., Travis, W.D., and Jen, J. 2002. A preliminary transcriptome map of non-small cell lung cancer. *Cancer Res. 62(12):*3340–3346.

Hunter, S., Young, A., Olson, J., Brat, D.J., Bowers, G., Wilcox, J.N., Jaye, D., Mendrinos, S., and Neish, A. 2002. Differential expression between pilocytic and anaplastic astrocytomas: Identification of apolipoprotein D as a marker for low-grade, noninfiltrating primary CNS neoplasms. *J. Neuropathol. Exp. Neurol. 61(3):*275–281.

Ishii, M., Hashimoto, S., Tsutsumi, S., Wada, Y., Matsushima, K., Kodama, T., and Aburatani, H, 2000. Direct comparison of GeneChip and SAGE on the quantitative accuracy in transcript profiling analysis. *Genomics 68:*136–143.

Kenzelmann, M., and Muhlemann, K. 1999. Substantially enhanced cloning efficiency of SAGE (Serial Analysis of Gene Expression) by adding a heating step to the original protocol. *Nucl. Acids Res. 27:*917–918.

Lal, A., Lash, A.E., Altschul, S.F., Velculescu, V., Zhang, L., McLendon, R.E., Marra, M.A., Prange, C, Morin, P.J., Polyak, K., Papadopoulos, N., Vogelstein, B., Kinzler, K.W., Strausberg, R.L., and Riggins, G.L. 1999. A public database for gene expression in human cancers. *Cancer Res. 59:*5403–5407.

Larsson, M., Stahl, S., Uhlen, M., and Wennborg, A. 2000. Expression profile viewer (ExProView): A software tool for transcriptome analysis. *Genomics 63(3):*341-353.

Lockhart, D.J., Dong, H., Byrne, M.C., Follettie, M.T., Gallo, M.V., Chee, M.S., Mittmann, M., Wang, C., Kobayashi, M., Horton, H., and Brown, E.L. 1996. Expression monitoring by hybridization to high-density oligonucleotide arrays. *Nat. Biotechnol. 14:* 1675–1680.

Madden, S.L., Wang, C.J., and Landes, G. 2000. Serial analysis of gene expression: From gene discovery to target identification *DDT 5:*415–425.

Man, M.Z., Wang, X., and Wang Y. 2000. POWER SAGE: Comparing statistical tests for SAGE experiments. *Bioinformatics 16(11):*953–959.

Margulies, E.H., and Innis, J.W. 2000. eSAGE: Managing and analysing data generated with serial analysis of gene expression (SAGE). *Bioinformatics 16(7):*650–651.

Myerowitz, R., Lawson, D., Mizukami, H., Mi, Y., Tifft, C.J., and Proia, R.L., 2002. Molecular pathophysiology in Tay-Sachs and Sandhoff diseases as revealed by gene expression profiling. *Hum. Mol. Genet. 11(11):*1343–1350.

Nacht, M., Ferguson, A.T., Zhang, W., Petroziello, J.M., Cook, B.P., Gao, Y.H., Maguire, S., Riley, D, Coppola, G., Landes, G.M, Madden, S.L., and Sukumar, S. 1999. Combining serial analysis of gene expression and array technologies to identify genes differentially expressed in breast cancer. *Cancer Res. 59:* 5464–5470.

Neilson, L., Andalibi, A., Kang, D., Coutifaris, C., Strauss III, J.F., Stanton, J.A.L., and Green, D.P. 2000. Molecular phenotype of the human oocyte by PCR-SAGE. *Genomics 63:*13–24.

Peters, D.G., Kassam, A.B., Feingold, E., Heidrich-O'Hare, E., Yonas, H., Ferrell, R.E., and Brufsky, A. 2001. Molecular anatomy of an intracranial aneurysm: coordinated expression of genes involved in wound healing and tissue remodeling. *Stroke 32:*1036–1042.

Peters, D.G., Kassam, A.M., Yonas, H., O'Hare, E.H., Ferrell, R.E., and Brufsky, A.M. 1999. Comprehensive transcript analysis in small quantities of mRNA by SAGE-Lite. *Nucl. Acids Res. 27:*e39(i–vi).

Polyak, K., and Riggins, G.J. 2001. Gene discovery using the serial analysis of gene expression technique: Implications for cancer research. *J. Clin. Oncol. 19:*2948–2958.

Powell, J. 1998. Enhanced concatemer cloning-a modification to the SAGE (Serial Analysis of Gene Expression) technique. *Nucl. Acids Res. 26(14):*3445–3446.

Riggins, G.J. 2001. Using serial analysis of gene expression to identify tumor markers and antigens. *Dis. Markers 17:*41–48.

Saha, S., Sparks, A.B., Rago, C., Akmaev, V., Wang, C.J., Vogelstein, B., Kinzler, K.W., and Velculescu, V.E. 2002. Using the transcriptome to annotate the genome. *Nat. Biotechnol. 20(5):*508–512.

Saha, S., Bardelli, A., Buckhaults, P., Velculescu, V.E., Rago, C., St. Croix, B., Romans, K.E., Choti, M.A., Lengauer, C., Kinzler, K.W., and Vogelstein, B. 2001. A phosphatase associated with metastasis of colorectal cancer. *Science 294:*1343–1346.

Schena, M., Shalon, D., Davis, R.W., and Brown, P.O. 1995. Quantitative monitoring of gene expression patterns with a complementary DNA microarray. *Science 270:*467–470.

Schulze, A., and Downward, J. 2001. Navigating gene expression using microarrays: A technology review. *Nat. Cell Biol. 3:*E190–E195.

Schwering, I., Brauninger, A., Klein, U., Jungnickel, B., Tinguely, M., Diehl, V., Hansmann, M.L., Dalla-Favera, R., Rajewsky, K., and Kuppers, R. 2003. Loss of the B-lineage-specific gene expression program in Hodgkin and Reed-Sternberg cells of Hodgkin lymphoma. *Blood 101(4):*1505–1512.

Velculescu, V.E., Madden, S.L., Zhang, L., Lash, A.E., Yu, J., Rago, C., Lal, A., Wang, C.J., Beaudry, G.A., Ciriello. K.M., Cook, B.P., Dufault, M.R., Ferguson, A.T., Gao, Y., He, T-C, Hermeking, H., Hiraldo, S.K., Hwang, P.M., Lopez, M.A.,

Luderer, H.F., Mathews, B., Petroziello, J.M., Polyak, K., Zawel, L., Zhang, W., Zhang, X., Zhou, W., Haluska, F.G., Jen, J., Sukumar, S., Landes, G.M., Riggins, G.J., Vogelstein, B., and Kinzler, K.W. 1999. Analysis of human transcriptomes. *Nat. Genet.* 23:387–388.

Velculescu, V.E., Zhang, L., Vogelstein, B., and Kinzlel, K.W. 1995. Serial analysis of gene expression. *Science* 270:484–487.

Velculescu, V.E., Zhang, L., Zhouj, W., Traverso, G., St.Croix, B., Vogelstein, B., and Kinzler, K.W. 2000. Serial analysis of gene expression Detailed Protocol (Ver.1.0e). Johns Hopkins Oncology Center and Howard Hughes Medical Institute, Baltimore, MD.

van Kampen, A.H., van Schaik, B.D., Pauws, E., Michiels, E.M., Ruijter, J.M., Caron, H.N., Versteeg, R., Heisterkamp, S.H., Leunissen, J.A., Bass, F., and van der Mee, M. 2000. USAGE.: A Web-based approach toward the analysis of SAGE data. Serial Analysis of Gene Expression. *Bioinformatics 16(10):* 899–905.

Virlon, B., Cheval, L., Buhler, J-M., Billon, E., Doucet, A., and Elalouf, J-M. 1999. Serial microanalysis of renal transcriptomes. *Proc. Natl, Acad. Sci. USA 96:*15286–15291.

Ye, S.Q., Zhang, L.Q., Zheng, F., Virgil, D., and Kwiterovich, P.O. 2000. miniSAGE: Gene expression profiling using serial analysis of gene expression from 1 microgram total RNA. *Anal. Biochem. 287:* 144–152.

Ye, S.Q., Usher, D.C., and Zhang, L.Q. 2002. Gene expression profiling of human diseases by serial analysis of gene expression. *J. Biomed. Sci. 9(5):*384–394.

III

Lung Carcinoma

1

Lung Carcinoma: An Introduction

M.A. Hayat

In the second half of the twentieth century, lung cancer became the leading cause of cancer death worldwide. Approximately 86% of the people who are diagnosed with lung cancer die of the disease within five years. About 169,000 new cases of lung cancer identified in the United States in 2002 will lead to more deaths than breast, colon, prostate, and cervical cancers combine. However, the incidence of these other cancers is more than half a million, which is far greater than the annual incidence of lung cancer. So, why is lung cancer so much more lethal than those other common cancers? One reason is that lung cancer is more frequently diagnosed at an advanced metastatic stage, which is generally not curable. Other types of cancers that develop deep within the body cavity, such as pancreas and ovaries, also present a challenge for early detection. They too have a higher probability of being detected in advanced stages, and therefore have low cure rates.

Lung carcinomas represent a heterogeneous category of tumors, including several entities with different prognosis and therapeutic approaches. From a clinical point of view, they are simply classified into two groups: small cell lung cancer (SCLC) and nonsmall cell cancer (NSCLC). The former accounts for 20–25% of all bronchogenic malignancies and follows a highly aggressive clinical course. Metastases (clinically apparent or occult) are usually present at diagnosis of SCLC, and the preferred treatment is usually combination chemotherapy. Less than 5% of patients currently survive

5 years past the initial diagnosis of SCLC, whereas the 5-year survival rate for patients diagnosed with NSCLC is 15%. Initially, SCLC is highly responsive to radiation and systemic chemotherapy regimens, whereas NSCLC is less sensitive to chemotherapy, and curative intent surgical resection is the preferred treatment.

Bronchial carcinoids and SCLC are neuroendocrine tumors, although their pathogenesis is very different. Carcinoids are tumors of low-grade malignancy, although a subtype, the atypical carcinoid, follows a more aggressive type. The overall survival of patients with carcinoids is far better than that of patients with SCLC. Carcinoids occur at an early age and are not smoking-related. In contrast to SCLC, they are highly resistant to chemotherapy, and most of them are cured by surgical resection. There are both similarities and differences in the molecular changes present in SCLC and carcinoid (Onuki et al., 1999). For example, mutations of the menin gene, characteristic of many carcinoids, are absent in high-grade neuroendocrine tumors, including SCLC.

Lung cancer is a highly aggressive neoplasm that is reflected by a multitude of genetic aberrations detectable at the chromosomal and molecular levels. To improve the poor prognosis of lung cancer, it is important to explore fundamental biological properties of this cancer and design new approaches based on the application of molecular genetics to therapy. A number of genes and proteins play a part in lung carcinomas. Several examples follow.

Handbook of Immunohistochemistry and in situ Hybridization of Human
Carcinomas, Volume 1: Molecular Genetics; Lung and Breast Carcinomas

101

Copyright © 2004 by Elsevier (USA)
All rights reserved.

Advances in molecular genetics are being applied to identify persons at high risk of lung cancer, including those at high risk of postoperative recurrence of the disease. There is a relationship between the expression of epidermal growth factor receptor family members (HER-1, HER-2, HER-3, and HER-4) and lung malignancy. For example, HER-2 is a marker of tumor progression in NSCLC, which can be observed at protein level by using immunohistochemistry (Kristiansen *et al.*, 2001); this information indicates chromosomal alterations at 17q21. HER-3 overexpression is a useful predictor of postoperative tumor recurrence in patients with NSCLC (Lai *et al.*, 2001).

BRAF mutations are also involved in lung adenocarcinoma, although such mutations are more abundant in colon carcinoma and melanomas. Activating mutations of the KRAS protooncogenes are also known to occur in lung adenocarcinomas. The presence of both BRAF and KRAS mutations in this adenocarcinoma suggests that other members of the receptor tyrosine kinase/RAS/Raf nitrogen-activated protein kinase cascade should be evaluated for mutations in these tumors (Naoki *et al.*, 2002).

Carbonic anhydrase-related protein has been reported to play a part in NSCLC (Akisawa *et al.*, 2003). Hiratsuka *et al.* (2002) report that distant primary tumors in organs other than the lung can elevate MMP9 through VEGFR-1/Flt-1 tyrosine kinase, promoting lung metastasis. Aryl hydrocarbon receptor has been reported to be up-regulated in lung adenocarcinoma (Lin *et al.*, 2003). Sturm *et al.* (2003) recommend that Ck34βE12 should be included in a routine diagnostic panel of antibodies used in the differential diagnosis of lung cancer, especially that of small cell proliferations such as basaloid carcinoma versus small cell carcinoma.

The role of mucin, MDM2, beta-catenin, thyroid transcription factor-1, caveolin-1, blood group antigen, angiogenesis, and global methylation of DNA in lung cancer is discussed later in this volume. It is concluded that genome-wide screening for gene alterations in lung carcinoma should provide new therapeutic opportunities for this disease.

Major classification of lung cancer is shown in Table 8. Genetic-based classification of human lung cancer has been presented by Petersen *et al.* (2001). The role of cigarette smoking in lung cancer is summarized next.

Lung Cancer and Cigarette Smoking

Lung cancer is the most important chemically induced tumor type in humans, for the vast majority of lung cancer is associated with long-term cigarette

Table 8 World Health Organization-Based Classification of Malignant Epithelial Lung Tumors

1. Squamous cell carcinoma (SCC)
   Variants: papillary, clear cell, small cell, basaloid SCC
2. Small cell carcinoma (SCLC)
   Variant: combined SCLC
3. Adenocarcinoma (Adeno)
   Variants: acinar, papillary, bronchioloalveolar, solid, mixed, others
4. Large cell carcinoma (LCC)
   Variants: large cell neuroendocrine carcinoma, basaloid, lympho-epithelioma–like clear cell with rhabdoid phenotype
5. Adenosquamous carcinoma
6. Carcinoma with pleomorphic, sarcomatoid or sarcomatous elements
7. Carcinoid tumors
   Variants: typical, atypical
8. Carcinomas of salivary gland type
9. Unclassified carcinomas

Spectrum of neuroendocrine lung tumors
   Typical carcinoid—atypical carcinoid—LCNEC—SCLC
Major clinical differentiation
   NSCLC (Adeno, SCC, LCC) ↔ SCLC

After Petersen *et al.*, (2001).

smoking, which is true for all major subtypes (Petersen *et al.*, 2001). The latency is usually several decades, and most carcinomas develop after the age of 50. Although a subgroup of smokers better tolerate this carcinogen, the risk steadily rises with age, beyond 50.

According to Piyathilake *et al.* (2001), although ~ 85% of lung cancers may be attributable to cigarette smoking, only 10–20% of smokers develop lung cancer during their lifetimes, suggesting that other factors, including environmental, genetic, and epigenetic, may be implicated in the pathogenesis of the disease. Mutational basis of cancer is well established. Epigenetic changes alter the way genes are expressed without affecting genetic coding. Alteration in global DNA methylation is thought to be an important epigenetic difference, which may contribute to the susceptibility to lung cancer. Other molecular mechanisms responsible for inducing lung cancer due to smoking are described here. For further details, see Part III, Chapter 10, by Piyathilake *et al.*, in this volume.

Tobacco smoke contains an array of potent chemical carcinogens and reactive oxygen species that may produce DNA bulky adducts, cross-links, oxidative or base DNA damage, and DNA strand breaks. Among several major DNA repair pathways that operate on specific types of damaged DNA by cigarette smoking, base excision repair is involved in repair of DNA base damage and single-strand breaks, and nucleotide excision repair is involved in the repair of bulky monoadducts, cross-links, and oxidative damages (Berwick *et al.*, 2000). Individuals with a reduced DNA repair capacity have a high level of carcinogen-DNA adducts in their tissues. Lung cancer patients may have lower DNA repair capacity.

X-ray cross-complementing group 1 (XRCC1) and excision repair cross-complementing group 2 (ERCC2) are two major DNA repair proteins. Polymorphisms of these two genes have been associated with altered DNA repair capacity and cancer risk (Zhou *et al.*, 2003). Cumulative cigarette smoking exposure is thought to play an important role in altering the direction and magnitude of the association between XRCC1 and ERCC2 polymorphisms, and the risk of lung cancer.

The increase in blood plasma nicotine level after smoking a cigarette (nicotine boost) is a measure of the absorbed dose of nicotine, which in turn reflects exposure to other smoke constituents. The dose of nicotine taken by a smoker is determined by a variety of factors, including the way in which the cigarette is smoked (the depth and speed of inhalation and the number of puffs taken), characteristics of the cigarette (tar and nicotine yield and additives such as menthol), and, possibly, individual factors such as gender, race, and pulmonary-related symptoms. Patterson *et al.* (2003) have examined demographic, smoking behavior and dependence, and psychological variables as predictors of nicotine boost-assessed in a nonlaboratory setting.

References

Akisawa, Y., Nishimori, I., Taniuchi, K., Okamoto, N., Takeuchi, T., Sonobe, H., Ohtsuki, Y., and Onishi, S. 2003. Expression of carbonic anhydrase-related protein CA-RP VIII in nonsmall cell lung cancer. *Virchows Arch. 442:*66–70.

Berwick, M., and Vineis, P. 2000. Markers of DNA repair and susceptibility to cancer in humans: An epidermiologic review. *J. Natl. Cancer Inst. 92:*874–897.

Hiratsuka, S., Nakamura, K., Iwai, S., Murakami, M., Itoh, T., Kijima, H., Shipley, J.M., Senior, R.M., and Shibuya, M. 2002. MMP9 induction by vascular endothelial growth factor receptor-1 is involved in lung-specific metastatis. *Cancer Cell 2:*289–300.

Kristiansen, G., Yu, Y., Petersen, S., Kaufmann, O., Schluns, K., Dietel, M., and Petersen, I. 2001. Overexpression of c-erbB2 protein correlates with disease-stage and chromosomal gain at the c-erbB2 locus in nonsmall cell lung cancer. *Eur. J. Cancer 37:*1089–1095.

Lai, S.L., Perng, R.P., and Hwang, J. 2000. p53 gene status modulates the chemosensitivity of nonsmall cell lung cancer cells. *J. Biomed. Sci. 7:*64–70.

Lin, P., Chang, H., Tsai, W.-T., Wu, M.-H., Liao, Y.-S., Chen, J.-T., and Su, J.-M. 2003. Overexpression of aryl hydrocarbon receptor in human lung carcinoma. *Toxicol. Pathol. 31:*22–30.

Naoki, K., Chen, T.-H., Richards, W.G., Sugarbaker, B.J., and Meterson, M. 2002. Missense mutations of the BRAF gene in human lung adenocarcinoma. *Cancer Res. 62:*7001–7003.

Onuki, N., Wistuba, I., Travis, W.D., Virmani, A.K., Yashima, K., Brambilla, E., Hasleton, P., and Gazdar, A.F. 1999. Genetic changes in the spectrum of neuroendocrine lung tumors. *Cancer 85:*600–607.

Patterson F., Benowitz, N., Shields, P., Kaufmann, V., Jespen, C., Wileyto, P., Kucharski, S., and Lerman, C. 2003. Individual differences in nico-tine intake per cigarette. *Cancer Epidemiol. Biomark. Prev. 12:*359–365.

Petersen, I., and Petersen, S. 2001. Toward a genetic-based classification of human lung cancer. *Anal. Cell Pathol. 22:*111–121.

Piyathilake, C.J., Frost, A.R., Bell, W.C., Oelschlager, D., Weiss, H., Johanning, G.L., Niveleau, A., Heimburger, D.C., and Grizzle, W.E. 2001. Altered global methylation of DNA: An epigenetic difference in susceptibility for lung cancer is associated with its progression. *Hum. Pathol. 32:*856–862.

Sturm, N., Rossi, G., Lantuejoul, S., Laverriere, M.-H., Papotti, M., Brichon, P.-Y., Brambilla, C., and Brambilla, E. 2003. 34BE12 expression along the whole spectrum of neuroendocrine proliferations of the lung from neuroendocrine cell hyperplasia to small cell carcinoma. *Histopathology 42:*156–166.

Zhou, W., Liu, G., Miller, D.P., Thurston, S.W., Xu, L.L., Wain, J.C., Lynch, T.J., Su, L., and Christiani, D.C. 2003. Polymorphisms in the DNA repair genes XRCC1 and ERCC2, smoking and lung cancer risk. *Cancer Epidemil. Biomark. Prev. 12:*359–365.

2

Histopathological Classification Phenotype and Molecular Pathology of Lung Tumors

Elisabeth Brambilla

Introduction

Lung cancer is the most common diagnosed cancer and the major cause of mortality (Greenlee *et al.*, 2001) worldwide. With 169,500 new cases per year and 157,400 cancer deaths (Greenlee *et al.*, 2001) in the United States and 182,000 new cases per year and 190,000 cancer deaths in the European Union in 2001, it is the only cancer where incidence and mortality are quite similar. The main risk factor is tobacco carcinogen. Although lung cancer incidence began to decline in men in the United States beginning in 1980, its rate is increasing in women, as a consequence of an increase in the number of women smoking (Travis *et al.*, 1996).

This review focuses on histopathological classification and phenotypical characteristics of lung cancer. Histologic evaluation for lung cancer diagnosis is based on several types of biopsy specimens, including bronchoscopy or fine-needle biopsies and video-assisted thoracoscopic biopsy, as well as wedge resection, lobectomy, or pneumonectomy. Light microscopy is sufficient for most of the diagnosis of lung cancer types and subtypes, rendering the need for histochemical stains or immunohistochemistry to a few histologic types. The international standard for histologic classification of lung tumors is that proposed by the World Health Organization (WHO) and the International Association for the Study of Lung Cancer (Travis *et al.*, 1999). The four major histologic types of lung cancer are squamous cell carcinoma, adenocarcinoma—the incidence of which is increasing at the expense of squamous cell carcinoma—small cell lung carcinoma (SCLC) and large cell lung carcinoma (LCLC). These major types have been subclassified into subtypes, the clinical significance of which might be extremely important such as the bronchioloalveolar carcinoma (BAC) as a variant of adenocarcinoma (Travis *et al.*, 1999).

Preinvasive Lesions

Evidence is accumulating that invasive lung cancer is the end result of a multistep and multifocal process in which molecular changes accompany or even precede histological changes. This fulfills the concept of "field cancerization," which reflects the fact that the entire epithelium is the target of tobacco carcinogens. Active mutagenesis may randomly affect any anatomical location in the bronchial tree. The genetic changes, because of their exquisite specificity or selectivity, are suited to be markers of premalignancy, providing they

Handbook of Immunohistochemistry and *in situ* Hybridization of Human
Carcinomas, Volume 1: Molecular Genetics; Lung and Breast Carcinomas

105

Copyright © 2004 by Elsevier (USA)
All rights reserved.

are maintained in malignancy. The pathology of preinvasive lesions for lung cancer has attracted increasing interest in recent years because of the importance of early detection of lung cancer to screen high-risk patients using fluorescence bronchoscopy and low-dose spiral and helical computerized tomodensitometry (CT). Early detection aims should therefore be to identify the field effect, evaluate the severity of the lesions in the field, and devise suitable methods of intervention according to individual potential for progression. Histological diagnosis of preneoplasia, as well as that of neoplasia, is currently the most reliable standard of diagnosis, although molecular pathology markers have value in indicating the severity of the cancerization field and the risk of progression into lung cancer.

Squamous Dysplasia and Carcinoma *in situ*

Morphological transformation of the normal bronchial mucosa occurs through a continuous spectrum of lesions, including basal cell hyperplasia; squamous metaplasia; mild, moderate, and severe dysplasia; and carcinoma *in situ* (CIS).

According to the thickness and severity of cytologic atypia within the bronchial epithelium, squamous dysplasia are subclassified as mild, moderate, and severe for the lower third, two-thirds, or all thickness of the bronchial epithelium involved, respectively (Travis *et al.*, 1999). Carcinoma *in situ* shows full-thickness involvement of the epithelium and marked cytologic atypia. It differs from severe dysplasia by lack of visible maturation and orientation of the cells from the basal to luminal part of the epithelium. Use of the term "microinvasive squamous cell carcinoma" is not recommended because these microinvasive tumors are true squamous cell carcinoma classified as T1 of the tumor size, node, and metastases (TNM) classification and should be separated from carcinoma *in situ* with involvement of submucosal glands. Pathologists, however, experience the frequency of invasive carcinoma characterized by disruption of basal lamina at less than 300-µ distance from a CIS on serial sections. Serial sections of these areas are thus recommended.

Atypical Alveolar Hyperplasia

This is a millimetric lesion that is considered as a preinvasive state for bronchioloalveolar carcinoma. Atypical alveolar hyperplasia (AAH) is characterized by type II cell proliferation that resembles but falls short of criteria for BAC, nonmucinous type. This is an incidental discovery on histologic examination of a lung cancer resection specimen, the incidence of which varies from 5.7–21.4%, depending on the extent of the search and the criteria used for this diagnosis as well as the range of ages in reported autopsy studies. Most AAH are less than 5 mm in diameter and are multiple. Histologically, AAH consists of a focal proliferation of slightly atypical cuboidal to low columnar epithelial cells along alveolar walls and respiratory bronchioles. Alveolar septa may present slight thickening and discrete lymphoid infiltration.

Diffuse Idiopathic Pulmonary Neuroendocrine Cell Hyperplasia

Diffuse idiopathic pulmonary neuroendocrine cell hyperplasia (DIPNECH) is thought to represent a precursor lesion for carcinoid tumors because a subset of these patients have one or more carcinoid tumors (Travis *et al.*, 1999). It is a rare condition involving peripheral airways at the level of terminal and respiratory bronchioles characterized by linear neuroendocrine cell hyperplasia and tumorlets.

Squamous Cell Carcinoma

Variants of squamous cell carcinoma include:

Papillary
Clear cell
Small cell
Basaloid

Squamous cell carcinoma accounts for approximately 30% of all lung cancers in the United States and 45% in Europe. Its incidence in Europe is progressively decreasing, whereas that of adenocarcinoma is increasing. Two-thirds of squamous cell carcinomas are present as central tumors with a primary site on main system or lobar, whereas one-third are present as peripheral tumors on segmental bronchi. The morphologic features that characterize squamous differentiation include intercellular bridging and keratinization (or individual cell keratinization or squamous pearl formation). These differentiated features are readily apparent in well-differentiated tumors and difficult to detect in poorly differentiated tumors. However, the degree of differentiation does not correlate with prognosis in lung squamous cell carcinoma. Variants described in the WHO classification include papillary, clear cell, small cell, and basaloid subtypes. This last variant has a dismal prognosis compared with poorly differentiated squamous cell carcinoma (Brambilla *et al.*, 1992; Moro *et al.*, 1994). Papillary squamous cell carcinoma often show a pattern of exophytic endobronchial growth.

Most squamous carcinoma express high molecular weight cytokeratins (recognized by antibody AE1/AE3,

KL1, 34βE12, anti-cytokeratine 5-6). Very few cases express cytokeratine 7 (Ck7), whereas thyroid transcription factor 1 (TTF-1) is never expressed.

Adenocarcinoma

Usual subtypes of adenocarcinoma include:

Acinar
Papillary
Bronchioloalveolar carcinoma
> Nonmucinous
> Mucinous
> Mixed mucinous and nonmucinous or indeterminate
Solid adenocarcinoma
Adenocarcinoma with mixed subtypes
Variants
> Well-differentiated fetal adenocarcinoma
> Mucinous "colloid" adenocarcinoma
> Mucinous cystadenocarcinoma
> Signet ring adenocarcinoma
> Clear cell adenocarcinoma

Adenocarcinomas account for about 40% of lung cancer in Europe and the United States. Because most adenocarcinomas are histologically heterogeneous, consisting of two or more of the histologic subtypes, 80% of lung adenocarcinomas diagnosed fall into the category of mixed subtype. The acinar and papillary subtypes are recognized by their architectural pattern of tumor cell growth and invasion. A substantially different definition has been given to bronchioloalveolar carcinoma (BAC subtype), which should be restricted to tumors that grow in a purely lepidic fashion. The solid-type adenocarcinoma is a poorly differentiated carcinoma presenting intracytoplasmic mucins that should be at least five mucin droplets in two different high-power fields. Mucin stains recommended are periodic acid–Schiff (PAS) with diastase digestion and Kreyberg staining with alcian blue.

Bronchioloalveolar carcinoma is uncommon and probably restricted to fewer than 3% of all lung malignancies. In the 1999 WHO/International Association for the Study of Lung Cancer (IASLC) classification, BAC received a restrictive definition: it is defined as a tumor showing lepidic growth along respected alveolar septa with intact elastic and basal lamina frames, without invasive growth. The lack of invasive growth is added as an essential criterion (Travis *et al.*, 1999) based on clinico-pathological data indicating that patients with less than 2-cm BAC may be curable by economic surgical resection (Noguchi *et al.*, 1995).

As a result of the definition, this tumor can be considered as a carcinoma *in situ* at alveolar site. As a consequence of this revised definition of BAC, the literature dealing with these tumors needs complete reevaluation. Indeed, previous to the last classification, BAC included tumors with obvious invasive growth. It is common to observe central scars in pulmonary adenocarcinoma that contains invasive components and a focal BAC-like pattern at the periphery of the tumor. More than 50% of previously called BACs present focal central desmoplastic scaring tissue or intra-alveolar complex papillary growth, whereas the lepidic growth starts around the edge of the scar. For tumors showing malignant tumor cell nests in a desmoplastic central stromal reaction, the diagnosis is adenocarcinoma *mixed subtype* and the various subtypes present should be mentioned (such as acinar, papillary, or BAC). These are not considered any longer as BAC.

According to the 1999 WHO/IASLC classification, a final diagnosis of BAC can only be achieved at examination of a surgical resection specimen. Small biopsies obtained by bronchoscopy or fine-needle sampling may show a lepidic growth pattern suggesting the possibility of BAC but are not sufficient to exclude the presence of an invasive growth. Several important clinicopathological studies have shown the clinical significance of BAC (Noguchi *et al.*, 1995; Suzuki *et al.*, 2000). Noguchi *et al.* (1995) reported that in a large series of 236 cases, the patients with less than 2-cm peripheral lung adenocarcinoma achieve 100% 5-year survival. Multiple pathologic factors for prognostic assessment associated with 100% 5-year survival include at least one of the following features: 1) a pattern of lepidic growth of more than 75% 2) central scar measuring 5 mm or less; and 3) lack of destruction of the elastic fiber framework by tumor cells (Yokose *et al.*, 2000).

The immunohistochemical characteristics of adenocarcinoma may vary according to their histological subtype. They all express epithelial markers (cytokeratins, epithelial membrane antigen (EMA), carcinoembry antigen (CEA). Among cytokeratins, Ck7 is almost constantly expressed, whereas Ck20 is infrequently expressed except for mucinous type BAC, which are often Ck20 positive and Ck7 negative. TTF-1 is expressed in 85% of adenocarcinomas, the most frequently negative ones being mucinous types.

Deregulation of oncogenes and tumor suppressor genes are similar to those observed in other non–small cell lung carcinoma except for Ras mutation, which is quite restricted to adenocarcinoma and more characteristic of peripheral type adenocarcinoma. The differential diagnosis of primary lung adenocarcinoma from metastatic colonic adenocarcinoma cannot rely on Ck20 positivity and CdX2 homeobox gene expression because both are commonly expressed in these two types of tumors (Rossi *et al.*, 2004). A distinction of pulmonary adenocarcinoma from epithelial malignant mesothelioma as a peripheral tumor rests on phenotypical differences. Although primary pulmonary adenocarcinomas express TTF-1, mesotheliomas do not. In contrast with adenocarcinoma, mesothelioma expresses calretinin and cytokeratin 5-6. A number of general adenocarcinoma markers (CEA, CD15, BerEP-4) also help in this distinction.

Small Cell Lung Carcinoma

Small cell lung carcinoma (SCLC) accounts for 25% of all lung cancers in the United States as well as in Europe. Two-thirds of SCLC are proximal and present as a perihilar tumor. The 1999 WHO/IASLC classification presents only two types of SCLC: SCLC (with pure SCLC histology) and combined SCLC (combined with any non–small cell type).

SCLC has a distinctive histological appearance. Tumor cells have a small size, not exceeding that of three lymphocytes. They have a round or fusiform shape, scant cytoplasm with a nuclear-to-cytoplasmic ratio of 9 to 10, a finely granular nuclear chromatin (salt-and-pepper appearance), and absent or inconspicuous nucleoli (Travis *et al.*, 1999). Owing to the scarcity of cytoplasm, nuclear molding and smearing of nuclear chromatin is frequent, caused by crush artifact. There is usually an extensive necrosis and mitotic rate exceeding 20 and reaching 100 mitoses per 2 mm^2 area. Most often, the growth pattern consists of diffuse sheets, although endocrine differentiations with rosettes, palisading, ribbons, and organoid nesting might be seen.

The immunohistochemical features of SCLC are not required for the diagnosis of SCLC. However, crush artifact is common in small biopsy specimens for immunohistochemistry for neuroendocrine differentiations and keratins, and common leucocytes antigens become more useful in marking SCLC versus lymphoid cells, respectively. Moreover, the most important differential diagnosis resides in the distinction between SCLC and non–small cell lung carcinoma (NSCLC) because of different therapeutic implications.

The most useful and specific neuroendocrine markers for distinction of SCLC are chromogranin A, synaptophysin, and neural cell adhesion molecule (NCAM) (clones 123C3 and CD56). The vast majority (95%) of SCLC are reactive with NCAM antibodies, with a specific membranous pattern. This is the most specific and sensitive marker to distinguish SCLC from NSCLC (Lantuejoul *et al.*, 1998). Twenty percent of small cell lung carcinoma may lack chromogranin A and/or synaptophysin expression. Keratin (AE1–AE3)

and EMA stain 85% of SCLC each, in contrast with a specific set of cytokeratin (CK1, 5, 10, 14) recognized by the antibody 34βE12, which is never expressed in pure SCLC. However, this reactivity with 34βE12 enlights the NSCLC component combined with SCLC in the variant of *SCLC* called *small cell lung carcinoma combined*. In 85% of SCLC, TTF-1 is expressed. Antigen Mic2 (CD99), characteristic of primitive neuroectodermal tumors (PNET) and small round cell sarcoma, is also present in a large proportion of SCLC. C-kit (CD117) is expressed in about 40% of SCLC.

Combined Small Cell Lung Cancer

A SCLC associated with at least 10% of another NSCLC component is diagnosed as combined SCLC. Therefore a tumor with at least 10% SCLC component is a SCLC combined. The frequency of combined SCLC depends on the extent of histologic sampling and the extent of the associated component. In a recent study on surgically treated SCLC and using a conservative estimate of 10% of tumors showing NSCLC associated for subclassifying a tumor as a combined variant of SCLC, 28% of the SCLC cases showed a combination with NSCLC, more commonly with large cell lung carcinoma followed by adenocarcinoma and squamous cell carcinoma (Nicholson *et al.*, 2002). When SCLC is associated with spindle cell carcinoma, giant cell carcinoma, or carcinosarcoma, the tumor is diagnosed as *SCLC combined*. Immunohistochemistry might help to recognize associated components such as cytokeratin antibody cocktails, which tend to stain non–small cell lung carcinoma components, a good example of which is CK1, 5, 10, 14 recognized by 34βE12. However, evidence is lacking that pure small cell lung carcinoma and combined small cell lung carcinoma behave differently in regard to prognosis and response to therapy. Following chemotherapy, a mixture of large cells, squamous cell, adenocarcinoma, or giant cells with SCLC may be seen in 15–45% of the cases (Brambilla *et al.*, 1991; Sehested *et al.*, 1986).

Large Cell Lung Carcinoma

Large cell carcinoma is a tumor that shows no differentiation pattern allowing classification into squamous cell carcinoma, adenocarcinoma, or small cell carcinoma. These poorly differentiated tumors most often arise in the lung periphery, although they may be located centrally. They frequently appear at gross examination as large, necrotic tumors. Histologically, these consist of sheets or nests of large polygonal cells with vesicular nuclei and prominent nucleoli. Although they are undifferentiated by light microscopy, features of squamous cell or adenocarcinoma might be found with electron microscopy examination.

Large Cell Neuroendocrine Carcinoma

Large cell neuroendocrine carcinoma (LCNEC) is a variant of large cell carcinoma (Travis *et al.*, 1999). It is a high-grade non–small cell neuroendocrine carcinoma that differs from atypical carcinoid and small cell carcinoma (Travis *et al.*, 1991, 1998). Histologic criteria include (Table 9): 1) neuroendocrine morphology (organoid, palisading, trabecular, or rosette-like growth patterns (Figure 12A); 2) non–small cell cytologic features (large size, polygonal shape, low nuclear-to-cytoplasmic (N/C) ratio, coarse or vesicular nuclear chromatin, and obvious nucleoli; 3) high mitotic rate (≥ 11 per 2 mm^2) with a mean of 60 mitoses per 2 mm^2; 4) frequent necrosis; and 5) at least one positive neuroendocrine immunohistochemical specific marker or neuroendocrine granules by electron microscopy (Travis *et al.*, 1991, 1999). It is difficult to diagnose LCNEC based on small biopsy specimens because of the frequent lack of neuroendocrine morphology without a substantial sampling of the tumors.

Combined LCNEC

The term *combined LCNEC* is used for tumors associated with other histologic types of NSCLC, such as adenocarcinoma or squamous cell carcinoma. Any combination of LCNEC with SCLC is diagnosed as

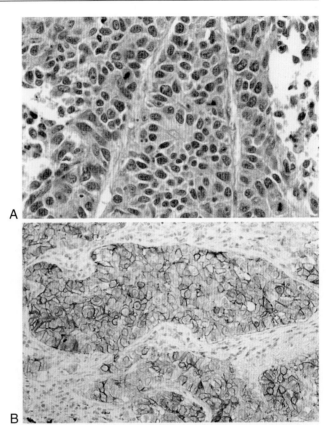

Figure 12 Large cell neuroendocrine carcinoma. **A:** Numerous rosettes give this tumor a neuroendocrine morphologic appearance. Mitotic rate is high (HES 400X). **B:** NCAM is expressed with a membrane pattern (200X).

Table 9 Light Microscopic Features for Distinguishing Small Cell Carcinoma and Large Cell Neuroendocrine Carcinoma[a]

Histologic Feature	Small Cell Carcinoma	Large Cell Neuroendocrine Carcinoma
Cell Size	Smaller (less than diameter of 3 lymphocytes)	Larger
Nuclear/cytoplasmic (N/C) ratio	Higher	Lower
Nuclear chromatin	Finely granular, uniform	Coarsely granular or vesicular, less uniform
Nucleoli	Absent or faint	Often (not always) present, may be prominent or faint
Nuclear molding	Characteristic	Less prominent
Fusiform shape	Common	Uncommon
Polygonal shape with ample pink cytoplasm	Uncharacteristic	Characteristic
Nuclear smear	Frequent	Uncommon
Basophilic staining of vessels and stroma	Occasional	Rare

[a]From Travis W.D. *et al.*, 1991. Neuroendocrine tumors of the lung with proposed criteria for large cell neuroendocrine carcinoma: An ultrastructural, immunohistochemical, and flow cytometric study of 35 cases. *Am. J. Surg. Pathol.* *15*:529–533. With permission.

SCLC combined (Travis *et al.*, 1999). A variety of criteria must be used to separate SCLC from LCNEC (Table 9).

There is no specific marker to distinguish large cell carcinoma from squamous cell carcinoma and adenocarcinoma. In contrast, there are specific features of LCNEC that distinguish them as a specific variant among large cell carcinomas. This distinction is supported by a worse survival observed for LCNEC compared with classical large cell carcinoma. Ninety percent of LCNECs express NCAM (clone 123C3, Cd56) (Figure 12B) in addition to the typical morphological neuroendocrinoid pattern, and most of them express in addition one or both chromogranin A and synaptophysin (Lantuejoul *et al.*, 1998). Half of LCNECs express TTF1, whereas CK 1, 5, 10, 14 (34βE12) is never expressed in pure LCNEC but is expressed in combined components.

Basaloid Carcinoma

Basaloid carcinoma is the most prominent variant of large cell carcinoma after LCNEC (Brambilla *et al.*, 1992; Moro *et al.*, 1994; Travis *et al.*, 1999). Basaloid carcinomas represent 3–4% of NSCLC in Europe and always occurs in males and smokers. Most of these tumors develop in proximal bronchi, where they frequently have an endobronchial component. Two-thirds of these tumors arise from long areas on bronchial mucosa and show prolonged and laterally extended *in situ* carcinoma. About half of the tumors present with a pure basaloid pattern that belongs to a variant of large cell carcinoma. The remaining cases have minor (< 50%) components of squamous cell carcinoma or, more rarely, adenocarcinoma and are thus classified as squamous cell carcinoma (basaloid variant) or adenocarcinoma, respectively. These tumors consist of lobular, trabecular, or palisading gross pattern of relatively small monomorphic cuboidal to fusiform cells with moderately hyperchromatic nuclei, finely granular chromatin, absent or only focally conspicuous nucleoli, scant cytoplasm but an N/C ratio lower than that of SCLC, and high mitotic rate from 20 to 100 mitoses per 2 mm² (Figure 13A). Neither intercellular bridges nor individual cell keratinization are present, which allow them to be distinguished from poorly differentiated squamous cell carcinoma. Patients with basaloid carcinoma have a significantly shorter survival than those with poorly differentiated squamous cell carcinoma, which deserves this differential diagnosis (Brambilla *et al.*, 1992; Moro *et al.*, 1994).

The differential diagnosis between LCNEC and basaloid carcinoma is somewhat puzzling because both may have peripheral palisades and rosettes can be

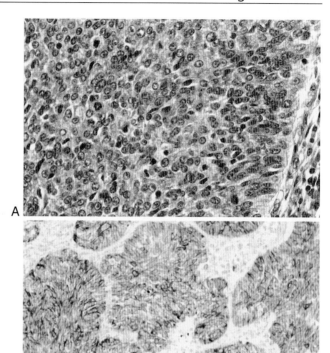

Figure 13 Basaloid carcinoma. **A:** This tumor consists of lobules of rather small uniform cells with peripheral palisading, and a high rate of mitosis (HES 400X). Chromatin is moderately dense and the nucleolus unconspicuous (400X). **B:** Cytokeratins 1, 5, 10, 14 stained by antibody 34βE12 are diffusely expressed.

observed in 30% of basaloid carcinoma. Fortunately, immunhistochemical features are quite distinctive. SCLC never expresses CK1, 5, 10, 14 (34βE12 reactively), whereas the vast majority (100% in our series) of basaloid carcinoma are reactive with 34βE12 (Figure 13B). In a quite opposite figure, TTF-1 is never expressed in basaloid carcinoma but is expressed in the majority of LCNEC (Sturm *et al.*, 2001).

Other large cell carcinoma variants include clear cell, rhabdoid type and lymphoepithelioma-like carcinoma. This last one displays Epstein-Barr virus genomic sequences at *in situ* hybridization (Chan *et al.*, 1995).

Adenosquamous Carcinoma

Adenosquamous carcinoma accounts for 0.6–2.3% of all lung cancers and is defined as a lung carcinoma having at least 10% of squamous cell or adenocarcinoma components. Adenosquamous carcinoma should not be confused with mucoepidermoid carcinoma, a malignant epithelial tumor characterized by the presence of squamoid cells, mucin secreting cells, and cells having intermediate type, identical to the same tumors encountered in the salivary glands. Mucoepidermoid carcinoma of high-grade malignancy is differentiated from adenocarcinoma by a variety of features, including a mixture of mucin-containing cells and squamoid cells, transition areas from classical low-grade mucoepidermoid carcinoma, and lack of keratinization.

The different immunohistochemical staining characteristics previously described for squamous cell carcinoma and adenocarcinoma are admixed according to the component in adenosquamous carcinoma.

Carcinomas with Pleomorphic Sarcomatoid or Sarcomatous Elements (Sarcomatoid Carcinoma)

The subtypes of these carcinomas include:

Carcinomas with spindle and/or giant cells
 Pleomorphic carcinoma
 Spindle cell carcinoma
 Giant cell carcinoma
Carcinosarcoma
Pulmonary blastoma
Other

This group of lung carcinomas is poorly differentiated and expresses the features and biological behavior of epithelial cells that adopt epithelial to mesenchymal transition in certain conditions of cultures *in vitro*. Pleomorphic carcinomas tend to be large peripheral tumors invading bronchial lumens, forming endobronchial growth. They often invade the chest wall and are associated with a poor prognosis (Fishback *et al.*, 1994). Because of the characteristic histologic heterogeneity of this tumor, adequate sampling is required and should consist of at least one section per centimeter of the tumor diameter. To enter in this category, a pleomorphic carcinoma should have at least a 10% component of a spindle or giant cells associated with other histological types, such as adenocarcinoma or squamous cell carcinoma (Travis *et al.*, 1999).

Pure giant cell or spindle cell carcinomas are extremely rare but cause an obvious challenge in distinguishing them from sarcoma. Pleomorphic and sarcomatoid carcinomas have a dismal prognosis even at stage I.

A recent large review of 78 cases of pleomorphic sarcomatoid carcinoma indicated that immunohistochemical features of these carcinomas reflect their epithelial lung epithelial cell origin with TTF-1 expression in 40% of them, as well as CK7 positivity and CK20 negativity (Rossi *et al.*, 2003).

This class of tumor also includes carcinosarcoma and pulmonary blastoma.

Carcinoid Tumor

Subtypes of carcinoid include

Typical carcinoid
Atypical carcinoid

Carcinoid tumors account for 1–2% of all invasive lung malignancies.

Typical and atypical carcinoids are characterized histologically by endocrinoid, organoid growth pattern, and uniform cytologic features, consisting of moderate eosinophilic, finely granular cytoplasm, a nuclear with a finely granular chromatin, and inconspicuous nucleoli that can be discretely more prominent in atypical carcinoid. A variety of histologic patterns may occur in atypical and typical carcinoids, including trabecular, palisading, rosette-like, papillary, sclerosing papillary, glandular, paragangliomatous, spindle cell, and follicular patterns.

The most distinguishing feature between typical carcinoid and atypical carcinoid is the rate of mitosis and the presence or absence of necrosis. Typical carcinoids show less than 2 mitoses per 2 mm^2 area of viable tumor (10× high power field) and no necrosis. The presence of mitosis between 2 to 10 per 2 mm^2 or the presence of necrosis (Travis *et al.*, 1998, 1999) define the diagnosis of atypical carcinoids. The presence of features such as cell pleomorphism, vascular invasion, and increased cellularity are of no help in separating typical carcinoid from atypical carcinoid and in allowing stratification of patients for prediction of survival (Travis *et al.*, 1998). Typical carcinoid may well show focal cytologic pleomorphism as do paraganglioma in head and neck area. The necrosis in atypical carcinoid usually consists of small foci centrally located within organoid nests of tumor cells.

Although neuroendocrine markers are not required for the diagnosis of carcinoids, they are positive in every case. Carcinoids do not express TTF-1 in contrast with the high-grade neuroendocrine tumors, LCNEC, and SCLC (Sturm *et al.*, 2002). Protein S100, which is a nonspecific antigen, might be present in the nuclei of any kind of neuroendocrine lung tumor. In carcinoids, they are specifically expressed in perilobular cells that have appearance and phenotype of sustentacular cells, as seen in paraganglioma. The differential diagnosis between carcinoids, meningothelial-like bodies, and paraganglioma resides in expression of cytokeratins in carcinoids but absence of expression in meningothelial-like bodies and in paraganglioma, and the absence of neuroendocrine markers in meningothelial-like bodies.

Molecular Pathology of Lung Carcinoma

Molecular pathology refers here to phenotypical markers used to detect abnormal expression of proteins, *in situ* overexpression of oncogenes, or loss of protein expression of tumor suppressor genes. Both p53 and retinoblastoma gene (Rb), as well as molecules from their pathways, belong to tumor suppressor genes involved in cell cycle regulation and apoptosis; Bcl2 and Bax as well as the Fragile Histidine Trial (FHIT) protein are members of apoptosis markers that influence the cell susceptibility to death; telomerase reactivation reflects cell immortalization; matrix degrading proteases and vascular endothelial growth factor (VEGF) are proteins reflecting angiogenic potential, migratory capacities, and survival of clonal proliferation. Overall, the clonal expansion (tumor growth) might be regarded as the net result of a gain of a cell numbers by increasing intrinsic proliferation (rate of cell division) and loss of cells by decreasing propensity to cell death (escape from apoptosis). The complex genetic and epigenetic changes that result in lung carcinoma are thought to begin before the occurrence of invasive carcinoma in a process referred to as multistep carcinogenesis. A set of phenotypical traits of lung cancer appears at the level of preinvasive lesions.

DNA Damage

Damaged DNA may give rise to allelic loss, a frequent early molecular finding in lung carcinogenesis. Loss of heterozygosity (LOH) refers to allelic loss. Chromosomal loci that normally harbor two different polymorphic alleles are assessed for loss of one (loss of heterozygosity, LOH) or both of these alleles (homozygous deletion). Remanant allelic loss at specific chromosomal loci, as a candidate for the presence of tumor suppressor genes, fosters search for mutation of the second allele. Regions that have received considerable attention are chromosomes 3p14-23 (5 loci), 8p21-23, 9p, **17p** (p53 locus), **and 13q** (Rb locus) because of the high density of loss and the probability that tumor suppressor genes are present. One of the best-studied chromosome 3 genes is FHIT. The FHIT gene is situated in a highly fragile histidine site, where it is particularly prone to partial deletion as a result of direct DNA damage by smoking associated carcinogens (Sozzi *et al.*, 1996). Candidate tumor suppressor genes on chromosome 9 include p15^{INK4A} (9p21) and p16^{INK4B} (9p21).

p16^{INK4} inhibits activation of cyclinD/Cdk4, 6 kinases and thus impairs progression of the cell through the mitotic cycle. Loss of p16^{ink4} would cause more rapid progression through the cell cycle. Allelic loss is observed in more than 70% of NSCLC and p16^{INK4} is inactivated by homozygous loss in nearly half of smoking-associated lung cancers (Gazzeri *et al.*, 1998). Genomic molecular damage accumulating during lung carcinogenesis eventually results in chromosomal rearrangements and aneusomy involving multiple chromosomes. Abnormalities have been described in every chromosome, but some chromosomes and chromosomal loci are more frequently unbalanced than others. Gains of chromosomes 6, 7, and 8 occur in approximately 50% of the NSCLC.

Ras Mutation

Ras mutations occur almost exclusively in adenocarcinoma, most frequently at Ki-Ras codon 12, and are strongly related to tobacco smoke nitrosamines. A single-base mutation at Ki-Ras codon 12, observed in 30% of adenocarcinoma, is responsible for the lack of intrinsic GTPase functions of the Ras protein, which is endowed with constitutive activity toward proliferations. Ras mutation is a relatively late event in the process of Clara-type II cell mutagenesis but precedes invasion since it occurs in a proportion of the preneoplastic lesion atypical alveolar hyperplasia. Adenocarcinoma shares similar molecular abnormalities with squamous cell carcinoma, except for the high frequency of Ki-Ras mutations that are not or extremely rarely found in other long tumor types. The frequency of Ki-Ras mutation is much lower in nonsmokers (5%) than in smokers (30%). The restricted number of potential mutants at Ras condon makes Ras mutation a candidate for molecular early detection of adenocarcinoma from exfoliated cells. There is no reliable change of protein expression related to Ras mutation.

p53 Mutations

Mutations of *TP53* tumor suppressor gene occurs in about 50% of NSCLC and more than 70% of SCLC. *TP53* mutations are the most extensively studied mutation in lung cancer. The database maintained at the International Agency for Research on Cancer (IARC) is a valuable resource to study the role of the gene in lung tumorigenesis (Hainaut *et al.*, 2001). Previous reports demonstrated that *TP53* mutational spectra of lung cancer was unique from other cancers. In particular, an excess of G:C to T:A transversions was characteristic of lung cancers and was related to smoke exposure. In those who never smoked, there is a reciprocal increase in G:C to A:T transitions.

Squamous cell carcinoma showed the highest frequency of p53 mutations (about 70%) among all histological types of lung carcinoma. p53 is a tumor suppressor gene with functions in G1 arrest and apoptosis in response to cytotoxic stress (stress includes DNA damage and several genetic abnormalities in the gene sequence) that stabilizes P53 and allows immunohistochemical detection. P53 immunoreactivity is fairly correlated with TP53 missense mutation. However, other types of TP53 mutation accounting for a maximum of 20% lead to an absence of functional P53 protein and are thus called null phenotype mutations, not recognized by immunohistochemistry. P53 protein overexpression and, less commonly, mutations may precede invasion. The preinvasive lesions of squamous cell carcinoma, dysplasia, and carcinoma *in situ* display P53 accumulation and immunoreactivity in an increasing proportion of mild dysplasia to CIS (20–60%).

Rb Gene Inactivation

Inactivation of the Rb pathway is frequent in NSCLC, but mechanisms of Rb functions inactivation is different from that seen in high neuroendocrine tumors. Although loss of Rb protein expression is detected in only 15% of NSCLC, RB is frequently inactivated through deregulation of its phosphorylation pathway. Both CdK inhibitor inactivation (p16^{INK4}) and cyclin D1 overexpression contribute to this indirect Rb inactivation and leading to overphosphorylation and loss of Rb function on G1 arrest. Immunohistochemistry is a straightforward method of detecting p16^{INK4} inactivation and cyclin D1 overexpression (both occur in 50% of NSCLC). There is a constant inverse relation between loss of Rb, loss of P16, and overexpression of cyclin D1 consistent with P16 and cyclin D1 being exclusively devoted to Rb phosphorylation pathway.

The tumor suppressor genes p53 and Rb are frequently mutated in high-grade neuroendocrine lung tumors, including SCLC and LCNEC. In 50–80% of SCLC and LCNEC, p53 mutation and/or p53 protein accumulation and immunoreactivity are observed. High-grade neuroendocrine tumors share a high frequency of Rb loss of protein expression (80–100%) detectable by immunohistochemistry, compared with nuclear staining of normal cells. p14^{ARF}, a protein encoded at the p16-9p21 locus, is induced by oncogenic stress and exerts its effect through p53 stabilization and transcriptional activation. However, both p14^{ARF} loss and TP53 mutation have been found in SCLC and LCNEC, enlighting p14^{ARF} functions independent of p53. E2F1, which includes p14^{ARF} is frequently overexpressed in high-grade neuroendocrine tumors.

Aberrant Methylation

Epigenetic silencing of the promoter regions of multiple genes are universal in cancers. The major mechanism studied is methylation, although histone deacetylation plays an important cooperative role and may, in fact, precede the onset of methylation. Most of the silenced genes are known or suspected tumor suppressor genes. The methylation profile varies with the tumor type. Carcinoids, SCLC, squamous cell carcinomas, and adenocarcinomas of the lung have unique profiles of aberrant methylation, and the methylation rates of *APC*, *CDH13*, and *RARβ* were significantly higher in adenocarcinomas than in squamous cell carcinomas (Toyooka *et al.*, 2001).

Neuroendocrine Lung Tumors Phenotype

The gradual increase of fractional allelic loss and molecular abnormalities for TP53, Rb, p14^{ARF}, E2F1 along the spectrum of neuroendocrine lung tumors strongly supports the grading concept of typical carcinoid as low-grade, atypical carcinoid as intermediate grade, and LCNEC and SCLC as high-grade neuroendocrine lung tumors. However, a continuous spectrum is challenged. MEN1 gene mutation and loss of heterozygosity at the MEN1 gene locus 11q13 was recently demonstrated in 65% of sporadic atypical carcinoids and was not found in high-grade neuroendocrine tumors; TTF-1 is not expressed in carcinoids, whereas it is expressed in high-grade SCLC and LCNEC; gene expression profiling using cDNA microarrays recently compared profiles of carconoids and SCLC showed similarities between carcinoids and central nervous tumors, whereas SCLC has more closely inherited the phenotype of basal bronchial cells.

CONCLUSIONS

Genetic and molecular abnormalities in lung cancer are the result of DNA damage, mostly reflecting tobacco carcinogenesis. Their full knowledge and validated tools for their detection are useful in designing new therapeutic strategies and establishing the risk of progression of any early stage into lung cancer.

The spectrum of neuroendocrine tumors has not been described as such in the WHO classification in 1999, with the epidemiological, clinical, and behavior as well as prognostic and response to treatment being extremely different. Although they belong to a morphological and biological spectrum sharing neuroendocrine properties and features, their genetic and molecular profiles have more differences than similarities.

Histologic subclassification of lung tumors is essentially based on light microscopy in order to achieve widest application throughout the world and assume comparability and consistency of data. However, techniques, including immunohistochemistry, electron microscopy, tissue culture, and molecular biology, might provide valuable information on carcinogenesis, histogenesis, and differentiation. It is highly recognized that immunohistochemistry or electron microscopy may detect differentiation, specifically regarding histological heterogeneity of lung cancer, that cannot be seen by routine light microscopy. However, these techniques are occasionally required for precise classification. An example of this is LCNEC and malignant mesothelioma that require appropriate immunohistochemical and/or electron microscopy findings to confirm the diagnosis.

References

Brambilla, E., Moro, D., Gazzeri, S., Morel, F., Jacrot, M., and Brambilla, C. 1991. Cytotoxic chemotherapy induces cell differentiation in small cell lung carcinoma. *J. Clin. Oncol. 9:*50–61.

Brambilla, E., Moro, D.,Veale, D., Brichon, P.Y., Stoebner, P., Paramelle, B., and Brambilla, C. 1992. Basal cell (basaloid) carcinoma of the lung: A new morphologic and phenotypic entity with separate prognostic significance. *Hum. Pathol. 23:*993–1003.

Chan, J.K., Hui, P.K., Tsang, W.Y., Law, C.K., Ma,. C.C., Yip, T.T., and Poon, Y.F. 1995. Primary lymphoepithelioma-like carcinoma of the lung: A clinicopathologic study of 11 cases. *Cancer 76:* 413–422.

Fishback, N.F., Travis, W.D., Moran, C.A., Guinee, D.G. Jr., McCarthy, W.F., and Koss, M.N. 1994. Pleomorphic (spindle/giant cell) carcinoma of the lung: A clinicopathologic correlation of 78 cases. *Cancer 73:*2936–2945.

Gazzeri, S., Della Valle, V., Chaussde, L., Brambilla, C., Larsen, C.J., and Brambilla, E. 1998b. The human p19ARF protein encoded by the b transcript of the p16INK4 gene is frequently lost in small cell lung tumors. *Cancer Res. 58:*3926–3931.

Gazzeri, S., Gouyer, V., Vour'ch, C., Brambilla, C., and Brambilla, E. 1998a. Mechanism of p16INK4A inactivation in non–small-cell lung cancers. *Oncogene 16:*497–505.

Greenlee R.T., Hill-Harmon, M.B., Murray, T., and Thun, M. 2001. Cancer Statistics, 2001. *CA Cancer J. Clin. 51:*15–36.

Hainaut, P., Olivier, M., and Pfeifer, G.P. 2001. TP53 mutation spectrum in lung cancers and mutagenic signature of components of tobacco smoke: Lessons from the IARC TP53 mutation database. *Mutagenesis 16(6):*551–553; *author reply* 555–556.

Lantuejoul, S., Moro, D., Michalides, R.J., Brambilla, C., and Brambilla, E. 1998. Neural cell adhesion molecules (NCAM) and NCAM-PSA expression in neuroendocrine lung tumors. *Am. J. Surg. Pathol. 22:*1267–1276.

Moro, D., Brichon, P.Y., Brambilla, E., Veale, D., Labat-Moleur, F., and Brambilla, C. 1994. Basaloid bronchial carcinoma: A histological group with a poor prognosis. *Cancer 73:*2734–2739.

Naylor, S.L., Johnson, B.E., Minna, J.D., and Sakaguchi, A.Y. 1987. Loss of heterozygosity of chromosome 3p markers in small-cell lung cancer. *Nature 329:*451–454.

Nicholson, S.A., Beasley, M.B., Brambilla, E., Hasleton, P.S., Colby, T.V., Sheppard, M.N., Falk, R., and Travis, W.D. 2002. Small cell lung carcinoma (SCLC): A clinicopathologic study of 100 cases. *Am. J. Surg. Pathol. 26(9):*1184–1197.

Noguchi, M., Morikawa, A., Kawasaki, M., Matsuno, Y., Yamada, T., Hirohashi, S., Kondo, H., and Shimosato, Y. 1995. Small adenocarcinoma of the lung. Histologic characteristics and prognosis. *Cancer 75:*2844–2852.

Ordonnez, N. 2000. Value of thyroid transcription factor-1 immunostaining in distinguishing small cell lung carcinomas from other small cell carcinomas. *Am. J. Surg. Pathol. 24:*1217–1223.

Rossi, G., Cavazza, A., Sturm, N., Migaldi, M., Facciolongo, N., Longo, L., Maiorana, A., and Brambilla, E. 2003. Pulmonary carcinomas with pleomorphic, sarcomatoid or sarcomatous elements: A clinicopathologic and immunohistochemical study of 75 cases. *Am. J. Surgical Pathol. 27(3):*311–324.

Rossi, G., Murer, B., Cavazza, A., Losi, L., Natalpi, P., Marchioni, A., Migaldi, M., Capitanio, G., and Brambilla, E. 2004, in press. Primary mucinous (so-called coloid) carcinomas of the lung: A clinicopathologic and immunohistochemical study with special reference to CDX-2 homeobox gene and MUC2 expression. *Am. J. Surg. Pathol.*

Sehested, M., Hirsch, F.R., Osterlind, K., and Olsen, J.E. 1986. Morphologic variations of small cell lung cancer: A histopathologic study of pretreatment and posttreatment specimens in 104 patients. *Cancer 57:*804–807.

Sozzi, G., Veronese, M.L., Negrini, M., Baffa, R., Cotticelli, M.G., Inoue, H., Tornielli, S., Pilotti, S., De Gregorio, L., Pastorino, U., Pierotti, M.A., Ohta, M., Huebner, K., and Croce, C.M. 1996. The FHIT gene at 3p14.2 is abnormal in lung cancer. *Cell 85:*17–26.

Sturm, N., Lantuejoul, S., Laverriere, M.H., Papotti, M., Brichon, P.Y., Brambilla, C., and Brambilla, E. 2001. Thyroid transcription factor-1 (TTF-1) and cytokeratin 1, 5, 10, 14 (34betaE12) expression in basaloid and large cell neuroendocrine carcinomas of the lung. *Hum. Pathol. 32(9):*918–925.

Sturm, N., Rossi, G., Lantuejoul, S., Papotti, M., Frachon, S., Claraz, C., Brichon, P.Y., Brambilla, C., and Brambilla, E. 2002. Expression of thyroid transcription factor-1 (TTF-1) in the spectrum of neuroendocrine cell lung proliferations with special interest in carcinoids. *Human Pathol. 33(2):* 175–182.

Suzuki, K., Yokose, T., Yoshida, J., Nishimura, M., Takahashi, K., Nagai, K., and Nishiwaki, K. 2000. Prognostic significance of the size of central fibrosis in peripheral adenocarcinoma of the lung. *Ann. Thorac. Surg. 69:*893–897.

Toyooka, S., Toyooka, K.O., Maruyama, R., Virmani, A.K., Girard, L., Miyajima, K., Harada, K., Ariyoshi, Y., Takahashi, T., Sugio, K., Brambilla, E., Gilcrease, M., Minna, J.D., and Gazdar, A.F. 2001. DNA methylation profiles of lung tumors. *Mol. Cancer Therapeut. 1(1):*61–67.

Travis, W.D., Linnoila, R.I., Tsokos, M.G., Hitchcock, C.L., Cutler, G.B. Jr., Nieman, L., Chrousos, G., Pass, H., and Doppman, J. 1991. Neuroendocrine tumors of the lung with proposed criteria for large-cell neuroendocrine carcinoma: An ultrastructural, immunohistochemical, and flow cytometric study of 35 cases. *Am. J. Surg. Pathol. 15:*529–553.

Travis, W.D., Lubin, J., Ries, L., and Devesa, S. 1996. United States lung carcinoma incidence trends: Declining for most histologic types among males, increasing among females. *Cancer 77:*2464–2470.

Travis, W.D., Rush, W., Flieder, D.B., Falk, R., Fleming, M.V., Gal, A.A., and Koss, M.N. 1998. Survival analysis of 200 pulmonary neuroendocrine tumors with clarification of criteria for atypical carcinoid and its separation from typical carcinoid. *Am. J. Surg. Pathol. 22:*934–944.

Travis, W.D., Colby, T.V., Corrin, B., Shimosato, Y., and Brambilla, E. 1999. *Histological Typing of Lung and Pleural Tumors.* Berlin, Springer.

Yokose, T., Suzuki. K., Nagai, K., Nishiwaki, Y., Sasaki, S., and Ochiai, A. 2000. Favorable and unfavorable morphological prognostic factors in peripheral adenocarcinoma of the lung 3 cm or less in diameter [In Process Citation]. *Lung Cancer 29:*179–188.

3

Immunohistochemistry and *in situ* Hybridization of Mucin in Lung Carcinoma

M.C. Copin, L. Devisme, J.P. Aubert, N. Porchet, and M.P. Buisine

Introduction

Mucins are glycoproteins synthesized by epithelial cells and thought to play a key role in cytodifferentiation and to promote tumor-cell invasion. Mucin gene expression was first studied by nonmorphological molecular techniques such as Northern Blot and more recently reverse transcriptase–polymerase chain reaction (RT-PCR). *In situ* hybridization and immunohistochemistry have provided data on the precise localization of mucins at the cellular level.

Mucus properties can be attributed largely to their constituent mucin O-glycoproteins that exhibit high density and viscoelasticity. Mucins are synthesized by a wide variety of epithelial tissues and are thought to promote tumor-cell invasion and metastasis. They have a characteristic amino acid composition, with a high content of threonine and serine carrying O-linked glycan chains and distributed in tandemly repeated motifs in the central part of the protein backbone. There are secretory or membrane-associated forms of mucins. Secretory mucins form the mucus gel and contain ~ 80% carbohydrates by weight. To date, eight human mucin genes (which encode the apomucin backbone)—*MUC1* to *MUC4*, *MUC5AC*, *MUC5B*, *MUC6*, and *MUC7*—have been well characterized from various epithelial tissues or mucosae, including airways, though not all have been completely sequenced (for review, see Moniaux *et al.*, 2001). Additional complementary deoxyribonucleic acid (cDNA) clones have been proposed as *MUC8*, *MUC9*, *MUC11*, and *MUC12*, but their patterns of expression are not so well-defined. The family of four genes (*MUC2*, *MUC5AC*, *MUC5B*, *MUC6*) maps to 11p15.5 (Pigny *et al.*, 1996) and encodes large secretory gel forming mucins containing cystein-rich subdomains. The other genes, *MUC1*, *MUC3*, *MUC4*, *MUC7*, *MUC8*, are scattered on different chromosomes (Porchet *et al.*, 1995) and encode membrane-bound or secreted mucins having other own peptide organizations, the first and the best known of them being *MUC1*. More recently, many human genes encoding mucins have been named MUC, despite the absence of sequence homology (Dekker *et al.*, 2002).

Several arguments suggest that mucins play a role in tumor-cell invasion and metastasis, resulting in prognostic implications. The mucin MUC1 is a transmembrane molecule with a large extracellular domain protruding high above the cell surface, which is thought

Handbook of Immunohistochemistry and *in situ* Hybridization of Human
Carcinomas, Volume 1: Molecular Genetics; Lung and Breast Carcinomas

115

Copyright © 2004 by Elsevier (USA)
All rights reserved.

to reduce cell-cell and extracellular matrix (ECM)-cell adhesion in cancer cells, but direct evidence for a role of specific mucin genes in tumor progression is lacking. One study shows that inactivation of *Muc2* causes intestinal tumor formation with spontaneous progression to invasive carcinoma in mice genetically deficient in *Muc2* (Velcich *et al.*, 2002). Sialomucin complex (SMC), a rat homologue of the human mucin MUC4 isolated from highly metastatic ascites 13762 mammary adenocarcinoma cells, is thought to potentiate metastasis by sterically disrupting molecular interactions for cell-cell and cell-ECM adhesions and by suppressing antitumor immunity by inhibiting interactions between cytotoxic lymphocytes and target tumor cells (Carraway *et al.*, 2000). One recent study shows that *in vivo*, subcutaneous injection of SMC-overexpressing cells in nude mice results in substantially greater lung metastasis than injection of SMC-repressed cells. Moreover, injection of A375 human melanoma cells followed by *in vivo* induction of SMC overexpression within the solid tumor results in spontaneous distant metastasis in nude mice (Komatsu *et al.*, 2000).

Physiological expression patterns of most of these genes have been studied in different laboratories using various techniques, including Northern Blot, Dot Blot, RNase protection assay, RT-PCR or *in situ* hybridization. With the differentiation state of epithelial cells and malignant transformation, the tissue and cell-specific expression of mucin genes becomes dysregulated. The patterns of mucin messenger ribonucleic acid (mRNA) and glycoproteins are frequently altered in adenocarcinomas compared with normal tissues. Alterations include increased, decreased, lost, or aberrant expression of mucin mRNA and peptides. Immunohistochemical studies using antibodies that recognize the tandem repeat sequences of apomucin have shown an increased immunoreactivity in carcinomas compared with normal tissues, most likely due to incomplete glycosylation and exposure of peptide epitopes.

Four of the mucin genes (*MUC4, MUC5AC, MUC5B,* and *MUC8*) have been isolated by screening a human tracheobronchial cDNA library. *MUC4, MUC5AC,* and *MUC5B* were isolated in our laboratory with a polyclonal antiserum raised against deglycosylated glycopeptides purified from human sputum (Aubert *et al.*, 1991; Porchet *et al.*, 1991). The latter three genes have been demonstrated encoding major airway mucins in normal specimens by Northern Blot analysis and *in situ* hybridization. The expression of seven mucin genes (*MUC1, MUC2, MUC4, MUC5AC, MUC5B, MUC7,* and *MUC8*) has been demonstrated in normal tracheobronchial mucosa by Northern Blot. In contrast to Northern or Slot Blot analysis, *in situ*

hybridization allows the cellular localization of the signal and the determination of the expression pattern of mucin genes in human adult mucosae. The tandem repeat organization and the choice of consensus oligonucleotide probes specific for the tandem repeat of each gene is a convenient way to obtain the amplification of the signal by hybridizing a maximum number of small probes all along the same mRNA molecule (Audié *et al.*, 1993).

Recent studies from our laboratory have characterized the pattern of apomucin expression in normal adult and fetal lung. These studies focused on the cellular localization and variations along the respiratory tract of mucin genes by *in situ* hybridization and MUC5AC and MUC5B protein by immunohistochemistry (Buisine *et al.*, 1999; Copin *et al.*, 2000). We showed that the mucin gene expression pattern is complex in adult normal airways and lung involving six genes *MUC1, MUC2, MUC4, MUC5AC, MUC5B,* and *MUC7*. They were expressed in an array of epithelial cells exhibiting various phenotypes: *MUC1, MUC2, MUC5B,* and *MUC7* in the submucosal glands, and *MUC1, MUC2, MUC4, MUC5AC,* and *MUC5B* were expressed in the surface epithelium. In distal bronchioles, Clara cells expressed *MUC1* and *MUC4* genes (data unpublished). The alveolar type II epithelial cells express the MUC1 glycoprotein (Jarrard *et al.*, 1998) detected by immunohistochemistry, but any other mucin genes are detectable by *in situ* hybridization only (Copin *et al.*, 2000). In our experience, the two types of alveolar cells express MUC1 by immunohistochemistry (data unpublished) using M8 antibody, which recognizes the tandem repeat sequences (McIlhinney *et al.*, 1985). Whereas we did not detect *MUC4* gene expression in normal type II pneumocytes by *in situ* hybridization, we found *MUC4* gene and protein expression in type II pneumocyte hyperplasia (Copin *et al.*, 2000).

MUC5AC and MUC5B are two secreted mucins, encoded by genes of the 11p15.5 mucin gene cluster, which are associated with mucus secretion in airways, MUC5AC in surface epithelium and MUC5B in submucosal glands (Copin *et al.*, 2000). *MUC5AC* mRNA and protein are located in the more specialized cells for mucus secretion (i.e., the goblet cells). *MUC5B* mRNA and protein are mostly detected in the mucous cells of submucosal glands and in gland ducts, serous cells being negative.

In contrast to *MUC5AC* and *MUC5B*, which are expressed in specialized cells, *MUC1, MUC2,* and *MUC4* are found in all epithelial cells of the surface epithelium, both in ciliated cells and goblet cells as well as in Clara cells, suggesting various functions of mucin gene besides mucus production (Copin *et al.*, 2000;

Lopez-Ferrer *et al.*, 2001). Moreover, *MUC4* is the earliest mucin gene to be expressed in the foregut (Buisine *et al.*, 1999). The primitive epithelial cells have the potential to differentiate in all epithelial cell types of the conducting airways and alveolar epithelium, suggesting that *MUC4* has a great importance in lung development and cell differentiation. Mucin gene expression is subjected to profound differential regulation during human fetal development from 6.5 to 27 weeks of gestation. The *MUC4* gene expression at 6.5 weeks is followed by *MUC1* and *MUC2* gene expression from 9.5 weeks of gestation before epithelial cytodifferentiation. In contrast, *MUC5AC*, *MUC5B*, and *MUC7* are expressed at later gestational ages concomitant with epithelial cytodifferentiation. The genes *MUC3* and *MUC6* have never been detected in normal adult and fetal respiratory tract by *in situ* hybridization. The latter results are confirmed for MUC6 at the protein level by immunohistochemistry (Bartman *et al.*, 1998). Only one study by Northern Blot has reported low levels of *MUC3* mRNA in nonneoplastic tissue adjacent to carcinomas (Nguyen *et al.*, 1996). The MUC3 antibodies are not yet available.

The alteration of the level and expression pattern of mucin mRNA in lung carcinoma or carcinoma cell lines compared with normal tissue has been suggested using Dot Blot, Northern Blot, or immunohistochemistry techniques (Nguyen *et al.*, 1996; Seregni *et al.*, 1996; Yu *et al.*, 1996). Yu *et al.* (1996) found overexpression of *MUC5AC* in 13 carcinomas (21.7%), eight adenocarcinomas, and five squamous cell carcinomas, and overexpression of *MUC5B* in 10 carcinomas (16.7%), 6 adenocarcinomas, and 4 squamous cell carcinomas by Slot Blot analysis. Seregni *et al.* (1996), studying *MUC1-MUC4* expression by Northern Blot analysis, showed that *MUC1* mRNA are most expressed in lung cancer, followed by *MUC4*, and that the highest reactivity was observed for *MUC1* and *MUC4* mainly in the mucus-secreting adenocarcinoma type. *MUC2* and *MUC3* were undetectable in the cancer specimens studied (10 adenocarcinomas and six squamous cell carcinomas). Nguyen *et al.* (1996), using the same technique, found that squamous carcinoma overexpressed *MUC4* only compared with nonneoplastic tissue adjacent to carcinomas. High levels of *MUC1*, *MUC3*, and *MUC4* expression could be detected in well-differentiated adenocarcinomas.

There is now a great interest in mucins as markers of differentiation and tumor progression. Moreover, the study of the expression of mucin genes may contribute to the determination of molecular basis of histopathological classification of lung adenocarcinomas. We analyzed the apomucin expression at the cell level in preinvasive lesions of the bronchus, squamous

cell carcinomas and adenocarcinomas of the lung (Copin *et al.*, 2000, 2001). We analyzed and compared the qualitative and semiquantitative pattern of expression of *MUC1-4*, *MUC5AC*, *MUC5B*, and *MUC6-7* genes using *in situ* hybridization. We completed the study by immunostaining the secreted mucins (MUC2, MUC5AC, MUC5B, and MUC6) with specific antibodies for MUC2, MUC5AC, MUC5B, and MUC6 apomucins.

MATERIALS AND METHODS

Tissues

The samples were quickly immersed in fresh 10% neutral formaldehyde solution (pH 7.4) in phosphate buffer. The amount used is 15–20 times the volume of tissue. They were processed for paraffin-embedding and cut under sterile conditions (3-μm-thick sections) and placed on gelatin-covered slides for *in situ* hybridization or silan-covered slides for immunohistochemistry. Slides were stored at 4°C until used. Morphological control of the same blocks was obtained by sections stained with hematoxylin-eosin-saffron (HES) and astra blue.

In situ Hybridization

The original protocol was first described by Audié *et al.*, (1993) and modified by Buisine *et al.*, (1999).

Reagents

1. Sterile deionized glass-distilled water.
2. 10X terminal desoxynucleotidyl transferase (TdT) buffer: 1.54 mg dithiothreitol, 1.97 mg MnCl$_2$, 0.95-mg MgCl$_2$, 2.14 g sodium cacodylate; bring vol to 10 ml with sterile water.
3. Diethyl pyrocarbonate (DEPC) water: 100 μl diethylpyrocarbonate in 100 ml sterile water; agitate vigorously and let stand for 24 hr before autoclaving.
4. Gelatin-covered slides: 5 g gelatin in 1 L sterile water; incubate in a waterbath at 55°C until complete dissolution; equilibrate the temperature at 43°C, add 4 g chromium (III) potassium sulfate. 2H$_2$O and incubate at 40°C until complete dissolution; immerse slides in the preparation and let them dry in a ventilated incubator at 37°C.
5. 30, 70, and 100% ethanol.
6. Glycin buffer: 24 g Tris-HCl, 7.5 g glycin; bring volume to 1 L with sterile water; adjust pH to 7.4 with HCl; autoclave.

7. 10X proteinase K buffer: 12.1 g Tris-HCl, 18.6 g EDTA; bring volume to 100 ml with sterile water (pH 8.0).

8. Phosphate buffered saline (PBS): 8-g NaCl, 200-mg KCl, 1.78 g $Na_2HPO_4 \cdot 2H_2O$, 240 mg KH_2PO_4; bring volume to 1 L with sterile water; adjust pH to 7.4 with HCl; autoclave.

9. 4% paraformaldehyde: 4 g paraformaldehyde in 100-ml PBS; incubate in a waterbath at 55°C until complete dissolution; filtrate on 0.45-μm filter units; add 0.5 ml M $MgCl_2$.

10. 20X SSPE: 174 g NaCl, 27.6 g $Na_2HPO_4 \cdot 2H_2O$, 7.4 g EDTA; bring volume to 1 L with sterile water; adjust pH to 7.4 with NaOH; autoclave.

11. Acetylation buffer: 20 ml 20X SSPE in 100 ml sterile water, 1.3 ml triethanolamine; adjust pH to 8.0 with HCl; add 250 μl acetic anhydride.

12. 50X Denhardt's: 1 g Ficoll 400, 1 g poly-vinylpyrrolidone, 1 g bovine serum albumin; bring volume to 100 ml with sterile water; fitrate on 0.22-μm filter units.

13. 1.2 M phosphate buffer: solution A: 5.34 g $Na_2HPO_4 \cdot 2H_2O$ in 25 ml sterile water; solution B: 1.87 g $Na_2HPO_4 \cdot 2H_2O$ in 10 ml sterile water; mix solutions A and B and adjust pH to 7.2.

14. 20X sarcosyl: 2 g N-lauroylsarcosine sodium salt in 100 ml sterile water; filtrate on 0.22-μm filter units.

15. Prehybridization mixture: 15 mg dithiothreitol, 300 μl DEPC water, 200 μl 20X SSPE, 500 μl deionized formamide, 100 μl 1.2 M phosphate buffer, 50-μl 20X sarcosyl, 20-μl 50X Denhardt's buffer, 30 μg calf thymus DNA, 30 μg yeast transfer ribonucleic acid (tRNA).

16. Hybridization mixture: 15 mg dithiothreitol, 300 μl DEPC water, 200 μl 20X SSPE, 500 μl deionized formamide, 100 μl 1.2 M phosphate buffer, 50 μl 20X sarcosyl, 20 μl 50X Denhardt's buffer, 3 μg yeast tRNA.

17. Developer: 500 mg p-methylaminophenol hemisulfite salt, 2.25 g hydroquinone, 18 g sodium sul-fite, 12 g sodium carbonate, 1 g potassium bromide in 250-ml water.

18. Fixative: 80 g sodium thiosulfate in 250 ml water.

Probes

The eight ^{35}S-labeled oligonucleotide probes corre-sponded to each tandem repeat domain of *MUC1*, *MUC2*, *MUC3*, *MUC4*, *MUC5AC*, *MUC5B*, *MUC6*, and *MUC7*. Nucleotide sequences of nucleotide probes were for *MUC1*: 5′ GTCCGGGGCCGAGGTGACA-CCGTGGGCTGGGGGGGGCGGTGGAGCCCGG

3′; *MUC2*: 5′ GGT CTGTGTGCCGGTGGGTGTTG-GGGTTGGGGTCACCGTGGTGGTGGT 3′; *MUC3*: 5′ GGTGGTCTCGGTGGTGGTGATGGAAGAAG-TGAAGCTGGGAGTACTGTG 3′; *MUC4*: 5′ GTCG-GTGACAGGAAGAGGGGTGGCGTGACCTGTGG ATGCTGAGGAAGT 3′; *MUC5AC*: 5′ AGGGGCA-GAAGTTGTGCTCGTTGTGGGAGCAGAG-GTTGTGCTGGTTGT 3′; *MUC5B*: 5′ TGTGGTCA G C T C T G T G A G G A T C C A G G T C G T C C C CGGAGTGGAGGAGGG 3′; *MUC6*: 5′ TTCAGGA TGGTGTGTGGAGGAAGCATGTGAGTGGAG-GATGTAGAAGT 3′; *MUC7*: 5′ CGGTGGAGCT GTGTAGTTGCAGAAGGTGTGGGTGGG-GCAGCTGTGGT 3′.

Labeling of the Probe

Labeling was performed using terminaldeoxynu-cleotidyltransferase using ^{35}S-dATP.

1. Heat denature 60 ng oligonucleotide in DEPC water (to final volume 15 μl) for 5 min in boiling water; rapidly cool on ice.

2. On ice, add 1.5 μl 10X TdT buffer, 35 μCi α-[^{35}S-dATP], 20 U terminaldeoxynucleotidyltransferase.

3. Incubate at 37°C for 2 hr.

4. Stop the reaction by adding 1 μl of 0.5 M EDTA.

5. Purify the probe by gel permeation chromatogra-phy using a Sephadex G-25 type spin column accord-ing to the manufacturer's instructions.

6. Determine the labeling efficiency of the labeled probe and store at −80°C until use.

In situ Hybridization

1. Immerse the slides in xylene at room tempera-ture for 10 min; repeat using new xylene (for deparaf-finization).

2. Rinse the slides in 100% ethanol for 5 min and rehydrate in an ethanol series (100%, 70%, 30%) and sterile water at room temperature for 3 min each.

3. Immerse the slides in glycin buffer at room tem-perature for 10 min.

4. Incubate the slides with 2 μg/ml proteinase K in X proteinase K buffer at 37°C for 20 min. Rinse in water.

5. Immerse the slides in fresh 4% paraformalde-hyde at room temperature for 15 min. Rinse in PBS.

6. Immerse the slides in fresh acetylation buffer at room temperature for exactly 10 min. Rinse in water.

7. Dehydrate the slides in an ethanol series (30%, 70%, 100%) and air-dry at 37°C.

8. Place the slides horizontally in a humidified chamber prewarmed to 42°C and apply prehybridiza-tion mixture on specimen area.

9. Cover and incubate the slides at 42°C for at least 45 min.

10. Dehydrate the slides in an ethanol series (30%, 70%, 100%) and air-dry at 37°C.

11. Replace the slides in the humidified chamber and apply between 20 and 120 µl (depending on specimen size) hybridization mixture containing $7.5 \cdot 10^3$ dpm/ml of ^{35}S-labeled probe previously denatured on specimen area.

12. Cover and incubate overnight the slides at 42°C.

13. Rinse the slides gently with 4X SSPE containing 2.5 mg/ml dithiothreitol.

14. Wash the slides in 4X SSPE containing 2.5 mg/ml dithiothreitol and in 4X SSPE at room temperature for 30 min each.

15. Wash the slides in X SSPE at room temperature and at 42°C for 30 min each.

16. Wash the slides in 0.1X SSPE at 42°C for 2×30 min.

17. Dehydrate the slides in an ethanol series (30%, 70%, 100%) and air-dry at 37°C.

18. Dip the slides in the dark in photographic emulsion prewarmed to 43°C and air-dry in a vertical position in a dark chamber for at least 2 hr.

19. Transfer the slides in a dark box and store at 4°C.

20. Develop the slides 1–3 weeks after exposure by immersion in developer and in fixative for 5 min each.

21. Rinse the slides in water and counterstain with methyl green pyronin.

Controls

They consisted of (a) competition studies with a fiftyfold excess of unlabeled relevant and irrelevant oligonucleotides and (b) careful examination of nonepithelial structures on the slide: vessels, muscle, and connective tissue (negative control) and (c) for carcinomas, a representative sample was chosen by morphological examination, including both carcinoma and a normal bronchiole or bronchus used as an internal positive control.

Scoring

The results were evaluated blindly by two pathologists with experience in pulmonary pathology. Scoring of reactions was performed semiquantitatively according to the intensity of labeling: +++, strong (visible at magnification X25); ++, moderate (visible at magnification X100); +, weak (visible at magnification X200); and −, absent. When the signal was too intense to identify the cell type, the exposure was shortened for 1 week. Anthracosic deposits were easily distinguished from silver grain deposition by their perivascular and peribronchiolar situation.

Immunohistochemistry

The following staining procedures were conducted using an automated immunostainer (ES, Ventana Medical Systems, Strasbourg, France).

Primary Antibodies

The following primary antiapomucin antibodies were applied to the sections: polyclonal MUC2 (LUM2-3) diluted at 1/1000 (Herrmann *et al.*, 1999), monoclonal MUC5AC (CLH2) diluted at 1/5 (Reis *et al.*, 1997), polyclonal MUC5B (LUM5B.2) diluted at 1/1000 (Wickström *et al.*, 1998) provided by Ingemar Carlstedt, monoclonal MUC1 (M8) diluted at 1/50 (McIlhinney *et al.*, 1985) provided by Dallas Swallow, and monoclonal MUC6 (CLH5) diluted at 1/250, Novocastra (Reis *et al.*, 2000). For MUC5AC and MUC5B, the antibodies stemmed from a different batch in comparison with relevant publications. For MUC1, MUC5AC, MUC5B, and MUC6, peptides corresponding to the tandem repeat sequences were used, whereas for MUC2 a peptide from the COOH-terminus was selected.

Materials

1. 10% neutral buffered formalin.
2. Silanized microscope slides.
3. Microwave oven.
4. Bar code labels.
5. Xylene.
6. 80%, 95%, and 100% solutions of ethanol.
7. Deionized/distilled water.
8. Ventana Automated Slide Staining System.
9. Primary antibodies.
10. Ventana Medical Systems DAB Detection Kit.
   ▲ Dispenser Inhibitor Solution, 3% H_2O_2 (hydrogen peroxide solution).
   ▲ Dispenser biotinylated Ig secondary antibody, affinity purified goat-anti-mouse and goat-anti-rabbit IgG in PBS with 0.05% ProClin 300 (preservative).
   ▲ Dispenser SA-HRP conjugate streptavidin horseradish peroxidase in protein stabilizer and 0.05% ProClin 300.
   ▲ Dispenser hydrogen peroxyde solution, 0.04%–0.08% H_2O_2 in a stabilizing solution.
   ▲ Dispenser substrate solution diaminobenzidine (DAB) 2 g/L in a stabilizer solution and preservative.

▲ Dispenser copper sulfate, copper sulfate 5 g/L, in a solution buffered solution and preservative.

11. Wash solution.
12. Liquid coverslip.
13. Mounting medium.
14. Cover glass.
15. Counterstain (hematoxylin).

Method

1. Each slide must be labeled with the appropriate bar code specifying the staining recipe and primary antibody.

2. Immerse the slides in two xylene baths at room temperature for 5 min each (for deparaffinization).

3. Rinse the slides for 3 min in two baths containing 100% ethanol, 3 min in 95% ethanol, 3 min in 80% ethanol, and 10 dips in water.

4. Immerse the slides in 500-ml citrate buffer (10 mM) (1/10) (pH 6.0).

5. Perform microwave pretreatment for two 10-min cycles and rinse the slides with deionized glass-distilled water.

6. Incubate sections for 32 min with normal goat serum to block the nonspecific antibody binding sites and rinse.

The immunohistochemistry method is a three-step indirect process based on the biotin-streptavidin complex.

7. The appropriate primary antibody dispenser and the DAB detection kit dispensers are loaded onto the reagent carousel and placed in the Ventana automated immunohistochemistry system.

8. Slides are incubated in 3% H_2O_2 for 4 min at 37°C to suppress endogeneous peroxidase activity and rinsed.

9. Slides are incubated for 32 min at 37°C in primary antibody diluted in PBS and rinsed.

10. Slides are incubated in universal biotinylated secondary antibody for 8 min at 37°C and rinsed.

11. Slides are incubated with streptavidin-peroxidase conjugate for 8 min at 37°C and rinsed.

12. The DAB solution is mixed with H_2O_2 solution on the specimen slide. Slides are incubated for 8 min at 37°C and rinsed.

13. Copper DAB enhancer is applied with mixing for 4 min at 37°C.

14. Slides are incubated with hematoxylin for 4 min at 37°C and mounted with cover glass.

Controls

Positive and negative controls were added in each automated immunohistochemistry run. Negative controls consisted of slides run without the primary antibody and negative tissues for *MUC5AC* and *MUC5B* by *in situ* hybridization (vessels, muscle). Normal bronchus was used as positive control for *MUC5B* and *MUC5AC*; *MUC5B* has been shown to be expressed predominantly in submucosal glands and *MUC5AC* in goblet surface cells (Buisine *et al.*, 1999; Copin *et al.*, 2000), normal gastric mucosae for *MUC6* (De Bolos *et al.*, 1995) and normal small intestinal mucosae for *MUC2* (Buisine *et al.*, 1998).

RESULTS AND DISCUSSION

Precursor Lesions of Squamous Cell Carcinoma, Invasive Squamous Cell Carcinoma

In hyperplasia (basal cell/goblet cell), metaplasia, and dysplasia, the pattern of qualitative expression of mucin genes is similar to that determined for normal mucosae by *in situ* hybridization. Nevertheless, quantitative variations of the *MUC5AC* and *MUC4* gene expression levels are observed. In goblet cell hyperplasia, *MUC5AC* gene and protein are overexpressed compared with normal mucosae, reflecting the increased number of goblet cells. Both *MUC5AC* gene and protein are overexpressed in the upper part of the epithelium in basal cell hyperplasia, metaplasia, and dysplasia, whereas in basal cells, *MUC5AC* is absent. Moreover, in our experience, *MUC5AC* gene and protein are overexpressed in normal bronchus epithelium adjacent to squamous cell carcinoma compared with the normal lobe by *in situ* hybridization and immunohistochemistry (Copin *et al.*, 2000). The overexpression of *MUC5AC* that is clearly related to mucous goblet cells in normal airways might be associated with mucous cell hyperplasia involving cell division and/or differentiation of small-granules mucous cells in precursor squamous lesions. These findings provide additional arguments in favor of a mucous cell origin of precursor lesions of squamous cell carcinoma, already suggested by experimental studies in hamsters (Becci *et al.*, 1978). Proliferative changes were induced experimentally by damaging the bronchial lining epithelium with a variety of physical or toxic agents. The earlier phase consisted of an increased number of small, basally located cells. These metaplastic cells in turn might be derived from proliferating mucous cells as suggested by the presence of many small mucous granules ultrastructurally in these metaplastic cells (Trump *et al.*, 1978). It may also be true in humans because mucous cells can persist in preinvasive carcinoma and detection of traces of mucins is common in carcinomas of various types. Moreover, focal expression of cytokeratins that are usually associated with glandular phenotype as cytokeratin 7 can be found in squamous cell carcinoma.

The pattern of expression of mucin genes was identical in squamous cell carcinomas whether they were *in situ* or invasive. The gene *MUC4* was heterogenously detected in 14 of 17 invasive carcinomas regardless of the differentiation grade (Figure 14). The expression was frequently moderate to strong in the center of the carcinomatous strands where the neoplastic cells are more differentiated and, in some cases, keratinized and weak or absent in the periphery. In the five carcinomas *in situ*, the expression of *MUC4* was diffuse, stronger in the upper part of the tumor as described in hyperplasia and weaker than in normal epithelium. For *MUC1* and *MUC2*, the expression was very weak and diffuse or absent. In all cases, *MUC5B* and *MUC5AC* were undetectable. Expression of *MUC3*, *MUC6,* and *MUC7* was not found in the normal respiratory surface epithelium.

Entrapment of bronchial glands and bronchioles is commonly found in lung carcinoma and can be recognized by careful evaluation of the cytologic detail. In contrast to Northern or Slot Blot analysis, *in situ* hybridization and immunohistochemistry allow the cellular localization of the signal. These techniques showed that many tumors contained numerous residual normal cells that could be readily identified by morphological examination and that mucin genes could be expressed strongly by these entrapped cells; for example, *MUC5AC* in goblet cells, *MUC5B* in submucosal glands, and *MUC4* in bronchioles. These data probably explain some conflicting results about

the pattern of mucin gene expression in lung carcinoma using Blot Analysis; for example, the frequent expression of *MUC5AC* in epidermoid carcinomas (Yu *et al.*, 1996).

In *in situ* or invasive squamous cell carcinoma, when epidermoid differentiation processes are achieved, consisting of the appearance of intracellular bridges and loss of ciliated and goblet cells, *MUC5AC* mRNA expression is decreased and then labeling completely disappears, whereas *MUC4* expression is maintained in all carcinomatous cells. The gene *MUC4* is expressed independently of mucus secretion both in normal airways or in various stages of epidermoid differentiation (Copin *et al.*, 2000; Lopez-Ferrer *et al.*, 2001). In normal airways, it is expressed by basal cells and probably ciliated cells as well as collecting ducts and goblet cells. Its expression is associated with squamous metaplasia even with complete squamous cell differentiation, dysplasia, and most of squamous cell carcinomas well differentiated and keratinized. Moreover, we have shown that *MUC4* is the earlier mucin gene expressed in the foregut, before epithelial differentiation into ciliated or secretory cells, and that *MUC5AC* is acquired concomitantly to epithelial cell differentiation. These findings in adult squamous lesions and embryonic respiratory tract might support the histogenetic theory of non–small cell bronchogenic carcinoma originating from an immature precursor. This perhaps is a pluripotent mucous cell and the distinctive role of *MUC4* in lung carcinoma histogenesis. The mucin phenotype of this pluripotent mucous cell might be *MUC5AC+, MUC4+*.

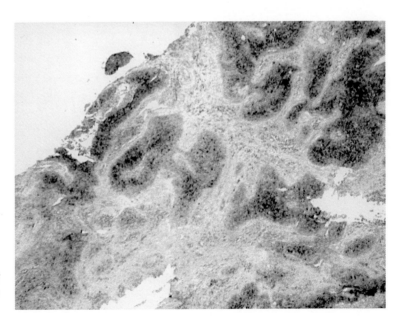

Figure 14 Intense and heterogeneous expression of *MUC4* mRNA in invasive epidermoid carcinoma by *in situ* hybridization (25X).

Lung Adenocarcinomas

Lung adenocarcinomas express mucin mRNA, which is expressed in normal respiratory mucosa (*MUC1, MUC2, MUC4, MUC5AC, MUC5B*), and *MUC3* and/or *MUC6* mRNA, which are not detected in normal lung by *in situ* hybridization (Copin *et al.*, 2001; Lopez-Ferrer *et al.*, 2001). Interestingly, the expression pattern of mucin genes differs according to the histological subtypes of adenocarcinomas as defined by the World Health Organization classification (Travis *et al.*, 1999). Nonmucinous type of bronchioloalveolar carcinomas and nonbronchioloalveolar carcinomas share the constant expression of *MUC1* and *MUC4*, the variable expression of *MUC2, MUC3, MUC5AC*, and *MUC5B,* and the absence of *MUC6* gene expression (Copin *et al.*, 2001).

Among adenocarcinomas, the mucinous type of bronchioloalveolar carcinoma has a particular pattern of mucin gene expression since all mucin genes are expressed except *MUC7* (Copin *et al.*, 2001). The expression of *MUC5AC* and *MUC5B* genes and proteins is the most intense and diffuse among all subtypes (Figures 15–16). Coexpression of *MUC1* and *MUC3* genes is constant. Coexpression of *MUC2, MUC4,* and *MUC6* genes is very frequent. MUC6 protein expression has been confirmed by immunohistochemistry. This complex but homogeneous expression pattern in mucinous bronchioloalveolar carcinoma is in agreement with the great cellular homogeneity of this type of adenocarcinoma. There is now limited but convincing evidence that bronchioloalveolar carcinoma is distinct from typical adenocarcinoma of the lung.

When restrictive criteria for diagnosis of bronchioloalveolar carcinoma are used, this tumor exhibits a better prognosis than the other types (Breathnach *et al.*, 1999). Whether the histological subtype is a prognostic factor in bronchioloalveolar carcinoma remains debated (Clayton, 1986). A characteristic feature of the biology of bronchioloalveolar carcinoma is the development of multifocal lesions within the lung parenchyma, the high frequency of diffuse pulmonary involvement with limited regional lymph node involvement, and rare brain metastases (Breathnach *et al.*, 1999). The different morphological and clinical patterns of the subtypes of bronchioloalveolar carcinoma suggest that their biological behavior may differ.

Clinically, the mucinous subtype is more strongly associated with diffuse pulmonary involvement than the nonmucinous subtype (Breathnach *et al.*, 1999). There are already some molecular data that support the biological difference of mucinous type of bronchioloalveolar carcinoma. A significantly higher frequency of K-Ras mutations in the mucinous form of bronchioloalveolar carcinoma than in the other subtypes has been demonstrated (Marchetti *et al.*, 1996). In mucinous carcinomas of the colon and prostate, altered expression of one or several mucin genes appears to be a marker of particularly aggressive tumors, whereas mucinous carcinomas of the breast, so-called colloid carcinomas, exhibit better prognosis than their nonmucinous breast counterparts (O'Connell *et al.*, 1998). Thus, mucous secretion may have a distinct biological significance in adenocarcinoma according to the organ involved. In lung, mucin gene

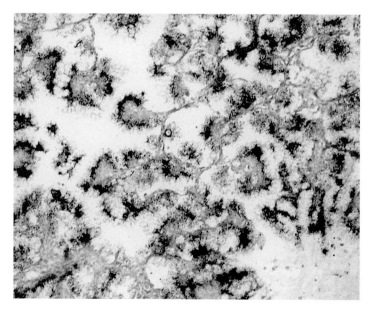

Figure 15 Intense and homogeneous expression of *MUC5AC* mRNA in mucinous type of bronchioloalveolar carcinoma by *in situ* hybridization (100X).

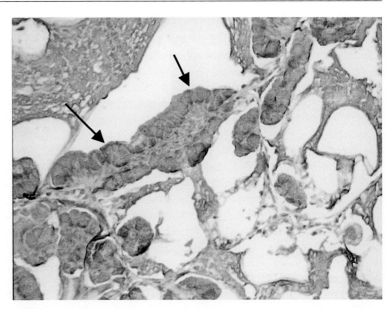

Figure 16 MUC5AC protein detected in mucinous type of bronchioloalveolar carcinoma by immunohistochemistry (arrows; 200X).

expression may serve as a marker of cytodifferentiation. We showed that *MUC5AC* is expressed in goblet cells in normal respiratory mucosa and disappears when complete epidermoid differentiation is achieved in metaplasia and squamous cell carcinoma (Copin *et al.*, 2000). The great amount and constant expression of this mucin in mucinous type of bronchioloalveolar carcinoma might indicate that this subtype of adenocarcinoma had sustained a well-differentiated phenotype similar to the goblet cell, correlated with a noninvasive pattern and a better prognosis than non-bronchioloalveolar carcinomas.

The expression of secretory mucins *MUC2*, *MUC5AC*, and *MUC5B* is associated both with the different histological subtypes of adenocarcinomas and mucous secretion (Copin *et al.*, 2001). Mucous-secreting adenocarcinomas maintain focal *MUC2*, *MUC5AC*, and *MUC5B* gene and protein expression, confined to mucous-secreting carcinomatous cells. Nonsecreting and poorly differentiated adenocarcinomas lost both *MUC5AC* and *MUC5B* expression and maintain *MUC2* mRNA expression. We could not detect MUC2 protein expression in this latter type. These results are in agreement with a previous report that investigated the mucociliary differentiation–dependent mucin gene expression in bronchial cells cultured in the presence of retinoic acid (Koo *et al.*, 1999). Following retinoic acid treatment of retinoid-deficient human tracheobronchial epithelial cell cultures, induction of mucin gene expression occurs sequentially: *MUC2*, *MUC5AC*, and *MUC5B* were up-regulated at 24, 48, and 72 hr, respectively. This study indicates that MUC2 mRNA expression is an early marker of mucous

differentiation, whereas *MUC5AC* and *MUC5B* mRNAs are expressed later, during more advanced stages of mucous differentiation.

MUC3 and *MUC6*, which are not expressed in normal adult and fetal lung, are expressed in lung adenocarcinomas (Copin *et al.*, 2001). The *MUC6* gene is clustered with *MUC2*, *MUC5AC*, and *MUC5B* (in that order) at chromosomal location 11p15.5 (Pigny *et al.*, 1996). All four of the mucin genes show different expression patterns *in vivo* both in normal tissues and carcinomas, suggesting that the regulation of this gene cluster is complex. However, the pattern of expression in mucinous type of bronchioloalveolar carcinoma (i.e., *MUC2*, *MUC5AC*, *MUC5B*, and *MUC6*) might support some coregulatory features for the 11p15.5 mucin gene cluster. *MUC6*, like *MUC5AC*, is primarily expressed in the stomach, whereas *MUC5AC* is associated with surface mucous epithelial cells and *MUC6* with pyloric mucous gland cells (De Bolos *et al.*, 1995). Honda *et al.* (1998) previously described a subtype of mucinous bronchioloalveolar carcinoma expressing gastric mucins by histochemistry—so-called gastric-type pulmonary carcinoma—morphologically indistinct from the other subtypes of mucinous bronchioloalveolar carcinoma. Histochemically, this subtype is characterized by carcinoma cells (located in the protruding portions of the papillary structures) that differentiate as gastric-type surface mucous cells and carcinoma cells in the indented portions that differentiate as pyloric gland-type mucous cells. Heterotopic gastric mucosa was considered to be the origin of these tumors, especially since respiratory and digestive tracts are foregut derivatives. The expression of gastric mucins has already been described in various sites, including mucous metaplasia of the pancreatic duct and pancreatic ductal adenocarcinoma (Matsuzawa *et al.*, 1992). This particular phenotype seems to be associated with malignant progression. We believe as others (Lopez-Ferrer *et al.*, 2001) that the expression pattern of mucin genes in lung carcinomas, particularly *MUC3* and *MUC6* genes in mucinous bronchioloalveolar carcinoma, supports that lung carcinomas originate from a common endodermal precursor cell with the potential for multicellular differentiation, including expression of mucins of gastric type.

We did not detect *MUC4* gene expression in normal type II pneumocytes, but it was found in type II pneumocyte hyperplasia (Copin *et al.*, 2000). The other mucin genes were not detectable in this cell type. Adenocarcinomas could be separated in two categories depending on mucin gene expression (Copin *et al.*, 2001). On the one hand, mucinous type of bronchioloalveolar carcinoma is a very differentiated adenocarcinoma with both constant expression of the 11p15 mucin genes and phenotypic characteristics of the goblet-cell type. On the other hand, nonmucinous type of bronchioloalveolar carcinoma and nonbronchioloalveolar carcinoma share constant expression of *MUC1* and *MUC4* genes, as type II pneumocyte hyperplasia and focal expression of *MUC2*, *MUC5AC,* and *MUC5B* genes and proteins only in mucous-secreting areas. Finally, histological subtyping of adenocarcinomas might be more related to the degree of glandular differentiation than to the cellular origin.

In conclusion, in contrast to Northern or Slot Blot analysis, *in situ* hybridization and immunohistochemistry allow the cellular localization of the signal. Several arguments could explain some conflicting results between Blot analysis and morphological techniques. *In situ* hybridization and immunohistochemistry showed that many carcinomas contain numerous residual normal cells that could be readily identified by morphological examination, and mucin genes (sometimes distinct from those expressed by carcinomatous cells), can be expressed strongly by these entrapped cells. Moreover, expression of apomucins confined to carcinomatous cells at a more differentiated state (such as MUC5AC and MUC5B proteins in mucus-secreting adenocarcinomas or a weak and diffuse expression such as *MUC2* mRNA in epidermoid carcinomas) can be easily distinguished by these morphological techniques.

In the same way, heterogeneity of the expression of mucin genes in some lung carcinomas reflect different states of cytodifferentiation (i.e., *MUC2*, *MUC5AC* and *MUC5B* in adenocarcinomas) or might suggest for *MUC4* a role of mucins in tumor progression.

The expression of mucin genes reflects a precise state of differentiation more complex than morphologic differentiation grade, which indicates only the similarity to normal glands. The study of the expression of mucin genes contributes to the determination of molecular basis of histopathological classification of lung carcinomas and supports the concept of a common cell origin from an immature and pluripotent mucous cell.

From a clinical point of view, mucin genes could serve as diagnostic and prognostic markers. For instance, in the management of patients with bronchioloalveolar carcinoma (BAC), detection of tumor markers in sputum or bronchial lavage fluid and bronchioalveolar lavage by immunohistochemistry could be of great importance since this type of carcinoma does not invade the bronchi. Histological diagnosis of BAC is often made using surgical specimens or samples obtained by transbronchial or needle biopsies. Further investigations are necessary to evaluate the potential utility of the aberrant expression of *MUC3* and *MUC6* in bronchial lavage fluid as a diagnostic argument of BAC recurrence. In the future, both studies about the regulation of mucin genes and the functional approach with knockout mice will allow the understanding of all the data accumulated with expression studies in cancers.

Acknowledgments

This work was supported by grants from the Association pour la Recherche sur le Cancer and the CHU de Lille (contract no 96/09.29/95,95) and the Ligue contre le Cancer du Pas de Calais. The authors would like to acknowledge the members of the European Union Consortium (contract No. CEE BMH4-CT98-3222), especially Ingemar Carlstedt, Dallas Swallow, and Leonor David, for providing antibodies. Contract grant sponsor: Association pour la Recherche sur le Cancer and CHU de Lille. Contract grant number: 96/09.29/95,95.

References

Aubert, J.P., Porchet, N., Crepin, M., Duterque-Coquillaud, M., Vergnes, G., Mazzuca, M., Debuire, B., Petitprez, D., and Degand, P. 1991. Evidence for different human tracheobronchial mucin peptides deduced from nucleotide cDNA sequences. *Am. J. Respir. Cell. Mol. Biol. 5:*178–185.

Audié, J.P., Janin, A., Porchet, N., Copin, M.C., Gosselin, B., and Aubert, J.P. 1993. Expression of human mucin genes in respiratory, digestive, and reproductive tracts ascertained by *in situ* hybridization. *J. Histochem. Cytochem. 41:*1479–1485.

Bartman, A.E., Buisine, M.P., Aubert, J.P., Niehans, G.A., Toribara, N.W., Kim, Y.S., Kelly, E.J., Crabtree, J.E., and Ho, S.B. 1998. The *MUC6* secretory mucin gene is expressed in a wide variety of epithelial tissues. *J. Pathol. 186:*398–405.

Becci, P.J., Mc Dowell, E.M., and Trump, B.F. 1978. The respiratory epithelium : IV. Histogenesis of epidermoid metaplasia and carcinoma *in situ* in the hamster. *J. Natl. Cancer Inst. 61:*577–586.

Breathnach, O.S., Ishibe, N., Williams, J., Linnoila, R.I., Caporaso, N., and Johnson, B.E. 1999. Clinical features of patients with stages IIIB and IV bronchioloalveolar carcinoma of the lung. *Cancer 86:*1165–1173.

Buisine, M.P., Devisme, L., Savidge, T.C., Gespach, C., Gosselin, B., Porchet, N., and Aubert, J.P. 1998. Mucin gene expression in human embryonic and fetal intestine. *Gut 43:*519–524.

Buisine, M.P., Devisme, L., Copin, M.C., Durand-Reville, M., Gosselin, B., Aubert, J.P., and Porchet, N. 1999. Developmental mucin gene expression in the human respiratory tract. *Am. J. Respir. Cell Mol. Biol. 20:*209–218.

Carraway, K.L., Price-Schiavi, S.A., Komatsu, M., Idris, N., Perez, A., Li, P., Jepson, S., Zhu, X., Carvajal, M.E., and Carraway, C.A. 2000. Multiple facets of sialomucin complex/MUC4, a membrane mucin and erbb2 ligand, in tumors and tissues (Y2K update). *Front. Biosci. 5:*D95–D107.

Clayton, F. 1986. Bronchioloalveolar carcinomas: Cell types, patterns of growth and prognostic correlates. *Cancer 57:*1555–1564.

Copin, M.C., Buisine, M.P., Leteurtre, E., Marquette, C.H., Porte, H., Aubert, J.P., Gosselin, B., and Porchet, N. 2001. Mucinous bronchioloalveolar carcinomas display a specific pattern of mucin gene expression among primary lung adenocarcinomas. *Human Pathol. 32:*274–281.

Copin, M.C., Devisme, L., Buisine, M.P., Marquette, C.H., Wurtz, A., Aubert, J.P., Gosselin, B., and Porchet, N. 2000. From normal respiratory mucosa to epidermoid carcinoma: Expression of human mucin genes. *Int. J. Cancer 86:*162–168.

De Bolos, C., Garrido, M., and Real, F.X. 1995. *MUC6* apomucin shows a distinct normal tissue distribution that correlates with Lewis antigen expression in the human stomach. *Gastroenterology 109:*723–734.

Dekker, J., Rossen, J.W., Buller, H.A., and Einerhand, A.W. 2002. The MUC family: An obituary. *Trends Biochem. Sci. 27:* 126–131.

Herrmann, A., Davies, J.R., Lindell, G., Martensson, S., Packer, N.H., Swallow, D.M., and Carlstedt, I. 1999. Studies on the insoluble glycoprotein complex from human colon. Identification of reduction-insensitive *MUC2* oligomers and C-terminal cleavage. *J. Biol. Chem. 274:*15828–15836.

Honda, T., Ota, H., Ishii, K., Nakamura, N., Kubo, K., and Katsuyama, T. 1998. Mucinous bronchioloalveolar carcinoma with organoid differentiation simulating the pyloric mucosa of the stomach: Clinicopathologic, histochemical, and immuno-histochemical analysis. *Am. J. Clin. Pathol. 109:*423–430.

Jarrard, J.A., Linnoila, R.I., Lee, H., Steinberg, S.M., Witschi, H., and Szabo, E. 1998. MUC1 is a novel marker for the type II pneumocyte lineage during lung carcinogenesis. *Cancer Res. 58:*5582–5589.

Komatsu, M., Tatum, L., Altman, N.H., Carothers, C.A., Carraway, K., and Carraway, L. 2000. Potentiation of metastasis by cell surface sialomucin complex (rat MUC4), a multifunctional antiadhesive glycoprotein. *Int. J. Cancer 87:*480–486.

Koo, J.S., Yoon, J.H., Gray, T., Norford, D., Jetten, A.M., and Nettesheim, P. 1999. Restoration of the mucous phenotype by retinoic acid in retinoid-deficient human bronchial cell cultures: Changes in mucin gene expression. *Am. J. Respir. Cell Mol. Biol. 20:*43–52.

Lopez-Ferrer, A., Curull, V., Barranco, C., Garrido, M., Lloreta, J., Real, F.X., and De Bolos, C. 2001. Mucins as differentiation markers in bronchial epithelium: Squamous cell carcinoma and adenocarcinoma display similar expression patterns. *Am. J. Respir. Cell Mol. Biol. 24:*22–29.

Marchetti, A., Buttitta, F., Pellegrini, S., Chella, A., Bertacca, G., Filardo, A., Tognoni, V., Ferreli, F., Signorini, E., Angeletti, C.A.,
and Bevilacqua, G. 1996. Bronchioloalveolar lung carcinomas: K-Ras mutations are constant events in the mucinous subtype. *J. Pathol. 179:*254–259.

Matsuzawa, K., Akamatsu, T., and Katsuyama, T. 1992. Mucin histochemistry of pancreatic duct cell carcinoma, with special reference to organoid differentiation simulating gastric pyloric mucosa. *Hum. Pathol. 23:*925–933.

McIlhinney, R.A.J., Patel, S., and Gore, M.E. 1985. Monoclonal antibodies recognizing epitopes carried on both glycolipids and glycoproteins of the human fat globule membrane. *Biochem. J. 227:*155–162.

Moniaux, N., Escande, F., Porchet, N., Aubert, J.P., and Batra, S.K. 2001. Structural organization and classification of the human mucin genes. *Front. Biosci. 6:*D1192–D1206.

Nguyen, P.L., Niehans, G.A., Cherwitz, D.L., Kim, Y.S., and Ho, S.B. 1996. Membrane-bound (*MUC1*) and secretory (*MUC2, MUC3,* and *MUC4*) mucin gene expression in human lung cancer. *Tumour Biol. 17:*176–192.

O'Connell, J.T., Shao, Z.M., Drori, E., Basbaum, C.B., and Barsky, S.H. 1998. Altered mucin expression is a field change that accompanies mucinous (colloid) breast carcinoma histogenesis. *Hum. Pathol. 29:*1517–1523.

Pigny, P., Guyonnet-Dupérat, V., Hill, A.S., Pratt, W.S., Galiègue-Zouitina, S., D'hooge, M.C., Laine, A., Van-Seuningen, I., Degand, P., Gum, J. R., Kim, Y.S., Swallow, D.M., Aubert, J.P., and Porchet, N. 1996. Human mucin genes assigned to 11p15.5: Identification and organization of a cluster of genes. *Genomics 38:*340–352.

Porchet, N., Nguyen, V.C., Dufossé, J., Audié, J.P., Guyonnet-Dupérat, V., Gross, M.S., Denis, C., Degand, P., Bernheim, A., and Aubert, J.P. 1991. Molecular cloning and chromosomal localization of a novel human tracheo-bronchial mucin cDNA containing tandemly repeated sequences of 48 base pairs. *Biochem. Biophys. Res. Commun. 175:*414–422.

Porchet, N., Pigny, P., Buisine, M.P., Debailleul, V., Degand, P., Laine, A., and Aubert, J.P. 1995. Human mucin genes: Genomic organization and expression of *MUC4, MUC5AC,* and *MUC5B. Biochem. Soc. Trans. 23:*800–805.

Reis, C.A., David, L., Nielsen, P.A., Clausen, H., Mirgorodskaya, K., Roepstorff, P., and Sobrinho-Simöes, M. 1997. Immuno-histochemical study of MUC5AC expression in human gastric carcinomas using a novel monoclonal antibody. *Int. J. Cancer 74:*112–121.

Reis, C.A., David, L., Carvalho, F., Mandel, U., De Bolos, C., Mirgorodskaya, E., Clausen, H., and Sobrinho-Simoes, M. 2000. Immunohistochemical study of the expression of MUC6 mucin and co-expression of other secreted mucins (MUC5AC and MUC2) in human gastric carcinomas. *J. Histochem. Cytochem. 48:*377–388.

Seregni, E., Botti, C., Lombardo, C., Cantoni, A., Bogni, A., Cataldo, I., and Bombardieri, E. 1996. Pattern of mucin gene expression in normal and neoplastic lung tissues. *Anticancer Res. 16:*2209–2213.

Travis, W.D., Colby, T.V., Corrin, B., Shimosato, Y., and Brambilla, E. 1999. *Histological Typing of Lung and Pleural Tumors: International Histological Classification of Tumours.* 3rd ed. Geneva: World Health Organization.

Trump, B.F., McDowell, E.M., Glavin, F., Barrett, L.A., Becci, P.J., Schurch, W., Kaiser, H.E., and Harris, C.C. 1978. The respiratory epithelium. III. Histogenesis of epidermoid metaplasia and carcinoma *in situ* in the human. *J. Natl. Cancer Inst. 61:*563–575.

Velcich, A., Yang, W.C., Heyer, J., Fragale, A., Nicholas, C., Viani, S., Kucherlapati, R., Lipkin, M., Yang, K., and Augenlicht, L. 2002. Colorectal cancer in mice genetically deficient in the mucin *Muc2. Science 295:*1726–1729.

Wickström, C., Davies, J., Eriksen, G., Veerman, E., and Carlstedt, I. 1998. MUC5B is a major gel-forming, oligomeric mucin from human salivary gland, respiratory tract, and endocervix: Identification of glycoforms and C-terminal cleavage. *Biochem. J. 334:*685–693.

Yu, C.J., Yang, P.C., Shun, C.T., Lee, Y.C., Kuo, S.H., and Luh, K.T. 1996. Overexpression of *MUC5* genes is associated with early postoperative metastasis in nonsmall cell lung cancer. *Int. J. Cancer 69:*457–465.

4

Immunohistochemical Expression of MDM2 in Lung Carcinoma

Sylvie Gazzeri

Introduction

The MDM2 gene was initially identified as an amplified gene on a murine double-minute chromosome in the spontaneously transformed BALB/c 3T3 cells (Cahilly-Snyder *et al.,* 1987). This oncogenic property was further demonstrated in *in vitro* experiments where overexpression of MDM2 was able to increase the tumorigenic potential (Finlay, 1993) and proliferative rate (Martin *et al.,* 1995) of cultured cells. The analysis of more than 3000 tumors samples shows that MDM2 is amplified in 7% of these tissues. The highest frequency of MDM2 amplification is observed in soft tissue tumors, osteosarcomas, and oesophageal carcinomas. Furthermore, many reports describe the overexpression of MDM2 in different types of tumors. The presence of a high level of MDM2 in these tumors might be an important element for their survival because it decreases their ability to activate p53.

MDM2 in the p53 Pathway

The function of MDM2 was unclear until it was demonstrated that it binds to the p53 tumor suppressor protein with high affinity and inhibits its function in a number of ways (Momand *et al.,* 2000). In fact, MDM2 is now believed to be the major physiological

antagonist of p53. Besides being a transcriptional target for p53, MDM2 can also antagonize p53-dependent transcriptional activation by direct binding via its N-terminal region, generating a negative feedback loop that probably serves as a pivotal mechanism for restraining p53 function in normal cells in the absence of stress. In line with this notion, overexpression of MDM2 is a mechanism, independent of gene mutation, by which wild-type p53 function can be inactivated, abrogating its tumor suppressor effects. In this respect, limitation of p53 apoptotic function by MDM2 is consistent with the lethal phenotype of MDM2 knockout mice rescued by p53 knockout.

Besides its negative control on p53 functions at the transcription level, MDM2 also regulates p53 protein stability. Binding of MDM2 to p53 can promote the degradation of p53, acting as an ubiquitin-protein ligase to ubiquitinate p53 and triggering its nuclear export and degradation in cytoplasmic proteasomes (Michael *et al.,* 2002). The inhibitory effect of MDM2 toward p53 is counteracted by human p14[ARF], a tumor suppressor gene that acts as a sensor of hyperproliferative signals emanating from oncoproteins (Sherr, 2001). Direct binding of p14[ARF] to MDM2 prevents p53 ubiquitination and degradation by blocking p53-MDM2 nuclear export, sequesters MDM2 into the nucleolus, and inhibits its ubiquitin ligase activity toward p53

Handbook of Immunohistochemistry and in situ Hybridization of Human Carcinomas, Volume 1: Molecular Genetics; Lung and Breast Carcinomas

127

Copyright © 2004 by Elsevier (USA)
All rights reserved.

(Zhang *et al.*, 2001). Because of this role, p14[ARF] prevents the negative-feedback regulation of p53 by MDM2 and leads to the activation of p53 in the nucleoplasm. Conversely, high levels of MDM2 relocalize endogenous p14[ARF] from nucleoli to nucleoplasm, suggesting that balance between both protein levels and their respective subcellular location might be important to regulate their effects.

Alternative Effector Pathways Regulated by MDM2

In addition to its clear role in the regulation of p53, MDM2 is part of a complex network of interactions through which MDM2 affects the cell cycle, apoptosis, and tumorigenesis. For example, overexpression of MDM2 in the mammary gland of transgenic mice induces the uncoupling of S-phase from mitosis, leading to cell cycle arrest independently of p53. On the other hand, MDM2 shows transforming activities in the absence of p53 binding and this may be related to the ability of MDM2 to affect transcription. These other mechanisms of MDM2 function are much less well-understood than those depending on p53, but other important cell growth–regulatory proteins have been shown to interact with MDM2. This is examplified by MDM2 that binds Rb and abrogates its ability to control G_1 arrest. It also induces E2F1/DP1 activities, encouraging G_1-S transition. Furthermore, MDM2 is able to interact with and inhibits CBP/p300 and the L5 ribosomal ribonucleoprotein particles.

Spliced Variants and Products of MDM2

The analysis of the MDM2 oncogene expression in tumors has revealed the existence of at least 40 alternatively and aberrantly spliced transcripts of MDM2 RNA, but it is currently unknown how many of these are actually expressed as protein products and what is their respective biological significance (Bartel *et al.*, 2002). In addition to the full-length protein (p85/90), p54/57 and p74/76 isoforms are the most studied as MDM2 gene products (Eymin *et al.*, 2002). Although the function of these MDM2 isoforms remains to be elucidated, it is noteworthy that the p74/76 isoforms lacking the N-terminal domain of the full-length protein do not bind to p53 and can antagonize the ability of p90 to target p53 destruction. In contrast, the p54/57 isoforms bind to p53 but lack C-terminal epitopes.

Activation of MDM2 Gene in Lung Cancer

Amplification of the MDM2 gene has been reported in human lung tumors in only a few cases of non–small cell lung cancer (NSCLC), and mutations were never found. In contrast, aberrant expression of MDM2 product has been variably appreciated because 24–70% of NSCLC and 40–70% of small cell lung carcinoma (SCLC) were reported to overexpress MDM2 proteins (Osada *et al.*, 2002). Interestingly, some studies reported an association between MDM2 overexpression and a favorable prognosis. Furthermore, a clear overexpression of MDM2 product has been observed in preneoplastic lung lesions, suggesting that alteration of MDM2 could be an early event during lung carcinogenesis (Rasidakis *et al.*, 1998).

Immunohistochemical Technique

MATERIALS

Preparation of Tissues Sections

1. Dextran solution: dextran 10% in H_2O.
2. Acetone.
3. Amino-propyltriethoxysilane (APES, Sigma A3648).
4. APES solution: APES 2% in acetone.
5. 150 mM phosphate-buffered saline (PBS) (pH 7.4).
6. Paraformaldehyde (PAF). Working solution at 3.7% prepared from a 20% stock solution and stored at 4°C.

Immunoperoxidase Technique

Incubation with Primary Antiserum

1. Blocking serum: 2% normal donkey serum in PBS (pH 7.4)/BSA 0.03%.
2. Primary antibody solution: primary antibodies are diluted in PBS (pH 7.4)/bovine serum albumin (BSA) 0.03%.
3. MDM2 (N20) Sc-813 Santa-Cruz polyclonal antibody is directed against residues 1-26. Used at 1/4000 dilution.
4. MDM2 (SMP14) Sc-965 Santa-Cruz monoclonal antibody recognizes residues 154-167. Used at 1/500 dilution.
5. MDM2 (2A10) (generous gift from A. Levine) monoclonal antibody is directed against residues 294-339. Used at 1/2000 dilution.
6. MDM2 (NCL-MDM2p) Novocastra polyclonal antibody recognizes a peptide site near the C-terminal of MDM2. Used at 1/2000 dilution.

Incubation With Secondary Antibody and ABC Reagent

1. Secondary antibody solution: secondary antibodies are diluted in PBS (pH 8.6)/BSA 0.03%.
2. Anti-rabbit biotinylated donkey F(ab′)2 (1/1000; The Jackson Laboratory, West Grove, PA).

3. Anti-mouse biotinylated donkey F(ab')2 (1/500; The Jackson Laboratory, West Grove, PA).

4. StreptABComplex/HRP (from Dako ref 0377). Add 10 μl streptavidine and 10-μl peroxidase (biotinylated horseradish peroxidase) to 2000 μl Tris 0.05 M (pH 7.6). Make up fresh each day, at least 30 min before using.

5. DAB working solution: Prepare fresh just prior to use in dark. Mix 2 ml DAB (3 g DAB in 50 ml Tris-HCl (hydrochloric acid) 0.05 M (pH 7.6)) and 2 ml imidazole (4 g imidazole in 50 ml Tris-HCl 0.05 M (pH 7.6) to 200 ml Tris-HCl 0.05 M (pH 7.6) and 35 μl H$_2$O$_2$ (Merck, ref 1.07209.0250).

6. Hematoxylin solution.

7. HCl solution: Prepare working solution by adding 10 drops of HCl (35%) to 200 ml of distilled water.

8. NH$_4$OH solution: Prepare working solution by adding 10 drops of NH$_4$OH to 200 ml of distilled water.

9. Merckoglas (Merck Eurolab).

METHODS

Preparation of Tissue Sections

1. Incubate slides overnight with dextran solution.
2. Rinse with warm water.
3. Rinse with distilled water.
4. Dry at 50°C in oven for 2 hr.
5. Immerse slides in APES solution for 20 sec.
6. Rinse three times with acetone.
7. Rinse with distilled water.
8. Dry at 50°C in oven for 2 hr.
9. Cut 6-μm-thick frozen samples on APES-coated slides.
10. Air-dry for 5 min.
11. Identify sections with Dako Pen.
12. Apply 3.7% PAF for 10 min to fix the cells.
13. Remove PAF and rinse twice in PBS/BSA 0.03% for 10 min.
14. Remove excess PBS/BSA.

Immunoperoxidase Technique

Incubation with Primary Antiserum

1. To block nonspecific binding, apply 100 μl of blocking solution and incubate for 30 min in humidity chamber at room temperature.

2. Remove the blocking serum and apply 100 μl of primary antibody solution.

3. Incubate overnight in humidity chamber at 4°C.

Incubation with Secondary Antibody and ABC Reagent

1. Remove slides from incubation racks and place in PBS (pH 8.6)/BSA 0.03% for 10 min. Repeat three times.

2. Remove excess PBS/BSA solution.

3. Dry the contour of the sections.

4. Apply secondary (biotinylated) antibody (100 μl) and incubate in humidity chamber racks at room temperature for 1 hr.

5. Remove from incubation racks and place in PBS/BSA for 10 min. Repeat three times.

6. Remove excess PBS/BSA solution.

7. Dry the contour of the sections.

8. Apply ABC reagent (100-μl per slide) and incubate in humidity chamber racks in dark at room temperature for 1 hr.

DAB Incubation

1. Remove slides from incubation racks and place in PBS (pH 8.6)/BSA 0.03% for 5 min. Repeat three times.

2. Rinse slides in Tris 0.05 M (pH 7.6) for 5 min.

3. Prepare fresh DAB working solution.

4. Immerse slides in DAB solution for 5 min. Keep away from light.

5. Rinse slides in distilled water for 5 min. Repeat three times.

6. Check the staining using microscope.

Counterstaining

1. Immerse slides in hematoxylin solution for 20 sec with shaking.

2. Rinse with tap water.

3. Immerse slides in HCl solution for 8 sec.

4. Rinse with tap water.

5. Immerse slides in NH$_4$OH solution for 8 sec.

6. Rinse with tap water.

7. Check counterstaining using microscope.

Slide Preparation for Storage

1. Dehydrate in three changes of 100% ethanol (5 min each) and three changes of 100% toluene.

2. Coverslip in Merckoglas.

Notes: Scores of immunostaining were calculated by multiplying the percentage of labeled cells with the intensity of staining (1+, 2+, 3+), and tumors were graded into four classes based on their score (class 0: no staining; class 1: <50; class 2: 50–100; class 3 ≥ 100). Tumors with more than 50% of tumor cells displaying moderate or strong nuclear staining were considered as tumors overexpressing MDM2 (score ≥ 100; class 3) when compared with normal tissues.

RESULTS AND DISCUSSION

Immunohistochemical Analysis of MDM2 Expression

In most studies, MDM2 overexpression was assessed by using immunohistochemical studies with one or two antibodies (IF2, Oncogene Sciences, NY; 1B10, Novocastra Laboratories, UK) mapping different epitopes onto MDM2 protein (N-Ter and C-Ter, respectively). However, analysis of MDM2 status is complicated by the existence of various isoforms, the detection of which depends on the antibody used, rendering hazardous the functional interpretation of immunostaining data without any concomittant biochemical analysis. We investigated the pattern of MDM2 expression by immunohistochemistry (IHC) using four antibodies mapping distinct epitopes onto the MDM2 protein, allowing detection of all MDM2 isoforms on a panel of human lung cancers comprising all histological types. Scores of immunostaining are recorded based on assessing the percentage of positive cells and the intensity of staining. Normal lung parenchyma adjacent to tumor on sections was considered as internal control, whereas three normal lung tissues taken for diagnosis in patients without history of cancer were taken as external controls.

Normal Lung Tissue

Nuclear staining only was considered to assess MDM2 immunoreactivity. Both epithelial and endothelial cells express MDM2 and serve as positive internal control, whereas lymphocytes do not express MDM2 and serve as negative internal control. These were the gold standards for comparison in our study. In normal lung that was present in the vicinity of the tumors, as well as in control normal lung tissues, a similar intensity of staining was observed on the entire target population of epithelial and endothelial cells using all antibodies.

Lung Tumors

Because MDM2 immunostaining is heterogeneous in lung tumors, differential score is ascribed to each case and for each antibody according to the intensity of staining and the percentage of stained cells. A score is calculated by multiplying the percentage of labeled cells (0–100%) by the intensity of staining (1–3), allowing a range of 0 to 300. The resulting scores are subdivided as described in the method section into the four following classes of staining: 0, undetectable; 1, faint; 2, moderate; and 3, high. According to this classification, MDM2 immunostaining in normal alveolar and bronchiolar epithelium and stromal cells (endothelial cells and fibroblasts) is uniformaly distributed between classes 1 and 2 (between 30 and 80 score values). Tumors displaying a mean score of class 3 are considered as MDM2 overexpressing cases since their level of staining was definitely higher than the maximum level reached by normal lung cells. Accordingly, MDM2 is overexpressed in 31 of 90 (34%) NSCLC and in

29 out of 102 (28%) neuroendocrine (NE) lung tumors (Figure 17). No significant difference in the pattern of MDM2 staining was observed between the distinct histological subclasses of lung tumors. In contrast, MDM2 overexpression was associated with an extended stage in adenocarcinomas ($P = 0.0248$).

Possible discordance between the reactivity of antibodies was considered according to the localization of their epitope onto the MDM2 protein. The SMP14 and 2A10 antibodies, both recognizing mid-region epitopes in the acidic zinc region of the molecule, consistently gave a similar score of MDM2 immunostaining that was lower than that obtained with the two other antibodies (N and C terminal domains) in 8% (16/192) of the cases. Absence or aberrantly low level of MDM2 staining was observed in 3% (5/192) and 7% (13/192) of the cases with N20 and C-Ter antibodies, respectively, compared with the high level of reactivity using the three other antibodies. Overall, 100% of the tumors analyzed displayed the same pattern of MDM2 immunostaining (according to intensity and percentage of positive tumor cells) with at least two antibodies and 88% (168/192) with at least three distinct antibodies, indicating that a good concordance existed between the reactivity of the antibodies.

Comparison of IHC and Western Blot Data

To validate the IHC data, a Western Blot analysis was performed on 28 of 192 lung tumors using the same four MDM2 antibodies. Results of Western Blotting were considered concordant with those of IHC when the same status of MDM2 was detected with at least two distinct antibodies, whatever the isoform. In these conditions, of the 21 samples considered as MDM2 nonoverexpressing cases using IHC, 17 (81%) displayed concordant results by Western Blotting. In the four other cases, overexpression of MDM2 was detected using Western Blot analysis. In two of these samples, however, discordances were observed between the antibodies used in the immunohistochemical analysis. In the last two cases, all antibodies gave concordant IHC data that were completely discordant from Western Blotting data.

On the other hand, of the seven samples considered as MDM2 overexpressing cases using IHC, six displayed concordant results by Western Blotting. Interestingly, nearly all these tumors overexpressed at least two isoforms as detected by Western Blot analysis. In the last case, overexpression was detected with only one antibody. However, this sample contained more than 50% of stromal cells, suggesting that overexpression of MDM2 in tumor cells was masked by the contamination with low-expressing stromal cells. Taken together,

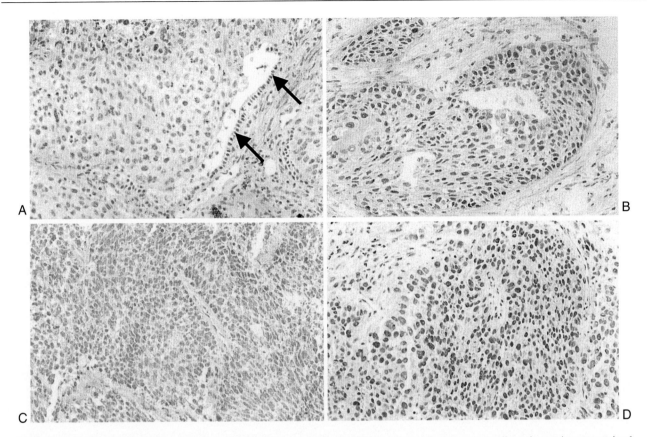

Figure 17 Immunohistochemical analysis of MDM2 in normal lung and lung tumors. **A:** Squamous cell carcinoma immunostained with C-Ter antibody showing a faint MDM2 nuclear expression (score 20) (200X). Note the moderate MDM2 expression in entrapped type II cells (arrows); **B:** Squamous cell carcinoma exhibiting a strong nuclear staining with 2A10 antibody (score 150) (200X); **C:** Small cell lung carcinoma immunostained with N20 antibody (score 50) (200X); **D:** Large cell neuroendocrine carcinoma immunostained with SMP14 antibody (score 180) (200X). This figure is part of Eymin *et al.*, 2002.

these data showed a good concordance (23/28, 82%) between both techniques.

References

Bartel, F., Taubert, H., and Harris L.C. 2002. Alternative and aberrant splicing of MDM2 mRNA in human cancer. *Cancer Cell 2:*9–15.

Cahilly-Snyder, L., Yang-Feng, T., Francke, U. and George, D.L. 1987. Molecular analysis and chromosomal mapping of amplified genes isolated from a transformed mouse 3T3 cell line. *Somat. Cell Mol. Genet. 13:*235–244.

Eymin, B., Gazzeri, S., Brambilla, C., and Brambilla, E. 2002. MDM2 overexpression and p14^{ARF} inactivation are two mutually exclusive events in primary human lung tumors. *Oncogene 21:*2750–2761.

Finlay, C.A. 1993. The MDM2 oncogene can overcome wild-type p53 suppression of transformed cell growth. *Mol. Cell Biol. 13:*301–306.

Martin, K., Trouche, D., Hagemeier, C., Sorensen, T.T., La Thangue, N.B., and Kouzarides, T. 1995. Stimulation of E2F1/DP1 transcriptional activity by MDM2 oncoprotein. *Nature 375:*691–694.

Michael, D., and Oren, M. 2002. The p53-MDM2 module and the ubiquitin system. *Semin. Cancer Biol. 13:*49–58.

Momand, J., Wu, H.H., and Dasgupta, G. 2000. MDM2-master regulator of the p53 tumor suppressor protein. *Gene 242:*15–29.

Osada, H., and Takahashi, T. 2002. Genetic alterations of multiple tumor suppressors and oncogenes in the carcinogenesis and progression of lung cancer. *Oncogene 21:*7421–7434.

Rasidakis, A., Orphanidou, D., Kalomenidis, J., Papamichalis, G., Toumbis, M., Lambaditis, J., Sacharidou, A., Papastamatiou, H., and Jordanoglou, J. (1998). Expression of MDM2 protein in neoplastic, preneoplastic, and normal bronchial mucosa specimens: Comparative study with p53 expression. *Hybridoma 17:*339–345.

Sherr, C. 2001. The INK4a/ARF network in tumor suppression. *Nat. Rev. Mol. Cell. Biol. 2:*731–737.

Zhang, Y., and Xiong Y, 2001. Control of p53 ubiquitination and nuclear export by MDM2 and ARF. *Cell Growth Diff. 12:*175–186.

5

Immunohistochemical Expression of E2F1 and p14^{ARF} in Lung Carcinoma

Sylvie Gazzeri and Elisabeth Brambilla

Introduction

The E2F family of transcription factors comprises six structurally related E2Fs (E2F1-6) that function as heterodimers with members of the DP family (DP-1 and DP-2). Initial studies implicated E2F/DP heterodimers mainly in transcriptional activation of genes required for cell cycle progression. However, it is currently clear that transcriptional activation of cell-cycle–related genes is only one facet of E2F activity. Data accumulated during the last few years demonstrate that E2Fs function in both transactivation and repression of gene expression. Furthermore, it is now clear that E2Fs also have important roles in regulating both cell proliferation and antiproliferative processes such as apoptosis and senescence.

E2F1-Dependent Regulation of Target Gene Expression and S-Phase Entry

By coordinating early cell cycle events with the transcription of genes required for S-phase entry, E2F1 plays a critical role in G$_1$/S transition. By so doing, E2F1 is a critical target of the action of the retinoblastoma protein (Rb) as a growth suppressor. Binding with an hypophosphorylated form of Rb sequesters E2F1 into inactive repressive complexes and inhibits its transcriptional capacity correlating with the ability of Rb to arrest cell growth in G1 (Harbour *et al.*, 2000). The phosphorylation of Rb by cyclinD-cdk4/6 complexes in late G$_1$ allows the release of a free active form of E2F1. Accordingly, impairing the Rb function (Rb loss or phosphorylation) leads to abnormal proliferation related to the release of active E2F1. Beside its capacity to activate gene transcription, E2F can associate with Rb to form an active transcriptional repressor complex at promoters that can block transcription by recruiting histone deacetylase (HDAC) and remodeling chromatin. The relative importance *in vivo* of these two potential mechanisms for inhibiting E2F1 transactivation is still unclear.

E2F1-Induced Apoptosis

It is now clear that beside its property to regulate cell proliferation, E2F1 is implicated in antiproliferative processes such as apoptosis (Ginsberg, 2002). Experiments using tissue culture cells as well as transgenic mice demonstrate that E2F1-induced apoptosis is mediated by both p53-dependent and p53-independent pathways. The existing data strongly suggest that E2F1

Handbook of Immunohistochemistry and *in situ* Hybridization of Human Carcinomas, Volume 1: Molecular Genetics; Lung and Breast Carcinomas

133

Copyright © 2004 by Elsevier (USA)
All rights reserved.

affects p53-dependent apoptosis through a number of pathways. One of these pathways involves activation of ARF, a direct E2F1 target gene (see the following sections), another relies on physical interaction with p53, while yet another is caffeine-sensitive.

Additional pathways also mediate E2F1-induced apoptosis independently of p53. They implicate the regulation of expression of several E2F1-target genes, among which are the p53 homologue p73, the apoptois protease-activating factor 1 Apaf-1, the proapoptotic members of the Bcl-2 family, including Bok, Bad, Bak, and Bidl, and several members of the caspase family (caspases -3 and -7).

E2F Family Members

The major member of the E2F family of transcription factors is E2F1 (DeGregori, 2002). Based on structure, transcriptional properties, and association with pocket proteins, the E2F family can be divided onto three distinct subgroups. The factors E2F1, E2F2, and E2F3 associate exclusively with pRb, are potent transcriptional activators, and can induce apoptosis. In contrast, E2F4, which associates with Rb, p107, and p130, and E2F5, which associates with p130, seem to be primarily involved in the active repression of E2F-responsive genes. Meanwhile, E2F6 does not interact with pocket proteins and functions as a negative regulator of E2F-dependent transcription via complexing with chromatin modifiers.

E2F1 and Carcinogenesis

Ectopic expression of E2F1 induces DNA synthesis in quiescent immortal rodent cells, confers neoplastic properties to immortalized cells, and increases tumorigenesis in transgenic mice that lack p53 function. Moreover, old transgenic mice overexpressing E2F1 develop spontaneous tumors in a variety of tissues. Altogether, these results confer oncogenic properties to E2F1. Paradoxically, mice lacking E2F1 also develop lymphoma or pulmonary adenocarcinoma, suggesting that E2F1 can function as a tumor suppressor, probably via its ability to induce p53-dependent or -independent apoptosis. Thus the ability of E2F1 to promote cell cycle progression and tumor growth or apoptosis and tumor suppression might be tissue-specific.

Connecting E2F1 to Alternative Reading Frame Pathways

The ARF tumor suppressor encodes a nucleolar protein (p19^{ARF} in mice, p14^{ARF} in humans) from an **A**lternative **R**eading **F**rame of the INK4a locus (Quelle et al., 1995). Although encoded at the same locus, p16^{INK4a} and p14^{ARF} proteins have no homology because of a one-base shift in the exon 2 of the reading frame. The ARF protein is part of an ARF/MDM2/p53 signaling pathway that negatively regulates cell growth through the capacity of the p53 transcription factor to induce cell cycle arrest and/or apoptosis (Zhang et al., 2001). Expression of ARF increases p53 activity by interacting with the MDM2 protein. MDM2 is encoded by a p53-responsive gene and antagonizes p53 function by inhibiting its transcriptional activity, catalyzing its ubiquitination and triggering its nuclear export and degradation in cytoplasmic proteasomes. Binding of ARF to MDM2 relocalizes it to the nucleolus and inhibits its E3 ubiquitin ligase activity and nuclear export to induce a p53 transcriptional response.

The ARF tumor suppressor acts as a sensor of hyperproliferative signals emanating from oncoproteins and inducers of S-phase entry such as E2F1. The transcription factor E2F1 is able to transactivate ARF expression, which in turn inhibits E2F1 transcriptional activity, generating a negative feed-back loop onto ARF expression. The protein ARF triggers p53-dependent growth arrest in the G$_1$ and G$_2$ phases of the cell cycle or, in the presence of appropriate collateral signals, sensitizes cells to apoptosis. Because ARF-null mice are highly tumor-prone and die early in life, ARF has been proposed to be most important in tumor surveillance. Interestingly, p14^{ARF} and p16^{INK4a} act both as tumor suppressor genes, one through intersection with the p53 pathway, the other intersecting with the Rb pathway.

Expanding Roles for ARF in p53-Independent Signaling Pathways

In murine systems, ARF can suppress colony formation of p53 $-/-$ cells by affecting the Rb pathway. Furthermore, mice lacking both p53 and ARF develop a wilder spectrum of tumors than animals lacking either gene alone, and reintroducing ARF in mouse embryo fibroblasts (MEF) lacking p53, MDM2, and ARF induces cell growth inhibition.

In human, p53-independent ARF activity has been recently ascribed to its ability to impede S-phase progression and induce p53 and Bax-independent apoptosis. Finally, we recently showed that p14^{ARF} can induce G$_2$ arrest and apoptosis independently of p53, leading to regression of tumors established in nude mice (Eymin et al., 2003).

Immunohistochemical Technique

E2F1

We describe here a technique optimized for use on the Discovery Ventana automated IHC system (Tucson, AZ).

MATERIALS

Preparation of Tissues Sections

1. Dextran solution: Dextran 10% in H_2O.
2. Acetone.
3. Amino-propyltriethoxysilane (APES, Sigma A3648).
4. APES solution: APES 2% in acetone.
5. 150 mM phosphate-buffered saline (PBS) (pH 7.4).
6. Paraformaldehyde (PAF). Working solution at 3.7% prepared from a 20% stock solution and stored at 4°C.

Immunostaining Technique

1. Primary antibody solution: Primary antibody is diluted in PBS (pH 7.4)/BSA 0.03%.
2. E2F1 (KH95) Pharmingen monoclonal antibody recognizes an epitope between amino acids 342 and 386. Used at 1/100 dilution.
3. APK wash 10X buffer (Ventana ref 250-6042).
4. Liquid coverslip (Ventana ref 250-009).
5. IVIEW DAB Detection Kit (Ventana ref 760-91).
6. Amplification system (Ventana ref 750-080).
7. Counterstain1 (Ventana ref 771-741).
8. Merckoglas (Merck Eurolab, Dorset, United Kingdom).
9. Bar code labels (Ventana ref 1 418 700).

Discovery Ventana Automated

Ventana Medical Systems, Inc. 1910 Innovation Park Drive, Tucson, AZ 85737.

METHODS

Preparation of the Automated

1. Switch on the Ventana automated (computer and command module).
2. Open the NexES software.
3. Select new program.
4. Select the procedure "Research IHC IVIEW DAB Nonparaffinized."
5. Select "Antibody."
6. Select "Manual Deposit."
7. Select "Standard Incubation for 60 min."
8. Select "Amplification" then "Monoclonal Antibody."
9. Select "Secondary Antibody," then "Secondary Ventana for 8 min."
10. Select "Counterstaining," then "Counterstain 1 for 2 min."
11. Save the program.
12. Print bar code labels and stick them on the slides.

Preparation of Tissue Sections

1. Incubate slides overnight with dextran solution.
2. Rinse with warm water.
3. Rinse with distilled water.
4. Dry at 50°C in oven for 2 hr.
5. Immerse slides in APES solution for 20 sec.
6. Rinse three times with acetone.
7. Rinse with distilled water.
8. Dry at 50°C in oven for 2 hr.
9. Cut 6-μm-thick frozen samples on APES-coated slides.
10. Air-dry for 5 min.
11. Identify sections with Dako Pen.
12. Apply 3.7% PAF for 10 min to fix the cells.
13. Remove PAF and rinse twice in PBS/BSA 0.03% for 10 min.
14. Remove excess PBS/BSA.

Tissue sections must be immunostained immediately after preparation.

Immunoperoxidase Technique

1. Fill the appropriate reservoirs with the APK wash buffer IX (from a 10X solution diluted with distilled water) and liquid coverslip.
2. Load the IVIEW DAB Detection Kit, amplification, and counterstain1 dispensers onto the reagent carousel and place on the Ventana automated IHC system.
3. Load bar-coded frozen sections onto Ventana automated IHC system.
4. Add 500 μl of APK wash buffer onto each slide.
5. Start the appropriate program.
6. When the first round is finished and the module rings, apply 100 μl of primary antibody solution onto each slide and press onto the module to continue the program.
7. After the program has finished, immerse slides in soapy water for 5 min.
8. Rinse extensively with tap water.
9. Dehydrate in three changes of 100% ethanol (5 min each) and three changes of 100% toluene.
10. Coverslip in Merckoglas.

p14^{ARF}

MATERIALS

Preparation of Tissue Sections

1. Dextran solution: Dextran 10% in H_2O.
2. Acetone.
3. Amino-propyltriethoxysilane (APES, Sigma A3648).
4. APES solution: APES 2% in acetone.
5. 150 mM phosphate-buffered saline (PBS) (pH 7.4).
6. Paraformaldehyde (PAF). Working solution at 3.7% prepared from a 20% stock solution and stored at 4°C.

Immunoperoxidase Technique

Incubation with Primary Antiserum

1. Blocking serum: 2% normal donkey serum in PBS (pH 7.4)/BSA 0.03% for p14^{ARF} antibody and 2% normal rabbit serum in PBS (pH 7.4)/BSA 0.03% for hp19^{ARF} antibody.
2. Primary antibody solution: Primary antibodies are diluted in PBS (pH 7.4)/BSA 0.03%.
3. hp19^{ARF} polyclonal antibody recovered from rabbits inoculated with a recombinant GST-hp19^{ARF} protein (Della Valle *et al.*, 1997). Used at 1/200 dilution.
4. p14^{ARF} (C-18) sc-8613 Santa-Cruz polyclonal antibody raised against a peptide mapping at the carboxyterminus of p14^{ARF} of human origin. Used at 1/750 dilution.

Incubation with Secondary Antibody and ABC Reagent

1. Secondary antibody solution: Secondary antibodies are diluted in PBS (pH 8.6)/bovine serum albumin (BSA) 0.03%.
2. Anti-goat biotinylated rabbit F(ab')2 (1/1250; The Jackson Laboratory; West Grove, PA).
3. Anti-rabbit biotinylated donkey F(ab')2 (1/300; The Jackson Laboratory; West Grove, PA).
4. StreptABComplex/HRP (from Dako ref 0377). Add 10 µl streptavidine and 10 µl peroxidase (biotinylated horseradish peroxidase) to 2000 µl Tris 0.05 M (pH 7.6). Make up fresh each day, at least 30 min before using.
5. DAB working solution: Prepare fresh just before use in dark. Mix 2 ml DAB (3 g DAB in 50 ml Tris-HCl 0.05 M [pH 7.6]) and 2 ml imidazole (4 g imidazole in 50 ml Tris-HCl 0.05 M (PH 7.6) to 200 ml Tris-HCl 0.05 M (pH 7.6) and 35 µl H$_2$O$_2$ (Merck, ref 1.07209.0250).
6. Hematoxylin solution.
7. HCl solution: Prepare working solution by adding 10 drops of HCl (35%) to 200 ml distilled water.
8. NH$_4$OH solution: Prepare working solution by adding 10 drops of NH$_4$OH to 200 ml distilled water.
9. Merckoglas (Merck Eurolab).

METHODS

Preparation of Tissues Sections

1. Incubate slides overnight with dextran solution.
2. Rinse with warm water.
3. Rinse with distilled water.
4. Dry at 50°C in oven for 2 hr.
5. Immerse slides in APES solution for 20 sec.

6. Rinse three times with acetone.
7. Rinse with distilled water.
8. Dry at 50°C in oven for 2 hr.
9. Cut 6-µm-thick frozen samples on APES-coated slides.
10. Air-dry for 5 min.
11. Identify sections with Dako Pen.
12. Apply 3.7% PAF for 10 min to fix the cells.
13. Remove PAF and rinse twice in PBS/BSA 0.03% for 10 min.
14. Remove excess PBS/BSA.

Immunoperoxidase Technique

Incubation with Primary Antiserum

1. To block nonspecific binding, apply 100 µl of blocking solution and incubate for 30 min in humidity chamber at room temperature.
2. Remove the blocking serum and apply 100 µl primary antibody solution.
3. Incubate overnight in humidity chamber at 4°C.

Incubation with Secondary Antibody and ABC Reagent

1. Remove slides from incubation racks and place in PBS (pH 8.6)/BSA 0.03% for 10 min. Repeat three times.
2. Remove excess PBS/BSA solution.
3. Dry the contour of the sections.
4. Apply secondary (biotinylated) antibody (100 µl) and incubate in humidity chamber racks at room temperature for 1 hr.
5. Remove from incubation racks and place in PBS/ BSA for 10 min. Repeat three times.
6. Remove excess PBS/BSA solution.
7. Dry the contour of the sections.
8. Apply ABC reagent (100 µl per slide) and incubate in humidity chamber racks in dark at room temperature for 1 hr.

DAB Incubation

1. Remove slides from incubation racks and place in PBS (pH 8.6)/BSA 0.03% for 5 min. Repeat three times.
2. Rinse slides in Tris 0.05 M (pH 7.6) for 5 min.
3. Prepare fresh DAB working solution.
4. Immerse slides in DAB solution for 5 min. Keep away from light.
5. Rinse slides in distilled water for 5 min. Repeat three times.
6. Check the staining using microscope.

Counterstaining

1. Immerge slides in hematoxylin solution for 20 sec with shaking.
2. Rinse with tap water.
3. Immerse slides in HCl solution for 8 sec.

4. Rinse with tap water.
5. Immerse slides in NH₄OH solution for 8 sec.
6. Rinse with tap water.
7. Check counterstaining using microscope.

Slide Preparation for Storage
1. Dehydrate in three changes of 100% ethanol (5 min each) and three changes of 100% toluene.
2. Coverslip in Merckoglas.

RESULTS AND DISCUSSION

E2F1 in Human Lung Tumors

The factor E2F1 was poorly analyzed in lung tumors, and all studies were carried out on non–small cell lung tumors. Up-regulation of E2F1 mRNA and protein expression has been reported. Amplification and mutation are very rare. In some studies, overexpression of E2F1 was associated with higher growth index and adverse prognosis (De Muth *et al.*, 1998; Gorgoulis *et al.*, 2002), whereas in others no prognostic value was found (Volm *et al.*, 1998).

We investigated E2F1 expression by IHC using an antihuman monoclonal antibody (clone KH95, Pharmingen) on a panel of 47 lung tumor tissue samples comprising neuroendocrine and non-neuroendocrine tumors (Eymin *et al.*, 2001). Normal lung parenchyma adjacent to tumors on sections were considered as internal controls, whereas normal lung tissues taken for diagnosis in patients without a history of cancer were taken as external controls. Scores of immunostaining were recorded based on an assessment of the percentage of positive cells and the intensity of staining. Normal lung cells, lymphocytes, and vessels were all negative. Cases were recovered as positive when at least 20% of tumor cells showed obvious nuclear staining.

When compared with normal lung cells, nuclear overexpression of E2F1 is observed in 9% of adenocarcinoma, 12.5% of squamous carcinoma, 55% of LCNEC, and 100% of SCLC. Overall, up-regulation of E2F1 is detected in 26/30 (86%) high-grade NE tumors, in 50% of LCNEC and 92% of SCLC, and in only 1/17 (6%) NSCLC (Figures 18A and 18B) demonstrating a distinct pattern of E2F1 protein expression in human lung tumors. Moreover, E2F1 overexpression in neuroendocrine lung tumors is significantly associated with a high proliferation index (p < 0.0001).

To validate the IHC data, we performed a Western Blot analysis on the same cases. IHC results were

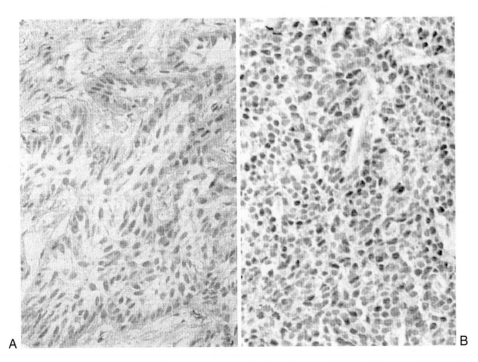

A   B

Figure 18 E2F1 immunostaining. **A:** E2F1 negative immunostaining in a squamous cell carcinoma. One percent of cells are positive. **B:** E2F1 positive immunostaining with 50% of cells showing nuclear immunostaining in a SCLC.

consistent with those of Western Blot analysis in 91% (43/47) of cases. Four cases displayed discordance: in three of them where necrosis was intense, E2F1 staining was positive using IHC analysis, whereas Western analysis was negative. In the last case, a faint band was observed by Western Blot but tumor cells were negative in IHC analysis. Taken together, these data showed a good concordance between both techniques.

ARF in Human Lung Cancer

NSCLC have been primarily studied. Homozygous deletions of both INK4A and ARF genes are detected in about 30% of cases (Park *et al.*, 2003). Mutations are rare, from 5% in human tumors to 12% in cell lines. They were never detected in exon1beta, suggesting that p14[ARF] intragenic mutations are rare or absent in lung cancer. Some mutations in exon2 common to p14[ARF] and p16[INK4A] were predicted to have some effects on both genes, whereas others affected either p16[INK4A] or p14[ARF] sequences. Methylation of the p14[ARF] gene was detected in 8% of cases in a large series of NSCLC, but it was also seen in some corresponding nonmalignant lung tissues from the same patients (Zöchbauer-Müller *et al.*, 2001).

We investigated p14[ARF] expression by IHC using two antibodies p14[ARF] C-18 Santa-Cruz TEBU and hp19[ARF] (Della Valle *et al.*, 1997) in a large panel of human lung tumor samples comprising neuroendocrine and non-neuroendocrine tumors (Gazzeri *et al.*, 1998). In normal lung, which was systematically present in the vicinity of the tumors, as well as in normal lung tissue serving as external control, hp14[ARF] immunostaining was found to be exclusively nuclear. All epithelial cells of bronchi, bronchioli, and alveoli, as well as endothelial cells, fibroblast, and smooth muscle cells appear to be stained. Lymphoid cells identified on the slides are only weakly stained or negative and serve as negative control.

Compared with normal lung structures and stromal cells, a loss of hp14[ARF] nuclear immunostaining is observed in 25 out of 101 nonsmall cell cancer (NSCLC) (Figures 19A and 19B). No significant difference in the distribution was observed between the different histological subclasses of tumors. The overall frequency of loss of hp14[ARF] protein expression was significantly higher in neuroendocrine (NE) lung tumors (55/100, 55%) than in NSCLC (25/101, 25%) ($p < 0.0001$) and specially in high-grade NE tumors (51/78, 65%) ($p < 0.0001$). With respect to each NE tumor subtype, hp14[ARF] loss of immunoreactivity occurred in 1 out of 15 (6%) typical carcinoid, 3 out of 7 (43%) atypical carcinoid, 13 out of 26 (50%) large cell neuroendocrine carcinoma (LCNEC), and 38 out of 52 (73%) small cell lung cancer (SCLC). Therefore, the frequency of loss of the protein expression increased significantly from typical carcinoid to LCNEC ($p < 0.0001$) and to the group of high-grade NE tumors LCNEC and SCLC ($p = 0.0001$).

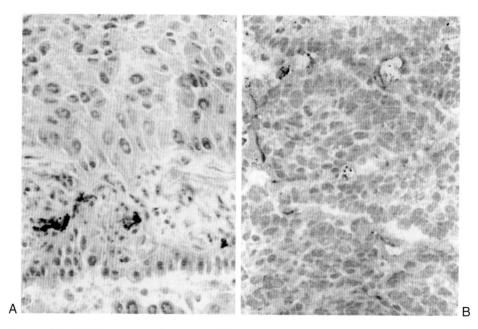

A   B

Figure 19 p14[ARF] immunostaining. **A:** p14[ARF] positive immunostaining in a squamous cell carcinoma (X400) showing a strong nuclear staining. **B:** p14[ARF] negative immunostaining in a large cell neuroendocrine carcinoma with positive staining of some stromal cells.

Western Blot analysis of the hp14^{ARF} product was also performed on 17 tumor samples. The results were consistent with those of immunohistochemistry (IHC), with the hp14^{ARF} protein being detected in all cases displaying a positive IHC and never detected in cases with negative IHC, except in one case (data not shown).

References

DeGregori, J. 2002. The genetics of the E2F family of transcription factors: Shared functions and unique roles. *Biochim. Biophys. Acta 1602:*131–150.

Della Valle, V., Duro, D., Bernard, O., and Larsen, C.J. 1997. The human protein p19^{ARF} is not detected in hemopoietic human cell lines that abundantly express the alternative beta transcript of the p16^{INK4a}/MTS1 gene. *Oncogene 15:*2475–2481.

DeMuth, J.P., Jackson, C.M., Weaver, D.A., Crawford, E.L., Durzinsky, D.S., Durham, S.J., Zaher, A., Phillips, E.R., Khuder, S.A., and Willey, J.C. 1998. The gene expression index c-myc x E2F1/p21 is highly predictive of malignant phenotype in human bronchial epithelial cells. *Am. J. Respir. Cell. Mol. Biol. 19:*18–24.

Eymin, B., Gazzeri, S., Brambilla, C., and Brambilla, E. 2001. Distinct pattern of E2F1 expression in human lung tumors: E2F1 is up-regulated in small cell lung carcinoma. *Oncogene 20:*1678–1687.

Eymin, B., Leduc, C., Coll, J.L., Brambilla, E., and Gazzeri, S. 2003. p14^{ARF} induces G$_2$ arrest and apoptosis independently of p53, leading to regression of tumors established in nude mice. *Oncogene 22:*1822–1835.

Gazzeri, S., Della Valle, V., Chaussade, L., Brambilla, C., Larsen, C.J., and Brambilla, E. 1998. The human p19^{ARF} protein encoded by the beta transcript of the p16^{INK4A} gene is frequently lost in small cell lung cancer. *Cancer Res. 58:* 3926–3931.

Ginsberg, D. 2002. E2F1 pathways to apoptosis. *FEBS Letters 529:*122–125.

Gorgoulis, V.G., Zacharatos, P., Mariatos, G., Kotsinas, A., Bouda, M., Kletsas, D., Asimacopoulos, P.J., Agnantis, N., Kittas, C., and Papavassiliou, A.G. 2002. Transcription factor E2F1 acts as a growth-promoting factor and is associated with adverse prognosis in non-small cell lung carcinomas. *J. Pathol. 198:* 139–141.

Harbour, J.W., and Dean, D.C. 2000. The Rb/E2F pathway: Expanding roles and emerging paradigms. *Genes Dev. 14:* 2393–2409.

Park, M.J., Shimizu, K., Nakano, T., Park, Y.B., Kohno, T., Tani, M., and Yokota, J. 2003. Pathogenetic and biologic significance of TP14^{ARF} alterations in nonsmall cell lung carcinoma. *Cancer Genet. Cytogenet. 141:*5–13.

Quelle, D.E., Zindy, F., Ashmun, R.A., and Sherr, C.J. 1995. Alternative reading frames of the INK4a tumor suppressor gene encode two unrelated proteins capable of inducing cell cycle arrest. *Cell 83:*993–1000.

Volm, M., Koomagi, R., and Rittgen, W. 1998. Clinical implications of cyclins, cyclin-dependent kinases, Rb, and E2F1 in squamous-cell lung carcinoma. *Int. J. Cancer 79:*294–299.

Zhang, Y., and Xiong, Y. 2001. Control of p53 ubiquitination and nuclear export by MDM2 and ARF. *Cell Growth Diff. 12:*175–186.

Zöchbauer-Müller, S., Fong, K.M., Virmani, A.K., Geradts, J., Gazdar, A.F., and Minna, J.D. 2001. Aberrant promoter methylation of multiple genes in nonsmall cell lung cancers. *Cancer Res. 61:*249–255.

6

Role of Immunohistochemical Expression of Beta-Catenin in Lung Carcinoma

Vassilis G. Gorgoulis, Panayotis Zacharatos, and Athanassios Kotsinas

Introduction

Cell adhesion is an important factor in the maintenance of tissue structure in multicellular organisms (Pokutta *et al.*, 2002). Adhesion receptors are not only intercellular transmembrane connectors, but also link extracellular matrix and adjacent cells to the intracellular cytoskeleton, and furthermore serve as signal transducers (Van Aken *et al.*, 2001). In addition, molecular components of these receptors that have not been incorporated in their structure, participate in gene transcriptional activities, whereas their aberrant forms have profound effects on cancer progression and metastasis (Conacci-Sorell *et al.*, 2002; Hajra *et al.*, 2002). The best-known such components are the alpha-, beta-, gamma- (or plakoglobin), and delta- (or p120^{ctn}) members of the catenin family (Figure 20).

The *CTNNB1* gene, encoding beta-catenin protein (β-cat), is located on the chromosomal region 3p21 (Bremnes *et al.*, 2002b). The primary structure of β-cat consists of a 149 amino acid N-terminal region, which contains a binding site for α-catenin, followed by a 515-residue domain that consists of 12 motifs that represent approximately 42 amino acid repeats known

as armadillo (arm) repeats, and a C-terminal 119 residue region (Pokutta *et al.*, 2002). The arm-repeat region binds to cadherins and also T cell factors (Tcfs).

In normal nonproliferative cells, the cellular pool of β-cat is sequestered mainly for incorporation and assembly of adhesion receptors. In developing tissues and/or tissues with continuous renewing capabilities, such as epithelia, binding of this pool is further antagonized by recruitment into intracellular signaling and gene transcriptional activation. In any case, the cytoplasmic surplus of β-cat is actively down-regulated by proteasome degradation (Figure 20). (For comprehensive reviews of the preceding, see Conacci-Sorell *et al.*, 2002; Polakis, 2001; Van Aken *et al.*, 2001).

Involvement of β-cat in cell adhesion requires E-cadherin (E-cad) binding. E-cad–sequestered β-cat is involved first in the assembly of cell adhesion receptors that are responsible for Ca$^+$-dependent homotypic and heterotypic cell interactions (Van Aken *et al.*, 2001), and second in connecting the receptor with the intracellular actin cytoskeleton via α-catenin, as a molecular bridge (Van Aken *et al.*, 2001). The function of these receptors is also strongly influenced by growth receptors, specifically with tyrosine kinase activity that

Handbook of Immunohistochemistry and *in situ* Hybridization of Human
Carcinomas, Volume 1: Molecular Genetics; Lung and Breast Carcinomas

141

Copyright © 2004 by Elsevier (USA)
All rights reserved.

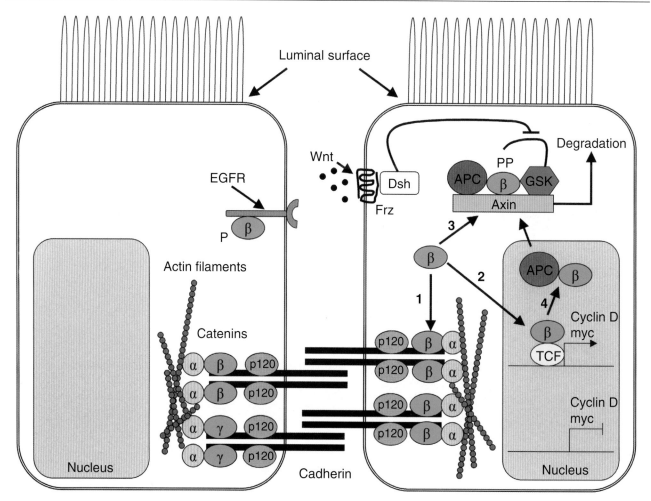

Figure 20 Schematic representation of β-cat (β) cellular functions. The cytoplasmic pool of β-cat is involved in: **1)** Cadherin binding and incorporation in cellular adhesion. β-Cat also links these receptors via α-cat (α) to the intracellular actin cytoskeleton. Growth factor receptors with tyrosine kinase activity (e.g., EGFR) destabilize cell adhesion through β-cat phosphorylation at tyrosine residues (P). **2)** Intracellular signaling and gene transcription activation. Wnt signals stabilize β-cat by blocking its intracellular degradation (see also point 3) and lead to nuclear translocation. In the nucleus, complexes with LEF/TCFs (TCF) activate a broad pattern of genes. **3)** A multi-component system (APC-Axin-GSK) in the absence of proliferating signals (e.g., Wnt) sequesters the surplus of β-cat and marks it for proteasome degradation through phosphorylation (PP). **4)** APC is also able to sequester β-cat in the nucleus and shuttle it in the cytoplasm to the multicomponent system (APC-Axin-GSK) (see point 3). Other abbreviations: gamma-catenin (γ), delta-catenin or p120^{ctn} (p120), frizzled receptors (Frz), disheveled (Dsh).

phosphorylate β-cat at tyrosine residues, thus promoting dissociation of the cell adhesion complex (Van Aken *et al.*, 2001) (Figure 20).

Proliferating signals, such as Wnt, stabilize the cytoplasmic β-cat by blocking its sequestration and lead to nuclear translocation. In the nucleus, dimmers are formed with members of the lymphoid enhancer factor (LEF)/T cell factor (TCF) transcriptional family that activate a broad pattern of genes. These genes are involved in G$_1$ to S-phase cell cycle transition, tissue development, extracellular matrix degradation, and so on (Bremnes *et al.*, 2002b; Conacci-Sorell *et al.*, 2002) (Figure 20).

B-cat that escapes from the preceding cellular functions is actively down-regulated by multicomponent proteasome degradation complexes (reviewed in Polakis, 2001). Two types of such complexes have been described, and both use adenomatous polyposis coli (APC) as a scaffold protein (Polakis, 2001). The first employs a phosphorylation-dependent mechanism that is mediated by glycogen synthase kinase (GSK), whereas the second one depends on p53 response. The former has been suggested to be the main degradation route of β-cat that regulates its excessive cytoplasmic levels, whereas the latter functions in response to cell cycle arrest, protecting cells from aberrant β-cat–dependent transcriptional activation. In both cases, APC binds and sequesters β-cat for destruction and also regulates its nucleocytoplasmic shuttling ability (Henderson *et al.*, 2002) (Figure 20).

Stabilized and/or high levels of β-cat that result from 1) mutations in genes coding for components of the degradation systems; 2) mutations in β-cat; or 3) deregulated Wnt signaling accumulate in the cytoplasm, leading to an increased β-cat nuclear translocation. As a consequence, β-cat exerts oncogenic activity through aberrant target gene expression (Conacci-Sorel *et al.*, 2002).

Results from cellular systems, animal models, and certain types of tumors have shown that aberrant β-cat expression leads to increased cellular proliferation (Orford *et al.*, 1999; Polakis, 2000; Wong *et al.*, 1998). In cell lines it has been shown that this is due to enhanced G_1 to S transition (Orford *et al.*, 1999). In normal cells this β-cat–augmented cellular proliferation impinges on several cell cycle regulators. First, E-cad modulates its transcriptional activity (Stockinger *et al.*, 2001) and regulates cell growth via p27[KIP1] (St Croix *et al.*, 1998), which is inversely expressed to β-cat (Orford *et al.*, 1999). Second, p53 not only inhibits β-cat–dependent cellular growth through p14[ARF] (reviewed in Oren *et al.*, 2002), but also promotes its degradation in response to its oncogenic cell cycle insults (Oren *et al.*, 2002; Polakis, 2001). Third, APC, apart from the previously mentioned down-regulating effects on β-cat, controls entry into S-phase, through inhibiting β-cat–dependent transcriptional activation of downstream targets, such as cyclin D and c-Myc, that in, turn promote E2F1 release from pRB (Yang, 2002).

Many of the β-cat downstream transactivation targets function as apoptosis inhibitors (e.g., cyclin D). Interestingly, recent data have shown that ectopic overexpression of β-cat has additional cellular effects (Kim *et al.*, 2000; Wong *et al.*, 1998). First, it induces apoptosis, but this effect is independent of its transactivation function or the involvement of major G_1-phase cell cycle regulators (Kim *et al.*, 2000). Second, it attenuates cell cycle blocks in response to genotoxic stress that might lead to accumulation of DNA damage (Orford *et al.*, 1999).

Aberrant expression of β-cat has been reported in a variety of cancers (reviewed in Polakis, 2000). However, given its dual cellular function, different immunohistochemical expression patterns have been reported not only among distinct types of cancer, but also between tumors of the same origin (Bremnes *et al.*, 2002b; Polakis, 2000). Membranous localization is frequently interpreted as an association with E-cad–dependent cell adhesion. In such cases loss or reduction in staining correlates with tumor dedifferentiation, infiltrative growth, metastasis, and poor prognosis (Van Aken *et al.*, 2001). On the other hand, nuclear and/or cytoplasmic presence has been correlated with mutations either in the β-cat N-terminus or in the mutation cluster region (MCR) mainly of APC, or in other components of the degradation complexes, respectively (Hajra *et al.*, 2002; Polakis, 2000). In some types of tumors the later pattern is related to increased proliferation, whereas in others the opposite was observed (Jung *et al.*, 2001; Kotsinas *et al.*, 2002). There are a few reports that examine the association of β-cat status with proliferation and at the same time in relation to cell cycle regulatory molecules. In some reports the membranous and the nuclear localization are inversely related (Kotsinas *et al.*, 2002; Pirinen *et al.*, 2001). It has been proposed that competition between different cellular partners for the cytoplasmic pool of β-cat influences its final incorporation either in cell adhesion or in signal transduction mechanisms (Conacci-Sorell *et al.*, 2002).

In lung cancer there is an increasing number of reports on the status of β-cat (Bremnes *et al.*, 2002b). Many of these studies describe β-cat association with E-cad and their mutual impact on clinicopathological parameters (Bremnes *et al.*, 2002b). Few reports provide data regarding β-cat correlation with tumor growth parameters (proliferation index [PI] and apoptosis index [AI]), but with conflicting results on proliferation (Hommura *et al.*, 2002; Pirinen *et al.*, 2001), whereas only one examines its relationship with apoptosis (Kotsinas *et al.*, 2002). Another important aspect concerns its relationship with key cell cycle regulators (Bremnes *et al.*, 2002a; Kotsinas *et al.*, 2002) and their concomitant influence on tumor growth parameters (PI versus AI) in lung cancer (Kotsinas *et al.*, 2002).

The aim of this chapter is to provide insights into β-cat expression in lung tumors and on the methodologies employed to evaluate these results. Specific emphasis is placed on its relationship with expression status of key cell cycle regulators, such as p53, MDM2, APC, p27[KIP1], pRB, and E2F1, and their mutual influence on tumor growth parameters (proliferation versus apoptosis) that could lead to a better understanding and interpretation of the results. Finally, correlations with clinicopathological parameters are discussed.

Materials and Methods

MATERIALS

Immunohistochemistry

1. Tris-buffered saline (TBS) 0.05 M Tris/HCl, 0.15 M NaCl: Dissolve 0.61 g Tris-base, 8.2 g NaCl per 1000 ml double distilled water (ddH$_2$O) and adjust pH to 7.6 with NaOH.

2. Sodium citrate buffer: Dissolve 2.1 g citric acid per 1000 ml ddH$_2$O and adjust pH to 6.0 with NaOH.

3. 30% H$_2$O$_2$.

4. Xylene.

5. Ethanol 100%.

6. 3,3-diaminobenzidine tetrahydrochloride (DAB): Dissolve 6 mg of DAB in 10 ml of 0.05 M TBS (pH 7.6). Add 0.1 ml of 3% H$_2$O$_2$.

7. StreptABC complex/HRP (Dako, Glostrup, Denmark; Santa Cruz, Santa Cruz, CA, USA or other).

8. Microscopy slides and coverslips.

9. DPX mounting.

10. Harris' hematoxylin solution.

11. Bovine serum albumin (BSA) stock solution 10% w/v: Dissolve 1 g BSA fraction V in 10 ml distilled water.

12. Triton X-100.

13. Rabbit and/or swine normal serum.

14. Primary antibodies (mouse and/or rabbit) (see Method).

15. Secondary antibodies (rabbit and/or swine) (Dako, Santa Cruz).

Tdt-mediated dUTP Nick End Labeling Assay (TUNEL)

1. Tris-buffered saline (TBS) 0.05 M Tris/HCl, 0.15 M NaCl (pH 7.6): Given previously.

2. 30% H$_2$O$_2$.

3. Xylene.

4. Ethanol 100%.

5. DAB: given previously.

6. StreptABC complex/HRP (Dako, Santa Cruz).

7. Microscopy slides and coverslips.

8. DPX mounting.

9. Harris' hematoxylin solution.

10. Terminal dideoxynucleotidil-transferase (Tdt) (New England Biolabs, Beverly, MA, USA, or other).

11. 1XTdt buffer: 25 mM Tris/HCl (pH 7.2), 200 mM potassium cacodylate, 0.25 mM CoCl$_2$, 250 mg/ml BSA (usually provided as a 10X concentrated solution with the enzyme from the supplier).

12. DNase (Promega, Madison, WI, USA; Invitrogen, Paisley, UK).

13. Proteinase K stock solution 20 mg/ml.

14. 0.4 mM biotin-dATP (Invitrogen).

15. Ethylenediamine tetraacetate (EDTA) 20 mM: Dissolve 7.45 g EDTA per 1000 ml double-distilled water and adjust pH to 8.0 with NaOH.

Western Blot Analysis

1. Tris-buffered saline (TBS) 0.05 M Tris/HCl, 0.15 M NaCl: Dissolve 0.61 g Tris-base, 8.2 g NaCl per 1000 ml ddH$_2$O and adjust pH to 7.6 with NaOH.

2. TBS-T: TBS with the addition of 0.05% Tween-20 (0.5-ml Tween-20 to 1000-ml TBS).

3. Nonfat dry milk.

4. Methanol.

5. Protein lysis buffer (RIPA buffer): 0.05 M Tris-base pH 8.0, 0.15 M NaCl, 0.5% NaDOC (sodium deoxycholate), 1% NP40 or Triton X-100, and 0.1% sodium dodecyl sulfate (SDS). Dissolve 0.061 g Tris-base, 0.82 g NaCl per 80 ml ddH$_2$O and adjust pH to 7.6 with NaOH. Add 1 ml 1% NP40 or Triton X-100, 0.5 g NaDOC, 0.1 g SDS and adjust with sterile ddH$_2$O to 100-ml final volume.

6. Nuclear and cytoplasmic extraction reagents (Pierce, Rockford, IL, USA).

7. Protein assay reagent (Biorad München, Germany).

8. Protease inhibitors cocktail [benzamidine, aprotinin, leupeptin, Phenly-Methly-Sulfonyl-Flouride (PMSF)] (Sigma Athens, Greece, or Pierce). Dilute in sterile distilled water to a 1 mg/ml final stock concentration. Store frozen in small aliquots and use at 50-fold final dilution. Phosphatase inhibitor cocktails (Sigma): i) imidazole, sodium molybdate, sodium orthovanadate, sodium tartrate for inhibition of tyrosine protein phosphatase; and ii) (−)-p-bromotetramisole, canthardin, microcystin LR. Store frozen in small aliquots and use at 100-fold final dilution.

9. Running buffer: 24.8 mM Tris-base, 192 mM glycine (pH 8.3), 0.1% SDS. Dissolve 3.03 g Tris-base, 18.8 g glycine and 1 g SDS in up to 1000 ml distilled water.

10. Transfer (Towbin) buffer: 24.8 mM Tris-base, 192 mM glycine (pH 8.3), 10% methanol. Dissolve 3.03 g Tris-base, 18.8 g glycine in up to 900 ml distilled water and add 100 ml methanol.

11. Protein loading buffer: Add 1.0 ml glycerol, 0.5 ml β-mercaptoethanol, 3.0 ml of 10% SDS (w/v), 1.25 ml Tris-HCl (pH 6.8), and 2.0 ml of 0.1% bromophenol blue (w/v) in distilled water to a final volume of 10 ml. Store frozen in small aliquots.

12. Protein Molecular Weight Marker (Pierce, New England Biolabs).

13. Nitrocellulose or polyvinylidene difluoride (PVDF) transfer membrane.

14. Ponceau-S: 0.1% Ponceau-S (w/v) in 5% acetic acid (v/v).

15. Chemiluminescent reagents (Pierce; Amersham, Athens, Greece).

16. Precast acrylamide gels (optional), or in-house prepared according to the electrophoresis system manufacturer's instructions.

17. Pestles for tissue homogenization.

18. Autoradiography films.

19. Photographic developer and fixer.

20. Whatman No. 3 blotting paper.
21. Primary antibody (see Methods).
22. Secondary antibody (see Methods).

METHODS

Immunohistochemistry

Antibodies

For immunohistochemical analysis, the following antibodies (Abs) were used:

1. Anti-beta-catenin (E-5) (Class: IgG1 mouse monoclonal; epitope: amino acids 680-781, carboxy terminus of beta-catenin, human origin) (Santa Cruz).

2. Anti-APC (C-20) (Class: IgG rabbit polyclonal: epitope: carboxy terminus of APC of human origin) (Santa Cruz).

3. Anti-E2F1 (KH95) (Class: IgG2a mouse monoclonal; epitope: Rb-binding domain of E2F-1 p60, human origin) (Santa Cruz).

4. Anti-p53 (DO7) (IgG2b) (mouse monoclonal; residues 1 to 45 of human p53) (Dako).

5. Anti-MDM2 (SMP14) (Class: IgG1 mouse monoclonal; epitope: residues 154-167 of MDM2, human origin) (Santa Cruz).

6. Anti-Ki-67 (MIB-1) (Class: IgG1, mouse monoclonal, epitope: Ki-67 nuclear antigen) (Dako).

Remark: The most often-used anti-beta-catenin antibody for its expression analysis is clone (14) from Transduction Laboratories (San Diego, CA, USA). However, in our hands, the Santa Cruz antibody performed very well.

Method

1. Paraffin sections, 5-μm-thick, mounted on poly-l-lysine slides are placed in an oven for 2 hr at 60°C.

2. Deparaffinize in xylene (2X 5 min) and rehydrate in graded ethanols.

2X 5 min 100% Ethanol
2X 5 min 96% Ethanol
2X 5 min 80% Ethanol
2X 5 min 70% Ethanol
2X 5 min 50% Ethanol
1X 5 min TBS

3. Immerse the slides in a 3% H_2O_2 (30% solution diluted in distilled water), for 15 min at room temperature in the dark to block the endogenous peroxidase activity.

4. Subsequently immerse the slides in 0.01-M sodium citrate buffer (pH 6.0) and process in a microwave oven with the power set at 700W for two cycles of 5 min each. The fluid level should be maintained during

the procedure by adding antigen retrieval solution. Alternatively, heat the slides for 1 hr immersed in 0.01 M sodium citrate buffer (pH 6.0) in a steamer (650W).

5. Allow sections to cool at room temperature for approximately 30 min and rinse in 1X TBS (pH 7.6).

6. Overlay the sections with 20% normal rabbit serum (RDS), if mouse monoclonal is used as 1[ry] antibody, or 20% normal swine serum (NSS), if rabbit polyclonal is used as 1[ry] antibody, and incubate for 20 min at room temperature.

7. The primary layer consists of the corresponding antiserum at a dilution of 1:100 in TBS containing 0.3% Triton X-100 (BDH, Dorset, United Kingdom), 0.1% bovine serum, albumin (BSA; Sigma) and 2.5% RDS or NSS, respectively, overnight at 4°C in a humidified chamber.

8. After washing in TBS (3X 10 min), the secondary antibody layer, consisting of either biotinylated rabbit anti-mouse or swine anti-rabbit, respectively, is applied for 2 hr at room temperature at a dilution of 1:250 in TBS containing 0.3% Triton X-100 and 0.1% BSA.

9. Sections are subsequently washed in TBS 3X 10 min and incubated in ABC complex for 2 hr at room temperature at a dilution of 1:100 in TBS containing 0.3% Triton X-100 and 0.1% BSA.

10. Visualization of the reaction product is carried out by incubation with diaminobenzidine (DAB) at a dilution of 0.03% in TBS containing 0.1% H_2O_2 (from a 30% stock solution) for 7 min at room temperature in the dark.

11. The sections are washed in TBS for 5 min, counterstained with hematoxylin for 1 min, reverse-serially dehydrated, and mounted under a coverslip with 100 μl of DPX for observation at the light microscope.

Evaluation

1. *Beta-catenin status.* The staining patterns were discerned based on the following criteria (Kotsinas *et al.*, 2002): 1) membranous pattern (M), if immunoreactivity was present solely at the cell membranes as a linear staining; 2) membranous-cytoplasmic (MC), if immunoreactivity was also present in the cytoplasm; 3) cytoplasmic (C), if immunoreactivity was restricted only to the cytoplasm; 4) cytoplasmic-nuclear (CN), if immunoreactivity was present in the cytoplasm and concomitantly in more than 20% of the nuclei; and 5) nuclear (N), if immunoreactivity was confined mainly in the nuclei. Cases with negative staining in tumor areas, but without positive signal in the surrounding stromal as well as normal areas, were considered as noninformatives and were not included in the analysis.

2. *APC status.* For APC three patterns were discerned in a similar manner as for β-cat (Kotsinas *et al.*, 2002): 1) cytoplasmic (C); 2) cytoplasmic-nuclear (CN); and 3) nuclear (N).

3. E2F1, p53, MDM2, and Ki-67 staining was evaluated as previously described (Gorgoulis *et al.*, 2000, 2002). Independent observers carried out slide examination. Inter-observer variability was minimal (P < 0.01).

Controls

Sections from human hepatocellular carcinomas are used as beta-catenin positive controls, as suggested by the manufacturer. The specificity of anti-APC antibody was tested by incubating the latter with the appropriate control peptide (Santa Cruz), against which it was raised. Elimination of immunostaining verified APC positivity. In addition, Jurkat cell line was used as a positive control. For E2F1 expression, the MCF-7 cell line (derived from breast cancer) was employed as a positive control (Gorgoulis *et al.*, 2002). Positive controls for the remaining antibodies have been previously described (Gorgoulis *et al.*, 2000). In each set of immunoreactions, antibody of the corresponding IgG fraction, but of unrelated specificity, was used as a negative control.

Tdt-mediated dUTP Nick End Labeling Assay (TUNEL)

Method

Double-strand DNA breaks were detected by TUNEL, as previously described (Gorgoulis *et al.*, 2000).

1. Paraffin sections, 5-μm-thick, mounted on poly-l-lysine slides are placed in an oven for 2 hr at 60°C.

2. Deparaffinize in xylene (2X 5 min) and rehydrate in graded ethanols.

$$2X \ 5 \ min \ 100\% \ Ethanol$$
$$2X \ 5 \ min \ 96\% \ Ethanol$$
$$2X \ 5 \ min \ 80\% \ Ethanol$$
$$2X \ 5 \ min \ 70\% \ Ethanol$$
$$2X \ 5 \ min \ 50\% \ Ethanol$$
$$1X \ 5 \ min \ TBS$$

3. Incubate slides for 15 min at 37°C with proteinase K solution diluted at 20 μg/ml with TBS. Wash for 2X 5 min in TBS.

4. Immerse the slides in 3% H_2O_2 (30% solution diluted in distilled water) for 15 min at room temperature in the dark to inactivate the endogenous peroxidase activity. Wash for 2X 5 min in TBS.

5. Preincubate sections for 5 min at 37°C in 1X Tdt buffer.

6. Incubate slides for 60 min at 37°C. Overlay slides with: 15 units Tdt, 24 μM biotin-ATP and 1X Tdt buffer/slide, and incubate for 60 min at 37°C.

7. Stop labeling reaction by immersing the sections 2X 5 min in 20 mM ethylenediamine tetra-acetic acid (EDTA) and wash for 2X 5 min in TBS.

8. Sections are subsequently incubated in ABC complex for 2 hr at room temperature at a dilution of 1:100 in TBS.

9. Visualization of the reaction product is carried out by incubation with DAB at a dilution of 0.03% in TBS containing 0.1% H_2O_2 (from a 30% stock solution) for 7 min at room temperature in the dark.

10. The sections are washed in TBS for 5 min, counterstained with hematoxylin for 1 min, reverse-serially dehydrated and mounted under a coverslip with 100 μl of DPX for observation with the light microscope.

Controls

We used as positive controls tissue sections incubated with DNAse I prior to treatment with Tdt and as negative ones sections incubated in Tdt buffer without the presence of the enzyme.

Evaluation

Cells are considered to undergo apoptosis when nuclear staining, without cytoplasmic background, is observed. The AI is estimated as the percentage of apoptotic cells in 10 high-power fields (HPFs) (counted cells: 900–10,000). In our analysis, independent observers performed slide examination. Intraobserver variability was minimal (P < 0.001).

Western Blot Analysis

Total Protein Extraction

1. Homogenize frozen tissue samples in 2.5 volumes of protein lysis buffer.

2. Centrifuge the homogenate at 1000X g at 4°C for 5 min.

3. Collect supernatants and determine their protein concentration spectrophotometrically.

4. Adjust the supernatants to 1 μg/ml aprotinin, 1 μg/ml leupeptin, and 1 μg/ml pepstatin A.

Nuclear and Membranous/Cytoplasmic Extracts

Note: Adapted from Pierce: NE-PER Nuclear and cytoplasmic extraction reagents.

1. Frozen tumor tissues (50 mg) are homogenized directly in CER I buffer (500 μl) and centrifuged at 500X g for 3 min. Supernatant is carefully discarded.

2. Add 200 μl of ice-cold CER I buffer (Pierce).

3. Vortex vigorously at high speed for 15 sec and incubate on ice for 10 min.

4. Add 27.5 μl CER II buffer (Pierce).

5. Vortex vigorously at high speed for 5 sec and incubate on ice for 1 min.

6. Vortex vigorously at high speed for 5 sec. Centrifuge for 5 min at maximum speed (16,000X g).

7. Transfer immediately supernatant fractions, representing membranous-cytoplasmic extract, to clean prechilled tubes and store on ice.

8. Resuspend the pellet fractions, corresponding to nuclei, in 100 μl of ice-cold NER buffer (Pierce), respectively.

9. Vortex vigorously at high speed for 15 sec. Return samples on ice and continue vortexing for 15 sec every 10 min for a total of 40 min.

10. Centrifuge for 10 min at maximum speed (16,000X g).

11. Immediately transfer supernatants (nuclear extracts) to clean, prechilled tubes and store on ice.

12. Determine protein concentration and either store all extracts at −80°C or load immediately on a gel.

Antibodies

1. Anti β-cat (E-5) mouse monoclonal antibody (Class: IgG1; epitope: amino acids 680-781, carboxy terminus of beta-catenin, human origin, [Santa Cruz]) was used as first antibody.

2. Anti APC (F3) mouse monoclonal antibody (Class: TgC; epitope: amino acids 2–289, human origin, Santa Cruz).

3. A goat-anti-mouse IgG (H+L) horseradish peroxidase–labeled one was employed as a secondary antibody (31430, Pierce).

4. Anti-actin C-2 mouse monoclonal antibody (Santa Cruz) was used to assess equal loading of total protein per sample.

Controls

Human tumor cell lines Hella and MCF7 are used as positive controls.

Gel Electrophoresis and Blotting

1. 50 μg of protein from total extracts or 70 μg of protein from subcellular extracts from each sample are adjusted with an equal volume of protein loading buffer and boiled for 5 min.

2. Samples are loaded on 4–20% gradient PAGEr Gold precast gels (Biowhittaker) or a custom 10% gel.

3. Gel electrophoresis is performed at 120 V with constant cooling, until the front dye reaches the end of the gel.

4. Gel, membrane, and six pieces of Whatman paper, cut to the size of the gel, are pre-equilibrated in transfer buffer and mounted for blotting. Protein transfer to nitrocellulose membranes is performed for 2 hr at 200 mA.

5. Immerse membrane in Ponceau-S to assess transfer efficiency and wash with several rinsing of distilled water.

Signal Development and Quantitation

1. Blots are blocked for 1 hr in 5% nonfat dry milk/TBS-T (TBS-T:TBS, 0.1% Tween-20) at room temperature.

2. Subsequently, membranes are incubated overnight with primary antibody (1:200 dilution) at 4°C.

3. Wash 4X 10 min in TBS-T, followed by 1 hr incubation with secondary peroxidase labeled antibody (1:50,000 dilution) in TBS-T at room temperature.

4. Wash 5X 10 min in TBS-T and develop signal with the chemiluminescent substrate, following the manufacturer's instructions.

5. Develop autoradiographs.

RESULTS AND DISCUSSION

The protein β-cat is one of the components of the E-cad associated adhesion complex that also plays a significant role in cell signal transduction, gene activation, apoptosis inhibition, increased cellular proliferation, and migration. These properties, as well as its frequent aberration in many types of tumors, including those of the lung, have made β-cat a target for intensive investigations. Immunohistochemical (IHC) analysis is one of the most widely used methods for assessing its expressional status. An increasing body of reports deal with its expression in non–small cell lung carcinomas (NSCLC) (Bremnes et al., 2002b), whereas very few concern neuroendocrine tumors (Clavel et al., 2001) or small cell lung carcinomas (SCLC) (Rodriguez-Salas et al., 2001).

β-cat Immunohistochemical Expression Evaluation

All reports that describe the staining reaction of β-cat in normal lung tissue support a predominant linear membranous localization in the pneumocytes of the alveoli, at the lateral cellular sides in the bronchial epithelium and in the seromucous glands of the submucosa that is occasionally accompanied by a faint and diffuse cytoplasmic presence (Bremnes et al., 2002a; Eberhart et al., 2001). A similar signal is found in the endothelial cells of the vessels in lung parenchyma (Kotsinas et al., 2002). Sporadically, type II pneumocytes show a diffuse cytoplasmic and nuclear staining (Bremnes et al., 2002a; Nakatani et al., 2002). These elements are usually used as positive internal controls for the specificity of β-cat staining.

Several scoring methods and classifications of β-cat IHC expression results are being used for the in situ interpretation in lung tumors (Figure 21). Scoring methods make use of the proportion of positive cells with a specific β-cat subcellular location (Clavel et al.,

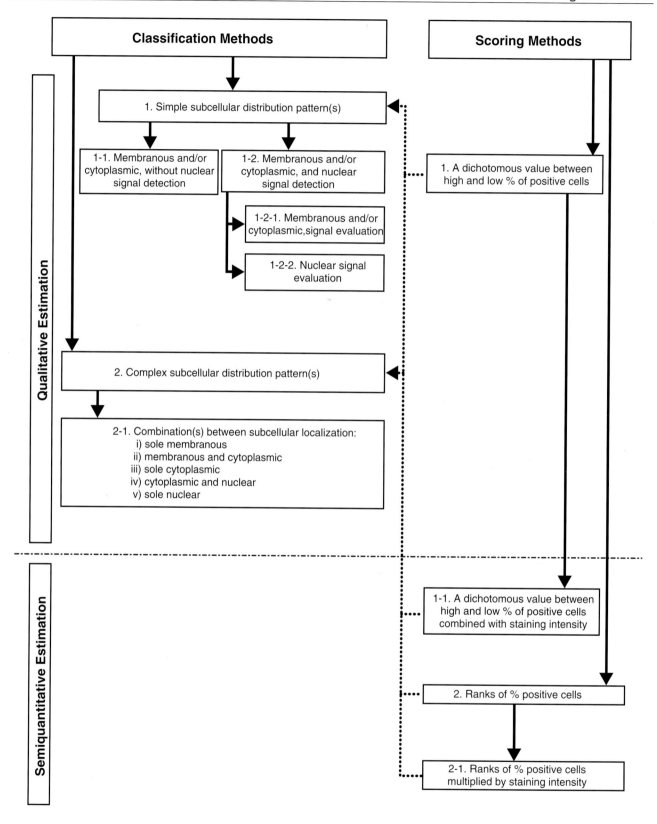

2001; Kase *et al.*, 2000; Pirinen *et al.*, 2001). They include either: 1) a dichotomous value between high and low percent of β-cat expression; or 2) a grading of low, moderate, and high number of immunoreactive cells, with arbitrary cutoff values (Figure 21). Often a combination of scoring methodologies with staining intensity is employed, aiming at a semiquantitative analysis of β-cat expression (Hommura *et al.*, 2002; Retera *et al.*, 1998). Classification methods lead to a description of expression patterns and rely on evaluation of its subcellular localization and/or percent of cells stained in particular subcellular compartments (Hommura *et al.*, 2002; Kotsinas *et al.*, 2002; Pirinen *et al.*, 2001; Retera *et al.*, 1998). Nevertheless, not all studies provide a uniform definition of expression profiles. The most common classification includes simple membranous, cytoplasmic, and nuclear positivity of β-cat presence (Pirinen *et al.*, 2001; Retera *et al.*, 1998). In more recent reports, the specific subcellular distribution comprises more complex patterns resulting from the combined IHC staining of β-cat in more than one intracellular compartment in each sample (Kotsinas *et al.*, 2002; Nakatani *et al.*, 2002). Scoring methods are often applied to a specific pattern emerging from a classification one (Figure 21).

Finally, less commonly reported is the IHC signal from elements in the tumor surrounding stroma. In one study, fibroblasts in this area display a weak cytoplasmic stain (Kotsinas *et al.*, 2002), whereas in another they exhibit a nuclear/cytoplasmic signal (Nakatani *et al.*, 2002).

Use of β-Cat Immunohistochemical Scoring Methods and Clinicopathological Evaluation

Several studies employing scoring methods primarily evaluate the percent of positive cells displaying a membranous signal or a combined membranous and cytoplasmic pattern (Bremnes *et al.*, 2002a; Pirinen *et al.*, 2001) (Figure 21). Either an arbitrary dichotomous value separates low from high (or preserved) expression or arbitrary ranking values are used to describe a gradual increase in the percent of immunoreactive cells (Figure 21). A similar approach is used for nuclear staining (Bremnes *et al.*, 2002a) and is considered only if more than 20% of cells are positive (Pirinen *et al.*, 2001). Sometimes this pattern is either not mentioned (Retera *et al.*, 1998) or not found (Kase *et al.*, 2000). Usually, membranous and/or membranous with cytoplasmic positivity represents normal staining, whereas nuclear presence alone or nuclear combined with cytoplasmic reactivity reflects abnormal staining, if the latter is found. However, in other reports a positive signal is considered if cytoplasmic and/or nuclear staining is present, whereas membranous localization is judged as negative (Hommura *et al.*, 2002). Nevertheless, for semiquantitative β-cat expression estimation, the proportion of positive cells multiplied by their staining intensity is predominantly used (Bremnes *et al.*, 2002a; Retera *et al.*, 1998; Winn *et al.*, 2002). In certain reports IHC intensity-scoring methods are

Figure 21 Summary of the β-cat IHC evaluation methods employed in the literature.

Scoring methods are used to estimate the proportion of positive cells. They include: **1)** A simple dichotomous value separating cases with positive (or preserved) staining and low (or abnormal) staining. The cutoff value is arbitrarily set: i) >10% (Ramasami *et al.*, 2000), ii) ≥ 50% (Lee *et al.*, 2000), iii) ≥ 70% (Kase *et al.*, 2000; Sugio *et al.*, 2002), iv) ≥ 80% (Kimura *et al.*, 2000), v) ≥ 87% (Shibanuma *et al.*, 1998). **1-1)** The staining intensity in each group can be further evaluated as weak, medium, or strong (Lee *et al.*, 2002) or combined with intracellular distribution (apical, basal, focal, circumferential/pericellular) (Ramasami *et al.*, 2000). **2)** Ranks of percent-positive cells separated by arbitrary values, representing groups of cases with absent to low, intermediate, and up to high (or preserved) levels of expression: i) < 20%, 20–50%, > 50% (Clavel *et al.*, 2001); ii) < 10%, 10–90%, ≥ 90% (Pirinen *et al.*, 2001); iii) < 25%, 25–75%, >75% (Hommura *et al.*, 2002; Retera *et al.*, 1998). **2-1)** These levels are further multiplied by the staining intensity (trace, weak, moderate, or intense) displayed by each case and produce a final semiquantitative score (Bremnes *et al.*, 2002a; Hommura *et al.*, 2002; Retera *et al.*, 1998; Winn *et al.*, 2002).

Classification methods are used to describe the topological distribution of the staining signal. They include: **1)** Assignment of the staining signal as a membranous, cytoplasmic, or nuclear one. Reports up to now have shown: **1-1)** Cases with either sole membranous or cytoplasmic distribution, or sometimes combined membranous with cytoplasmic one, but without the presence of cases with nuclear staining (Clavel *et al.*, 2001; Kase *et al.*, 2000; Nawrocki *et al.*, 1998; Sugio *et al.*, 2002). This pattern is usually further divided into "normal" or "abnormal" one by employing scoring methods. **1-2)** Cases that exhibit one of the sole membranous, cytoplasmic, or nuclear signal, respectively, or a combination of membranous with cytoplasmic staining. This pattern is usually further classified if: **1-2-1)** Membranous and/or cytoplasmic signal is evaluated as normal (or preserved) and nuclear localization is considered as abnormal (Bremnes *et al.*, 2002a; Pirinen *et al.*, 2001); and **1-2-2)** Cytoplasmic and/or nuclear signal is evaluated as positive and sole membranous distribution is considered as negative one (Hommura *et al.*, 2002; Shibanuma *et al.*, 1998). Both these classifications are often combined with scoring methods. **2)** Complex β-cat staining patterns that include sole membranous, cytoplasmic and nuclear signal, and subcellular combinations representing membranous with cytoplasmic staining and cytoplasmic with nuclear one. These patterns can be further combined into groups with biological significance that represent: i) the membranous associated patterns (sole membranous, sole cytoplasmic, and membranous with cytoplasmic); and ii) the nuclear associated ones (sole cytoplasmic, sole nuclear, and cytoplasmic with nuclear) (Kotsinas *et al.*, 2002).

also confirmed by β-cat Western Blot analysis (Kase *et al.*, 2000).

Most of the studies employing scoring methods, and more specifically those that evaluate semiquatitatively the expression of β-cat, have shown that its absence or reduction in lung tumor areas, compared with adjacent normal areas, highly associates with dedifferentiated tumors, aggressive and infiltrating carcinomas, promotion of metastasis, and dissemination of malignant cells (extensively reviewed in Bremnes *et al.*, 2002b). Furthermore, it correlates with reduced survival, and in univariate and multivariate analyses represents a significant independent unfavorable prognostic factor (Bremnes *et al.*, 2002b). It should be mentioned that in most of these studies β-cat is considered part of the adhesion complex and therefore is examined in parallel mainly with E-cad and/or other catenins (alpha- and gamma-) (Bremnes *et al.*, 2002a, b), and only in certain cases on its own (Hommura *et al.*, 2002; Pirinen *et al.*, 2001; Retera *et al.*, 1998).

Relationship of β-cat Immunohistochemical Scoring Methods with Tumor Kinetics and G_1 to S-Phase Cell Cycle Regulators

There are a few studies that address the expression status of β-cat with scoring methods and relate it with tumor proliferative activity. Results are controversial, since Pirinen and co-workers (Pirinen *et al.*, 2001) found that tumors with reduced β-cat staining had higher PI values (P = 0.0001), in contrast to that of Hommura *et al.* (2002) who showed that tumors with high or moderate scores of β-cat expression exhibit higher Ki-67 PI (P = 0.005). A possible explanation for this discrepancy is the different evaluation system for β-cat positivity employed in each report (see paragraph 5) (Figure 21). Taking into account that cyclin D is a β-cat transcriptional target, Hommura *et al.* (2002) also explored for a possible mutual relationship, but they failed to find one. Again the previously mentioned explanation could be the reason for this negative result.

Evaluation of β-cat Immunohistochemical Patterns

Several studies of β-cat in NSCLCs employ the classification of its subcellular distribution into patterns of expression, either alone or in combination with scoring methods (Kotsinas *et al.*, 2002; Winn *et al.*, 2002) (Figure 21). Those that use only IHC patterns have reported a strong and homogenous signal that does not reflect a grading in staining intensity. This in certain cases was confirmed by Western Blot analysis and did not permit further semiquantitative estimation of β-cat

levels of expression (Kotsinas *et al.*, 2002; Winn *et al.*, 2002; Toyoyama *et al.*, 1999). In two reports investigating neuroendocrine tumors and SCLCs, respectively, both methods of classification and scoring are being used (Clavel *et al.*, 2001; Rodriguez-Salas *et al.*, 2001).

As previously mentioned, sole cytoplasmic or nuclear staining, or a combined staining is considered as aberrant β-cat expression. These patterns are usually found inversely correlated with membranous-related ones (Kotsinas *et al.*, 2002; Pirinen *et al.*, 2001) and support the view that competition between different cellular partners for the cytoplasmic pool of β-cat influences its final incorporation either in cell adhesion or in signal transduction mechanisms (Conacci-Sorell *et al.*, 2002). In view of its nucleo-cytoplasmic shuttling ability, we considered it necessary to further verify these IHC patterns by Western Blot analysis on subcellular fractions consisting of membranous-cytoplasmic and nuclear ones (Kotsinas *et al.*, 2002).

The aberrant pattern has led to investigations for alterations that provide: 1) β-cat stabilization, through activating point mutations in its NH_2 terminal domain, allowing it to escape from phosphorylation-dependent proteasomal degradation; and 2) inactivation of components that regulate β-cat turnover (e.g., APC or Axin). An increasing number of reports have shown that in lung tumors, point mutations either in β-cat (Clavel *et al.*, 2001, Hommura *et al.*, 2002) or in components of the down-regulating mechanisms are infrequent or absent, except for protein phosphates 2A (PP2A) (Polakis, 2000).

The most common studied β-cat regulator is APC (Hajra *et al.*, 2002; Polakis, 2000). Recently we have shown by a combined IHC analysis and Western Blot analysis that APC expression levels in a set of NSCLCs are generally lower in tumor areas than in normal ones (Kotsinas *et al.*, 2002). Although APC has nucleo-cytoplasmic shuttling ability that allows sequestration of aberrant β-cat (Henderson *et al.*, 2002), we did not find any correlation between APC IHC expression patterns and β-cat ones (Kotsinas *et al.*, 2002).

In many tumors APC is inactivated by point mutations (Polakis, 2000). The most frequent ones occur in the central third, mainly in a region termed mutation cluster region (MCR) roughly defined by codons 1250–1500. This region contains β-cat binding and down-regulating domains. Mutations in the MCR usually lead to expression of truncated APC proteins that retain the ability to bind but are unable to suppress β-cat activity (Polakis, 2000). However, in lung tumors APC inactivating point mutations are also not frequent (Cooper *et al.*, 1996; Kotsinas *et al.*, 2002). Nevertheless, loss of heterozygosity, which is common in lung tumors (Cooper *et al.*, 1996), can partly explain the diminished APC expression levels. Recently, methylation suppression of APC expression has been documented in lung cancer (Brabender *et al.*, 2001).

Association of β-cat Immunohistochemical Patterns with Tumor Kinetics and G_1 to S-Phase Cell Cycle Regulators

There are a few reports that examine the expression status of β-cat in combination with cell cycle regulators and their mutual impact on tumor kinetics (PI versus AI) in different types of tumors (Jung *et al.*, 2001). In lung cancer two reports also take into account other cell cycle regulators (Bremnes *et al.*, 2002a; Kotsinas *et al.*, 2002), but only one further associates these relationships with tumor kinetics (Kotsinas *et al.*, 2002). Most of these studies employ expression pattern classification that also reflects its disparate cellular functions.

In NSCLCs, two reports have shown that nuclear-associated β-cat IHC patterns correlate with increased PI (Kotsinas *et al.*, 2002; Pirinen *et al.*, 2001), confirming previous findings in cellular and animal systems (Orford *et al.*, 1999; Wong *et al.*, 1998). A third report was able to show a relationship only between staining intensity and increased PI (Hommura *et al.*, 2002). However, the reason behind this misleading discrepancy is that Hommura *et al.* (2002) have assigned the previously mentioned patterns as positive staining, whereas membranous ones were disregarded as negative ones (Figure 21). Therefore, staining intensity is compared with the PI within the nuclear-associated patterns of expression, actually agreeing with the previous reports. In other types of tumors, there are controversial findings that probably reflect a cell-type– and tumor-type–dependent activity (Jung *et al.*, 2001). We have found that the significant association between β-cat nuclear localization and increased PI is accompanied by overexpression of E2F1, deregulation of p53 and MDM2 regulatory loop, and underexpression of p27[KIP] (Kotsinas *et al.*, 2002).

In NSCLCs, E2F1 is strongly associated with tumor growth, mainly due to increased proliferation (Gorgoulis *et al.*, 2002). Release of E2F1 from pRb sequestration forces cells to enter the S-phase (Lundberg *et al.*, 1999). This effect may be mediated by several downstream β-cat transcriptional targets (e.g., Myc and cyclin D1). However, we did not find any correlation between β-cat and pRb expression status (Kotsinas *et al.*, 2002), implying that additional factors, such as inactivation of p16[INK4a] that is frequently observed in NSCLCS (Sekido *et al.*, 1998), may be involved in E2F1 activation. It is interesting to note that despite low APC levels, its nuclear localization was related to

aberrant pRb expression ($P = 0.04$) (Kotsinas *et al.*, 2002), suggesting a fail-safe protection mechanism of APC against defected pRb pathway from β-cat oncogenic effects. This finding is in accordance with the recently proposed role of APC as a checkpoint gene (Yang, 2002), because of its ability to control entry into S-phase through regulation of the pRb pathway.

Oncogenic β-cat effects, like those of other oncogenes (e.g., Ras, Myc), impinge on the p53 protective response. This effect is exerted via p14[ARF] on the p53/MDM2 regulatory loop (Oren *et al.*, 2002). In such a case, p53 is also able to induce its degradation (Polakis, 2001). In NSCLCs this protective loop is ablated and highly associated with tumor growth (Gorgoulis *et al.*, 2000). Taking into consideration that p53 is the main E2F1 mediator of programmed cell death, loss of this checkpoint correlates with E2F1-driven proliferation and concomitant nuclear (aberrant) presence of β-cat.

E-cad is known to sequester cytoplasmic β-cat (Conacci-Sorell *et al.*, 2002) and also to modulate its transcriptional activity in a cell adhesion–independent manner (Stockinger *et al.*, 2001). In turn, it regulates cell growth via p27[KIP]. Correlation between nuclear β-cat presence, low p27[KIP] expression level, and increased PI could be a result of loss of E-cad expression (Bremnes *et al.*, 2002b; St. Croix *et al.*, 1998) combined with reduced or absent β-cat turnover mechanisms (Kotsinas *et al.*, 2002) and with repression of p27[KIP] expression through β-cat-induced Myc-dependent transcription (Yang *et al.*, 2001).

Finally, no correlation was found either between β-cat with AI or with tumor growth (PI/AI) (Kotsinas *et al.*, 2002). A possible explanation is that β-cat functions upstream of final downstream-effectors like E2F1 and p27[KIP], while itself represents an intermediate one that exerts its cellular effects indirectly, through other downstream mediators (Oren *et al.*, 2002). These mediators receive and integrate signals from other modulators too, transducing them further downstream to final ones, like E2F1 and p27[KIP] (Lundberg *et al.*, 1999). Thus, β-cat influences cell proliferation, but disturbance of the overall cell growth probably requires the additive deregulation(s) of other cell cycle modulator(s). As it has been postulated, excessive rates of cell division, although necessary for neoplastic development, are not sufficient to ensure this process (Bernstein *et al.*, 2002). Mutations in DNA repair and/or apoptosis mechanisms that lead to genomic instability, which in turn is allowed to escape through decreased rates of apoptosis, appears to be the engine for both tumor progression and tumor heterogeneity (Bernstein *et al.*, 2002). According to our findings, nuclear accumulation of β-cat in NSCLCs is accompanied by an increase in proliferation that is related to overexpression of E2F1. This effect is facilitated by a selective advantage from inactivation of the p53 and p27^{KIP} cell cycle checkpoints and the presence of diminished APC levels.

Conclusions and Further Prospects

Either scoring methods or classification of β-cat expression have provided insights into its role in lung tumorigenesis. However, both of them have pros and cons. Scoring methods seem to cover adequately the loss of β-cat associated adhesion and its impact on the clinicopathological patient's outcome (Bremnes *et al.*, 2002b). On the other hand, this is highly dependent also on the status of E-cad, because most of the studies examine both of them in parallel (Bremnes *et al.*, 2002b). Furthermore, not all studies obtain or provide a consistent grading intensity that in turn leaves no other choice but to employ classification of expression patterns, which are always visible. Actually, all works employing scoring methods either report or consider a simplified expression pattern that is usually oriented toward membranous and/or cytoplasmic staining on which intensity grading is applied (Kase *et al.*, 2000; Pirinen *et al.*, 2001; Retera *et al.*, 1998). A more in-depth analysis could reveal expression patterns. Technical problems due to the antibodies used also can be part of the problem. For example, although most of the reports on β-cat IHC expression in lung cancer employ clone 14 from Transduction Laboratories, the staining intensity results are not uniformly reproduced between these studies. On the other hand, not all classification-employing studies adopt or report a common format of expression patterns.

However, what seems to be important in evaluating β-cat IHC results, regardless of the methodology, is the grouping of findings into biologically relevant patterns. These are the membranous-associated patterns of β-cat expression and the nuclear ones (Kotsinas *et al.*, 2002). The first one is related to cell adhesion integrity, whereas the second one reflects its aberrant transcriptional form. We should keep in mind the cellular functions of β-cat and explore its relationship with clinicopathological parameters considering the molecular background in each specific lung tumor to obtain meaningful biological results. Such an approach could provide a more appropriate management of patients with lung cancer.

Acknowledgments

Panayotis Zacharatos and Athanassios Kotsinas are recipients of a postdoctoral scholarship from S.S.F., Greece.

The authors of this chapter apologize for any references omitted unintentionally.

References

Bernstein, C., Bernstein, H., Payne, CM., and Garewal, H. 2002. DNA repair/proapoptotic dual-role proteins in five major DNA repair pathways: Fail-safe protection against carcinogenesis. *Mut. Res. 511:*145–178.

Brabender, J., Usadel, H., Danenberg, K.D., Metzger, R., Schneider, P.M., Lord, R.V., Wickramasinghe, K., Lum, C.E., Park, J., Salonga, D., Singer, J., Sidransky, D., Holscher, A.H., Meltzer, S.J., and Danenberg, P.V. 2001. Adenomatous polyposis coli gene promoter hypermethylation in nonsmall cell lung cancer is associated with survival. *Oncogene 20:*3528–3532.

Bremnes, R.M., Veve, R., Gabrielson, E., Hirsch, F.R., Baron, A., Bemis, L., Gemmill, R.M., Drabkin, H.A., and Franklin, W.A., 2002a. High-throughput microarray analysis used to evaluate biology and prognostic significance of the E-cadherin pathway in nonsmall cell lung cancer. *J. Clin. Ocol. 20:*2417–2428.

Bremnes, R.M., Veve, R., Hirsch, F.R., and Franklin, W.A. 2002b. The E-cadherin cell-cell adhesion complex and lung cancer invasion, metastasis, and prognosis. *Lung Cancer 36:*115–124.

Clavel, C.E., Nollet, F., Berx, G., Tejpar, S., Nawrocki-Rabi, B., Kaplan, H.H., van Roy, E.M., and Birembaut, P.L. 2001. Expression of the E-cadherin-catenin complex in lung neuroendocrine tumours. *J. Pathol. 194:*20–26.

Conacci-Sorell, M., Zhurinsky, J., and Ben Ze'ev, A. 2002. The cadherin-catenin adhesion system in signaling and cancer. *J. Clin. Invest. 109:*987–991.

Cooper, C.A., Bubb, V.J., Smithson, N., Carter, R.L., Gledhill, S., Lamb, D., Wyllie, A.H., and Carey, F.A. 1996. Loss of heterozygosity at 5q21 in nonsmall cell lung cancer: A frequent event but without evidence of APC mutation. *J. Pathol. 180:* 33–37.

Eberhart, C.G., and Argani, P. 2001. Wnt signaling in human development: Beta-catenin nuclear translocation in fetal lung, kidney, placenta, capillaries, adrenal, and cartilage. *Ped. Develop. Pathol. 4:*351–357.

Gorgoulis, V.G., Zacharatos, P., Kotsinas, A., Mariatos, G., Liloglou, T., Vogiatzi, T., Foukas, P., Rassidakis, G., Garinis, G., Ioannides, T., Zoumpourlis, V., Bramis, J., Michail, P.O., Asimacopoulos, P.J., Field, J.K., and Kittas, C. 2000. Altered expression of the cell cycle regulatory molecules pRb, P53, and MDM2 exert a synergetic effect on tumor growth and chromosomal instability in nonsmall cell lung carcinomas (NSCLC). *Mol. Med. 6:*208–237.

Gorgoulis, V.G., Zacharatos, P., Mariatos, G., Kotsinas, A., Bouda, M., Kletsas, D., Asimacopoulos, P.J., Agnantis, N., Kittas, C., and Papavassiliou, A. 2002. Transcription factor E2F1 acts as a growth-promoting factor and is associated with adverse prognosis in nonsmall cell lung carcinomas (NSCLCs). *J. Pathol. 198:*142–156.

Hajra, K.M., and Fearon, F.R. 2002. Cadherin and catenin alterations in human cancer. *Genes Chrom. Cancer 34:*255–268.

Henderson, B.R., and Fagotto, F. 2002. The ins and outs of APC and beta-catenin nuclear transport. *EMBO. Rep. 3:*834–839.

Hommura, F., Furuuchi, K., Yamazaki, K., Ogura, S., Kinoshita, I., Shimi, M., Moriuchi, T., Katoh, H., Nishimura, M., and Dosaka-Akita, H. 2002. Increased expression of beta-catenin predicts better prognosis in nonsmall cell lung carcinomas. *Cancer 94:*752–758.

Jung, A., Schrauder, M., Oswald, U., Knoll, C., Sellberg, P., Palmqvist, R., Niedobitek, G., Brabletz, T., and Kirchner, T.

2001. The invasion front of human colorectal adenocarcinomas shows co-localization of nuclear beta-catenin, cyclin D1, and p16INK4A and is a region of low proliferation. *Am. J. Pathol. 159:*1613–1617.

Kase, S., Sugio, K., Yamasaki, K., Okamoto, T., Yano, T., and Sugimachi, K. 2000. Expression of E-cadherin and β-catenin in human nonsmall cell lung cancer and the clinical significance. *Clin. Cancer Res. 6:*4789–4796.

Kim, K., Pang, K.M., Evans, M., and Hay, E.D. 2000. Overexpression of beta-catenin induces apoptosis independent of its transactivation function with LEF-1 or the involvement of major G1 cell cycle regulators. *Mol. Biol. Cell 11:*3509–3523.

Kimura, K., Endo, Y., Yonemura, Y., Heizmann, C.W., Schafer, B.W., Watanabe, Y., and Sasaki, T. 2000. Clinical significance of S100A4 and E-cadherin-related adhesion molecules in nonsmall cell lung cancer. *Int. J. Oncol. 16:*1125–1131.

Kotsinas, A., Evangelou, K., Zacharatos, P., Kittas, C., and Gorgoulis, V.G. 2002. Proliferation, but not apoptosis, is associated with distinct beta-catenin expression pattern in nonsmall cell lung carcinomas: Relationship with adenomatous polyposis coli and G1- to S-phase cell cycle regulators. *Am. J. Pathol. 161:*1619–1634.

Lee, Y.C., Wu, C.T., Chen, C.S., Hsu, H.H., and Chang, Y.L. 2002. The significance of E-cadherin and alpha-, beta-, gamma-catenin expression in surgically treated nonsmall cell lung cancers of 3-cm or less in size. *J. Thorac. Cardiovasc. Surg. 123:*502–507.

Lundberg, S.A., and Weinberg, R.A. 1999. Control of the cell cycle and apoptosis. *E. J. Cancer 35:*1886–1894.

Nakatani, Y., Masudo, K., Miyagi, Y., Inayama, Y., Kawano, N., Tanaka, Y., Kato, Y., Ito, T., Kitamura, H., Nagashima, Y., Yamanaka, S., Nakamura, N., Sano, J., Ogawa, N., Ishiwa, N., Notohara, K., Resl, M., and Mark, EJ. 2002. Aberrant nuclear localization and gene mutation of beat-catenin in low-grade adenocarcinoma of fetal lung type: Up-regulation of the Wnt signaling pathway may be a common denominator for the development of tumors that form morules. *Mod. Pathol. 15:*617–624.

Nawrocki, B., Polette, M., Van Hengel, J., Tournier, J.M., Van Roy, F., and Birembaut, P. 1998. Cytoplasmic redistribution of E-cadherin-catenin adhesion complex is associated with down-regulated tyrosin phosphorylation of E-cadherin in human bronchopulmonary carcinomas. *Am. J. Pathol. 153:*1521–1530.

Oren, M., Damalas, A., Gottlieb, T., Michael, D., Taplick, J., Leal, J.F.M., Maya, R., Moas, M., Seger, R., Taya, Y., and Ben-Ze'ev, A. 2002. Regulation of p53: Intricate loops and delicate balances. *Biochem. Pharmacol. 64:*865–871.

Orford, K., Orford, C.C., and Byers, S.W. 1999. Exogenous expression of the β-catenin regulates contact inhibition, anchorage-independent growth, anoikis, and radiation-induced cell cycle arrest. *J. Cell. Biol. 146:*855–867.

Pirinen, R.T., Hirvikoski, P., Johansson, R.T., Hollmén, S., and Kosma, V.-M. 2001. Reduced expression of α-catenin, β-catenin, and γ-catenin is associated with high cell proliferative activity and poor differentiation in nonsmall cell lung cancer. *J. Clin. Pathol. 54:*391–395.

Pokutta, S., and Weis, W.I. 2002. The cytoplasmic face of cell contact sites. *Curr. Opin. Struct. Biol. 12:*255–262.

Polakis, P. 2000. Wnt signaling and cancer. *Genes Dev. 14:* 1837–1851.

Polakis, P. 2001. More than one way to skin a catenin. *Cell 105:* 563–566.

Ramasami, S., Kerr, K.M., Chapman, A.D., King, G., Cockburn, J.S., and Jeffrey, R.R. 2000. Expression of CD44v6 but not E-cadherin or beta-catenin influences prognosis in primary pulmonary adenocarcinoma. *J. Pathol. 192:*427–432.

Retera, J., Leers, M., Sulzer, M., and Theunisen, P. 1998. The expression of β-catenin in nonsmall cell lung cancer: A clinicopathological study. *J. Clin. Pathol. 51:*891–894.

Rodriguez-Salas, N., Palacios, J., de Castro, J., Moreno, G., Gonzalez-Baron, M., Gamallo, C. 2001. Beta-catenin expression pattern in small cell lung cancer: Correlation with clinical and evolutive features. *Histol. Histopathol. 16:*353–358.

Sekido, Y., Fong, K.M., and Minna, J.D. 1998. Progress in understanding the molecular pathogenesis of human lung cancer. *Biochim. Biophys. Acta 1378:*F21–F59.

Shibanuma, H., Hirano, T., Tsuji, K., Wu, Q., Shrestha, B., Konaka, C., Ebihara, Y., and Kato, H. 1998. Influence of E-cadherin dysfunction upon local invasion and metastasis in nonsmall cell lung cancer. *Lung Cancer 22:*85–95.

St, Croix, B., Sheehan, C., Rak, J.W., Florenes, V.A., Slingerland, J.M., and Kerbel, R.S. 1998. E-cadherin-dependent growth suppression is mediated by the cyclin-dependent kinase inhibitor p27^{KIP1}. *J. Cell Biol. 142:*557–571.

Stockinger, A., Eger, A., Wolf, J., Beug, H., and Foisner, R. 2001. E-cadherin regulates cell growth by modulating proliferation-dependent β-catenin transcriptional activity. *J. Cell Biol. 154:* 1185–1196.

Sugio, K., Kase, S., Sakada, T., Yamazaki, K., Yamaguchi, M., Ondo, K., and Yano, T. 2002. Micrometastasis in the bone marrow of patients with lung cancer associated with a reduced expression of E-cadherin and beta-catenin: Risk assessment by immunohistochemistry. *Surgery 131:*S22–S31.

Toyoyama, H., Nuruki, K., Ogawa, H., Yanagi, M., Matsumoto, H., Nishijima, H., Shimotakahara, T., Aikou, T., and Ozawa, M. 1999. The reduced expression of E-cadherin, α-catenin, and γ-catenin but not β-catenin in human lung cancer. *Oncol. Rep. 6:*81–85.

Van Aken, E., De Wever, O., Correia da Rocha, A.S., and Mareel, M. 2001. Defective E-cadherin/catenin complexes in human cancer. *Virchows Arch. 439:*725–751.

Winn, R.A., Bremnes, R.M., Bemis, L., Franklin, W.A., Miller, Y.E., Cool, C., and Heasley, L.E. 2002. Gamma-Catenin expression is reduced or absent in a subset of human lung cancers and re-expression inhibits transformed cell growth. *Oncogene 21:*7497–7506.

Wong, M.H., Rubinfeld, B., and Gordon, J.I. 1998. Effects of forced expression of an NH2-terminal truncated beta-catenin on mouse intestinal epithelial homeostasis. *J. Cell Biol. 141:* 765–777.

Yang, V.W. 2002. APC as a checkpoint gene: The beginning or the end? *Gastroenterology 123:*935–939.

Yang, W., Shen, J., Wu, M., Arsura, M., FitzGerald, M., Suldan, Z., Kim, D.W., Hofman, C.S., Pianetti, S., Romieu-Mourez, R., Freedman, L.P., and Sonenshein, G.E. 2001. Repression of transcription of the p27 (Kip1) cyclin-dependent kinase inhibitor gene by C-Myc. *Oncogene 20:*1688–1702.

7

Immunohistochemistry of Laminin-5 in Lung Carcinoma

Toshiro Niki

Introduction

Laminins are a family of glycoproteins that consist of one heavy α chain and two light β and γ chains (Timple, 1989). The laminin molecule is a major component of the basement membrane and plays important roles in cell differentiation, adhesion, and migration (Timple, 1989). Currently, 5 α chain (α1–α5), 3 β chain (β1–β3), and 3 γ chain (γ1–γ3) variants have been identified, and these chains assemble into at least 12 different laminin isoforms (Lohi, 2001). These different isoforms show overlapping but distinct tissue distribution and are thought to perform different functions (Lohi, 2001).

Laminin-5 (nicein, kalinin, epiligrin, ladsin) was identified as a new adhesion ligand for integrin α3β1 and α6β4 in epithelial basement membrane (Carter *et al.*, 1991; Miyazaki *et al.*, 1993; Rousselle *et al.*, 1991; Verrando *et al.*, 1988). The laminin-5 molecule consists of α3, β3, and γ2 chains, the latter two being unique to this isoform. As a component of anchoring filaments, laminin-5 plays crucial roles in the attachment of epithelial cells to the basement membrane (Carter *et al.*, 1991; Rousselle *et al.*, 1991; Verrando *et al.*, 1988). Inactivating mutation of laminin-5 chains is one of the causes of epidermolysis bullosa, a skin-blistering disease. Interestingly, laminin-5 was

also identified as a cell-adhesive scatter factor secreted by keratinocytes and gastric carcinoma cell lines (Miyazaki *et al.*, 1993). The dual roles of laminin-5 (i.e., cell attachment and migration) are probably attributable to differential use of the integrins (Nguyen *et al.*, 2000). In normal skin, laminin-5 is localized in the basement membrane and stabilizes hemidesmosomes via integrin α6β4 (Carter *et al.*, 1991; Rousselle *et al.*, 1991). During tissue injury, however, laminin-5 is up-regulated in the keratinocytes of the migrating front and is thought to promote cell migration and wound closure (Nguyen *et al.*, 2000). The results of *in vitro* studies have suggested that laminin-5 promotes cell migration and scattering through interaction with integrin α3β1 (Verrando *et al.*, 1994).

Epithelial induction of laminin-5 is not restricted to the epidermis; studies show that laminin-5 is induced in mucosal injury of gastrointestinal tract (Vlaamo *et al.*, 1998). Also, laminin-5 seems to be expressed in type II pneumocytes in certain forms of lung injury and inflammation as well (Niki, unpublished). With regard to laminin-5 expression in cancer, Pyke *et al.* (1994) first demonstrated that laminin-5 is frequently localized in the cytoplasm of invading cancer cells, characteristically located in the margin of cancer cell nests. The expression pattern of laminin-5 suggested that laminin-5 may be involved in tumor invasion as

Handbook of Immunohistochemistry and *in situ* Hybridization of Human
Carcinomas, Volume 1: Molecular Genetics; Lung and Breast Carcinomas

155

Copyright © 2004 by Elsevier (USA)
All rights reserved.

well as cell migration during tissue injury. It has been reported that laminin-5 is strongly expressed at the invasive front of colorectal (Aoki *et al.*, 2002; Pyke *et al.*, 1994; Sordat *et al.*, 1998), gastric (Koshikawa *et al.*, 1999; Pyke *et al.*, 1994), pancreatic (Fukushima *et al.*, 2001; Takahashi *et al.*, 2002), breast (Pyke *et al.*, 1994), endometrial (Lundgren *et al.*, 2003), ovarian (Kohlberger *et al.*, 2002) adenocarcinomas, urothelial carcinomas (Hindermann *et al.*, 2003), uterine cervical (Pyke *et al.*, 1994; Skyldberg *et al.*, 1999), esophageal (Yamamoto *et al.*, 2001) and oral (Katoh *et al.*, 2002; Ono *et al.*, 1999) squamous cell carcinomas, and malignant melanoma (Pyke *et al.*, 1994). Studies suggest that laminin-5 may serve as an invasion marker and/or a useful prognostic indicator for certain types of cancer (Fukushima *et al.*, 2001; Kohlberger *et al.*, 2002; Moriya *et al.*, 2001; Ono *et al.*, 1999; Skyldberg *et al.*, 1999; Takahashi *et al.*, 2002).

Several studies demonstrated that laminin-5 is frequently expressed in lung cancer (Kagesato *et al.*, 2001; Maatta *et al.*, 1999; Moriya *et al.*, 2001; Niki *et al.*, 2002; Terasaki *et al.*, 2003). These studies show that 1) laminin-5 is characteristically expressed at the invasive front of cancer (i.e., at the interface between cancer cell nests and fibrous stroma) (Kagesato *et al.*, 2001; Maatta *et al.*, 1999; Moriya *et al.*, 2001; Niki *et al.*, 2002); 2) overexpression of laminin-5 is associated with poor patient prognosis for adenocarcinoma of small size (Moriya *et al.*, 2001); and 3) expression levels of laminin-5 may vary according to the four major histologic subtypes of lung cancer (Maatta *et al.*, 1999).

MATERIALS

1. 50 mM Tris-buffered saline (TBS), pH 7.6.
2. 10 mM citrate buffer, pH 6.0.
3. Primary antibody to laminin-5 γ2 chain.[a]
4. Biotinylated anti-mouse secondary antibody (Dako Cytomation, Dako, Glostrup, Denmark).
5. 3,3'-Diaminobenzidine tetrahydrochloride (DAB).
6. ABC kit (Vectastain, Vector Laboratories, Burlingame, CA).

METHODS

1. Slice paraffin blocks into 3–5 micron thick sections.
2. Bring the sections onto silan-coated glass slides and heat-dry overnight at 50°C.

[a]To detect laminin-5 γ2 chains, two monoclonal antibodies 1) clone 1-97 (Ono *et al.*, 1999) or 2) clone D4B5 (Koshikawa *et al.*, 1999) are widely used. The latter is commercially available from Chemicon International, Temecula, CA.

3. Rehydrate the sections through graded xylene and ethanol.
4. Autoclave the sections for 10 min in citrate buffer (pH 6.0) at 121°C.
5. Allow to cool down to room temperature.
6. Block endogenous peroxidase activity by incubating the sections in 0.3% hydrogen peroxide/methanol for 30 min at room temperature.
7. Block nonspecific binding by incubating the sections in 10% horse serum for 10 min at room temperature.
8. Incubate for 2 hr at room temperature in primary antibody diluted in TBS (final concentration; 1 μg/ml for clone 1-97 or 2 μg/ml clone D4B5).
9. Rinse the sections three times in TBS for 5 min at room temperature.
10. Incubate for 30 min at room temperature in biotinylated secondary antibody diluted in TBS.
11. Rinse as in **step 9**.
12. React with ABC complex 20 min at room temperature.
13. React with 0.02% DAB solution in Tris-HCl buffer (pH 7.6) containing 0.007% hydrogen peroxide.
14. Rinse in water and counterstain with hematoxylin.

RESULTS AND DISCUSSION

Because lung carcinomas are generally classified into four major histologic subtypes, the results are described for each histologic subtypes accordingly. To date, the available data is mainly derived from laminin-5 γ2 chain expression in adenocarcinomas.

Laminin-5 Expression in Adenocarcinomas

Moriya *et al.* investigated the expression of laminin-5 γ2 chain in 102 cases of lung adenocarcinomas of small size (maximum dimension 2 cm or less) (Moriya *et al.*, 2001). A significant proportion of peripheral lung adenocarcinomas of small size consists of two components: 1) tumor cells with bronchioloalveolar growth pattern in which tumor cells spread along the pre-existing alveolar walls in a noninvasive manner; and 2) tumor cells with invasive growth pattern, often accompanied with fibroblastic proliferation (desmoplasia) (Noguchi *et al.*, 1995; Terasaki *et al.*, 2003). The invasive component could be papillary or acinar adenocarcinoma, solid adenocarcinoma with mucin, or combinations of more than one of these histologic subtypes. Typically, the invasive component is located in the center of the tumor surrounded by the bronchioloalveolar component.

In these mixed bronchioloalveolar and invasive tumors, immunoreactive laminin-5 was frequently

(90.2%, 37/41 cases) found in the cytoplasm of cancer cells at the cancer–stroma interface (Figure 22A); only minimal immunoreactivity for laminin-5 was found in the bronchioloalveolar carcinoma component, except in cancer cells that showed budding into the fibrous stroma (Figure 22B). In contrast, intense staining for laminin-5 was typically observed in cancer cells that infiltrated the fibrous stroma in a scattered manner (Figure 22C). The same patterns of laminin-5 expression were seen in 38 of 43 cases (88.4%) of invasive adenocarcinoma. Although laminin-5 was preferentially distributed at the cancer–stroma interface, there were exceptional cases; laminin-5 was expressed by the bulk of tumor cells, in some cases of solid adenocarcinoma with mucin (Figure 22D). In 18 pure bronchioloalveolar carcinomas without vascular or stromal invasion (Travis *et al.,* 1999), by contrast, staining for laminin-5 was absent or minimal in 11 of

18 cases (61.1%) studied. In the remaining cases, however, laminin-5 immunoreactivity was occasionally observed in cancer cells budding into fibrous stroma (as shown in Figure 22B).

Laminin-5 Expression in Squamous Cell and Other Histologic Types of Carcinomas

Maatta *et al.* compared the expression of laminin-5 γ2 chain in four histologic subtypes of lung cancer (Maatta *et al.,* 1999). They demonstrated laminin-5 expression in all cases of non–small cell carcinomas (21 squamous cell carcinomas, 19 adenocarcinomas, and 5 large cell carcinomas). Expression of laminin-5 was strongest in squamous cell carcinomas, followed by adenocarcinomas and larger cell carcinomas. In squamous cell carcinomas, expression pattern of laminin-5 was similar to that in adenocarcinomas; laminin-5 was

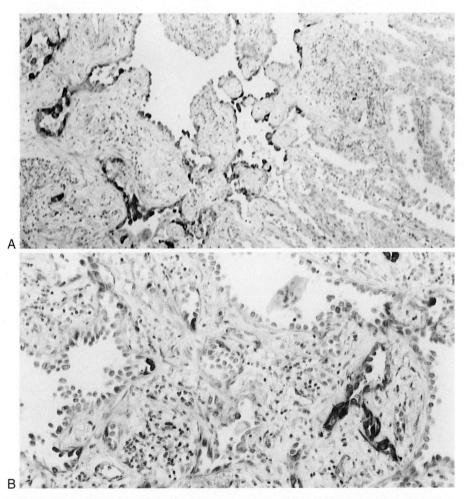

Figure 22 A: Laminin-5 is characteristically localized in the cytoplasm of cancer cells at the cancer–stroma interface. **B:** Only minimal immunoreactivity for laminin-5 is found in the bronchioloalveolar carcinoma component, except in cancer cells that showed budding into the fibrous stroma.   *(Continued)*

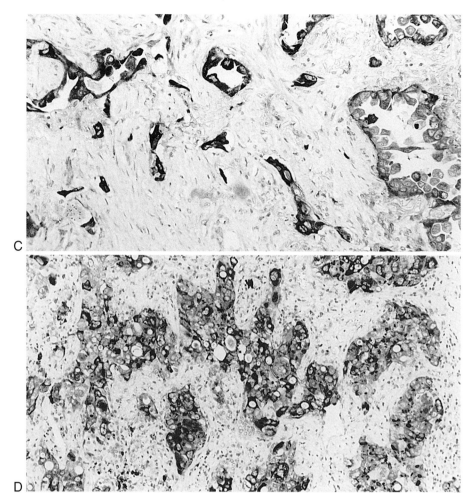

Figure 22 Cont'd C: Intense staining for laminin-5 is typically observed in cancer cells that infiltrated the fibrous stroma in a scattered manner. **D:** Laminin-5 is expressed by the bulk of tumor cells in some cases of solid adenocarcinoma with mucin.

localized in the tumor–stroma interface of the tumor cell nests or individual tumor cells scattered in the fibrous stroma. Only limited data are available concerning laminin-5 expression in other histologic types of lung carcinomas; Maatta *et al.* reported that laminin-5 was only weakly expressed in 10 of 19 small cell carcinomas (Maatta *et al.,* 1999). To date, there are no data about laminin-5 expression in large cell neuroendocrine carcinomas.

Clinical Significance of Laminin-5 Expression in Lung Carcinomas

To investigate the significance of laminin-5 expression, Moriya *et al.* graded laminin-5 expression levels into three categories based on the area showing the highest level of laminin-5 expression (Moriya *et al.,* 2001). First, sections were scanned at low magnification to

identify the area showing the highest level of laminin-5 expression. Then, that area was viewed with 10X objective, and laminin-5 expression was graded on a scale of 0 to 2 as follows: 0, either no laminin-5–positive cancer cells present or only a few laminin-5–positive cancer cells observed in a scattered manner; 1, cluster(s) of laminin-5-positive cancer cells present, but accounting for less than 30% of the tumor cells within the visual field; 2, cluster(s) of laminin-5-positive cancer cells that accounted for more than 30% of the tumor cells within the visual field. This grading method was based on the reasoning that the prognosis of patients would be influenced by the most aggressive subpopulation of tumor cells present.

By this grading method, overexpression of laminin-5 (grade 2, 24 cases, 23.5%) was shown to be associated with vascular invasion (p = 0.0210) and stromal fibroblastic reaction (p = 0.0046), but not with pathologic stage (p = 0.3592), nodal involvement (p = 0.7564), pleural invasion (p = 0.1697), or lymphatic invasion (p = 0.3152), indicating that laminin-5 overexpression may occur at a relatively early stage of the disease, and that laminin-5 is unrelated to metastasis via lymphatic channels (Moriya et al., 2001). Survival analyses with the Kaplan-Meyer method demonstrated that prognosis of patients became less favorable as the laminin-5 grade increased from 0 to 2 (p = 0.0018, log-rank test). Moreover, prognostic significance of laminin-5 overexpression was retained on multivariate analyses (Moriya et al., 2001).

Although the study by Moriya et al. (2001) demonstrated the prognostic value of laminin-5 expression in lung adenocarcinomas of small size (maximum dimension 2 cm or less), whether the results can be extrapolated to lung adenocarcinoma in general remains to be determined. It is currently unknown whether laminin-5 immunostaining is also useful for other histologic types of lung cancer (e.g., squamous cell carcinoma and large cell carcinoma).

Usefulness of Laminin-5 as an Invasion Marker

The preferential expression of laminin-5 at the invasive front of cancer indicates that laminin-5 may serve as a marker for stromal invasion of cancer cells. In fact, Skyldberg et al. suggested the possibility of using laminin-5 immunostaining to identify microinvasion of cervical cancer in routine histologic diagnosis (Skyldberg et al., 1999). In contrast, however, laminin-5 expression may be found in a subset of bronchioloalveolar carcinomas (Moriya et al., 2001; Terasaki et al., 2003), for which no evidence of stromal invasion was found on routine histologic examination using hematoxylin and eosin stain and/or elastica stains. It may be that, although laminin-5 immunostaining identifies bronchioloalveolar carcinoma cells with some potential to cause remodeling of alveolar structures (minimal invasion), it may not be necessarily associated with vascular invasion or nodal involvement. It has already been shown that patients with pure bronchioloalveolar carcinomas show no nodal involvement and excellent prognosis when standard surgical resection is performed (Noguchi et al., 1995). Thus, additional studies will be required to determine the clinical significance of laminin-5–positive cells in bronchioloalveolar carcinomas.

Role of Laminin-5 in Tumor Cell Invasion

In vitro studies have shown that laminin-5 promotes cell migration and scattering (Miyazaki et al., 1993; Verrando et al., 1994), indicating that laminin-5 has a causal effect on cancer cell invasion. However, it is yet to be determined whether expression of laminin-5 has a causal relationship with cancer cell invasion and poor prognosis of patients. In fact, the predominantly cytoplasmic localization of laminin-5 γ2 chains presents a paradox regarding the role of laminin-5 in cell invasion: to facilitate cell migration via integrins, laminin-5 should be secreted into extracellular space and incorporated into basement membranes. In this context, it is noteworthy that conflicting results have been reported concerning 1) localization of laminin-5 in the basement membranes, and 2) coexpression of laminin α3, β3, and γ2 chains.

In several studies, laminin-5 was localized in the basement membrane surrounding cancer cell nests (Koshikawa et al., 1999; Sordat et al., 1998), as well as in the cytoplasm of invading cancer cells. Although discrepancies concerning laminin-5 localization in basement membrane may be explained by the use of different monoclonal antibodies that may recognize different epitopes of the laminin γ2 chain, it appears that cytoplasmic accumulation of laminin-5, rather than its localization in basement membrane, is associated with poor prognosis of patients (Takahashi et al., 2002). Recently, Koshikawa et al. reported that laminin-5 γ2 chain was strongly expressed at the invasive margin of gastric cancer cells without significant signal for laminin-5 β3 or α3 (Koshikawa et al., 2000). These authors also demonstrated the secretion of the laminin γ2 monomer, as well as the laminin-5 heterotrimer, by two-dimensional sodium dodecyl sulfate polyacrylamide gel electrophoresis (SDS-PAGE). The authors speculated that the monomeric form of the γ2 chain may have a function distinct from the laminin-5 trimer. In contrast, colocalization of laminin β3 and γ2 chains have been shown in lung and colorectal adenocarcinomas (Niki et al., 2002; Sordat et al., 1998). The reason for these discrepancies is not yet clear, but it may also be attributable to the use of different antibodies and/or different types of cancer specimens investigated. Certainly, to establish the role of laminin-5 in tumor cell invasion, these issues need to be addressed by further investigations.

It has been reported that laminin-5 may perform two opposite functions, promotion of cell migration and assembly of hemidesmosomes. Quaranta and colleagues reported that the cleavage of the laminin γ2 chain by MMP-2 elicits cell migration on laminin-5 (Giannelli et al., 1997). Conversely, after cleavage of the laminin α3 chain by plasmin, laminin-5 impedes cell motility and promotes hemidesmosome assembly (Goldfinger et al., 1998). Thus, different functions of laminin-5 could be explained by differential processing of the subunits that comprise laminin-5. More recently, Koshikawa et al. found that MT1-MMP, which cleaves laminin γ2 chain more efficiently than MMP-2, plays essential roles in cell migration on laminin-5 (Koshikawa *et al.*, 2000). These authors found that cell migration on laminin-5 was significantly reduced by metalloproteinase inhibitors and MT1-MMP antisense oligonucleotides.

Wound Healing and Cancer

Laminin-5 is induced at the ulcer edge of skin and gastrointestinal tract (Nguyen *et al.*, 2000; Vaalamo *et al.*, 1998) and invasive front of cancer (Lohi, 2001), probably through activation of growth factor receptors (Katoh *et al.*, 2002; Mizushima *et al.*, 1996; Niki *et al.*, 2002; Olsen *et al.*, 2003; Ono *et al.*, 2002). However, the exact mechanisms underlying up-regulation of laminin-5 at the invasive front of cancer are still unknown.

Laminin-5 is yet another molecule involved in wound healing and cancer invasion. Given the wide range of molecules implicated in these two processes (Dano *et al.*, 1999; Rowley, 1998), it is tempting to speculate that invading cancer cells are using a genetic program normally used for wound healing. The essential difference between these two processes is that while wound healing is a self-limiting process, cancer invasion is a continuous process that will eventually kill the host unless the patient is treated. In this regard, it is noteworthy that wild-type p53 has a suppressive effect on the expression of genes involved in inflammation and tissue remodeling, including vascular endothelial growth factor (VEGF) (Mukhopadhyay *et al.*, 1995), COX-2 (Subbaramaiah *et al.*, 1999), and the inducible isoform of nitric oxide synthase (Forrester *et al.*, 1996). The promoter activity of laminin-5 γ2 chain is also suppressed by wild-type, but not by mutant-type p53 (Niki, unpublished). In primary resected lung adenocarcinomas, laminin-5 overexpression is associated with p53 abnormality (Niki *et al.*, 2002). Thus, accumulations of genetic alterations may be responsible for the dysregulation of wound-related gene expression and induction of invasion in cancer. This hypothesis needs to be verified by future studies.

CONCLUDING REMARKS

The characteristic expression pattern of laminin-5 strongly indicates that it plays important roles in the invasion of lung cancer cells. The prognostic value of laminin-5 in lung cancer needs further investigation for both adenocarcinomas and squamous cell carcinomas, and maybe for large cell carcinomas as well. The mechanistic details concerning how the laminin-5 molecule facilitates cancer cell invasion and how its overexpression occurs at the invasive front is still an open question that warrants further study.

Acknowledgment

I would like to thank Dr. Setsuo Hirohashi for giving me the opportunity to contribute to this chapter and Dr. Yukiko Ono and Dr. Yukihiro Nakanishi for the laminin-5 γ2 antibody. I would also like to thank Yuko Yamauchi, Sanae Iba, and Miyuki Saito for excellent technical assistance.

References

Aoki, S., Nakanishi, Y., Akimoto, S., Moriya, Y., Yoshimura, K., Kitajima, M., Sakamoto, M., and Hirohashi, S. 2002. Prognostic significance of laminin-5 gamma2 chain expression in colorectal carcinoma: Immunohistochemical analysis of 103 cases. *Dis. Colon Rectum 45:*1520–1527.

Carter, W.G., Ryan, M.C., and Gahr, P.J. 1991. Epiligrin, a new cell adhesion ligand for integrin alpha3 beta1 in epithelial basement membranes. *Cell 65:*599–610.

Dano, K., Romer, J., Nielsen B.S., Bjorn, S., Pyke, C., Rygaard, J., and Lund, L.R. 1999. Cancer invasion and tissue remodeling: Cooperation of protease systems and cell types. *APMIS 107:* 120–127.

Forrester, K., Ambs, S., Lupold, S.E., Kapust, R.B., Spillare, E.A., Weinberg, W.C., Felley-Bosco, E., Wang, X.W., Geller, D.A., Tzeng, E., Billiar, T.R., and Harris, C.C. 1996. Nitric oxide-induced p53 accumulation and regulation of inducible nitric oxide synthase expression by wild-type p53. *Proc. Natl. Acad. Sci. USA. 93:*2442–2447.

Fukushima, N., Sakamoto, M., and Hirohashi, S. 2001. Expression of laminin-5-gamma2 chain in intraductal papillary-mucinous and invasive ductal tumors of the pancreas. *Mod. Pathol. 14:* 404–409.

Giannelli, G., Falk-Marzillier, J., Schiraldi, O., Stetler-Stevenson, W.G., and Quaranta, V. 1997. Induction of cell migration by matrix metalloproteinase-2 cleavage of laminin-5. *Science 277:*225–228.

Goldfinger, L.E., Stack, M.S., and Jones, J.C.R. 1998. Processing of laminin-5 and its functional consequences: Role of plasmin and tissue-type plasminogen activator. *J. Cell Biol. 141:*255–265.

Hindermann, W., Berndt, A., Haas, K.M., Wunderlich, H., Katenkamp, D., and Kosmehl, H. 2003. Immunohistochemical demonstration of the gamma2 chain of laminin-5 in urinary bladder urothelial carcinoma: Impact for diagnosis and prognosis. *Cancer Detect. Prev. 27:*109–115.

Kagesato, Y., Mizushima, H., Koshikawa, N., Kitamura, H., Hayashi, H., Ogawa, N., Tsukuda, M., and Miyazaki, K. 2001. Sole expression of laminin gamma-2 chain in invading tumor cells and its association with stromal fibrosis in lung adenocarcinomas. *Jpn. J. Cancer Res. 92:*184–192.

Katoh, K., Nakanishi, Y., Akimoto, S., Yoshimura, K., Takagi, M., Sakamotom, M., and Hirohashi, S. 2002. Correlation between laminin-5 gamma2 chain expression and epidermal growth factor receptor expression and its clinicopathological significance in squamous cell carcinoma of the tongue. *Oncology 62:*318–326.

Kohlberger, P., Muller-Klingspor, V., Heinzl, H., Obermair, A., Breitenecker, G., and Leodolter, S. 2002. Prognostic value of laminin-5 in serous adenocarcinomas of the ovary. *Anticancer Res. 22:*3541–3544.

Koshikawa, N., Giannelli, G., Cirulli, V., Miyazaki, K., and Quararnta, V. 2000. Role of cell surface metalloprotease MT1-MMP in epithelial cell migration over laminin-5. *J. Cell Biol.* 148:615–624.

Koshikawa, N., Moriyama, K., Takamura, H., Mizushima, H., Nagashima, Y., Yanoma, S., and Miyazaki, K. 1999. Overexpression of laminin gamma2 chain monomer in invading gastric carcinoma cells. *Cancer Res.* 59:5596–5601.

Lohi, J. 2001. Laminin-5 in the progression of carcinomas. *Int. J. Cancer* 94:763–767.

Lundgren, C., Frankendal, B., Silfversward, C., Nilsson, B., Tryggvason, K., Auer, G., and Nordstrom, B. 2003. Laminin-5 gamma2-chain expression and DNA ploidy as predictors of prognosis in endometrial carcinoma. *Med. Oncol.* 20:147–156.

Maatta, M., Soini, Y., Paakko, P., Salo, S., Tryggvason, K., and Autio-Harmainen, H. 1999. Expression of the laminin gamma2 chain in different histological types of lung carcinoma: A study by immunohistochemistry and *in situ* hybridization. *J. Pathol.* 188:361–368.

Miyazaki, K., Kikkawa, Y., Nakamura, A., Yasumitsu, H., and Umeda, M. 1993. A large cell-adhesive scatter factor secreted by human gastric carcinoma cells. *Proc. Natl. Acad. Sci. USA.* 90:11767–11771.

Mizushima, H., Miyagi, Y., Kikkawa, Y., Yamanaka, N., Yasumitsu, H., Misugi, K., and Miyazaki, K. 1996. Differential expression of laminin-5/ladsin subunits in human tissues and cancer cell lines and their induction by tumor promoter and growth factors. *J. Biochem. (Tokyo)* 120:1196–1202.

Moriya, Y., Niki, T., Yamada, T., Matsuno, Y., Kondo, H., and Hirohashi, S. 2001. Increased expression of laminin-5 and its prognostic significance in small-sized lung adenocarcinoma: An immunohistochemical analysis of 102 cases. *Cancer* 91:1129–1141.

Mukhopadhyay, D., Tsiokas, L., and Sukhatme, V.P. 1995. Wild-type p53 and v-Src exert opposing influences on human vascular endothelial growth factor gene expression. *Cancer Res.* 55:6161–6165.

Nguyen, B.P., Ryan, M.C., Gil, S.G., and Carter, W.G. 2000. Deposition of laminin-5 in epidermal wounds regulates integrin signaling and adhesion. *Curr. Opin. Cell Biol.* 12:554–562.

Niki, T., Kohno, T., Iba, S., Moriya, Y., Takahashi, Y., Saito, M., Maeshima, A., Yamada, T., Matsuno, Y., Fukayama, M., Yokota, J., and Hirohashi, S. 2002. Frequent colocalization of COX-2 and laminin-5 gamma2 chain at the invasive front of early-stage lung adenocarcinomas. *Am. J. Pathol.* 160:1129–1141.

Noguchi, M., Morikawa, A., Kawasaki, M., Matsuno, Y., Yamada, T., Hirohashi, S., Kondo, H., and Shimosato, Y. 1995. Small adenocarcinoma of the lung: Histologic characteristics and prognosis. *Cancer* 75:2844–2852.

Olsen, J., Kirkeby, L.T., Brorsson, M.M., Dabelsteen, S., Troelsen, J.T., Bordoy, R., Fenger, K., Larsson, L.I., and Simon-Assman, P. 2003. Converging signals synergistically activate the LAMC2 promoter and lead to accumulation of the laminin gamma 2 chain in human colon carcinoma cells. *Biochem. J.* 371: 211–221.

Olsen, J., Lefebvre, O., Fritsch, C., Troelsen, J.T., Orian-Rousseau, V., Kedinger, M., and Simon-Assmann, P. 2000. Involvement of activator protein 1 complexes in the epithelium-specific activation of the laminin gamma2-chain gene promoter by hepatocyte growth factor (scatter factor). *Biochem. J.* 347:407–417.

Ono, Y., Nakanishi, Y., Gotoh, M., Sakamoto, M., and Hirohashi S. 2002. Epidermal growth factor receptor gene amplification is correlated with laminin-5 γ2 chain expression in oral squamous cell carcinoma cell lines. *Cancer Lett.* 175:197–204.

Ono, Y., Nakanishi, Y., Ino, Y., Niki, T., Yamada, T., Yoshimura, K., Saikawa, M., Nakajima, T., and Hirohashi, S. 1999. Clinocopathologic significance of laminin-5 gamma2 chain expression in squamous cell carcinoma of the tongue: Immunohistochemical analysis of 67 lesions. *Cancer 85:* 2315–2321.

Pyke, C., Romer, J., Kallunki, P., Lund, L.R., Ralfkiaer, E., Dano, K., and Tryggvason, K. 1994. The gamma2 chain of kalinin/laminin-5 is preferentially expressed in invading malignant cells in human cancer. *Am. J. Pathol.* 145:782–791.

Rousselle, P., Lunstrum, G.P., Keene, D.R., and Burgeson, R.E. 1991. Kalinin: An epithelium-specific basement membrane adhesion molecule that is a component of anchoring filaments. *J. Cell Biol.* 114:567–576.

Rowley, D.R. 1998. What might a stromal response mean to prostate cancer progression? *Cancer Metastasis Rev.* 17:411–419.

Skyldberg, B., Salo, S., Eriksson, E., Aspenblad, U., Moberger, B., Tryggvason, K., and Auer, G. 1999. Laminin-5 as a marker of invasiveness in cervical lesions. *J. Natl. Cancer Inst.* 91: 1882–1887.

Sordat, I., Bosman, F.T., Dorta, G., Guillou, L., Mazzucchelli, L., Saraga, E., Benhattar, J., Tran-Thang, C., Blum, A.L., Dorta, G., and Sordat, B. 1998. Differential expression of laminin-5 subunits and integrin receptors in human colorectal neoplasia. *J. Pathol.* 185:44–52.

Subbaramaiah, K., Altorki, N., Chung, W.J., Mestre, J.R., Sampat, A., and Dannenberg, A.J. 1999. Inhibition of cyclooxygenase-2 gene expression by p53. *J. Biol. Chem.* 274:10911–10915.

Takahashi, S., Hasebe, T., Oda, T., Kinoshita, T., Konishi, M., Ochiai, T., and Ochiai, A. 2002. Cytoplasmic expression of laminin gamma2 chain correlates with postoperative hepatic metrastasis and poor prognosis in patients with pancreatic ductal adenocarcinoma. *Cancer* 94:1894–1901.

Terasaki, H., Niki, T., Matsuno, Y., Yamada, T., Maeshima, A., Asamura, H., Hayabuchi, N., and Hirohashi, S. 2003. Lung adenocarcinoma with mixed bronchioloalveolar and invasive components: Clinicopathological features, subclassification by extent of invasive foci, and immunohistochemical characterization. *Am. J. Surg. Pathol.* 27:937–951.

Timple, R. 1989. Structure and biological activity of basement membrane proteins. *Eur. J. Biochem.* 180:487–502.

Travis, W., Colby, T., Corrin, B., Shimosato, Y., and Brambilla, E. 1999. *Histological Typing of Lung and Pleural Tumors.* 3rd. ed. World Health Organization. Berlin: Springer.

Vaalamo, M., Karjalainen-Lindsberg, M.L., Puolakkainen, P., Kere, J., and Saarialho-Kere, U. 1998. Distinct expression profiles of stromelysin-2 (MMP-10), collagenase-3 (MMP-13), macrophage metalloelastase (MMP-12), and tissue inhibitor of metalloproteinases-3 (TIMP-3) in intestinal ulcerations. *Am. J. Pathol.* 152:1005–1014.

Verrando, P., Lissitzky, J.C., Sarret, Y., Winberg, J.O., Gedde-Dahl Jr., T. Schmitt, D., and Bruckner-Tuderman, L. 1994. Keratinocytes from junctional epidermolysis bullosa do adhere and migrate on the basement membrane protein nicein through alpha 3 beta 1 integrin. *Lab. Invest.* 71:567–574.

Verrando, P., Pisani, A., and Ortonne, J.P. 1988. The new basement membrane antigen recognized by the monoclonal antibody GB3 is a large-size glycoprotein: Modulation of its expression by retinoic acid. *Biochim. Biophys. Acta* 942:45–56.

Yamamoto, H., Itoh, F., Iku, S., Hosokawa, M., and Imai, K. 2001. Expression of the gamma 2 chain of laminin-5 at the invasive front is associated with recurrence and poor prognosis in human esophageal squamous cell carcinoma. *Clin. Cancer Res.* 7:896–900.

8

Role of Immunohistochemical Expression of Caveolin-1 in Lung Carcinoma

Chao-Chi Ho, Pei-Hsin Huang, Pan-Chyr Yang, and Su-Ming Hsu

Introduction

Caveolae, "cave-like" invaginations of the cell surface, were first described in the middle of the twentieth century (Palade *et al.*, 1968). Caveolae exist in most mammalian cell types and are especially abundant in adipocytes, endothelial cells, fibroblasts, smooth-muscle cells, and type I pneumocytes (Couet *et al.*, 1997). Caveolae were thought to be involved in potocytosis and transcytosis (Simionescu, 1983). Recent evidence has indicated that caveolae are a concentrated pool of signaling molecules and play an important role in regulating signaling cascades (Liu *et al.*, 2002).

The principal component of caveolae is caveolin-1, a 21–24 kDa integral membrane protein (Rothberg *et al.*, 1992). Caveolin-1 forms a homo-oligomer and is distributed in the membranes of caveolae. Expression of caveolin-1 in cells lacking caveolae produces bona fide caveolae, and depletion of caveolin-1 causes loss of caveolae (Fra *et al.*, 1995). Caveolin-1 interacts with a number of signaling molecules through its "scaffolding" domain (Li *et al.*, 1996). Recent studies have also highlighted the role of a caveolin-1 in lipid regulation and the relationship between lipid regulation and signaling pathways. Caveolin-1 can bind cholesterol and fatty acid within caveolae and is involved in cholesterol transport to the cell membrane (Fielding *et al.*, 2001). Caveolin-1 maintains the cholesterol-rich lipid-raft domains where specific signaling pathways are dependent (Feron *et al.*, 1999; Roy *et al.*, 1999).

The gene encoding caveolin-1 is localized in the chromosome 7q31.1 region that is frequently deleted in a wide variety of human epithelial tumors (Razani *et al.*, 2001). It is noteworthy that expression of caveolin-1 mRNA and protein is frequently lost in human cancer cell lines. Furthermore, recombinant expression of caveolin-1 in transformed NIH3T3 cells or cell lines derived from human cancers suppresses their transformed phenotype (Engelman *et al.*, 1997). Targeted down-regulation of caveolin-1 expression promotes anchorage-independent cell growth in soft agar and drives tumorigenesis in nude mice (Galbiati *et al.*, 1998). All of this evidence suggests that caveolin-1 functions as a tumor suppressor. Caveolin-1 may exert its tumor-growth inhibition by contact inactivation of signaling molecules such as v-src, Ha-Ras, protein kinase A, protein kinase C, and p42/44 MAP kinase within caveolae (Carman *et al.*, 1999; Galbiati *et al.*, 1998).

Handbook of Immunohistochemistry and in situ Hybridization of Human Carcinomas, Volume 1: Molecular Genetics; Lung and Breast Carcinomas

163

Copyright © 2004 by Elsevier (USA)
All rights reserved.

In addition, caveolin-1 can inhibit tumorigenesis by lowering the local nitric oxide concentration through its participation in degradation of nitric oxide synthase via the proteosome pathway (Felley-Bosco *et al.*, 2000).

Caveolin-1 may also function as a tumor metastasis-promoting molecule, which is unrelated to its obvious function of cell growth inhibition (Hulit *et al.*, 2000). Using cell lines derived from primary mouse prostate cancer, Yang *et al.* (1998) reported that caveolin-1 cDNA, along with a number of metastasis-related sequences, was identified in the associated metastastic cancer cells. Elevated expression of caveolin-1 is also found to be associated with progression of esophageal, pancreatic, and breast carcinoma (Kato *et al.*, 2002; Suzuoki *et al.*, 2002; Yang *et al.*, 1998). Inhibition of C-Myc–induced apoptosis by caveolin-1 was recently proposed to promote progression of prostate cancer (Timme *et al.*, 2000), and it may serve as a prognostic indicator for these patients (Yang *et al.*, 1999).

MATERIALS

1. Dulbecco's phosphate-buffered saline (PBS): 100 mg anhydrous calcium chloride, 200 mg potassium chloride, 200 mg monobasic potassium phosphate, 100 mg magnesium chloride · 6 H_2O; 8 g sodium chloride, and 2.16 g dibasic sodium phosphate · 7 H_2O; bring vol to 1 L with deionized glass-distilled water (pH 7.4). Tris-buffered saline (TBS): 8 g sodium chloride, 200 mg potassium chloride, and 3 g Tris base, bring vol to 1 L with deionized glass-distilled water, adjust to pH 7.4 HCl.

2. Primary antibody reactive against the desired antigen (anti-human caveolin-1 antibody, C37120, a mouse monoclonal antibody, 1:1000 dilution; BD Transduction Laboratories, Lexington, KY).

3. A secondary antibody labeled with biotin, which is reactive against the species of immunoglobulin used for the primary reagent.

4. 10% normal serum in PBS from the species from which the secondary antibody was generated (for this example: horse serum).

5. Avidin-biotin complex horseradish peroxidase reagent (Vectastain ABC kit, Vector Laboratories, Inc., Burlingame, CA).

6. Chromogen: 0.01% hydrogen peroxide (H_2O_2) and 0.05% diaminobenzidine tetrachloride (DAB) in 0.05 M Tris buffer (pH 7.2).

7. Counterstain (e.g., hematoxylin or methyl green).

8. 6 M ammonium hydroxide diluted in deionized water to 3.7 mM.

9. Xylene.

10. 30% H_2O_2 diluted in deionized water to 3%.

11. 100% ethanol diluted in deionized water to 95%, 85%, 75%, and 50% ethanol.

12. 0.01 N sodium citrate.

METHOD

For Paraffin Tissue Section

1. Place paraffin sections onto glass slides.

2. Deparaffinize the sections in xylene for 30 min or soak the tissue section in a container filled with xylene for 5 min. Repeat and soak in fresh xylene four more times.

3. Hydrates sections in graded ethanol and water by incubating the sections in 100%, 95%, 85%, 75%, and 50% ethanol every 4 min and shaking gently.

4. Rinse the sections twice with TBS or PBS.

5. Place the sections in 0.01 N sodium citrate and autoclave at 121°C for 5 min to retrieve the antigen. Alternatively, sections can be heated in a microwave oven (700–800W) for 5–10 min and cooling at room temperature.

6. Treat the sections with 3% H_2O_2 for 15 min to abolish endogenous peroxidase activity.

7. Rinse the sections with TBS or PBS.

8. Incubate sections in a moist chamber with 1:20 dilution of normal preimmune serum from species, providing the biotinylated secondary antibody for at least 10 min.

9. Blot off excess serum from sections and rinse again with TBS or PBS.

10. Incubate sections with primary antiserum (dilute as suggested) at room temperature for 1–2 hr. Alternative choices for incubating include 37°C incubation for 30 min or overnight at 4°C.

11. Rinse the sections with TBS or PBS three times for 5 min each.

12. Incubate sections with biotin-labeled bridge antiserum (dilute as suggested) for 20–30 min.

13. Rinse the sections with TBS or PBS three times for 5 min each.

14. Incubate sections for 30 min with avidin-biotin complex formed by mixing solutions of avidin and biotin-conjugated peroxidase. The complex should be prepared 30 min before use.

15. Rinse the sections with TBS or PBS three times for 5 min each.

16. Incubate sections in 0.01% H_2O_2 and 0.05% DAB in 0.05 M Tris buffer, pH 7.2, for 2–10 min. Monitor development of color under the microscope.

17. Stop the reaction by rinsing the sections with distilled water for 5–10 min.

18. Counterstain with hematoxylin or methyl green. Rinse in large amount of TBS. Soak in 3.7 mN ammonia for 3–5 sec and then wash with a lot of water.
19. Dehydrate sections in graded ethanol and water by incubating the sections in 50%, 75%, 85%, 95%, and 100% ethanol every 4 min and shaking gently.
20. Expose to xylene and add entellan for mounting.

RESULTS AND DISCUSSION

In this study, we analyzed paraffin-embedded specimens of clinically well-defined and pathologically proved lung adenocarcinoma obtained from the Department of Pathology, National Taiwan University Hospital. Among the 95 lung cancer specimens, 35 cases (labeled as "P/N+") already had tumor cells metastasizing to the ipsilateral hilar or peribronchial lymph nodes. The other 60 cases (labeled as "P/N−") had tumor cells in the lung without evidence of tumor metastasis to regional lymph nodes (Ho *et al.*, 2002).

Caveolin-1 immunoreactivity was normally localized to fibroblasts, type I pneumocytes, and endothelial cells of blood vessels in all tissue specimens examined, which could serve as an internal quality control of immunohistochemistry. Caveolin-1 immunoreactivity, however, was not seen in primary lung adenocarcinoma cells in the great majority of cases. In contrast, immunohistochemical examination of ipsilateral hilar/peribronchial lymph nodes (35 cases) with metastatic lung adenocarcinoma revealed moderate to intense caveolin-1 immunoreactivity in the cytoplasm of varying percentages of cancer cells in all lymph nodes except one (Figure 23).

Altogether, when we artificially chose 30% caveolin-1 immunoreactivity as a cutoff value for the assignment of positivity (or negativity) of caveolin-1 staining, only four cases had caveolin-1 positivity at the primary lesion site, and all of them had metastatic tumor cells in lymph nodes, which also had caveolin-1 positivity. Six cases had caveolin-1 positivity in the nodal metastasis site, but not at the primary site. Twenty-five cases were caveolin-1–negative both at the primary lung lesion site and at nodal metastases. No cases had caveolin-1 positivity at the primary lung lesion site, but were caveolin-1-negative at the nodal metastasis site. Therefore, there was a trend toward caveolin-1 expression in regional nodal metastases rather than in the primary lung lesion site.

Analysis of the association of caveolin-1 expression with the patients' survival rate was performed on 35 patients who had lung adenocarcinoma with regional nodal metastases (P/N+). The survival time after surgery for these four caveolin-1^{+ve} and 31 caveolin-1^{-ve} cases was calculated by the Kaplan-Meier method. Although the survival time was not significantly different (P = 0.0704) between the two groups, a trend toward a poor prognosis for caveolin-1^{+ve} patients was noted. When all variables were evaluated, the multiple Cox regression model suggested that the expression of caveolin-1 was an independent factor for prediction of poor survival in patients with pulmonary adenocarcinoma (hazard ration 7.189; P = 0.0041).

We also investigated the expression pattern of caveolin-1 in a series of lung carcinoma cell lines (CL) with varying invasive/metastatic ability. Caveolin-1 protein was undetectable in low-invasive cells. In contrast, it was abundantly expressed in highly invasive cells (Ho *et al.*, 2002). To assess whether the presence of caveolin-1 in CL cells may potentiate cellular invasive capability, we transfected plasmids harboring a full-length human *caveolin-1*-encoding cDNA into less invasive cells. Tet-Off induced cells and constitutively

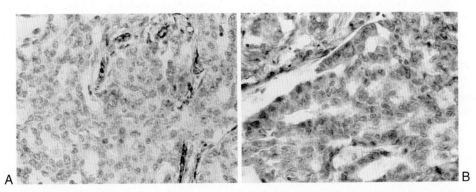

Figure 23 Caveolin-1 immunoreactivity in lung adenocarcinoma and ipisilateral hilar/peribronchial lymph nodes with tumor metastases. **A:** Negative or faint staining of caveolin-1 in the primary lung adenocarcinoma. **B:** Positive caveolin-1 staining was detected in cancer cells in a lymph node with metastatic lung adenocarcinoma.

expressing cells were established. Both showed enhanced invasive capability when caveolin-1 was expressed. The most interesting finding is that expression of caveolin-1 in the non–caveolin-1 expressing cells induced filopodia formation in these CL cells. Cells cultured under delipidated condition lost the ability to form filopodia despite the presence of caveolin-1 expression (Ho *et al.*, 2002).

Our results are consistent with the dual function of *caveolin-1* both as a tumor suppressor gene and as a metastasis-promoting gene. The same observation was also noted from varying invasive/metastatic cell lines. In a low-invasive lung cancer cell line, caveolin-1 expression was absent or extremely low, which is consistent with the fact that down-regulated caveolin-1 expression facilitates cell transformation. However, in more invasive cell lines, caveolin-1 expression was abundant, which is consistent with the proposed metastasis-promoting function of caveolin-1 (Kato *et al.*, 2002; Suzuoki *et al.*, 2002; Yang *et al.*, 1998). When we introduced caveolin-1 expression into less-invasive cells, the ability of cell invasiveness was enhanced.

Several mechanisms have been elucidated to explain how down-regulated caveolin-1 helps cell transformation during tumorigenesis. The reciprocal negative transcriptional regulation between caveolin-1 and other growth factor or signal transducers like α-folate receptor, p42/44 MAP kinase, and the neu oncogene has been proposed to account for cell transformation in various cell types (Li *et al.*, 1996; Razani *et al.*, 2001). In addition, transcriptional silencing by methylation of CpG islands in the 5 promoter region of the *caveolin-1* gene was found in two human breast cancer cell lines that failed to express the caveolin-1 protein (Engleman *et al.*, 1999).

In summary, our study clearly revealed that caveolin-1 in CL cells is necessary for mediating filopodia formation, which may enhance cell migration and, in part, the invasive ability of lung adenocarcinoma cells. Caveolin-1, the principal component of caveolae, binds cholesterol and sphingolipids to form platforms and serves to support membrane traffic and signal transduction. Thus, it is not totally unexpected that the formation of filopodia mediated by up-regulated caveolin-1 in cells requires the presence of lipid in culture medium. Recently, Kanzaki *et al.* (2002) reported that cholesterol depletion effectively disrupted the colocalization of F-actin with caveolae (Kanzaki *et al.*, 2002). When a yeast two-hybrid screen was used, the actin-binding filamin was identified as a ligand for caveolin-1 (Stahlhut *et al.*, 2000). Possibly, the interaction between caveolin-1 and actin or filamin provides a physical link for caveolae and the cytoskeleton, and,

thus, participation in filopodia formation. How caveolin-1 interacts with the intracellular cytoskeleton (especially that in filopodia) in cells with various transformation or differentiation states will be another interesting issue for understanding the role of caveolin-1 as a tumor metastasis-promoting molecule.

References

Carman, C.V., Lisanti, M.P., and Benovic, J.L. 1999. Regulation of G protein-coupled receptor kinases by caveolin. *J. Biol. Chem.* 247:8858–8864.

Couet, J., Li, S., Okamoto, T., Scherer, P.S., and Lisanti, M.P. 1997. Molecular and cellular biology of caveolae: Paradoxes and plasticities. *Trends Cardiovasc. Med.* 7:103–110.

Engelman, J.A., Wycoff, C.C., Yasuhara, S., Song, K.S., Okamoto, T., and Lisanti, M.P. 1997. Recombinant expression of caveolin-1 in oncogenically transformed cells abrogates anchorage-independent growth. *J. Biol. Chem.* 272:16374–16381.

Engelman, J.A., Zhang, X.L., and Lisanti, M.P. 1999. Sequence and detailed organization of the human caveolin-1 and 2 genes located near the D7S522 locus (7q31.1). Methylation of a CpG island in the 5′ promoter region of the caveolin-1 gene in human breast cancer cell lines. *FEBS Lett.* 448:221–230.

Felley-Bosco, E., Bender, F.C., Courjault-Gautier, F., Bron, C., and Quest, A.F. 2000. Caveolin-1 down-regulates inducible nitric oxide synthase via the proteosome pathway in human colon carcinoma cells. *Proc. Natl. Acad. Sci. USA* 97:14334–14339.

Feron, O., Dessy, C., Moniotte, S., Desager, J.P., and Balligand, J.L. 1999. Hypercholesterolemia decreases nitric oxide production by promoting the interaction of caveolin and endothelial nitric oxide synthase. *J. Clin. Invest.* 103:897–905.

Fielding, C., and Fielding, P.E. 2001. Caveolae and intracellular trafficking of cholesterol. *Adv. Drug Deliv. Rev.* 49:251–264.

Fra, A.M., Williamson, E., Simons, K., and Parton, R.G. 1995. De novo formation of caveolae in lymphocytes by expression of VIP21-caveolin. *Proc. Natl Acad. Sci. USA* 92:8655–8659.

Galbiati, F., Volonte, D., Engelman, J.A., Watanabe, G., Burk, R., Pestell, R.G., and Lisanti, M.P. 1998. Targeted down-regulation of caveolin-1 in sufficient to drive cell transformation and hyperactivate the p42/44 MAP kinase cascade. *EMBO J.* 17:6633–6648.

Ho, C.C., Huang, P.H., Huang, H.Y., Chen, Y.H., Yang, P.C., and Hsu, S.M. 2002. Up-regulated caveolin-1 accentuates the metastasis capability of lung adenocarcinoma by inducing filopodia formation. *Am. J. Pathol.* 161:1647–1656.

Hulit, J., Bash, T., Fu, M., Galbiati, F., Albanese, C., Sage, D.R., Schlegel, A., Zhurinsky, J., Shtutman, M., Ben-Ze'ev, A., Lisanti, M.P., and Pestell, R.G. 2000. The cyclin D1 gene is transcriptionally repressed by caveolin-1. *J. Biol. Chem.* 275:21203–21209.

Kanzaki, M., and Pessin, J.E. 2002. Caveolin-associated filamentous actin (Cav-actin) defines a novel F-actin structure in adipocyte. *J. Biol. Chem.* 277:25867–25869.

Kato, K., Hida, Y., Miyamoto, M., Hashida, H., Shinohara, T., Itoh, T., Okushiba, S., Kondo, S., and Katoh, H. 2002. Overexpression of caveolin-1 in esophageal squamous cell carcinoma correlates with lymph node metastasis and pathologic stage. *Cancer 94:* 929–933.

Li, S., Couet, J., and Lisanti, M.P. 1996. Src tyrosine kinases, G_x subunits, and H-Ras share a common membrane-anchored scaffolding protein, Caveolin. Caveolin binding negatively regulates the auto-activation of src tyrosine kinases. *J. Biol. Chem.* 271:29182–29190.

Liu, P., Rudick, M., and Anderson, R.G.W. 2002. Multiple functions of caveolin-1. *J. Biol. Chem. 277:*41295–41298.

Palade, G.E., and Bruns, R.R. 1968. Structural modulations of plasmalemmal vesicles. *J. Cell Biol. 37:*633–649.

Razani, B., Schlegel, A., Liu, J., and Lisanti M.P. 2001. Caveolin-1, a putative tumor suppressor gene. *Biochem. Soc. Trans. 29:* 494–499.

Rothberg, K., Heuser, J.E., Donzell, W.C., Ying, Y.S., Glenney, J.R., and Anderson, R.G.W. 1992. Caveolins, a protein component of caveolae membrane coats. *Cell 68:*673–682.

Roy, S., Leutterforst, R., Harding, A., Apolloni, A., Etheridge, M., Stang, E., Rolls, B., Hancock, J.F., and Parton, R.G. 1999. Dominant-negative caveolin inhibit H-Ras function by disrupting cholesterol-rich plasma membrane domains. *Nature Cell Biol. 1:*98–105.

Simionescu, N. 1983. Cellular aspects of transcapillary exchange. *Physiol. Rev. 63:*1563–1560.

Stahlhut, M., and Deurs, B.V. 2000. Identification of filamin as a novel ligand for caveolin-1: Evidence for the organization of caveolin-1-associated membrane domains by the actin cytoskeleton. *Mol. Biol. Cell 11:*325–337.

Suzuoki, M., Miyamoto, M., Kato, K., Hiraoka, K., Oshikiri, T., Nakakubo, Y., Fukunaga, A., Shichinohe, T., Shinohara, T., Itoh, T., Kondo, S., and Katoh, H. 2002. Impact of caveolin-1 expression on prognosis of pancreatic ductal adenocarcinoma. *Br. J. Cancer 87:*1140–1144.

Timme, T.L., Goltsov, A., Tahir, S., Li, L., Wang, J., Ren, C., Johnston, R.N., and Thompson, T. C. 2000. Caveolin-1 is regulated by C-Myc and suppresses C-Myc-induced apoptosis. *Oncogene 19:*325–326.

Yang, G., Truong, L.D., Timme, T.L., Ren, C., Wheeler, T.M., Park, S.H., Nasu, Y., Scardino, P.T., and Thompson, T.C. 1998. Elevated expression of caveolin is associated with progression in prostate and breast cancer. *Clin. Cancer Res. 4:*1873–1880.

Yang, G., Truong, L.D., Wheeler, T.M., and Thompson, T.C. 1999. Caveolin-1 expression in clinically confined human prostate cancer: A novel prognostic marker. *Cancer Res. 59:*5719–5723.

9

Role of Thyroid Transcription Factor-1 in Pulmonary Adenocarcinoma

Yasushi Yatabe

Introduction

Tissue-specific gene expression is mediated largely by transcription factors, and a master regulatory gene is thus a potential marker of cellular lineage. The myoD gene family is an example of such a lineage marker (Weintraub *et al.*, 1991). The expression of these genes is restricted to the skeletal muscle cell lineage, and ectopic expression switches the cell type to a myogenic phenotype. Thus, the myoD family fulfills most criteria for a lineage marker due to its restricted expression and its function as a master regulatory molecule (Weintraub *et al.*, 1991). It is noted that the gene is not expressed after cell differentiation is complete. Using this property, the expression of myoD in tumor cells indicates both neoplastic nature and differentiation into skeletal muscle cells, implying the diagnosis of rhabdomyosarcoma. In this manner, the lineage markers decide the fate of the cells, and the cancer is also likely to be characterized by the features of the originating cells, acquired through the lineage marker.

Thyroid transcription factor-1 (TTF-1) also known as Nkx2.1 or thyroid-specific enhancer-binding protein, (Korfhagen *et al.*, 1997; Mendelson, 2000) is a homeodomain-containing transcription factor that regulates tissue-specific expression of the surfactant apoprotein A (SPA) (Bruno *et al.*, 1995), surfactant apoprotein B (Yan *et al.*, 1995), surfactant apoprotein C (Kelly *et al.*, 1996), Clara cell antigen (Toonen *et al.*, 1996), and T1α (Ramirez *et al.*, 1997) by directly binding to these promoters. The expression of TTF-1 is initiated at a very early stage of lung morphogenesis. In the developing mouse lung, TTF-1 expression first appears at emergence of the laryngotracheal diverticulum and is localized primarily in the branching bronchial epithelium for the next 6 days (Kimura *et al.*, 1996; Lazzaro *et al.*, 1991; Zhou *et al.*, 1996). Once peripheral airway tubes develop, the expression shifts to the peripheral airway epithelium, and this pattern is retained until death (Kimura *et al.*, 1996; Lazzaro *et al.*, 1991; Zhou *et al.*, 1996). TTF-1$^{-/-}$ knockout mice show a tracheoesophageal fistula and severe pulmonary hypoplasia, suggesting TTF-1 also plays a crucial role in lung morphogenesis (Kimura *et al.*, 1996; Minoo *et al.*, 1999). TTF-1 is indispensable for lung function.

Several reports have already described the differences in TTF-1 expression between histologic types.

Handbook of Immunohistochemistry and *in situ* Hybridization of Human Carcinomas, Volume 1: Molecular Genetics; Lung and Breast Carcinomas

169

Copyright © 2004 by Elsevier (USA)
All rights reserved.

The expression of TTF-1 appears in 71–76% (mean, 72%) of adenocarcinomas and 81–92% (mean, 89%) of small cell carcinomas (SCLC), whereas it was not found, or only found at very low frequency, in other types of non–small cell carcinoma (NSCLC) and carcinoma of the other organs, except thyroid cancers (Di Loreto *et al.*, 1997; Fabbro *et al.*, 1996; Kaufmann *et al.*, 2000; Ordonez, 2000b; Puglisi *et al.*, 1999; Yatabe *et al.*, 2002). Therefore, most of the reports emphasized the practical usefulness of TTF-1 in the differential diagnosis of lung cancer from nonpulmonary cancers, as a result of the tissue specific expression (Di Loreto *et al.*, 1997; Fabbro *et al.*, 1996; Katoh *et al.*, 2000). In contrast, few studies referred to the possible significance of TTF-1 expression in lung carcinogenesis, despite its crucial roles in lung development and maintenance of pulmonary function by means of SPA and Clara cell-specific 10-KD protein (CC10) induction.

Based on the recent development of molecule-targeted drugs, tumor classification schemes are shifting to that based on the molecular mechanisms of carcinogenesis. Imatinib mesylate (Demetri, 2001; Joensuu *et al.*, 2001) elucidated the molecular basis of relatively rare gastrointestinal tumors and proposed new subclassification within the ordinary classification schema (Berman *et al.*, 2001). Because of its multistep derivation, the tumors should be quite heterogeneous. To understand the underlying mechanism, it is important to recognize what categorization is biologically significant and what discriminator is used for the categorization. In this chapter, we attempt to describe the biological significance of TTF-1 expression in lung adenocarcinomas and show that TTF-1 sheds light on a subtype of lung adenocarcinomas, in terms of cellular lineage and molecular carcinogenesis (Yatabe *et al.*, 2002).

MATERIALS

1. Phosphate-buffered saline (PBS) (for 20 L).
   a. 9 g of $NaH_2PO_42H_2O$, 64.5g of $NaH_2PO_412H_2O$ and 160 g of NaCl.
   b. Place in the carboy/plastic container, pour 40 L of distilled water, and mix thoroughly.
   c. Adjust pH to 7.4, if necessary (unnecessary most times).
2. McIlvaine citrate buffer for antigen retrieval (for 500 mL).
   a. 3.76g of citrate acid monohydrate and 23g of $NaH_2PO_412H_2O$.
   b. Dissolve well with 500 mL of distilled water, and adjust pH to 6.4.
3. Antibody dilution buffer.
   a. Filtered PBS with 0.001% thimerosal (for an antiseptic) and 0.01% bovine serum albumin.

4. Aminobenzidine tetrachloride (DAB) solution: DAB tablets are commercially available (Dako Copenhagen, Denmark, or Novocastra, Newcastle upon Tyne, U.K.).
5. Mayer's hematoxylin.
6. Vextastain elite ABC kit (Vector Laboratories, Inc., Burlingame, CA); follow the manufacturer's instructions.

METHODS

The staining procedure is the same as that used for common immunohistochemical analysis, except for the use of freshly prepared sections. This is critical for TTF-1 staining, and the staining must be performed within a week of preparation. Non-neoplastic type II pneumocytes served as internal controls for antigen preservation.

1. Deparaffinize with four changes of xylene for 5 min each, followed by four changes of 100% ethanol for 4 min each.
2. Blocking of endogenous peroxidase in methanol with 0.1% H_2O_2 for 20 min.
3. Rinse sections in three changes of prechilled PBS.
4. Antigen retrieval:
   a. Place sections in a heat-stable plastic Coplin jar.
   b. Fill with an excess amount of citrate buffer (pH 6.4).
   c. Autoclave for 10 min.
   d. Keep the Coplin jar in the processor until the solution is cooled down to room temperature (~ 60 min).
5. Rinse sections in three changes of prechilled PBS.
6. Remove the excess PBS, place slides horizontally in a humidity incubation chamber, and cover specimens with 3–4 drops of 10% normal serum for 30 min (swine serum is commonly used) to avoid nonspecific antibody binding.
7. Remove the excess normal serum, place slides horizontally in a humidity incubation chamber, and cover specimens with 3–4 drops of TTF-1 antibody (8G7G3, Dako, Copenhagen, Denmark; 1:150 diluted with the antibody dilution buffer) for 1 hr at room temperature or overnight at 4°C.
8. Wash sections in three changes of prechilled PBS.
9. Remove the excess PBS, place slides horizontally in a humidity incubation chamber, and cover specimens with 3–4 drops of anti-mouse antibody for 30 min.
10. Wash sections in three changes of prechilled PBS.
11. Remove the excess PBS, place slides horizontally in a humidity incubation chamber, and cover specimens with 3–4 drops of ABC complex for 30 min.

12. Wash sections in three changes of prechilled PBS.
13. Develop the color with DAB solution (0.6 mg/mL PBS + 0.1% H_2O_2) for 5–10 min.
14. Wash sections with tap water.
15. Counterstain with hematoxylin.
16. Dehydrate using four changes of ethanol, and then clear with three changes of xylene, 3 min each.
17. Coverslip sections with Permount (Fisher Scientific, Pittsburgh, PA).

The positive signals are seen as a brown color that is localized in the nuclei.

RESULTS AND DISCUSSION

TTF-1 Expression as a Marker of Terminal Respiratory Unit in the Normal Lung

It has been reported that TTF-1 expression during rat development is restricted to the thyroid, lung, and forebrain. In the former two organs, the expression is observed at the earliest stage of development and is preserved throughout lifetime (Kimura et al., 1996; Minoo et al., 1999; Stahlman et al., 1996). Expression of TFF-1 in the human lung is also restricted to the peripheral portion in both fetal and adult lungs (Stahlman et al., 1996, Yatabe, 2002), and expressing cells include pneumocytes (both type I and type II) and a part of small-sized bronchioles (Figure 24). These epithelia stained very uniformly and consistently, and TTF-1 appeared to label a series of cells that represented a certain functional unit or common lineage. Taken together with transcriptional activity of TTF-1 for functional molecules, such as surfactant apoprotein and Clara cell antigen, we therefore considered these epithelia as a functional or lineage unit, termed the terminal respiratory unit (TRU). It is of note that the morphological difference between TTF-1 positive and negative bronchioles was obscured, and that not all, but only a part of the bronchioles, (more peripheral portion of bronchioles) included the TRU.

TTF-1 Expression in the Neoplasia

To examine whether the restricted expression in normal tissue is reflected in the tumors, we applied a tissue microarray method in addition to regular examination with whole sections. The results are summarized in Table 10. The distribution of the TTF-1 expressing tumors is quite similar to that of the pattern in normal tissues, and the expression is restricted to lung adenocarcinoma and thyroid tumors, with a few exceptions. One of the exceptions is SCLC (Figure 24). Because none of the carcinoid tumors expressed TTF-1

(Fabbro et al., 1996; Sturm et al., 2002), the expression is not related to neuroendocrine differentiation. Recent evidences in stem cell research indicate that stem cells expressed a broad range of genes (Akashi et al., 2003; Shamblott et al., 2001). Morphologically, SCLCs appear as very primitive or undifferentiated cells, and TTF-1 expression in SCLCs is suggested to have some association with the multilineage gene expression of the stem cells. Indeed, ectopic expression of c-Kit, stem cell factor (Hibi et al., 1991; Sekido et al., 1991) and some cancer testis antigens (Sugita et al., 2002) are more common in SCLC than in NSCLC. Although expression of TTF-1 in nonpulmonary small cell carcinomas is controversial (Agoff et al., 2000; Kaufmann et al., 2000; Oliveira et al., 2001; Ordonez, 2000b), the hypothesis is supported by our data showing frequent positivity in nonpulmonary small cell carcinoma. Of the nine cases, six of the nonpulmonary small cell carcinoma expressed TTF-1, and in the combined small cell and ordinary carcinomas, only portions showing the small cell carcinoma morphology were positive for TTF-1. As for the tumors with small cell carcinoma morphology, the TTF-1 expression may represent undifferentiated features of the cancer cells.

In thyroid carcinomas, TTF-1 is invariably expressed in follicular neoplasm and papillary carcinoma (Bejarano et al., 2000; Katoh et al., 2000), reflecting the normal expression pattern. However, negative expression of TTF-1 in anaplastic carcinoma is rather common. This was explained by "dedifferentiation" of the follicle-derived tumors (Bejarano et al., 2000; Heldin et al., 1991).

It is of note that one case of breast cancer was shown to express TTF-1. The case is a 43-year-old female without prior history of any cancers. Two cancer nodules (25 mm and 5 mm in the largest dimensions) in the right breast were surgically resected. Although histology showed an ordinary infiltrating lobular carcinoma (Figure 24). TFF-1 was expressed in both tumors. The presence of in situ lesions and no other cancers during 8 years of follow-up indicates the cancers are actually of breast origin.

TTF-1 Expression in the Lung Adenocarcinoma

The expression of TTF-1 was observed in 75.7% of lung adenocarcinomas in our series, and about 70–75% of the adenocarcinomas have been reported to be positive in the literature (Kaufmann et al., 2000; Ordonez, 2000a; Pelosi et al., 2001; Puglisi et al., 1999). The expression pattern is characterized by quite uniform staining; when the tumor cells are positive, almost all cancer cells express TTF-1, regardless of malignancy potential, differentiation, and cytologic atypia. For example,

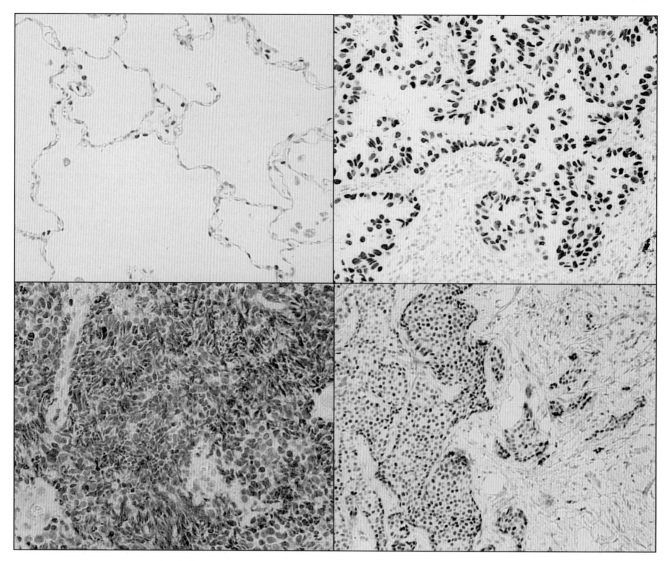

Figure 24 TTF-1 expression in normal peripheral lung (upper left), lung adenocarcinoma (TRU-type, upper right), small cell lung cancer (lower left), and an unusual case of breast cancer (lower right).

adenocarcinoma frequently presents as a low-grade lesion in the periphery and high-grade in the center; TTF-1 is expressed in both the low-grade peripheral lesion and central high-grade lesion in the individual TTF-1 positive tumors. This feature is maintained even in the metastatic sites. This is contrasted to the expression pattern of surfactant apoprotein A, a functional marker of the lung, which was frequently lost in the metastatic site. In this sense, it is quite reasonable to use TTF-1 to detect micrometastasis in this lymph nodes and malignant effusion of lung adenocarcinoma.

Conversely, ~75% of TTF-1 positivity indicated that lung adenocarcinomas could be subdivided by their TTF-1 expression status, which is a functionally important transcription factor in the lung and a marker of

TRU in the normal lung. This raised two questions: 1) whether TTF-1 is useful as a marker of adenocarcinoma derived from the TRU, and 2) what is the significance of subdividing by TTF-1 status.

Morphologic Characteristics of TTF-1 Expressing Adenocarcinoma

To obtain the morphologic characteristics of TTF-1 expressing tumors, we focused on the similarity to normal counterparts of the airway epithelium. Based on the morphological resemblance to type II pneumocytes, Clara cells, and small-sized bronchioles, we simply categorized the tumor as TRU or non-TRU type (Figure 25). This morphological classification is very

Table 10 TTF-1 Expression in Various Tumors

Organ	Subtype	n=	Positive
		Nonneuroendocrine Tumors	
Breast	AD	14	1(7.1%)
Cervix uteri	AD	5	0(0%)
	SQ	10	0(0%)
Colon	AD	16	0(0%)
Endometrium	AD	10	0(0%)
Esophagus	AD	1	0(0%)
	SQ	7	0(0%)
Head and neck	SQ[a]	20	0(0%)
Lung	AD	296[b]	224(75%)
	SQ	74	2(2.7%)
	LA	25	7(28.0%)
	AS	11	2(18.2%)
Ovary	Adenoma	4	0(0%)
	AD	10	0(0%)
Pancreas	AD	3	0(0%)
Parotid glands	Benign tumors	38	0(0%)
	AD	60	0(0%)
Stomach	AD	19	0(0%)
Thyroid	Adenomatous goiter	6	6(100.0%)
	Follicular adenoma	18	18(100.0%)
	Follicular carcinoma	13	12(92.3%)
	Papillary carcinoma	20	19(95.0%)
	Anaplastic carcinoma	2	0(0%)
		Neuroendocrine Tumors	
Lung	Typical carcinoid	2	0(0%)
	LCNEC	15	13(86.7%)
	SCLC	26	25(96.2%)
Esophagus	Small cell carcinoma	6	4(66.7%)
Stomach	Carcinoid	3	0(0%)
Colon	Carcinoid	4	0(0%)
Bladder	Small cell carcinoma	2	1(50%)
Uterine cervix	Small cell carcinoma	1	1(100%)
Thymus	Carcinoid	1	0(0%)

[a]Including three head and neck cancer and one esophageal cancer cases that were weakly positive.

[b]64 were examined with regular whole sections; the other results were from tissue microarrays.

AD, adenocarcinoma; *SQ*, squamous carcinoma; *LA*, large cell carcinoma; *LCNEC*, large cell neuroendocrine tumor; *SCLC*, small cell lung cancer.

close to Shimosato's cytological classification of lung adenocarcinoma (Shimosato, 1989), except dealing with adenocarcinoma resembling bronchioles. We include the adenocarcinoma resembling bronchioles into TRU type. This results from the unexpected findings that TTF-1 labeled small-sized bronchioles in addition to pneumocytes, like a unique functional or lineage unit. Because bronchial-surface type and bronchiole-like adenocarcinoma mimic each other, we therefore paid most attention to distinguishing between them. Major differential characteristics of bronchiole-like adenocarcinoma includes low columnar cells with a dome-shaped

protrusion of each luminal cellular border, but not showing a smooth line at luminal border (Figure 25).

With this simple classification of lung adenocarcinomas, 48 of 64 (75%) lung adenocarcinomas in our series were categorized as the TRU subtype and 16 cases (25%) as the non-TRU subtype (Table 11). This classification is justified by the frequent expression of the surfactant apoprotein A (SPA) in the TRU subtype, and also by the fact that SPA is expressed in half of the TRU adenocarcinomas, in contrast to none of the non-TRU adenocarcinomas (Table 11). The majority of TTF-1 positive cases showed TRC morphology (42/46, 91%),

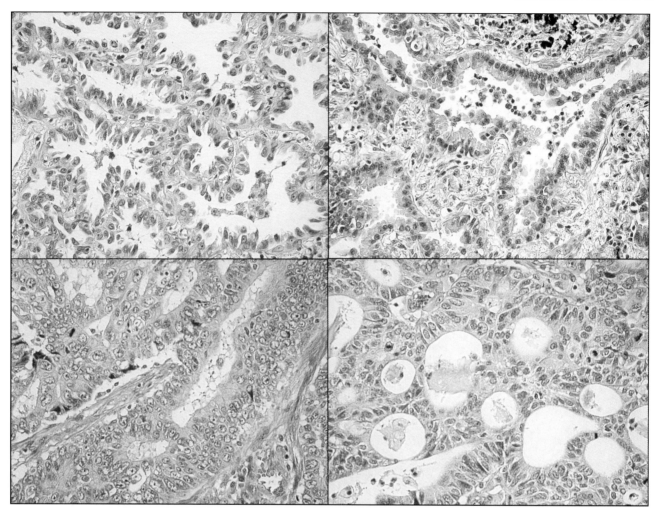

Figure 25 Histologic features of TRU (upper right and left) and non-TRU (lower left and right) types of adenocarcinomas. Upper left shows bronchiole-like adenocarcinoma (TTF-1 positive), which mimics bronchial surface epithelium-like adenocarcinoma (TTF-1 negative, lower left).

whereas a major proportion of TTF-1 negative adenocarcinomas (12/18, 66.7%) belong to the non-TRU subtype. Conversely, 88% (42/48) of adenocarcinomas with the TRU morphology were TTF-1 positive, whereas only 25% (4/16) of non-TRU tumors expressed TTF-1 (Table 12). These results implied that TTF-1 expression is largely maintained even after malignant transformation of TRU cells. Therefore, TTF-1 is capable of being used as a lineage marker for TRU cells.

Difference in Carcinogenetic Mechanism by TTF-1 Expression Status

Susceptibility to a particular carcinogen is likely to differ among cell types due to intrinsic cellular features or anatomic features, such as carcinogen accessibility and clearance rates. For example, an erythroid precursor cell is particularly susceptible to benzene toxicity (Corti *et al.*, 1998), whereas the size of the particles inhaled as carcinogen determines the anatomic sites to which it will be deposited. Therefore, it is quite reasonable to suggest that cellular lineage affects the carcinogenetic mechanism significantly. Accordingly, clinicopathologic features and cancer-associated genes were compared between TTF-1 positive and negative lung adenocarcinomas. The results are summarized in Table 12. Tumors with TTF-1 expression were significantly more prevalent in females and nonsmokers. Because gender and smoking status were tightly correlated with each other (i.e., 4 of 31 females were smokers and only 4 of 33 men were nonsmokers) it was unclear which of them affected TTF-1 status more significantly in the cohort. The close relationship reflected cultural background, and recently this has been

Table 11 Clinicopathologic Features by TTF-1 Expression Status

$n=$	TTF-1 (+) 46(72%)	TTF-1(-) 18(28%)	p-value[a]
Morphological Characteristics			
SPA (<1+/2≥)	22/24	2/16	<0.01
TRC/non-TRC morphology	42/4	6/12	<0.01
Clinicopathologic Characteristics			
Female/male ratio	27/19	4/14	<0.01
Smoker/nonsmoker	26/20	13/5	0.04
pStage (I/II/IIIA)	31/2/13	11/2/5	0.60
pN (0/1/2)	31/2/13	12/2/4	0.57
Cancer-Associated Genes			
p53 accumulation (+/−)	12/34	12/6	<0.01
p53 mutation/wildtype[b]	21/51	17/16	0.03
G:T > C:A transversion[b]	3/14(18%)	6/9(40%)	0.24
K-Ras mutation/wildtype[b]	4/32	3/19	1.00
CCND1 (+/−)	29/17	7/11	0.100
Rb (+/−)	41/5	11/7	0.02
p27 (1+/2+/3+/4+)	0/10/14/22	3/7/5/3	<0.01
COX2 (+/−)	35/10	15/3	0.74

[a]Each value represents the results from Fisher's exact test or chi-square test.

[b]Analysis using a different cohort from that of the other analysis.

Note: Underlined values indicate statistically significant differences.

SPA, surfactant apoprotein A; TRC, terminal respiratory unit.

changing. Using another recent cohort, the logistic regression analysis suggested that smoking status was more highly associated with TTF-1 expression than gender. In contrast to this relationship, pathologic stage as well as local tumor-factor (pT) and nodal status are not associated with TTF-1 expression status, supporting the idea that TTF-1 is not a marker of malignant potential or advanced disease, but a lineage marker of the originating cellular feature. The prognostic imact of TTF-1 status further confirmed the idea, and the Cox-proportional hazard model resulted in no prognostic significance of TTF-1 expression in lung adenocarcinoma.

Cancer-associated molecules, p53 accumulation, p53 mutation, K-Ras mutation, and expression of cyclin D1 (CCND1), Rb, p27[KIP1], and COX2 were compared between TTF-1 positive and negative lung adenocarcinomas. The well-known tumor suppressor gene p53 is frequently involved in a variety of human cancers. Approximately 50% of NSCLC and 90% of SCLCs exhibit the p53 mutation. Meanwhile K-Ras is also a well-known oncogene, and the mutation of K-Ras is frequently detected in adenocarcinoma but is quite rare in SCLC. In some reports (Kobayashi et al., 1990; Marchetti et al., 1996), the K-Ras is mutated preferentially in mucinous bronchioloalveolar carcinomas, which morphologically resemble pancreatic cancer.

We reported that the lack of the expression in the NSCLCs (61%) was significantly associated with shorter survival by the Cox-proportional hazard model (Nishio et al., 1997). This tendency was particularly remarkable in adenocarcinoma. One of the causative genes for retinoblastoma is Rb, and nearly all SCLCs and ~20% of NSCLCs lack Rb expression (Nishio et al., 1997) due to inactivation of the Rb gene. The CDK inhibitor p27[KIP1] is considered responsible for the onset and/or maintenance of the quiescent state (Rivard et al., 1996; Zhang et al., 2000), and only differentiated cells expressed this inhibitor in vivo (Endl et al., 2001). Whereas the majority of SCLCs show increased staining when compared with normal epithelium, p27[KIP1] was reduced in 72% of NSCLCs and its reduction was correlated with poor prognosis (Yatabe et al., 1998b). COX2 is an enzyme involved in the conversion of arachidonic acid to prostanoids and plays a well-known role in inflammatory reactions. In additions, recent reports suggest that it has various other functions, including an association with carcinogenesis (Vane, 1994; Williams et al., 1997). The molecule was initially investigated in association with colon carcinogenesis, because COX2 was up-regulated in ~50% of adenoma and 80–85% of adenocarcinoma of the colon (Eberhart et al., 1994; Sano et al., 1995), and mice without a functional COX2 gene had a significantly reduced incidence of intestinal polyps (Oshima *et al.*, 1996). In the lung, we have reported, increased expression of COX2 was detected in about one-third of precancerous and carcinoma *in situ*, and in 70% of invasive adenocarcinomas (Yatabe *et al.*, 1998a). Interestingly, we found that the expression is heterogeneous, and the up-regulation was remarkable, especially in the invasive portion of the adenocarcinoma. This was evidenced by homogeneous up-regulation in the metastatic sites and the invasion portion in the corresponding primary tumor. This finding was subsequently confirmed (Niki *et al.*, 2002) and further studies from our group demonstrated that a COX2-specific inhibitor suppressed the invasive phenotype of LNM35, which was established as a cell line with consistent lymphogenous metastasis (Kozaki *et al.*, 2001). These findings suggest that COX2-is associated with invasive growth of lung adenocarcinomas.

In comparison with these cancer-associated molecules, TTF-1 status was associated with p53 accumulation, p53 mutation, lack of Rb, and level of $p27^{KIP1}$ expression. The TTF-1 positive TRU type of adenocarcinoma demonstrated less p53 alteration, infrequent Rb inactivation, and frequent $p27^{KIP1}$ expression, preserved. As previously mentioned, alteration of p53 and Rb is a hallmark of SCLC and squamous cell carcinoma, both of which are suggested to be associated with cigarette smoking. In terms of the association with p53, Rb, and smoking, TTF-1 negative adenocarcinomas are more closely aligned to SCLCs and squamous cell carcinoma. A difference in the carcinogenetic mechanism between the two subtypes was also supported by frequent G->T transversion in TTF-1 negative adenocarcinoma, although the difference did not reach statistical significance, possibly because of the small number of cases. Taken together with all of these findings,

TTF-1 status divides lung adenocarcinoma into two subtypes from the aspect of molecular carcinogenesis. The characteristics are summarized in Table 13.

Correlation with the Other Subclassifications of Lung Adenocarcinomas

Several subclassification schemes for lung adenocarcinomas have been proposed, and thus the question how the distinction by TTF-1 status is related with the previous subclassification may arise. In this chapter, we select three subclassifications of lung adenocarcinomas and compare them with TTF-1 positive and negative adenocarcinoma. First, Dr. Shimosato, of the National Cancer Center in Japan, has proposed a subclassification system based on cytologic features (Shimosato, 1989). As previously mentioned, TTF-1 expressing adenocarcinomas correspond to their classification of type II pneumocyte type, Clara cell type, and their mixed type. Most of the other types, including bronchial surface types and bronchial gland types, do not express TTF-1 in our experience. Second, in comparison with the World Health Organization (WHO) classifications (Travis *et al.*, 1999), bronchioloalveolar, acinar, and papillary (Clara/type II pneumocyte type) subtypes are a major source of TTF-1 positive tumors, whereas most of the TTF-1 negative tumors are composed of acinar (bronchial gland-like or bronchial lining cell-like) and solid tumors with mucin. Classifications by the WHO are mostly based on growth structure but not cellular features, and thus it is difficult to fit the subtypes to TTF-1 expression status. Finally, Noguchi *et al.* (1995) reported prognostic differences based on their own histologic classification. They focused on the structure and divided adenocarcinoma into six subtypes: localized bronchioloalveolar carcinoma (LBAC), LBAC with foci of

Table 12 Characteristics of TTF-1 Positive and Negative Lung Adenocarcinoma

	TTF-1 Positive Adenocarcinoma	TTF-1 Negative Adenocarcinoma
Putative Originating Cells	Terminal respiratory unit (TRU-deriving carcinoma)	Central bronchial epithelium (bronchogenic carcinoma)
Location in the Lung	Periphery	Hilum, sites associated with larger bronchi
Prevalent Population	Nonsmoker	Smoker
Carcinogenesis	Probably unique	Similar to SCLCs or SQ
Associated Molecules	Unknown	p53,Rb
Precancerous Lesions	Atypical adenomatous hyperplasia	Not known (*de novo?*)
Characteristic Morphology	Lepidic growth, central scar	Solid, homogenous appearance Necrosis, frequent high-grade tumor
Frequency in Lung Adenoca	70–75%	25–30%

TRU, terminal respiratory unit; *SCLC*, small cell lung cancer; S2, squamous cell carcinoma.

collapse (type B), LBAC with foci of active fibroblastic proliferation (type C), poorly differentiated adenocarcinoma (type D), tubular adenocarcinoma (type E), and papillary adenocarcinoma with destructive growth (type F). The first three were grouped as "replacement type" and are characterized by lepidic growth. In addition, the study emphasized that patients with the LBAC (type A) and LBAC with foci of collapse (type B) showed an excellent prognosis, and thus suggested that these subtypes represent a preinvasive lesion. Adenocarcinoma with TTF-1 expression is equivalent to the replacement type and a part of papillary adenocarcinoma with destructive growth. They further subdivided the replacement type/TTF-1 positive subgroup into an invasive and noninvasive group, by means of the presence or absence of active fibroblastic foci, representing a stromal reaction because of invasion. It is of note that TTF-1 is not a marker of aggressiveness. Indeed, micropapillary adenocarcinoma, a recently suggested aggressive form of adenocarcinoma, frequently expressed TTF-1 (Amin et al., 2002). Almost all atypical adenomatous hyperplasia, a putative precancerous lesion of adenocarcinoma. Expression profiling analysis gave insight into classification based on molecular signatures of tumors. One of the earliest reports on breast cancers categorized three major subtypes of cancers (i.e., luminal, basal, and HER-2/neu amplification subtypes). Interestingly, the expression profiles of each subtype were characterized by cellular lineage and a feature represented by a peculiar gene alteration. In regard to lung cancers, the expression profiling analysis addressed the issue on diversity of lung adenocarcinoma. Two excellent reports had been published in tandem. Using a nonsupervised clustering method, both results identified particular subtypes of adenocarcinoma, some of which were equivalent to the TTF-1 positive TRU type of adenocarcinoma. In the article by Garber et al., (2001), lung cancers were divided into six types according to the molecular signature-adenocarcinoma type 1 to 3, squamous cell carcinoma, small cell carcinoma, and large cell carcinoma. Hierarchical clustering of the relationship among them largely grouped them into two arms; one was composed of the latter four groups (adenocarcinoma type 3, squamous cell carcinoma, small cell carcinoma, and large cell carcinoma), and the other included adenocarcinoma types 1 and 2, of which the profile was close to that of normal lung. Conversely, there were two distinct groups in lung adenocarcinoma (i.e., adenocarcinoma type 1 and 2 versus adenocarcinoma type 3). To distinguish the two groups, molecular hallmarks were extracted among thousands of genes examined, and TTF-1 was listed as one of the hallmarks for a highly expressed gene in adenocarcinoma type 1 and 2.

On the other hand, Bhattacharjee et al., (2001) described four distinct subtypes (types 1–4) of adenocarcinoma, in addition to squamous cell carcinoma, small cell carcinoma, and metastatic carcinoma, based on their expression profiles. Type 1 was poorly differentiated adenocarcinoma, and this type showed high levels of cell division genes, similar to squaous and small cell carcinomas. Type 2 is characterized by the expression of neuroendocrine molecules and demonstrated poor prognosis, confirming the previous report (Carnaghi et al., 2001; Graziano et al., 1994). Type 3 and 4 shared high expression of the surfactant apoproteins, and morphologically, type 4 are largely bronchioloalveolar carcinoma. Therefore, types 3 and 4 appeared to be consistent with TTF-1 positive adenocarcinoma. Both reports were summarized to suggest the existence of distinct subtypes of adenocarcinoma, correspond to TTF-1 positive adenocarcinoma, and that again, TTF-1 positive adenocarcinoma could be further subdivided, probably reflecting invasive growth.

In conclusion, we indicated that TTF-1 is expressed consistently throughout the life stages and uniformly in the TRU. The TTF-1 positive adenocarcinomas, which are suggested to derive from the TRU, share distinct clinicopathologic and molecular characteristics, suggesting different mechanisms of molecular carcinogenesis from the TTF-1 negative adenocarcinomas. We suspect that TTF-1 is a marker that discriminates the molecular mechanism of carcinogensis, based on the originating cellular lineage.

Acknowledgments

The author thanks Yoshitsugu Horio for critical review of the manuscript, Takashi Takahashi and Tetsuya Mitsudomi for insightful discourse, and Kaori Hayashi for technical assistance.

References

Agoff, S.N., Lamps, L.W., Philip, A.T., Amin, M.B., Schmidt, R.A., True, L.D., and Folpe, A.L. 2000. Thyroid transcription factor-1 is expressed in extrapulmonary small cell carcinomas but not in other extrapulmonary neuroendocrine tumors. *Mod. Pathol.* 13:238–242.

Akashi, K., He, X., Chen, J., Iwasaki, H., Niu, C., Steenhard, B., Zhang, J., Haug, J., and Li, L. 2003. Transcriptional accessibility for genes of multiple tissues and hematopoietic lineages is hierarchically controlled during early hematopoiesis. *Blood* 101:383–389.

Amin, M.B., Tamboli, P., Merchant, S.H., Ordonez, N.G., Ro, J., Ayala, A.G., and Ro, J.Y. 2002. Micropapillary component in lung adenocarcinoma: A distinctive histologic feature with possible prognostic significance. *Am. J. Surg. Pathol.* 26: 358–364.

Bejarano, P.A., Nikiforov, Y.E., Swenson, E.S., and Biddinger, P.W. 2000. Tyroid transcription factor-1, thyroglobulin, cytokeratin 7, and cytokeratin 20 in thyroid neoplasms. *Appl. Immunohistochem. Mol Morphol. 8:*189–194.

Berman, J. and O' Leary, T.J. 2001. Gastrointestinal stromal tumor workshop. *Hum. Pathol. 32:* 578–582.

Bhattacharjee, A., Richards, W.G., Staunton, J., Li, C., Monti, S., Vasa, P., Ladd, C., Beheshti, J., Bueno, R., Gillette, M., Loda, M., Weber, G., Mark, E.J., Lander, E.S., Wong, W., Johnson, B.E., Golub, T.R., Sugarbaker, D.J., and Meyerson, M. 2001. Classification of human lung carcinomas by mRNA expression profiling reveals distinct adenocarcinoma subclasses. *Proc. Natl. Acad. Sci. USA 98:*13790–13795.

Bruno, M.D., Bohinski, R.J., Huelsman, K.M., Whitsett, J.A., and Korfhagen, T.R. 1995. Lung cell-specific expression of the murine surfactant protein A (SP-A) gene is mediated by interactions between the SP-A promoter and thyroid transcription factor-1. *J. Biol. Chem. 270:*6531–6536.

Carnaghi, C., Rimassa, L., Garassino, I., and Santoro, A. 2001. Clinical significance of neuroendocrine phenotype in nonsmall cell lung cancer. *Ann. Oncol. 12 (Suppl. 12):*S119–S123.

Corti, M. and Snyder, C.A. 1998. Gender- and age-specific cytotoxic susceptibility to benzene metabolites *in vitro. Toxicol. Sci. 41:*42–48.

Demetri, G.D. 2001. Targeting c-kit mutations in solid tumors: Scientific rationale and novel therapeutic options. *Semin. Oncol. 28:*19–26.

Di Loreto, C., Di Lauro, V., Puglisi, F., Damante, G., Fabbro, D., and Beltrami, C.A. 1997. Immunocytochemical expression of tissue-specific transcription factor-1 in lung carcinoma. *J. Clin. Pathol. 50:*30–32.

Eberhart, C.E., Coffey, R.J., Radhika, A., Giardiello, F.M., Ferrenbach, S., and DuBois, R.N. 1994. Up-regulation of cyclooxygenase 2 gene expression in human colorectal adenomas and adenocarcinomas. *Gastroenterology 107:*1183–1188.

Endl, E., Kausch, I., Baack, M., Knippers, R., Gerdes, J., and Scholzen, T. 2001. The expression of Ki-67, MCM3, and p27 defines distinct subsets of proliferating, resting, and differentiated cells. *J. Pathol. 195:*457–462.

Fabbro, D., Di Loreto, C., Stamerra, O., Beltrami, C.A., Lonigro, R., and Damante, G. 1996. TTF-1 gene expression in human lung tumours. *Eur. J. Cancer 32A:*512–517.

Garber, M.E., Troyanskaya, O.G., Schluens, K., Petersen, S., Thaesler, Z., Pacyna-Gengelbach, M., van de Rijn, M., Rosen, G.D., Perou, C.M., Whyte, R.I., Altman, R.B., Brown, P.O., Botstein, D., and Petersen, I. 2001. Diversity of gene expression in adenocarcinoma of the lung. *Proc. Natl. Acad. Sci. USA 98:*13784–13789.

Graziano, S.L., Tatum, A.H., Newman, N.B., Oler, A., Kohman, L.J., Veit, L.J., Gamble, G.P., Coleman, M.J., Barmada, S., and O' Leary, S. 1994. The prognostic significance of neuroendocrine markers and carcinoembryonic antigen in patients with resected stage I and II nonsmall cell lung cancer. *Cancer Res. 54:*2908–2913.

Heldin, N.E., and Westermark, B. 1991. The molecular biology of the human anaplastic thyroid carcinoma cell. *Thyroidology 3:*127–131.

Hibi, K., Takahashi, T., Sekido, Y., Ueda, R., Hida, T., Ariyoshi, Y., and Takagi, H. 1991. Coexpression of the stem cell factor and the c-kit genes in small cell lung cancer. *Oncogene 6:*2291–2296.

Joensuu, H., Roberts, P.J., Sarlomo-Rikala, M., Andersson, L.C., Tervahartiala, P., Tuveson, D., Silberman, S., Capdeville, R., Dimitrijevic, S., Druker, B., and Demetri, G.D. 2001. Effect of the tyrosine kinase inhibitor STI571 in a patient with a metastatic gastrointestinal stromal tumor. *N. Engl. J. Med. 344:*1052–1056.

Katoh, R., Kawaoi, A., Miyagi, E., Li, X., Suzuki, K., Nakamura, Y., and Kakudo, K. 2000. Thyroid transcription factor-1 in normal, hyperplastic, and neoplastic follicular thyroid cells examined by immunohistochemistry and nonradioactive *in situ* hybridization. *Mod. Pathol. 13:*570–576.

Kaufmann, O., and Dietel, M. 2000. Thyroid transcription factor-1 is the superior immunohistochemical marker for pulmonary adenocarcinomas and large cell carcinomas compared to surfactant proteins A and B. *Histopathology 36:*8–16.

Kelly, S.E., Bachurski, C.J., Burhans, M.S., and Glasser, S.W. 1996. Transcription of the lung-specific surfactant protein C gene is mediated by thyroid transcription factor 1. *J. Biol. Chem. 271:* 6881–6888.

Kimura, S., Hara, Y., Pineau, T., Fernandez-Salguero, P., Fox, C.H., Ward, J.M., and Gonzalez, F.J. 1996. The T/ebp null mouse: Thyroid-specific enhancer-binding protein is essential for the organogenesis of the thyroid, lung, ventral forebrain, and pituitary. *Genes Dev. 10:*60–69.

Kobayashi, T., Tsuda, H., Noguchi, M., Hirohashi, S., Shimosato, Y., Goya, T., and Hayata, Y. 1990. Association of point mutation in c-Ki-Ras oncogene in lung adenocarcinoma with particular reference to cytologic subtypes. *Cancer 66:*289–294.

Korfhagen, T.R., and Whitsett, J.A. 1997. Transcriptional control in the developing lung: The Parker B. Francis lectureship. *Chest 111:*83S–88S.

Kozaki, K., Koshikawa, K., Tatematsu, Y., Miyaishi, O., Saito, H., Hida, T., Osada, H., and Takahashi, T. 2001. Multifaceted analyses of a highly metastatic human lung cancer cell line NCI-H460-LNM35 suggest mimicry of inflammatory cells in metastasis. *Oncogene 20:*4228–4234.

Lazzaro, D., Price, M., de Felice, M., and Di Lauro, R. 1991. The transcription factor TTF-1 is expressed at the onset of thyroid and lung morphogenesis and in restricted regions of the foetal brain. *Development 113:*1093–1104.

Marchetti, A., Buttitta, F., Pellegrini, S., Chella, A., Bertacca, G., Filardo, A., Tognoni, V., Ferreli, F., Signorini, E., Angeletti, C.A., and Bevilacqua, G. 1996. Bronchioloalveolar lung carcinomas: K-Ras mutations are constant events in the mucinous subtype. *J. Pathol. 179:*254–259.

Mendelson, C.R. 2000. Role of transcription factors in fetal lung development and surfactant protein gene expression. *Annu. Rev. Physiol. 62:*875–915.

Minoo, P., Su, G., Drum, H., Bringas, P., and Kimura, S. 1999. Defects in tracheoesophageal and lung morphogenesis in Nkx2.1(−/−) mouse embryos. *Dev. Biol. 209:*60–71.

Niki, T., Kohno, T., Iba, S., Moriya, Y., Takahashi, Y., Saito, M., Maeshima, A., Yamada, T., Matsuno, Y., Fukayama, M., Yokota, J., and Hirohashi, S. 2002. Frequent colocalization of COX2 and laminin-5 gamma2 chain at the invasive front of early-stage lung adenocarcinomas. *Am. J. Pathol. 160:*1129–1141.

Nishio, M., Koshikawa, T., Yatabe, Y., Kuroishi, T., Suyama, M., Nagatake, M., Sugiura, T., Ariyoshi, Y., Mitsudomi, T., and Takahashi, T. 1997. Prognostic significance of cyclin D1 and retinoblastoma expression in combination with p53 abnormalities in preimary, resected nonsmall cell lung cancers. *Clin. Cancer Res. 3:*1051–1058.

Noguchi, M., Morikawa, A., Kawasaki, M., Matsuno, Y., Yamada, T., Hirohashi, S., Kondo, H., and Shimosato, Y. 1995. Small adenocarcinoma of the lung: Histologic characteristics and prognosis. *Cancer 75:*2844–2852.

Oliveira, A.M., Tazelaar, H.D., Myers, J.L., Erickson, L.A., and Lloyd, R.V. 2001. Thyroid transcription factor-1 distinguishes metastatic pulmonary from well-differentiated neuroendocrine tumors of other sites. *Am. J. Surg. Pathol. 25:*815–819.

Ordonez, N.G. 2000a. Thyroid transcription factor-1 is a marker of lung and thryroid carcinomas. *Adv. Anat. Pathol. 7:*123–127.

Ordonez, N.G. 2000b. Value of thyroid transcription factor-1 immunostaining in distinguishing small cell lung carcinomas from other small cell carcinomas. *Am. J. Surg. Pathol. 24:* 1217–1223.

Oshima, M., Dinchuk, J.E., Kargman, S.L., Oshima, H., Hancock, B., Kwong, E., Trzaskos, J.M., Evans, J.F., and Taketo, M.M. 1996. Suppression of intestinal polyposis in Apc delta716 knockout mice by inhibition of cyclooxygenase 2 (COX-2). *Cell 87:*803–809.

Pelosi, G., Fraggetta, F., Pasini, F., Maisonneuve, P., Sonzogni, A., Iannucci, A., Terzi, A., Bresaola, E., Valduga, F., Lupo, C., and Viale, G. 2001. Immunoreactivity for thyroid transcription factor-1 in stage I nonsmall cell carcinomas of the lung. *Am. J. Surg. Pathol. 25:*363–372.

Puglisi, F., Barbone, F., Damante, G., Bruckbauer, M., Di Lauro, V., Beltrami, C.A., and Di Loreto, C. 1999. Prognostic value of thyroid transcription factor-1 in primary, resected, nonsmall cell lung carcinoma. *Mod. Pathol. 12:*318–324.

Ramirez, M.I., Rishi, A.K., Cao, Y.X., and Williams, M.C. 1997. TGT3, thyroid transcription factor I, and Sp1 elements regulate transcriptional activity of the 1.3-kilobase pair promoter of T1 alpha, a lung alveolar type I cell gene. *J. Biol. Chem. 272:* 26285–26294.

Rivard, N., L Allemain, G., Bartek, J., and Pouyssegur, J. 1996. Abrogation of p27Kip1 by cDNA antisense suppresses quiescence (G0 state) in fibroblasts. *J. Biol. Chem. 271:*18337–18341.

Sano, H., Kawahito, Y., Wilder, R.L., Hashiramoto, A., Mukai, S., Asai, K., Kimura, S., Kato, H., Kondo, M., and Hla, T. 1995. Expression of cyclooxygenase-1 and –2 in human colorectal cancer. *Cancer Res. 55:*3785–3789.

Sekido, Y., Obata, Y., Ueda, R., Hida, T., Suyama, M., Shimokata, K., Ariyoshi, Y., and Takahashi, T. 1991. Preferential expression of c-kit protooncogene transcripts in small cell lung cancer. *Cancer Res. 51:*2416–2419.

Shamblott, M.J., Axelman, J., Littlefield, J.W., Blumenthal, P.D., Huggins, G.R., Cui, Y., Cheng, L., and Gearhart, J.D. 2001. Human embryonic germ cell derivatives express a broad range of developmentally distinct markers and proliferate extensively *in vitro. Proc. Natl. Acad. Sci. USA 98:*113–118.

Shimosato, Y.1989. Pulmonary neoplasms. In Sternberg S.S., (ed). *Diagnostic Surgical Pathology.* New York: Raven Press, 785–827.

Stahlman, M.T., Gray, M.E. and Whitsett, J.A. 1996. Expression of thyroid transcription factor-1 (TTF-1) in fetal and neonatal human lung. *J. Histochem. Cytochem. 44:*673–678.

Sturm, N., Rossi, G., Lantuejoul, S., Papotti, M., Franchon, S., Claraz, C., Brichon, P.Y., Brambilla, C., and Brambilla, E. 2002.

Expression of thyroid transcription factor-1 in the spectrum of neuroendocrine cell lung proliferations with special interest in carcinoids. *Hum. Pathol. 33:*175–182.

Sugita, M., Geraci, M., Gao, B., Powell, R.L., Hirsch, F.R., Johnson, G., Lapadat, R., Gabrielson, E., Bremnes, R., Bunn, P.A., and Franklin, W.A. 2002. Combined use of oligonucleotide and tissue microarrays identifies cancer/testis antigens as biomarkers in lung carcinoma. *Cancer Res. 62:*3971–3979.

Toonen, R.F., Gowan, S., and Bingle, C.D. 1996. The lung enriched transcription factor TTF-1 and the ubiquitously expressed proteins Sp1 and Sp3 interact with elements located in the minimal promoter of the rat Clara cell secretory protein gene. *Biochem. J. 316:*467–473.

Travis, W.D., Colby, T.V., Sobin, S.H., Corrin, B., Shimosata, Y., and Brambilla, E. 1999. Histological Typing of Lung and Pleural Tumors, 3rd ed. New York: Springer-Verlag.

Vane, J. 1994. Towards a better aspirin. *Nature 367:*215–216.

Weintraub, H., Davis, R., Tapscott, S., Thayer, M., Krause, M., Benezra, R., Blackwell, T.K., Turner, D., Rupp, R., Hollenberg, S. *et al.* 1991. The myoD gene family: Nodal point during specification of the muscle cell lineage. *Science 251:* 761–766.

Williams, C. S., Smalley, W., and DuBois, R.N. 1997. Aspirin use and potential mechanisms for colorectal cancer prevention. *J. Clin. Invest. 100:*1325–1329.

Yan, C., Sever, Z., and Whitsett, J.A. 1995. Upstream enhancer activity in the human surfactant protein B gene is mediated by thyroid transcription factor 1. *J. Biol. Chem. 270:*24852–24857.

Yatabe, Y., Hida, T., Achiwa, H., Muramatsu, H., Kozaki, K., Nakamura, S., Ogawa, M., Mitsudomi, T., Sugiura, T., and Takahashi, T. 1998a. Increased expression of cyclooxygenase 2 occurs frequently in human lung cancers, specifically in adenocarcinomas. *Cancer Res. 58:*3761–3764.

Yatabe, Y., Masuda, A., Koshikawa, T., Nakamura, S., Kuroishi, T., Osada, H., Takahashi, T., and Mitsudomi, T. 1998b. $p27^{KIP1}$ in human lung cancers: Differential changes in small cell and nonsmall cell carcinomas. *Cancer Res. 58:*1042–1047.

Yatabe, Y., Mitsudomi, T., and Takahashi, T. 2002. TTF-1 expression in pulmonary adenocarcinomas. *Am. J. Surg. Pathol. 26:* 767–773.

Zhang, X., Wharton, W., Donovan, M., Coppola, D., Croxton, R., Cress W.D., and Pledger, W.J. 2000. Density-dependent growth inhibition of fibroblasts ectopically expressing p27(kip1). *Mol. Biol. Cell 11:*2117–2130.

Zhou, L., Lim, L., Costa, R.H., and Whitsett, J.A.1996. Thyroid transcription factor-1, hepatocyte nuclear factor-3beta, surfactant protein B, C, and Clara cell secretory protein in developing mouse lung. *J. Histochem. Cytochem. 44:*1183–1193.

10

Role of Global Methylation of DNA in Lung Carcinoma

Chandrika J. Piyathilake, Alain Niveleau, and William Grizzle

Introduction

DNA methylation in mammals occurs at the cytosine residues of cytosine guanine dinucleotide (CpG) dinucleotides by an enzymatic reaction that produces 5-methylcytosine (5-mc). One of the first alterations of DNA methylation to be recognized in neoplastic cells was a decrease in overall 5-mc content, referred to as genome-wide or global DNA hypomethylation. Despite the frequently observed cancer-associated increases of regional hypermethylation, the prevalence of global DNA hypomethylation in many types of human cancers (Gama-Sosa et al., 1983; Kliasheva, 1990) suggests that such hypomethylation plays a significant and fundamental role in tumorigenesis. Global DNA hypomethylation has been implicated in chromosome instability, loss of imprinting, abnormal chromosomal structures (Lewis et al., 1991), and activation of oncogenes and reactivation of transposons of retroviruses; all these changes have been proposed to contribute to carcinogenesis (Baylin et al., 1998; Feinberg et al., 1983a, b).

Although the methylation of DNA is increasingly recognized as an important epigenetic change influencing the process of cellular neoplastic transformation, the exact mechanisms of hypomethylation of DNA are unknown. Because folate is an essential cofactor in the synthesis of S-adenosyl-methionine (SAM), the primary

donor of methyl groups for methylation of DNA, we evaluated methylation of DNA in lung tissues of smokers, where folate is likely to be deficient due to biological inactivation by exposure to cigarette smoke (Dastur et al., 1972; Khaled et al., 1986; Linnell et al., 1986; Stedman, 1968). In our initial studies we used a radio-labeled methyl incorporation (RMI) assay to assess the methylation status in tissues of the lung. In the RMI assay, DNA is methylated with Sss I methylase in the presence of ^{3}H-labeled S-adenosylmethionine (source of methyl groups) (Balaghi et al., 1993). The methylation status of DNA within a tissue sample containing a mixed population of cells (fibroblasts, lymphocytes, etc., in addition to cancer and normal epithelial cells) is evaluated with the RMI assay. The advantage of the RMI assay is that it allows for evaluation of methylation in the same frozen tissue specimens used for folate and other vitamin assays. Although we have observed that the RMI assay has some day-to-day variability, the relative methylation values for groups of samples done on the same day are comparable, especially when matched cancer and normal samples are used. With this technique, we documented that lower levels of tissue folate and vitamin C are associated with global DNA hypomethylation in neoplastic tissues of the lung (Piyathilake et al., 2000a, b).

We recently developed an immunohistochemical assay (using monoclonal antibodies against 5-methyl

Handbook of Immunohistochemistry and in situ Hybridization of Human
Carcinomas, Volume 1: Molecular Genetics; Lung and Breast Carcinomas

Copyright © 2004 by Elsevier (USA)
All rights reserved.

cytosine [5-mc]) for global methylation of DNA that has several advantages over previous methylation assays (Piyathilake *et al.*, 2000c). The ability to evaluate global methylation in intact and specific types of cells involved in carcinogenesis is a unique feature of this newly developed assay. The immunohistochemical evaluation of global DNA methylation using monoclonal antibodies against 5-mc is described below.

MATERIALS

Buffers

1. 4 L of *Tris*-HCl buffer: 50 m*M Tris*-HCl base (24.23 g), 150 m*M* NaCl (35.06 g), 0.01% Triton X-100 (16 drops). Bring solution to 4 L with deionized H_2O and adjust pH to 7.6 with concentrated HCl.

2. 1 L of phosphate-buffered saline (PBS): 137 m*M* NaCl (8.0 g), 2.7 m*M* KCl (0.2 g), 8.1 m*M* Na_2HPO_4 (1.5 g), 1.5 m*M* KH_2PO_4 (0.2 g). Bring volume to 1 L with deionized H_2O.

3. 100 mL of PBS with 1% bovine serum albumin (BSA) fraction V (1 g), 15 m*M* NaN_3 (9.75 mg), 1 mM ethylenediamine tetra-acetic acid (EDTA) (29.2 mg). This buffer is referred to as PBE.

Antigen-Retrieval Solution and Acid Treatment Solution to Open DNA Coils

1. 0.01 *M* citric acid: 1.92 g anhydrous citric acid in 1 L of deionized H_2O; adjust pH to 6.0 with NaOH.

2. 3.5 N HCl (HCl treatment is to open up DNA).

Reagent for Quenching Endogenous Peroxidase Activity and for Blocking Nonspecific Binding of the Antibodies

1. 3% H_2O_2 in deionized water.
2. Goat serum diluted to 1% with PBE buffer.

Primary Antibody

Antibodies specific for 5-mc were obtained from Dr. Alain Niveleau. This antibody has been well-characterized (Reynaud *et al.*, 1991) and used with success in several laboratories (Bensaada *et al.*, 1998; Miniou *et al.*, 1994). Appropriate dilutions with PBE were determined for this antibody for tissue sections from different organs used in our laboratory.

Detection System

Biotin-streptavidin detection system (Signet, Dedham, MA).

Chromogen

3,3′-Diaminobenzidine tetrahydrochloride (DAB) kit from Biogenex. Add 0.5 mL of substrate buffer to 4.5 mL of deionized H_2O. Add four drops of DAB solution and two drops of H_2O_2 from the kit.

Other Materials

1. Superfrost slides.
2. Xylene.
3. 70%, 95% and 100% ethanol.
4. PAP pen.
5. Plastic microwaveable container.
6. Plastic Coplin jars.
7. Racks for holding slides.
8. Humidity chambers.
9. Glass staining dishes with glass slide holders.
10. 700–800W microwave oven with carousal.
11. Counterstain: hematoxylin.

METHODS

Slide Preparation

1. Cut two 5-μm sections from a paraffin block of a 10% neutral buffered formalin-fixed specimen and mount on plus slides (one section is for negative control, commonly referred to as the "delete" because of the deletion of the primary antibody).

2. Heat the slides for 1 hr at 58°C.

Preparation for Immunostaining

1. Remove the paraffin with three changes of xylene for 2 min each. Rehydrate tissues by placing slides in 100%, 95%, and 75% ethanol (3 min each).

2. Place the slides in Tris-buffer for 3–30 min.

3. Dry the area around the tissues with a laboratory wipe and make a hydrophobic ring around the tissues with a PAP pen.

4. Add 50–200 μL of Tris-buffer to cover the specimen (the specimens are stable for several hours at this point as long as they are covered with the buffer and not allowed to dry).

Antigen Retrieval and Opening of DNA

1. Drain Tris-HCl buffer from the slides and place them in plastic Coplin jars filled with deionized water (H_2O) for 1–3 min.

2. Preheat an 8 × 8-inch microwave oven–safe dish, filled with 1 inch of H_2O in a microwave oven for 8 min.

3. Taking great care, remove very hot H_2O from the Coplin jar and completely fill with the antigen retrieval solution (0.01 M citric acid).

4. Place Coplin jar in the 8 × 8-inch previously pre-heated dish, which now has 1 inch of hot H_2O in the bottom.

5. Heat the slides in the microwave oven on high and continue heating for 5 min after the solution begins to boil. Make sure that the antigen retrieval solution has not evaporated to the point that the tissue sections dry. Add more antigen retrieval solution, if needed, and heat the slides for an additional 5 min of boiling.

6. With great care, remove the Coplin jar from the microwave oven and allow the slides to cool for 10–15 min.

7. Rinse the slides twice with deionized H_2O and place the slides in 3.5 N HCl for 15 min. Rinse again with deionized H_2O and place the slides in Tris-HCl buffer.

8. Mark the slides as described with a PAP pen.

Quenching Endogenous Peroxidase

Drain the Tris-HCl buffer from the slides and add 50–200 µl of 3% H_2O_2 for 3 min to quench the endogenous peroxidase activity. Rinse in Tris-HCl buffer for 3 min.

Immunostaining Procedure

1. Drain the Tris-HCl buffer from the slides and place them in a humidity chamber. Add 50–200 µl of 1% goat serum to the tissue for 10 min to reduce non-specific staining.

2. Drain the blocking agent from the slides (except the deletes, which are left in the chamber as they are) and add the primary antibody diluted in PBE buffer. Place the slides in a humidity chamber for 1 hr at room temperature.

3. Rinse the slides in Tris-HCl buffer for 10 min (two 5-min rinses). Drain the excess Tris-HCl buffer from the slides and place them on level slide racks.

4. Add the secondary antibody diluted in PBE buffer to the slides for 10 min.

5. Rinse the slides for 10 min (two 5-min rinses) in Tris-HCl buffer and drain the excess buffer from the slides.

6. Add the streptavidin-peroxidase diluted in PBE buffer to the slides for 5 min.

7. Rinse the slides as in **step 5**.

8. Add the DAB mixture to the slides for 7 min or until the desired intensity of staining is achieved. This may take 2–15 min. Rinse the slides with deionized water.

Counterstain Protocol

Hematoxylin Counterstain

1. Place slides in Mayers hematoxylin for 1 min. If hematoxylin is freshly prepared, 5 sec of staining in hematoxylin will provide a light nuclear counterstain that will not interfere with the detection of immuno-staining. Do not overstain with hematoxylin.

2. Rinse slides with tap water for 1 min.

3. Dehydrate slides through graded alcohols (3 min each in 70%, 95%, and 100% ethanol) and three changes of xylene (3 min each).

4. Mount over slips with permount.

Evaluation of Immunostaining

Three observers independently grade the intensity of immunostaining on a scale of 0 (no staining) to 4+ (intense staining) in cervical cells (Grizzle *et al.*, 1998). The percentage of cells at each intensity is estimated by scanning the entire tissue section. In the example shown in Table 13, the percentage of normal cells staining at any intensity is 70. The weighted average of the immunostaining score is derived by multiplying the percentage of cells at each intensity with the intensity of staining (0–4) and by adding the scores as shown in Table 13. The final results reported are the average of the three observers. If the degree of staining is different in basal and luminal cells of normal or hyperplastic tissues, we score the staining in these compartments separately. We have observed a highly significant correlation (R > 0.85) between the total number of cells stained and weighted average of the score, suggesting that either one can be used in data analysis.

RESULTS AND DISCUSSION

The purpose of the first study was to compare the results of global DNA methylation evaluated by radio-labeled methyl incorporation (inversely related to the degree of DNA methylation *in vivo*) with immunohis-tochemical staining of the same tissue sections with the monoclonal antibody developed against 5 mc (Piyathilake *et al.*, 2000c). We used archival specimens of squamous cell cancer (SCC) of the human lung with a matched uninvolved specimen (n = 18 pairs) and 18 lung specimens from subjects without lung cancer (noncancer specimens) to make this comparison. The radio-labeled methyl incorporation (CPM/µg of DNA) was significantly higher in the SCCs compared with matched uninvolved tissues (p = 0.002). The difference in radio-labeled methyl incorporation between SCCs and the tissues from patients without cancer was not

Table 13 Evaluation Form for Immunohistochemical Assays

Case #					Marker	DNA Methylation	
		Intensity of Staining					
Cell Type	0	1	2	3	4		
Normal	30	50	10	10	0	70	% Cells staining
		0.5	0.2	0.3		1	Score
Dysplasia							
Cancer							

significant (p = 0.79), apparently due to high interpatient variability in the radio-labeled technique. The immunostaining for 5-mc, localized in the nuclei of cells, was reported as a percentage of cells positive for staining as well as a weighted average of the intensity score. The percentage of bronchial epithelial cells positive for immunostaining evaluated either on a section containing SCC (close to SCC) or on a section from matched uninvolved specimen (away from SCC) was significantly higher compared with SCCs (p = 0.03, p = 0.005, respectively).

The percentage of bronchial epithelial cells positive for 5-mc in noncancer specimens was not significantly different from the percentage of positive bronchial cells evaluated either close or away from SCCs, but was significantly higher compared with the percentage of SCC cells positive for 5-mc. the weighted average of the intensity scores followed a similar pattern. The results suggested that both radio-labeled methyl incorporation assay and immunostaining for 5-mc can be used to demonstrate hypomethylation of DNA in SCC tissues compared with matched uninvolved tissues. An advantage of immunostaining, however, was its ability to demonstrate hypomethylation of SCC compared with adjacent bronchial mucosa, precluding the need to use sections from both SCC and matched uninvolved tissues. Only the immunostaining technique was able to document a statistically significant difference in DNA methylation between SCC and tissues from patients without cancer. We concluded that the immunostaining technique has advantages over the radio-labeled methyl incorporation assay and may be best suited for the evaluation of global DNA methylation when the methylation status of cancer cannot be normalized by methylation of normal tissues or when the number of samples available for evaluation is small.

To evaluate whether altered global DNA methylation in specific types of cells represents an epigenetic difference in susceptibility for SCC, we examined the status of global DNA methylation by using a monoclonal antibody specific for 5-mc in randomly selected lung specimens of 60 cigarette smokers that developed SCC and 30 cigarette smokers that did not (Piyathilake et al., 2001). We reported that 5-mc immunostaining scores of DNA of SCC (0.61 ± 0.42) and associated hyperplastic lesions (0.82 ± 0.27) were significantly lower compared with DNA of histologically normal bronchial epithelial cells (0.99 ± 0.52) and hyperplastic lesions (1.2 ± 0.22) of noncancer specimens. The ratio of 5-mc scores between SCC and matched uninvolved bronchial epithelial cells was significantly associated with advanced stage and size of the tumor. These results suggested that alteration in global DNA methylation is an important epigenetic difference in susceptibility for the development of lung cancer. The reduced global DNA methylation in SCC compared with epithelial hyperplasia and its association with tumor size and stage of the disease is suggestive of its involvement in the progression of SCC. The results also indicate that normal methylation of DNA in epithelial hyperplastic lesions may prevent the transformation of these lesions to invasive cancer. Figure 26 shows the 5-mc staining of normal bronchial epithelium, epithelial hyperplasia, dysplasia, and SCC.

To evaluate the race and age-dependent alterations in global DNA methylation on the development and progression of SCCs of the lung, we assessed the global methylation status in SCC and in the associated uninvolved bronchial mucosa and epithelial hyperplasia of 53 Caucasians and 23 African Americans by using an antibody specific for 5-mc (Piyathilake et al., 2003). 5-mc scores of SCC (0.59 ± 0.06) were significantly lower compared with 5-mc scores of uninvolved bronchial mucosa (UBM) (0.87 ± 0.07) and epithelial hyperplasia (EH) (0.82 ± 0.07) in Caucasians (p < 0.05). In African Americans, 5-mc scores of SCC (0.55 ± 0.09) were not significantly different from 5-mc scores of UBM (0.60 ± 0.09) and EH (0.54 ± 0.14), suggesting an involvement of methylation in the development of SCCs in Caucasians, but not in African Americans. 5-mc scores were lower in younger (< 65 years) subjects compared with older (> 65 years) subjects in Caucasians. Because cancers in younger subjects tend to be more aggressive than those in older subjects, these observations

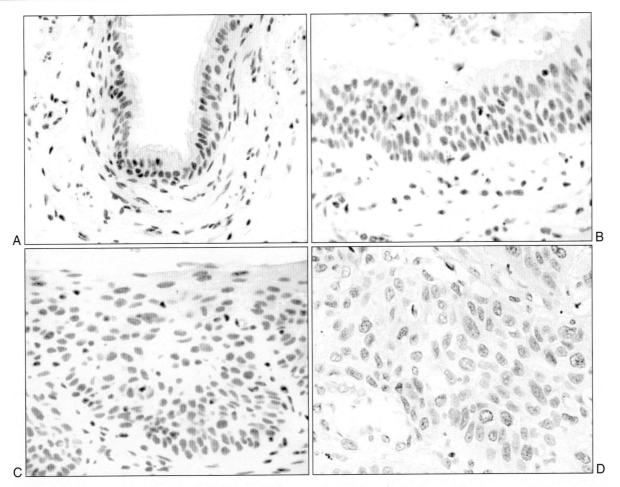

Figure 26 5-mc staining of normal bronchial epithelium. **A:** epithelial hyperplasia; **B:** dysplasia; **C:** and SCC; **D:** final magnification (400X).

may suggest that hypomethylation may have contributed to the aggressiveness of cancers in younger Caucasians. Hypomethylation of SCCs of the lung in Caucasian men was associated with shorter survival from the disease. These preliminary results suggest that the methylation status of DNA may affect the development, aggressiveness, and prognosis of SCCs of the lung in Caucasians. Our results in the study of colorectal cancer have demonstrated similar racial differences in the prognostic usefulness of specific molecular markers (Grizzle *et al.*, 2002; Talley *et al.*, 2002). Because it was unclear whether the inconsistencies across race and gender subgroups are an effect of selection and size of the study groups, replication of this study in other populations is necessary to increase the scientific credibility of the observed results.

As discussed, an important advantage of immunohistochemical evaluation of global DNA methylation is its ability to assess the status of methylation in specific types of cells. Another advantage is that it allows

associating methylation of DNA with the expression of other biomarkers in the same tissue. We believe that an evaluation of relationships between DNA methylation and expression of intermediate endpoint biomarkers may reveal associations that are not readily apparent from studies with specific biomarkers alone. To generate preliminary data in this area, we assessed the global DNA methylation by the RMI assay and by immunohistochemistry. In the same tissues, the expression of epidermal growth factor receptor (EGFR) was assessed by immunohistochemistry (Piyathilake *et al.*, 2002). We selected EGFR to evaluate these associations because we identified it as an important marker in our studies of lung cancer because of its significant stepwise increase in expression observed from noncancer tissues to tissues of preneoplastic lesions to invasive cancer, and also because of the observed association between higher expression in SCCs of the lung and poor survival. We observed, using the immunohistochemical method, that lower global methylation in malignant cells was associated with a higher expression of EGFR in these tissues (p < 0.05). In this study, there was no association between global DNA methylation assessed by the RMI assay and the expression of EGFR in SCC. These results suggest that evaluation of methylation in specific cells rather than in a mixed population of cells may provide more information with regard to its association with the expression of intermediate endpoint biomarkers that are important in carcinogenesis.

In conclusion, in a study of patients with SCCs of the lung using a biochemical as well as an immunohistochemical method to analyze hypomethylation, the immunohistochemical method demonstrated that global methylation significantly decreases from uninvolved bronchial mucosa to epithelial hyperplasia and to squamous cell cancers. Hypomethylation as determined by the biochemical or immunohistochemical methods demonstrated the same general pattern, with the differences in methylation between uninvolved tissue and matched SCCs of the lung being statistically significant by both methods; however, when SCCs of the lung were compared with tissues from the lungs of patients without cancer of the lung, differences in the biochemical assay were not statistically significant, whereas immunohistochemical method was. This probably is because so many and variable numbers of nonmalignant cells are included in the biochemical analysis, increasing its variability. The most important factor is that the immunohistochemical method permitted discrimination between both normal and uninvolved and hyperplastic as well as malignant lesions. Similarly, using the immunohistochemical method but not the biochemical method, we found statistically significant racial differences in the importance of global methylation to clinical outcome and significant associations between DNA methylation and the expression of other biomarkers that are important in lung carcinogenesis.

Acknowledgments

Supported in part by K07 CA70160 and R 03 CA83094 (Piyathilake) and the Early Detection Research Network 1U24 CA86359-03 (Grizzle).
Collaborators: Drs. Douglas Heimburger, Gary Johanning, Phillip Cornwell, Walter Bell, Andra Frost, Heidi Weiss, Santosh Niwas, Maurizio Macaluso, and Upender Manne.
Technical assistance: Denise Oelschlager, Jennifer Jones, Martin Whiteside, and Li-Juan Fan.

References

Balaghi, M., and Wagner, C. 1993. DNA methylation in folate deficiency: Use of CpG methylase. *Biochem. Biophys. Res. Commun. 193:*1184–1190

Baylin, S.B., Herman, J.G., Graff, J.R., Vertino, P.M., and Issa, J.P. 1998. Alterations in DNA methylation: A fundamental aspect of neoplasia. *Adv. Cancer Res. 72:*141–196.

Bensaada, M., Kiefer, H., Tachdjian, G., Lapierre, J.M., Cacheux, V., Niveleau, A., and Metezeau, P. 1998. Altered patterns of DNA methylation on chromosomes from leukemia cell lines: Identification of 5-methylcytosines by indirect immunodetection. *Cancer Genet. Cytogen. 103:*101–109.

Dastur, D.K., Quadros, E.V., Wadia, N.H., Desai, M.M., and Bharucha, E.P. 1972. Effect of vegetarianism and smoking on vitamin B-12, thiocyanate, and folate levels in the blood of normal subjects. *Br. Med. J. 3:*260–263.

Feinberg, A.P., and Vogelstein, B. 1983a. Hypomethylation of Ras oncogenes in primary human cancers. *Biochem. Biophys. Res. Commun. 111:*47–54.

Feinberg, A.P., and Vogelstein, B. 1983b. Hypomethylation distinguishes genes of some human cancers from their normal counterparts. *Nature 301:*89–92.

Gama-Sosa, M.A., Slagel, V.A., Trewyn, R.W., Oxenhandler, R., Kuo, K.C., Gehrke, C.W., and Ehrlich, M. 1983. The 5-methylcytosine content of DNA from human tumors. *Nucleic Acids Res. 11:*6883–6894.

Grizzle, W.E., Manne, U., Weiss, H.L., Jhala, N., and Talley, L.I. 2002. Molecular staging of colorectal cancer in African-American and Caucasian patients using phenotypic expression of p53, Bcl-2, MUC-1, and p27[kip-1]. *Int. J. Cancer 97:* 403–409.

Grizzle, W.E., Myers, R.B., Manne, U., and Srivastava, S. 1998. Immunohistochemical evaluation of biomarkers in prostatic and colorectal neoplasia. In John Walker's *Methods in Molecular Medicine—Tumor Marker Protocols.* Hanausek, M., and Walaszek, Z. (eds). Totowa, NJ: Humana Press, Inc.: 143–160.

Khaled, M.A., Watkins, C.L., and Krumdieck, C.L. 1986. Inactivation of B-12 and folate coenzymes by butyl nitrite as observed by NMR: Implications on one-carbon transfer mechanism. *Biochem. Biophys. Res. Commun. 135:*201–207.

Kliasheva, R.I. 1990. DNA methylation in human lung tumors. *Voprosy. Onkologii. 36:*186–189.

Lewis, J., and Bird, A. 1991. DNA methylation and chromatin structure. *FEBS Lett. 285:*155–159.

Linnell, J.C., Smith, A.D.M., Smith, C.L., Wilson, J., and Matthews, D.M. 1986. Effects of smoking on metabolism and excretion of vitamin B-12. *Br. Med. J. 2:*215–216.

Miniou, P., Jeanpierre, M., Blanquet, V., Sibella, V., Bonneau, D., Herbelin, C., Fischer, A., Niveleau, A., and Viegas-Pequignot, E. 1994. Abnormal methylation pattern in constitutive and facultative (X inactive chromosome) heterochromatin of ICF patients. *Hu. Mol. Genet. 3:*2093–2102.

Piyathilake, C.J., Bell, W.C., Johanning, G.L., Cornwell, P.E., Heimburger, D.C., and Grizzle, W.E. 2000b. The accumulation of ascorbic acid by squamous cell carcinomas of the lung and larynx is associated with global methylation of DNA. *Cancer 89:*171–176.

Piyathilake, C.J., Frost, A.R., Bell, W.C., Oelschlager, D., Weiss, H., Johanning, G.L., Niveleau, A., Heimburger, D.C., and Grizzle, W.E. 2001. Altered global methylation of DNA: An epigenetic difference in susceptibility for lung cancer is associated with its progression. *Hum. Pathol. 32:*856–862.

Piyathilake, C.J., and Johanning, G.L. 2002. Cellular vitamins, DNA methylation, and cancer risk. *J. Nutr. 132(8 Suppl):*2340S–2344S.

Piyathilake, C.J., Johanning, G.L., Frost, A.R., Whiteside, M.A., Manne, U., Grizzle, W.E., Heimburger, D.C., and Niveleau, A. 2000c. Immunohistochemical evaluation of global DNA methylation: Comparison with *in vitro* radiolabeled methyl incorporation assay. *Biotech. Histochem. 75:*251–258.

Piyathilake, C.J., Johanning, G.L., Macaluso, M., Whiteside, M.A., Heimburger, D.C., and Grizzle, W.E. 2000a. Localized deficiencies of folate and vitamin B-12 in lung tissues are associated with global DNA methylation. *Nutr Cancer* *37*:99–107.

Piyathilake, C.J., Macaluso, M., Henao, O., Frost, A.R., Bell, W.C., Johanning, G.L., Heimburger, D.C., Niveleau, A., and Grizzle, W.E. 2003. Race and age-dependant differences in global methylation of DNA. *Cancer Causes Control 14:*37–42.

Reynaud, C., Bruno, C., Boullanger, P., Grange, J., Barbesti, S., and Niveleau, A. 1991. Monitoring of urinary excretion of modified nucleosides in cancer patients using a set of six monoclonal antibodies. *Cancer Lett. 61:*255–262.

Stedman, R.L., 1968. The chemical composition of tobacco and tobacco smoke. *Chem. Rev. 68:*153–207.

Talley, L.I., Grizzle, W.E., Waterbor, J.W., Brown, K., Weiss, H., and Frost, A.R. 2002. Hormone receptors and proliferation in breast carcinomas of equivalent histologic grades in pre- and postmenopausal women. *Int. J. Cancer. 98:*118–127.

11

Immunohistochemical and Molecular Pathology of Angiogenesis in Primary Lung Adenocarcinoma

Oichi Kawanami

Introduction

Angiogenesis or neovascularization is required for an excess blood supply to the tumors (Folkman *et al.*, 1992). It is known that the number of capillaries in primary adenocarcinoma generally exceeds those in squamous cell carcinoma of the lung, and the microvessel counts suggest a positive correlation with the development of metastasis in primary lung adenocarcinomas. A higher vascularization in lung cancer might reflect poorer prognosis of patients. Meanwhile, four different patterns of tumor angiogenesis were proposed in primary lung adenocarcinomas (Pezzella *et al.*, 1997). They included alveolar, basal, papillary, and diffuse types. In their definition, they indicated that no angiogenesis occurred in the alveolar pattern. However, these results might not reflect the details of alveolar capillary angiogenesis at ultrastructural level (Jin *et al.*, 2001a,b).

In this chapter, techniques of immunohistochemical, ultrastructural, and molecular pathology are described in detail. Reverse transcription-polymerase chain reaction (RT-PCR) was applied incorporating

laser capture microdissection that enabled selective collection of mRNAs from alveolar walls alone. The morphological characteristics of normal alveolar capillaries and their neighboring microvessels in the human lungs would be defined. Our study focused especially on the process of phenotypic and functional alterations of capillary endothelial cells of the alveolar walls when neoplastic cells spread on their surface. Morphologic details of neovascularization was further demonstrated by electron microscopy. Significant expression of mRNAs related to the growth of endothelial cells were obtained, and analyses of the basic factors, including vascular endothelial growth factor (VEGF) (Dvorak *et al.*, 1999) and their receptors (KDR and Flt-1), thrombin, trypsin (Koshikawa *et al.*, 1998), and protease-activated receptor (PAR)-1 and PAR-2 (Dery *et al.*, 1998), were made.

Protease-activated receptors (PAR-1 to 4) are expressed in many cells of mammalian tissues (Dery *et al.*, 1998). The PARs receptors are not elicited by conventional ligand–receptor interaction. They require partial proteolytic cleavage by serine proteases such as

Handbook of Immunohistochemistry and *in situ* Hybridization of Human Carcinomas, Volume 1: Molecular Genetics; Lung and Breast Carcinomas

189

Copyright © 2004 by Elsevier (USA)
All rights reserved.

thrombin, trypsin, and mast cell tryptase to expose tethered ligand that actually stimulates G-proteins of the cells leading to promote biological functions. The PARs are elicited also by activating peptides, such as SFLLRN for PAR-1 and SLIGKV for PAR-2, as substitution of serine proteases. The consequence of the activations will lead to proliferation of endothelial cells, an increased synthesis of procollagen in smooth muscle cells, and airway contraction/relaxation in bronchial epithelial cells. There is increasing evidence that PAR family members play pivotal roles in wound healing (Dery *et al.*, 1998). It is intriguing to examine possible roles of PARs in angiogenesis of the alveolar walls in primary lung adenocarcinoma.

Tissues were obtained from solitary nodules of primary lung adenocarcinomas. Diagnosis of papillary or bronchiolalveolar cell carcinoma was established according to the World Health Organization (WHO) classification. Data obtained were presented as means ± SD. Analysis was performed using the student's t test for unpaired observations (two-tailed), and significance was considered at p < 0.05.

Immunohistochemistry and Immunofluorescence Studies

Fixation, Antibodies, and Immunohistochemical Procedures

Normal and tumor tissues were fixed in 4% buffered paraformaldehyde for immunofluorescence staining, or in 10% buffered formalin for embedding in paraffin. Deparaffinized sections were stained by an avidin-biotin complex (ABC) immunoperoxidase method. The sections were treated with 0.3% hydrogen peroxide in methanol for 15 min at 20°C to suppress endogenous peroxidase activity (Jin *et al.*, 2003).

The primary polyclonal rabbit antibody to VEGF (1:100 dilution, Santa Cruz Biotechnology, Inc., Santa Cruz, CA) was used for 1 hr at room temperature. For the proliferating cell nuclear antigen (PCNA) staining, sections were pretreated with 0.01 M buffer (pH 6) at 100°C for 10 min, followed by cooling at room temperature, and incubated with the primary monoclonal mouse antibody to PCNA (1:100 dilution, PC10, Dako, Glostrup, Denmark). Incubation with the primary antibody for PAR-1 (mouse monoclonal; mm IgG, WEDE15, ImmunoTech, Marseille, France, 1:100) was performed overnight at 4°C, and the sections were reacted with 3,3′-diaminobenzidine (DAB) and counterstained with hematoxylin.

Indirect double immunofluorescence method was applied on 4–6 μm thick frozen sections, and cancer cells and endothelial cell lines were fixed with acetone.

They were incubated overnight at 4°C with a combination of mm IgG antibody or rabbit polyclonal (rp) IgG antibody against different antigens, including thrombomodulin (mmIgG, TM1009, Dako, Carpinteria, CA, 1:100), von Willebrand factor (vWf: rpIgG, Dako, Glostrup, 1:400), Ki67 (mmIgG, Ki = S5 Dako, Glostrup, 1:100): trypsin (mmIgG, Chemicon International Temecula, CA, 1:150), thrombin (mmIgG, T4, Biogenesis, England, 1:100) and PAR-2 (rpIgG, Teijin Co. Ltd, Tokyo, Japan, 1:150) as the primary antibodies.

After washing, the sections were incubated with FITC-labeled goat-anti-rabbit IgG (Vector Laboratories, Burlingame, CA) for rpIgG antibodies and Texas Red-labeled horse-anti-mouse IgG for mmIgG antibodies in a double stain; they were incubated with secondary antibodies for 60 min at room temperature in the dark. Nuclear counterstaining was done at room temperature with 0.01% TOTO-3 iodide (Molecular Probes, Inc., Eugene, OR).

Immunohistochemical Control Procedures and Observations

Negative control procedures consisted of both the omission of the primary antibody from the staining procedure and the substitution of the primary antibody with corresponding amounts of normal immunoglobulin from the same animal species. Both control procedures consistently gave negative results.

The sections were examined by light microscopy or confocal laser scanning microscopy (Model TC-SP, Leica, Heidelberg, Germany) equipped with argon and argon-krypton laser sources. In the resulting preparations, red fluorescence represented thrombomodulin or trypsin, and green fluorescence depicted von Willebrand factor or PAR-2, and the nuclei appeared as blue color.

Characteristics of Normal Pulmonary Microvessels
Normal pulmonary vessels are under a dual-blood circulation system, including the pulmonary and bronchial circulation. It is nearly impossible to histologically identify the origin of microvessels in the lung parenchyma except for the capillaries. It is also impossible to accurately localize microvessels that are committed to physiologic communications between the bronchial and pulmonary circulation. Nevertheless, carbon particles infused through the bronchial circulation simultaneously filled alveolar capillaries via juxta-alveolar microvessels (Kawanami, 1997). Confocal microscopy revealed normal alveolar capillary endothelium and linear expression of thrombomodulin along the plasma membrane (Figure 27A), but their cytoplasm was not reactive for vWf. Such thrombomodulin-dominant pattern was clearly recognizable by a three-dimensional image construction

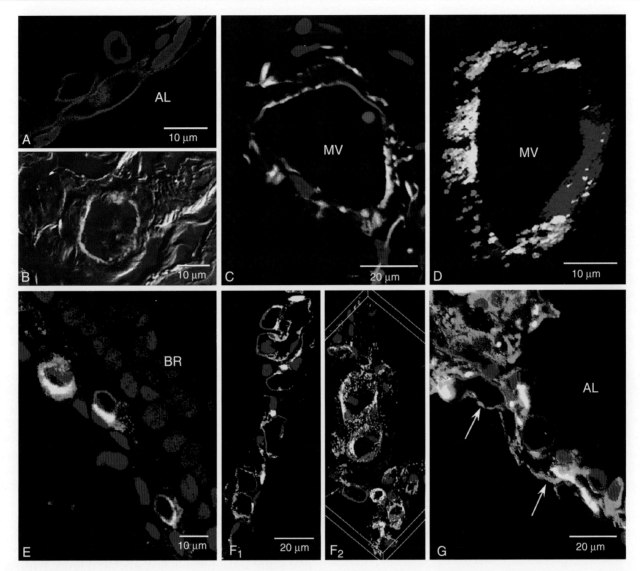

Figure 27 Dual immunofluorescence stainings for thrombomodulin (TM) *(red)* and von Willebrand factor (vWf) *(green)* by confocal laser microscopy in the normal lung microvessels. DAPI-stained nuclei appear *blue*. **A:** The plasma membrane of alveolar capillary endothelial cells is fully covered with TM, and thin cytoplasmic segments do not contain vWf at all (TM-dominant pattern). (AL = alveolar lumen.) **B:** Endothelial cells of a connective tissue microvessel (vasa vasorum in a pulmonary artery) express only vWf, and TM is scarcely expressed (vWf-dominant pattern). **C:** The endothelial cell lining of a juxta-alveolar microvessel (*MV*) consists of cytoplasmic segments alternately expressing TM and vWf (mosaic-like pattern). **D:** Three-dimensional image of a microvessel (*MV*) located at juxta-alveolar zone. The endothelial cells show a mosaic-like distribution pattern of TM and vWf along endothelial cell linings. **E:** Note that both antigens of TM and vWf in submucosal bronchial capillaries are expressed in the same cytoplasmic segments as a mixed pattern. **F$_1$:** Alveolar capillaries are fully distended when neoplastic cells of primary lung adenocarcinoma spread on the walls. vWf-positive masses accumulate along TM-positive cell membranes. Some are possibly platelets aggregations in the vascular lumens. **F$_2$:** At tumor areas, a number of small particulates positive for vWf (Weibel-Palade bodies) occurred in cytoplasmic segments of capillary endothelial cells (three-dimensional image of alveolar capillaries). **G:** Along an alveolar wall, groups of isolated neoplastic cells (left upper corner) are reactive for PAR-2 *(green fluorescence)*. Note the lack of continuity for TM reaction along an alveolar capillary loop. Instead, remaining capillary segments clearly demonstrate a gain of immunohistochemical property to PAR-2 *(arrows)*. (AL = alveolar lumen.)

(Jin *et al.*, 2001a). Endothelial cells of microvessels (> ~10-µm diameter) located in the connective tissue zones, including peribronchial, interlobular, and pleural areas, and vasa vasorum in large vessel walls, demonstrated band-like reaction for vWf alone, and their

plasma membrane lacked the reaction for thrombomodulin, as shown in vasa vasorum (Figure 27B) of a large pulmonary artery (vWf-dominant pattern). The juxta-alveolar microvessels were located along the borders between the alveolar and connective tissue zones.

They include the interlobular pulmonary venules, peri-bronchial, and subpleural microvessels (< ~ 40-µm diameter). They similarly demonstrated the alternate expression of thrombomodulin (red fluorescence) in the plasma membrane and vWf (green) along their thin cytoplasmic segments, respectively (MV: microvessel in Figure 27C). The mosaic-like pattern was obvious in a three-dimensional image (Figure 27D). Capillaries of the bronchial circulation located in submucosa of the bronchial airways are characterized as the mixed expression pattern of both antigens, demonstrating them in a single-cell compartment (Figure 27E).

The von Willebrand factor is localized in the cytoplasmic inclusions (Weibel-Palade bodies) of endothelial cells, and it is widely accepted as one of the biological markers of endothelial cells. However, normal alveolar capillary endothelial cells lack immunohistochemical reactivity for vWf. Nevertheless, the expression of vWf became extremely prominent in alveolar walls that underwent fibrotic changes in humans (Kawanami *et al.*, 1992) and animals (Kawanami *et al.*, 1995).

Vascular endothelial cells are consistently exposed to procoagulant and anticoagulant factors, the function of which is balanced under normal circumstances. A mosaic-like distribution pattern in the juxta-alveolar microvessels is suggestive of the transition sites between two types of endothelial cells, comprising thrombomodulin-dominant (pulmonary circulation) and vWf-dominant types (mostly bronchial circulation). This heterogeneous distribution of the antigens suggests topographic functional differences of endothelial cells to maintain blood coagulation and anticoagulation balance in the normal human lung (Kawanami *et al.*, 2000).

Phenotypic Alteration and Neovascularization of Alveolar Capillaries in Primary Lung Adenocarcinoma

In the tumor-bearing alveolar walls, nuclei of the capillary endothelium were hypertrophic and occasionally reactive for proliferating cell markers such as PCNA and Ki67. The PCNA reaction was also seen in endothelial cell nuclei of microvessels located in tumor scars (Jin *et al.*, 2001b). The number of nuclei in the tumor capillaries showed a 1.5-fold increase per unit length of the vessel circumference. The luminal width, shorter, and greater diameters of capillaries showed 3.8-, 1.8-, and 2-fold increases compared with the normal, respectively. These changes indicate proliferating capacity of endothelial cells and luminal dilatation of tumor capillaries in primary lung adenocarcinoma.

Concerning endothelial cell renewals, Thurston *et al.* (1998) showed unique characteristics of tracheal microvessels in mice under the experimental mycoplasma infection. They found that submucosal capillaries changed into venule-like structures with marked dilatation in C3H mice. The number of capillaries in C57Bl mice were markably increased, thus indicating different types of neovascularization taking place in each species. Species-dependant structural alterations of microvessels were coincidently demonstrated in alveolar capillaries of patients with primary lung adenocarcinomas. Both reactions were commonly initiated by plasma fluid extravasation as an early event, probably because of VEGF$_{121}$ function, as shown later in these patients (Dvorak *et al.*, 1999).

Proliferation of neoplastic cells often coincided with a gradual increase of interstitial fibrosis in alveolar walls. At areas of neoplastic cell proliferation, the plasma membrane of capillary endothelial cells tended to lose the reactivity for thrombomodulin. The cytoplasm in turn developed a number of tiny spots reactive for vWf (Figure 27F$_1$). These spots (Weibel-Palade bodies) increased in number to wide ranges and accumulated in the cytoplasm while alveolar fibrosis advanced; this phenomenon is well-demonstrated in a three-dimensional image (Figure 27F$_2$). In relatively early stages of tumor cell metastasis, cytoplasmic segments of alveolar capillaries demonstrated mosaic-like mixed patterns of the two antigens, as seen in the normal bronchial capillary endothelium. These findings are in good accord with the concept that blood coagulation activity is augmented in the primary lung adenocarcinoma, because an anticoagulant factor, thrombomodulin, is lost in contrast to a new gain of a coagulant vWf in the endothelium.

The reaction for VEGF was frequently intense in the neoplastic cell cytoplasm. Microvessel endothelial cells were scarcely reactive for Flt-1 and KDR in alveolar walls at the immunohistochemical level. With the exception of type II alveolar epithelial cells, PAR-1 and PAR-2 were hardly expressed in any cell located in the normal alveolar walls. Neoplastic cells of adenocarcinoma often appeared reactive for both antigens. In the capillary endothelium of tumor-bearing alveolar walls, PAR-2 tended to occur in segments deficient in thrombomodulin expression. Fine expression of PAR-2 (arrows) was demonstrated along the cell membranes of capillary endothelial cells that lacked the reaction to thrombomodulin (Figure 27G). Microvessels deep in the tumor scar were well reactive for PAR-1 and PAR-2, and they totally lost the reaction to thrombomodulin.

Induction of vWf can be promoted by the activation of endothelial PAR-2 that is generally known to be generated by inflammatory mediators such as tumor necrosis factor (TNF). Cytokines released from neoplastic cells suppress transcription of the thrombomodulin gene, which results in a loss of its expression in vascular endothelial cells (Conway *et al.*, 1988).

Without thrombomodulin expression, endothelial cells show efficient mitogenic response to thrombin via activation of PAR-1. PAR-2 activation in human vascular endothelial cells mediate their mitogenic response *in vitro* (Mirza *et al.*, 1996). Both PAR-1 and PAR-2 are closely related to cell growth of endothelial cells. It is interesting to note that the loss of thrombomodulin might suggest a functional transition of the capillaries from quiescent to alert state by unveiling new antigens such as vWf, PAR-1, and PAR-2.

Higher expressions of PAR-1 and PAR-2 mRNAs were found in the microdissected tissues of alveolar walls when neoplastic cells were spread. PAR-2 expression is localized in the normal human bronchial epithelial cells and peribronchial smooth muscle cells. Bronchial epithelial PAR-2 can exert a protective role against bronchoconstriction in humans and animals, as proved by the *in vitro* activations of PARs. Furthermore, the elicitation of PAR-2 induces growth activity of lung fibroblasts. This process might cause promotion of procollagen deposit in alveolar walls (Jin *et al.*, 2003).

Trypsin was immunohistochemically demonstrated in the normal bronchial basal cells, but barely in any alveolar wall cells (Jin *et al.*, 2003). Trypsin (-ogen) was originally found in pancreatic acinar cells. However, the distribution of trypsin is widely recognized throughout the body, including skin, esophagus, small intestine, lung, kidney, liver, extrahepatic bile duct, and neuronal cells (Koshikawa *et al.*, 1998). Wide distribution of trypsin might suggest its physiologic function in the tissue cells that express PAR-2. Neoplastic cells of primary lung adenocarcinoma were reactive for trypsin in 44% of patients and far less frequently (0–12%) in other types of lung tumors in our study (Jin *et al.*, 2003). The neoplastic cell cytoplasm positive for trypsin did not usually show a positive reaction to PAR-2, suggesting unfavorable tendency in the tumor cell growth by autocrine manner. Thrombin was not retrieved in any neoplastic cells or cells in the alveolar walls in paraffin or frozen sections, in contrast to an increased expression of PAR-1 in microvessel endothelial cells in lung adenocarcinomas. Thrombin was not expressed in any cancer and endothelial cell lines.

Evidence of Alveolar Wall Angiogenesis by Electron Microscopy

Fixation and Staining for Electron Microscopy

Small pieces of tissue fixed in 2.5% glutaraldehyde in 0.1 M phosphate buffer (pH 7.4) were postfixed in 1% OsO_4 in the same buffer at 4°C for 1 hr and embedded in Quetol resin (Nissin EM) (Kawanami *et al.*, 1992). Ultrathin sections were stained with uranyl acetate and lead citrate and examined with a transmission electron microscope (Hitachi H-7000) at an accelerating voltage of 75KV.

Capillary Endothelium in Alveolar Walls With or Without Neoplastic Cell Spread

Ultrastructural changes (Jin *et al.*, 2001b) of alveolar capillary endothelial cells in tumor areas consisted of:

1. Opening of the intercellular junctions (open gap junction)

2. Remarkable luminal dilatation in existing capillaries

3. Cytoplasmic multilayering or bridging at the elongated intercellular junctions of endothelial cells

4. Sprouting of cytoplasmic segments for angiogenesis

5. Development of organelles such as Weibel-Palade bodies and vesiculo-vacuolar organelles

6. Fenestrae formation at the cytoplasmic segments with extreme attenuation.

At areas of neoplastic cells' spreading on alveolar surface lining, cytoplasmic segments of capillary endothelium became round and thick and contained a large number of free ribosomes, pinocytotic vesicles, vesiculo-vacuolar organelles, and Weibel-Palade bodies (arrowheads, Figure 28A). Such segments of endothelial cells showed a slit-like lumen and occasionally extended to sprout some portions (stars) into the alveolar interstitium. These ultrastructural images strongly suggest a new vessel formation from the proper alveolar capillaries. On the process of such angiogenesis, the cytoplasm developed extremely thin segments that contained a number of foci of diaphragm-like fenestrae with ~70 nm diameter (Figure 28B). Such fenestrae (arrows) were found in the alveolar capillaries beneath neoplastic cells of 65% of patients with bronchioloalveolar or papillary type adenocarcinoma examined. Most-frequently developed were the new vessels extending from preexisting alveolar capillaries as intraalveolar papillary stalks of papillary adenocarcinoma (Jin *et al.*, 2001a,b). However, these ultrastructural remodelings with cytoplasmic fenestrae were not specific in the alveolar capillaries of primary lung adenocarcinomas. They were occasionally found in fibrotic alveolar walls without cancer cells in humans, as well as in experimental animals (Kawanami *et al.*, 1983, 1992, 1995).

In our results, the fenestrae formation of endothelial cells was limited to the capillaries of alveolar walls overlaid by VEGF-positive tumor cells of primary lung adenocarcinomas. Interestingly, fenestrated capillaries could be induced *in vitro*. Tumors stably transfected with human $VEGF_{165}$ and cell pellets containing basic fibroblast growth factor developed vessels with

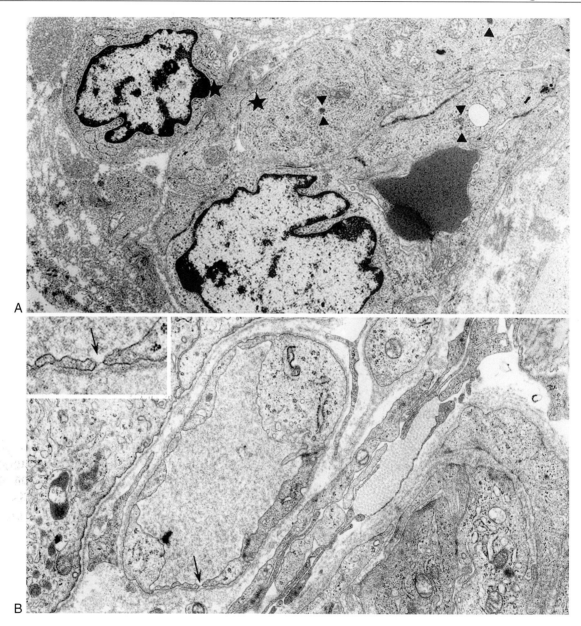

Figure 28 Ultrastructures of capillary sprouting and fenestrae formation. **A:** An alveolar capillary containing a red blood cell in a narrow lumen comprises several cytoplasmic segments of endothelial cells. They develop a large number of Weibel-Palade bodies (arrowheads), free ribosomes, and endoplasmic reticulum. Two cytoplasmic segments (stars) of endothelial cells extend into the interstitium and show long linear densities of the junction between adjacent segments. Such sprouting segments hardly develop a lumen. **B:** Among thick segments of endothelial cells (right lower corner), they occasionally develop cytoplasmic portions of extreme attenuation. These segments further develop diaphragm-like fenestrae 70-nm diameter (arrows indicate fenestrae), just beneath neoplastic cells of adenocarcinoma (left corner).

fenestrated endothelium in the host microvasculature (Roberts *et al.*, 1997). Basic fibroblast growth factor showed a least effectiveness to induce fenestrae in the new vessels. In contrast, chronic exposure of $VEGF_{165}$ was highly effective for the fenestration formation in nonfenestrated endothelium.

The RT-PCR revealed four isoforms of VEGF mRNA comprising 121 and 165, 189 and 206 amino acids, and they were generally up-regulated in the tumor-bearing alveolar walls. In our microdissection method, $VEGF_{165}$ mRNA appeared generated from neoplastic cells and caused new fenestrae formation in the alveolar capillaries. This is in good accord with a significant expression of the receptor, KDR, in the corresponding endothelial cells for altering endothelial cell phenotype.

Analyses of Reverse Transcription-Polymerase Chain Reaction

Microdissection of the Alveolar Walls and cDNA Synthesis Formation

Fresh tumor tissues were embedded in OCT compound, snapped frozen in acetone-dry ice, and stored at $-80°C$ until use. Using laser capture microdissection method (LCM 100 Image Archiving Workstation, Arcturus Engineering, Inc. Mountain View, CA), we microdissected unfixed 6-μm-thick sections containing alveolar walls that included capillary endothelial cells, pericytes, fibroblasts, and alveolar epithelial cells with or without neoplastic cells. Total RNA was extracted from the tissues by acid guanidinium thiocyanate-phenol-chloroform. Complementary DNA (cDNA) was synthesized from 2 μl of total RNA primed with oligo (dT) using superscript II reverse transcriptase (Gibco BRL Rockville, MD).

Primers of VEGF and Its Receptors (Flt-1 and KDR) and RT-PCR Method

For VEGF mRNA, nested PCR was carried out using primers that span the variable splice regions of VEGF mRNA: (a) 5′-GCT ACT GCC ATC CAA TCG AGA CC-3′ (sense) (exon 3); (b) 5′-GTT TCT GGA TTA AGG ACT GTT CTG TCG-3′ (anti-sense) (exon 8); and (c) 5′-AAT CCAATT CCAAGA GGG ACC GTG C-3′ (anti-sense) (exon 8). First amplification was carried out using primers (a) and (c) for 15 cycles (1 min at 94°C, 2 min at 62°C, and 3 min at 72°C). Then 1 μl of the first PCR products was used for amplification with the nested primers (a) and (b). Amplification was for 30 cycles under the same conditions as in the first amplification. The expected PCR products for each VEGF variant—440, 572, 644, and 695 bp—are encoding the isoforms of $VEGF_{121}$, $VEGF_{165}$, $VEGF_{189}$, and $VEGF_{206}$, respectively. For KDR, PCR was done with primers of 5′-ACGCTGACATGTACGG TCTATG-3′ (sense) and 5′-TTCCCAT-TTGCTGGCATCATA-3′ (anti-sense) for 40 cycles (1 min at 94°C, 2 min at 56°C, and 3 min at 72°C; the product size, 405 bp). For Flt-1, VEGF receptor, single PCR was done with primers of 5′-GCAACCTG TGACTTTTGTTCC-3′ (sense) and 5′-GAGGATTTCTTCCCCTGTGTA-3′ (anti-sense) for 40 cycles (1 min at 94°C, 2 min at 56°C, and 3 min at 72°C; the product size, 512 bp).

The PCR products were electrophoresed on 2% agarose gels and visualized by ethidium bromide staining. A glyceraldehyde 3-phosphate dehydrogenase (GAPDH) was used as an internal control. Semiquantitative measurements were done based on the standard curves constructed for the products and GAPDH. The analyses were done using NIH Image Software.

Messenger RNA Expression of VEGF Isoforms and Their Receptors

Both normal and tumor-containing alveolar walls demonstrated four isoforms of VEGF ($VEGF_{121}$, $VEGF_{165}$, $VEGF_{189}$ and $VEGF_{206}$). The expression of VEGF mRNAs was normalized against GAPDH. All isoforms of VEGF mRNA in tumor-bearing alveolar walls exceeded those in the normal walls. The expressions of $VEGF_{165}$ and KDR mRNAs were significantly higher in tumor-bearing walls than those in the normal walls ($p < 0.05$). The expression of Flt-1 mRNA in the tumor-bearing walls was not significantly different from that in the normal walls. A significant up-regulation was consistently evident of KDR and $VEGF_{165}$ mRNAs in cancerous alveolar walls (Jin et al., 2001a,b).

Tsopanoglou and Maragoudakis (1999) showed that KDR mRNA was up-regulated in human umbilical cord vein endothelial cell (HUVEC) following the exposure to thrombin, thus mitogenic activity of VEGF was sufficiently promoted. This result might imply a possible link of thrombin to PAR-1 activation that results in the up-regulation of KDR mRNA, thus playing a role as an accelerator of VEGF function. Taken together, alveolar capillary endothelial cells could stimulate angiogenesis in relation to serine proteases and protease-activated receptors.

Serine Proteases and Protease-Activated Receptors
Primers of PAR-1, PAR-2, Thrombin, and Trypsin

The primer sequences for thrombin were: forward, 5′-TGGGTACTGCGACCTCAACTAT-3′; reverse, 5′-CAGACACACAGGGTGAATGTAGTC-3′; the product size, 601 bp, and for trypsin: forward, 5′-CTCCT-GATCCTTACCTTTGTGG-3′; reverse, 5′-AGGGTAG-GAGGCTTCACACTTAG-3′; the product size, 519 bp, respectively. The primer sequences for PAR-1 were: forward, 5′-CAGTTTGGGTCTGAATTGTGTCG-3′; reverse, 5′-TGCACGAGCTTATGCTGCTGAC-3′; the product size, 586 bp, and for PAR-2: forward, 5′-TGGATGAGTTTTCTGCATCTGTCC-3′; reverse 5′-CGTGATGTTCAGGGCAGGAATG-3′; the product size, 491 bp, respectively (Jin et al., 2003). Following one cycle denaturation at 94°C for 10 min, PCR was performed at 94°C for 45 sec, 60°C for 45 sec for thrombin and trypsin (for PAR-1 and PAR-2, annealing was done at 58°C for 45 sec), and extension at 72°C for 2 min for all primers. Following 18 to 25 cycles of PCR for endothelial cell and human lung adenocarcinoma cell lines or 40 cycles for the others, the reactions were stopped by chilling to 4°C.

In the microdissected normal alveolar wall tissues, PAR-1, PAR-2, and trypsin mRNAs were detectable in 28% (mean/GAPDH ± SE; 0.011 ± 0.006), 36% (0.063 ± 0.023) and 7% of the patients examined, respectively (Jin et al., 2002). Compared with normal alveolar walls, mean expressions of the mRNAs of PAR-1, PAR-2, and trypsin indicated 9.6- (p < 0.05), 15.6- (p < 0.01), and 4.7-fold increase in the neoplastic cell-bearing alveolar walls, respectively. Major contribution for the higher expressions of PAR-1 and PAR-2 mRNA would rely on the neoplastic cells spreading on alveolar walls. Alveolar capillary endothelial cells should confer some contribution to the levels because the receptor proteins were newly expressed in the cancerous tissues.

As thrombin was not retrieved immunohistochemically in any cell types of the alveolar wall, RT-PCR analysis failed to detect its mRNA. However, thrombin is supposed to be well-generated in hypercoagulability condition under primary adenocarcinoma. Thus, PAR-1 in alveolar capillary endothelial cells might be elicited by thrombin for their own cell growth. These results confirmed a potential growth activity of the endothelial cells, via elicitation of G-protein coupled receptors, in the human alveolar capillaries that were terminal branches of the pulmonary arteries. The relation of thrombin, trypsin, and PARs will be further discussed with the results of *in vitro* studies.

In Vitro Proliferation Assay with Thrombin and Activating Peptides: SFLLRN and SLIGKV

Human Lung Adenocarcinoma Cell Lines

Four human lung cancer cell lines (Riken Cell Bank, Tokyo, Japan) were used for the experiments:

1. A549; alveolar epithelial cells (Dulbecco's MEM + 10% FBS).
2. HLC-1; well-differentiated bronchogenic adenocarcinoma (HamF12 + 10% FBS).
3. LC-2/ad; moderately differentiated adenocarcinoma (HamF12 + RPMI1640 + 15% FBS + 25mM HEPES).
4. PC-14; poorly differentiated adenocarcinoma (RPMI1640 + 10% FBS).

Each cancer cell line expressed PAR-1 and PAR-2 mRNAs. Cells derived from less-differentiated adenocarcinomas (LC-2 and PC-14) expressed trypsin mRNA, but neither A549 nor HLC-1 (well-differentiated cells) expressed it. The immunohistochemical examination of tumor tissues also showed a marked variability in

the expression of trypsin, even in the same patient. Only selected tumor cells could exert proteolytic activity for the activation of PAR-2, which would lead to cell growth in adjacent capillary endothelial cells.

Human Endothelial Cell Lines and Their Growth Activity via PARs

Cells used in the growth activity test were obtained from Clonetics Corporation (Walkersville, MD), including:

1. Human pulmonary artery endothelial cell (HPAEC).
2. Human umbilical vein endothelial cell (HUVEC).
3. Human aortic endothelial cells (HAEC).

They were cultured on gelatin-coated dishes in endothelial cell basal medium (EBM) (Clonetics) supplemented with EGM SingleQuots (BioWhittaker, Walkersville, MD). Cells were maintained in a tissue culture incubator under a 5% CO_2-95% air, 95% humidity, and 37°C temperature. Only 4–8 passages of the cell lines were used for the experiments (Jin et al., 2003; Fujiwara et al., 2001). Compared with the mRNA levels in HPAEC, the expression ratios of mean PAR-1 mRNA in HUVEC and HAEC were 108% and 122% and of mean PAR-2 mRNA were 105% and 53%, respectively.

Materials prepared were recombinant human tumor necrosis factor alpha, (TNFα) (R&D Systems, Minneapolis, MN), human plasma thrombin (Sigma Bachem, St. Louis, MO), synthetic activating peptides; SFLLRN for PAR-1 and SLIGKV for PAR-2 (Bachem, Bubendorf, Switzerland). Cells were treated for 12 hr at indicated concentration of activating peptide and further incubated for 12 hr with the presence of BrdU (10 μM/L). Then, cells were fixed in 70% ethanol (0.5M HCl) for 30 min at −20°C and processed for enzyme-linked immunosorbent assay (ELISA) using peroxidase (POD) conjugated antibody against BrdU according to the manufacturer's instructions (ELISA using BrdU labeling and detection kit III; Boehringer Mannheim, Mannheim, Germany). To compare the results with those after up-regulating PAR-2 mRNA expression, a 10 ng/ml TNF-α pretreatment (Fujiwara et al., 2001) was done for 24 hr, and SLIGKV was added to the cells of HPAEC and HUVEC. The DNA-uptake was measured in the same way as previously described.

Alpha-thrombin and its corresponding activating peptides, SFLLRN, induced a significant growth activity in the cells of HPAEC and HUVEC in a dose-dependant manner (up to 230% in HPAEC compared with nonstimulated cells). The activating peptide SFLLRN for PAR-1 induced 1.7-fold growth activity in HPAEC, but it did not exert significant effect in HUVEC (Jin et al., 2003). Both HPAEC and HUVEC with pretreatment by TNF-α resulted in a significant cell growth activity, up to 143% and 124%, respectively, due to the elicitation with SLIGKV, the activating peptide for PAR-2. However, the growth activity in these cells was not significant unless pretreatment was done.

In conclusion, proliferation of alveolar capillary endothelial cells could be initiated in part by PAR-1 activations with serum thrombin and by PAR-2 with neoplastic cell-derived trypsin. With this background, alveolar angiogenesis might be synergistically promoted with the function of VEGF, especially with VEGF$_{165}$ generated from the tumor. This hypothesis of alveolar wall angiogenesis was well-supported by morphology of microvessel endothelial cells, and by *in vitro* studies in which DNA-synthesis was further promoted via elicitation of protease–activated receptors in collaboration with serine proteases.

Acknowledgment

I appreciate the help offered by Enjing Jin, Masakazu Fujiwara, Mohammad Ghazizadeh, Hajime Shimizu, and Seiko Egawa.

References

Cocks, T.M., Fong, B., Chow, J.M., Anderson, G.P., Frauman, A.G., Goldie, R.G., Henry, P.J., Carr, M.J., Hamilton, J.R., and Moffatt, J.D. 1999. A protective role for protease-activated receptors in the airways. *Nature (letters). 398:*156–160.

Conway, E.M., and Rosenberg, R.D. 1988. Tumor necrosis factor suppresses transcription of the thrombomodulin gene in endothelial cells. *Mol. Cell. Biol. 8:*5588–5592.

Dery, O., Corvera, C.U., Steinhoff, M., and Bunnett, N.W. 1998. Proteinase-activated receptors: Novel mechanisms of signaling by serine proteases. (Invited review) *Am. J. Physiol. (Cell Physiol.)* 274:C1429–C1452.

Dvorak, H.F., Nagy, J.A., Brown, L.F., and Dvorak, A.M. 1999. Vascular permeability factor/vascular endothelial growth factor and the significance of microvascular hyperpermeability in angiogenesis. *Cur. Topics Microbiol. Immunol. 237:*97–132.

Folkman, J., and Shing, Y. 1992. Angiogenesis. *J. Biol. Chem. 267:* 10,931–10,934.

Fujiwara, M., Jin E., Ghazizadeh, M., and Kawanami, O. 2001. An *in vitro* model to evaluate regulatory mechanisms of antigen expression by normal pulmonary vessel endothelial cells. *Microvasc. Res. 61:*215–219.

Jin, E., Ghazizadeh, M., Fujiwara, M., Nagashima, M., Shimizu, H., Ohaki, Y., Arai, S., Gomibuchi, M., Takemura, T., and Kawanami, O. 2001a. Angiogenesis and phenotypic alteration of alveolar capillary endothelium in areas of neoplastic

cell spread in primary lung adenocarcinoma. *Pathol. Int. 51:*691–700.

Jin, E., Ghazizadeh, M., Fujiwara, M., Nagashima, M., Shimizu, H., Ohaki, Y., Arai, S., Gomibuchi, M., Takemura, T., and Kawanami, O. 2001b. Aerogenous spread of primary lung adenocarcinoma induces ultrastructural remodeling of the alveolar capillary endothelium. *Hum. Pathol. 32:*150–158.

Jin, E., Fujiwara, M., Xin, P., Ghazizadeh, M., Arai, S., Ohaki, Y., Kajiwara, K., Takemura, T., and Kawanami, O., 2003. Protease-activated receptors (PAR)-1 and PAR-2 participate in the cell growth of alveolar capillary endothelium in primary lung adenocarcinomas. *Cancer* 97:703–713.

Kawanami, O. 1997. The endothelium of the pulmonary microvessels. *J. Nippon. Med. Sch.* 64:1–17.

Kawanami, O., Basset, F., Barrios, R., Lacronique, J.G., Ferrans, V.J., and Crystal, R.G. 1983. Hypersensitivity pneumonitis in man: Light- and electron-microscopic studies of 18 lung biopsies. *Am. J. Pathol. 110:*275–289.

Kawanami, O., Jiang, H.X., Mochimaru, H., Yoneyama, H., Kudoh, S., Ohkuni, H, Ooami, H., and Ferrans, V.J. 1995. Alveolar fibrosis and capillary alteration in experimental pulmonary silicosis in rats. *Am. J. Respir. Crit. Care Med. 151:*1946–1955.

Kawanami, O., Jin, E., Ghazizadeh, M., Fujiwara, M., Jiang, L., Ohaki, Y., Gomibuchi, M., and Takemura, T. 2000. Mosaic-like distribution of endothelial cell antigens in capillaries and juxta-alveolar microvessels in normal human lung. *Pathol. Int. 50:*136–141.

Kawanami, O., Matsuda, K., Yoneyama, H., Ferrans, V.J., and Crystal, R.G. 1992. Endothelial fenestration of the alveolar capillaries in interstitial fibrotic lung diseases. *Acta Pathol. Jpn. 42:*177–184.

Koshikawa, N., Hasegawa, S., Nagashima, Y., Miyata, S., Miyagi, Y., Yasumitsu, H., and Miyazaki, K. 1998. Expression of trypsin by epithelial cells of various tissues, leukocytes, and neurons in human and mouse. *Am. J. Pathol. 153:*937–944.

Mirza, H., Yatsula, V., and Bahou, W. 1996. The proteinase activated receptor-2 (PAR-2) mediates mitogenic responses in human vascular endothelial cells. Molecular characterization and evidence for functional coupling to the thrombin receptor. *J. Clin. Invest.* 97:1705–1714.

Pezzella, F., Pastorino, U., Tagliabue, E., Andreola, S., Sozzi, G., Gasparini, G., Menard, S., Gatter, K.C., Harris, A.L., Fox, S., Buyse, M., Pilotti, S., Pierotti, M., and Rilke, F. 1997. Nonsmall cell lung carcinoma tumor growth without morphological evidence of neoangiogenesis. *Am. J. Pathol* 151:1417–1423.

Roberts, W.G., and Palade, G.E. 1997. Neovasculature induced by VEGF is fenestrated. *Cancer Res.* 57:765–772.

Thurston, G., Murphy, T.J., Baluk, P., Lindsey, J.R., and McDonald, D.M. 1998. Angiogenesis in mice with chronic airway inflammation. Stain-dependent differences. *Am. J. Pathol. 153:*1099–1112.

Tsopanoglou, N.E., and Maragoudakis, M.E. 1999. On the mechanism of thrombin-induced angiogenesis. Potentiation of vascular endothelial growth factor activity on endothelial cells by up-regulation of its receptors. *J. Biol. Chem.* 274: 23,969–23,976.

12

Immunohistochemistry of Human Leukocyte Antigen Expression in Lung Carcinoma

Mirjana Urosevic and Beatrix Müller

Introduction

The delineation of molecular events that lead to recognition of tumor cells by the immune system has helped us to understand that human leukocyte antigen (HLA) molecules play a fundamental role in these interactions. The HLA class I molecules bind antigenic peptides, which are mostly derived from endogenous proteins (tumor-associated antigens [TAA] in tumor cells) and present them on the cell surface to the T-cell receptors (TCRs). The recognition of these peptides by cytotoxic T-cell lymphocytes (CTLs) triggers the series of events that can eventually lead to tumor cell lysis. The HLA class I proteins are also involved in the interaction with natural killer (NK) cells. Through the receptors that they express, NK cells can sense the cells that are not expressing one or more HLA alleles and lyse them. On the other hand, HLA class II molecules present exogenous antigens to CD4+ T helper 1 (Th1) cells that subsequently prime antigen-specific CD8+ T-cell responses (Wang, 2001). CD4+ T cells have the central role in initiating and maintaining antitumor immunity (Wang, 2001). There is strong circumstantial evidence on the association of malignant transformation with alterations in HLA expression and/or function in a way that these abnormalities enable tumor cells to escape from host immunosurveillance (Gilboa, 1999; Hicklin *et al.*, 1999; Hiraki *et al.*, 1999; Korkolopoulou *et al.*, 1996).

HLA Class I Molecules

Among HLA class I antigens, two families are indistinguishable, both sharing homology at the sequence level consistent with the broadly similar secondary structure (O'Callaghan and Bell, 1998). These two groups consist of ubiquitously expressed, highly polymorphic class Ia (classical) HLA-A, HLA-B, and HLA-C molecules and selectively expressed, rather monomorphic class Ib (nonclassical), principally HLA-E, -F, and -G molecules, whose distinct features we will discuss later in this chapter. Both classical and nonclassical HLA molecules are cell surface, transmembrane glycoproteins that are assembled as heterodimers of a polymorphic 45-kDa heavy chain and a nonpolymorphic 12-kDa β_2-microglobulin (β_2m) light chain. The class I heavy chains are encoded by genes located within the major histocompatibility complex (MHC)

Handbook of Immunohistochemistry and *in situ* Hybridization of Human Carcinomas, Volume 1: Molecular Genetics; Lung and Breast Carcinomas

199

Copyright © 2004 by Elsevier (USA)
All rights reserved.

complex on chromosome 6, whereas β_2m is encoded by a gene mapped to chromosome 15 (Hicklin *et al.*, 1999; O'Callaghan and Bell, 1998).

With the development of monoclonal antibodies specific for different HLA class I determinants (Table 14), it became possible to investigate HLA class I expression *in situ* by immunohistochemistry in surgically removed tumors. This allows analysis of HLA class I antigen expression within an entire, heterogeneous tumor-cell population, and then comparison of these patterns to surrounding normal tissue. Distinct phenotypes of HLA class I down-regulation have been identified in malignant lesions that include: 1° total HLA class I loss; 2° total HLA class I down-regulation; and 3° selective loss or down-regulation of HLA class I allospecificities (Hicklin *et al.*, 1999; Seliger *et al.*, 2002). Total HLA class I antigen loss is generally caused by structural defects in one of the β_2m gene copies associated with the loss of functional β_2m expression (Hicklin *et al.*, 1999; Seliger *et al.*, 2002). Mutations of the β_2m gene, described in lung cancer as well (Chen *et al.*, 1996b), usually affect β_2m expression at the post-translational level (i.e., β_2m mRNA is present whereas the functional protein is not). Total HLA class I down-regulation on the other side can be corrected by IFN-γ and may be caused by different mechanisms, such as altered binding of the regulatory factors to the HLA class I heavy chain gene enhancer element, or concomitant with transporter associated with antigen processing (TAP) down-regulation (Hicklin *et al.*, 1999; Seliger *et al.*, 2002). The basis of TAP down-regulation lies in the structural defects in the TAP1 gene, resulting in the dysfunctional TAP1 protein (Chen *et al.*, 1996a). Different mechanisms underlie

selective HLA class I allele loss (down-regulation) (Hicklin *et al.*, 1999; Seliger *et al.*, 2002). Even though they have been described in a number of different cancer types, there is a lack of studies delineating these mechanisms in lung cancer (Korkolopoulou *et al.*, 1996). Several phenotypes can be represented within a given tumor-cell population, resulting in the heterogeneous pattern of HLA class I expression within a lesion (examples shown in Figure 29). The majority of the HLA class I expression studies in lung cancer have been performed by using frozen (cryostat) sections with a monoclonal antibody W6/32 (Chen *et al.*, 1996b; Korkolopoulou *et al.*, 1996; Passlick *et al.*, 1996). The frequency of the detected total HLA class I loss (down-regulation) ranges from 27–38% for all types of lung cancer (Hicklin *et al.*, 1999; Korkolopoulou *et al.*, 1996; Passlick *et al.*, 1996). Depending on the lung cancer histology and antibody used, the percentage of HLA class I expression varies, as represented in Table 16. Several studies have shown that presence versus the absence of HLA class I antigens does not have a significant impact on overall survival (Korkolopoulou *et al.*, 1996; Passlick *et al.*, 1994), as well as on clinical stage of the disease (Korkolopoulou *et al.*, 1996; Passlick *et al.*, 1996). However, Passlick *et al.* showed deficient HLA class I expression in 60% of patients with micrometastatic spread of tumor cells to regional lymph nodes (Passlick *et al.*, 1996).

The HLA class Ib molecule HLA-G is implicated in immunescape of tumor cells in lung cancer (Urosevic *et al.*, 2001). An HLA-G molecule differs from other HLA class I molecules—hence the prefix "nonclassical"—by its quasimonomorphism, tissue-restricted distribution, and unique processing of HLA-G transcript

Table 14 Antibodies Detecting Different HLA Antigens Used for Immunohistochemistry

Antibody	Specificity	Subclass	Tissue Sections	Concentration/Dilution for IHC	Reference
W6/32	HLA-A, -B, -C, -G, -E+B2m	IgG2a	Cryostat	Manufacturer recommendation	(Dako[a])
HCA2	HLA-A, -G	IgG1	Cryostat	1:10	(Blaschitz *et al.*, 2000; Stam *et al.*, 1990)
TP25.99	HLA-A, -B, -C, -E ± B2m	IgG1	Paraffin	1:10	(D'Urso *et al.*, 1991)
4H84	HLA-G	IgG1	Paraffin	1:500–2000	(McMaster *et al.*, 1998)
87G	HLA-G	IgG2a	Cryostat	5–10 μg/ml	(Blaschitz *et al.*, 2000; Lee *et al.*, 1995)
MEM-G/1	HLA-G	IgG1	Paraffin	Manufacturer recommendation	(Serotec[b])
HLA class II	HLA-DP, DQ, DR	IgG1	Paraffin	Manufacturer recommendation	(Dako)
HLA-DR	HLA-DR	IgG1	Paraffin	Manufacturer recommendation	(Dako)

HLA, human leukocyte antigen.

[a]DakoCytomation, Glostrup, Denmark.

[b]Serotec Ltd., Oxford, U.K.

Table 15 The Frequency of Altered HLA Antigen Expression in Lung Cancer

Antibody	Total in Lung Cancer	Squamous Cell Carcinoma	Adenocarcinoma	Large Cell Carcinoma	Small Cell Lung Cancer
W6/32	38% (Korkolopoulou et al., 1996)	47.4% (Korkolopoulou et al., 1996), 25% (Passlick et al., 1996)	34.5% (Korkolopoulou et al., 1996), 23.1% (Passlick et al., 1996)	75% (Passlick et al., 1996)	(Doyle et al., 1985; Funa et al., 1986)
HCA2	8.3% (Korkolopoulou et al., 1996)				
TP25.99	94% (Urosevic et al., 2001)	90% (Urosevic et al., 2001)	94% (Urosevic et al., 2001)	100% (Urosevic et al., 2001)	
4H84	26% (Urosevic et al., 2001)	1/9 (Urosevic et al., 2001)	2/9 (Urosevic et al., 2001)	5/9 (Urosevic et al., 2001)	(Urosevic et al., 2001)
87G	Only on infiltrating cells 29% (Passlick et al., 1996), 18%	(Onno et al., 2000; Pangault et al., 2002)			
HLA Class II	(Redondo et al., 1991), 22% (Foukas et al., 2001)	29% (Passlick et al., 1996)	34.6% (Passlick et al., 1996)	25% (Passlick et al., 1996)	

HLA, human leukocyte antigen.

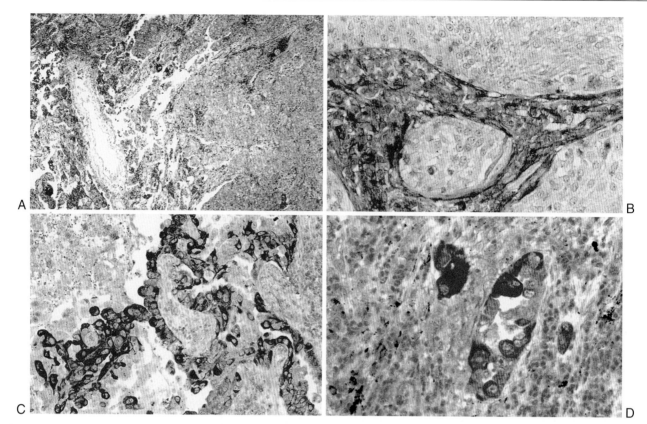

Figure 29 The expression of different HLA molecules detected by immunohistochemistry in lung cancer. **A:** Low magnification overview of HLA class I expression using TP25.99 antibody in large-cell carcinoma of the lung; **B:** Expression of HLA class I antigens in lung squamous cell carcinoma; **C** and **D:** Expression of HLA-G visualized using 4H84 antibody on a large-cell carcinoma.

(Urosevic *et al.*, 2002). The HLA-G protein is not expressed in the normal lung tissue (Onno *et al.*, 1994; Pangault *et al.*, 2002; Urosevic *et al.*, 2001). Because HLA-G mainly provides inhibitory signals to immune cells, it has been hypothesized that through up-regulation of HLA-G the tumors cells could evade host immuno-surveillance (Cabestre *et al.*, 1999; Seliger *et al.*, 2003; Urosevic *et al.*, 2003). We have shown the up-regulation of HLA-G in lung cancer (Urosevic *et al.*, 2001). The HLA-G immunoreactivity correlated with high-grade histology, with HLA-G being preferentially expressed on large-cell carcinomas (Urosevic *et al.*, 2001) (Table 14 and Figure 29). One of the HLA-G antibodies (MEM-G/1) recently became commercially available, which will enable more thorough screening of different cancer types, including lung cancer.

HLA Class II Molecules

The HLA class II molecules are cell-surface glyco-proteins consisting of noncovalently bound homo-dimeric complex of two chains, α and β (Janeway *et al.*, 1999). In contrast to ubiquitously expressed HLA class I

molecules, these molecules are constitutively expressed on limited number of cell types such as B lymphocytes, macrophages, and dendritic cells, but HLA class II mol-ecules are inducible on many other cell types with apparent immune (i.e., T cells) or other function (i.e., endothelial cells, tumor cells) (Blanck, 1999; Foukas *et al.*, 2001; Janeway *et al.*, 1999). Various cytokines such as IFN-γ can induce the expression of class II antigens (Blanck, 1999; Foukas *et al.*, 2001). Human tumor cell lines display one of three phenotypes with regard to HLA class II inducibility: noninducible, inducible, or constitutive (Blanck, 1999). It is gener-ally assumed that only the inducible phenotype is nor-mal and that the noninducible and constitutive phenotypes are the result of tumorigenesis (Blanck, 1999). Because the tumor-antigen pool presented by HLA class II molecules is yet to be well-defined, the expression of these molecules has not been so thor-oughly investigated (Table 15 shows representative studies in lung cancer). Yano *et al.*, detected class II expression in 45% of their lung adenocarcinoma spec-imens that correlated with both histological grade and the tumor stage (Yano *et al.*, 1998). Similar findings were shown by Redondo *et al.* (1991) where 18% of the tumors expressing class II antigens were preferentially well-differentiated tumors. Passlick and co-workers detected MHC class II on primary lung tumors in 10% of patients with micrometastates in regional lymph nodes, and in 30% of patients without tumor spread (Passlick *et al.*, 1996). Even though it has been shown that immunization with virus-derived MHC class II T-cell peptide or adoptive transfer of tumor-reactive CD4+ T cells results in antitumor immunity against MHC-negative tumors (Wang, 2002), biological significance of class II expression on tumor cells still remains elusive.

MATERIALS

1. Tris-buffered saline with bovine serum albumin (BSA) (Sigma, Division of Fluka Chemie GmbH, Buchs, Switzerland): Containing, when dissolved in 1 L of distilled or deionized water, 0.05 M Tris buffered saline (NaCl −0.138 M; KCl −0.0027 M); 1% BSA (w/v), pH 8.0 at 25°C.

2. Tris-buffered saline with Tween-20 (Sigma): Containing, when dissolved in 1 L distilled or deionized water, 0.05 M Tris buffered saline (NaCl −0.138 M; KCl −0.0027 M); 0.05% Tween-20, pH 8.0 at 25°C.

3. Primary antibody diluted to appropriate concentration (Table 14) in antibody diluent (DakoCytomation, Glostrup, Denmark).

4. ChemMate alkaline phosphatase anti-alkaline phosphatase (APAAP) mouse Detection Kit (code No. K5000 from DakoCytomation). The kit contains ready-to-use link (secondary) antibody and ready-to-use APAAP immunocomplexes (alkaline phosphatase and mouse monoclonal antialkaline phosphatase). Substrate provided with the kit is a five-component naphthol phosphate/new fuchsin chromogen.

5. Mayer's hematoxylin (ready-to-use aqueous solution, code No. S3309 from DakoCytomation).

6. Eukitt mounting medium (ProSciTech, Thuringowa, Australia).

METHOD

Alkaline phosphatase antialkaline phosphatase method using ChemMate APAAP (mouse) Detection Kit.

Processing of Formalin-Fixed, Paraffin-Embedded Tissue Sections Prior to Immunohistochemistry
1. Deparaffinize the slides.
2. Proceed to antigen retrieval. Heat the buffer (1 mM ethylenediaminetetraacetic acid (EDTA),

pH 8.0) in the microwave oven to boiling, then place the slide in the buffer and repeat the boiling process.

Processing of Cryostat Tissue Sections Prior to Immunohistochemistry
1. Thaw the cryostat sections.
2. Fix in acetone for 10 min.

Immunohistochemistry Protocol
3. Incubate the slides in Tris-buffered saline containing BSA for 10 min (to reduce the nonspecific binding).
4. Add primary antibody and incubate for 60 min.
5. Rinse the slides in Tris-buffered saline containing BSA for 5 min.
6. Add link (secondary) antibody and incubate for 30 min.
7. Rinse the slides in Tris-buffered saline containing Tween-20 for 5 min.
8. Add APAAP immunocomplexes and incubate for 30 min.
9. Rinse as in **step 7**.
10. Add link (secondary) antibody and incubate for 10 min.
11. Rinse as in **step 7**.
12. Add APAAP immunocomplexes and incubate for 10 min.
13. Rinse as in **step 7**.
14. Repeat the **step 10**.
15. Rinse as in **step 7**.
16. Repeat the **step 12**.
17. Rinse as in **step 7**.
18. Add the chromogen substrate and incubate according to the speed of developing color between 10–30 min. As soon as the color develops proceed with the next step.
19. Rinse the slides in distilled water.
20. Immerse the slides in hematoxylin solution for 1 min.
21. Rinse as in **step 19**.
22. Rinse the slides under running tap water.
23. Rinse as in **step 19**.
24. Immerse the slides in increasing concentration of ethanol (70%, 95%, and 100%) and finally in xylene.
25. Mount the slides using the Eukitt mounting medium.

References

Blanck, G. 1999. HLA class II expression in human tumor lines. *Microbes Infect.* 1:913–918.

Blaschitz, A., Hutter, H., Leitner, V., Pilz, S., Wintersteiger, R., Dohr, G., and Sedlmayr, P. 2000. Reaction patterns of monoclonal antibodies to HLA-G in human tissues and on cell lines: A comparative study. *Hum. Immunol. 61:*1074–1085.

Cabestre, F.A., Lefebvre, S., Moreau, P., Rouas Friess, N., Dausset, J., Carosella, E.D., and Paul, P. 1999. HLA-G expression: Immune privilege for tumour cells? *Semin. Cancer Biol. 9:* 27–36.

Chen, H.L., Gabrilovich, D., Tampe, R., Girgis, K.R., Nadaf, S., Carbone, D.P., 1996a. A functionally defective allele of TAP1 results in loss of MHC class I antigen presentation in a human lung cancer. *Nat. Genet. 13:*210–213.

Chen, H.L., Gabrilovich, D., Virmani, A., Ratnani, I., Girgis, K.R., Nadaf Rahrov, S., Fernandez Vina, M., and Carbone, D.P. 1996b. Structural and functional analysis of beta 2-microglobulin abnormalities in human lung and breast cancer. *Int. J. Cancer 67:*756–763.

Doyle, A., Martin, W.J., Funa, K., Gazdar, A., Carney, D., Martin, S.E., Linnoila, I., Cuttitta, F., Mulshine, J., Bunn, P., and Minna, J. 1985. Markedly decreased expression of class I histocompatibility antigens, protein, and mRNA in human small cell lung cancer. *J. Exp. Med. 161:*1135–1151.

D'Urso, C.M., Wang, Z.G., Cao, Y., Tatake, R., Zeff, R.A., and Ferrone, S. 1991. Lack of HLA class I antigen expression by cultured melanoma cells FO-1 due to a defect in B2m gene expression. *J. Clin. Invest. 87:*284–292.

Foukas, P.G., Tsilivakos, V., Zacharatos, P., Mariatos, G., Moschos, S., Syrianou, A., Asimacopoulos, P.J., Bramis, J., Fotiadis, C., Kittas, C., and Gorgoulis, V.G. 2001. Expression of HLA-DR is reduced in tumor infiltrating immune cells (TIICs) and regional lymph nodes of nonsmall cell lung carcinomas. A putative mechanism of tumor-induced immunosuppression? *Anticancer Res. 21:*2609–2615.

Funa, K., Gazdar, A.F., Minna, J.D., and Linnoila, R.I. 1986. Paucity of beta 2-microglobulin expression on small cell lung cancer, bronchial carcinoids, and certain other neuroendocrine tumors. *Lab. Invest. 55:*186–193.

Gilboa, E. 1999. How tumors escape immune destruction and what we can do about it. *Cancer Immunol. Immunother. 48:*382–385.

Hicklin, D.J., Marincola, F.M., and Ferrone, S. 1999. HLA class I antigen down-regulation in human cancers: T-cell immunotherapy revives an old story. *Mol. Med. Today 5:*178–186.

Hiraki, A., Kaneshige, T., Kiura, K., Ueoka, H., Yamane, H., Tanaka, M., and Harada, M. 1999. Loss of HLA haplotype in lung cancer cell lines: Implications for immunosurveillance of altered HLA class I/II phenotypes in lung cancer. *Clin. Cancer Res. 5:*933–936.

Janeway, C.A., Travers, P., Walport, M., and Capra, J.D. 1999. Antigen recognition by T lymphocytes. In *Immunobiology: The Immune System in Health and Disease.* Edinburgh: UK: Churchill Livingstone, pp 4:1–4:20.

Korkolopoulou, P., Kaklamanis, L., Pezzella, F., Harris, A.L., and Gatter, K.C. 1996. Loss of antigen-presenting molecules (MHC class I and TAP-1) in lung cancer. *Br. J. Cancer 73:* 148–153.

Lee, N., Malacko, A.R., Ishitani, A., Chen, M.C., Bajorath, J., Marquardt, H., and Geraghty, D.E. 1995. The membrane-bound and soluble forms of HLA-G bind identical sets of endogenous peptides but differ with respect to TAP association. *Immunity 3:*591–600.

McMaster, M., Zhou, Y., Shorter, S., Kapasi, K., Geraghty, D., Lim, K.H., and Fisher, S. 1998. HLA-G isoforms produced by placental cytotrophoblasts and found in amniotic fluid are due to unusual glycosylation. *J. Immunol. 160:*5922–5928.

O'Callaghan, C.A., and Bell, J.I. 1998. Structure and function of the human MHC class Ib molecules HLA-E, HLA-F, and HLA-G. *Immunol. Rev. 163:*129–138.

Onno, M., Guillaudeux, T., Amiot, L., Renard, I., Drenou, B., Hirel, B., Girr, M., Semana, G., Le Bouteiller, P., and Fauchet, R. 1994. The HLA-G gene is expressed at a low mRNA level in different human cells and tissues. *Hum. Immunol. 41:*79–86.

Onno, M., Le Friec, G., Pangault, C., Amiot, L., Guilloux, V., Drenou, B., Caulet-Maugendre, S., Andre, P., and Fauchet, R. 2000. Modulation of HLA-G antigens expression in myelomonocytic cells. *Hum. Immunol. 61:*1086–1094.

Pangault, C., Le Friec, G., Caulet-Maugendre, S., Lena, H., Amiot, L., Guilloux, V., Onno, M., and Fauchet, R. 2002. Lung macrophages and dendritic cells express HLA-G molecules in pulmonary diseases. *Hum. Immunol. 63:*83–90.

Passlick, B., Izbicki, J.R., Simmel, S., Kubuschok, B., Karg, O., Habekost, M., Thetter, O., Schweiberer, L., and Pantel, K. 1994. Expression of major histocompatibility class I and class II antigens and intercellular adhesion molecule-1 on operable nonsmall cell lung carcinomas: Frequency and prognostic significance. *Eur. J. Cancer 30A:*376–381.

Passlick, B., Pantel, K., Kubuschok, B., Angstwurm, M., Neher, A., Thetter, O., Schweiberer, L., and Izbicki, J.R. 1996. Expression of MHC molecules and ICAM-1 on nonsmall cell lung carcinomas: Association with early lymphatic spread of tumour cells. *Eur. J. Cancer 32A:*141–145.

Redondo, M., Concha, A., Oldiviela, R., Cueto, A., Gonzalez, A., Garrido, F., and Ruiz-Cabello, F. 1991. Expression of HLA class I and II antigens in bronchogenic carcinomas: Its relationship to cellular DNA content and clinical-pathological parameters. *Cancer Res. 51:*4948–4954.

Seliger, B., Cabrera, T., Garrido, F., and Ferrone, S. 2002. HLA class I antigen abnormalities and immune escape by malignant cells. *Semin. Cancer Biol. 12:*3–13.

Seliger, B., Abken, H., and Ferrone, S. 2003. HLA-G and MIC expression in tumors and their role in anti-tumor immunity. *Trends Immunol. 24:*82–87.

Stam, N.J., Vroom, T.M., Peters, P.J., Pastoors, E.B., and Ploegh, H.L. 1990. HLA-A- and HLA-B-specific monoclonal antibodies reactive with free heavy chains in Western Blots, in formalin-fixed, paraffin-embedded tissue sections, and in cryo-immuno-electron microscopy. *Int. Immunol. 2:*113–125.

Urosevic, M., and Dummer, R. 2002. HLA-G—An ace up the sleeve? *ASHI Q 26:*106–109.

Urosevic, M., and Dummer, R. 2003. HLA-G and IL-10 expression in human cancer—different stories with the same message. *Semin. Cancer Biol. 13:*337–342.

Urosevic, M., Kurrer, M.O., Kamarashev, J., Mueller, B., Weder, W., Burg, G., Stahel, R.A., Dummer, R., and Trojan, A. 2001. HLA-G up-regulation in lung cancer associates with high-grade histology, HLA class I loss, and IL-10 production. *Am. J. Pathol. 159:*817–824.

Wang, R.F. 2001. The role of MHC class II-restricted tumor antigens and CD4+ T cells in antitumor immunity. *Trends Immunol. 22:*269–276.

Wang, R.F. 2002. Enhancing antitumor immune responses: Intracellular peptide delivery and identification of MHC class II-restricted tumor antigens. *Immunol. Rev. 188:*65–80.

Yano, T., Fukuyama, Y., Yokoyama, H., Kuninaka, S., Asoh, H., Katsuda, Y., and Ichinose, Y. 1998. HLA class I and class II expression of pulmonary adenocarcinoma cells and the influence of interferon gamma. *Lung Cancer 20:*185–190.

13

Immunohistochemistry and *in situ* Hybridization of Telomerase Expression in Lung Carcinoma

Fumiyuki Kumaki

Introduction

Lung carcinoma is one of the most common fatal malignancies throughout the world and is the number one cause of cancer death in the United States, in both males and females. Histologically, lung carcinoma is mainly divided into four types: adenocarcinoma, squamous cell carcinoma, large cell carcinoma, and small cell carcinoma (Travis *et al.*, 1999). These four types account for more than 95% of diagnosed cases. The increase of death by lung carcinoma is thought to be caused by both an increase of spontaneous generation of lung carcinoma due to smoking and/or environmental pollution and a delay of detection at an early stage in spite of the technological improvements of radiography and better medical treatment. To improve the outcome, biological methods would also be needed for early detection.

Telomeres, which represent the ends of eukaryotic cell chromosome arms, are specific DNA-protein complexes. Telomeric DNA usually consists of multiple repeating G-enriched nucleotide sequences that are directed toward the 3′-end of the chromosome (Zakien, 1995). In human chromosome, telomeres are composed of the hexanucleotide units, TTAGGG, with a total length of 10 to 15 kilobase (kb) (Blackburn, 1991; Moyzis *et al.*, 1988; Meyne *et al.*, 1989). Telomeres seem to play a key role in the stabilization of chromosomes on replication (Blackbu *et al.*, 1991; Greider, 1994) by protecting their ends from exonucleases and ligases (Rhyu *et al.*, 1995). In addition, telomeres prevent degradation and unwanted recombination (Counter *et al.*, 1992; Greider, 1994; Harley, 1991), such as fusion of the terminal regions of broken chromosomes.

In all normal somatic cells, each cycle of cell division and DNA replication results in a loss of 50–200 terminal nucleotides from each chromosomes (Allsopp *et al.*, 1992; Harley *et al.*, 1990; Vaziri *et al.*, 1993), because DNA polymerase cannot completely replicate the ends of linear DNA molecules (Olovnikov *et al.*, 1973; Watson *et al.*, 1972). It has been suggested that this shortening of telomeres may function as a mitotic clock (Hayflick *et al.*, 1961), by which normal cells count their divisions (Harley, 1991) and eventually signal their senescence. In contrast, immortalized cell lines and cancer cells show no net loss of telomere length with cell divisions (Counter *et al.*, 1992), suggesting that telomere maintenance is essential for cellular immortality.

Copyright © 2004 by Elsevier (USA)
All rights reserved.

The ribonucleoprotein-telomerase complex present in most cancer cell lines and in certain germline and stem cells (Eisenhauer *et al.*, 1997; Hiyama *et al.*, 1995; Kim *et al.*, 1994; Shay *et al.*, 1997; Taylor *et al.*, 1996; Wright *et al.*, 1996; Yui *et al.*, 1998) is a specialized reverse transcriptase that synthesizes telomeric repeats (Lingner *et al.*, 1997) using a segment of its RNA moiety as a template (telomerase RNA component [TERC]) on chromosome 3q26.3, and the protein moiety as the site of the catalytic activity (telomerase reverse transcriptase [TERT]) on 5p15.33 (Feng *et al.*, 1995; Harrington *et al.*, 1997; Kilian *et al.*, 1997; Meyerson *et al.*, 1997; Nakamura *et al.*, 1997). The TERC is present in all cells, whereas the expression of TERT is confined to cells that express telomerase activity. Recent reconstitution experiments show that TERT is the major determinant of human telomerase activity, and its expression is indicative of activation of telomerase, strongly suggesting that cells showing this activity undergo unlimited replication (Nakayama *et al.*, 1998; Weinrich *et al.*, 1997).

Information on telomerase activity in tumors has been obtained almost exclusively by the use of the telomeric repeat amplification protocol (TRAP) assay, which shows that telomerase activity may be present in greater than 80% of tumor biopsies yet absent or reduced in normal somatic tissue (Breslow *et al.*, 1997; Kim *et al.*, 1994; Raymond *et al.*, 1996; Shay *et al.*, 1996). However, it is unlikely that TRAP assay alone will show the true complexities of telomerase regulation, and not only molecular approaches but also histologic approaches will be required to understand telomerase activity.

We investigated the expression of telomerase activity not only in formalin-fixed, paraffin-embedded tissues but also stamp cytological tissues and pleural effusion of lung carcinoma using immunohistochemical staining and *in situ* hybridization. Based on our findings, we describe these methods of detecting telomerase expression by using immunohistochemistry and *in situ* hybridization techniques.

MATERIALS

Sample Preparations

Formalin-Fixed Paraffin-Embedded Tissues

A total of 115 tissues were examined. Eighty-five men and 30 women ranging in age from 33 to 86 years (mean 65.8 years). The patients underwent surgery at the National Defense Medical College Hospital, Tokorozawa, Japan, between January 1986 and February 1998, without prior chemotherapy or radiotherapy. The tumors were classified histologically as squamous cell carcinoma in 45 patients, adenocarcinoma in 54, large cell carcinoma in 12, and small cell carcinoma in 4.

The tissues were fixed with 10% formalin, embedded in paraffin, sectioned at a thickness of 5 μm and collected on 3-amino-propyltriethoxysilane coated slides for immunohistochemical staining and *in situ* hybridization. The TRAP assay mentioned in this chapter refers to the report by the author (Kumaki *et al.*, 2001); this report should be consulted for more details.

Stamp Cytological Tissues

A total of 35 of the 115 stamp cytological tissues were examined. They were 10 squamous cell carcinoma, 19 adenocarcinoma, 5 large cell carcinomas, and 1 small cell carcinoma. The tumors were cut and stamped on the coating slides without any fixation. After drying, the slides were stored at −80°C.

Pleural Effusion

A total of seven samples of pleural effusion from patients diagnosed with pulmonary adenocarcinoma were examined. Fluid was centrifuged at 500 g for 15 min and cells were pulled up from the border of red blood cells and plasma. They were then collected on the coated slides and dried. They were stored at −80°C and dried.

Reagents

Common Reagents

1. Target retrieval solution, 10X (Dako Corporation, Carpinteria, CA).
2. Hydrogen peroxide, 30% (Fisher Scientific, Springfield, NJ).
3. Phosphate-buffered saline (PBS), pH 7.4 (Invitrogen Life Technologies, Carlsbad, CA).
4. Normal goat serum (Vector Laboratories, Burlingame, CA).
5. Saline-sodium citrate (SSC), 20X (Invitrogen).

Antibodies and Amplification Reagents

1. Rabbit polyclonal TERT antibody (Novus Biologicals, Catalog #NB100-141, Littleton, CO).
2. Anti-rabbit antibody labeled with horseradish peroxidase (HRP) polymer (EnVision + systems, Dako).
3. Avidin-biotin complex (ABC) labeled alkarynphosphatase (ALP) system (Dako).
4. 5-Bromo-4-chloro-3-indolyl phosphate (BCIP; Sigma Chemical, St Louis, MO).
5. Nitro blue tetrazolium (NBT, Sigma).
6. Fast Red TR/Naphthol AS-MX tablet sets (Sigma).
7. Fluorescein isothiocyanate (FITC) conjugated streptavidin (Vector).

8. 4', 6-Diamidino-2-phenylindole (DAPI, Vector).

9. Propidium iodide (PI, Vector).

10. DAB (Vectrastain kit, Vector).

Reagents for *in situ* Hybridization

1. Biotin-labeled oligonucleotide probe (Figure 30 shows the sequences of probes of TERC and TERT), 100 p*M*/μL in distilled water.

2. Hybridization solution (1000 μL for 20 slides).

Oligonucleotide probe 20 μL.
Formamide (Sigma) 500 μL.
50% Dextran sulphate (Sigma) 100 μL.
Denhardt's solution (Sigma) 100 μL.
Salmon sperm DNA, 10 mg/mL (Sigma) 20 μL.
20X SSC 100 μL.
EDTA pH 8.0 10 μL.
DEPC 150 μL.

3. Proteinase K (Roche Applied Science, Indianapolis, IN).

4. Paraformaldehyde (Sigma).

5. Glycine (Sigma).

6. Triethanolamine (Fisher).

METHODS

Immunohistochemical Staining of Paraffin Sections

1. Deparaffinization.

2. Microwave for 10 min at 1000W in 1X target retrieval solution and cool down to room temperature.

3. 0.3% hydrogen peroxide in methanol for 30 min.

4. PBS wash (10 min each, two times).

5. 10% normal goat serum in PBS for 30 min.

6. Rabbit polyclonal primary antibody, 1:1000 for 2 hr at room temperature.

7. PBS wash (15 min each, four times).

8. Anti-rabbit antibody labeled with HRP polymer for 1 hr at room temperature.

9. PBS wash (10 min each, two times).

10. Development of color by DAB for 5–7 min.

11. Counterstaining with hematoxylin and mount.

In situ Hybridization of Paraffin Sections

1. Deparaffinization.

2. 0.2-N HCl for 20 min.

3. 2X SSC wash for 5 min (two times) at 37°C.

4. PBS wash for 5 min (two times) at 37°C.

5. Proteinase K (20 mg/ml in PBS) for 10 min at 37°C.

6. 4°C DEPC water wash for 5 min (two times).

7. Fixation with 4% paraformaldehyde for 5 min at 4°C.

8. 0.2% glycine in distilled water for 10 min.

9. 0.1 M triethanolamine for 10 min.

10. Dehydration by 100% ethanol and dry.

11. Drop 50 μL hybridization solution on each slide and incubation overnight at 42°C.

12. 2X SSC wash for 10 min (two times) at 42°C.

13. 2X SSC wash for 10 min at room temperature.

14. 0.2X SSC wash for 10 min at room temperature.

15. 4% normal goat serum for 20 min.

16. ABC labeled ALP system for 1 hr at room temperature.

17. Detection by BCIP in NBT solution (pH 9.5) for 1 hr at most with a shield for the light.

1. TERC Probes (48 Oligonucleotides)

(A) Sense
5'-tgg tgg cca ttt ttt gtc **taa ccc t**ta ctg aga agg gcg tag gcg ccg-3'

(B) Antisense
5'-cgg cgc cta cgc cct tct cag tta ggg tta gac aaa aaa tgg cca cca-3'

2. TERT Probes (60 Oligonucleotides)

(A) Sense
5'-gca aag cat tgg aat cag aca gca ctt gaa gag ggt gca gct gcg gga gct gtc gga agc-3'
      (1)   (2)

(B) Antisense
5'-gct tcc gac agc tcc cgc agc tgc acc ctc ttc aag tgc tgt ctg att cca atg ctt tgc-3'

Figure 30 Sequences of sense and antisense mRNA oligonucleotide probes used for TERC and TERT. Biotin is labeled to the ends of each 5' site. The TERC sense probe contains the telomerase template for telomere (bold). TERT sense probe was chosen to include both telomerase specific motif (1) and the conserved RT motif shown in (2).

18. Wash in running water for 5 min and dry.
19. Mount by water-soluble medium.

Fluorescent *in situ* Hybridization of Frozen Stamp Cytological Tissues

1. Defrost and dry at room temperature.
2. Fixation in 4% paraformaldehyde.
3. 0.2 N HCl for 20 min.
4. 2X SSC wash for 5 min (two times) at 37°C.
5. PBS wash for 5 min (two times) at 37°C.
6. Proteinase K in PBS for 5 min at 37°C.
7. 4°C DEPC water wash for 5 min (two times).
8. Fixation with 4% paraformaldehyde for 5 min at 4°C.
9. 0.2% glycine in distilled water for 10 min.
10. 0.1 M triethanolamine for 10 min.
11. Dehydration in 100% ethanol and dry.
12. Drop 50 µL hybridization solution on each slide and incubation overnight at 42°C.
13. 2X SSC wash for 10 min (two times) at 42°C.
14. 2X SSC wash for 10 min at room temperature.
15. 0.2X SSC wash for 10 min at room temperature.
16. 4% normal goat serum for 20 min.
17. FITC conjugated streptavidin for 2 hr at room temperature.
18. Counterstain with DAPI or PI and mount.
19. Examination with a confocal microscope (Leica model TCS-4D/DMIRBE, Heidelberg, Germany).

In situ Hybridization of Pleural Effusion

1. Defrost and dry at room temperature.
2. Fixation in 4% paraformaldehyde.
3. 0.2 N HCl for 20 min.
4. 2X SSC wash for 5 min (two times) at 37°C.
5. PBS wash for 5 min (two times) at 37°C.
6. Proteinase K (20 mg/mL in PBS) for 5 min at 37°C.
7. DEPC water (4°C) wash for 5 min (two times).
8. Fixation with 4% paraformaldehyde for 5 min at 4°C.
9. 0.2% glycine in distilled water for 10 min.
10. 0.1 M triethanolamine for 10 min.
11. Dehydration in 100% ethanol and dry.
12. Drop 50-µL hybridization solution on each slide and incubate overnight at 42°C.
13. 2X SSC wash for 10 min (two times) at 42°C.
14. 2X SSC wash for 10 min at room temperature.
15. 0.2X SSC wash for 10 min at room temperature.
16. 4% normal goat serum for 20 min.
17. Avidin-biotin complex system labeled ALP.
18. Development of the color with Fast Red TR.
19. Counterstaining with hematoxylin and mounted.

RESULTS

Immunohistochemical Staining of Paraffin Sections

Prior to immunohistochemical staining for TERT, telomerase activity was detected in 107 (93%) of 115 lung carcinomas by TRAP assay and was negative in non-neoplastic tissues (Kumaki *et al.*, 2001).

Immunoreactivity for TERT protein was detected in 108 (93.9%) of 115 cases that were positive for TRAP assay. Only one case of adenocarcinoma was discrepant. The TERT protein was mainly localized in the nuclei of cancer cells and some activated lymphocytes. Stromal cells in carcinoma and normal adjacent tissues showed no reaction for TERT (Figures 31A and 31B).

In situ Hybridization of Paraffin Section

First, it is necessary to confirm the presence of intact RNA by a housekeeping gene and absence of reaction with sense probes in all cases. The expression of TERC mRNA was recognized in cancer cells in all of the 115 lung carcinomas. The signals were mainly detected in the cytoplasm and weaker in the nuclei of carcinoma cells and inflammatory cells but not in stroma cells (Figure 31C).

The expression of TERT mRNA was found in 108 (93.9%) of 115 cases that were the same cases with positive reaction for TERT protein. It was mainly detected in the cytoplasm just like TERC. There were no signals in stroma cells (Figure 31D).

Fluorescent *in situ* Hybridization of Frozen Stamp Cytological Tissues

The expression of TERC and TERT mRNA was 100% and 91% (32 of 35 cases), respectively. All TERT mRNA negative cases (n = 3) were adenocarcinoma, and these results were parallel to those of paraffin tissues. Figures 31E and 31F show the expression of TERC mRNA in squamous cell carcinoma (E) and TERT mRNA in adenocarcinoma (F). Both signals (FITC, showing green) were recognized mainly in the cytoplasm of cancer cells. The nuclei were stained by PI (red) or DAPI (blue). Non-neoplastic cells (fibroblasts, red blood cells, etc.) showed no or very weak signals.

In situ Hybridization of Pleural Effusion

The expression of TERC and TERT was seen in all seven cases of adenocarcinoma cells. Figures 31G and 31H show the expression of TERC mRNA (G) and TERT mRNA (H) in adenocarcinoma of pleural effusion.

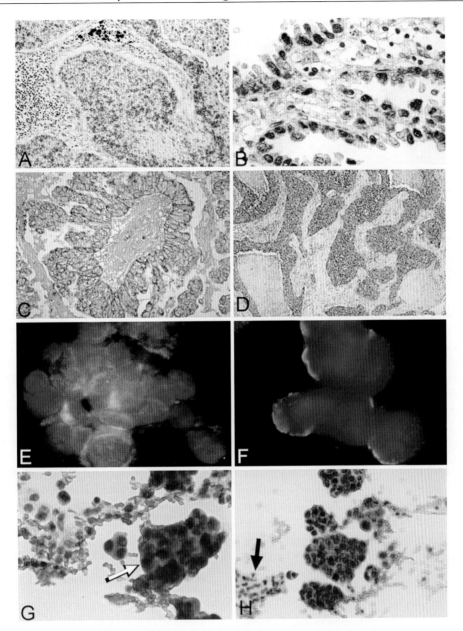

Figure 31 Photomicrographs showing immunohistochemical staining of TERT protein and *in situ* hybridization of TERC and TERT mRNA. A positive reaction of TERT protein is shown in the nuclei of squamous cell carcinoma **A:** and adenocarcinoma cells **B:** TERT is also recognized in activated lymphocytes. However, it is negative in any other non-neoplastic tissues. *In situ* hybridization of TERC mRNA in adenocarcinoma **C:** and of TERT mRNA in squamous cell carcinoma **D:** the reactions are localized in the cytoplasm of the cancer cells and inflammatory cells; however, the signal is negative in stroma cell. TERC and TERT mRNAs are shown in the stamp cytological tissues of adenocarcinoma **E** and **F**. In both, the reaction is localized in the cytoplasm of the cancer cells and visualized with FITC (green). The nuclei are stained either by PI (red) or DAPI (blue). TERC and TERT mRNA in the adenocarcinoma cells of pleural effusion (G and H). The positive reaction (red) is recognized in the cytoplasm of the cancer cells (white arrow); however, negative in the reactive mesothelial cells (black arrow) and red blood cells. (A and D, 100X; H, 200X; B, C, and G, 400X; E, 1000X; F, 1600X.)

To identify the eukaryotic cells, hematoxylin was used for counterstaining. The positive signals colored red with PI were recognized in the cytoplasm of cancer cells (white arrow). However, they were not recognized in the reactive mesothelial cells that were not malignant cells (black arrow). Red blood cells that had no nuclei were also negative signals.

DISCUSSION

During the past few years, numerous studies have reported that telomerase is expressed and active in various types of tumors, including carcinomas and/or sarcomas of the lung (Hiyama et al., 1995), liver (Tahara et al., 1995), colon (Chadeneau et al., 1995), breast (Sugino et al., 1996), uterus (Zheng et al., 1997), brain (Langford et al., 1995), bone (Schwartz et al., 1995), and soft tissues (Taylor et al., 1996) by the use of TRAP assay. However, the cellular localization of this enzyme in paraffin sections of formalin-fixed tissues and cytological tissues have not been shown. We successfully demonstrated the presence of telomerase in lung cancer cells using in situ hybridization and immunohistochemical staining (Kumaki et al., 2001).

Using in situ hybridization methods, we detected mRNAs for TERC and TERT in the cytoplasm of cancer cells. In all cases TERC mRNA was recognized, whereas TERT mRNA was detected in 93.9% cases (108 of 115 cases). The TERT protein was mainly expressed in the nuclei of cancer cells in all TERT mRNA-positive cases. We interpret these results as indicating that TERT protein produced by transcription of TERT mRNA in the cytoplasm becomes localized in the nucleus and acts as an internuclear protein. In these studies, most of the cases that were positive for TERT staining and in situ hybridization also gave a positive reaction with the TRAP assay. These results also suggest that TERT protein binds to TERC mRNA after returning to the nucleus, forming a ribonucleoprotein complex. Activation of this complex increases telomerase activity. Therefore, expression of TERT reflects telomerase activity.

In recent years, surgical therapy for lung cancer is decided based on prior testing (i.e., radiography and biopsy by bronchoscopy). However, sometimes surgeons have a difficult time making decisions based on rapid diagnosis during the operation due to unclear diagnosis of the biopsy specimen. These cases are very difficult to diagnose even with the use of hematoxylin-eosin (H-E) stain because of the small lesions involved. Therefore, we examined whether the expression of telomerase would be useful for the diagnosis in stamp cytological tissues by in situ hybridization. The expression of TERC and TERT was positive in 100% and 91% of cancer cells and correlated to that in paraffin tissues,

respectively. Naritoku et al. (1999) showed that the rate of expression of telomerase in the fine-needle aspiration specimens was the same as that of biopsies using TRAP assay. These results suggest that in situ hybridization methods and TRAP assay could be useful for the diagnosis of telomerase in stamp cytological tissues.

Despite the development of cytology, it is difficult to distinguish pulmonary adenocarcinoma from benign reactive mesothelial cells in pleural effusion. Electron microscopy is useful but tissue preparation requires a long time (more than 1 day) for diagnosis. Although several reports indicated that telomerase activity by TRAP assay was useful in cytology tissues of bronchial washings (Yahata et al., 1998), colonic luminal washings (Yoshida et al., 1997), pancreatic juice (Suehara et al., 1998), and urine (Muller et al., 1998), this method cannot distinguish severe pleuritis, which shows high telomerase activity by activated lymphocytes from malignant pleural effusion. Therefore, it is necessary to examine the expression of telomerase in the cytology tissues from pleural effusion using in situ hybridization. The expression of TERC and TERT was positive in all cancer cells of pleural effusion. In contrast, it was not recognized in reactive mesothelial cells at all. These results suggest that the use of in situ hybridization may be useful for the differential diagnosis between pulmonary adenocarcinoma and reactive mesothelial cells in pleural effusion.

In summary, the determination of telomerase using immunohistochemical staining and in situ hybridization may be useful for the diagnosis of lung cancer on cytological tissues as well as TRAP assay.

Acknowledgment

This paper is dedicated to Dr. Victor J. Ferrans, former chief of Pathology Section, National Heart, Lung and Blood Institute (NHLBI), National Institutes of Health (NIH). Some of these studies were performed in his laboratory. The author wishes to acknowledge Dr. Zu-Xi Yu, Pathology Section, NHLBI, NIH, and Dr. Toshiaki Kawai and Dr. Yuichi Ozeki, Departments of Pathology and Thoracic Surgery, National Defense Medical College, Japan, for their support of this research and comments on the manuscript.

References

Allsopp, R.C., Vaziri, H., Patterson, C., Goldstein, S., Younglai, E.V., Futcher, A.B., Greider, C.W., and Harley, C.B. 1992. Telomere length predicts replicative capacity of human fibroblasts. Proc. Natl. Acad. Sci. USA 89:10114–10118.

Blackburn, E.H. 1991. Structure and function of telomeres. Nature 350:569–573.

Breslow, R.A., Shay, J.W., Gazdar, A.F., and Srivastava, S. 1997. Telomerase and early detection of cancer: A National Cancer Institute workshop. J. Natl. Cancer Inst. 89:618–623.

Chadeneau, C., Hay, K., Hirte, H.W., Gallinger, S., and Bacchetti, S. 1995. Telomerase activity associated with acquisition of malignancy in human colorectal cancer. Cancer Res. 55:2533–2536.

Counter, C.M., Avilion, A.A., LeFeuvre, C.E., Stewart, N.G., Greider, C.W., Harley, C.B., and Bacchetti, S. 1992. Telomere shortening associated with chromosome instability is arrested in immortal cells which express telomerase activity. *EMBO J. 11*:1921–1929.

Eisenhauer, K.M., Gerstein, R.M., Chiu, C.P., Conti, M., and Hsueh, A.J. 1997. Telomerase activity in female and male rat germ cells undergoing meiosis and in early embryos. *Biol. Reprod. 56*:1120–1125.

Feng, J., Funk, W.D., Wang, S.S., Weinrich, S.L., Avilion, A.A., Chiu, C.P., Adams, R.R., Chang, E., Allsopp, R.C., Yu, J., Le, S., West, M.D., Harley, C.B., Andrews, W.H., Greider, C.W., and Villeponteau, B. 1995. The RNA component of human telomerase. *Science 269*:1236–1241.

Greider, C.W. 1994. Mammalian telomere dynamics: Healing, fragmentation shortening and stabilization. *Curr. Opin. Genet. Dev. 4*:203–211.

Harley, C.B., Futcher, A.B., and Greider, C.W. 1990. Telomeres shorten during aging of human fibroblasts. *Nature 345*:458–460.

Harley, C.B. 1991. Telomeres loss: Mitotic clock or genetic time bomb? *Mutat. Res. 256*:271–282.

Harrington, L., Zhou, W., McPhail, T., Oulton, R., Yeung, D.S., Mar, V., Bass, M.B., and Robinson, M.O.1997. Human telomerase contains evolutionarily conserved catalytic and structural subunits. *Genes Dev. 11*:3109–3115.

Hayflick, L., and Moorhead, P.S. 1961. The serial cultivation of human diploid cell strains. *Exp. Cell Res. 25*:585–621.

Hiyama, K., Hiyama, E., Ishioka, S., Yamakido, M., Inai, K., Gazdar, A.F., Piatyszek, M.A., and Shay, J.W. 1995. Telomerase activity in small cell and nonsmall cell lung cancers. *J. Natl. Cancer Inst. 87*:895–902.

Kilian, A., Bowtell, D.D., Abud, H.E., Hime, G.R., Venter, D.J., Keese, P.K., Duncan, E.L., Reddel, R.R., and Jefferson, R.A. 1997. Isolation of a candidate human telomerase catalytic subunit gene, which reveals complex splicing patterns in different cell types. *Hum. Mol. Genet. 6*:2011–2019.

Kim, N.W., Piatyszek, M.A., Prowse, K.R., Harley, C.B., West, M.D., Ho, P.L., Coviello, G.M., Wright, W.E., Weinrich, S.L., and Shay, J.W. 1994. Specific association of human telomerase activity with immortal cells and cancer. *Science 266*: 2011–2015.

Kumaki, F., Kawai, T., Hiroi, S., Shinomiya, N., Ozeki, Y., Ferrans, V.J., and Torikata, C. 2001. Telomerase activity and expression of human telomerase RNA component and human telomerase reverse transcriptase in lung carcinomas. *Hum. Pathol. 32*: 188–195.

Langford, L.A., Piatyszek, M.A., Xu, R., Schold, S.C. Jr., and Shay, J.W. 1995. Telomerase activity in human brain tumors. *Lancet 346*:1267–1268.

Lingner, J., Hughes, T.R., Shevchenko, A., Mann, M., Lundblad, V., and Cech, T.R. 1997. Reverse transcriptase motifs in the catalytic subunit of telomerase. *Science 276*:561–567.

Meyerson, M., Counter, C.M., Eaton, E.N., Ellisen, L.W., Steiner, P., Caddle, S.D., Ziaugra, L., Beijersbergen, R.L., Davidoff, M.J., Liu, Q., Bacchetti, S., Haber, D.A., and Weinberg, R.A. 1997. hEST2, the putative human telomerase catalytic subunit gene, is up-regulated in tumor cells and during immortalization. *Cell 90*:785–795.

Meyne, J., Ratliff, R.L., and Moyzis, R.K. 1989. Conservation of the human telomere sequence (TTAGGG)n among vertebrates. *Proc. Natl. Acad. Sci. USA 86*:7049–7053.

Moyzis, R.L., Buckingham, J.M., Cram, L.S., Dani, M., Deaven, L.L., Jones, M.D., Meyne, J., Ratliff, R.L., and Wu, J.R. 1988. A highly conserved repetitive DNA sequence, (TTAGGG)n,

present at the telomeres of human chromosomes. *Proc. Natl. Acad. Sci. USA 85*:6622–6626.

Muller, M., Krause, H., Heicappell, R., Tischendorf, J., Shay, J.W., and Miller, K. 1998. Comparison of human telomerase RNA and telomerase activity in urine for diagnosis of bladder cancer. *Clin. Cancer Res. 4*:1949–1954.

Nakamura, T.M., Morin, G.B., Chapman, K.B., Weinrich, S.L., Andrews, W.H., Lingner, J., Harley, C.B., and Cech, T.R. 1997. Telomerase catalytic subunit homologs from fission yeast and human. *Science 277*:955–959.

Nakayama, J., Tahara, H., Tahara, E., Saito, M., Ito, K., Nakamura, H., Nakanishi, T., Tahara, E., Ide, T., and Ishikawa, F. 1998. Telomerase activation by hTRT in human normal fibroblasts and hepatocellular carcinomas. *Nat. Genet. 18*:65–68.

Naritoku, W.Y., Datar, R.H., Li, P., Groshen, S.L., Taylor, C.R., and Imam, S.A. 1999. Telomerase activity: Comparison between fine-needle aspiration and biopsy specimens for the detection of tumor cells. *Cancer 87*:210–215.

Olovnikov, A.M. 1973. A theory of marginotomy: The incomplete copying of template margin in enzymic synthesis of polynucleotides and biological significance of the phenomenon. *J. Theor. Biol. 41*:181–190.

Raymond, E., Sun, D., Chen, S.F., Windle, B., and Von Hoff, D.D. 1996. Agents that target telomerase and telomeres. *Curr. Opin. Biotechnol. 7*:583–591.

Rhyu, M.S. 1995. Telomeres, telomerase, and immortality. *J. Natl. Cancer Inst. 87*:884–894.

Schwartz, H.S., Juliao, S.F., Sciadini, M.F., Miller, L.K., and Butler, M.G. 1995. Telomerase activity and oncogenesis in giant cell tumor of bone. *Cancer 75*:1094–1099.

Shay, J.W., and Wright, W.E. 1996. Telomerase activity in human cancer. *Curr. Opin. Oncol. 8*:66–71.

Shay, J.W., and Bacchetti, S. 1997. A survey of telomerase activity in human cancer. *Eur. J. Cancer 33*:787–791.

Suehara, N., and Tanaka, M. 1998. Telomerase activity detected in pancreatic juice 19 months before a tumor is detected in a patient with pancreatic cancer. *Am. J. Gastroenterol. 93*:1967–1971.

Sugino, T., Yoshida, K., Bolodeoku, J., Tahara, H., Buley, I., Manek, S., Wells, C., Goodison, S., Ide, T., Suzuki, T., Tahara, E., and Tarin, D. 1996. Telomerase activity in human breast cancer and benign breast lesions: Diagnostic applications in clinical specimens including fine-needle aspirates. *Int. J. Cancer 69*:301–306.

Tahara, H., Nakanishi, T., Kitamoto, M., Nakashio, R., Shay, J.W., Tahara, E., Kajiyama, G., and Ide, T. 1995. Telomerase activity in human liver tissues: Correlation between chronic liver disease and hepatocellular carcinoma. *Cancer Res. 55*:2734–2736.

Taylor, R.S., Ramirez, R.D., Ogoshi, M., Chaffins, M., Piatyszek, M.A., and Shay, J.W. 1996. Detection of telomerase activity in malignant and nonmalignant skin conditions. *J. Invest. Dermatol. 106*:759–765.

Travis, W.D., Colby, T.V., Corrin, B., Shimosato, Y., and Branhamera, E. 1999. *Histologic typing of lung and pleural tumors. World Health Organization International Histological Classification of Tumors.* Berlin, Germany: Springer.

Vaziri, H., Schachter, F., Uchida, I., Wei, L., Zhu, X., Effros, R., Cohen, D., and Harley, C.B. 1993. Loss of telomeric DNA during aging of normal and trisomy 21 human lymphocytes. *Am. J. Hum. Genet. 52*:661–667.

Watson, J.D. 1972. Origin of concatemeric T7 DNA. *Nat. New Biol. 239*:197–201.

Weinrich, S.L., Pruzan, R., Ma, L., Ouellette, M., Tesmer, V.M., Holt, S.E., Bodnar, A.G., Lichtsteiner, S., Kim, N.W., Trager, J.B., Taylor, R.D., Carlos, R., Andrews, W.H., Wright, W.E., Shay, J.W., Harley, C.B., and Morin, G.B. 1997. Reconstitution of human telomerase with the template RNA component hTR and the catalytic protein subunit hTRT. *Nat. Genet. 17:*498–502.

Wright, W.E., Piatyszek, M.A., Rainey, W.E., Byrd, W., and Shay, J.W. 1996. Telomerase activity in human germline and embryonic tissues and cells. *Dev. Genet. 18:*173–179.

Yahata, N., Ohyashiki, K., Ohyashiki, J.H., Iwama, H., Hayashi, S., Ando, K., Hirano, T., Tsuchida, T., Kato, H., Shay, J.W., and Toyama, K. 1998. Telomerase activity in lung cancer cells obtained from bronchial washings. *J. Natl. Cancer Inst. 90:* 684–690.

Yoshida, K., Sugino, T., Goodison, S., Warren, B.F., Nolan, D., Wadsworth, S., Mortensen, N.J., Toge, T., Tahara, E., and Tarin, D. 1997. Detection of telomerase activity in exfoliated cancer cells in colonic luminal washings and its related clinical implications. *Br. J. Cancer 75:*548–553.

Yui, J., Chiu, C.P., and Landsdorp, P.M. 1998. Telomerase activity in candidate stem cells from fetal liver and adult bone marrow. *Blood 91:*3255–3262.

Zakien, V.A. 1995. Telomeres: Beginning to understand the end. *Science 270:*1601–1607.

Zheng, P.S., Isawaka, T., Yamasaki, F., Ouchida, M., Yokoyama, M., Nakao, Y., Fukuda, K., Matsuyama, T., and Sugimori, H. 1997. Telomerase activity in gynecologic tumors. *Gynecol. Oncol. 64:*171–175.

14

Use of Fluorescence *in situ* Hybridization in Detecting Lung Cancer Cells

Irina A. Sokolova and Larry E. Morrison

Introduction

The idea that carcinogenesis is related to structural changes of nuclei and chromosomes appeared at the end of the nineteenth century and was clearly formulated as the somatic mutations theory by T. Bovery in 1914 (Balmain, 2001). This theory postulated that tumor growth is based on incorrect chromosome combinations that result in the abnormal growth characteristics of tumors. It took almost half a century to obtain the first direct proof of that theory, which was the discovery of the Philadelphia chromosome associated with chronic myeloid leukemia (Nowell *et al.*, 1960). From the early 1960s, investigations of hematological malignancies revealed relatively simple correlations between transformed phenotype and chromosomal aberrations. However, investigations of solid tumors of epithelial origin revealed considerably more complex relationships between carcinogenesis and chromosomal aberrations. In general, hematological tumors possessed more specific chromosomal aberrations than tumors of epithelial origin (Heim, *et al.*, 1995). Solid tumors showed multiple chromosomal aberrations and it was difficult to understand how these aberrations related to specific tumor type.

In the 1970s, two techniques were introduced that were designed to resolve the complexity of tumor karyotyping. One set of techniques involved the preparation and microscopic detection of metaphase chromosome banding patterns, and the other set involved *in situ* hybridization of labeled probes with metaphase chromosomes (Caspersson *et al.*, 1970; Hennig, 1973; Pardue *et al.*, 1975). Chromosome banding provided a unique ability to identify every human chromosome in normal and transformed cells, and allowed the identification of multiple chromosome aberrations in solid tumors (Heim *et al.*, 1995). *In situ* hybridization techniques provided the ability to further increase resolution and permit the localization of specific genes or repetitive sequences on the metaphase chromosomes (Pardue, *et al.*, 1975; Hennig *et al.*, 1973). Taken together, these two technologies provided the basis for systematic investigations of chromosomal aberrations associated with cancer.

The early *in situ* hybridization work used radiolabeled nucleic acid probes, but this laid the foundation for all subsequent *in situ* hybridization techniques. In particular, from a technical point of view these early studies showed that *in situ* hybridization is a complex

213

Copyright © 2004 by Elsevier (USA)
All rights reserved.

process, depending on the complexity and concentration of the labeled probes, the number of target sequences in the genome, the method of sample treatment, and the method of hybridization. However, the ability to use and detect radiolabeled probes was restricted by the chemistry of isotope incorporation, the ability to detect isotopes *in situ*, and the lifetimes of radioactive decay of isotopes. The breakthrough in the field was provided in the 1980s with the introduction of fluorescently labeled probes and their visualization using fluorescent microscope (Bauman *et al.*, 1981, Pinkel *et al.*, 1986). Since then fluorescence *in situ* hybridization (FISH) has proved to be the most reliable and high-resolution technique for the *in situ* detection of chromosomal aberrations in transformed cells. Introduction of this technique allowed investigation of individual regions of chromosomes involved in the pathogenesis of different types of cancer, with resolution on the order of 10^5 base pairs, and sometimes as low as several thousand base pairs (Patel *et al.*, 2000; Popescu *et al.*, 1996; Roylance, 2002; Teixeira, 2002). By comparison, chromosome banding alone provided considerably lower resolution (on the order of 10^6–10^7 nucleotide base pairs) (Heim *et al.*, 1995).

The combination of FISH with molecular cloning of rearranged chromosome sequences revealed four basic mechanisms of tumorigenesis, associated with chromosomal changes:

1. Deletion of chromosomal segments containing tumor suppressor genes (for example, allelic losses on the long arm of chromosome 13, containing both BRCA2 and retinoblastoma protein (Rb) are associated with sporadic breast carcinomas (Tsukamoto *et al.*, 1996).

2. Amplification of oncogenes, for example, amplification of MYCN (2p24) in neuroblastomas or amplification of ERBB2 (17q11-12) in breast cancer (Pauletti *et al.*, 1996; Schwab, 1998).

3. Formation of fusion proteins as a result of translocations, most frequently connected with the formation of aberrant transcriptional factors, like the MLL-1/AF4 rearrangement in ALL (Borkhardt *et al.*, 1994).

4. Transfer of potential oncogenes via translocations to regions where they can be activated by promotor elements. For example, the t(8;14) translocation in Burkitt's lymphoma that activates C-myc through proximally located enhancer elements of the immunoglobulin heavy chain locus at chromosome 14q32 (Gauwerky *et al.*, 1993).

Therefore, the physical location data afforded by FISH analysis can be directly related to the mechanisms of carcinogenesis. From the late 1980s through the 1990s, different variations of FISH methodology arrived, including multicolor FISH (mFISH), multiplex FISH (M-FISH), spectral karyotyping (SKY), rainbow cross-species FISH (R-FISH), and comparative genomics hybridization (CGH) (Morrison *et al.*, 1999; Patel *et al.*, 2000; Teixeira, 2002). Also, FISH techniques applied to interphase nuclei were described (Hopman *et al.*, 1988; Popescu *et al.*, 1997). The ability to stain interphase nuclei changed the entire paradigm and allowed investigators to define gain or losses of genetic material without preparation of the metaphase chromosomes.

In this chapter, we will describe the application of FISH to interphase nuclei of lung carcinoma. Lung cancer is a leading cause of death among both men and women in the United States. The American Cancer Society estimates there are 164,000 new lung cancer cases each year in the United States alone. The mortality rate associated with lung cancer diagnosis approaches 90%, despite intensive efforts to improve diagnosis and patient care. Patients with the best prognosis have early-stage disease with no local involvement beyond the primary tumors (Patz *et al.*, 2000). The current belief is that early detection offers the best chance of survival. Thus, new methods of early lung cancer diagnosis are greatly needed to detect the disease while it is still curable. During the past decade or so, the genetic events that promote lung cancer tumorigenesis have begun to be elucidated. As for most solid tumors, lung cancer tumorigenesis is a multistep process. For example, it is known that squamous cell carcinoma (SCC) of the lung follows the morphologic sequence from normal bronchial epithelium to squamous metaplasia to low-grade squamous dysplasia to high-grade squamous dysplasia to carcinoma *in situ* to microinvasive SCC to invasive SCC and finally to metastatic SCC (Hirsch *et al.*, 2001). Each of these steps is due to the accumulation of additional genetic alterations. Tumors at their earliest stages have relatively few genetic alterations, whereas those at later stages have numerous genetic alterations. FISH analysis allows investigators to examine the specimen directly, on a cell-by-cell basis, without prior cell culturing. Chromosome alterations associated with lung cancer have been described in the literature based on CGH, FISH, and loss of heterozygosity (LOH) studies, as well as other molecular studies (Balsara *et al.*, 2002; Petersen *et al.*, 2000; Wistuba *et al.*, 2002). Multiple genetic changes identified in lung tumor samples involve tumor suppressor genes such as TP53 (17q13), RB (13q14), p16 (9p21), FHIT gene (3p14), RASFF1 (3p21), oncogenes such as the family of RAS protooncogenes, the family of myc oncogenes, and the AIS oncogene, as well as growth factor receptors, such as c-erb-2 (Hibi *et al.*, 2000; Mao, 2002; Sekido, 2003).

Our approach to improving the diagnosis of non–small cell lung cancer (NSCLC) has been to select a group of the most consistently reported chromosome aberrations, develop FISH probes to these loci, screen a number of NSCLC tissues and normal adjacent tissues with this collection of probes, and select a subset of four probes that best detected the presence of cancer cells. These probes are then combined into a single four-color FISH probe set and their effectiveness in detecting lung cancer established in a collection of bronchial wash specimens (Sokolova *et al.*, 2002). The FISH methods required for these studies are reviewed in detail, including probe hybridization to a variety of specimen types, and data analysis required to identify the best probes set.

MATERIALS

1. Antifade solution DAPI II (Vysis, Inc., Downers Grove, IL) – 125 ng/ml 4,6-diamidino-2-phenylindole in aqueous buffer solution with 1-4-phenylenediamine and glycerol. Store at −20°C protected from light.

2. 20X Saline-sodium citrate (SSC) solution: Mix thoroughly 132 g of SSC (3 M NaCl, 0.3 M sodium citrate) in 400 ml purified water. Measure pH and adjust to pH 5.3 with HCl. Add purified water to bring final volume to 500 ml. Store at ambient temperature. Discard stock solution after 6 months or sooner if solution appears cloudy or contaminated.

3. 2X SSC solution: Mix thoroughly 100 ml of 20X SSC (pH 5.3) with 850 ml purified water. Measure pH and adjust to 7.0 ± 0.2 with 1 N NaoH. Add purified water to bring final volume to 1 L. Store at ambient temperature. Discard stock solution after 6 months or sooner if solution appears cloudy or contaminated.

4. Posthybridization wash solution 2X SSC/0.3% NP-40: Mix 100 ml 20X SSC, pH 5.3, with 3 ml NP-40 and 847 ml purified water. Adjust pH to 7.0–7.5 with 1 N NaOH. Adjust to 1000 ml final volume by adding purified water. Filter through 0.45-μm filtration unit. Store in a covered container at room temperature for up to 6 months. Discard solution that was used in the assay at the end of each day.

5. Posthybridization wash solution 0.4X SSC/0.3% NP-40: Mix 40 ml 20X SSC, pH 5.3, with 3 ml NP-40 and 907 ml purified water. Adjust pH to 7.0–7.5 with 1 N NaOH. Adjust to 1000 ml final volume by adding purified water. Filter through 0.45-μm filtration unit. Solution can be stored in a covered container at room temperature for up to 6 months. Discard solution that was used in the assay at the end of each day.

6. 2X SSC/0.1% NP-40 post-hybridization wash solution: Mix 100 ml 20X SSC, pH 5.3, with 1 ml NP-40 and 849 ml purifier water. Adjust pH to 7.0–7.5 with 1 N NaOH if necessary. Adjust to 1000 ml final volume by adding purified water. Filter through 0.45 μm filtration

unit. The solution can be stored in a covered container at room temperature for up to 6 months. Discard solution that was used in the assay at the end of each day.

7. Denaturation solution (70% formamide/2X SSC): Mix thoroughly 7 ml 20X SSC, pH 5.3, with 49 ml formamide and 14 ml purified water. Final pH of solution should be 7.0–8.0. Between uses, store covered at 2–8°C, discard after 7 days.

8. Formaldehyde solution: Mix together 12.5 ml of 10% neutral buffered formalin, 37 ml 1X phosphate buffer saline (PBS) and 0.5 ml 100X MgCl₂. Use at room temperature, discard after using 1 day. Store between 2–8°C when not in use.

9. Ethanol wash solutions: Prepare volume to volume dilutions of 70% and 85% ethanol, using 100% ethanol and purified water. Dilutions may be used for 1 week unless evaporation occurs or the solution becomes diluted because of excessive use. Store at room temperature in tightly capped containers when not in use.

10. Protease solution: Dissolve pepsin in diluted HCl at concentrations suggested in Table 16 for different sample types. Discard after using 1 day.

11. *In situ* hybridization probes: Many FISH probes can be obtained from commercial sources. However, if not commercially available, FISH probes can be prepared by culturing bacteria or yeast containing the human DNA sequence of interest. Vectors that accept large inserts are preferred, including P1 and bacterial chromosome vectors PAC and BAC, because FISH probe targets are typically greater in size than 40,000 base pairs, often greater than 100,000 base pairs, to yield good signals. After isolating DNA, the DNA can be labeled by any of several techniques, most commonly involving DNA polymerases and fluorophore-labeled nucleoside triphosphates, including nick translation, random priming, and DOP-PCR (see Morrison *et al.*, 2003, for a review of FISH probe labeling methods).

METHODS

In general, the FISH procedure consists of the following major steps:

1. Specimen fixation.

2. Specimen pretreatment with proteases to liberate tissue DNA from protein.

3. Denaturation of specimen and denaturation of fluorescently labeled DNA probes.

4. Hybridization of specimen with fluorescently labeled DNA probes.

5. Posthybridization washing to remove unbound probes.

6. Counterstaining.

7. Visualization of probes using fluorescent microscopy.

Table 16 Recommendations for Protease Concentration and Post-Hybridization Wash Solutions for Different Types of Lung Specimens

Condition	Lung Sample Type					
	Nonstained Bronchial Samples	Prestained Bronchial Samples	Sputum Samples	Tissue Touch Prep Samples	Paraffin-Embedded Tissues	Lung Cell Lines
Pepsin 0.05 mg/mL in 0.01 N HCl	√			√		√
Pepsin 0.5 mg/ml in 0.01 N HCl		√	√			
Pepsin 4 mg/ml in 0.2 N HCl					√	
Post-Hyb Wash 2X SSC/0.3% NP-40	√	√	√		√	
Post-Hyb Wash 0.4X SSC/0.3% NP-40				√		√

However, depending on the type of specimen, these steps must be adjusted and optimized to provide the best hybridization results. In the following section, we provide detailed protocols for several types of bronchial specimens, which allowed us to obtain high-quality signal intensity of fluorescently labeled DNA probes, as well as low background staining, while still maintaining good morphology of tissues and cells.

Sample Pretreatment Procedure

NOTE: Sample slides should be prepared according to a clinical lab protocol for a specific lung specimen type.

1. Immerse slides in 2X SSC, 73°C for 2 min.
2. Incubate slides in protease solution at 37°C for 10 min (see details provided in Table 16 for appropriate protease solution concentration).
3. Incubate slides in 1X PBS for 5 min at room temperature.
4. Incubate slides in formaldehyde fixative for 5 min at room temperature.
5. Incubate slides in 1X PBS for 5 min at room temperature.
6. Dehydrate slides through an ethanol series: 70% ethanol for 1 min, 85% ethanol for 1 min, 100% ethanol for 1 min. Air-dry slides.

Denaturation of Probe and Target DNA

1. Prepare denaturation solution and ensure that the temperature is 73 ± 1°C.
2. Mark hybridization areas with a diamond-tipped scribe on the sample slide.

3. Immerse the slides in the denaturation solution for 5 min.

NOTE: Immerse no more than four slides in the coplin jar simultaneously.

4. Dehydrate slides for 1 min in 70% ethanol, followed by 1 min in 85% ethanol and 1 min in 100% ethanol. Keep the slides in 100% ethanol until you are ready to dry all slides and apply the probe mixture.

Hybridization of the Probe to the Sample Target

1. Air-dry the slides to evaporate remaining ethanol, or place them on a slide warmer at 40–45°C for 1–2 min.
2. Apply 10 µl of probe mixture to one target area and immediately apply coverslip. Repeat for additional target areas.
3. Seal coverslip with rubber cement.
4. Place slides in the prewarmed humidified box and place box in a 37°C incubator for 16–24 hr.
5. Proceed to post-hybridization wash step.

Codenaturation of Probe and Target DNA and Subsequent Hybridization

In some cases it appears advantageous to perform codenaturation of probe and target DNA in one step. This can be done using a hot plate or hybridization oven, such as a HYBrite automated codenaturation oven (Vysis, Inc., Downers Grove, IL) that allows programming of the denaturation (Melt) and hybridization (Hyb) temperatures.

1. Set the Melt temp to 72°C and the Melt time to 2 min. Set the Hyb temp to 37°C and the Hyb time to 16–24 hr.

2. Mark hybridization areas with a diamond-tipped scribe on the sample slide and place slides on the HYBrite.

3. Apply 10 µl of probe mixture to the slide and immediately apply coverslip.

4. Seal coverslip with rubber cement.

5. Begin the cohybridization program.

6. When the hybridization step is completed, proceed to post-hybridization wash step.

Post-Hybridization Wash

1. Pour 70 ml of post-hybridization wash solution into a Coplin jar and put it in water bath set at 73°C. See details in Table 16 for appropriate post-hybridization wash solution concentration.

2. Pour 70 ml of 2X SSC/0.1% NP-40 into a second Coplin jar and keep at room temperature. The solutions may be used for 1 day, then discard.

3. Remove rubber cement and coverslips from each slide and immerse in appropriate post-hybridization wash solution. Let stand for 2 min.

4. Remove slides after 2 min and rinse in ambient 2X SSC/0.1% NP-40.

Visualizing the Hybridization

1. Air-dry slide(s) in darkness.

2. Apply 10 µl DAPI II counterstain to the target area of slide and apply coverslip.

3. View slides using a suitable filter set on an optimally performing fluorescence microscope.

RESULTS

Investigation of Genomic Changes in Lung Tumor Tissues

Using the FISH protocols previously described, we studied the occurrence of a number of chromosomal aberrations in lung tumor tissues compared to normal lung tissues. The chromosomal loci selected for this study were derived from a number of publications that used a variety of techniques to identify chromosomal abnormalities, including FISH, CGH, and LOH. The ultimate goal of this work was to identify several FISH probes that could be used together in a multicolor panel to detect lung cancer with high sensitivity and specificity. For this purpose we selected 13 centromere enumeration (CEP) and 13 locus-specific indicator (LSI) probes (Vysis, Inc./Abbott Laboratories), and labeled each probe with one of the following fluorophores: SpectrumAqua, SpectrumGreen, SpectrumOrange,

SpectrumGold, or SpectrumRed, and combined these probes into eight probe set mixes. When possible, CEP and LSI probes targeting the same chromosome were combined into the same probe set mix to enable the determination of whether locus-specific gains or losses were due to chromosomal aneusomy or were true allelic gains or losses at these loci. The following probes were tested: CEP 1, CEP 3, CEP 4, CEP 6, CEP 7, CEP 8, CEP 9, CEP 10, CEP 11, CEP 12, CEP 16, CEP 17, CEP 18, LSI 17p13, LSI 17q21, LSI 8q24, LSI 13, LSI 9p21, LSI 21, LSI 5p15, LSI 5q31, LSI 7p12, LSI 3p14, LSI 3q26, LSI 10q23, and LSI 20q13. These probes were hybridized to 27 lung tumor touch preparations and 10 normal lung touch preparations. Hybridized specimen slides were viewed on a fluorescence microscope using single band pass filter sets specific for each of the four fluorescent labels and the DAPI counterstain. Each touch preparation was analyzed by counting the number of signals of each fluorescent color in 100 consecutive noninflammatory cells, and the copy number of each probe target recorded.

Gains and Losses in Normal and Tumor Specimens

Cells were considered to show gains of a chromosome (CEP probes) or locus (LSI probes) if there were three or more signals for that probe in a cell. Cells were considered to show loss of a chromosome or chromosome locus if the cell had only a single signal for that probe. Cells were considered to show relative gain or loss of a locus if there were greater or fewer signals, respectively, for that locus than for the corresponding chromosome. For example, a cell having three 9p21 signals and four CEP 9 signals shows relative loss of the 9p21 locus. In contrast, a cell with three copies of CEP 8 and five copies of 8q24 shows relative gain of the 8q24 locus.

In the normal touch preparations, the average percentage of cells with gains ranged from 1.0% for CEP 8 to 7.8% for LSI 21. The average percentage of cells with losses ranged from 1.3% for 7p12 to 10.9% for CEP 17. The average percentage of cells showing gains for each of the LSI probes relative to the corresponding CEP probes in the normal touch preparations ranged from 6.2% for LSI 8q24 to 11.4% for LSI 17q21. The average percentage of cells showing losses for each of the LSI probes relative to the corresponding CEP probes in the normal touch preparations ranged from 1.1% for LSI 7p12 to 6.2% for LSI 17q21.

In the tumor touch preparations, the average percentage of cells with gains ranged from 13.4% for LSI 13 to 34% for LSI 5p15 (Figure 32). The average percentage of cells with losses in the tumor touch preparation specimens ranged from 1.3% for LSI 7p12 to 10.9% for CEP 17. The average percentage of cells showing gains of the LSI probes relative to the corresponding CEP probes in the tumor touch preparations ranged from 9.2% for LSI 9p21 to 16.7% for LSI 17q21. The high standard deviations reflect the fact that some tumors exhibited high levels of gains or losses for a certain probe whereas others did not.

Sensitivity and Specificity for Detecting Malignancy Based on Gains and Losses of Individual Probes

Each probe can be evaluated for its ability to effectively identify lung tumor tissue by calculating the sensitivity and specificity of tumor detection relative to a threshold value of abnormal cells per specimen. The threshold value is often set relative to the number of abnormal cells found in a collection of normal tissue of the same tissue type. In initial investigations the threshold value is typically set at three standard deviations above the mean of the percentage of abnormal cells found in the normal specimen set to assure good specificity. Figure 33 shows sensitivity values calculated in this manner for the most sensitive probes. Single probes showing consistently high sensitivity with the three standard deviation threshold were gain of LSI probes 5p15, 8q24, 7p12, and 3q26, and gain of centromeric probes CEP 1, CEP 6, CEP 7, CEP 8, and CEP 9. Highest sensitivities based on relative gain or loss (locus-to-corresponding CEP or opposite arm probe) were achieved by gain of the 5p locus relative to the 5q locus, and loss of 3p14, gain of 7p12, loss of 7p21, loss of 9p21, and gain of 8q24 relative to the corresponding CEP probes. All of the probes mentioned and probe ratios showed the maximum specificity of 100% at the three standard deviation threshold.

Complementation of Probes

To make the FISH assay more sensitive and specific in detecting malignant cells, it is possible to combine several probes into a single probe set. The reason the FISH assay can become more sensitive is because tumors are heterogeneous, and one lung tumor cell population may have alterations of a locus that another lung tumor may not, and visa versa. In addition, the FISH assay can be made more specific because with multiple probes the requirement for positivity can be set at two or more abnormal loci per cell, thereby decreasing the chance of false positive calls. For example, gains of two or more chromosomes in a single cell are very rare in normal cell populations and essentially pathogenomic of neoplasia. The one possible exception of this is the finding of a cell that shows tetrasomy for all signals. Tetrasomic cells are occasionally observed in normal specimens and are probably not indicative of neoplasia unless present in significant numbers.

When designing a multiple probe assay, it is important to select probes that provide complementation. That is, when the probes are used in combination, they identify more tumors than when each of the probes is used separately. This can be seen for the probe combinations listed in Table 17. Each pair of probes listed shows a combined sensitivity that is greater than that of either probe alone. For example, CEP 1 when used alone showed a sensitivity for detecting tumor tissue of 79.3%. When CEP 1 was combined with LSI 3q26 the sensitivity increased to 89.5%, and when combined with either CEP 7, 8, or 9 the sensitivity increased to 86.2%.

Selection of a Probe Panel

The 26 probes in the selection study were narrowed down to four probes to form a multicolor probe panel that could be used in a single *in situ* hybridization. The four probes were selected based on their individual performances, as well as the ability to complement one another. We did not look beyond a two-probe complementation because the size of the data set did not warrant

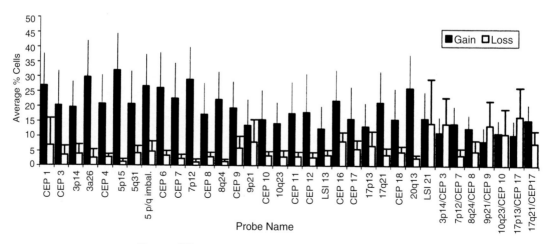

Figure 32 Frequency of gains and losses in lung tumors.

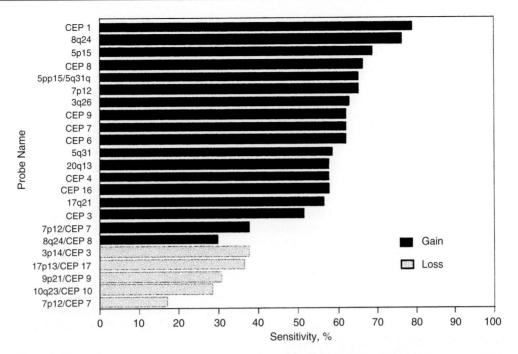

Figure 33 Sensitivity for detecting lung cancer for single probes across all lung tumor types versus normal lung tissues.

a three- or four-probe complementation analysis. However, four probes were put in the panel to provide redundancy and further enable the use of a two-locus abnormality minimum to call a cell abnormal, thereby improving specificity. The probe panel included a repetitive sequence centromeric probe to chromosome 1,

and three unique-sequence probes to the loci 5p15, 8q24 (containing the C-myc gene), and 7p12 (containing the epidermal growth factor receptor [EGFR] gene). These probes were labeled respectively with Spectrum-Aqua, SpectrumGreen, SpectrumGold, and Spectrum-Red and formulated into a single probe set.

Application of the FISH Probe Panel for Detecting Malignant Cells in Bronchial Washing Specimens

To test the feasibility of the four-probe FISH panel to improve the diagnosis of lung cytology specimens, FISH was performed on a collection of bronchial washing specimens (Sokolova *et al.*, 2002). Bronchial washing specimens were selected from the cytopathology archives of the Institute of Pathology in Basel, Switzerland. These were cytology specimens prestained with PAP stain and permanently mounted under coverslips. The specimens had been archived for a period of time ranging from a few months to 2 years. Forty-eight specimens were selected with a histologically confirmed diagnosis of lung cancer, and 26 specimens were selected with a negative histological diagnosis.

Based on previous experience (Halling *et al.*, 2000), we suggested the following criteria for cell abnormality: a cell is abnormal if it shows copy number gains for

Table 17 Sensitivity for Combinations of Two Probes Across All Lung Tumor Types versus Normal Lung Tissues

Probe 1 (Gain)	Probe 2 (Gain)	Sensitivity (%)
3q26	CEP 1	89.5
8q24	CEP 7	86.2
8q24	CEP 1	86.2
CEP 7	CEP 1	86.2
CEP 8	CEP 1	86.2
CEP 9	CEP 1	86.2
5p15	3q26	84.2
8q24	CEP 8	83.3
8q24	5p15	82.8
CEP 6	CEP 1	82.8
CEP 8	5p15	82.8
CEP 9	CEP 8	82.8
5p15	CEP 1	82.8
8q24	7p12	79.3
8q24	CEP 6	79.3

at least two probes included in the probe mixture. Therefore, each sample was analyzed after FISH hybridization and the number of "abnormal" cells in each specimen was tabulated. The optimal point in this assay corresponded to a cutoff value of 5–6 "abnormal" cells per specimen, as determined from receiver operating characteristic curves. An example of bronchial washing specimens hybridized with the panel of FISH probes is presented in Figure 34. Our results showed that the sensitivity of the FISH assay was 82% in this selected set of specimens, compared with the 54% sensitivity of cytology (p = 0.007). This indicates that the lung FISH assay provided a substantial improvement in the sensitivity for detecting lung cancer in bronchial washing specimens, and therefore can be potentially used as an adjunct to cytology in examining bronchial specimens for evidence of malignancy. The specificity of FISH

was 82%, compared with a specificity for cytology that was 100% by design. This difference, however, was not statistically significant (p = 0.993). Furthermore, "false positive" FISH specimens may represent a group of patients at higher risk for cancer development.

The purpose of this study was to determine if malignant cells are present in bronchial specimens at a level sufficient to permit effective detection of lung cancer. However, it is possible that the finding of amplification of one or more of the loci assessed with this probe set (i.e., 5p15, 8q24, 7p12) may also have prognostic and therapeutic significance. For example, some studies suggest that C-myc (8q24 region) amplification or overexpression is associated with a worse prognosis (Volm et al., 2000). Also, increased expression of the EGFR (7p12 locus) has been shown to correlate with increased responsiveness to genistein and tyrphostin in non–small lung cancer cell lines (Lei et al., 1999). Further properly controlled prospective studies should be performed on large cohorts of lung cancer patients to determine if these markers may provide additional prognostic and therapeutic information.

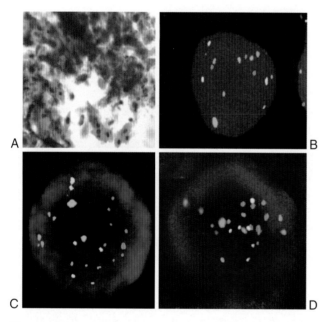

Figure 34 Example of abnormal cells in bronchial washing specimens. Panel **A**: PAP staining of a bronchial washing specimen from the patient with the clinical diagnosis squamous cell carcinoma of the lung. Panel **B**: FISH assay results on the same specimen as in **A**. Probes used in the FISH assay are as follows: CEP 1 Aqua, 5p15 Green, 8q24 (MYC) Gold, 7p12 (EGFR) Red. Note that abnormal cell demonstrates three Aqua signals, three Red signals, four Gold signals and four Green signals. Panel **C**: Example of FISH abnormal cells found in a patient diagnosed with lung adenocarcinoma. Cytology results were negative for this patient, however, the FISH assay clearly shows that the cell is not disomic, demonstrating six Gold signals, eight Red signals, six Green signals, and two Aqua signals. Panel **D**: Example of FISH abnormal cells found in a patient diagnosed with lung squamous cell carcinoma but negative cytology results. FISH shows that the cell contains six Gold signals (8q24 MYC locus), seven Red signals (7p12 EGFR locus), six Green signals (5p15), and four Aqua signals (CEP 1).

Conclusion

Tumor progression in lung cancer is associated with numerous genetic alterations detectable by FISH. We provided a number of FISH protocols that can be used to assess the level of abnormal cells in various specimens relevant to the analysis of lung cancer. With these protocols we studied abnormalities at 26 chromosomal loci of lung tissue. FISH probes targeting a number of these loci can provide good sensitivity and specificity for detecting lung tumor specimens and distinguishing them from normal tissue.

Overall, FISH studies are becoming important not only for improved diagnosis of tumors, but also for improved cancer prognosis, selection of appropriate therapy, and detection of minimal residual disease. Also, due to the increasing sensitivity of FISH techniques there is hope that FISH can better identify premalignant lesions, and, in turn, better identify individuals at higher risk for developing disease (Hittelman, 2001).

References

Balmain, A. 2001. Cancer genetics: From Boveri and Mendel to microarrays. *Nat. Rev. 1:*77–82.

Balsara, B.R., and Testa, J.R. 2002. Chromosomal imbalances in human lung cancer. *Oncogene 21:*6877–6883.

Bauman, F.G.J., and van Duijn, P. 1981. Hybrido-cytochemical localization of specific DNA sequences by fluorescence microscopy. *Histochem. J. 13:*723–733.

Borkhardt, A., Repp, R., Haupt, E., Brettreich, S., Buchen, U., Gossen, R., and Lampert, F. 1994. Molecular analysis of MLL-1/ AF4 recombination in infant acute lymphoblastic leukemia. *Leukemia 8:*549–553.

Caspersson, T., Zech, L., and Johansson, C. 1970. Different binding of alkylating fluorochromes in human chromosomes. *Exp. Cell Res. 60:*315–319.

Gauwerky, C.E., and Croce, C.M. 1993. Chromosomal translocations in leukaemia. *Sem. Cancer Biol. 4:*333–340.

Halling, K.C., King, W., Sokolova, I.A., Meyer, R.G., Burkhardt, H.M., Halling, A.C., Cheville, J.C., Sebo, T.J., Ramakumar, S., Stewart, C.S., Pankratz, S., O'Kan, D.J., Seelig, S.A., Lieber, M.M., and Jenkins, R.B. 2000. A comparison of cytology and fluorescence *in situ* hybridization for the detection of urothelial carcinoma. *J. Urol. 164:*1768–1775.

Heim, S., and Mitelman, F. 1995. *Chromosomal and Molecular Genetic Aberrations of Tumor Cells.* 2nd. ed. New York: Wiley-Liss. John Wiley & Sons, Inc.

Hennig, W. 1973. Molecular hybridization of DNA and RNA *in situ*. *Int. Rev. Cytol. 36:*1–44.

Hibi, K., Trink, B., Patturajan, M., Westra, W.H., Caballero, O.L., Hill, D.E., Ratovitski, E.A., Jen, J., and Sidransky, D. 2000. AIS is an oncogene amplified in squamous cell carcinoma. *Proc. Natl. Acad. Sci. USA 97:*5462–5467.

Hirsh, F.R., Franklin, W.A., Gazdar, A.F., and Bunn, P.A. Jr. 2001. Early detection of lung cancer: Clinical perspectives of recent advances in biology and radiology. *Clin. Cancer Res. 7:*5–22.

Hittelman, W.N. 2001. Genetic instability in epithelial tissues at risk for cancer. *Ann. N. Y. Acad. Sci. 952:*1–12.

Hopman, A.H.N., Ramaekers, F.C.S., Raap, A.K., Beck, J.L.M., Devilee, P., van der Ploeg, M., and Vooijs, G.P. 1988. *In situ* hybridization as a tool to study numerical chromosome aberrations in solid bladder tumors. *Histochemistry 89:*307–316.

Lei, W., Mayotte, J.E., and Levitt, M.L. 1999. Enhancement of chemosensitivity and programmed cell death by tyrosine kinase inhibitors correlates with EGFR expression in nonsmall cell lung cancer cells. *Anticancer Res. 19:*221–228.

Mao L. 2002. Recent advances in the molecular diagnosis of lung cancer. *Oncogene 21:*6960–6969.

Morrison, L.E., and Legator, M.S. 1999. Multicolor fluorescence *in situ* hybridizations techniques. In Andreeff, M. and Pinkel, D. (eds.) *An Introduction to Fluorescence In Situ Hybridization: Principles and Clinical Applications.* New York: Wiley-Liss, 77–118.

Morrison, L.E., Ramakrishnan, R., Ruffalo, T.M., and Wilber, K.A. 2002. Labeling fluorescence *in situ* hybridization probes for genomic targets. In: Fan, Y-S (ed). *Molecular Cytogenetics: Protocols and Applications* (Methods of Molecular Medicine Series). Totowa, NJ: Humana Press, 21–40.

Nowell, P.C., and Hungerford, D.A. 1960. A minute chromosome in human chronic granulocytic leukemia. *Science 132:* 1497.

Pardue, M.L., and Gall, J.G. 1975. Nucleic acid hybridization to the DNA of cytological preparations. *Methods Cell Biol. 10:*1–16.

Patel, A.S., Hawkins, A.L., and Griffin, C.A. 2000. Cytogenetics and cancer. *Curr. Opin. Oncol. 12:*62–67.

Patz, E.F. Jr., Goodman, P.C., and Bepler, G. 2000. Screening for lung cancer. *N. Engl. J. Med. 343:*1627–1633.

Pauletti, G., Godolphin, W., Press, M.F., and Slamon, D.J. 1996. Detection and quantitation of HER-2/neu gene amplification in human breast cancer archival material using fluorescence *in situ* hybridization. *Oncogene 13:*63–72.

Petersen, S., Aninat-Meyer, M., Schluns, K., Gellert, K., Dietel, M., and Petersen, I. 2000. Chromosomal alterations in the clonal evolution to the metastatic stage of squamous cell carcinomas of the lung. *Br. J. Cancer 82:* 65–73.

Pinkel, D., Straume, T., and Gray, J.W. 1986. Cytogenetic analysis using quantitative, high-sensitivity, fluorescence hybridization. *Proc. Natl. Acad. Sci. USA 83:*2934–2938.

Popescu, N.C., and Zimonjic D.B. 1997. Molecular cytogenetic characterization of cancer cell alterations. *Cancer Genet. Cytogenet. 93:*10–21.

Roylance, R. 2002. Methods of molecular analysis: Assessing losses and gains in tumours. *Mol. Pathol. 55:*25–28.

Schwab, M. 1998. Amplification of oncogenes in human cancer cells. *Bioessays 20:*473–479.

Sekido, Y., Fong, K.M., and Minna, J.D. 2003. Molecular genetics of lung cancer. *Annu. Rev. Med. 54:*73–87.

Sokolova, I.A., Bubendorf, L., O'Hare, A., Legator, M.S., Jacobson, K.K., Grilli, B., Dalquen, P., Halling, K.C., Tamm, M., Seelig, S.A, and Morrison, L.E. 2002. A fluorescence *in situ* hybridization-based assay for improved detection of lung cancer cells in bronchial washing specimens. *Cancer 96:*306–315.

Teixeira, M.R. 2002. Combined classical and molecular cytogenetic analysis of cancer. *Eur. J. Cancer 38:*1580–1584.

Tsukamoto, K., Ito, N., Yoshimoto, M., Iwase, T., Tada, T., Kasumi, F., Nakamura, Y., and Emi, M. 1996. Two distinct commonly deleted regions on chromosome 13q suggest involvement of BRCA2 and retinoblastoma genes in sporadic breast carcinomas. *Cancer 78:*1929–1934.

Volm, M., and Koomagi, R. 2000. Prognostic relevance of C-myc and caspase-3 for patients with nonsmall cell lung cancer. *Oncol. Rep. 7:*95–98.

Wistuba, I.I., and Gazdar, A.F. 2002. Characteristic genetic alterations in lung cancer. *Methods Mol. Med. 74:*3–28.

15

Immunohistochemistry of BCL-2 Gene Expression in Lung Carcinoma

Valeria Masciullo and Antonio Giordano

Introduction

Tumor growth and proliferation strictly depend on an imbalance between cellular proliferation and cell death. Alterations in different genes, controlling the proliferation/death ratio, are frequently involved in the development and progression of several types of human cancers.

Lung Cancer

Lung cancer is one of the leading causes of cancer death in the world and its incidence is steadily increasing in women (Shopland et al., 1991). Despite diagnostic and therapeutic improvements (Wiest et al., 1997), the prognosis of lung cancer has changed little during the last 20 years, and the 5-year overall survival is still less than 15%. The only curable tumors are those diagnosed in the early stages and treated surgically. Because early detection and effective chemoprevention may represent the most promising clinical approach, the identification of biomarkers to stratify patients into different risk groups could be a potential strategy for clinical trials.

Early studies indicated that several distinct chromosomal loci (3p, 9p, 13q, 17p, and others) are implicated, suggesting that sequential genetic events possibly occur during initiation and progression of lung carcinogenesis (Shimizu et al., 1997).

However, recent studies have indicated that allelic loss of several other chromosomal regions could be involved in the pathogenesis of lung cancer. These chromosomal regions include 1p, 1q, 2q, 5p, 6p, 8p, 8q, 10q, 14q, 17q, 18q, and 22q (Virmani et al., 1998). The malignant transformation of pulmonary epithelial cells is the result of a multistep process in which accumulation of genetic and molecular alterations, involving key regulatory elements of the cell cycle and mechanisms of proliferation and apoptosis, accompany or even precede histological changes (Salgia et al., 1998). Oncogene activation (Ras, Myc, and autocrine growth factor loops) or more importantly tumor suppressor gene inactivation (p53, pRb family, and cyclin-dependent kinase inhibitor p16 and p27) at a genetic, epigenetic, or post-translational level removes crucial regulatory constraints on the cell cycle at the G1 checkpoint, accelerating cell division (Esposito et al., 1997; Sekido et al., 1998). p53 inactivation is one of

Handbook of Immunohistochemistry and in situ Hybridization of Human
Carcinomas, Volume 1: Molecular Genetics; Lung and Breast Carcinomas

223

Copyright © 2004 by Elsevier (USA)
All rights reserved.

the most common alterations in lung cancer (75% of genetic alterations). In fact, mutations of p53 have been reported with frequencies up to 50% in non–small cell cancer (NSCLC) (Mitsudomi *et al.*, 1993) and 70–80% in small cell lung cancer (SCLC) (Greenblatt *et al.*, 1994). On the other hand, some authors have reported mutations or deletions of the Rb gene in NSCLCs in more than 90% of the cases (Reissmann *et al.*, 1993). Mutation of p53 missense is highly concordant with p53 stabilization and immunoreactivity; other gene products, like pRb and Ras, are either rapidly degraded or not detectable at the immunohistochemical level if mutated. The expression of p53, pRb, and Ras was investigated by immunohistochemistry (IHC) in panel of 65 samples of preneoplastic lesions of the bronchial epithelium (Ferron *et al.*, 1997). The frequency of p53-positive and pRb-negative microscopic fields was directly related to the morphological grading of the lesions. One of the main patterns found to be correlated with the severity of histopathological features was characterized by combined p53 hyperexpression and pRb hypoexpression.

Interestingly, some authors could correlate the prognostic significance of the loss of Rb protein either alone or combined with Ras or with p53 in patients with NSCLC. Individuals with theoretically the best pattern of protein expression in their tumors versus those with theoretically the worst pattern of gene expression (i.e., Rb1/Ras2 versus Rb2/Ras1 and Rb1/p532 versus Rb2/p531) showed a longer period of survival. The correlation between Rb and Ras appeared to be a better prognostic factor in NSCLCs compared with Rb/p53 status. On the other hand, in patients affected by squamous cell carcinoma, neither Rb/Ras nor Rb/p53 status was a significant prognostic factor (Dosaka-Akita *et al.*, 1997).

In an immunohistochemical study on 77 lung cancer specimens (Baldi *et al.*, 1996), pRb2/p130, another member of the retinoblastoma family, was shown to be undetectable in a higher percentage of patients compared with pRb and p107. In an additional study (Baldi *et al.*, 1997), we found a negative correlation between histological grading and pRb2/p130 staining in 158 specimens of human lung cancer. Moreover, we recently demonstrated the role of pRb2/p130 in lung cancer patients as an independent prognostic predictor of clinical outcome (Caputi *et al.*, 2002). Patients with lung cancer that show loss of pRb2/p130 expression are at a higher risk of death from disease and may eventually benefit from more aggressive adjuvant therapy. Finally, in support of the involvement of pRb2/p130 as a tumor suppressor gene in lung cancer, we showed that *in vivo* retroviral transduction of pRb2/p130 in established tumors, derived from injection of the lung

adenocarcinoma cell line H23 grown in nude mice, reduced the mass 12-fold with respect to the control viruses (Claudio *et al.*, 2000).

Clonal expansion to reach a critical size necessary for tumor growth and progression could also be influenced by the level of cell death. Therefore, apoptosis susceptibility factors and their antagonists of the Bcl-2 family of proteins are potential regulators of clonal expansion and tumor progression in lung cancer.

The Bcl-2 gene family is composed of a divergent group of proteins that regulate programmed cell death and stress-induced apoptosis. One arm of this family includes antiapoptotic members such as mammalian Bcl-2, Bcl-xl, Bcl-w, A1, Mcl-1, and their *C. Elegans* homologue CED-9 and are required for cell survival (Adams *et al.*, 1998). On the other hand, the Bax/Bak-like proteins, which are structurally related to Bcl-2, as well as the BH3-only proteins (Bid, Bad, Bim/Bod, and PUMA) are both essential initiators of apoptotic cell death (Huang *et al.*, 2000). The BH3-only proteins are activated in response to various apoptotic stimuli and are able to induce cell death by interacting directly with Bax and Bak (Wei *et al.*, 2000) or binding to the hydrophobic groove of antiapoptotic members such as Bcl-2 or Bcl-xL and removing the inhibition of these antiapoptotic molecules on Bax and Bak. The ratio of death antagonists to agonists determines whether a cell will respond to an apoptotic signal.

The Bcl-2 family controls programmed cell death during embryogenesis and tissue differentiation as well as in response to cytotoxic stimuli (Hockenbery *et al.*, 1990; Le Brun *et al.*, 1993). Abnormalities in cell death control can cause a variety of disease, including cancer.

The Bcl-2 gene (B cell lymphoma gene-2) was originally identified in follicular B cell lymphoma, where a chromosomal translocation t(14;18) moves the Bcl-2 gene into juxtaposition with transcriptional enhancer elements of the immunoglobulin heavy chain locus (Aisemberg *et al.*, 1988; Tsujimoto *et al.*, 1984, 1986). This translocation results in deregulated high-level expression of the Bcl-2 gene in B lymphocytes. Transgenic mice overexpressing Bcl-2 in B lymphocytes show a tumor incidence of 5–10% (Aisemberg *et al.*, 1988; McDonnel *et al.*, 1991; Strasser *et al.*, 1993). Those tumors were classified as either pro-B/pre-B lymphomas or plasmacytoma/large B cell lymphomas and were carrying rearrangements of the C-Myc protooncogene, suggesting a model in which Bcl-2 promotes neoplastic transformation by increasing the lifespan of normally short-lived cells, thus allowing them to accumulate oncogenic mutations.

Experiments with transgenic mice overexpressing Bcl-2 under the control of different promoters showed that it can also promote transformation of T lymphocytes (Linette *et al.*, 1995), myeloid cells (Traver *et al.*, 1998), and mammary epithelial cells (Jager *et al.*, 1997).

The Bcl-2 protooncogene is encoded by a 230-kb gene. Its product, a 26-kDa protein, is located in the inner mitochondrial membrane, and to a lesser extent in cell membranes (Jong *et al.*, 1994). The major function of Bcl-2 appears to be to inhibit programmed cell death by preventing caspase activation and to prolong cell survival by arresting cells in the G_1/G_2-phase of the cell cycle (Green *et al.*, 1998).

Besides being controlled through transcription, phosphorylation and proteolytic cleavage, Bcl-2 is also regulated by ubiquitination and proteasome degradation systems (Breitschopf *et al.*, 2000; Dimmeler *et al.*, 1999). However, transregulatory mechanisms appear to be responsible for most cases of Bcl-2 overexpression that occurs in various types of human epithelial malignancies, including prostate (Colombel *et al.*, 2000), breast (Silvestrini *et al.*, 1994), thyroid (Pilotti *et al.*, 1994), gastrointestinal (Bronner *et al.*, 1995) and non–small cell lung cancer (Fontanini *et al.*, 1995; Pezzella *et al.*, 1993). One of the most common transregulators of Bcl-2 in cancer is the tumor suppressor protein p53. Negative responsive elements for wild-type p53, but not mutant p53, have been found in the 5′ untranslated regions of the Bcl-2 gene through which wild-type, but not mutant p53, is able to repress the Bcl-2 gene (Miyashita *et al.*, 1994). Thus, p53 mutations could result in elevated production of Bcl-2. Moreover, experiments with normal lymphocytes and lymphomas have clearly demonstrated that Bcl-2 overexpression does not only inhibit radiation and anticancer drug-induced apoptosis in short-term assays but promotes long-term survival and continued clonogenic growth (Schmitt *et al.*, 2000; Strasser *et al.*, 1994). In solid cancer it is less clear whether apoptosis regulators determine sensitivity to anticancer therapy because overexpression of Bcl-2-like molecules often had only minimal effects (Brown *et al.*, 1999).

Although there are now a large number of studies on the immunohistochemical expression of Bcl-2 in lung cancer, their value in predicting the prognosis of patients with lung cancer remains controversial. In this chapter, we present an extensive review of the literature to assess how Bcl-2 expression relates to lung tumors and their clinical outcome.

MATERIALS

1. Tissue fixation: Formalin-fixed, paraffin-embedded tissue sections. Poly-L-lysine coated slides.
2. Deparaffinization: Xylene, 100% alcohol, 90% alcohol, 85% alcohol, 70% alcohol, and 50% alcohol.

3. Block of endogenous peroxidase activity: 3% hydrogen peroxide. To prepare 100 ml: 10 ml of 30% H_2O_2 in 90 ml of PBS or methanol. Mix well.
4. Dulbecco's 1X phosphate buffer saline (PBS) (0.01 M) without calcium and magnesium: 0.32 g monobasic sodium phosphate, 9 g sodium chloride, and 1.09 g dibasic sodium phosphate *7H$_2$O; bring vol to 1 L with deionized-glass distilled water, (pH 7.4).
5. Blocking solution: 2% normal serum in PBS.
6. Primary antibody: Mouse anti-Bcl-2 antibody (clone 124 or clone 100).
7. Secondary antibody: Biotinylated horse anti-mouse IgG.
8. ABC reagent: HRP-streptavidin diluted 1:400 or avidin-biotin-complex (ABC) kit.
9. DAB reagent: 0.05% 3,3′-diaminobenzidine tetrahydrochloride (DAB) and 0.03% H_2O_2 in PBS: To prepare 100 ml, stir 50 mg of DAB in 100 ml of PBS. Add 100 µl of 30% H_2O_2 and filter. Use promptly.
10. Harris' hematoxylin.

METHODS

Immunohistochemical Staining Protocol for Paraffin Sections

1. Deparaffinize sections in xylene for 3×5 min.
2. Hydrate with 100% ethanol for 2×5 min.
3. Hydrate with 95% ethanol for 5 min followed by 90%, 85%, 70%, and 50% alcohol for 5 min.
4. Rinse in distilled water.
5. Hydrogen peroxide: Incubate sections in 3% H_2O_2 in PBS (or methanol) for 15 min to block endogenous peroxidase activity.
6. Rinse in PBS for 2×5 min.
7. Antigen retrieval: Immerse sections in 10 mM citrate buffer (in distilled water, pH 6.0) and irradiate them with the microwave for two cycles of 8 and 5 min, separated by 5 min of pause at 700W.
8. Cool slides 20 min at room temperature and then rinse in PBS for 5 min.
9. Blocking: Incubate sections with 2% normal horse serum in PBS for 30 min to block non-specific binding of secondary immunoglobulin.
10. Primary antibody: Incubate sections with mouse anti-Bcl-2 antibody diluted in PBS overnight at 4°C at the concentration suggested by the manufacturer.
11. Rinse in PBS for 3×5 min.
12. Secondary antibody: Incubate sections with biotinylated horse anti-mouse IgG diluted in PBS for 30 min at room temperature.
13. Rinse in PBS for 3×5 min.
14. Avidin-biotin-complex (ABC): Incubate sections with HRP-strepatavidin reagent diluted 1:400 in PBS

for 30 min at room temperature (or use ABC kit following instructions of the kit).

15. Rinse in PBS for 3×5 min.

16. DAB: Incubate sections with DAB solution for 2–10 min.

17. Rinse in distilled water 2×5 min.

18. Counterstain with hematoxylin.

19. Rinse in running tap water for 5 min.

20. Dehydrate through 50% ethanol for 5 min followed by 70%, 85%, 90%, and 95% ethanol for 5 min.

21. Clear in xylene for 2×5 min.

22. Coverslip with mounting medium.

Control tissue: normal human tonsil

RESULTS AND DISCUSSION

Bcl-2 Expression in Normal Lung Tissues and Preinvasive Bronchial Lesions

Bcl-2 immunostaining is cytoplasmic and granular and is restricted in normal bronchial epithelium to the basal epithelial layer and to some epithelial cells that are perpendicularly oriented to the basal lamina and represent another variant of reserve cells, in addition to those basally located.

Bronchial epithelial cell transformation proceeds through stepwise morphological changes, including hyperplasia, metaplasia, dysplasia of progressive severity (mild, moderate, and severe), and carcinoma *in situ* (CIS) (Auerbach *et al.*, 1979). Basal hyperplasia and squamous metaplasia are common reactive lesions, whereas dysplasia and CIS are preinvasive lesions that could regress spontaneously (Hammond *et al.*, 1991). Hyperplasia shows the same staining pattern of normal bronchial epithelium, whereas Bcl-2 overexpression occurred frequently in metaplasia, dysplasia, and CIS (Brambilla *et al.*, 1998). Specifically, the prominent basal staining pattern seen in histologically normal epithelium was lost in preinvasive lesions from metaplasia to CIS, whereas a heterogeneous Bcl-2 expression was observed in suprabasal layers (Walker *et al.*, 1995).

Bcl-2 Expression in Cancer SCLC

Several immunohistochemical data have demonstrated Bcl-2 expression in 60–90% of cases of SCLC (Ben-Ezra *et al.*, 1994; Jiang *et al.*, 1997; Stefanaki *et al.*, 1998). This is consistent with experiments indicating Bcl-2 overexpression in SCLC cell lines (Ben-Ezra *et al.*, 1994; Ikegaki *et al.*, 1994).

It has been hypothesized that Bcl-2 overexpression in SCLC may promote cell survival in response to chemotherapeutic drugs and account for SCLC chemoresistance, particularly after tumor recurrence. This idea is supported by experimental data showing that Bcl-2 can modulate the apoptosis induced by anticancer drugs by inhibiting the biochemical pathway activated in response to DNA damage (Fisher *et al.*, 1993).

Bcl-2 Expression in NSCLC

Despite several studies, the role of Bcl-2 in NSCLC is still controversial. The first study showed Bcl-2 overexpression in 25% of squamous cell carcinomas and 12% of adenocarcinomas (Pezzella *et al.*, 1993), demonstrating that the 5-year survival was higher among patients with Bcl-2–positive tumors both in the whole group and in the group with squamous cell carcinoma alone. However, in a series of 126 T1N0M0 NSCLCs, no significant difference in clinical outcome was observed between Bcl-2–positive (63%) and -negative (59%) cases (Ritter *et al.*, 1995). Finally, another study (Pastorino *et al.*, 1997) conducted on 515 cases of stage I NSCLC failed to disclose any significant prognostic role of Bcl-2 in the whole series. However, although not significant, the risk of recurrence within the subset of pT1N0 patients was eight times lower in Bcl-2–positive cases.

Bcl-2 as a Prognostic Factor in Lung Cancer

A prognostic factor is a variable measured in individual patients that, alone or in combination with other factors, explains part of the heterogeneous population, and is at the time of diagnosis able to provide information about clinical outcome (Yip *et al.*, 2000). In lung cancer, the prognostic factor currently used are clinical variables such as performance status or disease extent (Paesmans *et al.*, 2000). Analysis and characterization of proteins involved in cancer development may help identify novel prognostic factor more closely related to tumor biology. Moreover, the histopathological biomarkers might be useful in risk assessment and provide intermediate end points for chemopreventive trials (Jeanmart *et al.*, 2003). A recent review of the literature with meta-analysis revealed that Bcl-2 overexpression is a good prognostic factor for survival in patients with NSCLC (Martin *et al.*, 2003). This result is encouraging and supports the development of properly designed prospective studies to demonstrate the usefulness of markers like Bcl-2 in lung cancer, as assessed by immunohistochemistry. Unfortunately, the possibility of carrying out large studies is strongly limited by several biases. First, the technique used to identify overexpression of Bcl-2 status may vary in different studies. The IHC is not always performed with the same antibody and the reaction of epitope unmasking is not always conducted. Moreover, the cutoff for the number of positive cells defining a tumor with Bcl-2 overexpression is often arbitrary and varies according to investigators, leading to seriously biased conclusions.

No conclusion has been drawn until now on the role of Bcl-2 in SCLC and neuroendocrine lung cancer. This is due to the small number of patients included in these trials; thus, further studies are needed.

Concluding Remarks

Bcl-2 overexpression, as assessed by immunohisto-chemistry, may represent an independent prognostic parameter of clinical outcome in patients with NSCLC. Patients with Bcl-2-positive tumors had significantly better survival than those with Bcl-2-negative tumors. This result is potentially important and opens perspectives for identifying high-risk patients that may benefit from a more specific therapy or could be differently stratified in randomized trials.

The mechanism underlying the effect of Bcl-2 expression on prognosis remains unknown. The process of apoptosis may involve pro- and antiapoptotic proteins that interact to balance the entire process of programed cell death. Thus, the analysis of only Bcl-2 in tissue should be performed with caution and studies evaluating the combination of these proteins should be conducted to assess their impact on clinical outcome.

References

Adams, J.M., and Cory, S. 1998. The Bcl-2 protein family: Arbiters of cell survival. *Science 281:*1322–1326.

Aisemberg, A.C., Wilkes, D.M., and Jacobson, J.O. 1988. The Bcl-2 gene is rearranged in many diffuse B cell lymphomas. *Blood 71:*969–972.

Auerbach, O., Hammond, E.C., and Garfinkel, I. 1979. Changes in bronchial epithelium in relation to cigarette smoking 1955–1960 vs 1970–1977. *N. Engl. J. Med. 300:*381–386.

Baldi, A., Esposito, V., De Luca, A., Howard, C.M., Mazzarella, G., Baldi, F., Caputi, M., and Giordano, A. 1996. Differential expression of the Retinoblastoma gene family members. Rb/p105, p107, Rb2/p130 in lung cancer. *Clin. Cancer Res. 2:*1239–1245.

Baldi, A., Esposito, V., De Luca, A., Fu, Y., Meoli, I., Giordano, G.G., Caputi, M., Baldi, F., and Giordano, A. 1997. Differential expression of Rb2/p130 and p107 in normal human tissues and in primary lung cancer. *Clin. Cancer Res. 3:*1691–1697.

Ben-Ezra, J.M., Kornstein, M.J., Grimes, M.M., and Krystal, G. 1994. Small cell carcinomas of the lung express the Bcl-2 protien. *Am. J. Pathol. 145:*1036–1040.

Brambilla, E., Gazzeri, S., Lantuejoul, S., Coll, J.L., Moro, D., Negoescu, A., and Brambilla, C. 1998. p53 mutant immuno-phenotype and deregulation of p53 transcription pathway (Bcl-2, Bax and Waf1) in precursor bronchial lesion of lung cancer. *Clin. Can. Res. 4:*1609–1618.

Breitschopf, K., Haendeler, J., Malchow, J., Zeiher, A.M. and Dimmeler, S. 2000. Posttranslational modification of Bcl-2 facilitates its proteasome-dependent degradation. Molecular characterization of the involved signaling pathway. *Mol. Cell Biol. 20:*1886–1896.

Bronner, M.P., Culin, C., Reed, J.C., and Furth, E.E. 1995. The Bcl-2 protooncogene and the gastrointestinal epithelial tumor progression model. *Am. J. Pathol. 146:*20–26.

Brown, J.M., and Wouters, B.G. 1999. Apoptosis, p53, and tumor cell sensitivity to anticancer agents. *Cancer Res. 59:*1391–1399.

Caputi, M., Groeger, A.M., Esposito, V., De Luca, A., Masciullo, V., Mancini, A., Baldi, F., Wolner, E., and Giordano, A. 2002. Loss of pRb2/p130 expression is associated with unfavorable clinical outcome in lung cancer. *Clin. Can. Res. 8:*3850–3856.

Claudio, P.P., Howard, C.M., Pacilio, C., Cinti, C., Romano, G., Minimo, C., Maraldi, N.M., Minna, J.D., Gelbert, L., Leoncini, L., Tosi, G.M., Micheli, P., Caputi, M., Giordano, G.G., and Giordano, A. 2000. Mutations in the Retinoblastoma-related gene Rb2/p130 in lung tumors and suppression of tumor growth *in vivo* by retrovirus-mediated gene transfer. *Cancer Res. 60:*372–382.

Colombel, M., Symmans, F., Gil, S., O'Toole, K.M., Chopin, D., Benson, M., Olsson, C.A., Korsmeyer, S., and Byttyan, R. 2000. Detection of the apoptosis-suppressing oncoprotein Bcl-2 in hormone-refractory human prostate cancers. *Am. J. Pathol. 143:*390–400.

Dimmeler, S., Breitschopf, K., Haendeler, J., and Zeiher, A.M. 1999. Dephosphorylation targets Bcl-2 for ubiquitin-dependent degradation: A link between the apoptosome and the protea-some pathway. *J. Exp. Med. 189:*1815–1822.

Dosaka-Akita, H., Hu, S., Fujino, M., Harada, M., Kinoshita, I., Xu, H.J., Kuzumaki, N., Kawakami, Y., and Benedict, W.F. 1997. Altered Retinoblastoma protein expression in non-small cell lung cancer: Its synergistic effect with altered Ras and p53 protein status on prognosis. *Cancer 79:*1329–1337.

Esposito, V., Baldi, A., De Luca, A., Groger, A.M., Loda, M., Giordano, G.G., Caputi, M., Baldi, F., Pagano, M., and Giordano, A. 1997. Prognostic role of the cyclin-dependent kinase inhibitor p27 in nonsmall cell lung cancer. *Cancer Res. 57:*3381–3385.

Fisher, T.C., Milner, A.E., Gregory, C.D., Jackman, A.L., Aberne, W., Hartley, J.A., Dive, C., and Hickman, J.A. 1993. Bcl-2 modulation of apoptosis induced by anticancer drugs: Resistance to thymidi-late stress is independent of classical resistance pathways. *Cancer Res. 53:*3321–3326.

Ferron, P.E., Bagni, I., Guidoboni, M., Beccati, M.D., and Nenci, I. 1997. Combined and sequential expression of p53, Rb, Ras, and Bcl-2 in bronchial preneoplastic lesions. *Tumori 83:*587–593.

Fontanini, G., Vignati, S., Bigini, D., Mussi, A., Lucchi, M., Angeletti, C.A., Basolo, F., and Bevilacqua, G. 1995. Bcl-2 pro-tein: A prognostic factor inversely correlated to p53 in nonsmall cell lung cancer. *Br. J. Cancer 71:*1003–1007.

Green, D.R., and Reed, J.C. 1998. Mithocondria and apoptosis. *Science 281:*1309–1311.

Greenblatt, M.S., Bennett, W.P., Hollstein, M., and Harris, C.C. 1994. Mutations in the p53 tumor suppressor gene: Clues to cancer etiology and molecular pathogenesis. *Cancer Res. 54:*4855–4878.

Hammond, W.G., Teplitz, R.L., and Benfield, J.R. 1991. Variable regression of experimental bronchial preneoplasia during carcinogenesis. *J. Thorac. Cardiovasc. Surg. 101:*800–806.

Hockenbery, D., Numez, G., Milliman, C., Schreiber, R.D., and Korsmeyer, S. 1990. Bcl-2 is an inner mitochondrial membrane protein that blocks programmed cell death. *Nature 348:* 334–336.

Huang, D.C.S., and Strasser, A. 2000. BH3-only proteins-essential initiators of apoptotic cell death. *Cell 103:*839–842.

Ikegaki, N., Katsumata, M., Minna, J., and Tsujimoto, Y. 1994. Expression of Bcl-2 in small cell lung carcinoma cells. *Cancer Res. 54:*6–8.

Jager, R., Herzer, U., Schenkel, J., and Weiher, H. 1997. Overexpression of Bcl-2 inhibits alveolar cell apoptosis during involution and accelerates C-Myc-induced tumorigenesis of the mammary gland in transgenic mice. *Oncogene 15:*1787–1795.

Jeanmart, M., Lantuejoul, S., Fievet, F., Moro, D., Sturm, N., Brambilla, C., And Brambilla, E. 2003. Value of immunohistochemical markers in preinvasive bronchial lesions in risk assessment of lung cancer. *Clin. Can. Res. 9:*2195–2203.

Jiang, S.X., Kuwao, S., Kameya, T., Sato, Y., Yanase, N., Yoshimura, H., and Kodama, T. 1997. Bcl-2 protein expression in lung cancer and close correlation with neuroendocrine differentiation. *Am. J. Pathol. 148:*837–846.

Jong, D., Prins, F.A., Masson, D.Y., Reed, J.C., Van Omen, G.B., and Kluin, P.M. 1994. Subcellular localization of the Bcl-2 protein in malignant and normal lymphoid cells. *Cancer Res. 54:*256–260.

LeBrun, D.P., Warnke, R.A., and Cleary, M.L. 1993. Expression of Bcl-2 in fetal tissues suggests a role in morphogenesis. *Am. J. Pathol. 142:*743–753.

Linette, G.P. Hess, J.L., Sentman, C.L., Korsmeyer, S.J. 1995. Peripheral T-cell lymphomas in lckpr-Bcl-2 transgenic mice. *Blood 86:*1255–1260.

Martin, B., Paesmans, M., Berghmans, T., Branle, F., Ghisdal, L., Mascaux, C., Meert, A.P., Steels, E., Vallot, F., Verdebout, J-M., Lafitte, J-J., and Sculier, J-P. 2003. Role of Bcl-2 as a prognostic factor for survival in lung cancer: A systematic review of the literature with meta-analysis. *Br. J. Cancer 89:*55–64.

McDonnell, T.J., and Korsemeyer, S.J. 1991. Progression from lymphoid hyperplasia to high-grade malignant lymphoma in mce transgenic for the t(14;18). *Nature 349:*254–256.

Mitsudomi, T., Oyama, T., Kusan, T., Osali, T., Nakanishi, R., and Shirakusa, T. 1993. Mutations of the p53 gene as a predictor of poor prognosis in patients with nonsmall cell lung cancer. *J. Natl. Cancer Inst. 85:*2018–2023.

Miyashta, T., Harigai, M., Hanada, M., and Reed, J.C. 1994. Identification of a p53-dependent negative response element in the Bcl-2 gene. *Cancer Res. 54:*3131–3135.

Paesmans, M., Sculier, J-P., Lecomtre, J., Thiriaux, J., Libert, P., Sergysels, R., Bureau, G., Dabouis, G., Van Custem, o. Mommen, p., Ninane, V., Klastersky, J. For the European Lung Cancer working party. 2000. Prognostic factors in patients with small cell lung cancer: Analysis of a series of 763 patients included in four consecutive prospective trials and with a minimal 5-year follow-up duration. *Cancer 89:*523–533.

Pastorino, U., Andreola, S., Tagliabue, E., Pezzella, F., Incarbone, M., Sozzi, G., Buyse, M., Menare, S., Pienotti, M., and Rilke, M. 1997. Immunocytochemical markers in stage I lung cancer: Relevance to prognosis. *J. Clin. Oncol. 15:*2858–2865.

Pezzella, F., Turley, H., Kuzu, I., Tungekar, M.F., Dunnil, M.S., Pierce, C.B., Harris, A., Gatter, K.C., and Mason, D.Y. 1993. Bcl-2 protein in nonsmall cell lung carcinoma: Immunohistochemical evidence for abnormal expression and correlation with survival. *N. Engl. J. Med. 329:*690–694.

Pilotti, S., Collini, P., Rilke, F., Cattoretti, G., Del Bo, R., Pienotti, M.A. 1994. Bcl-2 Protein expression in carcinomas originating from the follicular epithelium of the thyroid gland *J. Pathol. 172:*337–342.

Reissmann, P.T., Koga, H., Takahashi, R., Figlin, R.A., Holmes, E.C., Piantadosi, S., Cordon-Cardo, C., and Slamon, D.J. 1993. Inactivation of the Retinoblastoma susceptibility gene in nonsmall cell lung cancer. *Oncogene 8:*1913–1919.

Ritter, J.H., Dresler, C.M., and Wick, M.R. 1995. Expression of Bcl-2 protein in stage T1N0M0 nonsmall cell lung carcinoma. *Hum. Pathol. 26:*1227–1232.

Salgia, R., and Skarin, A.T. 1998. Molecular abnormalities in lung cancer. *J. Clin. Oncol. 16:*1207–1217.

Schmitt, C.A., Rosenthal, C.T., and Lowe, S.W. Genetic analysis of chemoresistance in primary murine lymphomas. *Nat. Med. 6:*1029–1035.

Sekido, Y., Fong, K.M., and Minna, J.D. 1998. Progress in understanding the molecular pathogenesis of human lung cancer. *Biochem. Biophys. Acta. 1378:*F21–F59.

Shimizu, E., and Sone, S., 1997. Tumor suppressor genes in human lung cancer. *J. Med. Investig. 44:*15–24.

Shopland, D.R., Eyre, H.J., and Pechacek, T.F., 1991. Smoking attributable cancer mortality in 1991: Is lung cancer now the leading cause of death among smokers in the United States? *J. Natl. Cancer Inst. 83:*1142–1148.

Silvestrini, R., Veneroni, S., Daidone, M.G., Benini, E., Boracchi, P., Mezzetti, M., Di Fronzo, G., Rilke, F., and Veronesi, U. 1994. The Bcl-2: A prognostic indicator strongly related to p53 protein in lymph-node-negative breast cancer patients. *J. Natl. Cancer Inst. 86:*499–504.

Stefanaki, K., Rontogiannis, D., Vamvouka, C., Bolioti, S., Chaniotis, V., Sotsiou, F., Vlychou, M., Delidis, G., Kakolyris, S., Georgoulias, V., and Kanavaros, P. 1998. Immunohistochemical detection of Bcl-2, p53, mdm-2, and p21/wafl proteins in small cell lung carcinomas. *Anticancer Res. 18:* 1689–1696.

Strasser, A., Harris, A.W., and Cory, S. 1993. Eμ-Bcl-2 transgene facilitates spontaneous transformation of early pre-B and immunoglobulin secreting cells but not T cells. *Oncogene 8:*1–9.

Strasser, A., Harris, A.W., Jacks, T., and Cory, S. 1994. DNA damage can induce apoptosis in proliferating lymphoid cells via p53-independent mechanisms inhibitable by Bcl-2. *Cell 79:*329–339.

Traver, D., Akashi, K., Weissman, I.L., and Lagasse, E. 1998. Mice defective in two apoptotic pathways in the myeloid lineage develop acute myeloblastic leukemia. *Immunity 9:*47–57.

Tsujimoto, Y., Finger, L.R., Yunis, J., Nowell, P.C., and Croce, C.M. 1984. Cloning of the chromosome breakpoint of neoplastic B cells with the t(14;18) chromosome translocation. *Science 226:*1097–1099.

Tsujimoto, Y., and Croce, C.M. 1986. Analysis of the structure, transcripts, and protein products of Bcl-2 the gene involved in human follicular lymphoma. *Proc. Natl. Acad. Sci. USA 83:*5214–5218.

Virmani, A.K., Fong, K.M., Kodagoda, D., McIntire, D., Hung, J., Tonk, V., Minna, J.D., and Gazdar, A.F. 1998. Allelotyping demonstrates common and distinct patterns of chromosomal loss in human lung cancer types. *Genes Chromosom. Cancer 21:*308–319.

Walker, C., Robertson, L., Myskow, M., and Dixon, G. 1995. Expression of the Bcl-2 protein in normal and dysplastic bronchial epithelium and in lung carcinomas. *Br. J. Cancer 72:*164–169.

Wei, M.C., Lindsten, T., Mootha, V.K., Weiler, S., Gross, A., Ashiya, M., Thompson, C.B., and Korsmeyer, S.J. 2000. tBID, a membrane-targeted death ligand, oligomerizes BAK to release cythocrome c. *Genes Dev. 14:*2060–2071.

Wiest, J.S., Franklin, W.A., Drabkin, H., Gemmill, R., Sidransky, D., and Anderson, M.W. 1997. Genetic markers for early detection of lung cancer and outcome measures for response to chemoprevention. *J. Cell Biochem. 29:*64–73.

Yip, D., and Harper, P.G. 2000. Predictive and prognostic factors in small cell lung cancer: Current status. *Lung Cancer 28:*173–185.

IV

Breast Carcinoma

1

Breast Carcinoma: An Introduction

M.A. Hayat

Breast Carcinoma: An Introduction

Breast cancer is the leading cause of cancer among women worldwide in the age group of 35–55 years. This cancer occurs at high frequency and affects one in nine Western women; it comprises 18% of all cancers in females. There are ~200,000 new cases of breast cancer diagnosed every year, and ~50,000 women die annually from this cancer in the United States. The majority of these women are diagnosed with invasive breast cancer, and the remaining with *in situ* carcinoma. Despite important advances in therapy, more than half of the affected patients suffer from relapse. Although breast cancer is infrequent in men, in the United States, it is estimated that 1500 men develop this cancer yearly, and ~400 may die from this disease. The lifetime risk of being diagnosed with breast cancer in men is ~0.11% compared with ~13% in women.

The primary reason for the high mortality rate is the highly heterogeneous nature of this disease. Various pathological breast cancer subclasses have markedly different clinical courses and treatment responses. Thus, breast cancer subclasses need to be further defined by genetic markers to improve diagnosis and therapy and follow-up strategies. Although limited information is available on the genetic events implicated in tumor development and progression, common hallmarks of cancer cells include oncogene activation and/or loss of tumor suppressor gene function, as well as karyotypic

mutations. Some of these genetic markers are discussed in this volume. Table 20 shows most of the important biomarkers.

Approximately 8–10% of breast cancer cases are inheritable, which may be caused by mutation of tumor suppressor genes. Gene mutations result in altered mRNA and protein levels. A number of such genes have been identified; examples are C-Myc, HER-2/neu, p53, BRCA1, and *MUC1* genes. The oncoprotein products of these genes have become the target for novel immunotherapy approaches in the treatment of breast cancer. These proteins have been extensively studied using immunohistochemistry (IHC). It should be noted that all genes previously mentioned also show expression in some normal tissues. Moreover, breast cancer is a disease with extreme complexity. Identifying additional genes that may be up- or down-regulated in breast tumors will help us better understand the process of breast tumorigenesis and provide additional markers for diagnosis and treatment of the disease. These genes are briefly summarized in this chapter; human homolog of the rat neuroglioblastomas (HER-2/neu) is discussed in detail.

p53

The tumor suppressor wild-type p53 gene is called the "guardian of the genome" because it suppresses carcinogenesis *in vivo*. It is located on chromosome 17p and encodes for a nuclear transcription factor.

Handbook of Immunohistochemistry and *in situ* Hybridization of Human
Carcinomas, Volume 1: Molecular Genetics; Lung and Breast Carcinomas

233

Copyright © 2004 by Elsevier (USA)
All rights reserved.

p53 gene mutations or expression alterations are commonly associated with human carcinogenesis, which have been reported in approximately one-third of breast cancers. Expression of the p53 gene is activated in response to DNA damage. The wild-type p53 protein prevents the cell cycle from proceeding from G1- to S-phase in cells with DNA damage, allowing DNA repair. The p53 protein itself also plays a role in DNA repair and in apoptosis. When the p53 gene undergoes mutation, the cell loses these three important controls of the cell life cycle as the p53 protein becomes nonfunctional. Usually loss of one allele and mutation of the other inactivate the p53 tumor suppressor gene. Wild-type p53 protein is rapidly eliminated by virtue of its short half-life, whereas mutant p53 protein has a half-life of many hours.

Immunohistochemical expression of p53 protein is associated with p53 mutation, although p53 alterations may not be detectable using immunoassays. Mutant p53 protein accumulates in the nucleus of neoplastic cells. p53 protein immunoreactivity is associated with poor clinical outcome in breast cancer patients, especially in those with node negative disease. In fact, mutation of p53 has been reported in 20–40% of invasive breast cancer. In invasive breast cancer, the presence of p53 autoantibodies is associated with parameters of poor prognosis such as high nuclear grade and absence of hormone receptors. For patients with breast, colorectal, and lung tumors, the specificity of p53 autoantibodies is close to 100%.

The biochemical activity of p53 required for antiproliferation relies on its ability to bind to specific DNA sequences and function as a transcription factor. The importance of the activation of transcription by p53 is underscored by the fact that the majority of p53 mutations found in tumors are located within the domain required for sequence-specific DNA binding. The p53 functions as transcriptional activator by interacting with coactivators such as the CREB-binding protein. Also, p53 increases the expression of multiple genes, including the cyclin-dependent kinase inhibitor, p21. Increased expression of the p21 gene by p53 correlates with cell cycle by inducing G1 arrest or apoptosis. Immunohistochemical studies indicate that p21 expression is increased with breast epithelial proliferation, hyperplasia, dysplasia, and carcinogenesis. p53 and p21 immunoreactivity is generally inverse, although there are exceptions.

The ability of p53 to induce cell cycle arrest or apoptosis is closely regulated under normal conditions. Some oncogenes and stress signals regulate p53 activity through MDM2. As a negative regulator of p53, MDM2 functions in two ways: 1) it binds to the activation domain of p53 and inhibits its activity to stimulate transcription; and 2) it mediates the degradation of p53.

C-Myc

Amplification of the C-Myc gene and overexpression of its nuclear phosphoprotein has been detected in 25% and 45%, respectively, of primary breast carcinomas. This information suggests that amplification of this gene and overexpression of its protein play an important role in tumor progression from noninvasive to invasive and that it has the potential as a marker of poor prognosis of breast carcinoma.

Human Breast Cancer Gene 1/2 (BRCA1/2)

BRCA1 is a breast cancer susceptibility gene, the mutant form of which predisposes to both breast and ovarian cancers. BRCA1 functions as a classical tumor suppressor gene, and loss of the wild-type allele is required for tumorigenesis in mutation carriers. This gene encodes a multifunctional protein that together with other proteins, contributes to homologous recombination, DNA damage response, and transcriptional regulation. Tumors associated with BRCA1 display aggressive pathological features. A better understanding of the molecular mechanism underlying the functions of BRCA1 and HER-2/neu will lead to the rational design of effective therapeutic targets against biologically aggressive, estrogen receptor-negative breast cancer, which disproportionally affect young women (see Part IV, Chapter 15 in this volume). Unlike BRCA1 mutations, germline mutation of BRCA2 are involved in the development of a considerable number of male breast cancers (see Part IV, Chapters 13, 14, and 15 in this volume).

Mucin 1 Glycoprotein (MUC1)

Mucin production and secretion by specialized epithelial cells is a common mechanism used by mammals to protect the underlying mucosa against various injuries. The expression of mucin genes is cell- and tissue-specific but is subjected to variation during cell differentiation and the inflammatory process and is altered during carcinogenesis. Mucins can be divided into secreted forms (e.g., *MUC2*) and membrane-bound (*MUC1*). Only *MUC1* is discussed here.

MUC1 is a high molecular weight transmembrane glycoprotein with a large extracellular domain. It is found on the luminal side of epithelia in a variety of tissues, including breast, pancreas, sweat glands, lung, and ovaries. A limited amount of *MUC1* enters the blood under normal conditions, but this is often increased in breast cancer patients. In breast cancer, *MUC1* is overexpressed and aberrantly glycosylated (underglycosylated). Such underglycosylation of the core protein in cancer tissues exposes new epitopes on the cell surface that are unique to cancer tissues.

Epithelial cells in normal tissues express membrane-bound *MUC1* at their apical surface facing the lumen of the duct. In carcinoma the localization at the apical surface is lost, and high concentrations of this protein spread out over the whole cell surface. This aberrant expression mediates the initial step in the metastatic cascade of tumor cells by the antiadhesive effects of *MUC1*, which shields the cell surface sterically and electrostatically. Overexpression of *MUC1* interferes cell-substrate and cell-cell adhesion by masking cell-surface integrins and E-cadherin.

MUC1 glycoprotein on breast cancer cells carries shortened carbohydrate chains. The partially deglycosylated *MUC1* structure is recognized by the monclonal antibody SM3 that is being tested for its diagnostic utility.

Role of HER-2/neu Oncogene in Breast Cancer and Other Cancers

The HER-family tyrosine kinases play a central role in the proliferation, differentiation, and development of cells. The family consists of four members: HER-1 (epidermal growth factor receptor), HER-2, HER-3, and HER-4. Each of these receptors has an extracellular region, a single transmembrane region, and a cytoplasmic sequence containing a tyrosine kinase domain and a C-terminal tail (Walker, 1998). The extracellular region has ~40% homology among the four family members and can be further divided into four domains (I-IV). The N-terminal domain I has a sequence similarity to domain III, which is flanked by two cysteine-rich domains, II and IV.

The HER-2/neu (erbB-2) protoonocogene was discovered in the early 1980s and subsequently localized on the long arm of chromosome 17q 21, encoding a 185 kDa transmembrane glycoprotein with tyrosine kinase activity and homology to the epidermal growth factor receptor (EGFR) (Schechter *et al.*, 1984). This gene is present in normal cells as a single copy, is involved in the regulation of normal cell growth and division, and is expressed at low levels in many normal epithelial cells. Amplified HER-2/neu gene and its overexpressed protein product are found in many types of cancers, including breast, ovary, lung, pancreas, stomach, and renal.

Amplification of this gene occurs in ~15–30% of invasive breast carcinomas. This results in the overexpression of the corresponding protein receptor (p185), leading to proliferative responses in the tumor cells via a trosine-kinase–mediated signal. Clinically, such overexpression and/or amplification portends a worse prognosis for the patients in terms of both a shorter disease-free interval and overall survival. HER-2/neu abnormalities are not only a prognostic factor but also a resistance factor against hormone therapy. The role of this gene in breast cancer is discussed further in this chapter and its role in other cancer types is also reviewed.

HER-2/neu protein is considered to be a ligand orphan receptor because none of the soluble ligands bind to this receptor HER-2/neu protein amplifies the signal provided by other receptors of the HER family by heterodimerizing with them. Ligand-dependent activation of HER-1, HER-3, and HER-4 by EGF or heregulin results in heterodimerization and thereby HER-2 activation. Dimerization is followed by receptor autophosphorylation, a process in which one receptor molecule phosphorylates the other in the dimer. Receptor autophosphorylation creates binding sites for a series of cytosolic proteins endowed with src homology 2 domains, which are recruited to phosphorylated tyrosine residues on the activated receptor (Bhargava *et al.*, 2001). Other cytosolic proteins include a series of adapter proteins, which couple the receptor to the Ras signaling pathway, components of the phosphoinositide 3-kinase pathway, and phospholipase c-γ in the protein kinase C pathway (Marshall, 1995). These systems in concert generate a cascade of responses that ultimately signal irreversible commitment of the cell to enter the S-phase of the cell cycle.

Sixteen years ago, Slamon *et al.* (1987) first identified HER-2/neu gene amplification in breast cancer using Southern hybridization of frozen breast cancer specimens. Subsequently, a correlation was shown between gene amplification and overexpression of its protein product. Although initially controversial, it is established now that HER-2/neu gene amplification and/or protein overexpression predicts shorter disease-free survival or shorter overall survival in both axillary lymph node-negative and node-positive breast cancer (Gullick *et al.*, 1991). In addition, HER-2/neu protein overexpression is associated with responsiveness to cytotoxic chemotherapy and resistance to tomoxifen antiestrogen therapy. The amplification/overexpression of this gene is an independent prognostic and predictive marker of response to therapy with Herceptin. However, the exact mechanism by which HER-2/neu is selectively amplified/overexpressed in some cancers remains poorly understood. Also, the chromosomal machinery through which the oncogene amplification is generated is largely unknown.

Detection of HER-2/neu Amplification and Protein Expression

The status of an individual gene, its mRNA, and protein product should be routinely used in the rational practice of molecular medicine. Such information should be the basis for diagnosis and therapy. Methods used to determine HER-2/neu status include assays based on the extraction of nucleic acids (Southern and Northern Blots, quantitative polymerase chain reaction), protein (Western Blots), and overexpression of message (RNA mRNA *in situ* hybridization). Although Southern and Western Blotting are being used, they are not well-suited for routine diagnostic studies. These techniques have been mostly replaced with IHC and fluorescence *in situ* hybridization (FISH). Overexpression of the HER-2/neu protein is identified by IHC, and FISH is applied to detect HER-2/neu gene amplification. These two methods are used most commonly to diagnose breast cancer microscopically.

The main advantages of the immunohistochemical method compared with the FISH procedure are its rapidity, simplicity, and lower cost. In addition, its results can be compared with those obtained routinely in other laboratories. Another advantage of this method is that the analyte is studied *in situ* and visualized in the context of cell morphology. It can also be used with fresh, frozen, or paraffin-embedded tissue samples, including archival specimens. In the case of using Herceptin, this method is the ideal approach because this antibody targets the plasma membrane protein and not DNA or RNA.

However, IHC has certain technical problems, including sensitivity differences between different antibodies used and tissue pretreatments (Hayat, 2002). Lack of quantification in conventional immunohistochemistry is another problem. Although standardized reagent kits (HercepTest) are commercially available, the results are not always consistent. The subject of IHC is also discussed in Part I of this volume.

An alternative to IHC is FISH, the advantages of which are presented in Part I of this volume. FISH is extensively used for detecting HER-2/neu gene amplification. According to some studies, this technique yields better sensitivity and specificity than those provided by IHC (Press *et al.*, 2002). The FISH technique provides quantitation of the number of gene copies in the cancer cell nucleus. Both freshly frozen and paraffin-embedded tissues can be used with this technique. It can be carried out either using single-color (HER-2 probe only) or dual-color hybridization (using HER-2 and chromosome 17 centromere probes) simultaneously. The two-color protocol makes it easier to distinguish true HER-2/neu amplification from chromosomal aneuploidy.

Study of cultured cells, tissue fragments, and imprint touch specimens from tumors can be conducted with FISH, but use of tissue sections might create difficulty in obtaining the quantification because of nuclear truncation as a result of sectioning. Another practical limitation is the unavailability of the epifluorescence microscope in most diagnostic laboratories.

Although a vast majority of studies indicate that HER-2/neu gene amplification is closely related to the overexpression of its protein product, the detection rates for IHC reported by clinical laboratories have not been uniform. Such lack of agreement among various studies carried out in different laboratories is related to the differing sensitivities and specificities of the assay types used. Whether assays were carried out with or without antigen retrieval will affect immunohistochemical detection. Differences in the subjective interpretation of immunohistochemical images will play a role in the final assessment. Whether the assessment was carried out by one or more than one person may also influence the accuracy of the results. Such striking variability in interpretation of FISH images has not been demonstrated.

Discordance between FISH results and IHC with respect to HER-2/neu gene amplification and protein overexpression is not uncommon (Persons *et al.*, 1997). A recent study has demonstrated that HER-2 protein staining was scored as 2+ or 3+ positive by immunohistochemistry, whereas the tumors demonstrated no gene amplification (Grushko *et al.*, 2002). In some such studies, false-positive staining cannot be ruled out. In addition, besides gene amplification, other mechanisms to induce protein overexpression might be present. However, in the absence of HER-2/neu gene amplification, the expression of its protein product is low and difficult to detect with immunohistochemistry.

To resolve this discrepancy, Press *et al.* (2002) used four antibodies for IHC and two antibodies for FISH to test for HER-2/neu alterations in breast cancer tissues. Two protocols approved by the Food and Drug Administration (FDA) for FISH are 1) PathVysion (Vysis, Inc., Downers Grove, IL) and 2) Ventana InformHer (Ventana Medical Systems, Inc., Tucson, AZ). Four antibodies used for IHC are 1) R60 rabbit polyclonal antibody, 2) 10H8 mouse monoclonal antibody, 3) Herceptin, and 4) CB11 monoclonal antibody. According to this study, the FISH assays were highly sensitive and specific for detecting HER-2/neu gene amplification, whereas immunohistochemical assays were not significantly different from the FISH assays for identifying protein overexpression. However, two commercially available, FDA-approved immunohistochemical assays (CB11 and Herceptin) were less sensitive at detecting protein overexpression. Studies by Tubbs *et al.* (2001) also report that determination of HER-2/neu gene copy number by FISH may be a more accurate and reliable method for selecting patients eligible for Herceptin therapy. It should be noted that normal cells have only one or two copies of the HER-2/neu gene.

Although FISH is an excellent method for profiling gene amplification *in situ*, correlation with tissue morphology is difficult because of dark-field visualization. Recently, a new, bright-field gold-based autometallographic method (GOLDFISH) was introduced for detecting amplified HER-2/neu oncogene in sections of paraffin-embedded breast cancer tissues (Tubbs *et al.*, 2002). The advantage of this method is that simultaneous gene amplification and morphological changes can be examined with a conventional bright-field microscope. Additional studies are awaited to evaluate the reproducibility of this novel method.

Herceptin (Trastuzumab)

The FDA has approved an immunotherapeutic antibody for women with metastatic, HER-2/neu overexpressing breast carcinoma. This is a recombinant, humanized anti-HER-2/neu monoclonal antibody Herceptin (trastuzumab). Clinical trials in various laboratories have shown the potential benefits of Herceptin therapy for patients with invasive breast carcinoma that overexpresses the protein product of amplified HER-2/neu oncogene. Herceptin is specifically directed against the extracellular domain of this protein and down-regulates the expression of this cell surface protein. This therapy inhibits the growth of HER-2/neu gene amplifying tumor cells, resulting in tumor regression. In other words, Herceptin directly targets the protein and indirectly targets the oncogene.

The HercepTest (Dako) yields significantly more accurate data compared with those achieved by using other antibodies such as CB11 and TAB250 clones. Also, the level of agreement in evaluations by different laboratories using the HercepTest is excellent on cell lines as well as tumor tissues. This test yields the highest level of reproducibility in assay sensitivity. Clinical laboratory testing for HER-2/neu gene and/or protein status largely determines which cancer patients will receive Herceptin. Another advantage of using Herceptin is that it may enhance the antitumor activity of drugs such as Taxol and anthracyclines. Treatment with Herceptin is most effective when used in combination with chemotherapy or in sequence after chemotherapy.

However, stringent quality control and an ongoing quality assurance program using a standard reference material improve the reliability of immunohistochemical results for HER-2/neu, irrespective of the antibody used. Rhodes *et al.* (2002a, b) have developed a quality control system consisting of four cell lines with differing levels of HER-2/neu expression and with known HER-2/neu gene status. This system provides a biologic standard against which HER-2/neu immunohistochemical assay sensitivity can be accurately evaluated, irrespective of the antibody used. This approach is important

in determining the level of interlaboratory agreement when conducting the same assay on the same specimens, ensuring the same quality of service to patients with breast cancer, regardless of where the patient is treated.

HER-2 Status in Breast Carcinoma Using CISH, FISH, and IHC

The HER-2/neu oncogene is a member of the EGFR family, and its amplification is known to be one of the most common genetic alterations associated with human breast cancer and some other cancers (Sahin, 2000). Identification of HER-2 status is important for determining prognosis of patients that have invasive breast cancer as well as selecting subgroups that have HER-2 overexpression in metastasis for therapy with trastuzumab (Herceptin, Genetech, Inc., South San Francisco, CA) (Shak, 1999), a humanized monoclonal antibody to the HER-2 protein. Trastuzumab has been found to be effective only in patients whose tumors show HER-2 gene amplification and/or HER-2 protein overexpression. Therefore, accurate, consistent, and straightforward methods for evaluating HER-2 status have become increasingly important.

Currently, IHC and FISH of formalin-fixed tissues are the two methodologies approved by the FDA for use in HER-2 testing. To accurately identify gene status by bright-field microscopy and overcome FISH limitations, chromogenic *in situ* hybridization (CISH) using the HER-2 DNA probe generated by SPT HER-2 DNA probe (Zymed) was first reported to evaluate against FISH (PathVision, Vysis, Downers Grove, IL) and a HER-2 antibody CB11 (Novocastra, Newcastle, UK) in 157 breast cancer tissue specimens (Tanner *et al.*, 2000). The CISH method was performed on fixed sections, whereas FISH was used on imprints of whole tumor nuclei prepared from fresh tissue samples. Chr.17cen imprints of whole tumor nuclei were prepared from fresh tissue samples. A chr.17cen probe was used in the FISH assay but not in the CISH assay.

The CISH method allows detection of the gene copies through an immunoperoxidase reaction. This shows a good concordance with FISH with regard to HER-2 status in breast carcinoma. The CISH approach can be used as an alternative to FISH for determining gene amplification status in 2+ breast tumors. In addition, CISH is a reliable method to calibrate the IHC procedure, or as a quality control test, to check that the IHC signal is in agreement with the gene status. Figure 35 shows comparative results of CISH and FISH in the breast tumor. For further interpretation of CISH results, see Part I in this volume.

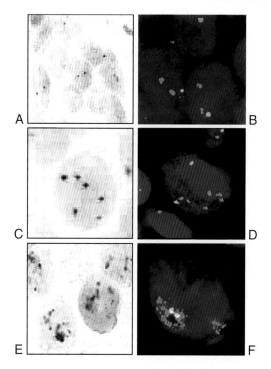

Figure 35 A: CISH, only one or two signals are present in the nucleus of tumor cells. **B:** FISH, only one or two pink dots correspond to HER-2 probe and green dots correspond to centromere 17 probe in similar cells as in A. **C:** CISH, six signals are present in the nucleus. **D:** FISH also shows six pink dots; ratio of gene signals with centromere 17 is 3:1. **E:** CISH, large gene copy clusters are present in the nucleus. **F:** FISH, the nucleus also shows a cluster of gene copies. CISH was performed according to the protocol provided by the supplier (Zymed Inc., South San Francisco, CA), whereas FISH was carried out based on the instructions given by the supplier (PathVysion kit, Vysis, Downers Grove, IL). Arnould *et al.* (2003).

Which is the Best Method to Assess HER-2/neu Status?

HER-2/neu protein is overexpressed in 25–30% of breast carcinomas and the HER-2/neu gene is amplified in ~20% of invasive ductal carcinomas of the breast. Both gene amplification and protein overexpression are associated with decreased disease-free survival and overall survival in node-positive and node-negative patients. Both methods are good predictors of response to Herceptin treatment. Also, protein overexpression is thought to be an important predictor of response to different chemotherapies and resistance to hormonal therapy (Allred *et al.*, 1998). Which method, IHC or FISH, provides better assessment of HER-2/neu status, and thus better diagnosis and prognosis? This question is discussed in this section.

The IHC and FISH methods measure two different targets; the former assesses the overexpression of the plasma membrane receptor protein, whereas the latter allows counting the amplified gene signal numbers. Although both methods are currently being implemented in pathology laboratories, IHC is used much more commonly. However, standardization of immunohistochemical results is difficult. Although FISH is technically more difficult and expensive and requires a special microscope, signal counting is easier to standardize. An agreement between the results of these two methods is common, but discordance may occur in the case of moderately positive (2+) immunohistochemical results.

A recent comparative study reported that it is more important to assess HER-2/neu amplification using FISH than to analyze protein overexpression with IHC (Sauer *et al.*, 2003). Based on this report, failure to detect FISH-amplified but immunohistochemical negative cases would result in diagnostic and prognostic errors, which would have an adverse effect on the survival of patients. On the other hand, tumors showing protein overexpression with IHC without gene amplification belong to a better prognostic group, and failure to detect them is less likely to have a negative effect on the survival of these women. This and some other studies advocate the use of FISH as the primary method of investigating HER-2/neu status. In conclusion, both methods should be used to ensure accurate diagnosis, prognosis, and therapy.

Standard Reference Material for HER-2 Testing

Despite a large number of published studies on using IHC or FISH to measure the HER-2/neu gene and HER-2 protein for selecting patients for therapy, the reliability of the results lacks consensus. Comparative advantages and limitations of these two methods are discussed elsewhere in this chapter. The main reason for this problem is differences in both the processing conditions and the interpretation of results. The impact of methodologic differences cannot be dealt with logically (objectively) to determine which method most reliably selects breast cancer patients that will respond to treatment with Herceptin or other drugs. To address this problem, a workshop was sponsored by the National Institute of Standards and Technology, the FDA, the NCI, and the CAP for recommending a standardized reference material for HER-2 protein testing (Hammond *et al.*, 2003).

The need for such reference material becomes apparent when one considers that to qualify a patient for treatment with Herceptin, a diagnostic test is required. The test must not only have consistent characteristics regardless of the method and testing platform but also produce quantitative results. The availability of such reference material will encourage pathologists to interpret results consistently. Also, concordance between laboratories will improve. At the workshop, the following characteristics of HER-2 reference material were agreed (Hammond *et al.*, 2003).

1. It can be used for all assay methods, including detection of DNA, RNA, and protein expression.
2. It has known DNA/protein content.
3. An infinite supply of identical material is available.
4. It can be easily produced and distributed.
5. Rigorous quality control is carried out in manufacturing and use.
6. The same fixation and other processing protocols are used as for the test.
7. Range of positivity from negative to positivity can be carried out.
8. It mimics test tissue as closely as possible.

The materials that could provide these characteristics include tumor tissue blocks with known HER-2 gene and its protein product, artificial tissues produced from known tissue components (faux tissue), tumor xenografts, and cell lines with differential HER-2 gene protein expression.

Bispecific Antibody

Carcinoembryonic antigen (CEA) and HER-2 are expressed in ~50% and 30% of breast cancers, respectively. Targeting these two antigens simultaneously by a bispecific antibody (BAb) may provide efficient tumor uptake and prolonged tumor residence duration. It is known that multivalent binding can enhance the apparent affinity of an antibody to its target (Hayat, 2002; Terskikh *et al.*, 1997). Although the use of Herceptin has demonstrated encouraging therapeutic results (discussed in this volume), its success is limited by the relatively low percentage of breast cancers expressing HER-2 (Mass, 2000). In fact, Herceptin immunotargetting strategy requires the presence of a specific target highly expressed in the tumor and little or no expression in the non-malignant tissues.

Besides HER-2, other frequently used target antigens in breast cancers include CEA, *MUC1*, and TAG-72. None of these targets can be considered a universal target antigen in breast cancers. To develop a new diagnostic and/or therapeutic tool that could be used in almost all breast cancers, Dorvillius *et al.* (2002) have prepared a BAb that is directed against HER-2 and CEA. By using BIACORE technology, they showed that this BAb recognized both CEA and HER-2. Thus, targeting two distinct tumor-associated antigens on the same cell could improve tumor localization. This BAb has potential as a new candidate for breast cancer targeting.

Breast Cancer Therapy

The association of HER-2/neu gene amplification and protein overexpression with human tumors, its extracellular accessibility, as well as its involvement in tumor aggressiveness are all factors that make this receptor an appropriate target for tumor-specific therapies. A number of approaches are being investigated as possible therapeutic strategies that target HER-2: 1) growth inhibitory antibodies that can be used alone or in combination with standard chemotherapeutics; 2) tyrosine kinase inhibitors (TKI) that have been developed in an effort to block receptor activity because phosphorylation is the key event leading to activation and initiation of the signaling pathway; and 3) active immunotherapy because the HER-2 oncoprotein is immunogenic in some breast cancer patients. Some of the hormonal inhibitors are summarized in this section.

In many patients the progress of breast cancer is dependent on estrogens, which has led to an intensive search for estrogen antagonists. Tamoxifen, a competitive estrogen inhibitor, binds to the ligand-binding site of the estrogen receptors, $ER\alpha$ and $ER\beta$. Since its introduction more than three decades ago, tamoxifen has accumulated more than 20 million patient years of experience and has become widely used in the treatment of estrogen receptor-positive breast cancer, both as adjuvant therapy and for the treatment of advanced disease. Tamoxifen is also thought to be effective for the prevention of breast cancer (Wickerham, 2002). However, although this treatment has been successful, tamoxifen may be detrimental in some patients with HER-2 postive tumors. The women who take tamoxifen are at increase risk for endometrial cancer and blood coagulation disorders and eventually may develop resistance to this drug. These facts have prompted research to find inhibitors that interfere with early stages of steroid biosynthesis as possible alternatives to tamoxifen.

So far, trilostane is the only clinically available inhibitor of this class. Its action is attributed, at least in part, to a noncompetitive inhibition of estrogen receptor functions (Puddefoot *et al.*, 2002). Trilostane reversibly inhibits 3β-hydroxysteroid dehydrogenase in the adrenal cortex. However, this inhibition does not fit with certain observations in breast cancer patients. CI-1033 is another orally available 3-chloro, 4-fluoro, 4-anilinoguinazoline that is a receptor inhibitor. It posseses a morpholine solubilizing side chain at the 7 position and an acryllamide at the 6 position (Allen *et al.*, 2002). In contrast to the majority of receptor inhibitors that target only one or two of the HER-2/neu family members, CI-1033 is a small molecule that inhibits transmembrane receptor tyrosine kinases (RTK). It blocks signal transduction through all four members of the HER-2/neu receptor family with essentially identical potency. Receptor inhibition is irreversible and signaling through this pathway can only be reestablished by the synthesis of new HER-2/neu receptors. Because CI-1033 is a highly selective inhibitor of these receptors, it is expected to have limited nonspecific toxicities. It may have clinical activity in solid tumors that overexpress a member of the HER-2/neu receptors. Although according to Shin *et al.* (2001) and Allen *et al.* (2002), CI-1033 has acceptable side effects at potentially therapeutic dose levels, it requires further testing.

Raloxifene is another estrogen receptor modifier used for estrogen- and progesterone-positive breast cancer patients. Breast cancer patients are also responsive to various degrees to doxorubicin, cyclophosphamide, methotrexate, fluorouracil, and paclitaxel. However, as stated earlier, resistance develops against these hormone inhibitors, a phenomenon that limits the duration of positive response to this type of hormonal therapy. Approximately, one-third of all patients with advanced breast cancer will initially benefit from the antiestrogen therapy, but the majority of these patients will relapse (Gao *et al.*, 2002). It should be noted that the role of HER-2 protein overexpression in determining the sensitivity of cancer cells to drugs is complex, and molecules involved in its signaling pathway are probably the actual protagonists of the sensitivity to drugs.

Role of HER-2/neu Gene Expression in Endometrial Carcinoma

Endometrial carcinoma is the most common female gynecologic malignancy, accounting for nearly half of all cancers of the female genital tract. Although 5-year survival rates up to 80% have been reported, some women still relapse. Recognition of prognostic factors should help to define high-risk patients and subsequently could have an impact on adjuvant therapy. A factor that may influence endometrial cancer patient prognosis is overexpression of the HER-2/neu gene product. As many as 48% of endometrial carcinomas demonstrate elevated HER-2 expression (Rolitsky *et al.*, 1999). Some other studies also report correlation between increased HER-2 expression and endometrial metastasis, decreased survival of these patients, and depth of myometrial invasion (Berchuck *et al.*, 1991; Seki *et al.*, 1998). Unlike, for example, in the case of Barrett carcinoma (Geddert *et al.*, 2002) HER-2/neu gene amplification does not seem to be a late event in the development of endometrial cancer because this gene is found in all clinical stages (Kohlberger *et al.*, 1996). However, some other studies did not find any prognostic

significance of HER-2 overexpression in the endometrial carcinoma (Gassel *et al.*, 1998).

HER-2 and the CD44 transmembrane glycoproteins interact with each other in numerous cell types. This interaction helps to maintain HER-2 activity that contributes to tumor progression. CD44 proteins play roles in cell-cell and cell matrix adhesion, regulation of growth factor signaling, and metastasis. Aberrant CD44 splice variant expression often occurs in human carcinomas, including endometrial tumors (Ayhan *et al.*, 2001). It is also known that CD44 is colocalized and coimmunoprecipitated with HER-2 in ovarian, cervical, and mammary cell carcinoma cell lines (Bourguignon *et al.*, 1997; Wobus *et al.*, 2002). In contrast to other carcinomas, although CD44 and HER-2 are robustly expressed in endometrial carcinoma cells *in vitro* and *in situ*, these proteins do not interact with each other (Wobus *et al.*, 2002). Lack of CD44-HER-2 interactions may reduce the contribution of HER-2 to endometrial carcinoma progression. It is possible that other proteins are required to mediate the association between CD44 and HER-2, and that such accessory proteins are not present in endometrial carcinoma cells. Alternatively, endometrial carcinoma cells may express a protein(s) that prevents CD44-HER-2 interactions. In immunohistochemical studies, mouse-anti-human CD44 antibody Hermes-3 (1:25) and a rabbit-anti-HER-2 antibody (1:50) (Santa Cruz Biotechnology, Santa Crux, CA) can be used overnight at 4°C.

Role of HER-2/neu Gene Expression in Barrett Carcinoma

A metaplasia-dysplasia sequence usually precedes Barrett carcinoma (Haggitt, 1994). Because of chronic gastroesophageal reflux, the initial step in this development is the replacement of normal stratified squamous epithelium in the lower esophagus by metaplastic specialized columnor epithelium, the so-called Barrett esophagus. In patients with Barrett esophagus, the risk of developing adenocarcinoma is ~30–125-fold higher than in the general age-matched population. Therefore, early detection of patients at risk will improve the clinical surveillance of patients with this disease. Reliable criteria for the individual patient's risk or stage of tumor development will also contribute to our understanding of tumor evolution. To accomplish these goals, the best approach is to study amplification of the HER-2/neu gene and overexpression of its protein product. Although distinct histologic discrimination of metaplastic, dysplastic, and invasive stages is of great clinical interest, considerable interobserver and intraobserver disagreement in the assessment of histologic alterations in Barrett esophagus has been reported (Reid *et al.*, 1988).

Only a few studies have compared HER-2/neu amplification and its protein overexpression within the same series of tissue specimens. Al Kasspooles et al. (1993) studied 10 Barrett carcinomas and found HER-2/neu gene amplification in 15.4% and protein overexpression in 60% of cases. Walch et al. (2000) found gene amplification in five of six Barrett carcinomas and protein expression in three of four carcinomas. Two adjacent high-grade dysplasias were negative for amplification and overexpression. In another study from the same laboratory, 25 patients with this disease showed gene amplification and simultaneous protein overexpression in 35% of carcinomas and in 31% of high-grade dysplasia, but not in low-grade dysplasia or intestinal metaplasias (Walch et al., 2001).

Recently, Geddert et al. (2002) examined 39 specimens of metaplastic specialized epithelium (SE), 27 of low-grade dysplasia (LGD), 27 of high-grade dysplasia (HGD), and 46 of adenocarcinoma (CA) derived from Barrett esophagus for HER-2/neu gene amplification using differential polymerase chain reaction and for overexpression of HER-2/neu protein using IHC. Amplification of the gene was: SE, 0%; LGD, 0%; HGD, 11.1%; and CA, 13.6%. Protein overexpression was: SE, 0%; LDG, 7.4%; HGD, 18.5%; and CA, 21.7%. In eight (89%) of nine specimens, the amplification correlated with overexpression. The reverse was true in eight (47%) of 17 specimens, whereas overexpression of the protein appeared more often and earlier. These results show that HER-2/neu gene amplification is a late event in the metaplasia-dysplasia-carcinoma sequence of Barrett esophagus. Gene amplification is regularly accompanied by protein overexpression, but the latter also occurs in the absence of the former. It is suggested that besides gene amplification, other mechanism(s) to induce protein overexpression may exist.

Role of HER-2/neu Gene Expression in Pancreatic Carcinoma

Approximately 28,000 pancreatic cancer cases are diagnosed in the United States every year. It is one of the most aggressive tumors and has an overall five-year survival rate of only 2% (Warshaw et al., 1992), which exists only for a minority of patients with locally limited and surgically resectable tumor. However, 70–80% of patients whose tumor can be surgically extirpated will suffer an incurable local relapse, distant metastases, or peritoneal carcinosis (Henne-Brums et al., 1998). Information on the molecular basis of pancreatic cancer may provide new therapeutic targets and improved clinical outcomes. In this respect, the role of HER-2/neu oncogene in pancreatic cancer is summarized in the following.

The assessment of the prognostic influence of HER-2/neu gene amplification and protein overexpression in the case of pancreatic cancer has produced conflicting data. One study indicated that the overexpression of this protein in pancreatic cancer correlated with shortened survival of patients (Lei et al., 1995).

Thybusch-Bernhardt et al. (2001) have analyzed HER-2 expression and function in pancreatic cancer cell growth in vivo and in vitro and indicated a rate-limiting role for HER-2 in cell proliferation. They suggest that HER-2/neu targeting has a potential role in pancreatic cancer therapy.

Role of HER-2/neu Gene Expression in Prostate Carcinoma

Prostate cancer is one of the most common noncutaneous malignancies in the United States. Approximately 12% of all men have prostate specific antigen (PSA) in the range of 2.5 to 4.0 ng/ml, and presumably the vast majority of them do not have this disease and even fewer would have clinically significant prostate cancer. The disease occurs predominantly in older men and shows a variable course. In some patients the cancer has an indolent course even in the presence of widespread metastases, whereas in others the disease progresses aggressively, resulting in significant morbidity and mortality. Its mortality is surpassed only by that of lung cancer. Despite technical improvements in both surgery and radiotherapy, progression-free 5-year survival rates do not exceed 20–40% in patients with extracapsular disease or with unfavorable prognostic variables, such as high-grade tumors or PSA 10 ng/ml at diagnosis (Lilleby et al., 2003).

The development of the prostate is determined mainly by multiple steroid hormones, polypeptide growth factors, and cytokines. In the regulation of prostatic growth and differentiation, several interactions between epithelial and stromal cells are involved. These interactions also play a crucial role in the development of adenocarcinoma of the prostate. An objective parameter is needed to predict the biological behavior of early stages of prostatic cancer.

Is assessment of the HER-2/neu oncogene amplification to detect early stages of prostate cancer relevant? Has the detection of such amplification any prognostic value for prostatic cancer? An agreement on these questions is lacking. According to Sadashivan et al. (1993), HER-2/neu may be a prognostic marker in this cancer. In contrast, Kuhn et al. (1993) reported that although HER-2/neu oncoprotein was overexpressed in the archival prostate cancer tissues, gene amplification was less than significant. A number of other protocols, such as PSA serum test (Fowler et al., 2001), comparative genomic hybridization (Paris *et al.*, 2003), and digital rectal examination, with their advantages and limitations, are available for prostate cancer diagnosis. One of the more important goals of research on prostate cancer is the identification of the molecular basis for therapeutic resistance by prostate cancer cells. Genetic markers of prostate cancer progression are highly desirable and necessary for improved treatment and survival (see below and Volume 2 of this series of handbooks).

Role of HER-2/neu Expression in Lung Carcinoma

Neoplastic malignancy of the lung is a major cause of death in the Western world. In 1998 in the United States, 172,000 new cases of lung cancer were diagnosed, and 160,000 of patients died of the disease. The mortality rate of lung cancer is among the highest of all cancer types. During the last two decades, the mortality rate was lowered by only ~ 6% as a result of improved therapies. In lung NSCLC, the most common form of lung cancer, HER-2/neu protein overexpression has been observed, although to a variable extent (Giatromanolaki *et al.*, 1996). According to Kristiansen *et al.* (2001), HER-2/neu oncoprotein up-regulation is associated with tumor progression in lung cancer. On the other hand, an immunohistochemical study by Pfeiffer *et al.* (1996) reported a lack of prognostic significance of the HER-2/neu oncoprotein in patients with systemically untreated NSCLC. Cumulative evidence indicates that the importance of HER-2/neu oncoprotein overexpression for prognosis in NSCLC is still controversial. The role of this oncogene in lung cancer biology clearly necessitates further studies.

Breast Cancer Biomarkers

In the United States breast cancer is the second leading cause of cancer deaths in women. In 2002, ~ 205,000 women were diagnosed with this cancer, and ~ 40,000 patients died from the disease. Women have a lifetime risk of ~1 in 8. Both genetic and environmental factors play a role in the development of breast cancer. Understanding the basis for tumor development, progression, the identification of biomarkers for assessment of prognosis, and prediction of therapy outcome are integral parts of current research efforts. Strategies for early detection and breast cancer prevention can be carried out because biomarkers of disease susceptibility are available.

Women at increased genetic risk of breast cancer have long been identified on the basis of their family history. For example, an unaffected woman having a family history of breast cancer is associated with an increased risk for the disease. The identification of several breast cancer predisposition genes (e.g., BRCA1/2; Table 18) has clarified cancer risk assessment for women within families with a history of breast cancer. Genetic testing now makes it possible for some women to be more certain of their risk of developing breast cancer. Optimally, genetic testing can be used, along with a careful pathologic examination of breast cancer specimens, to guide surgical decision making, such as appropriate surgical techniques, prior to prophylactic mastectomy in women with high familial risk of breast cancer.

A large number of studies have examined the relationship between genetic features, including markers and prognosis. Such markers are enumerated in Table 18. Considering the wide variety of these markers it is clear that a number of different pathways exist for the development of breast cancer, and some molecular characteristics correlate with clinicopathological features.

Publication bias is a serious problem in assessing the utility of any given marker because positive associations have a far higher probability of being published than do negative ones. Results that disagree with accepted views may, in part, reflect heterogeneity in breast cancers. Such contradictory results deserve to be accepted, provided that the results were obtained by studying multiple biopsies, DNA sampling, and by careful histological/immunohistochemical examination. Some of the important markers are summarized in the following sections.

The HER-2 receptor is involved with signal transduction and cell proliferation. Approximately 20–30% of breast cancer patients show amplification and/or overexpression of HER-2. Women with HER-2–overexpressing tumors have a significantly lower overall survival rate and a shorter time to relapse than patients whose tumors do not show amplification of this gene. The former patients also show less sensitivity to chemotherapy and endocrine therapy. HER-2 alterations occur in approximately one-third of invasive ductal carcinoma of the breast. Mammary Paget disease also shows a high expression of HER-2 (Anderson *et al.*, 2003). Most patients with advanced breast cancer develop osteolytic bone metastases, which are common causes of morbidity and sometimes mortality.

Primary intratumoral heterogeneity for both HER-2 gene amplification and protein overexpression is not uncommon. This heterogeneity may arise via random genetic alterations with clonal progression, resulting in genetic subclones of cells within the primary tumor (Edgerton *et al.*, 2003). A recent study reveals a complex expression pattern of the four members of the EGF receptor family, suggesting that it is this pattern, rather than the expression of individual family members, that

Table 18 Breast Cancer Biomarkers

Akt	Knuefermann *et al.* (2003)
*Androgen receptor	Selim *et al.* (2002)
Angiogenesis	Shirakawa *et al.* (2002)
*Bag-1	Townsend *et al.* (2002)
*Bcl-2	Callagy *et al.* (2003)
*Bone morphogenetic protein 7	Schwalbe *et al.* (2003)
BRCA1/BRCA2 genes	Osorio *et al.* (2002)
C-KIT receptor	Palmu *et al.* (2002)
*Carbonic anhydrase-9	Span *et al.* (2003)
Carcinoembryonic antigen	Dorvillus *et al.* (2003)
*CD34	Colpaert *et al.* (2003)
*CD44	Wobus *et al.* (2002)
*Chromosome 1 (polysomy)	Nakopoulo *et al.* (2002)
Chromosome 8	Charape-Jauffret *et al.* (2003)
*Chromosome 17 (aneusomy)	Wang *et al.* (2002)
Chromosome 20	Hodgson *et al.* (2003)
*CLDN-7	Kominsky *et al.* (2003)
*Collagen types	Amenta *et al.* (2003)
Cyclin D1	Ormandy *et al.* (2003)
Cyclin D2 (methylation)	Lehmann *et al.* (2002)
*Cyclooxygenase 2	Spizzo *et al.* (2003)
Cytokeratins	Denoux *et al.* (2003)
CYP17 and HSD17B1	Wu *et al.* (2003)
DLC-1 gene	Plaumann *et al.* (2003)
*E-cadherin	Colpaert *et al.* (2003)
EDG1 protein	Wittmann *et al.* (2003)
*Epidermal growth factor receptor	Santini *et al.* (2002)
Epstein-Barr virus gene	Xue *et al.* (2003)
*Estrogen receptor alpha	Suo *et al.* (2001)
*Estrogen receptor beta	Saunders *et al.* (2002)
*Fibroblast growth factor-binding—protein	Kagan *et al.* (2003)
*Fos B protein	Milde-Langosch *et al.* (2003)
Galectin-3	Song *et al.* (2002)
*HER-2/neu	Gupta *et al.* (2003)
*Hyaluronan	Wernicke *et al.* (2003)
Isolated tumor cells	Wiedswang *et al.* (2003)
*Ki-67	Yang *et al.* (2003)
*Maspin	Umekita and Yoshida (2003)
*Met	Ocal *et al.* (2003)
MTI-MMP	Deryugina *et al.* (2003)
MUC1 gene	Porowska *et al.* (2002)
MUC4/5MC	Price-Schiavi *et al.* (2002)
Nuclear Syk	Wang *et al.* (2003)

(Continued)

Table 18 Breast Cancer Biomarkers—cont'd

*P 27	De Paola *et al.* (2002)
*P 53 protein	Tsutsui *et al.* (2002)
Parathyroid hormone–related protein	Hoey *et al.* (2003)
*P-Cadherin	Paredes *et al.* (2002)
*Pdef protein	Feldman *et al.* (2003)
Plasminogen activator inhibitor-1	Offersen *et al.* (2003)
Polymorphonuclear leukocyte elastase	Foekens *et al.* (2003)
PPAR	Suchanek *et al.* (2002)
*Progesterone	Van Poznak *et al.* (2002)
RECK	Span *et al.* (2003)
Sialyl Le ^{x}antigen	Nakagoe *et al.* (2002)
*Stromelysin–3	Nakopoulou *et al.* (2002)
TACC	Lauffart *et al.* (2003)
*Tenacin-C	Tsunoda *et al.* (2003)
*Topoisomerase II α	Järvinen and Liu (2003)
TMS1 (ASC) gene	Virmani *et al.* (2003)
Vascular endothelial growth factor	Pegram and Resse (2002)
*Zonula occludens-1	Bell *et al.* (2003)

*These biomarkers have been identified with IHC or ISH.

should be taken into account when evaluating antitumoral drugs designed to target these receptors (Biéche *et al.*, 2003). It is interesting to note that two hypothetical genes, GRB7 and MLN64, are coamplified and overexpressed with HER-2/neu, contributing to the development and progression of breast cancer (Kauraniemi *et al.*, 2003).

Other important biomarkers are estrogens. They cause proliferation of breast epithelial cells through estrogen-receptor–mediated processes. Rapidly proliferating cells are susceptible to genetic errors during DNA replication; uncorrected, these can ultimately lead to malignancy. In addition, specific oxidative metabolites and conjugates may serve as biomarkers to predict a patient's risk of breast cancer. Quantitation of estrogen and progesterone receptors in breast tumors can be used as prognostic factors and as targets for endocrine therapy.

Two major genes, BRCA1 and BRCA2, are also implicated in inherited predisposition to female breast and ovarian cancers. Germline mutations in these two genes are responsible for 5–10% of all breast cancer cases. The BRCA1 protein may affect the cell cycle and DNA repair by its ability to modulate gene expression at the level of transcription. BRCA1 is known to regulate the expression of many genes, including p21, Gadd 45, cyclin B1, DBB2, and XPC at the level of transcription. The heterogeneity of breast cancer risk among women who carry the same BRCA1 mutation

suggests the possibilities of modifying environmental and genetic factors. For example, the product of the RAD51 gene functions with BRCA1/2 in the repair of DNA breaks (Jakubowska *et al.*, 2003). Thus, RAD51 may be the strongest genetic modifier of breast cancer risk in BRCA1 carriers.

Phosphatase and tensin homologue (PTEN) deleted on chromosome 10 in human cancers is regarded as a candidate tumor suppressor gene. The tumor suppressive effects of this gene are mediated via cell survival, G1 cell cycle entry, and apoptosis. Mutations involving PTEN have been identified in several types of human cancers, including breast, prostate, brain, endometrium, and colorectal cancers, and in cancer cell lines. In addition to structural alterations of the PTEN gene, PTEN may be inactivated by pathways other than mutation or genetic deletion. Notably, epigenetic mechanisms (e.g., methylation) may be involved in breast cancer. Using immunohistochemistry, loss of PTEN expression in primary ductal adenocarcinomas of the breast has been reported (Perren *et al.*, 1999).

Higher levels of circulating insulin-like growth factor-1 (IGF-1) are associated with higher risks for premenopausal breast cancer. The increased levels of the free form of IGF-1 may promote proliferation in the breast epithelium (Dabrosin, 2003). The aforementioned discussion clearly indicates that breast cancer is a multigenic disease, substantiating its profound complexity. The protocols for IHC, ISH, and FISH of important breast cancer diagnostic markers are detailed in Part 4 of this volume.

References

Al-Kassapodes, M., Moore, J.H., Orringer, M.B., and Beer, D.G., 1993. Amplification and overexpression of the EGFR and erbB-2 genes in human esophageal adenocarcinomas. *Int. J. Cancer 54:*213–219.

Allen, L.F., Lenehan, P.F., Eiseman, I.A., Elliot, W.L., and Fry, D.W. 2002. Potential benefits of the irreversible pan-erbB inhibitor, CI-1033, in the treatment of breast cancer. Ann Arbor, MI: Pfizer Global Res. Develop.

Allred, D.C., Harvey, J.M., Berardo, M., and Clark, G.M. 1998. Prognostic and predictive factors in breast cancer by immunohistochemical analysis. *Mod. Pathol. 11:*155–168.

Amenta, D.C., Hadad, S., Lee, M., Barnard, N., Li, D., and Myers, J. 2003. Loss of types XV and XIX collagen precedes basement membrane invasion in ductal carcinoma of the female breast. *J. Pathol. 199:*298–308.

Anderson, J.M., Ariga, R., Govil, H., Bloom, K.J., Francescatti, D., Reddy, V.B., Gould, V.E., and Gattuso, P. 2003. Assessment of HER-2/neu status by immunohistochemistry and fluorescence in situ hybridization in mammary Paget disease and underlying carcinoma. *Appl. Immunohistochem. Mol. Morphol. 11:*120–124.

Arnould, L., Denoux, Y., MacGrogan, G., Penault-Lloorca, F., Fiche, M., Treilleux, I., Mathieu, M.C., Vincent-Salomon, A., Vilain, M.O., and Couturier, J. 2003. Agreement between chromogenic *in situ* hybridization (CISH) and FISH in the determination of HER-2 status in breast cancer. *Br. J. Cancer 88:*1587–1591.

Ayhan, A., Taskiran, C., Celik, C., Aksu, T., and Yuce, K. 2001. Surgical stage III endometrial cancer: Analysis of treatment outcomes, prognostic factors, and failure patterns. *Eur. J. Gynaecol. Oncol. 23:*553–556.

Bièche, I., Onody, P., Tozlu, S., Driouch, K., Vidaud, M., and Lidereau, R. 2003. Prognostic value of ERBB family mRNA expression in breast carcinomas. *Int. J. Cancer 106:*758–765.

Bell, J., Walsh, S., Nusrat, A., and Cohen, C. 2003. Zonula occludens-1 and HER-2/neu expression in invasive breast carcinoma. *Appl. Immunohistochem. Mol. Morphol. 11:*125–129.

Berchuck, A., Rodriguez, G., Kinney, R.B., Soper, J.T., Dodge, R.K., Clarke-Pearson, D.L., and Bast, R.C. 1991. Overexpression of HER-2/neu in endometrial cancer is associated with advanced stage disease. *Am. J. Obstet. Gynecol. 164:*15–21.

Berner, H., Suo, Z., Risberg, B., Villman, K., Karlsson, M., and Nesland, J. 2003. Clinicopathological associations of CD44 mRNA and protein expression in primary breast carcinomas. *Histopathology 42:*546–554.

Bhargava, R., Neem, R., Marconi, S., Luszcz, J., Garb, J., Gasparini, R., and Otis, C.N. 2001. Tyrosine kinase activation in breast carcinoma with correlation to HER-2/neu gene amplification and receptor overexpression. *Hum. Pathol. 32:*1344–1350.

Bourguignon, L.Y., Zhu, H., Chu, A., Lida, N., Zhang, L., and Hung, M.C. 1997. Interaction between the adhesion receptor CD44 and the oncogene product p185 HER-2 promotes human ovarian tumor cell activation. *J. Biol. Chem. 272:* 27,913–27,918.

Callagy, G., Cattaneo, E., Daigo, Y., Happerfield, L., Bobrow, L., Pharoah, P., and Caldas, C. 2003. Molecular classification of breast carcinomas using tissue microarrays. *Diagn. Mol. Pathol. 12:*27–34.

Charafe-Jauffret, E., Moulin, J., Ginestier, C., Bechlian, D., Conte, N., Geneix, J., Adelaide, J., Noguchi, T., Hassoun, J., Jacquemier, J., and Birnbaum, D. 2003. Loss of heterozygosity at microsatellite markers from region p11-21 of chromosome 8 in microdissected breast tumor but not in peritumoral cells. *Int. J. Oncol. 21:*989–996.

Colpaert, C., Vermeulen, P., Benoy, I., Soubry, A., Van Roy, F., van Beest, P., Goovaerts, G., Dirix, L., Van Dam, P., Fox, S., Harris, A., and Van Marck, E. 2003. Inflammatory breast cancer shows angiogenesis with high endothelial proliferation rate and strong E-cadherin expression. *Brit. J. Cancer 88:*718–725.

Dabrosin, C. 2003. Increase of free insulin-like growth factor-1 in normal human breast in vivo late in the menstrual cycle. *Breast Cancer Res. Treat. 80:*193–198.

Denoux, Y., Lebeau, C., Michels, J., and Chasle, J. 2003. Double immunohistochemical labeling technique applied to different types of cytokeratins in epithelial proliferations of the breast. *Biotech. Histochem. 78:*23–26.

DePaola, F., Vecci, A., Granato, A., Liverani, M., Monti, F., Innoceta, A., Gianni, L., Saragoni, L., Ricci, M., Falcini, F., Amadori, D., and Volpi, A. 2002. p27/kip I expression in normal epithelium, benign and neoplastic breast lesions. *J. Pathol. 196:*26–31.

Deryugina, E., Ratnikov, B., and Strongin, A. 2003. Prinomastat, a hydroxamate inhibitor of matrix metalloproteinases, has a complex effect on migration of breast carcinoma cell. *Int. J. Cancer 104:*533–541.

Dorvillus, M., Garambois, V., Pourquier, D., Gutowski, M., Rouanet, P., Mani, J.C., Pugniere, M., Hynes, N.E., and Pelegrin, A. 2002. Targeting of human breast cancer by a bispecific antibody directed against two tumor-associated antigens: erbB-2 and carcinoembryonic antigen. *Tumor Biol. 23:*337–347.

Edgerton, S.M., Moore, D., Merkel, D., and Thor, A.D. 2003. erbB-2 (HER-2) and breast cancer progression. *Appl. Immunohistochem. Mol. Morphol. 11:*214–221.

Feldman, R., Sementchenko, V., Gayed, M., Fraig, M., and Watson, D. 2003. Pdef expression in human breast cancer is correlated with invasive potential and altered gene expression. *Cancer Res. 63:*4626–4631.

Foekens, J., Ries, C., Look, M., Gippner-Steppart, C., Klijn, J., and Jochum, M. 2003. Elevated expression of polymorphonuclear leukocyte elastase in breast cancer tissue is associated with tamoxifen failure in patients with advanced disease. *Br. J. Cancer 88:*1084–1090.

Fowler, J.E., Bigler, S.A., and Farabaugh, P.B. 2001. Prospective study of cancer detection in black and white men with normal digital rectal examination but prostate-specific antigen equal or greater than 4.0 ng/ml. *Cancer 94:*1661–1667.

Gao, Z.O., Gao, Z.P., Fields, J.Z., and Boman, B.M. 2002. Development of cross-resistance to tamoxifen in faloxifene-treated breast carcinoma cells. *Antican. Res. 22:*1379–1384.

Gassel, A.M., Backe, J., Kreb, S., Caffier, H., and Muller-Hermelink, H.K. 1998. Endometrial carcinoma: Immunohistochemically detected proliferation index is a prognosticator of long-term outcome. *J. Clin. Pathol. 51:*25–29.

Geddert, H., Zeriouh, M., Wolter, M., Heise, J.W., Gabbeert, H.E., and Sarbia, M. 2002. Gene amplification and protein overexpression of c-erbB-2 in Barrett carcinoma and its precursor lesions. *Am. J. Clin. Pathol. 118:*60–66.

Giatromanolaki, A., Gorgoulis, V., Chetty, R., Koukourakis, M.I., Whitehouse, R., Kittas, C., Veslemes, M., Gatter, K.C., and Lordanoglou, I. 1996. C-erbB-2 oncoprotein expression in operable nonsmall cell lung cancer. *Anticancer Res. 16:* 987–994.

Grushko, T.A., Blackwood, M.A., Schumm, P.L., Hagos, F.G., Adeyanju, M.O., Feldman, M.D., Sanders, M.O., Weber, B.L., and Olopade, O.I. 2002. Molecular-cytogenetic analysis of HER-2/neu gene in BRCA1-associated breast cancers. *Cancer Res. 62:*1481–1488.

Gullick, W., Love, S., Wright, C., Barnes, D.M., Gusterson, B., Harris, A.L., and Altman, D.G. 1991. C-erbB-2 protein in breast cancer is a risk factor in patients with involved and uninvolved lymph nodes. *Br. J. Cancer 63:*434–438.

Gupta, D., Middleton, L., Whitaker, M., and Abrams, J. 2003. Comparison of fluorescence and chromogenic *in situ* hybridization for detection of HER-2/neu oncogene in breast cancer. *Anat. Path. 119:*381–387.

Haggitt, R.C. 1994. Barrett's esophagus, dysplasia, and adenocarcinoma. *Hum. Pathol. 25:*982–993.

Hammod, M.E.H., Barker, P., Taube, S., and Gutman, S. 2003. Standard reference material for HER-2 testing. Report of a National Institute of Standards and Technology-sponsored consensus workshop. *Appl. Immunohistochem. Mol. Morphol. 11:*103–106.

Hayat, M.A. 2002. *Microscopy, Immunohistochemistry and Antigen Retrieval Methods.* New York: Kluwer Academic/ Plenum Publishers.

Henne-Bruns, D., Vogel, I., Luttges, J., Kloppel, G., and Kremer, B. 1998. Ductal adenocarcinoma of the pancreas head: Survival after regional versus extended lymphandectomy. *Hepato-Gastroenterol. 102:*S13–S24.

Hogsdon, J., Chin, K., Collins, C., and Gray, J. 2003. Genome amplification of chromosome 20 in breast cancer. *Breast Cancer Res. Treatment 78:*337–345.

Hoey, R., Sanderson, C., Iddon, J., Brady, G., Bundred, N., and Anderson, N. 2003. The parathyroid hormone-related protein receptor is expressed in breast cancer bone metastases and promotes autocrine proliferation in breast carcinoma cells. *Brit. J. Cancer 88:*567–573.

Järvinen, T., and Liu, E. 2003. HER-2/neu and topoisomerase II in breast cancer. *Breast Cancer Res. Treat. 78:*299–311.

Jakubowska, A., Narod, S.A., Goldgar, D.E., Mierzejewski, M., Masoje, B., Nej, K., Huzarska, J., Byrski, T., Gorski, B., and Lubinski, J. 2003. Breast cancer risk reduction associated with the RAD51 polymorphism among carriers of the BRCA1 5382insC mutation in Poland. *Cancer Epidemiol. Biomark Prev. 12:* 457–459.

Kagan, B.L., Henke, R.T., Cabal-Manzano, R., Stoica, G.E., Nguyen, Q., Wellstein, A., and Riegel, A.T. 2003. Complex regulation of the fibroblast growth factor-binding protein in MDAMB-468 breast cancer cells by CCAAT/enhancer-binding protein B[1]. *Cancer Res. 63:*1696–1705.

Kato, T., Kameoka, S., Kimura, T., Nishikawa, T., and Kobayashi, M. 2003. The combination of angiogenesis and blood vessel invasion as a prognostic indicator in primary breast cancer. *Br. J. Cancer 88:*1900–1908.

Kauraniemi, P., Kuukasjarvi, T., Sauter, G., and Kallioniemi, A. 2003. Amplification of a 280-kilobase core region at the ERBB2 locus leads to activation of two hypothetical proteins in breast cancer. *Am. J. Pathol. 163:*1979–1984.

Knuefermann, C., Lu, Y., Lu, B., Jin, W., Liang, K., Wu, L., Schmidt, M., Mills, G.B., Mendelsohn, D., and Fan, Z. 2003. HER2/P1-3K/Akt activation leads to a multidrug resistance in human breast adenocarcinomas cells. *Oncogene 22:*3205–3212.

Kohlberger, P., Loesch, A., Koelbl, H., Breitenecker, G., Kainz, C., and Gitsch, G. 1996. Prognostic value of immunohistochemically detected HER-2/neu oncoprotein in endometrial cancer. *Cancer Lett. 98:*151–155.

Kominsky, S., Argani, P., Korz, D., Evron, E., Raman, V., Garrett, E., Rein, A., Sauter, G., Kallioniemi, O., and Sukumar, S. 2003.

Loss of the tight junction protein claudin-7 correlates with histological grade in both ductal carcinoma *in situ* and invasive ductal carcinoma of the breast. *Oncogene 22:*2021–2033.

Kristiansen, G., Yu, Y., Petersen, S., Kaufmann, O., Schluns, K., Dietel, M., and Petersen, I. 2001. Overexpression of c-erbB-2 protein correlates with disease-stage and chromosomal gain at the c-erbB-2 locus in nonsmall cell lung cancer. *Eur. J. Cancer 37:*1089–1095.

Kuhn, E.J., Kurnot, R.A., Sesterhenn, I.A., Chang, E.H., and Moul, J.W. 1993. Expression of the c-erbB-2 (HER-2/neu) oncoprotein in human prostatic carcinoma. *J.Urol. 150:*1427–1433.

Lauffart, B., Gangisetty, O., and Still, I. 2003. Molecular cloning, genomic structure and interactions of the putative breast tumor suppressor TACC2. *Genomics 81:*192–201.

Lehmann, U., Länger, F., Feist, H., Glöckner, S., Hasemeier, B., and Kreipe, H. 2002. Quantitative assessment of promoter hypermethylation during breast cancer development. *Am. J. Pathol. 160:*605–612.

Lei, S., Appert, H.E., Nokata, B., Domenico, D.R., Kim, K., and Howard, J.M. 1995. Overexpression of HER-2/neu oncogene in pancreatic cancer correlates with shortened survival. *Int. J. Pancreatol. 17.1:*15–21.

Lilleby, W., Dale, E., Olsen, D.R., Gude, U., and Fossa, S.D. 2003. Changes in treatment volume of hormonally treated and untreated cancerous prostate and its impact on rectal dose. *Acta. Oncol. 42:*10–14.

Marshall, C.J. 1995. Specificity of receptor tyrosine kinase signaling: Transient versus sustained extracellular signal regulated kinase activation. *Cell 80:*179–185.

Mass, R. 2000. The role o HER-2 expression in predicting response to therapy in breast cancer. *Semin. Oncol. 27:*46–52; discussion 92–100.

Milde-Langosch, K., Kappes, H., Riethdorf, S., Löning, T., and Bamberger, A.-M. 2003. FosB is highly expressed in normal mammary epithelia, but down-regulated in poorly differentiated breast carcinomas. *Breast Cancer Res. Treat. 77:*265–275.

Nakagoe, T., Fukushima, K., Itoyanagi, N., Ikuta, Y., Oka, T., Nagayasu, T., Ayabe, H., Hara, S., Ishikawa, H., and Minami, H. 2002. Expression ABH/Lewis-related antigens as prognostic factors in patients with breast cancer. *J. Cancer Res. Clin. Oncol. 128:*257–264.

Nakopoulou, L., Giannopoulou, I., Trafalis, D., Gakiopoulou, H., Keramopoulos, A., and Davaris, P. 2002. Evaluation of numeric alterations of chromosomes 1 and 17 by *in situ* hybridization in invasive breast carcinoma with clinicopathologic parameters. *App. Immunohistochem. Mol. Morph. 10:*20–28.

Nakopoulou, L., Panayotopoulou, E., Giannopoulou, I., Alexandrou, P., Katsarou, S., Athanassiadou, P., and Keramopoulos, A. 2002. Stromelysin-3 protein expression in invasive breast cancer: Relation to proliferation, cell survival and patients' outcome. *Mod. Pathol. 15:*214–219.

Offersen, B.V., Nielsen, B.S., Høyer-Hansen, G., Rank, F., Hamilton-Dutoit, S., Overgaard, J., and Andreasen, P.A. 2003. The myofibroblast is the predominant plasminogen activator inhibitor-1-expressing cell type in human breast carcinomas. *Am. J. Pathol. 163(5):*1887–1899.

O'Grady, A., Flahavan, C., Kay, E., Barrett, H., and Leader, M. 2003. HER-2 analysis in tissue microarrays of archival human breast cancer. *App. Immunohistochem. Mol. Morph. 11:*177–182.

Ormandy, C., Musgrove, E., Hui, R., Daly, R., and Sutherland, R. 2003. Cylin D1, EMS1 and 11q13 amplification in breast cancer. *Breast Cancer Res. Treat. 78:*323–335.

Osorio, A., De La Hoya, M., Rodríguez-López, R., Martínez-Ramírez, A., Cazorla, A., Granizo, J.J., Esteller, M., Rivas, C., Caldéz, T., and Benítez, J. 2002. Loss of heterozygosity analysis at the BRCA loci in tumor samples from patients with familial breast cancer. *Int. J. Cancer 99:*305–309.

Palmu, S., Söderström, K.O., Quazi, K., Isola, J., and Salminem, E. 2002. Expression of C-KIT and HER-2 tyrosine kinase receptors in poor-prognosis breast cancer. *Anticancer Res. 22:*411–414.

Paredes, J., Milanezi, F., Reis-Filho, J.S., Leitão, D., Athanazio, D., and Schmitt, F. 2002. Aberrant P-Cadherin expression: Is it associated with estrogen-independent growth in breast cancer? *Pathol. Res. Pract. 198:*795–801.

Paris, P.L., Albertson, K.G., Alers, J.C., Andaya, A., Carroll, P., Fridlyand, J., Jain, A.N., Kamkar, S., Kowbel, K., Krijtenburg, P.J., Pinkel, D., Schroder, F.H., Vissers, K.J., Watson, M.J.E., Wildhagen, M.F., Collins, C., and Dekken, H.V. 2003. High-resolution analysis of paraffin-embedded and formalin-fixed prostate tumors using comparative genomic hybridization to genomic microarrays. *Am. J. Pathol. 162:*763–770.

Pegram, M. and Reese, D. 2002. Combined biological therapy of breast cancer using monoclonal antibodies directed against HER-2/neu protein and vascular endothelial growth factor. *Semin. in Oncol. 29(3):*29–37.

Perren, A., Weng, L.P., Boag, A.H., Ziebold, U., Thakore, K., Dabia, P.L., Komminoth, P., Lees, J.A., Mulligan, L.M., Mutter, G.L., and Eng, C. 1999. Immunohistochemical evidence of loss of PTEN expression in primary ductal adenocarcinomas of the breast. *Am. J. Pathol. 155:*1253–1260.

Persons, D.L., Borelli, K.A., and Hsu, P.H. 1997. Quantitation of HER-2/neu and C-Myc gene amplification in breast carcinoma using fluorescence *in situ* hybridization. *Pathology 10:*720–727.

Pfeiffer, P., Clausen, P.P., Andersen, K., and Rose, C. 1996. Lack of prognostic significance of epidermal growth factor receptor and the oncoprotein p185HER-2 in patients with systematically untreated small cell lung cancer: An immunohistochemical study on cryosections. *Br. J. Cancer 74:*86–91.

Plaumann, M., Seitz, S., Frege, R., Estevez-Schwarz, L., and Scherneck, S. 2003. Analysis of DLC-1 expression in human breast cancer. *J. Cancer Res. Clin. Oncol. 129:*349–354.

Porowska, H., Paszkiewicz-Gadek, A., Wolczyński, S., and Gindzienski, A. 2002. MUC1 expression in human breast cancer cells is altered by the factors affecting cell proliferation. *Neoplasma 49:*104–109.

Press, M.F., Slamon, K.J., Flom, K.J., Park, J., Zhou, J.Y., and Bernstein, L. 2002. Evaluation of HER-2/neu gene amplification and overexpression: Comparison of frequently used assay methods in a molecularly characterized cohort of breast cancer specimens. *J. Clin. Oncol. 20:*3095–3105.

Price-Schiavi, S., Jepson, S., Li, P., Arango, M., Rudland, P., Yee, L., and Carraway, K. 2002. Rat MUC4 (sialomucin complex) reduces binding of anti-ERBB2 antibodies to tumor cell surfaces, a potential mechanism for herceptin resistance. *Int. J. Cancer 99:*783–791.

Puddefoot, J.R., Barker, S., Glover, H.R., Malouitre, S.D., and Vinsow, G.P. 2002. Non-competitive steroid inhibition of oestrogen receptor functions. *Int. J. Cancer 101:*17–22.

Reid, B.J., Haggitt, R.C., Rubin, C.E., Roth, G., Surawicz, C.M., Van Belle, G., Lewin, K., Weinstein W.M., Antonioli, D.A., and Goldman, H. 1988. Observation variation in the diagnosis of dysplasia in Barrett's esophagus. *Hum Pathol. 19:*166–178.

Rhodes, A., Jasani, B., Anderson, E., Dodson, A.R., and Balaton, A.J. 2002b. Evaluation of HER-2/neu immunohisto-chemical assay sensitivity and scoring on formalin-fixed and paraffin–processed cell lines and breast tumors. *Am. J. Clin. Pathol. 118:*408–417.

Rhodes, A., Jasani, B., Couturier, J., McKinley, M.J., Morgan, J.M., Dodson, A.R., Navabi, H., Miller, K.D., and Balaton, A.J. 2002a. A formalin-fixed and paraffin–processed cell line standard for

quality control of immunohistochemical assay of HER-2/neu expression in breast cancer. *Am. J. Clin. Pathol. 117:*81–89.

Rolitsky, C.D., Theil, K.S., McGaughy, V.R., Copeland, L.J., and Neimann, T.H. 1999. HER-2/neu amplification and overexpression in endometrial carcinoma. *Int. J. Gynecol. Pathol. 18:* 138–143.

Sadashivan, R., Morgan, R., Jennings, S., Austenfeld, M., Van Veldhuizen, P., Stephens, R., and Noble, M. 1993. Overexpression of HER-2/neu may be an indicator of poor diagnosis in prostate cancer. *J. Urol. 150:*126–131.

Sahin, A.A. 2000. Biologic and clinical significance of HER-2/neu (c-erbB-2) in breast cancer. *Adv. Anat. Pathol. 7:*158–166.

Santini, D., Ceccarelli, C., Tardio, M., Taffurelli, M., and Marrano, D. 2002. Immunocytochemical expression of epidermal growth factor receptor in myoepithelial cells of the breast. *App. Immunohistochem. Mol. Morph. 10:*29–33.

Sauer, T., Wiedswang, G., Boudjema, G., Christensen, H., and Karesen, R. 2003. Assessment of HER-2/neu overexpression and/or gene amplification in breast carcinomas: Should *in situ* hybridization be the method of choice? *APMIS 111:*444–450.

Saunders, P.T.K., Millar, M.R., Williams, K., Macpherson, S., Bayne, C., O'Sullivan, C., Anderson, T.J., Groome, N.P., and Miller, W.R. 2002. Expression of oestrogen receptor beta (ERβI) protein in human breast cancer biopsies. *Br. J. Cancer 86:*250–256.

Schechter, A.L., Stern, D.F., Vaidyanathan, L., Decker, S.J., Drebin, J.A., Green, M.I., and Weinberg, R.A. 1984. The neu oncogene: An erbB-related gene encoding a 185,000-Mr tumor antigen. *Nature 312:*513–516.

Schroeder, J.A., Adriance, M.C., Thompson, M.C., Camenisch, T.D., and Gendler, S.J. 2003. MUC1 alters β-catenin-dependent tumor formation and promotes cellular invasion. *Oncogene 22:* 1324–1332.

Schwalbe, M., Sänger, J., Eggers, R., Naumann, A., Schmidt, A., Höffken, K., and Clement, J. 2003. Differential expression and regulation of bone morphogenetic protein 7 in breast cancer. *Int. J. Oncol. 23:*89–95.

Seki, A., Nakamura, K., Kodama, J., Miyagi, Y., Yoshinouchi, M., and Kudo, T. 1998. A close correlation between c-erbB-2 gene amplification and local progression in endometrial adenocarcinoma. *Eur. J. Gynaecol. Oncol. 19:*90–92.

Selim, A.-G., El-Ayat, G., and Wells, C.A.2002. Androgen receptor expression in ductal carcinoma *in situ* of the breast: Relation to oestrogen and progesterone receptors. *J. Clin. Pathol. 55(1):* 14–16.

Shak, S. 1999. Overview of the trastuzumab (Herceptin) anti-HER-2 monoclonal antibody clinical program in HER-2-overexpressing metastatic breast cancer. Herceptin Multinational Investigator Study Group. *Semin. Oncol. 26:*71–77.

Shin, D.M. 2001. A phase I clinical and biomarker study of CI-1033, a novel pan-erbB tyrosine kinase inhibitor in patients with solid tumors. *Proc. Am. Soc. Clin. Oncol. 20:*82a.

Shirakawa, K., Shibuya, M., Heike, Y., Takashima, S., Watanabe, I., Konishi, F., Kasumi, F., Goldman, C., Thomas, K., Bett, A., Terada, M., and Wakasugi, H. 2002. Tumor-infiltrating endothelial cells and endothelial precursor cells in inflammatory breast cancer. *Int. J. Cancer 99:*344–351.

Slamon, D.J., Clark, G.M., Wong, S.G., Levin, W.J., Ullrich, A., and McGuire, W.L. 1987. Human breast cancer: Correlation of relapse and survival with amplification of the HER-2/neu oncogene. *Science 235:*177–182.

Song, Y., Billiar, T., and Lee, Y. 2002. Role of Galectin-3 in breast cancer metastasis. *Am. J. Pathol. 160:*1069–1075.

Span, P., Bussink, J., Manders, P., Beex, L., and Sweep, C.G.J. 2003. Carbonic anhydrase-9 expression levels and prognosis in human breast cancer: Association with treatment outcome. *Br. J. Cancer 89:* 271–276.

Span, P., Sweep, C.G.J., Manders, P., Beek, L., Leppert, D., and Lindberg, R. 2003. Matrix metalloproteinase inhibitor reversion-inducing cysteine-rich protein with kazal motifs. *Cancer 97(11):* 2710–2715.

Spizzo, G., Gastl, G., Wolf, D., Gunsilius, E., Steurer, M., Fong, D., Amberger, A., Margreitr, R., and Obrist, P. 2003. Correlation of COX-2 and Ep-CAM overexpression in human invasive breast cancer and its impact on survival. *Br. J. Cancer 88:*574–578.

Stathopoulou, A., Mavroudis, D., Perraki, M., Apostolaki, S., Vlachonikolis, I., Lianidou, E., and Georgoulias, V. 2003. Molecular detection of cancer cells in the peripheral blood of patients with breast cancer: Comparison of CK-19, CEA, and maspin as detection markers. *Anticancer Res. 23:*1883–1890.

Suchanek, K.M., May, F.J., Robinson, J.A., Lee, W.J., Holman, N.A., Monteith, G.R., and Roberts-Thomson, S.J. 2002. Peroxisome proliferator-activated receptor α in the human breast cancer cell lines MCF-7 and MDA-MB-231. *Mol. Carcinogen 34:* 165–171.

Suo, Z., Bjaamer, A., Med, S., Ottestad, L., and Nesland, J.M. 2001. Expression of EGFR family and steroid hormone receptors in ductal carcinoma in situ of the breast. *Ultrastruct. Pathol. 25:* 349–356.

Tanner, M., Jarvinem, T.A.H., and Kawaniemi, P. 2000. Topoisomerase 11α amplification and deletion predict response to chemotherapy in breast cancer. *Proc. Am. Assoc. Cancer Res. 41:*803-abstract.

Terskikh, A.V., Le Doussal, J.M., Crameri, R., Fisch, I., Mach, J.P., and Kajava, A.V. 1997. 'Peptabody': A new type of high avidity binding protein. *Proc. Natl. Acad. Sci. USA. 94:* 1663–1668.

Thybusch-Bernhardt, A., Aigner, A., Beckman, S., Czubayko, F., and Juhl, H. 2001. Ribozyme targetting of HER-2 inhibits pancreatic cancer cell growth *in vivo. Eur. J. Cancer 37:*1688–1694.

Tolgay, O.I., Dolled-Filhart, M., D'Aquila, T.G., Camp, R.L., Rimm, D.L. 2003. Tissue microarray-based studies of patients with lymph node negative breast carcinoma show that met expression is associated with worse outcome but is not correlated with epidermal growth factor family receptors. *Cancer 97(9):*1841–1848.

Townsend, P., Dublin, E., Hart, I., Kao, R.-H., Hanby, A., Cutress, R., Poulsom, R., Ryder, K., Barnes, D., and Packham, G. 2002. BAG-1 expression in human breast cancer: Interrelationship between BAG-I RNA, protein, HSC70 expression and clinico-pathological data. *J. Pathol. 197:*51–59.

Tsunoda, L., Inada, H., Kalembeyi, I., Imanaka-Yoshida, K., Sakakibara, M., Okada, R., Katsuta, K., Sakakura, T., Majima, Y., and Yoshida, T. 2003. Involvement of large tenascin-C splice variant in breast cancer progression. *Am. J. Pathol. 162:*1857–1867.

Tsutsui, S., Ohno, S., Murakami, S., Hachitanda, Y., and Oda, S. 2002. DNA aneuploidy in relation to the combination of analysis of estrogen receptor, progesterone receptor, p53 protein and epidermal growth factor receptor in 498 breast cancers. *Oncology 63:*48–55.

Tubbs, R.T., Pettay, J.D., Roche, P.C., Stoler, M.H., Jenkins, R.B., and Grogan, T.M. 2001. Discrepancies in clinical laboratory testing of eligibility for trastuzumab therapy: Apparent immunohistochemical false-positives do not get the message. *J. Clin Oncol. 19:*2714–2721.

Tubbs, R., Pettay, J.D., Skacel, M., Powell, R., Stoler, M., Roche, P., and Hainfeld, J. 2002. Gold-facilitated in situ hybridization. *Am. J. Pathol. 160:*1589–1595.

Umekita, Y. and Yoshida, H. 2003. Expression of maspin gene is up-regulated during the progression of mammary ductal carcinoma. *Histopathology 42:*541–545.

Van Poznak, C., Tan, L., Panageas, K., Arroyo, C., Hudis, C., Norton, L., and Seidman, A. 2002. Assessment of molecular markers of clinical sensitivity to single-agent taxane therapy for metastatic breast cancer. *J. Clin. Oncol. 20(9):*2319–2326.

Virmani, A., Rathi, A., Sugio, K., Sathyanarayana, U., Toyooka, S., Kischel, F., Tonk, V., Padar, A., Takahashi, T., Roth, J., Euhus, D., Minna, J., and Gazdar, A. 2003. Aberrant methylation of *TMS1* in small cell, non small cell lung cancer and breast cancer. *Int. J. Cancer 106:*198–204.

Walch, A., Bink, K., Gais, P., Stangl, S., Hutzler, P., Aubele, M., Mueller, J., Hofler, H., and Werner, M. 2000. Evaluation of c-erbB-2 overexpression and HER-2/neu gene copy number heterogeneity in Barrett's adenocarcinoma. *Anal. Cell Pathol. 20:*25–32.

Walch, A., Specht, K., Bink, K., Zitzelsberger, H., Braselmann, H., Bauer, M., Aubele, M., Stein, H., Siewert, J. R., Hofler, H., and Werner, M. 2001. HER-2/neu gene amplification, elevated mRNA expression, and protein overexpression in the metaplasia-dysplasia-adenocarcinoma sequence of Barrett's esophagus. *Lab. Invest. 81:*791–801.

Walker, R.A. 1998. The erb-B/HER type 1 tyrosine kinase receptor family. *J. Pathol. 185:*234–235.

Wang, L., Duke, L., Zhang, P., Arlinghaus, R., Symmans, W., Sahin, A., Mendez, R., and Dai, J. 2003. Alternative splicing disrupts a nuclear localization signal in spleen tyrosine kinase that is required for invasion suppression in breast cancer. *Cancer Res. 63:*4724–4730.

Wang, S., Saboorian, M., Frenkel, E., Haley, B., Siddiqui, M., Gokaslan, S., Hynan, L., Ashfaq, R. 2002. Aneusomy 17 in breast cancer: Its role in HER-2/neu protein expression and implication for clinical assessment of HER-2/neu status. *Mod. Pathol. 15:*137–145.

Warshaw, A.L., and Castillo, F.D. 1992. Pancreatic carcinoma. *N. Engl. J. Med. 26:*455–465.

Wernicke, M., Piñeiro, L., Caramutti, D., Dorn, V., Raffo, M., Guixa, H., Telenta, M., and Morandi, A. 2003. Breast cancer stromal myxoid changes are associated with tumor invasion and metastasis: A central role for hyaluronan. *Mod. Pathol. 16:* 214–219.

Wickerham, L. 2002. Tamoxifen-an update on current data and where it can now be used. *Breast Cancer Res. Treat. 75:*S7–S12.

Wiedswang, G., Borgen, E., Karesen, R., Kvalheim, G., Nesland, J.M., Qvist, H., Schlichting, T., Sauer, T., Janbu, J., Harbitz, T., and Naume, B. 2003. Detection of isolated tumor cells in bone marrow is an independent prognostic factor in breast cancer. *J. Clin. Oncol. 21(18):*3469–3478.

Wittmann, B.M., Wang, N., and Montano, M.M. 2003. Identification of a novel inhibitor of breast cell growth that is down-regulated by estrogens and decreased in breast tumors. *Cancer Res. 63:* 5153–5158.

Wobus, M., Rangwala, R., Sheyn, I., Hennigan, R., Coila, B., Lower, E.E., Yassin, R.S., and Sherman, L.S. 2002. CD44 associates with EGFR and erbB-2 in metastasizing mammary carcinoma cells. *Appl. Immunohistochem. Mol Morph. 10:* 34–39.

Wu, A., Seow, A., Arakawa, K., Van Der Beek, D., Lee, H., and Yu, M. 2003. HSD17B1 and CYP17 polymorphisms and breast cancer risk among Chinese women in Singapore. *Int. J. Cancer 104:*450–457.

Xue, S., Lampert, I., Haldane, J., Bridger, J., and Griffin, B. 2003. Epstein-Barr virus gene expression in human breast cancer: Protagonist or passenger? *Br. J. Cancer 89:*113–119.

Yang, M., Moriya, T., Oguma, M., De La Cruz, C., Endoh, M., Ishida, T., Hirakawa, H., Orita, Y., Ohuchi, N., and Sasano, H. 2003. Microinvasive ductal carcinoma (T1mic) of the breast. The clinicopathological profile and immunohistochemical features of 28 cases. *Pathol. Intern. 53:*422–428.

2

Expression of Vascular Endothelial Growth Factor Receptor-2/Flk-1/ KDR in Breast Carcinoma

Lydia Nakopoulou

Introduction

Angiogenesis, the formation of new blood vessels from existing vasculature, is a tightly regulated event playing an important role in embryonic development, wound healing, and pathological conditions like tumor growth and progression (Feige *et al.*, 2000; Plate *et al.*, 1994). Neovascularization is essential for the development of tumors, including breast cancer, because it supplies not only nutrients such as growth factors and oxygen requirements for the growing tumor but also a vascular route for metastasis (Hanahan *et al.*, 1996). Several growth factors produced by tumor cells and the surrounding stroma have been identified as possible regulators of angiogenesis (Klagsbrun *et al.*, 1991). Among these factors, vascular endothelial growth factor (VEGF) and its receptors are thought to play a key role in tumor angiogenesis (Frelin *et al.*, 2000; McMahon, 2000; Neufeld *et al.*, 1999).

The VEGFs are a family of related dimeric glycoproteins, six of which (VEGF-A/VEGF, VEGF-B, VEGF-C, VEGF-D, placental growth factor, and the orf parapoxvirus VEGF) have been identified to date (Cross *et al.*, 2001). The VEGFs family has multiple and diverse functions, including promotion of endothelial cell mitogenesis, promotion of endothelial cell survival, increased vascular permeability, vasodilation, increased expression of proteolytic enzymes (e.g., collagenases) involved in stromal degradation, and chemotactic and immune effects via inhibition of maturation of antigen-presenting dendritic cell (Cross *et al.*, 2001; Ferrara, 2001). Gene targeting has indicated that VEGF-A is perhaps the most critical regulator of the development of the vascular system (Ferrara, 2002). The human VEGF-A gene, localized on chromosome 6, is organized into eight exons, and alternative splicing produces at least five different molecular species, ranging in size from 121 to 206 amino acids. The VEGF-A variant containing 165 amino acids ($VEGF_{165}$) is the predominant and biologically most active variant (Ferrara, 2002). The glycoprotein VEGF regulates vasculogenesis during embryonic development showing a progressive decrease postnatally, being minimal in most adult tissues (Gerber *et al.*, 1999). However, VEGF expression is reinduced during pathological angiogenesis such as in ischemic myocardium or retina and in inflamed tissues (Carmeliet *et al.*, 2000).

Handbook of Immunohistochemistry and in situ Hybridization of Human Carcinomas, Volume 1: Molecular Genetics; Lung and Breast Carcinomas

249

Copyright © 2004 by Elsevier (USA)
All rights reserved.

The VEGFs modulate their activities through several receptors: VEGF-R1/Flt-1 (fms like tyrosine kinase 1), VEGF-R2/Flk-1 (fetal liver kinase 1)/KDR (kinase domain region), VEGF-R3/Flt-4, and neuropilin-1 (Cross *et al.*, 2001). All VEGF-A isoforms interact with Flt-1 (VEGF-R1) and Flk-1/KDR (VEGF-R2). In contrast, VEGF-B and placental growth factor (PLGF-1 and PLGF-2) bind Flt-1 but not Flk-1/KDR. The factors VEGF-C and VEGF-D bind Flt-4 and Flk-1/KDR but not Flt-1. Neuropillin-1, a receptor for semapharins in the nervous system, is also a receptor for the heparin-binding isoforms of VEGF and PlGF (Ferrara, 2002). The functional effects of VEGF depend on which ligand-receptor complex is undergoing activation. In both normal physiology and tumor pathophysiology, VEGF appears to have its most potent effects via the Flk-1/KDR/VEGF-R2 receptor, where it potently stimulates angiogenesis (Sledge, 2002).

The receptors Flt-1 and Flk-1/KDR both consist of an extracellular domain that includes seven immunoglobulin-like regions, a transmembrane region, and an intracellular domain that contains the consensus tyrosine kinase sequence (Ferrara, 1999). Both receptors show an amino acid sequence homology of approximately 44% (DeVries *et al.*, 1992; Millauer *et al.*, 1993). The murine homologue of KDR is known as Flk-1 (fetal liver kinase 1) and shares 85% sequence identity with human KDR. Both receptors are predominantly expressed in vascular endothelial cells (Breier *et al.*, 1995; Millauer *et al.*, 1993). Binding of VEGF to its receptors causes receptor dimerization and activation of the intrinsic kinase, followed by autophosphorylation of the receptor and subsequent signal transduction. There are significant differences between Flk-1/KDR and Flt-1. Targeted homozygous null mutations of Flk-1 are essential for endothelial organization during vascular development, whereas KDR/Flk-1 is required for the formation of blood islands and hemopoiesis (Millauer *et al.*, 1993). Experiments in gene knockout mice suggest that Flt-1 regulates endothelial cell-cell or cell-matrix interactions (Ferrara *et al.*, 1996), whereas the Flk-1 receptor is important for endothelial cell differentiation and mitogenesis (Shalaby *et al.*, 1995).

Numerous studies have shown that overexpression of VEGF and Flk-1/KDR is strongly associated with invasion and metastasis in human malignant disease (Dvorak *et al.*, 1997). The levels of VEGF expression are high in various types of tumors, and newly sprouting capillaries are clustered around VEGF-producing tumor cells (Dvorak *et al.*, 1997). The VEGF/VEGF receptor signaling has been implicated in angiogenesis that occurs in many human solid tumors, including bladder (O'Brien *et al.*, 1992), colon (Elllis *et al.*, 1998), gastrointestinal (Brown *et al.*, 1993), renal (Takahashi

et al., 1999) carcinomas, gliomas (Millauer *et al.*, 1994), and neuroblastomas (Davidoff *et al.*, 2001).

Although previous data about the expression of VEGF and its receptors in breast cancer is available (Brown *et al.*, 1995; DeJong *et al.*, 1998; Yoshiji *et al.*, 1996), there is limited evidence regarding the clinical significance of VEGF receptors in breast cancer. Vascular endothelial growth factor, both by virtue of its general importance in angiogenesis and its specific importance in breast cancer, represents a reasonable therapeutic target (Sledge, 2002). Targeting of VEGF has been the subject of several approaches, including ligand sequestration, attacks on the external membrane receptor, inhibition of the internal (tyrosine kinase) domain of the receptor, inhibition of VEGF receptor message, inhibition of downstream intermediates, and indirect inhibition through upstream regulators of VEGF (Shinkaruk *et al.*, 2003; Sledge, 2002). More specifically, as far as targeting of VEGF receptors is concerned, monoclonal antibodies have been developed against the external membrane domain of the Flk-1 (Zhu *et al.*, 1999; Kozin *et al.*, 2001) and clinical trials have recently been initiated with these agents, though there are no published reports to date. One of these agents, DC101, when combined with low-dose vinblastine (so-called "metronomic therapy") in preclinical models, has shown synergistic activity, resulting in sustained remissions of large tumors in a murine neuroblastoma model (Klement *et al.*, 2000). Moreover, several companies have developed receptor tyrosine kinase inhibitors directed against the internal membrane tyrosine kinase domain of VEGF receptors 1 and/or 2 (Flt-1 and/or Flk-1) (Drevs *et al.*, 2000; Laird *et al.*, 2000). Several of these compounds target more than one receptor. For instance, SU6668 targets the Flk-1 as well as the PDGF and FGF-1 receptors (Laird *et al.*, 2000). PTK787/ZK 222584 targets both Flt-1 and Flk-1 (Drevs *et al.*, 2000). Several of these compounds have shown *in vitro* and *in vivo* preclinical evidence of antiangiogenic activity. SU5416 has recently entered the breast cancer arena in a phase I trial in inflammatory breast cancer (Overmoyer *et al.*, 2001). In this trial, it was combined with doxorubicin as preoperative therapy, with acceptable toxicity and no evidence that doxorubicin modulated the pharmacokinetics of SU5416. Trials with other receptor tyrosine kinase inhibitors of the VEGF receptors are in progress in several centers. Recently, Flk-1 was used as a model antigen to explore the feasibility of the immunotherapy with a vaccine based on a xenogeneic homologous protein (Liu *et al.*, 2003). Immunotherapy with a quail homologous KDR/Flk-1 protein vaccine was effective at both protective and therapeutic antitumor immunity in several solid and hematopoietic tumor models in mice (Liu *et al.*, 2003).

It is well-established that balanced proteolysis of the extracellular matrix (ECM) is essential for angiogenesis. The tissue inhibitor of metalloproteinases type-1 (TIMP-1) is a 28,000 dalton sialoglycoprotein that specifically inhibits matrix metalloproteinases (MMPs) (Gomez et al., 1997). The TIMPs are multifunctional proteins capable of influencing the cellular microenvironment both physically and physiologically (Fata et al., 1999). Beyond their direct influence on the turnover of structural matrix molecules, TIMPs can also affect soluble extracellular factors (Gomez et al., 1997). This is through the inhibition of at least three other MMP-mediated processes: the processing of cytokines, the degradation of growth factor binding proteins, and the release of ECM-bound growth factors (Fata et al., 1999; Gomez et al., 1997). Inhibition of the bioavailability of growth-promoting factors suggests that TIMPs can limit cellular proliferation and this may be the basis of TIMPs, ability to suppress primary tumor growth (Khokha, 1994). Moreover, it has been demonstrated that TIMP-1 can act as an inhibitor of angiogenesis in a variety of angiogenic models (Guedez et al., 2001; Takigawa et al., 1990). These data prompted us to investigate Flk-1 expression in relation to TIMP-1 mRNA expression.

Moreover, prompted by the hypothesis that VEGF/Flk-1 system may have another regulatory role in breast carcinogenesis apart from angiogenesis, we examined the expression of Flk-1 in invasive breast carcinomas in correlation with the expression of proliferation indices such as Ki-67 and topoisomerase IIα (TopoIIα), as well as TIMP-1 mRNA localization. In addition, we investigated any correlation of Flk-1 expression with clinical and immunohistochemical prognostic parameters and survival.

MATERIALS AND METHODS

Patients and Tumor Specimen

One hundred and forty-one paraffin blocks with tumor samples were available from patients with resectable breast cancer. Patient data were analyzed for various clinicopathological factors (menopausal status, tumor size, histologic type, histologic grade, nuclear grade, lymph node status, stage, estrogen and progesterone receptor [ER, PR] status, Ki-67, p53, Bcl-2, c-erbB-2, TIMP-1 mRNA in intratumoral fibroblasts, and TIMP-1 mRNA at the tumor margin).

In this study, all carcinomas were classified according to the criteria of the World Health Organization (WHO) (Association of Directors of Anatomic and Surgical Pathology Recommendations, 1996) and were recorded as invasive ductal and lobular. All invasive ductal carcinomas were of the not-otherwise-specified type and were graded according to the modified Scarff-Bloom-Richardson histological grading system, with guidelines suggested by Nottingham City Hospital pathologists (Association of Directors of Anatomic and Surgical Pathology Recommendations, 1996). Nuclear grading was separately assessed, based on the Scarff-Bloom-Richardson scheme (Association of Directors of Anatomic and Surgical Pathology Recommendations, 1996). Staging at the time of diagnosis was based on the tumor nodes metastasis (TNM) system. Tumor size (<2 cm, 2–5 cm, >5 cm) and lymph node status were evaluated separately. All women have been followed-up after surgical treatment at 6-month intervals for a mean period of 70.34 months (range: 7–94 months).

Immunohistochemistry

For the detection of Flk-1/KDR, the avidin-biotin indirect immunoperoxidase method was performed:

1. Paraffin-embedded tissue sections, 4-μm thick were cut, dewaxed, rehydrated, and incubated with 0.3% hydrogen peroxide (H_2O_2) for 30 min to block endogenous peroxidase activity and nonspecific binding.
2. To enhance antigen retrieval, sections were microwave-treated in 0.01 M citrate buffer pH 6.0 at 700W for 5 min (one cycle).
3. After rinsing with 0.01 M phosphate-buffered saline (PBS), normal horse serum was applied for 20 min to block nonspecific antibody binding.
4. Subsequently, sections were incubated overnight at 4°C with the primary antibody (mouse/A-3, SC-6251; Santa Cruz Biotechnology, Santa Cruz, CA) at a dilution of 1:80.
5. After additional rinsing in PBS (0.01 M), sections were incubated with biotinylated horse anti-mouse for anti-rabbit secondary antibody (Vector Laboratories, Burlingame, CA) for 30 min, at room temperature, and then incubated with avidin-biotinylated peroxidase complex (Vectastain Elite ABC Kit, Vector Laboratories) for 30 min.
6. The peroxidase reaction was developed with a 0.5 mg/ml solution of 3,3'-diaminobenzidine tetrahydrochloride (Sigma Chemical Co., St. Louis, MO) supplemented with 0.01% H_2O_2.
7. Finally, sections were counterstained with Harris' hematoxylin and mounted.

The other immunomarkers assessed in this study in combination with Flk-1 had been previously detected with the following antibodies:

1. Anti-ER clone 1D5 and anti-PR clone 1A6 (Dako, Glostrup, Denmark) at dilutions 1:450 and 1:150, respectively.

2. Anti-c-erbB-2 clone CB11 (Biogenex, San Ramon, CA) at a dilution of 1:150.

3. Rabbit anti-human Ki-67 (Dako, Glostrup, Denmark) at a dilution of 1:50.

4. Anti-Bcl-2 clone 124 (Dako, Glostrup, Denmark) at a dilution of 1:100.

5. Anti-p53 clone BP53.12.1 (Oncogene, Cambridge, MA) at a dilution of 1:50.

6. Anti-topoisomerase IIα (TopoIIα) clone H2.7 (Biocare Medical, Walnut Creek, CA) at a dilution 1:100 overnight. A standard avidin-biotin-peroxidase complex (ABC) method (Hsu *et al.*, 1981) (Vectastain Elite, Vector Laboratories) was used for visualization with diaminobenzidine as a chromogen. Sections were counterstained with hematoxylin and mounted.

To enhance antigen retrieval for ER, PR, Ki-67, Bcl-2, p53, and TopoIIα, sections were microwave-treated, 5 cycles of 5 min for ER and PR and two cycles of 5 min for Ki-67, Bcl-2, p53, and TopoIIα in 0.01 M citrate buffer (pH 6.0) at 700W. Positive controls included breast cancer tissue with known immunoreactivity for these markers. Negative controls had the antibody replaced by PBS.

Evaluation of Immunohistochemistry

The extent of Flk-1 was evaluated using a score of 0 to 3 as follows: score 0 < 10% positive cells per 10 high-power fields (400X); score 1: 11–30%, positive cells; score 2: 31–50%, score 3: >50%. The extent of Ki-67 expression was evaluated using a score of 0 to 2 as follows: score 0: < 10% positive cells per 10 high-power fields (400X); score 1: 11–30% positive cells, score 2: >30% positive cells.

Staining for ER and PR was evaluated semiquantitatively using the H score system [H = p1 + p2 × 2 + p3 × 3, where p1 corresponds to the percentage of positive nuclei with mild staining intensity, p2 represents the percentage of positive nuclei with moderate staining intensity, and p3 corresponds to the percentage of positive nuclei with intense staining intensity] (McClelland *et al.*, 1991). Carcinomas with H = 0–50 were considered as negative, with H = 51–100 considered to be of mild reactivity, H = 101–200 of moderate reactivity, and H > 200 graded as highly immunoreactive cancers. The fraction of c-erbB-2–positive stained cells was scored from 0 to 3 according to the guidelines for scoring HercepTest (Dako) [0: negative (no staining is observed, or membrane staining in less than 10% of the tumor cells); 1+: negative (a faint/barely perceptible membrane staining is detected in more than 10% of the tumor cells. The cells are only stained in part of the membrane); 2+: positive (a weak to moderate

complete membrane staining is observed in more than 10% of the tumor cells); 3+: positive (a strong complete membrane staining is observed in more than 10% of the tumor cells)]. For purposes of statistical analysis, negative (0, 1+) and positive (2+, 3+) c-erbB-2 staining was grouped in two categories. Bcl-2 expression was scored as negative (score 0) if less than 10% of tumor cells were positive, slightly positive (score 1) if 10–50% of tumor cells were positive, and strongly positive (score 2) if more than 50% of neoplastic cells showed cytoplasmic staining. For purposes of statistical analysis as far as ER, PR, Ki-67, p53, and Bcl-2 protein are concerned, cases with scores 1, 2, 3 (ER, PR) and cases with 1, 2 (Ki-67, Bcl-2) were included in the same group of positive protein expression.

The evaluation of TopoIIα immunopositivity was performed in areas with a notable number of immunoreactive cells, because this marker was generally expressed in relatively few neoplastic cells within the tumor mass. The ratio, expressed as a percentage of the number of immunohistochemically positive neoplastic nuclei in the total number of 500 (stained and unstained) was calculated automatically.

In situ Hybridization

In situ hybridization was performed for the localization of TIMP-1 mRNA transcripts. From the full-length human TIMP-1 cDNA, a 782-bp EcoRI-Xhol fragment coding for a unique portion of the 3′ untranslated region was subcloned into Bluscript KSII+plasmid (Stratagene, La Jolla, CA). Antisense cRNA probe was made by T7 transcription and digoxigenin-UTP labeling (Promega, Madison WI). The sense probe was made using T3 polymerase on a Bam HI linearized fragment and used as negative control.

1. Seven-μm-thick sections were rehydrated and digested with proteinase K (1 μg/ml) at 37°C for 30 min.

2. Sections were treated with 0.25% acetic anhydrite in 0.1 M triethanolamine for 10 min, washed in 2X SSC 5 min, and then rehydrated.

3. Overnight hybridization at 42°C was carried out using a digoxigenin-labeled RNA probe at a dilution of 1:1000.

4. Sections were stringently washed (2X SSC, 50% formamide, 30 min, 50°C) and incubated with RNase A (50 μg/ml RNase A in TES: 100-mM Tris-HCl, 1 mM ethylenediamine tetra-acetic acid (EDTA), 500 mM NaCl, pH 8.0) for 10 min at 37°C.

5. Stringent washings (2X SSC at 50°C for 15 min twice) followed.

6. Nonradioisotopic detection of hybridized probe was performed using an alkaline-phosphatase conjugated

antidigoxigenin antibody at a dilution of 1:750 for 1 hr at room temperature.

7. Sections were incubated overnight with nitroblue tetrazolium salt (NBT), 175 ng/mL and 5-bromo-4-chloro-3 indolyl phosphate (BCIP), 340 ng/mL in tris magnesium sulfate (TSM) buffer (100 mM Tris-HCl, pH 9.5, 100 mM NaCl, 50 mM MgCl$_2$).

8. Finally, sections were counterstained with Mayer's hematoxylin.

As an additional negative control, one slide in each experimental procedure was treated with ribonuclease A prior to hybridization to deplete the sample of mRNA.

Semiquantitative estimation based on the staining intensity and relative abundance of immunoreactive cells was performed independently by two pathologists. The fraction of TIMP-1 mRNA stained cells was scored after having examined 10 high-power fields (400X) of one section for each sample, and the percentage of positive cells was the average of the positive cells of 10 fields (Nakopoulou et al., 2002).

Statistical Analysis

Pearson's chi-square test with continuity correction was employed to assess any significant correlation between Flk-1 expression in tumor cells and major prognostic variables for breast carcinoma. The effect of Flk-1 expression on postoperative survival rates was assessed by multivariate analysis using stepwise-forward Cox's proportional hazard regression model. TopoIIα expression failed to fit the Gaussian distribution. Therefore, nonparametric analysis of variance with ranks was employed to assess TopoIIα with Flk-1. A p value of ≤ 0.05 was considered statistically significant.

RESULTS AND DISCUSSION

Immunoreaction of Flk-1/VEGFR$_2$ was detected in 64.5% (91 of 141) of 141 invasive breast carcinomas included in our study showing a widespread cytoplasmic expression in cancer cells (Figure 36). Although Flk-1 was not detected in the adjacent normal breast tissue observed in some cases of invasive carcinomas, it was expressed in the majority of the included in situ component. Moreover, Flk-1 was detected in endothelial cells of a few intratumoral small vessels. It is interesting that Flk-1 expression in invasive breast carcinomas showed a suggestive correlation with the menopausal status of the patients (p = 0.051). On the other hand, there was no significant correlation between Flk-1 expression and ER (p = 0.108) and PR (p = 0.413) detection in our invasive breast carcinomas. The expression of Flk-1 in invasive breast carcinomas was significantly

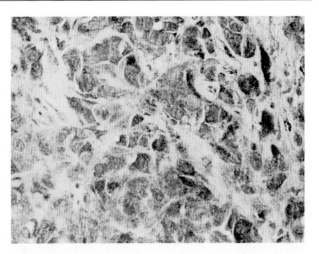

Figure 36 Flk-1 expression in the cytoplasm of breast cancer cells (ABC/HRP, 250X).

correlated with nuclear grade (p = 0.003), whereas its expression demonstrated no correlation with histological type (p = 0.999) and grade (p = 0.274), tumor size (p = 0.332), lymph node status (p = 0.492), and stage (p = 0.381).

The expression of Flk-1 demonstrated a significant correlation with two well-established proliferation indices: Ki-67 (p = 0.037) and topoisomerase IIα (p = 0.009). Positive staining of Flk-1 was detected in 76.4% of cases (42 of 55) with a high Ki-67 proliferation index but only in 57.6% (49 of 85) of cases with a low Ki-67 index.

On the other hand, there was no significant correlation between Flk-1 expression and p53 (p = 0.607), Bcl-2 (p = 0.745), or c-erb-B2 detection (p = 0.325). It is interesting that Flk-1 expression in invasive breast carcinomas showed a significant inverse correlation with TIMP-1 mRNA expression in intratumoral stromal cells (p = 0.013). Localization of TIMP-1 mRNA was evaluated as previously described (Nakopoulou et al., 2002). Thus, Flk-1 was detected in 73.5% (50 of 68) of invasive breast carcinomas with low TIMP-1 mRNA expression in intratumoral stromal cells, compared to 48.9% (23 of 47) of cases with high intratumoral TIMP-1 mRNA expression. However, no correlation was observed between Flk-1 expression and TIMP-1 mRNA localization in stromal cells at the tumor margin (p = 0.999).

Multivariate analysis applying the Cox's hazard regression model did not prove any significant correlation between Flk-1 expression and clinical outcome (p = 0.096).

Proliferation and angiogenesis are important prognostic variables in breast cancer (Eppenberger et al., 1998; Nakopoulou et al., 1999; Nakopoulou et al., 2000).

Although the intrinsic and extrinsic factors controlling these variables are not fully determined, VEGF-1 and its receptors Flk-1 and Flt-1 may play an important role in tumor growth having a parallel effect on neovascularization (Association of Directors of Anatomic and Surgical Pathology Recommendations, 1996).

In the current study, $VEGFR_2$/Flk-1 was detected in 64.5% of 141 invasive breast carcinomas showing a widespread expression in cancer cells. Concomitant with our findings, recent studies have demonstrated that VEGF receptors have been detected in several types of nonendothelial cells such as melanona (Gitay-Goren et al., 1993), ovarian (Boocock et al., 1995) and leukemic cell lines (Shibaya, 1995), Hela cells, macrophages (Barleon et al., 1996), and retinal progenitor cells (Yang and Cepko, 1996), suggesting that VEGF receptors are not restricted to endothelial cells. However, the biologic significance of nonendothelial VEGFR expression remains elusive, because Flk-1 positive ovarian cancer cell lines fail to respond to exogenous VEGF (Boocock et al., 1995). Moreover, it has been noted that the binding of VEGF to its receptors on nonendothelial cells could induce motility of monocytes, differentiation of osteoblasts, and production of insulin by beta cells (Enomoto et al., 1994). These findings suggest that VEGF could be multifunctional growth factor in different types of cells. Conflicting data have been reported regarding Flk-1 expression in various types of human malignancies. Although Von Marschall et al. (2001) have reported Flk-1 expression only in endothelial cells of human hepatocellular carcinomas, other studies have demonstrated variable expression of Flk-1 in tumor cells of pancreatic carcinomas (Von Marschall et al., 2000), neuroblastomas (Davidoff et al., 2001), and AIDS-related Kaposi's sarcomas (Mascood et al., 1997).

As far as breast carcinoma is concerned, Xie et al., (1999) have reported that cancer cells of hormone-induced mammary carcinoma in the Noble rat expressed not only high levels of VEGF but also of its receptors Flt-1 and Flk-1. Moreover, previous studies have shown variable expression of VEGF, Flt-1, and Flk-1 in tumor cells of invasive breast carcinomas (Brown et al., 1995; De Jong et al., 1998; Yoshiji et al., 1996).

The variable data regarding Flk-1 expression in neoplastic cells of different tumor types may be partly explained by experimental data indicating that endothelia from different tissues vary in their ability to express VEGF receptors given identical stimuli (Roberts et al., 1998). In this view, the previous notion could be further applied to tumor cells with VEGFR expression being determined by cell type and microenvironment.

In addition, the absence of Flk-1 expression in the adjacent normal breast tissue included in some of our invasive breast carcinomas is compatible with previous data regarding normal breast and pancreatic ductal cells (De Jong et al., 1998; Von Marschall et al., 2001). Therefore, de novo expression of VEGF and its receptor Flk-1 in breast and pancreatic carcinoma compared with normal tissue suggests that up-regulation of the VEGF/VEGFR system may be a feature restricted to malignant transformation.

It is interesting that Flk-1 expression in the invasive breast carcinomas of our series was significantly correlated with both proliferation indices Ki-67 ($p = 0.037$) and TopoIIα ($p = 0.009$). To the best of our knowledge, the correlation of Flk-1 expression with proliferation indices like Ki-67 and especially TopoII has not been investigated in human breast carcinomas despite previous limited experimental studies (Xie et al., 1999).

Our findings in a considerable number of breast carcinomas are compatible with the reported experimental data indicating a significant correlation of high VEGF and Flk-1 activity with increased Ki-67 expression in hormone-induced mammary cancer in the Noble rat (Xie et al., 1999), suggesting that VEGF may act as an autocrine growth factor for mammary cancer cells in vivo and this autocrine regulatory role may be mediated through Flk-1. To further confirm the autocrine growth stimulatory action of VEGF in mammary cancer cells, Xie et al. (1999) treated the cells with exogenous synthetic VEGF and observed an apparently increased growth rate in the treated cells. According to other experimental studies, the proliferative effects of VEGF after binding to Flk-1 on endothelial cells are mediated at least in part by activation of the mitogen activated protein kinase (MAPK) signaling pathway (D'Angelo et al., 1995), a notion also demonstrated for pancreatic cell lines (Von Marschall et al., 2000). Von Marschall et al. (2000) found that the KDR/Flk-1–activated intracellular signaling cascade is functional in pancreatic tumor cells and tested the mitogenic responsiveness of cell lines expressing KDR/Flk-1 to recombinant human $VEGF_{165}$. They observed that treatment with $VEGF_{165}$ results in a significant stimulation of cell proliferation in the pancreatic cell lines. This data, in combination with our observations, are further supported by strong evidence that Flk-1 mediates the mitogenic stimulus in response to VEGF, which may act as a growth factor for mammary cancer cells (Waltenberger et al., 1994).

The clinical and therapeutic significance of the correlation of Flk-1 expression with proliferation indices as demonstrated in our study is further undermined by recent experimental studies demonstrating that anti–Flk-1 (DC-101) treatment significantly suppressed the growth of mouse mammary and colonic carcinomas by decreasing microvessel density and tumor cell proliferation as reflected by decreased proliferating cell nuclear antigen (PCNA) expression (Prewett *et al.*, 1999; Shaheen *et al.*, 2001).

Although previous studies have shown that overexpression of VEGF and Flk-1 is associated with invasion and metastasis in human malignant disease, including breast cancer (Dvorak *et al.*, 1997), the clinical significance of VEGF receptor Flk-1 has not been fully elucidated.

Interestingly, Flk-1 expression in our breast carcinomas demonstrated a suggestive correlation with menopausal status but no relation with ER/PR detection in the invasive carcinomas. Conflicting data have been reported regarding the correlation between ER/PR and angiogenic factors, especially VEGF. Experimental studies have indicated that ERa overexpression in cancer cell lines like Ishikawa cells caused down-regulation of VEGF, whereas others have proved that $VEGF_2$ is ERa indicible (Ali *et al.*, 2000; Bausero *et al.*, 2000). Similar discrepancies have been reported regarding regulation of VEGF by progestins, because some investigators have demonstrated increased VEGF levels in response to synthetic progestins (Classen-Linke *et al.*, 2000; Hyder *et al.*, 1998). On the other hand, clinical studies have demonstrated an inverse correlation between VEGF levels estimated by immunoassays and ER detection in human breast carcinomas (Eppenberger *et al.*, 1998; Linderholm *et al.*, 1998).

Concomittant with the well-established notion that balanced proteolysis of the extracellular matrix is essential for angiogenesis, an inverse correlation was observed between Flk-1 expression in breast cancer cells and TIMP-1 mRNA localization in intratumoral stromal cells (p = 0.013). In this view, the low levels of Flk-1 detection in our cases with high TIMP-1 mRNA expression could be due to the inhibitory effect of TIMP-1 on endothelial proliferation and invasion as it has been reported in a variety of angiogenic models (Takigawa *et al.*, 1990) and in Burkitt's lymphoma cell lines (Guedez *et al.*, 2001). In addition, the role of TIMPs as an angiogenesis inhibitor is further supported by *in vivo* angiogenesis assays, indicating that TIMP-1 inhibited the endothelial response induced by basic fibroblast growth factor (bFGF) (Johnson *et al.*, 1994).

In conclusion, the significant correlation of Flk-1 expression in neoplastic cells of invasive breast carcinomas with proliferation indices like Ki-67 and TopoIIα suggests that VEGF may exert a growth factor activity on mammary cancer cells through its receptor Flk-1. Moreover, the inverse correlation between Flk-1 expression in breast cancer cells and TIMP-1 mRNA localization in intratumoral stromal cells further supports the notion that TIMP-1 may play an inhibitory role on angiogenesis. Further investigation is necessary to elucidate the interactions between extracellular matrix regulators (like MMPs and TIMPs) and angiogenesis.

References

Ali S.H., O'Donell, A.L., Balu, D., Pohl, M.B., Seyler, M.J., Mohamed S., Mousa, S., and Dandona, P. 2000. Estrogen receptor alpha in the inhibition of cancer growth and angiogenesis. *Cancer Res.* 60:7094–7098.

Association of Directors of Anatomic and Surgical Pathology Recommendations for life reporting of breast carcinoma. 1996. *Hum. Pathol.* 27:220–224.

Barleon, B., Sozzani, S., Zhou, D., Weich, H.A., Mantovani, A., and Marme, D. 1996. Migration of human monocytes in response to vascular endothelial growth factor is mediated via the VEGF receptor Flt-1. *Blood* 87:3336–3343.

Bausero, P., Ben-Mahdi, M., Mazucafelli, J., Bloy, C., and Perrot-Applanat, M. 2000. Vascular endothelial growth factor is modulated in vascular muscle cells by estradiol, tamoxifen, and hypoxia. *Am. J. Physiol. Heart Circ. Physiol.* 279:H2033–H2042.

Boocock, C.A., Charnock-Jones, S., Sharkey, A.M., McLaren, J., Barker, P.J., and Wright, K.A. 1995. Expression of vascular endothelial growth factor and its receptors Flt and KDR in ovarian carcinoma. *J. Natl. Cancer Inst.* 7:506–516.

Breier, G., Clauss, M., and Risau, W. 1995. Coordinate expression of vascular endothelial growth factor receptor-1 (Flt-1) and its ligand suggests a paracrine regulation of murine vascular development. *Dev. Dyn.* 204:228–239.

Brown, L.F., Berse, B., Jackman, R.W., Tognazzi, K., Manseau, E.J., Dvorak, H.F., and Senge, D.R. 1993. Expression of vascular permeability factor (vascular endothelial growth factor) and its receptors in adenocarcinomas of the gastrointestinal tract. *Cancer Res.* 53:4727–4735.

Brown, L.F., Berse B., Jackman, R.W., Tognazzi, K., Guidi, A.J., Dvorak, H.F., Senger, D.R., Connolly, J.L. and Schmitt, S.J. 1995. Expression of vascular permeability factor (vascular endothelial factor) and its receptors in breast cancer. *Hum. Pathol.* 55:510–513.

Carmeliet, P., and Collen, D. 2000. Transgenic mouse models in angiogenesis and cardio-vascular disease. *J. Pathol.* 190:387–405.

Classen-Linke, I., Alfer, J., Krusche, C.A., Chwalisz, K., Rath, W., and Beier, H.M. 2000. Progestins, progesterone receptor modulators, and progesterone antagonists change VEGF release of endothelial cells in culture. *Steroids* 65:763–771.

Cross, M., and Claesson-Welsh, L. 2001. FGF and VEGF function in angiogenesis: Signalling pathways, biological responses, and therapeutic inhibition. *Trends Pharmacol. Sci.* 22:201–207.

D'Angelo, G., Struman, I., Martial, J., and Weiner, R.T. 1995. Activation of mitogen-activated protein kinases by vascular endothelial growth factor and basic fibroblast growth factor in capillary endothelial cells is inhibited by the anti-angiogenic factor 16 kDA-N-terminal fragment of prolactin. *Proc. Natl. Acad. Sci. USA* 92:6374–6378.

Davidoff, A.M., Leary, M.A., Ng, C.Y., and Vanin, E.F. 2001. Gene therapy-mediated expression by tumor cells of the angiogenesis inhibitor Flk-1 results in inhibition of neuroblastoma growth *in vivo*. *J. Pediatr. Surg.* 36:30–36.

De Jong, J.S., van Diest, P.J., Van der Valk, P., and Baak, J.P.A. 1998. Expression of growth factors, growth inhibiting factors, and their receptors in invasive breast cancer I: An inventory in search of autocrine and paracrine loops. *J. Pathol.* 184:44–52.

DeVries, C., Escobedo, J.A., Ueno, H., Houck, K.A., Ferrara, N., and Williams, L.T. 1992. The fms-like tyrosine kinase, a receptor for vascular endothelial growth factor. *Science* 255:989–991.

Drevs, J., Hofmann, I., Hugenschmidt, H., Wittig, C., Madjar, H., Muller, M., Wood, J., Martiny-Baron, G., Unger, C., and Marme, D. 2000. Effects of PTK787/ZK222584, a specific inhibitor of vascular endothelial growth factor receptor tyrosine kinases, on primary tumor, metastasis, vessel density, and blood flow in a murine renal cell carcinoma model. *Cancer Res.* 60:4819–4824.

Dvorak, H., Nagy, J.A., Feng, D., Brown, L. F., and Dvorak, A.M. 1997. Vascular permeability factor/vascular endothelial growth factor and the significance of microvascular hyperpermeability in angiogenesis. *Curr. Top. Microbiol. Immunol.* 237:97–132.

Ellis, L.M., Staley, C.A., Liu, W., Fleming, R.Y., Parish, N.U., Bucana, C.D., and Gallick, G.E. 1998. Down-regulation of vascular endothelial growth factor in a human colon carcinoma cell line transfected with an antisense expression vector specific for src. *J. Biol. Chem.* 273:1052–1057.

Enomoto, T., Okamoto, T., and Sato, J.D. 1994. Vascular endothelial growth factor induces the disorganization of actin stress fibers accompanied by protein phosphorylation and morphological change in Balb/C3T3 cells. *Biochem. Biophys. Res. Commun.* 202:1716–1723.

Eppenberger, U., Kueng, W., Schlaeppi, J.M., Roesel, J.L., Benz, C., Mueller, H., Matter, A., Zuber, M., Luescher, K., Litschgi, M., Schmitt, M. Foekens, J.A., and Eppenberger-Castori, S. 1998. Markers of tumor angiogenesis and proteolysis independently define high- and low-risk subsets of node-negative breast cancer patients *J. Clin. Oncol.* 16:3129–3136.

Fata, J.E., Leco, K.J., Moorehead, R.A., Martin, D.C., and Kokha, R. 1999. TIMP-1 is important for epithelial proliferation and branching morphogenesis during mouse mammary development. *Devel. Biol.* 211:238–254.

Feige, J.J., and Bailly, S. 2000. Molecular bases of angiogenesis. *Bull. Acad. Natl. Med.* 184:537–544.

Ferrera, N. 1999. Molecular and biological properties of vascular endothelial growth factor. *J. Mol. Med.* 77:527–543.

Ferrara, N. 2001. Role of vascular endothelial growth factor in regulation of physiological angiogenesis. *Am. J. Physiol. Cell Physiol.* 280:C1358–C1366.

Ferrara, N. 2002. Role of vascular endothelial growth factor in physiologic and pathologic angiogenesis: Therapeutic implications. *Semin. Oncol.* 29 (Suppl. 16):10–14.

Ferrara, N., Carver-Moore, K., Chen, H., Dowd, M., Lu, L., O'Shea, K.S., Powell-Braxton, L., Hillan, K.J., and Moore, M.W. 1996. Heterozygous embryonic lethality induced by targeted inactivation of the VEGF gene. *Nature* 380:439–442.

Frelin, C., Ladoux, A., and D'Angelo, G. 2000. Vascular endothelial growth factors and angiogenesis. *Ann. Endocrinol. (Paris)* 61: 70–74.

Gerber, H.P., Hillan, K.J., Ryan, A.M., Kowalski, J., Keller, G.A. Rangell, L., Wright, B.D., Radtke, F., Aguet, M., and Ferrara, N. 1999. VEGF is required for growth and survival in neonatal mice. *Development* 126:1149–1159.

Gitay-Goren, H., Halaban, R., Neufeld, G., McLaren, J., Barker, P.J., Wright, K.A., Twentyman, P.R., and Smith, S.K. 1993. Human melanoma cells but not normal melanocytes express vascular endothelial growth factor receptors. *Biochem. Biophys. Res. Commun.* 190:702–708.

Gomez, D.E., Alonso, D.F., Yoshiji, H., and Thorgeirsson, U.P. 1997. Tissue inhibitors of metallo-proteinases: Structure, regulation, and biological functions. *Eur. J. Cell Biol.* 74:111–112.

Guedez, L., McMarlin, A.J., Kingma, D.W., Bennett, T.A., Stetler-Stevenson, M., and Stetler-Stevenson, W.G. 2001. Tissue inhibitor of metalloproteinase-1 alters the tumorigenicity of

Burkitt's lymphoma via divergent effects on tumor growth and angiogenesis. *Am. J. Pathol.* 158:1207–1215.

Hanahan, D., and Folkman, J. 1996. Patterns and emerging mechanisms of the angiogenic switch during tumorigenesis. *Cell* 86:353–364.

Hsu, S.M., Raine, L., and Fanger, H. 1981. The use of avidin-biotin peroxidase complex (ABC) in immunoperoxidase technique. A comparison between ABC and unlabeled antibody (PAP) procedures. *J. Histochem. Cytochem.* 29:577–580.

Hyder, S.M., Murthy, L., and Stancel, G.M. 1998. Progestin regulation of vascular endothelial growth factor in human breast cancer cells. *Cancer Res.* 58:392–395.

Johnson, M.D., Kim, H.C., Chesler, L., Tsao-Wu, G., Bouck, N., and Polverini, P.J. 1994. Inhibition of angiogenesis by tissue inhibitor of metalloproteinase. *J. Cell Physiol.* 160: 194–202.

Khokha, R. 1994. Suppression of the tumorigenic and metastatic abilities of murine B16-F10 melanoma cells *in vivo* by the over-expression of the tissue inhibitor of metalloproteinases-1. *J. Natl. Cancer Inst.* 86:299–304.

Klagsbrun, M., and D'Amore, P.A. 1991. Regulators of angiogenesis. *Annun. Rev. Physiol.* 53:217–239.

Klement, G., Baruchel, S., Rak, J., Man, S., Clark, K., Hicklin, D.J., Bohlen, P., and Kerbel, R.S. 2000. Continuous low-dose therapy with vinblastine and VEGF receptor-2 antibody induces sustained tumor regression without overt toxicity. *J. Clin. Invest.* 105:R15–R24.

Kozin, S., Boucher, Y., Hicklin, D., Bohlen, P., Jain, R.K., and Suit, H.D. 2001. Vascular endothelial growth factor receptor-2-blocking antibody potentiates radiation-induced long-term control of human tumor xenografts. *Cancer Res.* 61:39–44.

Laird, A., Vajkoczy, P., Shawver, L., Thurnher, A., Liang, C., Mohammadi, M., Schlessinger, J., Ullrich, A., Hubbard, S.R., Blake, R.A., Fong, T.A., Strawn, L.M., Sun, L., Tang, C., Hawtin, R., Tang, F., Shenoy, N., Hirth, K.P., McMahon, G., Cherrington. 2000. SU6668 is a potent antiagiogenic and antitumor agent that induces regression of established tumors. *Cancer Res.* 60:4152–4160.

Linderholm, B., Tavelin, B., Grankvist, K., and Henriksson, R. 1998. Vascular endothelial growth factor is of high prognostic value in node-negative breast carcinoma. *J. Clin. Oncol.* 16: 3121–3128.

Liu, J.Y., Wei, Y.Q., Yang, L., Zhao, X., Tian, L., Hou, J.M., Niu, T., Liu, F., Jiang, Y., Hu, B., Wu, Y., Su, J.M., Lou, Y.Y., He, Q.M., Wen, Y.J., Yang, J.L., Kan, B., Mao, Y.Q., Luo, F., and Peng, F. 2003. Immunotherapy of tumors with vaccine based on quail homologous vascular endothelial growth factor receptor-2. *Blood* 102:1815–1823.

Mascood, R., Cai, J., Zheng, T., Smith D.I., Naidu, Y., and Gill, P.S. 1997. Vascular endothelial growth factor/vascular permeability factor is an autocrine growth factor for AIDS-Kaposi sarcoma. *Proc. Natl. Acad. Sci. USA* 94:979–984.

McClelland, R.A., Wilson, D., and Leake, R. 1991. A multicentre study into the reliability of steroid receptor immunocytochemical assay quantification. *Eur. J. Cancer* 27:711–716.

McMahon, G. 2000. VEGF receptor signaling in tumor angiogenesis. *Oncologist 5* (Suppl. 1):3–10.

Millauer, B., Shawer, L.K., Plate, K.H., Risau, W., and Ullrich, A. 1994. Glioblastoma growth inhibited *in vivo* by a dominant-negative Flk-1 mutant. *Nature* 367:576–579.

Millauer, B., Wizigmann-Voos, S., Schuurch, H., Martinez, R., Moller, N.P.H., Risau, W., and Ullrich, A. 1993. High-affinity VEGF binding and developmental expression suggest Flk-1 as a major regulator of vasculogenesis and angiogenesis. *Cell* 72:835–846.

Nakopoulou, L., Lekkas, N., Lazaris, A.C., Athanassiadou, P., Giannopoulou, I., Mavrommatis, J., and Davaris, P. 1999. An immunohistochemical analysis of angiogenesis in invasive breast cancer with correlations to clinicopathologic predictors. *Anticancer Res.* 19:4547–4554.

Nakopoulou, L., Lazaris, A.C., Kavantzas, N., Alexandrou, P., Athanassiadou, P., Keramopoulos, A., and Davaris, P. 2000. DNA Topoisomerase II-Alpha immunoreactivity as a marker of tumor aggressiveness in invasive breast cancer. *Pathobiology* 68:137–143.

Nakopoulou, L., Giannopoulou, I., Stefanaki, K., Panayotopoulou, E., Tsirmpa, I., Alexandrou, P., Mavrommatis, J., Katsarou, S., and Davaris, P. 2002. Enhanced mRNA expression of tissue inhibitor of metalloproteinase-1 (TIMP-1) in breast carcinoma is correlated with adverse prognosis. *J. Path.* 197:307–313.

Neufeld, G., Cohen, T., Gengrinovitch, S., and Poltorak, Z. 1999. Vascular endothelial growth factor (VEGF) and its receptors. *FASEB J.* 13:9–22.

O'Brien, T., Cranston, D., Fuggle, S., Bicknell, R., and Harris, A.L. 1992. Different angiogenic pathways characterize superficial and invasive bladder cancer. *Cancer Res.* 55:510–513.

Overmoyer, B., Robertson, K., and Persons, M. 2001. A phase I pharmacokinetic and pharmacodynamic study of SU5416 and doxorubicin in inflammatory breast cancer. *Proc. Am. Soc. Clin. Oncol.* 20:99a (Abstract 391).

Plate, K.H., Breier, G., and Risau, W. 1994. Molecular mechanisms of development and tumor angiogenesis. *Brain Pathol.* 4:207–218.

Prewett, M., Huber, J., Li, Y., Santiago, A., O'Connor, W., King, K., Overholser, J., Hooper, A., Pytowski, B., Witte, L., Bohlen, P., and Hicklin, D.J. 1999. Antivascular endothelial growth factor receptor (fetal liver kinase 1) monoclonal antibody inhibits tumor angiogenesis and growth of several mouse and human tumors. *Cancer Res.* 59:5209–5218.

Roberts, G.W., Delaat, J., Nagane, M., Huang, S., Cavenee, W.K., and Palade, G.E. 1998. Host microvasculature influence on tumor vascular morphology and endothelial gene expression. *Am. J. Pathol.* 153:1239–1248.

Shaheen, R.M., Tseng, W.W., Vellagas, R., Liu, W., Ahmad, S.A., and Jung, Y.D. 2001. Effects of an antibody to vascular endothelial growth factor receptor-2 on survival, tumor vascularity, and apoptosis in a murine model of colon carcinomatosis. *Int. J. Oncol.* 18:221–226.

Shalaby, F., Rossant, J., Yamaguchi, T.P., Gernenstein, M., Wu, X.-F., Breitman, M.L., and Schuh, A.C. 1995. Failure of blood island formation and vasculogenesis in Flk-1 deficient mice. *Nature* 376:62–66.

Shibaya, M. 1995. Role of VEGF-Flt receptor in normal and tumor angiogenesis. *Adv. Cancer Res.* 67:281–316.

Shinkaruk, S., Bayle, M., Lain, G., and Deleris, G. 2003. Vascular endothelial cell growth factor (VEGF), an emerging target for cancer chemotherapy. *Cur. Med. Chem. Anti-Canc. Agents* 3:95–117.

Sledge, G.W. 2002. Vascular endothelial growth factor in breast cancer: Biologic and therapeutic aspects. *Semin. Oncol.* 29 (Suppl. 11):104–110.

Takahashi, A., Sasaki, H., Jinkim, S., Kakizoe, T., Miyao, N., Sugimura, T., Terada, M., and Tsukamoto, T. 1999. Identification of receptor genes in renal cell carcinoma associated with angiogenesis by differential hybridization technique. *Biochem. Biophys. Res. Commun.* 257:855–859.

Takigawa, M., Nishida, Y., Suzuki, F., Kishi, J., Yamashita, K., and Hayakawa, T. 1990. Induction of angiogenesis in chick yolk-sac membrane by polyamines and its inhibition by tissue inhibitors of metalloproteinases (TIMP-1 and TIMP-2). *Biochem. Biophys. Res. Commun.* 171:1264–1271.

Von Marschall, Z., Cramer, T., Hocker, M., Burde, R., Plath, T., Schirner, M., Heidenreich, R., Breier, G., Riecken, E.O., Wiedermann, B., and Rosewicz, S. 2000. De novo expression of vascular endothelial growth factor in human pancreatic cancer: Evidence for an autocrine mitogenic loop. *Gastroenterology* 119:1358–1372.

Von Marschall, Z., Cramer, T., Hocker, M., Finkenzeller, G., Wiedenmann, B., and Rosewicz, S. 2001. Dual mechanism of vascular endothelial growth factor up-regulation by hypoxia in human hepatocellular carcinoma. *Gut* 48:87–96.

Waltenberger, J., Claessou-Welsh, L., Siegbahn, A., Shibuya, M., and Heldin, C.H. 1994. Different signal transduction properties of KDR and Flt-1, two receptors for vascular endothelial growth factor. *J. Biol. Chem.* 269:26988–26995.

Xie, B., Tam, N.N.C., Tsao, S.W., and Yong, Y.C. 1999. Coexpression of vascular endothelial growth factor (VEGF) and its receptors (Flk-1 and Flt-1) in hormone-induced mammary cancer in the Noble rat. *Br. J. Cancer* 81:1335–1343.

Yang, K., and Cepko, C.L. 1996. Flk-1, a receptor for vascular endothelial growth factor (VEGF), is expressed in retinal progenitor cells. *J. Neurosci.* 16:6089–6099.

Yoshiji, H., Gomez, D.E., Shibuya, M., and Thorgeirsson, U.P. 1996. Expression of vascular endothelial growth factor, its receptors, and other angiogenic factors in human breast cancer. *Cancer Res.* 56:2013–2016.

Zhu, Z., and Witte, L. 1999. Inhibition of tumor growth and metastasis by targeting tumor-associated angiogenesis with antagonists to the receptors of vascular endothelial growth factor. *Invest. New Drugs* 17:195–212.

3

HER-2/neu Gene Amplification and Protein Overexpression in Breast Carcinoma: Immunohistochemistry and Fluorescence *in situ* Hybridization

Akishi Ooi

Introduction

The HER-2/neu gene (also known as c-erbB-2) is a protooncogene located on chromosome 17 (17q12-21.32). This gene encodes a plasma membrane glyco-protein with a molecular weight of 185,000 daltons that is composed of an extracellular ligand-binding domain at the N-terminal region, a transmembrane lipophilic segment, and a C-terminal intracellular region containing a tyrosine kinase domain. After high-affinity ligand binding, HER-2 dimerization occurs, which results in activation of the intrinsic protein tyrosine kinase and tyrosine autophosphorylation. These events initiate a cascade of biochemical and physiological responses that are relayed to transcription factors such as activator protein-1. These, in turn, influence gene expression and hence a multiplicity of cellular events. The introduction of vectors that cause overexpression of HER-2 results in cellular transformation. Many types of epithelial malignancies display increased HER-2 on their surface membranes with the gene amplification.

Overexpression of the HER-2 protein product is closely related to gene amplification in breast cancer. Since the first report by Slamon *et al.* (1987), many authors have reported that the amplification and/or overexpression of HER-2 in breast cancer is significantly correlated with time to relapse and overall survival. In addition, HER-2 abnormalities have attracted a great deal of attention due to the emergence of a new adjuvant therapy that uses a humanized monoclonal antibody against the HER-2 gene product, trastuzumab (Herceptin: Genentech, Inc, South San Francisco, CA). To select breast cancers that are likely to be responsive to the therapy, it is necessary to precisely evaluate HER-2 abnormalities. Two techniques are currently readily applicable: immunohistochemistry (IHC) or fluorescence *in situ* hybridization (FISH), which estimates protein overexpression and gene amplification, respectively.

Handbook of Immunohistochemistry and *in situ* Hybridization of Human Carcinomas, Volume 1: Molecular Genetics; Lung and Breast Carcinomas

259

Copyright © 2004 by Elsevier (USA)
All rights reserved.

HER-2 is significant because it is overexpressed in the plasma membrane, and thus a direct measurement of its expression level would be an ideal diagnostic assay. However, lack of standardization of these IHC methods accounts, in large part, for the fact that a wide range of overexpression rates, varying from 5–85% is reported. In addition, errors relating to antibody selection, fixation, and interpretation have been widely documented (Bartlett *et al.*, 2003). FISH is an alternative laboratory test that can be carried out using paraffin-embedded breast tissues to detect HER-2 gene aberrations. Although IHC and FISH have been approved by the U.S. Food and Drug Administration (FDA) as clinical tests for breast carcinoma, the question remains as to which is the better method of assessing the HER-2 status of patients with this disease. These two methods are described in detail next.

MATERIALS

Immunohistochemistry

1. 20% buffered formalin: 20% formalin in phosphate buffer (0.067 M, pH 7.4).

2. Matsumami adhesive silan (MAS) coated glass slides: Matsunami, Co. Ltd., Tokyo.

3. 0.01 M phosphate buffered saline (PBS, pH 7.2): dissolve 4.5 g $NaH_2PO_4 \cdot 2H_2O$, 32.27 g $Na_2HPO_4 \cdot 12H_2O$, 80 g NaCl in 10 L of distilled water.

4. 3% hydrogen peroxide in water: add 12 ml 30% H_2O_2 to 108 ml of distilled water.

5. 10X normal goat serum: 1 ml normal goat serum and 9 ml PBS.

6. Anti-HER-2 antibody: polyclonal antibody against the internal domain of the human HER-2 protein (Nichirei, Tokyo, Japan), diluted 1:100 in PBS.

7. Link Antibody (Dako Co., Carpinteria, CA): Biotin-labeled, affinity-isolated goat-anti-rabbit and goat-anti-mouse immunoglobulins in PBS, containing carrier protein and 0.015 M sodium azide.

8. Streptavidin-HRP (Dako): Streptavidin conjugated to horseradish peroxidase in PBS containing carrier protein and antimicrobial agents.

9. DAB-hydrogen peroxidase solution: 20-μg 3:3′-diaminobenzidine tetrachloride (DAB) and 20 μl of 30% H_2O_2 in 100 ml of PBS.

Fluorescence *in situ* Hybridization

1. 20X SSC: Dissolve 175.3 g NaCl, 88.2 g *tri*-sodium citrate dihydrate ($Na_3C_6H_5O_7 \cdot 2H_2O$) in 800 ml distilled water, adjust the pH to 7.0 with 1 N HCl, and adjust to 1 L with distilled water.

2. 20% sodium bisulfate/2X SSC: Dissolve 8 g sodium bisulfate in 40 ml of 2X SSC.

3. Proteinase K stock solution (25 mg/ml): Dissolve 500 mg proteinase K (Rosh Diagnostics Co, Ltd.; Cod. No. 1 000 144) in 20 ml distilled water, and divide into 800 μl aliquots. Store at −20°C.

4. Proteinase K (0.5 mg/ml)/10 mM Tris (pH 7.8)/ 5 mM ethylenediamine tetra-acetic acid (EDTA)/0.5% sodium dodecyl sulfate (SDS): Dissolve 800 μl proteinase K stock solution (25 mg/ml) in 40 ml of 10 mM Tris (pH 7.8) containing 5 mM EDTA and 0.5% SDS.

5. 10 mM Tris (pH 7.8)/5 mM EDTA/0.5% SDS: dissolve 1.21 g of Tris (hydroxymethyl)-aminomethane, 1.86 g of ethylenediamine tetra-acetic acid disodium salt, and 5.0 g of SDS in distilled water, adjust the pH to 7.8 with 1 N HCl, and adjust to 1 L with distilled water. Autoclave at 121°C for 20 min.

6. Locus Specific Identifier DNA Probe (LSI) HER-2/neu SpectrumOrange/Chromosome Enumeration DNA Probe (CEP), 17 SpectrumGreen Dual Color Probe (Vysis Inc., Downers Grove, IL).

7. Rubber cement (Paper Bond, Kokuyo Co. Ltd., Tokyo, Japan).

8. 70% formamide in 2X SSC: 28 ml deionized formamide (Nakarai Tesque, Co. Ltd.; COD. No. 16345-65), 4 ml 20X SSC and 8 ml distilled water.

9. 10% (W/V) Nonidet P-40: Dissolve 10 g of Nonidet P-40 in distilled water, and adjust to 100 ml with distilled water.

10. 0.1% NP-40/2X SSC: 0.4 ml 10% Nonidet P-40 and 39.6 ml of 2X SSC.

11. DAPI II (Vysis Inc., Downers Grove, IL): 4′, 6-diamidine-2′-phenylindole dihydrochloride and p-phenylenediamine in phosphate-buffered saline and glycerol.

12. 50% formamide in 2X SSC: 20 ml deionized formamide, 4 ml 20X SSC and 16 ml distilled water.

METHODS

Immunohistochemistry

Resected breast tissue was immediately immersed in 20% buffered neutral formalin, fixed overnight, and embedded in paraffin according to standard procedures. Because the immunoreactivity diminishes with time in formalin-fixed, paraffin-embedded sections stored on MAS-coated glass slides at room temperature, sections (3 μm) were stained within 6 weeks of being cut.

1. Deparaffinate paraffin-embedded 3-μm-thick sections stored on MAS-coated glass slides by five successive 10-min washes in xylene, and hydrate with graded ethanol.

2. Immerse slides in 3% hydrogen peroxide in water for 10 min.

3. Rinse slides by immersing them three times in PBS for 5 min each.

4. Incubate slides with 10X normal goat serum for 30 min.

5. Rinse slides by immersing them three times in PBS for 5 min each.

6. Apply the anti-HER-2 antibody solution and incubate slides overnight in a moist chamber at 4°C.

7. Wash slides in PBS three times for 5 min each.

8. Incubate with link antibody for 20 min.

9. Wash slide in PBS three times for 5 min each.

10. Incubate slides with streptavidin-HRP for 20 min.

11. Wash slides by immersing them three times in PBS for 5 min each.

12. Develop the peroxidase reaction by immersing slides in DAB-hydrogen peroxidase solution for 5 min.

13. Wash slides by immersing them three times in distilled water for 5 min each.

14. Counterstain with hematoxylin.

15. Wash slides in tap water, dehydrate using graded ethanol, clear in xylene, and mount.

For evaluation of HER-2 positivity, each tumor was scored according to the criteria recommended by Dako HercepTest, but the number of positive cells was not considered. The criteria were: negative, no discernible staining or background type staining; 1+, definite cytoplasmic staining and/or equivocal discontinuous membrane staining; 2+, unequivocal membrane staining with moderate intensity; 3+ strong and complete plasma membrane staining.

Fluorescence *in situ* Hybridization

1. Mark hybridization areas with a diamond-tipped scriber on the bottom of the specimen slide.

2. Aliquot the probe mixture into microcentrifuge tubes. Ten μl of the probe mixture will be applied per 24 × 24 mm area of sample. Place the tube in a 75°C water bath for 5 min. Immediately chill the tube in ice water, and keep it away from light until it is used.

3. Deparaffinate the paraffin-embedded 3-μm-thick sections stored on MAS-coated glass slides by three successive 10-min washes in xylene, followed by three 5-min washes in absolute ethanol and two 1-min washes in distilled water.

4. Incubate slides in 20% sodium bisulfate/2X SSC at 43°C for 20 min.

5. Wash slides in 2X SSC for 5 min.

6. Incubate slides in a glass Coplin jar containing 40 ml of proteinase K (0.5 mg/ml)/ at 37°C for 20 min.

7. Wash slides in 2X SSC three times for 5 min each.

8. Prewarm slides to 37°C on a slide warmer. Prepare 40 ml of denaturation solution (70% deionized formamide/2X SSC) and prewarm the solution to 73°C.

Denature slides by immersing in a glass Coplin jar at 73°C for 5 min.

9. Immediately transfer slides to a Coplin jar containing 40 ml of ice-cold 70% ethanol and rinse for 2 min. Then carry out successive rinses in ice-cold 80%, 90% and 100% ethanol for 2 min each.

10. Allow slides to dry, aided by cool air flow.

11. Apply the DNA probe solution to each sample. Cover sample with a cover glass and seal with rubber cement.

12. Heat slides at 69°C for 5 min on a slide warmer.

13. Incubate the slides in a prewarmed moist chamber at 37°C overnight for hybridization.

14. Remove rubber cement with a rubber stick. Carefully remove the cover glass. Wash slides in a Coplin jar containing 50% formamide/2X SSC at 45°C three times for 10 min each.

15. Rinse slides in 2X SSC at 45°C for 10 min.

16. Rinse slides in 0.1% NP-40/2X SSC at 45°C for 5 min.

17. Rinse slides in 2X SSC at room temperature for 5 min.

18. Add 10 μl DAPI II (Vysis) to each section. Cover the section with a glass coverslip and seal using nail cement.

19. Examine slides with a fluorescence microscope (Olympus, Tokyo, Japan) equipped with Triple Bandpass Filter sets (Vysis) for DAPI II, SpectrumOrange, and SpectrumGreen.

A cell is considered to show HER-2 amplification when a definite cluster or more than 10 orange signals are found, as in a previous FISH study. The FISH images were taken using a photographic camera and records were stored as film slides.

RESULTS AND DISCUSSION

There is general consensus that the frequency of HER-2 overexpression in breast cancer is between 20–30%, although this figure used to be much more controversial (Bartlett *et al.*, 2003). There are still discrepancies concerning frequencies of HER-2 overexpression in cancers involving other organs. An accurate comparison of the extent of HER-2 reactivity as indicated by IHC in cancers has been hampered by differences in HER-2 detection techniques, different choices of antibodies used in various techniques, and different criteria for evaluating the experimental results (Bartlett *et al.*, 2003). Antibodies that can reliably detect HER-2 expression in fixed and paraffin-embedded pathological materials have only recently become available. In our studies, overexpression was defined as staining of the cellular membrane. However, some previous studies reporting high frequencies of overexpression adopted cytoplasmic staining as an additional positive indicator. Furthermore, because epithelial cells usually express HER-2, it is important to assess whether there is relative overexpression of HER-2 in carcinoma cells, compared with expression levels in healthy epithelial cells, in the cancerous organs.

Comparisons of the specificities and characteristics of different antibodies used in IHC are readily found (Press *et al.*, 1994). Concerning breast carcinomas, standardization of IHC methods to detect HER-2 overexpression has been recently attempted by the introduction of tests such as the HercepTest by Dako. We examined HER-2 overexpression in carcinomas of the breast, stomach, lung, colon, and ovary by conventional IHC using a polyclonal antibody directed against the internal domain of HER-2 (Kobayashi *et al.*, 2002; Takehana *et al.*, 2002) and obtained good results: equivocal stainings were seen only occasionally and little background staining was observed. In these studies we did not employ any antigen retrieval procedures as described in the IHC protocol. However, in our recent study of bile duct carcinomas, we found that the sensitivity of the antibody is enhanced by antigen retrieval with autoclaving the section in 0.01 M citrate buffer (pH 7.0) at 110°C for 10 min between step 1 and 2.

To detect a numerical aberration of a gene, dual-color FISH was applied: a centromeric probe was used as a reference probe to assist in distinguishing real gene amplification from an increased gene number resulting from chromosomal polysomy at gene located. We previously performed dual-color FISH using a digoxigenin-labeled HER-2 specific probe and a biotin-labeled centromere 17 probe (Hara *et al.*, 1998; Ishikawa *et al.*, 1997). In those studies, multistep immunohistochemical procedures were necessary for signal detection, the procedure was rather time-consuming, and it was occasionally difficult to obtain an optimal balance of the two fluorescent signals (from fluorescein isothiocyanate, and rhodamine).

Later, directly labeled fluorescent probes became available from Vysis, Inc., and when these were used the FISH procedure became less time-consuming, as described in the protocol. A fluorescence microscope with a triple band filter is probably the only equipment purchase necessary for a nonresearch laboratory. The fluorescence microscope should be equipped with two additional single-band filters that are specific to each fluorescent wavelength used, because occasionally these fluorescence-specific filters produce clearer signals than the triple band filter.

The FISH protocol has also been modified to incorporate an intermittent, short-term microwave treatment of the sample during the initial period of hybridization. This was exploited by Kitayama *et al.* (2000), who demonstrated a significant improvement in hybridization results in several cases, although several additional steps, including use of a microwave processor with a temperature sensor, was necessary. In our experience, the critical step in FISH using paraffin-embedded tissue is removal of nuclear proteins by enzymatic digestion. The optimal digestion conditions may be modified to accommodate each section, because fixation conditions were different for various individual specimens. PathVysion is the most recently developed dual-color FISH detection kit available from Vysis. Vysis also provides its original protocol for using PathVysion and a kit with digestive agents (Paraffin Pretreatment Kit.)

FISH allows determination of three gene amplification parameters: the fraction of cells with an amplified gene, the amplification levels within the subpopulations, and the amplification patterns, which may appear as clustered signals or as multiple scattered signals indicating gene amplification. Different investigators have used different criteria for HER-2 amplification detected by FISH. The most standardized method is to score tumors according to the ratio of the HER-2 signal to the number of copies of chromosome 17 for each cancer cell, calculating the mean ratio for each tumor. A mean ratio greater than two is generally taken to indicate amplification positive, whereas signals near a cutoff point of 1.8–2.2 indicate low-level amplification.

Alternatively, in previous studies (Kobayashi *et al.*, 2002; Kunitomo *et al.*, 2002), we defined cells that had a greater number of HER-2 than chromosome 17 signals as amplification positive, and cells in which both centromere 17 and HER-2 signals were increased equally as polysomy 17. Cells with positive HER-2 amplification were further divided into high- and low-level amplification categories according to the ratio of HER-2 signal to chromosome 17 number, with a greater ratio of 4 indicating high-level amplification. Some investigators, however, used single-color FISH and did not correct the HER-2 signal by the number of copies of chromosome 17. In those studies, more than four signals per nucleus were considered as positive amplification (Couturier *et al.*, 2000; Jacobs *et al.*, 1999; Wang *et al.*, 2000), more than two but less than eight signals as low-level amplification, and more than eight as high-level amplification (Xing *et al.*, 1996). In our own study all the tumors with high-level amplification (greater than 4 of mean ratio HER-2 to chromosome 17 by dual-color FISH) showed the mean number of more than 10 HER-2 signals. Thus, alternatively, in single-color FISH, by using 10 as the cutoff point, high-level amplification could be clearly discriminated from low-level amplification, polysomy 17, or no amplification.

It is usually not difficult to identify the breast cancer cells with high-level amplification, because most of them can be recognized qualitatively by large or small clusters of signals. On the other hand, to detect low-level amplification, which is defined quantitatively as cells having two or three additional signals of HER-2 to centromeric 17 signal, or cells with HER-2 to centromeric 17 signal ratios less than 4, precise enumeration of both signals is necessary. In breast cancer, cells with polysomy 17 in which HER-2 signals are greater than two are not unusual, and cells in S- or G2M- phase have doubled HER-2 signals due to the sister chromatids. Because of the nuclear truncation that is inevitable in FISH carried out on tissue sections, only portion of the signals per nucleus can be observed. It is therefore very difficult to differentiate polysomy from low-level amplification, even when centromere 17 signals are used for reference (Kunitomo et al., 2002).

In our FISH evaluation of 235 consecutive resected breast carcinomas (Kunitomo et al., 2002), we found that in 14 tumors the cancer cells typically had one or occasionally two large clusters (LC) of HER-2 signals (Figure 37A). Precise quantitation was impossible in most cancer cells, due to the tightly coalesced signals, but image analysis of each LC showed a fluorescence signal that was enhanced ~ 20-fold over that of single-copy signals. In 17 tumors, the cancer cells had smaller miniclusters (CL) and numerous scattered signals (SS), as shown in Figure 37B. In 12 cases, the cancer nuclei had homogeneous multiple scattered signals (MS) (Figure 37C). In mammalian cells, highly amplified DNA is found in two distinct structures: homogeneously staining regions (HSRs) that are located within a chromosomal site in the form of expanded chromosomal structures, and double minute chromosomes (DM), which are centromere-free circular structures (Fried et al., 1991). Interestingly, it has been shown in several

cell lines that amplified genes in the DMs were integrated to HSRs, but HSR has not been observed to be broken down to generate DM (Ruiz et al., 1990).

It is generally accepted that LC signals found by FISH correspond to amplified signals in HSR, and MS signals correspond to DMs, excluding a few exceptional cases. In fact, in our previous comparative studies imaging interphase nuclei and metaphase spreads of cell lines with known amplification types, the two types of FISH images, LC and MS signals, that were found in the interphase nuclei corresponded to HSRs and DMs, respectively (Hara et al., 1998; Ishikawa et al., 1997). Most FISH studies reporting HER-2 amplification in breast cancers were LC-type signals, and thus the amplicon was probably in an HSR. Although the exact karyotypic changes corresponding to FISH images of mini-CL and SS are unknown, they may represent complex translocations of the amplified genes to other chromosomes. Coene et al. (1997) demonstrated that breast cancer cell lines having no HSRs produce similar FISH images in interphase nuclei. MSs reminiscent of the amplicon signal in DM were rare, although there have been a few case reports. In Mitelman's tabulation of conventional karyotype analysis of 337 breast cancers, five cases of DM chromosomes were reported in addition to 35 breast cancers with HSRs (Mitelman, 1994). To establish the status of the amplicons, FISH analyses of metaphase spreads are needed.

There have been a limited number of studies directly comparing the IHC and FISH methods in evaluating paraffin-embedded breast cancer tissues (Couturier et al., 2000; Jacobs et al., 1999; Wang et al., 2000). In these studies, the concordance between IHC and FISH ranged from 73–98%. Most studies used four-tier systems (no staining or 1+ to 3+ staining), with 2+ and 3+ staining generally accepted as positive staining. The concordance rates for 3+ staining were very high, but

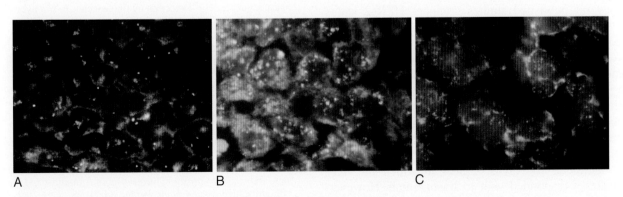

A   B   C

Figure 37 Dual-color FISH of HER-2 gene amplification. Amplified HER-2 genes forms large clusters **A:** small clusters and scattered signals **B:** or multiple scattered signals **C:** Amplification 570×.

those for 2+ staining were variable. In our study, the concordance for 2+ and 3+ staining was 85.9%. However, all of the 3+ staining tumors had high-level gene amplification and only 4 of 27 (14.8%) 2+ tumors had high-level amplification. If only 3+ staining was considered a definite indicator of overexpression, the concordance rate for our study was 95.9% (Kobayashi *et al.*, 2002).

In September 1998, the FDA approved trastuzumab, a recombinant monoclonal antibody targeting the HER-2 protein, for the treatment of metastatic breast cancer. There have been a number of clinical reports indicating that this therapy is effective for metastatic breast cancer (Slamon *et al.*, 2001). At the present time, the indication of trastuzumab therapy is for metastatic cancers. However, in the majority of cases, HER-2 status is determined for the primary tumors. One question that arises is whether HER-2 assays carried out on primary tumors reliably predict the HER-2 status of the metastatic foci. Thus, there are concerns regarding possible molecular heterogeneity of HER-2 amplification within a tumor, or between primary tumors and metastatic cancer foci.

Intratumoral heterogeneity of HER-2 overexpression, especially between the ductal carcinoma *in situ* and invasive components, has been noted by some authors. However, because of technical difficulties, studies focusing on intratumoral heterogeneity of gene amplification are sparse. In our study, IHC intratumoral heterogeneity was found in 33% (18 of 52) of tumors that overexpressed HER-2 and in the two tumors with confirmed genetical heterogeneity by FISH. Glockner *et al.*, (2002), using laser-assisted microdissection and quantitative polymerase chain reaction (PCR), found heterogeneity occurring in 36% of HER-2 amplified samples and concluded that HER-2 amplification is the most stable amplification in individual tumors compared with C-myc or cyclin D1. The FISH method may be the best way to reveal genetic heterogeneity of HER-2 amplification, because it enables the detection of gene amplification on a cell-by-cell basis.

Several published studies compare HER-2 status between primary and metastatic sites using IHC and/or FISH. Most have reported a high, though not complete, level of consistency in HER-2 status between primary tumors and metastases (Simon *et al.* 2001; Thor, 2001). Thor *et al.*, (2001) found a nearly 20% discordance in HER-2 expression between primary tumors and distant visceral metastases, and raised the possibility that treatment with trastuzumab caused the selection of resistant subclones lacking this alteration. In fact, we reported a case of metastatic breast cancer in which the liver metastases responded well to trastuzumab. However, FISH examination of the pleuritis carcinomatosa that was refractory to the agent revealed an outgrowth of HER-2-negative cancer cells (Kunitomo *et al.*, in press). Future studies will be required to clarify the genetic heterogeneity of HER-2 expression between metastatic foci, which may be able to distinguish between cancer cells that are responsive and refractory to trastuzumab.

In conclusion, FISH is an excellent tool to clarify the status of HER-2 amplification in breast cancers. Most investigators recommend that the optimal approach to detect tumors with HER-2 aberrations is to combine both the IHC and FISH assays. The IHC assay can be used for initial screening, followed by FISH to analyze the cases for which IHC was inconclusive. The FISH method may also prove useful to evaluate the therapeutic effects of trastuzumab in the course of treatment (see also Part IV, Chapters 4 and 5).

References

Bartlett, J., Mallon, E., and Cooke, T. 2003. The clinical evaluation of HER-2 status: Which test to use? *J. Pathol. 199:*411–417.

Coene, E.D., Schelfhout, V., Winkler, R.A., Schelfout, A.M., Van Roy, N., Grooteclaes, M., Speleman, F., and De Potter, C.R. 1997. Amplification units and translocation at chromosome 17q and c-erbB-2 overexpression in the pathogenesis of breast cancer. *Virch. Arch. 430:*365–372.

Couturier, J., Vincent-Salomon, A., Nicolas, A., Beuzeboc, P., Mouret, E., Zafrani, B., and Sastre-Garau, X. 2000. Strong correlation between results of fluorescent *in situ* hybridization and immunohistochemistry for the assessment of the ERBB2 (HER-2/neu) gene status in breast carcinoma. *Mod. Pathol. 13:*1238–1243.

Fried, M., Feo, S., and Heard, E. 1991. The role of inverted duplication in the generation of gene amplification in mammalian cells. *Biochim. Biophys. Acta 1090:*143–155.

Glockner, S., Buurman, H., Kleeberger, W., Lehmann, U., and Kreipe, H. 2002. Marked intramoral heterogeneity of c-myc and cyclinD1 but not of c-erbB2 amplification in breast cancer. *Lab. Invest. 82:*1419–1426.

Hara, T., Ooi, A., Kobayashi, M., Mai, M., Yanagihara, K., and Nakanishi, I., 1998. Amplification of C-Myc, K-sam, and c-met in gastric cancers: Detection by fluorescence *in situ* hybridization. *Lab. Invest. 78:*1143–1153.

Ishikawa, T., Kobayashi, M., Mai, M., Suzuki, T., and Ooi, A. 1997. Amplification of the c-erbB-2 (HER-2/neu) gene in gastric cancer cells. Detection by fluorescence *in situ* hybridization *Am. J. Pathol. 151:*761–768.

Jacobs, T.W., Gown, A.M., Yaziji, H., Barnes, M.J., and Schnitt, S.J. 1999. Comparison of fluorescence *in situ* hybridization and immunohistochemistry for the evaluation of HER-2/neu in breast cancer. *J. Clin. Oncol. 17:*1974–1982.

Kitayama, Y., Igarashi, H., and Sugimura, H. 2000. Initial intermittent microwave irradiation for fluorescence *in situ* hybridization analysis in paraffin-embedded tissue sections of gastrointestinal neoplasia. *Lab. Invest. 80:*779–781.

Kobayashi, M., Ooi, A., Oda, Y., and Nakanishi, I. 2002. Protein overexpression and gene amplification of c-erbB-2 in breast carcinomas: A comparative study of immunohistochemistry and fluorescence *in situ* hybridization of formalin-fixed, paraffin-embedded tissues. *Hum. Pathol. 33:*21–28.

Kunitomo, K., Takehana, T., Inoue, S., Fujii, H., Suzuki, S., Matsumoto, Y., and Ooi, A. 2002. Detection of c-erbB-2

(HER-2/neu) amplification in breast carcinoma by fluorescence *in situ* hybridization on tissue sections and imprinted cells. *Pathol. Int.* 53:451–457.

Kunitomo, M., Inoue, S., Ichihara, F., Kono, K., Fujii, H., Matsumoto, Y., and Ooi, A. 2003. A case of metastatic breast cancer with outgrowth of HER-2-negative cells after eradication of HER-2-positive cells by humanized anti-HER-2 monoclonal antibody (trastuzumab) combined with docetaxel. *Hum. Pathol. In Press.*

Mitelman, F. 1994. *Catalog of Chromosome Aberrations in Cancer.* 5th ed. New York: Wiley-Liss.

Press, M.F., Hung, G., Godolphin, W., and Slamon, D.J. 1994. Sensitivity of HER-2/neu antibodies in archival tissue samples: Potential source of error in immunohistochemical studies of oncogene expression. *Cancer Res.* 54: 2771–2777.

Ruiz, J.C., and Wahl, G.M. 1990. Chromosomal destabilization during gene amplification. *Mol. Cell. Biol. 10:* 3056–3066.

Simon, R., Nocito, A., Hubscher, T., Bucher, C., Torhorst, J., Schraml, P., Bubendorf, L., Mihatsch, M.M., Moch, H., Wilber, K., Schotzau, A., Kononen, J., and Sauter, G. 2001. Patterns of Her-2/neu amplification and overexpression in primary and metastatic breat cancer. *J. Natl. Cancer Inst. 93:* 1141–1146.

Slamon, D.J., Clark, G.M., Wong, S., Levin, W.J., Ullrich, A., and McGuire, W.L. 1987. Human breast cancer: Correlation of relapse and survival with amplification of the HER-2/neu oncogene. *Science 235:*177–182.

Slamon, D.J., Leyland-Jones, B., Shak, S., Fuchs, H., Paton, V., Bajamonde, A., Fleming, T., Eiermann, W., Wolter, J., Pegram, M., Baselga, J., and Norton, L. 2001. Use of chemotherapy plus a monoclonal antibody against HER-2 for metastatic breast cancer that overexpresses HER-2. *N. Engl. J. Med.* 344:783–792.

Takehana, T., Kunitomo, K., Kono, K., Kitahara, F., Iizuka, H., Matsumoto, Y., Fujino, M.A., and Ooi, A. 2002. Status of c-erbB-2 in gastric adenocarcinoma: A comparative study of immunohistochemistry, fluorescence *in situ* hybridization, and enzyme-linked immuno-sorbent assay. *Int. J. Cancer* 98:833–837.

Thor, A. 2001. Are patterns of HER-2/neu amplification and expression among primary tumors and regional metastases indicative of those in distant metastases and predictive of Herceptin response? *J. Natl. Cancer Inst.* 93:1120–1121.

Wang, S., Saboorian, M.H., Frenkel, E., Hynan, L., Gokaslan, S.T., and Ashfaq, R. 2000. Laboratory assessment of the status of HER-2/neu protein and oncogene in breast cancer specimens: Comparison of immunohistochemistry assay with fluorescence *in situ* hybridisation assays. *J. Clin. Pathol. 53:* 374–381.

Xing, W.R., Gilchrist, K.W., Harris, C.P., Samson, W., and Meisner, L.F. 1996. FISH detection of HER-2/neu oncogene amplification in early onset breast cancer. *Breast Cancer Res. Treat 39:*203–212.

4

HER-2/neu Amplification Detected by Fluorescence *in situ* Hybridization in Cytological Samples from Breast Cancer

Cecilia Bozzetti

Introduction

The proto-oncogene HER-2/neu (also called c-erbB-2) is located on chromosome 17q and encodes a 185-kDa transmembrane glycoprotein (p185) with tyrosine kinase activity that has partial homology with members of the epidermal growth factor (EGF) receptor family (Coussens *et al.*, 1985). Amplification of the HER-2/neu gene and/or p185 overexpression has been reported in 20–30% of primary breast cancer. Both alterations have been correlated with a more aggressive phenotype and worse disease-free and overall survival (Allred *et al.*, 1992; Gullick, 1990).

In light of this evidence, an innovative therapeutic approach targeting the receptor by monoclonal antibodies (MAb) to the HER-2/neu protein product has been promoted. Trastuzumab (Herceptin) is a high-affinity anti-HER-2/neu MAb developed by Genentech (San Francisco, CA) for the treatment of advanced breast disease. The results of the initial trastuzumab trials have shown clearly that this agent represents a

new specific therapy active in patients with advanced HER-2-positive breast cancer either alone or in combination with anthracycline and taxane-based chemotherapy (Vogel *et al.*, 2002).

Despite these results, the method of choice to determine whether a tumor is HER-2/neu positive is still under debate. Laboratory methods for assessing HER-2/neu amplification or HER-2/neu protein overexpression include Southern, Northern, and Western Blot techniques, fluorescence *in situ* hybridization (FISH), and immunohistochemistry (IHC). Amplification of the HER-2/neu gene is usually analyzed by FISH and HER-2/neu protein overexpression by IHC on tissue sections from the primary tumor. Assessment of HER-2/neu by IHC is relatively simple and inexpensive, but the use of different antibodies as well as interpretative difficulties do not make this technique highly reproducible. Gene-based techniques, such as FISH, have less variability, but require special equipment and are more expensive. Nevertheless, FISH seems to be a more powerful technique. In some clinical situations,

267

Copyright © 2004 by Elsevier (USA)
All rights reserved.

such as preoperative chemotherapy in primary disease or unavailability of archival samples in metastatic patients, the determination of HER-2/neu status by FISH on cytological material is helpful. This approach allows the selection of the best treatment. The biology of HER-2/neu and its use as a target for antibody-based therapeutics, as well as current methodologies for HER-2/neu testing and its evolving role in the management of breast cancer has been reviewed by Ravdin (2000). Fine-needle aspiration, a well-established method in the diagnosis and biological characterization of breast cancer, has been successfully combined with molecular and cytogenetic analysis. Only a few studies concerning the feasibility of dual-color FISH on cytological material from primary breast cancer fine-needle aspirates (FNAs) has been published (e.g., Bozzetti et al., 2002).

Though metastases are the target for trastuzumab-based therapy, HER-2/neu status is usually evaluated on the primary tumor, since metastatic lesions are rarely removed or biopsied to obtain a biological characterization. It has been postulated that breast cancer is a heterogeneous disease where, among different clones, an early stem line clone might evolve independently in the primary tumor and its metastasis, suggesting that clonal diversification and heterogeneity may account for cancer progression and metastatic dissemination (Kuukasjarvi et al., 1997). Tumor heterogeneity may explain why biomarkers of prognosis or of therapy responsiveness, evaluated exclusively from primary tumors, may not completely reflect all the biological properties of metastatic breast cancer. Amplification or overexpression of HER-2/neu may confer biological advantage to tumor cells, and genetic instability induced by amplification might generate a more aggressive cancer phenotype. Theoretically, this could explain the high proportion of patients that ultimately do not respond to trastuzumab-based therapies.

A number of studies have shown a high level of concordance, although not complete, between HER-2/neu status evaluated on primary tumor and lymph node metastases by means of both IHC (Masood and Bui, 2000) and FISH (Xu et al., 2002). However, few studies have compared HER-2/neu status between primary tumors and paired distant metastases, showing an heterogeneous pattern of results (Gancberg et al., 2002).

To our knowledge, none of the studies that evaluated HER-2/neu status on metastatic sites from breast cancer patients have been carried out on cytological material. In our experience, fine-needle aspiration biopsy (FNAB), possibly coupled with ultrasound for the sampling of deep lesions, is a relatively safe and less-invasive alternative to surgical biopsy to obtain cellular material for an updated definition of HER-2/neu status by FISH. This protocol circumvents many of the drawbacks related to immunocytochemistry of cytological samples (Nizzoli et al., 2003).

In a previous publication (Bozzetti et al., 2002) we reported that HER-2/neu gene amplification can be reliably estimated by FISH on FNAs from primary breast cancer, and a good correlation was found between FISH cytology and FISH and/or IHC results from the corresponding histological sections. More recently, we demonstrated the feasibility of HER-2/neu evaluation by FISH on cytological samples obtained from distant metastatic lesions of breast cancer (Bozzetti et al., 2003).

Our study emphasizes the feasibility and advantages of the two rapid and very informative techniques (FNAB and FISH). These techniques were performed to ascertain the malignant nature of a suspicious lesion and obtain predictive markers for response. Since the advent of trastuzumab, the characterization of the molecular profile in metastatic breast disease is becoming increasingly important for targeted therapies selection.

MATERIALS

Breast Cancer Specimens

A first series of samples included FNAs from 66 primary breast cancer patients, which were stained with May-Grunwald Giemsa (MGG) stain for routine diagnostic cytology as well as HER-2/neu evaluation by FISH. After surgery, the corresponding sections of archival formalin-fixed, paraffin-embedded tissue were evaluated for HER-2/neu by FISH and IHC. A second series of samples included cytological samples from metastatic lesions obtained from 22 patients that presented, at different times after primary treatment, metastatic relapses in liver (12 patients), in pleura (three patients), in peritoneum (three patients), and in skin (four patients). Cytologic smears from metastases were submitted for routine diagnostic cytology as well as for HER-2/neu evaluation by FISH. Moreover, their corresponding primary breast tumors were evaluated by FISH either on paraffin histological sections, when available, or on destained archival cytological smears.

Laboratory Reagents

1. PathVysion HER-2 DNA Probe Kit. Components: Locus Specific Identifier (LSI) HER-2/neu SpectrumOrange/CEP 17 SpectrumGreen DNA Probe.
   20X saline-sodium citrate (SSC) salts.
   NP-40.

DAPI counterstain (4,6-diamidino-2-phenylindole) and 1,4-phenylenediamine in phosphate-buffered saline and glycerol.

2. 20X SSC (3 M sodium chloride, 0.3 M sodium citrate, pH 5.3): 66 g 20X SSC, 200 ml distilled water; adjust pH to 5.3 with concentrated HCl. Bring volume to 250 ml with distilled water. Filter through a 0.45-μm pore filtration unit. Can be stored at room temperature for up to 6 months.

3. Wash buffer (2X SSC/0.3% NP-40, pH 7.0–7.5): 100 ml 20X SSC, pH 5.3, 847 ml distilled water, 3 ml NP-40; adjust pH to 7.0–7.5 with 1 N NaOH. Bring volume to 1 L with distilled water. Filter through a 0.45-μm pore filtration unit. Discard used solution at the end of each day. Unused solution can be stored at room temperature for up to 6 months.

4. 70%, 85%, and 100% ethanol.

5. Proteinase K solution: 10 mg/ml proteinase K in PBS. Store 150-μl aliquots at −10°C to −20°C for up to 6 months.

6. Xylene.

7. Carnoy's solution: 60 ml ethanol 100%, 30 ml chloroform, 10 ml glacial acetic acid.

8. Rubber cement.

9. Immersion oil for microscopy.

Laboratory Equipment

1. HYBrite Denaturation/Hybridization System for FISH (Vysis).

2. Water bath (37°C ± 1°C, 72°C ± 1°C).

3. Vortex mixer.

4. Microcentrifuge.

5. pH meter.

6. Calibrated thermometer.

7. Coplin jars: vertical staining jars.

8. 22 mm × 22 mm glass coverslip.

9. 0.45-μm pore filtration unit.

METHOD

Fluorescence *in situ* Hybridization—Principles of the Method

The FISH technique is based on the ability of single-stranded DNA to anneal to complementary DNA. Detailed procedures for FISH have been reviewed (Le Beau, 1993). The target DNA is the nuclear DNA of interphase cells smeared on a glass microscope slide. The test probe is directly labeled with a fluorochrome. To allow hybridization of complementary sequences to occur, a formamide solution containing the predenaturated labeled DNA probe is applied to the microscope slide. The slide is coverslipped, sealed, and the cellular

DNA is denatured by heating to form single-stranded DNA. Hybridization is allowed to occur by overnight incubation at 37°C. Thereafter, the unbound labeled probe is removed by wash buffer and cells are counterstained with DAPI. The hybridized probe is detected by fluorescence microscopy.

Our procedure is based on the application of a dual-color technique, using a probe for HER-2/neu gene together with a probe that hybridizes to a pericentromeric region of chromosome 17. The LSI HER-2/neu DNA probe is directly labeled with Spectrum-Orange fluorophore and targets the DNA sequences ~ 190 Kb in size spanning the HER-2/neu gene within band 17q11.2-q12 on long arm of chromosome 17. The Chromosome Enumerator Probe (CEP) 17 (alpha satellite) DNA probe is a 5.4-Kb probe targeting to the centromeric region of chromosome 17 (17p11.1-q11.1) and is directly labeled with SpectrumGreen fluorophore. Path-Vysion HER-2 DNA Probe Kit provides LSI HER-2/neu SpectrumOrange and CEP 17 Spectrum-Green DNA probes premixed and predenaturated in hybridization buffer. The probe mixture allows for the simultaneous determination of the HER-2/neu gene locus copy number and chromosome 17 copy number in interphase cells. Determination of the ratio of HER-2/neu to chromosome 17 copy number is useful in the discrimination of aneusomy of chromosome 17 from true gene amplification of the HER-2/neu. This allows for proper assessment of amplification levels, especially low levels of amplification.

In our procedure, both denaturation and hybridization take place in a HYBrite Denaturation/Hybridization System for FISH (Vysis). This system is optimized for performing FISH assay by a reproducible denaturation/hybridization temperature control and an accurate timing of events. Slides are placed on a heating surface where denaturation and hybridization successively occur, simplifying standard FISH steps in particular by eliminating formamide-denaturation baths.

HER-2/neu FISH on Cytologic Samples

Collection of Samples

Samples from primary breast cancer and superficial metastases are obtained by multidirectional FNAB using a 22-gauge needle and 20-ml syringe. Samples from liver and other deep metastases are obtained by ultrasound-guided FNAB. Cell preparations from metastatic pleural and ascitic fluid are cytospins. Proceed as follows:

1a. Samples obtained by FNAB from superficial or deep lesion: smear the aspirated material on at least two glass slides, and air-dry.

1b. Samples obtained from pleural and ascitic fluid: cytocentrifuge the cellular suspension for 5 min at 600 rpm (45 g) onto glass slides, and air-dry.

2. Stain one or more slide with MGG stain for routine cytology, and keep the remaining slides unstained.

3. After cytologic diagnosis of malignancy, choose one unstained slide, representative of the lesion, for FISH assessment.

4. Mark the area to be hybridized with a diamond-tipped scribe. The area should cover approximately a 22 mm × 22 mm portion of the slide. Store the slides in a container at room temperature until assay.

Pretreatment of Cytologic Slides

Pretreatment is a necessary procedure for cytologic samples in the prehybridization steps. It helps increase the accessibility of the target DNA of the smeared cells to the DNA probe. Unstained slides are usually pretreated with wash buffer only, except for smears showing considerable thickness and overlapping cells, for which an alternative pretreatment with proteinase K is recommended. The latter treatment should be applied in case of a lack of hybridization, after that FISH procedure has been carried out on slides pretreated with wash buffer only. In this case, if more than one unstained cytologic slide is available for FISH procedure, choose a new slide to submit to proteinase K pretreatment.

The steps for the pretreatment of MGG stained slides are described next.

Pretreatment of no more than six slides at one time per Coplin jar is recommended to avoid decrease of baths temperature.

Pretreatment with Wash Buffer

1. Fix the unstained smears twice in absolute methanol at room temperature for 5 min each, and air-dry.

2. Verify that the pH of the wash buffer is 7.0–7.5 at room temperature before use.

3. Add wash buffer (2X SSC/0.3% NP-40, pH 7.0–7.5) to a Coplin jar.

4. Prewarm the wash buffer by placing the Coplin jar in the water bath at 37°C until solution temperature has reached 37°C.

5. Incubate the samples in wash buffer for 30 min, and air-dry.

6. Dehydrate the samples in 70%, 85%, and 100% ethanol at room temperature for 2 min each, and air-dry.

Pretreatment with Proteinase K

1. Dehydrate the samples in 70%, 85%, and 100% ethanol at room temperature for 2 min each, and air-dry.

2. Add 70 ml PBS to a Coplin jar.

3. Prewarm PBS by placing the Coplin jar in the water bath at 37°C.

4. When PBS temperature has reached 37°C, thaw an aliquot of the 10 mg/ml proteinase K solution and add 140 μl to 70 ml prewarmed PBS (20 μg/ml proteinase K).

5. Incubate the samples in 20 μg/ml proteinase K at 37°C for 5 min.

6. Rinse the samples in distilled water, and air-dry.

7. Dehydrate the samples in 70%, 85%, and 100% ethanol at room temperature for 2 min each, and air-dry.

Pretreatment of Slides Stained with MGG Stain

1. Place the prestained slides in a Coplin jar with xylene until the coverslip could be removed easily.

2. After removing the coverslip, immerse the slides twice in clean xylene for 10 min each to remove the mounting medium.

3. Fix in Carnoy's solution for 10 min at room temperature. Repeat twice using clean Carnoy's solution.

4. Dehydrate the samples in 70%, 85%, and 100% ethanol at room temperature for 2 min each, and air-dry.

5. Pretreat slides with wash buffer or alternatively with proteinase K as previously described.

FISH Procedure

Denaturation and Hybridization

1. Allow the hybridization solution (LSI HER-2/neu SpectrumOrange/CEP 17 SpectrumGreen DNA Probe) to warm at room temperature for ~10 min so that the viscosity decreases sufficiently to allow accurate pipetting. Vortex to mix. Centrifuge the tube for 2–3 sec in a bench-top microcentrifuge to bring the content to the bottom of the tube.

2. Apply 10 μl of the hybridization solution to the target area of the slide.

3. Immediately place a 22 mm × 22 mm glass coverslip over the probe mix and allow it to spread evenly under the coverslip. Gently tap the coverslip with forceps to ensure its adhesion to the slide and to avoid air bubbles. The remaining probe solution should be refrozen immediately after use.

4. To prevent drying of the specimen during hybridization, seal coverslip by ejecting a small amount of rubber cement around the perimeter of the coverslip, creating an air-tight seal. Keep slides in the dark long enough to allow the rubber cement to dry.

5. Denaturate samples at 70°C for 5 min and hybridize at 37°C overnight in a HYBrite Denaturation/ Hybridization System for FISH (Vysis).

Posthybridization Washes

1. Verify that the pH of the wash buffer is 7.0–7.5 at room temperature before use.

2. Add 70 ml of wash buffer to a Coplin jar.

3. Prewarm the wash buffer by placing the jar in the water bath until solution temperature has reached 73°C.

4. Note: Wash no more than six slides at once. For subsequent washes, ensure that the temperature of the wash solution is 73°C before adding slides.

5. Add 70 ml of wash buffer to a second Coplin jar and use at room temperature.

6. Remove the rubber cement seal from the slides by gently scraping with forceps.

7. Immerse slides in wash buffer at room temperature for 1–2 min until coverslip floats off and, if necessary, carefully use forceps to remove it. Keep slides in wash buffer until all coverslips have been removed.

8. Remove slides from wash buffer and drain excess liquid by touching the bottom edge of the slides to a blotter.

9. Verify with a calibrated thermometer that the temperature of the prewarmed wash buffer is 73°C.

10. Incubate slides in wash buffer at 73°C for 2 min.

11. Remove each slide from the wash bath and drain excess liquid by touching the bottom edge of the slides to a blotter and wiping the underside of the slide with a paper towel.

12. Air-dry the slides in the dark on a slide rack in an oblique position.

13. Apply 10 µl of DAPI counterstain to the target area of the slide and apply a 22 mm × 22 mm coverslip. Gently press the slide between two sheets of blotting paper, allowing DAPI to spread evenly under the coverslip.

14. Seal coverslip with nail enamel around the perimeter of the coverslip, creating an air-tight seal. Air-dry in the dark in a horizontal position.

15. Store the slides in the dark before signal enumeration. If slides are not viewed at once, store them at −20°C in the dark. After removing from −20°C storage, allow slides to reach room temperature prior to viewing, using fluorescence microscopy.

Slide Evaluation

1. View samples using a 100X oil immersion fluorescence objective on a fluorescence microscope equipped with a 100 W mercury lamp and a triple excitation/emission filter for simultaneous detection of SpectrumOrange, SpectrumGreen, and DAPI.

2. Visually scan the entire slide to find out areas where the signals are bright, distinct, and easily evaluable. Skip signals with weak intensity. There should be a complete, distinct separation between signals in order for them to be considered more than one signal.

The same slide could show areas where hybridization successfully occurs, areas lacking signals, or where only red or green signals are present; this could be ascribed in case of fresh smears to particularly thick areas or in case of destained smears to areas where destaining unsuccessfully occurs. The background should appear dark or black and relatively free of fluorescent particles or debris. Avoid fields with clusters of cells exhibiting overlapping of nuclei, and avoid areas of necrosis and where the nuclear border are ambiguous. Counterstain with DAPI should allow for distinguishing between tumor cells and lymphocytes or normal epithelial cells that might be present in the smears, on the base of shape and dimension of cells. If any doubts arise about the nature of cells, verify it viewing the corresponding MGG stained slide.

3. Isolated cells or monolayered clusters are best for evaluation. Focus up and down to find all the signals present in the nucleus. Count split centromere signals as one. Exclude from the evaluation nuclei lacking hybridization signals. Score only nuclei with both green and red signals. As the level of gene amplification is heterogeneous among the nuclei of the same specimen, determine an average HER-2/neu gene copy number and an average centromere 17 copy number for each preparation, on the basis of the distribution of signals over the entire slide, or scoring at least 50 evaluable nuclei for each case.

4. Express results as a ratio of the number of copies of the HER-2/neu gene to the number of chromosome 17 centromeric markers. Classify the samples as:

▼ *Unamplified:* When two copies of HER-2/neu and two copies of chromosome 17 are found in the majority of cells and when the HER-2/neu to chromosome 17 ratio is lower than 2.0.

▼ *Amplified:* When HER-2/neu copy number per centromere 17 is greater than 2. On the basis of the average of HER-2/neu gene copy number, different levels of amplification can be defined as follows:

   Low level of amplification: When the number of HER-2/neu signals range from 5 to 10.
   Medium level of amplification: When the number of signals range from 11 to 20.
   High level of amplification: When there is a consistent presence of signal clusters or > 20 signals/cell.

▼ *Polysomic:* When an equal number of HER-2/neu and centromere 17 signals but greater than two in > 10% of cells is present.

▼ *Deleted:* When HER-2/neu copies are fewer than centromere 17 copies.

Aneuploidy of chromosome 17 was excluded as a source of increased HER-2/neu copy number. A positive control slide is included in each run and consists of a cytologic slide known to be amplified for the HER-2/neu gene because it was assayed by FISH in a previous run.

HER-2/neu FISH on Paraffin Sections

Formalin-fixed, paraffin-embedded tissue was cut into 4-µm-thick sections that were heated overnight at 56°C. Deparaffinization, pretreatment, enzyme digestion, and fixation of slides were performed using the Vysis Paraffin Pretreatment Kit (Vysis) according to the manufacturer's recommended protocol. Denaturation and hybridization were carried out in a HYBrite Denaturation/Hybridization System for FISH (Vysis). Ten microliters of HER-2/neu probe mix were applied to the tissue sections that were denatured at 72°C for 2 min and hybridized overnight at 37°C. The slides were then washed in posthybridization wash buffer at 72°C for 2 min and counterstained with DAPI.

For each specimen, at least 100 cells were scored for both HER-2/neu and chromosome 17 signals by image analysis. Images were processed at 1250X magnification using an Olympus MX60 fluorescence microscope with a 100W mercury lamp. Separate narrow band pass filters were used to detect SpectrumOrange, SpectrumGreen, and DAPI. Images were processed with Software Quips (Applied Imaging, Newcastle, UK; Olympus Distributor). Amplification of the HER-2/neu gene was indicated by a ratio of HER-2/neu to chromosome 17 copy number greater than two. For polysomic and deleted cases the same criteria of cytologic samples were applied.

HER-2/neu Immunohistochemistry

Sections of archival formalin-fixed, paraffin-embedded tissue (5 µm) were placed on slides coated with poly-L-lysine. After deparaffinization and blocking of endogenous peroxidase, HER-2/neu immunostaining was performed using rabbit anti-human c-erbB-2 oncoprotein as primary antibody (Dako, Copenhagen, Denmark) at 1/100 dilution. Binding of the primary antibody was revealed by means of the Dako Quick-Staining, Labeled Streptavidin-Biotin System (Dako LSAB), followed by the addition of diaminobenzidine (DAB) as a chromogen. The immunohistochemical expression of HER-2/neu was evaluated in a semiquantitative way. The tumor samples were scored as 3+ when > 10% of the cells showed a specific dark-brown border associated with the cell membrane. Scores of 0, 1+, and 2+ were assigned to negative, weak, or moderate membrane staining, respectively.

RESULTS

HER-2/neu on Primary Breast Cancer

Amplification of the HER-2/neu gene was evaluated by FISH on 66 primary breast cancer FNAs. Twenty-three paired paraffin sections were tested by FISH and 36 by IHC. Figure 38 shows an example of a highly amplified primary breast cancer FNA.

HER-2/neu FISH on Cytologic Smears

Forty-eight (73%) of the 66 primary breast carcinomas evaluated for HER-2/neu on FNAs were unamplified and 18 cases (27%) were amplified. In unamplified

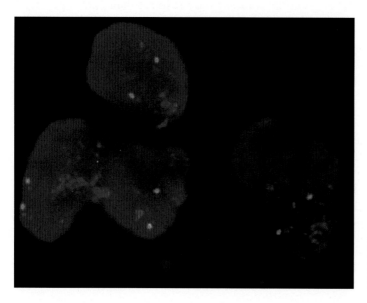

Figure 38 Photomicrograph of fine-needle aspirate from primary breast cancer hybridized with a HER-2/neu oncogene probe; multiple red signals in a cluster pattern indicate HER-2/neu amplification. Original magnification 1250X.

cases, the most frequent pattern consisted of two red and two green signals; one case (1.5%) was polysomic and seven cases (11%) showed a HER-2/neu deletion. Of the 18 amplified tumors, one, six, and 11 cases showed low, medium, and high levels of amplification, respectively. Highly amplified tumors showed HER-2/neu spot numbers ranging from 20 to 100, and often distributed in clusters. Samples with gene amplification showed centromere 17 disomy in nine cases and chromosome 17 polysomy in nine cases, with a number of centromere 17 signals up to 10.

With regard to the feasibility of FISH on cytology, hybridization was successful in 80% of the samples pretreated with wash buffer only. Twenty percent of samples were submitted to a further digestion with proteinase K.

HER-2/neu FISH on Paraffin Sections

Among 66 breast cancer FNAs, 23 also had HER-2/neu FISH evaluated on paired paraffin-embedded sections out of the 66 previously mentioned primary breast carcinomas. One case was not evaluable because of the lack of hybridization. Five of 22 evaluable cases (23%) were amplified and 17 (77%) unamplified. Among the unamplified cases, four were classified as polysomic. Amplified tumors showed a percentage of amplified cells ranging between 50–100%, and often distributed in clusters.

HER-2/neu Immunohistochemistry

Immunohistochemistry was performed on 36 paired paraffin-embedded sections out of the 66 primary breast carcinomas. Nine cases (25%) showed overexpression: eight cases had a percentage of intensively stained (3+) tumor cells ranging between 90–100% and one case showed moderate (2+) staining intensity. The remaining 27 cases (75%) showed normal HER-2 protein expression.

FISH Cytology versus FISH Histology

Table 19 shows the relationship between HER-2/neu results evaluated by FISH on cytologic specimens and by both FISH and immunohistochemistry on corresponding histologic sections. Among 66 cases examined, matched results from FISH cytology and FISH histology were obtained in 22 cases. Five cases were amplified and 15 unamplified, according to FISH on both cytological and histological samples. Four of five amplified cytological specimens showed a high level and one a medium level of amplification. Two cases, moderately amplified on FISH cytology, were classified as polysomic on FISH histology. Concordance between FISH cytology and histology was 91%.

FISH Cytology versus Immunohistochemistry

Paired results according to cytology by FISH and histology by immunohistochemistry were obtained in 36 cases (Table 19). Eight paired cases were both amplified and overexpressed. All were strongly immunostained; four were moderately and four highly amplified. Twenty-five cases were negative for both amplification and overexpression. Three of 36 cases were discordant. Two of them, moderately amplified on FISH cytology, were not overexpressed and were polysomic by FISH histology. The other discordant case, unamplified on FISH cytology, showed a medium level of overexpression. Concordance between FISH cytology and immunohistochemistry was 92%.

HER-2/neu on Metastatic Lesions

Table 20 shows the HER-2/neu results obtained by FISH in primary breast tumor and in corresponding distant metastatic lesion. The time period between primary tumor diagnosis and cytologic diagnosis of metastatic lesion ranged from 1 month to 13 years. HER-2/neu was assessed on 22 cytologic samples from metastatic sites: 14 were unstained and eight were MGG destained smears. Seven out of 22 (32%) metastases were amplified. Amplification was observed in 4 of 12 liver metastases, in 1 of 3 ascitic fluids and in 2 of 4 skin metastases. Among amplified cases, one

Table 19 HER-2/neu Results Comparing Fluorescence *in situ* Hybridization on Cytological Samples With FISH and Immunohistochemistry on Corresponding Histological Sections

| | Histology | | | |
| | FISH | | IHC | |
FISH Cytology	Amplified	Unamplified	Overexpressed	Normal
Amplified	5	2	8	2
Unamplified	0	15	1	25
Concordance	20 of 22 (91%)		33 of 36 (92%)	

Table 20 HER-2/neu Status Assessed by Fluorescence *in situ* Hybridization (FISH) on Primary Breast Cancer and on Cytological Samples from Metastatic Sites

Case	HER-2/neu Primary Tumor	Date Primary/metastasis	Metastatic Site	HER-2/neu Metastatic Site
1	*Not evaluable*	May 1986/June 1998	Liver	Amplified* (low level)
2	Amplified	May 1996/Aug 1998	Liver	Amplified*
3	Amplified	June 1997/March 2002	Liver	Amplified*
4	Polysomic	March 2000/Aug 2002	Liver	Polysomic
5	Amplified*	Nov 2001/Oct 2002	Liver	Amplified
6	Unamplified	March 2000/Feb 2001	Liver	Unamplified*
7	Unamplified	May 1995/June 2000	Liver	Unamplified
8	Unamplified	March 2000/May 2002	Liver	Unamplified
9	Unamplified*	Sept 2002/Oct 2002	Liver	Unamplified
10	Unamplified	April 1989/June 2001	Liver	Unamplified
11	*Unavailable*	Feb 1994/June 2001	Liver	Unamplified
12	*Unavailable*	June 1988/Feb 2002	Liver	Unamplified
13	*Unavailable*	June 1997/April 2002	Pleura	Unamplified
14	*Unavailable*	Nov 1995/July 2002	Pleura	Unamplified
15	*Not evaluable**	Feb 1998/Jan 2002	Pleura	Unamplified*
16	Amplified	Jan 1993/Dec 2002	Peritoneum	Amplified
17	*Not evaluable*	Dec 1997/Dec 2002	Peritoneum	Unamplified*
18	Unamplified	Sept 1999/Jan 2003	Peritoneum	Unamplified*
19	Unamplified*	April 1995/May 2001	Skin	Unamplified
20	Unamplified	May 1997/July 2002	Skin	Polysomic
21	*Unavailable*	June 1998/March 2002	Skin	Amplified
22	Amplified (low level)	Feb 1993/Nov 2002	Skin	Amplified (low level)

*FISH on destained cytological smears.

liver and one skin metastasis resulted poorly amplified. In all three pleural fluids, HER-2/neu was unamplified. One out of 12 liver and 1/4 of skin metastases were polysomic.

Seventeen out of 22 metastatic cases also had HER-2/neu assessed on the primary tumor. In 5 of 22 primary tumors, neither histologic nor cytologic specimens were retrieved because patients had been surgically treated elsewhere. Thirteen out of 17 cases were evaluated by FISH on archival histologic sections; 4 cases had HER-2/neu evaluation on MGG destained cytologic smears obtained at the time of diagnosis. In 3 of 17 (18%) cases, HER-2/neu was not evaluable: one histologic sample because of technical pitfalls and two samples, one cytologic and one histologic, because of lack of hybridization. Among the 14 evaluable primary tumors, five were amplified, eight were unamplified, and one polysomic. Among the five amplified cases, one resulted poorly amplified. Paired FISH results on primary and corresponding metastatic tumor were obtained in 14 cases. All five matched amplified metastatic lesions showed amplification also in their primary tumor. Of the two polysomic metastases, one had the primary tumor polysomic and one unamplified. In the remaining seven cases, no amplification was detected at metastatic site and in the corresponding primary tumor. Overall, HER-2/neu evaluation on primary and metastatic lesions yielded concordant results in all cases submitted for comparison.

DISCUSSION

Overexpression of the HER-2/neu protooncogene is associated with breast cancer progression and poor patient prognosis (Press *et al.*, 1997; Slamon *et al.*, 1997). However, until now the prognostic role of this marker had limited value in clinical decisions, mainly because most studies did not find HER-2/neu to be a prognostic indicator in the case of lymph node-negative patients (Ottestad *et al.*, 1993). On the other hand, HER-2/neu has an increasingly established role in the area of predictive molecular markers. Apparently, a uniform population of breast cancers can be divided, as far as their sensitivity to medical treatments is concerned, into subgroups based on their likely or unlikely response to given treatments. Patients that overexpress HER-2/neu were repeatedly reported to be less responsive to the cyclophosphamide, methotrexate, fluorouracil (CMF)-containing adjuvant chemotherapy regimens (Miles *et al.*, 1999), although with a few contrasting results (Menard *et al.*, 2001), but they might preferentially benefit from the anthracycline-based adjuvant chemotherapy regimens (Thor *et al.*, 1998). As far as endocrine therapies are concerned, only a limited proportion of HER-2/neu positive breast cancers are estrogen receptor positive. In these patients, the efficacy of adjuvant tamoxifen seems to be adversely influenced, according to some authors (Carlomagno *et al.*, 1996), whereas response to oophorectomy plus tamoxifen adjuvant therapy in premenopausal women may be favored (Love *et al.*, 2003). Considering the neoadjuvant endocrine therapy model, treatment with tamoxifen in a randomized clinical trial was confirmed as unfavorably influenced by HER-2/neu positivity, whereas treatment with the new aromatase inhibitor letrozole was reported as more likely associated with objective response (Ellis *et al.*, 2001).

The meaning of the predictive value of HER-2/neu molecular marker was greatly enhanced by the availability of trastuzumab, a high-affinity humanized anti-HER-2/neu antibody. The specific efficacy of this agent was confirmed in a few pivotal trials in metastatic disease in HER-2/neu-overexpressing patients. Administered as single agent in patients pretreated with 1–2 lines of chemotherapy, a limited but definite proportion of them responded to treatment (Cobleigh *et al.*, 1999). Administered as a single agent in non-pretreated patients, more than one-third showed an objective response (Vogel *et al.*, 2002). Addition of this agent to chemotherapy increased response rates, time to disease-progression, and survival duration (Slamon *et al.*, 2001).

Difficulties in interpreting and comparing data from different series aimed at assessing the prognostic and predictive value of HER-2/neu amplification/overexpression in clinical specimens might be due to the variety of reagents and techniques that had been employed. The availability of a specific treatment such as trastuzumab increases the need to select patients that will get the maximum benefit from this treatment and avoid any undue toxicity either in metastatic disease and, possibly in the future, in the adjuvant setting. In fact, although generally well-tolerated, trastuzumab-based therapy is correlated with a significant risk of cardiotoxicity. In any case, in the next years there is likely to be an expanding role of HER-2/neu assessment, either in the selection of type of chemotherapy (anthracycine or not), or in the selection of patients for trastuzumab treatment. The possibility of determining HER-2/neu status not only from histological samples but also on cytological specimens may help clinicians in treatment decisions.

It has been reported that HER-2/neu positivity of primary tumor incompletely predicts the response to trastuzumab, possibly because of a change of HER-2/neu status during metastatic progression. It has been previously demonstrated that recurrent breast cancer is actually an heterogeneous disease where a biologically dominant clone could eventually overcome the others and evolve independently in both primary cancer and metastases (Kuukasjarvi *et al.*, 1997). In this light, it would be worth ascertaining the concordance of HER-2/neu status between primary and metastatic lesion.

Nevertheless, few reports concern HER-2/neu assessment on distant metastatic lesions, whereas the majority compares HER-2/neu status in primary tumor and in regional concurrent lymph node metastases. However, concurrent regional lymph node metastases should not be assumed to be equivalent to distant ones; there is actually a genetic heterogeneity that underlies the development of a distant lesion, years after the primary cancer. Moreover, cells that metastatize via lymphatics could display different biological properties from those traveling to distant sites, due to the invasion by blood vessels.

Up to now, few studies have evaluated the feasibility of detecting HER-2/neu amplification by FISH in breast cancer FNAs (McManus *et al.*, 1999). Results obtained by FISH on FNAs from surgical samples or on FNAs from breast cancer patients have been correlated with results obtained by immunohistochemistry on the corresponding frozen sections or cytological smears. To our knowledge, only one report has correlated the detection of HER-2/neu amplification by FISH both on breast cancer FNAs and on the corresponding frozen section (Klijanienko *et al.*, 1999).

In one study (Bozzetti *et al.*, 2002), we evaluated HER-2/neu by FISH on breast cancer FNAs and by both FISH and IHC on the corresponding paraffin sections. A good concordance was obtained between FISH cytology and FISH histology (91%) as well as between FISH cytology and IHC (92%). In our opinion, the discrepancies between FISH cytology and FISH histology could mainly be ascribed to the fact that FISH spots on cytological preparation are more easily visualized than those on the corresponding tissue section, as previously reported by Klijanienko *et al.* (1999). The presence of monolayered and isolated cells allows for a more accurate signal enumeration on FNAs than on histological sections where fluorescence signal clusters are often prevalent. For this reason, although cytological samples are classified by visual evaluation only, the use of image analysis is recommended for histological samples. On the other hand, difficulties in detecting and counting the signals on cytological samples that showed considerable thickness and overlapping cells could be overcome by proteinase K pretreatment. The FISH method on FNAs allows for the visualization of HER-2/neu on a cell-by-cell basis. Counterstaining with DAPI permits the recognition of some nuclear details for the differentiation between epithelial cells and host elements. The application of dual-color FISH technique, using a probe for the HER-2/neu gene together with a probe for a pericentromeric region of chromosome 17, has the advantage of providing an accurate evaluation of gene copy number alterations, allowing one to distinguish between tumors with normal HER-2/neu gene content, tumors with a gain of only a few extra copies of the HER-2/neu gene, tumors highly amplified, and tumors with a HER-2/neu deletion.

At present, few studies have compared HER-2/neu status in primary breast cancer with paired distant metastases. In a preliminary report by Edgerton et al. (2000), overall 25% discordance was found in HER-2/neu status between 193 primary tumors and 68 paired local recurrence, 32 lymph nodal, and 93 distant metastases. Nevertheless, a note of caution should be applied due to the lack of further details regarding this work. On the contrary, other studies based on FISH or IHC assessment, and carried out on small series of patients, have reported a stable, although not always complete, HER-2/neu status congruence between primary and distant metastases (Tanner et al., 2001). A wider series, recently published by Gancberg et al. (2002), found 94% and 93% of concordance between paired primary tumors and distant metastatic lesions when analyzed by IHC or FISH, respectively.

Actually, our data suggest that HER-2/neu status is mostly stable in primary breast cancer and in the corresponding distant metastatic sites. In our analysis, one case was found polysomic on skin metastasis and unamplified on the paired primary histologic sample. As discussed by Wang et al. (2002), in primary breast cancer, aneusomy 17 is a common feature occurring in absence of HER-2/neu amplification; even high polysomy 17 is not sufficient to produce a significant increase in gene transcription, eventually leading to HER-2/neu protein overexpression. Interestingly, a low-level gain of only a few extra copies of HER-2/neu gene was found in one case, both in primary and in the metastatic lesion. This finding must be distinguished from extra gene copies due to the formation of sister chromatids in S- or G2-phase cells, mostly arranged in pairs. Criteria for a low-level gene copy number increase are the presence of extra signals in a major subpopulation and their random distribution in the nuclei. The biologic and clinical significance of a gain of a few gene copies is still unclear.

In our opinion, testing HER-2/neu status exclusively on primary breast cancer specimen could be a safe policy for using trastuzumab in metastatic disease, especially when results are obtained by FISH methodology. Reasons other than a possible change in HER-2/neu status could be advocated to explain the lack of response to trastuzumab therapy. Many methodological limitations may interfere with HER-2/neu assessment either by IHC or FISH in primary breast cancer, particularly when dealing with archival samples collected years before. Previous studies have pointed out that, in terms of feasibility and accuracy, FISH provides a tempting alternative to HER-2/neu evaluation by IHC, whose specificity and sensitivity problems have been described (Press et al., 1994). On the other hand, hybridization may be eventually compromised by Bouin fixation in archival histological samples subsequently assessed by FISH; this may account for the fact that, in the study of Gancberg et al. (2002), a high rate of samples was not evaluable by FISH. Similarly, in our experience on HER-2/neu evaluation on distant metastases, among archival series, two of the 13 histological samples and one of the 12 cytological destained smears were not assessable by FISH.

To our knowledge, none of the studies that evaluated HER-2/neu status on metastases from breast cancer patients have been carried out on cytological material. In our experience, FNAB, coupled with ultrasound methodology for hepatic lesions, is a relatively easy and safe method to obtain cellular material for FISH analysis, which has been proved to circumvent many of the shortcomings due to molecular and immunocyto/histochemical techniques both on frozen tissue sections and fresh aspirates (Nizzoli et al., 2003; Pauletti et al., 2000). The FISH technique on cytological specimens, including smears, FNAs, and imprint preparations, allows for a rapid and reproducible quantitative analysis of DNA alterations in single cells (Wolman, 1997), thus avoiding many of the drawbacks of tissue sections. We suggest that FNAB, performed on a suspicious breast cancer metastatic lesion, may thus provide fresh cytological material for an updated characterization of relevant predictive factors and a real-time HER-2/neu assessment for trastuzumab-based therapy. Furthermore, retesting of HER-2/neu status on metastatic lesions may be worthwhile when a negative score is obtained by IHC performed on the primary tumor sample resected many years before.

In conclusion, our results underline the feasibility and advantages of two reliable, rapid, and informative techniques, such as FNAB and FISH. The FISH technique is suitable on unstained smears for patient candidates for preoperative chemotherapy, as well as on smears already stained for patients whose paraffin blocks of primary tumor are not available. Moreover, FISH, performed on cytologic smears obtained by FNAB from metastatic sites, as well as on cytologic preparation from pleural and ascitic fluids, allows not only a better definition of HER-2/neu status but also aids in subsequent treatment decisions. Since the advent of trastuzumab, the characterization of the molecular profile in metastatic disease has been becoming increasingly important for targeted therapies selection (see also Part IV, Chapters 3 and 5).

Acknowledgment

I am grateful to Dr. Giorgio Cocconi for his expert review of this manuscript.

References

Allred, D.C., Clark, G.M., Molina, R., Tandon, A.K., Schnitt, S.J., Gilchrist, K.W., Osborn, C.K., Tormey, D.C., and McGuire, W.L. 1992. Overexpression of HER-2/neu and its relationship with other prognostic factors change during the progression of in situ to invasive breast cancer. *Hum. Pathol. 23:*974–979.

Bozzetti, C., Nizzoli, R., Guazzi, A., Flora, M., Bassano, C., Crafa, P., Naldi, N., and Cascinu, S. 2002. HER-2/neu amplification detected by fluorescence in situ hybridization in fine-needle aspirates from primary breast cancer. *Ann. Oncol. 13:*398–403.

Bozzetti, C., Personeni, N., Nizzoli, R., Guazzi, A., Flora, M., Bassano, C., Negri, F., Martella, E., Naldi, N., Franciosi, V., and Cascinu, S. 2003. HER-2/neu amplification by fluorescence in situ hybridization in cytologic samples from distant metastatic sites of breast carcinoma. *Cancer Cytopathol. 99:* 310–315.

Carlomagno, C., Perrone, F., Gallo, C., De Laurentiis, M., Lauria, R., Morabito, A., Pettinato, G., Panico, L., D'Antonio, A., Bianco, A.R., and De Placido, S. 1996. C-erbB-2 overexpression decreases the benefit of adjuvant tamoxifen in early-stage breast cancer without axillary lymph node metastases. *J. Clin. Oncol. 14:*2702–2708.

Cobleigh, M.A., Vogel, C.L., Tripathy, D., Robert, N.J., Scholl, S., Fehrenbacher, L., Wolter, J.M., Paton, V., Shak, S., Lieberman, G., and Slamon, D.J. 1999. Multinational study of the efficacy and safety of humanized anti-HER-2 monoclonal antibody in women who have HER-2-overexpressing metastatic breast cancer that has progressed after chemotherapy for metastatic disease. *J. Clin. Oncol. 17:*2639–2648.

Coussens, L., Yang-Feng, T.L., Liao, Y.C., Chen, E., Gray, A., McGrath, J., Seeburg, P.H., Libermann, T.A., Schlessinger, J., Francke, U., Levinson, A., and Ullrich, A. 1985. Tyrosine kinase receptor with extensive homology to EGF receptor shares chromosomal location with neu oncogene. *Science 230:*1132–1139.

Edgerton, S.M., Merkel, D., Moore, D.H., and Thor, A.D. 2000. HER-2/neu/erbB-2 status by immunohistochemistry and FISH: Clonality and progression with recurrence and metastases. *Breast Cancer Res. Treat. 64:*55 (Abstract 180).

Ellis, M.J., Coop, A., Singh, B., Mauriac, L., Llombert-Cussac, A., Janicke, F., Miller, W.R., Evans, D.B., Dugan, M., Brady, C., Quebe-Fehling, E., and Borgs, M. 2001. Letrozole is more effective neoadjuvant endocrine therapy than tamoxifen for erbB-1- and/or erbB-2-positive, estrogen receptor-positive primary breast cancer: Evidence from a phase III randomized trial. *J. Clin. Oncol. 19:*3808–3816.

Gancberg, D., Di Leo, A., Cardoso, F., Rouas, G., Pedrocchi, M., Paesmans, M., Verhest, A., Bernard-Marty, C., Piccart, M.J., and Larsimont, D. 2002. Comparison of HER-2 status between primary breast cancer and corresponding distant metastatic sites. *Ann. Oncol. 13:*1036–1043.

Gullick, W. J. 1990. The role of the epidermal growth factor receptor and the c-erbB-2 protein in breast cancer. *Int. J. Cancer Suppl. 5:*55–61.

Klijanienko, J., Couturier, J., Galut, M., El-Naggar, A.K., Maciorowski, Z., Padoy, E., Mosseri, V., and Vielh, P. 1999. Detection and quantitation by fluorescence in situ hybridization (FISH) and image analysis of HER-2/neu gene amplification in breast cancer fine-needle samples. *Cancer 87:*312–318.

Kuukasjarvi, T., Karhu, R., Tanner, M., Kahkonen, M., Schaffer, A., Nupponen, N., Pennanen, S., Kallioniemi, A., Kallioniemi, O.P., and Isola, J. 1997. Genetic heterogeneity and clonal evolution underlying development of asynchronous metastasis in human breast cancer. *Cancer Res. 57:*597–604.

Le Beau, M.M. 1993. Fluorescence in situ hybridization in cancer diagnosis. In DeVita, V.T., Hellman, S., and Rosemberg S.A. (eds). *Important Advances in Oncology.* Philadelphia: J.B. Lippincott Company, 29–45.

Love, R.R., Duc, N.B., Havighurst, T.C., Mohsin, S.K., Zhang, Q., DeMets, D.L., and Allred D.C. 2003. HER-2/neu overexpression and response to oophorectomy plus tamoxifen adjuvant therapy in estrogen receptor-positive premenopausal women with operable breast cancer. *J. Clin. Oncol. 21:*453–457.

Masood, S., and Bui, M.M. 2000. Assessment of HER-2/neu overexpression in primary breast cancers and their metastatic lesions: An immunohistochemical study. *Ann. Clin. Lab. Sci. 30:*259–265.

McManus, D.T., Patterson, A.H., Maxwell, P., Humphrey, M.W., and Anderson N.H. 1999. Fluorescence in situ hybridization detection of erbB-2 amplification in breast cancer fine-needle aspirates. *Mol. Pathol. 52:*75–77.

Menard, S., Valagussa, P., Pilotti, S., Gianni, L., Biganzoli, E., Boracchi, P., Tomasic, G., Casalini, P., Marubini, E., Colnaghi, M.I., Cascinelli, N., and Bonadonna, G. 2001. Response to cyclophosphamide, methotrexate, and fluorouracil in lymph node-positive breast cancer according to HER-2 overexpression and other tumor biologic variables. *J. Clin. Oncol. 19:*329–335.

Miles, D.W., Harris, W.H., Gillett, C.E., Smith, P., and Barnes, D.M. 1999. Effect of c-erbB-2 and estrogen receptor status on survival of women with primary breast cancer treated with adjuvant cyclophosphamide/methotrexate/fluorouracil. *Int. J. Cancer 84:*354–359.

Nizzoli, R., Bozzetti, C., Crafa, P., Naldi, N., Guazzi, A., Di Blasio, B., Camisa, R., and Cascinu, S. 2003. Immunocytochemical evaluation of HER-2/neu on fine-needle aspirates from primary breast carcinomas. *Diagn. Cytopathol. 28:*42–46.

Ottestad, L., Andersen, T.I., Nesland, J.M., Skrede, M., Tveit, K.M., Nustad, K., and Borresen, A.L. 1993. Amplification of c-erbB-2, int-2 and C-Myc genes in node-negative breast carcinomas. Relationship to prognosis. *Acta Oncol. 32:*289–294.

Pauletti, G., Dandekar, S., Rong, H., Ramos, L., Peng, H., Seshadri, R., and Slamon, D.J. 2000. Assessment of methods for tissue-based detection of the HER-2/neu alteration in human breast cancer: A direct comparison of fluorescence in situ hybridization and immunohistochemistry. *J. Clin. Oncol. 18:*3651–3664.

Press, M.F., Bernstein, L., Thomas, P.A., Meisner, L.F., Zhou, J.Y., Ma, Y., Hung, G., Robinson, R.A., Harris, C., El-Naggar, A., Slamon, D.J., Phillips, R.N., Ross, J.S., Wolman, S.R., and Flom, K.J. 1997. HER-2/neu gene amplification characterized by fluorescence in situ hybridization: Poor prognosis in node-negative breast carcinomas. *J. Clin. Oncol. 15:*2894–2904.

Press, M.F., Hung, G., Godolphin, W., and Slamon, D.J. 1994. Sensitivity of HER-2/neu antibodies in archival tissue samples: Potential source of error in immunohistochemical studies of oncogene expression. *Cancer Res. 54:*2771–2777.

Ravdin, P. The use of HER-2 testing in the management of breast cancer. 2000. *Semin. Oncol. 27:*33–42.

Slamon, D.J., Clark, G.M., Wong, S.G., Levin, W.J., Ullrich, A., and McGuire, W.L. 1997. Human breast cancer: Correlation of relapse and survival with amplification of the HER-2/neu oncogene. *Science 235:*177–182.

Slamon, D.J., Leyland-Jones, B., Shak, S., Fuchs, H., Paton, V., Bajamonde, A., Fleming, T., Eiermann, W., Wolter, J., Pegram, M., Baselga, J., and Norton, L. 2001. Use of chemotherapy plus a monoclonal antibody against HER-2 for metastatic breast cancer that overexpresses HER-2. *N. Engl. J. Med. 344:* 783–792.

Tanner, M., Jarvinen, P., and Isola, J. 2001. Amplification of HER-2/neu and topoisomerase II alpha in primary and metastatic breast cancer. *Cancer Res. 61:*5345–5348.

Thor, A.D., Berry, D.A., Budman, D.R., Muss, H.B., Kute, T., Henderson, I.C., Barcos, M., Cirrincione, C., Edgerton, S., Allred, C., Norton, L., and Liu, E.T. 1998. ErbB-2, p53, and efficacy of adjuvant therapy in lymph node-positive breast cancer. *J. Natl. Cancer Inst. 90:*1346–1360.

Vogel, C.L., Cobleigh, M.A., Tripathy, D., Gutheil, J.C., Harris, L. N., Fehrenbacher, L., Slamon, D.J., Murphy, M., Novotny, W.F., Burchmore, M., Shak, S., Stewart, S.J., and Press, M. 2002. Efficacy and safety of trastuzumab as a single agent in first-line treatment of HER-2-overexpressing metastatic breast cancer. *J. Clin. Oncol. 20:*719–726.

Wang, S., Hossein Saboorian, M., Frenkel, E.P., Haley, B.B., Siddiqui, M.T., Gokaslan, S., Hynan, L., and Ashfaq, R. 2002. Aneusomy 17 in breast cancer: Its role in HER-2/neu protein expression and implication for clinical assessment of HER-2/neu status. *Mod. Pathol. 15:*137–145.

Wolman, S.R. 1997. Applications of fluorescence *in situ* hybridization techniques in cytopathology. *Cancer 81:*193–197.

Xu, R., Perle, M.A., Inghirami, G., Chan, W., Delgado, Y., and Feiner, H. 2002. Amplification of HER-2/neu gene in HER-2/neu-overexpressing and-nonexpressing breast carcinomas and their synchronous benign, premalignant, and metastatic lesions detected by FISH in archival material. *Mod. Pathol. 15:*16–24.

5

Detection of HER-2 Oncogene in Human Breast Carcinoma Using Chromogenic *in situ* Hybridization

Nadia Dandachi and Cornelia Hauser-Kronberger

Introduction

A large number of characteristic chromosome aberrations have now been implicated in the development and progression of human breast cancer. Among these, HER-2 gene amplification has been the focus of intense study during the past two decades and determination of HER-2 status has become an important tool in the clinical management of patients with breast cancer.

Oncogenic amplification of the HER-2 gene has been observed in approximately 20–30% of breast cancers and is associated with more frequent relapse and poorer prognosis and may be predictive of response to certain anticancer therapies (Slamon *et al.*, 1987, 1989). Furthermore, and fundamental to clinician and patient, a positive HER-2 status is an eligibility requirement for Herceptin therapy in women with meta-static breast cancer. Consequently, issues relating to accurate and reliable laboratory assessment of HER-2 status have become one of the central debating points in current breast cancer diagnosis and biology.

Out of a wide range of assays that have been used in research for detecting HER-2 status, only two techniques are now predominant and readily applicable in the routine clinical pathology laboratory: determination of HER-2 overexpression by immunohistochemistry (IHC) and HER-2 gene amplification by fluorescence *in situ* hybridization (FISH).

Although two FISH (Inform Ventana, Tucson, AZ and PathVysion Vysis, Downer's Grove, Il) and two IHC (Dako Herceptest, Glostrup, Denmark and Pathway, Ventana) assays have been approved by the U.S. Food and Drug Administration (FDA) as clinical tests for breast carcinomas, the question of which test should be performed for assessing HER-2 status in practice remains a hotly debated and controversial issue, especially in the context of Herceptin therapy (Bartlett *et al.*, 2003; Bilous *et al.*, 2003; Schnitt *et al.*, 2001).

Whereas the FISH procedure has been proven to be reliable and reproducible, a colorimetric or chromogenic modification of this assay would be highly desirable for most practicing pathologists. This chromogenic *in situ* hybridization (CISH) uses conventional enzymes to generate bright-field gene copy signals and enables the assessment of HER-2 gene amplification using regular light microscopes.

Handbook of Immunohistochemistry and in situ Hybridization of Human Carcinomas, Volume 1: Molecular Genetics; Lung and Breast Carcinomas

279

Copyright © 2004 by Elsevier (USA)
All rights reserved.

Immunohistochemistry and Fluorescence *in situ* Hybridization

Two methods—IHC (HercepTest, Dako and Pathway, Ventana) and FISH (PathVysion, Vysis and Inform, Ventana)—are currently approved by the FDA for use in HER-2 testing in breast cancer samples. However, these assays have not been approved for the same clinical applications. The FDA approval for gene amplifications tests was originally for identifying women with node-negative disease at high risk for recurrence or disease-related death or for selection to doxorubicin chemotherapy. A recently approved application by FDA broadened the indications for FISH testing to include selection of women with metastatic breast cancer for trastuzamab therapy. Both IHC assays are approved for selection of women with metastatic breast cancer to receive trastuzamab treatment (Schnitt *et al.*, 2001). In addition, the recent FDA approval for the first anti-HER-2 therapy Herceptin has intensified the debate as to which of these methods provides the most accurate, reliable, and reproducible results.

The IHC analysis offers advantages over FISH because it is relatively inexpensive and fast, uses a light microscope, and can be performed in most pathology laboratories. Moreover, results have been linked to prognosis and response to treatment. Furthermore, from the biological aspect, it seems more logical to determine HER-2 protein overexpression than the level of gene amplification when treatments such as Herceptin are specifically targeted toward the HER-2 protein on the cell surface. Published data have supported a link between FISH results and prognosis; however, data linking FISH assay results and response to therapy are more limited (Schnitt *et al.*, 2001). Assay reliability has been questioned, because specimen handling, fixation, cell conditioning, and subjective scoring can affect the quality of results (Jacobs *et al.*, 1999a; van de Vijver, 2001). In addition, major concerns have been raised with respect to possible oversensitivity of Dako HercepTest, resulting in false-positive results (Jacobs *et al.*, 1999b; Tubbs *et al.*, 2001).

A major advantage of FISH over IHC is the relative stability of DNA for hybridization-based technologies (Nuovo *et al.*, 1989) as well as circumventing antigenic changes that occur in formalin-fixed and paraffin-embedded tissues, a major limitation inherent to IHC. Moreover, because FISH quantifies the number of gene copies in the cancer cell, which objectively reflects the HER-2 gene status of tumors, it is less arbitrary and subjective than the IHC scoring system. However, FISH methodology also has its disadvantages. Evaluation of FISH requires a modern and expensive fluorescence microscope equipped with high-quality 60X or 100X oil immersion objectives and multiband-pass fluorescence filters, which are not used in most routine diagnostic laboratories. Furthermore, relatively few diagnostic laboratories currently have expertise in FISH analysis. The FISH method requires more technologist time and more interpretation time by the pathologist than does IHC, because gene copy number aberration needs to be evaluated in individual cells one at a time, unlike IHC, in which an area of tumor cells is assessed collectively. Other FISH limitations include fluorescent signal fading and bleaching, which limits slide archiving.

Overall, the FISH procedure is more complicated, time-consuming, and expansive than IHC analysis. Yet it is possible that the development of automated methods to perform the assay and screen the slides may make FISH more practical for routine determination of HER-2 status.

Concerns related to a significant incidence of apparent false-positive cases using the HercepTest were expressed by several studies (Jacobs *et al.*, 1999b; Roche *et al.*, 1999; Tubbs *et al.*, 2001). Although the HercepTest provides the opportunity to strictly adhere to a unified methodology and explicit scoring criteria by the users, the clinical significance of the 2+ positive group has been questioned. This was further substantiated by Tubbs *et al.* (2001), who showed that 15% of cases that were 2+ positive by HercepTest did not show gene amplification by either FISH or messenger ribonucleic acid (mRNA) detection. These results indicate that the HercepTest is specific but too sensitive and that the 2+ score includes a subset of tumors that does not have HER-2 gene amplification (Hanna *et al.*, 2001).

This becomes particularly important since reliable clinical laboratory data regarding the HER-2 status are essential in making the management decision to administer Herceptin. Furthermore, recently it has become apparent that patients with +2 HercepTest score did not seem to benefit significantly from trastuzumab therapy (Vogel *et al.*, 2001). In addition, the available evidence indicates that HER-2 amplification might provide more meaningful prognostic information than HER-2 overexpression in breast cancer patients. However, it is not clear why tumors that are FISH positive for HER-2 but IHC negative should respond to an antibody-based treatment. One explanation could be that this might result from epitope loss during fixation and consequently the IHC result being a false negative.

In most studies, the concordance rates between IHC and FISH results are in the 80–95% range, with the highest levels of concordance in cases that are either completely negative or strongly positive by IHC. Much lower levels of agreement between IHC analysis and FISH results are seen in cases in which the IHC result is less than strongly positive (Schnitt *et al.*, 2001).

When performing gene-specific FISH, the use of a reference probe (usually the centromere probe of the respective chromosome) has been considered very important in distinguishing gene amplification from deoxyribonucleic acid (DNA) aneuploidy. Recent results clearly indicate, however, that inclusion of the centromere copy number counts affects the gene amplification classification in only a very small fraction of tumors (Rummukainen *et al.*, 2001; Tubbs *et al.*, 2001).

Many authors recently have advocated the use of a diagnostic algorithm for assessing HER-2 status. The IHC analysis is used as the initial screening step for all cases, and, subsequently, all IHC-positive cases should be confirmed by FISH (Lebeau *et al.*, 2001; Tubbs *et al.*, 2001; Wang *et al.*, 2000).

MATERIALS

One hundred and seventy-three cases of invasive breast carcinomas diagnosed between 1993 and 1999 were analyzed in this study (Dandachi *et al.*, 2002). Routinely formalin-fixed, paraffin-embedded tumor blocks were available for all patients.

METHODOLOGY

Tissue Preparation

Optimal fixation and tissue processing are the initial key steps for successful ISH. Tissue preparation, including storage and fixation, should be optimized to detect intracellular nucleic acids. It is a fact that nucleic acids are better preserved in frozen sections than in paraffin-embedded tissues. In general, cross-linking fixatives such as formaldehyde, paraformaldehyde, or glutaraldehyde are preferable over precipitating fixatives because of the excellent retention of nucleic acids in cross-linked tissues. Formalin-fixed archived tissue can be used for CISH after storage for several years. Our own archived paraffin-embedded tissue specimens have been stored for 9 years before use, without any noticeable reduction of sensitivity and specificity.

Overfixation can decrease the availability of nucleic acids for hybridization and cause higher background binding. In our opinion, neutral buffered formalin (NBF) is the most favorable fixative for ISH on archival paraffin-embedded tissue in general and for the specified purpose of performing CISH. The main aspects are the broad distribution in histopathology use, relatively wide fixation times, and excellent preservation of morphology. However, fixation times in buffered formalin should be standardized to a maximum of 24 hr to minimize staining variability. The effect on the intensity of the hybridization signal seems to be reduced with other fixatives (e.g., Hollande, Zenker, and Bouin).

Preparing the Target

In working with paraffin-embedded tissue, one must remove 1) the paraffin (deparaffinization), 2) the protein cross-linked to the DNA, and 3) DNA cross-links. Several manufacturers of DNA probes recommend not baking the sections at 60°C. In contrast, others suggest that it may be helpful to bake for about 15 min and to use positively charged silane-treated slides to improve adhesion of the section to the slide. In our study, we baked sections at 60°C for 1 hr and used silanized slides.

Antigen Retrieval

Most fixatives link cellular proteins to macromolecules such as nucleic acids or nuclear proteins themselves. These effects may hinder penetration of the probe to the target nucleic acid. Target retrieval using proteases and/or heat-induced antigen retrieval are therefore essential for ISH in general to provide the hybridization probe access to target nucleic acid sequences. The appropriate fixation and pretreatment conditions must be determined by the user for each type of specimen.

It should be noted that if microwave pretreatment is added, the length of proteinase K treatment should be reduced. Protease treatment is considered to be another important step to increase the accessibility of the target nucleic acid, especially for paraffin sections with labeled probes. Determination of the proper concentration of the protease and digestion time requires some testing and is dependent on the individual laboratory fixation and embedding procedure. In our lab, proteinase K pretreatment (1 mg/ml, see step-by-step protocol) for 8 min works well for breast cancer tissue fixed for 6–36 hr in 7% neutral buffered formalin. Insufficient target retrieval may lead to a diminished or completely absent hybridization signal. Pretreatment of slides with microwaving to increase ISH sensitivity, especially for old archival paraffin-embedded tissue, has been reported using 10 mM citrate buffer (pH 6.0). In our protocol, the sequential use of high-temperature target retrieval and proteolytic digestion are required to obtain optimal results with the used HER-2–probe. Target retrieval was performed first using a waterbath.

Hybridization

The probes used are commercially available (Zymed, South Francisco, CA) and provided with a suitable hybridization buffer. Generally, both the probe and target DNA require denaturation, but this may depend on the manufacturer's instructions. Some probes require no denaturation and some tolerate codenaturation.

Hybridization between the labeled probe and target DNA is defined by hydrogen bonding and the hydrophobic interactions in equilibrium. Annealing and separation of the two hybrid strands depend on various factors, including temperature, salt concentration, pH, the nature of the probes and target molecules, and the composition of the hybridization and washing solution. High stringency can lead to detection of small differences or changes such as a change in a single base. The optimal temperature for hybridization is 15–25°C below the melting temperature, which can be easily calculated with a standard formula.

Hybridization buffers contain reagents to maximize nucleic acid duplex and inhibit nonspecific binding of probe to tissue. After an overnight incubation, the coverslips are removed and the posthybridization washing is performed. The posthybridization wash is an important step to minimize background and nonspecific binding. The use of low salt concentrations, relatively high temperature, and formamide breaks the weaker unspecifically bound hydrogen bonds between the probe and nontarget molecules. Increased stringent wash temperature may reduce cross-hybridization with target DNA; however, wash temperature higher than 55°C will be accompanied by a decrease of signal intensity.

Visualization

Detection systems for biotin and digoxigenin are readily available and easy to use and generate a permanent signal in combination with peroxidase and alkaline-phosphatase conjugated antibodies. In our test system, we combined the digoxigenin-labeled HER-2 probe with a sandwich technique using an anti-digoxigenin antibody followed by a biotinylated anti-mouse antibody to increase staining sensitivity. As a final step, a complex of streptavidin and biotinylated HRP was added. The chromogenic visualization was performed using the substrate chromogen solution from the Dako ABC Kit composed of diaminobenzidine (DAB) and hydrogen peroxide, resulting in a brown signal. To ease evaluation of morphology, counterstaining with hematoxylin was done.

Controls

Control tissue sections were hybridized in the absence of probe. Positive controls were cases of breast carcinoma shown to have high levels of HER-2 amplification by Southern Blot analysis. To validate

sensitivity and intensity of staining, breast cancer cell lines with defined copy numbers of HER-2 were used (BT474 13 copies, SKBR3 6–8 copies of HER-2/ nucleus; MD-MB 361 4–6 copies, MCF-7 1–2 copies).

Scoring

Hybridization signals from at least 60 tumor cells were scored to assess oncogene copy number. The probe displayed a single distinct small dark brown signal at the location of each copy of the HER-2 gene. The expected number of signals in a normal and in an unamplified tumor cell varied from two to four, depending on the phase of the cell cycle, and was classified as negative. Precise signal enumeration was not possible in some sections because tumors with high levels of gene amplification often exhibit coalescing signal clusters. Thus, amplification was defined as more than six signals per nucleus or when gene copy clusters were seen in > 50% of cancer cells. Because of truncation of nuclei in thin paraffin-embedded tissue sections, some nuclei did not contain target sequence, and therefore did not display hybridization signals.

Limitations

Overdigestion

Negative results may be caused by improper incubation times, dilutions, temperatures, or suboptimal fixation or pretreatment conditions. Overpretreatment is visible in destruction of morphology, and underdigestion may result in no signal at all.

High Background

The most common problem encountered with chromogenic or fluorescence systems is background. This is due to nonspecific binding of the probe to nontarget molecules (e.g., cellular membranes and nontarget nucleic acids) and can even lead to render the slide uninterpretable. Tissue sections should not be permitted to dry out once the hybridization procedure has begun because sections that are allowed to dry after the hybridization step may exhibit high levels of background staining. Other causes for high background may be the incomplete blocking of endogenous peroxidase or nonspecific binding of the bridging antibody. Biotinylated probes are known to cause background staining because of nonspecific binding to endogenous biotin, which is present in tissues such as liver and kidney. This can result in high background or even false-positive results and is clearly a disadvantage (Hayat, 2002).

In our lab the use of a biotinylated HER-2 probe showed increased background in invasive breast cancer tissue. In contrast, digoxigenin-labeled probes have higher sensitivity and less background staining than biotinylated probes. The fact that digoxigenin is not present in mammalian cells and that the anti-digoxigenin antibody does not bind to other biologic material is a particular advantage.

Scoring of Clusters

Precise signal enumeration of high-level amplification was frequently not possible because of coalescing signal clusters, but at least in routine diagnostics enumeration of gene copies exceeding 10 is not needed. The most difficult category in CISH is the low level amplification (six to eight copies), in which accurate enumeration of the gene copies is necessary.

STEP-BY-STEP PROTOCOL

Step 1: Deparaffinization

1. Immerse slides in fresh xylene for 10 min. Repeat.
2. Dehydrate slides in 100% EtOH for 2 min. Repeat.
3. Dehydrate slides in 96% EtOH for 2 min. Repeat.
4. Rinse slides in distilled water for 3 min. Repeat.

Step 2: Pretreatment

1. Immerse slides in preheated Dako target retrieval solution (pH 6.0) at 95°C for 20 min.
2. Allow slides to cool down in the solution at room temperature for 20 min.
3. Rinse slides several times in distilled water.
4. Incubate slides in a bath of proteinase K solution (Dako, diluted 1:2000 in 50 mM Tris-HCl, pH 7.6) for 8 min at room temperature.
5. Rinse slides several times in distilled water.

Step 3: Background Quenching

1. Immerse slides in 3% H_2O_2 in methanol for 20 min at room temperature.
2. Rinse slides in distilled water for 10 min.
3. Air-dry slides.

Step 4: Hybridization

1. Remove slides from distilled water; carefully wipe excess water from around the specimen.
2. Apply 15 μl of dig-labeled probe (Zymed) to the section and cover with coverslip (avoid air bubbles).
3. Heat slides on a heating block for denaturing DNA at 94°C for 5 min.
4. Incubate slides in a moist chamber at 37°C overnight.

Step 5: Posthybridization Wash

1. Remove coverslips by soaking in 1X TBST (Tris-buffered saline/Tween-20; 0.1%) for 5–10 min.
2. Immerse slides in stringent wash solution (Dako, GenPoint) at 55°C for 20 min.

Step 6: Signal Amplification

1. Immerse slides in 1X TBST for 5 min.
2. Eliminate nonspecific background by incubating tissue with normal goat serum for 10 min.
3. Apply 150 μl of a 1:200 diluted mouse-anti-dig antibody (Boehringer Mannheim) on the slide and incubate in a moist chamber at room temperature for 60 min.
4. Wash slides in 1X TBST for 5 min. Repeat twice.
5. Apply 150 μl of a 1:200 diluted biotinylated goat-anti-mouse (Dako ABC kit) on the slide and incubate in a moist chamber at room temperature for 30 min.
6. Wash slides in 1X TBST for 5 min. Repeat twice.
7. Apply 150 μl of a 1:300 diluted complex of streptavidin and biotinylated HRP complex (Dako ABC kit) to the slide and incubate in a moist chamber at room temperature for 30 min.
8. Wash slides in 1X TBST for 5 min. Repeat twice.

Step 7: Chromogen Reaction

1. Incubate slides with 100 μl of the substrate chromogen solution (diaminobenzidine and 0.1% hydrogen-peroxide in PBS) for 3 min.
2. Rinse slides several times in distilled water.
3. Immerse slides in a bath of hematoxylin and incubate for 2–5 min.
4. Rinse slides in distilled water, dehydrate in graded alcohols, and mount slides.

RESULTS AND DISCUSSION

Despite the recognized potential advantages of chromogenic detection over fluorescence detection in DNA ISH and its availability as an invaluable molecular tool in research and clinical diagnosis for years, the practical use of this technique has been limited for several reasons. The major disadvantage of conventional CISH is its severely restricted sensitivity, which is essential for many clinical applications (Speel *et al.*, 1999).

Although the FISH procedure has been proven to be reliable and reproducible (Bartlett *et al.*, 2001; Pauletti *et al.*, 1996; Tubbs *et al.*, 2001), a colorimetric or chromogenic modification of this assay that allows the use of a standard bright-field microscope would be highly desirable for most pathologists, in particular in a setting where routine analysis has to be performed.

Moreover, a method that combines the advantages of both IHC analysis and FISH might be an alternative in terms of reliability and practicability to routinely assess HER-2 status. A chromogenic *in situ* hybridization (CISH) has been introduced by Tanner and co-workers (Tanner *et al.*, 2000) as an alternative to FISH, in which the DNA probe was detected using a peroxidase reaction. Others (Kumamoto *et al.*, 2001; Sharma *et al.*, 1999; Smith *et al.*, 1994; Vos *et al.*, 1999; Zhao *et al.*, 2002) have also successfully used nonfluorescent ISH for detecting HER-2 oncogene.

Herein we describe such a system for bright-field detection of amplification of the HER-2 oncogene in archival human breast cancers. This CISH assay uses conventional peroxidase to generate bright-field gene copy signals and enables the assessment of HER-2 gene amplification using regular light microscopes.

In our study (Dandachi *et al.*, 2002), we used the digoxigenin (DIG) labeled Spot-Light HER-2 DNA probe (Zymed, South Francisco, CA), which has been generated by Subtracted Probe Technology (SPT). With SPT, repetitive DNA sequences, such as *Alu* and LINE elements, which may consist of up to 40% of template and cause unspecific hybridization, are quantitatively removed. Therefore, the final probe is very specific, and the need for blocking nonspecific hybridization with Cot-1 DNA in traditional FISH probes is eliminated (Zhao *et al.*, 2002).

The current CISH procedure is based on a single-color detection of one DIG-labeled HER-2 DNA probe, similar to the FDA-approved FISH test (Inform, Ventana). The theoretical advantage of two-colored FISH (PathVysion, Vysis) is its ability to distinguish chromosomal amplification from aneuploidy using a chromosome 17 centromere as a differentially labeled reference probe. The use of chromosome 17 correction has been suggested as an important means to correct for HER-2 pseudo-amplification due to chromosome 17 polysomy (Press *et al.*, 1997; Watters *et al.*, 2003). However, although aneusomy 17 is a common alteration in breast cancer, several studies recently showed that except in a certain subset of cases, aneusomy 17 probably is not a significant factor for HER-2 protein expression or for clinical assessment of HER-2 status and makes analysis even more time-consuming (Tubbs *et al.*, 2001; Wang *et al.*, 2002).

Using a cohort of 173 archival breast cancer specimens, we showed in our study (Dandachi *et al.*, 2002) that CISH is an assay that is simple to use and can be easily integrated into routine testing. CISH has several fundamental advantages over FISH. For example CISH does not require an expensive fluorescence microscope and thus can be easily accommodated in any routine histopathology laboratory. More important, CISH allows

interpretation of gene amplification in the context of morphological and histological changes, which has important practical value. By contrast, in FISH, tissue morphology and gene amplification are primarily disconnected because selection of tumor cells for signal evaluation is based on nuclear 4′6-diamidino-2-phenylindole-2HCL (DAPI) staining, which does not always allow sufficient histopathological evaluation of the cells. Using adjacent hematoxylin and eosin (H&E) stained slides can overcome this problem, but it makes analysis even more time-consuming and tedious. Such a colorimetric detection system presents a convenient format for pathologists who are more experienced and familiar with a standard light microscope. In addition, unlike FISH, cases treated by CISH can be stored for long periods because the reaction product is permanent, a particularly important and useful property when assessing clinical specimens. Finally, FISH takes much longer to analyze and the cost of the procedure is substantially higher than that of the CISH assay.

Confirmation of nonamplification is just as important as recognizing the amplification of the HER-2 gene. Thus, the absence of gene amplification can only be verified if normal endogenous HER-2 gene signals are detectable, allowing the discrimination between a true negative and the failure of the hybridization procedure. Therefore, our study criteria for successful CISH analysis included identification of at least one copy of HER-2 gene per nucleus in most cancer cells and appropriate high-temperature target retrieval and proteinase K digestion as evidenced by well-preserved cell morphology.

The signals obtained with CISH are optically well-defined, sharp, and dark brown signals. In general, gene amplification by CISH was easy to identify either as individual scattered signals (Figure 39A) or consistent clusters (Figure 39B) in a microscope with a 25X or 40X objective magnification. Tumors with normal HER-2 gene showed typically one to two dots per nucleus (Figure 39C) or two to four spots in cases of chromosomal aneuploidy (Figure 39D). Signal distribution as groups or clusters has been described previously by others using FISH for detecting HER-2 (Pauletti *et al.*, 1996; Press *et al.*, 1997). Even though double minute chromosomes have been shown to have a tendency to aggregate, this pattern is more likely due to intrachromosomal amplification, namely homogeneously staining regions (Cowell, 1982). Precise signal enumeration of high-level amplification was frequently not possible because of coalescing signal clusters, but at least in routine diagnostics enumeration of gene copies exceeding 10 is not needed. The most difficult category in CISH is the low-level amplification (six to eight copies), in which accurate enumeration of the gene copies is necessary.

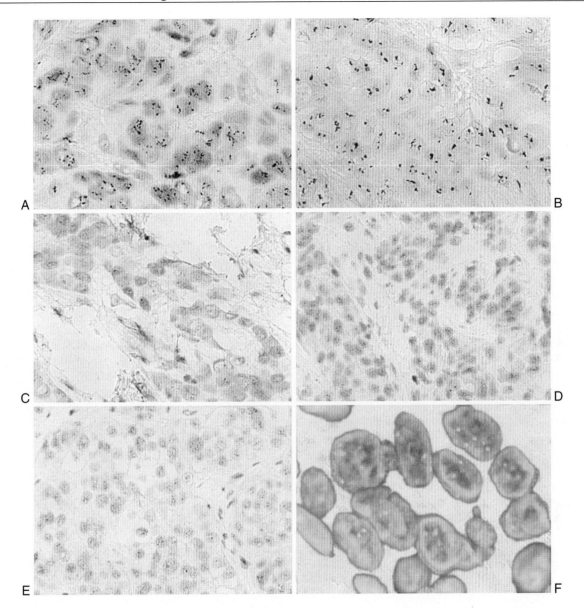

Figure 39 Representative photomicrographs of HER-2 oncogene detection determined by CISH in archival invasive breast carcinomas. Tumor with high level of HER-2 gene amplification showing either individual scattered brown reaction products (**A**) or numerous signals, arranged in clusters (**B**). Tumor with no amplification shows typically one to two signals per nucleus (**C**) or two to four spots in a chromosomally aneuploidy case (**D**). An invasive ductal carcinoma with HER-2 low-level amplification by CISH (**E**) and by FISH (**F**). Counterstained with hematoxylin (**A–E**) and DAPI (**F**), magnification 405X (**A–E**) and 1013X (**F**).

Rummukainen and co-workers (Rummukainen *et al.*, 2001) modified the conventional three-step detection method used by Tanner *et al.* (Tanner *et al.*, 2000) (anti-digoxigenin-fluorescein isothiocyanate (FITC) and anti-FITC-HRP + DAB chromogen). The key reagent that allowed higher sensitivity was the anti-mouse-HRP polymer (Powervision), which has high peroxidase enzyme content compared with the small size of the polymer molecule. With the HRP-polymer–based

detection, the authors successfully detected even low-level amplification of C-Myc.

A disadvantage of performing CISH as described in our protocol might be the handling of carcinogenic diaminobenzidine (DAB), but this shortcoming may be overcome using alkaline phosphatase instead of peroxidase enzyme. However, the use of DAB as a chromogen allows the application of conventional counterstains and the mounting of slides in xylene-based mounting solutions, making interpretation for the pathologist more acceptable.

Although FISH represents an excellent method for profiling HER-2 gene amplification, correlation with tissue morphology is difficult or even impossible because of dark-field visualization. This is of particular significance in breast carcinomas, where distinguishing between *in situ* and invasive components is essential for correctly identifying HER-2 status, especially because up to 80% of ductal carcinoma *in situ* (DCIS) overexpresses HER-2 in contrast to 15–30% of invasive breast carcinomas (Dowsett *et al.*, 2003).

In FISH, the DAPI counterstaining makes it difficult or even impossible to restrict the copy number counting only on cells that are truly malignant by morphologic criteria (using hematoxylin counterstain). In contrast, CISH allows morphology-guided HER-2 evaluation and can be restricted to invasive carcinoma cells. This drawback of FISH could also explain in part the discrepancies between FISH and CISH found by Tanner and co-workers (2000); 10 tumors (6.4%) that were amplified by FISH were classified as negative by CISH.

In our previously published study, we evaluated IHC and CISH concurrently in a cohort of 173 invasive breast carcinomas. A total of 42 (24.3%) of 173 investigated tumors displayed positive (+2, +3) immunostaining and 131 (75.7%) cases were scored as negative (0, +1) with IHC. In comparison, 33 (19.1%) of 171 patients in this same cohort demonstrated HER-2 gene amplification with CISH and 138 (79.8%) cases were negative with CISH. The HER-2 overexpression/ amplification rates determined by IHC (24.3%) and CISH (19.1%) are in concordance with established measures in breast cancers (Press *et al.*, 1997; Slamon *et al.*, 1987, 1989).

In our series, the overall concordance between CISH and IHC was 95.9%. Furthermore, κ statistics revealed an excellent agreement between IHC and CISH (κ = 0.878). Finally, discrepant cases between CISH and HercepTest (7/171) and all IHC positive cases (+2 and +3), a total of 42 cases, were analyzed with the FISH (PathVysion) assay. With respect of those tumors defined as nonamplified by CISH, seven cases (4.1%) of 171 tumors showed HER-2 overexpression in the absence of gene amplification. These seven IHC-positive cases were also classified as nonamplified using FISH. However, it is important to mention that HER-2 is a growth factor and enhanced transcription in the absence of gene amplification is a well-recognized mechanism for cellular function. This is due to enhanced production of mRNA by phosphorylation of tyrosine kinase acting on growth factors and regulators of cell growth and proliferation (Earp *et al.*, 1995; Pauletti *et al.*, 1996). Thus, discordances between

HER-2 overexpression and amplification may simply reflect this fact.

Concordance between CISH and FISH was 100% for the 38 cases analyzed (Figure 39E and 39F). Others have indicated a similar high level of concordance between CISH and FISH (Tanner *et al.*, 2000; Zhao *et al.*, 2002). The complete agreement of FISH and CISH results indicates that this chromogenic ISH technique appears to be sensitive and specific for detecting HER-2 amplification in human archival tumor samples. Pathologists may use this assay to simultaneously evaluate HER-2 gene alteration and morphological changes by conventional bright-field microscopy. These advantages, together with the possibility that CISH could be automated to study large numbers of cases, make this assay an interesting and promising alternative to FISH and IHC in the clinical laboratory setting.

Few data are available currently to verify the accuracy of oncogene amplification detection by CISH compared with FISH: therefore, CISH has not yet been widely recommended for routine use in clinical decision making. Although CISH has been tested with outstanding results, further studies are necessary to verify the applicability of this new method. In particular, studies that assess the clinical value of CISH are needed.

Future Outlook

The CISH method offers an ideal approach that allows detection of HER-2 amplification in the context of morphology and thus might have great potential practical value.

Further comparative studies are under investigation that need to evaluate the interobserver reproducibility of this assay, its portability to automated instrumentation, its relevance to clinical therapeutic outcome, and the utility of CISH for profiling other gene amplification applications in various carcinomas.

In the future, the application of expert teleconsultation in pathology will expand using the Internet as the communication medium. The efficiency and accuracy of expert teleconsultations in telepathology have been demonstrated in numerous studies (Kayser, 2002; Leong *et al.*, 2003). This tool will become widely used and teleconsultation with CISH may overcome technical difficulties as they occur if fading fluorescent labels are digitized.

Several investigators have successfully used a combination of two non–isotopic-labeled probes, mainly biotin and digoxigenin, conjugated to different enzymes (alkaline phosphatase or horseradish peroxidase), metals (nanogold), or fluorescence. This allows simultaneous localization of multiple genomic DNA or mRNA in the same tissue section or cell. In combination with tyramide signal amplification (TSA) or silver-autometallography leading to improvement of ISH sensitivity, multicolor CISH might become an interesting assay technique for detecting multiple DNA targets in routine histopathology (Frater *et al.*, 2001; Hopman *et al.*, 1997). The CISH localization of target nucleic acids could also easily be combined with a pre-embedding electronmicroscopy technique for gene localization at the ultrastructural level or with conventional IHC methods.

Recently, automated immunostainers have become available for routine use in the clinical laboratory that have the ability to control the temperature of each slide. These instruments can perform deparaffinization and antigen retrieval in a fully automated way and seem to be an ideal platform to carry out CISH (Beck *et al.*, 2003). The fast development of automated immuno-stainers, and new supersensitive chromogenic detection systems, will likely enhance the utility of automated CISH for detecting HER-2. These potential improvements save the lab technician time and allow for more flexibility in the staining protocol.

Acknowledgment

The authors are very grateful to Dr. Raymond Tubbs for his support and assistance with the development of this technique. This work was supported in part by the Austrian National Bank Fund (grant number 7645).

References

Bartlett, J., Mallon, E., and Cooke, T. 2003. The clinical evaluation of HER-2 status: Which test to use? *J. Pathol. 199:*411–417.

Bartlett, J.M., Going, J.J., Mallon, E.A., Watters, A.D., Reeves, J.R., Stanton, P., Richmond, J., Donald, B., Ferrier, R., and Cooke, T.G. 2001. Evaluating HER-2 amplification and overexpression in breast cancer. *J. Pathol. 195:*422–428.

Beck, R.C., Tubbs, R.R., Hussein, M., Pettay, J., and Hsi, E.D. 2003. Automated colorimetric *in situ* hybridization (CISH) detection of immunoglobulin (Ig) light chain mRNA expression in plasma cell (PC) dyscrasias and non-Hodgkin lymphoma. *Diagn. Mol. Pathol. 12:*14–20.

Bilous, M., Dowsett, M., Hanna, W., Isola, J., Lebeau, A., Moreno, A., Penault-Llorca, F., Ruschoff, J., Tomasic, G., and Van De Vijver, M. 2003. Current perspectives on HER-2 testing: A review of national testing guidelines. *Mod. Pathol. 16:*173–182.

Cowell, J.K. 1982. Double minutes and homogeneously staining regions: Gene amplification in mammalian cells. *Annu. Rev. Genet. 16:*21–59.

Dandachi, N., Dietze, O., and Hauser-Kronberger, C. 2002. Chromogenic *in situ* hybridization: A novel approach to a practical and sensitive method for the detection of HER-2 oncogene in archival human breast carcinoma. *Lab Invest. 82:*1007–1014.

Dowsett, M., Bartlett, J., Ellis, I., Salter, J., Hills, M., Mallon, E., Watters, A., Cooke, T., Paish, C., Wencyk, P., and Pinder, S. 2003. Correlation between immunohistochemistry (HercepTest) and fluorescence *in situ* hybridization (FISH) for HER-2 in 426 breast carcinomas from 37 centres. *J. Pathol. 199:*418–423.

Earp, H.S., Dawson, T.L., Li, X., and Yu, H. 1995. Heterodimerization and functional interaction between EGF receptor family members: A new signaling paradigm with implications for breast research. *Breast Cancer Res. Treat. 35:*115–132.

Frater, J.L., and Tubbs, R.R. 2001. Tyramide amplification methods for *in situ* hybridization. In Lloyd, R.V. (ed) *Morphology Methods. Cell and Molecular Biology Techniques.* Totowa, NY: Humana Press, 113–128.

Hanna, W.M., Kahn, H.J., Pienkowska, M., Blonda, J., Seth, A., and Marks, A. 2001. Defining a test for HER-2/neu evaluation in breast cancer in the diagnostic setting. *Mod. Pathol. 14:*677–685.

Hayat, M.A. 2002. *Microscopy, Immunohistochemistry, and Antigen Retrieval Methods.* New York: Kluwer Academic/Plenum Publishers.

Hopman, A.H., Claessen, S., and Speel, E.J. 1997. Multicolor bright-field *in situ* hybridization on tissue sections. *Histochem. Cell. Biol. 108:*291–298.

Jacobs, T.W., Gown, A.M., Yaziji, H., Barnes, M.J., and Schnitt, S.J. 1999a. Comparison of fluorescence *in situ* hybridization and immunohistochemistry for the evaluation of HER-2/neu in breast cancer. *J. Clin. Oncol. 17:*1974–1982.

Jacobs, T.W., Gown, A.M., Yaziji, H., Barnes, M.J., and Schnitt, S.J. 1999b. Specificity of HercepTest in determining HER-2/neu status of breast cancers using the United States Food and Drug Administration-approved scoring system. *J. Clin. Oncol. 17:*1983–1987.

Kayser, K. 2002. Interdisciplinary telecommunication and expert teleconsultation in diagnostic pathology: Present status and future prospects. *J. Telemed. Telecare 8:*325–330.

Kumamoto, H., Sasano, H., Taniguchi, T., Suzuki, T., Moriya, T., and Ichinohasama, R. 2001. Chromogenic *in situ* hybridization analysis of HER-2/neu status in breast carcinoma: Application in screening of patients for trastuzumab (Herceptin) therapy. *Pathol. Int. 51:*579–584.

Lebeau, A., Deimling, D., Kaltz, C., Sendelhofert, A., Iff, A., Luthardt, B., Untch, M., and Lohrs, U. 2001. Her-2/neu analysis in archival tissue samples of human breast cancer: Comparison of immunohistochemistry and fluorescence *in situ* hybridization. *J. Clin. Oncol. 19:*354–363.

Leong, F.J., and Leong, A.S. 2003. Digital imaging applications in anatomic pathology. *Adv. Anat. Pathol. 10:*88–95.

Nuovo, G.J., and Richart, R.M. 1989. Buffered formalin is the superior fixative for the detection of HPV DNA by *in situ* hybridization analysis. *Am. J. Pathol. 134:*837–842.

Pauletti, G., Godolphin, W., Press, M.F., and Slamon, D.J. 1996. Detection and quantitation of HER-2/neu gene amplification in human breast cancer archival material using fluorescence *in situ* hybridization. *Oncogene 13:*63–72.

Press, M.F., Bernstein, L., Thomas, P.A., Meisner, L.F., Zhou, J.Y., Ma, Y., Hung, G., Robinson, R.A., Harris, C., El-Naggar, A., Slamon, D.J., Phillips, R.N., Ross, J.S., Wolman, S.R., and Flom, K.J. 1997. Her-2/neu gene amplification characterized by fluorescence *in situ* hybridization: Poor prognosis in node-negative breast carcinomas. *J. Clin. Oncol. 15:*2894–2904.

Roche, P.C., and Ingle, J.N. 1999. Increased HER-2 with U.S. Food and Drug Administration-approved antibody. *J. Clin. Oncol. 17:*434.

Rummukainen, J.K., Salminen, T., Lundin, J., Joensuu, H., and Isola, J.J. 2001. Amplification of C-Myc oncogene by chromogenic and fluorescence *in situ* hybridization in archival breast cancer tissue array samples. *Lab Invest. 81:*1545–1551.

Schnitt, S.J., and Jacobs, T.W. 2001. Current status of HER-2 testing: Caught between a rock and a hard place. *Am. J. Clin. Pathol. 116:*806–810.

Sharma, A., Pratap, M., Sawhney, V.M., Khan, I.U., Bhambhani, S., and Mitra, A.B. 1999. Frequent amplification of c-erbB2 (HER-2/neu) oncogene in cervical carcinoma as detected by nonfluorescence *in situ* hybridization technique on paraffin sections. *Oncology 56:*83–87.

Slamon, D.J., Clark, G.M., Wong, S.G., Levin, W.J., Ullrich, A., and McGuire, W.L. 1987. Human breast cancer: Correlation of relapse and survival with amplification of the HER-2/neu oncogene. *Science 235:*177–182.

Slamon, D.J., Godolphin, W., Jones, L.A., Holt, J.A., Wong, S.G., Keith, D.E., Levin, W.J., Stuart, S.G., Udove, J., and Ullrich, A. 1989. Studies of the HER-2/neu proto-oncogene in human breast and ovarian cancer. *Science 244:*707–712.

Smith, K.L., Robbins, P.D., Dawkins, H.J., Papadimitriou, J.M., Redmond, S.L., Carrello, S., Harvey, J.M., and Sterrett, G.F. 1994. c-erbB-2 amplification in breast cancer: Detection in formalin-fixed, paraffin-embedded tissue by *in situ* hybridization. *Hum. Pathol. 25:*413–418.

Speel, E.J., Hopman, A.H., and Komminoth, P. 1999. Amplification methods to increase the sensitivity of *in situ* hybridization: Play card(s). *J. Histochem. Cytochem. 47:*281–288.

Tanner, M., Gancberg, D., Di Leo, A., Larsimont, D., Rouas, G., Piccart, M.J., and Isola, J. 2000. Chromogenic *in situ* hybridization: A practical alternative for fluorescence *in situ* hybridization to detect HER-2/neu oncogene amplification in archival breast cancer samples. *Am. J. Pathol. 157:*1467–1472.

Tubbs, R.R., Pettay, J.D., Roche, P.C., Stoler, M.H., Jenkins, R.B., and Grogan, T.M. 2001. Discrepancies in clinical laboratory testing of eligibility for trastuzumab therapy: Apparent immunohistochemical false-positives do not get the message. *J. Clin. Oncol. 19:*2714–2721.

van de Vijver, M.J. 2001. Assessment of the need and appropriate method for testing for the human epidermal growth factor receptor-2 (HER-2). *Eur. J. Cancer 37(Suppl. 1):*11–17.

Vogel, C.L., Cobleigh, M.A., Tripathy, D., Gutheil, J.C., Harris, L.N., Fehrenbacher, L., Slamon, D.J., Murphy, M., Novotny, W.F., Burchmore, M., Shak, S., and Stewart, S.J. 2001. First-line Herceptin monotherapy in metastatic breast cancer. *Oncology 61(Suppl. 2):*37–42.

Vos, C.B., Ter Haar, N.T., Peterse, J.L., Cornelisse, C.J., and van de Vijver, M.J. 1999. Cyclin D1 gene amplification and overexpression are present in ductal carcinoma *in situ* of the breast. *J. Pathol. 187:*279–284.

Wang, S., Hossein Saboorian, M., Frenkel, E.P., Haley, B.B., Siddiqui, M.T., Gokaslan, S., Hynan, L., and Ashfaq, R. 2002. Aneusomy 17 in breast cancer: Its role in HER-2/neu protein expression and implication for clinical assessment of HER-2/neu status. *Mod. Pathol. 15:*137–145.

Wang, S., Saboorian, M.H., Frenkel, E., Hynan, L., Gokaslan, S.T., and Ashfaq, R. 2000. Laboratory assessment of the status of HER-2/neu protein and oncogene in breast cancer specimens: Comparison of immunohistochemistry assay with fluorescence *in situ* hybridization assays. *J. Clin. Pathol. 53:*374–381.

Watters, A.D., Going, J.J., Cooke, T.G., and Bartlett, J.M. 2003. Chromosome 17 aneusomy is associated with poor prognostic factors in invasive breast carcinoma. *Breast Cancer Res. Treat. 77:*109–114.

Zhao, J., Wu, R., Au, A., Marquez, A., Yu, Y., and Shi, Z. 2002. Determination of HER-2 gene amplification by chromogenic *in situ* hybridization (CISH) in archival breast carcinoma. *Mod. Pathol. 15:*657–665.

6

Immunohistochemical Evaluation of Sentinel Lymph Nodes in Breast Carcinoma

Cynthia Cohen

Introduction

In breast cancer patients, lymph nodal status, tumor size, histologic grade and type of carcinoma, hormone receptor status, and HER-2/neu expression/amplification are significant prognostic factors. However, the axillary nodal status is the single most important predictor of overall and disease-free survival, with the absolute number of nodes involved by carcinoma being of prognostic significance (Reynolds *et al.*, 2003).

Although axillary dissection has become an integral part of breast cancer management because of its prognostic importance, the resultant excellent local control with only rare axillary recurrences, and the debatable improvement in long-term survival, it does have several disadvantages. It is an expensive procedure with substantial morbidity (lymphedema in up to 30% of patients in some series and major vessel or motor nerve damage) and is of no benefit to 70–75% of patients who are node-negative. Hence, the introduction of sentinel lymph node (SLN) biopsy for breast cancer patient management in 1993 by Krag *et al.* (1993). This procedure is based on the principles that

there is an orderly and predictable pattern of drainage to the regional lymph node basin, and that the first, or sentinel, lymph node is an effective filter for tumor cells. Several reported studies have shown that only ~ 1% of complete axillary lymph node (ALN) dissections will be positive for metastatic breast carcinoma, if the SLN is negative (Cohen *et al.*, 2002). On the other hand, there are reported higher frequencies (< 5–16%) of positive ALN in the face of negative SLN. Conversely, up to 70% of patients with SLN positive for metastatic carcinoma have non-SLN involved by tumor (Cohen *et al.*, 2002).

Thus, SLN status can be used to determine which patients will benefit from ALN dissection and which require adjuvant therapy or can be spared the morbidity and expense of such therapy. If the SLN is negative for metastatic carcinoma, no ALN dissection is performed. If the SLN is positive, the surgeon continues with an ALN dissection, if possible, during the initial surgery. Thus, in 20–25% of patients that have positive SLNs, a second operation to complete the ALN dissection will not be required. This necessitates that every effort be taken to diagnose the SLN metastasis

Handbook of Immunohistochemistry and *in situ* Hybridization of Human
Carcinomas, Volume 1: Molecular Genetics; Lung and Breast Carcinomas

289

Copyright © 2004 by Elsevier (USA)
All rights reserved.

intraoperatively by imprint cytology, frozen section (FS), and possibly rapid immunohistochemistry (RIHC). Fresh tissue is sometimes sent for reverse transcription-polymerase chain reaction (RT-PCR), and levels and immunohistochemistry (IHC) are performed on permanent sections. However, if results indicate metastatic carcinoma in the SLN, a second operation for an ALN dissection is required. As yet, there are many unanswered issues regarding intraoperative lymphatic mapping for the localization of SLN, the method of pathologic assessment, the method of reporting results, and the significance of micrometastases identified initially only by IHC.

This chapter reports on our technique of handling and immunostaining SLN from breast cancer patients, the results we obtained with intraoperative scrape touch preparations (TP), FS, and RIHC, and the subsequent seven levels (L_{1-7}) of hematoxylin and eosin (H&E) stained permanent sections, with cytokeratin immunostaining on levels 1-2 (L_{1-2}) and 4-5 (L_{4-5}). The literature regarding the controversial areas will be discussed. Recommendations, based on our and others' experience, will be given.

MATERIALS

I. Rapid Immunohistochemistry (EPOS) on Frozen Sections
  A. Equipment
    1. Slide warmer capable of 37°C.
    2. Tissue-Tek container.
    3. Slide holder.
    4. Plus microscope slides (FisherBrand Superfrost, Cat. No. 12-550-15, Fisher Scientific, Pittsburgh, PA).
  B. Materials
    1. Tris-buffered saline (TBS) (Dako, Cat. No. S1968, Dako Corporation, Carpinteria, CA). Dissolve entire contents of one packet in deionized water and bring final volume to 5 l. Add 2.5-ml of Dako Tween (Cat. No. S1966). Good for 5 days at room temperature. Otherwise, store TBS buffer at 2–8°C.
    2. Dako Liquid 3,3′ diaminobenzidine (DAB). Add 1 drop (or 20 μl) of the liquid DAB chromogen per ml of buffered substrate.

DAB chromogen is a suspected carcinogen. Wear gloves when preparing and aliquoting.

    3. EPOS, Cytokeratin (Dako, Cat. No. U7022).
    4. Aqueous hematoxylin (Richard-Allan, Cat. No. 7221, Kalamazoo, MI).
    5. Mounting medium (Richard-Allan, Cat. No. 4111).

II. Automated Immunohistochemistry (LSAB2) on Permanent Formalin-Fixed, Paraffin-Embedded Sections
  A. Equipment
    1. Dako autostainer (Dako).
    2. Electric pressure cooker, decloaking chamber (Biocare Medical, Cat. No. DC2000, Walnut Creek, CA).
    3. Plus microscope slides (FisherBrand).
    4. Coverslipper, Tissue-Tek SCA (Sakura, Torrance, CA).
  B. Materials
    1. 1X target retrieval solution (Dako, Cat. No. S1699).
       200 ml of 10X target retrieval solution. Quantity sufficient with deionized water to 2 L.
    2. 100% absolute ethanol.
       Store stock in flammable cabinet.
    3. Xylene.
       Store stock in flammable cabinet.
    4. 50% ethanol.
       Use equal amounts of deionized water and 100% ethanol.
       Store stock in flammable cabinet.
       Stable for 1 year.
    5. 3% hydrogen peroxide in water.
       10 ml of 30% hydrogen peroxide.
       Quantity sufficient to 100 ml.
    6. TBS.
       Dissolve entire contents of one packet in distilled water and bring final volume to 5 L. Add 2.5 ml of Dako Tween (Cat. No. S1966). Good for 5 days at room temperature. Otherwise, store TBS solution at 2–8°C.
    7. Dako LSAB2 HRP kit for the autostainer, which contains the following:
       a. Biotinylated link: Biotin-labeled affinity isolated goat-anti-rabbit and goat-anti-mouse immunoglobulins in phosphate-buffered saline (PBS), containing carrier protein and 0.015 M sodium azide.
       b. Streptavidin-HRP: Streptavidin conjugated to horseradish peroxidase in PBS containing carrier protein and anti-microbial agents.
    8. Dako liquid DAB.
       Add 1 drop of the liquid DAB chromogen per ml of buffered substrate.
    9. Aqueous hematoxylin.
    10. Primary antibodies.
        Prediluted and concentrated primary antibodies titrated on known positive controls and used at appropriate dilutions. Primary

antibodies are diluted with an antibody diluent, Zymed antibody diluent reagent solution (Zymed, South San Francisco, CA).
    11. Mounting medium.
III. Manual Immunohistochemistry on Permanent Formalin-Fixed, Paraffin-Embedded Sections
  A. Equipment
    1. Electric pressure cooker (Biocare Medical).
    2. Tissue-Tek container.
    3. Slide holder.
    4. Plus microscope slides (FisherBrand).
  B. Materials
    1. 1X target retrieval solution (Dako Cat. No. S1699).
       200 ml of 10X target retrieval solution.
       Quantity sufficient with deionized water to 2 L.
    2. 100% absolute ethanol.
       Store stock in flammable cabinet.
    3. Xylene.
       Store stock in flammable cabinet.
    4. 50% ethanol.
       Use equal amounts of deionized water and 100% ethanol.
       Store stock in flammable cabinet.
       Stable for 1 year.
    5. 95% ethanol.
       Store stock in flammable cabinet.
    6. 3% hydrogen peroxide in water.
       10 ml of 30% hydrogen peroxide.
       Quantity sufficient to 100 ml.

Caution: Hydrogen peroxide is corrosive and causes burns. Wear gloves when handling.

    7. TBS.
       Dissolve entire contents of one packet in distilled or deionized water and bring final volume to 5 L. Add 2.5 ml of Dako Tween. Good for 5 days at room temperature. Otherwise, store TBS solution at 2–8°C.
    8. Dako LSAB2 HRP, which contains the following:
       a. Biotinylated-link: Biotin-labeled affinity isolated goat-anti-rabbit and goat-antimouse immunoglobulins in PBS, containing carrier protein and 0.015 M sodium azide.
       b. Streptavidin-HRP: Streptavidin conjugated to horseradish peroxidase in PBS containing carrier protein and antimicrobial agents.
    9. Dako liquid DAB, large volume kit.
       Add 1 drop (or 20 µl) of the liquid DAB chromogen per ml of buffered substrate.

    10. Aqueous hematoxylin.
    11. Primary antibodies.
        Prediluted and concentrated antibodies are titrated on known positive controls and used at appropriate dilutions. Primary antibodies are diluted with an antibody diluent, Zymed antibody diluent reagent solution.
    12. Mounting medium.

METHODS

I. Rapid Immunohistochemistry (EPOS) on Frozen Sections
  A. Cut two 6-µm-thick sections and place them on Superfrost plus slides.
  B. Fix in cold acetone for 2–3 min.
  C. Allow slides to come to room temperature and place in TBS buffer. (One slide is the patient negative control, which will have primary antibody replaced with TBS).
  D. Place slides on slide warmer, remove excess buffer, but do not allow tissue to dryout.
     Cover section with EPOS and incubate for 10 min.
     TBS buffer only on negative control slide.
  E. Rinse well with TBS buffer.
  F. Remove excess buffer, cover slide with DAB, and incubate for 5 min.
  G. Rinse well with water.
  H. Counterstain with hematoxylin for 1 min and rinse well with water.
  I. Dehydrate in 70%, 95%, and 100% alcohol, and in two changes of xylene.
  J. Coverslip in mounting medium.

Note: At the same time, also immunostain a known positive control (a frozen section of a breast or lung adenocarcinoma previously used to optimize the procedure) with a negative control slide (TBS buffer instead of EPOS cytokeratin, the specific antibody).

II. Automated Immunohistochemistry (LSAB2) on Permanent Formalin-Fixed, Paraffin-Embedded Sections
  A. All fixed, paraffin-embedded tissue should be mounted on Superfrost plus slides.
  B. Paraffin sections should be mounted on slides from preheated water baths containing distilled or deionized water. The water bath should contain no additives (e.g., gelatin, polylysin).
  C. Stain slides of patient carcinomas and appropriate positive control tissue such as breast carcinoma (each with a negative TBS control) simultaneously in the same run.
  D. Fill out the slide layout map to indicate where each slide will be placed on the slide holder.

Place all slides in a microwavable Tissue-Tek slide holder. Microwave on high for 1 min, add 1 additional min for each rack up to three racks (two racks for 2 min, etc.). This melts the paraffin.

E. Turn the dial on the pressure cooker to "Keep Warm" while deparaffinizing slides.
   1. Fill pan with 500 ml of deionized water.
   2. Place from 1–4 Tissue-Tek containers filled with 250 ml of 1X target retrieval solution.

F. Deparaffinize and rehydrate slides by running them through the following solutions:
   1. Formalin-fixed or B5-fixed tissue:
      a. Xylene—5 min
      b. Xylene—5 min
      c. 100% ethanol—2 min
      d. 100% ethanol—2 min
      e. 50% ethanol—2 min
      f. 50% ethanol—2 min
      g. Deionized water twice for 2 min each

G. Transfer deparaffinized slides from water to 1X target retrieval solution in the pressure cooker. Put lid on securely by aligning the "open" arrow with the white dot on the pan's handle. Grip the lid handle, and rotate clockwise to the "close" position. When the lid is locked in the proper position, the *vent lever* will lower the *weight* on the *vent nozzle*.

H. Set timer for 3–5 min by turning "Timer Switch" above 20, then back down to 3–5 min. The "Heat-on" light will turn on. The pressure will rise from 0 to 15–20 pounds per square inch (PSI), and temperature will rise to 120°C. Upon reaching full temperature and pressure, the timer starts counting down.

I. When the timer reaches zero and the pressure has dropped to zero, it is safe to remove the pressure cooker lid. Slightly push down on the brown lid handle and rotate counter clockwise to the "open" position. Remove the lid slowly and allow the steam to escape. Use heat-resistant gloves to remove Tissue-Tek containers from pan. Allow slides to cool for 5–10 min.

J. After the cool down, rinse slides with deionized water and place in TBS for 5 min.

K. Set up Autostainer for your run according to the Dako user guide.
   1. Click on "Program."
   2. Click on "Slides," enter the number of slides and hit "Enter Key."
   3. Fill in programming grid according to protocol being used, click "Next."
   4. Compare your slide layout map to the screen's slide layout map to be sure you

have programmed the computer correctly. Click "Next."
   5. Save the program with the current date (example: 030503am), print "Load Reagent" map.
   6. Set up reagent racks according to map.
   7. Make sure Autostainer has appropriate buffer and water.
      Load slides (see example labeled "Loading Slides") on Autostainer according to slide layout map and start your run (see example labeled "Starting a Run").

L. When the run has ended (see example labeled "Completing a Staining Run"), do not print run log.

M. Transfer slides from stainer to a slide holder for the coverslipper, place in a Tissue-Tek container filled with water.

N. Dehydrate slides by dipping them several times through two changes each of 70%, 95%, and 100% alcohol, and two changes of xylene.

O. Coverslip in mounting medium and label slides.

III. Manual Immunohistochemistry on Permanent Formalin-Fixed, Paraffin-Embedded Sections

A. All fixed, paraffin-embedded tissue should be mounted on Superfrost plus slides.

B. Paraffin sections should be mounted from pre-heated water baths containing distilled or deionized water. The water bath should contain no additives (such as gelatin, polylysin).

C. Stain 5 micron slides of patient carcinomas and appropriate positive control tissue, such as breast carcinoma (each with a negative TBS control) simultaneously in the same run.

D. Organize patient slides and appropriate positive control tissue slides. Place all slides in a microwavable Tissue-Tek slide holder. Microwave on high for 1 min, then add 1 addtional min for each rack up to three racks. This melts the paraffin.

E. Deparaffinize and rehydrate slides by running them through the following solutions:
   1. Formalin-fixed or B5-fixed tissue:
      a. Xylene—5 min
      b. Xylene—5 min
      c. 100% ethanol—2 min
      d. 100% ethanol—2 min
      e. 95% ethanol—2 min
      f. 95% ethanol—2 min
      g. 50% ethanol—2 min
      h. 50% ethanol—2 min
      i. Deionized water-running 1–2 min each

F. Transfer slides from water to warm 1X target retrieval solution in the electric pressure cooker. Put lid on securely by aligning the "open" arrow with the white dot on the pan's handle. Grip the lid handle and rotate clockwise to the "close" position. When the lid is locked in the proper position, the *vent lever* will lower the *weight* on the *vent nozzle*.

G. Set timer for 3–5 min by turning "Timer Switch" above 20, then back down to 3–5 min. The "Heat-on" light will turn on. The pressure will rise from 0 to 15–20 PSI, and the temperature will rise to 120°C. Upon reaching full temperature and pressure, the timer starts counting down.

H. When the timer reaches zero and the pressure has dropped to zero, it is safe to remove the lid. Slightly push down on the brown handle and rotate counter clockwise to the "open" position. Remove the lid slowly and allow the steam to escape. Use heat-resistant gloves to remove Tissue-Tek slide containers from pan. Allow slides to cool for 5–10 min.

I. Rinse cooled slides with deionized water and place in TBS for 5 min.

Note: Do not allow slides to become dry at any time.

J. For blocking of endogenous peroxidase activity, incubate slides in 3% hydrogen peroxide for 5 min, rinse with 1X TBS.

K. To apply primary antibodies, wipe carefully around the tissue with gauze and apply 100 µl of the appropriate antibody to the tissue, for both patient and known positive controls. For negative controls, add TBS buffer or appropriate isotype antibody. Incubate slides for 25 min in a humidity chamber at room temperature.

L. Rinse slides individually with 1X TBS from a wash bottle. Wash slides in 1X TBS for two washes. Place slides in wash buffer (1X TBS).

M. Wipe slides carefully around tissue sections with gauze. Cover tissue with several drops of biotinylated link. Incubate for 15 min in a humidity chamber at room temperature.

N. Remove slides from humidity chamber and place in a slide holder in staining dish containing 1X TBS. Agitate slides briefly and pour off buffer. Wash slides in 2 changes of 1X TBS.

O. Wipe slides carefully around tissue sections with gauze. Cover tissue with several drops of streptavidin-HRP. Incubate for 15 min in a humidity chamber at room temperature.

P. Remove slides from humidity chamber and place in slide holder in staining dish containing 1X TBS. Agitate slides briefly and pour off buffer. Wash slides in two changes of 1X TBS.

Q. Wipe slides carefully around tissue sections with gauze. Cover tissue with several drops of prepared DAB substrate-chromogen solution. Incubate for 5 min in a humidity chamber at room temperature.

R. Remove slides from humidity chamber and place in slide holder in staining dish containing deionized water. Agitate slides briefly and pour off water. Wash slides in 2 changes of water.

S. Counterstain for 45 sec to 1 min in an aqueous hematoxylin. Rinse well in tap water.

T. Dehydrate slides by dipping them several times through the following solutions:
  1. Two changes of:
    a. 50% alcohol
    b. 95% alcohol
    c. 100% alcohol
    d. Xylene

U. Coverslip in mounting medium.

RESULTS AND DISCUSSION

In our laboratory, we examined SLNs from 91 patients. Sixty-six percent were negative and 34% were positive for metastatic carcinoma. The size of the metastasis was > 2 mm in 10 (32%) (macrometastases), and < 2 mm in 21 (68%) (micrometastases), with clusters of cells in 8, and isolated cells in 3 (Figure 40). Of the 241 SLNs from the 91 patients (270 sections), 56 nodes from 31 patients were positive. In TP, 8.9% were positive, in FS 9.7%, in H&E L_1 11.7%, in H&E L_{2-7} 10.4%, in IHC L_{1-2} 16.6%, and in IHC L_{4-5} 13.2%. Of the positive SLNs in 31 patients, 13 (15.5%) were positive in TP, 18 (20.4%) in FS, 27 (29.7%) in H&E L_1, 24 (27%) in H&E L_{2-7}, 35 (39%) in IHC L_{1-2}, and 28 (31.4%) in IHC L_{4-5}. Thus, IHC L_{1-2} was the most sensitive method of detection, with an additional six levels and one extra IHC L_{4-5} resulting in no increase in sensitivity.

By RIHC, 14 of 72 SLN (19%) were positive for metastatic carcinoma in 13 of 32 patients studied (Figure 41). The 57% sensitivity of RIHC was less than the 69% for TP, 86% for FS, and 100% for H&E L_1 and IHC L_{1-2}. The six false-negatives (FN) by RIHC were micrometastases (four) or undetected (two) by intraoperative TP and FS. Seven of the eight true-positive RIHC were macrometastases.

The technique of SLN biopsy was initially used for management of penile cancer in 1977. Since then, its use has been extended to melanoma and breast carcinoma patients. The localization of the SLN was initially by radioactive isotope and gamma probe, then with the use of blue dye alone, and subsequently a

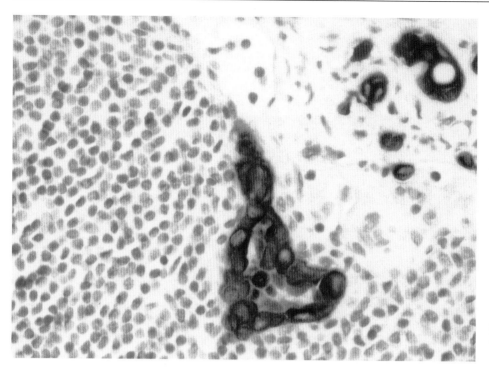

Figure 40 Cytokeratin AE$_{1-3}$ immunostained isolated and clusters of breast carcinoma cells (≤ 0.2 mm in diameter) (pNO[i+]), in the peripheral sinus of a sentinel lymph node, subsequently confirmed in the hematoxylin and eosin stained section 400 X.

combination of radioisotope and blue dye has given the best results. Success rates range from 66–98% and are operator- and technique-dependent (Reynolds, 2003). The accuracy is 95–100%, and the FN rate 0–15%.

At present, there is no consensus as to the most sensitive and cost-effective pathologic method of handling SLNs. Lymph nodes may be bivalved if < 1 cm, and breadloaved at 2 mm intervals if > 1 cm. The entire SLN is usually examined. Intraoperative cytology as an imprint or scraping, of all or most cut surfaces, is useful in alcohol-fixed H&E or air-dried Dif-Quick preparations. Frozen sections (1–3 levels, or multiple serials) are

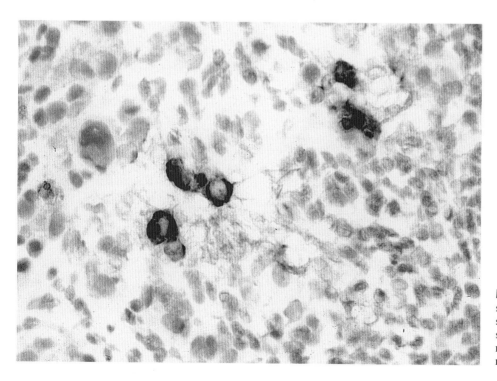

Figure 41 Small clusters and single cells (≤ 0.2 mm) of metastatic breast carcinoma immunostained for cytokeratin AE$_{1-3}$ by rapid immunohistochemistry 400 X, reported as pNO(i+).

usually performed and H&E stained. Both RIHC and RT-PCR are still research tools. The permanent H&E sections are examined in one to multiple levels and/or serial sections; IHC is usually performed on one level, but sometimes on more than one level.

Cytologic examination is simple, rapid, and inexpensive. It does not use up diagnostic tissue (as might multiple FS levels or serial sections, RIHC, or RT-PCR). Good cytologic detail is obtained, especially for lobular carcinoma. However, especially with micrometastases, too few cells may be obtained, resulting in FN or indeterminate diagnoses. Experience in cytopathology is also useful for the pathologist. In the literature, intraoperative TP and/or FS are reported accurate in 83–100% of SLNs (Cohen *et al.*, 2002). They are concordant with subsequent H&E sections in 82–99%, with a sensitivity of 29–96% and a specificity of 100% for TP. For TP, the positive predictive value (PPV) is 100%, the negative predictive value (NPV) 93–99%, and the FN rate 6–62% per patient or 0–71% per SLN (Cohen *et al.*, 2002; Reynolds *et al.*, 2003).

With FS, the nodal architecture and location of metastases are visualized, and fewer indeterminate diagnoses are obtained (Reynolds, 2003). Disadvantages include the 10–15 minutes preparation and slide review time per SLN, resulting in lengthier operative procedures. Our surgeons, however, do the lumpectomy while waiting for SLN FS results, and then, if necessary, perform complete ALN dissections, so that there is no increase in operative time. FS do have artifacts that are then present in permanent sections. Diagnostic tissue may be consumed (with a positive diagnosis, it is hoped), and the FS procedure is more expensive than cytology. Again, accuracy, with or without TP, is reported at 83–100% for FS, with concordance with permanent sections in 82–99%, a sensitivity of 86–87%, specificity of 100%, PPV of 72%, NPV of 100%, and FN rate of 14–33% compared with permanent H&E, 28–52% compared with permanent IHC, and 13–48% compared with permanent H&E and IHC (Cohen *et al.*, 2002; Reynolds *et al.*, 2003). With serial FS, at times of the whole SLN, sometimes with RIHC, the FN rate is reported as 0% (Viale *et al.*, 1999; Veronesi *et al.*, 1999). This labor-intensive, time-consuming procedure (particularly if on multiple SLNs), may be less cost-effective than obtaining an FN result, necessitating a second operative procedure.

By facilitating detection of small foci, single, or small clusters of metastatic carcinoma < 2 mm (micrometastases) and lobular carcinoma intraoperatively, RIHC (IHC performed on FS) would enable the complete ALN dissection to be performed during the same procedure for ~ 25% of patients. For the remaining 75% not requiring a complete lymphadenectomy, unnecessary cost of the RIHC would have to be met, although this is much less than the cost of a second

operative procedure. The RIHC technique is felt to be highly sensitive, despite inherent technical and interpretative problems. Results are, however, mixed. The reported accuracy of RIHC is 95% (Krishnamurthy *et al.*, 2001), sensitivity 57% (Beach *et al.*, 2003), specificity 100% (Beach *et al.*, 2003), and NPV 94% (Viale *et al.*, 1999). Our sensitivity of 57% with RIHC is less than that obtained with TP (69%), FS (86%), permanent H&E (100%), or permanent IHC (100%) (Beach *et al.*, 2003). Thus, intraoperative RIHC detected no additional micrometastases than were demonstrated by TP and/or FS, whether ductal or lobular carcinoma (Beach *et al.*, 2003; Galimberti *et al.*, 2000; Viale *et al.*, 1999). This appeared to be due to sampling error and inherent difficulties with FS (folding of sections, higher background, weaker stain of metastatic carcinoma). Positive controls of breast and lung adenocarcinoma, used to optimize the technique and run with each batch stained, were cut by a histotechnologist and results (clean background, strong immunostain) were excellent. The SLNs are usually fatty and difficult to cut FS, especially for less-experienced residents sometimes on call for FS. Results should be markedly improved where FS are routinely cut by technical staff, who could also do the RIHC.

In the literature, time taken to perform RIHC with various techniques has ranged from 10–65 min (Galimberti *et al.*, 2000; Hyjek *et al.*, 2000; Krishnamurthy *et al.*, 2001; Viale *et al.*, 1999). We were able to perform RIHC in 16 min with the Dako EPOS method (Chilosi *et al.*, 1994; Richter *et al.*, 1999), because incubation times at 37°C are less than at room temperature. Staining times will be increased as the number of SLNs (range 1–11) submitted per patient increases. Our surgeons feel that this delay is no problem, as they perform the SLN biopsy, then the lumpectomy while awaiting results, and then the complete ALN dissection, where appropriate.

Recently, using the Zymed Sentinel Lymph Node Rapid IHC Kit for pancytokeratin (Zymed, South San Francisco, CA), in which the antibodies are conjugated with horseradish peroxidase polymer, we obtained a sensitivity and accuracy of 92% and 94%, respectively. The technique took 8 min and was easy to perform and interpret because of minimal background stain.

The RIHC method has also been performed intraoperatively on TP (Krishnamurthy *et al.*, 2001; Ku, 1999; Reintgen *et al.*, 2000) and is suggested to improve diagnostic accuracy, especially for low-grade, lobular, and low-volume metastatic carcinoma (Ku, 1999). As with RIHC on FS, experience in interpretation of the immunostained TP is essential to avoid potential errors (Ku, 1999).

With a variety of mRNA markers (keratin 19, *MUC1*, β human chorionic gonadotrophin, mammoglobin, carcinoembryonic antigen transcripts) (Hoon *et al.*, 1996; Min *et al.*, 1998; Noguchi *et al.*, 1996), RT-PCR has been undertaken on fresh frozen tissue. Frequency of detection of SLN metastasis occurs in 25–31% of patients with subsequent upstaging (Hoon *et al.*, 1996; Min *et al.*, 1998; Noguchi *et al.*, 1996). Using a method of RT-PCR, amplification of mammoglobin gene transcripts on frozen or formalin-fixed, paraffin-embedded cell pellets from lymph nodes, sensitivity was increased by 9.5% and 50%, respectively (Javonic *et al.*, 2003). Sampling error or lack of mammoglobin expression in the primary tumor seems to be the cause of FNs in 15.4% of fresh frozen nodes and 12.5% of fixed lymph nodes (Javonic *et al.*, 2003). Some authors feel that submission of half of a bisected SLN for RT-PCR (Smith *et al.*, 1999) or even the cutting of FS for intraoperative diagnosis (Ku, 1999) results in loss of micrometastases in up to 40% of nodes (Smith *et al.*, 1999). The use of cytokeratin mRNA markers in RT-PCR may yield false-positive results because of the presence of cytokeratin-positive dendritic cells in lymph nodes (Yaziji *et al.*, 2000).

In permanent sections, serial sectioning with H&E staining is reported by some to increase detection of metastatic carcinoma in SLN by 7–33% (Cohen *et al.*, 2002; Reynolds *et al.*, 2003). IHC on permanent sections is shown to increase detection of micrometastatic carcinoma in SLNs by 8–50%, with subsequent upstaging (Cohen *et al.*, 2002; Reynolds *et al.*, 2003). Others report no improvement in detection with serial sectioning and IHC (Cohen *et al.*, 2002). In our hands, IHC on one level of a paraffin section demonstrated metastases in 16.6% of SLN sections and in 39% of patients, a much higher frequency than in TP (8.9%, 15.5%, respectively), in FS (9.7%, 20.4%), one level H&E stained (11.7%, 29.7%), six extra H&E stained levels (10.4%, 27%), or an extra IHC stained level (L_{4-5}) (13.2%, 31.4%) (Cohen *et al.*, 2002).

Different cytokeratin (CK) antibodies have been used in SLN IHC. The antibody cocktail AE_{1-3} has been found by us and others (Cohen *et al.*, 2002) to be most sensitive and specific compared with panCK and CAM 5.2. Others have found a CK cocktail of seven antibodies to be most sensitive compared with AE_{1-3} and CAM 5.2, or MAK-6 to be the antibody of choice over AE_1 and CAM 5.2. Epithelial membrane antigen (EMA) and *MUC1* antisera have been used, with EMA shown to have low specificity and sensitivity (Cohen *et al.*, 2002; Reynolds *et al.*, 2003). Besides demonstrating CK-positive dendritic cells, IHC may also stain axillary lymph node epithelial inclusions, or carcinoma or normal breast epithelium mechanically transported to the SLN after surgical or needle biopsy. In the latter, hemosiderin-laden macrophages and damaged red blood cells are often also present. Epithelial cells and cell clusters < 1 mm in diameter and not associated with features

of established metastases are associated with prebiopsy breast massage, suggesting benign mechanical transport (Diaz *et al.*, 2003). Immunostained plasma cells have been mistaken for isolated tumor cells.

There are many controversial issues pertaining to SLN biopsy in breast cancer patients. The prognostic implications of micrometastases, particularly of isolated and clusters of tumor cells highlighted only by IHC, remain to be shown. It is not known which patients with SLN micrometastases will benefit from ALN dissections. Also, as systemic adjuvant therapy is increasingly being recommended for patients without axillary nodal metastases, and as no local recurrences were found in patients with small, estrogen receptor-positive carcinomas with SLN macro- and micrometastases (the majority identified by IHC only) (Guenther *et al.*, 2003), SLN biopsy and ALN dissection may no longer be required management. In axillary non-SLN, micrometastases detected by IHC have been shown to correlate significantly with poor outcome (decreased overall and disease-free survival, increased recurrence rate) (Cohen *et al.*, 2002); others fail to show any relationship to prognosis. Patients with IHC-detected micrometastases in SLN are reported to have significantly worse overall and disease-free survival than patients without them (Hsueh *et al.*, 2000; Teng *et al.*, 2000). The pNO (i+) patients with IHC-positive isolated cancer cells had a similar frequency of non-SLN metastases as patients staged as pN1 with micrometastases < 2 mm (Mathieu *et al.*, 2003). The larger size of the SLN metastasis (mean 9 mm) and the presence of markers of increased biologic aggressiveness (angiogenesis, proliferation, histologic grade, extracapsular extension [Changsri *et al.*, 2003; Dabbs *et al.*, 2003]) usually predict the presence of non-SLN metastases, so that ALN dissection would be recommended in these patients. Yet reports vary from finding metastases in ALN with SLN micrometastases < 1 mm, to no positive non-SLN when SLN micrometastases were detected only by IHC. A lower frequency of positive SLN is reported with smaller breast carcinomas, but correlation between size of SLN metastasis and carcinoma size, histologic grade, ductal, or lobular-type of the primary breast carcinoma is not always shown (Cohen *et al.*, 2002). Presence of peritumoral angiolymphatic invasion and extranodal hilar tissue invasion have also been associated with non-SLN involvement (Reynolds *et al.*, 2003).

Whether additional serial sections or levels from frozen and/or paraffin-embedded blocks should be H&E stained and/or immunostained is questionable. The more techniques used and the more tissue examined, the greater the likelihood of demonstrating micrometastases. However, keeping in mind time constraints, labor, and economic factors, and the questionable necessity to detect single cells or clusters of micrometastatic carcinoma, an optimized SLN examination is needed. The College of American Pathologists (Fitzgibbons *et al.*, 2000) recommends an intraoperative gross examination with imprint cytology, preferable to FS, if needed, followed by entire submission of the SLN sectioned at 2-mm intervals to be examined in one H&E section. Routine serial sections and CK IHC are not recommended. The Association of Directors of Anatomic and Surgical Pathology (ADASP, 2001) recommends intraoperative TP and FS if results will affect immediate surgical management. If nodes are negative, they suggest more than one section, but no number of sections is indicated. It is stated that IHC studies may or may not be clinically relevant, but results should be included in the pathology report if tumor cells are identified by CK immunostain only and not visible in standard H&E sections. In 1971, micrometastases were defined as metastatic foci ≤ 2 mm in greatest dimension and macrometastases as tumor foci > 2 mm in diameter (Huvos *et al.*, 1971). More recently, the American Joint Commission on Cancer (AJCC, 2002) defined micrometastases as > 0.2 mm–2.0 mm with extravasation, proliferation, and stromal reaction, whereas isolated tumor cell groups were ≤ 0.2 mm in diameter, typically without evidence of metastatic activity (proliferation, stromal reaction), usually detected by IHC or molecular methods, and subsequently sometimes verified in the H&E stain. Multiple foci of isolated or clusters of tumor ≤ 0.2 mm in one SLN, are not defined. The AJCC Cancer Staging Manual and International Union Against Cancer (IUAC) TNM Classification 2002 uses pNO for negative regional lymph nodes, without IHC. pNO (i–) and pNO (i+) indicate regional lymph nodes negative and positive by IHC, respectively, with no IHC cluster > 0.2 mm. Micrometastasis (> 0.2 mm–2.0 mm) are reported as pN1mi.

We use and recommend examination of the entire SLN. An SLN < 1 cm has three H&E stained FS levels, without TP, because they are not bisected and have no cut surface. SLNs larger than 1 cm are bisected, or breadloafed, at 2-mm intervals for the larger nodes. All cut surfaces are scraped (on one or more slides) and H&E stained for cytologic examination. All FS profiles are cut at three levels, in one or more blocks as appropriate, and H&E stained. No routine intraoperative RIHC is performed, and tissue is not sent for RT-PCR. Patients with positive SLNs immediately undergo complete ALN dissection. All SLN profiles are routinely fixed, paraffin-embedded, and H&E stained at one level. Lymph nodes with macrometastases > 2 mm previously diagnosed by TP and FS are not immunostained. SLN with micrometastases > 0.2 mm–2 mm or negative by TP and FS are immunostained with AE_{1-3}

antibody on one level. Reporting is as previously indicated as pNO (i–), pNO (i+), or pN1mi, with subsequent completion lymphadenectomy in the latter group, although this management is controversial.

Acknowledgments

The author thanks the Histology and Immunohistochemistry Laboratories (Debbie Sexton) for the innumerable levels and immunostains, Judy Dunbar for her secretarial assistance, and Robert Santoianni for his photographic expertise.

This chapter was presented in part at the International Academy of Pathology biannual meeting, Nagoya, Japan, in October 2000; and, in part, at the 90th annual meeting of the United States and Canadian Academy of Pathology, Atlanta, GA, in March 2001.

References

AJCC Cancer Staging Manual. American Joint Committee on Cancer, Sixth Edition. 2002. New York: Springer-Verlag.

Association of Directors of Anatomic and Surgical Pathology. 2001. ADASP recommendations for processing and reporting lymph node specimens submitted for evaluation of metastatic disease. *Am. J. Surg. Pathol.* 25:961–963.

Beach, R.A., Lawson, D., Waldrop, S.M., and Cohen, C. 2003. Rapid immunohistochemistry for cytokeratin in the intraoperative evaluation of sentinel lymph nodes for metastatic breast carcinoma. *Appl. Immunohistochem. Mol. Morphol.* 11:45–50.

Changsri, C., Prakash, S., Sandweiss, L., and Bose, S. 2003. Prediction of additional axillary metastasis of breast cancer following sentinel lymph node surgery. *Mod. Pathol.* 16:25A (Abstract).

Chilosi, M., Lestani, M., Pedron, S., Montagna, L., Benedetti, A., Pizzolo, G., and Menestrina, F. 1994. A rapid immunostaining method for frozen sections. *Biotech. Histochem.* 69:235–239.

Cohen, C., Alazraki, N., Styblo, T., Waldrop, S.M., Grant, S.F., and Larsen, T. 2002. Immunohistochemical evaluation of sentinel lymph nodes in breast carcinoma patients. *Appl. Immunohistochem. Mol. Morphol.* 10:286–303.

Dabbs, D.J., Kessinger, R.L., McManus, K., and Johnson, R. 2003. Sentinel lymph node micrometastatic disease has an aggressive biological immunophenotype in patients with axillary micrometastases. *Mod. Pathol.* 16:26A–27A (Abstract).

Diaz, N.M., Vrcel, V., Ebert, M.D., Clark, J.D., Stowell, N., Sharma, A., Jakub, J.W., Cantor, A.B., Centeno, B.A., and Cox, C.E. 2003. Epithelial cells in sentinel lymph nodes (SLNs) associated with breast massage prior to SLN biopsy: A mode of benign mechanical transport. *Mod. Pathol.* 16: 28A (Abstract).

Fitzgibbons, P.L., Page, D.L., Weaver, D., Thor, A.D., Allred, D.C., Clark, G.M., Ruby, S.G., O'Malley, F., Simpson, J.F., Connolly, J.L., Hayes, D.F., Edge, S.B., Lichter, A., and Schnitt, S.J. 2000. Prognostic factors in breast cancer: College of American Pathologists consensus statement 1999. *Arch. Pathol. Lab. Med.* 24:966–978.

Galimberti, V., Zurrida, S., Intra, M., Monti, S., Arnone, P., Pruneri, G., and DeCicco, C. 2000. Sentinel node biopsy interpretation: The Milan experience. *Breast J.* 6:306–309.

Guenther, J.M., Hansen, N.M., DiFronzo, L.A., Giuliano, A.E., Collins, J.C., Gruber, B.L., and O'Connell, T.X. 2003. Axillary dissection is not required for all patients with breast cancer and positive sentinel nodes. *Arch. Surg.* 138:52–56.

Hoon, D.S., Sarantou, T., Doi, F., Chi, D.D., Kuo, C., Conrad, A.J., Schmid, P., Turner, R., and Giuliano, A. 1996. Detection of metastatic breast cancer by beta-hCG polymerase chain reaction. *Int. J. Cancer 69:*369–374.

Hsueh, E.C., Hansen, N., and Giuliano, A.E. 2000. Intraoperative lymphatic mapping and sentinel lymph node dissection in breast cancer. *CA Cancer J. Clin. 50:*279–291.

Huvos, A.G., Hutter, R.V.P., and Berg, J.W. 1971. Significance of axillary macrometastases and micrometastases in mammary cancer. *Ann. Surg. 173:*44–46.

Hyjek, E., Chiu, A., Chadburn, A., Swistel, A., Hoda, S., and DeLellis, R. 2000. Rapid intraoperative immunostaining for cytokeratin of sentinel lymph nodes in breast cancer. *Mod. Pathol. 13:*223A (Abstract).

Jovanic, I., Woodrick, R., Goldschmidt, R., and Kaul, K.L. 2003. Detection of tumor cells in lymph nodes of breast cancer patients: Comparison of histology with RT-PCR. *Mod. Pathol. 16:*35A (Abstract).

Krag, D.N., Weaver, D.L., Alex, J.C., and Fairbanks, J.T. 1993. Surgical resection and radiolocalization of the sentinel node in breast cancer using a gamma probe. *Surg. Oncol. 2:*335–340.

Krishnamurthy, S., Tarco, E., Hunt, K., Kuerer, H., Ross, M., Ames, F., Singletary, E., and Sneige, N. 2001. Utility of imprint cytology and rapid cytokeratin immunostaining for the intraoperative evaluation of axillary sentinel lymph nodes in breast cancer. *Mod. Pathol. 14:*29A (Abstract).

Ku, N.N. 1999. Pathologic examination of sentinel lymph nodes in breast cancer. *Surg Oncol. Clin. North Am. 8:*469–479.

Mathieu, M.C., Suciu, V., Garbay, J.R., Rouzier, R., Lumbroso, J., and Travagli, J.P. 2003. Isolated tumor cells in the sentinel node and risk of axillary nonsentinel node involvement in patients with breast carcinomas. *Mod. Pathol. 16:*40A (Abstract).

Min, C.J., Tafra, L., and Verbanac, K.M. 1998. Identification of superior markers for polymerase chain reaction detection of breast cancer metastases in sentinel lymph nodes. *Cancer Res. 58:*4581–4584.

Noguchi, S., Aihara, T., Motamura, K., Inaji, H., Imaoka, S., and Koyama, H. 1996. Detection of breast cancer micrometastases in axillary lymph nodes by means of reverse transcriptase-polymerase chain reaction: Comparison between *MUC1* mRNA and keratin 19 mRNA amplification. *Am. J. Pathol. 148:*649–656.

Reintgen, D., Giuliano, R., and Cox, C.E. 2000. Sentinel node biopsy in breast cancer: An overview. *Breast J. 6:*299–305.

Reynolds, C., and Visscher, D.W., March 2003. Recent developments in diagnostic and therapeutic approaches to breast diseases. Annual meeting of the United States and Canadian Academy of Pathology, Washington, DC. Short course #9 (syllabus, supplemental handout).

Richter, T., Nahrig, J., Komminoth, P., Kowolik, J., and Werner, M. 1999. Protocol for ultrarapid immunostaining of frozen sections. *J. Clin. Pathol. 52:*461–463.

Smith, P.A., Harlow, S.P., Krag, D.N., and Weaver, D.L. 1999. Submission of lymph node tissue for ancillary studies decreases the accuracy of conventional breast cancer axillary node staging. *Mod. Pathol. 12:*781–785.

Teng, S., Dupont, E., McCann, C., Wang, J., Bolano, M., Durand, K., Peltz, E., Bass, S.S., Cantor, A., Ku, N.N., and Cox, C.E. 2000. Do cytokeratin-positive-only sentinel lymph nodes warrant complete axillary lymph node dissection in patients with invasive breast cancer? *Am. Surg. 66:*574–578.

Veronesi, U., Paganelli, G., Viale, G., Galimberti, V., Luini, A., Zurrida, S., Robertson, C., Sacchini, V., Veronesi, P., Orvieto, E., de Cicco, C., Intra, M., Tosi, G., and Scarpa, D. 1999. Sentinel lymph node biopsy and axillary dissection in breast cancer: Results in a large series. *J. Natl. Cancer Inst. 91:*368–373.

Viale, G., Bosari, S., Mazzorol, G., Galimberti, V., Luini, A., Veronesi, P., Paganelli, G., Bedoni, M., and Orvieto, E. 1999. Intraoperative examination of axillary sentinel lymph nodes in breast carcinoma patients. *Cancer 85:*2433–2438.

Yaziji, H., and Gown, A.M. 2000. Detection of pelvic node metastases. *Am. J. Clin. Pathol. 114:*149–154 (Letter).

7

CD10 Expression in Normal Breast and Breast Cancer Tissues

Keiichi Iwaya and Kiyoshi Mukai

Introduction

CD 10 is a 90–110-kDa cell surface, zinc-dependent metalloprotease that has been called "common acute lymphoblastic leukemia antigen." This antigen has also been referred to as "neutral metalloendopeptidase" in the kidney, and "enkephalinase" in the brain, and is expressed on the cell surface of most acute lymphoblastic leukemias (Mumford et al., 1981; Patey et al., 1981). The enzyme, present in a variety of cell types, is now known to be virtually identical to CD10 or neutral endopeptidase 24.11 (Shipp et al., 1989). CD10 has been very useful for classifying acute leukemias and subclassifying malignant lymphomas (Greaves et al., 1983), and it is also used to determine the primary site of metastatic tumors. CD10 is expressed at high rates in renal cell carcinoma (Avery et al., 2000; Chu et al., 2000), hepatocellular carcinoma (Chu et al., 2000), trophoblastic tumor (Ordi et al., 2003), and carcinoma of the urinary bladder and prostate, and at much lower rates in breast cancer, stomach cancer, and mullerian epithelial tumors of the female genital tract.

CD10 expression by tumor cells of colon carcinoma and melanoma is significantly correlated with metastasis (Kanitakis et al., 2002; Yao et al., 2002); however, a study using cell migration assays revealed an inverse correlation between CD10 expression and cell migration in prostatic carcinoma cell lines (Sumitomo et al., 2000). Our previous study showed an association between CD10 expression by the stromal cells of invasive ductal carcinoma of the breast and aggressive clinical behavior (Iwaya et al., 2002), but expression in breast carcinoma cells was neither associated with aggressive clinical behavior nor correlated with clinical outcome. Because different materials and methods were used in different studies, the results cannot be compared. Whether CD10 promotes or inhibits tumor aggressiveness is still a matter of controversy. Nevertheless, these data suggest that CD10 is to some extent involved in tumor invasion or metastasis.

CD10 is clearly detectable in myoepithelial cells, and so they are used as an internal positive control in routine immunohistochemical studies of the human breast tissue (Kaufmann et al., 1999). In this chapter, the routine immunohistochemical method used in our laboratory is described. The Results and Discussion sections describe the types of cells in breast tissue (including benign and malignant lesions), which are stained by CD10 immunohistochemistry, and how CD10 immunochemistry can be used for pathological diagnosis.

Handbook of Immunohistochemistry and in situ Hybridization of Human Carcinomas, Volume 1: Molecular Genetics; Lung and Breast Carcinomas

299

Copyright © 2004 by Elsevier (USA)
All rights reserved.

MATERIALS

1. Phosphate-buffered saline (PBS), pH 7.4: Dissolve 8 g NaCl, 28.7 g $Na_2HPO_4 \cdot 12H_2O$, and 3.3 g $NaH_2PO_4 \cdot 2H_2O$ in deionized glass-distilled water to a final volume of 1 L.

2. 1 M Tris buffer, pH 8.0: Dissolve Tris 121.1 g in deionized glass-distilled water to a final volume of 1 L; adjust to pH 8.0 with HCl.

3. 10% buffered neutral formalin solution: Dissolve 4 g $NaH_2PO_4 \cdot 2H_2O$, 6.5 g Na_2HPO_4, and 100-mL formalin in deionized glass-distilled water to a final volume of 1 L.

4. Paraffin, pastilles, solidification point 56–58°C (Merck, Darmstadt, Germany).

5. Silane-coated glass slides (Muto-Glass, Tokyo, Japan).

6. Primary antibody diluted in PBS with 1% normal swine serum.

7. Biotinylated secondary antibody diluted in PBS (Dako Cytomation California Inc., Carpinteria, CA).

8. Horseradish peroxidase (HRP)-labeled streptavidin diluted in PBS (Dako Cytomation California Inc.).

9. 0.01 M sodium citrate buffer, pH 6.0:18 mL of 0.1 M citric acid, 82 mL of 0.1 M trisodium citrate dihydrate. Add deionized glass-distilled water to a final volume of 1 L.

10. 0.02% 3,3′-diaminobenzidine, (Dojin, Kumamoto, Japan) and 0.02% H_2O_2 in Tris buffer, pH 8.0: Dissolve 30 mg 3,3′-diaminobenzidine in 0.05 M Tris buffer, pH 8.0 to a volume of 150 mL. Add 30 μL of H_2O_2 immediately before use.

11. Mayer's hematoxylin.

12. Coverslip.

13. Malinol.

METHOD

Immunohistochemistry (Labeled Streptavidin Biotinylated Antibody Method)

1. Cut tissue into small blocks about 1–1.5 $cm^2 \times$ 0.4 cm.

2. Fix the tissue blocks overnight in 10% buffered neutral formalin solution.

3. Follow standard paraffin-embedding procedures.

4. Cut paraffin blocks into 4 μm-thick sections and collect onto silane-coated glass slides.

5. Heat the sections at 37°C overnight.

6. Dewax the sections in xylene (six incubations, 3 min each).

7. Rehydrate by passing through a graded alcohol series (two washes in absolute ethanol, 3 min each; followed by 95% ethanol, 90% ethanol, 80% ethanol, 70% ethanol, 3 min each).

8. Rinse in water.

9. Heat in 0.01 M sodium citrate buffer (pH 6.0) for 10 min in a conventional pressure sterilizer.

10. After cooling, incubate the sections with 10% normal swine serum in PBS for 10 min, then with an anti-CD10 monoclonal antibody (56C6, Novocastra, Newcastle-upon-Tyne, U.K.) diluted to 1:100 overnight in a humidified chamber.

11. Wash the slides in PBS three times for 5 min each.

12. Apply the labeled secondary reagent. Biotin-labeled rabbit-anti-mouse immunoglobulin antibody can be purchased from several suppliers.

13. Incubate with the labeled secondary reagent for 30 min at room temperature in the humidified chamber.

14. Wash the slides in PBS three times for 5 min each.

15. Incubate with HRP-labeled streptavidin. The color reaction is developed in 0.02% 3,3′-diaminobenzidine and 0.02% H_2O_2 in Tris buffer, pH 8.0. Incubate for 5–15 min. The incubation time should be determined by the intensity of the stain.

16. Wash the slides gently in water.

17. Counterstain with Mayer's hematoxylin. The incubation time should be determined by the intensity of the stain.

18. Wash the slides gently in water.

19. Dip the slides in 0.5% glacial acetic acid/99.5% ethanol for 10 sec.

20. Wash the slides gently in water.

21. Dehydrate sections by passing through a graded alcohol series, then clean sections in xylene. Each step requires at least 3 min.

22. Mount the sections in Malinol.

23. Remove any excess mounting material with a paper towel.

RESULTS AND DISCUSSION

Because CD10 is constantly expressed in the myoepithelial cells of the normal breast and no expression is detected in other cells, such as luminal epithelial cells, fibroblasts, myofibroblasts, smooth muscle cells, or endothelial cells, CD10 immunostaining by this method clearly highlights the distribution of myoepithelial cells in normal breast (Figure 42). The myoepithelial cells of the breast express CD10 during its development and after its maturation (Atherton *et al.*, 1994), making CD10 a stable and specific marker of myoepithelial cells in normal breast (Atherton *et al.*, 1994). Although comparisons of CD10 with other well-established markers of myoepithelial cells have shown that its expression pattern is almost the same as that of α-smooth muscle actin (α-SMA) (Mukai *et al.*, 1981), myofibroblasts and vascular smooth muscle cells also express α-SMA. High molecular weight keratins and S-100 protein are often expressed in other elements of breast tissue, including some luminal epithelial cells and peripheral nerve fibers (Gillett *et al.*, 1990; Guelstein *et al.*, 1993), and thus CD10 is more specific than these conventional markers of myoepithelial cells.

Two biological markers of myoepithelial cells have recently been reported. One is p63, a member of the p53 gene family that is expressed in the nuclei of myoepithelial cells (Barbareschi *et al.*, 2001), and while antibodies to p63 offer excellent sensitivity and increased specificity, nuclear localization is often difficult to interpret because the nucleus of a given myoepithelial cell is not directly juxtaposed to the nuclei of its nearest neighbors, and thus the pattern of reactivity of p63 is discontinuous even in distinct cases with continuous myoepithelial cell layers (Werling *et al.*, 2003). CD10, on the other hand, reveals the border of the myoepithelial cell very clearly, and the distribution of CD10 on the membrane of myoepithelial cells has been demonstrated by Gusterson *et al.* electron-microscopically (1986). The other new marker is maspin, a novel proteinase inhibitor of a serpin superfamily member that is consistently expressed in the nucleus and cytoplasm of myoepithelial cells (Maass *et al.*, 2001; Reis-Filho *et al.*, 2001; Zou *et al.*, 1994). However, a recent study demonstrated maspin expression in 34.4% of early-stage breast cancer (Kim *et al.*, 2003),

whereas the frequency of expression of CD10 on cancer cells is very low. Our unpublished data, obtained in an immunohistochemical study of CD10 in 151 invasive ductal carcinomas, no CD10 expression was detected in the cancer cells except for one case, in which the majority of cancer cells were positive for CD10. Although it is preferable to use several markers to identify myoepithelial cells, CD10 is especially useful for detecting myoepithelial cells in normal breast tissue and cancer tissue.

The presence of a myoepithelial cell layer is a pathological hallmark of benign breast disease. There are epithelial proliferative lesions in benign breast disease that mimic invasive cancer, and intraductal papillomatosis and sclerosing adenosis, for example, are sometimes difficult to distinguish from invasive ductal carcinoma. CD10 immunohistochemistry is useful for detecting the myoepithelial cell layer in these benign lesions (Moritani *et al.*, 2002). We confirmed CD10 expression in the outer layer of papillary growth and tubular growth in all four cases of intraductal papillomatosis and in all five cases of sclerosing adenosis, although other elements of those lesions, such as proliferating stromal cells and capillary vessels, were negative for CD10. These findings indicate that CD10 is a practical marker for determining whether a myoepithelial cell layer is present when making the differential diagnosis between a benign proliferative epithelial lesion and ductal carcinoma (Figures 43–46).

No stromal expression of CD10 is detected in normal breast tissue, benign epithelial proliferating lesions, or fibrocytic disease, although CD10 is detected on stromal cells in granulation tissue and

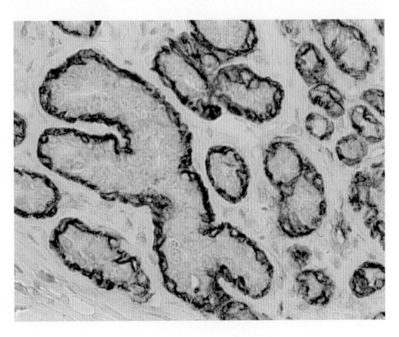

Figure 42 CD10 expression in normal breast tissue. CD10 is strongly expressed by the myoepithelial cell. 400×.

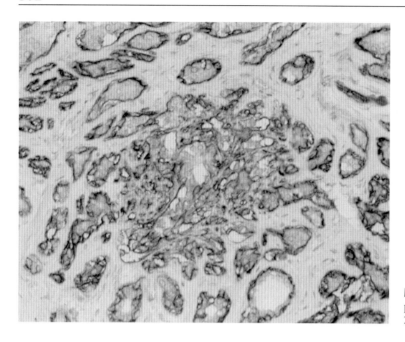

Figure 43 CD10 expression in adenosis. CD10 is positive in the outer layer of the proliferated tubules. 200×.

suture granulomas. Stromal CD10 was detected in granulation tissue or suture granuloma in 8 of 10 breast tissue specimens obtained after needle or incisional biopsy, but it was not detected in an old scar or in a case with prominent fresh bleeding. These findings are consistent with a report that CD10 is expressed during wound healing in a rat model (Olerud *et al.*, 1999).

Although most human breast tumors are negative for stromal CD10 expression, proliferating stromal cells in fibroadenoma and phyllodes tumors have been reported to express CD10 in a study of frozen tissue specimens (Mechtersheimer *et al.*, 1990). In our preliminary study, CD10 was detected in paraffin sections of all 10 phyllodes tumors (eight benign, one borderline, and one malignant). CD10 was also found expressed by stromal cells in 13 of 15 (86.7%) fibroadenomas. Stromal cellularity was low, and no nodular myxoid change was seen in two fibroadenomas that were negative for stromal CD10 expression.

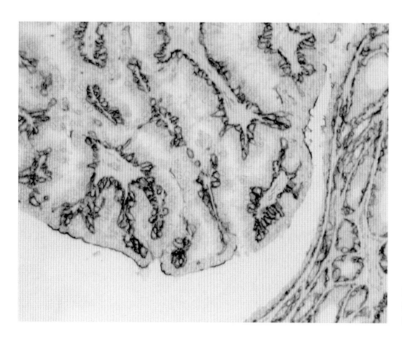

Figure 44 CD10 expression in intraductal papillomatosis. CD10 is positive in the cells beneath the epithelial cells that line the papillary structures. 400×.

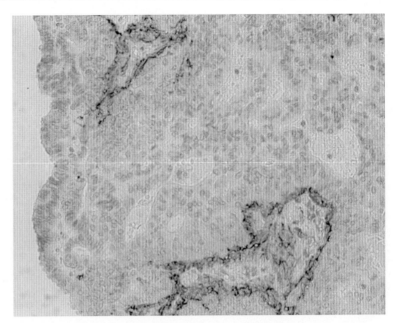

Figure 45 CD10 expression in intracystic papillary carcinoma. CD10-positive myoepithelial cells are absent in some parts of the area of papillary growth. 400×.

No stromal expression of CD10 is detected in normal breast tissue, benign proliferative lesions, or intraductal carcinoma. Nine predominantly intraductal carcinomas were examined for expression of CD10. The CD10-positive myoepithelial layer was focally lost around the cancer nest in the microinvasive area, but a small number of CD10-positive stromal cells was present in all nine cases. (Figure 47). In our previous study, CD10 was expressed in more than 10% of stromal cells around carcinoma cells in 20 of 110 invasive ductal carcinomas (Iwaya *et al.*, 2002). The CD10-positive stromal cells were diffusely distributed among invasive ductal cancer cells, and the CD10 staining was more intense in the stromal areas close to the cancer cells.

Stromal CD10 expression is a new and strong predictor of tumor recurrence and death. Multivariate analysis by Cox's proportional hazards regression model revealed that stromal CD10 expression is an independent prognostic factor for patient outcome.

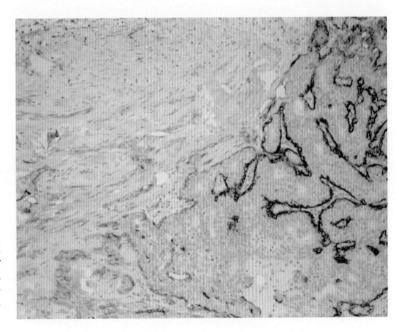

Figure 46 CD10 expression in invasive ductal carcinoma. A CD10-positive myoepithelial cell layer is present around the cancer nests on the right side, but CD10 immunostaining is weak in the myoepithelial cell layer in the area of the invasion on the left. 200×.

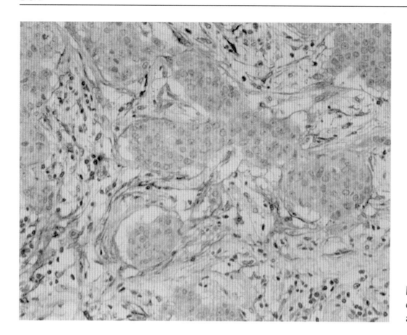

Figure 47 CD10 expression in invasive ductal carcinoma. CD10-positive stromal cells are present around the cancer nests. 400×.

Lymph node status was another independent factor. Tumor size, histological grade, and clinical stage were not independent prognostic factors (Iwaya *et al.*, 2002). Although further retrospective and prospective studies are necessary before clinical application, our findings strongly suggest that stromal CD10 expression is closely linked to aggressive biological behavior of breast carcinoma, including invasion and metastasis.

We found that the emergence of CD10-positive stromal cells is a late event in colorectal carcinogenesis, and that their presence was significantly associated with invasive growth (Ogawa *et al.*, 2002). We believe that CD10-positive stromal cells are recruited by the cancer cells, which acquire the capacity for invasive growth by interacting with CD10-positive stromal cells. How CD10-positive stromal cells increase the aggressiveness of breast carcinoma is unknown, but CD10 is structurally similar to matrix metalloproteinase and stromelysin, and may therefore create a microenvironment that facilitates cancer cell invasion or metastasis (Basset *et al.*, 1994; Sternlicht *et al.*, 1999; Talvensaari-Mattila *et al.*, 1998).

In summary, CD10 immunohistochemistry clearly detects the presence of myoepithelial cells in the human breast in both benign and malignant lesions. No other cells in normal breast express CD10. The stromal cells expressing CD10 are seen around invasive ductal carcinoma cells. Stromal CD10 expression is an independent prognostic factor for invasive ductal carcinoma.

References

Atherton, A.J., Anbazhagan, R., Monaghan, P., Bartek, J., and Gusterson, B.A. 1994. Immunolocalisation of cell surface peptidases in the developing human breast. *Differentiation* 56:101–106.

Atherton, A.J., O'Hare, M.J., Buluwela, L., Titley, J., Monaghan, P., Paterson, H.F., Warburton, M.J., and Gusterson, B.A. 1994. Ectoenzyme regulation by phenotypically distinct fibroblast sub-populations isolated from the human mammary gland. *J. Cell Sci. 107:*2931–2939.

Avery, A.K., Beckstead, J., Renshaw, A.A., and Corless, C.L. 2000. Use of antibodies to RCC and CD10 in the differential diagnosis of renal neoplasms. *Am. J. Surg. Pathol. 24:*203–210.

Barbareschi, M., Pecciarini, L., Cangi, M.G., Macri, E., Rizzo, A., Viale, G., and Doglioni, C. 2001. p63, a p53 homologue, is a selective nuclear marker of myoepithelial cells of the human breast. *Am J. Surg. Pathol. 25:*1054–1060.

Basset, P., Wolf, C., Rouyer, N., Bellocq, J.P., Rio, M.C., and Chambon, P. 1994. Stromelysin-3 in stromal tissue as a control factor in breast cancer behavior. *Cancer 74:*1045–1049.

Chu, P., and Arber, D.A. 2000. Paraffin-section detection of CD10 in 505 nonhematopoietic neoplasms. Frequent expression in renal cell carcinoma and endometrial stromal. *Am. J. Clin. Pathol. 113:*374–382.

Chu, P.G., Ishizawa, S., Wu, E., and Weiss, L.M. 2002. Hepatocyte antigen as a marker of hepatocellualr carcinoma: An immunohistochemical comparison to carcinoembryonic antigen, CD10, and alpha-fetoprotein. *Am. J. Surg. Pathol. 26:*978–988.

Gillett, C.E., Bobrow, L.G., and Millis, R.R. 1990. S100 protein in human mammary tissue: Immunoreactivity in breast carcinoma, including Paget's disease of the nipple and value as a marker of myoepithelial cells. *J. Pathol. 160:*19–24.

Greaves, M.F., Hariri, G., Newman, R.A., Sutherland, D.R., Ritter, M.A., and Ritz, J. 1983. Selective expression of the common acute lymphoblastic leukemia (gp 100) antigen on immature lymphoid cells and their malignant counterparts. *Blood 61:*628–639.

Guelstein, V.I., Tchypysheva, T.A., Ermoilova, V.D., and Ljubimov, A.V. 1993. Myoepithelial and basement membrane antigen and malignant human breast tumors. *Int. J. Cancer 53:*269–277.

Gusterson, B.A., Monaghan, P., Mahendran, R., Ellis, J., and O'Hare, M.J. 1986. Identification of myoepithelial cells in human and rat breasts by anti-common acute lymphoblastic leukemia antigen antibody A12. *J. Natl. Cancer Inst. 77:*343–349.

Iwaya, K., Ogawa, H., Izumi, M., Kuroda, M., and Mukai, K. 2002. Stromal expression of CD10 in invasive breast carcinoma: A new predictor of clinical outcome. *Virchows Arch. 440:*589–593.

Kanitakis, J., Narvaez, D., and Claudy, A. 2002. Differential expression of the CD10 antigen (neutral endopeptidase) in primary versus metastatic malignant melanomas of the skin. *Melanoma Res. 12:*241–244.

Kaufmann, O., Flath, B., Spath-Schwalbe, E., Possinger, K., and Dietel, M. 1999. Immunohistochemical detection of CD10 with monoclonal antibody 56C6 on paraffin sections. *Am. J. Clin. Pathol. 111:*117–122.

Kim, D.H., Yoon, D.S., Dooley, W.C., Nam, E.S., Ryu, J.W., Jung, K.C., Park, H.R., Sohn, J.H., Shin, H.S., and Park, Y.E. 2003. Association of maspin expression with the high histological grade and lymphocyte-rich stroma in early-stage breast cancer. *Histopathology 42:*37–42.

Maass, N., Teffner, M., Rosel, F., Pawaresch, R., Jonat, W., Ngasaki, K., and Rudolph, P. 2001. Decline in the expression of the serine proteinase inhibitor maspin is associated with tumor progression in ductal carcinomas of the breast. *J. Pathol. 195:*321–326.

Mechtersheimer, G., Kruger, K.H., Born, I.A., and Moller, P. 1990. Antigenic profile of mammary fibroadenoma and cytosarcoma phyllodes. A study using antibodies to estrogen- and progesterone receptors and to a panel of cell surface molecules. *Pathol. Res. Pract. 186:*427–438.

Moritani, S., Kushima, R., Sugihara, H., Bamba, M., Kobayashi, T.K., and Hattori, T. 2002. Availability of CD10 immunohistochemistry as a marker of breast myoepithelial cells on paraffin sections. *Mod. Pathol. 15:*397–405.

Mukai, K., Schollmeyer, J.V., and Roasai, J. 1981. Immunohistochemical localization of actin: Applications in surgical pathology. *Am. J. Surg. Pathol. 5:*91–97.

Mumford, R.A., Pierzchala, P.A., Strauss, A.W., and Zimmerman, M. 1981. Purification of a membrane-bound metalloendopeptidase from porcine kidney that degrades peptide hormones. *Proc. Natl. Acad. Sci. USA 78:*6623–6627.

Ogawa, H., Iwaya, K., Izumi, M., Kuroda, M., Serizawa, H., Koyanagi, Y., and Mukai, K. 2002. Expression of CD10 by stromal cells during colorectal tumor development. *Hum. Pathol. 33:*806–811.

Olerud, J.E., Usui, M.L., Seckin, D., Chiu, D.S., Haycox, C.L., Song, I.S., Ansel J.C., and Bunnett, N.W. 1999. Neutral endopeptidase expression and distribution in human skin and wounds. *J. Invest. Dermatol. 112:*873–881.

Ordi, J., Romagosa, C., Tavassoli, F.A., Nogales, F., Palacin, A., Condom, E., Torne, A., and Cardesa, A. 2003. CD10 expression in epithelial tissues and tumors of the gynecologic tract: A useful marker in the diagnosis of mesonephric, trophoblastic, and clear cell tumors. *Am J. Surg. Pathol. 27:*178–186.

Patey, G., De La Baume, S., Schwartz, J.C., Gros, C., Roques, B., Fournie-Zaluski, M.C., and Soroca-Luca, E. 1981. Selective protection of methionine enkephalin released from brain slices by enkephalinase inhibition. *Science 212:*1153–1155.

Reis-Filho, J.S., Milanezi, F., Silva, P., and Schmitt, F.C. 2001. Maspin expression in myoepithelial tumors of the breast. *Pathol. Res. Pract. 197:*817–821.

Shipp, M.A., Vijayaraghavan, J., Schmidt, E.V., Masteller, E.L., D'Adamio, L., Hersh, L.B., and Reinherz, E.L. 1989. Common acute lymphoblastic leukemia antigen (CALLA) is active neutral endopeptidase 24.11 ("enkephalinase"): Direct evidence by cDNA transfection analysis. *Proc. Natl. Acad. Sci. USA 86:*297–301.

Sternlicht, M.D., Lochter, A., Sympson, C.J., Huey, B., Rougier, J.P., Gray, J.W., Pinkel, D., Bissel, M.J., and Werb, Z. 1999. The stromal proteinase MMP3/stromelysin-1 promotes mammary carcinogenesis. *Cell 98:*137–146.

Sumitomo, M., Shen, R., Walburg, M., Dai, J., Geng, Y., Navarro, D., Boileau, G., Papandreou, C.N., Giancotti, F.G., Knudsen, B., and Nanus, D.M. 2000. Neutral endopeptidase inhibits prostate cancer cell migration by blocking focal adhesion kinase signaling. *J. Clin. Invest. 106:*1399–1407.

Talvensaari-Mattila, A., Paakko, P., Hoyhtya, M., Blanco-Sequeiros, G., and Turpeenniemi-Hujanen, T. 1998. Matrix metalloproteinase-2 immunoreactive protein: A marker of aggressiveness in breast carcinoma. *Cancer 83:*1153–1162.

Yao, T., Takata, M., Tustsumi, S., Nishiyama, K., Taguchi, K., Nagai, E., and Tsuneyoshi, M. 2002. Phenotypic expression of gastrointestinal differentiation markers in colorectal adenocarcinomas with liver metastasis. *Pathology 34:*556–560.

Werling, R.W., Hwang, H., Yaziji, H., and Gown, A.M. 2003. Immunohistochemical distinction of invasive from noninvasive breast lesions: A comparative study of p63 versus calponin and smooth muscle myosin heavy chain. *Am. J. Surg. Pathol. 27:*82–90.

Zou, Z., Anisowicz, A., Hendrix, M.J., Thor, A., Neveu, M., Sheng, S., Rafidi, K., Seftor, E., and Sager, R. 1994. Maspin, a serpin with tumor-suppressing activity in human mammary epithelial cells. *Science 265:*1893–1894.

8

Role of Immunohistochemical Expression of AKT Protein in Breast Carcinoma

Bradley L. Smith, Debbie Altomare, Neil L. Spector, and Sarah S. Bacus

Introduction

Protein phosphorylation is a major regulatory mechanism for cellular processes that determine growth, development, and death. Phosphorylation modulates protein activation, intracellular localization, expression, and degradation. Changes in phosphorylation of signaling proteins that make up these control pathways may be experimentally detected by using antibodies to specific proteins and phosphorylation-specific antibodies. In this way, the activation of the pI3K/AKT pathway may be determined in cell lines, animal model tissues, or patient samples. As detected using phosphorylation-specific AKT antibodies, AKT activation has been reported in a number of tumors (Gupta *et al.*, 2002; Itoh *et al.*, 2002; Malik *et al.*, 2002; Roy *et al.*, 2002; Stal *et al.*, 2003; Sun *et al.*, 2001a). These studies, however, cannot distinguish which isoforms are activated because the two primary regulatory phosphorylation sites (Threonine 308 and serine 473) are conserved among the three isoforms. Nevertheless, cell line, xenograft, and human studies have demonstrated that AKT phosphorylation can be used to monitor the biological activity of a drug targeting an upstream receptor such as the epidermal growth factor receptor (EGFR) (Rusnak *et al.*, 2001; Ng *et al.*, 2001). Similarly, AKT phosphorylation may be used to select patients that are most likely to respond to such a therapeutic. As new targeted therapeutics enter the clinic, it has become clear that resistance may arise in patients, such as that seen in CML patients treated with Gleevec, a BCR/ABL inhibitor (Gore *et al.*, 2001). The AKT activation, through loss of the upstream regulator PTEN, may be one cause for such resistance to arise in breast cancer patients (Bianco *et al.*, 2003). Previous studies, as well as the results presented here, suggest that an AKT assay may be especially useful for the diagnosis and treatment of breast carcinomas.

The multiple functions of AKT in the cell is further complicated by the potential expression of three isoforms: AKT1, AKT2, and AKT3. One cell line-based study suggested that the three isoforms may be differentially activated by EGFR (Okano *et al.*, 2000). Initial studies on human samples have shown that expression of each isoform may be tissue- or tumor-specific. For example, AKT2 expression has been reported in a number of cancers (Bacus *et al.*, 1993; Bellacosa *et al.*, 1995; Cheng *et al.*, 1996; Ruggeri *et al.*, 1998).

Handbook of Immunohistochemistry and in situ Hybridization of Human Carcinomas, Volume 1: Molecular Genetics; Lung and Breast Carcinomas

307

Copyright © 2004 by Elsevier (USA)
All rights reserved.

Overexpression of AKT3 has been reported in estrogen receptor-negative breast tumors (Nakatani *et al.*, 1999) and AKT1 activation was observed in approximately 40% of breast carcinomas in one study (Sun *et al.*, 2001b) and 24% in another (Stal *et al.*, 2003). The AKT2 isoform has also been reported to be activated in many breast cancers through an interaction of estrogen receptor (ER) alpha with PI-3-kinase (Sun *et al.*, 2001b). These authors also report that cells with activated AKT2 were ER alpha-independent and were resistant to Tamoxifen. Interestingly, overexpression of HER-2/ neu also has been associated with resistance to Tamoxifen (Stal *et al.*, 2000; Wright *et al.*, 1992). In addition, AKT2 overexpression has been linked to an up-regulation of intergrins and increased invasiveness in breast cancer cells (Arboleda *et al.*, 2003). Overall, these results suggest another dimension to the complexity of the AKT pathway.

Breast carcinomas may be characterized by the overexpression of growth factor receptors such as the erbB family (HER-2/neu, EGFR, erbB-3, and erbB-4), insulin-like growth factor receptors (IGF-IR) and others (Nahta *et al.*, 2003). Activation of these receptors leads to the activation of downstream signaling pathways, including the AKT and MAP kinase pathways. Of these pathways, the AKT pathway may be most important in protecting the cancer cell from growth arrest and death (apoptosis). The AKT pathway functions to promote cell survival by inhibiting apoptosis by means of its ability to phosphorylate and regulate several targets, including Bad (Cardone *et al.*, 1998), Forkhead transcription factors (Brunet *et al.*, 1999) and caspase-9 (Rommel *et al.*, 1999). In addition, AKT maintains protein production through regulation of p70 S6 kinase and other proteins that regulate protein translation (Polakiewicz *et al.*, 1998; Pullen *et al.*, 1997). The AKT pathway is capable of directly regulating cell cycle through phosphorylation of meitotic inhibitors such as p27 and p21 (Li *et al.*, 2002; Viglietto *et al.*, 2002). The AKT pathway further prevents cell cycle arrest by phosphorylating DNA damage response proteins such as MDM2 (Mayo *et al.*, 2001; Zhou *et al.*, 2001). Activation of AKT influences other pathways such as the ERK, JNK, and NfkB pathways through cross-talk via Raf, ASK1 (Kim *et al.*, 2001), and IKK (Romashkova, *et al.*, 1999), respectively. In breast carcinomas, AKT activation may occur through multiple mechanisms involving complex mechanisms such as heterodimerization of the erbB receptors. For example, HER-3 contains a major docking site for phosphoinositide-3-kinase/AKT (pI3K) (Hellyer *et al.*, 1998). Therefore, NDF/ heregulin ligand stimulation of HER-2/neu and HER-3 receptor heterodimers causes activation of the pI3K pathway and phosphorylation of AKT (Altiok *et al.*, 1998; Hellyer, 1998; Liu *et al.*, 1999; Xing *et al.*, 2000). These observations implicate pI3K/AKT in the signaling cascade that results from HER-3 heterodimerization with overexpressed HER-2/ neu receptors in breast cancer cells. Amplification and/or overexpression of the HER-2/neu gene in breast cancers is associated with aggressive behavior and resistance to therapeutic regimens (Alaoui-Jamali *et al.*, 1997; Ross *et al.*, 1999; Toikkanen *et al.*, 1992). The molecular mechanisms that contribute to therapeutic resistance/survival of HER-2/neu–overexpressing tumor cells have not been well defined. However, activation of pI3K/AKT promotes cell survival and may enhance tumor aggressiveness (results herein). Therefore, it is clear that AKT plays a central role in determining whether a cell will become malignant and whether a breast carcinoma will respond to a given therapy.

One example of a targeted therapeutic whose development was based on a knowledge of the role of HER-2/neu in breast cancer is Herceptin. Herceptin is an antibody therapeutic that targets the HER-2/neu receptor tyrosine kinase. HER-2/neu, the receptor, is overexpressed or activated in a number of cancers, including breast cancer. In the case of Herceptin, if the patients selected were not chosen on the basis of their HER-2/neu expression, then the number of patients required and the length of the study would have been much greater (Baselga *et al.*, 2000). Likewise, the overall response rate to Herceptin would be much lower if the patients were not first selected based on HER-2/neu expression. However, the response rate to Herceptin as a single agent therapy in breast cancer patients is significantly lower than 50% (Hortobagyi, 2001). This response rate suggests that additional cellular factors are affecting the ability of the drug to inhibit the survival and proliferation of cancer cells. Initial attempts to identify these factors, as reported herein, suggest that expression and activation of other signaling proteins that also result in the activation of AKT may determine response to Herceptin as well as other drugs targeting the erbB family of growth factor receptors. Thus, determination of the activation state of the AKT pathway is central to these alternative oncogenic events.

To better understand the role of AKT in breast carcinomas, we undertook three studies, one focusing on expression of AKT isoforms (AKT2 in particular), another focusing on AKT phosphorylation and patients' response to Herceptin therapy, and the third a test of cancer biopsies for response to HER-2/neu tyrosine kinase inhibitors. All studies examined the roles these distinct mechanisms of protein regulation play in breast tumorigenesis, patient prognosis, and potential patient response to targeted therapy such as Herceptin inhibitors or EGFR and HER-2/neu.

MATERIALS

For the AKT isoform studies; MCF7 cells were obtained from the Michigan Cancer Foundation (Detroit, MI). A variant of MCF7 cells, MCF7/HER-2, with 5 ± 8-fold elevated expression of HER-2/neu, was generated earlier by transfection of MCF7 cells with an erbB-2 cDNA expression vector (Bacus et al., 2001). AU-565 cells (Bacus et al., 1992) were obtained from the Cell Culture Laboratory, Navy Biosciences Laboratory (Navy Supply Center, Oakland, CA). The MDA-MB-435 and MDA-MB-435/HER-2 cell lines were obtained from Dr. D. Yu (UTMD Anderson Cancer Center, Houston, TX). Cell lines T47D, BT474, MDA-MB-361, MDA-MB-231, and HLB100 were purchased from the American Type Culture Collection (ATCC) (Rockville, MD). All of the cells were cultured in RPMI 1640 (Gibco, Grand Island, NY) supplemented with 10% fetal bovine serum, penicillin (100 units/ml), and streptomycin (100 mg/ml) in a humidified incubator with 7% CO_2 at 37°C. For the AKT2 studies, antibodies used consisted of an anti-phospho-AKT antibody (Dr. R. Seger, Weizmann Institute, Rehovat, Israel, or Ser 473 antibody from Cell Signaling Technology, Beverly, MA), anti-AKT1 (D-17, Santa Cruz Biotechnology, Santa Cruz, CA), anti-AKT2 (D-17, Santa Cruz), or anti-PTEN monoclonal antibody (A2B1, Santa Cruz). Detection of horseradish peroxidase–conjugated secondary antibodies was carried out with Renaissance Chemiluminescence Reagent Plus (NEN Life Science, Boston, MA). To detect HER-3/HER-2 heterocomplexes, anti-HER-3 and anti-HER-2 antibodies (NeoMarkers, Fremont, CA) were used for immunoprecipitation and Western Blotting, respectively.

For the Herceptin tissue microarray studies, EGFR and HER-2 antibodies were obtained from Ventana Medical Instruments, Inc. (VMSI, Tucson, AZ). The antibodies HER-3 (1:10) and heregulin (1:25) were obtained from NeoMarkers (Fremont, CA). The TGF-α and IGF-IR antibodies were obtained from Oncogene Sciences (San Diego, CA) and NeoMarkers (Fremont, CA), respectively. Phospho-specific ERK, phospho-AKT, and phospho-S6 ribosomal protein antibodies were obtained from Cell Signaling Technology (Beverly, MA) and the LSAB2 kit (Dako, Carpinteria, CA) was used as the detection chemistry. DAB (Dako) was used as the chromogen. Tissue microarrays of 250 breast cancer patients that received chemotherapy together with Herceptin were obtained from Clinomics Biosciences (Pittsfield, MA). The

histology of the tumors varied, with infiltrating ductal carcinoma being most common. All patients received radiotherapy post-surgery. The tissue samples in the array were taken before treatment. HER-2/neu expression had been determined by the Herceptest on the original biopsies for all patients. Patient response was based on the case histories at last follow-up, as decided by an independent pathologist provided by Clinomics.

For the erbB inhibitor studies, we used GW572016, a reversible inhibitor of EGFR and erbB-2 tyrosine kinases. The cell lines used were the erbB-2 overexpressing human breast adenocarcinoma line BT474, which was obtained from the American Type Culture Collection (Rockville, MD). The EGFR expressing LICR-LON-HN5 head and neck carcinoma cell line (HN5) was kindly provided by Helmout Modjtahedi at the Institute of Cancer Research, Surrey, U.K. Cells were cultured as previously described. GW572016, N-{3-chloro-4-[(3-fluorobenzyl)oxy]phenyl}-6-[5-({[2-(methylsulfonyl)ethyl]amino}methyl)-2-furyl]-4-quinazolinamine, was synthesized as previously described and dissolved in DMSO for cell culture work. For immunohistochemistry (IHC), antibodies to EGFR, erbB-2, and cyclin D were from Ventana Medical Scientific Instruments VMSI, (Tucson, AZ); anti- p-AKT (Ser 437) and p-Erk1/2 were from Cell Signaling Technology, Inc. (Beverly, MA); anti p-EGFR and p-erbB2 were from Chemicon (Temecula, CA) and NeoMarkers (Fremont, CA), respectively; antibodies to TGF-α, erbB-3, heregulin, IGF-IR-1, and phospho-erbB-2 were from NeoMarkers; and anti-Erk1/2 antibodies were from Santa Cruz Biotechnology (Santa Cruz, CA). For Western Blot, anti-phosphotyrosine was purchased from Sigma Chemical (St. Louis, MO); anti-EGFR (Ab-12) and anti-erbB-2 (Ab-11) were from NeoMarkers; antibodies to p-AKT (Ser 437), p-Erk1/2, and Erk1/2 were purchased from Santa Cruz Biotechnology; and anti-cyclin D1 and 2 from Upstate Biotechnology (Lake Placid, NY).

METHODS

Cell Treatments

Wortmannin and LY294002 (Calbiochem, San Diego, CA) were dissolved in dimethylsulfoxide at a concentration of 1 mM for worthmannin and 50 mM for LY294002 and stored at −70°C. Stock solutions were diluted immediately before use in medium to a final concentration of 1 mM and 50 mM, respectively. Heregulin (Neomarkers, Fremont, CA) was stored as a frozen, concentrated stock solution at −70°C and was diluted in medium immediately before use. Herceptin was stored at −70°C at a concentration of 100 mg/ml in 50% glycerol. Where indicated, the cells were exposed to heregulin (150 ng/ml) or Herceptin (100 mg/ml). Treatment with wortmannin, LY294002, heregulin, or Herceptin was initiated 7 ± 48 hr after cell plating. The final concentration of dimethylsulfoxide (maximum 0.25%) in growth medium did not affect cell growth. To simulate hypoxic conditions, cells were also treated with deferoxamine (DFO) for 2 ± 4 days in concentrations ranging from 12–50 mM. For UV irradiation, cells were treated for 16 sec with 2500 mJoules/cm$_2$ light energy and harvested after 17 hr. Cell cycle analysis/apoptosis assay of DNA content was determined by flow cytometry, using an EPICS-753 flow cytometer, as previously described (Dolbeare *et al.*, 1983). All experiments involving FACS analysis of the cell cycle were repeated at least three times, and representative results are presented. Apoptosis was assessed by DAPI staining to identify apoptotic cells with condensed or fragmented nuclei, and by flow cytometry to measure the fraction of subgenomic (sub-G1) cells.

Immunoprecipitation and Western Blot Analyses

Cells were extracted with lysis buffer (50 mM Tris-HCl (pH 7.4), 150 mM sodium chloride, 1 mM EDTA, 1% Nonidet P-40, 1 mM sodium orthovanadate, 1 mM sodium fluoride, 1 mM phenylmethylsulfonyl fluoride, 1 mg/ml pepstatin, 1 mg/ml leupeptin, 1 mg/ml aprotinin). Insoluble material was removed by centrifugation at 48°C for 5 min at 6500 g. Protein concentration was determined with a protein assay kit (BioRad, Hercules, CA). Protein lysates (65 ± 75 mg per lane) were subjected to electrophoresis using a Bis-Tris NuPAGE system with 16 MOPS running buffer (Novex, San Diego, CA), and Western Blotted onto Hybond C-extra (Amersham Pharmacia Biotech, Piscataway, NJ). Membranes were blocked with bovine serum albumin or nonfat milk and incubated with anti-phospho AKT antibody (Dr. R. Seger, Weizmann Institute, Rehovat, Israel, or Ser 473 antibody from Cell Signaling Technology, Beverly, MA), anti-AKT1 (D-17, Santa Cruz Biotechnology, Santa Cruz, CA), anti-AKT2 (D-17, Santa Cruz) or anti-PTEN monoclonal antibody (A2B1, Santa Cruz). Detection of horseradish peroxidase–conjugated secondary antibodies was carried out with Renaissance Chemiluminescence Reagent Plus (NEN Life Science, Boston, MA). To detect HER-3/HER-2 heterocomplexes, anti-HER-3 and anti-HER-2 antibodies (NeoMarkers, Fremont, CA) were used for immunoprecipitation and Western Blotting, respectively. The phospho-specific, pan-AKT (Ser 473)

antibodies recognize equivalent phosphorylation sites in all AKT isoforms. Because AKT1, 2, and 3 comigrate, the individual activated AKT isoforms cannot be distinguished. The isoform specificity of the anti-AKT1 and anti-AKT2 antibodies was tested by transfecting cells with HA-tagged AKT1 or HA-tagged AKT2 expression constructs, immunoprecipitating with anti-HA antibody, and then the isoform specificity of the antibodies was confirmed by Western Blotting using individual AKT isoform-specific antibodies. Similar specificity of these antibodies has been reported by other groups (Sun *et al.*, 2001a,b).

FISH Analysis

Fluorescence *in situ* hybridization (FISH) analysis was performed using the Path-Vysion HER-2 DNA Probe Kit (Vysis, Downers Grove, IL), according to the manufacturer's instructions.

Immunohistochemical Staining

Cells and tissues were fixed in 10% neutral buffered formalin for 60 min at room temperature, washed with H$_2$O, rinsed with Tris-buffered saline (0.05 M Tris, 0.15 M NaCl, pH 7.6), and blocked with 10% goat serum (for HER-2) in 0.1% bovine serum albumin/Tris-buffered saline for 15 min. Primary antibodies used for IHC included AKT1, AKT2, phospho-AKT, and HER-2/neu (previously described), and a monoclonal AKT2 antibody (Ruggeri *et al.*, 1998). Primary, secondary, and tertiary reagents were incubated for 30, 20, and 15 min, respectively, 37°C and washed with Tris-buffered saline between steps. Final detection was achieved using CAS red chromagen (Quantitative Diagnostic Laboratory, Elmhurst, IL). Slides for p-S6 ribosomal protein, p-ERK, and p-AKT were antigen-retrieved using 0.1 M citrate buffer, pH 6.0 in the "decloaker" (Biocare Corp.), and the sections incubated overnight with the primaries at 4°C. The next day, the slides were placed onto the autostainer (Dako Corp.) and the LSAB2 kit (Dako) was used as the detection chemistry. DAB (Dako) was used as the chromagen. After immunostaining, all slides were counterstained manually with 4% ethyl green (Sigma). For paraffin-embedded sections, 4-m sections were placed onto charged slides and dried at 60°C for 1 hr. The sections were deparaffinized and hydrated in water. Paraffin sections were stained for AKT1 and AKT2, as described, for the fresh cells, whereas HER-2/neu staining required antigen retrieval with citrate buffer before immunostaining. Antigen retrieval was required for immunostaining of estrogen and progesterone receptor, which were visualized with monoclonal antibodies for ER and progesterone receptor (PR) (Clones 6F11 and 636, respectively, Ventana Medical Instruments, Tucson, AR) and a streptavidin peroxidase reaction using DAB as the chromogen (Ventana Medical Instruments). The paraffin sections were stained for ER and PR according to the manufacturer's instructions using the Ventana Nexus automated staining platform, and ethyl green (Sigma) was employed as a nuclear counterstain. Phospho-AKT was detected following antigen retrieval using polyclonal antibodies against phosphorylated Ser473 (Cell Signaling Technology, Beverly, MA) and stained with a streptavidin alkaline phosphatase reaction using CAS Red (Jackson Research Laboratories, West Grove, PA).

Image and Data Analysis

Protein expression or phosphorylation levels were quantified using alkaline phosphatase or peroxidase techniques and microscope-based image analysis of immunohistochemically stained slides (Bacus *et al.*, 1993). Quantification was by means of a CAS 200 image analyser, as previously described (Bacus *et al.*, 1997). For the purpose of analysis, scores of 0 and 1 were classified as negative and scores of 2 and 3 were classified as positive for all antibodies. Statistical analysis was performed using Systat to quantify frequencies and calculate Pearson Chi-squared tests of significance for interactions between variables. Comparisons were performed only on samples for which all relevant data was available.

RESULTS

AKT Isoform Expression in Breast Carcinoma

Up-regulation of AKT2 in Breast Cancer Correlates with Overexpression of HER-2/neu

The HER-2/neu-overexpressing breast cancers exhibit increased resistance to some forms of cytotoxic therapy compared with breast cancers lacking overexpression of this oncoprotein (Alaoui-Jamali *et al.*, 1997; Ross *et al.*, 1999; Toikkanen *et al.*, 1992). Activation of the anti-apoptotic AKT signaling pathway has been found in some HER-2/neu-overexpressing cell lines (Zhou *et al.*, 2000, 2001). To extend these findings, we tested patient tumor samples for HER-2/neu expression and found a correlation with expression of AKT2, but not AKT1. The AKT3 kinase was not tested, because commercial anti-AKT3 antibody for IHC was not available. For this study, we selected tissues from 26 tumors that expressed normal levels of HER-2/neu and 26 tumors with known overexpression of HER-2/neu (Figure 48A–C). FISH analysis showed increased HER-2/neu copy numbers in tumor specimens

overexpressing HER-2/neu (Figure 48 B). We observed that AKT2 levels were significantly higher in the HER-2/neu overexpressing group (Figure 48C). The AKT2 levels varied from a low of 0.01 optical density units (OD) to a high of 2.5. Samples without overexpression of HER-2/neu (level of +1 or less) had AKT2 levels ranging from 0.01–0.27 OD. Most samples (25 of 26) with HER-2/neu overexpression (+3) had AKT2 levels between 0.26–1.0 OD, although some were even higher. Statistical analysis by the Spearman method revealed a significant correlation between HER-2/neu overexpression and elevated levels of AKT2 protein. Statistical analysis was as follows: Number of pairs $= 52$; $r_s = 0.76$; 95% confidence interval (0.61–0.86); P value (one tailed) $= P < 0.0001$. The tumor tissues available for our study were paraffin-embedded clinical specimens, and it was not possible to obtain parallel snap-frozen material for *in vitro* kinase assays. As an alternative, we used a phospho-specific, pan-AKT antibody for the IHC analysis of tumors with or without overexpression of AKT2. The results demonstrated that increased AKT2 expression correlated with increased total AKT activity.

AKT2 Protein Is Up-regulated by Ectopic Expression of HER-2/neu and Mediates Response to Hypoxia

To determine if AKT2 can be up-regulated by overexpression of HER-2/neu, we used MDA-MB-435 cells and a MDA-MB-435/HER-2 derivative that stably overexpresses HER-2/neu as a result of transfection with a HER-2/neu expression vector (a gift from Dr. D. Yu,

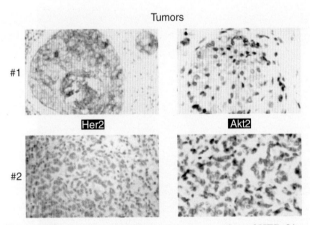

Figure 48 A: Correlation between overexpression of HER-2/neu and overexpression of AKT2 in human breast cancer cell lines and tumors. **A:** Western blot analysis of MDA-MB-435 and MDA-MB-435/HER-2 cells using antibodies against HER-2/neu and AKT2. Paraffin sections of breast cancer tissues stained immunohistochemically for HER-2/neu and AKT2, detected by alkaline phosphatase (red stain) and counter stained by Feulgen. Tumor #1 has high HER-2/neu and high AKT2 expression, whereas tumor #2 has low HER-2/neu and low AKT2 expression.

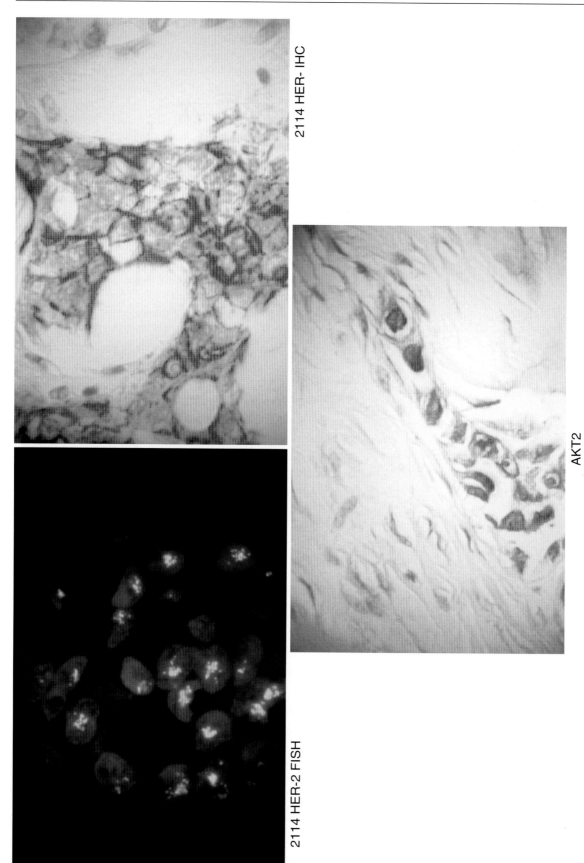

2114 HER- IHC

2114 HER-2 FISH

AKT2

Figure 48 B: Fluorescent *in situ* hybridization detection of HER-2/neu amplification, and corresponding HER-2/neu and AKT2 immunohistochemical analyses showing over-expression in a breast cancer specimen.

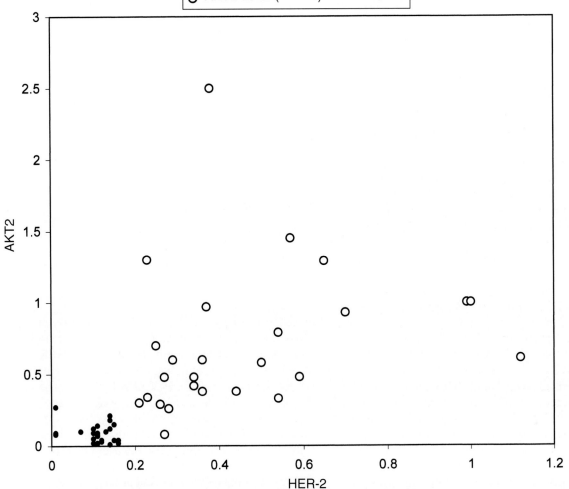

Figure 48 C: Graphic representation of AKT2 expression as it relates to HER-2/neu expression in a total of 52 tumor specimens. HER-2/neu results are expressed in picograms of HER-2/neu protein per cell, whereas AKT2 results are shown in arbitrary units of optical density, following image quantitation. Cells that do not overexpress HER-2/neu are indicated by solid black dots, and cells that overexpress HER-2/neu are indicated by open circles. Note that HER-2/neu overexpression correlates with high AKT2 expression.

UTMD Anderson Cancer Center, Houston, TX). Western Blot analysis revealed that AKT2 expression in MDA-MB-435/HER-2 was approximately tenfold greater than in the parental cell line (data not shown), consistent with our findings for breast tumor tissues. Up-regulation of AKT is associated with resistance to hypoxia and correlates with HER-2/neu overexpression. To determine the biological implications of AKT up-regulation in connection with overexpression of HER-2/neu, we investigated the growth of MCF7 and MCF7/HER-2 cells under hypoxic conditions in the presence or absence of the pI3K inhibitors wortmannin and LY294002. The receptor AKT2 kinase

was overexpressed and activated in MDA-MB-435/HER-2 cells when compared with parental MDA-MB-435 cells (data not shown). Similarly, AKT2 was overexpressed and activated in MCF7/HER-2 cells compared with parental MCF7 cells. The MCF7 cells were sensitive to hypoxia-induced apoptosis, whereas MCF7/HER-2 cells showed increased resistance to programmed cell death induced by hypoxic conditions (data not shown). Viability decreased dramatically in the presence of wortmannin and LY294002 (viability dropped from an average of 80–30% following inhibition of the PI3K pathway and induction of hypoxia in the MCF7/HER-2 cells), indicating that resistance to hypoxia in HER-/neu-overexpressing cells is associated with activation of the pI3K pathway and that inhibition of this pathway restores sensitivity to hypoxia-induced apoptosis. These results indicate that HER-2/neu overexpression is associated with cell survival under stress conditions such as hypoxia and that inhibition of the pI3K/AKT pathway restores cell sensitivity to these conditions.

Activation of AKT Occurs in Cells That Overexpress HER-2/neu

To investigate the functional significance of elevated AKT expression in HER-2/neu overexpressing cells and cancers, we evaluated AKT activity in lysates prepared from the HER-2/neu-overexpressing cell line BT474. This cell line was chosen because of its high level of endogenous HER-2/neu expression and resistance to hypoxia. Cells were treated with 150 ng/ml heregulin or 100 mg/ml Herceptin, a monoclonal antibody directed against the extracellular domain of human HER-2/neu, and tested for activation of AKT. Treatment with heregulin resulted in increased AKT phosphorylation, as shown by Western Blot analysis with a phospho-specific AKT antibody (data not shown), indicating that activation of HER-2/neu by heregulin activates AKT signaling. Treatment with Herceptin diminished AKT activity. As expected, protein expression levels of AKT1, AKT2, and PTEN (data not shown) remain unchanged during this short time course. To investigate whether HER-2/neu and HER-3 heterodimerization play a role in AKT activation, we immunoprecipitated HER-3 from untreated and treated cells and assayed for HER-2/neu by Western Blot analysis. Treatment with heregulin resulted in increased heterodimerization and increased activation of AKT, whereas treatment with Herceptin inhibited the formation of HER-2/HER-3 heterodimers and abolished AKT enzymatic activity.

Analysis of Breast Cancer Tissue Arrays

Because Herceptin down-regulates the phosphorylation of AKT *in vitro* on cell lines (results not shown), we further examined the role of AKT phosphorylation in breast carcinoma and, in particular, patient response to Herceptin and chemotherapy. We analyzed tissue microarrays of patient tumor samples taken before treatment. From the original tissue arrays of 250 patients, 43 samples were not used in this analysis for lack of clinical data or because the sections did not contain useable tumor tissue. The demographics of the remaining 207 patients are depicted in Table 21. Sixty-five percent of patients expressed HER-2/neu at the +3 level,

Table 21 Demographics of Breast Cancer Patient Samples Used in Final Analysis of Tissue Microarray. (Analysis on Tissue Array Samples for Which Clinical and HercepTest Data Was Available.)

	Number of Patients
Patients Included in Final Analysis	207
Histology	
Infiltrating ductal carcinoma	178
Lobular carcinoma	11
Medullary carcinoma	4
Tubular carcinoma	4
Scirrhous carcinoma	7
Papillary carcinoma	3
Treatment (plus Herceptin)	
Anthracycline plus cyclophosphamide	140
Doxorubicin	63
Paclitaxel	4

as determined by the HercepTest at the time of diagnosis. Thirty percent of patients expressed HER-2/neu at the +2 level. The remaining 5% expressed HER-2/neu at +1 or 0 levels and were administered Herceptin on the basis of serum HER-2/neu levels. The great majority of patients had infiltrating ductal carcinomas and received anthracycline plus cyclophosphamide. All patients had 4 mg/kg Herceptin loading dosage and 2 mg/kg weekly maintenance dosage. Overall, 15% of patients were disease-free or had stable disease after therapy, whereas 85% relapsed at last followup.

The patient tissue microarrays were first analyzed with phospho-specific antibodies to key downstream proteins that mediate the major pathways activated by the erbB receptors. The results of this analysis was statistically correlated with patient status following therapy. Table 22 presents the results of this analysis. Both AKT and ERK phosphorylation were found to statistically affect patient status. Of these two, AKT was most highly predictive of patient status; patients that have activated AKT were very likely to be deceased at the time this study was ended. In contrast, phosphorylation of STAT3, another protein that is downstream of erbB receptors, did not statistically affect patient survival. Phosphorylation of p38 also did not affect patient survival (results not shown). The importance of AKT in patient survival following Herceptin treatment and chemotherapy is not surprising, given AKT's role in determining cell survival following exposure to these two treatments.

Table 22 Analysis of Downstream Protein Activation and Patient Survival Following Herceptin/chemotherapy

Patient Group (IHC Score)	n	%Deceased	%Alive	Chi^2 Value/ P Value
p-AKT-0	42	37	63	64.6/0.000
1	51	84	16	
2	56	86	14	
3	41	85	15	
p-ERK-0	27	37	63	18.0/0.000
1	82	77	23	
2	49	78	22	
3	37	76	24	
p-SAT3-0	9	44	56	8.4/0.04
1	25	56	44	
2	54	80	20	
3	101	74	26	

Inhibition of the pI3K-AKT Survival Pathway Predicts Clinical Response to GW572016

As we have shown from the Herceptin tissue microarrays, increased expression of p-AKT in breast tumors is a poor prognostic factor. We therefore investigated the effects of GW572016, a reversible dual kinase inhibitor of HER-1 and HER-2/neu therapy, on the pI3K/-AKT survival pathway in sequential tumor biopsies from patients entered on clinical trial for this dual inhibitor (GW572016). Inhibition of p-AKT expression varied, but nonetheless predicted a favorable clinical response to therapy (Table 23). Two patients with metastatic breast cancer that had failed to progress on prior therapy with Herceptin and multiple chemotherapy regimens achieved documented partial responses following GW572016 treatment. Intratumor inhibition of p-AKT was observed by comparing day 0 and day 21 biopsies from these two patients (372 and 361; Table 23). Profiling the erbB receptors on these two patients that achieved partial responses revealed that both overexpressed target receptors EGFR/erbB-2 (361) or erbB-2/erbB-3 (372) (Table 23).

Inhibition of p-AKT was associated with disease stabilization, notably in patient 369 with metastatic head and neck carcinoma that had radiographic evidence of tumor regression though not sufficient to qualify as a partial response (Table 23). Inhibition of p-AKT was also found to be associated with increased tumor apoptosis. The tumor cell prosurvival factor AKT correlates with a poor clinical outcome. It was reported previously that inhibition of AKT phosphorylation in GW572016 treated tumor cell lines induced apoptosis (Xia *et al.*, 2002). To further investigate the therapeutic consequences of inhibiting AKT phosphorylation in patients tumors, we performed TUNEL assays on biopsies to assess tumor cell apoptosis. Consistent with its role in promoting tumor cell survival, inhibition of p-AKT expression following GW572016 therapy was associated with induction of

Table 23 Intra-tumor Biological Effects of GW572016 Therapy According to Clinical Response

Patient and Tumor Type	Dose (mg/d)	EGFR	erbB-2	p-ERK Index	p-AKT	cyclin D1	erbB-3	TGF-α	Heregulin
Patients Achieving a Partial Remission at 8 Weeks									
361 **Breast**	**1200**								
d 0		20	43	4015	36	31	0	35	20
d 21				0	24	3			
372 **Breast**	**1200**				28				
d 0		7	50	378	48	20	60	38	10
d 21				10	30	4			
425 **Breast**	**1200**								
d 0		23	62	0	85	19		32	5
d 21		24	68	7	26	12		22	8

OD values of ≤ 10, 10–15, and ≥ 15 roughly correspond to HercepTest standards of 1+, 2+, 3+, respectively.

P-ERK index is the product of the percentage of cells staining positive for p-ERK 1/2 in the tissue section and the OD value for p-ERK 1/2 immunoreactivity.

tumor cell apoptosis, as indicated by increased percentages of TUNEL + cells in day 21 biopsies (results not shown).

DISCUSSION

Breast cancer patients whose tumor cells overexpress HER-2/neu have a poor prognosis (Toikkanen et al., 1992). Up-regulation of HER-2/neu enhances the metastatic potential of cancer cells and is associated with resistance to various therapeutic regimens (Alaoui-Jamali et al., 1997, Ross et al., 1999). The mechanism of HER-2/neu-mediated resistance in cancer cells is currently unclear. Unlike the other EGF receptors, HER-2/neu has intrinsic tyrosine kinase activity that activates receptor-mediated signaling in the absence of ligand. Ligand binding by EGF receptors induces receptor dimerization and activates pI3K. Although HER-3 signals via pI3K, it is not known whether HER-2/neu homodimers activate pI3K in the absence of extracellular activation. The pI3K/AKT signaling pathway plays a critical role in apoptosis inhibition, which may contribute to the pathogenesis of cancer and the development of a hormone-independent phenotype.

To explore the relationship between the pI3K/AKT pathway components and HER-2/neu in breast cancer, we first evaluated AKT1 and AKT2 protein levels in a series of breast cancers in which the HER-2/neu status was known. HER-2/neu-positive tumors exhibited significantly more AKT2 protein than HER-2/neu-negative breast cancers (Figure 48A). In a recent report using similar antibodies against phospho-AKT (Ser 473) and AKT2, AKT2 activation was associated with activation of ERα (Sun et al., 2001b). These data, together with our findings, suggest that AKT2 expression and activity correlate with potentially important prognostic markers for breast cancer, and that HER-2/neu overexpression and AKT overexpression/activation may result in a hormone-independent phenotype and resistance to anti-hormone therapy such as Tamoxifen. Another report of IHC with phospho-specific, pan-AKT antibodies found 7 of 10 HER-2/neu-positive tumors that showed strong phospho-AKT staining. We suggest that activation of AKT2 in particular plays an important role in breast tumorigenesis based on the immunohistochemical analyses showing overexpression of AKT2 in the breast tumors. Because the phospho-specific, pan-AKT antibodies recognize equivalent phosphorylation sites in AKT1, AKT2, and AKT3, we acknowledge that our positively stained paraffin-embedded tumor specimens may have activation of multiple AKT isoforms. Indeed, a recent report has described increased AKT1 kinase activity in 38% (19 out of 50)

of snap-frozen breast tumors. We observed overexpression of AKT1 in a significant subset of our tumors, although we did not detect a significant correlation between AKT1 protein expression and HER-2/neu expression. It is likely that any of our specimens that stained positively with the phospho-specific, pan-AKT antibody would have activation of all AKT isoforms expressed in that tumor, because of shared upstream signaling via pI3K.

We further examined the expression of AKT2 in breast cancer cell lines to elucidate the role of pI3K/AKT signaling in HER-2/neu-associated resistance to cellular stress. Transfection of breast cancer cells with HER-2/neu resulted in increased expression of AKT1 (data not shown), as well as AKT2, which could reflect differences in cell culture versus in situ conditions. Nevertheless, we show that pI3K/AKT activity is required for resistance to hypoxia-induced apoptosis in breast cancer cell lines overexpressing HER-2/neu, and these cells become sensitive to apoptosis-inducing conditions when the pI3K/AKT pathway is blocked by wortmannin or LY294002.

Several lines of evidence presented here suggest that there is cross talk between the heterodimers of HER-2/neu and HER-3 and the pI3K/AKT pathway. Heregulin, a ligand that causes phosphorylation of HER-2 through heterodimerization with other HER receptors (Nahta et al., 2003), stimulated activation of AKT. Moreover, inhibition of the heterodimerization between HER-2 and HER-3 by a monoclonal antibody to HER-2/neu (Herceptin) inhibited AKT activation (results not shown, Yakes et al., 2002). Although AKT activation by HER-2/neu overexpression or heregulin previously has been reported (Hellyer et al., 2001), our results further suggest that AKT2 plays a significant role in HER-2/neu-overexpressing breast cancer cells to acquire resistance to cellular stress induced by hormonal therapy, chemotherapeutic agents, and/or radiation therapy. Furthermore, in our tissue microarrays of patients treated with Herceptin and chemotherapy, high levels of activation of AKT correlated with worse outcomes. Moreover, this investigation suggests that agents that down-regulate AKT2 expression or AKT activation may be of therapeutic value in breast cancers that overexpress HER-2/neu.

In our studies, overexpression of AKT2, but not AKT1 or AKT3, was sufficient to duplicate the invasive effects of phosphoinositide 3-OH kinase (pI3-K) transfected in breast cancer cells. Furthermore, expression of kinase-dead AKT2, but not kinase-dead AKT1 or AKT3, was capable of blocking invasion induced by either human epidermal growth factor receptor-2 (HER-2) overexpression or by activation of pI3K. Taken together, these data indicate that AKT2 mediates pI3K-dependent effects on adhesion, motility, invasion, and metastasis *in vivo*.

To further understand the importance of activation of the downstream signals of MAPK and AKT, we tested biopsies of patients treated with GW572016 dual kinase inhibitor of HER-1 and HER-2. Downstream regulation of AKT phosphorylation, a key prosurvival factor, was associated with tumor cell apoptosis and regression of metastases. In addition, inhibition of ERK 1 and 2 (extracellular signal-regulated kinases 1 and 2) phosphorylation, regulators of tumor cell growth and survival, predicted favorable clinical responses to therapy. Conversely, disease progression tended to occur in the absence of inhibitory effects of GW572016 therapy on phosphor-AKT and phosphor-ERK1/2. Moreover, lately localization of phosphorylated AKT has shown to play a role in response to stress of chemotherapy (Heliez *et al.*, 2003). Thus, determining the biological consequences of therapeutics targeting EGFR and erbB-2 on signaling pathways in clinical tumor samples treated with targeted therapy to HER-1 and HER-2 provides tools for establishing molecular profiles of tumors and may be used to maximize therapeutic efficacy.

Our studies demonstrate that identifying target populations is more complex than merely quantifying EGFR or erbB-2 expression, or erbB-2 gene amplification. It requires establishing a tumor profile, whereby the predominant growth and survival pathways operative in tumors can be identified. The EGFR/erbB-2 inhibitors warrant rational clinical development. The EGFR/erbB-2 receptors are expressed in normal tissue (e.g., myocardium, hepatocytes), so that dose selection should be based on the identification of a biologically effective dose for inhibition of growth and survival pathways rather than maximum tolerated doses to enable safer chronic administration. Although correlations between biological effects and clinical response will require confirmation in larger studies, identifying and using the biological markers of cell signaling pathways will help identify a tumor profile of patients likely to respond to targeted therapeutics.

In summary, the origins of cancer, including breast carcinomas, have been shown to lie in the inappropriate activation of oncogenes, resulting in abnormal activation or inhibition of cell signaling pathways. These events lead to unregulated cell proliferation, failure to respond to genetic damage resulting in genetic instability, and lack of programmed cell death, the result of which is the accumulation of malignant cells. Basic research has identified a select group of key cellular proteins that regulate these cellular functions and are most often disrupted during oncogenesis. The protein kinase AKT is one such protein, and AKT activation has been found in most cancers studied (Vivanco *et al.*, 2002). This activation has been shown to lead to unregulated cell survival and proliferation. An understanding of these oncogenic events has provided the foundation for the development of new therapeutics targeted at specific signaling molecules that mediate cellular transformation and malignancy. The clinical testing and patient use of this new class of drugs requires that the clinician be aware of the expression and activation of the relevant signaling pathways or molecules in the individual patient being considered for inclusion in a trial or for therapy with targeted drugs. Patients most likely to respond to the targeted drug will be those in whom the target is activated and is driving the continued presence or growth of the tumor. In this review, we concentrated on the unregulated activation of the pI3K/AKT pathway.

References

Alaoui-Jamali, M.A., Paterson, J., Al Moustafa, A.E., and Yen, L. 1997. The role of erbB-2 tyrosine kinase receptor in cellular intrinsic chemoresistance: Mechanisms and implications. *Biochem. Cell. Biol. 75*:315–325.

Altiok, S., Batt, D., Altiok, N., Papautsky, A., Downward, J., Roberts, T.M., and Avraham, H. 1999. Heregulin induce phosphorylation of BRCA1 through phosphatidylinositol 3-kinase/AKT in breast cancer cells. *J. Biol. Chem. 274*:32,274–32,278.

Arboleda, M.J., Lyons, J.F., Kabbinavar, F.F., Bray, M.R., Snow, B.E., Ayala, R., Danino, M., Karlan, B.Y., and Slamon, D.J. 2003. Overexpression of AKT2/protein kinase beta leads to up-regulation of beta1 integrins, increased invasion, and metastasis of human breast and ovarian cancer cells. *Cancer Res. 63*:196–206.

Bacus, S.S., Altomre, D.A., Lyass, L., Chin, D.M., Farrell, M.P., Gurova, K., Gudkov, A., and Testa, J.R. 2002. AKT2 is frequently up-regulated in HER-2/neu-positive breast cancers and may contribute to tumor aggressiveness by enhancing cell survival. *Oncogene 21*:3532–3540.

Bacus, S.S., Chin, D., Stewart, J., Zelnick, C., Mahvi, D., and Gilchrist, K. 1997. Potential use of image analysis for the evaluation of cellular predicting factors for therapeutic response in breast cancers. *Cytol. Histol. 19*:316–328.

Bacus, S.S., Gudkov, A.V., Lowe, M., Lyass, L., Yung, Y., Komarov, A.P., Keyomarsi, K., Yarden, Y., and Seger, R. 2001. Taxol-induced apoptosis depends on MAP kinase pathways (ERK and p38) and is independent of p53. *Oncogene 20*:147–155.

Bacus, S.S., Gudkov, A.V., Zelnick, C.R., Chin, D., Stern, R., Stancovski, I., Peles, E., Ben-Baruch, N., Farbstein, H., Lupu, R., Wen, D., Sela, M., and Yarden, Y. 1993. Neu differentiation factor (heregulin) induces expression of intercellular adhesion molecule 1: Implications for mammary tumors. *Cancer Res. 53*:5251–5261.

Bacus, S.S., Huberman, E., Chin, D., Kiguchi, K., Simpson, S., Lippman, M., and Lupu, R. 1992. A ligand for the erbB-2 oncogene product (gp30) induces differentiation of human breast cancer cells. *Cell Growth Different. 3*:401–411.

Bacus, S.S., and Ruby, S.G. 1993. Application of image analysis to the evaluation of cellular prognostic factors—in breast carcinoma. *Pathol. Ann. 28*:179–204.

Baselga, J. 2002. Combined anti-EGF receptor and anti-HER-2 receptor therapy in breast cancer: A promising strategy ready for clinical testing. *Ann. Oncology 13:*8–9.

Bellacosa, A., de Feo, D., Godwin, A.K., Bell, D.W., Cheng, J.Q., Altomare, D.A., Wan, M., Dubeau, L., Scambia, G., Masciullo, V., Ferrandina, G., Benedetti, P., Mancuso, S., Neri, G., and Testa, J.R. 1995. Molecular alterations of the AKT2 oncogene in ovarian and breast carcinomas. *Int. J. Cancer 64:*280–285.

Bianco, R., Shin, I., Ritter, C.A., Yakes, F.M., Basso, A., Rosen, N., Tsurutani, J., Dennis, P.A., Mills, G.B., Arteaga, C.L., 2003. Loss of PTEN/MMAC1/TEP in EGF receptor-expressing tumor cells counteracts the antitumor action of EGFR tyrosine kinase inhibitors. *Oncogene 22:*2812–2822.

Brunet, A., Bonni, A., Zigmond, M.J., Lin, M.Z., Juo, P., Hu, L.S., Anderson. M.J., Arden, K.C., Blenis, J., Greenberg, M.E. 1999. AKT promotes cell survival by phosphorylating and inhibiting a Forkhead transcription factor. *Cell 96:*857–868.

Cardone, M.H., Roy, N., Stennicke, H.R., Salvesen, G.S., Franke, T.F., Stanbridge, E., Frisch, S., and Reed, J.C.1998. Regulation of cell death protease caspase-9 by phosphorylation. *Science 282:*1318–1321.

Cheng, J.Q., Ruggeri, B., Klein, W.M., Sonoda, G., Altomare, D.A., Watson, O.K., and Testa, J.R. 1996. Amplification of AKT2 in human pancreatic cells and inhibition of AKT2 expression and tumorigenicity by antisense RNA. *Proc. Natl. Acad. Sci. USA 93:*3636–3641.

Dolbeare, F., Gratzner, H., Pallavicini, M.G., Gray, J.W. 1983. Flow cytometric measurement of total DNA content and incorporated bromodeoxyuridine. *Proc. Natl. Acad. Sci. USA. 80(18):*5573–5577.

Gore, M.E., Mohammed, M., Ellwood, K., Hsu, N., Paquette, R., Rao, P.N., and Sawyers, C.L. 2001. Clinical resistance to STI-571 cancer therapy caused by BCR-ABL gene mutation or amplification. *Science 293:*876–880.

Gupta, A.K., McKenna, W.G., Weber, C.N., Feldman, M.D., Goldsmith, J.D., Mick, R., Machtay, M., Rosenthal, D.I., Bakanauskas, V.J., Cerniglia, G.J., Bernhard, E.J., Weber, R.S., and Muschel, R.J. 2002. Local recurrence in head and neck cancer: Relationship to radiation resistance and signal transduction. *Clin. Cancer Res. 8:*885–892.

Hellyer, N.J., Cheng, K., and Koland, J.G. 1998. ErbB3 (HER-3) interaction with the p85 regulatory subunit of phosphoinositide 3-kinase. *Biochem J. 333:*757–763.

Hellyer, N.J., Kim, M.S., and Koland, J.G. 2001. Heregulin-dependent activation of phosphoinositide 3-kinase and AKT via the erbB-2/erbB-3 co-receptor. *J. Biol. Chem. 276:*42153–42161.

Heliez, C., Baricault, L., Barboule, N., and Valette, A. 2003. Paclitaxel increases p21 synthesis and accumulation of its AKT-phosphorylated form in the cytoplasm of cancer cells. *Oncogene 22:*3260–3268.

Hortobagyi, G.N. 2001. Overview of treatment results with trastuzumab (Herceptin) in metastatic breast cancer. *Semin. Oncol. 28:*43–47.

Itoh, N., Semba, S., Ito, M., Takeda, H., Kawata, S., and Yamakawa, M. 2002. Phosphorylation of AKT/PKB is required for suppression of cancer cell apoptosis and tumor progression in human colorectal carcinoma. *Cancer 94:*3127–3134.

Kim, A.H., Khursigara, G., Sun, X., Franke, T.F., and Chao, M.V. 2001. AKT phosphorylates and negatively regulates apoptosis signal-regulating kinase 1. *Mol. Cell. Biol.* 893–901.

Li, Y., Dowbenko, D., and Lasky, L.A. 2002. AKT/PKB phosphorylation of p21Cip/WAF1 enhances protein stability of p21Cip/WAF1 and promotes cell survival. *J. Biol. Chem. 277:*11352–11361.

Liu, W., Li, J., and Roth, R.A. 1999. Heregulin regulation of AKT/protein kinase B in breast cancer cells. *Biochem. Biophys. Res. Comm. 261:*897–903.

Malik, S.N., Brattain, M., Ghosh, P.M., Troyer, D.A., Prihoda, T., Bedolla, R., and Kreisberg, J.I. 2002. Immunohistochemical demonstration of phospho-AKT in high Gleason grade prostate cancer. *Clin. Cancer Res. 8:*1168–1171.

Mayo, L.D., and Donner, D.B. 2001. A phosphatidylinositol 3-kinase/AKT pathway promotes translocation of MDM2 from the cytoplasm to the nucleus. *Proc. Natl. Acad. Sci. USA 98:*11598–11603.

Nahta, R., Hortobagyi, G.N., and Esteva, F.J., 2003. Growth factor receptors in breast cancer: Potential for therapeutic intervention. *Oncologist 8:*5–17.

Nakatani, K., Thompson, D., Barthel, A., Sakaue, H., Liu, W., Weigel, R.J., and Roth, R.A. 1999. Up-regulation of AKT3 in estrogen receptor-deficient breast cancers and androgen-independent prostate cancer lines. *J. Biol. Chem. 274:*21528–21532.

Ng, S.S., Tsao, M.S., Nicklee, T., and Hedley, D.W. 2001. Wortmannin inhibits pkb/AKT phosphorylation and promotes gemcitabine antitumor activity in orthotopic human pancreatic cancer xenografts in immunodeficient mice. *Clin. Cancer Res.7:*3269–3275.

Okano, J., Gaslightwala, I., Birnbaum, M.J., Rustgi, A.K., Nakagawa, H. 2000. Akt/protein kinase B isoforms are differentially regulated by epidermal growth factor stimulation. *J. Biol. Chem. 275(40):*30934–30942.

Polakiewicz, R.D., Schieferl, S.M., Gingras, A.C., Sonenberg, N., and Comb, M.J. 1998. mu-Opioid receptor activates signaling pathways implicated in cell survival and translational control. *J. Biol. Chem. 273:*23534–23541.

Pullen, N., and Thomas, G. 1997. The modular phosphorylation and activation of p70s6k. *FEBS Lett. 410:*78–82.

Romashkova, J.A., and Makarov, S.S. 1999. NFkappa B is a target of AKT in anti-apoptotic PDGF signaling. *Nature 401:*86–90.

Rommel, C., Clarke, B.A., Zimmermann, S., Nunez, L., Rossman, R., Reid, K., Moelling, K., Yancopoulos, G. D., and Glass, D.J. 1999. Differentiation stage-specific inhibition of the Raf-MEK-ERK pathway by AKT. *Science 286:*1738–1741.

Ross, J.S., and Fletcher, J.A. 1999. HER-2/neu (c-erbB-2) gene and protein in breast cancer. *Am. J. Clin. Pathol. 112:*S53–S67.

Roy, H.K., Olusola, B.F., Clemens, D.L., Karolski, W.J., Ratashak, A., Lynch, H.T., and Smyrk, T.C. 2002. AKT proto-oncogene overexpression is an early event during sporadic colon carcinogenesis. *Carcinogenesis 23:*201–205.

Ruggeri, B.A., Huang, L., Wood, M., Cheng, J.Q., and Testa, J.R. 1998. Amplification and overexpression of the AKT2 oncogene in a subset of human pancreatic ductal adenocarcinomas. *Mol. Carcinog. 21:*81–86.

Rusnak, D.W., Lackey, K., Affleck, K., Wood, E.R., Alligood, K.J., Rhodes, N., Keith, B.R., Murray, D.M., Knight, W.B., Mullin, R.J., and Gilmer, T.M. 2001. The effects of the novel, reversible epidermal growth factor receptor/erbB-2 tyrosine kinase inhibitor, GW2016, on the growth of human normal and tumor-derived cell lines *in vitro* and *in vivo. Mol. Cancer Ther. 1:*85–94.

Stal, O., Borg, A., Ferno, M., Kallstrom, A-C., Malmstrom, P., and Nordenskjold, B. 2000. ErbB2 status and the benefit from two or five years of adjuvant tamoxifen in postmenopausal early stage breast cancer. *Ann. Oncol. 11:*1545–1550.

Stal, O., Perez-Tenorio, G., Akerberg, L., Olsson, B., Nordenskjold, B., Skoog, L., and Rutqvist, L.E. 2003. AKT kinases in breast cancer and the results of adjuvant therapy. *Breast Cancer Res. 5:*R37–R44.

Sun, M., Paciga, J.E., Feldman, R.I., Yuan, Z., Coppola, D., Lu, Y.Y., Shelley, S.A., Nicosia, S.V., and Cheng, J.Q. 2001b.

Phosphatidylinositol-3-OH kinase (pI3K)/AKT2, activated in breast cancer, regulates and is induced by estrogen receptor alpha (ERalpha) via interaction between ERalpha and PI3k. *Cancer Res. 61:*5985–5991.

Sun, M., Wang, G., Paciga, J.E., Feldman, R.I., Yuan, Z.Q., Ma, X.L., Shelley, S.A., Jove, R., Tsichlis, P.N., Nicosia, S.V., and Cheng, J.Q. 2001a. AKT1/PKBalpha kinase is frequently elevated in human cancers and its constitutive activation is required for oncogenic transformation in NIH3T3 cells. *Am. J. Pathol. 159:*431–437.

Toikkanen, S., Helin, H., Isola, J., and Joensuu, H.J. 1992. Prognostic significance of HER-2 oncoprotein expression in breast cancer: A 30-year follow-up. *Clin. Oncol. 10:*1044–1048.

Viglietto, G., Motti, M.L., Bruni, P., Melillo, R.M., D'Alessio, A., Califano, D., Vinci, F., Chiappetta, G., Tsichlis, P., Bellacosa, A., Fusco, A., and Santoro, M. 2002. Cytoplasmic relocalization and inhibition of the cyclin-dependent kinase inhibitor p27(Kip1) by PKB/AKT-mediated phosphorylation in breast cancer. *Nat. Med. 8:*1136–1144.

Vivanco, I., and Sawyers, C.L. 2002. The phosphatidylinositol 3-kinase AKT pathway in human cancer. *Nat. Rev. Cancer 2:*489–501.

Wright, C., Nicholson, S., Angus, B., Sainsbury, J.R., Farndon, J., Cairns, J., Harris, A., and Horne, C.H. 1992. Relationship between c-erbB-2 protein product expression and response to endocrine therapy in advanced breast cancer. *Br. J. Cancer 65:*118–121.

Xia, W., Mullin, R.J., Keith, B.R., Liu, L.H., Ma, H., Rusnak, D.W., Owens, G., Alligood, K.J., and Spector, N.L. 2002. Antitumor activity of GW572016: A dual tyrosine kinase inhibitor blocks EGF activation of EGFR/erbB-2 and downstream Erk1/2 and AKT pathways. *Oncogene 21:*6255–6263.

Xing, X., Wang, S.C., Xia, W., Zou, Y., Shao, R., Kwong, K.Y., Yu, Z., Zhang, S., Miller, S., Huang, L., and Hung, M.C. 2000. The ets protein PEA3 suppresses HER-2/neu overexpression and inhibits tumorigenesis. *Nat. Med. 6:*189–195.

Yakes, F.M., Chinratanalab, W., Ritter, C.A., King, W., Seelig, S., and Arteaga, C.L. 2002. Herceptin-induced inhibition of phosphatidylinositol-3 kinase and AKT is required for antibody-mediated effects on p27, cyclin D1, and antitumor action. *Cancer Res. 62:*4132–4141.

Zhou, B.P., Hu, M.C., Miller, S.A., Yu, Z., Xia, W., Lin, S.Y., and Hung, M.C. 2000. HER-2/neu blocks tumor necrosis factor-induced apoptosis via the AKT/NF-kappaB pathway. *J. Biol. Chem. 275:*8027–8031.

Zhou, B.P., Liao, Y., Xia, W., Spohn, B., Lee, M.H., and Hung, M.C. 2001. Cytoplasmic localization of p21Cip1/WAF1 by AKT-induced phosphorylation in HER-2/neu-overexpressing cells. *Nat. Cell Biol. 3:*245–252.

9

Expression of Extracellular Matrix Proteins in Breast Cancer

Anna Kádár, Janina Kulka, and Anna-Mária Tõkés

Introduction

Role of Extracellular Matrix Proteins

It has been recognized that the composition and organization of the extracellular matrix (ECM) highly affects the behavior of cells and the organism. The mammary gland comprises stromal and epithelial cells that communicate with each other via the ECM. According to classical theory, the development of breast cancer involves epithelial hyperplasia, premalignant changes, in situ carcinoma, and invasive carcinoma. The alterations in the structure of the mammary epithelium are accompanied by a reorganization in the composition of the periductal stroma compartment and ECM. The ECM proteins, including collagen, elastin, basement membrane proteins, structural glycoproteins, and proteoglycans, not only provide a physical support for cells but also play a role as an informational system in mammary gland branching. The ECM proteins and integrins collaborate and profoundly influence gene expression and major cellular programs, including growth, migration, differentiation, and survival. Many biological activities require specific interactions with ECM proteins including development, morphogenesis, growth, and tumor progression. Thus, there exists a dynamic relationship between cells and the ECM production. The ECM is composed of a number of different structures and molecules; the most common are the structural glycoproteins: *tenascin*, fibronectin, osteonectin, osteopontin, thrombospondin, vitronectin, and WNT proteins.

Role of Extracellular Matrix Protein Molecules in Tumorigenic Breast Morphology

During the normal mammary gland development, the gland displays several properties associated with breast cancer. Many of the factors known to act in breast cancer formation are also vital for mammary development. The developing mammary gland shows many properties associated with or similar to tumor progression, such as cell proliferation, invasion, reinitiation of proliferation, angiogenesis, and response to apoptosis (Wiseman et al., 2002). Many of the factors implicated in mammary gland development are also associated with carcinogenesis, so it is desirable to distinguish those molecules that are associated with tumor formation only.

Interaction between breast epithelium and ECM plays a major role in mammary gland branching formation. It is the mammary epithelium that proliferates,

Handbook of Immunohistochemistry and in situ Hybridization of Human
Carcinomas, Volume 1: Molecular Genetics; Lung and Breast Carcinomas

321

Copyright © 2004 by Elsevier (USA)
All rights reserved.

invades, and has tumorigenetic potential; however, the neighboring stromas give permissive or instructive signals. The origin of breast tumorigenesis is attributed to genetic alterations in the breast epithelial cell compartments, but fibroblasts displaying abnormal phenotypes may produce an altered stroma that may influence the progression of tumorigenesis in the breast epithelium. The altered stroma may contain high amounts of collagen, fibronectin (FN), and proteoglycans. Some consequences of modified synthesis of ECM molecules related to tumorigenesis include changes in proliferation, induction of angiogenesis, altered cell adhesion, and cell migration, as well as tumor metastasis. A combination of autocrine and paracrine growth factors and cytokines, and ECM molecules lead to deregulation of cell cycle control, which results in increased cell proliferation and malignant transformation.

It has been recognized that the stroma around a malignant tumor differs from that of the normal ECM. One of the changes seen in the stroma of the breast is the overexpression of one of the ECM glycoproteins, tenascin-C (Howeedy *et al.*, 1990).

Tenascin Family Proteins

Tenascin-C (TN-C) is an ECM glycoprotein isolated from human glioblastoma cells (Bourdon *et al.*, 1989), chicken embryos, embryonic mouse, and chicken brain. Four other forms of tenascin (TN-R, TN-W, TN-X, TN-Y) have the same consecutive arrangement of domains found in TN-C. The EMC glycoprotein TN-R, named also as J1 160/180 molecule, or janusin or restrictin, exhibits a structure similar to TN-C and is synthesized by oligodendrocytes during myelinization in late gestation, and its expression partially overlaps with that of TN-C. As described for TN-C, the isoform 160/180 is generated by alternative splicing of the A1 fibronectin type III homologous repeat (Fuss *et al.*, 1993). The EMC glycoprotein TN-X is a prominent component of connective tissues. It is the largest member of the tenascin family; the gene consisting of 39 exons (Bristow *et al.*, 1993). Deletion within the region of the TN-X gene results in Ehlers-Danlos syndrome, a connective tissue disorder involving vascular fragility and poor wound healing.

The EMC glycoprotein TN-W was identified in the zebrafish and its expression colocalizes with TN-C in several developmental contexts (Weber *et al.*, 1998). The EMC glycoprotein TN-Y is expressed in connective tissues and brain, modulating muscle cell growth. In the adult chicken, it is expressed in skeletal muscles and heart. It is a multidomain protein and when compared with other members of the tenascin family, its fibronectin (FN) type III-like repeats are interrupted by

a domain consisting of repeats of serines, prolines, or other amino acids (Hagios *et al.*, 1996). The mostly characterized member of the tenascin family and known to be implicated in breast cancer tumorigenesis is TN-C.

Tenascin-C Structure

Ericson *et al.*, (1989) demonstrated the stellate, six-armed structure of the TN-C having subunits that range between 190–300 kDa and thus called the molecule hexabranchion. In the TN-C hexamer, six arms emanate from a central knob. It exists as multiple isoforms and the precise structure of these isoforms determines the ultimate effect of the protein. The oligomer has a linear arrangement of domains, with a cysteine-rich NH2 terminus termed the tenascin assembly domain, followed by a series of epidermal growth factor-like (EGF-like) repeats, 8–17 FN-III–like repeats, and a fibrinogen-like globular domain at the carboxyl terminus. The structure and size of TN-C vary as a result of alternative splicing of exons within the FN-III repeat domain. A number of biologically active sites have been mapped to the FN-III repeat domain, including recognition sites for cell surface receptors such as integrins, domains that bind to other ECM proteins such as glycosaminoglicans, cell adhesion molecules of the immunoglobulin superfamily, and annexin II, as well as sites susceptible to proteolytic cleavage by matrix metalloproteinase. The latter function allows TN-C containing matrices to be selectively remodeled. Some of the TN-C splice variants occur at distinct frequencies during development of the nervous system, in the kidney, cornea, and smooth muscle. Alternative splicing of FN-III repeats occurs in two other members of the tenascin family, TN-R and TN-X. There are several studies in splice variants but the mechanism that determines the preferential splice site in the FN-III repeats is unknown (Adams *et al.*, 2002).

The Tenascin-C Gene

The human TN-C gene is located on chromosome 9, bands q32–q34. Structurally, the coding region of the human TN-C gene is ~80 kb of DNA. The gene consists of 27 exons separated by 26 introns. The fibrinogen-like domain has been reported to be encoded by five exons, the EGF-like domain by a single exon and each FN-III repeats are encoded either by a single exon or by two exons separated by an intron (Rocchi *et al.*, 1991). The promoter region of TN-C gene contains a variety of potential regulatory sequences. Most expressive are the binding site for homeodomain proteins and the TRE/AP1 phorbol response element. The strongest activation properties reside in the most proximal 220 nucleotides, whereas a silencer element, located between 220 and 2300, is active in both TN-C producing and nonproducing cell lines. Near the multiple binding site in the promoter region a collagen responsive element has been mapped within the TN-C promoter between positions 570 and 469. This finding suggests a rather important functional interpretation of TN-C.

Expression and Degradation of Tenascin-C in Different Tissues

Studies indicate that TN-C, like other large matrix proteins, are assembled into hexamers during translation. A variety of cells have the ability to synthesize TN-C but the expression pattern varies in normal development and in pathological conditions. This property suggests that TN-C containing matrices may influence tissue morphogenesis by determining whether cells adhere to substrate or to other cells in the same tissue, or provide a provisional matrix conductive for cellular migration, division, or apoptosis. It is important to study the molecules and regulatory pathways that govern local expression of TN-C.

At first, TN-C expression is observed at gastrulation and somite formation in a series of rostral-caudal waves in the chicken embryo. At a later stage of somitogenesis, expression of TN-C is restricted to the rostral half of the somite during invasion of neural crest cells. In the later stage of development, TN-C is produced by glial elements in the central nervous system. Its expression is particularly eminent on radial and Bergmann glial fibers during neuronal differentiation and migration in the cerebellum and cortex. Outside of the nervous system, TN-C is expressed during morphogenesis in the skeleton, connective tissue, and vasculature. In adults, TN-C is observed at sites of neovascularization and wound healing, and its expression is also induced by mechanical stress. The TN-C protein is up-regulated during pathological conditions, including vascular hypertension and in stroma surrounding tumor formation and metastasis. Other members of the TN family are coexpressed with TN-C in several tissues, raising the possibility that various forms of TNs may be complemental in several biological processes.

Degradation of TN-C is executed by members of the matrix metalloproteinase (MMP) family. The MMP-7 cleaves all TN-C isomers, separating the amino-terminal knob from the remainder subunit. The MMP-2, -3, and -7 cleave large TN-C isoforms within the additional fibronectin type III-like region. It is of importance that the expression of MMPs often occurs under the same pathological circumstances as does expression of tenascin, for example, in tumorigenesis.

Induction of Tenascin-C Expression

The TN-C promoter contains structural elements that may serve as potential targets for growth factor signaling transduction pathways. Available evidence demonstrates the induction of increased TN-C mRNA expression by growth factors. A large series of studies demonstrates that growth factors, cytokins, and other soluble factors are the paracrine acting factors in epithelial mesenchymal tissue recombinants and are stimulating agents inducing an increased expression of TN-C mRNA. Various growth factors have been identified as inducers of TN-C expression. Early studies showed that synthesis of TN-C by embryonic fibroblasts could be up-regulated by fetal bovine serum and TGFβ-1 (Vollmer, 1997). Fibroblast growth factor-2 (FGF-2) stimulates TN-C expression in Ewing's sarcoma–derived cell line. The FGF selectively stimulates an increased expression of the larger splice variants of TN-C.

There is evidence that TN-C gene expression is induced not only by growth factors but also by mechanical regulation. Mechanical strain induces TN-C in neonatal rat cardiac myocytes in an amplitude-dependent manner. *In vivo* studies examining TN-C expression in fibroblasts and chondrocytes within the osteotendineous junction showed that removal of mechanical stress from the junction suppresses TN-C expression. Recent studies have indicated that the control of TN-C expression by mechanical factors occurs at the level of the gene promoter, and that different promoter motifs and transcription factors may be used, depending on cell type and the culture conditions used.

Tenascin-C Interaction with Ligands and Its Effect on Cells

Tenascin possesses adhesive and antiadhesive activities that coexist on the native molecule. These opposing activities arise as a consequence of TN-C binding to other components of the ECM and to cell surface receptors. TN-C binds with high affinity to several proteins and carbohydrates. Cell surface receptors for TN-C include members of the integrin family of heterodimers, cell adhesion molecules (CAMs) of the immunoglobulin superfamily, a transmembrane chondroitin sulfate proteoglycans receptor, and annexin II. TN-C and TN-R also bind with other extracellular matrix proteins, including fibronectin, and a class of extracellular chondroitin sulfate proteoglycan called lecticans (aggrecan, versican, brevican, and neurocan).

In vitro, tenascin binds to immobilized fibronectin and to cell surface proteoglycans. There is some tissue specificity in the interaction of heparan sulfate proteoglycans and tenascin because the integral membrane heparan sulfate proteoglycan syndecan interacts with immobilized tenascin in cell-free binding assays, whereas preparations of syndecan isolated from mammary gland do not. Tenascin interacts with and is cleaved by proteases and MMPs. Differential expression of one or more alternatively spliced forms of TN-C determines the binding to a cellular receptor or ECM component. For example, the smallest TN-C variant binds to fibronectin with higher affinity than larger forms. Larger variants of TN-C isoforms have been correlated with cell migration and invasiveness of carcinomas.

Tenascin-C and Cell Adhesion

Tenascin-C in Relation to Other Extracellular Matrix Molecules

Immunodetection of TN-C in close proximity to proteoglycans suggests a close interaction of these two families of molecules in many experimental models. It has been noted that TN-C strongly interacts with the proteoglycan syndecan, versican, neuran, phosphacan, and keratan sulfate. It was demonstrated that syndecan isolated from embryonic tooth mesenchyme binds TN-C, the binding being mediated by the heparan sulfate side chains of syndecan (Salmivirta et al., 1991). Preparations of syndecan isolated from mammary gland do not interact with immobilized TN-C in cell-free binding assays.

Tenascin-C and Fibronectin

After the identification of TN-C it was demostrated that TN-C and fibronectin not only share FN type III-like repeats as common structural part, but TN-C has been isolated from cell surface fibronectin preparation (Ericson et al., 1984). In fact, fibronectin and TN-C colocalization was observed in a variety of developing tissues or developmental processes such as cell migration in some tumors and cell cultures. The interaction of fibronectin and TN-C molecules is documented, and it has been demonstrated that TN-C inhibits adhesion of cells to intact fibronectin or fibronectin fragments that contain the classical ARG-GLY-ASP (RGD) sequence (Chiquet-Ehrismann et al., 1988). The interaction of TN-C with cells and substrate is undoubtedly more complex because adjacent regions in the alternative-spliced type III repeats alter the interaction of TN-C with fibronectin substrates.

It is questioned whether the role of TN-C is to inhibit cell binding to fibronectin or if TN-C and fibronectin act synergically. An important result was published a few years ago: plating fibroblasts on a mixed substrate of fibronectin and TN-C but not fibronectin alone up-regulated the transcription of four genes: c-fos, collagenase, stromelysisin, and the 92 kDa gelatinase (Tremble et al., 1994). This observation may suggest that the alteration of the composition of the ECM by TN-C expression may regulate the expression of genes involved in cell migration or in tissue invasion.

Tenascin-C as a Basement Membrane Component

In our study of breast tissues, TN-C immunoreactivity was often detected in a membrane-like basement, raising the question whether TN-C is a constituent element of the basement membrane. This localization of TN-C was described in skin (Lightner et al., 1989), and teeth (Thesleff et al., 1987). High-resolution immune-electron microscopy of human skin tissue demonstrated that TN-C located in the basement membrane is not otherwise associated with collagen and elastic fibers. In developing lung it appears punctate and different from the uniform staining of laminin (Young et al., 1994).

In conclusion, it appears unlikely that TN-C is an integral constituent of the basement membrane; however, functionally it may be implicated.

Effect of Tenascin-C Substrates and Soluble Tenascin

The current question is whether TN-C represents an adhesive or anti-adhesive protein. Some authors who added soluble TN-C to cell cultures detected that TN-C reduces focal adhesion integrity in confluent endothelial and smooth muscle cell cultures (Hahn et al., 1995). It was also found that TN-C causes a dose-dependent reduction in the number of focal adhesion positive cells to ~50% of albumin treated controls. A maximum response was reported with 20–60 µg/ml of soluble TN-C and the reduction in focal adhesion in TN-C-treated cells was observed after 10 min (Murphy-Ulrich et al., 1991). It has also been described that the proportion between TN-C and FN used to coat the substrate determines the shape of the cells (Chiquet Ehrismann et al., 1988). Cells adhere to FN and TN-C in part via interaction of certain integrins with the RGD sequence. Interaction between FN type III domains and beta 1 integrins mediate focal adhesion kinase phosphorylation and focal contact adhesion (Chi-Rosso et al., 1997). How the number of FN type III domains of TN-C influences the binding capacity of TN-C to FN is questionable.

Using TN-C as substrate, the fibroblast-like cells characteristically adhere to the TN-C substrate and cells remain rounded and do not spread, in contrast to fibronectin-coated substrate (Chiquet-Ehrismann et al., 1988). Like fibroblasts, neurons adhere to TN-C but remain rounded. Faissner et al., (1990) reported that

TN-C is repulsive for neurons of the central nervous system. To control the mechanisms by which TN-C inhibits cell spreading, different fragments of TN-C were used as substrate. When a recombinant form of the FN type III repeat containing the RGD sequence was used as substrate, endothelial cells adhere strongly and start spreading (Joshi et al., 1993). Using the fibrinogen-like terminal knob as substrate for endothelial cells, the cells were demonstrated to adhere without any spreading.

An experiment with a series of tumor cell lines demonstrated that those producing either fibronectin only or coexpressing fibronectin and TN-C adhere to substrate or to tissue culture plastic. Cell lines producing TN-C only remained unattached (Kawakatsu et al., 1992).

Tenascin in Adult Tissues Other Than Breast Lesions

Tenascin-C in Normal Adult Tissues

Early reports suggested that normal adult tissues do not express TN-C. However, since then TN-C expression has been detected in many normal tissues. In the endometrium during the first part of the menstrual cycle, the high proliferative activity is correlated with higher TN-C expression. In the breast, the highest proliferative activity is found in the second half of the menstrual cycle and is accompanied by high TN-C levels. These observations suggest a possible relationship of TN-C expression with proliferative events.

Tissue injury mediates TN-C expression in in vivo and experimental conditions. This is a common feature in cornea and skin wounds, ruptured ligament, and nerve injury. Depending on the extent and site of the damage, TN-C seems to be involved in wound healing in the skin, regeneration, muscles, and reinnervation. In the skin a strong immunostaining appears at the dermal/epidermal junction beneath migrating and proliferating epidermal cells. Following nerve injury it has been observed that fibroblasts proliferating close to denervated synaptic sites synthesize TN-C. In the neuromuscular system peripheral nerve damage results in the modulation of TN-C expression both in nerves and muscles.

Tenascin-C in Benign and Precancerous Proliferative Lesions

In malignant tumors, there is strong evidence of TN-C up-regulation and overexpression. Increased proliferation is also associated with numerous benign or precancerous lesions (Iskaros et al., 2000). Immunohistochemical studies have demonstrated an increased TN-C expression in benign and precancerous proliferative lesion of the skin, endometrial tissue, and breast. However, TN-C immunohistochemistry cannot discriminate between TN-C-isoforms that appear in benign and presumably precancerous lesions.

Expression of Tenascin-C in Tumors

There is an increased expression of TN-C in a variety of tumors. A final conclusion or a consequent role of TN-C in oncogenesis has not been clearly discussed. A precise and profound examination of TN-C during normal mammary gland development and tumor formation indicates that it is differentially regulated at each stage of development and during the course of oncogenesis.

In adenocarcinoma in humans and in animal organs examined, TN-C immunoreactivity was found to be over most of the extracellular space of the stromal mesenchyme. In mesenchymal tumors, such as leiomyosarcoma of the uterus, immunostaining was detected around islands of tumor cells or encircling single tumor cells. In oral squamous cell carcinoma with metastasis the expression of TN-C was found to be increased when compared with nonmetastatic cases (Harada et al., 1994). A profound analysis of TN-C in salivary gland tumors has shown a strong TN-C immunoreactivity in Warthin's tumors in the basement membrane zone beneath the epithelial cell layer. In adenoid carcinomas and mucoepidermoid carcinomas, TN-C showed uniform stromal distribution. No TN-C immunoreaction was detected in the normal serous or mucous gland (Soini et al., 1992).

Tenascin-C Expression in Breast Diseases

Tenascin-C Expression in Normal Human Breast Tissue

On the 14th postgestational day, TN-C appears in the dense embryonic mammary mesenchyme, then expression decreases in the postnatal and adult stages. Small amounts of TN-C can be detected in the mature, normal mammary gland. A thin, focally discontinuous band of TN-C immunoreactivity can be consistently noted around ducts and acini. No TN-C is present beyond the immediate periductal and periacinar stromal regions, except for occasional thin bands around vessels. During pregnancy, because of the increased amounts of progestins, estrogen, and placental lactogen, the mammary epithelium proliferates from a branching ductal system to form a known structure of the breast. In this phase the TN-C production is suppressed. During lactation, a continued suppression of TN-C is observed. Following weaning, loss of milk production and massive apoptosis is observed in the female breast. This process is associated with myoepithelial cell-derived TN-C positivity (Jones et al., 2000).

Tenascin-C Expression in Fibrocystic Breast

In fibrocystic changes, continuous bands of TN-C immunoreactivity of variable thickness were consistently noted around ductal and acinar structures in all variants of fibrocystic disease (FCD). In dilated ducts with or without apocrine metaplasia, and with variable degrees of stromal fibrosis, a notable layer of TN-C reactivity was seen. In samples of ductal hyperplasia, variably thick bands of intense TN-C immunostaining were noted by Howeedy et al., (1990). In all variants of FCD, TN-C reactivity was not entirely restricted to the immediate periepithelial regions; its presence also extended further into the stroma. In fibroadenomas, TN-C immunoreactivity was observed around ductal structures. The TN-C in the periepithelial regions was prominent and extended unevenly into more distant stromal regions. The stroma of fibroadenomas in general showed irregularly distributed areas of TN-C immunoreactivity.

Tenascin-C Expression in *in situ* Ductal Breast Carcinomas (DCIS)

Around ducts distended by *in situ* carcinoma of all types, prominent bands of TN-C appeared either as concentric rings or irregularly. Stromal TN-C locates focally, often forming bridges between the ducts affected by DCIS. When DCIS was associated with microinvasion, stromal tenascin seemed to accompany the invading cells (Jahkola et al., 1998a). Howeedy et al., (1990) reported that the stromal region beyond ducts containing *in situ* carcinomas presented little TN-C reactivity.

Tenascin-C Expression in Different Types of Invasive Breast Carcinomas

In infiltrating ductal carcinomas not otherwise specified (IDC NOS), TN-C expression involves a large part of the stroma. Strong and generally diffuse TN-C immunostaining can be noted around clusters of carcinoma cells or around individual cells. In small tumors, TN-C expression is similar in the central and peripheral regions, whereas in larger tumors the peripheral regions show more intensive immunoreactivity.

In infiltrating lobular carcinomas, extensive TN-C immunoreactivity has been described in areas showing the typical "Indian file" pattern. However, this was not fully elucidated, though it was described that the TN-C positivity is more intense at the invasive edge of the tumor (Howeedy et al., 1990), and thus the presence of TN-C positivity may indicate the site of active cancer spread (Jahkola et al., 1998a, b). The expression of TN-C was also described in carcinoma cells. TN-C appears to be secreted by both epithelial and stromal cells close to the basement membrane zone. In certain malignant-tumor cell lines, TN-C was secreted by tumor cells, especially those of epithelial origin (Jessen et al., 1996).

In our study, our aim was to systematically investigate the distribution of TN-C with characterized monoclonal antibodies by immunohistochemistry. Primary, second primary, and recurrent breast carcinomas were analyzed. The effect of TN-C on breast cancer cell lines was also investigated.

MATERIALS

Tenascin-C Immunodetection in Primary, Secondary, and Recurrent Tumors

1. Samples taken for diagnostic purposes from breast specimens of 62 female patients (mean age 58, between 38 and 84 years of age) were examined immunohistochemically for TN-C expression. None of the patients had received chemotherapy or irradiation before surgery. All 62 cases were primary invasive ductal breast carcinomas not otherwise specified.

2. Samples of breast biopsies from 20 female patients diagnosed with breast cancer were examined (mean age 61.5, between 37 and 86 years of age). Primary tumors, and their recurrences and second primary tumors were analyzed in this part of the study. The mean follow-up period of the patients was 2.9 years. During this period, 14 out of 20 patients had local recurrences and 3 had a second recurrence, while 6 developed a second primary breast carcinoma.

3. Cell culture. MDA-MB-435 human breast cancer cell line was used to study the effect of TN-C on focal adhesion plaques.

METHODS

Materials for TN-C Immunodetection

1. 4% neutral-buffered formalin (4% formaldehyde in 0.1 M phosphate buffer [pH 7.2]).

2. Ethyl alcohol 100%, 70%, and 50%.

3. Phosphate-buffered saline (PBS) 10X, (pH 7.4) (87.5 g NaCl, 15.33 g $Na_2HPO_4 \cdot 2H_2O$, 1.92 g KH_2PO_4 and adjusted to 1000 ml with distilled water. Dilute to 1:10 before use.

4. Endogenous peroxidase blocking solution (3% H_2O_2 in distilled water).

5. Normal bovine serum albumin (BSA) in PBS 1%, (Sigma, No. A-9647, St. Louis, MO) supplemented with 0.1% sodium azide (NaN_3, Sigma, S-2002).

6. Pepsin (0.1% in 0.01 M HCl).

7. Primary antibodies: Antibodies derived from two different clones, both of mouse origin:

    a. DB7 (BioGenex, San Ramon, CA), the immunogen of which was the supernatant of a human fibroblast cell culture; and

    b. TN2 (Dako, Glostrup, Denmark), which was raised against tenascin purified from the supernatant of a glioma cell line (U 251). Both diluted to 1:40 and 1:25 in PBS/BSA/azide.

8. Biotinilated secondary antibody conjugated with streptavidin-biotin complex Dako LSAB 2 System, HRP ready to use, K0675.

9. Secondary antibody Dako LSAB+ Kit alkaline phosphatase, K 0689.

10. Dako 3.3′-diaminobenzidin (DAB) chromogen tablets S 3000.

Step-by-step protocol for DAB preparation: Place 10 ml of buffer diluent in test tube, transfer 1 tablet of DAB to the 10-ml buffer. Transfer 2 ml of the diluted DAB chromogen to a clean tube, add 15 μl of 3% H_2O_2 and mix.

11. Dako New Fuchsin substrate system K 698.

Step-by-step protocol for chromogen substrate preparation: Place two drops of Tris buffer concentrate into a test tube. Bring up volume to 2 ml with distilled water, then add one drop of substrate concentrate and mix gently. Place one drop from new fuchsin chromogen and one drop of activating agent into an empty tube and mix gently. Allow to stand for 3 min. Transfer the substrate reagent into the test tube containing the chromogen and mix.

12. Mayer's hemalaun (hematoxylin 1 g, distilled water 1000 ml, natrium sodium iodate 0.2 g, ammonium or potassium-aluminium sulfate 50 g, citric acid 1 g, chloral hydrate 50 g).

13. Dako ultramount aqueous permanent mounting medium 003181.

14. Pertex xylol-based mounting medium PER 30000 (Burgdorf, Germany).

Immunohistochemistry Protocol for Tenascin-C Detection

1. Fix the tissue without delay with 4% neutral-buffered formalin for 24 hr.

2. Dehydrate in a series of ascending concentrations of ethanol.

3. Infiltrate and embed in paraffin.

4. Cut 3–5 μm thick sections with microtome and float them on a water bath kept at room temperature to stretch the sections.

5. Mount sections on silane-treated glass slides.

6. Deparaffinize with three changes of 5 min each slides in xylene, followed by three changes of 100% ethanol, two changes of 75% ethanol, and two changes of 50% ethanol.

7. Place the slides in three changes of distilled water.

8. Treat the sections with pepsin (0.1% in 0.01 M HCl) for 50 min at 37°C (use only freshly prepared enzymatic solution, as the enzyme activity decreases with age).

9. Place the slides in PBS (pH 7.4).

10. When DAB is used as chromogen, inhibit endogenous peroxidase activity by immersing the sections for 30 min in a solution of 3% H_2O_2 in distilled water.

11. Rinse the slides in PBS.

12. Block the background staining by treating the sections for 30 min at room temperature with BSA.

13. Incubate the slides with monoclonal antibodies diluted to 1:40 and 1:25 in PBS/BSA/azide for 90 min at 37°C in humid chamber.

14. Use negative controls. Incubate the slides in absence of primary antibody.

15. Use positive control. (3-5 μm sections of human tonsils, removed during tonsillectomy, may serve as positive control.)

16. Rinse slides three times for 5 min each in PBS further processed for second antibody.

17. Incubate for 30 min with biotinylated anti-mouse secondary antibody at room temperature.

18. Rinse slides three times for 5 min each in PBS.

19. Incubate with streptavidin horseradish peroxidase **or** with streptavidin alkaline phosphatase for 30 min at room temperature.

20. Rinse slides three times for 5 min each in PBS.

21. Develop the color reaction with DAB for 5–10 min. Use fumehood at this step of reaction. If using the alkaline phosphatase as enzyme, use the New Fuchsin substrate system.

22. Wash in distilled water for 10 min.

23. Counterstain for 20 sec with hemalaun.

24. Wash in running tap water for 10 min.

25. Dehydrate in ascending ethanol.

26. Clear in xylene and coverslip with mounting media when DAB chromogen is used.

27. **Or** use aqueous permanent mounting medium without dehydration when alkaline phosphate method is applied.

The positive TN-C reaction using New Fuchsin as chromogen was observed as purple staining, and brown when DAB was used as chromogen substrate. Independent of the chromogen substrate used, the positivity appeared as a continuous band of variable thickness around ducts, as tumor cell nests, as a delicate fibrillar component in the stroma of carcinomas, or as a cytoplasmic reaction in some of the carcinoma cells.

Immunostaining was analyzed by 2 independent observers (A-M. T., E. H.). Immunoreaction for TN-C was evaluated in the stroma close (< 0.5 mm) to epithelial elements of the tumor. An average of 2 observers' results were determined for each slide. If the interobserver difference was greater than 20%, slides were reevaluated using discussion microscopy (Olympus, BH2). The positivity in ducts, carcinoma cells, and vessels was also noted and recorded.

The Effect of Tenascin-C on Focal Adhesion Plaques

Immunohistochemistry for Actin and Vinculin Detection

Materials for Immunohistochemistry

1. MDA-MB-435 human breast cancer cell line.

2. Cell culture medium (Dulbeco modified eagle medium (DMEM) supplemented with 10% fetal calf serum (FCS), 1 mM pyruvate and antibiotics-free).

3. Purified TN-C preparations obtained from Sigma and Chemicon.

4. 1% paraformaldehyde for cell fixation (1 g paraformaldehyde in 100 ml distilled water, mix at 70°C in water bath. Add normal natrium hydroxide to adjust the pH to 7.4.

5. PBS 10X, (pH 7.4) (87.5 g NaCl, 15.33 $Na_2HPO_4 \cdot 2H_2O$, 1.92 g KH_2PO_4) adjusted to 1000 ml with distilled water. Dilute to 1:10 before use.

6. Endogenous peroxidase blocking solution (3% H_2O_2 in distilled water).

7. Triton-X 100 for cell membrane permeabilization use in 0.04%.

8. Use Phalloidin rhodamin (Molecular Probes) for actin staining.

9. Monoclonal anti-vinculin (clone No. HVIN-1) from Sigma.

10. Biotin conjugated secondary antibody (Amersham, Buckinghamshire, U.K.).

11. Streptavidin-FITC (Fluorescein izo-thiocyanate) (Amersham).

12. Fluorescent mounting medium (Dako).

Immunodetection of Actin and Vinculin in Focal Adhesion Plaques

1. Grow the MDA-MB-435 human breast cancer cells in DMEM supplemented with 10% FCS, 1 mM pyruvate and antibiotics-free for 24 hr on glass coverslips.

2. Examine the attachments of cells to glass coverslips.

3. Rinse the cells with PBS three times for 5 min each.

4. Treat the cells for 10 min, 30 min, 1 hr and 2 hr with 10, 20, 100, 200 µg/ml soluble tenascin, respectively.

5. Fix the cells with cold 1% paraformaldehyde/PBS for 10 min.

6. Rinse the cells with PBS three times.

7. Permeabilize the cells with 0.04% Triton-X.

8. Rinse the cells with PBS three times.

9. Stain the permeabilized cells for actin staining with phalloidin rhodamin for 1 hr at room temperature. Use humified chamber and dark conditions.

10. Use negative controls. Incubate the slides in absence of primary antibody.

11. Rinse the cells with PBS three times.

12. Incubate the cells with anti-vinculin monoclonal antibody (1:100 in PBS) for 1 hr at room temperature in dark conditions.

13. Wash the cells in PBS three times.

14. Incubate with goat-anti-mouse IgG-biotin conjugate, diluted in 1:100 in PBS for 30 min at room temperature and dark conditions.

15. Wash the cells in PBS three times.

16. Incubate with streptavidin-FITC diluted 1:100 in PBS for 1 hr at room temperature and in dark conditions.

17. Wash the cells in PBS three times.

18. Mount the coverslips with fluorescent mounting medium on glass slides.

19. Analyze the double-stained specimens on laser scanning confocal microscopy.

RESULTS

Tenascin-C Expression in Primary Breast Carcinomas

Tumor grade was assessed on hematoxylin-eosin (H&E) stained sections. Tubule formation, cellular pleiomorphism and mitotic rate were evaluated and scored according to the Elston-Bloom-Richardson scheme: grade 1, grade 2, and grade 3 correspond to well, moderately, and poorly differentiated invasive carcinoma respectively of the breast (Elston, 1991). Areas of tumor-free breast parenchyma showed variable degrees of TN-C positivity. In normal or fibrocystic breast close to the carcinoma, TN-C immunoreaction was found in the vicinity of basement membrane of the ducts, meanwhile the intra- and interlobular stroma was negative (Figure 49A). The TN-C expression of the *in situ* component of carcinomas varied even within one slide. In the majority of cases, the *in situ* component showed an increased periductal expression of TN-C in the form of irregular concentric rings. However, not every duct in the *in situ* component exhibited periductal TN-C positivity. The stromal regions beyond the ducts containing *in situ* carcinoma displayed little or no TN-C reactivity. Infiltrating ductal carcinoma NOS, of 62 cases displayed strong and generally diffuse stromal TN-C immunostaining around tumor cells nests, around clusters of carcinoma cells, or around individual cells (Figure 49B and Figure 50).

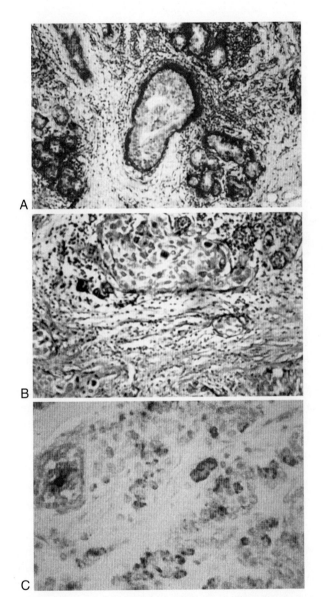

Figure 49 **A:** Fibrocystic breast. Tenascin-C immunoreaction is shown in the basement membrane zone, however, not in the intra- and interlobular stroma. Chromogen substrate New Fuchsin. Magnification 100X. **B:** Invasive ductal carcinoma of the breast. Tenascin-C positivity in the stroma, around tumor cell nests and in the cytoplasms of cancer cells. Chromogen substrate DAB. Magnification 200X. **C:** Invasive ductal carcinoma of the breast. Tenascin-C positivity in the cytoplasm of cancer cells. Chromogen substrate New Fuchsin. Magnification 200X.

In small tumors, the TN-C expression was similar in the central and peripheral stromal regions, whereas in larger tumors the peripheral regions were more intensely reactive. Strong TN-C expression was also observed in thin septa around tumor cell nests. Tumors with different grades did not differ in the distribution of stromal TN-C staining. In 10 out of 62 cases (16.1%), TN-C expression was detected in the cytoplasms of some of the neoplastic epithelial cells as well. These positive cells were scattered in the tumor and rarely formed groups (see Figure 49C). In 12 out of 62 cases (19.3%), increased stromal TN-C immunoreaction was observed in areas infiltrated by chronic inflammatory cells. Areas of normal breast parenchyma within the tumorous breast showed variable degrees of TN-C positivity. An occasional TN-C expression was also found around vascular channels and in the smooth muscle cells.

Tenascin-C Expression in Primary, Recurrent, and Second Primary Breast Carcinomas

Samples of 20 primary breast carcinomas, their recurrence, or second primary tumors occurring in the same or opposite breasts were investigated in our laboratory. Fifteen of 20 primary breast carcinomas were invasive ductal carcinomas, 3 of 15 contained DCIS components, 2 were of mixed type, 1 was invasive lobular carcinoma, 1 was neuroendocrine, and another invasive ductal carcinoma with Paget's disease. The recurrences occurred between 1–10 years following breast conserving surgery, but in 9 cases the local recurrences were noted in the first 2 years after the operation. Typically, the stromal TN-C expression was focal. In 5 cases, TN-C was not detected around all the ducts. TN-C was also observed around tumor cell nests as well as around the DCIS component. No differences were found in the intensity of the stromal TN-C expression in the primary, second primary, or recurrent tumors. Weaker stromal TN-C positivity in some of the recurrent tumors may be explained by the reduced stromal component in the second primary or recurrent tumors. In 4 cases, TN-C was present in the primary breast carcinoma in the apical layer of the epithelium in some normal ducts. There was no evidence whether these cells were going to proliferate or not in a later phase.

Ten of 62 cases (see Material in Part 1 of this chapter) of invasive ductal carcinomas of NOS exhibited TN-C positivity in the carcinoma cells, whereas in the selected group of 20 recurring cases 7 revealed intracellular positivity in the tumor cells. These positive cells were scattered in 1 case of invasive ductal carcinoma with mucinous component; the TN-C positive cells formed

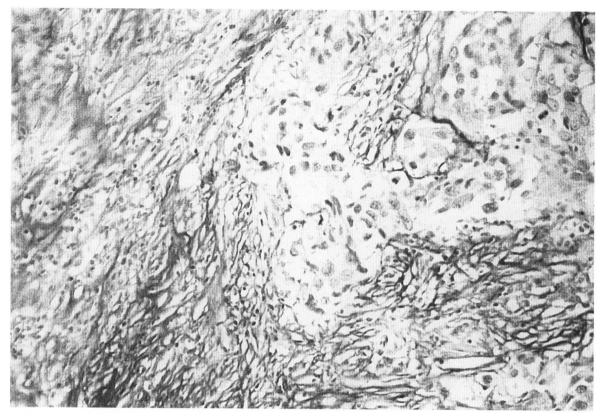

Figure 50 Invasive ductal carcinoma of the breast. Tenascin-C immunoreaction in the stroma of the tumor. Chromogen substrate New Fuchsin. Magnification 200X.

groups at the periphery of the mucinous tumor cell nests. In 1 case, TN-C positivity was found around a duct affected by *in situ* ductal carcinoma exhibiting microinvasion. No TN-C was present in the only case of primary invasive lobular carcinoma. It should be noted that there was no follow-up study in the 62 cases of invasive ductal carcinoma NOS retrieved from our archives. In the 20 cases of primary breast carcinomas, follow-up studies were available, so the recurrences and/or second primary tumors could be followed. It should also be mentioned that no differences were observed between the retrieved and relatively "fresh" formalin-fixed material as far as TN-C immunostaining was concerned. No differences were observed between the two antibodies used.

The Effect of Tenascin-C on Focal Adhesion Plaques

The high stromal expression of TN-C in invasive ductal breast carcinoma, and the TN-C positivity in tumor cells and in the apical epithelial layer of normal ducts provided us with the rationale to investigate the effect of TN-C on the organization of filamentous actin cytoskeleton and focal adhesion complexes in the

highly invasive MDA-MB-435 breast carcinoma cell line. The addition of soluble TN-C to the cell cultures did not cause a general rounding of tumor cells. Soluble TN-C added to the well-spread breast carcinoma cell line in the amount of 10–20 μg/ml had no significant effect on focal adhesion plaques nor on phalloidin stained F-actin stress fibers. No significant effect was observed following longer incubation. A slight tendency for reduction in the number of vinculin plaques was observed when high concentration (100 μg/ml of TN-C) was administered. The question arises as to which extent could the experimental results be applied to human conditions? The extreme high concentrations of TN-C added to adhesion plaques cannot be compared with the small amount of TN-C present in breast carcinomas. The examination was carried out simultaneously using two different TN-C products provided by two companies (Chemicon, El Segundo, CA; Sigma), with both reactions leading to the same result.

DISCUSSION

Immunodetection may have many technical hazards. Fixation with high concentration or more than 24 hr of formaldehyde may alter the conformation of the protein molecule to be analyzed, making it unrecognizable by the antibody. Antigen retrieval methods may renature the protein structure that was altered during fixation (Hayat, 2002). Several antigen retrieval methods are currently in use, depending on the tissue and the protein structure. The presence of endogenous biotin is a known source of nonspecific staining in immunohistochemical methods based on the avidin-biotin systems. Background staining is 1 of the common problems in immunohistochemistry and has numerous causes. The best approach to significantly reduce background staining is to use a blocking protein before the application of primary antibody. In our methodology, the previously described setbacks were taken into consideration.

Tenascin-C Expression in Primary Breast Carcinomas

The progressive alteration in the structure of the mammary epithelium during the development of invasive breast cancer is accompanied by reorganization in the composition of the epithelial cells, periductal, and stromal extracellular matrix. One of the ECM components that has been the subject of increasing investigations as a potentially prognostic and predictive marker for human breast cancer and some other disorders, is TN-C. Expression of this large, hexameric glycoprotein is suppressed in the normal, resting, adult mammary gland and is induced in breast cancer. One of the most prominent features of TN-C reactivity is its restricted location in the normal and fibrocytic breast stroma. The distribution of TN-C in other organs of benign and malignant *in situ* or invasive epithelial tumors is similar to that seen in breast lesions. In colon adenomas, TN-C expression in the basal lamina is increased compared with the normal mucosa (Riedl *et al.*, 1992). The role of TN-C in normal breast ducts remains to be elucidated. In the breast the highest proliferative activity is found in the second half of the menstrual cycle and is accompanied by the highest TN-C levels (Ferguson *et al.*, 1990). Whether TN-C acts as a mitogen in *in vivo* situations, or whether its expression parallels with proliferative events is questionable. In ductal carcinoma *in situ*, TN-C expression was predominantly noted around the ducts containing DCIS.

Jahkola *et al.*, (1998a) investigated 89 DCIS for TN-C expression and reported positive correlation of increased periductal expression of TN-C in comedo type with development of microinvasion. It was shown with *in situ* hybridization that TN-C is produced by the carcinoma cells of early invasive nests of DCIS and that TN-C mRNA positive cells are particularly frequent at the margins of the carcinoma cell nests (Lightner *et al.*, 1994). Studies by Jahkola *et al.*, (1998a) concluded that stromal or moderate to strong periductal TN-C expression in DCIS may be related to early invasion. A weak or absent tenascin-C expression in DCIS represents a subgroup of tumors with indolent benign behavior. In invasive ductal carcinomas, TN-C expression was found predominantly in the stroma. The TN-C in invasive carcinomas may be the product of cells like fibroblasts or produced by malignant or normal epithelial cells. It has been proposed that the stromal TN-C in carcinomas inhibits tumor growth by creating cell boundaries, inhibiting cell migration, and regulating cancer cell growth (Natali *et al.*, 1991). The presence of TN-C in the tumor cells observed in 10 of 62 primary tumors, especially at the margins of tumor cell nests, indicates that TN-C may be involved in cancer cell spreading, thus resulting in unfavorable prognosis.

Based on the antiadhesive properties of TN-C, it has been suggested that TN-C produced by tumor cells in breast carcinoma might stimulate cell movement and subsequent metastasis. (Shoji *et al.*, 1993). Researchers investigating the capacity of malignant tumor cell lines to secrete TN-C have shown that tenascin was secreted by some tumor cells, specially those of epithelial origin (Jessen *et al.*, 1996). Based on both other studies and our own findings, it is controversial that the intracellular and stromal TN-C acts differently (Kádár *et al.*, 2002).

Previous studies emphasized the association between inflammation and TN-C expression (Jahkola *et al.*, 1996). In 12 out of 62 cases of IDC NOS, increased stromal TN-C immunoreaction was found in areas infiltrated by chronic inflammatory cells. The relationship between these phenomena is not fully understood. It is questionable whether there is a causal relationship between the TN-C overexpression and chronic inflammatory cell infiltration or if they are different manifestations of the same process. The close association between stromal TN-C expression and vascularity shown in this study raises the question whether endothelial or smooth muscle cells are capable of producing TN-C (Tõkés *et al.*, 1999). It was found that endothelial cells may also be responsible for TN-C synthesis; however, the mechanism is not fully elucidated (Webersinke *et al.*, 1992). Sharifi *et al.* (1992) reported that smooth muscle cells are capable of producing TN-C under *in vitro* conditions.

How do cellular interactions with TN-C affect signaling networks and downstream cellular processes? The TN-C binds and interacts with integrin receptors, such as β1 and αvβ3 integrins, which are receptors known to control the growth of breast tumor cells and microvessels. In addition TN-C modulates the activity of EGFR and up-regulates the expression of growth-associated genes such as cyclin D1 and C-Myc in mammary epithelial cells (Jones *et al.*, 2000). This coexpression is important because TN-C expression in breast carcinoma is incompatible with the presence of an intact basement membrane. Destruction of the basement membrane by stromelysin results in the appearance of TN-C in the stromal ECM. It was also demonstrated that TN-C up-regulates the expression of ECM degrading proteinases in fibroblasts (Jones *et al.*, 2000).

Tenascin-C Expression in Primary, Recurrent, and Second Primary Carcinomas

The presence of TN-C in tumor cells and in certain epithelial cells of the ducts, around tumor cell nests, and around *in situ* carcinomas in our cases indicates that TN-C might be involved in cancer cell spreading, resulting in unfavorable prognosis. TN-C positivity in the apical layer of normal ducts is of questionable importance. Should TN-C be secreted by normal epithelial cells or are these cells in the earlier stage of transformation into cancerous cells? Are these tumors of higher risk or not? Studies on higher numbers of cases (143) suggest that TN-C positivity in the invasion zone is the only prognostic marker predicting local recurrences (Jahkola *et al.*, 1998b). Yoshida *et al.*, (1995) found that cytoplasmic TN-C positivity indicates poor prognosis. Jahkola *et al.*, (1998b) found that the expression of TN-C in the invasion border was the only marker that predicted local recurrence. It was suggested that patients with carcinomas that do not express TN-C at the invasion border do not need adjuvant treatment. In a study by Ishihara *et al.*, (1995) the analyzed breast carcinomas were classified into three groups: cancer cell tenascin+/stromal tenascin+; cancer cells tenascin negative/stromal tenascin+; and cancer cells tenascin negative/stromal tenascin negative. They found that the group expressing TN-C positivity in both cancer cells and stroma exhibited an increased frequency of lymph node metastasis and poor outcome.

The Effect of Soluble Tenascin-C on Focal Adhesion Plaques in Breast Cancer Cell Culture

One of the consequences of modified ECM in tumors may be the increased invasive potential of tumor cells (Jessen *et al.*, 1996). TN-C and thrombospondin are considered to favor invasion and modulate sprouting of endothelial cells (Canfield *et al.*, 1995). The number of focal adhesion in bovine aortic endothelial cells is diminished by TN-C,

secreted protein, acidic and rich in cystein (SPARC), and thrombospondin (Sage *et al.*, 1991). However, the effect of TN-C is undoubtedly more complex, and TN-C was proven to control cell migration. Cell culture experiments demonstrated that both epithelial and fibroblasts cells spread less easily on a substrate coated with TN-C (Sage *et al.*, 1991; Sriramarao *et al.*, 1993). Alteration in cell shape after TN-C treatment was reported in several studies and seemed to be related directly to the number and distribution of focal adhesions on the cell surface (Sage *et al.*, 1991). This structure links ECM glycoproteins and proteoglycans to the cytoskeleton and its associated components via several types of integrins. Interaction of TN-C with a cell surface molecule might trigger a cytoplasmatic signal that alters the cytoskeleton. Because TN-C alters the integrity of the actin cytoskeleton, it is possible that this molecule modulates epidermal growth factor signaling through its effect on the actin cytoskeleton and focal adhesion molecules (Murphy-Ullrich *et al.*, 1991). Publications regarding the attachment behavior of some cell cultures have shown that cells secreting TN-C alone are round and float in suspension, whereas cells secreting FN only attach and spread quickly (Chiquet-Ehrismann *et al.*, 1995). Some authors reported that TN-C reduces focal adhesion integrity in confluent endothelial and smooth muscle cell cultures (Hahn *et al.*, 1995).

We observed a slight tendency for reduction in the number of vinculin plaques at a concentration of 100 µg/ml of TN-C treatment on MDA-MB-435 cell lines. A general rounding of cells was not observed in our experiment. Murphy-Ullrich *et al.* (1991) found that TN-C causes a dose-dependent reduction in the number of focal adhesion positive cells to approximately 50% of albumin-treated controls. They reported a maximum response with 20–60 µg/ml of soluble TN-C and the reduction in focal adhesion in tenascin-treated cells was observed after 10 min. Cells adhere to the ECM proteins FN and TN-C in part via the interaction of certain integrins with the RGD sequence. Interaction between FN type III domains and beta 1 integrins mediates focal adhesion kinase phosphorylation and focal contact adhesion (Chi-Rosso *et al.*, 1997). How the number of FN type III domains of TN-C influences the binding capacity of TN-C is questionable. One or more of the FN type III repeats bind to FN and thereby inhibits attachment of primary fibroblasts and T lymphocytes to immobilized FN (Chung *et al.*, 1995). The alternatively spliced region, Tn FN-III A-D, induces loss of focal adhesion and inhibits proliferation of endothelial cells through interaction with annexin II (Chung *et al.*, 1995), and promotes proliferation of hemopoetic precursor cells in the bone marrow

(Seiffert *et al.*, 1998). According to our findings, administration of TN-C in cell culture does not contribute to the reduction of focal adhesion plaques. Consequently, TN-C on its own in *in vitro* conditions does not promote cell spreading.

In conclusion, although a large number of studies have been completed regarding the structure, function, and regulation of TN-C, many questions still remain open. With respect to the TN gene, it might be fruitful to look for other members of the tenascin family that may considerably increase our knowledge of this interesting gene family. Little is known about the molecular mechanisms involved in TN-C expression. Studying the dynamics of the TN-C promoter has led to the identification of several distinct functional elements. For TN-C, it has been established that there is an AP1 site and that promoter activation can be triggered by Hox-genes (Jones *et al.*, 1990). Although the regulation of TN-C by homeobox proteins appears to be a general theme, responses of ATTA (nucleic acids) elements at the TN-C promoter will depend on homeobox proteins in a given cell type. The signal transduction pathways and responses initiated by TN-C binding to different cell surfaces are complex and diverse, reflecting the existence of numerous receptors that produce different signal inputs. These signals result in different gene expression. The permanent open question is whether TN-C or other ECM molecules interact with the same receptor that may initiate identical signal transduction pathways? Efforts are also made to identify soluble factors that stimulate an increased TN-C messenger ribonucleic acid expression.

A rapidly emerging theme in tenascin biology is its regulation by biomechanical factors. There are *in vivo* evidences that support the idea that TN-C helps the tissue adapt to physical stress. Localization and identification of the TN-C glycoprotein may represent an important approach in identifying functions of TN-C. Tremendous efforts have been made to recreate three-dimensional tissue microenvironments composed of multiple ECM proteins, including TN proteins. For example, colocalization of tenascin-R with members of the aggrecan family results in the formation of ECM architecture that is biologically significant for neuronal conductivity. It appears very promising to focus on investigations on the expression of specific TN-C splice variants. Although there have been studies on the organ-specific expression of TN-C splice variants, up to now little is known about the function of the different splice variants. Irrespective of how TN-C is regulated and acts in the neoplastic mammary gland, the observation that TN-C is expressed in human breast cancer indicates that it may represent a tumor-associated antigen with future promise as a target for immunotherapy. Phase I clinical trials have already been conducted with 131 iodine labeled monoclonal antibodies to TN-C in patients with malignant glioma (Cokgor *et al.*, 2000). The early results indicate that ablation of TN-C-positive tumor cells rapidly stabilizes the disease without causing significant toxic effects.

References

Adams, M., Jones, J.L., Walker, R.A., Pringle, J.H., and Bell, S.C. 2002. Changes in tenascin-C isoform expression in invasive and preinvasive breast disease. *Cancer Res. 62:*3289–3297.

Bourdon, M.A., and Rouslahti, E. 1989. Tenascin mediates cell attachment through an RGD-dependent receptor. *J. Cell Biol. 108:*1149–1155.

Bristow, J., Tee, M.K., Gittelman, S.E., Mellon, S.H., and Miller, W.L. 1993. Tenascin-X: A novel extracellular matrix protein encoded by the human XB gene overlapping P450c21B. *J. Cell Biol. 122:*265–268.

Canfield, A.E., and Schor, A.M. 1995. Evidence that tenascin and thrombospondin-1 modulate sprouting of endothelial cells. *J. Cell Sci. 108:*797–809.

Chiquet-Ehrismann, R., Hagios, C., and Schenk, S. 1995. The complexity in regulating the expression of tenascins. *BioEssays 17:*873–878.

Chiquet-Ehrismann, R., Kalla, P., Pearson, C.A., Beck, K., and Chiquet M. 1988. Tenascin interferes with fibronectin action. *Cell 53:*383–390.

Chi-Rosso, G., Gotwals, P.J., Yang, J., Ling, L., Jiang, K., Chao, B., Baker, D.P., Burkly, L.C., Fawell, S.E., and Koteliansky, V.E. 1997. FN type III repeats mediate RGD-independent adhesion and signaling through activated beta1 integrins. *J. Biol. Chem. 272:*31447–31452.

Chung, C.Y., Murphy-Ullrich, J.E., and Erickson, H.P. 1996. Mitogenesis, cell migration, and loss of focal adhesion induced by TN-C interacting with its cell surface receptor, annexin II. *Mol. Biol. Cell 7:*883–892.

Chung, C.Y., Zardi, L., and Erickson, H.P. 1995. Binding of TN-C to soluble fibronectin and matrix fibrils. *J. Biol. Chem. 270:*29012.

Cokgor, I., Akabani, G., Kuan, C.T., Friedman, H.S., Friedman, A.H., Coleman, R.E., McLendon, R.E., Bigner, S.H., Zhao, X.G., Garcia-Turner, A.M., Pegram, C.N., Wikstrand, C.J., Shafman, T.D., and Herndon, J.E. 2000. Phase I trial results of iodine-131-labeled antitenascin monoclonal antibody 81C6 treatment of patients with newly diagnosed malignant gliomas. *J. Clin. Oncol. 18:*3862–3872.

Elston, C.W., and Ellis, I.O. 1991. Pathological prognostic factors in breast cancer. I. The value of histological grade in breast cancer: Experience from a large study with long-term follow up. *Histopathology 19:*403–410.

Ericson, H.P., and Bourdon, M.A. 1989. Tenascin: An extracellular matrix protein prominent in specialized embryonic tissues and tumors. *Ann. Rev. Cell Biol. 5:*71–91.

Ericson, H.P., and Iglesias, J.L. 1984. A six-armed oligomer isolated from cell surface fibronectin preparations. *Nature 311:*267–269.

Faissner, A., and Kruse, J. 1990. J1/tenascin is a repulsive substrate for central nervous system neurons. *Neuron 5:*627–637.

Ferguson, J.E., Schor, A.M., Howell, A., and Ferguson, M.W. 1990. Tenascin distribution in the normal breast is altered during the menstrual cycle and in carcinomas. *Differentiation* *42:*199–297.

Fuss, B., Wintergerst, E.S., Bartsch, U., and Schachner, M. 1993. Molecular characterization and *in situ* mRNA localization to the neural recognition molecule J1 160/180, a modular structure similar to tenascin. *J. Cell Biol. 120:*1237–1249.

Hagios, C., Koch, M., Spring, J., Chiquet, M., and Chiquet-Ehrisman, R. 1996. Tenascin-Y: A protein of novel domain structure is secreted by differentiated fibroblasts of muscle connective tissue. *J. Cell Biol. 134:*1499–1512.

Hahn, A.W., Kern, F., Jonas, U., John, M., Buhler, F.R., and Resink T.J. 1995. Functional aspect of vascular tenascin-C expression. *J. Vasc. Res. 32:*162–174.

Hahn, A.W., Kern, F., Jonas, U., John, M., Buhler, F.R., and Resink, T.J. 1995. Functional aspect of vascular TN-C expression. *J. Vasc. Res. 32:*162–174.

Harada, T., Shinohara, M., Nakamura, S., and Oka, M. 1994. An immunohistochemical study of the extracellular matrix in oral squamous cell carcinoma and its association with invasive and metastatic potential. *Virch. Arch. 425:*257–266.

Hayat, M.A. 2002. Microscopy, Immunohistochemistry, and Antigen Retrieval Methods. New York: Kluwer Academic/Plenum Publishers.

Howeedy, A.A., Virtanen, I., Laitinene, L., Gould, N.S., Koukoulis, G.K., and Gould, V.E. 1990. Differential distribution of tenascin in the normal, hyperplastic, and neoplastic breast. *Lab. Invest. 63:*798–806.

Ishihara, A., Yoshida, T., Tamaki, H., and Sakakura, T. 1995. Tenascin expression in cancer cells and stroma of human breast cancer and its prognostic significance. *Clin. Cancer Res. 1:*1035–1041.

Iskaros, B.E., Sison, C.P., and Hajdu, S.I. 2000. Tenascin patterns of expression in duct carcinoma in situ of breast. *Ann. Clin. Lab. Sci. 30:*267–271.

Jahkola, T., Toivonen, T., Smitten, K., Blomqvist, C., and Virtanen, I. 1996. Expression of tenascin in invasion border of early breast cancer correlates with higher risk of distant metastasis. *Int. J. Cancer (Pred. Oncol.) 69:*445–447.

Jahkola, T., Toivonen, T., Nordling, S., Smitten, K., and Virtanen, I. 1998a. Expression of tenascin-C in intraductal carcinoma of human breast: Relationship to invasion. *Eur. J. Cancer 34:*1687–1692.

Jahkola, T., Toivonen, T., Virtanen, I., von Smitten, K., Nordling, S., von Boguslawski, K., Haglund, C., Nevanlinna, H., and Blomqvist, C. 1998b. Tenascin-C expression in invasion border of early breast cancer: A predictor of local and distant recurrence. *Brit. J. Cancer 78:*1507–1513.

Jessen, L.R., Petersen, W.O., and Bissell, M.J. 1996. Cellular changes involved in conversion of normal to malignant breast: Importance of the stromal reaction. *Physiol. Rev. 76:*69–125.

Jones, F.S., Crossin, K.L., Cunningham, B.A., and Edelman, G.M. 1990. Identification and characterization of the promoter for the cytotactin gene. *Proc. Natl. Acad. Sci. USA 89:*6497–6501.

Jones, F.S., and Jones, P.L. 2000a. The tenascin family of ECM glycoproteins: Structure, function, and regulation during embryonic development and tissue remodeling. *Dev. Dyn. 218:*235–259.

Jones, P.L., and Jones, F.S. 2000b. Tenascin-C in development and disease: Gene regulation and cell function. *Matrix Biol. 19:*581–596.

Joshi, P., Chung, C.Y., Aukhil, I., and Erikson, H.P. 1993. Endothelial cells adhere to RGD domain and the fibrinogen-like terminal knob of tenascin. *J. Cell Sci. 106:*389–400.

Kádár, A., Tõkés, A-M., Kulka, J., and Robert, L. 2002. Extracellular matrix components in breast carcinomas. *Semin. Cancer Biol. 12:*243–257.

Kawakatsu, H., Shiurba, R., Obara, M., Hiraiwa, H., Kusakabe, M., and Sakakura, T. 1992. Human carcinoma cells synthesize and secrete tenascin in vitro. *Jpn. J. Cancer Res. 83:*1073–1080.

Lightner, V.A., Gumkowski, F., Bigner, D.D., and Erickson, H.P. 1989. Tenascin/hexabranchion in human skin: Biochemical identification and localization by light and electron microscopy. *J. Cell Biol. 108:*2483–2493.

Lightner, V.A., Marks, J.R., and McCachren, S.S. 1994. Epithelial cells are an important source of tenascin in normal and malignant human breast tissue. *Exp. Cell Res. 210:*177–184.

Murphy-Ullrich, J.E., Lightner, V.A., Aukhil, I., Yan, Y.Z., Ericson, H.P., and Höök, M. 1991. Focal adhesion integrity is downregulated by alternatively spliced domain of human tenascin. *J. Cell Biol. 115:*1127–1136.

Natali, P.G., Nicotra, M.R., Bigotti, A., Castellani, P., Risso, A.M., and Zardi, L. 1991. Comparative analysis of the expression of the extracellular matrix protein tenascin in normal human fetal, adult, and tumor tissues. *Int. J. Cancer 47:*811–816.

Riedl, S., Faissner, A., Schlag, P., Herbay, A.V., Koretz, M., and Möller, P. 1992. Altered content and distribution of tenascin in colitis, colon adenoma, and colorectal carcinoma. *Gastroenterology 103:*400–406.

Rocchi, M., Archidiacono, N., Romeo, G., Saginati, M., and Zardi, L. 1991. Assignment of the gene for human tenascin in the region q32-q34 of chromosome 9. *Hum. Genet. 86:*621–623.

Sage, E.H., and Bornstein, P. 1991. Extracellular proteins that modulate cell-matrix interaction. SPARC, tenascin, and thrombospondin. *J. Biol. Chem. 266:*14831–14834.

Salmivirta, M., Elenius, K., Vainio, S., Hofer, U., Chiquet-Ehrismann, R., Thesleff, I., and Jalkanen, M. 1991. Syndecan from embryonic tooth mesenchyme binds tenascin. *J. Biol. Chem. 266:*7733–7739.

Seiffert, M., Beck, S.C., Schermutzi, F., Müller, C.A., Erickson, H.P. and Klein, G. 1998. Mitogenetic and adhesive effects of TN-C on human hematopoietic cells are mediated by various functional domains. *Matrix Biol. 17:*47–63.

Sharifi, B.G., LaFleur, D.W., Pirola, C.J., Forrester, J.S., and Fagin, J.A. 1992. Angiotensin II regulates tenascin gene expression in vascular smooth muscle cells. *J. Biol. Chem. 267:*23910–23915.

Shoji, T., Kamiya, T., Tsubura, A., Hamada, Y., Hatano, T., Hioki, K., and Morii, S. 1993. Tenascin staining positivity and the survival of patients with invasive breast carcinoma. *J. Surg. Res. 55:*295–297.

Soini, Y., Paakko, P., Virtanen, I., and Lehto, V.P. 1992. Tenascin in salivary tumours. *Virchows Arch. A. Pathol. Anat. 421:*217–222.

Sriramarao, P., Mendler, M., and Bourdon, M.A. 1993. Endothelial cell attachment and spreading on human tenascin is mediated by $\alpha 2 \beta 1$ and $\alpha v \beta 3$ integrins. *J. Cell. Sci. 105:*1001–1002.

Thesleff, I., Mackie, E.J., Vaino, S., and Chiquet-Ehrismann, R. 1987. Changes in the distribution of tenascin during tooth development. *Development 101:*289–296.

Tõkés, A.-M., Hortoványi, E., Kulka, J., Jäckel, M., Kerényi, T., and Kádár, A. 1999. Tenascin expression and angiogenesis in breast cancers. *Pathol. Res. Pract. 195:*821–828.

Tremble, P., Chiquet-Ehrismann, R., and Werb, Z. 1994. The extracellular matrix ligands fibronectin and tenascin collaborate in regulating collagenase gene expression in fibroblasts. *Mol. Biol. Cell 5:*439–453.

Vollmer, G. 1997. Biologic and oncologic implications of tenascin-C/hexabranchion proteins. *Crit. Rev. Oncol. Hematol. 25:*187–210.

Weber, P., Montag, D., Schachner, M., and Bernhardt, R.R. 1998. Zebrafish tenascin-W: A new member of the tenascin family. *J. Neurobiol. 35:*1–16.

Webersinke, G., Bauer, H., Amberger, A., Zach, O., and Bauer, H.C. 1992. Comparison of gene expression of extracellular matrix molecules in brain microvascular endothelial cells and astrocytes. *Biochem. Biophys. Res. Comm. 189:*877–884.

Wiseman, B.S., and Werb, Z. 2002. Stromal effects on mammary gland development and breast cancer. *Science 296:* 1046–1049.

Yoshida, T., Ishihara, A., Hirokawa, Y., Kusakabe, M., and Sakakura, T. 1995. Tenascin in breast cancer development: Is epithelial tenascin a marker for poor prognosis? *Cancer Lett. 90:*65–73.

Young, S.L., Chang, L.Y., and Ericson, H.P. 1994. Tenascin-C in rat lung: Distribution, ontogeny, and role in branching morphogenesis. *Dev. Biol. 161:*615–625.

10

Immunohistochemistry of Adhesion Molecule CEACAM1 Expression in Breast Carcinoma

Ana-Maria Bamberger and Christoph M. Bamberger

Introduction

Cell adhesion molecules are important regulators of tissue architecture and cellular polarity, which can also modulate proliferation, differentiation, and invasion processes in virtually all tissues of the human body (Crossin, 2002). They can function either as cell-cell adhesion molecules or as cell-matrix adhesion molecules, governing proper cell-cell and cell-matrix interactions, respectively. In contrast to normal tissues, malignant tumors are characterized by a disruption of these processes. Thus, cell adhesion molecules have been shown to be expressed and/or to function abnormally in many types of tumors (Crossin, 2002).

Carcinoembryonic antigen cell adhesion molecule 1 (CEACAM, synonyms: CD66a, biliary glycoprotein [BGP]) is a member of the carcinoembryonic antigen (CEA) family, which in turn is part of the immunoglobulin superfamily (Thompson et al., 1991). The adhesion molecule CEACAM1 is the human homologue of rat C-CAM (Lin et al., 1989). It can bind to other members of the CD66 cluster (Rojas et al., 1990), as well as to the cell adhesion molecule integrin beta3 (Brümmer et al., 2001) and the actin-based cytoskeleton (Ebrahimnejad et al., 2000). Furthermore, it has been suggested as a ligand for E-selectin (Kuijpers et al., 1992). It contains intracellular motifs potentially involved in signal transduction (Obrink, 1997) and has been shown to associate with pp60 c-src (Brümmer et al., 1995). The physiological function of CEACAM1 appears to be of a pleiotropic kind (e.g., it can act as a receptor for bacterial surface molecules, such as the opacity-associated adhesins of *Neisseria gonorrhoea*) (Booth et al., 2003). It is also involved in insulin degradation and, thus, maintenance of insulin sensitivity (Najjar, 2002), and it can both stimulate and inhibit angiogenesis in an apparently tissue-dependent manner (Ergün et al., 2000; Volpert et al., 2002). To date, CEACAM1 has been shown to be expressed in epithelium and cells of the myeloid lineage in numerous tissues, including liver, colon, bladder, prostate, breast, and endometrium (Prall et al., 1996; Thompson et al., 1991). In addition, we have demonstrated specific expression of CEACAM1 in the invasive part of the human placenta (i.e., the extravillous trophoblast) (Bamberger et al., 2000).

The role of CEACAM1 in tumor growth, invasion, and metastasis has not been fully elucidated as yet. This role seems to be rather complex and most probably tissue-dependent. Loss of CEACAM1 expression

337

Copyright © 2004 by Elsevier (USA)
All rights reserved.

seems to occur early in the development of colon carcinoma, (i.e., at the premalignant adenoma stage) (Neumaier *et al.*, 1993; Nollau *et al.*, 1997). Zhang *et al.* (1997) demonstrated that CEACAM1 is, in fact, among the 20 genes most frequently down-regulated in colon cancer. Overexpression of CEACAM1 inhibits the growth of this type of cancer, with the Ser503 residue of the cytoplasmic domain being crucial for this effect (Fournes *et al.*, 2001). Similarly, in prostate cancer, CEACAM1 can act as a tumor suppressor protein. As in colonic carcinomas, down-regulation of CEACAM1 occurs early in the pathogenesis of prostate cancer (Lin *et al.*, 1999) and is associated with a loss of cell polarity (Busch *et al.*, 2002). The CEACAM1 overexpression reduces the tumorigenic potential of prostate carcinomas (Lin *et al.*, 1999), and again, the cytoplasmic domain is necessary and sufficient for growth inhibition (Estrera *et al.*, 1999). Finally, down-regulation of CEACAM1 and loss of its polar expression pattern has also been reported for endometrial cancer (Bamberger *et al.*, 1998).

Although these examples indicate an overall tumor suppressive role of CEACAM1, there are other tissues in which this adhesion molecule appears to have quite the opposite effect. As mentioned, CEACAM1 is specifically expressed in the invasive part of the human trophoblast, indicating a proinvasive role in this semi-tumorous tissue (Bamberger *et al.*, 2000). In lung cancer, CEACAM1 expression correlates with a poor prognosis (Laack *et al.*, 2002), and in malignant melanoma it predicts a higher rate of metastasis (Thies *et al.*, 2002). This chapter summarizes immunohistochemistry data on CEACAM1 expression in benign and malignant breast tissue and focuses on our published studies on this issue (Bamberger *et al.*, 2002; Riethdorf *et al.*, 1997).

MATERIALS

1. Fixative: 4% formalin in phosphate-buffered saline (PBS).
2. Slides coated with APES (3-aminopropyltriethoxysilane).
3. Xylene.
4. Graded alcohol: Ethanol at different concentrations in Tris-buffered saline (TBS).
5. TBS, 50 mM Tris, 150 mM NaCl, pH 7.4.
6. 10 mM citrate (pH 6.0).
7. Normal goat serum (Dako, Carpinteria, CA), diluted 1:20 in TBS.
8. Monoclonal mouse anti-CEACAM1 antibody Mab 4D1/C2, 4 μg/ml in TBS (Stoffel *et al.*, 1993).
9. Normal (nonimmune) murine serum (Dako).

10. Anti-mouse, alkaline phosphatase-linked secondary antibody (Dako).
11. Universal alkaline phosphatase anti–alkaline phosphatase (APAAP) detection kit (Dako).
12. Hematoxylin.

METHODS

1. Cut serial sections (4–6 μm thick) from paraffin blocks.
2. Mount on APES-coated slides.
3. Deparaffinize in xylene.
4. Rehydrate in graded alcohol to TBS.
5. Microwave for 5 × 2 min to 10 mM citrate (pH 6.0).
6. Cool down for 20 min to room temperature.
7. Wash in TBS.
8. Incubate for 30 min with normal goat serum (1:20 in TBS) at room temperature.
9. Incubate overnight at 4°C with Mab 4D1/C2 (4 μg/ml).
10. For controls, incubate overnight at 4°C with nonimmune murine serum.
11. Wash with TBS and carry out APAAP detection protocol according to the manufacturer's instructions (Dako).
12. Counterstain with hematoxylin.

RESULTS AND DISCUSSION

In the studies summarized here, CEACAM1 expression was analyzed by immunohistochemistry in normal mature breast tissue, fibrocystic lesion, adenosis, radial scar, ductal and lobular hyperplasia, atypical ductal and lobular hyperplasia, noninvasive ductal and lobular carcinomas, and invasive carcinomas of different types and gradings (Riethdorf *et al.*, 1997). The CEACAM1 expression was correlated with the expression of different molecules previously shown to play a role in cell cycle regulation and/or in the development and progression of breast cancer (Bamberger *et al.*, 2002) (i.e., cyclin D1, cyclin E, p16, p21, p27, Retinoblastoma [Rb] protein, Rb2, estrogen receptor, progesterone receptor (PR)-A, PR-B, Ki-67, and HER-2/neu). In these studies, anti-CEACAM1 monoclonal antibody, Mab D1/C2, was employed (Stoffel *et al.*, 1993). This antibody has also been extensively used by our group to study the expression of CEACAM1 in other tissues (Bamberger *et al.*, 1998, 2000, 2001; Brümmer *et al.*, 2001).

In normal tissue and benign lesions of the breast, CEACAM1 was consistently expressed at the apical sites of epithelial cells and in myoepithelia, whereas it was absent or restricted to some apical membranes within the ductal tree (Riethdorf *et al.*, 1997) (Figure 51A). The specific staining of myoepithelia was most evident in pseudoinfiltrative radial scars and sclerosing adenosis. However, the apical expression of CEACAM1 disappeared with the development of the malignant phenotype in noninvasive and invasive carcinomas, and changed gradually from low- to high-grade noninvasive carcinomas into a predominant uniform membrane location around the atypical cells (Figure 51B). The CEACAM1 expression was irregular in intensity and distribution. The native apical CEACAM1 staining was partially preserved in some highly differentiated invasive carcinomas with a better prognosis, such as tubular and papillary carcinomas. Reduction or loss of

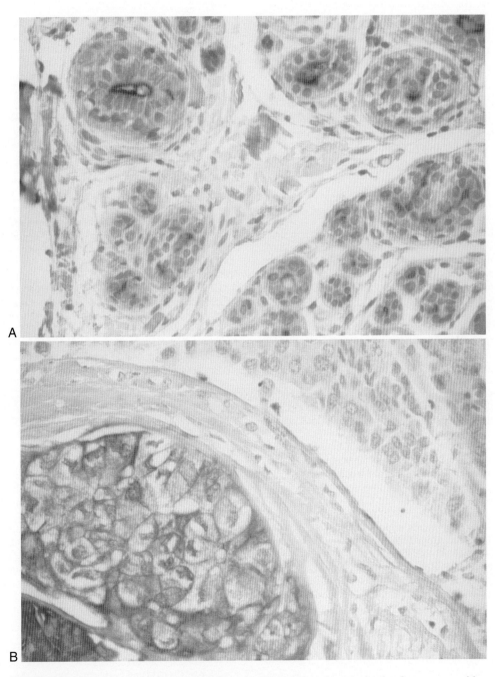

Figure 51 A: Immunohistochemical detection of CEACAM1 at the apical pole of mammary epithelial cells 40×. **B:** Loss of polar expression pattern of CEACAM1 in a mammary carcinoma. Note the membrane location around the atypical cells (40×).

CEACAM1 expression was observed in only 30% of the tumors. These findings indicate that a change in expression patterns rather than loss of CEACAM1 expression coincides with the development of the malignant phenotype (Riethdorf *et al.*, 1997). These findings were later confirmed by Huang *et al.* (1998), who extended them to the RNA level. Consistent with these findings, the same group showed that reinstitution of CEACAM1 expression in CEACAM1-negative MCF-7 breast cancer cells confers the ability to form luminar structures (Huang *et al.*, 1999), apparently by inducing apoptosis in the central cells within the alveolar structures (Kirshner *et al.*, 2003). We demonstrated that CEACAM1 expression in human breast cancer samples significantly correlated with the expression levels of the cell cycle inhibitors Rb, Rb2, and p27 (Bamberger *et al.*, 2002). Therefore, as in colon, prostate, and endometrium, and as opposed to lung, melanoma, and trophoblast, the role of CEACAM1 in the human breast appears to be basically that of a tumor suppressor protein.

References

Bamberger, A.-M., Kappes, H., Methner, C., Rieck, G., Brümmer, J., Wagener, C., Löning, L., and Milde-Langosch, K. 2002. Expression of the adhesion molecule CEACAM1 (CD66a, BGP, C-CAM) in breast cancer is associated with the expression of the tumor-suppressor genes Rb, Rb2, and p27. *Virchows Arch.* 440:139–144.

Bamberger, A.-M., Riethdorf, L., Nollau, P., Naumann, M., Erdmann, I., Götze, J., Brümmer, J., Schulte, H.M., Wagener, C., and Löning, Y. 1998. Dysregulated expression of CD66a (BGP, C-CAM), an adhesion molecule of the CEA family, in endometrial cancer. *Am. J. Pathol.* 152:1401–1406.

Bamberger, A.-M., Sudahl, S., Löning, T., Wagener, C., Bamberger, C.M., Drakakis, P., Coutifaris, C., and Makrigiannakis, A. 2000. The adhesion molecule CEACAM1 (CD66a, C-CAM, BGP) is specifically expressed by the extravillous intermediate trophoblast. *Am. J. Pathol.* 156:1165–1170.

Bamberger, A.M., Sudahl, S., Wagener, C., and Löning, T. 2001. Expression pattern of the adhesion molecule CEACAM1 (C-CAM, CD66a, BGP) in gestational trophoblastic lesions. *Int. J. Gynecol. Pathol.* 20:160–165.

Booth, J.W., Telio, D., Liao, E.H., Mc Caw, S.E., Matsuo, T., Grinstein, S., and Gray-Owen, S.D. 2003. Phosphatidylinositol 3-kinases in carcinoembryonic antigen-related cellular adhesion molecule-mediated internalization of *Neisseria gonorhoeae*. *J. Biol. Chem.* 278:14037–14045.

Brümmer, J., Ebrahimnejad, A., Flayeh, R., Schumacher, U., Löning, T., Bamberger, A.-M., and Wagener, C. 2001. *Cis* interaction of cell adhesion molecule CEACAM1 with integrin beta3. *Am. J. Pathol.* 159:537–546.

Brümmer, J., Neumaier, M., Göpfert, C., and Wagener, C. 1995. Association of pp60c-src with biliary glycoprotein (CD66a), an adhesion molecule of the carcinoembryonic antigen family down-regulated in colonic carcinomas. *Oncogene* 11:1649–1655.

Busch, C., Hanssen, T.A., Wagener, C., and Obrink, B. 2002. Down-regulation of CEACAM1 in human prostate cancer: Correlation

with loss of cell polarity, increased proliferation rate, and Gleason grade 3 to 4 transition. *Hum. Pathol.* 33:290–298.

Crossin, K.L. 2002. Cell adhesion molecules activate signaling networks that influence proliferation, gene expression, and differentiation. *Ann. N.Y. Acad. Sci.* 961:159–160.

Ebrahimnejad, A., Flayeh, R., Unteregger, G., Wagener, C., and Brümmer, J. 2000. Cell adhesion molecule CEACAM1 associates with paxillin in granulocytes and epithelial and endothelial cells. *Exp. Cell. Res.* 260:365–373.

Ergün, S., Kilic, N., Ziegeler, G., Hansen, A., Nollau, P., Götze, J., Wurmbach, J.H., Horst, A., Weil, J., Fernando, M., and Wagener, C. 2000. CEA-related cell adhesion molecule 1: A potent angiogenic factor and a major effector of vascular endothelial growth factor. *Mol. Cell* 5:311–320.

Estrera, V.T., Luo, W., Phan, D., Earley, K., Hixson, D.C., and Lin, S.H. 1999. The cytoplasmic domain of C-CAM1 tumor suppressor is necessary and sufficient for suppressing the tumorigenicity of prostate cancer cells. *Biochem. Biophys. Res. Commun.* 263:797–803.

Fournes, B., Sadekova, S., Turbide, C., Letourneau, S., and Beauchemin, N. 2001. The CEACAM1-L Ser503 residue is crucial for inhibition of colon cancer cell tumorigenicity. *Oncogene* 20:219–230.

Huang, J., Hardy, J.D., Sun, Y., and Shively, J.E. 1999. Essential role of biliary glycoprotein (CD66a) in morphogenesis of the human mammary epithelial cell line MCF10F. *J. Cell Sci.* 112:4193–4205.

Huang, J., Simpson, J.F., Glackin, C., Riethdorf, L., Wagener, C., and Shively, J.E. 1998. Expression of biliary glycoprotein (CD66a) in normal and malignant breast epithelial cells. *Anticancer Res.* 18:3203–3212.

Kirshner, J., Chen, C.-J., Liu, P., Huang, J., and Shively, J.E. 2003. CEACAM1-4S, a cell-cell adhesion molecule, mediates apoptosis and reverts mammary carcinoma cells to a normal morphogenic phenotype in a 3D culture. *Proc. Natl. Acad. Sci. USA* 100:521–526.

Kuijpers, T.W., Hoogerwerf, M., Van der Laan, L.J., Nagel, G., Van der Schoot, C.E., Grunert, F., and Roos, D. 1992. CD66 nonspecific cross-reacting antigens are involved in neutrophil adherence to cytokine-activated endothelial cells. *J. Cell Biol.* 118:457–466.

Laack, E., Nikbakht, H., Peters, A., Kugler, C., Jasiewicz, Y., Edler, L., Brümmer, J., Schumacher, U., and Hossfeld, D.K. 2002. Expression of CEACAM1 in adenocarcinoma of the lung: A factor of independent prognostic significance. *J. Clin. Oncol.* 20:4279–4284.

Lin, S.H., and Giudotti, G. 1989. Cloning and expression of a cDNA coding for a rat liver plasma membrane ecto-ATPase. The primary structure of the ecto-ATPase is similar to that of the human biliary glycoprotein I. *J. Biol. Chem.* 264:14,408–14,414.

Lin, S.H., and Pu, Y.S. 1999. Function and therapeutic implications of C-CAM cell-adhesion molecule in prostate cancer. *Semin. Oncol.* 26:227–233.

Najjar, S.M. 2002. Regulation of insulin action by CEACAM1. *Trends Endocrinol. Metab.* 13:240–245.

Neumaier, M., Paululat, S., Chan, A., Matthes, P., and Wagener, C. 1993. Biliary glycoprotein, a potential human cell adhesion molecule, is down-regulated in colorectal carcinomas. *Proc. Natl. Acad. Sci. USA* 90:10,744–10,748.

Nollau, P., Scheller, H., Kona-Horstmann, M., Rohde, S.F.H., Wagener, C., and Neumaier, M. 1997. Expression of CD66a (human C-CAM) and other members of the carcinoembryonic antigen family of adhesion molecules in human colorectal adenomas. *Cancer Res.* 57:2354–2357.

Obrink, B. 1997. CEA adhesion molecules: multifunctional proteins with signal-regulatory properties. *Curr. Opin. Cell Biol.* 9:616–626.

Prall, F., Nollau, P., Neumaier, M., Haubeck, H.-D., Drzerniek, Z., Helmchen, U., Löning, T., and Wagener, C. 1996. CD66a (BGP), an adhesion molecule of the carcinoembryonic antigen family, is expressed in epithelium, endothelium, and myeloid cells in a wide range of normal human tissues. *J. Histochem. Cytochem. 44:*35–41.

Riethdorf, L., Lisboa, B.W., Henkel, U., Naumann, M., Wagener, C., and Löning, T. 1997. Differential expression of CD66a (BGP), a cell adhesion molecule of the carcinoembryonic antigen family, in benign, premalignant, and malignant lesions of the human mammary gland. *J. Histochem. Cytochem. 45:*957–963.

Rojas, M., Fuks, A., and Stanners, C.P. 1990. Biliary glycoprotein, a member of the immunoglobulin supergene family, functions *in vitro* as a Ca2+-dependent intercellular cell adhesion molecule. *Cell Growth Diff. 1:*527–533.

Stoffel, A., Neumaier, M., Gaida, F.-J., Fenger, U., Drzeniek, Z., Haubeck, H.-D., and Wagener, C. 1993. Monoclonal, anti-domain and anti-peptide antibodies assign the molecular weight 160,000 granulocyte membrane antigen of the CD66 cluster to a mRNA species encoded by the biliary glycoprotein gene, a member of the carcinoembryonic antigen family. *J. Immunol. 150:*4978–4984.

Thies, A., Moll, I., Berger, J., Wagner, C., Brümmer, J., Schulze, H.J., Brunner, G., and Schuhmacher, U. 2002. CEACAM1 expression in cutaneous malignant melanoma predicts the development of metastatic disease. *J. Clin. Oncol. 20:*2530–2536.

Thompson, J., Grunert, F., and Zimmermann, W. 1991. Carcinoembryonic antigen family: Molecular biology and clinical perspectives. *J. Clin. Lab. Anal. 5:*344–366.

Volpert, O., Luo, W., Liu, T.J., Estrera, V.T., Logothetis, C., and Lin, S.H. 2002. Inhibition of prostate tumor angiogenesis by the tumor suppressor CEACAM1. *J. Biol. Chem. 277:*35,696–35,702.

Zhang, L., Zhou, W., Velculescu, V.E., Kern, S.E., Hruban, R.H., Hamilton, S.R., Vogelstein, B., and Kinzler, K.W. 1997. Gene expression profiles in normal and cancer cells. *Science 276:*1268–1272.

11

Role of Cadherins in Breast Cancer

Aaron C. Han

Introduction

Cadherins are important cell adhesion proteins that have a role in organogenesis and cell sorting during morphogenesis (Geiger *et al.*, 1992; Takeichi, 1991). These proteins are cell surface molecules that mediate homotypic adhesion. There are three classical cadherins that are defined by the organs in which they were initially described: E (epithelial) cadherin, N (neural) cadherin, and P (placental) cadherin. E-cadherin is the predominant cadherin in simple epithelium, including glandular, squamous, and transitional types. N-cadherin is found in neural tissue, but also expressed in cells of mesenchymal and mesodermal origin. In addition to its presence in placental tissue, P-cadherin is also found in basal epithelium.

Diagnostic Utility of Cadherin Expression

Cadherins have been useful in the study of normal, benign, and malignant tumor tissues (Furukawa *et al.*, 1997; Hashizume *et al.*, 1996; Inoue *et al.*, 1992). Because of their restricted tissue-specific expression, the pattern of cadherin expression is useful in determining and assigning origin of tissue (Peralta Soler *et al.*, 1995, 1997a, 1997b). In neoplasia, this tissue-specific pattern of cadherin expression has been observed to be preserved in certain tumor types, and has been useful in discriminating tumors of similar histology but distinct subtypes or cell type of origin. One area in which this has been useful is distinguishing tumors of simple epithelial from mesenchymal origin that may have similar histologic phenotypes. Examples of this include tumors in the lung and pleura, where lung adenocarcinoma and pleural mesothelioma can have similar histology (Han *et al.*, 1997; Peralta Soler *et al.*, 1995). Because the cadherin phenotypes are different, these tumors can be discriminated on the basis of their cadherin profile. Specifically, adenocarcinomas are derived from simple epithelia and predominantly express E-cadherin and rarely N-cadherin. In contrast, mesothelial cells are known expressors of N-cadherin, and the corresponding tumors are N-cadherin positive, but E-cadherin negative. Similar distinct cadherin profiles have been documented for prostatic tumors compared with epithelia of seminal vesicle (Peralta Soler *et al.*, 1997a) and among ovarian epithelial tumors (Peralta Soler *et al.*, 1997b). Serous tumors of the ovary express N-cadherin but not E-cadherin, whereas mucinous tumors have the opposite phenotype. This would be expected, because serous tumors are Mullerian tumors thought to arise from modified mesothelium, whereas mucinous tumors arise from endodermal origin.

Prognostic Utility of Cadherin Expression

Cadherins are also known tumor suppressor genes (Frixen *et al.*, 1991; Hedrick *et al.*, 1993; Zarka *et al.*, 2003). This has been best demonstrated in E-cadherin

Handbook of Immunohistochemistry and *in situ* Hybridization of Human
Carcinomas, Volume 1: Molecular Genetics; Lung and Breast Carcinomas

343

Copyright © 2004 by Elsevier (USA)
All rights reserved.

loss mutants, which demonstrate lack of cell-cell adhesion, but can be restored to an adhesive phenotype with reintroduction of the E-cadherin gene (Oka *et al.*, 1993; Vleminckx *et al.*, 1991). Individuals of families harboring E-cadherin mutations have been shown to be at increased risk of developing adenocarcinoma of the stomach (Guilford *et al.*, 1998; Huntsman *et al.*, 2001) consistent with a tumor suppressor role. Clinical studies have shown that tumors that have lost or reduced expression of, or mutated genes for E-cadherin, have a more aggressive natural history (Charpin *et al.*, 1997, 1998; Gamallo *et al.*, 1993; Lipponen *et al.*, 1994; Siitonen *et al.*, 1996). Some of these tumors that have lost E-cadherin are due to promoter hypermethylation (Miyakis *et al.*, 2002).

The E-cadherin gene is located on chromosome 16q21-22.1, and loss of heterozygosity (LOH) of this region has been shown in several tumor types (Miyakis *et al.*, 2002). Specifically, lobular carcinoma shows a high frequency of LOH in this gene region (Huiping *et al.*, 1999; Miyakis *et al.*, 2002).

Breast Cancer and Tumor Markers

Breast cancer is a devastating tumor that affects approximately one in 10 women. The diagnosis and prognosis of breast cancer depend on well-defined clinical and pathologic parameters, including histologic type, tumor histologic grade, tumor size, margin status, lymph node status, and presence or absence of distant metastasis (Rosen, 1997; Tavassoli, 1999). In addition, biomarkers have been shown to predict prognosis, but more importantly guide choices of available treatment options, such as hormone treatment based on estrogen/progesterone receptor status and Herceptin therapy based on HER-2/neu gene expression status (Slamon *et al.*, 2001). Known inherited breast cancer genes and phenotypes have also been described (Miki *et al.*, 1994). The search for new genes and proteins that may be predictive of prognosis or response to targeted therapies remains a goal in breast cancer research.

The purposes of our studies were to examine the role of cadherin expression in breast cancer to determine if there are distinct patterns of cadherin expression in certain breast tumors that may be useful for tumor classification or diagnosis, and also to determine if cadherin expression was correlated with clinical behavior.

MATERIALS

1. 10% neutral buffered formalin for tissue fixation, generally fixed for up to 24 hr.

2. Paraffin-embedded sections cut at 3–4 microns placed on Fisher-brand Superfrost slides for immunohistochemistry.

3. Xylene and graded alcohols for clearing of paraffin.

4. 10X Citra antigen retrieval fluid.

5. Steamer or microwave oven for heat-induced antigen retrieval.

6. Proteases for pretreatment.

7. Inhibitor for blocking of endogenous peroxidase activity.

8. Primary antibodies.

9. Biotinylated immunoglobulins.

10. Wash solution (buffered saline) for rinses between steps.

11. Developing reagents, including avidin-HRPO, diaminobenzidine (DAB), hydrogen peroxide.

12. Coverslips.

METHODS

1. 3–4 μm sections are cut from paraffin blocks of tissue and placed on the slides. Appropriate positive and negative controls are cut.

2. Slides are placed on a metal plate and heated/dried in the oven at 60°C for at least 30 min.

3. Staining is performed on the automated machine by Ventana systems (Nexxus). In our protocol, the machine is turned on at this step in the procedure. Slides are removed from the oven and placed in slide holders for deparaffinization step.

4. Slides are placed in xylene twice for 10 min each time.

5. Slides are hydrated through graded alcohols 100%, then 95%.

6. Slides are hydrated in distilled water.

7. Slides are placed with distilled water in steamer for antigen retrieval step.

8. 1X Citra buffer is added to the steamer. Certain antibodies that do not require antigen retrieval skip to step 15.

9. After water reaches a vigorous boil in the steamer, the chamber and slides are placed on the base of the steamer.

10. Steam antigen retrieval for 30 min.

11. Appropriate primary and secondary antibodies are placed on the Nexxus instrument.

12. Caps and fluids are checked on the machine.

13. After antigen retrieval step is complete, remove the top and allow the buffer to cool prior to removing the slides.

14. Remove the slides and attach the appropriate label for the primary antibody. The order and bar code are determined and printed out by the Nexxus stainer.

15. Cover slides with appropriate wash solution.

16. Once the automated stainer is loaded, press "Run" and follow appropriate instructions from the computer screen.

17. The automated steps include the following (18–26) starting with tissue incubation with primary antibody (Colvin et al., 1995; Dabbs, 2002).

18. Rinse slides with wash solution or buffered saline three times.

19. Incubate with biotinylated secondary antibody.

20. Rinse slide with wash solution or buffered saline three times.

21. Incubate with avidin-biotin-peroxidase complex.

22. Rinse slide with wash solution or buffered saline three times.

23. Incubate slide with DAB.

24. Rinse slide with wash solution or buffered saline three times.

25. Counter stain with hematoxylin.

26. When the stainer is finished, slides can be removed and placed in a rack.

27. Slides should be rinsed in warm running water with dish detergent until the water runs clear. This removes the liquid coverslip.

28. Slides are dehydrated through graded alcohols (95%, then 100%) and cleared in xylene.

29. Coverslips are applied. This can be manually performed or by an automated system.

RESULTS AND DISCUSSION

Role of Cadherins in Defining Breast Cancer Subtypes

Studies of cadherin expression in breast cancer have shown that E-cadherin is the predominant classical cadherin expressed in both invasive and *in situ* ductal carcinomas (Acs *et al.*, 2001; Gupta *et al.*, 1997; Peralta Soler *et al.*, 1999). This is not surprising, as these cancers recapitulate duct epithelium in which E-cadherin is the prevalent cadherin molecule. Studies examining different subtypes of epithelial primary breast tumors demonstrated that while the majority of ductal tumors express E-cadherin, lobular carcinoma typically is negative for E-cadherin (Moll *et al.*, 1993; Vos *et al.*, 1997). This is interesting from a biologic standpoint as histologically lobular tumors are less cohesive and do not form epithelial groups. They are often described histologically and microscopically as forming single-file tumor cells. Teleologically, one could say that lobular carcinoma has lost adhesive properties as a result of lack of cadherin adhesion protein expression. Thus, the histologic phenotype shows a lack of cellular cohesion and adhesion.

At the genetic level, the loss of E-cadherin expression in lobular carcinoma has been shown to be due to deletions and can be seen in genetic studies as LOH (Huiping *et al.*, 1999; Miyakis *et al.*, 2002). In addition, E-cadherin point mutations and hypermethylation phenotype has been shown to account for loss of E-cadherin expression in lobular carcinoma.

This differential expression of E-cadherin in ductal versus lobular tumors is a useful diagnostic adjunct in lesions that present in anatomic pathology practice. Not infrequently, when these lesions present as *in situ* cancers, they can be difficult to discriminate based on routine stains. E-cadherin is a useful adjunctive diagnostic stain because duct carcinoma *in situ* (DCIS) is E-cadherin positive, whereas lobular carcinoma *in situ* (LCIS) is E-cadherin negative (Acs *et al.*, 2001; Bratthauer *et al.*, 2002). Hybrid lesions including both DCIS and LCIS have a staining pattern that is heterogeneous. This diagnostic distinction based on the presence or absence of E-cadherin expression in *in situ* cancer is important because the clinical implication and treatment for DCIS is often different from LCIS.

Special breast tumor types have also been examined for their classical cadherin profiles (Han *et al.*, 1999). Interestingly, although the predominant cadherin in epithelial tumors is E-cadherin, for medullary carcinoma, which is a specific type of epithelial breast cancer, there is a high frequency of N-cadherin and P-cadherin expression. This raises an issue regarding the histogenesis of the tumor cells in medullary carcinoma (Rosen, 1997; Tavassoli, 1999). Possibly, the cell of origin is different from the ductal or lobular cells, and is a cell that has origin from basal epithelium, which is typically P-cadherin positive. We have also studied biphasic tumors (mixed epithelial mesenchymal tumors) that show distinct cadherin profiles. These tumors often recapitulate the cadherin profile of the benign counterpart, with E-cadherin being dominant in the epithelial components, and N-cadherin being dominant in the mesenchymal component.

This suggests that cadherins, in addition to other tumor markers, may be best used in a panel to define the phenotype of tumors. The customization of each individual patient's tumor type based on a panel of protein and genetic markers is very likely to be the direction in which we characterize individual patient tumors, with the aim to customize patient treatment in the near future (Callagy *et al.*, 2003; Miller *et al.*, 1999).

Role of Cadherin in Predicting Tumor Behavior: Loss of E-Cadherin in Ductal Cancer

Early research focused on the role of E-cadherin on tumor behavior in breast cancer. Cell culture studies showed the role of E-cadherin in carcinoma cell adhesiveness (Frixen *et al.*, 1991; Furukawa *et al.*, 1997). Loss of E-cadherin has been demonstrated correlating with increased biologic aggressiveness in carcinomas from several sites (Bailey *et al.*, 1998; Charpin *et al.*, 1998; Denk *et al.*, 1997; Gamallo *et al.*, 1993; Rimm *et al.*, 1995; Sakuragi *et al.*, 1994; Tada *et al.*, 1996). Possibly, the best worked-out model is in certain inherited gastric cancers, where E-cadherin mutations are a predisposition gene to cancer development in these families (Guilford *et al.*, 1998; Huntsman *et al.*, 2001). In breast cancer there are reports of E-cadherin loss correlating with poor clinical behavior; however, these have not been universally confirmed in other studies, nor have we been able to reproduce these results (Charpin *et al.*, 1998; Gamallo *et al.*, 1993; Lipponen *et al.*, 1994; Oka *et al.*, 1993; Peralta Soler *et al.*, 1999; Siitonen *et al.*, 1996). Some of the discrepancies may be due to use of different antibody clones recognizing different epitopes or comparing immunohistochemical versus genetic mutation studies. Possibly, there are minor changes in E-cadherin resulting in minor epitope loss that is not readily identified in all immunohistochemical assays.

Aberrant Expression of P-cadherin as a Marker of Aggressive Biology

Our study of approximately 200 breast cancers showed a frequency of aberrant P-cadherin expression in more than half of the cases (Figure 52) (Peralta Soler *et al.*, 1999). We classified tumors based on the intensity and quantity of P-cadherin expression and showed that there was a correlation of increasing aberrant P-cadherin expression with poor survival. This deduction was valid irrespective of other parameters such as receptor status and node status. Thus suggesting that P-cadherin may be an independent prognostic factor and potential marker for poor survival in breast cancer.

Other studies have also demonstrated aberrant P-cadherin expression in carcinomas from the breast, cervix, endometrium, esophagus, and skin (Han *et al.*, 2000, Moreno-Bueno *et al.*, 2003, Palacios *et al.*, 1995; Pizarro *et al.*, 1995). It will be interesting to see the extent to which aberrant expression of P-cadherin is a marker of malignancy or aggressive tumor behavior; for example, in carcinomas from various primary sites.

Role of Cadherin in Tumor Biology (Paget Disease)

Cadherins are known mediators of cell sorting in organogenesis and formation of organs (Furukawa *et al.*, 1997; Takeichi, 1991). The ability to mediate cell-cell adhesion plays a critical role in normal tissue development, and its disruption presumably leads to tumor dispersion and probably metastatic potential (Koukoulis *et al.*, 1998; Takeichi, 1993). Whether cadherin expression is involved positively to mediate cell trafficking in cancer is not known. However, it is

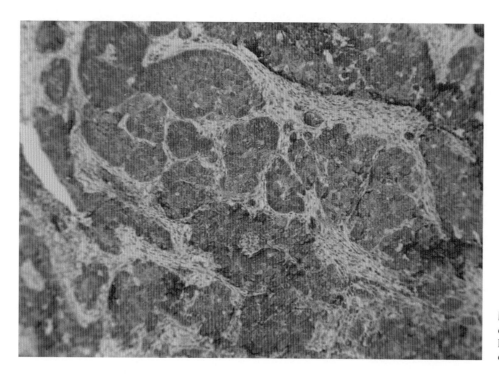

Figure 52 Breast cancer with expression of P-cadherin (immunohistochemical stain 100× magnification).

known that some tumors have a marked tropism for specific tissue types. In the breast, Paget disease is a specific adenocarcinoma that affects the nipple-areolar complex (Rosen, 1997; Tavassoli, 1999). Tumor cells are preferentially localized to the benign nipple epithelium of this region. One possible mechanism is that receptors or cell surface proteins mediate this specific localization of tumor cells to the nipple (Kondo *et al.*, 2002; Kothari *et al.*, 2002; Kuan *et al.*, 2001).

We examined this mechanism by analyzing cases of Paget disease of the breast. As the basal epithelium in the nipple predominantly expresses P-cadherin, we studied the frequency and intensity of P-cadherin expression in Paget disease. We observed that virtually all cases of Paget disease of the breast expressed high levels of P-cadherin, which is significant as P-cadherin is only seen in high intensity in approximately 20% of ductal carcinomas (Peralta Soler *et al.*, 1999). Furthermore, the P-cadherin positive Paget tumor cells are in contact with the P-cadherin expressing cells of the nipple basal epithelium. Interestingly, in some areas within the clusters and nests of tumor cells there is less P-cadherin expression in the tumor cell-cell junctions, compared with tumor cell-benign basal epithelial cell junction.

One possible mechanism in Paget disease is that P-cadherin is more critical for tumor cell migration and adhesion to the nipple epithelium than it is in tumor cell-tumor cell adhesion. This would also imply that there may be transient or varying levels of P-cadherin expression during different stages of the life history of a tumor cell, and trafficking of Paget tumor cells requires acquisition of higher levels of P-cadherin expression than that required for the nest of cells to remain in the nipple area. Transient expression of cadherins as a mechanism for metastasis (loss of cadherin) and then adhesion and growth of metastatic cells (with re-expression of cadherin) has been suggested in studies of E-cadherin (Graff *et al.*, 2000). This would have implications regarding the ability to target anti-cadherin therapy, and the need to identify specific stages of oncogenesis in which certain proteins and molecules are operative, targeting therapy at appropriate clinical junctures (Corada *et al.*, 2002; Kashima *et al.*, 2003; Mareel *et al.*, 1996; Schuhmacher *et al.*, 1999).

CONCLUSION

In conclusion, cadherins are useful markers in the study of benign and malignant disease processes. In breast cancer, cadherins are useful for differentiating tumor subtypes, as a marker of prognosis, and may be useful in defining tumor biology and trafficking. More studies of these molecules will hopefully yield additional insights into tumor biology, and also identify potential molecular and cell surface targets for earlier diagnosis and more specific biologic therapy.

References

Acs, G., Lawton, T.J., Rebbeck, T.R., LiVolsi, V.A., and Zhang, P.J. 2001. Differential expression of E-cadherin in lobular and ductal neoplasms of the breast and its biologic and diagnostic implications. *Am. J. Clin. Pathol. 115*:85–98.

Bailey, T., Biddlestone, L., Shepherd, N., Barr, H., Warner, P., and Jankowski, J. 1998. Altered cadherin and catenin complexes in the Barrett's esophagus-dysplasia-adenocarcinoma sequence. Correlation with disease progression and dedifferentiation. *Am. J. Pathol. 152*:135–144.

Bratthauer, G.L., Moinfar, F., Stamatakos, M.D., Mezzetti, T.P., Shekitka, K.M., Man, Y.G., and Tavassoli, F.A. 2002. Combined E-cadherin and high molecular weight cytokeratin immunoprofile differentiates lobular, ductal, and hybrid mammary intraepithelial neoplasia. *Hum. Pathol. 33*:620–627.

Callagy, G., Cattaneo, E., Daigo, Y., Happerfield, L., Bobrow, L.G., Pharoah, P.D., and Caldas, C. 2003. Molecular classification of breast carcinomas using tissue microarrays. *Diagn. Mol. Pathol. 12*:27–34.

Charpin, C., Garcia, S., Bonnier, P., Martini, F., Andrac, L., Choux, R., Lavaut, M., and Allasia, C. 1998. Reduced E-cadherin immunohistochemical expression in node-negative breast carcinomas correlates with 10-year survival. *Am. J. Clin. Pathol. 109*:431–438.

Charpin, C., Garcia, S., Bouvier, C., Devictor, B., Andrac, L., Choux, R., and Lavaut, M. 1997. E-cadherin quantitative immunocytochemical assays in breast carcinomas. *J. Pathol. 181*:294–300.

Colvin, R.B., Bhan, A.K., and McCluskey, R.T. 1995. *Diagnostic Immunopathology*. New York: Raven Press.

Corada, M., Zanetta, L., Orsenigo, F., Breviario, F., Lampugnani, M.G., Bernasconi, S., Liao, F., Hicklin, D.J., Bohlen, P., and Dejana, E. 2002. A monoclonal antibody to vascular endothelial-cadherin inhibits tumor angiogenesis without side effects on endothelial permeability. *Blood 100*:905–911.

Dabbs, D.J. 2002. *Diagnostic Immunohistochemistry*. Philadelphia, PA: Churchill Livingstone.

Denk, C., Hulsken, J., and Schwarz, E. 1997. Reduced gene expression of E-cadherin and associated catenins in human cervical carcinoma cell lines. *Cancer Lett. 120*:185–193.

Frixen, U., Behrens, J., Sachs, M., Eberle, G., Voss, B., and Warda, A. 1991. E-cadherin mediated cell-cell adhesion prevents invasiveness of human carcinoma cell lines. *J. Cell. Biol. 113*:173–185.

Furukawa, F., Fujii, K., Horiguchi, Y., Matsuyoshi, N., Fujita, M., Toda, K.I., Imamura, S., Wakita, H., Shirahama, S., and Takigawa, M. 1997. Roles of E- and P-cadherin in the human skin. *Microsc. Res. Tech. 38*:343–352.

Gamallo, C., Palacios, J., Suarex, A., Pizarro, A., Navarro, P., Quintanilla, M., and Cano, A. 1993. Correlation of E-cadherin expression with differentiation grade and histological type in breast carcinoma. *Am. J. Pathol. 142*:987–993.

Geiger, B., and Ayalon, O.C. 1992. *Cadherins. Annu. Rev. Cell. Biol. 8*:307–332.

Graff, J.R., Gabrielson, E., Fujii, H., Baylin, S.B., and Herman, J.G. 2000. Methylation patterns of the E-cadherin 5′CpG island are unstable and reflect the dynamic, heterogeneous loss of E-cadherin expression during metastatic progression. *J. Biol. Chem. 275*:2727–2732.

Guilford, P., Hopkins, J., Harraway, J., McLeod, M., McLeod, N., Harawira, P., Taite, H., Soular, R., Miller, A., and Reeve, A.E. 1998. E-cadherin germline mutations in familial gastric cancer. *Nature 392:*402–405.

Gupta, S.K., Douglas-Jones, A.G., Jasani, B., Morgan, J.M., Pignatelli, M., and Mansel, R.E. 1997. E-cadherin (E-cad) expression in duct carcinoma *in situ* (DCIS) of the breast. *Virchows Arch. 430:*23–28.

Han, A.C., Edelson, M.I., Peralta Soler, A., Knudsen, K.A., Lifschitz-Mercer, B., Czernobilsky, B., Rosenblum, N.G., and Salazar, H. 2000. Cadherin expression in glandular tumors of the cervix. Aberrant P-cadherin expression as a possible marker of malignancy. *Cancer 89:*2053–2058.

Han, A.C., Peralta Soler, A., Knudsen, K.A., and Salazar, H. 1999. Distinct cadherin profiles in special variant carcinomas and other tumors of the breast. *Hum. Pathol. 30:*1035–1039.

Han, A.C., Peralta-Soler, A., Knudsen, K.A., Wheelock, M.J., Johnson, K.R., and Salazar, H. 1997. Differential expression of N-cadherin in pleural mesotheliomas and E-cadherin in lung adenocarcinomas in formalin-fixed, paraffin-embedded tissues. *Hum. Pathol. 28:*641–645.

Hashizume, R., Koizumi, H., Ihara, A., Ohta, T., and Uchikoshi, T. 1996. Expression of β-catenin in normal breast tissue and breast carcinoma: A comparative study with epithelial cadherin and β-catenin. *Histopathology 29:*139–146.

Hedrick, L., Cho, K.R., and Vogelstein, B. 1993. Cell adhesion molecules as tumour suppressors. *Trends Cell Biol. 3:*36–39.

Hirohashi, S. 1998. Inactivation of the E-cadherin-mediated cell adhesion system in human cancers. *Am. J. Pathol. 153:*333–339.

Huiping, C., Sigurgeirdottir, J.R., Jonasson, J.G., Eiriksdottir, G., Johannsdottir, J.T., Egilsson, V., and Ingvarsson, S. 1999. Chromosome alterations and E-cadherin gene mutations in human lobular breast cancer. *Br. J. Cancer 81:*1103–1110.

Huntsman, D.G., Carneiro, F., Lewis, F.R., MacLeod, P.M., Hayashi, A., Monaghan, K.G., Maung, R., Seruca, R., Jackson, C.E., and Caldas, C. 2001. Early gastric cancer in young, asymptomatic carriers of germ-line E-cadherin mutations. *New Engl. J. Med. 344:*1904–1909.

Inoue, M., Ogawa, H., Miyata, M., Shiozaki, H., and Tanizawa, O. 1992. Expression of E-cadherin in normal, benign, and malignant tissues of female genital organs. *Am. J. Clin. Pathol. 98:*76–80.

Kashima, T., Nakamura, K., Kawaguchi, J., Takanashi, M., Ishida, T., Aburatani, H., Kudo, A., Fukayama, M., and Grigoriadis, A.E. 2003. Overexpression of cadherins suppresses pulmonary metastasis of osteosarcoma *in vivo. Int. J. Cancer 104:*147–154.

Kondo, Y., Kashima, K., Daa, T., Fujiwara, S., Nakayama, I., and Yokoyama, S. 2002. The ectopic expression of gastric mucin in extramammary and mammary Paget's disease. *Am. J. Surg. Pathol. 26:*617–623.

Kothari, A.S., Beechey-Newman, N., Hamed, H., Fentiman, I.S., D'Arrigo, C., Hanby, A.M., and Ryder, K. 2002. Paget disease of the nipple. A multifocal manifestation of higher-risk disease. *Cancer 95:*1–7.

Kuan, S.F., Montag, A.G., Hart, J., Krausz, T., and Recant, W. 2001. Differential expression of mucin genes in mammary and extramammary Paget's disease. *Am. J. Surg. Pathol. 25:*1469–1477.

Koukoulis, G.K., Patriarca, C., and Gould, V.E. 1998. Adhesion molecules and tumor metastasis. *Hum. Pathol. 29:*889–892.

Lipponen, P., Saarelainen, E., Ji, H., Aaltomaa, S., and Syrjanen, K. 1994. Expression of E-cadherin (E-CD) as related to other prognostic factors and survival in breast cancer. *J. Pathol. 174:*101–109.

Mareel, M., Berx, G., Van Roy, R., and Bracke, M. 1996. Cadherin/catenin complex: A target for anti-invasive therapy? *J. Cell Biochem. 61:*524–530.

Miki, Y., Swensen, J., Shattuck-Eidens, D., Futreal, P.A., Harshman, K., Tavtigian, S., Liu, Q., Cochran, C., Bennett, L.M., Ding, W., Bell, R., Rosenthal, J., Hussey, C., Tran, T., McClure, M., Frye, C., Hattier, T., Phelps, R., Haugen-Strano, A., Katcher, H., Yakumo, K., Gholami, Z., Shaffer, D., Stone, S., Bayer, S., Wray, C., Bogden, R., Dayananth, P., Ward, J., Tonin, P., Narod, S., Bristow, P.K., Norris, F.H., Helvering, L., Morrison, P., Rosteck, P., Lai, M., Barrett, J.C., Lewis, C., Neuhausen, S., Cannon-Albright, L., Goldgar, D., Wiseman, R., Kamb, A., and Skolnick, M.H. 1994. A strong candidate for the breast and ovarian cancer susceptibility gene BRCA1. *Science 266:*66–71.

Miller, K.D., and Sledge Jr., G.W. 1999. Toward checkmate: Biology and breast cancer therapy for the new millennium. *Invest. New Drugs 17:*417–427.

Miyakis, S., and Spandidos, D.A. 2002. Allelic loss in breast cancer. *Cancer Detect. Prevent. 26:*426–434.

Moll, R., Mitze, M., Frixen, U.H., and Birchmeier, W. 1993. Differential loss of E-cadherin expression in infiltrating ductal and lobular breast carcinomas. *Am. J. Pathol. 143:*1731–1742.

Moreno-Bueno, G., Hardisson, D., Sarrio, D., Sanchez, C., Cassia, R., Prat, J., Herman, J.G., Esteller, M., Matias-Guiu, X., and Palacios, J. 2003. Abnormalities of E- and P-cadherin and catenin (beta-, gamma-catenin, and p120ctn) expression in endometrial cancer and endometrial atypical hyperplasia. *J. Pathol. 199:*471–478.

Oka, H., Shiozaki, H., Kobayashi, K., Inoue, M., Tahara, H., Kobayashi, T., Takasuka, Y., Matsuyoshi, N., Hirano, S., Takeichi, M., and Mori, T. 1993. Expression of E-cadherin cell adhesion molecules in human breast cancer tissues and its relationship to metastasis. *Cancer Res. 53:*1696–1701.

Palacios, J., Benito, N., Pizarro, A., Suarez, A., Espada, J., Cano, A., and Gamallo, C. 1995. Anomalous expression of P-cadherin in breast carcinoma. Correlation with E-cadherin expression and pathological features. *Am. J. Pathol. 146:*605–612.

Peralta Soler, A., Harner, G.D., Knudsen, K.A., McBrearty, F.X., Grujic, E., Salazar, H., Han, A.C., and Keshgegian, A.A. 1997a. Expression of P-cadherin identifies prostate-specific-antigen-negative cells in epithelial tissues of male sexual accessory organs and in prostatic carcinomas. Implications for prostate cancer biology. *Am. J. Pathol. 151:*471–478.

Peralta Soler, A., Knudsen, K.A., Jaurand, M.-C., Johnson, K.R., Wheelock, M.J., Klein-Szanto, A.J.P., and Salazar, H. 1995. The differential expression of N-cadherin and E-cadherin distinguishes pleural mesotheliomas from lung adenocarcinomas. *Hum. Pathol. 26:*1363–1369.

Peralta Soler, A., Knudsen, K.A., Recson-Miguel, A., McBrearty, F.X., Han, A.C., and Salazar, H. 1997b. Expression of E-cadherin and N-cadherin in surface epithelial-stromal tumors of the ovary distinguishes mucinous from serous and endometrioid tumors. *Hum. Pathol. 28:*734–739.

Peralta Soler, A., Knudsen, K.A., Salazar, H., Han, A.C., and Keshgegian, A.A. 1999. P-cadherin expression in breast carcinoma indicates poor survival. *Cancer 86:*1263–1272.

Pizarro, A., Gamallo, C., Benito, N., Palacios, J., Qunitanilla, M., Cano, A., and Contreras, F. 1995. Differential patterns of placental and epithelial cadherin expression in basal cell carcinoma and in the epidermis overlying tumours. *Br. J. Cancer 72:*327–332.

Rimm, D.L., Sinard, J.H., and Morrow, J.S. 1995. Reduced β-catenin and E-cadherin expression in breast cancer. *Lab. Invest. 72:*506–512.

Rosen, P.P. 1997. *Rosen's Breast Pathology*. Philadelphia, PA: Lippincott-Raven Publishers.

Sakuragi, N., Nishiya, M., Ideda, K., Ohkouch, T., Furth, E.E., Hareyama, H., Satoh, C., and Fujimoto, S. 1994. Decreased E-cadherin expression in endometrial carcinoma is associated with tumor dedifferentiation and deep myometrial invasion. *Gynecol. Oncol. 53:*183–189.

Schuhmacher, C., Becker, K., Reich, U., Schenk, U., Mueller, J., Siewert, J.R., and Hofler, H. 1999. Rapid detection of mutated E-cadherin in peritoneal lavage specimens from patients with diffuse-type gastric carcinoma. *Diagn. Mol. Pathol. 8:*66–70.

Siitonen, S.M., Kononen, J.T., Helin, J.H., Rantala, I.S., Holli, K.A., and Isola, J.J. 1996. Reduced E-cadherin expression is associated with invasiveness and unfavorable prognosis in breast cancer. *Am. J. Clin. Pathol. 105:*394–402.

Slamon, D.J., Leyland-Jones, B., Shak, S., Fuchs, H., Paton, V., Bajamonde, A., Fleming, T., Eiermann, W., Wolter, J., Pegram, M., Baselga, J., and Norton, L. 2001. Use of chemotherapy plus a monoclonal antibody against HER-2 for metastatic breast cancer that overexpresses HER-2. *New Engl. J. Med. 344:*783–792.

Tada, H., Hatoko, M., Muramatsu, T., and Shirai, T. 1996. Expression of E-cadherin in skin carcinomas. *J. Dermatol. 23:* 104–110.

Takeichi, M. 1991. Cadherin cell adhesion receptors as a morphogenetic regulators. *Science 251:*1451–1455.

Takeichi, M. 1993. Cadherins in cancer: Implications for invasion and metastasis. *Curr. Opin. Cell Biol. 5:*806–811.

Tavassoli, F.A. 1999. *Pathology of the Breast.* Stamford CT: Appleton & Lange.

Vleminckx, K., Vakaet, L., Mareel, M.M., Fiers, W., and Van Roy, F. 1991. Genetic manipulation of E-cadherin expression by epithelial tumour cells reveals an invasion suppressor role. *Cell 66:*107–119.

Vos, C.B.J., Cleton-Jansen, A.M., Berx, G., de Leeuw, W.J., ter Haar N.T., van Roy, F., Cornelisse, C.J., Peterse, J.L., and van de Vijver, M.J. 1997. E-cadherin inactivation in lobular carcinoma *in situ* of the breast: An early event in tumorigenesis. *Br. J. Cancer 76:*1131–1133.

Zarka, T.A., Han, A.C., Edelson, M.I., and Rosenblum, N.G. 2003. Expression of cadherins, p53, and BCL2 in small cell carcinomas of the cervix: Potential tumor suppressor role for N-cadherin. *Int. J. Gynecol. Cancer. 13:*240–243.

12

Immunohistochemical Expression of Erythropoietin and Erythropoietin Receptor in Breast Carcinoma

Geza Acs

Introduction

Hypoxic microregions are a characteristic pathophysiologic property of solid tumors and are thought to result from inadequate perfusion due to severe structural and functional abnormalities of the tumor microcirculation and from tumor-related anemia (Vaupel *et al.*, 1991, 2001). Tumor hypoxia is believed to make solid tumors resistant to radiation and chemotherapy (Höckel *et al.*, 1996; Vaupel *et al.*, 2001). However, recent studies suggest that sustained hypoxia can additionally enhance local and systemic malignant progression and may increase aggressiveness through clonal selection and genomic and proteomic changes (Dachs *et al.*, 1998; Höckel and Vaupel, 2001; Vaupel *et al.*, 2001). Multivariate analysis has shown that tumor hypoxia is indeed one of the most powerful prognostic factors in cancers independent of other variables (Colpaert *et al.*, 2001; Höckel *et al.*, 1999).

Many hypoxia-regulated genes, such as vascular endothelial growth factor, are known to play a key role

in carcinogenesis and tumor progression (Semenza, 2001; Shweiki *et al.*, 1992). The best-known hypoxia-regulated gene is erythropoietin (Epo), a glycoprotein hormone stimulator of erythropoiesis (Jelkmann, 1994). Epo gene expression is primarily modulated by tissue hypoxia, and this regulation occurs mainly at the mRNA level mediated by hypoxia-inducible transcription factor-1 (HIF-1) (Ebert *et al.*, 1999; Lacombe and Mayeux, 1999). During adult life, Epo is normally produced by the kidney and liver (Jelkmann, 1994; Koury *et al.*, 1992). The Epo receptor (EpoR) belongs to the cytokine receptor type I superfamily (Ihle *et al.*, 1995; Tilbrook *et al.*, 1999; Wojchowski *et al.*, 1999). The signaling mechanisms following receptor activation include the Jak/STAT and the Ras/MAP kinase pathways (Ihle *et al.*, 1995; Lacombe *et al.*, 1999; Tilbrook *et al.*, 1999; Wojchowski *et al.*, 1999). Stimulation of EpoR in erythroblasts promotes their proliferation and differentiation, and leads to increased expression of the antiapoptotic proteins Bcl-2 and Bcl-X_L (Silva *et al.*, 1996, 1999), and inhibition of

Handbook of Immunohistochemistry and *in situ* Hybridization of Human
Carcinomas, Volume 1: Molecular Genetics; Lung and Breast Carcinomas

351

Copyright © 2004 by Elsevier (USA)
All rights reserved.

apoptosis (Lacombe and Mayeux, 1999; Wojchowski et al., 1999).

Recently other sites of Epo production have been reported, including brain astrocytes (Masuda et al., 1994) and human female reproductive organs, including the uterus (Masuda et al., 2000; Yasuda et al., 1998). Considerable amounts of Epo are also present in human milk, the source of which has recently been shown to be lactating breast glands (Acs et al., 2002; Juul et al., 2000). In recent years it has also become clear that EpoR is expressed by a variety of cell types, including endothelial cells (Anagnostou et al., 1994), neurons (Juul et al., 1998), and mammary epithelial cells (Acs et al., 2001, 2002; Juul et al., 2000). Although the specific function(s) of EpoR in these nonhematopoietic sites is not fully understood, the EpoR expressed in these tissues appears to be functional, thus suggesting a wider biological role for Epo signaling unrelated to erythropoiesis (Masuda et al., 1999). There appears to be a paracrine Epo/EpoR system in the brain, where neurons express EpoR and astrocytes produce Epo (Masuda et al., 1994; Morishita et al., 1997). Evidence suggests that the signaling cascades that have been characterized in hematopoietic cells (Ihle et al., 1995; Wojchowski et al., 1999) are also functional in neurons and can be modulated by Epo. It has been demonstrated in vitro and in vivo that Epo is a potent inhibitor of neuronal apoptosis induced by ischemia and hypoxia (Siren et al., 2001). Endothelial cells also express EpoR mRNA, and Epo stimulates proliferation and migration of human endothelial cells and angiogenesis (Anagnostou et al., 1990; Yasuda et al., 1998).

Elevated Epo levels have long been recognized in patients with renal cell carcinomas, Wilms' tumors, hepatomas, and cerebellar hemangioblastomas, all tumors arising in anatomic sites in which Epo is normally expressed in low levels (Ebert et al., 1999). Moreover, ectopic Epo expression in erythroleukemia cells was found to mediate their autonomous growth (Mitjavila et al., 1991). The expression of EpoR was also reported in cases of renal cell carcinoma, and a potential paracrine or autocrine role for Epo signaling for promoting growth of renal carcinomas has been suggested (Westenfelder et al., 2000). We recently described that human breast cancer cell lines and human breast carcinomas express Epo and EpoR mRNA and protein and that their expression is enhanced by hypoxia (Acs et al., 2001). Furthermore, we demonstrated that Epo signaling is biologically active and stimulates tyrosine phosphorylation, DNA synthesis, and proliferation in breast cancer cells. We also characterized the expression of Epo and EpoR by immunohistochemistry in a large series of in situ and invasive breast carcinomas (Acs et al., 2002).

MATERIALS

Two methods are presented for immunohistochemical detection of Epo and EpoR. Compared with the traditional avidin-biotin peroxidase technique (Method B), Method A uses an enzyme (horseradish peroxidase or alkaline phosphatase) labeled dextran polymer conjugated to the secondary antibody as a detection system. Although more expensive, this method is faster and results in a cleaner background, especially in tissues rich in endogenous biotin.

Materials for Method A

1. Dulbecco's phosphate-buffered saline (PBS): 100 mg anhydrous calcium chloride, 200 mg potassium chloride, 200 mg monobasic potassium phosphate, 100 mg magnesium chloride. $6 H_2O$, 8 g sodium chloride, and 2.16 g dibasic sodium phosphate. $7 H_2O$; bring vol to 1 L with deionized water, pH 7.4.

2. Fixative: 10% formaldehyde in PBS.

3. Silanized glass microscope slides (to avoid detachment of tissue sections during processing).

4. Xylene.

5. 100% ethanol.

6. Citrate buffer (0.01 mol/L sodium citrate): 3.84-g anhydrous citric acid, bring vol to 2 L with deionized water, pH 6.0 (or 10X citrate buffer for heat-induced epitope retrieval, Labvision Corp., Fremont, CA).

7. H_2O_2 solution to block endogenous peroxidase activity: 3% H_2O_2 in 100% methanol.

8. Tris-buffered saline (TBS): 2.4 g Tris-HCl and 8.76 g sodium chloride; bring vol to 1 L with deionized water, pH 7.4.

9. TBS-Tween-20 buffer (TBST): 1% Tween-20 in TBS (or 10X TBST from Dako Corp., Carpinteria, CA, Cat. No. S3306).

10. Primary antibody for Epo (rabbit polyclonal, H-162, Santa Cruz Biotechnologies Inc., Santa Cruz, CA, 1:200 dilution, 1 µg/ml), and EpoR (rabbit polyclonal, C-20, Santa Cruz Biotechnologies Inc., 1:400 dilution, 0.5 µg/ml) diluted in Dako antibody diluent (Cat. No. S0809).

11. Hoseradish peroxidase (HRP) labeled dextran polymer conjugated to goat-anti-rabbit secondary antibodies (Dako Envision + System, Rabbit, HRP, Cat. No. K4002).

12. Diaminobenzidine chromogen (e.g., Dako Liquid DAB+, Cat. No. K3467).

13. Hematoxylin counterstain (Gill III formula).

14. 4% glacial acetic acid.

15. Saturated lithium carbonate solution in deionized water.

16. Permanent mounting medium, low viscosity (e.g., Cytoseal 60, Richard-Allan Scientific, Kalamazoo, MI).

17. Glass coverslips.

18. Humidified incubation chamber (water-tight plastic box with close-fitting lid, fitted with a damp paper towel to maintain high humidity).

Materials for Method B

1. Dulbecco's PBS (see step 1 in Method A).

2. Fixative: 10% formaldehyde in PBS.

3. Silanized glass microscope slides (see **step 3** in Method A).

4. Xylene.

5. 100% ethanol.

6. Citrate buffer (see step 6 in Method A).

7. H_2O_2 solution to block endogenous peroxidase activity: 3% H_2O_2 in 100% methanol.

8. PBS/BSA buffer: PBS containing 1% bovine serum albumin (BSA).

9. Dako Biotin Blocking System (solutions 1 and 2, Cat. No. X0590).

10. 1.5% normal goat serum in PBS/BSA.

11. Primary antibody for Epo (rabbit polyclonal, H-162, Santa Cruz Biotechnologies Inc. 1:200 dilution, 1 μg/ml), and EpoR (rabbit polyclonal, C-20, Santa Cruz Biotechnologies Inc., 1:400 dilution, 0.5 μg/ml) diluted in PBS/BSA.

12. Biotinylated goat-anti-rabbit immunoglobulin G secondary antibody (e.g., Vector Laboratories, Inc., Burlingame, CA), diluted 1:200 in PBS/BSA.

13. Horseradish peroxidase-conjugated streptavidin (e.g., Streptavidin HP Detection System, Research Genetics, Huntsville, AL).

14. Diaminobenzidine chromogen (see **step 12** in Method A).

15. Hematoxylin counterstain (Gill III formula).

16. 4% glacial acetic acid.

17. Saturated lithium carbonate solution in deionized water.

18. Permanent mounting medium, low viscosity.

19. Glass coverslips.

20. Humidified incubation chamber (water-tight plastic box with close-fitting lid, fitted with a damp paper towel to maintain high humidity).

METHODS

Method A

1. Fix tissue samples in 10% formaldehyde for 24 hr at 4°C.

2. Process fixed tissue sections for paraffin embedding according to standard histologic protocols.

3. Cut tissue sections 4–5 μm thick onto charged microscope slides and let them dry completely.

4. Heat slides in an oven at 70°C for 2 hr, then let them cool to room temperature.

5. Deparaffinize slides by immersing into xylene for 5 min, repeat two times.

6. Immerse slides in 100% ethanol for 5 min, repeat two times.

7. Place slides in running deionized water for 5 min.

8. Place slides into citrate buffer and perform heat-induced epitope retrieval by microwaving them at 700 W for 4 min.

9. Replace evaporated buffer and repeat microwave treatment as discussed.

10. Let slides cool for 20 min in citrate buffer.

11. Place slides in running deionized water for 5 min.

12. Block endogenous peroxidase activity by treating the slides with 3% H_2O_2 in methanol for 20 min.

13. Place slides in running deionized water for 5 min.

14. Place slides in TBST buffer for 5 min.

15. Dry excess buffer by blotting the sides and wiping the bottom of the slide (do not let the specimen dry out between solutions). Place slides into humidified chamber.

16. Apply primary antibody onto slides (100–200 μl per slide, depending on size of tissue section) to cover entire tissue.

Note: Covering the tissue section with a plastic coverslip can reduce the amount of primary antibody solution needed.

17. Incubate slides with the primary antibody in a humidified chamber at 4°C overnight (16 hr).

18. Let slides warm to room temperature (30 min).

19. Rinse slides four times with TBST.

20. Dry excess buffer (as in **step 15**).

21. Apply HRP-labeled dextran polymer conjugated to secondary antibody.

22. Incubate slides in humidified chamber at room temperature for 30 min.

23. Rinse slides four times with TBST.

24. Dry excess buffer (as in **step 15**).

25. Apply DAB chromogen and incubate 5–10 min.

26. Place slides into tap water to stop reaction.

27. Counterstain slides with hematoxylin for 1 min, rinse in water.

28. Place slides into glacial acetic acid for 20 sec, rinse in water.

29. Place slides into lithium solution, rinse in water.

30. Dehydrate slides in 100% ethanol for 5 min three times, followed by xylene for 5 min three times.

31. Coverslip slides with permanent mounting medium.

Method B

1–13. Perform steps as described for method A.

14. Place slides in PBS/BSA for 5 min.

15. Dry excess buffer by blotting the sides and wiping the bottom of the slide (do not let the specimen dry out between solutions). Place slides into humidified chamber.

16. Incubate slides with Dako Biotin Blocking System solution 1 at 37°C for 10 min.

17. Rinse slides two times in PBS/BSA.

18. Incubate slides with Dako Biotin Blocking System solution 2 at 37°C for 10 min.

19. Rinse slides two times in PBS/BSA.

20. Block slides by incubating with 1.5% normal goat serum in PBS/BSA at 37°C for 30 min.

21. Dry excess blocking solution (as in **step 15**); do not rinse slides.

22. Apply primary antibody onto slides (100–200 μl per slide, depending on size of tissue section) to cover entire tissue.

Note: Covering the tissue section with a plastic coverslip can reduce the amount of primary antibody solution needed.

23. Incubate slides with the primary antibody in a humidified chamber at 4°C overnight (16 hr).

24. Let slides warm to room temperature (30 min).

25. Rinse slides three times with PBS/BSA.

26. Dry excess buffer (as in **step 15**).

27. Apply biotinylated secondary antibody.

28. Incubate slides in humidified chamber at 37°C for 30 min.

29. Rinse slides three times with PBS/BSA.

30. Dry excess buffer (as in **step 15**).

31. Apply horseradish peroxidase-conjugated streptavidin.

32. Incubate slides in humidified chamber at 37°C for 40 min.

33. Rinse slides three times with PBS/BSA.

34. Dry excess buffer (as in **step 15**).

35. Apply DAB chromogen and incubate 5–10 min.

36. Place slides into tap water to stop reaction.

37. Counterstain, dehydrate, and coverslip slides as described (**steps 27–31** in Method A).

RESULTS AND DISCUSSION

In benign mammary epithelial cells we found weak to moderate granular cytoplasmic Epo and EpoR immunostaining in 92% and 96% of 184 samples, respectively. In normal ducts and lobules Epo staining was most prominent in the luminal aspect of lobular epithelial cells. Strong Epo immunostaining was found in lobules showing secretory change; however, EpoR staining was not increased in such secretory lobules. Juul *et al.* (2000) also reported strong Epo immunoreactivity in mammary epithelial cells from lactating breast tissue, with less reactivity noted during gestation and in nonlactating breast. These authors also found weak EpoR immunoreactivity in benign mammary epithelial cells regardless of lactational state. These findings and our results support the concept that Epo present in breast milk is synthesized in the lactating mammary gland epithelium. Although the physiological role of Epo in the breast is not yet clear, the presence of EpoR suggests a specific role for Epo signaling in the breast. In benign epithelial lesions, including usual hyperplasia without atypia, sclerosing adenosis, and intraductal papillomas, Epo and EpoR staining was similar to that in normal epithelial cells.

Immunostaining for Epo was found in 92% (112 of 122 cases) of *in situ* ductal carcinomas (DCIS), 94% (34 of 36 cases) of *in situ* lobular carcinomas (LCIS) and in 95% (174 of 184 cases) of invasive breast carcinomas. Although Epo staining was usually weak to moderate and heterogeneous, strong, prominent reactivity was present in viable tumor cells adjacent to necrotic areas and at the infiltrating edge of carcinomas (Figure 53), sites thought to be the most hypoxic parts of tumors (Colpaert *et al.*, 2001; Vaupel *et al.*, 1989). Expression of Epo was similar in the *in situ* and invasive components of the tumors.

Diffuse, moderate to strong cytoplasmic and membrane EpoR immunostaining was present in 92% (121 of 122 cases) of DCIS, 100% (36 of 36 cases) of LCIS, and 96% (183 of 184 cases) of invasive carcinomas (Figure 54). Although EpoR immunostaining was usually uniform throughout the tumor, reactivity was further increased in tumor cells adjacent to necrotic areas. The *in situ* and invasive components of the tumors showed similar EpoR immunoreactivity. In addition, strong EpoR expression was found in the endothelial and smooth muscle cells of the tumor vasculature.

To compare the immunoreactivity in benign and neoplastic mammary epithelium and correlate the immunostaining pattern with clinicopathologic features of the tumors, immunoreactivity for Epo and EpoR was scored semiquantitatively as follows: Cytoplasmic and/or membrane immunoreactivity was considered positive. First, the total percentage of positive tumor cells and benign ductal and lobular epithelial cells was assessed. Then the percentage of weakly, moderately, and strongly staining cells was determined, so that the sum of these categories equated with the overall percentage of positivity. A staining score was then calculated as follows: score (out of maximum of 300) = sum of $1 \times$ percentage of weak, $2 \times$ percentage of moderate, and $3 \times$ percentage of strong staining.

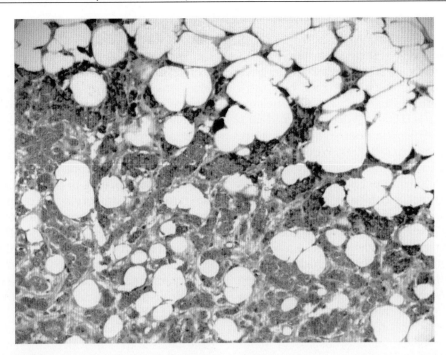

Figure 53 Expression of erythropoietin (Epo) in invasive ductal carcinoma of the breast. Note the heterogenous, weak Epo immunostaining in the tumor with strong reactivity for the hormone in cells at the infiltrating edge of the tumor (original magnification 100X, immunohistochemical stain for Epo with hematoxylin counterstain).

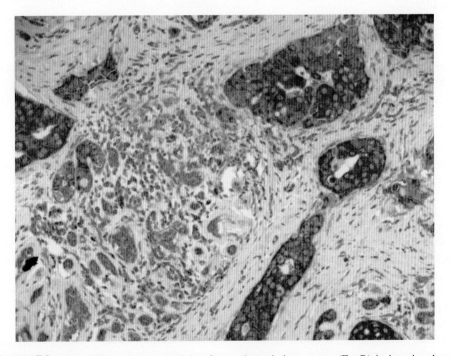

Figure 54 Strong, diffuse immunostaining for erythropoietin receptor (EpoR) in invasive ductal carcinoma of the breast, whereas the benign lobule in the center shows only weak reactivity (original magnification 100X, immunohistochemical stain for EpoR with hematoxylin counterstain).

Compared with the benign epithelial cells, Epo immunostaining was significantly increased in DCIS (p < 0.0001, Wilcoxon signed rank test), LCIS (p < 0.005, Wilcoxon signed rank test), and invasive carcinomas (p < 0.0001, Wilcoxon signed rank test). Similarly, EpoR immunostaining was significantly increased in DCIS (p < 0.0001, Wilcoxon signed rank test), LCIS (p < 0.0001, Wilcoxon signed rank test), and invasive carcinomas (p < 0.0001, Wilcoxon signed rank test) compared with benign mammary epithelial cells. We found no difference in Epo or EpoR immuno-staining between ductal and lobular carcinomas and carcinomas with mixed features.

Immunostaining for Epo did not show correlation with any of the clinico-histopathologic features of the tumors examined, including tumor size, grade, presence of lymphovascular invasion, presence of axillary lymph node metastasis, estrogen and progesterone receptor status, and HER-2/neu expression. In previous studies these clinico-histopathologic features were shown not to be correlated with the presence or degree of intra-tumoral hypoxia (Vaupel et al., 1991). Interestingly, Epo immunostaining did not show statistical correlation with the presence of tumor necrosis despite the prominent Epo staining observed near necrotic areas. This finding was likely due to the fact that areas surrounding necrotic regions usually represented small portions of the tumors and the prominent immunostaining restricted to these small areas did not increase the overall staining score of the tumors significantly.

No correlation was found between EpoR immunoreactivity and tumor size. In contrast, in DCIS cases, a significant correlation was seen between EpoR immunostaining and tumor grade and the presence of tumor necrosis. Similarly, in invasive carcinomas EpoR immunoreactivity was significantly increased in carci-nomas with high combined histologic grade (p = 0.026, Kruskal-Wallis test), in carcinomas showing tumor necrosis (p = 0.027, Mann-Whitney test), presence of lymphovascular invasion (p = 0.0069, Mann-Whitney test), and in tumors associated with axillary lymph node metastases (p = 0.0075, Mann-Whitney test). Immunoreactivity for EpoR was also significantly higher in breast carcinomas negative for estrogen recep-tor (p = 0.0375, Mann-Whitney test) and progesterone receptor (p = 0.0404, Mann-Whitney test) expression compared with hormone receptor positive tumors. No correlation was found between EpoR expression and HER-2/neu overexpression in the tumors.

Until recently, the biologic effects of Epo was con-sidered to be restricted to erythroid progenitor cells. There is now evidence that certain nonerythroid cells express EpoR and respond to Epo in vitro and in vivo. The expression of EpoR in neuronal tissues appears to

be functional, and evidence suggests that Epo may have mechanisms of action in the central nervous system similar to those described in erythroid cells, such as decreasing apoptosis of neurons during normal brain development, or exerting a neuroprotective effect under adverse conditions such as hypoxia (Juul et al., 1998; Sakanaka et al., 1998). Endothelial cells also express EpoR, and Epo stimulates proliferation and migration of human endothelial cells and angiogenesis (Anagnostou et al., 1990; Yasuda et al., 1998). Recent studies indi-cate that Epo signaling in endothelial cells is mediated via tyrosine phosphorylation of proteins, including phosphorylation of transcription factor STAT-5, which is similar to that in erythroid cells (Haller et al., 1996). We found that the vasculature of solid tumors, includ-ing breast carcinomas, also express EpoR, suggesting that a paracrine mechanism of Epo signaling may also play a role in the vascularization of these tumors (Acs et al., 2001).

An autocrine role for Epo, whereby the hormone contributes to the survival, proliferation, and differenti-ation of trophoblast cells (Fairchild Benyo et al., 1999), appears to be analogous to its role in erythroleukemia (Stage-Marroquin et al., 1996) and renal cell carcinoma cells (Westenfelder et al., 2000). Our demonstration that Epo signaling stimulates proliferation of human breast cancer cells suggests that an autocrine mecha-nism of Epo signaling may also play a role in the proliferation and hypoxic survival of breast cancers (Acs et al., 2001).

In the uterus and oviduct Epo expression has been shown to be regulated by estrogen in addition to hypoxia (Yasuda et al., 1998). Interestingly, even though the importance of increased estrogen effect in breast car-cinogenesis is well-established, we found no correla-tion of Epo expression with the hormone receptor status of the tumors. On the other hand, we found a correlation of increased EpoR expression by the tumors with negative hormone receptor status, a feature asso-ciated with poor differentiation and adverse prognosis.

Regions of low oxygen and necrosis are common features of solid tumors (Vaupel et al., 1989). Hypoxic conditions in vitro as well as in vivo result in elevated levels of HIF-1, a transcription factor that in turn stim-ulates the expression of a number of genes important for tumor cell survival, including Epo (Wang et al., 1995). Overexpression of HIF-1 protein was found in 53% of all primary malignant tumors, including 29% of primary breast cancers and 69% of breast cancer metastases (Zhong et al., 1999). In colonic adenocar-cinoma, cancer cells at the leading edge of infiltrating carcinoma showed the most intense HIF-1 expression; this pattern of staining is similar to the expression of Epo we found in breast (Acs et al., 2002) and cervical carcinomas (Acs *et al.*, 2003), supporting the hypothesis that increased Epo expression in these tumors is stimulated by tissue hypoxia.

Hypoxia can also induce EpoR expression (Acs *et al.*, 2001, 2003) and viable tumor cells adjacent to necrotic regions typically show increased EpoR immunostaining. However, hypoxic regions are usually heterogeneously distributed within tumors (Vaupel *et al.*, 1991). Although the heterogeneous pattern of Epo immunostaining is consistent with this phenomenon, EpoR immunostaining was diffuse and uniform throughout the tumor tissue, with further accentuation near necrotic areas. These findings suggest that while hypoxia induces EpoR expression, other oncogenic mechanisms are also likely to play a role in increased EpoR expression in neoplastic mammary epithelial cells. Furthermore, HIF-1 is not known to regulate EpoR gene expression. Thus, the biochemical mechanisms underlying elevated EpoR expression in breast cancer and its hypoxic enhancement remain unknown. Nevertheless, elevated EpoR expression might increase the sensitivity of the neoplastic cells to available Epo.

Experimental data suggest that one of the major modes of action of most anticancer treatment methods, including radiation and chemotherapy, may be via activation of apoptosis in sensitive cells, and the tendency of a cancer cell to undergo apoptosis may have important implications for tumor progression and response to treatment (Kerr *et al.*, 1994; Makin *et al.*, 2001; Zhivotovsky *et al.*, 1999). Overexpression of the anti-apoptotic protein Bcl-2 has been shown to protect cancer cells from apoptotic cell death induced by a variety of stimuli, including radiation and most cytotoxic drugs (Datta *et al.*, 1995; Krajewski *et al.*, 1995; Reed *et al.*, 1996). In erythroblasts and neurons the major action of Epo is inhibition of apoptosis (Lacombe *et al.*, 1999; Siren *et al.*, 2001; Wojchowski *et al.*, 1999). We have recently shown that Epo dose dependently inhibits the cytotoxic effect and apoptosis induced by the chemotherapeutic drug cisplatin in cervical carcinoma cells (Acs *et al.*, 2003). Since human cervical carcinoma cells also express functional EpoR, we hypothesize that the action of Epo in cancer cells is likely mediated by mechanisms similar to those described in erythroblasts and neurons.

Tumor hypoxia has been traditionally considered to be a therapeutic problem, as it makes solid tumors resistant to radiation and chemotherapy (Vaupel *et al.*, 2001). Patients with hypoxic tumors have a significantly shorter recurrence-free and overall survival, and tumor oxygenation has been shown to be one of the strongest independent prognostic factors (Höckel *et al.*, 1994, 1996). Importantly, disadvantage in outcome for hypoxic tumors of the uterine cervix appears to be independent of the mode of primary treatment (radiation or radical surgery). Thus, it was suggested that tumor hypoxia may not simply counteract oxygen-dependent therapy forms, but, through genomic and proteomic changes, it may also increase aggressiveness and advance tumor progression per se (Höckel *et al.*, 1996, 2001). Hypoxia is thought to mediate the selection of neoplastic cells with diminished apoptotic potential by providing a growth advantage to cells with genetic alterations that impair the process of apoptosis (Graeber *et al.*, 1996). This hypoxia-mediated clonal selection of tumor cells with diminished apoptotic potential has been suggested as an important biological mechanism of tumor progression. Thus, the presence of hypoxia may also be involved in the development of a more aggressive phenotype and contribute to metastasis and treatment resistance (Beavon, 1999; Cairns *et al.*, 2001; Teicher, 1994). Recent data suggest that stimulation of EpoR signaling by hypoxia may be an important mechanism, leading to increased Bcl-2 expression and decreased apoptotic potential and increased aggressiveness in cancer cells (Acs *et al.*, 2003). Further studies are needed to confirm this hypothesis.

Anemia has been a well-recognized complication of cancer and cancer treatment (Littlewood, 2001), and one important consequence of anemia is the resulting tumor hypoxia (Kelleher *et al.*, 1996). Although the relationship among anemia, hypoxia, blood transfusion, and treatment outcome is complex, anemia has been traditionally considered to be one of the most powerful prognostic factors in patients with cancer (Fyles *et al.*, 2000). Recently, recombinant human erythropoietin (rHuEpo) has become available for treatment of anemic patients. In fact, rHuEpo has been shown to effectively increase hemoglobin levels and is often used in patients receiving radio- and chemotherapy (Littlewood, 2001). However, Epo is a potent growth factor, which may stimulate proliferation and inhibit apoptosis of EpoR-bearing tumor cells (Acs *et al.*, 2001; Westenfelder *et al.*, 2000). In addition, Epo also stimulates proliferation and migration of vascular endothelial cells and augments angiogenesis (Anagnostou *et al.*, 1990; Jaquet *et al.*, 2002). Because the vasculature of solid tumors express EpoR (Acs *et al.*, 2001, 2003), the potential detrimental effects of Epo on tumor growth may be further aggravated by its known angiogenic activity. Previous studies in renal cell carcinoma patients suggested that increased serum Epo levels in the absence of polycythemia carried a worse prognosis and indicated a higher incidence of progressive metastatic disease (Ljungberg *et al.*, 1992). Furthermore, Epo administration to a patient with multiple myeloma may have caused further malignant transformation resulting in plasma cell leukemia, calling into question the safety of Epo treatment for patients with EpoR expressing myeloma cells (Olujohungbe *et al.*, 1997).

The available data raise the possibility that cellular responses to Epo may collectively promote growth of EpoR-bearing tumors, and these actions may be further enhanced by either high endogenous Epo production by the tumor (stimulated by hypoxia), or by exogenous Epo administration. Thus, until it is demonstrated that pharmacologic doses of Epo lack such trophic effects *in vivo*, treatment of cancer patients with rHuEpo should probably be carried out with some degree of caution. On the other hand, specific cell surface receptors on tumor cells may provide an ideal therapeutic target for anticancer treatment. Since cells of breast cancer and other solid tumors express EpoR, it may be possible to use EpoR or other targets in its signaling pathway in novel therapeutic approaches aimed at the eradication of tumor cells (Lappin *et al.*, 2002).

In summary, evidence indicates that benign mammary epithelial cells and cells of *in situ* and invasive breast carcinomas express both Epo and EpoR. Increased immunostaining of EpoR in breast carcinomas shows a positive correlation with clinico-histopathologic features suggestive of an adverse prognosis. Increased autocrine/paracrine Epo signaling may represent a novel mechanism by which hypoxia promotes a malignant phenotype in breast cancer.

References

Acs, G., Acs, P., Beckwith, S.M., Pitts, R.L., Clements, E., Wong, K., and Verma, A. 2001. Erythropoietin and erythropoietin receptor expression in human cancer. *Cancer. Res. 61:*3561–3565.

Acs, G., Zhang, P.J., McGrath, C.M., Acs, P., McBroom, J., Mohyeldin, A., Liu, S., Lu, H., and Verma, A. 2003. Hypoxia-inducible erythropoietin signaling in squamous dysplasia and squamous cell carcinoma of the uterine cervix and its potential role in cervical carcinogenesis and tumor progression. *Am. J. Pathol. 162:*1789–1806.

Acs, G., Zhang, P.J., Rebbeck, T.R., Acs, P., and Verma, A. 2002. Immunohistochemical expression of erythropoietin and erythropoietin receptor in breast carcinoma. *Cancer 95:* 969–981.

Anagnostou, A., Lee, E.S., Kessimian, N., Levinson, R., and Steiner, M. 1990. Erythropoietin has a mitogenic and positive chemotactic effect on endothelial cells. *Proc. Natl. Acad. Sci. USA 87:*5978–5982.

Anagnostou, A., Liu, Z., Steiner, M., Chin, K., Lee, E.S., Kessimian, N., and Noguchi, C.T. 1994. Erythropoietin receptor mrna expression in human endothelial cells. *Proc. Natl. Acad. Sci. USA 91:*3974–3978.

Beavon, I.R. 1999. Regulation of E-cadherin: Does hypoxia initiate the metastatic cascade? *Mol. Pathol. 52:*179–188.

Cairns, R.A., Kalliomaki, T., and Hill, R.P. 2001. Acute (cyclic) hypoxia enhances spontaneous metastasis of kht murine tumors. *Cancer Res. 61:*8903–8908.

Colpaert, C., Vermeulen, P., van Beest, P., Goovaerts, G., Weyler, J., Van Dam, P., Dirix, L., and Van Marck, E. 2001. Intratumoral hypoxia resulting in the presence of a fibrotic focus is an independent predictor of early distant relapse in lymph node-negative breast cancer patients. *Histopathology 39:*416–425.

Dachs, G.U. and Chaplin, D.J. 1998. Microenvironmental control of gene expression: Implications for tumor angiogenesis, progression, and metastasis. *Semin. Rad. Oncol. 8:*208–216.

Datta, R., Manome, Y., Taneja, N., Boise, L.H., Weichselbaum, R., Thompson, C.B., Slapak, C.A., and Kufe, D. 1995. Overexpression of Bcl-xl by cytotoxic drug exposure confers resistance to ionizing radiation-induced internucleosomal DNA fragmentation. *Cell. Growth Diff. 6:*363–370.

Ebert, B.L. and Bunn, H.F. 1999. Regulation of the erythropoietin gene. *Blood 94:*1864–1877.

Fairchild Benyo, D. and Conrad, K.P. 1999. Expression of the erythropoietin receptor by trophoblast cells in the human placenta. *Biol. Reprod. 60:*861–870.

Fyles, A.W., Milosevic, M., Pintilie, M., Syed, A., and Hill, R.P. 2000. Anemia, hypoxia and transfusion in patients with cervix cancer: A review. *Radiother. Oncol. 57:*13–19.

Graeber, T.G., Osmanian, C., Jacks, T., Housman, D.E., Koch, C.J., Lowe, S.W., and Giaccia, A.J. 1996. Hypoxia-mediated selection of cells with diminished apoptotic potential in solid tumours. *Nature 379:*88–91.

Haller, H., Christel, C., Dannenberg, L., Thiele, P., Lindschau, C., and Luft, F.C. 1996. Signal transduction of erythropoietin in endothelial cells. *Kidney Int. 50:*481–488.

Höckel, M., Knoop, C., Schlenger, K., Vorndran, B., Knapstein, P.G., and Vaupel, P. 1994. Intratumoral po2 histography as predictive assay in advanced cancer of the uterine cervix. *Adv. Exp. Med. Biol. 345:*445–450.

Höckel, M., Schlenger, K., Aral, B., Mitze, M., Schaffer, U., and Vaupel, P. 1996. Association between tumor hypoxia and malignant progression in advanced cancer of the uterine cervix. *Cancer Res. 56:*4509–4515.

Höckel, M., Schlenger, K., Höckel, S., and Vaupel, P. 1999. Hypoxic cervical cancers with low apoptotic index are highly aggressive. *Cancer Res. 59:*4525–4528.

Höckel, M., and Vaupel, P. 2001. Tumor hypoxia: Definitions and current clinical, biologic, and molecular aspects. *J. Natl. Cancer Inst. 93:*266–276.

Ihle, J.N., Witthuhn, B.A., Quelle, F.W., Yamamoto, K., and Silvennoinen, O. 1995. Signaling through the hematopoietic cytokine receptors. *Annu. Rev. Immunol. 13:*369–398.

Jaquet, K., Krause, K., Tawakol-Khodai, M., Geidel, S., and Kuck, K. 2002. Erythropoietin and VEGF exhibit equal angiogenic potential. *Microvasc. Res. 64:*326.

Jelkmann, W. 1994. Biology of erythropoietin. *Clin. Invest. 72:* S3–10.

Juul, S.E., Anderson, D.K., Li, Y., and Christensen, R.D. 1998. Erythropoietin and erythropoietin receptor in the developing human central nervous system. *Pediatr. Res. 43:*40–49.

Juul, S.E., Zhao, Y., Dame, J.B., Du, Y., Hutson, A.D., and Christensen, R.D. 2000. Origin and fate of erythropoietin in human milk. *Pediatr. Res. 48:*660–667.

Kelleher, D.K., Mattiensen, U., Thews, O., and Vaupel, P. 1996. Blood flow, oxygenation, and bioenergetic status of tumors after erythropoietin treatment in normal and anemic rats. *Cancer Res. 56:*4728–4734.

Kerr, J.F., Winterford, C.M., and Harmon, B.V. 1994. Apoptosis: Its significance in cancer and cancer therapy. *Cancer 73:*2013–2026.

Koury, M.J., and Bondurant, M.C. 1992. The molecular mechanism of erythropoietin action. *Eur. J. Biochem. 210:*649–663.

Krajewski, S., Blomqvist, C., Franssila, K., Krajewska, M., Wasenius, V.M., Niskanen, E., Nordling, S., and Reed, J.C. 1995. Reduced expression of proapoptotic gene bax is associated with poor response rates to comtination chemotherapy and shorter survival in women with metastatic breast adenocarcinoma. *Cancer Res. 55:*4471–4478.

Lacombe, C., and Mayeux, P. 1999. The molecular biology of erythropoietin. *Nephrol. Dial. Transplant 14:*22–28.

Lappin, T.R., Maxwell, A.P., and Johnston, P.G. 2002. Epo's alter ego: Erythropoietin has multiple actions. *Stem Cells 20:* 485–492.

Littlewood, T.J. 2001. The impact of hemoglobin levels on treatment outcomes in patients with cancer. *Semin. Oncol. 28:* 49–53.

Ljungberg, B., Rasmuson, T., and Grankvist, K. 1992. Erythropoietin in renal cell carcinoma: Evaluation of its usefulness as a tumor marker. *Eur. Urol. 21:*160–163.

Makin, G., and Dive, C. 2001. Apoptosis and cancer chemotherapy. *Trends Cell Biol. 11:*S22–26.

Masuda, S., Kobayashi, T., Chikuma, M., Nagao, M., Sasaki, R. 2000. The oviduct produces erythropoietin in an estrogen- and oxygen-dependent manner. *Am. J. Physiol. Endocrin. Metab. 278:*E1038–1044.

Masuda, S., Nagao, M., and Sasaki, R. 1999. Erythropoietic, neurotrophic, and angiogenic functions of erythropoietin and regulation of erythropoietin production. *Int. J. Hematol. 70:* 1–6.

Masuda, S., Okano, M., Yamagishi, K., Nagao, M., Ueda, M., and Sasaki, R. 1994. A novel site of erythropoietin production: Oxygen-dependent production in cultured rat astrocytes. *J. Biol. Chem. 269:*19488–19493.

Mitjavila, M.T., Le Couedic, J.P., Casadevall, N., Navarro, S., Villeval, J.L., Dubart, A., and Vainchenker, W. 1991. Autocrine stimulation by erythropoietin and autonomous growth of human erythroid leukemic cells *in vitro. J. Clin. Invest. 88:*789–797.

Morishita, E., Masuda, S., Nagao, M., Yasuda, Y., and Sasaki, R. 1997. Erythropoietin receptor is expressed in rat hippocampal and cerebral cortical neurons, and erythropoietin prevents *in vitro* glutamate-induced neuronal death. *Neuroscience 76:*105–116.

Olujohungbe, A., Handa, S., and Holmes, J. 1997. Does erythropoietin accelerate malignant transformation in multiple myeloma? *Postgrad. Med. J. 73:*163–164.

Reed, J.C., Miyashita, T., Takayama, S., Wang, H.G., Sato, T., Krajewski, S., Aime-Sempe, C., Bodrug, S., Kitada, S., and Hanada, M. 1996. Bcl-2 family proteins: Regulators of cell death involved in the pathogenesis of cancer and resistance to therpy. *J. Cell Biochem. 60:*23–32.

Sakanaka, M., Wen, T.C., Matsuda, S., Masuda, S., Morishita, E., Nagao, M., and Sasaki, R. 1998. *In vivo* evidence that erythropoietin protects neurons from ischemic damage. *Proc. Natl. Acad. Sci. USA 95:*4635–4640.

Semenza, G.L. 2001. Hypoxia-inducible factor 1: Control of oxygen homeostasis in health and disease. *Pediatr. Res. 49:* 614–617.

Shweiki, D., Itin, A., Soffer, D., and Keshet, E. 1992. Vascular endothelial growth factor induced by hypoxia may mediate hypoxia-initiated angiogenesis. *Nature 359:*843–845.

Silva, M., Benito, A., Sanz, C., Prosper, F., Ekhterae, D., Nunez, G., and Fernandez-Luna, J.L. 1999. Erythropoietin can induce the expression of Bcl-x(1) through stat5 in erythropoietin-dependent progenitor cell lines. *J. Biol. Chem. 274:*22,165–22,169.

Silva, M., Grillot, D., Benito, A., Richard, C., Nunez, G., and Fernandez-Luna, J.L. 1996. Erythropoietin can promote erythroid progenitor survival by repressing apoptosis through Bcl-xl and Bcl-2. *Blood 88:*576–582.

Siren, A.L., Knerlich, F., Poser, W., Gleiter, C.H., Bruck, W., and Ehrenreich, H. 2001. Erythropoietin and erythropoietin receptor in human ischemic/hypoxic brain. *Acta. Neuropathol. 101:*271–276.

Stage-Marroquin, B., Pech, N., and Goldwasser, E. 1996. Internal autocrine regulation by erythropoietin of erythroleukemic cell proliferation. *Exp. Hematol. 24:*1322–1326.

Teicher, B.A. 1994. Hypoxia and drug resistance. *Cancer Met. Rev. 13:*139–168.

Tilbrook, P.A., and Klinken, S.P. 1999. Erythropoietin and erythropoietin receptor. *Growth Factor 17:*25–35.

Vaupel, P., Kallinowski, F., and Okunieff, P. 1989. Blood flow, oxygen and nutrient supply, and metabolic microenvironment of human tumors: A review. *Cancer Res. 49:*6449–6465.

Vaupel, P., Keleher, D.K., and Höckel, M. 2001. Oxygen status of malignant tumors: Pathogenesis of hypoxia and significance for tumor therapy. *Semin. Oncol. 28:*29–35.

Vaupel, P., Schelenger, K., Knoop, C., and Höckel, M. 1991. Oxygenation of human tumors: Evaluation of tissue oxygen distribution in breast cancers by computerized O_2 tension measurements. *Cancer Res. 51:*3316–3322.

Wang, G.L., Jiang, B.H., Rue, E.A., and Semenza, G.L. 1995. Hypoxia-inducible factor 1 is a basic-helix-loop-helix-pas heteriodimer regulated by cellular o2 tension. *Proc. Natl. Acad. Sci. USA 92:*5510–5514.

Westenfelder, C. and Baranowski, R.L. 2000. Erythropoietin stimulates proliferation of human renal carcinoma cells. *Kidney Int. 58:*647–657.

Wojchowski, D.M., Gregory, R.C., Miller, C.P., Pandit, A.K., and Pircher, T.J. 1999. Signal transduction in the erythropoietin receptor system. *Exp. Cell Res. 253:*143–156.

Yasuda, Y., Masuda, S., Chikuma, M., Inoue, K., Nagao, M., and Sasaki, R. 1998. Estrogen-dependent production of erythropoietin in uterus and its implication in uterine angiogenesis. *J. Biol. Chem. 273:*25,381–25,387.

Zhivotovsky, B., Joseph, B., and Orrenius, S. 1999. Tumor radiosensitivity and apoptosis. *Exp. Cell. Res. 248:*10–17.

Zhong, H., De Marzo, A.M., Laughner, E., Lim, M., Hilton, D.A., Zagzag, D., Buechler, P., Isaacs, W.B., Semenza, G.L., and Simons, J.W. 1999. Overexpression of hypoxia-inducible factor 1alpha in common human cancers and their metastases. *Cancer Res. 59:*5830–5835.

13

Loss of *BRCA1* Gene Expression in Breast Carcinoma

Wen-Ying Lee

Introduction

BRCA1 (BReast CAncer 1) is a putative tumor suppressor gene located on chromosome 17q21, which encodes a nuclear protein of 220 kD consisting of 1863 amino acids (Chen *et al.*, 1996; Miki *et al.*, 1994). Many tumors with germline BRCA1 mutations display loss of heterozygosity at this locus, suggesting that *BRCA1* is a tumor suppressor gene (Smith *et al.*, 1992). Breast cancer occurs in familial (hereditary) and sporadic (nonhereditary) forms. Germline BRCA1 mutation accounts for 50% of breast cancer families (Ford *et al.*, 1998), whereas none has currently been reported in sporadic breast cancers (Futreal *et al.*, 1994).

BRCA1 protein has been implicated in several important cellular functions, including regulation of transcription, repair of DNA damage, and cell cycle control. It has been reported that the *BRCA1* gene has RING finger domain in its NH2 terminus (Miki *et al.*, 1994) and BRCT motif in its COOH terminus (Koonin *et al.*, 1996) that functions as a transactivator (Monteiro *et al.*, 1996). *BRCA1* interacts directly with p53 and transcriptionally activates p21^{waf1} (Zhang *et al.*, 1998), suggesting a role in cell cycle control. *BRCA1* also interacts with Rad51, a recA homologue of *Escherichia coli* that is involved in repair of double-strand breaks in

DNA, suggesting a role in maintaining genomic stability (Koonin *et al.*, 1996; Scully *et al.*, 1997a, b). BRCA1 expression has shown to be associated with functional differentiation in multiple tissues (Marquis *et al.*, 1995). It has been reported that BRCA1 transcripts are expressed in both alveolar and ductal epithelial cells of normal mammary gland, and breast cancer is frequently associated with loss of BRCA1 expression (Lane *et al.*, 1995). This evidence suggests that the *BRCA1* gene provides an important growth regulatory function in mammary epithelial cells.

Although no somatic mutation of the *BRCA1* gene has been detected (Futreal *et al.*, 1994; Merajver *et al.*, 1995), loss of heterozygosity at chromosome 17q21 (Jacobs *et al.*, 1993) and decreased levels of the BRCA1 mRNA have been demonstrated in some sporadic cases of breast cancer (Thompson *et al.*, 1995), indicating the involvement of *BRCA1* even in the sporadic form. BRCA1 mRNA levels have been demonstrated to be markedly decreased during the transition from carcinoma *in situ* to invasive cancer in sporadic cases (Thompson *et al.*, 1995). Furthermore, Chen *et al.* (1995) have demonstrated mislocalization of BRCA1 in the cytoplasm in breast cancer cells and cell lines from sporadic cases. Previous experiments have suggested that compromised BRCA1 function plays an

Handbook of Immunohistochemistry and in situ Hybridization of Human Carcinomas, Volume 1: Molecular Genetics; Lung and Breast Carcinomas

361

Copyright © 2004 by Elsevier (USA)
All rights reserved.

important role in the pathogenesis and progression of sporadic breast cancer.

However, controversy has existed during the last few years concerning the subcellular localization of BRCA1 (Chen *et al.*, 1995; Scully *et al.*, 1996; Jensen *et al.*, 1996). Chen *et al.* (1995) reported that BRCA1 protein is localized in the nuclei of normal cells, but is aberrantly localized in the cytoplasm in breast cancer cell lines. They also demonstrated immunohistochemical staining of this protein in 50 fixed tissue sections of breast carcinomas and revealed a variable subcellular location, including 8 in the nuclei, 6 in the cytoplasm, 34 in both the nuclei and cytoplasm, and its absence in 2 cases.

The observations reported by Chen *et al.* (1995) have been challenged by Scully *et al.* (1996), who stated that BRCA1 is exclusively nuclear regardless of cell type. Scully *et al.* (1996) reexamined the subcellular location of BRCA1 by using an affinity purified polyclonal antibody and a panel of monoclonal antibodies. Immunostaining with the various antibodies showed the BRCA1 nuclear signal in normal breast cells and breast cancer cell lines. Furthermore, all antibodies demonstrated a BRCA1 nuclear dot pattern in cells fixed with neutral paraformaldehyde, methanol, or 70% ethanol. Thus, the nuclear location of BRCA1 is not the result of a fixation artifact. BRCA1 concentrated in the nuclear but not in the cytoplasmic fraction was observed by biochemical extraction analysis in unfixed cells, indicating that BRCA1 nuclear subcellular location is independent of cell fixation conditions. Alcoholic, formalin-fixed, paraffin-embedded sections of breast cancer showed a variety of staining patterns, ranging from predominantly nuclear to mainly cytoplasmic. However, when these sections were treated with microwave heating, the staining was predominantly cytoplasmic. In cells fixed in neutral-buffered formalin and pretreated with microwave heating, the BRCA1 staining pattern was predominantly nuclear. Conversely, in neutral-buffered, formalin-fixed cells without microwave treatment, the signal was exclusively cytoplasmic. Thus, the subcellular location of BRCA1 was shown to be nuclear. The discrepancies of BRCA1 staining can be the result of variations in fixation, or staining conditions, or both (Scully *et al.*, 1996).

Studies by Chen *et al.* (1995) and Scully *et al.* (1996) indicate that BRCA1 is a 220-kD protein. However, Jensen *et al.* (1996) reported that BRCA1 as a 190-kD protein was located in both the cytoplasm and the cell membrane, and also suggested that BRCA1 with a granin motif functions as a secreted growth inhibitor but not a tumor suppressor gene product as previously described. Later studies reject the suggestion that BRCA1 is a granin (Koonin *et al.*, 1996; Wilson *et al.*, 1996).

Currently, there has been no independent corroboration to suggest that BRCA1 is secreted. Jensen *et al.* (1996) used C-20 antibody (Santa Cruz Biotechnology, Santa Cruz, CA) raised against residues 1843–1862, which has been proven to cross-react with epidermal growth factor receptor (EGFR) and HER-2 (Bernard-Gallon *et al.*, 1997; Wilson *et al.*, 1996). Wilson *et al.* (1996) have confirmed the reactive band of C-20 antibody migrated at 190 kD, the size expected for EGFR. Interestingly, Wilson *et al.* (1997) and Thakur *et al.* (1997) have demonstrated BRCA1 splice variants that lack exon 11 including the nuclear localization signal (NLS) motifs. As a consequence of splice elimination of NLS, the proteins of splice variants cannot autonomously translocate to the nucleus. Thus, both studies have shown nuclear localization of full-length BRCA1, but cytoplasmic localization of splice variants (Thakur *et al.*, 1997; Wilson *et al.*, 1997).

Wilson *et al.* (1999) have evaluated the specificity and use of 19 anti-BRCA1 antibodies directed against several different epitopes using immunoblotting, immunoprecipitation, immunocytochemical, and immunohistochemical techniques. Sixteen antibodies have identified BRCA1 220 kD in nuclear and total extracts, and several of these antibodies were highly specific (SD118, Ab-1, Ab-2, 17F8, and AP-16). Ab-1 (MS110) and Ab-2 (MS13) are commercially available (Oncogene Research Products, San Diego, California). SD118, AP16, and 17F8 are available from the academic laboratory (Livingston laboratory, Dana-Farber Cancer Institute and Harvard Medical School, Boston, MA and Lee laboratory Cancer for Molecular Medicine/Institute of Biotechnology, University of Texas Health Science Center, San Antonio, TX) (Chen *et al.*, 1995; Scully *et al.*, 1996). Other antibodies (C-20, D-20, and I-20; Santa Cruz Biotechnology) recognized additional proteins that are unlikely to be BRCA1 products of alternative splicing. Furthermore, only Ab-1 antibody (MS110), when combined with antigen retrieval, yielded consistent nuclear staining in formalin-fixed and paraffin-embedded tissues (Wilson *et al.*, 1999). The descriptions of five highly specific antibodies are summarized in Table 24. Perez-Valles *et al.* (2001) have compared the immunohistochemistry of four commercially available anti-BRCA1 antibodies (D-20, I-20, K-18, and MS110) on paraffin-embedded tissue sections from breast cancers. All positive cases showed predominantly cytoplasmic staining with the polyclonal antibodies D-20, I-20, and K-18. After heating pretreatment, both nuclear and cytoplasmic stainings were found with the I-20 antibody. Only monoclonal antibody MS110 showed predominantly nuclear staining after microwave treatment. The results are consistent with the previous report (Wilson *et al.*, 1999).

Table 24 Descriptions of Five Highly Specific BRCA1 Antibodies

Name	Reference	Antigen	Type	Source	Applications				
					IP	IB	ICC	IHC/F	IHC/P
N-terminus									
MS110 (Ab-1)	Scully *et al.* (1996)	r 1-304	Monoclonal	Oncogene Research Products[a]	+++	+++	+++	+++	+++
MS13 (Ab-2)	Scully *et al.* (1996)	r 1-304	Monoclonal	Oncogene Research Products[a]	+++	+++	+++	−	−
Exon-11									
17F8	Chen *et al.* (1995)	r 762-1315	Monoclonal	Lee laboratory[b]	+++	+++	++	+	−
SD118	Unpublished	r 1005-1313	Monoclonal	Livingston laboratory[c]	+++	+++	+++	+++	−
C-terminus AP-16	Scully *et al.* (1996)	r 1313-1863	Monoclonal	Livingston laboratory[c]	+++	+++	+++	−	−

IB, immunoblotting; ICC, immunocytochemistry on cultured cells; IHC/F, immunohistochemistry with frozen cryostat tissue sections; IHC/P, immunohistochemistry with formalin-fixed, paraffin-embedded tissue samples; IP, immunoprecipitation; −, no signal; +, ++, weak signals; +++, strong signals.

[a]San Diego, California.

[b]Center for Molecular Medicine/Institute of Biotechnology, University of Texas Health Science Center, San Antonio, TX.

[c]Dana-Farber Cancer Institute and Harvard Medical School, Boston, MA.

Thus, the confusing observations of BRCA1 subcellular localization may have resulted from the specificity of the antibodies used to detect BRCA1 protein. Certain technical treatment and splice variants of BRCA1 protein may further complicate the situation.

MATERIALS

1. Primary antibody: BRCA1 antibody (Ab-1, clone MS110; Oncogene Research Products).

2. Antibody diluent (Code No. S0809; Dako, Glostrup, Denmark), which contains 0.05 M Tris-HCl buffer (pH 7.2–7.6) and 1% bovine serum albumin.

3. Wash buffer solution: 0.02 M phosphate buffer saline (PBS) (Code No. S3024; Dako), which makes 1 liter of 0.02 mol/L sodium phosphate buffer, 0.15 mol/L NaCl (pH 7.0).

4. Dako LSAB2 system, peroxidase (Code No. K0675).

5. 3% hydrogen peroxide.

6. Methanol.

7. Chromogen: 3-amino-9-ethylcarbazole (AEC) or diaminobenzidine (DAB).

8. Antigen retrieval solution (Code No. S1699; Dako), a modified citrate buffer (pH 6.1).

9. Positive control tissue: normal breast tissue.

10. Counterstain: Mayer's hematoxylin.

11. Ethanol, absolute and 95%.

12. Xylene.

13. Distilled water.

14. Mounting media, such as Dako Faramount (Code No. S3025) or Dako Glycergel (Code No. C0563).

15. Poly L–lysine–coated slides.

16. Coverslips.

17. Humid chamber.

18. Timer.

19. Staining jars.

20. Absorbent wipes.

21. Drying oven, capable of maintaining 60°C or less.

METHOD

1. Cut 4-μm-thick sections from neutral-buffered, formalin-fixed and paraffin-embedded tissue blocks on Poly-L-lysine–coated slides.

2. Place tissue slides in drying oven at 56°C for 20–30 min.

3. Store tissue slides at 2–25°C before immunohistochemical staining.

4. Place tissue slides in a xylene bath and incubate for 5 min. Change baths and repeat once.

5. Tap off excess liquid and place slides in absolute ethanol for 3–5 min. Change baths and repeat once.

6. Tap off excess liquid and place slides in 95% ethanol for 3–5 min. Change baths and repeat once.

7. Change xylene and alcohol solutions after 40 slides.

8. Tap off excess liquid and place slides in distilled water for 5–10 min.

9. Tap off excess liquid and place slides in PBS for 5 min.

10. Tap off excess liquid and place slides in 3% H$_2$O$_2$ in methanol for 10 min.

11. Rinse gently with distilled water or PBS.

12. Place in PBS bath for 5 min.

13. Tap off excess liquid. Using a lintless tissue, carefully wipe around the specimen to remove any remaining liquid.

14. Place slides in antigen retrieval solution and heat in microwave oven at 750 W for 10 min.

15. Place slides at room temperature for 20 min.

16. Place slides in PBS bath for 5 min.

17. Tap off excess liquid and wipe slides as done previously (**step 13**).

18. Dilute primary antibody (Ab-1 antibody) at 1:50 with Dako Antibody Diluent.

19. Apply enough primary antibody to cover specimen.

20. Incubate slides with primary antibody overnight at 4°C.

21. Rinse gently with distilled water or PBS.

22. Place in PBS bath for 5 min.

23. Tap off excess liquid and wipe slides as done previously (**step 13**).

24. Apply enough drops of biotinylated link (Bottle of Dako LSAB2 System) to cover specimen.

25. Incubate for 10 min.

26. Rinse gently with distilled water or PBS.

27. Place in PBS bath for 5 min.

28. Tap off excess liquid and wipe slides as done previously (**step 13**).

29. Apply enough drops of streptavidin–horseradish peroxidase (HRP) (Bottle2 of Dako LSAB2 System) to cover specimen.

30. Incubate for 10 min.

31. Rinse gently with distilled water or PBS.

32. Place in PBS bath for 5 min.

33. Tap off excess liquid and wipe slides as done previously (**step 13**).

34. Apply substrate-chromogen solution (AEC or DAB) to cover specimen.

35. Incubate for 5–10 min.

36. Rinse gently with distilled water or PBS.

37. Place in PBS bath for 5 min.

38. Immerse slides in a bath of hematoxylin.

39. Incubate for 1–3 min, depending on the strength of hematoxylin used.

40. Rinse gently in a distilled water bath.

41. Mount coverslips with an aqueous-based mounting medium.

42. Include positive control slide (normal breast tissue) on each run.

43. Include negative control slide (omitting the primary antibody in normal breast tissue) on each run.

44. Interpret tumors as BRCA1-positive staining if more than 10% of tumor cells have distinct nuclear staining.

RESULTS AND DISCUSSION

Using monoclonal antibody Ab-1 for BRCA1 (MS110) on formalin-fixed and paraffin-embedded tissue sections, BRCA1 was exclusively localized in the nuclei of normal ductal (Figure 55) and lobular epithelia, including the normal breast tissue adjacent to breast cancer. Using the same antibody, loss of BRCA1 nuclear expression has been reported in 20% (Lee *et al.*, 1999), 27% (Yoshikawa *et al.*, 1999), 34.3% (Yang *et al.*, 2001), and 43.5% (Wilson *et al.*, 1999) of sporadic breast carcinoma. Lee *et al.* (1999) reported that in 108 consecutive cases of sporadic breast cancer (invasive ductal carcinoma, not otherwise specified type), 72% had BRCA1 expression exclusively in the nucleus (Figure 56), 1.9% had BRCA1 expression exclusively in the cytoplasm, 2.8% had both nuclear and cytoplasmic expression, and 18.5% were without BRCA1 expression. Complete loss of nuclear BRCA1 expression was demonstrated in 20.4% of invasive ductal carcinomas. In 86 cases with BRCA1 nuclear expression, 96.5% showed more than 50% positive cells and 63.9% showed more than 75% positive cells. BRCA1 nuclear expression could be considered to represent the normal or physiologic phenotype. Altered BRCA1 protein expression may play an important role in sporadic breast carcinoma. A similar conclusion was suggested using other antibodies in other studies (Rio *et al.*, 1999; Taylor *et al.*, 1998; Yoshikawa *et al.*, 1999).

The lack of somatic mutations of BRCA1 in sporadic breast carcinoma (Futreal *et al.*, 1994) suggests that BRCA1 expression might be down-regulated by epigenetic changes other than point mutations. One epigenetic alteration of BRCA1 inactivation that has been reported in sporadic breast carcinoma is hypermethylation of the BRCA1 promoter region (Rice *et al.*, 1998; Catteau *et al.*, 1999, 2002). It has been reported that aberrant methylation of the BRCA1 CpG island promoter is associated with decreased BRCA1 mRNA in sporadic breast cancer cells (Rice *et al.*, 1998). Hypermethylation of CpG-rich areas located within the promoter of genes may be a common mechanism of silencing tumor suppressor genes. However, because methylation was detected only in 11% of breast cancer cases (Catteau *et al.*, 1999), methylation cannot be the sole mechanism mediating the loss of BRCA1 expression in sporadic breast carcinoma. On the other hand, Baldassarre *et al.* (2003) reported that HMGA1b protein binds to and inhibits the activity of

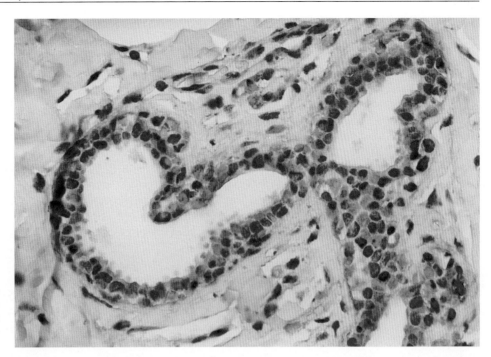

Figure 55 BRCA1 immuno-staining. Strong nuclear immuno-staining of normal ductal epithelia is visible (immunoperoxidase, original magnification X400).

the BRCA1 promoters both *in vitro* and *in vivo*. An inverse correlation between HMGA1 and BRCA1 mRNA and protein expression was also found in breast cancer cell lines and tissue. These results suggest that HMGA1 protein can negatively regulate BRCA1 gene expression in sporadic breast carcinoma. Thus, loss of BRCA1 expression in sporadic breast carcinoma might result from nonmutational mechanisms, such as

changes in upstream regulatory pathways or altered expression of the regulatory gene.

Although the mechanism of BRCA1-mediated tumor suppression and growth regulation is not fully elucidated, BRCA1 has been considered as a caretaker in maintaining genomic stability by cell cycle arrest and DNA repair (Scully *et al.*, 1997b; Somasundaram *et al.*, 1997). BRCA1-associated hereditary breast cancer

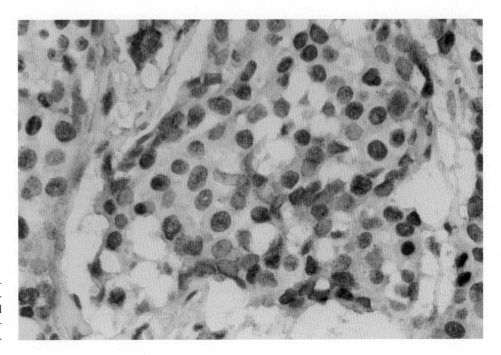

Figure 56 BRCA1 immuno-staining. Strong nuclear immuno-staining of an invasive ductal carcinoma, grade I (immunoperoxidase, original magnification X400).

has been reported to have greater tumor cell proliferation rates (Marcus *et al.*, 1996) and significant association with grade III tumor, indicating poor prognosis (Jacquemier *et al.*, 1995). Loss of BRCA1 nuclear expression has also been demonstrated to be significantly associated with high Ki-67 index (Jarvis *et al.*, 1998; Lee *et al.*, 2002). Furthermore, complete loss of BRCA1 nuclear expression was significantly more frequent in high histologic grade tumors (3.2% in grade I, 26.2% in grade II, and 31.3% in grade III) in sporadic breast cancers (Lee *et al.*, 1999). This observation is supported by other researchers (Taylor *et al.*, 1998; Wilson *et al.*, 1999; Yang *et al.*, 2001). On the basis of these observations, it is suggested that loss of nuclear BRCA1 expression contributes to the pathogenesis and progression of sporadic breast carcinoma. Loss of BRCA1 nuclear expression in sporadic breast carcinomas causes highly proliferative tumor growth and genetic instability, which may trigger further genetic alterations and lead to a more aggressive behavior.

Seery *et al.* (1999) have investigated 42 breast cancers and have reported that decreased BRCA1 mRNA expression is significantly associated with acquisition of metastatic capabilities, suggesting that BRCA1 expression might predict distant metastasis of sporadic breast cancers. Furthermore, BRCA1 expression has been correlated to disease-free survival (DFS) in a cohort of 175 sporadic breast carcinomas using immunohistochemical staining (Yang *et al.*, 2001). Univariate analysis focusing on DFS has revealed axillary lymph node status, histologic grade, and BRCA1 expression as significant prognostic factors. Multivariate Cox regression analysis also demonstrated that axillary lymph node status, histologic grade, and BRCA1 expression are independent prognostic factors. The odds ratio for BRCA1 expression is 5.724, indicating that the risk for patients with BRCA1 negativity dying within a specific time is 5.7 times greater than that for patients with BRCA1 positivity (Yang *et al.*, 2001). Thus, loss of BRCA1 nuclear expression appears to be a poor prognostic biomarker.

It has been reported that patients with BRCA1 mutation have a much worse DFS rate compared with those with wild-type BRCA1 tumors (Chappuis *et al.*, 2000a). Loss of BRCA1 expression by epigenetic mechanisms might confer a similar prognostic outcome to that observed as a consequence of BRCA1 mutation. Consistent with this hypothesis, van't Veer *et al.* (2002) reported a sporadic breast cancer without BRCA1 mutation that had a BRCA1-mutant genetic profile and shared a similar prognosis. Conversely, the extent of apoptosis in tumors can be enhanced by the administration of tamoxifen and almost any kind of

chemotherapeutic drug (Barry *et al.*, 1990; Cameron *et al.*, 2000), but loss of the genes inducing apoptosis or overactivation of the genes blocking apoptosis may make tumors resistant to anticancer agents. It has been reported that BRCA1 facilitates stress-induced apoptosis in breast and ovarian carcinoma cell lines (Thangaraju *et al.*, 2000). Loss of BRCA1 expression might up-regulate the threshold for drug-induced apoptosis, which results in worse prognosis. It is interesting to speculate that BRCA1 expression may be useful in the treatment of breast carcinoma as a decision-making biomarker for aggressive treatment after surgery. However, studies involving large population-based cohorts of patients with breast cancer are needed to definitively determine the prognostic impact of BRCA1 expression in sporadic breast carcinoma.

Young women (< 35 or 40 years old) with breast cancer have a more advanced disease and worse prognosis than older women (Bonnier *et al.*, 1995). It has been reported that breast cancer in women younger than 35 years has a significantly greater incidence of large tumor, high proliferative rate, high histologic grade, loss of BRCA1 nuclear expression, and Bcl-2 negativity than older women (Lee, 2002). No difference is found in lymph node status and *HER-2* and *p53* gene expression between these age-groups. Bogdani *et al.* (2002) have also reported that nuclear BRCA1 expression is significantly less in young women with breast cancer. These findings suggest that young women with breast cancer have frequent loss of BRCA1 nuclear expression, which may be responsible for the specific tumor biology different from older women. However, the frequency of *BRCA1* gene mutations in young women with breast cancer is low, from 6.2% to 10% in white women (Langston *et al.*, 1996; Malone *et al.*, 1998) and 8% in Chinese women (Tang *et al.*, 2001), and is not significantly greater than in a general group of patients with breast cancer. Although BRCA1 mutation cannot be completely responsible for the more aggressive tumors in young women, loss of BRCA1 nuclear expression by epigenetic mechanisms might confer a similar prognostic outcome to the patients with breast cancer with BRCA1 mutation who have an aggressive cancer phenotype (Chappuis *et al.*, 2000a).

BRCA1-associated hereditary breast cancers are more frequently estrogen receptor (ER) negative (Chappuis *et al.*, 2000b). Similarly, a significant correlation between expression of BRCA1 and ER mRNA has been observed (Seery *et al.*, 1999) in sporadic breast carcinoma. Furthermore, loss of BRCA1 nuclear expression has been frequently found in tumors with negative ER in sporadic form (Lee *et al.*, 1999), suggesting a functional relationship between these two genes. The positive correlation between BRCA1 and ER expression is consistent with the results reported by Gudas *et al.* (1995), whose data suggest that BRCA1 expression is regulated by the steroid hormones in the human breast cancer cell. The increase in BRCA1 expression on stimulation with estrogen is not coordinated with the early induction of the estrogen-dependent p53 gene, but closely parallels the delayed increase in the S-phase–dependent marker cyclin A. Steroid hormones might induce BRCA1 transcription indirectly by altering the proliferative status of the cells rather than acting directly on DNA sequences in the *BRCA1* gene itself (Gudas *et al.*, 1995).

Apoptosis is believed to act as a counterbalance to proliferation and is a critical factor in tissue homeostasis. Dysregulation of the apoptosis process may therefore play a crucial role in oncogenesis. It has been reported that BRCA1 is associated with the p53 tumor suppressor protein and functions as a transcriptional coactivator for p53 (Zhang *et al.*, 1998). BRCA1 can induce apoptosis, which is enhanced by the coexpression of p53 (Zhang *et al.*, 1998). Yang *et al.* (2002) have investigated the relationship between alterations of BRCA1 expression and the apoptosis-related proteins p53, Bcl-2, and bax in sporadic breast carcinomas. It has shown that BRCA1 expression was positively correlated with Bcl-2 expression, but not with BRCA1, p53, or bax expressions. These results are consistent with the report by Lee *et al.* (1999). These authors have demonstrated that loss of BRCA1 expression is significantly correlated with Bcl-2 negativity, but there was no correlation between BRCA1 and p53 expression. Furthermore, loss expression of Bcl-2 is also observed in BRCA1-associated breast cancers (Freneaux *et al.*, 2000).

Most BRCA1-associated breast carcinomas are characterized by high mitotic and apoptotic rates and a decrease of Bcl-2 expression. Therefore, the regulating relationship between *Bcl-2* and *BRCA1* genes may exit in both sporadic and hereditary breast carcinomas. The alterations of BRCA1 protein might be responsible for the down-regulation of Bcl-2, and thus for the high proliferative rate and increased apoptosis. However, the down-regulation of Bcl-2 in tumors with loss of BRCA1 expression might merely reflect the down-regulation of negative ER. Because BRCA1-associated breast carcinomas are frequently ER negative (Chappuis *et al.*, 2000b) and a positive association exists between the expression of ER and Bcl-2 (Bhargava *et al.*, 1994), there remain many questions to be addressed in the regulation mechanism between BRCA1 and Bcl-2.

CONCLUSION

It is concluded that *BRCA1* is located in nuclear foci of all cell types, including normal and malignant breast cells. BRCA1 is a tumor suppressor gene and acts as caretaker in maintaining genetic stability by cell cycle arrest and DNA repair. Although somatic BRCA1 mutation has not been reported in sporadic breast cancer, loss of nuclear BRCA1 expression may contribute to the pathogenesis and progression of sporadic breast carcinoma. Loss of BRCA1 protein expression in sporadic breast cancers causes highly proliferative tumor growth and genetic instability, which may trigger further genetic alterations and lead to a more aggressive behavior. Only monoclonal antibody Ab-1 for BRCA1 (MS110) (commercially available) is highly specific and can be applied on formalin-fixed, paraffin-embedded tissue sections that show a predominantly nuclear staining after microwave treatment.

References

Baldassarre, G., Battista, S., Belletti, B., Thakur, S., Pentimalli, F., Trapasso, F., Fedele, M., Pierantoni, G., Croce, C.M., and Fusco, A. 2003. Negative regulation of BRCA1 gene expression by HMGA1 proteins accounts for the reduced BRCA1 protein levels in sporadic breast carcinoma. *Mol. Cell. Biol. 23:*2225–2238.

Barry, M.A., Behnke, C.A., and Eastman, A. 1990. Activation of programmed cell death (apoptosis) by cisplatin, other anticancer drugs, toxins and hyperthermia. *Biochem. Pharmacol. 40:*2353–2362.

Bernard-Gallon, D.J., Crespin, N.C., Maurizis, J.C., and Bignon, Y.J. 1997. Cross-reaction between antibodies raised against the last 20 C-terminal amino acids of BRCA 1 (C-20) and human EGF and EGF-R in MCF 10a human mammary epithelial cell line. *Int. J. Cancer 71:*123–126.

Bhargava, V., Kell, D.L., van de Rijn, M., and Warnke, R.A. 1994. Bcl-2 immunoreactivity in breast carcinoma correlates with hormone receptor positivity. *Am. J. Pathol. 145:*535–540.

Bogdani, M., Teugels, E., De Greve, J., Bourgain, C., Neyns, B., and Pipeleers-Marichal, M. 2002. Loss of nuclear BRCA1 localization in breast carcinoma is age dependent. *Virchows Arch. 440:*274–279.

Bonnier, P., Romain, S., Charpin, C., Lejeune, C., Tubiana, N., Martin, P.M., and Piana, L. 1995. Age as a prognostic factor in breast cancer: Relationship to pathologic and biologic features. *Int. J. Cancer 62:*138–144.

Cameron, D.A., Keen, J.C., Dixon, J.M., Bellamy, C., Hanby, A., Anderson, T.J., and Miller, W.R. 2000. Effective tamoxifen therapy of breast cancer involves both antiproliferative and pro-apoptotic changes. *Eur. J. Cancer 36:*845–851.

Catteau, A., Harris, W.H., Xu, C.F., and Solomon, E. 1999. Methylation of the BRCA1 promoter region in sporadic breast and ovarian cancer: Correlation with disease characteristics. *Oncogene 18:*1957–1965.

Catteau, A., and Morris, J.R. 2002. BRCA1 methylation: A significant role in tumour development? *Semin. Cancer Biol. 12:*359–371.

Chappuis, P.O., Kapusta, L., Begin, L.R., Wong, N., Brunet, J.S., Narod, S.A., Slingerland, J., and Foulkes, W.D. 2000a. Germline

BRCA1/2 mutations and p27(Kip1) protein levels independently predict outcome after breast cancer. *J. Clin. Oncol.* *18:*4045–4052.

Chappuis, P.O., Nethercot, V., and Foulkes, W.D. 2000b. Clinicopathological characteristics of BRCA1- and BRCA2-related breast cancer. *Semin. Surg. Oncol. 18:*287–295.

Chen, Y., Chen, C.F., Riley, D.J., Allred, D.C., Chen, P.L., Von Hoff, D., Osborne, C.K., and Lee, W.H. 1995. Aberrant subcellular localization of BRCA1 in breast cancer. *Science 270:*789–791.

Chen, Y., Farmer, A.A., Chen, C.F., Jones, D.C., Chen, P.L., and Lee, W.H. 1996. BRCA1 is a 220-kDa nuclear phosphoprotein that is expressed and phosphorylated in a cell cycle-dependent manner. *Cancer Res. 56:*3168–3172.

Ford, D., Easton, D.F., Stratton, M., Narod, S., Goldgar, D., Devilee, P., Bishop, D.T., Weber, B., Lenoir, G., Chang-Claude, J., Sobol, H., Teare, M.D., Struewing, J., Arason, A., Scherneck, S., Peto, J., Rebbeck, T.R., Tonin, P., Neuhausen, S., Barkardottir, R., Eyfjord, J., Lynch, H., Ponder, B.A., Gayther, S.A., and Zelada-Hedman, M. 1998. Genetic heterogeneity and penetrance analysis of the BRCA1 and BRCA2 genes in breast cancer families: The Breast Cancer Linkage Consortium. *Am. J. Hum. Genet. 62:*676–689.

Freneaux, P., Stoppa-Lyonnet, D., Mouret, E., Kambouchner, M., Nicolas, A., Zafrani, B., Vincent-Salomon, A., Fourquet, A., Magdelenat, H., and Sastre-Garau, X. 2000. Low expression of bcl-2 in Brca1-associated breast cancers. *Br. J. Cancer 83:*1318–1322.

Futreal, P.A., Liu, Q., Shattuck-Eidens, D., Cochran, C., Harshman, K., Tavtigian, S., Bennett, L.M., Haugen-Strano, A., Swensen, J., and Miki, Y. 1994. BRCA1 mutations in primary breast and ovarian carcinomas. *Science 266:*120–122.

Gudas, J.M., Nguyen, H., Li, T., and Cowan, K.H. 1995. Hormone-dependent regulation of BRCA1 in human breast cancer cells. *Cancer Res. 55:*4561–4565.

Jacobs, I.J., Smith, S.A., Wiseman, R.W., Futreal, P.A., Harrington, T., Osborne, R.J., Leech, V., Molyneux, A., Berchuck, A., and Ponder, B.A. 1993. A deletion unit on chromosome 17q in epithelial ovarian tumors distal to the familial breast/ovarian cancer locus. *Cancer Res. 53:*1218–1221.

Jacquemier, J., Eisinger, F., Birnbaum, D., and Sobol, H. 1995. Histoprognostic grade in BRCA1-associated breast cancer. *Lancet 345:*1503.

Jarvis, E.M., Kirk, J.A., and Clarke, C.L. 1998. Loss of nuclear BRCA1 expression in breast cancers is associated with a highly proliferative tumor phenotype. *Cancer Genet. Cytogenet. 101:*109–115.

Jensen, R.A., Thompson, M.E., Jetton, T.L., Szabo, C.I., van der Meer, R., Helou, B., Tronick, S.R., Page, D.L., King, M.C., and Holt, J.T. 1996. BRCA1 is secreted and exhibits properties of a granin. *Nat. Genet. 12:*303–308.

Koonin, E.V., Altschul, S.F., and Bork, P. 1996. BRCA1 protein products ... Functions motifs... *Nat. Genet. 13:*266–268.

Lane, T.F., Deng, C., Elson, A., Lyu, M.S., Kozak, C.A., and Leder, P. 1995. Expression of Brca1 is associated with terminal differentiation of ectodermally and mesodermally derived tissues in mice. *Genes Dev. 9:*2712–2722.

Langston, A.A., Malone, K.E., Thompson, J.D., Daling, J.R., and Ostrander, E.A. 1996. BRCA1 mutations in a population-based sample of young women with breast cancer. *N. Engl. J. Med. 334:*137–142.

Lee, W.Y. 2002. Frequent loss of BRCA1 nuclear expression in young women with breast cancer: An immunohistochemical study from an area of low incidence but early onset. *Appl. Immunohistochem. Mol. Morphol. 10:*310–315.

Lee, W.Y., Jin, Y.T., Chang, T.W., Lin, P.W., and Su, I.J. 1999. Immunolocalization of BRCA1 protein in normal breast tissue and sporadic invasive ductal carcinomas: A correlation with other biological parameters. *Histopathology 34:*106–112.

Malone, K.E., Daling, J.R., Thompson, J.D., O'Brien, C.A., Francisco, L.V., and Ostrander, E.A. 1998. BRCA1 mutations and breast cancer in the general population: Analyses in women before age 35 years and in women before age 45 years with first-degree family history. *JAMA 279:*922–929.

Marcus, J.N., Watson, P., Page, D.L., Narod, S.A., Lenoir, G.M., Tonin, P., Linder-Stephenson, L., Salerno, G., Conway, T.A., and Lynch, H.T. 1996. Hereditary breast cancer: Pathobiology, prognosis, and BRCA1 and BRCA2 gene linkage. *Cancer 77:*697–709.

Marquis, S.T., Rajan, J.V., Wynshaw-Boris. A., Xu, J., Yin, G.Y., Abel, K.J., Weber, B.L., and Chodosh, L.A. 1995. The developmental pattern of BRCA1 expression implies a role in differentiation of the breast and other tissues. *Nat. Genet. 11:*17–26.

Merajver, S.D., Pham, T.M., Caduff, R.F., Chen, M., Poy, E.L., Cooney, K.A., Weber, B.L., Collins, F.S., Johnston, C., and Frank, T.S. 1995. Somatic mutations in the BRCA1 gene in sporadic ovarian tumours. *Nat. Genet. 9:*439–443.

Miki, Y., Swensen, J., Shattuck-Eidens, D., Futreal, P.A., Harshman, K., and Tavtigian, S. 1994. A strong candidate for the breast and ovarian cancer susceptibility gene BRCA1. *Science 266:*66–71.

Monteiro, A.N., August, A., and Hanafusa, H. 1996. Evidence for a transcriptional activation function of BRCA1 C-terminal region. *Proc. Natl. Acad. Sci. USA 93:*13595–13599.

Perez-Valles, A., Martorell-Cebollada, M., Nogueira-Vazquez, E., Garcia-Garcia, J.A., and Fuster-Diana, E. 2001. The usefulness of anitbodies to the BRCA1 protein in detecting the mutated BRCA1 gene: An immunohistochemical study. *J. Clin. Pathol. 54:*476–480.

Rice, J.C., Massey-Brown, K.S., and Futscher, B.W. 1998. Aberrant methylation of the BRCA1 CpG island promoter is associated with decreased BRCA1 mRNA in sporadic breast cancer cells. *Oncogene 17:*1807–1812.

Rio, P.G., Maurizis, J.C., Peffault de Latour, M., Bignon, Y.J., and Bernard-Gallon, D.J. 1999. Quantification of BRCA1 protein in sporadic breast carcinoma with or without loss of heterozygosity of the BRCA1 gene. *Int. J. Cancer 80:*823–826.

Scully, R., Chen, J., Ochs, R.L., Keegan, K., Hoekstra, M., Feunteun, J., and Livingston, D.M. 1997a. Dynamic changes of BRCA1 subnuclear location and phosphorylation state are initiated by DNA damage. *Cell 90:*425–435.

Scully, R., Chen, J., Plug, A., Xiao, Y., Weaver, D., Feunteun, J., Ashley, T., and Livingston, D.M. 1997b. Association of BRCA1 with Rad51 in mitotic and meiotic cells. *Cell 88:*265–275.

Scully, R., Ganesan, S., Brown, M., De Caprio, J.A., Cannistra, S.A., Feunteun, J., Schnitt, S., and Livingston, D.M. 1996. Location of BRCA1 in human breast and ovarian cancer cells. *Science 272:*123–126.

Seery, L.T., Knowlden, J.M., Gee, J.M.W., Robertson, J.F.R., Kenny, F.S., Ellis, I.O., and Nicholson, R.I. 1999. BRCA1 expression levels predict distant metastasis of sporadic breast cancers. *Int. J. Cancer 84:*258–262.

Smith, S.A., Easton, D.F., Evans, D.G., and Ponder, B.A. 1992. Allele losses in the region 17q12-21 in familial breast and ovarian cancer involve the wild-type chromosome. *Nat. Genet. 2:*128–131.

Somasundaram, K., Zhang, H., Zeng, Y.X., Houvras, Y., Peng, Y., Zhang, H., Wu, G.S., Licht, J.D., Weber, B.L., and El-Deiry, W.S. 1997. Arrest of the cell cycle by the tumour-suppressor

BRCA1 requires the CDK-inhibitor p21WAF1/CiP1. *Nature* 389:187–190.

Tang, N.L., Choy, K.W., Pang, C.P., Yeo, W., and Johnson, P.J. 2001. Prevalence of breast cancer predisposition gene mutations in Chinese women and guidelines for genetic testing. *Clin. Chim. Acta. 313*:179–185.

Taylor, J., Lymboura, M., Pace, P.E., A'hern, R.P., Desai, A.J., Shousha, S., Coombes, R.C., and Ali, S. 1998. An important role for BRCA1 in breast cancer progression is indicated by its loss in a large proportion of non-familial breast cancers. *Int. J. Cancer 79*:334–342.

Thakur, S., Zhang, H.B., Peng, Y., Le, H., Carroll, B., Ward, T., Yao, J., Farid, L.M., Couch, F.J., Wilson, R.B., and Weber, B.L. 1997. Localization of BRCA1 and a splice variant identifies the nuclear localization signal. *Mol. Cell Biol. 17*:444–452.

Thangaraju, M., Kaufmann, S.H., and Couch, F.J. 2000. BRCA1 facilitates stress-induced apoptosis in breast and ovarian cancer cell lines. *J. Biol. Chem. 275*:33487–33496.

Thompson, M.E., Jensen, R.A., Obermiller, P.S., Page, D.L., and Holt, J.T. 1995. Decreased expression of BRCA1 accelerates growth and is often present during sporadic breast cancer progression. *Nat. Genet. 9*:444–450.

van't Veer, L.J., Dai, H., van de Vijver, M.J., He, Y.D., Hart, A.A., Mao, M., Peterse, H.L., van der Kooy, K., Marton, M.J., Witteveen, A.T., Schreiber, G.J., Kerkhoven, R.M., Roberts, C., Linsley, P.S., Bernards, R., and Friend, S.H. 2002. Gene expression profiling predicts clinical outcome of breast cancer. *Nature 415*:530–536.

Wilson, C.A., Payton, M.N., Elliott, G.S., Buaas, F.W., Cajulis, E.E., Grosshans, D., Ramos, L., Reese, D.M., Slamon, D.J., and Calzone, F.J. 1997. Differential subcellular localization, expression and biological toxicity of BRCA1 and the splice variant BRCA1-delta11b. *Oncogene 14*:1–16.

Wilson, C.A., Payton, M.N., Pekar, S.K., Zhang, K., Pacifici, R.E., Gudas, J.L., Thukral, S., Calzone, F.J., Reese, D.M., and Slamon, D.I. 1996. BRCA1 protein products: Antibody specificity. *Nat. Genet. 13*:264–265.

Wilson, C.A., Ramos, L., Villasenor, M.R., Anders, K.H., Press, M.F., Clarke, K., Karlan, B., Chen, J.J., Scully, R., Livingston, D., Zuch, R.H., Kanter, M.H., Cohen, S., Calzone, F.J., and Slamon, D.J. 1999. Localization of human BRCA1 and its loss in high-grade, non-inherited breast carcinomas. *Nat. Genet. 21*:236–240.

Yang, Q., Sakurai, T., Mori, I., Yoshimura, G., Nakamura, M., Nakamura, Y., Suzuma, T., Tamaki, T., Umemura, T., and Kakudo, K. 2001. Prognostic significance of BRCA1 expression in Japanese sporadic breast carcinomas. *Cancer 92*:54–60.

Yang, Q., Yoshimura, G., Nakamura, M., Nakamura, Y., Suzuma, T., Umemura, T., Mori, I., Sakurai, T., and Kakudo, K. 2002. Correlation between BRCA1 expression and apoptosis-related biological parameters in sporadic breast carcinomas. *Anticancer Res. 22*:3615–3619.

Yoshikawa, K., Honda, K., Inamoto, T., Shinohara, H., Yamauchi, A., Suga, K., Okuyama, T., Shimada, T., Kodama, H., Noguchi, S., Gazdar, A.F., Yamaoka, Y., and Takahashi, R. 1999. Reduction of BRCA1 protein expression in Japanese sporadic breast carcinomas and its frequent loss in BRCA1-associated cases. *Clin. Cancer Res. 5*:1249–1261.

Zhang, H., Somasundaram, K., Peng, Y., Tian, H., Zhang, H., Bi, D., Weber, B.L., and El-Deiry, W.S. 1998. BRCA1 physically associates with p53 and stimulates its transcriptional activity. *Oncogene 16*:1713–1721.

14

Role of Immunohistochemical Detection of BRCA1 in Breast Cancer

Marika Bogdani

Introduction

A small proportion of breast cancers, in particular those diagnosed in young patients, are attributable to hereditary predisposition. *BRCA1* was the first gene linked to breast cancer. Its discovery opened a new area of research into the function of this gene. A few studies used immunochemistry and cell fractionation analysis to examine the subcellular localization of the BRCA1 gene product, mainly in normal and tumor cell lines, and their conclusions are often in contradiction. In addition, little was known about the application of antibodies for immunohistochemistry (IHC). We examined the expression of the *BRCA1* gene at the protein level and described its staining pattern in breast cancers, hoping to contribute to the understanding of the cellular localization of the BRCA1 gene product and its role in sporadic, especially early-onset, breast cancer.

First, a short description of breast morphology and physiology, breast carcinoma, tumor suppressor genes, and the BRCA1 gene and protein is given in the following sections.

Breast Morphology and Physiology

The breast or mammary gland tissue comprises parenchyma and stroma. The parenchyma is composed of 15 to 20 segments. Each segment contains a branching system that is labeled in the following order from the nipple opening: collecting duct, lactiferous sinus, segmental duct, subsegmental duct, ductule, and lobule. The last two structures are often described as terminal duct lobular unit. The whole branching system is lined by two cell layers: the inner cells are epithelial cells and the outer cells are myoepithelial cells, wedged between the bases of two adjacent inner cells. The specialized and interlobular stroma contains connective tissue, blood vessels, nerves, and lymphatics (Tavassoli, 1992).

Development and function of the breast are regulated by hormonal factors. The ovarian hormones regulate the proliferation of the epithelial cells of the mammary gland in women during the menstrual cycle. Proliferation of the mammary epithelial cells is maximal in the luteal phase of the cycle with an increase in the number of cells per terminal duct lobular unit

Handbook of Immunohistochemistry and *in situ* Hybridization of Human Carcinomas, Volume 1: Molecular Genetics; Lung and Breast Carcinomas

371

Copyright © 2004 by Elsevier (USA)
All rights reserved.

(Ferguson *et al.*, 1981). In the postmenopausal period, when estrogen levels are low, rates of breast epithelial cell proliferation are very low.

Breast Cancer

Breast cancer is one of the most common malignancies in Western countries. Approximately 1 in 8 to 10 women will be diagnosed with this disease (Kelsey *et al.*, 1993). A positive family history and hormonal and environmental factors have been identified as major contributors to the risk for development of breast cancer (Berg, 1984). The large majority of breast cancer cases are thought to be sporadic. Approximately 5% of breast cancers, however, are believed to arise from the inheritance of mutations in genes that confer an increased risk for the disease (Claus *et al.*, 1991). In such cases, breast cancer usually presents at a younger age, is often bilateral, and tends to cluster in families along with other cancer types, mainly ovarian cancer.

The breast carcinoma originates from the epithelial lining of the terminal mammary ducts (Azzopardi, 1979; Tavassoli, 1992). A major distinction is made between invasive (or infiltrating) carcinoma and non-invasive (*in situ*) carcinoma. The most frequently observed histologic type is infiltrating ductal carcinoma, whereas the other invasive subtypes are found in 1–10% of patients with breast cancer.

Cancer and Tumor Suppressor Genes

To evolve to a cancer phenotype, a cell has to escape from several surveillance mechanisms. The tightly regulated mechanisms that control cell division have to be circumvented, and the "protective" mechanisms responsible for the elimination of "unwanted" cells by apoptosis or immune response have to be inactivated. Also, to acquire invasive or metastatic properties, cellular interaction properties have to be modified. It is known that these cellular mechanisms are disregulated in cancer cells through the accumulation of mutations in specific genes: the cancer genes (proto-oncogenes and tumor suppressor genes). Tumor suppressor genes differ from proto-oncogenes in that both alleles have to be mutated to contribute to the cancer phenotype. It has been estimated that about three to seven mutations in cancer genes are necessary for a cell to become cancerous (Bruce *et al.*, 1994). In the majority of cases, all those mutations occur in somatic cells. They arise spontaneously or can be induced (e.g., by radiation, food, tabacco). Usually, the resulting tumors show no obvious clustering (familial, regional) and are therefore called sporadic tumors.

In a small proportion of families, however, there is a clear clustering of specific cancer types. Familial predisposing factors for colon cancer, breast cancer, and ovarian cancer have been identified. In such families, a mutation in a particular tumor suppressor gene in the germ cell line, and therefore present in all cells, is responsible for the high likelihood that cancer will develop. This first inactivating mutation is recessive and can be passed through the germline, giving rise to an inherited predisposition. A second mutation later in the life in the remaining normal allele will result in the complete absence of the normal protein product. This second mutation always occurs at the somatic level. The inactivation of the second allele of a tumor suppressor gene makes a functional contribution to the development of cancer.

BRCA1 Gene and BRCA2 Protein

BRCA1 Gene: Discovery, Structure, and Regulation of Expression

BRCA1, the first gene found to be associated with susceptibility to breast cancer, was localized to chromosome 17q21 by genetic linkage analysis in families with many cases of early-onset breast cancer (Hall *et al.*, 1990). Subsequent analysis by an international consortium of groups indicated that mutations in *BRCA1* account for disease in 80–90% of families with both breast and ovarian cancer, and in about 50% of families with breast cancer alone (Ford *et al.*, 1998). *BRCA2* (Wooster, *et al.*, 1994), *p53* (Lane, 1994), *PTEN* (Li *et al.*, 1997) and other undiscovered cancer predisposing genes are responsible for causing disease in the remaining families.

Currently, about 700 different germline mutations occurring throughout the *BRCA1* gene have been described (Breast Cancer Information Core database). These mutations are highly penetrant, conferring a risk of 59% by age 50 years and 82% by age 70 years. Only a few somatic mutations have been detected in sporadic breast or ovarian cancer (Merajver *et al.*, 1995; van der Looij *et al.*, 2000).

The *BRCA1* gene had been isolated by positional cloning in the latter half of 1994 (Miki *et al.*, 1994). It is composed of 22 coding exons with 5592 nucleotides distributed over roughly 100 kb of genomic DNA (Miki *et al.*, 1994). The gene encodes a 7.8-kb transcript. This transcript, presumed to be the full-length mRNA, is most abundant in testis and thymus. Low quantities are found in breast, ovary, uterus, and spleen. Several forms of alternatively spliced transcripts have been described. The *BRCA1* promoter regulates transcription initiation at two distinct transcription start sites (Xu *et al.*, 1997). It was speculated that estrogen stimulates *BRCA1* transcription (Romagnolo *et al.*, 1998), but the activation of *BRCA1* promoter is thought to be indirect, secondary to hormone-induced cell cycle regulation (Gudas *et al.*, 1995; Marks *et al.*, 1997).

BRCA1 Protein: Structure, Function, Localization, and Regulation

The BRCA1 gene product is a protein of 1863 amino acids. It contains in its NH2-terminal a single C3HC4-type Zinc finger domain, which is found in a wide variety of proteins of diverse origin and function and appears to interact with DNA either through direct binding or indirectly by mediating protein–protein interactions (Freemont, 1993). Although more than 80 RING finger–containing proteins have been currently identified, a specific cellular function has not yet been associated with the motif (Saurin *et al.*, 1996). Nevertheless, there is evidence to support the importance of the RING finger to BRCA1 function (Meza *et al.*, 1999; Wu *et al.*, 1996). The BRCA1 C-terminal region is negatively charged and contains two tandem copies of BRCT (BRCA1 C-terminus) region—a newly identified domain of unknown function, which is found in proteins involved in DNA repair, recombination, and cell cycle control (Callebaut and Mornon, 1997). The central portion of the protein bears two putative nuclear localization signals (Chen *et al.*, 1996). A putative granin consensus, typical for granins (proteins that are precursors of several secreted peptides), has been reported (Jensen *et al.*, 1996).

Although implicated in different cellular pathways, the mechanism by which inactivation of BRCA1 might lead to malignant transformation is unknown. Evidence to support the view that *BRCA1* functions as a tumor suppressor (Miki *et al.*, 1994) was apparent even before the gene had been cloned. Loss of heterozygosity has been observed in both sporadic and familial tumors (Borg *et al.*, 1994; Cropp *et al.*, 1994; Futreal *et al.*, 1992) and was almost always found to affect the wild-type allele in the familial tumors (Smith *et al.*, 1992). Constitutive mutations in the *BRCA1* gene were found in families that had shown evidence of linkage of breast or ovarian cancer, or both, to *BRCA1*. Most of these mutations lead to the synthesis of a nonfunctional truncated protein as a result of a nonsense mutation or by the insertion/deletion of short sequences resulting in a frameshift (Breast Cancer Information Core database; Miki *et al.*, 1994). Complete somatic deletion of one normal allele of *BRCA1* occurs in sporadic breast cancer (Cropp *et al.*, 1994). These observations and experimental data from studies of BRCA1 mRNA level (Magdinier *et al.*, 1998; Ozçelik *et al.*, 1998; Thompson *et al.*, 1995)

support a role of *BRCA1* not only in inherited, but also (although an apparently minor role) in sporadic breast cancer formation. However, the exact biochemical and cellular functions of *BRCA1* remain elusive, as do some basic characteristics, such as the subcellular localization. Nevertheless, evidence for important roles of *BRCA1* in normal functioning of cells is accumulating (Scully *et al.*, 2000).

BRCA1 in DNA Repair

The BRCT domain has been recognized as common to BRCA1, a p53 binding protein, 53BP1, and a number of other proteins (Koonin *et al.*, 1996). It appears to participate in either DNA binding or in protein–protein interactions (Koonin *et al.*, 1996; Scully *et al.*, 1997a). The BRCT domain may be involved directly in DNA repair and metabolism because a high proportion of the characterized proteins that contain it participate in such functions (Callebaut *et al.*, 1997). BRCA1 is shown to have a specific interaction with Rad51 protein (Scully *et al.*, 1997c). Rad51 is a member of a protein family involved with the homologous pairing of DNA during the process of recombination and recombinational repair (Baumann *et al.*, 1996; Sung, 1994), supporting the suggestion that BRCA1 is involved in DNA repair processes. Additional evidence supporting the participation of BRCA1 in DNA damage response comes from interactions observed between BRCA1 and p53 pathways. It is found that BRCA1 can specifically stimulate p53 response element, acting as a coactivator of *p53*-driven gene expression, and may play an important role in cell growth arrest, apoptosis, and DNA damage repair processes (Ouchi *et al.*, 1998). A p53 null background partially rescues the BRCA1-null phenotype (Hakem *et al.*, 1997). This suggests that BRCA1 may be responsible for DNA sensing and repair. When its function is lost, DNA damage accumulates and p53-responsive pathways cause cellular arrest and apoptosis. Loss of p53 function in these cells then allows the cells to continue dividing with unrepaired damage until genomic stability is sufficiently compromised to result in cell death (Brugarolas *et al.*, 1997). Futhermore, mutations in the *p53* gene were detected in the half of breast tumor DNA of patients with a family history of breast cancer (Glebov *et al.*, 1994) and in all BRCA1-associated tumors (Crook *et al.*, 1997), suggesting that loss of p53 function may be an absolute requirement for the progression of cancer. Large, multisubunit protein complexes containing the BRCA1 protein (BASC, or BRCA1-associated genome surveillance complex), and also tumor suppressors and DNA repair proteins such as MSH2, MSH6, MLH1, ATM, BLM, and the RAD50-MRE11-NBS1 protein complex, have been isolated from nuclear extracts and were ascribed a role in sensoring abnormal DNA structures and in repair processes (Wang *et al.*, 2000). More recently, it has been shown that cells mutated for BRCA1 and BRCA2 display defects in DNA repair by homologous recombination, which further support the view of BRCA1 participation in DNA repair pathway (Jasin, 2002).

BRCA1 in Cell Cycle Control

There is evidence that BRCA1 affects the cell cycle: reduced expression is associated with highly proliferative cells *in vivo* (Sobol *et al.*, 1996). Experimental inhibition of BRCA1 expression produce accelerated growth of normal and malignant mammary cells (Thompson *et al.*, 1995), and high expression of BRCA1 is toxic to cells (Holt *et al.*, 1996; Wilson *et al.*, 1997). Various cell cycle regulated and regulatory proteins interact with BRCA1 (Chen *et al.*, 1996b; Gudas *et al.*, 1996; Jin *et al.*, 1997; Ruffner and Verma *et al.*, 1997), and BRCA1 protein changes its subnuclear localization during the cell cycle (Scully *et al.*, 1996, 1997b). It has been proposed that BRCA1 is a DNA-binding transcription activator (Miki *et al.*, 1994), as supported by the demonstration of transcriptional activation *in vitro* by the BRCA1 carboxyl-terminal domain (Chapman *et al.*, 1996; Haile *et al.*, 1999; Monteiro *et al.*, 1996) and the association of BRCA1 with the RNA polymerase II holoenzyme (Chiba *et al.*, 2002; Scully *et al.*, 1997a, 2000). It is suggested that BRCA1 can contribute to cell cycle arrest and growth suppression through the induction of p21. The protein p21 is a universal cell cycle inhibitor, serving as a potent growth inhibitor and effector of cell cycle checkpoints. BRCA1 may transcriptionally induce p21 expression, and thus negatively regulate cell cycle progression (Somasundaram *et al.*, 1997).

BRCA1 in Development and Differentiation

Several studies suggest a role of BRCA1 in development and differentiation. The nullizygous mouse models show that BRCA1 is essential for cellular survival during embryogenesis (Hakem, 1997). The protein is expressed throughout all tissues of the developing animals, particularly those that are rapidly proliferating and undergoing differentiation (Korhonen *et al.*, 2003; Marquis *et al.*, 1995).

As for the subcellular localization of the BRCA1 protein, three different situations have been described: BRCA1 has been found in the nuclei of the normal cells of different types, in tumor cells other than breast and ovary, and in the cytoplasm of breast and ovarian carcinoma cells (Chen Y *et al.*, 1995). Another study locates BRCA1 exclusively in the nuclei of normal and tumor

cells (Scully *et al.*, 1996), whereas a third study describes BRCA1 in cytoplasmic secretory vesicles (Jensen *et al.*, 1996).

Because the subcellular localization of the BRCA1 protein may provide essential information with regard to its function and the potential role of immunohistochemistry for mutation screening, the immunohistochemical pattern of BRCA1 in breast tissue was analyzed. Therefore, we developed a reliable immunohistochemical method using commercially available antibodies to obtain more insight in the role of BRCA1 in breast cancer. Using this method, we analyzed the BRCA1 expression in normal and tumor specimens obtained from young and old breast cancer patients.

MATERIALS

Patient and Tissue Selection

Tissue was obtained from patients with breast cancer who underwent breast surgery at the Academic Hospital (VUB, Brussels, Belgium). Details of the selected cases were as follows: breast reduction specimens (12) for esthetic purposes from patients aged 20–23 years, consecutive tumorectomy and mastectomy specimens (52) from patients aged 24–40 years ("young patients"), consecutive tumorectomy and mastectomy specimens (50) taken from patients aged 60–93 years ("old patients").

For this study, all patients were checked for a possible family history of breast or ovarian carcinoma, or both. We could identify 11 "young" patients who had one or more relatives (mother, grandmother, or aunt) with breast cancer and one "old" patient, 86 years at time of diagnosis, whose mother had breast carcinoma.

Blood from six young patients bearing a breast cancer negative for BRCA1 immunostaining was tested by the protein truncation test and by single strand conformation polymorphism analysis to search for possible mutations in exon 11 and exons 2, 5, and 20. The staining control tissues were fixed either in formalin or in Bouin's solution. The patient material used in the study were fixed in Bouin's solution.

Antibodies and Peptides

Primary Antibodies

Four commercial antibodies from Santa Cruz, Inc. (Santa Cruz, CA) are affinity-purified rabbit polyclonal antibodies:

▲ D-20 (Cat. No. sc-641) is raised against a peptide corresponding to amino acids 2–21,

▲ K-18 (Cat. No. sc-1021) is raised against a peptide corresponding to amino acids 70–89,

▲ I-20 (Cat. No. sc-646) recognizes the amino acids 1823–1842, and

▲ C-20 (Cat. No. sc-642) reacts with amino acids 1843–1862.

Secondary Antibody

The biotinylated goat anti-rabbit Ig (code RPN 1004; Amersham, UK) was used as the secondary antibody.

Synthetic Blocking Peptides

D-20 P (Cat. No. sc-641 P), K-18 P (Cat. No. sc-1021 P), I-20 P (Cat. No. sc-646 P), and C-20 P (Cat. No. sc-642 P) from Santa Cruz were used as blocking peptides for the Santa Cruz polyclonal antibodies.

Additional Materials

1. Fixatives:

    3.7% neutral-buffered formaldehyde: 100 ml formaldehyde 37%, 900 ml phosphate buffer Bouin's fixative: 750 ml saturated picric acid, 250 ml formaldehyde 37%, 50 ml glacial acetic acid.

2. Toluol

3. Alcohol 70%, 90%, 100%

4. 0.3% Hydrogen peroxide (H_2O_2): 10 ml 3% H_2O_2, 90 ml methanol

5. Citrate buffer: 18 ml stock solution A (4.2 g citric acid, 200 ml distilled water), 82 ml stock solution B (14.7 g sodium citrate, 500 ml distilled water), add 900 ml distilled water; adjust pH at 6.0

6. 0.1% Trypsin solution: 100 mg trypsin, 200 mg calcium chloride, 100 ml distilled water; pH is adjusted at 7.8

7. Phosphate buffer saline (PBS): 7.52 g K_2HPO_4, 7.32 g NaH_2PO_4, 7.2 g NaCl in 1000 ml distilled water

8. 10% Normal goat serum (NGS): 0.8 g BSA^+ 10 ml normal goat serum, PBS added to 100 ml

9. Streptavidin-biotin complex (ready to use provided from Dako, Glostrup, Denmark).

10. 0.15% 3.3′-diaminobenzidine (DAB) tetrachloride: 15 mg DAB + 0.1 ml H_2O_2 + 10 ml PBS

11. Harris hematoxylin

12. Acrytol mounting medium

METHODS

Immunohistochemical Staining

1. Deparaffinize the sections by passing them in toluol 2 times for 5 min and rehydrate them through a series of graded alcohols—100%, 100%, 90%, and 70%—for 4 min.

2. Rinse the sections in tap water.

3. Perform antigen retrieval. Treat the slides in citrate buffer for 9 min at 750 W in a microwave oven or incubate them with trypsin solution at 37°C for 5 min.

4. Block the endogenous peroxidase activity with 3% hydrogen peroxide (H_2O_2) in methanol for 30 min at room temperature.

5. Rinse the sections with PBS for 5 min.

6. Incubate the sections with NGS for 30 min.

7. Incubate the sections overnight with the primary antibodies at 4°C.

8. Rinse the sections with PBS 3 to 5 times over a period of 15 min.

9. Incubate the sections with biotinylated secondary for 30 min at room temperature.

10. Rinse as in **step 8**.

11. Incubate the sections with horseradish peroxidase labeled streptavidin for 30 min at room temperature.

12. Rinse as in **step 8**.

13. Visualize the reaction product by incubation in DAB for 5 min at room temperature.

14. Rinse with tap water for 5 min.

15. Counterstain the sections with Harris hematoxylin.

16. Mount the sections in mounting medium

Evaluation of Staining

Any staining produced by the antibodies and present in epithelial or nonepithelial cells was recorded. The weak staining of collagen, when present, was considered background staining.

The localization pattern was evaluated at nuclear, cytoplasmic, and cell membrane level in epithelial and stromal cells. The staining intensity was graded as follows: no staining, weak, moderate, and strong. The immunoreactivity was an estimation of the percentage of stained cells and was classified as: <5%, 5–25%, 25–50%, 50–75%, or >75% of the cells.

Controls for Immunohistochemical Staining

To test the specificity of the immunostaining, the following experiments were performed on all control tissues, with or without antigen retrieval:

▲ Omission of the primary antibody and substitution with normal rabbit serum or PBS,

▲ Omission of the primary and secondary antibodies,

▲ Omission of the primary and secondary antibodies and ABC, and

▲ Absorbtion of the antibodies with excess peptide: each antibody solution was preincubated with 10-fold by weight excess of the appropriate peptide antigen, as recommended in the data sheets, at room temperature for 3 hr. This mixture was used as primary antibody. A parallel incubation of the tissue section, from the same block, with the nonabsorbed primary antibody was performed.

RESULTS AND DISCUSSION

Currently, the function and subcellular localization of the BRCA1 protein and its role in normal cells and in oncogenesis are under active investigation.

The putative motifs found in BRCA1 protein suggest that the full-length protein is synthetized in cytoplasm and then transported to the nucleus, where it would function as a transcriptional regulator and be involved in a DNA integrity-keeping mechanism. The subcellular localization of BRCA1 is controversial; three different views exist. Scully *et al.* (1996) reported the BRCA1 protein to be located exclusively in the nuclei of the normal and tumor cells. Chen *et al.* (1995) proposed that BRCA1 protein is primarily located in the nucleus in a variety of normal cell lines and in tumor cell lines derived from organs other than breast and ovary, whereas in most breast and ovarian cancer cell lines and in paraffin-embedded tissue sections of sporadic breast carcinoma, BRCA1 was present mainly in the cytoplasm or had a dual localization to both cytoplasm and nucleus. Jensen *et al.* (1996) have suggested that BRCA1 belongs to the granin family of proteins, is present in secretory vesicles in the cytoplasm, and functions as a secreted growth inhibitor. Finding out the reasons for the discrepancies in subcellular localization is difficult. Because most of the studies have been performed in cell lines mostly using self-made monoclonal or polyclonal antibodies, differences in antibody specificity and fixation technique may be responsible for the apparent incompatibilities.

We examined the expression of the BRCA1 protein in archival breast tumor specimens to determine its staining and localization pattern using commercially available polyclonal antibodies. We also investigated if breast tumors occurring in patients possibly belonging to breast cancer families show different staining properties.

The following sections first discuss the more technical part of this work that determined our choice for particular BRCA1 antibodies and the staining protocol; subsequently, the different types of staining observed are analyzed.

Technique and Antibody Specificity

Technique

The tissue specimens that are submitted to routine pathologic analysis are proceeded through fixation and paraffin embedding. Fixation will cross-link the antigens to larger macromolecules, thereby immobilizing them in the tissue. During conventional fixation, many antigenic epitopes may be destroyed or "masked" because of conformational changes or steric hindrance, by immobilization on nearby macromolecules. Obviously, an ideal fixative applicable to all antigens does not exist. Moreover, paraffin embedding may affect the immunoreactivity (Hayat, 2002).

The control tissue specimens used in our study were fixed in Bouin's fixative or formalin and paraffin embedded. Because only a weak or almost no immunoreactivity was detected with all antibodies, we decided to apply antigen retrieval methods to unmask the antigen. Without application of antigen retrieval, the formalin-fixed breast tissue sections showed almost no staining with all antibodies. The Bouin's fixed breast tissue sections showed a weak diffuse nuclear staining present in less than 10% of cells and a weak to moderate cytoplasmic staining.

In the breast epithelial cells of formalin or Bouin's fixed tissues treated with microwave heating before immunostaining, the four antibodies produced no nuclear staining and a weak to strong cytoplasmic staining, whereas a cell membrane staining was seen in formalin-fixed breast tissue. With use of trypsin before immunohistochemistry on formalin-fixed breast tissue sections, a moderate to strong cytoplasmic staining appeared; but no nuclear staining was detected. In Bouin's fixed breast tissue sections, trypsin digestion markedly enhanced the nuclear and cytoplasmic staining intensity and immunoreactivity produced by all antibodies. Therefore, we found that trypsinization was effective for epitope unmasking and staining with Santa Cruz BRCA1 rabbit polyclonal antibodies in Bouin's, but not in formalin-fixed tissues.

The duration of fixation also is of great importance. Overfixation of the specimens will result in destruction of the tissue antigens. A fixation time of 8–24 hr is desirable. In breast tissue fixed for more than 48 hr, the nuclear or cytoplasmic BRCA1 signal detectable by immunohistochemistry decreases, even when trypsin pretreatment is used. With increase in the fixation time, immunoreactivity also changes leading to variations in staining intensity among groups of cells within the same section.

D-20 antibody has been used on paraffin-embedded tissue sections in a few studies (Jarvis *et al.*, 1998;

Perez-Valles *et al.*, 2001), but the results and, consequently, the conclusions are different from ours (Bogdani *et al.*, 2002). This is because of the inappropriate protocol used in these studies (Jarvis *et al.*, 1998; Perez-Valles *et al.*, 2001) for the BRCA1 immunostaining with the Santa Cruz polyclonal antibodies. We would like to stress the importance of establishing an appropriate working protocol to obtain reliable results and draw valid conclusions.

Antibody Specificity

To analyze the presence and cellular localization of a protein, immunohistochemical techniques are useful, but the specificity of an antibody can never be totally proven. Nevertheless, antibody specificity controls should be carried out. If an immunohistochemical staining with a BRCA1 antibody disappears after preincubation of the primary antibody with the corresponding blocking peptide, this means that the recognized molecule is BRCA1 or BRCA1-like. In our study, with omission of the primary antibody, no immunostaining was present in any tissue component in experiments with the four BRCA1 antibodies. Immunoabsorbtion test with excess peptide for each corresponding antibody yielded different results. The immunopositivity observed on tissue sections with D-20 and K-18 antibodies entirely disappeared. I-20 immunoreactivity in cells completely disappeared, but in one case, stained secreted material in duct lumina was observed. The immunoreactivity obtained with C-20 was notably diminished, but small vesicular bodies were seen scattered all over the epithelial cell cytoplasm and a strong partial staining of the nuclear membrane (present only in some cases) was visible.

In all antibody preabsorbtion experiments, the four antibodies that we tested produced no nuclear staining at all. Moreover, all of them recognized about a 220-kD protein in Western Blots of normal breast tissue extracts (data not shown). This strongly suggests that all of them recognize the same protein. The additional staining still present in the preincubation experiments by the I-20 and C-20 suggests that these antibodies also recognize other proteins, which makes them unsuitable for an immunohistochemical study. Because the BRCA1 epitopes recognized by the D-20 and K-18 antibodies are localized very closely to each other on the NH2-terminal part of the protein, we may expect that those two antibodies generate very similar staining types. In all control and study tissue sections, D-20 and K-18 antibodies gave similar results. Therefore, only the results obtained with the D-20 antibody will be reported.

Several authors have already used the D-20 antibody for immunoprecipitation (Jensen *et al.*, 1996; Thakur *et al.*, 1997; Thomas *et al.*, 1996). They all found that this antibody clearly recognizes the full-length BRCA1 protein of the expected size of about 220 kD in total cell lysate. Some authors report other different lengths of BRCA1 (Coene *et al.*, 1997). Taking into account the diversity of the cell lines and antibodies used by different authors to detect the BRCA1 protein, these various bands may represent different alternative forms of the protein, may be because of differences in technical conditions, or may not even be related to BRCA1.

Localization and Staining Patterns of BRCA1 Protein in Normal Breast Epithelial Cells from Breast Reduction Specimens and from Patients with Breast Carcinoma

Localization

BRCA1 immunostaining was present in the nucleus and the cytoplasm of most of our study specimens and was in keeping with data of Chen *et al.* (1995). However, our finding of a cytoplasmic and a nuclear immunoreactivity in the normal breast is in contrast with their data on normal breast cell lines. The presence of cytoplasmic BRCA1 immunoreactivity, however, is consistent with the findings of Jensen *et al.* (1996), except that in their study no concurrent BRCA1 nuclear expression was present.

There also is controversy about the BRCA1 subcellular localization in studies on breast tissue sections. Two studies (Jarvis *et al.*, 1998; Taylor *et al.*, 1998) describe BRCA1 immunostaining to be present in the nuclear and cytoplasmic compartments, whereas other studies (Bernard-Gallon *et al.*, 1999; Perez-Valles *et al.*, 2001) report only cytoplasmic staining in a part of normal breast tissue specimens. Diversity of normal and tumor breast cell lines and tissues used and also different technical conditions and antibodies applied can affect the results.

In our study, in young patients the nuclear staining was mostly seen as fine spots and diffuse, whereas in older patients it was mostly diffuse. In 18 cases (11 young and 8 old patients), a mixed nuclear pattern composed of spots and diffuse staining was seen. Nuclear immunoreactivity showed a slight variation in young and old patients. It was absent in three cases: one young and two old patients.

The pattern of the cytoplasmic staining was granular, sometimes of fine appearance. In both groups, the cytoplasmic staining was strong in most of the cases. The cytoplasmic immunoreactivity was present in the majority of cells in both groups. A variation in cytoplasmic staining intensity was seen in the same tissue section in a small number of cases (four young and two old patients).

BRCA1 immunoreactivity also was observed in myoepithelial and stromal cells. Because of their location and their shape, the myoepithelial cells were not always easily visible, especially when epithelial cells showed a strong cytoplasmic staining. Nuclear staining was seen as either diffuse or spotlike. The cytoplasmic staining for both antibodies was too weak to evaluate a staining pattern. Endothelial cells showed a nuclear diffuse or spotlike staining pattern, whereas the cytoplasmic staining was granular or uniform when the staining intensity was weak. Smooth muscle cells showed a nuclear and cytoplasmic staining similar to that of the myoepithelial cells. The nuclei of the stromal cells revealed a diffuse staining or spots in most of the cases. The cytoplasmic staining was weak in most cells.

The presence of BRCA1 immunoreactivity in the nucleus and the cytoplasm can be explained in the findings of Wilson et al. (1997). Different alternatively spliced recombinant BRCA1 proteins bearing the nuclear localization signal (NLS) go to the nucleus, whereas those lacking it will be retained in the cytoplasm. The full-length protein is also found to be retained in cytoplasm. Three splice variants missing the major part of the sequence and NLS described by Miki et al. (1994) would generate obligate cytoplasmic BRCA1 variants. It is not possible, however, to discriminate the full-length and the described spliced forms of BRCA1 by our immunohistochemical approach because the NLS is located in exon 11 (in the middle of the protein).

Staining Pattern

In most reports describing the BRCA1 staining patterns, only immunofluorescence techniques were applied (Chen et al., 1995, 1996b; Jin et al., 1997; Scully et al., 1996; Thomas et al., 1996). Those studies were restricted to normal and tumor cell lines.

Chen et al. (1996b) report that immunostaining of BRCA1 changes during the cell cycle. At the G1-phase, BRCA1 is detected as a homogenous nuclear staining. As the cell progresses into the S-phase, staining intensifies and nuclear spots appear. During mitosis, BRCA1 staining tends to surround chromosomes as they align on the metaphase plate and then move apart. As cells reenter the G1-phase, a weak, homogeneous, nuclear staining returns. Scully et al. (1997c) suggest that the BRCA1 containing S-phase foci (or spots) are dynamic physiologic elements, responsive to the DNA damage, and these BRCA1-containing multiprotein complexes participate in a replication checkpoint response. The nature and the composition of these

subnuclear domains have not been fully clarified yet. It has been suggested that BRCA1 is associated with tumor suppressors and DNA damage repair proteins MSH2, MSH6, MLH1, ATM, BLM, and the RAD50-MRE11-NBS1 to form a large protein complex, named BASC, which may serve as a sensor for DNA damage (Wang et al., 2000). The mechanism underlying the accumulation of BRCA1 polypeptides into these nuclear dots is still unclear, but it was found that BRCA1 also undergoes hyperphosphorylation during the S-phase of the cell cycle (Chen et al., 1996b).

In our study, immunostaining with BRCA1 showed different staining patterns according to localization. In the nucleus there were coarse spots, dense round structures, easily visible at low magnification, whereas fine spots, smaller round structures were detected at high magnification; in some instances, the nucleus was homogenously stained. The cell membrane was stained in a chaplet-like or homogenous fashion. In the cytoplasm, granules or vesicles were visible. Cytoplasmic staining was strong in most cases and present in all cells. In most of the tissues, the pattern was represented by one of these three variants, but we could also detect the simultaneous presence of either fine and coarse spots, or spots and diffuse staining. The difference in the size of the nuclear spots is probably because of different amounts of the BRCA1 protein within these structures in the nuclei in different cell cycle phases. A cyclic variation in the frequency of mitosis and apoptosis in normal breast epithelial cells has been demonstrated (Anderson et al., 1982; Ferguson et al., 1981). Cancer cells are also proliferating cells, with a shorter G_1-phase than normal cells. Because BRCA1 protein is found to be expressed and phosphorylated in a cell cycle–dependent manner, this can explain the variation in the nuclear pattern. Finding different nuclear patterns simultaneously in one tissue section, in normal and tumor cells, could be because of variation in response of cells to different stimuli, resulting in an asynchronous cell cycle and, probably, variable BRCA1 expression.

Expression of BRCA1 in Tumor Breast Tissue

The involvement of BRCA1 in familial breast and ovarian cancer is well established. However, the role of BRCA1 in the development of the much more common sporadic tumors is still unknown. Several studies have reported that loss of heterozygosity (LOH) is a common event at the 17q21 region in sporadic breast cancer (Cropp et al., 1994; Futreal et al., 1992); therefore, it has been postulated that BRCA1 may have a role not only in hereditary, but also in sporadic breast cancer. But, BRCA1 somatic mutations rarely have been observed in sporadic breast tumors. Nevertheless, it is possible that BRCA1 plays a role in these tumors through other alterations. One of the proposed mechanisms was the reduction of BRCA1 mRNA expression levels in breast carcinoma cells in comparison with adjacent normal mammary epithelial cells (Magdinier *et al.*, 1998; Ozçelik *et al.*, 1998; Thompson *et al.*, 1995). But the expression levels still remained greater than those in breast cancer from patients with BRCA1 mutations (Kainu *et al.*, 1996). Inhibition of BRCA1 expression with anti-sense oligonucleotides indicated that reduced expression of BRCA1 increases the proliferative rate of the benign and malignant epithelial cells (Thompson *et al.*, 1995). Reduction in BRCA1 expression is also seen at the protein level, not only in the tumors with LOH, but also in a part of the tumors without detectable LOH (Rio *et al.*, 1999). By immunohistochemistry, loss of nuclear BRCA1 was observed in a part of sporadic breast tumors (Jarvis *et al.*, 1998; Taylor *et al.*, 1998).

When investigating the normal breast tissue adjacent to a neoplastic lesion, we obtained results very similar to those seen in breast tissue of nonaffected women: most patients showed nuclear immunopositivity in almost all the normal epithelial cells, whereas in the remaining specimens, a nuclear staining was observed in less than 75% of the cells. Because the decrease in the number of immunopositive normal epithelial cells could be observed in young and in old patients, we assume that this is not caused by age-dependent hormonal factors. Other physiologic factors must be responsible for the observed differences in BRCA1 expression levels.

Tumor tissue can be easily distinguished morphologically from adjacent normal tissue; therefore, we could estimate the proportion of the nuclear positive cells in both compartments for each patient. The staining pattern was similar to that seen in the normal tissue from both young and old patients. Compared with normal epithelial cells, cancer cells showed variability in BRCA1 staining. In a significant fraction of the breast cancers from young and old patients, the BRCA1 immunostaining was comparable in intensity and number of stained cells to that observed in the adjacent normal tissue (Figure 57). The tumors and their corresponding adjacent normal tissue obtained an equal score for nuclear immunoreactivity in 24 (46%) young and 25 (50%) old patients; an equal scoring for cytoplasmic staining intensity was observed in 19 (36%) young and 25 (50%) old patients. Apparently, in these breast cancers, the molecular pathway leading to the nuclear localization of BRCA1 was not disrupted.

In the other specimens we found a difference in nuclear BRCA1 expression between normal and tumor tissues, this difference being more marked in the young than in the old patients (Figure 57). This is reflected as a decrease in nuclear BRCA1 expression and nuclear immunoreactivity in the tumor specimens. In these specimens, the tumor tissue, even though positive for BRCA1, showed less nuclei stained and less intense cytoplasmic staining than adjacent normal tissue. In the tumors of 10 (19%) young and 21 (42%) old patients, the relative amount of the cells with unstained nuclei ranged between 4% and 75%. In 14 (27%) tumors of young and 3 (6%) tumors of old patients, BRCA1 nuclear staining was totally absent. The cells with no nuclear staining were clustered or mixed with cells with BRCA1 nuclear staining. The decrease in nuclear immunoreactivity and cytoplasmic staining intensity in the tumor was statistically significant when compared with normal tissue (Wilcoxon signed ranks test, $P < 0.01$). In old patients, this decrease was seen as a reduction in the nuclear immunoreactivity, whereas in young patients, this decrease mainly resulted from the occurrence of a high number of cases devoid of any nuclear staining in the cancer cells. Nuclear BRCA1 expression was significantly decreased in the tumors of the young patients compared with the old patients (logistic regression, $P = 0.0097$).

Thus, our results indicate that in tumor cells, BRCA1 expression, at protein level, could be the same as in the normal tissue, less, or absent. We also found a difference in nuclear BRCA1 expression in tumor tissues between young and old patients. The same proportion of tumors from young and old patients showed a decrease in the percentage of the cells with nuclear BRCA1 staining, but this decrease was more drastic in tumors of young patients. These results support a possible role of the BRCA1 gene product in sporadic, and in particular in early-onset breast cancer.

A significant fraction of the tumors from young patients showed no nuclear staining. It remains an open question whether the total absence of nuclear staining in these tumors either results from age-specific physiologic parameters (i.e., cyclic endocrine conditions) or is a property of a particular type of breast cancer occurring almost exclusively among young patients. When considering the "mixed" tumors in which cells with stained and unstained nuclei coexist, it is again unclear whether the tumor cells respond differently because they are (epi)genetically different or because an extracellular modulating factor does not reach the required concentration throughout the tumor tissue. In these tumors, nuclear negative cells are often clustered; but, although this observation could suggest a clonal proliferation, in some other tumors, the BRCA1 nuclear negative cells were mixed with the positive cells. Possible explanations for the absence

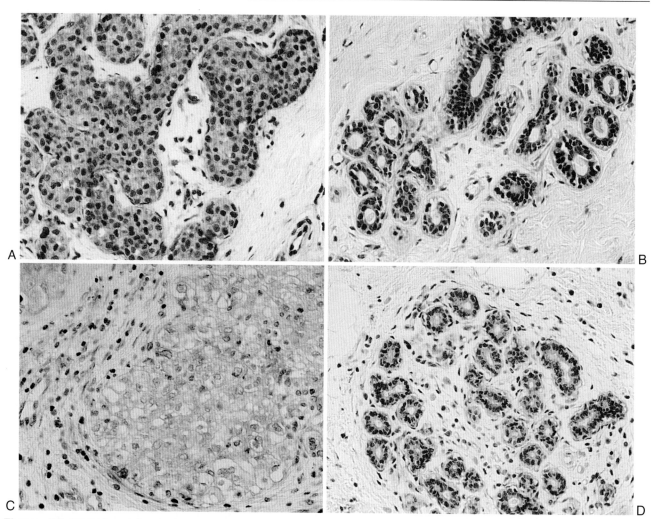

Figure 57 BRCA1 staining in breast cancer (**A** and **C**) and adjacent normal tissue (**B** and **D**). Tumor tissue shows either a BRCA1 staining comparable to that of normal tissue (**A** and **B**) or a faint to absent BRCA1 staining (**C** and **D**) (original magnification X280).

of nuclear BRCA1 expression in some tumors in our study include the following:

- ▲ An alteration in upstream BRCA1 regulatory mechanisms; BRCA1 mRNA is found to be expressed at five- to ten-fold greater levels in normal mammary tissue than in invasive, sporadic breast cancer samples (Thompson *et al.*, 1995).
- ▲ BRCA1 may be functionally inactivated by its mislocation from the nuclear to cytoplasmic compartment because of a defect in the cellular machinery involved in the receptor-mediated pathway of nuclear transport (Chen *et al.*, 1996).
- ▲ An epigenetic change, such as promoter methylation (Dobrovic *et al.*, 1997; Mancini *et al.*, 1998; Rice *et al.*, 2000). In the latter study, in tumor samples, nonmethylated promotor sequences were still detected at relatively high levels. Because

these nonmethylated promotor sequences are presumably not inactivated, we may assume that the BRCA1 protein is still synthesized, although at lower levels, in such tumors, which fits well with our immunohistochemical results.

Another explanation may be that patients displaying an absence of nuclear BRCA1 protein do have inherited cancer.

A specific BRCA1 granular cytoplasmic staining has been detected in almost all cells of normal and tumor specimens from all patients, indicating that the absence of the nuclear staining does not necessarily coincide with a total absence of the BRCA1 protein in the cell. However, a significant decrease in cytoplasmic staining intensity in tumor cells was observed mainly in these tumors showing a decrease in the percentage of cells with nuclear immunoreactivity, and especially in those tumors that showed a decrease in nuclear staining intensity. These observations led us to suggest that the decrease in the BRCA1 mRNA and protein level described by several authors and detected in a significant fraction of breast tissue extracts from sporadic tumors (Magdinier *et al.*, 1998; Sourvinos *et al.*, 1998, Thompson *et al.*, 1995) is restricted to those tumors presenting a smaller percentage of nuclear positive cells. We therefore believe that in those cells where BRCA1 mRNA is expressed at a low level, the total amount of the BRCA1 protein available is insufficient to be efficiently targeted into the nuclear complexes where it is supposed to participate in a DNA repair mechanism (Scully *et al.*, 2000), giving somatic mutations the possibility to accumulate.

A possible explanation for decreased BRCA1 expression in sporadic breast cancer could be a mutation in a gene(s) that regulates its expression. Alternatively, BRCA1 may be regulated during the differentiation of mammary epithelial cells and exhibit greater levels in the differentiated state that in the less differentiated phenotype present in breast cancer. Decreased expression of *BRCA1* supports the tumor-suppressive function of the gene, suggesting that a certain level of normally functioning *BRCA1* is needed for the maintenance of the negative growth regulatory mechanisms, and that cancer may arise because of the reduction in its relative level of expression. However, the exact molecular mechanism of altered BRCA1 phenotype in the sporadic breast cancer needs further investigation.

BRCA1 Expression in Cases with a Familial History of Breast Cancer

BRCA1 was originally identified as a breast cancer predisposing gene in about 50% of the patients with a strong family history of the disease (Ford, 1998). Familial tumors represent only 5% of all diagnosed breast cancers, suggesting that the tissue samples we analyzed in this study should be merely of sporadic origin. Because most of the reported BRCA1 mutations (about 90%) are truncating mutations and because the mRNA coding for truncated proteins is often unstable (Perrin-Vidoz *et al.*, 2002), we do not expect to obtain a signal with NH2-terminal–directed antibodies in BRCA1-associated tumors.

To investigate the hypothesis that tumors of patients with a BRCA1 mutation should not be stained, blood samples from six young patients with a breast cancer that did not show any nuclear staining after immuno-histochemistry were analyzed for the presence of a constitutive BRCA1 mutation. In one of these young patients who reported a family history of breast and ovarian carcinomas, we found a splice mutation at the exon5/intron 5 boundary (IVS5+3A > G) leading to a truncated protein. The RNA study in this patient demonstrated that an alternative splice site located 22bp within exon 5 was used.

These results suggest that tumors not stained for BRCA1 do not necessarily belong to BRCA1 mutation carriers. Because mutations in BRCA1 rarely have been reported in sporadic breast cancers, this could mean that somatic or constitutive mutations in genes other than *BRCA1* lead to the (regulatory) inactivation of the *BRCA1* gene. Alternatively, an epigenetic mechanism may also lead to the inactivation of this gene.

BRCA1 Mutation and Morphologic Features of Breast Cancer

Determination of estrogen receptors, progesteron receptors, and Neu oncoprotein expression on biopsies of invasive carcinomas before therapeutic manipulations has become standard practice in the management of patients with breast carcinoma. The determination of estrogen and progesteron receptors is considered a major feature of the tumor to be assessed; the results are used in making therapeutic decisions and are regarded as independent prognostic indicators. HER-2/neu, also known as c-erbB-2, is a member of a erbB oncogene family. This gene is found to be amplified in about 30% of human breast carcinomas and is an independent predictor of overall survival and time to relapse in patients with lymph node–positive breast cancer (Rilke *et al.*, 1991). BRCA1-associated breast cancers exhibit certain morphologic characteristics when compared with sporadic breast cancers. BRCA1-associated tumors display a greater proliferative potential, high histologic grade, more frequent estrogen and progesteron receptor negativity, and more frequent p53 mutations, which might have an effect on their clinical prognosis (Lakhani *et al.*, 2002; Lidereau *et al.*, 2000). However, it has not been possible until now to define a different subset of breast cancers from young women that are biologically more aggressive than the breast cancers in postmenopausal women. It has been suggested that morphologic characteristics of the BRCA1-associated cancers can be used to predict the risk for a young patient harboring a BRCA1 germline mutation. The use of these tumor features might be helpful, but they will still provide indirect information.

However, identification in our study of a patient with a *BRCA1* mutation among those with a breast tumor lacking a BRCA1 immunoreactivity strongly indicates that BRCA1 immunohistochemistry, using a well-established and reliable method, could enable us to identify at least a fraction of the women with a high risk for the disease. Although it can not discriminate the sporadic breast tumors from those associated with *BRCA1*, absence of BRCA1 nuclear immunoreactivity in tumors from "old patients" could be a strong indicator for a BRCA1 mutation.

CONCLUSIONS

The current study indicates the following conclusions:

1. Among the antibodies tested (Santa Cruz), only the aminoterminal antibodies can be successfully applied for immunohistochemistry. They work best on Bouin's fixed and paraffin-embedded breast tissue sections digested with trypsin before immunostaining.

2. The BRCA1 protein staining pattern is represented by spots or a diffuse staining in nuclei and by granules in cytoplasm. The BRCA1 protein is located in the nuclei and cytoplasm of normal epithelial and stromal cells and in breast cancer cells.

3. By BRCA1 immunohistochemistry, the breast cancers can be divided into two main categories according to nuclear BRCA1 immunostaining. In about half of the patients, young and old, the staining is present in almost all the normal and cancer cells. We have no reason to believe that BRCA1 is involved in the genesis of these cancers. In the other half of the patients, a variable but significant fraction of the tumor cells show no nuclear staining. The reason for this difference in BRCA1 expression among the cancer cells is currently unclear. In a subfraction of these patients, no nuclear BRCA1 staining could be observed in the tumor cells. These tumors are found almost exclusively among young patients and include those with a hereditary *BRCA1* mutation.

4. BRCA1 immunohistochemistry on tumor breast tissue might be used as a screening tool for identification of patients with a constitutive *BRCA1* mutation.

References

Anderson, T.J., Ferguson, D.J., Raab, G.M. 1982. Cell turnover in the "resting" human breast: Influence of the parity, contraceptive pill, age and laterality. *Br. J. Cancer 46:*376–382.

Azzopardi, J.G. 1979. *Problems in breast pathology.* Philadelphia: WB Saunders, 11–13.

Baumann, P., Benson, F.E., West, S.C. 1996. Human Rad51 protein promotes ATP-dependent homologous pairing and strand transfer reactions *in vitro. Cell 87:*757–766.

Berg, J.W. 1984. Clinical implications of risk factors for breast cancer. *Cancer 53:*589–591.

Bernard-Gallon, D.J., Peffault De Latour, M., Rio, P.G., Penault-LLorca, F.M., Favy, D.A., Hizel, C., Chassagne, J., Bignon, Y.J. 1999. Subcellular localization of BRCA1 protein in sporadic breast carcinoma with or without allelic loss of BRCA1 gene. *Int. J. Oncol. 14:*653–661.

Bogdani, M., Teugels, E., De Greve, J., Bourgain, C., Neyns, B., Pipeleers-Marichal, M. 2002. Loss of nuclear BRCA1 localization in breast carcinoma is age dependent. *Virchows Arch. 440:*274–279.

Borg, A., Zhang, Q.X., Johannsson, O., Olsson, H. 1994. High frequency of allelic imbalance at the BRCA1 region on chromosome 17q in both familial and sporadic ductal breast carcinomas. *J. Natl. Cancer. Inst. 86:*792–794.

Breast Cancer Information Core database. URL http://www.nhgri.nih.gov/Intramural_research/Lab_transfer/Bic

Breast Cancer Linkage Consortium. 1997. Pathology of familial breast cancer: Differences between breast cancers in carriers of BRCA1 and BRCA2 mutations and sporadic cases. *Lancet 349:*1505–1510.

Bruce, A., Bray, D., Lewis, J., Raff, M., Roberts, K., Watson, J. 1994. *Molecular biology of the cell* (3rd ed), New York: Garland Publishing, pp 1260–1266.

Brugarolas, J., Jacks, T. 1997. Double indemnity: p53, BRCA1 and cancer. *Nat. Med. 3:*721–722.

Callebaut, I., Mornon, J.P. 1997. From BRCA1 to RAP1: A widespread BRCT module closely associated with DNA repair. *FEBS Lett. 400:*25–30.

Chapman, M., Verma, I. 1996. Transcriptional activation by BRCA1. *Nature 382:*678–679.

Chen, C.F., Li, S., Chen, Y., Chen, P.L., Lee, W.H. 1996. The nuclear localization sequences of the BRCA1 protein interact with the importin-α subunit of the nuclear transport signal receptor. *J. Biol. Chem. 271:*32863–32868.

Chen, Y., Chen, C.F., Riley, D.J., Craig, A., Chen, P.L., Von Hoff, D., Kent, O., Lee, W.H. 1995. Aberrant subcellular localization of BRCA1 in breast cancer. *Science 270:*789–791.

Chen, Y., Chen, P.L., Riley, D.J., Lee, W.H., Allred, D.C., Osborne, C.K. 1996a. Location of BRCA1 in human breast and ovarian cancer cells. *Science 272:*125–126.

Chen, Y., Farmer, A., Chen, C.F., Jones, D., Chen, P.L., Lee, W.H. 1996b. BRCA1 is a 220-kDa protein that is expressed and phosphorylated in a cell cycle-dependent manner. *Cancer Res. 56:*3168–3172.

Chiba, N., Parvin, J.D. 2002. The BRCA1 and BARD1 association with the RNA polymerase II holoenzyme. *Cancer Res. 62:*4222–4228.

Claus, E.B., Risch, N., Thompson, W.D. 1991. Genetic analysis of BRCA1 in the cancer and steroid hormone study. *Am. J. Hum. Genet. 48:*232–242.

Coene, E., Van Oostveldt, P., Willems, K., Van Emmelo, J., De Potter, Ch. 1997. BRCA1 is localized in cytoplasmic tube-like invaginations in the nucleus. *Nat. Genet. 16:*122–124.

Crook, T., Crossland, S., Crompton, M.R., Osin, P., Gusterson, B. 1997. P53 mutations in BRCA 1-associated familial breast cancer. *Lancet 350:*638–639.

Cropp, C.S., Nevanlinna, H., Pyrhönnen, S., Stenman, U.H., Salmikangas, P., Albertsen, H., White, R., Callahan, R. 1994. Evidence for the involvement of BRCA1 in sporadic breast carcinoma. *Cancer Res. 54:*2548–2551.

Dobrovic, A., Simpfendorfer, D. 1997. Methylation of the BRCA1 gene in sporadic breast cancer. *Cancer Res. 57:*3347–3350.

Ferguson, D.J.P., Anderson, T.J. 1981. Morphological evaluation of cell turnover in relation to the menstrual cycle in the "resting" human breast. *Br. J. Cancer 44:*177–181.

Ford, D., Easton, D.F., Stratton, M., Narod, S., Goldgar, D., Devilee, P., Bishop, D.T., Weber, B., Lenoir, G., Chang–Claude, J., Sobul, H., Teare, M.D., Struewing, J., Arason, A., Scherneck, S., Peto, J., Rebbeck, T.R., Tonin, P., Neuhausen, S., Barkardottir, R., Eyfjord, J., Lynch, H., Ponder, B.A., Gayther, S.A., Zelada–Hedman, M., *et al.*, 1998. Genetic heterogeneity and penetrance analysis of the BRCA1 and BRCA2 genes in breast cancer families. The Breast Cancer Linkage Consortium. *Am. J. Hum. Genet.* 62:676–689.

Freemont, P.S., 1993. The RING finger: A novel protein sequence motif related to the Zinc finger. *Ann. NY Acad. Sci.* 684:174–192.

Futreal, A., Soderkvist, P., Marks, J.R., Iglehart, J.D., Cochran, C., Barrett, J.C, Wiseman, R.N. 1992. Detection of the frequent allelic loss on proximal chromosome 17q in sporadic breast carcinoma using microsatellite length polymorphism. *Cancer Res.* 52:2624–2627.

Glebov, O.K., McKenzie, K.E., White, C.A., Sukumar, S. 1994. Frequent p53 gene mutations and novel alleles in familial breast cancer. *Cancer Res.* 54:3703–3709.

Gudas, J., Li, T., Nguyen, D., Jensen, D., Rauscher, F.J., Cowan, K.M. 1996. Cell-cycle regulation of BRCA1mRNA in human breast epithelial cells. *Cell Growth Diff.* 7:717–723.

Gudas, J., Nguyen, H., Li, T., Cowan, K. 1995. Hormone-dependent regulation of BRCA1 in human breast cancer cells. *Cancer Res.* 55:4561–4565.

Haile, D.T., Parvin, J.D. 1999. Activation of transcription *in vitro* by the BRCA1 carboxyl-terminal domain. *J. Biol. Chem.* 274:2113–2117.

Hakem, R., de la Pompa, J.L., Elia, A., Potter, J., Mak, T.W. 1997. Partial rescue of BRCA1 (5-6) early embryonic lethality by p53 or p21 null mutation. *Nat. Genet.* 16:298–302.

Hall, J., Lee, M., Newman, B., Morrow, J., Anderson, L., Huey, B., King, M.C. 1990. Linkage of early-onset familial breast cancer to chromosome 17q21. *Science* 250:1684–1689.

Hayat, M.A. 2002. *Microscopy, Immunohistochemistry, and Antigen Retrieval Methods.* New York: Kluwer Academic/Plenum Publishers.

Holt, J., Thompson, M., Szabo, C., Robinson-Benion, C., Arteaga, C., King, M.C., Jensen, R. 1996. Growth retardation and tumor inhibition by BRCA1. *Nat. Genet.* 12:298–302.

Jarvis, E.M., Kirk, J.A., Clarke, C.L. 1998. Loss of nuclear BRCA1 expression in breast cancers is associated with a highly proliferative tumor phenotype. *Cancer Genet. Cytogenet.* 101:109–115.

Jasin, M. 2002. Homologous repair of DNA damage and tumorigenesis: The BRCA connection. *Oncogene 2002* 21:8981–8993.

Jensen, R., Thompson, M., Jetton, T., Szabo, C., Van Der Meer, R., Helow, R., Tronick, S., Page, D., King, M.C., Holt, J. 1996. BRCA1 ia secreted and exhibits properties of a granin. *Nat. Genet.* 12:303–308.

Jin, Y., Xu, X., Yang, M.-C.W., Wei, F., Ayi, T.C., Boxcock, A., Baer, R. 1997. Cell cycle-dependent colocalization of BRCA1 and BARD1 proteins in discrete nuclear domains. *Proc. Natl. Aca. Sci. USA* 94:12075–12080.

Kainu, T., Kononen, J., Johannssonn, O., Olsson, H.T., Borg, A., Isola, J. 1996. Detection of germline BRCA1 mutations in breast cancer patients by quantitative messenger RNA *in situ* hybridization. *Cancer Res.* 56:2912–2915.

Kelsey, J., Horn-Ross, P. 1993. Breast cancer: Magnitude of the problem and descriptive epidemiology. *Epidemiol. Rev.* 15(1):7–16.

Koonin, E., Altschul, S. 1996. BRCA1 protein products: Antibody specificity, functional and secreted tumor suppressors. *Nat. Genet.* 13:266–267.

Korhonen, L., Brannvall, K., Skoglosa, Y., Lindholm, D. 2003. Tumor suppressor gene BRCA-1 is expressed by embryonic and adult neural stem cells and involved in cell proliferation. *J. Neurosci. Res.* 71:769–776.

Lakhani, S.R., Van De Vijver, M.J., Jacquemier, J., Anderson, T.J., Osin, P.P., McGuffog, L., Easton, D.F. 2002. The pathology of familial breast cancer: Predictive value of immunohistochemical markers estrogen receptor, progesterone receptor, HER-2, and

p53 in patients with mutations in BRCA1 and BRCA2. *J. Clin. Oncol.* 20:2310–2318.

Lane, D.P. 1994. P53 and human cancers. *Br. Med. Bull.* 50:582–599.

Li, J., *et al.* 1997. PTEN, a putative protein phosphatase gene mutated in human brain, breast and prostate cancer. *Science* 275:1943–1947.

Lidereau, R., Eisinger, F., Champeme, M.H., Nogues, C., Bieche, I., Birnbaum, D., Pallud, C., Jacquemier, J., Sobol, H. 2000. Major improvement in the efficacy of BRCA1 mutation screening using morphoclinical features of breast cancer. *Cancer Res.* 60:1206–1210.

Magdinier, F., Ribieras, S., Lenuar, G.M., Frappart, L., Dante, R. 1998. Down-regulation of BRCA1 in human sporadic breast cancer: Analysis of DNA methylation patterns of the putative promoter region. *Oncogene* 17:3169–3176.

Mancini, D.N., Rodenhiser, D.I., Ainsworth, P.J., OiMalley, F.P., Singh, S.M., Xing, W., Archer T.K. 1998. CpG methylation within the 5'regulatory region of the BRCA1 gene is tumor specific and includes a putative CREB binding site. *Oncogene* 16:1161–1169.

Marks, J., Huper, G., Vaughn, J., Davis, P., Norris, J., McDonnell, D.P., Wiseman, R.W., Futreal, P.A., Iglehart, J.D. 1997. BRCA1 expression is not directly responsive to estrogen. *Oncogene* 14:115–121.

Marquis, S., Rajan, J., Wynshaw-Boris, A., Xu, J., Yin, G.Y., Abel, K., Weber, B., Chodosh, L. 1995. The developmental pattern of BRCA1 expression implies a role in differentiation of the breast and other tissues. *Nat. Genet.* 11:1726.

Merajver, S.D., Pham, T.M., Caduff, R.F., Chen, M., Poy, E.L., Cooney, K.A., Weber, B.L., Collins, F.S., Johnston, C., Frank, T.S. 1995. Somatic mutations in the BRCA1 gene in sporadic ovarian tumors. *Nat. Genet.* 9:439–443.

Meza, J.E., Brzovic, P.S., King, M.C., Klevit, R.E. 1999. Mapping the functional domains of BRCA1. *J. Biol. Chem.* 274:5659–5665.

Miki, Y., Swensen, J., Shattuck-Eidens, D., Futreal, P.A., Harshman, K., Tavtigian, S., Liu, Q., Cochran, C., Bennett, L.M., Ding, W., et al. 1994. A strong candidate for the breast and ovarian cancer susceptibility gene BRCA1. *Science* 266:66–71.

Monteiro, A., August, A., Hanafusa, H. 1996. Evidence for a transcriptional activation function of BRCA1 C-terminal region. *Proc. Natl. Acad. Sci. USA* 93:13595–13599.

Ouchi, T., Monteiro, A., August, A., Aaronson, S.A., Hanafusa, H. 1998. BRCA1 regulates p53-dependent gene expression. *Proc. Natl. Acad. Sci. USA* 95:2302–2306.

Ozçelilk, H., To, M.D., Couture, J., Bull, S.B., Andrulis, I.L. 1998. Preferential allele expression can lead to reduced expression of BRCA1 in sporadic breast cancers. *Int. J. Cancer* 77:1–6.

Perez-Valles, A., Martorell-Cebollada, M., Nogueira-Vazquez, E., Garcia-Garcia, J.A., Fuster-Diana, E. 2001. The usefulness of antibodies to the BRCA1 protein in detecting the mutated BRCA1 gene: An immunohistochemical study. *J. Clin. Pathol.* 54:476–780.

Perrin-Vidoz, L., Sinilnikova, O.M., Stoppa-Lyonnet, D., Lenoir, G.M., Mazoyer, S. 2002. The nonsense-mediated mRNA decay pathway triggers degradation of most BRCA1 mRNAs bearing premature termination codons. *Hum. Mol. Genet.* 11:2805–2914.

Rice, J.C., Futscher, B.W. 2000. Transcriptional repression of BRCA1 by aberrant cytosine methylation, histone hypoacetylation and chromatin condensation of the BRCA1 promoter. *Nuclei Acids Res.* 28:3233–3239.

Rilke, F., Colnaghi, M.I., Cascinelli, N., Andreola, S., Baldini, M.T., Bufalina, R., Della Porta, G., Menard, S., Pierotti, M.A.,

Testori, A. 1991. Prognostic significance of Her2/Neu expression in breast cancer and its relationship to other prognostic factors. *Int. J. Cancer 49*:44–49.

Rio, P., Maurizis, J.C., Peffault De Latour, M., Bignon, Y.J., Bernard-Gallon, D. 1999. Quantification of BRCA1 protein in sporadic breast carcinoma with or without loss of heterozygosity of the BRCA1 gene. *Int. J. Cancer 80:*823–826.

Romagnolo, D., Annab, L.A., Thompson, T.E., Risingner, J.I., Terry, L.A., Barrett, J.C., Afshari, C.A. 1998. Estrogen upregulation of BRCA1 expression with no effect on localization. *Mol. Carcinog. 22:*102–109.

Ruffner, H., Verma, I. 1997. Brca1 is a cell cycle-regulated nuclear phosphoprotein. *Proc. Natl. Acad. Sci. USA 94*:7138–7143.

Saurin, A.J., Borden, K.L.B., Boddy, M.N., Freemont, P.S. 1996. Does this have a familiar RING? *Trends Biochem. Sci. 21:*208–213.

Scully, R., Anderson, S., Chao, D., Wei, W., Ye, L., Young, R., Livingston, D., Parvin, J. 1997a. BRAC1 is a component of the RNA polymerase II holoenzyme. *Natl. Acad. Sci. USA 94:*5605–5610.

Scully, R., Chen, J., Ochs, R., Keegan, K., Hoekstra, M., Feunteun, J., Livingston, D. 1997b. Dynamic changes of BRCA1 subnuclear location and phosphorylation state are initiated by DNA damage. *Cell 90:*425–435.

Scully, R., Chen, Y., Plug, A., Xiao, Y., Weaver, D., Feunteun, J., Ashley, T., Livingston, D. 1997c. Association of BRCA1 with Rad51 in mitotic and meiotic cells. *Cell 88:*265–275.

Scully, R., Ganesan, S., Brown, M., De Caprio, J., Cannistra, S., Feunteun, J., Schnitt, S., Livingston, D. 1996. Location of BRCA1 in human breast and ovarian cancer cells. *Science 272:*123–125.

Scully, R., Livingston, D.M. 2000. In search of the tumour-suppressor functions of BRCA1 and BRCA2. *Nature 408:*429–432.

Smith, S., Easton, D., Evans, D.G.R., Ponder, B.A.J. 1992. Allele losses in the region 17q12-21 in familial breast and ovarian cancer involve the wild-type chromosome. *Nat. Genet. 2:*128–131.

Sobol, H., Stoppa-Lyonnet, D., Bressac-de-Paillerets, B., Peyrat, J.P., Keranguevan, F., Janin, N., Noguchi, T., Eisinger, F., Guinebretiere, J.M., Jacquemier, J., Birnbaum, D. 1996. Truncation at conserved terminal regions of BRCA1 protein is associated with highly proliferating hereditary breast cancers. *Cancer Res. 56:*3216–3219.

Somasundaram, K., Zhang, H., Zeng, Y.X., Houvras, Y., Peng, Y.I., Zhang, H., Wu, G.S., Licht, J.D., Weber, B., EI-Defry, W.S. 1997. Arrest of the cell cycle by the tumor-suppressor BRCA1 requires the CDK-inhibitor p21$^{WAF1/CIP1}$. *Nature 389:*187–190.

Sourvinos, G., Spandidos, D.A. 1998. Decreased BRCA1 expression levels may arrest the cell cycle through activation of p53 checkpoint in human sporadic breast tumors. *Biochem. Biophys. Res. Commun. 245:*75–80.

Sung, P. 1994. Catalysis of ATP-dependent homologous DNA pairing and strand exchange by yeast Rad51 protein. *Science 265:*1241–1243.

Tavassoli, F.A. 1992. *Pathology of the breast.* Norwalk, CT: Appleton & Lange, 1–23.

Taylor, J., Lymboura, M., Pace, P.E., Hern, R.P., Desai, A.J., Shousha, S. 1998. An important role for BRCA1 in breast cancer progression is indicated by its loss in a large proportion of non-familial breast cancers. *Int. J. Cancer 79:*334–342.

Thakur, S., Zhang, H.B., Peng, Y., Le, H., Carroll, B., Ward, T., Yao, J., Farid, L., Couch, F., Wilson, C., Weber, B. 1997. Localization of BRCA1 and a splice variant identifies the nuclear localization signal. *Mol. Cell. Biol. 17:*444–452.

Thomas, J., Smith, M., Rubinfeld, B., Gutowski, M., Beckmann, R., Polakis, P. 1996. Subcellular localization and analysis of apparent 180-kDa and 220-kDa proteins of the breast cancer susceptibility gene, BRCA1. *J. Biol. Chem. 271:*28630–28635.

Thompson, M., Jensen, R., Obermiller, P., Page, D., Holt, J. 1995. Decreased expression of BRCA1 accelerates growth and is often present during sporadic breast cancer progression. *Nat. Genet. 9:*444–450.

van der Looij, M., Cleton-Jansen, A.M., van Eijk, R., Morreau, H., van Vliet, M., Kuipers-Dijkshoorn, N., Olah, E., Cornelisse, C.J., Devilee, P. 2000. A sporadic breast tumor with a somatically acquired complex genomic rearrangement in BRCA1. *Genes Chromosomes Cancer 27:*295–302.

Wang, Y., Cortez, D., Yazdi, P., Neff, N., Elledge, S.J., Qin, J. 2000. BASC, a super complex of BRCA1-associated proteins involved in the recognition and repair of aberrant DNA structures. *Genes Dev. 14:*927–939.

Wilson, C., Payton, M., Elliott, G., Buaas, F.W., Cajulis, E.E., Grosshans, D., Ramos, L., Reese, D.M., Slamon, D.J., Calzone, F.L. 1997. Differential subcellular localization, expression and biological toxicity of BRCA1 and the splice variant BRCA1-11b. *Oncogene 14:*1–16.

Wooster, R., Neuhausen, S.L., Mangion, J., Quirk, Y., Ford, D., Collins, N., Nguyen, K., Seal, S., Tran, T., Averill, D., *et al.* 1994. Localization of a breast cancer susceptibility gene, BRCA2, to chromosome 13q12-13. *Science 265:*2088–2090.

Wu, L.C., Wang, Z.W., Tsan, J.T., Spillman, M., Phung, A.P., Xu, X., Yang, M.-C.W., Hwang, L.-Y., Bowcock, A.M., Baer, R. 1996. Identification of a Ring protein that can interact in vivo with the BRCA1 gene product. *Nat. Genet. 14:*430–440.

Xu, C.-F., Chambers, J., Solomon, E. 1997. Complex regulation of the BRCA1 gene. *J. Biol. Chem. 272:*20994–20997.

15

Fluorescence *in situ* Hybridization of *BRCA1* Gene in Breast Carcinoma

Tatyana A. Grushko, Karin K. Ridderstråle, and Olufunmilayo I. Olopade

Introduction

BRCA1 is a breast cancer susceptibility gene; the mutant form of *BRCA1* predisposes carriers to both breast and ovarian cancers (Arver *et al.*, 2000; Miki *et al.*, 1994). *BRCA1* functions as a classical tumor suppressor gene on 17q12-21, and loss of the wild-type allele (loss of heterozygosity [LOH]) is required for tumorigenesis in mutation carriers. *BRCA1* encodes a multifunctional protein, which together with other proteins contributes to homologous recombination, DNA damage response, and transcriptional regulation, and serves to maintain genomic stability (Venkitaraman, 2002). Although germline mutations in this gene account for approximately 4–10% of all breast cancers, about 50% of familial breast cancer cases and the majority (80%) of the breast/ovarian cancer families are caused by *BRCA1* mutations (Ford *et al.*, 1998). Breast cancers arising in *BRCA1* mutation carriers are usually high-grade, aneuploid, highly proliferative, and estrogen receptor–negative (ER⁻) (Arver *et al.*, 2000). Moreover, *BRCA1*-mutated tumors have unique gene expression profiles (Hedenfalk *et al.*, 2001; van't Veer *et al.*, 2002). However, the molecular mechanism(s)

contributing to tumor initiation and progression in women with germline *BRCA1* mutations are largely unknown. Although somatic mutations of the *BRCA1* gene are rare (Futreal *et al.*, 1994), hypermethylation of the *BRCA1* promoter may be an important mechanism for functionally inactivating *BRCA1* in nonhereditary forms of breast cancer (Catteau *et al.*, 2002), because 7–31% of sporadic breast tumors are reported to be *BRCA1*-methylated. It appears that *BRCA1*-methylated sporadic tumors may display pathologic features and gene expression profiles similar to those of *BRCA1*-mutated hereditary breast cancers (Catteau *et al.*, 2002; Esteller *et al.*, 2001; Hedenfalk *et al.*, 2001). However, the mechanisms of tumor initiation and progression in *BRCA1*-like sporadic breast cancers are unknown. The best approaches for reducing breast cancer mortality in young women have not yet been developed, mainly because of lack of knowledge of the mechanisms of tumor aggressiveness. Therefore, an understanding of the molecular mechanisms underlying tumor initiation and progression in cells lacking *BRCA1* is a critical step toward the development of more effective strategies for prevention, early detection, and treatment of aggressive breast cancer that disproportionately affects young women.

Handbook of Immunohistochemistry and *in situ* Hybridization of Human Carcinomas, Volume 1: Molecular Genetics; Lung and Breast Carcinomas

385

Copyright © 2004 by Elsevier (USA)
All rights reserved.

Methods for *BRCA1* analysis are available and include mutation screening, DNA methylation techniques, DNA, RNA, and protein expression by Southern, Northern, Western Blotting and immunohistochemistry (IHC). However, the large size of the *BRCA1* gene (220 kD) and the corresponding protein (1863aa), insufficient sensitivity of commonly used techniques, and the low specificity of the commercially available BRCA1 antibodies cause difficulties in studying *BRCA1* (Ford *et al.*, 1998; Perez-Valles *et al.*, 2001). Fluorescence *in situ* hybridization (FISH) is a high-throughput technology, which is becoming important in oncology for evaluation and characterization of genetic anomalies in complex biological specimens, including solid tumors, and for the study of disease mechanisms. FISH allows analysis of DNA in individual cells *in situ*, and thus provides the most sensitive and reliable means of assessing chromosomal aneuploidy and chromosome translocations, gene amplifications, and deletions. Thus, FISH is especially useful in cancer studies. Loss of function of the *BRCA1* tumor suppressor gene is accompanied by deletions in the *BRCA1* locus or even loss of the whole chromosome 17, which can be detected by FISH. In this chapter we describe protocols for performing *BRCA1/CEP17* FISH on both fresh-frozen and paraffin-embedded specimens from patients with breast cancer. Pretreatment of paraffin slides is a crucial step for successful FISH. However, the best conventional protocol based on acid and cheotrope treatment developed by Vysis/Abbott, Inc. (Downers Grove, Illinois; Vysis, 1998) is time-consuming and includes the use of harmful chemicals. Therefore, for pretreatment of paraffin slides, in addition to a detailed description of conventional protocol, we propose the alternative microwave-based method, which has been successfully used in our laboratory. We also describe two alternative methods of DNA labeling for FISH: direct DNA labeling using a fluorochrome-conjugated deoxyuridine-triphosphate (dUTP) such as *SpectrumGreen/Orange* dUTP, and indirect DNA labeling using biotin-conjugated dUTP and fluorescein isothiocyanate (FITC)-conjugated avidin for probe detection. Both procedures work well in our laboratory.

MATERIALS

Materials for Preparation of Slides from Frozen Tissues

1. Hypotonic solution: KCl 0.075 M (37°C).
2. Carnoy's fixative: 3 parts absolute methanol: 1 part glacial acetic acid (room temperature). Prepare immediately before use.

Materials for Pretreatment of Slides from Formalin-fixed, Paraffin-Embedded Tissue Sections

1. 20X SSC (standard saline citrate): Dissolve 175.3 g sodium chloride and 88.2 g sodium citrate in 800 ml deionized distilled water; adjust pH to 7.0, bring vol to 1 L and autoclave.
2. 2X SSC: Dilute 20X SSC 1:10 with deionized distilled water and adjust pH to 7.0.
3. 1 M sodium thiocyanate: 31.07 g sodium thiocynate (wear a facemask while handling the dry substance) and 1 L deionized distilled water; store at 4°C.
4. 0.01 M citrate buffer: Make stock solution of 0.1 M citric acid: 1.92 g citric acid anhydrous in 100 ml deionized distilled water, and 0.1 M sodium citrate: 14.7 g sodium citrate·2H$_2$O in 500 ml deionized distilled water. Working solution: 9 ml 0.1 M citric acid, 41 ml 0.1 M sodium citrate, 450 ml deionized distilled water; adjust pH to 6.0.
5. 0.2 M HCl: Dilute 2 M HCl 1:10.
6. 0.9% sodium chloride, pH 1.5: Dissolve 0.9 g sodium chloride in 100 ml deionized distilled water; adjust pH to 1.5.
7. Pepsin solution 0.5 mg/ml: Dissolve 25 mg pepsin (2500–3500 U/mg protein; Sigma, St. Louis, MO), in 50 ml 0.9% sodium chloride, pH 1.5; prewarm to 37°C. The same solution can be used ~3–5 times, store at −20°C between uses.
8. 10X PBS (phosphate-buffered saline): 80 g sodium chloride, 2 g potassium chloride, 14.4 g dibasic sodium phosphate, 2.4 g monobasic potassium phosphate, and 800 ml deionized distilled water; adjust pH to 7.4, bring vol to 1 L and autoclave.
9. 10% buffered formalin: 10 ml 10X PBS, 27 ml 37% formaldehyde solution (formalin), and 63 ml deionized distilled water.
10. Xylene
11. Ethanol: 70%, 80%, 90%, 100%
12. Coplin jars, one microwave-safe Coplin jar
13. Diamond-tipped glass scribe
14. Microwave

Materials for DNA Labeling Using Nick Translation

1. 1 M Tris-HCl (tris[hydroxymethyl]aminomethane): 60.55 g Tris base, 400 ml deionized distilled water, add concentrated HCl to adjust pH (~28 ml for pH 7.5 or ~21 ml for pH 8.0); bring vol to 500 ml with deionized distilled water and autoclave.
2. 0.5 M EDTA (ethylenediamine tetra-acetic acid): 93.06 g Na$_2$EDTA·2H$_2$O and 250 ml deionized distilled water, adjust pH to 8.0 with ~10 g NaOH pellets;

at this pH the EDTA will dissolve; bring vol to 500 ml and autoclave.

3. TE (Tris EDTA): 5 ml 1 M Tris pH 8.0, 1 ml 0.5 M EDTA pH 8.0; bring vol to 500 ml with deionized distilled water and autoclave.

4. 1 M magnesium chloride: Dissolve 20.33 g magnesium chloride·6H$_2$O in 70 ml deionized distilled water; bring vol to 100 ml and autoclave.

5. 1 M magnesium acetate: Dissolve 21.45 g magnesium acetate·4H$_2$O in 70 ml deionized distilled water; bring vol to 100 ml.

6. 10 mM phenylmethylsulfonyl fluoride (PMSF): 0.0174 g PMSF and 10 ml isopropanol; store at −20°C.

7. 10X tris-borate EDTA (TBE): 108 g Tris base, 55 g boric acid, 40 ml 0.5 M EDTA pH 8.0, and 1 L deionized distilled water; autoclave.

8. 1X TBE: Dilute 10X TBE 1:10 with deionized distilled water.

9. 1% agarose gel: 0.5 g agarose and 50 ml 1X TBE; microwave until the solution boils and all agarose is dissolved. Let it cool to ~60°C before pouring.

10. 5X loading dye: 3 ml glycerol, 0.025 g bromophenol blue, 0.025 g xylene cyanol; bring vol to 10 ml with sterile deionized distilled water; store at 4°C.

11. 100 basepair (bp) DNA ladder: The Invitrogen ladder has a 600-bp band, which is 2–3 times brighter than the other bands, making it easy to identify. Other DNA ladders can be used as well.

12. Ethidium bromide stock solution 10 mg/ml: 0.1 g ethidium bromide in 10 ml deionized distilled water; filter and store protected from light. Working solution 0.5 μg/ml: 12.5 μl stock solution in 250 ml 1X TBE.

13. 20% sodium dodecyl sulfate (SDS): 100 g SDS and 400 ml deionized distilled water; heat to 68°C to solute crystals; adjust pH to 7.2 with HCl and bring vol to 500 ml.

14. 10X PBS: For recipe, see materials for pretreatment.

15. 1X PBS: Dilute 10X PBS 1:10 with deionized distilled water; adjust pH to 7.4.

16. Biotin-l6-dUTP 0.3 mM: From Boehringer Mannheim (Indianapolis, IN). Other biotin-dUTP derivatives can be used.

17. Bovine serum albumin (BSA) 50 mg/ml: 0.05 g BSA in 1 ml deionized distilled water.

18. 0.2 mM *SpectrumOrange/Green*-dUTP: 40 μl 1 mM SpectrumOrange/Green-dUTP, 160 μl sterile TE pH 8.0; make 20 μl aliquots and store protected from light at −20°C.

19. Diethylnitrophenyl thiophosphate (dNTP), mix: 5 μl 100 mM deoxyadenosine triphosphate (dATP), 5 μl 100 mM deoxycytidine triphosphate (dCTP), 5 μl 100 mM deoxyguanosine triphosphate (dGTP) and 3 μl 100 mM deoxythymidine triphosphate (dTTP); bring

vol to 1 ml with sterile deionized distilled water, aliquot and store at −20°C.

20. 10X nick translation buffer (NTS) with BSA: 5 ml 1 M Tris-HCl pH 7.5, 0.5 ml autoclaved 1 M magnesium chloride, 70 μl 14.4 M β-mercaptoethanol and 20 μl BSA 50 mg/ml; bring vol to 10 ml with sterile deionized distilled water, aliquot, and store at −20°C.

21. Enzyme storage buffer: 500 μl 1 M Tris-HCl pH 7.5, 50 μl 1 M magnesium acetate, 0.7 μl 14.4 M β-mercaptoethanol, 100 μl 10 mM PMSF, 5 ml glycerol, and 20 μl BSA 50 mg/ml; bring vol to 10 ml with sterile deionized distilled water, aliquot, and store at −20°C.

22. 10X stop buffer: 2 ml 0.5 M EDTA pH 8.0, 0.5 ml 20% SDS, and 7.5 ml sterile deionized distilled water; store at room temperature.

23. DNAse I: Dilute to 0.2 U/μl.

24. DNA polymerase I: 10 U/μl Invitrogen.

25. 10X enzyme mix: 36.5 μl enzyme storage buffer, 1.5 μl 0.2 U/μl DNAse I, 2 μl 10. U/μl DNA polymerase I; make fresh each time and keep at −20°C until use.

Materials for Hybridization Procedure and Post-hybridization Washes

1. 20X SSC and 2X SSC: for recipe, see materials for pretreatment.

2. 4X SSC/1X SSC: dilute 20X SSC 1:5/1:20 with deionized distilled water and adjust pH to 7.0.

3. 10X RNAse A (1 mg/ml): 0.01 g RNAse A and 10 ml 2X SSC pH 7.0; store at −20°C.

4. Placental DNA 1 μg/μl (Sigma): 10 mg/ml placental DNA for hybridization (sonicated, denatured), dilute 1:10 with deionized distilled water; store at −20°C.

5. Salmon testis DNA 1 μg/μl (Sigma): 9.9 mg/ml salmon testis DNA for hybridization (sonicated, denatured), dilute 1:10 with deionized distilled water; store at −20°C.

6. Cot-1 DNA l μg/μl (Invitrogen) store at −20°C.

7. Avidin-conjugated fluorochrome 1 mg/ml: avidin conjugated to several different fluorochromes can be bought from Vector Laboratories (Burlingame, CA).

8. 3 M potassium acetate: 14.75 g potassium acetate and 50 ml deionized distilled water, pH 5.2.

9. Hybridization buffer: 2 g dextrose sulfate in 10 ml 4X SSC; before each use mix equal volumes of the 20% dextrose sulfate solution and formamide (total 10 μl per slide needed). Alternatively, use commercially available locus specific identifier/whole chromosome paint (LSI/WCP) hybridization buffer from Vysis/Abbott, Inc.

10. 10X PBS: for recipe, see materials for pretreatment.

11. 1X PBS: For recipe, see materials for DNA labeling using nick translation.

12. Denaturation solution: 70 ml formamide, 10 ml 20X SSC and 20 ml deionized distilled water; adjust pH to 7.0–7.5. The solution can be stored for up to 1 week at 4°C in an air-tight container.

13. 2X SSC/0.3% NP40: 66 ml 20X SSC, 20 ml 10% NP40, and 579.42 ml deionized distilled water; stir for 20 min and adjust pH to 7.0–7.5; store at room temperature.

14. 4X SSC/0.1% TritonX-100: 100 ml 20X SSC, 0.5 ml TritonX-100; bring vol to 500 ml and adjust pH to 7.0; store at room temperature.

15. 50/50 mix: 40 ml 20X SSC, 100 ml formamide, and 60 ml deionized distilled water; adjust pH to 7.0. The solution can be stored for up to 1 week at 4°C in an air-tight container.

16. 2% BSA: 0.4 g BSA in 20 ml 4X SSC; aliquot and store at −20°C.

17. DAPI (4'6' diamidino-2-phenylindole) solution: DAPI is light sensitive: Work fast shielded from light. Dissolve 10 mg DAPI in 10 ml sterile deionized distilled water and make 1-ml aliquots; store this stock solution at −20°C protected from light. Working solution: Add 10 μl stock solution to 5 ml 2X SSC; filter through 0.45 μm sterile filter and add 45 ml 2X SSC. Final concentration 200 ng/ml. The same solution can be used for ~2 months if stored protected from light at 4°C.

18. Mounting media, PDD (1,4-phenylenediamine) antifade: PDD is light sensitive: Work fast shielded from light. Wear a face mask when handling the dry substance. Dissolve 0.3 g PDD in 30 ml 1X PBS, adjust pH to 8.0 with NaOH. Pour the PDD solution into a beaker with 120 ml glycerol. Use a large magnetic stir bar at slow speed to mix the two layers. Store protected from light at −20°C. PDD anti-fade must be discarded after ~3 months (when it turns bluish/dark).

19. Ethanol: 70%, 80%, 90%, 100%; keep ~15 ml 100% ethanol in the freezer.

20. Coplin jars suitable for 10 slides and ~50 ml solution.

21. Rubber cement.

22. 10-ml syringe.

23. 22 × 22 mm cover glass.

24. Hybridization chamber: Opaque plastic box with a tight lid.

METHODS

Preparation of Slides from Frozen Tissues: Imprint Touch Preparation

1. Lightly press semi-thawed tumor pieces onto Superfrost Plus microscope slides (Fisher Scientific, Houston, Texas).

2. Treat slides with prewarmed hypotonic KCl (0.075 M) for 10 min at 37°C.

3. Fix slides for 10 min in Carnoy's fixative at room temperature.

4. Slides are ready for hybridization procedure. Store slides at −20°C.

Preparation of Slides from Formalin-Fixed, Paraffin-Embedded Tissues

Note: Use specimens that were fixed in formalin for 24–48 hr. Prolonged fixation may result in FISH assay failure (Vysis, 1998).

1. Cut 4–6-μm-thick paraffin sections using a microtome.

2. Float the sections on a protein-free (triple-distilled) waterbath at 40°C.

3. Mount a section on a positively charged slide; allow to air-dry and store at room temperature.

Pretreatment of Formalin-Fixed, Paraffin-Embedded Tissue Sections Using Acid and Chaotrope Treatment

Pretreatment is based on Vysis, Inc., Paraffin Pretreatment Kit Protocol (Vysis, 1998), which we modified and adjusted for use with archival material.

1. Bake slides in a 70°C warm oven, preferably overnight but at least for 2 hr.

2. Prepare/thaw the pepsin solution and warm to 37°C in a waterbath. Heat the sodium thiocyanate solution to 80°C.

3. Deparaffinize slides in xylene 3 times for 10 min each.

4. Immerse slides in 100% ethanol 2 times for 5 min each.

5. Dry slides on a slide warmer at 45°C and mark the area of interest with a glass marker on the back of the slide.

6. Immerse slides in 0.2 M HCl for 20 min.

7. Wash in deionized distilled water for 3 min.

8. Immerse slides in 2X SSC for 3 min.

9. Move slides to the 80°C 1 M sodium thiocyanate solution and incubate for 30 min.

10. Wash slides for 1 min in deionized distilled water.

11. Wash 2 times 5 min in 2X SSC; slides can be dried and steps 6–11 repeated for best results.

12. Wipe off excess liquid before placing slides in the pepsin solution. Digest tissue in 0.5 mg/ml pepsin solution at 37°C for 5–30 min. The correct time has to be evaluated empirically and may vary from one slide to another of the same tissue type. Digestion time is dependent on parameters such as initial fixation time, tissue type, and age of slide.

13. Wash in 2X SSC 2 times 5 min.

14. Check the digestion under a phase-contrast microscope. If necessary repeat steps 12 and 13, adjusting the digestion time.

15. Dry slides on a 45°C slide warmer.

16. Immerse slides in 10% buffered formalin for 10 min.

17. Wash two times 5 min in 2X SSC.

18. Dry slides on a 45°C slide warmer.

After this step the slides can be stored for a couple of weeks at room temperature. The procedure can be suspended after steps 5, 11 (dry slides on slide warmer), and 15, and can be continued later the same day or even the next day. Fixation (steps 16–18) can be omitted. If after hybridization the slides show a high degree of tissue loss, fixation step should be included.

Alternate Microwave Pretreatment of Formalin-Fixed, Paraffin-Embedded Tissue Sections

This procedure is faster than the conventional method described earlier. It is optimized for use with the 250 ml microwave buffer in Samsung MW1080STA microwave oven. If other brands of microwave oven and other volumes are used, the time, the power level, or both will need to be adjusted. There are several other reports that discuss different options and protocols using microwave oven (Bull *et al.*, 1999; Kitayama *et al.*, 2000).

1. Place slides in a 70°C warm oven, preferably overnight but at least for 2 hr.

2. Prepare/thaw pepsin solution and place it in a 37°C waterbath.

3. Deparaffinize slides in xylene for 10 min, then xylene:100% ethanol (1:1), for 10 min and in 100% ethanol for 10 min.

4. Hydrate slides in 90% and then 70% ethanol for 1 min each.

5. Dry slides on a slide warmer at 45°C and mark the area of interest with a glass marker on the back of the slide.

6. Immerse slides in 0.2 M HCl for 15 min.

7. Wash briefly (< 1 min) in deionized distilled water.

8. Put slides in a microwave safe jar filled with 0.01 M citrate buffer (pH 6.0).

9. Microwave for 1 min at 1000W immediately followed by 8 min at 200 W. The time and energy levels might have to be adjusted depending on the tissue used.

10. Immerse slides in 70% and 100% ethanol at 4°C for 3 min each.

11. Wash briefly in 2X SSC and wipe off excess liquid before placing slides in pepsin solution.

12. Digest tissue in 0.5 mg/ml pepsin solution at 37°C for 5–30 min. Digestion time has to be evaluated empirically and may vary from one slide to another.

13. Wash in 2X SSC 2 times 5 min.

14. Check digestion under a phase-contrast microscope. If necessary repeat steps 12 and 13, adjusting the digestion time.

15. Dry slides on a 45°C slide warmer.

16. Immerse slides in 10% buffered formalin for 10 min.

17. Wash two times 5 min in 2X SSC.

18. Dry slides on a 45°C slide warmer.

DNA Labeling Using Nick Translation

There is no commercially available *BRCA1* FISH probe. Use either BAC (283G2) (P1 library DMPC-HFF-1; Neuhausen *et al.*, 1994) or PAC (103014) (RPCI-1 library; Ioannou *et al.*, 1994) clones, which contain DNA sequences specific for the entire *BRCA1* gene locus on 17q12-q21. We and others have shown that both clones give successful FISH results (Staff *et al.*, 2000; Wei *et al.*, 2003). Follow the standard Qiagen phenol/chlorophorm extraction protocol to extract DNA from either clone and prepare for further fluorescent labeling. Use the control *CEP17* probe in dual hybridizations with the *BRCA1* probe to detect alterations of *BRCA1* copy number in relation to whole chromosome 17 gain or losses, or *BRCA1* gains or losses caused by structural chromosome rearrangements. The *CEP17* probe labeled with *SpectrumOrange* contains α-satellite DNA that hybridizes to the centromeric region of chromosome 17 (17p11.1-q11.1). *CEP17* is used as a marker to detect chromosome 17 aneusomy in cancer cells. This probe is commercially available from Vysis/Abbott, Inc., and is ready for hybridization.

BRCA1 DNA can be labeled either directly with a fluorochrome-conjugated dUTP such as *SpectrumGreen/Orange* dUTP or indirectly with, for example, biotin-conjugated dUTP (Espinosa *et al.*, 1997; Martin *et al.*, 2002). The procedures for both labeling techniques are similar, but the hybridization and washing procedures are markedly shorter for a directly labeled probe. The indirect BRCA1 DNA labeling was performed with biotin-16-dUTP and the *BRCA*1 probe was detected with FITC-conjugated avidin.

Note: We used ~5 μg DNA in the nick translation where most protocols recommend 1 μg (Espinosa *et al.*, 1997). The final concentration of DNAse I has to be determined empirically. The concentration of DNAse I described here has worked well for several of our probes with adjustment of the incubation time. During the procedure keep all tubes on ice. Calculate the volume of ddH_2O you need in the reaction mix to give a total volume of 50 μl. For direct labeling, keep the labeled dUTP and the reaction mix protected from light.

1. Prepare a 15°C waterbath by mixing ice and water in a Styrofoam box with a lid (Dow Chemical Company, Midland, MI).

2. Make the enzyme mix and store in the freezer until use. Follow step 3 for indirect labeling and steps 4 and 5 for direct labeling.

3. To a microcentrifuge tube add solutions in the following order: X ml deionized distilled water, 5 ml 10X NTS buffer, 1–5 mg DNA, 5 ml dNTP mix, 2.5 ml 0.3 mM biotin-16-dUTP, 1.5 ml 0.2 U/ml DNAse I, and 1 ml 10 U DNA polymerase I. Continue with step 6.

4. Make the 10X enzyme mix and store in the freezer until use.

5. In a microcentrifuge tube add solutions in the following order: X ml deionized distilled water, 5 ml 10X NTS buffer, 1 mg DNA, 5 ml dNTP mix, 5 ml 0.2 mM *SpectrumGreen/Orange*-dUTP and 5 ml enzyme mix. Save the remaining enzyme mix in the freezer (it might be needed later).

6. Vortex and centrifuge briefly.

7. Place tube in the 15°C waterbath, close the lid and incubate for 1–2 hr. The time must be determined experimentally; 1.5 hr is a good start.

8. Prepare and pour a 1% agarose gel into a gel tray with a comb suitable for a 10-μl sample.

9. When the incubation is finished, take a 5-ml sample from the nick translation mix, and immediately put the remaining part in the freezer.

10. Mix the reaction sample with 2 μl 5X loading dye and 3 μl TE. Load the sample on the gel using a 100-bp DNA ladder as a standard and run the gel at 100 V for 30–60 min.

11. Stain the gel with ethidium bromide (0.5 mg/ml) for 20 min (put the gel into a container with enough ethidium bromide solution to cover it) while gently rocking the container.

12. Analyze the gel under ultraviolet light. The nick translated DNA will appear as a smear. The largest part of the smear should be between 200 and 600 bp. For directly labeled probes the labeled dUTP will be seen as a bright fluorescent diffuse band on the gel.

13. If the DNA is the correct size, add 5 μl 10X stop buffer to the remaining reaction mix, vortex, and store at −20°C (the directly labeled probe should be protected from light).

14. If the DNA runs >600 bp, add 2 μl enzyme mix (or 1 μl of each enzyme if no mix was used), vortex, centrifuge briefly, and place the tube in the 15°C waterbath for 15–30 min, and then continue the procedure at step 6. For further nick translation of the same DNA, a higher concentration of DNAse I could be used.

15. If the DNA runs <200 bp, most fragments will be washed away during post-hybridization washes and no signal will be detected. Start from the beginning

with a shorter incubation time at 15°C or use more diluted DNAse I, or both.

Hybridization Procedure for Indirectly Biotin-Labeled Probe

Hybridization will take 16–24 hours, the best time to start is in the afternoon, allowing it to run overnight. It is recommended to use only eight slides at a time. The amounts of salmon testis, Cot-1, and placental DNA used have to be determined for each specific probe. Too much Cot-1/placental DNA will lead to a weakened probe signal, and too little will result in high background. A good starting point is 0.5 μg placental DNA, 0.5 μg Cot-1 DNA, and 3 μg salmon testis DNA. The following protocol is optimized for the *BRCA1* probe.

1. Prepare the RNAse A solution by mixing 5 ml 10X RNAse A with 45 ml 2X SSC in a Coplin jar and place the jar in a 37°C waterbath. Prepare the hybridization chamber by placing a wet piece of paper inside and let box float in a 37°C waterbath; prepare the denaturation solution and make sure all the waterbaths are turned on.

2. For each slide, mix (on ice): 0.5 μl placental DNA, 1.5 μl Cot-1 DNA, 3 μl salmon testis DNA, 1 μg (10 μl) labeled DNA, 3 μl 3 M potassium acetate, and 60 μl ice cold 100% ethanol.

3. Vortex and centrifuge briefly.

4. Place tubes at −80°C for 30 min.

5. Centrifuge for 30 min at 14,000 g at 4°C

6. Decant the supernatant carefully.

7. Add 500 μl 70% ethanol, rinsing the tube at the same time.

8. Centrifuge for 10 min at 14,000 g at 4°C. At this point start with slides (see step 14) so that the slides and the probe will be ready at approximately the same time.

9. Decant the supernatant and dry the DNA pellet in a speedvac for 10 min, at medium heat.

10. Resuspend the pellet in 10 μl hybridization buffer. Add 0.5 μl control *CEP17* probe diluted 1:10 in CEP hybridization buffer from Vysis/Abbott, Inc. (keeping the total volume of the hybridization mix ~10 μl).

11. Vortex for 30 min at low speed, centrifuge briefly.

12. Denature probe mixture by placing the tubes in a 74°C waterbath for 5 min.

13. Pre-anneal the probe mixture at 37°C (in a waterbath or on a slide warmer) until the slides are ready (at least 15 min).

14. Immerse slides in the 37°C RNAse A solution for 1 hr.

15. Wash slides 4 times 2 min in 2X SSC.

16. Dehydrate slides in a series of 70%, 80%, 90% and 100% ethanol soaks for 2 min each.

17. Air-dry slides.

18. Immerse tissue sections in the 75°C denaturation solution for 5 min (for cells in suspension or frozen touch preparations use 2 min). Keep slides separate from each other, putting only four slides per Coplin jar (50 ml solution).

19. Immediately immerse slides in 70% ethanol for 2 min followed by 80%, 90%, and 100% ethanol, 2 min each.

20. Transfer the slides from the ethanol to the slide warmer.

21. When the last drop of ethanol evaporate add 10 µl of probe solution to the area of interest and immediately cover with a 22 × 22 mm coverglass, avoiding air bubbles. When the liquid has spread under the whole glass, seal the edges with rubber cement using a syringe without needle.

22. Place slides in the prewarmed humidified hybridization chamber and close the lid. Return chamber to the 37°C waterbath and let hybridization proceed overnight (16–24 hr).

Washing Procedure and Detection of Indirectly Biotin-Labeled *BRCA1* Probe

1. Prewarm the following to 40°C: Four Coplin jars with 4X SSC, four Coplin jars with 50/50 mix, three Coplin jars with 4X SSC/0.1% TritonX-100, and the 2% BSA blocking solution. Prepare plastic coverslips by cutting pieces approximately 25 × 40 mm from a plastic bag.

2. Carefully remove rubber cement from each slide; make sure that the cover glass does not move and the tissue does not tear. Place slides in the first jar of 50/50 mix for 3 min or until the coverslips slide off. Wash slides 3 times for 2 min in the 50/50 mix at 40°C.

3. Wash slides 4 times for 2 min in 4X SSC at 40°C.

4. Drain and add 200 µl prewarmed blocking solution to each slide, cover with plastic coverslip and incubate for 45–60 min in a humid chamber at 37°C. In the meantime, for each slide prepare 200 µl blocking solution containing 5 µg/ml of avidin-fluorochrome (5 µl avidin-fluorochrome 1 mg/ml per ml blocking solution) and filter solution through a 0.2-µm filter. Keep shielded from light.

5. Gently remove coverslip and add 200 µl blocking solution containing the avidin-fluorochrome to each slide. Cover with plastic coverslip and incubate for 30 min in a humid chamber at 37°C. From this step forward protect slides from light (e.g., by coating clear glass jars with black tape and always using dark lids for jars).

6. Carefully remove coverslip and wash slides 3 times for 3 min in 4X SSC/0.1% TritonX-100 at 40°C.

7. Immerse slides in DAPI solution for 1 min.

8. Wash briefly in 2X SSC.

9. Apply ~2–3 drops of PDD anti-fade to each slide and add a coverglass, making sure no air bubbles are trapped under the glass. Gently squeeze out the excess PDD anti-fade and wipe off edges of slide.

10. Store slides at −20°C protected from light. Let slides come to room temperature before counting.

Hybridization Procedure for Directly Labeled Probes

The hybridization protocol is the same as for indirectly labeled probes except for the probe preparation. Use the protocol for indirectly labeled probes steps 1–22 but, in step 2 use the mixture described below (step 1). The amount of Cot-1 and labeled DNA must be determined empirically; some probes, especially those labeled with *SpectrumGreen;* will need up to 5 µg DNA per slide to show a bright signal. A good starting point for developing the probe preparation is described in step 1.

1. For each slide, mix (on ice): 12 µl (~1 µg) labeled DNA, 4 µl Cot-1 DNA 1 mg/ml, 5 µl salmon testis DNA 1 µg/ml, 2 µl (1/10 of total volume) 3 M potassium acetate, and 58 µl (2.5 times total volume) ice-cold 100% ethanol. Keep solutions containing the labeled DNA shielded from light at all steps.

2. Proceed with steps 3–22 from the protocol for indirectly labeled probes.

Washing Procedure for Directly Labeled Probes

1. Prepare Coplin jars with the post-hybridization solutions and place them in corresponding waterbaths. In the following steps protect the slides from light (e.g., by covering clear glass jars with black tape and by using dark lids for the jars).

2. Carefully remove the rubber cement from slides making sure the glass does not move and tear the tissue. Place slides in 2X SSC/0.3% NP40 until coverglasses slide off (2–4 min).

3. Immerse slides in 45°C 2X SSC/0.3% NP40 for 3 min; repeat once.

4. Place slides in 60°C 1X SSC for 5 min.

5. Immerse slides in 45°C 2X SSC/0.3% NP40 for 5 min.

6. Wash slides 3 times, briefly, in deionized distilled water.

7. Immerse slides in DAPI solution for 1 min.

8. Apply ~2–3 drops of PDD anti-fade and add coverslip. Make sure no air bubbles are trapped under

the coverslip. Wipe off excess PDD anti-fade from edges of slide.

9. Store slides at −20°C protected from light. Let slides come to room temperature before counting.

Troubleshooting

If the experiment fails, repeat it again before starting to troubleshoot. Remember, the pH of solutions and the temperatures of waterbaths and solutions are critical. Always confirm the pH and the temperature of the solutions before use; temperatures should be within ± 1°C of recommended. The most common problem of FISH failure is weak or no signal. Probable causes may include insufficient tissue pretreatment, tissue underdigestion, loss of DNA, or poor hybridization conditions. The following actions may be required: repeat pretreatment, increase digestion time, increase denaturation temperature of the slide (80°C) and denaturation time (10 min). For other problems and solutions we direct the reader to a troubleshooting guide (Vysis, 1998, 2001).

RESULTS AND DISCUSSION

Hybridization signals should be scored using the 100X objective of a Zeiss Axioplan epifluorescence photomicroscope (Carl Zeiss, Inc., Thornwood, NY) and multibandpass filters. Only well defined, non overlapping nuclei should be counted. In each tumor sample from imprint touch preparations score 100–200 cells. In each tumor sample from paraffin sections score 60–100 nuclei. In each normal sample an average of 50 (20–100) nonmalignant nuclei should be scored. Because the nuclei are usually present in slightly different planes of the section, the focus of the microscope has to be constantly adjusted to enable all positive signals to be visualized. Document counts in a two-way table as recommended by Vysis/Abbott, Inc. (Vysis, 1998). For each sample record the absolute number of *BRCA1* signals/cell, the number of *CEP17* signals/cell, and the ratio of *BRCA1* signals to chromosome 17 centromere signals. Alterations of the chromosome 17 copy numbers can be estimated by calculating the percentages of cells with reduction of *CEP17* signals to one copy (monosomy) or gain of *CEP17* signals to three or more copies (polysomy) for each sample. We also recommend a discrete classification scheme to assess the *BRCA1* copy number by determining the percentage of cells with one, two, three, and more than three copies of *BRCA1* gene per cell.

Control experiments are necessary for refinement and interpretation of FISH. To test the hybridization efficiency (probe and scoring efficiency) and to estimate the cutoff points for gene copy number gains and losses, use *BRCA1/CEP17* FISH on normal human lymphocytes, breast cancer cell lines, and nonmalignant breast samples. In our laboratory, for example, the hybridization efficiency of the *BRCA1* probe was confirmed by performing FISH on metaphase chromosomes and interphase nuclei from normal lymphocytes (two samples) and breast cancer cell line HCC1937 (one sample, data not shown), and on normal breast epithelium embedded in paraffin (two samples). We performed metaphase cell preparations according to routine protocols (Espinosa *et al.*, 1997). In normal lymphocytes the mean number (±SD) of *BRCA1* and *CEP17* signals per cell was 2.07±0.06 and 1.97±0.06, respectively. The mean *BRCA1/CEP17* ratio was 1.03±0.06. Scoring the chromosome 17 ploidy gave the following results (mean±SD): 94.3±3.9% disomic signals, 4.5±4.4% monosomic signals, and 0.9±0.7% polysomic signals. In nonmalignant breast samples the mean *BRCA1* and *CEP17* signals per cell were 1.9±0.0 and 1.9±0.07, respectively. The mean *BRCA1/CEP17* ratio was 1.1±0.07. The proportion of disomic cells was 81.7±7.0%, whereas 17.1±5.4% of cells were monosomic, and 1.2±1.6% of cells were polysomic for chromosome 17. Our results are consistent with previously published reports and indicate high hybridization efficiency (Grushko *et al.*, 2002; Staff *et al.*, 2000).

Analysis of *BRCA1* tumor suppressor gene using the FISH procedure is in its infancy stage. There are no published studies that correlate *BRCA1* gene copy number alterations with protein expression (Staff *et al.*, 2003). However, FISH analysis for detecting the gains and losses of *BRCA1* gene may shed light on mechanisms of *BRCA1* tumor suppressor gene inactivation and tumor initiation in *BRCA1* carriers and some sporadic tumors. *BRCA1* FISH has been used to investigate the genetic mechanisms that lead to the loss of the remaining wild-type allele of the gene, after inactivation of the first allele caused by germline mutation in breast tumors from *BRCA1*-mutation carriers (Staff *et al.*, 2000). In our laboratory we performed *BRCA1/CEP17* FISH on 13 sporadic *BRCA1*-methylated breast cancers and 37 control unmethylated tumors (Wei *et al.*, 2003). We found striking differences in mean copy number of *BRCA1* and *CEP17* between the two groups of tumors. The mean number of *BRCA1* copies was significantly less among the *BRCA1*-methylated cases (mean *BRCA1*/cell = 1.78) than among the unmethylated cases (mean = 2.30, *P* = 0.001). Similarly, the *CEP17* mean copy number was less among the *BRCA1*-methylated cases (mean = 1.85) as compared with the unmethylated cases (mean 2.29, *P* = 0.005). *BRCA1* methylation was frequently associated with loss of one copy of *BRCA1* gene and the occurrence of monosomy for chromosome 17 (Figure 58). Our data suggest that the

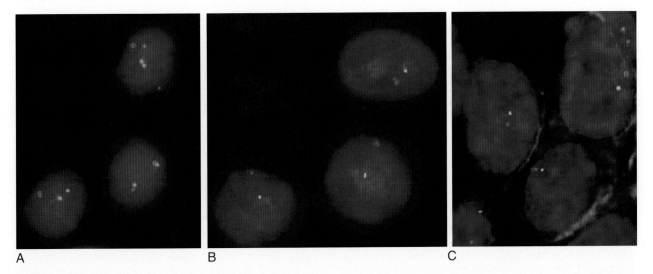

Figure 58 Representative photomicrographs *BRCA1/CEP17* fluorescence *in situ* hybridization (FISH) from patients with sporadic *BRCA1*-methylated (**A**) and unmethylated (**B**), and hereditary *BRCA1*-mutated (**C**) breast cancers. *BRCA1* gene is localized by green fluorescent signals, and chromosome 17 centromere (*CEP 17*) is localized by red fluorescent signals. The cells were counterstained with 4′6′diamidino-2-phenylindole (DAPI; blue). The tumor cells from the unmethylated case in **A** shows presumably normal *BRCA1* (2.06) and *CEP17* (2.05) copy numbers per cell. The majority of cells (93%) displayed normal two *BRCA1* and two *CEP17* signals per nucleus (ratio = 1.01). Slides from both cases were prepared from frozen tissues using imprint touch preparation procedure. The cells in **B** are a typical example of *BRCA1*-methylated sporadic cancer, in which the *BRCA1* methylation was correlated with chromosome 17 monosomy and loss of heterozygosity (LOH) in *BRCA1* locus. Of these nuclei, 76.5% displayed monosomy for chromosome 17 with corresponding loss of one copy of the *BRCA1* gene (*BRCA1/CEP17* ratio = 0.93). **C**, formalin-fixed, paraffin-embedded breast cancer tissue section from *BRCA1* mutation carrier is shown. The concomitant LOH in *BRCA1* locus (mean *BRCA1*/cell = 1.3) and loss of chromosome 17 (mean *CEP17*/cell = 1.1) was observed in 80% of tumor cells. Case was pretreated using conventional acid and chaotrope protocol. In all three cases the *BRCA1*-specific probe was labeled indirectly with biotin-l6-dUTP and detected with fluorescein isothiocyanate-conjugated avidin. Original magnification, 1200X.

loss of BRCA1 because of *BRCA1* promoter methylation is associated with LOH and may be an important mechanism for the inactivation of the *BRCA1* tumor suppressor gene in sporadic breast cancers. Our data also suggest that the detection of LOH in the *BRCA1* locus by FISH can be a useful method in the detection of sporadic breast cancers with BRCA1 dysfunction. The similarities between *BRCA1*-mutated hereditary and *BRCA1*-methylated sporadic tumors support a tumor progression model in which early loss of BRCA1 function causes defects in chromosome structure, cell division, and viability, so that a BRCA1-deficient cell must acquire additional alterations that overcome these problems. This presumably forces tumor evolution down a limited set of pathways (Venkitaraman, 2002), which may be similar in *BRCA1*-mutated hereditary and *BRCA1*-like sporadic breast cancers. A direct clinical application could be similar treatment strategies for BRCA1-deficient hereditary and sporadic breast cancers. FISH analysis of the *BRCA1* tumor suppressor gene described here should be expanded and could be applied to the other tumor suppressor genes.

Supported by the United States Army Department of Defense Grant DAMD17-99-1-9123.

References

Arver, B., Du, Q., Chen, J., Luo, L., and Lindblom, A. 2000. Hereditary breast cancer: A review. *Semin. Cancer Biol.* 10(4):271–288.

Bull, J.H., and Harnden, P. 1999. Efficient nuclear FISH on paraffin-embedded tissue sections using microwave pretreatment. *Biotechniques* 26(3):416–422.

Catteau, A., and Morris, J.R. 2002. *BRCA1* methylation: A significant role in tumour development? *Semin. Cancer Biol.* 12(5):359–371.

Espinosa, R., 3rd., and Le Beau, M.M. 1997. Gene mapping by FISH. *Methods Mol. Biol.* 68:53–76.

Esteller, M., Fraga, M.F., Guo, M., Garcia-Foncillas, J., Hedenfalk, I., Godwin, A.K., Trojan, J., Vaurs-Barriere, C., Bignon, Y.J., Ramus, S., Benitez, J., Caldes, T., Akiyama, Y., Yuasa, Y., Launonen, V., Canal, M.J., Rodriguez, R., Capella, G., Peinado, M.A., Borg, A., Aaltonen, L.A., Ponder, B.A., Baylin, S.B., and Herman, J.G. 2001. DNA methylation patterns in hereditary human cancers mimic sporadic tumorigenesis. *Hum. Mol. Genet. 10(26):*3001–3007.

Ford, D., Easton, D.F., Stratton, M., Narod, S., Goldgar, D., Devilee, P., Bishop, D.T., Weber, B., Lenoir, G., Chang-Claude, J., Sobol, H., Teare, M.D., Struewing, J., Arason, A., Scherneck, S., Peto, J., Rebbeck, T.R., Tonin, P., Neuhausen, S., Barkardottir, R., Eyfjord, J., Lynch, H., Ponder, B.A., Gayther, S.A., Zelada-Hedman, M., and the Breast Cancer Linkage Consortium. 1998. Genetic heterogeneity and penetrance analysis of the BRCA1 and BRCA2 genes in breast cancer families. The Breast Cancer Linkage Consortium. *Am. J. Hum. Genet. 62(3):*676–689.

Futreal, P.A., Liu, Q., Shattuck-Eidens, D., Cochran, C., Harshman K., Tavtgian, S., Bennett, L.M., Haugen-Strano, A., SwenSen, J., Miki, Y., Eddington, K., McClure, M., Frye, C., Weaver-Feldhous, J., Ding, W., Gholami, Z., SoederKvist, P., Terry, L., Jhanwar, S., Berchuck, A., Iglehart, J.D., Marks, J., Bailinger, D.G., Barrett, J.C., Skolnick, M.H., Kamb, A., Wiseman, R. 1994. BRCA1 mutations in primary breast and ovarian carcinomas. *Science 266(5182):*120–122.

Grushko, T.A., Blackwood, M.A., Schumm, P.L., Hagos, F.G., Adeyanju, M.O., Feldman, M.D., Sanders, M.O., Weber, B.L. and Olopade, O.I. 2002. Molecular-cytogenetic analysis of HER-2/neu gene in *BRCA1*-associated breast cancers. *Cancer Res 62(5):*1481–1488.

Hedenfalk, I., Duggan, D., Chen, Y., Radmacher, M., Bittner, M., Simon, R., Meltzer, P., Gusterson, B., Esteller, M., Kallioniemi, O.P., Wilfond, B., Borg, A., and Trent, J. 2001. Gene-expression profiles in hereditary breast cancer. *N Engl J Med 344(8):*539–548.

Ioannou, P.A., Amemiya, C.T., Garnes, J., Kroisel, P.M., Shizuya, H., Chen, C., Batzer, M.A., and de Jong, P.J. 1994. A new bacteriophage P1-derived vector for the propagation of large human DNA fragments. *Nat Genet 6(1):*84–89.

Kitayama, Y., Igarashi, H., and Sugimura, H. 2000. Initial intermittent microwave irradiation for fluorescence *in situ* hybridization analysis in paraffin-embedded tissue-sections of gastrointestinal neoplasia. *Lab Invest 80(5):*779–781.

Miki, Y., Swensen, J., Shattuck-Eidens, D., Futreal, P.A., Harshman, K., Tavtgian, S., Liu, Q., Cochran, C., Bennett, L.M., Ding, W., Bell, R., Rosenthal, J., Hussey, C., Tran, T., McClure, M., Frye, C., Hattier, T., Phelps, R., Haugen-Strano, A., Katcher, H., Yakumo, K., Gholami, Z., Shaffer, D., Stone, S., Bayer, S., Wray, C., Bogden, R., Dayananth, P., Ward, J., Tonin, P., Narod, S., Bristow, P.K., Norris, F.H., Helvering, L., Morrison, P., Rosteck, P., Lai, M., Barrett, J.C., Lewis, C., Neuhausen, S., Cannon-Albright, L., Goldgar, D., Wiseman, R., Kamb, A., and

Skolnick, M.H. 1994. A strong candidate for the breast and ovarian cancer susceptibility gene BRCA1. *Science 266(5182):*66–71.

Neuhausen, S.L., Swensen, J., Miki Y., Liu, Q., Tavtigian, S., Shattuck-Eidens, D., Kamb, A., Hobbs, M.R., Gingrich, J., and Shizuya, H. 1994. A P1-based physical map of the region from D17S776 to D17S78 containing the breast cancer susceptibility gene BRCA1. *Hum Mol Genet 3(11):*1919–1926.

Perez-Valles, A., Martorell-Cebollada, M., Nogueira-Vazquez, E., Garcia-Garcia, J.A., and Fuster-Diana, E. 2001. The usefulness of antibodies to the BRCA1 protein in detecting the mutated BRCA1 gene. An immunohistochemical study. *J. Clin. Pathol. 54(6):*4761–4780.

Staff, S., Isola, J., and Tanner, M. 2003. Haplo-insufficiency of *BRCA1* in sporadic breast cancer. *Cancer Res. 63(16):*4978–4983.

Staff, S., Nupponen, N.N., Borg, A., Isola, J.J., and Tanner, M.M. 2000. Multiple copies of mutant BRCA1 and BRCA2 alleles in breast tumors from germ-line mutation carriers. *Genes Chromosomes Cancer 28(4):*432–442.

van't Veer, L.J., Dai, H., van de Vijver, M.J., He, Y.D., Hart, A.A., Bernards, R., and Friend, S.H. 2002. Expression profiling predicts outcome in breast cancer. *Breast Cancer Res. 5(1):*57–58.

Venkitaraman, A.R. 2002. Cancer susceptibility and the functions of BRCA1 and BRCA2. *Cell 108(2):*171–182.

Vysis, I. 1998. *Pathvysion HER-2 DNA probe kit package insert.* Downers Grove, IL: Vysis, Inc., 12.

Vysis, I. 2001. *LSI locus specific identifie DNA probes.* Downers Grove, IL: Vysis; Abbott, Inc.

Wei, M., Grushko, T., Das, S., Dignam, J., Hagos, F., Sveen, L., Fackenthal, J., and Olopade, O.I. 2003. Methylation of the BRCA1 promoter in sporadic breast cancer related to BRCA1 copy number and pathologic features. In American Association for Cancer Research 94th Annual Meeting. July 11–14, 2003. Washington, D.C. *Proceedings of the AACR* 44 (2nd Ed.), Abstract 4950. 987–988.

Immunohistochemistry of c-*myc* Expression in Breast Carcinoma

Rakesh Naidu

Introduction

The *myc* family of protooncogenes is shown to play an important role in the development of many human cancers. The protooncogene c-*myc* was discovered as the cellular homologue to the transforming gene of the avian myelocytomatosis virus (Bishop, 1982). The chromosome 8q24 that maps the c-*myc* gene is frequently altered at the genetic and expression level. The length of the gene is 8.2 kbp.

The c-*myc* gene yields three different transcripts that can be translated into three nuclearphosphoproteins, namely c-*myc*1, c-*myc*2, and c-*myc*S. The protein c-*myc*2 has 439 amino acids with a molecular weight of 64 kD, whereas c-*myc*1 consists of 453 amino acids with a molecular weight of 67 kD. However, c-*myc*S protein is a 45-kD polypeptide that lacks approximately 100 amino acids at the N-terminal of c-*myc*2 (Spotts *et al.*, 1997). Both the c-*myc* mRNA and protein have short half-lives of less than 30 minutes. Among the three types of c-*myc* proteins, c-*myc*2, normally referred to as c-*myc*, is the major form that is expressed in various tissues and cell lines. The c-*myc* protein has several functional domains that include transactivational domain (TAD), which has Myc boxI (MbI) and Myc boxII (MbII), DNA binding basic region, nuclear localization sequence (NLS), the helix-loop-helix (HLH),

and the leucine zipper domain (LZ). As a transcription factor, c-*myc*1 and c-*myc*2 have the ability to activate and suppress transcription of target genes because they have both the MbI and MbII, which are required for transactivation and transsuppression, respectively. In contrast, c-*myc*S was shown to function as suppressor of transcription because it lacks MbI (Liao *et al.*, 2000).

The protein c-*myc* participates in various cellular activities involved in normal regulation of cell cycle progression, differentiation, and apoptosis. It initiates these functions through activation or repression, or both of various c-*myc* target genes (Facchini *et al.*, 1998). Expression of c-*myc* is tightly regulated in normal cells to ensure controlled proliferation. In normal cells, both transcriptional activation of growth stimulatory and transcriptional suppression of growth inhibitory of c-*myc* target genes are required to regulate cell cycle progression and monitor cellular growth (Dang, 1999). Up-regulation of c-*myc* protein in normal primary cells is required to stimulate quiescent cells to progress from G1- to S-phase promoting cell cycle progression. High levels of c-*myc* expression accelerate the growth rate of the cells. Autosuppression of c-*myc* is a negative feedback mechanism that allows the c-*myc* protein concentration to be retained within the normal physiologic range throughout the cell cycle progression.

Handbook of Immunohistochemistry and in situ Hybridization of Human Carcinomas, Volume 1: Molecular Genetics; Lung and Breast Carcinomas

395

Copyright © 2004 by Elsevier (USA)
All rights reserved.

Loss of autoregulation by c-*myc* because of mutations or genetic alteration may cause carcinogenesis (Facchini *et al.*, 1998). Studies of cell lines have reported that down-regulation of c-*myc* expression is required to induce differentiation. Expression of c-*myc* protein is greater in actively proliferating cells, and constitutive expression of c-*myc* protein was demonstrated to block differentiation in several cell types, suggesting that c-*myc* plays a key role in the switch from proliferation to differentiation (Freytag *et al.*, 1990).

Earlier studies have demonstrated that overexpression of c-*myc* could induce apoptosis. Studies on the functional interaction between p53 and c-*myc* have shown that c-*myc*–mediated apoptosis is dependent on p53 in some cell types, but not in others. However, p53-dependent and -independent mechanisms for c-*myc*–induced apoptosis have been shown to exist in epithelial cells (Sakamuro *et al.*, 1995). In addition, c-*myc* promotes apoptosis through other pathways by cooperating with cyclin D3, cdc25A, and ornithine decarboxylase (ODC) (Galaktionov *et al.*, 1996; Janicke *et al.*, 1996). However, overexpression of Bcl-2, Ras, platelet-derived growth factor, or other growth promoting factors has been implicated in inhibiting c-*myc*–induced apoptosis (Kauffmann-Zeh *et al.*, 1997; Marcu *et al.*, 1992). Inhibition of myc-induced apoptosis is essential to promote cell transformation and tumorigenesis.

Experimental evidence of c-*myc* molecular function suggests that c-*myc* exerts its transforming ability by activating and repressing transcription of critical sets of cellular genes (Dang, 1999; Facchini *et al.*, 1998). Uncontrolled regulation of the c-*myc* gene may cause alteration in expression of c-*myc* target genes. Studies have reported that c-*myc* requires cooperation of other oncogenes such as Ras, Bcl-2, viral proteins, and several peptide growth factors to promote transformation (Marcu *et al.*, 1992). Activation of c-*myc* oncogene leading to overexpression has been reported to occur through various mechanisms, such as proviral insertion, chromosomal translocation, gene amplification, and gene mutation, which may increase the rate of transcription or alter mRNA half-life (Marcu *et al.*, 1992). Constitutive expression of c-*myc* may result in shortening of G1-phase and facilitate progression from G1-to S-phase, or bypass Gl-phase growth arrest (Nass *et al.*, 1997). Overriding cell cycle arrest by ignoring G1-/S-phase checkpoint can cause genomic instability resulting in chromosomal aberrations and accumulation of somatic mutations in protooncogenes and tumor suppressor genes.

Each tumor mass consists of a heterogeneous population of tumor cells and the genetic change that occurs is not uniform throughout the entire tumor leading to different gene expression profiles. The role of normal and stromal cells that surround the tumor cannot be denied because they may facilitate tumor growth and invasion. Immunohistochemistry (IHC) allows us to visualize expression of a gene at a specific region of the tissue, which provides important clues to the role of the gene in the development of tumor. Currently, this technique has been given a lot of importance as a diagnostic tool to assist in patient treatment and management. Its ability to localize in detail the distribution of the various antigens in cells and tissues has become a tool in most pathology laboratories. Immunohistochemistry has been widely used in breast cancer studies to study the protein expression of various genes.

In this study, immunohistochemical technique was used to detect the presence of the c-*myc* oncoprotein in breast cancer cells. Immunoreactivity of c-*myc* is believed to represent the intracellular accumulation of c-*myc* oncoprotein. The c-*myc* gene is frequently overexpressed or amplified, or both, in human breast cancer, suggesting that overexpression of c-*myc* protein may play an important role in the development of breast cancer (Deming *et al.*, 2000; Liao *et al.*, 2000; Nass *et al.*, 1997). Increased levels of c-*myc* have been noted in benign breast lesions such as fibroadenoma and fibrocystic diseases, indicating that c-*myc* may be involved in the early development of the cancer (Spandidos *et al.*, 1987). Immunohistochemical analysis of various studies demonstrated that a greater percentage of breast cancer (50–100%) showed overexpression of c-*myc* protein than gene amplification (Deming *et al.*, 2000; Liao *et al.*, 2000). It is believed that up-regulation of c-*myc* may be associated with c-*myc* amplification, although other mechanisms such as increased rate of transcription and stability of c-*myc* mRNA or protein could be involved. Overexpression of c-*myc* was not correlated with clinicopathologic parameters and was not predictive of patient survival or disease-free survival (Locker *et al.*, 1989). Other authors have also reported lack of significant association between c-*myc* expression and prognosis of patients with breast cancer (Mizukami *et al.*, 1991; Spaventi *et al.*, 1994). Pietilainen *et al.* (1995) reported that c-*myc* has no independent prognostic value over standard prognostic factors, although strong expression was related to a long recurrence-free period. Bland *et al.* (1995) and Sierra *et al.* (1999) reported that c-*myc* may have prognostic value when coexpressed with other oncogenes. Although immunohistochemistry has been widely used to detect overexpression of c-*myc* protein in breast carcinomas, the number of studies published on c-*myc* immunohistochemistry were relatively small compared with those published on c-*myc* gene amplification. The reports on c-*myc* immunohistochemistry and its association with clinicopathologic parameters and prognosis from various studies are not consistent.

In this chapter, expression of c-*myc* protein detected by immunohistochemical methods is discussed in depth. Here, we report expression of c-*myc* protein in different histopathologic types and relate it to several clinicopathologic parameters such as estrogen receptor (ER) status, lymph node status, histologic grade, patient age, and proliferation index, as described earlier by Naidu *et al.* (2002). The role of c-*myc* in diagnosis and prognosis also is discussed.

MATERIALS

Reagents for c-*myc* Immunohistochemistry

1. Xylene
2. Absolute ethanol
3. 95% Ethanol: 5 ml absolute ethanol; bring volume to 100 ml with sterile distilled water
4. 90% Ethanol: 10 ml absolute ethanol; bring volume to 100 ml with sterile distilled water
5. 80% Ethanol: 20 ml absolute ethanol; bring volume to 100 ml with sterile distilled water
6. 75% Ethanol: 25 ml absolute ethanol; bring volume to 100 ml with sterile distilled water
7. 3% Hydrogen peroxide (H_2O_2): 3 ml of 30% H_2O_2; bring volume to 30 ml with sterile distilled water (keep in dark and refrigerate at 2–8°C)
8. 37 mM ammonia water: 2.5 ml 15 M ammonium hydroxide; bring volume to 1 liter with sterile distilled water
9. 3 M sodium hydroxide (NaOH): 24.0 g sodium hydroxide pellets; bring volume to 200 ml with sterile distilled water
10. Phosphate-buffered saline (PBS): 38.75 g sodium chloride, 7.50 g dibasic potassium phosphate, 1.00 g monobasic potassium phosphate; bring volume to 5 L with sterile distilled water; adjust pH to 7.6 with 3 M NaOH
11. 1% bovine serum albumin in PBS (PBS-BSA): 1 g BSA and 100 ml PBS; add BSA to buffer with stirring
12. 0.1% Trypsin containing 0.1% calcium chloride: 0.05 g trypsin and 0.05 g calcium chloride, and 50 ml PBS (pH 7.6); the trypsin solution is aliquoted into small quantities before frozen at −20°C
13. Mouse monoclonal antibody for c-*myc* (Clone 9E11) (Calbiochem, San Diego, CA): Primary antibody was diluted at 1:25 in PBS-BSA
14. Mouse monoclonal antibody (Clone DAK-G01) (Dako, Denmark): Negative control antibody was diluted at 1:25 in PBS-BSA
15. Detection kit: Universal Large Volume Dako 2 System Peroxidase contains prediluted biotinylated secondary antibody and streptavidin-conjugated peroxidase complex
16. Color detection kit: Large Volume Dako AEC Substrate System contains AEC chromogen (3-amino-9-ethylcarbazole) and substrate buffer
17. Mayer's hematoxylin (Fluka, France)
18. Aqueous mounting media (Dako Glycergel Mounting Medium)
19. Humidified chamber

METHODS

Indirect Immunoperoxidase Technique for Formalin-Fixed, Paraffin-Embedded Tissues

1. Place the slides with tissue sections of 5 µm in thickness in the oven at 56°C for 20 min to melt the paraffin.
2. Immediately transfer the slides into a xylene bath for 3 min to deparaffinize the sections.
3. Hydrate the tissue sections gradually in descending concentration of ethanol at 100% (absolute ethanol), 95%, 90%, 80%, and 75% for 2 min each.
4. Rinse the sections twice in sterile distilled water for 5 min to ensure complete removal of ethanol residues.
5. Flip the slides to remove excess water from the surface of the sections. Wipe the surrounding of the sections with tissue.
6. Incubate the sections in 3% hydrogen peroxide for 10 min at 37°C to block the endogenous peroxidase activity in a humidified chamber.
7. Rinse the sections in sterile distilled water for 5 min.
8. Rinse the sections in PBS for 5 min.
9. Flip the slides to remove excess PBS from the surface of the sections. Wipe the surrounding of the sections with tissue.
10. Incubate the sections with 0.1% trypsin in PBS containing calcium chloride for 12 min at room temperature in a humidified chamber.
11. Stop the digestion by rinsing the sections in cold PBS for 5 min.
12. Rinse the sections in PBS for 5 min at room temperature. Repeat **step 9**.
13. Incubate the positive control tissue sections and sample tissue sections with the c-*myc* primary antibody diluted in PBS containing 1% BSA for 2 hr at room temperature in a humidified chamber. The volume dispensed depends on the size of the tissue section.
14. For negative control, incubate one of the tissue sections with negative control antibody and the other with PBS-BSA diluent for 2 hr at room temperature in

a humidified chamber. The volume dispensed depends on the size of the tissue section.

15. Rinse the sections in PBS for 30 min at room temperature to remove unbound primary antibody and negative control antibody. Repeat **step 9**.

16. Incubate with biotinylated antibody for 10 min at 37°C in a humidified chamber. Cover the entire sections with the antibody.

17. Rinse the sections in PBS for 10 min at room temperature to remove unbound biotinylated antibody. Repeat **step 9**.

18. Incubate with streptavidin-conjugated horseradish peroxidase for 10 min at 37°C in a humidified chamber. Cover the entire sections with the streptavidin-conjugated horseradish peroxidase reagent.

19. Rinse the sections in PBS for 10 min at room temperature to remove unbound streptavidin-conjugated horseradish peroxidase. Repeat **step 9**.

20. Incubate with 70–100 μl of freshly prepared AEC substrate chromogen for 5–20 min at 37°C in a humidified chamber. The volume dispensed depends on the size of the tissue section. Stop the reaction until acceptable color intensity is reached.

21. Rinse the sections in distilled water for 1 min at room temperature. Flip the slides to remove excess of water from the surface of the sections.

22. Counterstain the tissue sections with Mayer's hematoxylin for 1–3 min at room temperature. Rinse the sections in distilled water for 1 min at room temperature. Flip the slides to remove excess of water from the surface of the sections.

23. Immerse the sections in 37 mM ammonia water for 10 sec.

24. Rinse the sections in distilled water for 1 min at room temperature.

25. While the sections are still wet, mount the sections with an aqueous mounting media.

26. The slides are allowed to dry for 1–2 hr at room temperature. The slides are now ready to be viewed under the light microscope.

RESULTS AND DISCUSSION

Optimization of c-*myc* Immunostaining

The staining method used was an indirect immunoperoxidase system using a standard biotin-streptavidin-peroxidase complex technique. This method is preferred and frequently used because of its greater sensitivity and it allows to amplify the signal in tissues with low concentration of the antigen. Therefore, precaution was taken in every step of the immunohistochemical method to optimize the conditions to produce satisfactory results. To obtain optimal, reproducible, and accurate results, the immunohistochemical technique requires assessment of several factors such as fixation, tissue handling, and processing. In most laboratories 10% neutral, buffered formalin is the most commonly used fixative for tissue preservation. All tissues obtained by our laboratory for immunohistochemistry were fixed in 4–10% buffered formalin for 12–18 hr immediately after surgery. These conditions were shown to preserve cell morphology and antigenicity. Before immunostaining, all the tissue sections were stained with hematoxylin and eosin (H&E) for histologic evaluation by consultant pathologist. Then the tissue was cut serially to obtain sections for immunohistochemical analysis. The first and last sections taken from each tissue were stained again with H&E to confirm the presence of tumor.

Heating tissues at 56°C for 20 min is sufficient to melt the wax. Sections were transferred to a xylene bath immediately for deparaffinization. This is an essential step for complete removal of the wax from the sections. Improper removal of paraffin will result in diffused nonspecific background staining. The xylene is cleared from the sections by absolute ethanol before rehydrating in descending concentrations of ethanol. This step may be repeated once to ensure that the xylene is completely removed to allow effective rehydration, otherwise the presence of xylene may hinder diffusion of water molecules into the tissues and reduce the ability of the antibody to reach and recognize the c-*myc* antigen. Checklist was maintained to make sure reagents for dewaxing and rehydration is changed periodically. Endogeneous peroxidases are present in many normal and neoplastic tissues. The enzyme activity was inhibited with 3% H_2O_2 for 10 min at 37°C before the application of the primary antibody. The temperature and incubation period should be sufficient for complete inhibition of the enzyme activity.

There has been considerable interest in antigen retrieval methods to enhance immunostaining by pretreating the paraffin-embedded, formalin-fixed sections with proteolytic enzymes or microwave heating. Optimization of c-*myc* immunostaining in formalin-fixed breast tissues using trypsin requires careful balance between cellular preservation and staining intensity of the nuclei. Before trypsinization, we noted that at higher dilution of the primary antibody, majority of the tissues showed absence of nuclear staining, whereas at lower dilution, there was no enhancement in the intensity of nuclear staining but there was an increase in the cytoplasmic and nonspecific background staining, irrespective of the incubation period. The primary concern of c-*myc* immunohistochemistry was the presence of cytoplasmic staining. We observed significant improvement in the nuclear staining intensity and reduction in the cytoplasmic and background staining when the tissues were trypsinized.

Although pretreatment was not recommended by the manufacturer, our experimental data suggest that trypsinization is required to facilitate the recognition of c-*myc* protein by the primary antibody. Trypsin treatment increases the permeability of the cell and the nuclear membrane of the tissue. It also exposes the epitopes by cleaving the cross-linking of the epitope due to formalin fixation for antibody–antigen binding. Standardization of trypsin digestion is important to obtain optimal immunohistochemical results. The nuclear staining intensity was improved by testing several dilutions of the primary antibody with varying incubation periods for both the primary antibody and trypsin. We found that incubation with trypsin for 10–20 min at room temperature increased the intensity of the nuclear staining. There was no cytoplasmic and background staining in majority of tissues. We showed that prolonged digestion decreases nuclear and cytoplasmic staining. Excessive treatment also increases nonspecific background staining and in some cases cytoplasmic staining. Our results demonstrated that there is an association between dilution of primary antibody and incubation period of primary antibody and trypsin. However, little improvement in the staining intensity was noticed after 20 min of trypsin treatment. Incubation with trypsin for 12 min and primary antibody with dilution of 1:25 for 2 hr at room temperature was sufficient to provide satisfactory nuclear staining in the absence of cytoplasmic staining. Appropriate antibody dilution and enzymatic digestion increases the antibody penetration into the cell and nucleus. Antibody dilution critically affects background staining, and antibody diluent (PBS) containing BSA is used to reduce the background staining.

The biotinylated antibody and streptavidin-conjugated, horseradish peroxidase enzyme was obtained commercially in a prediluted form. Incubation for 10 min at 37°C gave desirable results without any nonspecific background staining. The AEC substrate–chromogen reaction produces a red end product that precipitates at the antigen site. The incubation period for AEC substrate–chromogen varies between 5 and 20 min depending on the concentration of the c-*myc* antigen and nature of the tissue fixation. The optimal incubation for AEC substrate–chromogen reaction was determined by trials with different interval periods to avoid undersaturation or oversaturation of the chromogen. In general, the staining intensity result of AEC end product will reflect the concentration of the c-*myc* antigen present in the nucleus. However, oversaturation of the chromogen may result in very intense staining causing the result to be invalid. The color change is carefully

monitored and controlled. The sections are removed when desired color change is reached. Some tissue sections require a longer period to develop satisfactory color change, whereas some other tissues show immediate color reaction as soon as they are incubated. We noted that tissues incubated for more than 10 min irrespective of the rate of color change does not increase the intensity of the nuclear staining, but does increase the nonspecific background staining, which appears as diffused staining. After the color reaction, the sections are counterstained with Mayer's hematoxylin, which is a non–alcohol-containing counterstain, for 1–3 min. AEC forms an end product that is alcohol soluble. The counterstaining procedure is optimized to avoid overstaining of the nuclei with blue color. Excessive counterstaining causes dark coloration of the cell nuclei that may compromise proper interpretation of results. All the sections are mounted in aqueous-based mounting media and slides are stored in the dark because AEC is sensitive to excessive light and will fade in time.

The main limitation factor of the c-*myc* immunostaining presented here is the trypsinization process. Pretreatment of formalin-fixed, paraffin-embedded tissues with trypsin showed enhancement of nuclear staining with reduced cytoplasmic staining. Identifying the right method for enzymatic digestion is purely subjective and can be influenced by intralaboratory and interlaboratory variations. Careful monitoring of the incubation period and viewing under the microscope is required to assess whether the tissues are not undertreated or overtreated. We used both the weakly and strongly nuclear stained tissues as positive controls to monitor the variations and reproducibility of the results. Attempts were made to standardize the conditions for trypsinization. The incubation period for a particular concentration may vary between laboratories. Following a fixed scheme of enzymatic digestion may result in false-negative results. The effectiveness of trypsin digestion will depend on the fixation time, processing of tissues, embedding methods, and thickness of the tissues. Thus, the digestion time can be variable and must be optimized by the individual laboratories. Other factors that may influence the staining intensity of the trypsinized tissues are types of antibodies, source of antibodies, detection systems used, and experimental procedures.

Interpretation of c-*myc* Immunostaining

Positive and negative controls were included in each batch of immunostaining to validate immunohistochemical results. The negative control consists of commercially available negative control antibody, which has the same isotype (IgG1) and is from the same animal species as the primary antibody (mouse mono-clonal antibody), and PBS-BSA diluent. Breast control tissues that demonstrate strong and weak nuclear staining for c-*myc* were used as positive control tissues. In our laboratory, tissue sections that demonstrate consistently strong and weak nuclear staining without cytoplasmic and nonspecific staining with repeated runs were used as positive controls. The negative control antibody was obtained in a concentrated form and was diluted to the same concentration as the primary antibody using PBS-BSA diluent.

The tissues were carefully viewed under the microscope to observe for any nonspecific nuclear or cytoplasmic immunoreactivity in the nuclei and cytoplasm of the tumor and stromal cells. Positive control tissues with strong and weak nuclear staining were examined first to ensure that all the immunohistochemical reagents are functioning properly, staining procedures are well followed, and tissues are properly fixed and processed before examining the samples. Staining intensity of the strongly positive tissues was analyzed to ensure the staining is satisfactory. Weakly positive tissue was used to detect changes in the sensitivity of the antibody (e.g., overdiluted or sections that retain too much solution after rinsing) and working conditions of the immunohistochemical reagents, which might

not be apparent if only intensely stained positive control tissue was used. The c-*myc* nuclear oncoprotein was found predominantly in the nucleus (Figure 59). The presence of c-*myc* protein in the nucleus was visualized as red staining, which was indicative of positive reactivity. Staining intensity of the nuclei was noted in both the positive control tissues to ensure reproducibility of the staining intensity. The negative controls were examined for any nonspecific background staining caused by antibody cross-reactivity, endogenous peroxidase activity, precipitation of the AEC substrate chromogen or the end product, nonspecific binding of the biotinylated antibody, and streptavidin-conjugated peroxidase. Nonspecific staining was not encountered in these tissues. Throughout the immunohistochemical analysis the results observed in positive and negative control tissues were consistent.

We adapted a previously described method to score p53 immunostaining in breast tumors (Isola *et al.*, 1992) with some modification to assess the c-*myc* score in the study. This method is simple, reproducible, and useful to analyze the c-*myc* immunostaining. Using the scoring system we noted that c-*myc* overexpression was strongly correlated with the c-*myc* amplification detected in our previous study (Naidu *et al.*, 2002). It also reduces the intraobserver and interobserver variations,

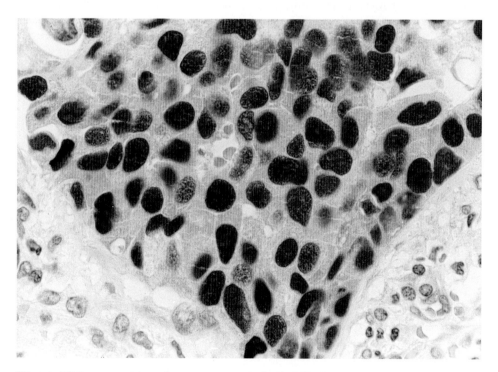

Figure 59 Immunostaining of c-*myc* oncoprotein in formalin-fixed, paraffin-embedded invasive ductal carcinoma tissue. Immunostaining of c-*myc* is confined predominantly in the nuclei of the malignant cells and the staining intensity is homogeneous in most of the stained cells. Magnification, 500X.

and the reproducibility was more than 98% in our laboratory. Immunostaining of c-*myc* was scored based on the intensity of the nuclear staining and the percentage of stained tumor cells. The staining intensity was classified into three categories on the basis of nuclear localization: negative represents no staining, weak represents faint nuclear staining, and strong staining represents moderate to darkly or densely stained nuclei with more widespread nuclear staining. A cutoff value of 20% of the entire tumor cells of the whole tissue section was chosen. The tumors were scored strongly positive if more than 20% of the nuclei were strongly stained. Immunostaining in more than 20% of the tumor cells represents overexpression of the c-*myc* protein. If a small proportion of the nuclei were stained (1–20%), irrespective of strong or weak nuclear staining, the tumor was scored as weakly positive.

Cell population in any tumor is heterogenous and heterogeneity in expression of c-*myc* protein is expected within an individual tumor and between different tumors. Staining intensity of the nuclei of the tumor cells varied from absence of staining to intense staining, but a marked heterogenous staining pattern was not observed in most of the tumors. Immunostaining of c-*myc* was confined predominantly in the nuclei of the malignant cells and no staining was seen in the nucleoli. In general, the staining appeared diffused throughout the nucleus. In very rare cases we noted weak cytoplasmic staining with strong nuclear staining in the same cell, but the majority of the tumor cells within the tumor showed staining restricted only to the nuclei. Presence of cytoplasmic staining was not a consistent feature; therefore, it could be a staining artefact and was not included for analysis. None of the tumor cells exhibited total cytoplasmic staining or weak nuclear staining accompanied with cytoplasmic immunoreactivity. Immunostaining of c-*myc* was not seen in the cytoplasm and nuclei of the normal and stromal cells.

Based on the scoring system, c-*myc* immunopositivity was detected in 252 (57%) of 440 primary breast carcinomas (Naidu *et al.*, 2002). Tumors strongly positive for c-*myc* were noted in 199 (45%) breast carcinomas. The percentage of tumor cells that exhibited, strong nuclear staining ranged from 70–80% for *in situ* and more than 90% for invasive tumors, and a similar pattern was observed when *in situ* and invasive lesions coexist. A small proportion of tumor cells demonstrated weak or absence of nuclear staining. Immunostaining of c-*myc* was evenly distributed throughout the cancer tissue in the majority of the strongly positive tumors, especially in the invasive carcinomas. In general, the staining intensity was homogeneous in most of the stained malignant epithelial cells. Weakly positive tumors were detected in 53 (12%) breast carcinomas. In these

tumors, approximately less than 20% of the tumor cells showed weak nuclear staining. However, a small number of tumors showed both weak and strong nuclear staining. All these tumors demonstrated a focal staining pattern and rarely the stained tumor cells were distributed throughout the tissue. Most of the tumor cells were negative for c-*myc* immunostaining.

Normal c-*myc* gene product accumulates or is located predominantly in the nuclei of the cells, irrespective of normal or tumor cells, because of the biological function of c-*myc* as DNA-binding protein. However, presence of c-*myc* protein in the cytoplasm has raised concerns regarding the localization of c-*myc* protein. Several reasons were suggested regarding the pattern of immunostaining. The majority of studies have reported the presence of c-*myc* immunoreactivity in the cytoplasm of the tumor cells (Mizukami *et al.*, 1991; Pavelic *et al.*, 1990, 1992a,b; Pietilainen *et al.*, 1995; Walker *et al.*, 1989). These studies have classified c-*myc* immunostaining into three patterns that include nuclear, nuclear and cytoplasmic, or cytoplasmic staining. Pavelic *et al.* (1990, 1991, 1992a,b) found c-*myc* immunostaining limited to the nuclear region. However, Pietilainen *et al.* (1995) and Mizukami *et al.* (1991) detected predominant cytoplasmic staining, although nuclear positivity was observed. Pavelic *et al.* (1990) have demonstrated that duration of fixation and types of fixatives may affect the staining intensity and intracellular localization of the c-*myc* protein, leading to decrease in nuclear staining and increase in cytoplasmic immunoreactivity with time. Furthermore, improper fixation procedures may promote autolysis because of the presence of endogenous hydrolytic enzyme causing the nuclear membrane to become porous and allow diffusion of the protein from the nucleus to the cytoplasm.

Expression of c-*myc*

Naidu *et al.* (2002) have demonstrated that c-*myc* oncoprotein was overexpressed in 45% of the primary breast carcinomas. Similarly, other investigators have detected overexpression of c-*myc* protein in about 50% to 100% of the breast carcinomas (Hehir *et al.*, 1993; Locker *et al.*, 1989; Mizukami *et al.*, 1991; Pavelic *et al.*, 1991, 1992a,b; Sierra *et al.*, 1999; Spandidos *et al.*, 1987, 1989; Spaventi *et al.*, 1994; Walker *et al.*, 1989). However, Bland *et al.* (1995) and Pietilainen *et al.* (1995) demonstrated c-*myc* immunopositivity in 17% and 12% of breast cancers, respectively, which is comparatively much less than the percentage described by Naidu *et al.* (2002). For immunohistochemistry, the differences among the studies could be because of several factors including types of antibodies used, dilution factor, differences in the detail of the experimental procedures used in different laboratories, nature of tissues (frozen or fixed tissues), types of fixatives, duration of fixation, number of cases and heterogeneity of patients studied, and scoring system that includes the staining intensity and proportion of positive cells to categorize c-*myc*–positive and -negative tumors.

Overexpression of c-*myc* protein detected by immunohistochemistry may reflect genetic alteration of the c-*myc* gene such as gene amplification. A recent meta-analysis performed on 29 studies indicated that the incidence of c-*myc* amplification varies greatly from 1% to 50% in different studies, but the average frequency appears to be 15.7% (Deming *et al.*, 2000). We have also reported that overexpression of the c-*myc* oncoprotein was significantly associated with amplification of the gene. The frequency of tumors overexpressing c-*myc* protein (45%) was much greater than c-*myc* amplification (24%) (Naidu *et al.*, 2002). However, it is not clear whether overexpression of c-*myc* protein is always associated with c-*myc* amplification (Deming *et al.*, 2000; Nass *et al.*, 1997). Although c-*myc* gene amplification is a frequent mechanism leading to the abnormal c-*myc* expression, it is unlikely to be the predominant mechanism of c-*myc* activation in breast carcinomas. Strong staining for c-*myc* protein also could indicate accumulation of the protein because of stabilization or increased half-life of the c-*myc* mRNA or the protein. Other mechanisms such as altered transcription rate or post-transcriptional modification may be responsible for sustained c-*myc* expression. Rearrangement of the c-*myc* gene has been reported in some breast tumors, but the incidence rate is less than gene amplification (Deming *et al.*, 2000; Liao *et al.*, 2000). Overexpression of c-*myc* mRNA also is frequently seen in tumors but is rarely associated with gene amplification or structural changes (Bieche *et al.*, 1999). Although the mechanism responsible for the transforming potential of overexpressed c-*myc* is not fully known, its ability to confer cellular growth advantage suggests that it may contribute to tumorigenesis.

Histopathologic Analysis

Overexpression of c-*myc* in various histopathologic types of primary breast carcinomas has been investigated (Naidu *et al.*, 2002). The study shows that 11 (27%) of 41 ductal carcinomas *in situ* (DCIS) were positive for c-*myc* immunoreactivity. Strong c-*myc* immunostaining was detected in 8 (35%) of the 23 comedo subtypes and 3 (17%) of the 18 noncomedo subtypes. Some investigators demonstrated c-*myc* immunopositivity in

noninvasive carcinomas but the authors did not classify the subtypes of these tumors (Hehir *et al.*, 1993; Mizukami *et al.*, 1991; Pavelic *et al.*, 1991, 1992b; Spandidos *et al.*, 1987). Sixty tumors with DCIS and adjacent invasive ductal carcinomas were analyzed independently. Strong c-*myc* immunopositivity was observed in 19 (32%) cases of *in situ* components and 27 (45%) cases of the adjacent invasive lesions. Of the 41 comedo DCIS-invasive tumors, 15 (37%) showed c-*myc* immunoreactivity in the comedo DCIS lesions, whereas 20 (49%) showed strong c-*myc* staining in the adjacent invasive components. Among the 19 noncomedo DCIS-invasive tumors, 4 (21%) demonstrated overexpression of c-*myc* in the noncomedo DCIS component and 7 (37%) in the adjacent invasive lesions. Overexpression of c-*myc* was noted more frequently in the invasive components than the *in situ* lesions, which confirms other findings (Mizukami *et al.*, 1991; Pavelic *et al.*, 1991, 1992a,b).

When comparison between comedo DCIS and other subtypes of DCIS was made, comedo DCIS was shown to be strongly associated with poor prognostic features and aggressive behavior. The ratio of recurrence in comedo to noncomedo DCIS was 10:1. In addition, presence of comedo DCIS not only predicts rate of recurrence, but it also has an increased cellular proliferation rate as measured by thymidine labeling index (TLI), as well as greater rate of progression to invasive breast carcinomas (Fisher *et al.*, 1995). According to the above data, enhanced c-*myc* expression occurs more often in comedo than noncomedo DCIS. A greater incidence of c-*myc* overexpression was noted in comedo DCIS-invasive than in noncomedo DCIS-invasive carcinomas. The invasive components of the adjacent comedo DCIS showed a greater frequency of c-*myc* overexpression compared with invasive lesions coexisting with the noncomedo DCIS. A study suggested that the biological and clinical behavior of invasive carcinomas might be determined by the genetic characteristics acquired during the preinvasive stage (Gupta *et al.*, 1997). It is possible that c-*myc* may be involved in the development of a highly aggressive phenotype of DCIS.

In situ tumors that express increased levels of c-*myc* are highly proliferative and may accumulate somatic genetic mutations during progression, leading to the development of the invasive phenotype. Therefore, up-regulation of c-*myc* may be an early pathogenic event in breast tumorigenesis. The greater frequency of c-*myc* overexpression in invasive tumors than in *in situ* tumors suggests that these abnormalities may be acquired by the cells during the development of invasive lesions or during the transition from *in situ* to invasive lesions. Greater incidence in invasive tumors indicates that c-*myc* overexpression can also be a late event in breast carcinogenesis. Increased levels of c-*myc* oncoprotein in *in situ* and invasive tumors suggest that up-regulation of c-*myc* may have a role in early and late stages of tumor progression in certain subgroups of breast tumors. Similarly, invasive tumors that possess c-*myc* abnormalities may progress to highly malignant tumors and may be more aggressive clinically than tumors without these alterations.

Strong statistical association was demonstrated between overexpression of c-*myc* protein in *in situ* and in adjacent invasive lesions. Fifteen (79%) of the 19 tumors showed strong c-*myc* immunostaining in both the *in situ* and invasive lesions. The staining pattern was similar in both lesion types of the tumor. These observations probably suggest that overexpression of c-*myc* acquired during the preinvasive stage may be sustained throughout the progression toward the invasive stage. Because up-regulation of c-*myc* occurs before the invasive process, it is possible that in some tumors c-*myc* overexpression may be a prerequisite to tumor invasion. Twelve (29%) of the 41 tumors demonstrated high levels of c-*myc* protein in the invasive lesions but absence in the *in situ* components. In certain subsets of tumors, up-regulation of c-*myc* expression may appear as a late event occuring at a later stage, which could have developed during the invasive growth or during the transition from preinvasive to invasive stage. The *in situ* tumors might have developed without c-*myc* abnormal expression. However, 4 (21%) of the 19 tumors showed overexpression of the c-*myc* protein in the *in situ* lesions but was not observed in the invasive components. In these tumors, the lesions may have a different origin or probably loss of c-*myc* expression during the invasive growth, or the invasive lesion may have acquired genetic abnormalities other than c-*myc*. Another possible reason could be a failure of a proportion of *in situ* tumors to progress to clinically detectable invasive tumors in the lifetime of the patients.

Overexpression of c-*myc* protein was analyzed in association with different histologic types of invasive breast carcinomas. Overexpression of c-*myc* protein was detected in 133 (49%) of 270 invasive ductal carcinomas, 11 (33%) of 33 invasive lobular carcinomas, 6 (29%) of 21 colloid carcinomas, and 7 (47%) of 15 medullary carcinomas. Using immunohistochemistry, Hehir *et al.* (1993) and Pavelic *et al.* (1992a) showed increased levels of c-*myc* protein in 56% and 100% of invasive ductal carcinomas, respectively. Mizukami *et al.* (1991) detected c-*myc* expression in 65% of invasive ductal carcinomas, 50% of invasive lobular carcinomas, 100% of colloid carcinomas, and 75% of medullary carcinomas. In another article by Pavelic *et al.* (1992b)

all of the 42 invasive carcinomas stained positive for c-*myc*, but no description of the types of invasive carcinomas were given. In these studies, the number of cases studied for each histologic type was insufficient for comparison purposes, particularly invasive lobular carcinomas, medullary carcinomas, and colloid carcinomas. Furthermore, there were not many studies that evaluate expression of c-*myc* in various histopathologic types of breast carcinomas. The data indicate that greater incidence of c-*myc* overexpression was detected in invasive ductal and medullary carcinomas, but a smaller incidence was detected in invasive lobular and colloid carcinomas. Each type of invasive carcinoma may have distinct biological and clinical features. The observation suggests that the expression of c-*myc* may be useful in understanding the biological behavior and the role of c-*myc* in different histopathologic types of invasive breast carcinomas. These tumors may have poor prognostic features, predictive of poor clinical outcome, and may have more aggressive behavior compared with tumors that do not overexpress c-*myc* protein. However, follow-up studies are required to understand the clinical outcome of these tumors in relation to c-*myc* expression, and c-*myc* might be useful to select tumors with greater malignancy potential among the invasive carcinomas.

Clinicopathologic Analysis

The relationship between c-*myc* overexpression and clinicopathologic factors such as estrogen receptor (ER) status, lymph node status, histologic grade, patient age, and proliferation index of invasive carcinomas has been reported (Naidu *et al.*, 2002). Ductal carcinoma *in situ* was not included in that analysis.

The data showed that overexpression of c-*myc* was not significantly associated with ER status and a similar association was noted by other authors (Locker *et al.*, 1989; Pavelic *et al.*, 1992b; Saccani-Jotti *et al.*, 1992; Spandidos *et al.*, 1989; Spaventi *et al.*, 1994). The experimental analysis showed that 50% of 241 ER-negative and 41% of 158 ER-positive tumors were strongly positive for c-*myc*. This indicates that increased levels of c-*myc* protein can be detected in both the estrogen-dependent and -independent tumors. Dubik and Shiu (1988) have shown that c-*myc* expression is critical in both the hormone-dependent and -independent breast cancer cell growth. In hormone-dependent human breast cancer cell lines, expression of c-*myc* was directly regulated by estrogen, suggesting that c-*myc* is an estrogen target gene. Treatment with estrogen stimulates but tamoxifen inhibits the expression of c-*myc* mRNA in hormone-responsive MCF-7 cells, indicating that estrogen is required for c-*myc* activation. Accumulation of c-*myc* mRNA and protein was observed in estrogen-treated cancer cell lines. Thus, c-*myc* is probably an important gene whose expression mediates the mitogenic effect of estrogen in hormone-responsive breast cancer cells. In contrast, the c-*myc* protein is constitutively expressed in hormone-independent cells (Shiu *et al.*, 1993). Estrogen or tamoxifen has no influence on c-*myc* expression in these cells. This observation showed that increased levels of c-*myc* expression confer estrogen independence and anti-estrogen resistance. High levels of c-*myc* in estrogen-independent cells may be achieved through gene amplification, growth factors, increased transcription rate, and increased stability of c-*myc* mRNA. In hormone-independent cells, c-*myc* mRNA was three times as stable as that in the hormone-responsive cells (Dubik *et al.*, 1988). Overexpression of c-*myc* alone is sufficient to provide growth stimulus to both the hormone-responsive and non–hormone-responsive breast cancer cells. A study using a c-*myc*–inducible human breast cancer cell model showed that aberrant expression of c-*myc* alone can induce resistance to anti-estrogen (Venditti *et al.*, 2002). These findings may suggest that overexpression of c-*myc* is one of the molecular changes that occur at the cellular level that supports the progression of breast tumors from the estrogen-dependent to nonestrogen-dependent stage or the resistance to anti-estrogen. With follow-up studies overexpression of c-*myc* protein in combination with ER status may be useful as a predictive marker for hormonal therapy and development of anti-estrogen resistance, and also to identify or to predict tumors most likely to progress from ER positivity to ER negativity.

The study also indicated that overexpression of c-*myc* was significantly associated with poorly differentiated tumors. Overexpression of c-*myc* was noted in 36% of 105 well differentiated, 40% of 157 moderately differentiated, and 61% of 137 poorly differentiated tumors. Other studies did not confirm this relationship (Mizukami *et al.*, 1991; Pietilainen *et al.*, 1995; Saccani-Jotti *et al.*, 1992; Spandidos *et al.*, 1989; Spaventi *et al.*, 1994; Tauchi *et al.*, 1989). In general, c-*myc* expression is down-regulated rapidly in cells that are differentiating. Constitutive expression of c-*myc* prevents withdrawal of the cell from the cell cycle and inhibits differentiation, as demonstrated in a number of cell lines (Freytag *et al.*, 1990). Expression of myc was low or undetectable in many differentiated adult tissues, consistent with the notion that its expression correlates with cell proliferation and cell differentiation. Decrease in c-*myc* mRNA and protein levels was observed with cell growth arrest and initiation of differentiation. Incidence of aberrant expression of c-*myc* protein increases from well differentiated to poorly differentiated tumors, suggesting that c-*myc* may be a contributing factor for loss of differentiation and increased aggressiveness.

Overexpression of c-*myc* protein was statistically associated with a high Ki67 proliferation index. Overexpression of c-*myc* was detected in 56% of 249 high Ki67 proliferation index tumors and 29% of 150 low Ki67 proliferation index tumors. Other authors reported significant correlation between rapidly proliferating tumors and c-*myc* overexpression (Pavelic *et al.*, 1992b). Conversely, Pietilainen *et al.* (1995) and Han *et al.* (2000) failed to show an association between c-*myc* immunopositivity and proliferation activity detected by S-phase fraction. Expression of c-*myc* is tightly regulated in order to have controlled cell proliferation. In normal cells, c-*myc* is essential in cell cycle regulation to stimulate quiescent cells to progress from the G1- to the S-phase. The c-*myc* protein is rarely detectable in nonproliferating or growth-arrested cells, but the protein is continuously expressed during the cell cycle. The concentration of c-*myc* protein is maintained at a constant level throughout all the phases of the cell cycle in continuously proliferating cells. Tumor cells constitutively expressing high levels of c-*myc* protein have increased growth rate, spend less or shorter time in the G1-phase, and become independent of external growth factor requirements. When c-*myc* expression is deregulated, the cells may grow at a faster rate and may be unable to withdraw from the cell cycle because of loss of growth suppression. This may allow epithelial cells within the mammary gland to survive. Such cells may lead to increased genetic instability and become targets for additional somatic mutations, which may potentially lead to tumor formation.

Although lymph node status is considered to be the strongest prognostic factor in breast cancer, the majority of studies found no relationship between c-*myc* immunopositivity and lymph node metastases (Bland *et al.*, 1995; Locker *et al.*, 1989; Pavelic *et al.*, 1992b; Pietilainen *et al.*, 1995; Saccani-Jotti *et al.*, 1992; Spandidos *et al.*, 1989; Spaventi *et al.*, 1994; Tauchi *et al.*, 1989), which was in agreement with the findings reported by Naidu *et al.* (2002). The results indicated that 49% of 238 lymph node–positive tumors and 42% of 161 lymph node–negative tumors were positive for c-*myc*. These studies showed that c-*myc* overexpression does not confer an advantage to the metastatic phenotype in breast cancer. Overexpression of c-*myc* may not be directly involved in promoting tumor metastases. In conjuction with Bcl-2 expression and loss of apoptosis, c-*myc* may facilitate metastases in breast tumors (Sierra *et al.*, 1999). Overexpression of Bcl-2 and c-*myc* increases life span and proliferative activity of tumor cells that may favor accumulation of genetic alteration that can promote nodal metastases. Pietilainen *et al.* (1995) noted that expression of c-*myc* protein was stronger at the invasive margin of the tumor, which indicates that c-*myc* may facilitate invasive growth. Overexpression of c-*myc* may be important for tumor cell invasion but is not essential for lymph node metastases.

On the basis of patient age, Naidu *et al.* (2002) found no correlation with c-*myc* expression. The study showed that 49% of 243 patients younger than 50 years and 42% of 156 patients 50 years and older were strongly positive for c-*myc*. Most of the studies found no association between age of the patient and c-*myc* immunopositivity (Pavelic *et al.*, 1992b). The above observation suggests that overexpression of c-*myc* is independent and is not influenced by patient age.

Currently, only a limited number of studies have used an immunohistochemical technique to evaluate c-*myc* protein expression. The majority of the publications have focused on the detection of c-*myc* gene amplification rather than c-*myc* immunohistochemistry. In a report by Deming and coinvestigators (2000), 29 studies were analyzed for an association between c-*myc* amplification and clinicopathologic parameters. The authors noted that c-*myc* amplification exhibited significant but weak association with histologic grade, lymph node metastases, absence of progesterone receptor, and postmenopausal status. The amplification of c-*myc* is significantly associated with risk for early relapse and shortened overall survival. The amplification of c-*myc* is also a stronger prognostic marker to predict early recurrence than HER-2, particularly in patients with node-negative breast cancer (Schlotter *et al.*, 2003). However, some studies were unable to find any relationship between c-*myc* amplification and poor prognosis (Deming *et al.*, 2000). According to Rummukianen *et al.* (2001a,b), the detection of c-*myc* amplification by fluorescence *in situ* hybridization (FISH) and chromogenic *in situ* hybridization (CISH) showed a significant correlation with established poor prognostic factors and poorer clinical outcome.

Reports on c-*myc* immunohistochemistry regarding prognosis are not consistent among investigators. The other concern was the presence of c-*myc* immunostaining in the nuclei and the cytoplasm of breast cancer cells. It is important to note whether sublocalization of c-*myc* protein has any prognostic or clinical significance. In three of their studies, Pavelic and coworkers (1991, 1992a,b) demonstrated that c-*myc* immunoreactivity was restricted in the nuclei of the tumor cells, but the prognostic significance was not reported. Other investigators have not found any association between c-*myc* expression and prognosis (Mizukami *et al.*, 1991; Spaventi *et al.*, 1994). In contrast, Pietilainen *et al.* (1995)

found that strong c-*myc* immunostaining is associated with better survival outcome. Mizukami *et al.* (1991) and Pietilainen *et al.* (1995) reported predominantly cytoplasmic localization, although nuclear staining was seen. The former found no relationship with the prognosis of the patients, but the latter have shown that nuclear staining was correlated with ER negativity but has no prognostic value, whereas cytoplasmic staining is correlated with low mitotic index and longer relapse-free survival. Another study reported that coexpression of c-*myc* with other oncogenes such as Ha-*ras* and c-*fos* showed a stronger association with reduced disease-free and overall survival than c-*myc* alone (Bland *et al.*, 1995). Sierra *et al.* (1999) reported that overexpression of Bcl-2 was significantly associated with lymph node metastases when coexpressed with c-*myc*, suggesting that c-*myc* has a role in promoting tumor metastases in the presence of Bcl-2. Both studies showed that cooperation of c-*myc* with other oncogenes may lead to a more unfavourable outcome than c-*myc* alone. Clinical studies that examine c-*myc* expression in conjunction with other genes are needed to provide new prognostic information for human breast cancer.

In summary, tumors with a high proliferation rate, high tumor grade, nodal involvement, and lack of ER are recognized as tumors with poor prognostic features. These are well established prognostic factors that are currently used to assess the prognosis of patients with breast cancer. The analysis on clinical parameters showed that overexpression of c-*myc* detected by immunohistochemistry was significantly associated with poorly differentiated and rapidly proliferating tumors. This suggests that c-*myc* may have the potential as a marker for poor prognosis and may be useful in identifying rapidly growing and aggressive tumors. Highly proliferative tumors are at increased risk for acquiring genetic alteration that favors disease progression toward higher state of malignancy and recurrence. Association with poor prognostic factors shows that overexpression of c-*myc* may predict poorer clinical outcome and greater risk for recurrence. Preferential expression of c-*myc* in poorly differentiated and highly proliferating epithelial tumor cells reflects the connection between c-*myc* expression and both cell cycle progression and dedifferentiation. In addition, overexpression of c-*myc* has been detected in preinvasive, preinvasive with adjacent invasive carcinomas, and invasive carcinomas with greater incidence in invasive than in preinvasive carcinomas. In some tumors, up-regulation of c-*myc* is present only in the *in situ* lesion and absent in the invasive component, whereas in other cases, c-*myc* overexpression was noted in the invasive but not detected in the *in situ* lesion.

However, the majority of tumors showed that increased levels of c-*myc* in the *in situ* components was consistently accompanied by overexpression in the adjacent invasive lesions and continues to present in the invasive component. This suggests that in some tumors c-*myc* overexpression represents an early and late event in breast cancer development. Up-regulation of c-*myc* could be one of the underlying molecular alterations that have a significant role in promoting tumor progression from the preinvasive to invasive stage. Also, c-*myc* may have the potential as a molecular marker to select tumors with greater malignancy potential and to identify *in situ* tumors that have the greatest risk for recurrence and are most likely to progress to the invasive stage. The knowledge and understanding of the molecular events of c-*myc* in *in situ* and invasive carcinoma might assist in individualized patient management and might prevent the progression to the invasive stage if the patient is diagnosed with having DCIS. Each type of *in situ* and invasive carcinoma is unique in terms of their genetic characteristics and may influence the clinical outcome of the patients. Understanding the role of c-*myc* in various histologic types of tumors may assist in understanding the biological characteristics of these tumors.

Overall analysis on c-*myc* immunohistochemistry suggests that c-*myc* has the potential as a molecular marker of poor prognosis for breast cancer. Furthermore, because of the lack of information on the patients' follow-up, we were not able to demonstrate the clinical significance of c-*myc* immunohistochemistry. Large scale clinical studies are required to determine the prognostic value and clinical importance of c-*myc*. Long-term follow-ups and standardization of experimental and tissue processing procedures for immunohistochemistry are warranted to achieve consistent and reproducible results before it can be proposed to be used routinely for diagnosis and to determine the prognosis of patients with breast cancer.

Acknowledgments

We acknowledge and thank the Ministry of Science, Technology and Environment for financial support (IRPA 06-02-03-0204) to conduct the study.

References

Bieche, I., Laurendeau, I., Tozlu, S., Olivi, M., Vidaud, D., Lidereau, R., and Vidaud, M. 1999. Quantitation of Myc gene expression in sporadic breast tumors with a real-time reverse transcription-PCR assay. *Cancer Res. 59:*2759–2765.

Bishop, J.M. 1982. Retroviruses and cancer genes. *Adv. Cancer Res. 37:*1–32.

Bland, K.I., Konstadoulakis, M.M., Vezeridis, M.P., and Wanebo, H.J. 1995. Oncogene protein co-expression: Value of Ha-ras, c-Myc, c-fos and p53 as prognostic determinants for breast carcinoma. *Ann. Surg. 221:*706–722.

Dang, C.V. 1999. C-Myc target genes involved in cell growth, apoptosis, and metabolism molecular and cellular biology. *Mol. Cell. Biol. 19:*1–11.

Deming, S.L., Nass, S.J., Dickson, R.B., and Trock, B.J. 2000. C-Myc amplification in breast cancer: A meta-analysis of its occurrence and prognostic relevance. *Br. J. Cancer 83:*1688–1695.

Dubik, D., and Shiu, R.P., 1988. Transcriptional regulation of c-Myc oncogene expression by estrogen in hormone-responsive human breast cancer cells. *J. Biol. Chem. 263:*12705–12708.

Facchini, L.M., and Penn, L.Z. 1998. The molecular role of myc in growth and transformation: Recent discoveries lead to new insights. *FASEB J. 12:*633–651.

Fisher, E.R., Costantino, J., Fisher, B., Paleker, A., Redmond, C., and Mamounas, E. 1995. Pathological findings from the National Surgical Adjuvant Breast Project (NSABP) protocol B-17, intraductal carcinoma (ductal carcinoma in situ). *Cancer 75:*1310–1319.

Freytag, S.O., Dang, C.V., and Lee, W.M. 1990. Definition of the activities and properties of Myc required to inhibit cell differentiation. *Cell Growth Differ. 1:*339–343.

Galaktionov, K., Chen, X., and Beach, D. 1996. Cdc25 cell-cycle phosphatase as a target of c-Myc. *Nature (Lond.) 382:*511–517.

Gupta, S.K., Douglas-Jones, A.G., Fenn, N., Morgan, J.M., and Mansel, R.E. 1997. The clinical behavior of breast carcinoma is probably determined at the preinvasive stage (ductal carcinoma in situ). *Cancer 80:*1740–1745.

Han, S., Park, K., Kim, H.Y., Lee, M.S., Kim, H.J., Kim, Y.D., Yuh, Y.J., Kim, S.R., and Suh, H.S. 2000. Clinical implication of altered expression of Mad 1 protein in human breast carcinoma. *Cancer 88:*1623–1632.

Hehir, D.J., Mcgreal, G., Kirwan, W.D., Kealy, W., and Brady, M.P. 1993. C-*Myc* oncogene expression: A marker for females at risk of breast carcinoma. *J. Surg. Oncol. 54:*207–210.

Isola, J., Visakorpi, T., Holli, K., and Kallioniemi, O-P. 1992. Association of overexpression of tumor suppressor protein p53 with rapid cell proliferation and poor prognosis in node-negative breast cancer patients. *J. Natl. Cancer Inst. 84:*1109–1114.

Janicke, R.U., Lin, X.Y., Lee, F.H.H., and Porter, A.G. 1996. Cyclin D3 sensitizes tumor cells to tumor necrosis factor-induced c-Myc-dependent apoptosis. *Mol. Cell. Biol. 16:*5245–5253.

Kauffmann-Zeh, A., Rodriguez-Viciana, P., Ulrich, E., Gilbert, C., Coffer, P., Downward, J., and Evan, G. 1997. Suppression of c-Myc-induced apoptosis by Ras signalling through PI(3)K and PKB. *Nature (Lond.) 385:*544–548.

Liao, D.J., and Dickson, R.B. 2000. C-Myc in breast cancer. *Endocr. Relat. Cancer 7:*143–164.

Locker, A.P., Dowle, C.S., Ellis, I.O., Elston, C.W., Blamey, R.W., Sikora, K., Evan, G., and Robbins, R.A. 1989. C-Myc oncogene product expression and prognosis in operable breast cancer. *Br. J. Cancer 60:*669–672.

Marcu, K.B., Bossone, S.A., and Patel, A.J. 1992. Myc function and regulation. *Annu. Rev. Biochem. 61:*809–860.

Mizukami, Y., Nonomura, A., Noguchi, M., Taniya, T., Koyasaki, N., Saito, Y., Hashimoto, T., Matsubara, F., and Yanaihara, N. 1991. Immunohistochemical study of oncogene product ras p21, c-Myc and growth factor EGF in breast carcinomas. *Anticancer Res. 11:*1485–1494.

Naidu, R., Norhanom, A.W., Yadav, M., Kutty, M.K., and Nair, S. 2002. Protein expression and molecular analysis of c-Myc gene in primary breast carcinomas using immunohistochemistry and differential polymerase chain reaction. *Int. J. Mol. Med.* 9:189–196.

Nass, S.J., and Dickson, R.B. 1997. Defining a role for c-Myc in breast tumorigenesis. *Breast Cancer Res. Treat.* 44:1–22.

Pavelic, K., Pavelic, Z.P., Denton, D., Reising, J., Khalily, M., and Preisler, H.D. 1990. Immunohistochemical detection of c-Myc oncoprotein in paraffin-embedded tissues. *J. Exp. Pathol.* 5:143–153.

Pavelic, Z.P., Steele, P., and Preisler, H.D. 1991. Evaluation of c-Myc proto-oncogene in primary human breast carcinomas. *Anticancer Res.* 11:1421–1427.

Pavelic, Z.P., Pavelic, K., Carter, C.P., and Pavelic, L. 1992a. Heterogeneity of c-Myc expression in histologically similar infiltrating ductal carcinomas of the breast. *J. Cancer Res. Clin. Oncol.* 118:16–22.

Pavelic, Z.P., Pavelic, L., Lower, E.E., Gapany, M., Gapany, S., Barker, E.A., and Preisler, H.D. 1992b. C-Myc, c-erbB2 and Ki-67 expression in normal breast tissue and in invasive and noninvasive breast carcinoma. *Cancer Res.* 52:2597–2602.

Pietilainen, T., Lipponen, P., Aaltomaa, S., Eskelinen, M., Kosma, V.M., and Syrjanen K. 1995. Expression of c-Myc proteins in breast cancer as related to established prognostic factors and survival. *Anticancer Res.* 15:959–964.

Rummukainen, J.K., Salminen, T., Lundin, J., Joensuu, H., and Isola, J.J. 2001a. Amplification of c-Myc oncogene by chromogenic and fluorescence in situ hybridization in archival breast cancer tissue array samples. *Lab. Invest.* 81:1545–1551.

Rummukainen, J.K., Salminen, T., Lundin, J., Kytola, S., Joensuu, H., and Isola, J.J. 2001b. Amplification of c-Myc by fluorescence in situ hybridization in a population-based breast cancer tissue array. *Mod. Pathol.* 14:1030–1035.

Saccani-Jotti, G., Fontanesi, M., Bombardieri, E., Gabrielli, M., Veronesi, P., Bianchi, M., Becchi, G., Bogni, A., and Tardini, A. 1992. Preliminary study on oncogene product immunohistochemistry (c-erbB2, c-Myc, ras p21, EGFR) in breast pathology. *Int. J. Biol. Markers* 7:35–42.

Sakamuro, D., Eviner, V., Elliott, K.J., Showe, L., White, E., and Prendergast, G.C. 1995. c-Myc induces apoptosis in epithelial cells by both p53-dependent and p53-independent mechanisms. *Oncogene* 11:2411–2418.

Schlotter, C.M., Vogt, U., Bosse, U., Mersch, B., and Wabetamann, K. 2003. C-myc, not HER-2/neu, can predict recurrence and mortality of patients with node-negative breast cancer. *Breast Cancer Res.* 5:30–36.

Shiu, R.P., Watson, P.H., and Dubik, D. 1993. C-Myc oncogene expression in estrogen-dependent and -independent breast cancer. *Clin. Chem.* 39:353–355.

Sierra, A., Castellsague, X., Escobedo, A., Moreno, A., Drudis, T., and Fabra, A. 1999. Synergistic cooperation between c-Myc and Bcl-2 in lymph node progression of T1 human breast carcinomas. *Breast Cancer Res. Treat.* 54:39–45.

Spandidos, D.A., Pintzas, A., Kakkanas, A., Yiagnisis, M., Mahera, H., Patra, E., and Agnantis, N.J. 1987. Elevated expression of the c-Myc gene in human benign and malignant breast lesions compared to normal tissue. *Anticancer Res.* 7:1299–1304.

Spandidos, D.A., Yiagnisis, M., Papadimitriou, K., and Field, J.K. 1989. Ras, c-Myc and c-erbB2 oncoproteins in human breast cancer. *Anticancer Res.* 9:1385–1393.

Spaventi, R., Kamenjicki, E., Pecina, N., Grazio, S., Pavelic, J., Kusic, B., Cvrtila, D., Danilovic, Z., and Spaventi, S. 1994. Immunohistochemical detection of TGF-alpha, EGFR, c-erbB2, c-Ha-ras, c-Myc, estrogen and progesterone in benign and malignant human breast lesions: A concomitant expression. *In vivo* 8:183–189.

Spotts, G.D., Patel, S.V., Xiao, Q., and Hann, S.R. 1997. Identification of downstream-initiated c-Myc proteins which are dominant-negative inhibitors of transactivation by full-length c-Myc proteins. *Mol. Cell. Biol.* 17:1459–1468.

Tauchi, K., Hori, S., Itoh, H., Osamura, R.Y., Tokuda, Y., and Tajima, T. 1989. Immunohistochemical studies on oncogene products (c-erbB2, EGFR, c-Myc) and estrogen receptor in benign and malignant breast lesions. *Virchows Arch. Pathol. Anat.* 416:65–73.

Venditti, M., Iwasiow, B., Orr, W.F., and Shiu, R.B.C. 2002. C-Myc gene expression alone is sufficient to confer resistance to antiestrogen in human breast cancer cells. *Int. J. Cancer* 99:35–42.

Walker, R.A., Senior, P.V., Jones, J.L., Critchley, D.R., and Varley, J.M. 1989. An immunohistochemical and in situ hybridization study of c-Myc and c-erbB2 expression in primary human breast carcinomas. *J. Pathol.* 158:97–105.

17

Immunohistochemical Localization of Neuropilin-1 in Human Breast Carcinoma

A Possible Molecular Marker for Diagnosis

Sushanta K. Banerjee

Introduction

Breast cancer is one of the most frequently occurring cancers in women and the third leading cause of cancer mortality, following lung and stomach cancer (Greenlee *et al.*, 2000; Parkin, 1998, 1999a,b). Currently, women face approximately a one in nine lifetime risk for the disease (American Cancer Society, 1994, 1999; Bryant *et al.*, 1994). Each year, an estimated 180,200 new cases are diagnosed and 43,900 women die of the disease (American Cancer Society, 1994). Breast cancer is the most commonly diagnosed cancer in women in the United States and worldwide. Treatment of this disease at advanced stages is often ineffective, as well as being disfiguring. Therefore, identification of genetic marker(s) for early detection of this disease is a high priority in medical management.

Genetic alterations (such as gene mutations, amplifications, or overexpression of genes) leading to the transformation of normal cells to malignant neoplastic cells are the ultimate cause of breast cancer (Desai *et al.*, 2002a,b; Squire *et al.*, 1992). Accordingly, gene expression profiling of human breast cancers using microarray techniques has increased our perceptive of the pathologic diversity of this disease (Desai *et al.*, 2002a; Perou *et al.*, 2000; Pollack *et al.*, 2002). These genomic analyses offer great promises for classification and identification of this disease's stages on the basis of variations in gene expression, protein levels, or both (Desai *et al.*, 2002a; Perou *et al.*, 2000). Recently, using cDNA microarray or similar gene

Handbook of Immunohistochemistry and *in situ* Hybridization of Human
Carcinomas, Volume 1: Molecular Genetics; Lung and Breast Carcinomas

409

Copyright © 2004 by Elsevier (USA)
All rights reserved.

profiling analysis technique (e.g., differential display method) identified several genes that were differentially expressed in breast tumor samples (Perou *et al.*, 1999, 2000; Saxena *et al.*, 2001). In addition, conventional Northern Blot or immuno–Western Blot assays and immunohistochemical analyses also identified several genes expressed differentially in human breast tumor samples. *Neuropilin-1* (*NRP-1*) is one of those genes that has been found to be overexpressed in the myoepithelial cells of the hyperplasic ducts and in ductal carcinoma *in situ* (DCIS) in human breast ductal carcinoma samples (Stephenson *et al.*, 2002).

Neuropilin-1 was first identified in the developing neuron and subsequently considered as receptor for the secreted proteins (i.e., Sema3A, Sema3B, Sema3C, and Sema3F) of the Semaphorin family, which are repellent factors of growing axon tips (Chen *et al.*, 1997b; He *et al.*, 1997; Kolodkin *et al.*, 1997). Neuropilin-1 is a 130- 140-kD, multidomain, cell surface glycoprotein (Kolodkin *et al.*, 1997a,b). The primary structure of this gene is highly conserved within vertebrate species, including Xenopus (Takagi *et al.*, 1987, 1991), chicken (Takagi *et al.*, 1995), mouse (Kawakami *et al.*, 1996), rat (Kolodkin *et al.*, 1997b), and human (Chen *et al.*, 1997a; Kolodkin *et al.*, 1997b). There are two subtypes of neuropilin protein, NRP-1 and NRP-2, which are products of genes on two different human chromosomes. The human *NRP-1* gene is located in chromosome 10p12 and *NRP-2* is located in chromosome 2q34 (Rossignol *et al.*, 1999, 2000). Both subtypes of neuropilin have similar structural features (Chen *et al.*, 1997a) and intimately interplay as neuron guidance factors (Giger *et al.*, 2000; Kolodkin *et al.*, 1997a,b).

Despite the neuron guidance, neuropilin proteins are also expressed in a variety of nonneural cells and mediate various intercellular signals to modulate diverse aspects of physiologic and pathophysiologic functions (Banerjee *et al.*, 1999; Gluzman-Poltorak *et al.*, 2001; Kawakami *et al.*, 2002; Neufeld *et al.*, 2002; Takashima *et al.*, 2002). Both NRP-1 and NRP-2 are overexpressed in endothelial cells and potentiate angiogenesis by binding with vascular endothelial growth factor-A (VEGF-A) isoforms (Gluzman-Poltorak *et al.*, 2000; Neufeld *et al.*, 2002; Soker *et al.*, 1998). In endothelial cells, NRP-1 acts as coreceptor or docking receptor that augments binding of VEGF-A$_{165}$, an isoform of VEGF-A, to its high-affinity receptor tyrosine kinase, KDR (Flk-1), regulating the mitogenic and chemotactic activity of VEGF (Neufeld *et al.*, 2002; Soker *et al.*, 1998). In addition, NRP-1 is able to bind with other subtypes of VEGF, such as VEGF-B and VEGF-C, and the heparin-binding form of placental growth factor, PIGF-2, and enhance angiogenesis (Gluzman-Poltorak

et al., 2000; Makinen *et al.*, 1999; Miao *et al.*, 2000a; Migdal *et al.*, 1998). The *in vitro* binding experiments have shown that NRP-1 binds with Flt-1, an additional receptor for VEGF, and the NRP-1–Flt-1 complex probably acts as a negative regulator of angiogenesis (Fuh *et al.*, 2000).

Neuropilin-1 is a cell adhesion receptor in fibroblast cells and interacts with ligand(s) of unknown characters (Shimizu *et al.*, 2000; Takagi *et al.*, 1995). The adhesion activity is mediated through the b1 and b2 extracellular domain of NRP-1 protein (Shimizu *et al.*, 2000). Moreover, abundant expression of NRP-1 has been perceived in various breast and prostate tumor cell lines lacking KDR and Flt-1 receptors (Miao *et al.*, 2000a). This tumor cell–derived NRP-1 is functionally active and possibly blocks the apoptotic death of tumor cells with or without involvement of VEGF (Miao *et al.*, 2000b). Recent studies from our laboratory have demonstrated that NRP-1 mRNA and protein expressions are differentially expressed in myoepithelial cells and vascular smooth muscle cells in preneoplastic and neoplastic human breasts. This study suggests that NRP-1 may be a multiple function protein in human breast and may be associated with local invasion and angiogenesis and, therefore, has direct relevance to the progression of breast cancer (Stephenson *et al.*, 2002). Thus, the NRP-1 protein can be considered a good prognostic marker for breast cancer that could help in early detection of this disease. This chapter outlines a detailed immunohistochemical method for the detection of NRP-1 in breast tumor samples. This procedure is similar to the one we reported previously elsewhere (Stephenson *et al.*, 2002).

MATERIALS

1. Phosphate buffer saline (1X PBS): 8.5 g sodium chloride, 1.43 g dibasic potassium phosphate, and 0.25 g monobasic potassium phosphate; bring volume to 1 L with deionized glass-distilled water (pH 7.4)

2. 4% Buffered-formalin (tissue fixative): 10.8 ml 37% formaldehyde (Cat. No. F79-500; Fisher Scientific, Houston, TX); bring to 100 ml with 1X PBS

3. Deparaffinization solutions: 100% xylene and 50% xylene + 50% ethanol solution

4. Graded ethanol: 50%, 70%, 90%, and 100% alcohol

5. 3% Hydrogen peroxide (H_2O_2): 5 ml 30% H_2O_2 (Cat. No. H325-500; Fisher Scientific) diluted in 45 ml methanol (prepare fresh before use)

6. Citrate buffer solution (1X): 5 ml 20X citrate buffer solution (Cat. No. 00-5000; Zymed Laboratories, Inc., San Francisco, CA), bring volume to 100 ml with deionized glass-distilled water

7. Immunohistochemical kits (Histostain-plus broad spectrum; Cat. No. 85-9643; Zymed Laboratories, Inc.) containing ready-to-use tissue blocker, biotinylated secondary antibody, peroxidase-conjugated linker, diaminobenzidine (DAB) tetrahydrochloride solution, and hematoxylin

8. Primary antibody diluted in 1X PBS

9. Hematoxylin solution

METHODS

1. Rinse breast tissue biopsy samples twice with 1X PBS, and fix in 4% buffered formalin overnight.

2. Dehydrate the tissue samples through increasing concentrations of ethanol (i.e., 50%, 70%, 90%, and 100%), 50% absolute alcohol 50% xylene, 100% xylene, and in paraffin.

3. Cut serial sections of 5 μm from the resulting blocks, mount on positively charged slides (Cat. No. 12-544-7; Fisher Scientific), and dry under slide warmer (Fisher Scientific).

4. Deparaffinize the tissue sections by incubating the slides in xylene for 15 min, followed by 10-min incubation in 50% xylene 50% ethanol solution.

5. Hydrate the tissue sections in 1X PBS through different concentration of decreasing concentrations of ethanol (i.e., 100%, 90%, 70%, and 50%).

6. Block endogenous peroxidase activity by incubating sections in 3% H_2O_2 solution for 5 min at room temperature.

7. Wash the slide 3 times (5 min each) in 1X PBS.

8. Incubate the slides in 1X citrate buffer in a microwave oven at high power for 45 sec and allow cooling for 15 min.

9. Rinse the slides 3 times in deionized glass-distilled water.

10. Wash the slides 3 times (5 min each) in 1X PBS.

11. Block nonspecific binding by incubating the sections in ready-to-use tissue blocker for 15 min at room temperature.

12. Incubate overnight in NRP-1 primary antibody ([rabbit polyclonal antibodies against human NRP-1; Zymed Laboratories, Inc.; or rabbit polyclonal antibodies against rat NRP-1]; 1:200 or 1:1200 dilutions, respectively, in 1X PBS) in a humid chamber at 4°C. Rabbit polyclonal antibodies against rat NRP-1 were generated by Kolodkin *et al.* (1997b).

13. Wash as in **step 9**.

14. Incubate the sections for 10 min at room temperature in biotinylated labeled second antibody.

15. Wash as in **step 9**.

16. Incubate the sections for 10 min at room temperature in peroxidase-conjugated linker.

17. Wash as in **step 9**.

18. Incubate the sections in DAB solution (Zymed Laboratories, Inc.) with their kit, for a period sufficient to yield dark brown color, usually 5 min.

19. Rinse in deionized glass-distilled water for 5 min.

20. Counterstain with hematoxylin for 2 min.

21. Rinse in running tap water for 5 min.

22. Dehydrate and mount in paramount for microscopic examination.

RESULTS AND DISCUSSION

Immunohistochemical localization and distribution of NRP-1 protein was determined in formalin-fixed, paraffin-embedded biopsy tissue samples. Normal ($n=16$), noninvasive ($n=15$) and invasive ($n=15$) human breast carcinoma tissues were assayed. Tissue samples were obtained from the Kansas Cancer Institute Tissue Repository Core Facilities (Kansas City, KS) and Cooperative Human Tissue Network (CHTN, NCI), respectively. The institution's Human Subjects Committee approved the study. The results of this study indicate significant increase of NRP-1 immunoreaction in the myoepithelial cells of ducts and lobules of invasive carcinoma tissue sections as compared with normal and noninvasive carcinoma of human breast (Figure 60). The NRP-1 immunoreaction in myoepithelial cells was minimal or undetectable in normal and noninvasive carcinoma tissue sections. Moreover, the number of ducts and lobules positive for NRP-1 in invasive carcinoma tissue samples increased by several fold when compared with samples of normal or noninvasive carcinoma tissue (Figure 61). Although the greater immunoreaction was observed in hyperplasic ducts in the invasive carcinoma samples, the reaction was minimal in DCIS, and the labeling intensity was virtually undetectable in areas of early invasion (Figure 60). The NRP-1 immunoreactivity in ductal epithelial cells cancer epithelial cells was sporadic and weak.

During cancer cell invasion in breast, along with several biochemical changes, destruction of myoepithelial cells surrounding the ducts and lobules is crucial (Lininger *et al.*, 2001; Tavassoli, 1999), because they are one of the important cell layers that line the entire duct system of the breast (Tavassoli, 1999) and are natural tumor suppressor cells that resist malignant transformation (Sternlicht *et al.*, 1997a,b). Lack of a myoepithelial layer surrounding the duct is an important diagnostic feature of DCIS and also an indication of the disappearance of a major obstacle for stromal invasion (Lininger *et al.*, 2001). However, the possible mechanisms involved in the removal of myoepithelial cells from around the duct to allow invasion of tumor cells into the stroma remain unknown. This and

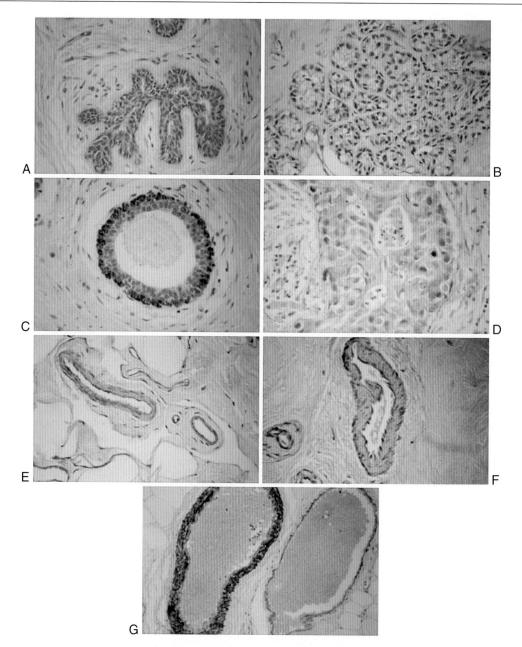

Figure 60 Immunohistochemical localization of neuropilin-1 antigen (*brown*) in myoepithelial cells of (**A**) ducts in normal breast section, (**B**) myoepithelial cells of ducts and lobules of noninvasive breast carcinoma section, (**C**) benign duct adjacent to the invasive carcinoma cells, (**D**) invasive ductal carcinoma section, and vascular smooth muscle cells of blood vessels from a (**E**) normal breast, (**F**) noninvasive breast carcinoma, and (**G**) invasive carcinoma of breast. Original magnification, (**A–C**) 300X; (**D**) 200X; (**E–G**) 150X.

previous studies (Stephenson *et al.*, 2002) together indicate an association between the overexpression of NRP-1 and the disappearance of myoepithelial cells, an event that is required for tumor epithelial cell invasion.

In addition to the myoepithelial cells, the immunoreaction of NRP-1 was detected in endothelial cells and was specifically prominent in vascular smooth muscle cells of the blood vessels in invasive breast carcinoma tissues (Figure 60). In normal and noninvasive tissues,

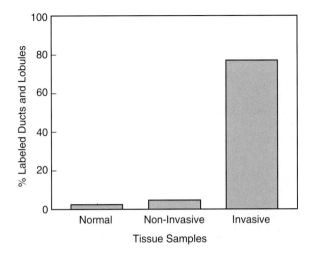

Figure 61 Immunohistochemical analyses of neuropilin-1 labeling index of myoepithelial cells in ducts and lobules in normal, noninvasive, and invasive carcinoma in breast.

NRP-1 immunoreaction was faint or almost undetected in both smooth muscle and endothelial cells. Although the functional role of this protein in smooth muscle cells is unknown, the current study and our previous findings suggest that NRP-1 may be associated with the migrations and proliferation of smooth muscle cells. In summary, using the procedures described in this chapter indicate, that NRP-1 may be a possible molecular marker, and its immunohistochemical detection in breast tumor specimens would be an ideal approach for the diagnosis and therapies of this disease.

Acknowledgments

This work was supported in part by NCI (National Institutes of Health) and Department of Veterans Affairs. This work is dedicated to the memory of my parents, Sakti Brata Banerjee and Bani Banerjee, and my teacher, Professor S.P. Ray Chaudhuri, Ph.D.

References

American Cancer Society. 1994. *Cancer Facts and Figures 1994.* Atlanta, GA: The Society.

American Cancer Society. 1999. *Cancer Facts and Figures 1999.* Atlanta, GA: The Society.

Banerjee, S.K., Zoubine, M.N., Trang, T.M., Weston, A.P., and Campbell, D.R. 1999. Overexpression of vascular endothelial growth factor164 and its co-receptor neuropilin-1 in estrogen-induced rat pituitary tumors and GH3 rat pituitary tumor cells. *Int. J. Oncol. 16:*253–260.

Bryant, H., and Brasher, P. 1994. Risks and probabilities of breast cancer: Short-term versus lifetime probabilities. *Can Med. Assoc. J. 150:*211–216.

Chen, H., Chedotal, A., He, Z., Goodman, C.S., and Tessier-Lavigne, M. 1997a. Neuropilin-2, a novel member of the neuropilin family, is a high affinity receptor for the semaphorins Sema E and Sema IV but not Sema III. *Neuron 19:*547–559.

Chen, H., Chedotal, A., He, Z., Goodman, C.S., and Tessier-Lavigne, M. 1997b. Neuropilin-2, a novel member of the neuropilin family, is a high affinity receptor for the semaphorins Sema E and Sema IV but not Sema III [published erratum appears in *Neuron* 1997 Sep; 19(3):559]. *Neuron 19:*547–559.

Desai, K.V., Kavanaugh, C.J., Calvo, A., and Green, J.E. 2002a. Chipping away at breast cancer: Insights from microarray studies of human and mouse mammary cancer. *Endocr. Relat Cancer 9:*207–220.

Desai, K.V., Xiao, N., Wang, W., Gangi, L., Greene, J., Powell, J.I., Dickson, R., Furth, P., Hunter, K., Kucherlapati, R., Simon, R., Liu, E.T., and Green, J.E. 2002b. Initiating oncogenic event determines gene-expression patterns of human breast cancer models. *Proc. Natl. Acad. Sci. U S A 99:*6967–6972.

Fuh, G., Garcia, K.C., and de Vos, A.M. 2000. The interaction of neuropilin-1 with vascular endothelial growth factor and its receptor flt-1. *J. Biol. Chem. 275:*26690–26695.

Giger, R.J., Cloutier, J.F., Sahay, A., Prinjha, R.K., Levengood, D.V., Moore, S.E., Pickering, S., Simmons, D., Rastan, S., Walsh, F.S., Kolodkin, A.L., Ginty, D.D., and Geppert, M. 2000. Neuropilin-2 is required *in vivo* for selective axon guidance responses to secreted semaphorins. *Neuron 25:*29–41.

Gluzman-Poltorak, Z., Cohen, T., Herzog, Y., and Neufeld, G. 2000. Neuropilin-2 is a receptor for the vascular endothelial growth factor (VEGF) forms VEGF-145 and VEGF-165. *J. Biol. Chem. 275:*18040–18045.

Gluzman-Poltorak, Z., Cohen, T., Shibuya, M., and Neufeld, G. 2001. Vascular endothelial growth factor receptor-1 and neuropilin-2 form complexes. *J. Biol. Chem. 276:*18688–18694.

Greenlee, R.T., Murray, T., Bolden, S., and Wingo, P.A. 2000. Cancer statistics, 2000. *CA Cancer J. Clin. 50:*7–33.

He, Z., and Tessier-Lavigne, M. 1997. Neuropilin is a receptor for the axonal chemorepellent Semaphorin III. *Cell 90:*739–751.

Kawakami, A., Kitsukawa, T., Takagi, S., and Fujisawa, H. 1996. Developmentally regulated expression of a cell surface protein, neuropilin, in the mouse nervous system. *J. Neurobiol. 29:*1–17.

Kawakami, T., Tokunaga, T., Hatanaka, H., Kijima, H., Yamazaki, H., Abe, Y., Osamura, Y., Inoue, H., Ueyama, Y., and Nakamura, M. 2002. Neuropilin 1 and neuropilin 2 co-expression is significantly correlated with increased vascularity and poor prognosis in nonsmall cell lung carcinoma. *Cancer 95:*2196–2201.

Kolodkin, A.L., and Ginty, D.D. 1997a. Steering clear of semaphorins: Neuropilins sound the retreat. *Neuron 19:*1159–1162.

Kolodkin, A.L., Levengood, D.V., Rowe, E.G., Tai, Y.T., Giger, R.J., and Ginty, D.D. 1997b. Neuropilin is a semaphorin III receptor. *Cell 90:*753–762.

Lininger, R.A., and Tavassoli, F.A. 2001. Breast. In Henson, D.E., and Albores-Saavedra, J., (eds) *Pathology of Incipient Neoplasia.* New York: Oxford University Press, 319–415.

Makinen, T., Olofsson, B., Karpanen, T., Hellman, U., Soker, S., Klagsbrun, M., Eriksson, U., and Alitalo, K. 1999. Differential binding of vascular endothelial growth factor B splice and proteolytic isoforms to neuropilin-1. *J. Biol. Chem. 274:*21217–21222.

Miao, H.Q., and Klagsbrun, M. 2000a. Neuropilin is a mediator of angiogenesis. *Cancer Metastasis Rev. 19:*29–37.

Miao, H.Q., Lee, P., Lin, H., Soker, S., and Klagsbrun, M. 2000b. Neuropilin-1 expression by tumor cells promotes tumor angiogenesis and progression. *FASEB J. 14:*2532–2539.

Migdal, M., Huppertz, B., Tessler, S., Comforti, A., Shibuya, M., Reich, R., Baumann, H., and Neufeld, G. 1998. Neuropilin-1 is a placenta growth factor-2 receptor. *J. Biol. Chem. 273:*22272–22278.

Neufeld, G., Cohen, T., Shraga, N., Lange, T., Kessler, O., and Herzog, Y. 2002. The neuropilins: Multifunctional semaphorin and VEGF receptors that modulate axon guidance and angiogenesis. *Trends Cardiovasc. Med. 12:*13–19.

Parkin, D.M. 1998. The global burden of cancer. *Semin. Cancer Biol. 8:*219–235.

Parkin, D.M., Pisani, P., and Ferlay, J. 1999a. Estimates of the worldwide incidence of 25 major cancers in 1990. *Int. J. Cancer 80:*827–841.

Parkin, D.M., Pisani, P., and Ferlay, J. 1999b. Global cancer statistics. *CA Cancer J. Clin. 49:*33–64.

Perou, C.M., Jeffrey, S.S., Van de, R.M., Rees, C.A., Eisen, M.B., Ross, D.T., Pergamenschikov, A., Williams, C.F., Zhu, S.X., Lee, J.C., Lashkari, D., Shalon, D., Brown, P.O., and Botstein, D. 1999. Distinctive gene expression patterns in human mammary epithelial cells and breast cancers. *Proc. Natl. Acad. Sci. USA 96:*9212–9217.

Perou, C.M., Sorlie, T., Eisen, M.B., Van de, R.M., Jeffrey, S.S., Rees, C.A., Pollack, J.R., Ross, D.T., Johnsen, H., Akslen, L.A., Fluge, O., Pergamenschikov, A., Williams, C., Zhu, S.X., Lonning, P.E., Borresen-Dale, A.L., Brown, P.O., and Botstein, D. 2000. Molecular portraits of human breast tumours. *Nature 406:*747–752.

Pollack, J.R., Sorlie, T., Perou, C.M., Rees, C.A., Jeffrey, S.S., Lonning, P.E., Tibshirani, R., Botstein, D., Borresen-Dale, A.L., and Brown, P.O. 2002. Microarray analysis reveals a major direct role of DNA copy number alteration in the transcriptional program of human breast tumors. *Proc. Natl. Acad. Sci. USA 99:*12963–12968.

Rossignol, M., Beggs, A.H., Pierce, E.A., and Klagsbrun, M. 1999. Human neuropilin-1 and neuropilin-2 map to 10p12 and 2q34, respectively. *Genomics 57:*459–460.

Rossignol, M., Gagnon, M.L., and Klagsbrun, M. 2000. Genomic organization of human neuropilin-1 and neuropilin-2 genes: Identification and distribution of splice variants and soluble isoforms. *Genomics 70:*211–222.

Saxena, N., Banerjee, S., Sengupta, K., Zoubine, M.N., and Banerjee, S.K. 2001. Differential expression of WISP-1 and WISP-2 genes in normal and transformed human breast cell lines. *Mol. Cell Biochem. 228:*99–104.

Shimizu, M., Murakami, Y., Suto, F., and Fujisawa, H. 2000. Determination of cell adhesion sites of neuropilin-1. *J. Cell Biol. 148:*1283–1293.

Soker, S., Takashima, S., Miao, H.Q., Neufeld, G., and Klagsbrun, M. 1998. Neuropilin-1 is expressed by endothelial and tumor cells as an isoform-specific receptor for vascular endothelial growth factor. *Cell 92:*735–745.

Squire, J., and Phillips, R.A. 1992. Genetic basis of cancer. In Tannock, I.F., and Hill, R.P. (eds) *The Basic Science of Oncology.* New York: McGraw-Hill, 41–60.

Stephenson, J.M., Banerjee, S., Saxena, N.K., Cherian, R., and Banerjee, S.K. 2002. Neuropilin-1 is differentially expressed in myoepithelial cells and vascular smooth muscle cells in preneoplastic and neoplastic human breast: A possible marker for the progression of breast cancer. *Int. J. Cancer 101:*409–414.

Sternlicht, M.D., and Barsky, S.H. 1997a. The myoepithelial defense: A host defense against cancer. *Med. Hypotheses 48:*37–46.

Sternlicht, M.D., Kedeshian, P., Shao, Z.M., Safarians, S., and Barsky, S.H. 1997b. The human myoepithelial cell is a natural tumor suppressor. *Clin. Cancer Res. 3:*1949–1958.

Takagi, S., Tsuji, T., Amagai, T., Takamatsu, T., and Fujisawa, H. 1987. Specific cell surface labels in the visual centers of Xenopus laevis tadpole identified using monoclonal antibodies. *Dev. Biol. 122:*90–100.

Takagi, S., Hirata, T., Agata, K., Mochii, M., Eguchi, G., and Fujisawa, H. 1991. The A5 antigen, a candidate for the neuronal recognition molecule, has homologies to complement components and coagulation factors. *Neuron 7:*295–307.

Takagi, S., Kasuya, Y., Shimizu, M., Matsuura, T., Tsuboi, M., Kawakami, A., and Fujisawa, H. 1995. Expression of a cell adhesion molecule, neuropilin, in the developing chick nervous system. *Dev. Biol. 170:*207–222.

Takashima, S., Kitakaze, M., Asakura, M., Asanuma, H., Sanada, S., Tashiro, F., Niwa, H., Miyazaki, J.J., Hirota, S., Kitamura, Y., Kitsukawa, T., Fujisawa, H., Klagsbrun, M., and Hori, M. 2002. Targeting of both mouse neuropilin-1 and neuropilin-2 genes severely impairs developmental yolk sac and embryonic angiogenesis. *Proc. Natl. Acad. Sci. USA 99:*3657–3662.

Tavassoli, F.A. 1999. Normal development and anaomalies. In *Pathology of the Breast.* New York: McGraw-Hill, 1–25.

18

Role of Epidermal Growth Factor Receptor in Breast Carcinoma

Careen K. Tang

Introduction

In the last decade, numerous studies have established the importance of growth factors and their receptors in cancer biology. Of these receptors, the epidermal growth factor receptor (EGFR) family is the most frequently implicated in human cancers. This class I subfamily of growth factor receptors is of four members: EGFR (Gullick, 1991), *HER-2*/p185*erbB-2*/neu (Hynes *et al.*, 1994), *HER-3*/p160*erbB-3* (Lemoine *et al.*, 1992), and *HER-4*/p180*erbB-4* (Plowman *et al.*, 1993). This EGFR family is one group of tyrosine kinases that have been widely implicated in promoting the proliferation and progression of malignant cells (Khazaie *et al.*, 1993; Salomon *et al.*, 1995). Because they are frequently overexpressed in a variety of carcinomas, they have received attention as targets for a variety of therapeutic interventions. The rationale for using EGFR-targeted approaches for cancer treatment is now firmly established and numerous clinical trials are in progress (Mendelsohn *et al.*, 2000).

Clinical Significance of Epidermal Growth Factor Receptor and Its Variance in Human Cancer

The epidermal *EGFR* gene is located on chromosome 7q12-13, which encodes a 170-kDa protein composed of three major domains: an extracellular ligand-binding domain, a transmembrane domain, and a cytoplasmic domain, which contains the catalytic protein tyrosine kinase and C-terminal regulatory domains (Downward *et al.*, 1984). The external ligand-binding region of the *EGFR* can be divided into four subdomains on the basis of amino acid homology. Subdomains I and III appear to be primarily responsible for ligand binding specificity (Bajaj *et al.*, 1987; Ward *et al.*, 1995), whereas subdomains II and IV are cysteine-rich domains containing 21 conserved cysteine residues. These domains may perform a role in the tertiary structure of the protein. The *EGFR* provided one of the first pieces of evidence for an activated oncogene to be associated with human tumor biology (Gullick, 1991; Khazaie *et al.*, 1993). Enhanced expression of *EGFR* is frequently detected in a variety of carcinomas, including breast, lung, head and neck, and glioblastoma (Berge *et al.*, 1987; Gullick, 1991; Harris *et al.*, 1992; Khazaie *et al.*, 1993; Libermann *et al.*, 1985; Ozanne *et al.*, 1986). Overexpression of *EGFR* in human malignancy has been extensively studied; cumulative evidence indicates the alteration of the *EGFR* gene to be as important as its amplification for eliciting oncogenic effects (Harris *et al.*, 1992). Many studies have shown *EGFR* expression and overexpression to correlate with higher disease stage, reduced survival, development of tumor metastasis, and the level of tumor

Handbook of Immunohistochemistry and *in situ* Hybridization of Human
Carcinomas, Volume 1: Molecular Genetics; Lung and Breast Carcinomas

415

Copyright © 2004 by Elsevier (USA)
All rights reserved.

dedifferentiation (Wells, 2000). Several groups have reported *EGFR* expression in human breast carcinomas, with an incidence ranging from 14% to 91% and a median value of 48% (Klijn *et al.*, 1992). Numerous studies have shown that the absence of estrogen receptor (ER) is often accompanied by the expression of high levels of EGFR and correlates with more aggressive states of the disease (Fitzpatrick *et al.*, 1984; Harris *et al.*, 1989; Nicholson *et al.*, 1989). Unlike *HER-2*/neu, the increased *EGFR* expression observed in primary breast tumors and in breast cancer cell lines is not caused by gene amplification; it results from an increased level of EGFR mRNA and protein expression.

Role of Epidermal Growth Factor–like Growth Factors in Human Cancers

The EGF superfamily includes EGF; transforming growth factor-alpha (TGF-α); heparin-binding epidermal growth factor–like factor (HB-EGF); amphiregulin (AR); betacellulin (BTC); the neuregulins (NRGs), also known as heregulin; neu differentiation factors; glial growth factors; acetylcholine receptor inducing activity; and epiregulin (EPR) (Riese II *et al.*, 1998). These growth factors share several common features. They are synthesized as large precursors with N-terminal single peptide sequences and possess single or multiple EGF-like domains. The EGF-like domain is characterized by the presence of six conserved cysteine residues, spaced within 40–50 amino acids at defined intervals. These cysteine residues pair to form three intramolecular disulfide bonds, which form a three-loop structure. The folded secondary structure of the EGF-like domain is involved in both the receptor-binding and biological activity of these molecules (Groenen *et al.*, 1994).

In normal tissues, EGF-like ligands are commonly expressed in cell types different from their receptors. These ligands are expressed in the stroma, whereas the receptor is expressed in the adjacent epithelial cells. Consequently, EGF-like ligands and their receptors are likely to regulate cell signaling in a paracrine-loop fashion. On malignant transformation and during tumor progression, this paracrine-loop regulation is often altered to an autocrine-regulated signaling pathway, with the ligands and their receptors being expressed on the same cells (De Miguel *et al.*, 1999; Scher *et al.*, 1995). This switch from paracrine to autocrine activity is likely to be a key factor in breast cancer progression, as well as in several other types of cancers, including non–small cell lung and prostate cancer (Scher *et al.*, 1995).

The *ErbB* family signaling network is very complex (Riese II *et al.*, 1998). This is the result of a large number

of ligands and their extensive cross-interactions within the receptors in this family. A single EGF-like ligand can bind and activate multiple receptors and a single *ErbB* family receptor can bind multiple ligands, with the exception of *ErbB-2*, which is an orphan receptor. For example, BTC HB-EGF and EPR activate both the EGFR and *ErbB-4*, whereas NRG binds to *ErbB-3* and *ErbB-4* (Riese II *et al.*, 1998). EGFR was the first receptor for which ligand-dependent tyrosine kinase activity was demonstrated (Ushiro *et al.*, 1980). A number of agonists can bind to and activate EGFR, including EGF, TGF- α, AR, BTC, and HB-EGF (Riese II *et al.*, 1998). Recent crystallographic studies have demonstrated the receptor-mediated dimerization in a dimeric EGF EGFR crystal structure, in which the two EGFR molecules are directly bound to each other by interactions between each subdomain II. Epidermal growth factor receptor–subdomains I–III are arranged in a C shape, and EGF is docked between subdomains I and III (Ogiso *et al.*, 2002). The same dimerization mode was found for an independently determined structure of the 2:2 complex of TGF-α and EGFR (Garrett *et al.*, 2002). Therefore, the mechanism of the EGFR dimerization is receptor-mediated and growth factor binding is thought to induce a conformational change that stabilizes this arrangement (Ferguson *et al.*, 2003; Garrett *et al.*, 2002; Ogiso *et al.*, 2002). Furthermore, on ligand binding, these receptors can heterodimerize with other members of the *ErbB* family to amplify and diversify their signaling pathways. The activation of *ErbB-2* can result in transphosphorylation through *ErbB-2* heterodimer with other members of the *ErbB* family (Riese II *et al.*, 1998). Recent crystallography of the extracellular region *HER-2*/neu structure reveals a fixed conformation that resembles a ligand-activated state, and demonstrates *HER-2*/neu poised to interact with other *ErbB* receptors in the absence of direct ligand binding (Cho *et al.*, 2003). Reports have shown that cancers cooverexpressing both EGFR and *ErbB-2* appear to have a poor prognosis compared with cancers with high expression of either receptors (Arteaga, 2002).

Epidermal Growth Factor Receptor vIII in Breast Cancer

Spontaneous deletions within the *EGFR* gene were first identified in primary human glioblastoma tumors (Ekstrand *et al.*, 1992; Wong *et al.*, 1992; Yamazaki *et al.*, 1988); and in nearly all cases, the alterations have been reported in tumors with EGFR amplification. Three mutated forms of EGFR have been identified— EGFRvI, vII, and vIII—formed by three different deletions in the extracellular domain (Batra *et al.*, 1995).

The most common of these is the type III EGF deletion-mutant receptor (EGFRvIII), which is characterized by the deletion of exons 2–7 in the EGFR mRNA. These deletions correspond to cDNA nucleotides 275–1075, which encode amino acids 6–276, presumably through alternative splicing or rearrangements (Ekstrand et al., 1992; Wong et al., 1992; Yamazaki et al., 1988). Deletion of 801 basepair (bp) within the extracellular domain of the EGFR gene causes frame truncation of the normal EGFR protein, resulting in a 145-kDa receptor (Ekstrand et al., 1992; Moscatello et al., 1995; Wong et al., 1992; Yamazaki et al., 1988). EGFRvIII is ligand-independent and is constitutively activated (Batra et al., 1995). Reports have demonstrated that 52–67% of primary human glioblastoma tumors detected EGFRvIII expression (Moscatello et al., 1995; Wikstrand et al., 1995). EGFRvIII is also frequently detected in various human cancers, including breast, prostate, ovarian, lung, and medulloblastoma tumors (Moscatello et al., 1995; Olapade-Olaopa et al., 2000; Wikstrand et al., 1995). The expression of EGFRvIII increased with dedifferentiation of prostatic epithelial cells with a concomitant decrease in wild-type EGFR expression (Olapade-Olaopa et al., 2000).

The EGFR gene is amplified in glioblastoma, but not in human breast cancer (Humphrey et al., 1990). Furthermore, EGFRvIII is expressed only in the tumor cells, but not in normal or stroma cells. The detection of EGFRvIII mRNA in breast cancer remains a challenge. Despite these hurdles, the laser capture microdissection (LCM) combined with reverse transcription-polymerase chain reaction (RT-PCR) detection was applied to detect EGFRvIII mRNA in primary invasive breast cancer specimens. As a result, a high incidence of EGFRvIII transcripts was observed by LCM in primary invasive breast cancers but not in normal breast tissue (Ge et al., 2002). Within a majority of breast cancer tissues, both mutant and wild-type EGFR mRNA were detected in the same tumor. The RT-PCR fragment derived from infiltrating breast carcinoma matched the published EGFRvIII sequence with a 801-bp deletion in the extracellular ligand binding domain, creating glycine at the splicing position as noted in human glioblastoma (Ge et al., 2002). These results confirmed the presence of EGFRvIII in human breast cancer. However, no tested normal adult human tissues, including those from the peripheral and central nervous system, the lymphoid system, skin, breast, liver, lung, ovary, placenta, endometrium, testes, and colon, have been found to express EGFRvIII by immunohistochemical or genetic analysis, or both (Wikstrand et al., 1995). Several functional differences between EGFRvIII and normal EGF receptor have been characterized.

Overexpression of EGFRvIII in NIH3T3 and NR6 cells results in transformed morphology, enhanced growth, and tumorigenicity in athymic mice (Batra et al., 1995; Moscatello et al., 1996). The EGFRvIII molecule appears to be unregulated by EGF or TGF-α (Batra et al., 1995; Moscatello et al., 1996; Tang et al., 2000) and is constitutively activated in various systems. Even though TGF-α cannot bind to EGFRvIII when expressed in Chinese hamster ovary (CHO) cells, these cells increase the levels of TGF-α DNA synthesis (Batra et al., 1995). Similar observations are seen in the EGFRvIII transfected 32D cell system, a nontumorigenic, IL-3–dependent murine hematopoietic cell, which does not express endogenous levels of EGF family receptors. Overexpressing EGFRvIII in 32D cells is capable of abrogating the IL-3–dependent pathway in the absence of ligands. In contrast, the parental cells—32D/EGFR, 32D/ErbB-4, and (32D/ErbB-2 + ErbB-3 cells)—all depend on IL-3 or EGF-like ligands for growth (Tang et al., 2000). Moreover, 32D cells expressing high levels of EGFRvIII formed large tumors in nude mice even when no exogenous EGF ligand was administered. In contrast, no tumors grew in mice injected with 32D/EGFR, 32D/ErbB-4, and (32D/ErbB-2+ ErbB-3 cells), or with low expressing clone 32D/EGFRvIII cells or the parental 32D cells (Tang et al., 2000). Such profound transforming activity has not been observed in any homodimer or heterodimer system of wild-type ErbB family receptors (Tang et al., 2000).

The biological significance of EGFRvIII in human cancer was explored by exogenous expression of EGFRvIII in gliomas, breast cancer, and a small cell lung cancer cell line. Exogenous expression of EGFRvIII in glioblastoma U87MG cells or MCF-7 cells, as well as lung cancer cells, are associated with self-phosphorylation and a pronounced enhancement of tumorigenicity in vivo (Damstrup et al., 2002; Nishikawa et al., 1994; Tang et al., 2000). Expression of EGFRvIII in MCF-7 cells increases ErbB-2 phosphorylation, presumably through heterodimerization and cross-talk (Tang et al., 2000).

Mechanisms of Epidermal Growth Factor Receptor Dysregulation

Epidermal growth factor receptor dysregulation can occur by many mechanisms. One mechanism is the high expression of EGFR, which mediates increased receptor signaling output. Another mechanism involves the overexpression of ligand, which results in increased EGFR signaling activity, despite normal or even low levels of receptor expression. EGFRs can also be mutated in cancers and confer constitutive signaling activity in the absence of ligand binding. Heterodimerization with *HER-2*/neu can potentiate EGFR function by increasing EGF binding affinity, resulting in the stabilization and recycling EGFR-*HER-2* heterodimers, and repertoire expansion of receptor-associated substrates and signaling responses (Arteaga, 2002). In addition, the EGFR can cross-talk with heterologous receptors activated by neurotransmitters, lymphokines, and stress inducers (Gschwind *et al.*, 2001; Prenzel *et al.*, 2000). Finally, EGFR dysregulation could be mediated by the inability to down-regulate receptors, which can be caused by a defect in an internalization mechanism or by dephosphorylation (Waterman *et al.*, 2001).

Targeting Epidermal Growth Factor Receptor

Substantial evidence supports a key role for EGFR in tumor cell motility, invasion, adhesion, angiogenesis, apoptosis, and cell cycle control. Clearly, EGFR is a promising target for various drugs (Kari *et al.*, 2003), resulting in an extremely dynamic field of EGFR-targeted cancer research. Ongoing investigations are continually improving our knowledge of the mechanisms involving EGFR regulation, activation, signal transduction, how they affect tumor cell function, and the progression of cancer. To date, our understanding of how the alteration of the *EGF* gene and the expression of mutated receptors contribute to tumor development is still limited and requires further study.

A variety of clinical approaches have been developed. Immunotherapy has long been considered a promising approach for the treatment of cancer (Baselga *et al.*, 1999; Mendelsohn *et al.*, 2000). Blockade of the TGF-α–EGFR autocrine pathway by using anti-EGFR blocking monoclonal antibodies, recombinant proteins containing TGF-α or EGF fused to toxins, or EGFR-specific tyrosine kinase inhibitors (TKIs) has been proposed as a potential therapeutic modality (Ciardiello *et al.*, 2002; Mendelsohn, 2002). Several blocking anti-EGFR monoclonal antibodies that inhibit the *in vitro* and *in vivo* growth of human cancer cell lines that express TGF-α and EGFR have been generated (Ciardiello *et al.*, 2002; Mendelsohn, 2002). Among these, monoclonal antibodies 528 and 225 are two mouse monoclonal antibodies that have been extensively characterized for their biological and pharmacological properties and represent the first anti-EGFR blocking agents that have entered clinical evaluation in patients with cancer (Ciardiello *et al.*, 2002; Mendelsohn, 2002).

IMC-C225 (cetuximab), a human-mouse chimeric anti-EGFR monoclonal antibody, is currently being investigated in several phase II and III clinical trials (Mendelsohn, 2002). Patients with head and neck, colorectal, breast, lung, prostate, and kidney carcinomas are being treated with IMC-C225 as a single therapeutic agent or in combination with other modalities (Mendelsohn, 2002).

During the past 5 years, significant progress has been made in the area of EGFR TKIs, and new structural classes have emerged that exhibit enormous improvements with regard to potency, specificity, and *in vitro* and *in vivo* activity (Ciardiello *et al.*, 2000; Fry *et al.*, 1994; Slichenmyer *et al.*, 2001; Yaish *et al.*, 1988). Very specific, irreversible inhibitors of the EGFR family have been synthesized to provide unique pharmacological properties with exceptional efficacy. Among them, ZD1839, OSI-774, CI-1033, and EKB-569 are currently being evaluated in clinical trials (Ciardiello *et al.*, 2000; Slichenmyer *et al.*, 2001; Torrance *et al.*, 2000). Preliminary data demonstrate that ZD1839 has promising clinical activity in patients with a wide range of tumor types (Baselga, 2000). CI-1033 is an irreversible EGFR TKI and is currently in phase II clinical trials targeting EGFR in different types of human cancer (Slichenmyer *et al.*, 2001; Torrance *et al.*, 2000).

Methods of Evaluating Epidermal Growth Factor Receptor Expression

Although the rationale for using EGFR-targeted approaches for cancer treatment is now firmly established and numerous clinical trials are in progress, how to reliably detect EGFR expression in heterogeneous cell populations and how to assess whether EGFR expression levels can predict response to therapy are still unclear (Arteaga, 2002). In addition, the situation with EGFR-targeted agents is less straightforward. No method of analysis of EGFR is consistently used in all laboratories, making the comparison of results from different studies difficult.

A variety of techniques can be used to evaluate EGFR at the DNA, RNA, and protein levels, as well as receptor activation *in situ*. Immunohistochemistry (IHC) is commonly used to evaluate EGFR protein levels and is arguably the most convenient method for analysis of clinical specimens. Unlike *ErbB-2*, detection of EGFR by IHC is much less formatted. The early literature indicates a wide range (15% to 90%) of EGFR overexpression detected in human breast cancers. This variation could be caused by any of the following factors: variability of tissue fixation protocols, different antibodies, lack of standardization of immunohistochemical assays, and different scoring methodologies (Arteaga, 2002). Of these factors, antibody specificity is most probably the cause for this variation. As is known, most commercially available antibodies to EGFR are able to recognize both wild-type and mutant EGFR (Ge *et al.*, 2002). Therefore, early publications describing EGFR overexpression in breast cancer may be questionable, because antibody detection of EGFR overexpression would not be able to discriminate the wild-type EGFR from the mutant EGFRvIII.

A series of highly promising EGFR signal inhibitors are moving rapidly through preclinical and clinical studies as anti-cancer agents. Despite a strong scientific foundation underlying development of EGFR as a valuable molecular target for cancer therapy, work to unravel the precise mechanisms by which EGFR inhibition exerts its anti-tumor effects is still ongoing. The establishment and agreement of a standard method for assessing EGFR levels is key to understanding the role of tumor EGFR expression in disease prognosis. When we can confidently measure the level of EGFR expressed in tumors, we will be in a better position to understand the response to EGFR-targeted therapy and whether overexpression of EGFR is correlated with prognosis. Answers to these questions would also help to determine if screening for tumor EGFR expression would be of benefit to patients with cancer. Therefore, characterization of EGFR antibody for IHC analysis with breast cancer specimens is crucial for further development of EGFR-targeting therapies. In this chapter, two of the specific EGFR antibodies that we have characterized are described. As shown in Table 26, EGFR (Ab-3) monoclonal antibody appears to recognize the wild-type EGFR only, but not the EGFRvIII receptor, whereas antibody EGFRvIII (Ab-18) specifically recognizes EGFRvIII, but not the wild-type EGFR protein (Table 26). Both EGFR (Ab-3) and EGFRvIII (Ab-18) antibodies do not recognize other EGF family receptors (Table 25). Figure 62 illustrates the IHC analysis of primary human infiltrated breast carcinoma specimens with specific antibodies against EGFR wild-type (Ab-3) and EGFRvIII (Ab-18).

MATERIALS

1. Xylene.
2. Ethyl alcohol (100%).
3. Citrate buffer (Dako, Carpinteria, CA).
4. Micro cover-glasses (VWR, West Chester, PA).
5. Slide Rack (Tissue Tek Slide holder) (VWR).
6. Dulbecco's phosphate-buffered saline (PBS) (1X, pH 7.4) without calcium and magnesium (BioFluids, Rockville, MD).
7. Diaminobenzidine (DAB) (Histostain plus Broad Spectrum kit; Zymed Laboratories, Inc., San Francisco, CA).
8. Hematoxylin, Gill 2X, 475 ml (VWR).
9. Permount (Fisher Scientific, Houston, TX).
10. Microwave.
11. Anti-EGFR (Ab-3) antibody (Lab Vision, Fremont, CA) diluted in 1X PBS (pH 7.4) in the presence of 1–3% bovine serum albumin (BSA).
12. Anti-EGFRvIII (Ab-18) antibody (Lab Vision) diluted in 1X PBS (pH 7.4) in the presence of 1–3% BSA.
13. H_2O_2 solution (Sigma, St. Louis, MO).
14. Microscope slides.

Detection of Epidermal Growth Factor Receptor with Frozen Tissue Sections by Epidermal Growth Factor Receptor (Ab–3) Monoclonal Antibody

This antibody can only be used for frozen sections, not for paraffin-embedded tissues.

1. Mount frozen sections on positively charged Plus slides (Fisher Scientific), and air dry for 30 min.
2. Fix frozen sections in cold acetone at −20°C for 15 min.

Table 25 Specificity of the EGFR and EGFRvIII Antibodies

Cell Line	EGFR	ErbB-2	ErbB-3	ErbB-4	EGFRvIII	Reactivity with αEGFRvIII (Ab-18)	Reactivity with αEGFR (Ab-3)
NIH/EGFR	++++	−	−	−	−	−	+
NIH/EGFRvIII	−	−	−	−	++++	+	−
MCF-7	−/+	++	+++	++++	−	−	−
MCF7/EGFRvIII	−/+	++	+++	++++	++++	+	−
MDA-MB-468	++++	++	+	−	−	−	+
MDA-MB-453	−	++++	+++	+	−	−	−

EGFR, epidermal growth factor receptor.

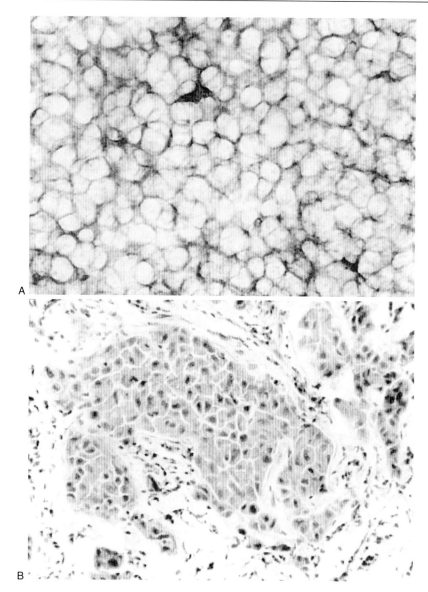

Figure 62 Representative immunohistochemical analysis of primary human infiltrated breast carcinoma specimens with specific antibodies against epidermal growth factor receptor (EGFR) or EGFRvIII. Counterstaining with hematoxylin was used for viewing negatively stained cells (*blue*). **A:** EGFR immunoreactivity shows in cellular membranes (*brown*). **B:** EGFRvIII positively stained cells shows in cytoplasms.

3. Wash with 1X PBS (pH 7.4) 3 times.

4. Dilute the primary antibody (anti-EGFR, Ab-3; NeoMarker) (1:20 dilution with 1X PBS (pH 7.4) in the presence of 1–3% BSA).

5. Add primary antibody (anti-EGFR, Ab-3; NeoMarker) prepared in **step 4** to the tissue section and incubate at room temperature for 1 hr.

6. Wash with 1X PBS (pH 7.4) three times.

7. Dilute second antibody (a horseradish peroxidase–conjugated goat anti-mouse IgG [H + L], Kirkegard & Perry Laboratories, Gaithersburg, MD) at a dilution of 1/250 with 1X PBS (pH 7.4) in the presence of 1–3% BSA and add to the sample and incubate for 1 hr at room temperature.

8. Wash with 1X PBS (pH 7.4) three times.

9. A positive immunoreaction was visualized by incubating the slides with diaminobenzidine (DAB) for 2–10 min.

10. Wash thoroughly with distilled water.

11. Counterstain with hematoxylin for 1 min.

12. Rinse thoroughly in distilled water.

Dehydration of Tissue Sections

1. 70% Ethanol (2–5 min)
2. 85% Ethanol (2–5 min)
3. 95% Ethanol (2–5 min)
4. Ethanol absolute (2–5 min)
5. Clear in xylene
6. Coverslip with Permount

Detection of Epidermal Growth Factor Receptor vIII Expression in Paraffin-embedded Human Primary Invasive Breast Carcinoma Tissues by Anti-Epidermal Growth Factor Receptor vIII (Ab-18) Deparaffinization Procedure (Pretreatment)

Deparaffinize immediately before immunostaining. Once deparaffinated you must go through the whole process and never let the slides dry.

In Coplin jar or in 50-ml Falcon tubes (two slides/tube, back to back):

1. Heat the sections of paraffin-embedded tissues at 56°C for 30 min.
2. Immediately deparaffinize by placing the slides in xylene (5 min each for three times; for thick sections warm the xylene to 37°C),
3. Follow by three changes in absolute ethanol (5 min each),
4. Two changes in 95% ethanol (5 min each),
5. 70% ethanol (5 min each),
6. Then three changes of distilled water (3 min each).
7. Rehydrate in 1X PBS (pH 7.4) for 10 min.

Blocking Endogenous Peroxidase

Incubate the slides in methanol containing 3% H_2O_2 in distilled water for 10 min or 0.3% H_2O_2 in distilled water for 30 min.

Antigen Retrieval Procedure for Detection of Epidermal Growth Factor Receptor vIII in Paraffin-Embedded Tissue

Microwave antigen retrieval produces improved staining for a wide range of monoclonal and polyclonal antibodies and reduces background staining (Hayat, 2002). It is essential for the detection of EGFRvIII in paraffin-embedded tissue sections. We modified the BioGenex microwave Antigen Retrieval procedure (BioGenex, San Ramon, CA). Using citrate-based (neutral pH) solutions works well in this case.

1. Reagents: Prepare 0.01 M citrate buffer (pH 6.0).
2. Stock solution A: 0.1 M citric acid (dissolve 21.01 g citric acid (anhydrous) in 1 L distilled water).
3. Stock solution B: 0.1 M sodium citrate (dissolve 29.41 g in 1 L distilled water).
4. The working solution is prepared just before use by mixing 18 ml solution A with 82 ml solution B plus sufficient distilled water to make 1000 ml and adjust the pH to 6.0.

Note: 10X H.I.E.R. buffer (HIER101) (BioTek Solutions, Inc., Santa Barbara, CA); or #H-3300 Antigen Unmasking Solution, 100X (Vector Laboratories, Burlingame, CA) also can be used.

Step-by-step procedure:

1. Place slides in Tissue Tek Slide holder; always fill any empty slots with blank slides if necessary to maintain the temperature.
2. Immerse the slide holder in 1X PBS (pH 7.4) for 5 min.
3. Place the holder in a slide bath or in a 1 L glass beaker filled with 0.01 M citrate buffer. Place the lid loosely, but secure with tape.
4. Position the slide holder in the center of the microwave oven on a paper towel to absorb any liquid run over.
5. Turn the microwave (800 W) on high for 30 min.* Closely watch the solution until it comes to a rapid boil. (**Note**: It usually takes 3–7 min before a boil is reached depending on the type of microwave, temperature, and other factors. It is important that a rapid boil is reached for every run before proceeding to the next step). Check the liquid level after 10–15 min; if buffer level is low, replenish with hot distilled water. The slides must remain fully submerged in the buffer at all times.
6. Remove the slide bath from the microwave oven. Allow slides to cool to room temperature (~30 min). Slowly cooling to room temperature is important.
7. Rinse with several changes in distilled water, 3–5 min each at room temperature.
8. Wash in 1X PBS (pH 7.4) for 10 min. Do not let the tissues dry out, and then continue with the immunostaining procedure.

Vectastain Immunostaining

1. Rinse the slides in distilled water for 5 min.
2. Wash in 1X PBS (pH 7.4) for 20 min.
3. Incubate with 5–10% normal goat serum in 1X PBS (pH 7.4) (serum from animals in which the

---

*Microwave times may need to be adjusted to compensate for differences between microwave brands and models; water should be boiling within about the first 3 min. If using the Vector Laboratories antigen unmasking solution, addition of 15–25 ml distilled water plus 25 ml of 1X unmasking solution is necessary after the initial 5 min heating, and the second 4 min microwaving must be changed to 2 × 2 min with an additional 25 ml unmasking solution added before the final 2 min heating. The power setting should be adjusted so that the oven cycles on and off every 20–30 sec and the solution boils about 5–10 sec for each cycle. This power setting should be noted and used for this step in all subsequent runs for the same antibody. Each antibody should be tested for the optimal time for this step.

secondary antibody was raised) for 20 min or 1–2% serum in 2% BSA/1X PBS (pH 7.4).

4. Drain the excess serum from the slide and incubate with primary antibody (anti-EGFRvIII antibody at 200 ng/ml) in 1X PBS (pH 7.4) for 2 hr at room temperature in a humidity chamber.

5. Rinse with 1X PBS (pH 7.4) for 3 min.

6. Wash in 1X PBS (pH 7.4) very carefully. Use a wash bottle of PBS and a tray for collecting the PBS, let PBS run over the section for several seconds, then put the slides in the humidity chamber and incubate in 1X PBS (pH 7.4) for 5 min two times.

7. Incubate the slides in secondary antibody (1:250) (a horseradish peroxidase–conjugated goat anti-mouse IgG ([H + L]; Kirkegard & Perry Laboratory) 30 min at room temperature.

8. Repeat the 1X PBS (pH 7.4) washes as above.

9. Develop the color by incubating the sections in DAB substrate solution for 2–8 min. (DAB and H_2O_2 is prepared without modification immediately before using). This results in a brown product. We do not use the optimal nickel solution, which will give a black product, but you can if desired).

10. Wash the slides thoroughly in running tap water for 10 min.

Counterstain

1. Counterstain with hematoxylin for 1 min.
2. Rinse thoroughly in the distilled water.

Dehydration of Tissue Sections

3. 70% Ethanol (2–5 min)
4. 85% Ethanol (2–5 min)
5. 95% Ethanol (2–5 min)
6. Ethanol absolute (2–5 min)
7. Clear in xylene
8. Let slides dry (20 min)
9. Coverslip with Permount

Notes:

1. Cultured cells grown directly on slides or coverslips and fixed with ice-cold methanol or with 2% formalin in 1X PBS (pH 7.4) also can be stained.

2. Instead of coating the slides with an adhesive, use SuperFrost Plus slides (Fisher Scientific). These slides are charged positively and the sections are charged negatively, thus preventing the sections from detaching from the slides while boiling.

3. To avoid the risk of drying the sections during microwave heating, it is necessary to heat them in multiple 4- to 5-min cycles and replenish the jars between heating periods.

RESULTS AND DISCUSSION

It has been well documented that the wild-type EGFR is a membrane receptor (Hackel et al., 1999), whereas EGFRvIII is mainly expressed in the cytoplasm of breast tumor cells. Similar observations have been reported in human glioblastoma tumors, where EGFRvIII is expressed exclusively in the cytoplasm (Ekstrand et al., 1992). As shown in Figure 62, the EGFRwt displays a membrane staining, whereas the EGFRvIII exhibits a predominantly cytoplasmic staining and some nuclear staining in cells derived from primary invasive breast cancer tissues. The altered subcellular location of EGFRvIII in breast cancer may suggest that the trafficking, signaling, recycling, and degradation of the mutant receptor may somehow differ from its wild-type form and may lead to cytoplasmic location. The changed subcellular location may attribute to a sustained tyrosine kinase activity, which manifests a more potent, aggressive, and oncogenic effect in human cancer. Indeed, one report has demonstrated that ZD1839 failed to inhibit EGFRvIII-expressing xenografts in nude mice, although it can sufficiently inhibit wild-type EGFR-expressing xenografts (Heimberger et al., 2002).

The *EGFR* overexpression in human breast cancer varies from 15% to 90% in reports from early studies. As now known, most commercially available antibodies to EGFR are able to recognize both wild-type and mutant EGFR, which raises a concern regarding the antibody specificity. Therefore, early publications describing EGFR overexpression in breast cancer may be questionable, because antibody detection of EGFR would not be able to discriminate the wild-type EGFR from the mutant EGFRvIII. In brain tumors, EGFR overexpression was associated with a mutant form of EGFR (Ekstrand et al., 1992; Wong et al., 1992), of which the most common mutant form is EGFRvIII (Frederick et al., 2000).

The current methods to measure EGFR levels in tumors are not quantitative and cannot be endorsed as predictive of either patient prognosis or response to treatment (Arteaga, 2002). However, antibodies used in this chapter provide useful information for the future development of standardized procedures to measure EGFR levels in breast cancers. This will serve to enhance our ability to best select patients, monitor drug activity, and assess tumor response in patient receiving EGFR inhibitory agents.

Acknowledgment

The author thanks Dr. Lee J. Romancyk for critically reviewing this manuscript.

References

Arteaga, C.L. 2002. Epidermal growth factor receptor dependence in human tumors: More than just expression? *Oncologist* 7(Suppl)4:31–39.

Bajaj, M., Waterfield, M.D., Schlessinger, J., Taylor, W.R., and Blundell, T. 1987. On the tertiary structure of the extracellular domains of the epidermal growth factor and insulin receptors. *Biochim. Biophys. Acta.* 16:220–226.

Baselga, J. 2000. New technologies in epidermal growth factor receptor targeted cancer therapy. *Signal 1:*12–21.

Baselga, J., Tripathy, D., Mendelsohn, J., Baughman, S., Benz, C.C., Dantis, L., Sklarin, N.T., Seidman, A.D., Hudis, C.A., Moore, J., Rosen, P.P., Twaddell, T., Henderson, I.C., and Norton, L. 1999. Phase II study of weekly intravenous trastuzumab (Herceptin) in patients with HER2/neu-overexpressing metastatic breast cancer. *Semin. Oncol.* 26(4 Suppl 12):78–83.

Batra, S.K., Castelino-Prabhu, S., Wikstrand, C.J., Zhu, X., Humphrey, P.A., Freidman, H.S., and Bigner, D.D. 1995. Epidermal growth factor ligand-independent, unregulated, cell-transforming potential of a naturally occurring human mutant EGFRvIII gene. *Cell Growth Differ.* 6:1251–1259.

Berge, M.S., Greenfield, C., Gullick, W.J., Haley, J., Downward, J., Neal, D.E., Harris, A.L., and Waterfield, M.D. 1987. Evaluation of epidermal growth factor receptors in bladder tumors. *Br. J. Cancer* 56:533–537.

Cho, H.S., Mason, K., Ramyar, K.X., Stanley, A.M., Gabelli, S.B., Denney, D.W. Jr, and Leahy, D.J. 2003. Structure of the extracellular region of HER2 alone and in complex with the Herceptin Fab. *Nature* 421:756–760.

Ciardiello, F., Caputo, R., Bianco, R., Damiano, V., Pomatico, G., De Placido, S., Bianco, A.R., and Tortora, G. 2000. Antitumor effect and potentiation of cytotoxic drugs activity in human cancer cells by ZD-1839 (Iressa), an epidermal growth factor receptor-selective tyrosine kinase inhibitor. *Clin. Cancer Res.* 6:2053–2063.

Ciardiello, F., and Tortora, G. 2002. Anti-epidermal growth factor receptor drugs in cancer therapy. *Exp. Opin. Investig. Drugs 11:*755–768.

Damstrup, L., Wandahl Pedersen, M., Bastholm, L., Elling, F., and Skovgaard Poulsen, H. 2002. Epidermal growth factor receptor mutation type III transfected into a small cell lung cancer cell line is predominantly localized at the cell surface and enhances the malignant phenotype. *Int. J. Cancer* 97:7–14.

De Miguel, P., Royuela Bethencourt, R., Ruiz, A., Fraile, B., and Paniagua, R. 1999. Immunohistochemical comparative analysis of transforming growth factor alpha, epidermal growth factor, and epidermal growth factor receptor in normal, hyperplastic and neoplastic human prostates. *Cytokine 11:*722–727.

Downward, J., Yarden, Y., Mayes, E., Scrace, G., Totty, N., Stockwell, P., Ullrich, A., Schlessinger, J., and Waterfield, M.D. 1984. Close similarity of epidermal growth factor receptor and v-erb-B oncogene protein sequences. *Nature 307:*521–527.

Ekstrand, A.J., Sugawa, N., James, C.D., and Collins, V.P. 1992. Amplified and rearranged epidermal growth factor receptor genes in human glioblastomas reveal deletions of sequences encoding portions of the N-and/or C-terminal tails. *Proc. Natl. Acad. Sci. USA* 89:4309–4313.

Ferguson, K.M., Berger, M.B., Mendrola, J.M., Cho, H-S., Leahy, D.J., and Lemmon, M.A. 2003. EGF activates its receptor by removing interations that autoinhibit ecodomain dimerization. *Mol. Cell 11:*507–517.

Fitzpatrick, S.L., Brightwell, J., Wittliff, J.L., Barrows, G.H., and Schultz, G.S. 1984. Epidermal growth factor binding by breast tumor biopsies and relationship to estrogen receptor and progestin receptor levels. *Cancer Res.* 44:3448–3453.

Frederick, L., Wang, X., Eley, G., and James, D.C. 2000. Diversity and frequency of epidermal growth factor receptor mutations in human glioblastomas. *Cancer Res.* 60:1383–1387.

Fry, D.W., Kraker, A.J., McMichael, A., Ambroso, L.A., Nelson, J.M., Leopold, W.R., Connors, R.W., and Bridges, A.J. 1994. A specific inhibitor of the epidermal growth factor receptor tyrosine kinase. *Science 265:*1093–1095.

Garrett, T.P., McKern, N.M., Lou, M., Elleman, T.C., Adams, T.E., Lovrecz, G.O., Zhu, H.J., Walker, F., Frenkel, M.J., Hoyne, P.A., Jorissen, R.N., Nice, E.C., Burgess, A.W., and Ward, C.W. 2002. Crystal structure of a truncated epidermal growth factor receptor extracellular domain bound to transforming growth factor alpha. *Cell 110:*763–773.

Ge, H., Gong, X., and Tang, C.K. 2002. Evidence of high incidence of EGFRvIII expression and coexpression with EGFR in human invasive breast cancer by laser capture microdissection and immunohistochemical analysis. *Int. J. Cancer* 98:57–61.

Groenen, L.C., Nice, E.C., and Burgess, A.W. 1994. Structure-function relationships for the EGF/TGF-alpha family of mitogens. *Growth Factors* 11:235–257.

Gschwind, A., Zwick, E., Prenzel, N., Leserer, M., and Ullrich, A. 2001. Cell communication networks: Epidermal growth factor receptor transactivation as the paradigm for interreceptor signal transmission. *Oncogene* 20:1594–1600.

Gullick, W.J. 1991. Prevalence of aberrant expression of the epidermal growth factor receptor in human cancer cells. *Br. Med. Bull.* 47:87–98.

Hackel, P.O., Zwick, E., Prenzel, N., and Ullrich, A. 1999. A epidermal growth factor receptors: Critical mediators of multiple receptor pathways. *Curr. Opin. Cell Biol.* 11:184–189.

Harris, A.L., Nicholson, S., Sainsbury, J.R., Farndon, J., and Wright, C. 1989. Epidermal growth factor receptors in breast cancer: Association with early relapse and death, poor response to hormones and interactions with neu. *J. Steroid. Biochem.* 34:123–131.

Harris, A.L., Nicholson, S., Sainsbury, R., Wright, C., and Farndon, J.J. 1992. Epidermal growth factor receptor and other oncogenes as prognostic markers. *Natl. Cancer Inst. Monogr.* 11:181–187.

Hayat, M.A., 2002. *Microscopy, Immunohistochemistry, and Antigen Retrieval Methods.* New York and London: Kluwer Academic/Plenum Publishers.

Heimberger, A.B., Learn, C.A., Archer, G.E., McLendon, R.E., Chewning, T.A., Tuck, F.L., Pracyk, J.B., Friedman, A.H., Friedman, H.S., Bigner, D.D., and Sampson, J.H. 2002. Brain tumors in mice are susceptible to blockade of epidermal growth factor receptor (EGFR) with the oral, specific, EGFR-tyrosine kinase inhibitor ZD1839 (Iressa). *Clin. Cancer Res.* 8: 3496–3502.

Humphrey, P.A., Wong, A.J., Vogelstein, B., Zalutsky, M.R., Fuller, G.N., Archer, G., Friedman, H.S., Kwatra, M.M., Bigner, S.H., and Bigner, D.D. 1990. Amplification and expression of the epidermal growth factor receptor gene in human glioma xenografts. *Proc. Natl. Acad. Sci. USA* 87:4207–4211.

Hynes, N.E., and Stern, D.F. 1994. The biology of erbB-2/ neu/HER-2 and its role in cancer. *Biochem. Biophys. Acta. 1198:*165–184.

Kari, C., Chan, T.O., Rocha de Quadros, M., and Rodeck, U. 2003. Targeting the epidermal growth factor receptor in cancer. *Cancer Res.* 6:1–5.

Khazaie, K., Schirrmacher, V., and Lichtner, R.B. 1993. EGF receptor in neoplasia and metastasis. *Cancer Metastasis Rev.* 12:255–274.

Klijn, J.G., Berns, P.M., Schmitz, P.I., and Foekens, J.A. 1992. The clinical significance of epidermal growth factor receptor (EGF-R) in human breast cancer: A review on 5232 patients. *Endocrin. Rev. 13:*3–17.

Lemoine, N.R., Barnes, D.M., Hollywood, D.P., Hughes, C.M., Smith, P., Dublin, E., Prigent, S.A., Gullick, W.J., and Hurst, H.C. 1992. Expression of the ErbB3 gene product in breast cancer. *Br. J. Cancer 66:*1116–1121.

Libermann, T.A., Nusbaum, H.R., Razon, N., Kris, R., Lax, I., Soreq, H., Whittle N., Waterfield M.D., Ullrich, A., and Schlessinger, J. 1985. Amplification, enhanced expression and possible rearrangement of EGF receptor gene in primary human brain tumors of glial origin. *Nature 313:*144–147.

Mendelsohn, J. 2002. Targeting the epidermal growth factor receptor for cancer therapy. *J. Clin. Oncol. 20(Suppl 18):*1S–13S.

Mendelsohn, J., and Baselga, J. 2000. The EGF receptor family as targets for cancer therapy. *Oncogene 19:*6550–6565.

Moscatello, D.K., Holgado-Madruga, Marina, Godwin, A.K., Ramirez, G., Gunn, G., Zoltick, P.W., Biegel, J.A., Hayes, R.L., and Wong, A.J. 1995. Frequent expression of a mutant epidermal growth factor receptor in multiple human tumors. *Cancer Res. 55:*5536–5539.

Moscatello, D.K., Montgomery, R.B., Sundareshan, P., McDanel, H., Wong, M.Y., and Wong, A.J. 1996. Transformational and altered signal transduction by a naturally occurring mutant EGF receptor. *Oncogene 13:*85–96.

Nicholson, S., Sainsbury, J.R., Halcrow, P., Chambers, P., Farndon, J.R., and Harris, A.L. 1989. Expression of epidermal growth factor receptors associated with lack of response to endocrine therapy in recurrent breast cancer. *Lancet 1:*182–185.

Nishikawa, R., Ji, X.D., Harmon, R.C., Lazar, C.S., Gill, G.N., Cavenee, W.K., and Huang, H.J. 1994. A mutant epidermal growth factor receptor common in human glioma confers enhanced tumorigenicity. *Proc. Natl. Acad. Sci. USA 91:*7727–7731.

Ogiso, H., Ishitani, R., Nureki, O., Fukai, S., Yamanaka, M., Kim, J.H., Saito, K., Sakamoto, A., Inoue, M., Shirouzu, M., and Yokoyama, S. 2002. Crystal structure of the complex of human epidermal growth factor and receptor extracellular domains. *Cell 110:*775–787.

Olapade-Olaopa, E.O., Moscatello, D.K., MacKay, E.H., Horsburgh, T., Sandhu, D.P., Terry, T.R., Wong, A.J., and Habib F.K. 2000. Evidence for the differential expression of a variant EGF receptor protein in human prostate cancer. *Br. J. Cancer 82:*186–194.

Ozanne, B., Richards, C.S., Hendler, F., Burns, D., and Gusterson, B. 1986. Overexpression of the EGF receptor is a hallmark of squamous cell carcinomas. *J. Pathol. 149:*9–14.

Plowman, G.D., Culouscou, J.-M., Whitney, G.S., Green, J.M., Carlton, G.W., Foy, L., and Neubauer, M.G. 1993. Ligand-specific activation of HER4/p180erbB-4, a fourth member of the epidermal growth factor receptor family. *Proc. Natl. Acad. Sci. USA 90:*1746–1750.

Prenzel, N., Zwick, E., Leserer, M., and Ullrich, A. 2000. Tyrosine kinase signalling in breast cancer. Epidermal growth factor receptor: Convergence point for signal integration and diversification. *Breast Cancer Res. 2:*184–190.

Riese II, D.J. and Stern, D.F. 1998. Specificity within the EGF family/ErbB receptor family signaling network. *BioEssays 20:*41–48.

Salomon, D.S., Brandt, R., Ciardiello, F., and Normanno, N. 1995. Epidermal growth factor-related peptides and their receptors in human malignancies. *Crit. Rev. Oncol. Hematol. 19:*183–232.

Scher, H.I., Sarkis, A., Reuter, V., Cohen, D., Netto, G., Petrylak, D., Lianes, P., Fuks, Z., Mendelsohn, J., and Cordon-Cardo, C. 1995. Changing pattern of expression of the epidermal growth factor receptor and transforming growth factor alpha in the progression of prostatic neoplasms. *Clin. Cancer Res. 1:*545–550.

Slichenmyer, W.J., Elliott, W.L., and Fry, D.W. 2001. CI-1033, a pan-erbB tyrosine kinase inhibitor. *Semin. Oncol. 28:*80–85.

Tang, C.K., Gong, X.Q., Moscatello, D.K., Wong, A.J., and Lippman, M.E. 2000. Epidermal growth factor receptor vIII enhances tumorigenicity in human breast cancer. *Cancer Res. 60:*3081–3087.

Torrance, C.J., Jackson, P.E., Montgomery, E., Kinzler, K.W., Vogelstein, B., Wissner, A., Nunes, M., Frost, P., and Discafani, C.M. 2000. Combinatorial chemoprevention of intestinal neoplasia. *Nat. Med. 6:*1024–1028.

Ushiro, H., and Cohen, S. 1980. Identification of phosphotyrosine as a product of epidermal growth factor-activated protein kinase in A-431 cell membranes. *J. Biol. Chem. 255:*8363–8365.

Ward, C.W., Hoyne, P.A., and Flegg, R.H. 1995. Insulin and epidermal growth factor receptors contain the cysteine repeat motif found in the tumor necrosis factor receptor. *Proteins 22:*141–153.

Waterman, H., and Yarden, Y. 2001. Molecular mechanisms underlying endocytosis and sorting of ErbB receptor tyrosine kinases. *FEBS Lett. 490:*142–152.

Wells, A. 2000. Tumor invasion: Role of growth factor-induced cell motility. *Adv. Cancer Res. 78:*31–101.

Wikstrand, C.J., Hale, L.P., Batra, S.K., Hill, L., Humphrey, P.A., Kurpad, S.K., McLendon, R.E., Moscatello, D., Pegram, C.N., Reist, C.J., Traweek, T., Wong, A.J., Zalutsky, M.R., and Bigner, D.D. 1995. Monoclonal antibodies against EGFRvIII are tumor specific and react with breast and lung carcinomas and malignant gliomas. *Cancer Res. 55:*3140–3148.

Wong, A.J., Ruppert, J.M., Bigner, S.H., Grzeschik, C.H., Humphrey, P.A., Bigner, D.S., and Vogelstein, B. 1992. Structural alterations of the epidermal growth factor receptor gene in human gliomas. *Proc. Natl. Acad. Sci. USA 89:*2965–2969.

Yaish, P., Gazit, A., Gilon, C., and Levitzki, A. 1988. Blocking of EGF-dependent cell proliferation by EGF receptor kinase inhibitors. *Science 242:*933–935.

Yamazaki, H., Fukui, Y., Ueyama, Y., Tamaoki, N., Kawamoto, T., Taniguchi, S., and Shibuya, M. 1988. A deletion mutation within the ligand binding domain is responsible for activation of epidermal growth factor receptor gene in human brain tumors. *Mol. Cell Biol. 8:*1816–1820.

19

Alterations of the Cell Cycle Regulating Proteins in Invasive Breast Cancer

Correlation with Proliferation, Apoptosis, and Clinical Outcome

N.J. Agnantis, A. Goussia, and P. Zagorianakou, M. Bai

Introduction

Tissue homeostasis is maintained by regulation of cell proliferation and apoptosis. This regulation may be achieved, in part, by coupling the process of cell cycle progression and apoptosis, using and controlling a shared set of factors including, among others, the tumor suppressor genes *p53* and *Rb* (Lundberg *et al.*, 1999; Sherr, 2000). The process of oncogenesis appears to be related to the impairment of various regulators of cell cycle apoptosis, and it has been a part of considerable progress in the understanding of how these disorders contribute to malignant phenotype (Evan *et al.*, 2001).

Once a quiescent cell has been stimulated to proliferate, there are a number of checkpoints that regulate its commitment to enter and progress through the cell cycle. There are two major checkpoints in the active cell cycle: one is in late G1-phase and the other one is at the G2/M transition. Passage through these checkpoints is controlled by a family of serine/threonine protein kinases, the cyclin-dependent kinases (CDKs), and temporal order is maintained by the successive accumulation of their regulatory subunits, the cyclins. The discovery that CDKs orchestrate key cell cycle transitions was recognized by the 2001 Nobel Prize in Physiology or Medicine, awarded to Leland Hartwell, Paul Nurse, and Tim Hunt (Enders, 2002). The periodic association of different cyclins with different CDKS has been shown to drive different phases of the cell cycle. The earliest players in this scheme are the D-type cyclins acting in conjunction with two related kinases, CDK4 and CDK6, and driving cells through the mid–G1-phase. Subsequent events at the

Handbook of Immunohistochemistry and *in situ* Hybridization of Human Carcinomas, Volume 1: Molecular Genetics; Lung and Breast Carcinomas

425

Copyright © 2004 by Elsevier (USA)
All rights reserved.

G1/S transition appear to be dictated by the cyclin E-CDK2 complex, whereas cyclin A-CDK2 controls entry into S-phase and cyclin B-CDK2 drives the G2/M transition. The understanding of control of the cell cycle has become more complete with the demonstration of specific inhibitors of several cyclin-CDKs complexes. Although the exact role of these inhibitors has not been fully explored, there is evidence that some of them are implicated in cell cycle arrest in response to deoxyribonucleic acid (DNA) damage and transforming growth factor-β (TGF-β) and are considered as potential tumor suppressor genes, because losing or reducing their inhibitory effects would presumably promote cell cycle progression (Datto *et al.*, 1995).

Cyclins are proteins that play a relevant role in cell cycle progression. According to their sequence, expression pattern, and their activity, they exert a regulatory function in the cell cycle. The cyclins bind to CDKs using a conserved 87-amino acid sequence, termed the *cyclin box* (Lees *et al.*, 1993). Currently, 10 different classes (designated as A–J) and a number of subclasses of cyclins exist. The D-type cyclins (D1, D2, and D3) form a distinct subgroup of the mammalian cyclin family, sharing ~60% pairwise identity throughout their amino acid sequences. Several lines of evidence indicate that deregulated expression of D-cyclins can contribute significantly in oncogenesis, both in model systems and in major human cancers (Notokura *et al.*, 1993).

Cyclin D1 is encoded by the *CCND1* gene on chromosome 11q13, which has been identified as the PRAD1 protooncogene and the most likely candidate for the *Bcl-1* protooncogene (Notokura *et al.*, 1991). It has been reported that cyclin D1 is a putative protooncogene because of its capacity to transform fibroblasts together with activated H-ras. Cyclin D1 activation was first observed in parathyroid adenomas, as a result of chromosomal inversion involving its locus at 11q13 under the control of the parathyroid hormone gene on 11p15 (Notokura *et al.*, 1991). However, the most frequent mechanism of cyclin D1 activation is gene amplification and protein overexpression. Indeed, cyclin D1 overexpression has been identified in a number of tumors, such as breast and thyroid cancers, as well as lymphomas. The cyclin D2 (*CCND2*) gene is located on chromosome 12p13. Accumulation of cyclin D2 contributes to sequestration of the cell cycle inhibitor P27[KIP1] and to cell cycle entry (Perez-Roger *et al.*, 1999). Aberrant expression of cyclin D2 has been noted in human ovarian granulosa cell tumors and testicular germ cell tumor cell lines, and recently it has been suggested that cyclin D2 is a direct target of c-myc in human fibroblasts (Bouchard *et al.*, 1999; Sicinski *et al.*, 1996). The cyclin D3 gene (*CCND3*) is

located on chromosome 6p21. An interaction between cyclin D3 and the cell death enzyme caspase 2 in yeast and mammalian fibroblasts has been reported, suggesting a connection between proliferation and apoptotic pathways (Mendelsohn *et al.*, 2002). At the transcriptional level, all three cyclin D genes become activated when quiescent cells are stimulated with cytokines, although the timing and nature of the responses seem to vary among different cell types. However, the general impression is that once the genes are maximally activated in the late G1-phase, their expression remains relatively constant throughout the remainder of the cycle.

The D-type cyclins can interact with several CDKs *in vitro* and many of the same complexes can be detected in cultured cells. However, the preferred partners, both *in vivo* and *in vitro*, are two closely related kinases, designated as CDK4 and CDK6, respectively (Bates *et al.*, 1994). They also appear to be associated only with the D-type and not with other members of the cyclin family. The major targets of cyclin D-CDK4 and cyclin D-CDK6 complexes are the members of the retinoblastoma protein (pRb) family: p105, p130, and p107. Cyclin D-CDK4 and cyclin D-CDK6 complexes bind to Rb and drive Rb inactivation through phosphorylation (Kato *et al.*, 1993). Retinoblastoma protein in the hyperphosphorylated form allows the progression of cell cycle through G1- to S-phase. It has been shown that pRb in turn has a regulatory effect on the levels of cyclin D1, as a part of a feedback mechanism. Hypophosphorylated pRb stimulates high cyclin D1 expression allowing pRb phosphorylation. It has been reported that cyclin D1 participation on pRb phosphorylation may be its only important physiologic function. It has been suggested that cyclin D1 is involved in cell cycle regulation through interactions with other cell cycle–related proteins, such as proliferating cell nuclear antigen (PCNA) and p21[CIP1] (de Jong *et al.*, 1999; Lallemand *et al.*, 1999).

Another G1 cyclin is cyclin E. In proliferating cells, the expression of cyclin E is maximal at the G1/S transition and cyclin E joins with CDK2 to regulate the above transition (Morgan, 1995). Knockin experiments in mice have shown that cyclin E can fully rescue cyclin D1–deficient phenotypes (functional redundancy) and skip the G1-phase to promote premature entry into the S-phase. Shortening of the G1-phase is one of the most prominent features of cyclin E overexpression and recent evidence suggests that premature entry into the S-phase, because of a shortened G1-phase, might cause chromosome segregation defects in mitosis. Cyclin E is both a regulator and a target of the E2F transcription family, and it has been suggested that the inactivation of Rb requires the sequential action of the cyclin D1-CDK4/CDK6 and cyclin E-CDK2 complexes (Lundberg *et al.*, 1999). Once cells enter the S-phase, cyclin E is degraded and the activation of CDK2 is taken over by cyclin A. Unlike G1 cyclins, cyclins A (A1, and A2 known as A) and cyclins B (B1 and B2) achieve maximal levels later in the cycle. Cyclin A is involved in DNA replication in the S-phase and G2/M transition. Cyclin A-CDK2 complex binds to the Rb regulator E2F and phosphorylates one of the heterodimeric components precluding DNA binding. It is evident that the activation of CDK2 by cyclin A is necessary for the continuation of S-phase, but toward the end of this stage cyclin A activates another kinase (probably a CDK). It signals the completion of S-phase and the onset of G2-phase. The activities of cyclin B are essential for G2 transition and mitosis. Overexpression of this cyclin may have an important role in G2 checkpoint control and immortalization of human fibroblasts.

The genes that encode CDK inhibitors have been identified and are designated as CDKI genes (Palazzo, 2001). Their function appears to be the formation of stable complexes that inactivate the catalytic units composed of a cyclin and a CDK, thereby arresting the cell cycle. Loss or inactivation of such inhibitors can result in uncontrolled growth; therefore, CDKIs are candidate tumor suppressor genes. The CDKIs identified in mammalian cells are classified into two major categories: those of the INK4 family (p15, p16, p18, and p19) and those of the CIP/KIP family which includes p21, p27, and p57. The INK4 proteins bind to CDK4 or CDK6 kinases and prevent D-type cyclin binding and activation. This binding blocks a cascade of events that leads to cell proliferation. The CIP/KIP family of inhibitors disrupt the function of all known cyclin-CDKs complexes *in vitro*.

The *p16* gene (CDKN2A, MTS1, or INK4A) maps to chromosome 9p21, a locus commonly deleted in a variety of human malignancies. Inactivation of the *p16^INK4a* gene, commonly by deletion, results in uninhibited phosphorylation of the pRb by the cyclin D-CDK4 and cyclin D-CDK6 complexes, and subsequently to uncontrolled cell growth (Elledge *et al.*, 1994). Thus, loss of cell cycle control may result from altered expression of any of these genes: loss of *p16^INK4a* expression, overexpression/amplification of CDK genes, or loss of Rb function. The *p15^INK4b* gene (CDKN2B, MTS2, or INK4B) is highly related to *p16^INK4a*, has a similar biochemical activity, and maps very closely to *p16* gene on chromosome 9p21. It has also been shown that the *p15^INK4b* gene is a potential effector of TGF-β that induces cell cycle arrest (Hannon *et al.*, 1994).

The *p21^CIP1* gene is an important intermediary molecule because it is under the transcriptional control of the *p53* gene. It has been found to be induced after γ-irradiation in a *p53*-dependent manner (Dulic *et al.*, 1994). This induced *p21^CIP1* protein was associated with cyclin E-CDK2 complex, inhibited its kinase activity, and thereby cell cycle progression. Increased binding of *p21^CIP1* to cyclin E-CDK2 complex contributes to the mechanism of action for the growth, suppression, or inhibition of *BRCA1* (a gene important in familial breast and ovarian cancer), of TGF-β in head and neck tumors, and of concanavalin A and phorbol 13-myristate 12-acetate in murine T lymphoma cell lines (Dennellan *et al.*, 1999). There is also a *p53*-independent activation of p21^CIP1, and this pathway appears to be inducible by a variety of growth factors and differentiating agents. Although p21^CIP1 induction has also been observed in p53-mediated apoptosis, there is no evidence that this protein is actually a mediator of apoptosis. Moreover, it has been found that p21^CIP1 associates with PCNA and that this association results in the inhibition of DNA polymerase δ processivity *in vitro*, thereby providing another pathway for cell cycle arrest mediated by p21^CIP1 (Li *et al.*, 1994). Furthermore, the association of p21^CIP1 expression with cyclin D1 overexpression suggests that p21^CIP1 could be involved in cyclin D1 modulation *in vivo*.

Current theories propose p27^KIP1 to be a central signal that coordinates the varied inputs from the extracellular environment and serves as a threshold for progression to S-phase or for exit from the cell cycle. p27^KIP1 product is present in quiescent cells and its levels are decreased when cells are stimulated by growth factors. It regulates progression through the cell cycle by binding to and inactivating the cyclin E-CDK2 complexes (Orend *et al.*, 1998). Recent evidence points out a role for p27^KIP1, as well as for p57^KIP2 protein inhibitor, in the inactivation of cyclin D-CDK4 and cyclin D-CDK6 complexes (Gillett *et al.*, 1999). Ancillary pathways using p27^KIP1, such as E-cadherin–induced contact inhibition, have been discovered (St. Croix *et al.*, 1998). Also of note is the demonstrated ability of p27^KIP1 to act not only as an inhibitor but also as a target of cyclin E-CDK2 complex. Specific mutations in the putative suppressor gene *p27^KIP1* have only rarely been reported in human cancer. Post-transcriptional control through the ubiquitin-proteasome pathway is thought to be the main process involved in the decreased expression observed in many cancers. An association between decreased p27^KIP1 product and worse outcome in various human malignancies, including breast cancer, has been shown (Chappuis *et al.*, 2000).

In normal cells, the transition through the G1 restriction point is negatively regulated by the pRb. The pRb exists in several phosphorylation states depending on the phase of the cell cycle. Its hypophosphorylated form (predominantly found in G0 cells and in cells that are arrested in G1) seems to act as a growth suppressor, whereas in its phosphorylated state it looses this activity. The action of pRb depends partly on its interaction with transcription factors. The best characterized of these interactions are those involving the E2F complexes. Hypophosphorylated pRb binds to E2F, which is important for the transition from G1- to S-phase. Both pRb and E2F form a complex in the G1-phase preventing the transactivation of some growth-regulating genes. As the cycle proceeds, pRb becomes hyperphosphorylated and looses its ability to produce a complex with E2F. E2F complexes are released allowing the transcription function of pRb to initiate DNA duplication machinery and G1/S transition. The phosphorylation of pRb is achieved by the cyclin D-CDK4/CDK6 kinase complex in mid–G1-phase, and probably by cyclin E-CDK2 in late G1-/S-phase (Kato *et al.*, 1993). Inactivation of pRb function can be achieved directly by mutational events or indirectly by disrupting the regulatory mechanisms that control cyclin D-CDK4/CDK6 kinase. Overexpression of cyclin D1 through amplification of 11q13 is documented in human cancer, and this overexpression could lead to an activation of the cyclin D1-CDK4/CDK6 kinase with subsequent phosphorylation and inactivation of pRb (Kato *et al.*, 1993). An additional level of regulation is provided by the INK4 family of CDKIs. These homologous proteins inhibit the kinase activity of CDK4/CDK6 by preventing its interaction with the activated cyclin D. The integrity of this regulatory cascade, termed the *Rb pathway* (INK4-CDK4/CDK6/cyclin D-Rb-E2F), appears to be compromised in most human cancers. This inactivation can be achieved by an increased activity of cyclin-CDKs complexes, by inactivation of the tumor suppressor genes INK4A or Rb or by both (Lundberg *et al.*, 1999).

The *p53* gene is the most commonly mutated gene in human cancers. The gene encodes a 53 kD nuclear protein that is a negative regulator of the G1-/S-phase transition in the cell cycle. The wild-type *p53* has a variety of biochemical activities, including transcriptional activation, DNA and ribonucleic acid (RNA) reannealing, and effects on DNA replication. A current model of *p53* function is that it is activated when DNA damage occurs and serves to block DNA replication, hence allowing time for DNA repair. Its presence is also able to induce a delay in the G1/S transition in cells with DNA damage, making DNA repair possible or inducing apoptosis in irreversibly damaged cells. A direct link with the cell cycle came with the demonstration that one of the main transcriptional targets of *p53* is the p21^{CIP1}. The target p21^{CIP1} is a wild-type

p53-inducible protein, which is involved in inhibition of the cell cycle progression through binding to cyclin E-CDK2 complexes. It seems that p53 is activated after DNA damage increases p21^{CIP1}. The transcription, which in turn inhibits cyclin E-CDK2, impedes progression along the cell cycle in the G1/S transition and S-phase. Therefore, one consequence of p53 inactivation is a dramatic reduction in the level of p21^{CIP1} and concomitant hyperactivity of cyclin E-CDK2 complex, which leads to uncontrolled cell division.

In tumor cells that have either lost the p53 protein or contain an altered form of p53, the p21^{CIP1} levels are reduced or the protein is absent. This could result in abnormal DNA replication control or loss of coordination between DNA replication and the cell cycle progression. Both could lead to genome instability, which is directly related to oncogenesis. Thus, p21^{CIP1} is an important downstream effector in the p53-specific pathway and its inactivation might potentially lead to tumor progression. Conversely, it has been shown that p21^{CIP1} expression also can be regulated independently of p53 expression, indicating that alternative mechanisms are involved in p21^{CIP1} inactivation. Interestingly, mutations of the *p21^{CIP1}*-encoded gene were not detected in many human malignancies. One of the target genes of *p53* is murine double minute 2 (*MDM2*) gene. The interaction of *MDM2* with *p53* was found to conceal its transcriptional activation domain and to inhibit *p53*-mediated transcription (Oliner *et al.*, 1993). In response to DNA damage, the *p53-MDM2* interaction is disrupted by phosphorylation or is antagonized by p14^{ARF} induction, respectively, resulting in the activation of *p53* (Pomerantz *et al.*, 1998). This "p53 pathway" is also frequently altered in most human cancers, either by inactivation of the tumor suppressor genes *p14^{ARF}* or *p53* or by hyperactivation of the *MDM2* gene (Sherr, 2000).

Many studies in human tumors indicate that genes that induce cell cycle progression also activate apoptosis. Thus, cells that enter the cell cycle die unless their program is commuted in some way (Lundberg *et al.*, 1999). It has been suggested that this obligatory coupling of cell proliferation with apoptosis provides a potent innate mechanism that suppresses neoplasia (Evan *et al.*, 2001).

Abnormalities of the positive and negative regulations of cell cycle are common in breast carcinomas. Among the most frequent alterations are *p53* mutation and overexpression, cyclin D1 and cyclin E overexpression, methylation, and decrease expression of *p16^{INK4a}* and *p27^{KIP1}* (Barbareschi, 2002; Geradts *et al.*, 2000). Currently, the roles of the major regulators in cell cycle progression in breast carcinomas remain unclear, and their prognostic significance is far from conclusive.

The aim of this chapter is to review the main findings of recent immunohistochemical and molecular studies, including our experience, concerning the alterations of the proteins involved in cell cycle regulation and their correlation with the proliferation profile, the apoptosis, the main prognostic indicators, and the clinical behavior in primary invasive breast cancer.

MATERIALS

1. Xylene.
2. Absolute alcohol.
3. Alcohol 96%: 480 ml absolute alcohol and 20 ml double-distilled water (DDW) to make 500 ml.
4. Tris-buffered saline (TBS): Contains Tris-base, Tris-HCl, and NaCl; bring vol to 1 L with DDW (pH 7.4).
5. 0.01 M citric acid antigen retrieval solution: 0.01 citric acid, 1N NaOH (pH 6.0).
6. 1% Bovine serum albumin (BSA): 1 g BSA and 100 ml TBS; add BSA to buffer with stirring.
7. 3% Hydrogen peroxidase (H_2O_2): 1 ml of 30% H_2O_2 and 99 ml methanol.
8. Primary antibodies diluted in TBS with 1% BSA: Anti-p53 protein (DO-7; dilution 1:50), p21/waf1 protein (EA-10; dilution 1:50; Calbiochem, San Diego, CA), anti-Rb (Ab-5, LM951; dilution 1:50; Oncogene, Carpinteria, CA), anti-cyclin D1 (DCS-6; dilution 1:20; Novocastra), anti-Ki67 (MIB-1; dilution 1:20; Immunotech), anti-cyclin A (6E6; dilution 1:10; Novocastra), cyclin B1 (7A9, dilution 1:10; Novocastra), anti-p27 (IB4; dilution 1:20; Novocastra), anti-Bcl-2 (M0887; dilution 1:40; Dako), anti-Bclx (M4512, A35-10, dilution 1:500; Dako), anti-bak (A3538, N/A, dilution 1:500; Dako), anti-bax (A3533, N/A, dilution 1:250; Dako).
10. Diaminobenzidine (DAB) tetrahydrochloride chromogen; add 20 µl DAB chromogen in 1 ml substrate buffer.
11. Mounting media.

METHOD

1. Place the sections in xylene for 5 min, twice for deparaffinazation.
2. Carry the sections through a series of graded alcohols (absolute alcohol; 96% alcohol) for 5 min, twice, for rehydration.
3. Rinse the sections in running tap water for 2 min.
4. Rinse the sections in DDW.
5. Place the sections, except those for PCNA detection, in a Coplin jar of microwave-compatible plastic, filled with preheated 0.01 M citric acid antigen retrieval solution.
6. Heat the jar in a microwave oven at 300W for 15 min.
7. Check the antigen retrieval fluid levels and add 0.01 M citric acid antigen retrieval solution, if necessary. Place the jar in the microwave oven for an additional heating cycle of 15 min.
8. Allow the sections to cool at room temperature for 20 min.
9. Rinse the sections in DDW.
10. Incubate the sections with 3% H_2O_2 for 30 min to block endogenous peroxidase.
11. Rinse the sections in DDW 3 times.
12. Place the sections in TBS for 10 min.
13. Incubate the sections with primary antibodies diluted in TBS containing 1% BSA, at 4°C overnight.
14. Rinse the sections with TBS and put them in TBS for 10 min.
15. Incubate the sections with EnVision/horseradish peroxidase (HRP) for 30 min at room temperature.
16. Place the sections in TBS, as in step 13.
17. Incubate the sections with DAB chromogen solution for color development. Check the slides under optic microscope for the staining color development.
18. Rinse the sections in DDW 3 times.
19. Place the sections in 10% Harris hematoxylin in DDW for 2 min for counterstaining.
20. Rinse the sections in running tap water.
21. Place the sections in 96% alcohol for 5 min, twice, and subsequently twice in absolute alcohol for 5 min each.
22. Mount the sections with mounting media. They are now ready for final examination in the light microscope.

RESULTS AND DISCUSSION

There is increasing evidence that immunohistochemical analysis of the complex molecular networks regulating the cell cycle in malignancies provide valuable information for the understanding of the impaired regulation of these networks and the biologic characteristics of the tumor (Bai *et al.*, 2001a; Kanavaros *et al.*, 2001).

Proliferation Profile in Breast Cancer

It is well recognized that the proliferative activity of the neoplastic cells correlates with the biological behavior of certain malignancies. In our study, the proliferation profile was determined by the expression of Ki67, cyclin A, and cyclin B1 proteins. Expression of these proteins was observed in all cases. A positive correlation between cyclin A and cyclin B1 was found, whereas Ki67 expression was positively associated with the expression of both cyclins. Greater levels of proliferation-related proteins were significantly more apparent in poorly differentiated tumors. Previous studies investigating cyclin A expression in breast carcinomas demonstrated similar associations of this protein with tumor proliferation (Collecchi *et al.*, 2000). Ki67 growth fraction has been associated with tumor size and ploidy, high histologic grade, lymph node involvement, and negative estrogen and progesterone status (Spyratos *et al.*, 2002; Steck *et al.*, 1999). In addition, an association between Ki67 expression and patient survival has also been demonstrated (Mirza *et al.*, 2002). Moreover, overexpression of cyclin A has been strongly related with early relapse and reduced survival (Bukholm *et al.*, 2001). In contrast, our study, confirming other reports (Pinto *et al.*, 2001), did not show similar correlations.

Alterations of the Proteins Involved in the Retinoblastoma Pathway in Breast Cancer

The tumor suppression function of Rb is inactivated by phosphorylation during cell cycle progression, mutation, sequestration by viral proteins, and degradation from caspases in response to apoptotic stimuli. In breast carcinomas and in other sporadic tumor types, Rb can be inactivated by targeting pathways that regulate the pRb phosphorylation, because Rb structural alterations rarely have been detected (T'Ang *et al.*, 1998). Loss of heterozygosity of Rb has been found in 25% of human breast carcinomas, but this abnormality does not always correlate with decreased protein expression (Borg *et al.*, 1992).

The distribution of Rb protein in normal breast tissue follows a pattern that is parallel to that of Ki67 expression. In our study, pRb expression was found in all cases. We observed a reduced expression of pRb in 13.9% of the cases, using parallel evaluation of proliferation activity (determined by Ki67 expression) because the absence or the low expression of pRb could be normal in cells that are in the G0- early G1-phase. Absence or low pRb expression has been observed in 10–45% of breast cancers, and this decreased expression has been considered to be important for the neoplastic progression, because it was associated with increased cell proliferation (Dublin *et al.*, 1998; Nielsen *et al.*, 1997). Ceccarelli *et al.* (1998) report a correlation of pRb expression with proliferative activity, suggesting that in breast carcinomas pRb behaves normally in regulating the cell cycle. Our study confirmed this finding, in that we revealed a significant positive correlation between pRb and Ki67 and cyclin A and B1 protein expression. Using immunohistochemistry, Nieslen *et al.*

(1997) reported a relation between loss of pRb immunoreactivity and highly proliferating tumors, and they raised the hypothesis that Ki67expression is related to hyperphosphorylated form of the protein. In the same study, using the Western Blot analysis, they demonstrated an accumulation of the hypophosphorylated protein. Most antibodies used in immunohistochemistry bind to pRb, independently of the phosphorylation status and the presence of point mutations. Therefore, a positive staining does not necessarily reflect a functional pRb.

The evaluation of the phosphorylation status has been proposed as a prognostic factor in breast carcinomas. The available results are contradictory regarding the relation of pRb expression with prognostic clinicopathologic parameters. Dublin *et al.* (1998) found that loss of pRb was significantly more frequent in tumors with negative ER status. In addition, a trend for tumors negative for pRb to be grade III was found. However, these authors failed to demonstrate a relation between pRb aberration and clinical outcome. In contrast, Ceccarelli *et al.* (1998) emphasized the finding of a high pRb expression in poorly differentiated tumors associated with loss of estrogen receptor (ER) and progesterone receptor (PR) content (The role of ER in breast carcinoma is also discussed in Part IV, Chapter 12). When cases were stratified by the number of metastatic lymph nodes, it was evident that all tumors with massive lymph node involvement were expressing pRb irrespective of their proliferative activity. Our study did not demonstrate any correlation between pRb and the histologic grade or the clinical outcome.

D-type cyclins are expressed differently in various cell types and some cells express all types of D-cyclins, whereas other cells express only one or two of them (Baldin *et al.*, 1993). These cyclins are implicated in cell cycle regulation and differentiation. Human mammary epithelial cell lines and normal breast epithelium express, at different levels, all three D-type cyclins (D1, D2, and D3) (Gillett *et al.*, 1996).

Overexpression of cyclin D1 has been found in 42–50% of invasive breast cancer cases (Barbareschi *et al.*, 1997; de Jong *et al.*, 1999; Geradts *et al.*, 2000; Gillett *et al.*, 1996; Michalides *et al.*, 1996). Similarly, overexpression of cyclin D1 was detected in 39.3% of our cases. However, the frequency of amplification of 11q13, where cyclin D1 is located, was only 10–24% (Barbareschi *et al.*, 1997). The discrepancies between the percentage of cases with cyclin D1 overexpression and cases with cyclin D1 gene amplification suggest that mechanisms other than gene amplification may be responsible for aberrant cyclin D1 expression in human breast carcinomas.

The biological effect of cyclin D1 overexpression seems to depend on how, when, and on what level this protein is induced. Overexpression of the protein promotes cell progression, which is generally observed as shortened G1 S transition. Cyclin D1 overexpression in late G1-phase can prolong the S-phase and inhibit DNA replication. Stable overexpression of exogenous cyclin D1 complementary DNA (cDNA) in human mammary epithelial cell line HBL-100 has been shown to inhibit cell growth. In our study, cyclin D1 expression was positively correlated with p21^{CIP1} expression and negatively with Ki67, suggesting that high levels of cyclin D1 cause inhibition of entry to S-phase by increasing the p21^{CIP1} protein in breast carcinomas. Previous studies report that cyclin D1 is associated with p21^{CIP1} expression and acts as a p21^{CIP1} inducer (de Jong et al., 1999; Lallemand et al., 1999). In addition, a trend for a positive correlation between cyclin D1 and pRb was revealed in our study. This finding is in keeping with previous data showing a positive correlation between pRb and cyclin D1 expression in breast carcinomas, which supports the finding that pRb can regulate cyclin D1 expression by creating a complex autoregulatory loop (Barbareschi et al., 1997). The cyclin D1 product has also been identified as important in mediating cell cycle growth arrest through the p53 pathway in murin fibroblast cell lines.

Cyclin D1 expression has been consistently correlated with positive ER status in breast carcinomas, and seems to activate transcription on ER-regulated genes (Barbareschi et al., 1997; Michalides et al., 1996). In vitro studies showed that cyclin D1 is up-regulated by estrogens and down-regulated by anti-estrogens (Altucci et al., 1996; Watts et al., 1995). Anti-estrogen-mediated G0/G1 arrest is associated with decreased cyclin D1 expression, inactivation of cyclin D1-CDK4 complexes, and decreased pRb phosphorylation (Carroll et al., 2000; Watts et al., 1995). Moreover, it has been shown that anti-estrogen inhibits the activity of cyclin E-CDK2 complexes and this inhibition is p21-dependent (Carroll et al., 2000). These experimental data showed that cyclin D1 expression can indeed be regulated by several proteins in breast carcinomas.

In our study, cyclin D1 expression was negatively associated with the histologic grade. This finding confirms previous results (Gillett et al., 1996). It is paradoxical that overexpression of cyclin D1 is associated with good prognostic parameters (low proliferative activity, positive ER status, and low histologic grade). A possible explanation is that cyclin D1 may exert its effect on functions other than cell proliferation, probably by the transcriptional activation of ER-regulated genes. As in our study, other reports failed to demonstrate a correlation between cyclin D1 overexpression and clinical outcome (Michalides et al., 1996). However, in some studies, it is actually associated with a good outcome, both in terms of prognosis and response to endocrine treatment (Barnes et al., 1998; Gillett et al., 1996).

The role of cyclins D2 and D3 was less extensively investigated in breast carcinomas. Courjal et al. (1996) analyzed messenger RNA (mRNA) expression of all three cyclins in 132 breast tumors and reported no evidence of cyclin D2 or D3 expression. Bukholm et al. (1998) observed expression of cyclins D2 and D3 in only 4% and 18% of breast cancer cases, respectively. Furthermore, it has been found that cyclin D2 mRNA and protein were absent in the majority of breast cancer cell lines, in contrast to the abundant expression in cultured normal breast epithelial cells (Sweeney et al., 1997), suggesting that cyclin D2 is involved in a vital tumor suppressor function in normal breast tissues and its loss may be related to tumorigenesis. Cyclins D2 and D3 rarely show abnormalities at the DNA level. Hypermethylation of the cyclin D2 promoter is an early and frequent event in breast carcinomas and may result in gene expression silencing and lack of protein expression (Courjal et al., 1996; Evan et al., 2001). In both univariate and multivariate analyses, expression of cyclin D3 was independent of tumor grade or metastases and did not influence the overall survival (Bukholm et al., 2001).

Molecular studies concerning the status of the CDK4/CDK6 and the E2F transcriptional factor in breast carcinomas are scarce. Amplification of the CDK4 gene has been described in 16% of breast cancer cases (An et al., 1999). E2F transcriptional factor is likely to function as tumor suppressor in breast cancer (Ho et al., 2001).

The cell cycle regulatory protein p16$^{INK4\alpha}$ is involved in tumor suppression in the Rb pathway. Little is known about the role of p16$^{INK4\alpha}$ in breast cancer. Decreased or low expression of p16$^{INK4\alpha}$ has been observed in 50% of human breast carcinomas (Geradts et al., 2000). Promoter methylation has been reported as the major mechanism by which its expression can be modulated (Hui et al., 2000; Silva et al., 2003). The methylated p16$^{INK4\alpha}$ gene has been demonstrated in ~20% of breast carcinoma cases (Hui et al., 2000). Homozygous deletion and mutation are rarely observed in breast carcinomas (Silva et al., 2003). Overexpression of p16$^{INK4\alpha}$ has been reported to be related with high tumor grade, lymph node involvement, ER negativity, increased risk for relapse, and poor outcome (Dublin et al., 1998; Hui et al., 2000). The observation of high levels of p16$^{INK4\alpha}$ expression in pRb-negative cells, which is caused by a feedback loop regulated by pRb, suggests that p16$^{INK4\alpha}$ overexpression may be a marker of Rb inactivation (Dublin et al., 1998; Nielsen et al., 1997). Further studies are required to clarify the role of p16$^{INK4\alpha}$ in breast carcinomas.

Alterations of the Proteins Involved in the p53 Pathway in Breast Cancer

That the *p53* gene is involved in breast cancer development was initially recognized in patients with Li Fraumeni syndrome, where germline mutations of the gene were found to be responsible for the disease. Patients with this syndrome had an increased risk for breast cancer development. Subsequent studies showed that *p53* was the most commonly mutated gene in breast carcinomas and pointed to an important role of the gene from the early stages of the carcinogenetic process, because these genetic alterations were also identified in preinvasive breast lesions (Borresen-Dale, 2003). Currently, data from transgenic animal studies, experimental cell *in vitro* studies, and studies in human cell carcinomas support the important role for *p53* in mammary carcinogenesis. The frequency of *p53* mutations in breast cancer is ~30% of cases (Beroud *et al.*, 2003). Mutations can lead to the synthesis of a stable malfunctional protein that is accumulated in tumor cells and can be detected by immunohistochemistry. The correlation between p53 mutation and immunohistochemical expression of the protein seems to be less than 75%, because not all mutations yield a stable protein; conversely, nonmutated protein also can be accumulated in tumors. Antibodies that are used for the immunohistochemical detection of p53 protein recognize an epitope shared by both wild and mutant type.

The protein p21$^{\text{CIP1}}$ is a critical downstream effector in the p53 pathway, and several studies have investigated the relationship and the clinical significance of both proteins in breast cancer. In our study, nuclear accumulation of p53 and p21$^{\text{CIP1}}$ was detected in 45.9% and 22.4% of cases, respectively, and the correlation between them was not significant. The correlation of p21$^{\text{CIP1}}$ with p53 expression is controversial (i.e., some studies show a positive association, whereas others show a negative association [Bankfalvi *et al.*, 2000; Barbareschi *et al.*, 1997]). Combined immunohistochemical evaluation of p53 and p21$^{\text{CIP1}}$ proteins could be helpful to obtain indirect information about the status of the *p53* gene, because only wild type *p53* can induce p21$^{\text{CIP1}}$ protein by arresting the cell cycle at the G1/S checkpoint. We found the following patterns: (a) p53$^+$/p21$^{\text{CIP1}}$– (33.6%); (b) p53$^+$/p21$^{\text{CIP1}}$+ (12.2%); (c) p53$^-$/p21$^{\text{CIP1}}$ + (10.2%); and (d) p53$^-$/p21$^{\text{CIP1}}$ – (43.8%). These findings suggest that in 33.6% of cases p53 protein is mutant, in 12.2% of cases the p53 protein is of wild type and could induce p21$^{\text{CIP1}}$ expression, whereas p21$^{\text{CIP1}}$ expression is induced independently of p53 protein in 10.2% of cases. It is also known that p21$^{\text{CIP1}}$ is induced independently of

p53 in normal and neoplastic tissues—that is, BRCA1-mediated growth arrest operates through p21$^{\text{CIP1}}$ expression.

One of the main functions of p53 uncoupled from the cell cycle regulatory effect is the proliferation of tumor cells. In our study, p53 overexpression was strongly associated with greater Ki67 expression, indicating that impairment of p53 cell cycle pathway is associated with tumor progression in breast carcinomas.

A number of studies have attempted to correlate p53 overexpression with clinical outcome. These data led to mixed conclusions, with most studies suggesting that p53 accumulation affects prognosis adversely, although other studies, like ours, have not confirmed these findings (Allred *et al.*, 1993; Rosen *et al.*, 1995). In a meta-analysis of more than 9000 patients with breast cancer, the prognostic role of p53 appeared weak when data were based on detecting p53 immunohistochemical accumulation (Barbareschi, 2002). However, a strong prognostic significance of p53 mutations detected by using sequencing has been reported in more than 30 studies to date (Pharoah *et al.*, 1999). All these studies have confirmed that p53 mutations confer a worse disease-free and overall survival in patients with breast cancer, and in several of the studies p53 mutations were the single most adverse prognostic indicator of recurrence and death.

Expression of p21$^{\text{CIP1}}$ has been detected in approximately 30–90% of breast cancer series (Caffo *et al.*, 1996; Wakasugi *et al.*, 1997) and this expression was related with large tumor size, high histologic grade, and positive nodal status (Caffo *et al.*, 1996). Contradictory results also exist in which p21$^{\text{CIP1}}$ seems to be associated with favorable parameters such as low histologic grade and positive ER status (Wakasugi *et al.*, 1997). In our study, we did not find correlation of p21$^{\text{CIP1}}$ expression with histologic grade. The prognostic role of p21$^{\text{CIP1}}$ in breast cancer is still unclear and contradictory results exist in the relative studies (Caffo *et al.*, 1996; Gohring *et al.*, 2001; Wakasugi *et al.*, 1997). Gohring *et al.* (2001) found no prognostic significance of p21$^{\text{CIP1}}$ expression in a large series of breast carcinomas; similarly, in our patient samples, p21$^{\text{CIP1}}$ expression does not seem to have a prognostic value. Analysis of the combined p21$^{\text{CIP1}}$ and p53 phenotypes showed that the p53$^+$/p21$^{\text{CIP1}+}$ and p53$^+$/p21$^{\text{CIP1}}$-phenotypes were associated with the worst outcome (Mathoulin-Portier *et al.*, 2000). In patients treated with systemic adjuvant therapy, an analysis showed that p53$^+$/p21$^{\text{CIP1}+}$ tumors were associated with long disease-free and overall survival. Multivariate analysis showed that the p53$^+$/p21$^{\text{CIP1}}$-phenotype was associated with short disease-free and overall survival and it has been suggested that p53/p21$^{\text{CIP1}}$ expression may be of clinical relevance for the therapeutic response to chemotherapy (Caffo *et al.*, 1996).

Accumulation of cyclin E protein in most cases reflects amplification of the gene, although mutations have been detected (Courjal *et al.*, 1996). However, cyclin E is sometimes modified by post-transcriptional mechanisms. The oncogenic role of this cyclin is not yet clear, although it has been reported that mammary carcinomas developed in more that 10% of trangenic female mice (Bortner *et al.*, 1997). The progression from normal and benign breast tissue through *in situ* to invasive carcinoma has been associated with accumulating levels of cyclin E. In invasive breast carcinomas, cyclin E is overexpressed in ~30% of the cases, suggesting that dysregulated expression of the gene may contribute to, rather than being a consequence of, increased cell division. Increased expression of cyclin E has been shown to correlate with high histologic grade, advanced stage, negative ER status, decreased cyclin D1 expression, and short overall survival (Kim *et al.*, 2001; Nielsen *et al.*, 1996).

p14^{INK4} is a cell cycle regulatory protein involved in tumor suppression in the p53 pathway. Because p14^{INK4} promoter is responsive to the E2F-1 transcription, it has been suggested that p14^{INK4} is a plexus between both p53 and Rb pathways. Overexpression of p14^{INK4} results in cell cycle arrest and apoptosis (Hemmati *et al.*, 2002). Alterations of p14^{INK4} are not well understood and most existing data are available from experimental animal or cell lines studies. Inactivation of p14^{INK4} has been reported in breast carcinomas and DNA methylation has been proposed as the most common mechanism of gene inactivation, whereas homozygous deletion and mutation were rare or absent (Silva *et al.*, 2003). In most of the tumors, p14^{INK4} is altered in concordance with p16^{INK4} (Silva *et al.*, 2003).

p27^{KIP1} Alterations in Breast Cancer

The protein p27 belongs to the family of KIP1 inhibitors, which has a broad spectrum of inhibitory activity on different CDKs and may act as a potential tumor suppressor gene. This protein decreases cyclin-CDK activity and induces cell cycle arrest in cell lines. Specific mutations in the *CDKN1/p27/KIP1* gene have rarely been reported in human breast carcinomas and breast carcinoma cell lines. Post-transcriptional mechanisms (ubiquitin-proteasome pathway, translational control, phosphorylation, and subcellular compartmentalization) are involved in decreased p27^{KIP1} protein expression observed in different tumors, including breast cancer (Slingerland *et al.*, 2000). In our study, decreased p27^{KIP1} protein expression was found in

32.6% of the cases. A negative correlation between p27^{KIP1} expression and histologic grade was revealed. In addition, a decreased p27^{KIP1} expression has been correlated with HER-2/neu overexpression (Spataro *et al.*, 2003). We also showed that decreased p27^{KIP1} expression was correlated with a shorter patient survival (Agnantis *et al.*, 2003). Previously, Chappuis *et al.* (2000) showed that low p27^{KIP1} expression was related to negative ER status, and it was a powerful and independent prognostic marker of poor clinical outcome. However, in a large series of patients, p27^{KIP1} protein expression did not predict the patient outcome (Barbareshi *et al.*, 2000).

Apoptosis in Breast Cancer: Correlation with Cell Cycle Proteins

Apoptosis is the most potent natural defence against cancer, because it eliminates premalignant cells that enter the S-phase inappropriately after alterations of restriction point controls. There is now evidence that cell proliferation and apoptosis are the result of a balanced interaction among multiple regulators. The proapoptotic and anti-apoptotic proteins act largely during the G0- and G1-phase of the cell cycle.

Apoptosis is a frequent phenomenon in breast cancer and it can be detected by light microscopy in conventional histopathologic sections. However, TUNEL technique (terminal deoxynucleotidyl transferase–mediated triphosphotase-biotin nick-end-label staining) is the most widely used method to study apoptotic activity, because it enables *in situ* detection of fragmented DNA at the single cell level. In some studies of breast cancer, high apoptotic index (AI) has been related to malignant cellular features and indicators of invasiveness; however, it was not confirmed by other studies. In breast cancer, a high AI has been correlated positively with large tumor size, high histologic grade, and negative hormone receptor content (Berardo *et al.*, 1998; Gonzalez-Campora *et al.*, 2000; Lipponen *et al.*, 1994; van Slooten *et al.*, 1998). A positive correlation between AI and aneuploidy has also been described (Berardo *et al.*, 1998; Lipponen *et al.*, 1994). Breast tumors that overexpress p53 protein exhibit high AI (Berardo *et al.*, 1998; Lipponen *et al.*, 1994). In addition, mitotic index and AI have been found to show a highly significant positive correlation (Lipponen *et al.*, 1994; van Slooten *et al.*, 1998). Other indicators of cell proliferation, such as Ki67, also have been found to be highly related to AI (Bai *et al.*, 2001b; van Slooten *et al.*, 1998). Bcl-2–positive tumors have generally low AI (Lipponen *et al.*, 1994). Multivariate analysis for survival has revealed that apoptosis is an independent factor for patient prognosis (Gonzalez-Campora *et al.*, 2000).

Overexpression of Bcl-2 alone is apparently not sufficient to induce a malignant phenotype. Expression of the protein is detectable in normal breast epithelium reaching its maximal levels at the end of the follicular phase. The exact mechanisms and the effect of the down-regulation of the Bcl-2 expression in breast cancer cells are not clearly defined. In breast carcinomas, Bcl-2 overexpression has been found in 65–80% of invasive ductal carcinomas, in 66% cases of micropapillary carcinomas, and in 2.9% cases of apocrine carcinomas (Krajewski *et al.*, 1995; Leal *et al.*, 2001; Luna-More *et al.*, 2000). In our study, Bcl-2 protein expression was detected in 83.3% of cases, and in 29.7% of them, Bcl-2 was undetectable or was detected at low level (Fig. 63).

In addition, our data showed that low Bcl-2 expression was associated with high proliferation activity as determined by Ki67, cyclin A, and cyclin B1 expression, suggesting that in breast cancer Bcl-2 protein not only mediates cell death but also cell proliferation. A negative correlation between Bcl-2 and Ki67 expression also has been observed (Castiglione *et al.*, 1999). It is evident that this anti-apoptotic protein prolongs survival of the noncycling cells and inhibits survival of the cycling cells. Cyclin cells overexpressing Bcl-2 undergo growth arrest in the G0-/G1- or G2-/M-phases promoting cell survival and oncogenic process (Knowlton *et al.*, 1998). Moreover, Bcl-2 overexpressing breast carcinoma cells showed increased doubling time, decreased S-phase fraction, and increased G0/G1 fraction (Knowlton *et al.*, 1998).

In breast cancer, Bcl-2 expression has been associated with various favorable prognostic factors that, in addition to low proliferative activity, include low histologic grade and high ER and PR levels (Krajewski *et al.*, 1999; Silvestrini *et al.*, 1994). In good accordance with the published data, we also observed greater Bcl-2 expression in better differentiated tumors.

The molecular connections between cell cycle and cell death are not entirely clear, but the p53 pathway seems to be involved. In some cell types, p53 induces apoptosis when overexpressed (Miyashita *et al.*, 1994). This apoptotic program does not depend on $p21^{CIP1}$ and may be partially mediated by Bcl-2. p53 is capable of down-regulating the transcription of Bcl-2 and up-regulating the apoptosis-promoting protein bax (Miyashita *et al.*, 1994). Studies in human breast tumors and cell lines revealed that expression of Bcl-2 and p53

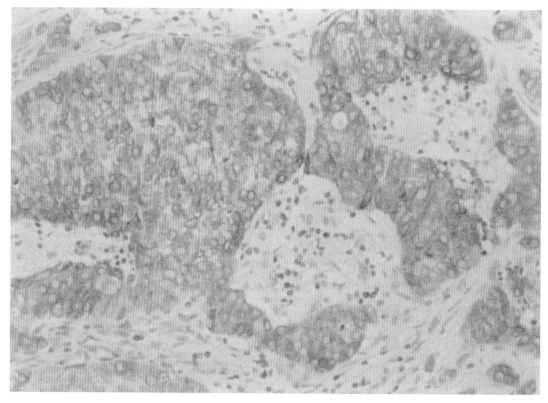

Figure 63 Invasive ductal breast carcinoma. Immunohistochemical expression of Bcl-2 in the majority of tumor cells. Original magnification, 400X.

is usually inversely correlated (Haldar *et al.*, 1994). Evaluating our data as continous variables, we found a trend for an inverse correlation between Bcl-2 and p53 overexpression. Wild-type p53 protein has been found to decrease Bcl-2 protein levels both *in vitro* and *in vivo*, and mutant p53 has been shown to inhibit Bcl-2 expression in some cancer cell lines (Miyashita *et al.*, 1994).

A positive correlation between Bcl-2 and cyclin D1 was found in our study. The role of cyclin D1 in the process of apoptosis in breast cancer is still being investigated. Cyclin D1 induction in breast cancer cells shortens G1-phase and is sufficient for cells arrested in the early G1 to complete the cell cycle. To complete the G1-phase, apoptotic mechanisms theoretically must be suppressed. In contrast, it has been suggested that cyclin D1 overexpression in late G1-phase prolongs the S-phase and inhibits DNA replication, pointing to a growth-restricting function and may be to a stimulatory effect on apoptosis. It is currently unclear which of the above actions of cyclin D1 on apoptosis contributes to breast cancer. However, it is difficult to evaluate and to produce results by studying the importance of individual factors in the apoptotic pathway without investigating many other regulators involved in this complex mechanism.

Data regarding the importance of other proteins of the Bcl-2 family are generally scarce. Expression of Bcl-x, bax, and bak proteins has been detected in all of our cases, and in the majority of them the total number of the cancer cell population was stained. Low levels of bax and high levels of Bcl-x expression were associated with high tumor grade. In addition, a linear positive correlation of bax and Bcl-2 was observed. A similar relation has been revealed by a previous study by Krajewski *et al.* (1995). Bcl-2 and bax are two functionally antagonistic proteins with a definite opposite effect on the apoptotic process. Relevant to the function of several members of the Bcl-2 family is their ability to homodimerize or heterodimerize with each other. Homodimerization or heterodimerization is important for the apoptosis regulating function and especially the dimerization that takes place between bax and Bcl-2. The protein bax promotes apoptosis in its homodimeric form; but after heterodimerizing with Bcl-2, it prevents apoptotic death.

It has been shown that Bcl-2 positivity is associated with a better patient prognosis in terms of a better relapse-free and overall survival (Krajewski et al., 1995, 1999). However, in multivariate analysis, Bcl-2 positivity does not seem to be an independent prognostic factor for either disease-free and overall survival (Lipponen *et al.*, 1995). Statistical analysis of the relationship between Bcl-2 expression and patient outcome showed that in our cases a high Bcl-2 expression detectable in more than 90% of tumor cells was associated with a better disease-free time and overall survival. None of the patients whose tumors overexpressed Bcl-2 died or even relapsed during the follow-up period (Briasoulis *et al.*, 2001).

These results appear paradoxical, because Bcl-2 protein, as an inhibitor of apoptosis, is expected to be associated with an aggressive tumor phenotype and, therefore, it should be a predictive factor of a poor clinical course. One possible explanation is the inhibitory effect of Bcl-2 on cell proliferation. Proliferating cells overexpressing Bcl-2 resist apoptosis and undergo growth arrest in G0-/G1- or G2-/M-phases, which promotes tumor cell survival but does not enhance cell proliferation (Knowlton *et al.*, 1998). The presence of anti-proliferative activity of Bcl-2, which would decrease the rate of cell proliferation, should be another explanation. The inverse correlation between Bcl-2 and Ki67 observed in breast carcinomas supports the latter hypothesis. In addition, it has been reported that estrogens can inhibit apoptosis by increasing the intracellular Bcl-2/bax ratio; therefore, estrogen may be an important regulator of Bcl-2 protein in breast cancer (Huang *et al.*, 1997).

In conclusion, abnormalities in many positive and negative regulators of the cell cycle and apoptosis are frequent in breast carcinomas. Currently, their role in the progression of the disease remains elusive and their prognostic significance is far from conclusive. The heterogeneous nature of breast cancer pathogenesis and the clinical course could be explained, in part, by different and distinctive sets of cell cycle and apoptosis defects. Immunohistochemical analysis of the complex molecular networks regulating the cell cycle and apoptosis in breast carcinomas provides valuable information for the understanding of the impaired regulation of these networks in breast carcinogenesis and progression. Moreover, combined analysis of the cell cycle and the apoptotic proteins may help to select subgroups of breast carcinoma cases with different biological characteristics.

References

Agnantis, N.J., Briasoulis, E., Zagorianakou, P., and Bai, M. 2003. Cyclin D1 and the cell cycle inhibitor p27^{Kip1} in human ductal breast cancer. A clinicopathological study with long term follow up. *Electronic J. Pathol. Histol.* (in press).

Allred, D.C., Clark, G.M., Elledge, R., Fuqua, S.A., Brown, R.W., Chamness, G.C., Osborne, C.K., and McGuire, W.L. 1993. Association of p53 protein expression with tumor or cell proliferation rate and clinical outcome in node-negative breast cancer. *J. Natl. Cancer Inst. 85*:200–206.

Altucci, L., Addeo, R., Cicatiello, L., Dauvois, S., Parker, M.G., Truss, M., Beato, M., Sica, V., Bresciani, F., and Weisz, A. 1996.

17beta-Estradiol induces cyclin D1 gene transcription, p36D1-p34cdk4 complex activation and p105Rb phosphorylation during mitogenic stimulation of G(1)-arrested human breast cancer cells. *Oncogene 12*:2315–2324.

An, H.X., Beckmann, M.W., Reifenberger, G., Bender, H.G., and Niederacher, D. 1999. Gene amplification and overexpression of CDK4 in sporadic breast carcinomas is associated with high tumor cell proliferation. *Am. J. Pathol. 154*:113–118.

Bai, M., Agnantis, N.J., Kamina, S., Demou, A., Zagorianakou, P., Katsaraki, A., and Kanavaros, P. 2001b. *In vivo* cell kinetics in breast carcinogenesis. *Breast Cancer Res. 3*:276–283.

Bai, M., Vlachonikolis, J., Agnantis, N.J., Tsanou, E., Dimou, S., Nicolaidis, C., Stefanaki, S., Pavlidis, N.J., and Kanavaros, P. 2001a. Low expression of p27 protein combined with altered p53 and Rb/p16 expression status is associated with increased expression of cyclin A and cyclin B1 in diffuse large B-cell lymphomas. *Mod. Pathol. 14*:1105–1113.

Baldin, V., Lukas, J., Marcote, M.J., Pagano, M., and Draetta, G. 1993. Cyclin D1 is a nuclear protein required for cell cycle progression in G1. *Genes Dev. 7*:812–821.

Bankfalvi, A., Tory, K., Kemper, M., Breeukelmann, D., Cubick, C., Poremba, C., Fuzesi, L., Lellee, R.J., and Bocker, W. 2000. Clinical relevance of immunohistochemical expression of p53-targeted gene products mdm-2, p211 and bcl-2 in breast carcinoma. *Pathol. Res. Pract. 196*:489–501.

Barbareschi, M. 2002. Prognostic value of the immunohistochemical expression of p53 in breast carcinomas: A review of the literature involving over 9,000 patients. *Appl. Immunohistochem. 4*:106–116.

Barbareschi, M., Pelosio, P., Caffo, O., Buttitta, F., Pellegrini, S., Barbazza, R., Dalla Palma, P., Bevilacqua, G., and Marchetti, A. 1997. Cyclin-D1-gene amplification and expression in breast carcinoma: Relation with clinicopathologic characteristics and with retinoblastoma gene product, p53 and p21WAF1 immunohistochemical expression. *Int. J. Cancer 74*:171–174.

Barbareschi, M., van Tinteren, H., Maurin, F.A., Veronese, S., Peterse, H., Maisonneure, P., Caffo, O., Scaioli, M., Doglioni, C., Galligioni, E., Dalla Palma, P., and Michalides, R. 2000. p27(kip1) expression in breast carcinomas: An immunohistochemical study on 512 patients with long-term follow-up. *Int. J. Cancer 89*:236–241.

Barnes, D.M., and Gillett, C.E. 1998. Cyclin D1 in breast cancer. *Breast Cancer Res. Treat. 52*:1–15.

Bates, S., Bonetta, L., MacAllan, D., Parry, D., Holder, A., Dickson, C., and Peters, G. 1994. CDK6 (PLSTIRE) and CDK4 (PSK-J3) are a dinstict subject of the cyclin-dependent kinases that associate with cyclin D1. *Oncogene 9*:71–79.

Berardo, M.D., Elledge, R.M., de Moor, C., Clark, G.M., Osborne, C.K., and Allred, D.C. 1998. Bcl-2 and apoptosis in lymph node positive breast carcinoma. *Cancer 82*:1296–1302.

Beroud, C., and Soussi, T. 2003. The UMD-p53 database: New mutations and analysis tools. *Hum. Mutat. 21*:176–181.

Borg, A., Zhang, Q.X., Olsson, H., and Wenngren, E. 1992. Chromosome 1 alterations in breast cancer: Allelic loss on 1p and 1q is related to lymphogenic metastases and poor prognosis. *Genes Chromosomes Cancer 5*:311–320.

Borresen-Dale, A.L. 2003. TP53 and breast cancer. *Hum. Mutat. 21*:292–300.

Bortner, D.M., and Rosenberg, M.P. 1997. Induction of mammary gland hyperplasia and carcinomas in transgenic mice expressing human cyclin E. *Mol. Cell Biol. 17*:453–459.

Bouchard, C., Thieke, K., Maier, A., Saffrich, R., Hanley-Hyde, J., Ansorge, W., Reed, S., Sicinski, P., Bartek, J., and Eilers, M. 1999. Direct induction of cyclin D2 by Myc contributes to cell

cycle progression and sequenstration of p27. *EMBO J. 18*:5321–5333.

Briasoulis, E., Agnantis, N.J., Zagorianakou, P., Kamina, S., Gorezi, M., Pavlidis, N., and Bai, M. 2001. Near-absolute expression of the bcl-2 protein identifies a subgroup of stage II breast cancer patients with a most favorable outcome. Results of a clinico-pathological study. *J. Exp. Clin. Cancer Res. 20*:341–344.

Bukholm, I.K., Berner, J.M., Nesland, J.M., and Borresen-Dale, A.L. 1998. Expression of cyclin Ds in relation to p53 status in human breast carcinomas. *Virchows Arch. 433*:223–228.

Bukholm, I.R., Bukholm, G., and Nesland, J.M. 2001. Overexpression of cyclin A is highly associated with early relapse and reduced survival in patients with primary breast carcinomas. *Int. J. Cancer 93*:283–287.

Caffo, O., Doglioni, C., Veronese, S., Bonzanini, M., Marchetti, A., Buttitta, F., Fina, P., Leek, R., Morelli, L., Palma, P.D., Harris, A.L., and Barbareschi, M. 1996. Prognostic value of p21(WAF1) and p53 expression in breast carcinoma: An immunohistochemical study in 261 patients with long-term follow-up. *Clin. Cancer Res. 2*:1591–1599.

Carroll, J.S., Prall, O.W., Musgrove, E.A., and Sutherland, R.L. 2000. A pure estrogen antagonist inhibits cyclin E-Cdk2 activity in MCF-7 breast cancer cells and induces accumulation of p130-E2F4 complexes characteristic of quiescence. *J. Biol. Chem. 275*:38221–38229.

Castiglione, F., Sarotto, I., Fontana, V., Destefanis, M., Venturino, A., Ferro, S., Cardaropoli, S., Orengo, M.A., and Porcile, G. 1999. Bc12, p53 and clinical outcome in a series of 138 operable breast cancer patients. *Anticancer Res. 19*:4555–4563.

Ceccarelli, C., Santini, D., Chieco, P., Taffurelli, M., Gamberini, M., Pileri, S.A., and Marrano, D. 1998. Retinoblastoma (RB1) gene product expression in breast carcinoma. Correlation with Ki-67 growth fraction and biopathological profile. *J. Clin. Pathol. 51*:818–824.

Chappuis, P.O., Kapusta, L., Begin, L.R., Wong, N., Brunet, J.S., Narod, S.A., Slingerland, J., and Foulkes, W.D. 2000. Germline BRCA1/2 mutations and p27(Kip1) protein levels independently predict outcome after breast cancer. *J. Clin. Oncol. 18*:4045–4052.

Collecchi, P., Santoni, T., Gnesi, E., Giuseppe Naccarato, A., Passoni, A., Rocchetta, M., Danesi, R., and Bevilacqua, G. 2000. Cyclins of phases G1, S and G2/M are overexpressed in aneuploid mammary carcinomas. *Cytometry 42*:254–260.

Courjal, F., Louason, G., Speiser, P., Katsaros, D., Zeillinger, R., and Theillet, C. 1996. Cyclin gene amplification and overexpression in breast and ovarian cancers: Evidence for the selection of cyclin D1 in breast and cyclin E in ovarian tumors. *Int. J. Cancer 69*:247–253.

Datto, M.B., Li, Y., Panus, J.F., Howe, D.J., Xiong, Y., and Wang, X.-F. 1995. Transforming growth factor beta induces the cyclin-dependent kinase inhibitor p21 through a p53-independent mechanism. *Proc. Natl. Acad. Sci. USA 92*:5545–5549.

de Jong, J.S., van Diest, P.J., Michalides, R.J., and Baak, J.P. 1999. Concerted overexpression of the genes encoding p21 and cyclin D1 is associated with growth inhibition and differentiation in various carcinomas. *Mol. Pathol. 52*:78–83.

Dennellan, R., and Chetty, R. 1999. Cyclin E in human cancers. *FASEB J. 13*:773–780.

Dublin, E.A., Patel, N.K., Gillett, C.E., Smith, P., Peters, G., and Barnes, D.M. 1998. Retinoblastoma and p16 proteins in mammary carcinoma: Their relationship to cyclin D1 and histopathological parameters. *Int. J. Cancer 79*:71–75.

Dulic, V., Kaufman, W.K., Wilson, S.J., Tlsty, T.D., Lee, E., Harper, J.W., Elledge S.J., and Reed, S.I. 1994. P53 dependent inhibition of cyclin-dependent kinase activities in human fibroblasts during radiation-induced G1 arrest. *Cell 76:*1013–1024.

Elledge, S.J., and Harper, J.W. 1994. Cdk inhibitors: On the threshold of checkpoints and development. *Curr. Opin. Cell Biol. 6:*847–852.

Enders, G.H. 2002. Cyclins in breast cancer: Too much of a good thing. *Breast Cancer Res. 4:*145–147.

Evan, G.I., and Vousden, K.H. 2001. Proliferation, cell cycle and apoptosis in cancer. *Nature 411:*342–348.

Geradts, J., and Ingram, C.D. 2000. Abnormal expression of cell cycle regulatory proteins in ductal and lobular carcinomas of the breast. *Mod. Pathol. 13:*945–953.

Gillett, C., Smith, P., Gregory, W., Richards, M., Millis, R., Peters, G., and Barnes, D. 1996. Cyclin D1 and prognosis in human breast cancer. *Int. J. Cancer 69:*92–99.

Gillett, C.E., Smith, P., Peters, G., Lu, X., and Barnes, D.M. 1999. Cyclin-dependent kinase inhibitor p27kip1 expression and interaction with other cell cycle-associated proteins in mammary carcinoma. *J. Pathol. 187:*200–206.

Gohring, U.J., Bersch, A., Becker, M., Neuhaus, W., and Schondorf, T. 2001. p21(waf) correlates with DNA replication but not with prognosis in invasive breast cancer. *J. Clin. Pathol. 54:*866–870.

Gonzalez-Campora, R., Galera Ruiz, M.R., Vazquez Ramirez, F., Rios Martin, J.J., Fernandez Santos, J.M., Ramos Martos, M.M., and Gomez Pascual, A. 2000. Apoptosis in breast carcinoma. *Pathol. Res. Pract. 196:*167–174.

Haldar, S., Negrini, M., Monne, M., Sabbioni, S., and Croce, C.M. 1994. Down-regulation of bcl-2 by p53 in breast cancer cells. *Cancer Res. 54:*2095–2097.

Hannon, G.J., and Beach, D. 1994. p^{15INK4B} is a potential effector of TGF-beta-induced cell cycle arrest. *Nature 371:*257–261.

Hemmati, P.G., Gillissen, B., von Haefen, C., Wendt, J., Starck, L., Guner, D., Dorken, B., and Daniel, P.T. 2002. Adenovirus-mediated overexpression of p14(ARF) induces p53 and Bax-independent apoptosis. *Oncogene 21:*3149–3161.

Ho, G.H., Calvano, J.E., Bisogna, M., and Van Zee, K.J. 2001. Expression of E2F-1 and E2F-4 is reduced in primary and metastatic breast carcinomas. *Breast Cancer Res. Treat. 69:*115–122.

Huang, Y., Ray, S., Reed, J.C., Ibrado, A.M., Tang, C., Nawabi, A., and Bhalla, K. 1997. Estrogen increases intracellular p26Bcl-2 to p21Bax ratios and inhibits taxol-induced apoptosis of human breast cancer MCF-7 cells. *Breast Cancer Res. Treat. 42:*73–81.

Hui, R., Macmillan, R.D., Kenny, F.S., Musgrove, E.A., Blamey, R.W., Nicholson, R.I., Robertson, J.F., and Sutherland, R.L. 2000. INK4a gene expression and methylation in primary breast cancer: Overexpression of p16INK4a messenger RNA is a marker of poor prognosis. *Clin. Cancer Res. 6:*2777–2787.

Kanavaros, P., Stefanaki, K., Rontogianni, D., Papalazarou, D., Sgantzos, M., Arvanitis, D., Vamvouka, C., Gorgoulis, V., Siatitsas, L., Agnantis, N.J., and Bai, M. 2001. Immunohistochemical expression of p53, p21/Walf, Rb, p16, cyclin D1, p27, Ki67, cyclin A, cyclin B1, bcl2, bax and bak proteins and apoptotic index in normal thymus. *Histol. Histopathol. 16:*1005–1012.

Kato, J.-Y., Matsushime, H., Hiebert, S.W., Ewen, M.E., and Sherr, C.J. 1993. Direct binding of cyclin D to the retinoblastoma product (pRb) and pRb phosphorylation by the cyclin D-dependent kinase, CDK4. *Genes 7:*331–342.

Kim, H.K., Park, I.A., Heo, D.S., Noh, D.Y., Choe, K.J., Bang, Y.J., and Kim, N.K. 2001. Cyclin E overexpression as an independent risk factor of visceral relapse in breast cancer. *Eur. J. Surg. Oncol. 27:*464–471.

Knowlton, K., Mancini, M., Creason, S., Morales, C., Hockenbery, D., and Anderson, B.O. 1998. Bcl-2 slows *in vitro* breast cancer growth despite its antiapoptotic effect. *J. Surg. Res. 76:*22–26.

Krajewski, S., Blomqvist, C., Franssila, K., Krajewska, M., Wasenius, V.M., Niskanen, E., Nordling, S., and Reed, J.C. 1995. Reduced expression of proapoptotic gene BAX is associated with poor response rates to combination chemotherapy and shorter survival in women with metastatic breast adenocarcinoma. *Cancer Res. 55:*4471–4478.

Krajewski, S., Krajewska, M., Turner, B.C., Pratt, C., Howard, B., Zapata, J.M., Frenkel, V., Robertson, S., Ionov, Y., Yamamoto, H., Perucho, M., Takayama, S., and Reed, J.C. 1999. Prognostic significance of apoptosis regulators in breast cancer. *Endocr. Relat. Cancer 6:*29–40.

Lallemand, F., Courilleau, D., Buquet-Fagot, C., Atfi, A., Montagne, M.N., and Mester, J. 1999. Sodium butyrate induces G2 arrest in the human breast cancer cells MDA-MB-231 and renders them competent for DNA replication. *Exp. Cell Res. 247:*432–440.

Leal, C., Henrique, R., Monteiro, P., Lopes, C., Bento, M.J., De Sousa, C.P., Lopes, P., Olson, S., Silva, M.D., and Page, D.L. 2001. Apocrine ductal carcinoma *in situ* of the breast: Histologic classification and expression of biologic markers. *Hum. Pathol. 32:*487–493.

Lees, E.M., and Harlow, E. 1993. Sequences within the conserved cyclin box of human cyclin A are sufficient for binding to and activation of cdc2 kinase. *Mol. Cell. Biol. 13:*1194–1201.

Li, R., Waga, S., Hannon, G.J., Beach, D., and Stillman, B. 1994. Differential effects by the p21 CDK inhibitor on PCNA-dependent DNA replication and repair. *Nature 371:*534–538.

Lipponen, P., Aaltomaa, S., Kosma, V.M., and Syrjanen, K. 1994. Apoptosis in breast cancer as related to histopathological characteristics and prognosis. *Eur. J. Cancer 30:*2068–2073.

Lipponen, P., Pietilainen, T., Kosma, V.M., Aaltomaa, S., Eskelinen, M., and Syrjanen, K. 1995. Apoptosis suppressing protein bcl-2 is expressed in well-differentiated breast carcinomas with favourable prognosis. *J. Pathol. 177:*49–55.

Luna-More, S., Casquero, S., Perez-Mellado, A., Rius, F., Weill, B., and Gornemann, I. 2000. Importance of estrogen receptors for the behavior of invasive micropapillary carcinoma of the breast. Review of 68 cases with follow-up of 54. *Pathol. Res. Pract. 196:*35–39.

Lundberg, A.S., and Weinberg, R.A. 1999. Control of the cell cycle and apoptosis. *Eur. J. Cancer 35:*1886–1894.

Mathoulin-Portier, M.P., Viens, P., Cowen, D., Bertucci, F., Houvenaeghel, G., Geneix, J., Puig, B., Bardou, V.J., and Jacquemier, J. 2000. Prognostic value of simultaneous expression of p21 and mdm2 in breast carcinomas treated by adjuvant chemotherapy with antracyclin. *Oncol. Rep. 7:*675–680.

Mendelson, A.R., Hamer, J.D., Wang, B.Z., and Brent, R. 2002. Cyclin D3 activates caspase 2, connecting cell proliferation with cell death. *Proc. Natl. Acad. Sci. USA 99:*6871–6876.

Michalides, R., Hageman, P., van Tinteren, H., Houben, L., Wientjens, E., Klompmaker, R., and Peterse, J. 1996. A clinicopathological study on overexpression of cyclin D1 and of p53 in a series of 248 patients with operable breast cancer. *Br. J. Cancer 73:*728–734.

Mirza, A.N., Mirza, N.Q., Vlastos, G., and Singletary, S.E. 2002. Prognostic factors in node-negative breast cancer: A review of studies with sample size more than 200 and follow-up more than 5 years. *Ann. Surg. 235:*10–26.

Miyashita, T., Krajewski, S., Krajewska, M., Wang, H.G., Lin, H.K., Liebermann, D.A., Hoffman, B., and Reed, J.C. 1994. Tumor suppressor p53 is a regulator of bcl-2 and bax gene expression *in vitro* and *in vivo*. *Oncogene 9:*1799–1805.

Morgan, D.O. 1995. Principles of Cdk regulation. *Nature* *374*:131–134.

Nielsen, N.H., Arnerlov, C., Emdin, S.O., and Landberg, G. 1996. Cyclin E overexpression, a negative prognostic factor in breast cancer with strong correlation to oestrogen receptor status. *Br. J. Cancer* 74:874–880.

Nielsen, N.H., Emdin, S.O., Cajander, J., and Landberg, G. 1997. Deregulation of cyclin E and D1 in breast cancer is associated with inactivation of the retinoblastoma protein. *Oncogene* *14*:295–304.

Notokura, T., Bloom, T., Kim, H.G., Juppner, H., Rudermans, J.V., Kronenberg, H.M., and Arnold, A. 1991. A novel cyclin encoded by a bell-linked candidate oncogene. *Nature* *350*:512–515.

Notokura, T., and Arnold, A. 1993. Cyclins and oncogenesis. *Biochim. Biophys. Acta. 1115*:63–78.

Oliner, J.D., Pietenpol, J.A., Thiagalingam, S., Gyuris, J., Kinzler, K.W., and Vogelstein, B. 1993. Oncoprotein MDM2 conceals the activation domain of tumour suppressor p53. *Nature* *362*:857–860.

Orend, G., Hunter, T., and Ruoslahti, E. 1998. Cytoplasmic displacement of cyclin E-cdk2 inhibitors p21Cip1 and p27Kip1 in anchorage-independent cells. *Oncogene 16*:2575–2583.

Palazzo, J.P. 2001. Cyclin-dependent kinase inhibitors-a novel class of prognostic indicators. *Hum. Pathol. 32*:769–770.

Perez-Roger, I., Kim, S.H., Griffiths, B., Sewing, A., and Land, H. 1999. Cyclins D1 and D2 mediate myc-induced proliferation via sequestration of p27 (kip1) and p21 (cip1). *EMBO J. 18*:5310–5320.

Pharoah, P.D., Day, N.E., and Caldas, C. 1999. Somatic mutations in the p53 gene and prognosis in breast cancer: A meta-analysis. *Br. J. Cancer 80*:1968–1973.

Pinto, A.E., Andre, S., Pereira, T., Nobrega, S., and Soares, J. 2001. Prognostic comparative study of S-phase fraction and Ki-67 index in breast carcinoma. *J Clin. Pathol. 54*:543–549.

Pomerantz, J., Schreiber-Agus, N., Liegeois, N.J., Silverman, A., Alland, L., Chin, L., Potes, J., Chen, K., Orlow, I., Lee, H.W., Cordon-Cardo, C., and De Pinho, R.A. 1998. The Ink4a tumor suppressor gene product, p19Arf, interacts with MDM2 and neutralizes MDM2's inhibition of p53. *Cell 92*:713–723.

Rosen, P.P., Lesser, M.L., Arroyo, C.D., Cranor, M., Borgen, P., and Norton, L. 1995. p53 in node-negative breast carcinoma: An immunohistochemical study of epidemiologic risk factors, histologic features, and prognosis. *J. Clin. Oncol. 13*:821–830.

Sherr, C.J. 2000. Cell cycle control and cancer. *Harvey Lect. 96*:73–92.

Sicinski, P., Donaher, J.L., Ceng, Y., Parker, S.B., Gardner, H., Park, M.Y., Robker, R.L., Richards, J.S., Mc Ginnis, L.K., Biggers, J.D., Eppig, J.J., Bronson, R.T., Elledge, S.J., and Weinberg, R.A. 1996. Cyclin D2 is an FSH-responsive gene involved in gonadal cell proliferation and oncogenesis. *Nature (Lond.) 384*:470–474.

Silva, J., Silva, J.M., Dominguez, G., Garcia, J.M., Cantos, B., Rodriguez, R., Larrondo, F.J., Provencio, M., Espana, P., and

Bonilla, F. 2003. Concomitant expression of p16INK4a and p14ARF in primary breast cancer and analysis of inactivation mechanisms. *J. Pathol. 199*:289–297.

Silvestrini, R., Veneroni, S., Daidone, M.G., Benini, E., Boracchi, P., Mezzetti, M., Di Fronzo, G., Rilke, F., and Veronesi, U. 1994. The Bcl-2 protein: A prognostic indicator strongly related to p53 protein in lymph node-negative breast cancer patients. *J. Natl. Cancer Inst. 86*:499–504.

Slingerland, J., and Pagano, M. 2000. Regulation of the Cdk inhibitor p27 and its deregulation in cancer. *J. Cell Physiol. 183*:10–17.

Spataro, V.J., Litman, H., Viale, G., Maffini, F., Masullo, M., Golouh, R., Martinez-Tello, F.J., Grigolato, P., Shilkin, K.B., Gusterson, B.A., Castiglione-Gertsch, M., Price, K., Lindtner, J., Cortes-Funes, H., Simoncini, E., Byrne, M.J., Collins, J., Gelber, R.D., Coates, A.S., and Goldhirsch, A. 2003. Decreased immunoreactivity for p27 protein in patients with early-stage breast carcinoma is correlated with HER-2/neu overexpression and with benefit from one course of perioperative chemotherapy in patients with negative lymph node status: Results from International Breast Cancer Study Group Trial V. *Cancer 97*:1591–1600.

Spyratos, F., Ferrero-Pous, M., Trassard, M., Hacene, K., Phillips, E., Tubiana-Hulin, M., and Le Doussal, V. 2002. Correlation between MIB-1 and other proliferation markers: Clinical implications of the MIB-1 cut off value. *Cancer 94*:2151–2159.

St Croix, B., Sheehan, C., Rak, J.W., Florenes, V.A., Slingerland, J.M., and Kerbel, R.S. 1998. E-Cadherin-dependent growth suppression is mediated by the cyclin-dependent kinase inhibitor p27(KIP1). *J. Cell Biol. 142*:557–571.

Steck, K., Hunt, K., Tucker, S., Singletary, S., and El-Naggar, A.K. 1999. Flow cytometric analysis of Ki-67 in invasive ductal carcinoma of the breast: Correlation with tumor and patient characteristics. *Oncol. Rep. 6*:835–838.

Sweeney, K.J., Sarcevic, B., Sutherland, R.L., and Musgrove, E.A. 1997. Cyclin D2 activates Cdk2 in preference to Cdk4 in human breast epithelial cells. *Oncogene 14*:1329–1340.

T'Ang, A., Varley, J.M., Chakraborty, S., Murphree, A.L., and Fung, Y.K. 1998. Structural rearrangement of the retinoblastoma gene in human breast carcinoma. *Science 242*:263–266.

van Slooten, H.J., van de Vijver, M.J., van de Velde, C.J., and van Dierendonck, J.H. 1998. Loss of Bcl-2 in invasive breast cancer is associated with high rates of cell death, but also with increased proliferative activity. *Br. J. Cancer 77*:789–796.

Wakasugi, E., Kobayashi, T., Tamaki, Y., Ito, Y., Miyashiro, I., Komoike, Y., Takeda, T., Shin, E., Takatsuka, Y., Kikkawa, N., Monden, T., and Monden, M. 1997. p21(Waf1/Cip1) and p53 protein expression in breast cancer. *Am. J. Clin. Pathol. 107*:684–691.

Watts, C.K., Brady, A., Sarcevic, B., deFazio, A., Musgrove, E.A., and Sutherland, R.L. 1995. Antiestrogen inhibition of cell cycle progression in breast cancer cells in associated with inhibition of cyclin-dependent kinase activity and decreased retinoblastoma protein phosphorylation. *Mol. Endocrinol. 9*:1804–1813.

20

Immunohistochemistry of Estrogen Receptor Expression in Breast Carcinoma

Philippa T.K. Saunders

Introduction

Estrogens modulate the growth and differentiation of mammary tissues and have also been implicated in the development of breast cancer. Estrogens are synthesized from androgens (testosterone and androstenedione) by the cytochrome P450 aromatase enzyme (Simpson *et al.*, 1994). In premenopausal women the major site of estrogen biosynthesis is the ovaries but after the menopause estrogen continues to be formed at locally high levels in extragonadal sites including the breast (reviewed by Simpson, 2002). In postmenopausal women use of hormone replacement therapy (HRT; estrogen alone or in combination with a progestagen) has been rising in Western countries over the past few decades. A recent review of four randomized trials that analyzed the effects of HRT on the health of over 20,000 women concluded that the excess incidence of breast cancer in 50- to 59- year-olds using HRT was 3.3 per 1000 users (Beral *et al.*, 2002).

Estrogen action is mediated via specific receptors (ER). The first estrogen receptor was cloned from human breast cancer cell line in 1986 (Green *et al.*, 1986). In 1996 a second ER was cloned, this time from rat prostate (Kuiper *et al.*, 1996), and its human homologue was cloned shortly thereafter (Mosselman *et al.*, 1996). The two receptors are known as estrogen receptor alpha (ERα, NR3A1) and estrogen receptor beta (ERβ, NR3A2), respectively. The ERα and ERβ receptors are the products of two genes located on different chromosomes (Enmark *et al.*, 1997). They share significant sequence homology and have been classified as subgroup A of the nuclear receptor subfamily 3 (Beato *et al.*, 2000). Key domains within ERα and β include the DNA binding domain (C) and two transactivation domains (AF-1 and AF-2, located in the N (A/B domain) and C terminal (E/F domain) portions of the protein, respectively (see reviews Pettersson *et al.*, 2001; Saunders, 1998). The crystal structure of the ligand binding domains of ERα and ERβ have been determined, and their overall arrangement has been shown to be similar, with both containing 12 helices (Pike *et al.*, 1999). *In vitro* studies have demonstrated that homodimers (ERα-ERα or ERβ-ERβ) or heterodimers (ERαERβ) can be formed when both isoforms are expressed in the same cell (Cowley *et al.*, 1997; Pettersson *et al.*, 1997). It has been suggested that one role played by ERβ may be to modulate ERα

Handbook of Immunohistochemistry and *in situ* Hybridization of Human Carcinomas, Volume 1: Molecular Genetics; Lung and Breast Carcinomas

439

Copyright © 2004 by Elsevier (USA)
All rights reserved.

transcriptional activity (Hall *et al.*, 1999; Weihua *et al.*, 2000).

In addition to the full-length receptors, a number of mRNAs encoding variant isoforms of both ERα and ERβ have been identified in breast cancers and breast cancer cell lines (Balleine *et al.*, 1999; Boccabella, 1963; Fuqua *et al.*, 1999; Huang *et al.*, 1999; McGuire *et al.*, 1991; Poola *et al.*, 2001). For example, Fuqua *et al.* (1991) have identified several exon deletion variants of ERα as well as a hypersensitive variant with a mutation within the hormone binding domain (Lemieux *et al.*, 1996). Variations of ERβ mRNAs lacking exons 5 and/or 6 have also been detected in normal human breast and in breast cancer (Lu *et al.*, 1998; Speirs *et al.*, 2000). In addition to these deletion variants, a number of other alternative splice forms have been described. Ogawa *et al.* (1998a,b) identified a novel human ERβ variant, which they named hERβcx. The open reading frame of the protein was identical to full-length hERβ for most of its length, but was truncated at the C-terminus leading to the loss of 61 amino acids and the addition of a unique 26 amino acid peptide. This protein (also known as ERβ2) (Moore *et al.*, 1998) is formed by alternative splicing of a novel eighth exon onto ERβ exon 7; this isoform is not found in rodents (Chu *et al.*, 2000). Moore *et al.* (1998) also identified four further ERβ splice forms in human tissue extracts.

Estrogen receptor status has been shown to correlate, with response of breast tumors to hormonal therapy such as treatment with the antiestrogen tamoxifen (Cuzick *et al.*, 2003). Tamoxifen acts by binding to estrogen receptors and is defined as a selective estrogen receptor modulator (SERM) because it acts as an estrogen antagonist in breast, whereas it has been shown to have agonist activity in the uterus and bone (Jensen *et al.*, 2003; van Leeuwen *et al.*, 1994). Studies *in vitro* have demonstrated that tamoxifen acts as a partial agonist on ERα but has a pure antagonist effect on ERβ (Barkhem *et al.*, 1998). Four trials have reported the use of tamoxifen as a prophylaxis for prevention of breast cancer, and the evidence shows that the risk of developing ER-positive tumors was reduced, although side effects included an increased risk of endometrial cancer and thromboembolism (Cuzick *et al.*, 2003). Aromatase inhibitors prevent synthesis of estrogens, reduce the amount of estrogens within tissues such as breast as well as in the circulation (Miller *et al.*, 2002; Simpson, 2002), resulting in the prevention of and therefore prevent estrogen-mediated gene activation via either ERα or ERβ. The most recently developed aromatase inhibitors have been reported to have improved efficacy compared with tamoxifen when used in patients with advanced breast cancer (Miller *et al.*, 2003).

Both tamoxifen and aromatase inhibitors have been given to women with ER-positive tumors preoperatively, and in the randomized trial reported by Dixon *et al.* (2002) the aromatase inhibitor letrozole was more effective in reducing tumor size and allowed for more conservative surgery.

The purpose of this chapter is to give details of immunohistochemical protocols that have been used successfully to achieve specific staining of breast cancer tissues for ERα and ERβ and to consider some of the data that have been published on the differential expression of these receptors and variant isoforms in the breast. The role of ER in conjunction with other proteins in breast cancer is discussed in Part IV, Chapter 19.

MATERIALS

Breast cancer tissues were fixed in 10% neutral buffered formaldehyde for 16 to 24 hr then stored in 70% (w/v) ethanol before processing using standard procedures. The patient samples used in this chapter are the same as those detailed in Saunders *et al.* (2002b).

A number of estrogen receptor antibodies have been described, and several are commercially available (see reviews in Kobayashi *et al.*, 2000; Pavao *et al.*, 2001). Table 26 contains the details of those that were used in the current study. All have been used previously on fixed sections from human tissues (Critchley *et al.*, 2001, 2002; Saunders *et al.*, 2000, 2002a; Scobie *et al.*, 2002). The location of the peptides used to generate the ERβ antibodies used in the current studies is shown in Figure 64. These antibodies are all specific to the ER to which they were raised when evaluated using Western blotting with either recombinant protein or nuclear tissue extracts (Saunders *et al.*, 2000, 2002a,b).

METHODS

All studies were conducted on fixed tissue sections cut to 5 µm. To enhance visualization of nuclear estrogen receptors, sections were mounted on charged slides (Superfrost, BDH, Poole, Dorset, UK) and subjected to heat-induced antigen retrieval (Hayat, 2002; Norton *et al.*, 1994) before immunohistochemical staining.

Antigen Retrieval and Blocking of Endogenous Peroxidase

Sections were dewaxed in xylene (2 × 5 min), rehydrated in alcohols (100% for 20 sec [repeat], 95% for 20 sec, 70% for 20 sec), and rinsed in distilled water. Citrate buffer (10 mM, pH 6) was added to a Tefal

Table 26 Estrogen Receptor Antibodies

Specificity	Supplier or Reference	Clone	Epitope	Raised in
ER alpha	Dako	1D5	Human recombinant	Mouse IgG1
ER alpha	Novocastra	6F11	Human recombinant	Mouse IgG1
ER beta	(1)	N/A	Hinge domain peptide	Sheep
ER beta	Serotec (2)	PPG5/10	C-terminus human ERβ1	Mouse IgG2a
ER beta	Novocastra	EMR02	Human ERβ (F domain)	Mouse IgG1
ER beta cx/beta2	(3)	M57	C-terminus ERβcx variant	Mouse IgG1

(1) Saunders *et al.* (2000) (peptide P4); (2) Saunders *et al.* (2000) (peptide P7); (3) Saunders *et al.* (2002a) (peptide P8).

Clypso pressure cooker and brought to boil on a halogen hot plate (the procedure was carried out inside a fume cupboard, and safety rules for handling pressure cookers were observed). Slides mounted in metal racks were added directly into the boiling buffer, and the lid closed, and the dial turned to setting 2. When full pressure was reached (steady release of steam, red button elevated), the timer was started, and the sections were left at full pressure for 5 min. The pressure cooker was then removed from the hot plate, the pressure was released, and the sections were left undisturbed for a further 20 min. The lid was removed carefully and cold water added until the solution cooled to room temperature. Slide racks were removed.

If a pressure cooker is not available, antigen retrieval can be carried out in a microwave oven by performing 4 × 5 min incubations on full power in 0.01 M citrate buffer (pH 6), which has been prewarmed for 5 min before addition of sections. For microwave retrieval, slides were mounted in plastic racks and placed in glass dishes covered with Saran wrap; the level of the buffer was noted and topped up with hot water if any loss of volume occurred during the procedure.

Endogenous peroxidases were blocked by incubating slides in a solution of 3% (v/v) hydrogen peroxide (in methanol) for 30 min on a rocker platform. Sections were washed in water and then in Tris buffered saline (TBS, 0.05 M Tris [pH 7.4], 0.85% NaCl) for 5 min before proceeding with incubation with appropriate primary antibodies (Table 27).

Sections of human endometrium (for ERα) or human adult testis (for ERβ) were included as positive controls. Negative controls were sections of tissue incubated either with normal mouse serum (monoclonal antibodies) or preimmune serum from the sheep (anti-P4 antibody) diluted to contain the same amount of IgG as the diluted primary antibodies.

Immunostaining for Estrogen Receptor Alpha

Overnight Method

Sections were blocked for 30 min in normal rabbit serum (NRS, Diagnostics Scotland) diluted 1:4 in TBS containing 5% bovine serum albumin (BSA) (NRS/TBS/BSA) and 4 drops of avidin (Vector Labs, Peter borough, UK) per ml buffer. Sections were rinsed in TBS for 5 min (twice), 4 drops of biotin were added

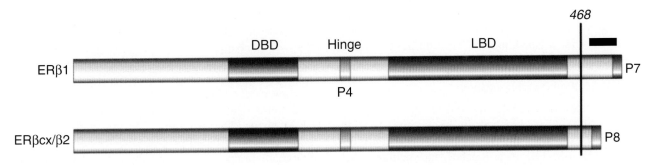

Figure 64 Diagram showing the relative positions of the epitopes recognized by ERβ antibodies. The sequences of the P4 and P7 peptides are given in Saunders *et al.* (2000) and P8 in Saunders *et al.* (2002a). The position of the black bar denotes the region of the protein recognized by the ERβ-specific antibody prepared by Novocastra (epitope mapping).

per ml TBS, and this was incubated on sections for 15 min, rinsed in TBS, and then washed in TBS for a further 15 min. Antibodies directed against ERα were diluted in NRS/TBS/BSA (6F11 1 in 20; 1D5 1 in 400), and sections were incubated overnight at 4°C. Sections were washed twice for 5 min each time in TBS and incubated with biotinylated rabbit anti-mouse (Dako, Dakocytomation, Glostrup, Denmark), diluted 1:500 in NRS/TBS/BSA. Incubations lasted for 1 hr, followed by two washes in TBS (5 min each). Thereafter, sections were incubated in ABC-HRP complex (Dako) for 1 hr, washed in TBS (2 × 5 min), and bound antibodies visualized by incubation with 3,3′-diaminobenzidine tetra-hydrochloride (liquid DAB cat K3468, Dako).

Rapid Method (Dako Antibody)

Anti-ERα was diluted 1:50 in biotin diluent (PBS, goat serum, and D-biotin), and sections were incubated in the sections for 60 min at room temperature. The secondary antibody, biotinylated anti-mouse IgG (Vector Laboratories) was diluted 1:2000, in background reducing diluent (Dako) and applied to sections for 30 min at room temperature. The tertiary system (ABC-HRP, Dako) was applied as per manufacturer's instructions for 30 min at room temperature. The tissue was visualized by immersing sections in DAB for 5 min.

Immunostaining for Estrogen Receptor Beta

Overnight Method

Sections were incubated for 30 min in normal rabbit serum (NRS, Diagnostics Scotland, Carluke) diluted 1:4 in TBS, containing 5% BSA (NRS/TBS/BSA), rinsed briefly in TBS, and an avidin biotin block performed using reagents (Vector). Anti-ERβ antibodies were diluted in NRS/TBS (ERβ anti P4 1 in 500; ERβ1 PPG5/10, 1 in 40; ERβ1 EMR02 1:40; ERβcx/β2 M57 1:50), and sections were incubated overnight at 4°C. Sections were washed twice for 5 min each in TBS and incubated with either biotinylated rabbit anti-sheep (Vector) or biotinylated rabbit anti-mouse (Dako), each of which was diluted 1:500 in NRS/TBS/BSA. Sections were incubated with ABC-HRP complex (Dako) for 1 hr, washed in TBS (2 × 5 min), and bound antibodies visualized by incubation with liquid DAB.

Rapid Method (Serotec antibody)

All incubations were carried out at room temperature. Sections were blocked in hydrogen peroxide (Dako Envision kit) for 5 min, washed in water followed by 50 mM TBS (5 min each), and incubated for 30 min in normal rabbit serum (NRS) diluted 1:4 in TBS containing 5% BSA (NRS/TBS/BSA). Slides were drained and excess solution wiped away before the addition of anti-ERβ antibody (PPG5/10) diluted 1:20 in NRS/TBS/BSA for 30 min. Slides were washed in TBS (2 × 5 min) and incubated with biotinylated rabbit anti-mouse IgG diluted 1:400 in NRS/TBS/BSA for 30 min before being washed twice in TBS (5 min each). Sections were incubated with streptavidin-HRP (Amdex, Amersham BioSciences, Piscataway, NJ) diluted 1:800 in TBS for 30 min, and washed twice in TBS, and bound antibodies were visualized by incubation with liquid DAB (cat K3468, Dako).

Counterstaining and Mounting

All sections were counterstained with hematoxylin for 3 min, washed in acid alcohol for 10–20 sec, washed in water, and blue color developed using Scotts tap water. Sections were dehydrated through a series of alcohols (70%, 95%, 100% 20 sec each), placed in Histoclear (National Diagnostics, Atlanta, GA) for 5 min, cleared in xylene for 5 min, and mounted under coverslips using Pertex (CellPath, Newtown, Wales, UK).

Evaluation of Results

Images were captured using an Olympus Provis microscope (Olympus Optical Co., London) equipped with a Kodak DCS330 camera (Eastman Kodak Co., Rochester, NY), stored on a Macintosh PowerPC computer and assembled using Photoshop 7 (Adobe, Mountain View, CA).

Quantitation of expression of estrogen receptors can be undertaken using an appropriate scoring system such as those reported in detail by others (Allred et al., 1998; Leake et al., 2000). This method is based on a composite additive score of intensity 0–3 and proportion of malignant epithelial cells staining 0–5.

RESULTS AND DISCUSSION

Antibodies were raised against estrogen receptors purified from calf uterine nuclei and MCF7 breast cancer cells, and these proved to be valuable tools for showing sites of expression of receptor proteins (Greene et al., 1980; King et al., 1984). Cloning of human ERα (Green et al., 1986) facilitated preparation of antibodies to specific peptides chosen from the protein sequence (Furlow et al., 1990). The identification of the human ERβ gene (Enmark et al., 1997; Mosselman et al., 1996) opened the possibility that this receptor might mediate some of the effects of estrogens on normal and malignant breast and has resulted in a large research effort to define the role(s) of ERα and ERβ in breast tumors (Fuqua et al., 2003; Jensen et al., 2001; Leygue et al., 1998; Palmieri et al., 2002; Speirs et al., 1999).

The existence of this second closely related protein also meant that it was possible that some antibodies raised previously to ERα might crossreact with ERβ and give misleading results. A further complication in interpreting immunohistochemical data from breast tissue arose from reports that mRNAs formed by alternative splicing of the ERα or ERβ genes were expressed in breast cancer tissue and breast cancer cell lines (Fuqua *et al.*, 1991, 1999; Hopp *et al.*, 1998; McGuire *et al.*, 1991; Poola *et al.*, 2001; Speirs *et al.*, 2000).

In light of these studies, questions that need to be addressed before interpreting the pattern of ER expression in breast cancer using immunohistochemistry are as follows:

1. Are the available antibodies specific?
2. Are the immunohistochemical methods reliable and can they be improved?
3. Are ERα or ERβ variant proteins expressed?

Figure 65 shows examples of the patterns of immunopositive staining obtained using each of the antibodies described earlier when methods for immunodetection of ERα and ERβ were applied.

Are the Currently Available Antibodies Specific?

Following identification of ERβ and cloning of the human cDNA (Ogawa *et al.*, 1998a) it has been possible to identify those regions of the estrogen receptors that have least sequence homology to each other and to other members of the nuclear receptor superfamily. These are located in the N terminal (A/B domain), the hinge (D domain), and at the C-terminus (F-domain). Furthermore, recombinant proteins ERα (~66 kDa) and ERβ (53 and 57 kDa) are commercially available (PanVera—Invitrogen Discovery Screening, Madison, WI), and therefore specificity can be checked using Western blot analysis.

The anti-ERα antibodies (ID5 and 6F11) used in the current study (Al Saati *et al.*, 1993) and 6F11 (Bevitt *et al.*, 1997) were generated using full-length recombinant proteins. The epitope recognized by ID5 has been mapped to the A/B domain of ERα (Al Saati *et al.*, 1993). Both antibodies react with recombinant ERα but not with recombinant ERβ on Western blots (Saunders *et al.*, 2000, 2002b). Using fixed tissue sections, anti-ERαID5 (Figure 65A) and anti ERα6F11 (Figure 65B,C) were localized on cell nuclei in the breast cancer. The proportion of immunopositive cells was variable between samples, and examples where most (Figure 65B) or very few (Figure 65C) cells were stained are shown. ERα immunopositive cells were not found within the surrounding stromal tissue.

In the current studies, ERβ proteins were immunolocalized on cell nuclei in the breast cancer and also in the stroma and epithelium of breast tissue adjacent to the malignant cells (Figure 65D–I). Expression of ERβ occurred both in ERα positive and in ERα negative (Figure 65E) breast tumors (Saunders *et al.*, 2002b). All antibodies used failed to recognize recombinant ERα protein on Western blots.

The monoclonal antibodies directed against the C-terminus of ERβ wild-type recognized both long and short forms of recombinant ERβ formed by the use of alternative ATG start sites. The antibodies used were raised to the C-terminus of full-length human ERβ1. Results using ERβ antibodies raised to the N-terminus (e.g., 14C8, Genetex; N19 Santa Cruz; and those described in Fuqua *et al.*, 2003; Mann *et al.*, 2001), the ligand binding domain (Choi *et al.*, 2001), or the C-terminus (D7N, Zymed; PA1313, Affinity Bioreagents) have been prepared, and some comparisons between the staining intensity have been reported (Skliris *et al.*, 2002). Mann *et al.* (2001) used a rabbit polyclonal antibody directed against the N-terminus of human ERβ on formalin-fixed samples: Using Western blot, multiple bands were present, the most prominent of which appeared shorter than the recombinant standard, and this may reflect degradation of protein in their extracts or nonspecific reactivity of the antibody used. We have detected cytoplasmic staining using some commercial polyclonal anti-ERβ antibodies, especially those that have not been affinity purified. Another source of nonspecific staining appears to arise from using poor secondary antibodies; this seems to be a particular problem with those raised in goats (unpublished observations). These findings may explain the presence of cytoplasmic staining seen in some published figures (Järvinen *et al.*, 2000; Mann *et al.*, 2001; Omoto *et al.*, 2001). Note that antibodies raised to epitopes other than those at the C-terminus of full-length wild-type human ERβ are likely to crossreact with variants such as ERβcx/β2 (Ogawa *et al.*, 1998b; Saunders *et al.*, 2002a) (see discussion following).

Are the Immunohistochemical Methods Reliable and Can They Be Improved?

Estrogen receptor status has been evaluated using a variety of assays including enzyme immunoassays (EIA) and immunohistochemistry. Studies comparing the results of EIA and immunohistochemistry using antibodies directed against ERα (Alberts *et al.*, 1996; Kurosumi, 2003) have concluded that immunohistochemistry is less expensive, more sensitive, and more specific than EIA (Zafrani *et al.*, 2000). However, several studies have also indicated that variable results can be

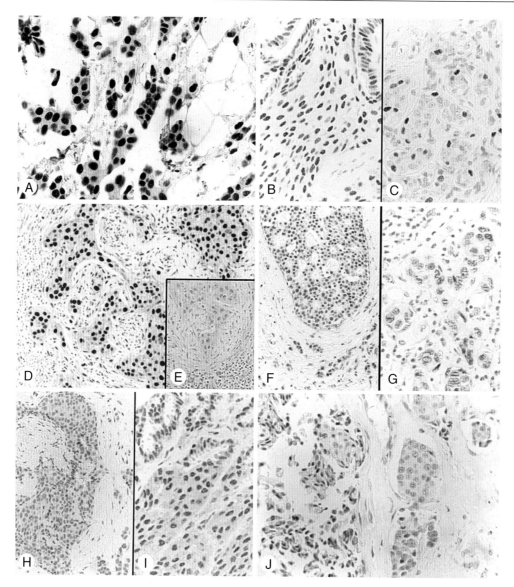

Figure 65 Immunopositive staining patterns obtained with selected antibodies specific for ERα or ERβ. Note that nuclei are immunopositive and cytoplasmic staining is low/absent. **A:** ERα immunopositive cells visualized using 1D5 and the rapid staining protocol. **B, C:** Variable staining of ERα positive cells in two different breast cancer samples stained with 6F11 and the overnight protocol. **D:** ERβ positive cell nuclei in malignant and surrounding stromal tissue visualized using a sheep polyclonal antibody. **E:** Parallel section to (D) showing that the tissue is immunonegative for ERα. **F, G:** ERβ1 monoclonal PPG5/10. **H, I:** ERβ1 monoclonal EMR02. **J:** ERβcx/β2 variant monoclonal M57 (Saunders *et al.*, 2002a).

obtained between laboratories, even when using the same sample blocks. False negative results are a problem if there is poor staining sensitivity (Rhodes *et al.*, 2000a,b; Rudiger *et al.*, 2002), endorsing the view that control positive samples should be included in all runs. Methods such as digital image analysis may improve reliability of assessments (Mofidi *et al.*, 2003).

Although there are several well-established antibodies directed against ERα, those available against ERβ have not been as rigorously tested. Some of the

published immunohistochemical data in the literature is thought to be of poor quality, and variations between the utility of different antibodies have been reported (Skliris *et al.*, 2002). Therefore, for assessment of ERβ expression there is clearly a need for standardization of methods and more quality controls.

In my experience there are two factors (other than the specificity of the primary antibodies) that can have an adverse effect on the quality and reliability of detection of ER proteins; these are poor tissue fixation and nonspecific staining by secondary antibodies. On Western blots false positive bands are often detected at ~66 kDa as a result of contamination of anti-IgG antibodies with those that bind to albumin. Based on results with different types of normal and malignant human tissues, I believe poor tissue preservation (e.g., underfixation seen in the center of large tissue fragments) and even overfixation (cross-linking of epitopes) can account for some of the variation in the detection of receptor proteins. Variations in staining intensity across the tissue section is a common observation when variable fixation has occurred. It is also important to check the cellular integrity of the tissue when heat-induced retrieval methods are used, as sometimes nuclear architecture may be destroyed by standard procedures, and in this instance immunostaining for nuclear proteins can be enhanced by reducing the duration of the retrieval procedure (e.g., from 20 min to 5 min). If cytoplasmic background staining is observed, this may also be reduced by either conducting retrieval in an alternative buffer (e.g., 10 mM glycine) (pH 3.5) or including Tween in buffer washes (0.01% w/v).

Are ERα and ERβ Variant Proteins Expressed?

There has been considerable debate over the role played by ERβ and the ERα and ERβ variants in cancer progression (Balleine et al., 1999; Huang et al., 1999; Lemieux et al., 1996; Leygue et al., 1999; Omoto et al., 2002; Speirs et al., 1999). Studies that have compared levels of expression of ERα and ERβ mRNAs have reported that the amount of ERβ mRNA does not appear to be correlated with that of ERα (Dotzlaw et al., 1997; Iwao et al., 2000; Vladusic et al., 2000). Similar findings have been reported from immunohistochemical evaluation of breast cancers (Jensen et al., 2001; Mann et al., 2001; Saunders et al., 2002b). Although several groups have reported evidence that mRNAs corresponding to alternatively spliced forms of ERα and ERβ have been detected in breast cancer tissues and cell lines (Iwao et al., 2000; Lu et al., 1998; Moore et al., 1998; Poola et al., 2001; Vladusic et al., 1998), there is far less data to show whether they are translated into functional proteins. Monoclonal antibodies directed against the N-terminus of ERβ have detected expression of proteins other than full-length ERβ in breast cancer cell lines (Fuqua et al., 1999) that might have been formed by translation of alternatively spliced mRNAs. Using the PPG5/10 monoclonal for Western blotting proteins corresponding in size to those that could be translated from mRNAs deleted in exons 5 or 6 (Brandenberger et al., 1999; Lu et al., 1998) were not detected (Saunders et al., 2002b). The most prominent proteins other than full-length ERβ1

migrated between 30 and 36 kDa. These could represent use of alternative start sites, translation from an exon 2 deleted mRNA (~35 kDa), or translation of protein from mRNA deleted for both exons 5 and 6 (AF074599), which is predicted to be ~43 kDa (short) or ~49 kDa (long) from the mRNA sequence.

Antibodies specific for ERβcx (ERβ2), ERβ4, and ERβ5 variants have been reported (Ogawa et al., 1998b; Saji et al., 2002b; Saunders et al., 2002a; Scobie et al., 2002). Expression of the ERβcx/β2 variant protein has been detected in breast cancer cells (Fig. 65J); in unpublished studies we did not find scoring of ERβcx expression correlated with that of ERβ1 (wild type) determined with the PPG5/10 antibody. In three different studies ERβcx immunopositive staining has been reported to be present in 54% (Omoto et al., 2002), 48% (Saji et al., 2002a), and 45% (Miller et al., unpublished) of breast tumor samples. Saji et al. (Saji, 2002a) reported that coexpression of ERα and ERβcx was associated with reduced levels of progesterone receptor positive immunostaining. Although studies in vitro suggest that ERβcx and the ERβdelta5 variants may have dominant negative activity (Inoue et al., 2000; Ogawa et al., 1998b) further investigations are required before we can be sure of the contribution of these variants to disease progression or response to endocrine therapy.

CONCLUSIONS

In conclusion, estrogen receptor status is an important predictor of response to therapies such as administration of tamoxifen or aromatase inhibitors. A role for splice variants of ERβ as modulators of the impact of response to estrogens has been claimed. Methods for evaluation of ERα expression using fixed tissue sections are well established, but the high variability in the quality of the published data using anti-ERβ antibodies is a cause for concern. Results with antibodies directed against regions of the receptors shared between full-length functional receptors and variant isoforms might give misleading results. At present ERα status is a well-established marker for breast cancer; however, there is an increasing interest in the potential that ERβ may mediate some estrogenic effects, and the use of ERα and ERβ selective antagonists is under investigation.

Acknowledgments

I am grateful for the skilled technical assistance of Michael Miller, Sheila MacPherson, Arantza Esnal, and Michelle Welsh, who undertook the immunostaining described in this chapter. My thanks to Professor William Miller (University of Edinburgh) for providing sections of breast cancer tissue.

References

Al Saati, T., Clamens, S., Cohen-Knafo, E., Faye, J.C., Prats, H., Coindre, J.M., Wafflart, J., Caveriviere, P., Bayard, F., and Delsol, G. 1993. Production of monoclonal antibodies to human estrogen receptor protein (ER) using recombinant ER (RER). *Int. J. Cancer 55:*651–654.

Alberts, S.R., Ingle, J.N., Roche, P.R., Cha, S.S., Wold, L.E., Farr, G.H., Jr., Krook, J.E., and Wieand, H.S. 1996. Comparison of estrogen receptor determinations by a biochemical ligand-binding assay and immunohistochemical staining with monoclonal antibody ER1D5 in females with lymph node positive breast carcinoma entered on two prospective clinical trials. *Cancer 78:*764–772.

Allred, D.C., Harvey, J.M., Berardo, M., and Clark, G.M. 1998. Prognostic and predictive factors in breast cancer by immunohistochemical analysis. *Mod. Pathol. 11:*155–168.

Balleine, R.L., Hunt, S.M.N., and Clarke, C.L. 1999. Coexpression of alternatively spliced estrogen and progesterone receptor transcripts in human breast cancer. *J. Clin. Endo. Metab. 84:*1370–1377.

Barkhem, T., Carlsson, B., Nilsson, Y., Enmark, E., Gustafsson, J., and Nilsson, S. 1998. Differential response of estrogen receptor alpha and estrogen receptor beta to partial estrogen receptor agonists/antagonists. *Mol. Pharmacol. 54:*105–112.

Beato, M., and Klug, J. 2000. Steroid hormone receptors: An update. *Hum. Reprod. Update 6:*225–236.

Beral, V., Banks, E., and Reeves, G. 2002. Evidence from randomised trials on the long-term effects of hormone replacement therapy. *Lancet 360:*942–944.

Bevitt, D.J., Milton, I.D., Piggot, N., Henry, L., Carter, M.J., Toms, G.L., Lennard, T.W., Westley, B., Angus, B., and Horne, C.H. 1997. New monoclonal antibodies to oestrogen and progesterone receptors effective for paraffin section immunohistochemistry. *J. Pathol. 183:*228–232.

Boccabella, A.V. 1963. Reinitiation and restoration of spermatogenesis with testosterone proprionate and other hormones after long-term and post-hypophysectomy regression period. *Endocrinology 72:*787–798.

Brandenberger, A.W., Lebovic, D.I., Tee, M.K., Ryan, I.P., Tseng, J.F., Jaffe, R.B., and Taylor, R.N. 1999. Oestrogen receptor (ER)-alpha and ER-beta isoforms in normal endometrial and endometriosis-derived stromal cells. *Mol. Human Reprod. 5:*651–655.

Choi, I., Ko, C., Park-Sarge, O.K., Nie, R., Hess, R.A., Graves, C., and Katzenellenbogen, B.S. 2001. Human estrogen receptor beta-specific monoclonal antibodies: Characterization and use in studies of estrogen receptor beta protein expression in reproductive tissues. *Mol. Cell Endocrinol. 181:*139–150.

Chu, S., Mamers, M., Burger, H.G., and Fuller, P.J. 2000. Estrogen receptor isoform gene expression in ovarian stromal and epithelial tumors. *J. Clin. Endo. Metab. 85:*1200–1205.

Cowley, S.M., Hoare, S., Mosselman, S., and Parker, S.G. 1997. Estrogen receptors alpha and beta form heterodimers on DNA. *J. Biol. Chem. 272:*19858–19862.

Critchley, H.O.D., Brenner, R.M., Drudy, T.A., Williams, K.A., Nayak, N.R., Millar, M.R., and Saunders, P.T.K. 2001. Estrogen receptor beta, but not estrogen receptor alpha, is present in the vascular endothelium of the human and nonhuman primate endometrium. *J. Clin. Endo. Metab. 86:*1370–1378.

Critchley, H.O.D., Henderson, T.A., Kelly, R.W., Scobie, G.S., Evans, L.R., Groome, N.P., and Saunders, P.T.K. 2002. Wild type estrogen receptor, ERβ1 and the splice variant (ERβcx/β2) are both expressed throughout the normal menstrual cycle. *J. Clin. Endo. Metab. 87:*5265–5273.

Cuzick, J., Powles, T., Veronesi, U., Forbes, J., Edwards, R., Ashley, S., and Boyle, P. 2003. Overview of the main outcomes in breast-cancer prevention trials. *Lancet 361:*296–300.

Dixon, J.M., Anderson, T.J., and Miller, W.R. 2002. Neoadjuvant endocrine therapy of breast cancer: A surgical perspective. *Eur. J. Cancer 38:*2214–2221.

Dotzlaw, H., Leygue, E., Watson, P.H., and Murphy, L.C. 1997. Expression of estrogen receptor-beta in human breast tumors. *J. Clin. Endo. Metab. 82:*2371–2374.

Enmark, E., Pelto-Huikko, M., Grandien, K., Lagercrantz, S., Lagercrantz, J., Fried, G., Nordenskjold, M., and Gustafsson, J.-A. 1997. Human estrogen receptor β-gene structure, chromosomal localization, and expression pattern. *J. Clin. Endo. Metab. 82:*4258–4265.

Fuqua, S.A., Fitzgerald, S.D., Chamness, G.C., Tandon, A.K., McDonnell, D.P., Nawaz, Z., O'Malley, B.W., and McGuire, W.L. 1991. Variant human breast tumor estrogen receptor with constitutive transcriptional activity. *Cancer Res. 51:*105–109.

Fuqua, S.A., Schiff, R., Parra, I., Friedrichs, W.E., Su, J.L., McKee, D.D., Slentz-Kesler, K., Moore, L.B., Willson, T.M., and Moore, J.T. 1999. Expression of wild-type estrogen receptor beta and variant isoforms in human breast cancer. *Cancer Res. 59:*5425–5428.

Fuqua, S.A., Schiff, R., Parra, I., Moore, J.T., Mohsin, S.K., Osborne, C.K., Clark, G.M., and Allred, D.C. 2003. Estrogen receptor beta protein in human breast cancer: Correlation with clinical tumor parameters. *Cancer Res. 63:*2434–2439.

Furlow, J.D., Ahrens, H., Mueller, G.C., and Gorski, J. 1990. Antisera to a synthetic peptide recognize native and denatured rat estrogen receptors. *Endocrinology 127:*1028–1032.

Green, S., Walter, P., Kumar, V., Krust, A., Bornert, J.-M., Argos, P., and Chambon, P. 1986. Human oestrogen receptor cDNA: Sequence, expression and homology to v-erb-A. *Nature 320:*134–139.

Greene, G.L., Sobel, N.B., King, W.J., and Jensen, E.V. 1980. Monoclonal antibodies to human estrogen receptor. *Proc. Natl. Acad. Sci. U S A 77:*5115–5119.

Hall, J.M., and McDonnell, D.P. 1999. The estrogen receptor β-isoform (ERβ) of the human estrogen receptor modulates ERα transcriptional activity and is a key regulator of the cellular response to estrogens and antiestrogens. *Endocrinology 140:*5566–5578.

Hayat, M.A., 2002. *Microscopy, Immunohistochemistry, and Antigen Retrieval Methods.* New York and London: Kluwer Academic/Plenum Publishers.

Hopp, T.A., and Fuqua, S.A. 1998. Estrogen receptor variants. *J. Mammary Gland. Biol. Neoplasia 3:*73–83.

Huang, A., Leygue, E., Dotzlaw, H., Murphy, L.C., and Watson, P.H. 1999. Influence of estrogen variants on the determination of ER status in human breast cancer. *Breast Cancer Res. Treat. 58:*219–225.

Inoue, S., Ogawa, S., Horie, K., Hoshino, S., Goto, W., Hosoi, T., Tsutsumi, O., Muramatsu, M., and Ouchi, Y. 2000. An estrogen receptor beta isoform that lacks exon 5 has dominant negative activity on both ERα and ERβ. *BBSRC 279:*814–819.

Iwao, K., Miyoshi, Y., Egawa, C., Ikeda, N., Tsukamoto, F., and Noguchi, S. 2000. Quantitative analysis of estrogen receptor-alpha and -beta messenger RNA expression in breast carcinoma by real-time polymerase chain reaction. *Cancer 15:*1732–1738.

Järvinen, T.A.H., Pelto-Huikko, M., Holli, K., and Isola, J. 2000. Estrogen receptor β is co-expressed with ERa, and PR and associated with nodal status, grade and proliferation rate in breast cancer. *Am. J. Pathol. 156:*29–35.

Jensen, E.V., Cheng, G., Palmieri, C., Saji, S., Makela, S., Van Noorden, S., Wahlstrom, T., Warner, M., Coombes, R.C., and Gustafsson, J.A. 2001. Estrogen receptors and proliferation markers in primary and recurrent breast cancer. *Proc. Natl. Acad. Sci. USA 98:*15197–15202.

Jensen, E.V., and Jordan, V.C. 2003. The estrogen receptor: A model for molecular medicine. *Clin. Cancer Res. 9:*1980–1989.

King, W.J., and Greene, G.L. 1984. Monoclonal antibodies localize oestrogen receptor in the nuclei of target cells. *Nature 307:* 745–747.

Kobayashi, S., Ito, Y., Ando, Y., Omoto, Y., Toyama, T., and Iwase, H. 2000. Comparison of five different antibodies in the histochemical assay of estrogen receptor alpha in human breast cancer. *Breast Cancer 7:*136–141.

Kuiper, G.G.J.M., Enmark, E., Pelto-Huikko, M., Nilsson, S., and Gustafsson, J.-A. 1996. Cloning of a novel estrogen receptor expressed in rat prostate. *Proc. Natl. Acad. Sci. USA 93:* 5925–5930.

Kurosumi, M. 2003. Significance of immunohistochemical assessment of steroid hormone receptor status for breast cancer patients. *Breast Cancer 10:*97–104.

Leake, R., Barnes, D., Pinder, S., Ellis, I., Anderson, L., Anderson, T., Adamson, R., Rhodes, T., Miller, K., and Walker, R. 2000. Immunohistochemical detection of steroid receptors in breast cancer: A working protocol. *J. Clin. Pathol. 53:*634–635.

Lemieux, P., and Fuqua, S.A. 1996. The role of the estrogen receptor in tumor progression. *J. Steroid Biochem. Mol. Biol. 56:*87–91.

Leygue, E., Dotzlaw, H., Watson, P.H., and Murphy, L.C. 1998. Altered estrogen receptor a and β messenger RNA expression during human breast tumorigenesis. *Cancer Res. 58:*3197–3201.

Leygue, E., Dotzlaw, H., Watson, P.H., and Murphy, L.C. 1999. Altered expression of estrogen receptor-alpha variant messenger RNAs between adjacent normal breast and breast tumor tissues. *Breast Cancer Res. 2:*64–72.

Lu, B., Leygue, E., Dotzlaw, H., Murphy, L.C., and Watson, P.H. 1998. Estrogen receptor-β mRNA variants in human and murine tissues. *Mol. Cell Endocrinol. 138:*199–203.

Mann, S., Laucirica, R., Carlson, N., Younes, P.S., Ali, N., Younes, A., Li, Y., and Younes, M. 2001. Estrogen receptor beta expression in invasive breast cancer. *Human Pathol. 32:*113–118.

McGuire, W.L., Chamness, G.C., and Fuqua, S.A.W. 1991. Estrogen receptor variants in clinical breast cancer. *Mol. Endocrinol. 5:*1571–1577.

Miller, W.R., and Jackson, J. 2003. The therapeutic potential of aromatase inhibitors. *Expert Opin. Investig. Drugs 12:*337–351.

Miller, W.R., Stuart, M., Sahmoud, T., and Dixon, J.M. 2002. Anastrozole ("Arimidex") blocks oestrogen synthesis both peripherally and within the breast in postmenopausal women with large operable breast cancer. *Br. J. Cancer 87:*950–955.

Mofidi, R., Walsh, R., Ridgway, P.F., Crotty, T., McDermott, E.W., Keaveny, T.V., Duffy, M.J., Hill, A.D., and O'Higgins, N. 2003. Objective measurement of breast cancer oestrogen receptor status through digital image analysis. *Eur. J. Surg. Oncol. 29:* 20–24.

Moore, J.T., McKee, D.D., Slentz-Kesler, K., Moore, L.B., Jones, S.A., Horne, E.L., Su, J.L., Kliewer, S.A., Lehmann, J.M., and Willson, T.M. 1998. Cloning and characterisation of human estrogen receptor beta isoforms. *BBSRC 247:*75–78.

Mosselman, S., Polman, J., and Dijkema, R. 1996. ERbeta: Identification and characterization of a novel human estrogen receptor. *FEBS lett 392:*49–53.

Norton, A.J., Jordan, S., and Yeomans, P. 1994. Brief, high-temperature heat denaturation (pressure cooking): A simple and effective method of antigen retrieval for routinely processed tissues. *J. Pathol. 173:*371–379.

Ogawa, S., Inoue, S., Watanabe, T., Hiroi, H., Orimo, A., Hosoi, T., Ouchi, Y., and Muramatsu, M. 1998a. The complete primary structure of human estrogen receptor beta (hER beta) and its heterodimerization with ER alpha *in vivo* and *in vitro*. *BBSRC 243:*122–126.

Ogawa, S., Inoue, S., Watanabe, T., Orimo, A., Hosoi, T., Ouchi, Y., and Muramatsu, M. 1998b. Molecular cloning and characterization of human estrogen receptor βcx: A potential inhibitor of estrogen action in human. *Nucl. Acids Res. 26:*3505–3512.

Omoto, Y., Inoue, S., Ogawa, S., Toyama, T., Yamashita, H., Muramatsu, M., Kobayashi, S., and Iwase, H. 2001. Clinical value of the wild type estrogen receptor β expression in breast cancer. *Cancer Let. 163:*207–212.

Omoto, Y., Kobayashi, S., Inoue, S., Ogawa, S., Toyama, T., Yamashita, H., Muramatsu, M., Gustafsson, J.A., and Iwase, H. 2002. Evaluation of oestrogen receptor beta wild-type and variant protein expression, and relationship with clinico-pathological factors in breast cancers. *Eur. J. Cancer 38:* 380–386.

Palmieri, C., Cheng, G., Saji, S., Zelada-Hedman, M., Warri, A., Weihua, Z., Van Noorden, S., Wahlstrom, T., Coombes, R.C., Warner, M., and Gustafsson, J.A. 2002. Estrogen receptor beta in breast cancer. *Endo. Rel. Cancer 9:*1–13.

Pavao, M., and Traish, A.M. 2001. Estrogen receptor antibodies: Specificity and utility in detection, localization and analyses of estrogen receptor a and β. *Steroids 66:*1–66.

Pettersson, K., Grandien, K., Kuiper, G.G.J.M., and Gustafsson, J.-A. 1997. Mouse estrogen receptor β forms estrogen receptor response element-binding heterodimers with estrogen receptor α. *Mol. Endocrinol. 11:*1486–1496.

Pettersson, K., and Gustafsson, J.-A. 2001. Role of estrogen receptor beta in estrogen action. *Ann. Rev. Physiology 63:*165–192.

Pike, A.C.W., Brzozowski, A.M., Hubbard, R.E., Bonn, T., Thorsell, A.-G., Engstrom, O., Ljunggren, J., Gustafsson, J.-A., and Carlquist, M. 1999. Structure of the ligand-binding domain of oestrogen receptor beta in the presence of a partial agonist and a full antagonist. *EMBO J. 18:*4608–4618.

Poola, I., and Speirs, V. 2001. Expression of alternatively spliced estrogen receptor alpha mRNAs is increased in breast cancer tissues. *J. Steroid Biochem. Mol. Biol. 78:*459–469.

Rhodes, A., Jasani, B., Balaton, A.J., and Miller, K.D. 2000a. Immunohistochemical demonstration of oestrogen and progesterone receptors: Correlation of standards achieved on in house tumours with that achieved on external quality assessment material in over 150 laboratories from 26 countries. *J. Clin. Pathol. 53:*292–301.

Rhodes, A., Jasani, B., Barnes, D.M., Bobrow, L.G., and Miller, K.D. 2000b. Reliability of immunohistochemical demonstration of oestrogen receptors in routine practice: Interlaboratory variance in the sensitivity of detection and evaluation of scoring systems. *J. Clin. Pathol. 53:*125–130.

Rudiger, T., Hofler, H., Kreipe, H.H., Nizze, H., Pfeifer, U., Stein, H., Dallenbach, F.E., Fischer, H.P., Mengel, M., von Wasielewski, R., and Muller-Hermelink, H.K. 2002. Quality assurance in immunohistochemistry: Results of an interlaboratory trial involving 172 pathologists. *Am. J. Surg. Pathol. 26:*873–882.

Saji, S., Omoto, Y., Shimizu, C., Horiguchi, S., Watanabe, T., Funata, N., Hayash, S., Gustafsson, J.A., and Yoi, M. 2002a. Clinical impact of assay of estrogen receptor betacx in breast cancer. *Breast Cancer 9:*303–307.

Saji, S., Omoto, Y., Shimizu, C., Warner, M., Hayashi, Y., Horiguchi, S.-i., Wantanabe, T., Hayashi, S.-i., Gustafsson, J.A., and Toi, M. 2002b. Expression of estrogen receptor (ER) βcx protein in ERa-positive breast cancer: Specific correlation with progesterone receptor. *Cancer Res. 62:*4849–4853.

Saunders, P.T.K. 1998. Oestrogen receptor beta (ERβ). *Rev. Reprod. 3:*164–171.

Saunders, P.T.K., Millar, M.R., Macpherson, S., Irvine, D.S., Groome, N.P., Evans, L.R., Sharpe, R.M., and Scobie, G.A. 2002a. Estrogen receptor beta (ERβ1), and the estrogen receptor beta 2 splice variant (ERβcx/2), are expressed in distinct cell populations in the adult human testis. *J. Clin. Endo. Metab. 87:* 2706–2715.

Saunders, P.T.K., Millar, M.R., Williams, K., Macpherson, S., Bayne, C., O'Sullivan C., Anderson, T.J., Groome, N.P., and Miller, W.R. 2002b. Expression of oestrogen receptor beta (ER beta 1) protein in human breast cancer biopsies. *Br. J. Cancer 86:*250–256.

Saunders, P.T.K., Millar, M.R., Williams, K., Macpherson, S., Harkiss, D., Anderson, R.A., Orr, B., Groome, N.P., Scobie, G., and Fraser, H.M. 2000. Differential expression of estrogen receptor-alpha and -beta and androgen receptor in the ovaries of marmoset and human. *Biol. Reprod. 63:*1098–1105.

Scobie, G.S., Macpherson, S., Millar, M.R., Groome, N.P., Romana, P.G., and Saunders, P.T.K. 2002. Human estrogen receptors: Differential expression of ER alpha and beta and the identification of ER beta variants. *Steroids 67:*985–992.

Simpson, E. 2002. Aromatization of androgens in women: Current concepts and findings. *Fertil. Steril. 77:*S6–S10.

Simpson, E.R., Mahendroo, M.S., Means, G.D., Kilgore, M.W., Hinshelwood, M.M., Graham-Lorence, S., Amarneh, B., Ito, Y., Fisher, C.R., and Michael, M.D. 1994. Aromatase cytochrome P450 the enzyme responsible for estrogen biosynthesis. *Endo. Rev. 15:*342–355.

Skliris, G.P., Parkes, A.T., Limer, J.L., Burdall, S.E., Carder, P.J., and Speirs, V. 2002. Evaluation of seven oestrogen receptor beta antibodies for immunohistochemistry, Western blotting, and flow cytometry in human breast tissue. *J. Pathol. 197:*155–162.

Speirs, V., Adams, I.P., Walton, D.S., and Atkin, S.L. 2000. Identification of wild-type and exon 5 deletion variants of estrogen receptor beta in normal human mammary gland. *J. Clin. Endo. Metab. 85:*1601–1605.

Speirs, V., Parkes, A.T., Kerin, M.J., Walton, D.S., Carleton, P.J., Fox, J.N., and Atkin, S.L. 1999. Coexpression of estrogen receptor a and β: Poor prognostic factors in human breast cancer. *Cancer Res. 59:*525–528.

van Leeuwen, F.E., van den Belt-Dusebout, A.W., Diepenhorst, F.W., van Tinteren, H., Coebergh, J.W.W., Kiemency, L.A.L.M., Gimbrère, C.H.F., Otter, R., Schouten, L.J., Damhuis, R.A.M., Benraadt, J., and Bontenbal, M. 1994. Risk of endometrial cancer after tamoxifen treatment of breast cancer. *Lancet 343:*448–452.

Vladusic, E.A., Hornby, A.E., Guerra-Vladusic, F.K., Lakins, J., and Lupu, R. 2000. Expression and regulation of estrogen receptor beta in human breast tumors and cell lines. *Oncol. Rep. 7:*157–167.

Vladusic, E.A., Hornby, A.E., Guerra-Vladusic, F.K., and Lupu, R. 1998. Expression of estrogen receptor beta messenger RNA variant in breast cancer. *Cancer Res. 58:*210–214.

Weihua, Z., Saji, S., Makinen, S., Cheng, G., Jensen, E.V., Warner, M., and Gustafsson, J.-A. 2000. Estrogen receptor (ER) β, a modulator of ERα in the uterus. *Proc. Natl. Acad. Sci. USA 97:*5936–5941.

Zafrani, B., Aubriot, M.H., Mouret, E., De Cremoux, P., De Rycke, Y., Nicolas, A., Boudou, E., Vincent-Salomon, A., Magdelenat, H., and Sastre-Garau, X. 2000. High sensitivity and specificity of immunohistochemistry for the detection of hormone receptors in breast carcinoma: Comparison with biochemical determination in a prospective study of 793 cases. *Histopathology 37:*536–545.

21

Immunofluorescence and Immunohistochemical Localization of Progesterone Receptors in Breast Carcinoma

P.A. Mote

Introduction

The Physiological Role of Progesterone

The ovarian steroid hormone progesterone plays a central role in many aspects of regulation of normal female reproductive function, and the predominant mode of mediation of progesterone effects is believed to be via classical steroid hormone action on its cognate receptor, the progesterone receptor (PR). The major physiological targets for progesterone action in the female mammals are the uterus, ovary, breast, and brain. Recent studies of growth and development in PR-null mice have significantly increased our understanding of the diversity and complexity of the role of progesterone (Lydon *et al.*, 1995, 1996). Many of the effects of progesterone may be through its ability to counteract the proliferative actions of estrogen (Clarke *et al.*, 1990; Lydon *et al.*, 1995). In the normal breast, progesterone has both proliferative and differentiating roles. It is essential for lobular-alveolar growth and development during pregnancy and also coordinates

with other hormones, such as prolactin, to inhibit lactation (Clarke *et al.*, 1990).

PR Expression in the Normal Breast

Because of difficulties in obtaining samples, there are few studies on PR expression in normal human breast tissue, in contrast to numerous reports available on receptor expression in malignant breast tissue. The PR expression analyzed by immunohistochemical techniques is observed only in the nucleus of breast epithelial cells, and there is no evidence of PR staining in stromal cells in this tissue (Battersby *et al.*, 1992). The PR expression remains constant during all phases of the menstrual cycle, unlike estrogen receptor (ER) expression, which declines during the luteal phase (Battersby *et al.*, 1992).

PR Expression in Breast Carcinoma

The PR is an important prognostic marker in breast cancer, and receptor expression is routinely assessed as

Handbook of Immunohistochemistry and in situ Hybridization of Human Carcinomas, Volume 1: Molecular Genetics; Lung and Breast Carcinomas

449

Copyright © 2004 by Elsevier (USA)
All rights reserved.

an integral part of disease management. Absence of PR in the primary breast tumor is associated with disease progression (Balleine *et al.*, 1999) and may be reflective of an aggressive tumor phenotype. The majority of studies of steroid receptor expression and its relationship to prognosis have reported a positive correlation with patient outcome (Pertschuk *et al.*, 1993). Receptor negativity is demonstrated to be strongly correlated with early disease recurrence (Pertschuk *et al.*, 1993), whereas receptor positive patients have a significantly longer recurrence-free survival time (Kommoss *et al.*, 1994). Although evaluation of PR expression is valuable in predicting disease-free and overall survival of the patient (Pertschuk *et al.*, 1993), it is probably most beneficial in establishing optimal therapy by determining the likelihood of a positive response to endocrine treatment (Osborne, 1998; Pichon *et al.*, 1980). For many years endocrine therapy was only a third-line treatment for breast cancer, due to its high failure rate and the inability to distinguish patients for which it might be effective (Pearce *et al.*, 1993). In the late 1960s, work from several laboratories provided a way of quantitating ER and PR in tumor cytosols by ligand binding assays, and measurements of these receptors soon became routine in patient management, greatly increasing the success rate of this form of treatment (Pearce *et al.*, 1993). Subsequent production of monoclonal antibodies permitted the development of enzyme immunoassays (EIA) based on direct recognition of receptor molecules, and more recently, immunohistochemical techniques have become popular, to the extent that they are now considered to be the method of choice for the detection of hormone receptors in tissue samples (Battifora, 1994; Taylor, 1994).

Immunohistochemical analysis of PR expression in breast tumors shows a marked heterogeneity of both staining intensity and positive cell distribution in malignant breast tissue, and areas within the same tumor, even adjacent areas, may range from intensely PR positive to clearly negative. Staining is confined to the nucleus of the epithelial cells, with no evidence of receptor expression in other cell types (Kommoss *et al.*, 1994; Zeimet *et al.*, 1994).

Progesterone Receptor Isoforms, PRA and PRB—*in vitro* Studies

The human PR belongs to the steroid-thyroid-retinoic acid receptor superfamily (Evans, 1988; Tsai *et al.*, 1994) and is a nuclear receptor encoded by a single-copy gene located on chromosome 11q22-23 (Rousseau-Merck *et al.*, 1987). PR is expressed as two proteins of dissimilar weight (PRB 100–120 kDa and PRA 79–94 kDa). PRA is identical to PRB except that

it lacks the first 164 amino acids. PR functions as a transcription factor that modulates target gene transcription in response to progesterone, and there have been several reviews published on the many studies of steroid receptor function (Conneely *et al.*, 2000; Gronemeyer, 1991).

The two isoforms, PRA and PRB, are both transcriptional activators of progesterone responsive target genes; however, recent *in vitro* evidence has suggested that the two PR isoforms may have different functions and that their relative expression may be important in determining cellular response to hormones (Meyer *et al.*, 1992; Vegeto *et al.*, 1993). In all cell types examined, PRB displays hormone-dependent transactivation and in general is more active than PRA. Conversely, in situations in which it is itself inactive, PRA can act as a transdominant repressor of PRB-mediated transcription under the influence of hormone (Tung *et al.*, 1993; Vegeto *et al.*, 1993). The PRA is also able to function as a repressor of mineralocorticoid receptor (MR), androgen receptor (AR), and glucocorticoid receptor (GR) activity when not active itself, suggesting it may play a major regulatory role in the overall cellular response to hormones (McDonnell *et al.*, 1994; Vegeto *et al.*, 1993). In support of this, overexpression of PRA in cultured breast cancer cells results in altered cell response to progestins (McGowan *et al.*, 1999). The results of *in vitro* studies have implications for the relationship between PR isoform expression and the potential clinical response to treatment *in vivo*. However, many *in vitro* findings are derived from transient transfections of each PR isoform in isolation, frequently into nontarget cell lines, and significant caution is therefore required to extrapolate significance *in vivo*, particularly as both PRA and PRB are frequently coexpressed in *vivo* in target cells (Graham *et al.*, 2002).

Progesterone Receptor Isoforms, PRA and PRB—*in vivo* Studies

Studies using transgenic mice have attempted to address the issue of PR isoform involvement in mouse mammary gland development. Mice expressing an excess of the PRA isoform, with a PRA:B ratio higher than the normal 3:1 for rodent reproductive tissue, demonstrated aberrations in mammary gland development such as ductal hyperplasia associated with basement membrane disorganization and decreased cell-to-cell adhesion. These data support the importance of a precise regulation of the relative expression of PRA:B for normal mammary gland development and suggest that an imbalance in PR isoform expression could result in inappropriate progesterone signaling, leading to aberrant mammary gland development (Shyamala, 1999), in addition to having significant implications for mammary carcinogenesis and treatment.

Recent studies on human breast and endometrial tissues have shown that in most normal cell types there are approximately equivalent levels of PRA and PRB expressed and that there is significant homogeneity in relative PRA and PRB expression in all PR positive cells (Arnett-Mansfield *et al.*, 2001; Mote *et al.*, 1999; Mote *et al.*, 2002). The coordinate regulation of PR isoform expression allows these cells to respond to systemic hormonal and other signals in unison and thereby supports a role for both PR isoforms in the normal response to progesterone.

In breast tissues, the cellular homogeneity of PRA: PRB expression observed in the normal state is replaced by an increasing frequency of cellular heterogeneity in benign and malignant breast lesions where adjacent tumor cells may display marked variability in their PRA:B levels, indicating a loss of regulated PR isoform expression and/or a predominance of one of the PR isoforms (Mote *et al.*, 2002). Loss of control of relative PRA:PRB expression is an early event in the development of breast cancer (Mote *et al.*, 2002). Breast cancers commonly express a predominance of one PR isoform (Mote *et al.*, 2002), and these changes in the ratio of PRA:PRB proteins within a cell are likely to result in aberrant hormonal responses, suggesting that expression of PRA or PRB predominance in breast carcinomas may alter hormone action in the breast and contribute to the evolution of the malignant phenotype. The clinical relevance of PRA:B expression is not yet fully understood, but given the proposed differential functions of PRA and PRB, as suggested by *in vitro* data, measurement of the relative expression of the individual PR isoforms may well become an important clinical marker in the future to facilitate optimal therapeutic decisions.

Immunohistochemical Detection of PR Proteins in Archival Breast Tissue

Reliable assessment of steroid receptor expression in malignant breast tissue is an important aid for successful patient management (Pertschuk *et al.*, 1993; Pichon *et al.*, 1996). This is most often performed using an immunoperoxidase technique because the stained slides are stable indefinitely at room temperature, easy to assess for PR content, and do not require the use of a fluorescent microscope.

Immunoperoxidase Staining

Determining hormone receptors by immunoperoxidase methods is now routine practice in many pathology laboratories to identify patients who are most likely to benefit from endocrine treatment. Rigorous standardization and quality assurance procedures are imposed to ensure accuracy and reproducibility, and laboratories performing immunoperoxidase staining techniques are encouraged to participate in proficiency testing programs (Regitnig *et al.*, 2002; Taylor, 1994). Immunohistochemical techniques have become more acceptable for the measurement of receptor proteins because they have certain advantages over biochemical methods and, importantly, correlate well with endocrine response (Page *et al.*, 1996). Success of immunohistochemical detection techniques is primarily due to the development of reliable antigen retrieval methods, providing a good correlation of results with early techniques (Cavaliere *et al.*, 1996; Katoh *et al.*, 1997) and to the widespread availability of commercial antibodies. Immunohistochemistry, using immunoperoxidase, is now the preferred method of routinely measuring PR expression in tumors for clinical assessment in most institutions (Hayat, 2002).

Dual Immunofluorescent Staining

The PR proteins can be visualized by fluorescent techniques, although for total PR content or evaluating PRA and PRB on adjacent sections there is no advantage compared to methods using immunoperoxidase. In fact immunofluorescence is generally a less-sensitive technique, storage of stained sections is limited, and interpretation of overall PR staining in a tissue is more complicated. Furthermore, most diagnostic laboratories have light microscopes, but the more expensive equipment required for analysis of fluorescent slides, such as a microscope fitted with the correct filters and digital imaging cameras for storage of data and semiquantitative analysis if needed, would be less likely to be readily available. Fluorescent techniques are essential, however, in dual staining to reveal both PR isoforms simultaneously within the same tissue section.

Advantages of Immunohistochemical Methods

Immunohistochemistry enables the visualization of circumscribed antigens in tissue sections by the binding of antibodies to their specific epitopes (Werner *et al.*, 1996). The PR can be detected in retrospective studies and in very small tumor samples when early tumor detection frequently results in reduced biopsy size, with insufficient material available for biochemical studies.

In addition to reduced tissue requirements, the most significant advantage of immunohistochemical detection of receptor proteins is the ability to demonstrate the exact subcellular location of the antigen, eliminating erroneous results introduced by the variable and unknown composition of tissue homogenates (Taylor, 1996).

Initial immunohistochemical analyses were performed predominantly on frozen tissue, due to problems of antigen masking encountered in archival, paraffin-embedded material. Measurement of PR by immunohistochemistry has been successful in frozen tissue (Tesch et al., 1993), but frozen material is not always readily available and has limitations in the use of retrospective studies, in ease of interpretation due to poor preservation of tissue morphology, and in reduced antigen stability during storage of sections before staining. In contrast, sections cut from paraffin-embedded tissue blocks can be stored at 4°C, for a period of several months with no loss of intensity in PR detection by immunoperoxidase staining, although storage for this length of time may not be satisfactory for all antigens (Jacobs et al., 1996). As a general rule, however, staining is best performed within three to four weeks of the sections being cut, and this is particularly important for fluorescent staining.

Antigen Retrieval

Formalin is the most widely used fixative for preserving tissues for pathological examination, as it is inexpensive, is easy to prepare, and conserves excellent morphological details without excessive tissue shrinkage. Unfortunately, though, it is by no means an ideal preservative for subsequent immunohistochemical analyses, as only a relatively few antibodies can be used satisfactorily, due to masking of antigenic epitopes during the fixation procedure (Werner et al., 1996). It was not until the fairly recent development of reliable antigen-retrieval techniques that immunohistochemistry (IHC) on archival or routinely processed tissues could become a viable option for receptor analysis.

Optimally revealing PR in archival tissues is a critical issue that, if not carried out correctly, could easily result in either no or inferior detection of PR proteins in a receptor positive tissue (Rhodes et al., 2001). The importance of antigen retrieval to reveal PR proteins by immunohistochemistry is shown (Figure 66A–B). Epitope unmasking is necessary to cleave bonds that are caused by fixation, and initial attempts to achieve this utilized enzymatic predigestion of the tissue. However, the cleavage sites of proteolytic enzymes are nonspecific and liable to affect the epitopes themselves, making timing and concentration critical for each tissue type and reproducibility difficult (Shi et al., 1995; Werner et al., 1996). These early techniques have now been largely replaced by heat-induced antigen-retrieval methods (Hayat, 2002).

Heat-Induced Methods of Antigen Retrieval

High-temperature preheating in the presence of a salt buffer, such as citrate, has been shown to unmask many antigens in formalin-fixed, paraffin-embedded sections for immunohistochemical staining (Taylor et al., 1994). Microwave techniques for antigen retrieval were first described by Shi and colleagues and have been in common use since 1991 (Shi et al., 1991). The microwave-retrieval method was found to be highly effective in archival material that frequently had prolonged formalin fixation (Cattoretti et al., 1993; Taylor et al., 1994). More recently, however, some authors have reported hydrated autoclaving methods to be superior to microwaving, both technically and in the staining results obtained (Mote et al., 1998). In addition to increasing retrieval of PR antigen, autoclaving has a number of technical advantages over microwaving. Firstly, the ability to uniformly heat a large number of slides without the need to monitor the sections carefully throughout the autoclaving process to prevent drying increases both ease and standardization of the method. Secondly, tissues are better preserved when pretreated by autoclaving, as microwaving frequently causes violent boiling of the buffer in which the sections are immersed, resulting in substantial tissue damage or loss. Antigen retrieval methods where sections are heated in a water bath to a temperature below boiling point are not recommended. Such methods will reveal PR in tissues where it is expressed abundantly, but where PR levels are low they are likely to produce a negative result. In PR positive tissues the amount of antigen detected is inferior to that detected after autoclaving (Figure 66C–D).

There are several different solutions for antigen retrieval that can be purchased from commercial sources, although sodium citrate (0.01 M, pH 6.0) remains the most popular because it is very simple to prepare in the laboratory and works well with most antibodies, including those to detect PR. Early work discussed in an excellent review by Shi and colleagues provides information on other solutions that can be used for antigen retrieval (Shi et al., 1995).

Mechanisms of Antigen Retrieval

Successful antigen retrieval, using high-heat methods, has challenged the original premise that fixation irreversibly alters the epitope, preventing formation of antigen–antibody complexes. It is now widely accepted that epitopes are merely concealed by fixation-related events and are not destroyed (Cattoretti et al., 1993). Masking is particularly acute when using monoclonal antibodies that recognize only a single epitope on the antigen (Bell et al., 1987). Formalin-modified epitopes are believed to be protected from denaturation during heating (Shi et al., 1995). The mechanism of heat-induced epitope unmasking is not clear, but it is hypothesized to be due to denaturation of cross-linked

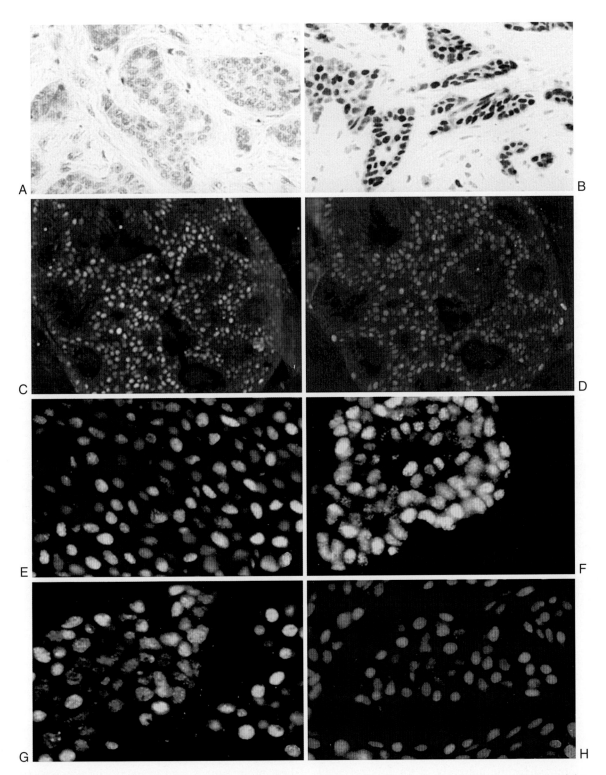

Figure 66 Expression of PR proteins in archival breast carcinomas determined either by immunoperoxidase staining (**A** and **B**) or by dual immunofluorescent histochemistry (**C–H**). PR expression in tissue without antigen retrieval pretreatment (**A**) versus tissue autoclaved in sodium citrate buffer (pH 6.0) for 30 min at 121°C (**B**). Dual immunofluorescent histochemistry of PR proteins unmasked by autoclaving in sodium citrate buffer (pH 6.0) at 121°C for 30 min (**C**), or by immersion in a 98°C water bath for 40 min (**D**). The PRA and B proteins viewed under dual-wavelength (TXR/FITC) excitation showing marked adjacent cell heterogeneity in the relative levels of PRA and B (**E**); similar levels of PRA and B (**F**); predominant expression of PRA (**G**), or predominant expression of PRB (**H**). Original magnification ×400 (**A, B,** and **E–H**), ×200 (**C** and **D**). The primary antibodies used were Novocastra Clone 16, catalogue number NCL-PGR-312 (Novocastra Laboratories Ltd, Newcastle, UK) for immunoperoxidase staining (**A** and **B**) or hPRa6 and hPRa7 (Neomarkers, Fremont, USA) for dual staining (**C–H**).

proteins (Cattoretti *et al.*, 1993; Werner *et al.*, 1996). Heat may supply the energy to break bonds between calcium ions or other divalent metal cations and proteins, and the buffer may subsequently precipitate or chelate released metal ions (Bankfalvi *et al.*, 1994; Morgan *et al.*, 1994; Werner *et al.*, 1996).

Factors Influencing Antigen Retrieval

The length of fixation in formalin is considered to be an important variable in the formation of formalin-induced cross-links, and cross-linking is suggested to be a progressive event, because longer retrieval times are correlated to longer fixation times (Munakata *et al.*, 1993; Werner *et al.*, 1996). Short (1–2 h) fixation times, preventing extensive cross-linking, have been used successfully to demonstrate PR in the absence of any pretreatment (Ozzello *et al.*, 1991; Taylor *et al.*, 1994). Unfortunately, the length of time that a sample remains in fixative is not easily controlled in a clinical setting and generally far exceeds 1–2 hr, with the consequence that there remains a need for reliable antigen-retrieval techniques (Rhodes *et al.*, 2001). Although newer fixatives especially designed for use in immunohistochemistry can now be obtained (Elias *et al.*, 1989), most pathologists still prefer to use formalin. The trend toward smaller biopsies and less-available material compounds this problem, because most pathologists would be unwilling to divide the specimen for formalin and alternative fixation.

Factors that are known to affect efficacy of antigen retrieval include duration and temperature of heating and pH, chemical composition, and molarity of the buffer solution used (Shi *et al.*, 1995; Taylor, 1996). All these factors are crucial for optimal epitope unmasking, and, furthermore, ideal conditions vary between antigens, as it has been clearly shown that demonstration of some antigens does not benefit, and may even be harmed by, antigen retrieval (Cattoretti *et al.*, 1993; Shi *et al.*, 1995; Werner *et al.*, 1996). The amount of pretreatment necessary for optimal antigen retrieval is thought to be dependent not only on fixation time and on the antigen to be demonstrated, but also on the actual epitope involved, with some epitopes requiring more vigorous unmasking than others (Werner *et al.*, 1996). For example, it has been shown that the PR epitope recognized by the antibody hPRa 7 is revealed effectively by no more than 5 min autoclaving time, whereas that recognized by the antibody hPRa 6 requires 30 min for optimal retrieval (Mote *et al.*, 1998). This lends weight to the view that cross-linking of epitopes as a consequence of formalin fixation may be more extensive for some antigen epitopes than others and establishes the importance of optimizing PR antigen retrieval before staining of routinely formalin-fixed

clinical specimens. A recent study of interlaboratory reliability and reproducibility of steroid receptor staining found too short an antigen retrieval time to be the principal contributory factor to variable results observed in immunohistochemical staining (Rhodes *et al.*, 2001).

The actual intensity of PR immunostaining achieved may not be of major concern if expression levels of the protein are high, and only information on presence or absence of receptor is required. However, low-level expression of PR may be undetected due to suboptimal antigen unmasking and could have important consequences in a therapeutic decision, particularly if there are a great number of cells expressing low levels of receptor, resulting in an overall moderate yet clinically important level of PR content. Furthermore, it is essential that antigen unmasking be equally effective for each PR epitope to be demonstrated in a dual-staining technique, especially if the levels of the two proteins expressed are to be subsequently compared. It is advisable that each individual antibody to be used in an immunohistochemical technique be first subjected to a rigorous optimization procedure that includes both method of antigen retrieval and length of time for pretreatment to ensure maximum demonstration of PR proteins.

Pretreatment is usually recommended for commercial anti-human PR antibodies in routine use. Newly developed antibodies to detect PR that have been selected prospectively for resistance to formalin fixation and paraffin embedding may eventually reduce the need for rigorous antigen retrieval in the future (Press *et al.*, 2002). However, most commercial antibodies currently available, including antibodies that recognize PRA or PRB individually, remain highly dependent on this procedure.

Section Cutting and Tissue Loss

One of the major disadvantages of rigorous heat-induced antigen retrieval is the problem of tissue loss or damage, and although sections are generally better preserved during autoclaving than microwaving, it can still be of concern, particularly with highly fatty and fibrous tissue such as breast (Mote *et al.*, 1998). Clearly not all tissue types are vulnerable during high-heat pretreatment, and in many situations section loss will not be considered a problem. However, breast tissue sections are very susceptible to loss during these rigorous procedures, particularly when prolonged heating times are necessary to achieve optimal PR immunostaining in formalin-fixed archival specimens (Mote *et al.*, 1998). The use of electrostatically charged slides results in a notable improvement in section adhesion and is strongly recommended. This improvement is further enhanced by the addition of Mayer Albumen adhesive (Mote *et al.*, 1998).

Section adhesion is also related to both section thickness and to slide drying time after cutting. It has been shown that fewer breast tissues will be lost when sections are cut at 2 μm compared to 4 μm (Mote *et al.*, 1998). Cutting tissue at 2 μm may prove to be technically demanding, especially with some samples, but because increased retention is obtained when using thinner sections in combination with an adhesive (Mote *et al.*, 1998), it should be considered, particularly if section loss is a problem. Drying sections at 37°C for 72 hr, rather than the routinely used 60°C for 1 hr, has also been shown to minimize section loss (Mote *et al.*, 1998).

Antibodies to Detect PR

There are several antibodies available commercially for use in immunohistochemical detection of PR—both monoclonal and polyclonal. Although the majority of suppliers provide detailed guidelines on their use in immunohistochemistry with respect to antigen retrieval required, antibody dilution, and incubation time, it is still advisable to optimize the conditions of new primary antibody use in each laboratory. In some breast tumor tissues background staining can be a problem, and optimization of the primary antibody will allow the best signal-to-noise ratio to be determined.

Most commercial antibodies for detection of PR are presumed to detect both isoforms, as they do so by Western immunoblotting. However, it has recently been reported that this may not be a correct assumption for formalin-fixed archival tissues (Mote *et al.*, 2001). Surprisingly, recent data show that, despite the similarity between PRA and PRB, not all available antibodies to human PR are able to recognize both proteins with equal efficacy by immunohistochemical methods, and that some antibodies fail to detect PRB proteins at all when they are in formalin-fixed conformations in tissue sections (Mote *et al.*, 2001). Detection of both PRA and PRB by immunoblot analysis when proteins are linearized, together with a lack of recognition of PRB when they are in formalin-fixed conformations in tissue sections, strongly implicates the involvement of protein folding. Alternatively, epitope masking could result from interaction of PRB, but not PRA, with another molecule *in vivo* prior to formalin fixation, concealing the critical antibody recognition site (Mote *et al.*, 2001).

For total PR expression, an antibody known to detect both PR isoforms can be used or, alternatively, a combination of two antibodies that recognize each isoform. For dual immunofluorescent staining of PRA and PRB proteins in the same section, antibodies recognizing individual isoforms must of course be used sequentially, as described in the methods section. Unfortunately, the PR antibodies currently available detect the two PR isoforms individually; these antibodies are raised in the same host animal, the mouse, with problems of potential cross-reactivity if utilized in a dual-staining technique. One way to avoid the problem of cross-reactivity because of both primary antibodies being raised in the same host species is to stain adjacent sections. This technique has been reported in some studies (Kudo *et al.*, 1996; Wang *et al.*, 1998), but is not a very satisfactory method as the direct comparison and analysis of two different sections stained using two different antibodies is not ideal, because it is remarkably difficult to accurately assess relative levels of antigen expression between two sections on a cell-to-cell basis. Demonstration of PRA and PRB proteins individually by immunohistochemistry is preferable by dual staining of the two antigens in the same tissue section, and this option is decidedly superior compared to staining adjacent sections. Recently, a dual-immunofluorescent staining technique has been established to visualize individual PR proteins within the same tissue section (Mote *et al.*, 1999). The dual-immunofluorescent staining method is described in detail in this chapter.

In conclusion, reliable demonstration of PR expression in breast tumors is clinically important to assess a patient's likely response to endocrine treatment, and given the heterogeneous nature of receptor expression in many tumors, immunohistochemistry is the ideal way to accomplish this. Careful determination of the best possible antigen retrieval method and staining protocol required for each antibody will ensure that optimal conditions are achieved to fully reveal PRA and PRB proteins in archival tissue sections. Two immunohistochemical staining methods to demonstrate PR proteins in archival tissues are described in this chapter:

1. An immunoperoxidase method to determine either total PR expression, or PRA and PRB proteins individually on adjacent sections.

2. A dual-immunofluorescent technique to simultaneously reveal PRA and PRB individually within the same tissue.

MATERIALS

1. Section cutting: Cryospray tissue coolant; disposable blades (*Feather*—S35 Arthur Bailey Surgical, Sydney) for use with a standard rotary microtome, electrostatically charged glass microscope slides (*Superfrost Plus*, Lomb Scientific, Sydney).

2. Mounting material: Mayer Albumen adhesive made with equal parts glycerine and egg white (Humason, 1979).

3. Antigen retrieval buffer: 0.1 M Tri-sodium citrate ($Na_3C_6H_5O_7 \cdot 2H_2O$) in distilled water, adjust to pH 6.0 using 1 M citric acid solution. Dilute 1/10 with distilled water and again adjust to pH 6.0 with 1 M citric acid solution immediately before use.

4. Dehydration and hydration: Xylene (dewaxing), 100% ethanol, 70% ethanol.

5. Phosphate-buffered saline (PBS): 137 mM sodium chloride (NaCl), 8.1 mM di-sodium hydrogen phosphate (Na_2HPO_4 [anhydrous]), 1.5 mM potassium di-hydrogen phosphate (KH_2PO_4), 2.7 mM potassium chloride (KCl), in distilled water (pH 7.3).

6. PBS with 0.5% (v/v) triton (PBT): 5 ml Triton-X 100 in 1l of PBS.

7. "Pap Pen"—use to form a fluid barrier around the tissues before staining.

8. Normal goat serum and biotinylated secondary antibodies diluted in PBS.

9. Primary mouse monoclonal antibodies to detect human PR diluted in PBT.

Immunoperoxidase Staining

10. Quenching: 3.0% (v/v) hydrogen peroxide—20 ml 30% hydrogen peroxide in 180 ml distilled water.

11. Streptavidin-biotin-horseradish peroxidase complex: Biotinylated-peroxidase and streptavidin reagents prepared, according to the manufacturer's instructions, in PBT.

12. Chromogen 3,3' diaminobenzidine (DAB) prepared by dissolving one tablet in 10 ml PBS. Activate immediately before use by adding 7.5 µl of 3.0% hydrogen peroxide per ml of DAB. (1 mg/ml DAB plus 0.02% (v/v) (final concentration) hydrogen peroxide, in PBS).

13. Harris hematoxylin as a counterstain.

14. Scott's alkaline blueing solution: 0.02 M sodium bicarbonate ($NaHCO_3$), 0.08 M magnesium sulfate ($MgSO_4$), in distilled water. Add a pinch of thymol to retard molds.

15. Xylene-miscible mountant.

Fluorescent Staining

16. Blocking antibody (Goat anti-mouse F(ab) immunoglobulins) diluted in 1% (w/v) bovine serum albumin (BSA): 1 g BSA in 100 ml PBS.

17. Texas red buffer: 0.1 M sodium bicarbonate ($NaHCO_3$), pH 8.2, 0.15 M sodium chloride (NaCl), 0.5% BSA, in distilled water, adjust pH to 8.2 if necessary.

18. Texas red (TXR)—avidin fluorochrome, diluted in Texas red buffer.

19. Fluorescein isothiocyanate buffer: 0.6 M sodium chloride (NaCl) pH 7.0, 0.06 M sodium citrate ($Na_3C_6H_5O_7 \cdot 2H_2O$), 1% (w/v) BSA, 0.2% (v/v) Tween-20, in distilled water.

20. Fluorescein isothiocyanate (FITC)—avidin fluorochrome diluted in fluorescein isothiocyanate buffer.

21. Fluorescent anti-fade water-miscible mountant.

METHODS

Section Cutting

1. Cool paraffin blocks at −20°C for a minimum of 2 hr before sectioning.

2. Cut sections at a thickness of 2 µm using a standard rotary microtome. A cryospray, to cool the block face during cutting, may be used.

3. Mount sections onto electrostatically charged microscope slides freshly smeared with a thin film of mounting adhesive (Mayer Albumen adhesive).

4. Dry sections in racks placed in a 37°C oven for 72 hr.

5. Store sections at 4°C for (preferably) no longer than 4 weeks.

Antigen Retrieval

6. Dewax sections by placing into two changes of xylene for a minimum of 15 min.

7. Rehydrate sections to distilled water by placing in two 5-min changes of 100% ethanol, followed by 70% ethanol for 5 min.

8. Wash sections 3 times in distilled water.

9. Place sections into 0.01 M sodium citrate buffer, freshly prepared and adjusted with 1 M citric acid to pH 6.0.

10. Position pairs of slides back to back into 50-ml polypropylene centrifuge tubes with sufficient 0.01 M sodium citrate buffer to cover the tissue.

11. Fit the tubes with loose-fitting screw caps and place vertically into a foil-covered 500-ml beaker.

12. Heat the slides in an autoclave at 121°C, 15 pounds per square inch (psi) for 30 min, using the liquid cycle.

13. Remove tubes from the autoclave and place into bench racks at room temperature.

14. Allow the slides to cool (in the tubes) for 30 min.

15. Remove slides and wash well in 3 changes of distilled water.

Detection of PR Proteins by Immunoperoxidase Staining

16. Following antigen retrieval, place sections into 3.0% (v/v) hydrogen peroxide for 5 min to reduce endogenous peroxidase activity.

17. Wash sections well in 3 changes of distilled water. Place slides into PBS.

18. All incubations are performed at room temperature (unless otherwise indicated by the supplier) in a moist chamber.

19. Include negative control sections in each staining run and treat in an identical way except for replacement of the primary antibody with preimmune serum, if available, or PBT.

20. Include a case or tissue known to be positive for PR in each staining run.

21. Outline a circle well clear of the tissue with a Pap pen, to retain reagents during incubation. 100 µl of fluid is usually sufficient to ensure that the section is completely covered during incubations.

22. Block nonspecific binding and reduce background staining by incubating slides in normal goat serum diluted in an equal volume of PBS for 30 min.

23. Do not wash slides. Tap off excess normal goat serum and incubate the sections in a primary mouse monoclonal antibody diluted in PBT. Follow the manufacturer's recommendations for the concentration of antibody and length of incubation. An initial small study using a range of conditions is strongly advised.

24. Wash sections well in a minimum of 4 changes of PBS—5 min for each wash.

25. Incubate sections in a goat anti-mouse biotinylated secondary antibody diluted in PBS for 30 min.

26. Wash sections as in **step 24**.

27. Incubate sections for 30 min in a streptavidin-biotin-horseradish peroxidase complex, prepared in accordance with the manufacturer's instructions.

28. Wash sections as in **step 24**.

29. PR proteins are visualized using DAB (1 mg/ml DAB plus 0.02% (v/v) (final concentration) hydrogen peroxide, in PBS).

30. Wash well in 2 changes of distilled water, 10 min each.

31. Lightly counterstain sections using Harris' hematoxylin for 5–10 sec.

32. Wash slides well in several changes of tap water.

33. Immerse slides in Scott's alkaline blueing solution for 1 min.

34. Wash as in **step 32**.

35. Dehydrate sections as follows: 70% ethanol (5 min), 2×100% ethanol (5 min each), and 2 changes of xylene (5 min each).

36. Using appropriately sized glass coverslips, mount sections with a xylene-based mountant.

37. The slides are now ready for examination by light microscopy. PR positive cells will have brown nuclear staining; the nuclei of PR negative cells will appear blue (Figure 66B).

Dual Immunofluorescent Staining to Simultaneously Detect PRA and PRB Proteins

38. Following antigen retrieval, place slides in PBS.

39. Incubations are performed at room temperature in a moist chamber.

40. Control sections must be included in each staining run and treated and stained in the same way as the test sections. Controls include adjacent sections to each sample stained using PBS/0.5% Triton-X 100 (1) in place of both primary antibodies to control for nonspecific staining and (2) to replace the second-sequence primary antibody to ensure no cross-reactivity between the two staining sequences.

41. Include a case or tissue known to be positive for PR in each staining run.

42. Outline a circle well clear of the tissue with a Pap pen to retain reagents during incubation. 100 µl of fluid is usually sufficient to ensure that the section is completely covered during incubations.

43. Block nonspecific binding and reduce background staining by incubating slides in normal goat serum diluted in an equal volume of PBS for 30 min.

44. *PRB staining:* Do not wash slides. Tap off excess normal goat serum and incubate the sections in a primary mouse monoclonal antibody to detect PRB proteins (hPRa6 [Mote *et al.*, 2001]), diluted optimally in PBT, overnight. It is recommended that to determine optimal concentrations, an initial small study, using a range of concentrations, be performed for each primary antibody.

45. Wash sections well in a minimum of 4 changes of PBS—5 min for each wash.

46. Incubate sections in a goat anti-mouse biotinylated secondary antibody, diluted in PBS, for 30 min.

47. Wash sections as in **step 45**.

48. For the remaining steps keep sections as dark as possible to minimize loss of fluorescence.

49. Incubate sections for 60 min in TXR-avidin diluted in TXR buffer.

50. Wash sections in 2×600 ml changes of PBS, using a magnetic stirrer, for a total of 30 min.

51. To block sites of potential cross-reactivity between the two staining sequences, incubate sections overnight with goat anti-mouse Ig F(ab) diluted in 1% (w/v) BSA in PBS.

52. Wash sections as in **step 45**.

53. Incubate sections in normal goat serum diluted 1:1 in PBS for 30 min.

54. *PRA staining:* Remove excess normal goat serum, as in **step 44,** and incubate sections for 2–4 hr in the second mouse primary monoclonal antibody, to detect PRA proteins (hPRa7).

55. Wash sections as in **step 45**.

56. The PRA protein is detected by incubation for 30 min with a biotinylated goat anti-mouse antibody diluted in PBS.

57. Wash sections as in **step 45**.

58. Incubate sections for 60 min with fluorescein isothiocyanate (FITC)-avidin diluted in FITC buffer.

59. Wash sections as in **step 50**.

60. Using appropriately sized glass coverslips, mount sections with a water-based antifade, fluorescent mountant and store in the dark at 4°C.

61. The slides are now ready for examination by fluorescent light microscopy using filters that detect the TXR and FITC fluorochromes either individually or together.

62. PR-positive staining is nuclear. Under dual-excitation fluorescence the colors observed will range from deep orange for cells expressing only PRB to bright green for cells expressing only PRA (Figure 66E). Cells expressing similar amounts of PRA and PRB will stain yellow (Figure 66F) showing colocalization of both isoforms. Cells in which both PRA and PRB are expressed, but where one isoform is predominant, appear green to green/yellow (Figure 66G) or orange to orange/yellow (Figure 66H), depending on the ratio of PRA:PRB expression.

RESULTS AND DISCUSSION

Immunoperoxidase Staining

Immunoperoxidase staining is based on the enzymatic detection of an antigen–antibody complex by a biotin-labeled secondary antibody and subsequent signal amplification using a biotin-peroxidase and strepavidin complex. The antigen is visualized by the reaction between peroxidase and diaminobenzidine forming a brown precipitate at the site of the PR protein (Figure 66B). The method described in this chapter is highly sensitive, and slides can be stored permanently at room temperature with no loss of signal. Immunoperoxidase staining is an excellent method for the detection of PR in archival breast tissues and is particularly useful when knowledge of only total PR expression is required, although levels of individual PR isoform expression can be determined by staining adjacent sections.

Control Sections for Immunoperoxidase Staining

The routine inclusion of standards and control sections is important, especially in clinical diagnostic work. Ideally, control sections should include material with a known range of intensities of PR staining, from low to high expression, plus a negative area. Such staining may occur within a single paraffin block, but if sufficient material is available, composite tissue arrays can be made for inclusion in each staining run (Kononen *et al.*, 1998). It has recently been reported that short peptides, simulating the portion of the native antigen to which the antibody binds, can be used effectively as controls (Sompuram *et al.*, 2002). The peptides can be attached directly to the same glass microscope slide as the patient sample and will detect subtle variations in immunohistochemical efficiency caused by analytical errors (Sompuram *et al.*, 2002).

In addition to the preceding, adjacent negative control slides, where the primary antibody is replaced by preimmune serum or PBT, should be included for each test case to ensure specificity of the antibody in detection of PR protein.

Interpretation of Immunoperoxidase Staining

Historically, researchers have evaluated steroid receptor expression in breast carcinomas by immunoperoxidase staining and have reported their findings as the percentage of cases examined that are PR positive. There is considerable variation between different studies, where the number of samples reported positive can range from 20–66% (Kommoss *et al.*, 1994; Luqmani *et al.*, 1993). The disparity of these data may be partially explained by methodological differences and the analysis criteria used to determine whether or not a section is receptor positive. In some studies the number of cells required to be stained for a case to be considered positive can be as high as 80%, whereas other studies may report a sample positive when only 5% staining is observed (Wishart *et al.*, 2002). The most frequently used cutoff level to determine whether a tumor is positive or negative is when 10% or more of the total cells in the section stain PR positive.

One of the disadvantages of measuring steroid receptor proteins by immunolocalization is that, unlike the biochemical assays, it cannot easily be accurately quantitated (Katoh *et al.*, 1997). However, for routine clinical decision making, a dependable and reproducible threshold value, obtained by semiquantitative scoring may be all that is required (Battifora, 1994), and for some years a scoring method (Histoscore) based on visual counting of positive cells within a section has been used quite successfully, being both cost effective and clinically adequate (Battifora, 1994). More recently the introduction of computerized image analysis methods have enabled more accurate quantitations to be made (Esteban *et al.*, 1993, 1994). Both image analysis and Histoscore analysis are based on parameters of the percentage of cells or nuclear area staining positive and on the intensity of staining within that area. In both methods the cutoff point, below which a cell is considered to be negative is important.

Concordance between image analysis and Histoscoring is shown to be quite good (Makkink-Nombrado *et al.*, 1995). Reports have indicated a good correlation between image analysis of immunolocalized PR and its corresponding biochemical analyses (Cavaliere *et al.*, 1996; Esteban *et al.*, 1993) and in overall and disease-free patient survival (Esteban *et al.*, 1993, 1994). Quantitation of immunohistochemical staining is based on the assumption that there is a direct relationship between cellular antigen content and the intensity of staining observed, and there is one report in the literature that questions whether this association can be presumed for all proteins (Watanabe *et al.*, 1996). They suggest that such factors as steric hindrance, antibody trapping, and a high antibody concentration causing unstable binding may all play a role in leading to a disproportionate decrease in antibody binding in areas where antigen content is actually high (Watanabe *et al.*, 1996). These data suggest caution when attempting to accurately quantitate immunohistochemical staining, but by no means devalue the technique when used as evidence of receptor positivity for potential endocrine treatment.

Dual Immunofluorescent Staining

The dual immunofluorescent staining technique is more complex and lengthy compared to the immunoperoxidase method. The fluorescent signals obtained are not permanent (although they will last for several weeks when stored at 4°C in the dark), necessitating prompt digital imaging or photography of representative areas of the tissue. Control sections, as described in the methods and discussed following, are critical to ensure that each PR protein is optimally visualized and that the PRA:B ratio within each cell is accurately determined.

The dual immunofluorescent method is based on an avidin-labeled fluorophore detecting a biotin-conjugated secondary antibody, but because there is no preformed complex in the tertiary step, the sensitivity of this method is lower when compared to immunoperoxidase staining. The greatest advantage of dual fluorescent staining is its ability to visualize the ratio of PRA and PRB proteins in individual cells and thus determine cell-to-cell heterogeneity (Figure 66E). Predominant expression of one PR isoform is also readily determined

by color under dual fluorescent signal excitation of both isoforms (Figure 66G–H).

The dual immunofluorescent staining technique described in this chapter has been optimized for use with the hPRa primary antibody series (Clarke *et al.*, 1987; Mote *et al.*, 1999). Briefly, the two PR isoforms are revealed sequentially as follows: the PRB proteins are detected first using a mouse monoclonal antibody that recognizes only this isoform (hPRa6), and labeled with TXR-avidin. Potentially vacant sites on the primary antibody are blocked with a small Fab fragment to mouse. This is followed by detection of the remaining PRA proteins, using an antibody that recognizes only PRA, and labeling with FITC fluorochrome. As both antibodies are raised in the same animal, there is a potential for cross-reactivity between the two detection protocols (Mote *et al.*, 1999). The different IgG isotypes of the primary antibodies (hPRA6 and hPRa7) used in the development of this technique and a blocking step employed before the second sequence have enabled a satisfactory protocol to be developed, although it is always advisable to include control slides, where the second primary antibody is omitted, to ensure no cross-reactivity.

Control Sections for Dual Immunofluorescent Staining

The inclusion of suitable control sections is critical for accurate interpretation of tissues stained for PR isoform expression by dual immunofluorescence. Staining of control sections should be (1) control sections with no primary antibodies should be negative for both fluorochromes; (2) control sections where the second sequence primary antibody (to detect PRA) has been omitted should reveal no staining under FITC wavelength excitation, despite the presence of PRA in the section, demonstrating no cross-reactivity between components of the sequential stains; and (3) positive control sections, where PRA and PRB expression is known, should show consistent signal in each staining run. Positive control sections can be obtained by using tissues with disparate PR isoform expression, where the relative levels of PRA:B have been previously determined by immunoperoxidase staining of adjacent sections. It is also recommended that, when first performing the dual-staining technique, each antigen of interest should be demonstrated individually by fluorescence in adjacent tissue sections. This determines the maximum signal expected for each antigen in a given tissue and ensures no significant loss of signal due to the lengthy dual-staining technique.

Interpretation of Dual Immunofluorescent Staining

By using a dual immunofluorescence staining technique the relative levels of PRA and PRB proteins can be determined within the same section of breast tumor by examining the tissue under dual wavelength excitation. A standard fluorescent microscope, fitted with fluorescent-quality (apochromatic) objectives, and narrow band pass filters suitable for detection of TXR (BP 545–580 nm), FITC (BP 450–480 nm), and TXR/FITC combined are essential for evaluation of PR proteins stained by this method.

The PRB proteins are labeled with TXR fluorochrome and are therefore visualized as a red color under TXR single wavelength excitation, whereas PRA proteins are labeled with FITC fluorochromes that reveal PRA to be green under FITC single wavelength excitation. When the same section is viewed with simultaneous excitation of both fluorochromes, nuclei that express predominantly PRA proteins appear green or green/yellow (Figure 66G), whereas nuclei that express primarily PRB proteins are orange or orange/yellow (Figure 66H). Nuclei co-expressing both PRA and PRB proteins in similar concentrations are yellow (Figure 66F), (Mote et al., 1999).

The relative levels of PRA and PRB proteins can be determined by merely observing the predominant color of PR positive cells under dual-wavelength fluorescent excitation. A more accurate method of analysis, however, can be achieved by capture of black-and-white digital images of the dual-stained slide under TXR and FITC wavelength excitations separately, followed by subsequent measurement of the amount of signal present for each protein using appropriate image analysis computer software. A PRA:PRB ratio can be calculated and significant deviations from 1.0 can be recorded as a predominance of one PR isoform. It is essential to evaluate several areas that are representative of the section as a whole, and given the limitations in quantitation of fluorescent images, cases should be scored conservatively with PRA:B ratios falling between 0.8 and 1.2 determined to reflect approximately equivalent PRA and PRB expression (Mote et al., 2002).

It must be emphasized that there are distinct limitations in quantifying proteins that are revealed by fluorescent immunostaining, because sections with abundant antigen expression may reach saturating levels of fluorescent signal. Furthermore, the relationship between the amount of protein expressed and the level of fluorescent intensity observed may not be linear. At best we can only report semiquantitatively in broad categories, such as low, moderate, high, and very high, to compare overall PR expression between tissues. The principal merit, however, of demonstrating PRA and PRB within the same tissue section by dual immunofluorescence lies in an ability for the technique to reveal differences, or changes, in relative expression of

these two isoforms, rather than to determine the absolute PRA and PRB protein levels present.

In conclusion, it has been well established that assessment of PR expression in breast carcinomas is of critical importance in determining a patient's likely response to endocrine treatment. Immunohistochemistry, using an immunoperoxidase staining technique, is now the preferred method of analysis of overall PR expression and is in routine use in most laboratories. It is a sensitive and reliable method, providing that care is taken to optimize the antigen-retrieval procedure and the primary antibody dilution and incubation times.

The dual-immunofluorescent staining technique has only recently been developed (Mote et al., 1999) and is not currently used routinely in diagnostic laboratories. Research has shown that the relative levels of the two PR isoforms change from similar expression of PRA and PRB in normal breast tissue to a frequent predominance of one isoform in the malignant state (Mote et al., 2002), although the clinical consequences of PR isoform predominance are not known. Any possible link between PRA:PRB ratios and tumor response to endocrine treatment has not yet been studied, nor has there been any evaluation of the potential prognostic value of relative PRA:PRB expression with patient outcome; however, such determinations in the future may well be of benefit to the patient. The relative expression of PRA and PRB in breast cancers may be a marker of abnormal tissue response to hormone stimulation and is, therefore, likely to have clinical consequences with respect to endocrine treatment.

References

Arnett-Mansfield, R.L., deFazio, A., Wain, G.V., Jaworski, R.C., Byth, K., Mote, P.A., and Clarke, C.L. 2001. Relative expression of progesterone receptors A and B in endometrioid cancers of the endometrium. *Cancer Res. 61:*4576–4582.

Balleine, R.L., Earl, M.J., Greenberg, M.L., and Clarke, C.L. 1999. Absence of progesterone receptor associated with secondary breast cancer in postmenopausal women. *Br. J. Cancer 79:*1564–1571.

Bankfalvi, A., Navabi, H., Bier, B., Bocker, W., Jasani, B., and Schmid, K.W. 1994. Wet autoclave pretreatment for antigen retrieval in diagnostic immunohistochemistry. *J. Pathol. 174:*223–228.

Battersby, S., Robertson, B.J., Anderson, T.J., King, R.J.B., and McPherson, K. 1992. Influence of menstrual cycle, parity and oral contraceptive use on steroid hormone receptors in normal breast. *Br. J. Cancer 65:*601–607.

Battifora, H. 1994. Immunocytochemistry of hormone receptors in routinely processed tissues. *Appl. Immunohistochem. 2:*143–145.

Bell, P.B., Rundquist, I., Svensson, I., and Collins, V.P. 1987. Formaldehyde sensitivity of a GFAP epitope, removed by extraction of the cytoskeleton with high salt. *J. Histochem. Cytochem. 35:*1375–1380.

Cattoretti, G., Pileri, S., Parravicini, C., Becker, M.H., Poggi, S., Bifulco, C., Key, G., D'Amato, L., Sabattini, E., Feudale, E.,

Reynolds, F., Gerdes, J., and Rilke, F. 1993. Antigen unmasking on formalin-fixed, paraffin-embedded tissue sections. *J. Pathol. 171:*83–98.

Cavaliere, A., Bucciarelli, E., Sidoni, A., Bianchi, G., Pietropaoli, N., Ludovini, V., and Vitali, R. 1996. Estrogen and progesterone receptors in breast cancer: Comparison between enzyme immunoassay and computer-assisted image analysis of immunocytochemical assay. *Cytometry 26:*204–208.

Clarke C.L., and Sutherland, R.L. 1990. Progestin regulation of cellular proliferation. *Endocr. Rev. 11:*266–300.

Clarke, C.L., Zaino, R.J., Feil, P.D., Miller, J.V., Steck, M.E., Ohlsson-Wilhelm, B.M., and Satyaswaroop, P.G. 1987. Monoclonal antibodies to human progesterone receptor: Characterization by biochemical and immunohistochemical techniques. *Endocrinology 121:*1123–1132.

Conneely, O.M., Lydon, J.P., De Mayo, F., and O'Malley, B.W. 2000. Reproductive functions of the progesterone receptor. *J. Soc. Gynecol. Investig. 7:*S25–S32.

Elias, J.M., Gown, A.M., Nakamura, R.M., Wilbur, D.C., Herman, G.C., Jaffe, E.S., Battifora, H., and Brigati, D.J. 1989. Quality control in immunohistochemistry: Report on a workshop sponsored by the Biological Stain Commission. *Am. J. Clin. Pathol. 92:*836–843.

Esteban, J.M., Ahn, C., Mehta, P., and Battifora, H. 1994. Biologic significance of quantitative estrogen receptor immunohistochemical assay by image analysis in breast cancer. *Am. J. Clin. Pathol. 102:*158–162.

Esteban, J.M., Kandalaft, P.L., Mehta, P., Odom-Maryon, T.L., Bacus, S., and Battifora, H. 1993. Improvement of the quantification of estrogen and progesterone receptors in paraffin-embedded tumors by image analysis. *Am. J. Clin. Pathol. 99:*32–38.

Evans, R.M. 1988. The steroid and thyroid hormone receptor superfamily. *Science 240:*889–895.

Graham, J.D., and Clarke, C.L. 2002. Expression and transcriptional activity of progesterone receptor A and progesterone receptor B in mammalian cells. *Breast Cancer Res. 4:*187–190.

Gronemeyer, H. 1991. Transcription activation by estrogen and progesterone receptors. *Annu. Rev. Genet. 25:*89–123.

Hayat, M.A. 2002. *Microscopy, Immunohistochemistry, and Antigen Retrieval Methods.* New York: Kluwer Academic/Plenum Publishers.

Humason, G. 1979. *Animal Tissue Techniques*, 4th ed. San Francisco: W.H. Freeman & Co.

Jacobs, T.W., Prioleau, J.E., Stillman, I.E., and Schnitt, S.J. 1996. Loss of tumor marker-immunostaining intensity on stored paraffin slides of breast cancer. *J. Natl. Cancer Inst. 88:*1054–1059.

Katoh, A.K., Stemmler, N., Specht, S., and D'Amico, F. 1997. Immunoperoxidase staining for estrogen and progesterone receptors in archival formalin fixed, paraffin embedded breast carcinomas after microwave antigen retrieval. *Biotechnic. Histochem. 72:*291–298.

Kommoss, F., Pfisterer, J., Idris, T., Giese, E., Sauerbrei, W., Schafer, W., Thome, M., and Pfleiderer, A. 1994. Steroid receptors in carcinoma of the breast. *Anal. Quant. Cytol. Histol. 16:*203–210.

Kononen, J., Bubendorf, L., Kallioniemi, A., Barlund, M., Schraml, P., Leighton, S., Torhorst, J., Mihatsch, M.J., Sauter, G., and Kallioniemi, O.P. 1998. Tissue microarrays for high-throughput molecular profiling of tumor specimens. *Nat. Med. 4:*844–847.

Kudo, A., Fukushima, H., Kawakami, H., Matsuda, M., Goya, T., and Hirano, H. 1996. Use of serial semithin frozen sections to evaluate the co-localization of estrogen receptors and progesterone receptors in cells of breast cancer tissues. *J. Histochem. Cytochem. 44:*615–620.

Luqmani, Y.A., Ricketts, D., Ryall, G., Turnbull, L., Law, M., and Coombes, R.C. 1993. Prediction of response to endocrine therapy in breast cancer using immunocytochemical assays for pS2, oestrogen receptor and progesterone receptor. *Int. J. Cancer 54:*619–623.

Lydon, J.P., DeMayo, F.J., Conneely, O.M., and O'Malley, B.W. 1996. Reproductive phenotypes of the progesterone receptor null mutant mouse. *J. Steroid Biochem. Molec. Biol. 56:*67–77.

Lydon, J.P., DeMayo, F.J., Funk, C.R., Mani, S.K., Hughes, A.R., Montgomery, C.A., Jr., Shyamala, G., Conneely, O.M., and O'Malley, B.W. 1995. Mice lacking progesterone receptor exhibit pleiotropic reproductive abnormalities. *Genes Dev. 9:*2266–2278.

Makkink-Nombrado, S.V., Baak, J.P., Schuurmans, L., Theeuwes, J.W., and van der Aa, T. 1995. Quantitative immunohistochemistry using the CAS 200/486 image analysis system in invasive breast carcinoma: A reproducibility study. *Anal. Cell. Pathol. 8:*227–245.

McDonnell, D.P., Shahbaz, M.M., Vegeto, E., and Goldman, M.E. 1994. The human progesterone receptor A-form functions as a transcriptional modulator of mineralocorticoid receptor transcriptional activity. *J. Steroid Biochem. Mol. Biol. 48:* 425–432.

McGowan, E.M., and Clarke, C.L. 1999. Alteration of the ratio of progesterone receptors A and B shows that both isoforms are active on endogenous progestin-sensitive endpoints in breast cancer cells. *Mol. Endocrinol. 13:*1657–1671.

Meyer, M.E., Quirin-Stricker, C., Lerouge, T., Bocquel, M.T., and Gronemeyer, H. 1992. A limiting factor mediates the differential activation of promoters by the human progesterone receptor isoforms. *J. Biol. Chem. 267:*10882–10887.

Morgan, J.M., Navabi, H., Schmid, K.W., and Jasani, B. 1994. Possible role of tissue-bound calcium ions in citrate-mediated high-temperature antigen retrieval. *J. Pathol. 174:*301–307.

Mote, P.A., Balleine, R.L., McGowan, E.M., and Clarke, C.L. 1999. Co-localization of progesterone receptors A and B by dual immunofluorescent histochemistry in human endometrium during the menstrual cycle. *J. Clin. Endocrinol. Metab. 84:*2963–2971.

Mote, P.A., Bartow, S., Tran, N., and Clarke, C.L. 2002. Loss of co-ordinate expression of progesterone receptors A and B is an early event in breast carcinogenesis. *Breast Cancer Res. Treat. 72:*163–172.

Mote, P.A., Johnston, J.F., Manninen, T., Tuohimaa, P., and Clarke, C.L. 2001. Detection of progesterone receptor forms A and B by immunohistochemical analysis. *J. Clin. Path. 54:*624–630.

Mote, P.A., Leary, J.A., and Clarke, C.L. 1998. Immunohistochemical detection of progesterone receptors in archival breast cancer. *Biotech. Histochem. 73:*117–127.

Munakata, S., and Hendricks, J.B. 1993. Effect of fixation time and microwave oven heating time on retrieval of the Ki-67 antigen from paraffin-embedded tissue. *J. Histochem. Cytochem. 41:*1241–1246.

Osborne, C.K. 1998. Steroid hormone receptors in breast cancer management. *Breast Cancer Res. Treat. 51:*227–238.

Ozzello, L., DeRosa, C., Habif, D.V., and Greene, G.L. 1991. An immunohistochemical evaluation of progesterone receptor in frozen sections, paraffin sections, and cytologic imprints of breast carcinomas. *Cancer 67:*455–462.

Page, D.L., and Simpson, J.F. 1996. Pathology of preinvasive and excellent-prognosis breast cancer. *Curr. Opin. Oncol. 8:*462–467.

Pearce, P.T., Myles, K.M., and Funder, J.W. 1993. Oestrogen and progesterone receptor assays in breast tumours. The Prince Henry's Hospital experience, 1983–1990. *Med. J. Aust. 159:*227–231.

Pertschuk, L.P., Feldman, J.G., Kim, D.S., Nayeri, K., Eisenberg, K.B., Carter, A.C., Thelmo, W.T., Rhong, Z.T., Benn, P., and Grossman, A. 1993. Steroid hormone receptor immunohisto-chemistry and amplification of c-*myc* protooncogene. Relationship to disease-free survival in breast cancer. *Cancer 71:*162–171.

Pichon, M.F., Pallud, C., Brunet, M., and Milgrom, E. 1980. Relationship of presence of progesterone receptors to prognosis in early breast cancer. *Cancer Res. 40:*3357–3360.

Pichon, M.F., Broet, P., Magdelenat, H., Delarue, J.C., Spyratos, F., Basuyau, J.P., Saez, S., Rallet, A., Courriere, P., Millon, R., and Asselain, B. 1996. Prognostic value of steroid receptors after long-term follow-up of 2257 operable breast cancers. *Br. J. Cancer 73:*1545–1551.

Press, M., Spaulding, B., Groshen, S., Kaminsky, D., Hagerty, M., Sherman, L., Christensen, K., and Edwards, D.P. 2002. Comparison of different antibodies for detection of progesterone receptor in breast cancer. *Steroids 67:*799–813.

Regitnig, P., Reiner, A., Dinges, H.P., Hofler, G., Muller-Holzner, E., Lax, S.F., Obrist, P., Rudas, M., and Quehenberger, F. 2002. Quality assurance for detection of estrogen and progesterone receptors by immunohistochemistry in Austrian pathology laboratories. *Virchows Arch. 441:*328–334.

Rhodes, A., Jasani, B., Balaton, A.J., Barnes, D.M., Anderson, E., Bobrow, L.G., and Miller, K.D. 2001. Study of interlaboratory reliability and reproducibility of estrogen and progesterone receptor assays in Europe. Documentation of poor reliability and identification of insufficient microwave antigen retrieval time as a major contributory element of unreliable assays. *Am. J. Clin. Pathol. 115:*44–58.

Rousseau-Merck, M.F., Misrahi, M., Loosfelt, H., Milgrom, E., and Berger, R. 1987. Localization of the human progesterone receptor gene to chromosome 11q22-q23. *Hum. Genet. 77:*280–282.

Shi, S.R., Key, M.E., and Kalra, K.L. 1991. Antigen retrieval in formalin-fixed, paraffin-embedded tissues: An enhancement method for immunohistochemical staining based on microwave oven heating of tissue sections. *J. Histochem. Cytochem. 39:*741–748.

Shi, S., Gu, J., Kalra, K.L., Chen, T., Cote, R.J., and Taylor, C.R. 1995. Antigen retrieval technique: A novel approach to immunohistochemistry on routinely processed tissue sections. *Cell Vision 2:*6–21.

Shyamala, G. 1999. Progesterone signaling and mammary gland morphogenesis. *J. Mam. Gland Biol. Neoplasia 4:*89–104.

Sompuram, S.R., Kodela, V., Zhang, K., Ramanathan, H., Radcliffe, G., Falb, P., and Bogen, S.A. 2002. A novel quality control slide for quantitative immunohistocytochemistry testing. *J. Histochem. Cytochem. 50:*1425–1434.

Taylor, C.R. 1994. An exaltation of experts: Concerted efforts in the standardization of immunohistochemistry. *Hum. Pathol. 25:*2–11.

Taylor, C.R., Shi, S.R., Chaiwun, B., Young, L., Imam, S.A., and Cote, R.J. 1994. Strategies for improving the immunohisto-chemical staining of various intranuclear prognostic markers in formalin-paraffin sections: Androgen receptor, estrogen receptor, progesterone receptor, p53 protein, proliferating cell nuclear antigen, and Ki-67 antigen revealed by antigen retrieval techniques. *Hum. Pathol. 25:*263–270.

Taylor, C.R. 1996. Paraffin section immunocytochemistry for estrogen receptor: The time has come. *Cancer 77:*2419–2422.

Tesch, M., Shawwa, A., and Henderson, R. 1993. Immuno-histochemical determination of estrogen and progesterone receptor status in breast cancer. *Am. J. Clin. Pathol. 99:*8–12.

Tsai, M.J., and O'Malley, B.W. 1994. Molecular mechanisms of action of steroid/thyroid receptor superfamily members. *Annu. Rev. Biochem. 63:*451–486.

Tung, L., Mohamed, M.K., Hoeffler, J.P., Takimoto, G.S., and Horwitz, K.B. 1993. Antagonist-occupied human progesterone B-receptors activate transcription without binding to progesterone response elements and are dominantly inhibited by A-receptors. *Mol. Endocrinol. 7:*1256–1265.

Vegeto, E., Shahbaz, M.M., Wen, D.X., Goldman, M.E., O'Malley, B.W., and McDonnell, D.P. 1993. Human progesterone receptor A form is a cell- and promoter-specific repressor of human progesterone receptor B function. *Mol. Endocrinol. 7:*1244–1255.

Wang, H., Critchley, H.O.D., Kelly, R.W., Shen, D., and Baird, D.T. 1998. Progesterone receptor subtype B is differentially regulated in human endometrial stroma. *Mol. Hum. Reprod. 4:*407–412.

Watanabe, J., Asaka, Y., and Kanamura, S. 1996. Relationship between immunostaining intensity and antigen content in sections. *J. Histochem. Cytochem. 44:*1451–1458.

Werner, M., Von Wasielewski, R., and Komminoth, P. 1996. Antigen retrieval, signal amplification and intensification in immunohistochemistry. *Histochem. Cell Biol. 105:*253–260.

Wishart, G.C., Gaston, M., Poultsidis, A.A., and Purushotham, A.D. 2002. Hormone receptor status in primary breast cancer—time for a consensus? *Eur. J. Cancer 38:*1201–1203.

Zeimet, A.G., Muller-Holzner, E., Marth, C., and Daxenbichler, G. 1994. Immunocytochemical versus biochemical receptor determination in normal and tumorous tissues of the female reproductive tract and the breast. *J. Steroid Biochem. Mole. Biol. 49:*365–372.

22

Immunohistochemical Expression of Cytosolic Thymidine Kinase in Patients with Breast Carcinoma

Qimin He, Yongrong Mao, and Jianping Wu

Introduction

Thymidine kinase (TK, ATP: thymidine 5′-phosphotransferase, EC.2.7.1.21), an enzyme of the pyrimidine salvage pathway, catalyzes the phosphorylation of thymidine to thymidine monophosphate. The TK enzyme in human cells appears in two isozyme forms, a cytosolic (TK1) and a mitochondrial (TK2) form. The TK1 enzyme plays the primary role in regulating intracellular thymidine pool throughout the cell cycle. The level of TK1 during the cell cycle rises at the G1/S boundary and increases dramatically from late G1 to late S-phase/early G2 phase in proliferating normal cells and tumor cells, but is virtually absent from quiescent cells. Thus, because TK1 enzyme is highly dependent on the growth stage of the cell, the enzyme is a useful indicator for cellular proliferation and, hence, for malignancy (He *et al.*, 1991; Sherley *et al.*, 1988).

Recent production of anti-TK1 antibodies has provided an attractive alternative in cancer research. We have developed and characterized an anti-TK1 rabbit polyclonal antibody (He *et al.*, 1996), two anti-TK1 monoclonal antibodies (Svanova Biotech, Sweden) raised against a 15 amino acid synthetic peptide (KPGEAVAARKLFAPQ), and an anti-TK1 chicken IgY antibody raised against a 31 amino acid synthetic peptide (GQPAG PDNKE NCPVP GKPGE AVAAR KLFAPQ), sequences corresponding to the part of the C-terminus of human TK1 (Wu *et al.*, 2003). Since 2000, an anti-TK1 monoclonal antibody raised against the purified human Raji TK1 (Zhang *et al.*, 2001), an anti-TK1 monoclonal antibody raised against the recombinant human HeLa TK1 (Kuroiwa *et al.*, 2001) and an anti-TK1 polyclonal rabbit antibody raised against a synthetic 15 amino acid peptide (PTVLPGSPSKTRGQI) corresponding to the part of the N-terminus of human TK1 (Voeller *et al.*, 2001) have also been reported. These antibodies are useful for probe in both serologic and immunohistochemic detection of TK1 in patients of all types of cancer diseases (He *et al.*, 2000; Kuroiwa *et al.*, 2001; Mao *et al.*, 2002; Wang *et al.*, 2001; Wu *et al.*, 2000, 2003; Zhang *et al.*, 2001; Zou *et al.*, 2002). In this chapter we discuss the immunohistochemic expression of TK1 in patients with breast carcinoma in comparison with PCNA and Ki-67. The role of Ki-67 in breast cancer is also discussed in Part IV, Chapter 19.

Handbook of Immunohistochemistry and in situ Hybridization of Human Carcinomas, Volume 1: Molecular Genetics; Lung and Breast Carcinomas

Copyright © 2004 by Elsevier (USA)
All rights reserved.

MATERIALS

1. Phosphate-buffered saline (PBS): pH 7.4. 11.5 g di-sodium hydrogen orthophosphate anhydrous (80 mM), 2.96 g sodium dihydrogen orthophosphate (20 mM), 8.76 g sodium chloride (150 mM). Dilute to 1000 ml with distilled water—check pH to 7.4.

2. Tris-buffered saline (TBS): 2.42 tris base (20 mM), 8 g sodium chloride (137 mM), 3.8 ml 1 M hydrochloride acid. Dilute to 1000 ml with distilled water check pH to 7.6.

3. Poly-L-lysine (Sigma, St. Louis, MO), 0.1% w/v with distilled water.

4. Ethanol solutions (anhydrous): 100%, and dilution to 95%, 70%, and 50% with distilled water, respectively.

5. Xylene (Sigma).

6. Trypsin (Zymed, CA). 0.5% (v/v) trypsin in PBS.

7. 0.3% hydrogen peroxide (H_2O_2).

8. Stored 10% bovine serum albumin (BSA, Immuno-degree Vector Laboratories, CA): add 10 g BSA in 100 ml TBS with stirring.

9. Primary antibody—anti-TK1 antibody diluted in PBS and add 1% BSA (from stored 10% BSA).

10. Blocking reagent—Normal serum (Included the VECTASTAN *Elite* ABC kit, Vector Laboratories, CA).

11. VECTASTAN *Elite* ABC kit (Vector Laboratories, CA). In accordance with the instruction of manufacter's 1) Biotinylated antibody: add two drops (150 µl) of normal blocking serum stock to 5 ml of PBS buffer in mixing bottle and then add two drops of biotinylated antibody stock. 2) VECTASTAN *Elite* ABC reagent (Avdin-HRP complex): add exactly two drops of reagent A to 5 ml of PBS buffer in the ABC reagent large mixing bottle. Then add exactly two drops of reagent B to the same mixing bottle, mix immediately, and allow VECTASTAN *Elite* ABC reagent to stand for about 30 min before use.

12. Diaminobenzidine (DAB). DAB 1.3 mM with H_2O_2 (0.02/ v/v) in 5 mM PBS buffer.

13. Hematoxylin (Vector Laboratories, Hematoxylin H-3401, 500 ml).

14. Distilled water.

METHOD

Paraffin-Embedded Sections

1. Sections cut from paraffin-embedded material should be cut at 4 µm and dried on slides, which are coated with Poly-L-lysine (0.1% w/v) and dried at 37°C.

2. The sections should be deparaffinized by taking them through 3 changes of xylene for 3 × 10 min and then rehydrated by passing through serial ethanol solutions from 100%, 95%, 70%, to 50%, ending finally in water. The slides should be air dried before further processing.

3. Rinse with distilled water for 5 min and wash in PBS for 5 min.

4. In order to unmask the TK1 activity in the tissue, incubate the sections with trypsin (Zymed, CA) for 10 min at room temperature. The trypsinized sections should be briefly rinsed in PBS.

5. Wash slides in PBS buffer for 3 × 5 min.

6. Sections are treated with H_2O_2 (0.3% v/v) in distilled water for 30 min in order to block the endogenous peroxidase.

7. Tap off H_2O_2 solution and wash slides in PBS buffer for 2 × 5 min.

8. Incubate the sections in normal serum diluted 1:67 in PBS for 30 min to block nonspecific binding sites.

9. Excess of normal serum is removed.

10. Add primary antibody—anti-TK1 antibody[a] at appropriate dilution with PBS and then incubate for 2 h at room temperature or overnight at 4°C.

11. Tap off antibody solution and wash slides in PBS buffer for 3 × 5 min.

12. Incubate with the diluted biotinylated second antibody[a] in PBS, 40 min at room temperature.

13. Tap off second antibody solution and wash slides in PBS buffer for 3 × 5 min.

14. Incubate with the diluted ABC reagent (HRP-streptavidin complex) for 40 min at room temperature.

15. Tap off the ABC reagent solution and wash slides in PBS buffer for 3 × 5 min.

16. Add enzyme substrate solution DAB and incubate at room temperature for 5–15 min until a chromogen is observed.

17. Stop enzyme reaction by washing with tap water for at least 5 min.

18. Counterstain with hematoxylin solution for 1–2 sec and then follow it by differentiation, dehydration, clearing, and mounting as for routine histological examination.

19. Examine under a light microscope.

Antibodies

Anti-TK1 rabbit polyclonal antibody (pAbl) (He *et al.*, 1996), anti-TK1 monoclonal antibodies—1E3 and 1D11 (Svanova Biotech, Sweden) and anti-TK1 chicken IgY antibody (TK1-IgY) (Wu *et al.*, 2003) were produced in our research group. Anti-PCNA antibody (PC10,

---

[a]For anti-TK1 monoclonal antibody, the biotinylated anti-mouse IgG should be chosen; for anti-TK1 polyclonal rabbit antibody, the biotinylated anti-rabbit IgG should be chosen; for anti-TK1 IgY antibody, the biotinylated anti-chicken IgY should be chosen (Vector Laboratories, selected biotinylated anti-immunoglobulins).

Santa Cruz, CA) and anti Ki-67 antibody (MIB-1, Dako, Denmark) are used for the comparison studies.

Patients

1. Comparative study of expression of TK1 and proliferating cell nuclear antigen (PCNA): Immunohistochemical detection of TK1 and PCNA in 52 patients with eight histologic types of breast carcinoma, histologic parts of normal tissues taken from the tumor-bearing breast of 16 patients, and histologic benign tumors from 20 patients were used for the comparative study. Anti-TK1 rabbit polyclonal antibody (pAbl) and anti-PCNA antibody (PC10) were used for this study (Mao *et al.*, 2002).

2. Comparative study of expression of TK1 and Ki-67: Immunohistochemical staining of TK1 and Ki-67 in 54 breast cancer patients with ductal infiltrated carcinoma and histologic parts of normal tissues taken from the tumor-bearing breast of 10 patients were used. To compare the immunohistochemical staining of TK1 expression, we used either one TK1 mAb (1D11) alone or a mixture of two TK1 mAbs (1D11 + 1E3).

3. Immunohistochemical staining of TK1 in four patients with advanced infiltrative ductal carcinoma using anti-TK1 IgY antibody. Four histologic normal parts of the tissues were taken from the same breast as bearing the tumor and identified as normal tissue (Wu *et al.*, 2003).

All patients were collected in Hubei Cancer Hospital, China, 1998–2000 and underwent radical mastectomy. These studies were conducted in accordance with the Helsinki Declaration in 1983.

Evaluation of Results of Immunohistochemistry

TK1-positive cells of the tissue section were counted among at least 100 cells in at least 10 microscopic fields at × 200–400 magnification. According to the numbers of positively stained cells in the tumor, we assess the expression of TK1 for each section as negative or weak staining <5% (−), or positive staining: 5–25% (+), 25–50% (++), or > 50% (+++).

Statistical Analysis

Chi-square test was applied for comparison of the results of immunohistochemical staining with tumor staging and grading as well as various histological types of tumors. A p value less than 0.05 (two tailed) was considered as statistically significant.

RESULTS AND DISCUSSION

Adenocarcinoma of breast (breast cancer) is one of the most common female cancers in developed countries (e.g., Europe, North America, and Australia). This carcinoma is five times higher in developed countries than that in the less-developed countries (e.g., Asia and Africa). The etiology of breast cancer is complex and is likely to involve the actions of genes at multiple levels along the multistage carcinogenesis process. The different types of cells in the normal breast and malignant breast tissues show different gene expression, making a true characterization of the gene expression by, for example, cDNA-array or proteomics difficult (Lakhani *et al.*, 2001; Perou *et al.*, 2000).

Understanding the role of an inherited genotype at different stages of carcinogenesis could improve our understanding of cancer biology, may identify specific exposures or events that correlate with carcinogenesis, or may target a relevant biochemical pathway for the development of preventive or therapeutic interventions.

The American Society of Clinical Oncology (ASCO, 2002) discussed breast tumor marker in terms of how it can help in screening (finding cancer early), diagnosis (Is it really cancer?), prognosis (How will the cancer behave?), monitoring (How is the treatment working?), predicting a response (Will a drug be effective?), and surveillance (Will the patient need more treatment?). Markers of proteins are few for use for histochemical staining. ASCO 2002 recommends only the use of steroid hormone receptors (estrogen [ER] and progesterone [PR]) in the diagnosis, prognosis, and treatment planning for women with breast cancer. The HER-2/neu (c-erbB-2) levels can also be used for diagnosis and prognosis of breast cancer (ASCO, 2002). However, the proportion of patients with ER-positive primary tumors is depended on age, with ~ 50% of premenopausal patients being positive compared to 75–85% among those menopausal patients. The HER2/neu is a 185-kDa growth factor receptor that is overexpressed only in ~25% of human breast cancers amplification of the HER2 gene (Pegram *et al.*, 2000). It indicates that such markers have limited use in clinical setting.

A recent study of breast cancer response to neoadjuvant chemotherapy, using Bcl-2, ER, p53, HER2/neu, and Ki-67 as predictive markers, reported that the proliferation decreased significantly after chemotherapy. The ER negativity and a high proliferation index, however, were associated with better response, and HER-2/neu status did not predict response (Faneyte *et al.*, 2003). Candidate prognostic biomarkers for breast cancer include elevated proliferation indices such as PCNA and Ki-67, which are still being used for clinical practice (Aziz *et al.*, 2003; Beenken *et al.*, 2002; Kato *et al.*, 2002). A recent study of pregnancy/lactation associated breast cancer reported that in spite of some significant differences in the expression of few prognostic markers (i.e., ER/PR, EGFR, and PCNA) there was no significant difference in the overall survived of breast carcinoma versus control group if compared stage for stage (Aziz *et al.*, 2003). The ER-2/neu and PCNA expressions were obviously important from a biological standpoint and were not recognized as independent molecular markers (Kato *et al.*, 2002). Ki-67 is the most widely used marker of proliferating cells (Molino *et al.*, 2002; Trihia *et al.*, 2003). Most biomarkers for breast cancer are discussed in Part I, Chapter 1.

Searching for new prognostic factors more closely related to the actual tumor biology has been advocated. Based on a previous study of cell cycle regulation of human TK1 (Kauffman *et al.*, 1991), the carboxyl-terminal (C-terminal) 40 residues are essential for regulation of TK1 activity during the cell cycle. Deletion of these residues completely abolishes cell cycle regulatory activity. Deletion of 10 residues at the C-terminal does not affect cell cycle regulation. This is why we chose this part of peptide from the C-terminus of human HeLa TK1 and developed anti-TK1 poly/monoclonal antibody. The epitope of our anti-TK1 polyclonal and monoclonal was found to be the crucial turn region: lysine 211, proline 212, glycine 213, and glutamic acid 214 (K211-P212-G213 -E214) at the C-terminus part of TK1 (residues 211-225) (He *et al.*, 1996). When P212 and E214 were substituted with alanine, the immunoresponse was reduced by 80%. Substitution of K211 reduced the immunoactivity by 55%. We used the anti-TK1 poly/monoclonal antibodies to detect the TK1 expression in patients with breast cancer.

Two other markers for proliferation, PCNA and Ki-67, were used for comparison.

Characterization of Immunohistostaining of TK1 in Breast Cancer Cells

The immunostaining for TK1 in malignant breast lesions appeared in the cytoplasm with a typically diffuse pattern (Figure 67). The positive staining was observed by spots of granules and concentrated mostly near the nuclei, previously described in proliferating cells (Wang *et al.*, 2001). The staining for PCNA and Ki-67 in malignant breast lesions was found in the nuclei with a diffuse or granular pattern or a mixture of both. In normal and benign tumors cells, only weak or no staining was found.

Expression of TK1 and PCNA in Normal Breast Tissues, Benign Breast Tumor, and Malignant Breast Lesions

Among the patients with benign tumor in Table 27, 20% and 25% showed positive staining for TK1 and PCNA, respectively, whereas the corresponding staining of normal tissues were 19% and 13% respectively. No significant differences in the expression of STK1 and PCNA were found between the specimens from benign tumor and normal tissues ($p = 1.0$, and $p = 0.40$, respectively). The TK1 expression in malignant lesions was higher (79%) as compared to PCNA expression (64.5%). Statistical analysis showed that the expressions

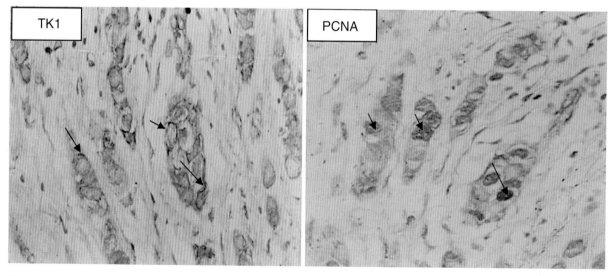

Figure 67 Demonstration of the cytosolic immunoreaction of TK1 and the nuclear immunoreaction of PCNA in breast carcinoma (brown color). Positive staining of 25–50% (++) for TK1 and PCNA in a large syringocarcinoma (original magnification 400X). Arrows in figures show examples of TK1 in cytoplasm and PCNA in nuclei, respectively. Slides were briefly counterstained with hematoxylin (light violet color).

Table 27 Immunohistochemical Staining of TK1 and Ki-67 in Normal Breast Tissues and Ductal Infiltrated Carcinomas

Type	TK1 Detected by TK1 mAb[1]								Ki-67 Detected by MIB-1[2]			
	1E3+1D11				1D11				MIB-1			
	−	+	++	+++	−	+	++	+++	−	+	++	+++
Ductal Infiltrated Carcinoma (n = 54)	23	20	8	3	28	21	5		32	20	2	
Normal Tissue (n = 10)	9	1			9	1			9	1		

[1]TK1 was detected using a mixture of two TK1 mAbs (1D11 + 1E3) and one TK1 mAb (1D11) alone.

[2]Ki-67 was detected using anti-Ki-67 mAb MIB-1.

of TK1 and PCNA were significantly higher in malignant lesions than in nonmalignant lesions ($p < 0.0001$ and $p < 0.0013$, respectively). Two types of the 20 benign tumors were identified, intraductal papilloma and fibroadenoma. No significant difference between the two types for expression of TK1 ($p = 0.65$) and PCNA ($p = 1.0$) was found. The immunohistochemic staining of eight various histological types of the breast cancers (noninfiltrating carcinoma: intraductal carcinoma and lobular carcinoma; infiltrating carcinoma: ductal carcinoma, lobular carcinoma, medullary carcinoma, mucinous adenocarcinoma, and large syringocarcinoma; Paget's disease of the nipple) was also determined. No difference was found between the TK1 and PCNA among different histologic types of malignant lesions. However, the strongest staining (+++) was found in the invasive group, such as ductal, lobular, and medullary carcinomas. Corresponding high staining for TK1 and PCNA was not found in the noninfitrating malignant lesions (Mao *et al.*, 2002).

Expression of TK1 and Ki-67 in Normal Tissues and Ductal Infiltrated Carcinomas

Expression of TK1 and Ki-67 in normal tissues showed only 10% positive staining for TK1 and Ki-67 (He *et al.*, 2003). The TK1 expression in ductal infiltrated carcinomas was higher (57.4%, 31/54) when two of our TK1 mAbs were combined (1D11 + 1E3), as compared to the TK1 expression using only one TK1 mAb (1D11) (48.1%, 28/54). In both cases TK1 expression was higher as compared to Ki-67 staining (40.7% 22/54) (Table 28). Although the expression of Ki-67 was lower than the expression of TK1, statistical

Table 28 Statistic Analysis of the Expression of TK1 Against the Expression of Ki-67 in Ductal Infiltrated Carcinomas

Ductal Infiltrated Carcinomas		Ki-67 (by mAbMIB-1)		TK1* (by mAbTK1 1D11)			TK1** (by mAbTK1 1E3 + 1D11)		TK1**/Ki-67
		Negative	Positive	Negative	Positive	*p*-value	Negative	Positive	*p*-value
Tumor Stage									
	I	6	1	7		0.28	7		—
	II	18	9	14	13	0.102	12	15	0.014
	III	8	12	7	13	0.684	4	16	0.068
Sum (*n*)		54	32	22	28	26	0.267	23	31	0.013
Tumor Grade									
	G1	11	5	14	2	0.106	14	2	0.105
	G2	16	7	9	14	0.002	8	15	< 0.01
	G3	5	10	5	10	0.751	1	14	0.028
Sum (*n*)		54	32	22	28	26	0.268	23	31	0.013

TK1**: TK1 was detected using a mixture of two TK1 mAbs (1D11 + 1E3). TK1*: TK1 was detected using one TK1 mAb (1D11). Ki-67 was detected using anti-Ki-67 mAb MIB-1.

Table 29 Expression of TK1 and Ki-67 in Relation to Tumor Stage and Grade in Ductal Infiltrated Carcinomas

Ductal Infiltrated Carcinomas		Ki-67 (by MIB-1)			TK1 (by mAbTK1 1D11)			TK1 (by mAbTK1 1E3 + 1D11)		
		Negative	Positive	*p* value	Negative	Positive	*p* value	Negative	Positive	*p* value
Tumor Stage										
	I–II	24	10	< 0.001	21	13	0.006	19	15	< 0.001
	III	8	12		7	13		4	15	
Tumor Grade										
	G1-2	27	12	< 0.001	23	16	< 0.001	22	17	< 0.001
	G3	5	10		5	10		1	14	
Sum (*n*)	54	32	22		28	26		23	31	

analysis of the expression of TK1 and Ki-67 in the malignant lesions was statistical by significantly higher as compared to normal tissues ($p < 0.001$ and $p < 0.001$, respectively, data not shown in Table 28).

The expression of TK1 in four patients with advanced infiltrative ductal carcinoma was also determined using anti-TK1 IgY antibody. A strong staining was observed in those four patients (+ + to + + +). No staining was found in normal tissues (Wu *et al.*, 2003).

Expression of TK1 and PCNA in Relation to Tumor Stage and Grade

The expression of TK1 in relation to tumor stage and tumor grade in 52 malignant breast patients compared with the expression of the PCNA in the same specimens. Our results showed that the TK1 expression showed a significant correlation with the increased tumor stages ($p = 0.023$) and tumor grades ($p = 0.009$). However, PCNA expression was neither significantly different in tumor stages ($p = 0.062$) nor in tumor grades ($p = 0.073$) (Mao *et al.*, 2002).

Expression of TK1 and Ki-67 in Relation to Tumor Stage and Tumor Grade

The TK1 expression was compared to Ki-67 expression (He *et al.*, 2003). No significant difference between the TK1 expression and Ki-67 expression was detected when using one TK1 mAb (1D11), but there was a statistical significant difference between the TK1 expression and Ki-67 when two TK1 mAbs were used (1D11 + 1E3). The results also indicated that the TK1 expression using two TK1 mAbs was more sensitive at increasing tumor stage and tumor grade as compared to that of Ki-67 expression. Concerning the tumor stage and the tumor grade, both TK1 and Ki-67 expression

showed a significant correlation with increased tumor stages and tumor grades.

We conclude that TK1 might be a more accurate and sensitive marker than PCNA and Ki-67 for estimation of cell proliferation and malignant potentials in breast carcinomas. Combining TK1 with other proliferating markers, such as Ki-67, will increase the prognostic index.

References

ASCO. 2002. Recommendations of the American Society of Clinical Oncology. A patient's guide: Understanding tumor markers for breast and colorectal cancers. *Am. Soc. Clin. Oncol.*:3–16.

Aziz, S., Pervez, S., Khan, S., Siddiqui., T., Kayani, N., Israr, M., and Rahbar, M. 2003. Case control study of novel prognostic markers and diseases outcome pregnancy/lactation-associated breast carcinoma. *Pathol. Res. Pract.* 199:15–21.

Beenken, S.W., and Bland, K.I. 2002. Biomarkers for breast cancer. *Minerva Chir.* 57:437–448.

Faneyte, I.F., Schrama, J.G., Peterse, J.L., Remijnse, P.L., Rodenhuis, S., and van de Vijver, M.J. 2003. Breast cancer response to neoadjuvant chemotherapy: Predictive markers and relation with outcome. *Br. J. Cancer* 88:406–412.

He, Q., Skog, S., and Tribukait, B. 1991. Cell cycle related studies on thymidine kinase and its isoenzymes in Ehrlich ascites tumors. *Cell Prolif.* 24:3–14.

He, Q., Wang, N., Skog, S., Eriksson, S., and Tribukait, B. 1996. Characterization of a peptide antibody against a C-terminal part of human and mouse cytosolic thymidine kinase, which is a marker for cell proliferation. *Eur. J. Cell Biol.* 70:117–124.

He, Q., Zou, L., Zhang, P.G., Lui, J.X., Skog, S., and Fornander, T. 2000. The clinical significance of thymidine kinase 1 measurement in serum of breast cancer patients using anti-TK1 antibody. *Int. J. Biol. Marker* 15:139–146.

He, Q., Mao, R.Y., Wu, J.P., Eriksson, S., and Skog, S. 2003. Cytosolic thymidine kinase (TK1) as a marker for cell proliferation in malignant tissues. Accepted by *AACC Annual Meeting D-35*.

Kato, T., Kameoka, S., Kimura, T., Nishikawa, T., and Kobayashi, M. 2002. C-erb-2 and PCNA as prognostic indicators of long-term survival in breast cancer. *Anticancer Res.* 22:1097–1103.

Kauffman, M.G., and Kelly, T.J. 1991. Cell cycle regulation of thymidine kinase: Residues near the carboxyl terminus are essential for the specific degradation of the enzyme at mitosis. *Mol. Cell. Biol. 11*:2538–2589.

Kuroiwa, N., Nakayama, M., Fufuda, T., Fufui, H., Ohwada, H., Hiwasa, T., and Fujimura, S. 2001. Specific recognition of cytosolic thymidine kinase in the human lung tumor by monoclonal antibody raised against recombinant human thymidine kinase. *J. Immunol. Methods 253*:1–11.

Lakhani, S.R., Ashworth, A. 2001. Microarray and histopathological analysis of tumours: The future and the past? *Nat. Rev. Cancer 2*:15–17.

Mao, Y.R., Wu, J.P., Wang, N., He, J.X., Wu, C.J., He, Q., and Skog, S. 2002. A comparison study: Immunohistochemical detection of cytosolic thymidine kinase and proliferating cell nuclear antigen in breast cancer. *Cancer Invest. 20*:922–931.

Molino, A., Pedersini, R., Micciolo, R., Frisinghelli, M., Giovannini, M., Pavarana, M., Nortilli, R., Santo, A., Manno, P., Padovani, M., Piubello, Q., and Cetto, G.L. 2002. Relationship between the thymidine labeling and Ki-67 proliferative indices in 126 breast cancer patients. *Appl. Immunohistochem. Mol. Morphol. 10*:304–309.

Pegram, M., and Slamon, D. 2000. Biological relation for Her/neu (c-erbB2) as a target for monoclonal antibody therapy. *Sim. Oncol. 27 (Suppl. 9)*:13–19.

Perou, C.M., Sørlie, T., Elsen, M.B., van de Rijn, M., Jeffrey, S.S., Rees, C.A., Pollack, J.B., Ross, D.T., Johnsen, H., Akslen, L.A., Fluge, Ø., Pergamenschlkov, A., Williams, C., Zhu, S.X., Lonning, P.E., Børresen-Dale, A-L., Brown, P.O., and Botstein, D. 2000. Molecular portraits of human breast tumors. *Nature 406*:747–752.

Sherley, J.L., and Kelly, T.J. 1988. Regulation of human thymidine kinase during the cell cycle. *J. Biol. Chem. 26*:8350–8358.

Trihia, H., Murray, S., Price K., Gelber, R.D., Golouh, R., Goldhirsh, A., Coates, A.S., Collins, J., Castiglione-Gertsch, M., and Gusterson, B.A. 2003. Ki-67 expression in breast carcinoma. *Cancer 97*:1321–1331.

Voeller, D.M., Parr, A., and Allegra, C.J. 2001. Development of human anti-thymidine kinase antibodies. *Anti-Cancer Drugs 12*:555–559.

Wang, N., He, Q., Skog, S., Eriksson, S., and Tribukait, B. 2001. Investigation on cell proliferation with a new antibody against thymidine kinase 1. *Analyt. Cell. Pathol. 23*:11–19.

Wu, J.P., Mao, Y.R., He, L.X., Wang, N., Wu, C.J., He, Q., and Skog, S. 2000. Comparison study on the expression of cytosolic thymidine kinase and proliferating cell nuclear antigen in colorectal carcinoma. *Anticancer Res. 6*:4867–4872.

Wu, C.J., Yang, R.J., Zhou, J., Bao, S., Zou, L., Zhang, P.G., Mao, Y.R., Wu, J.P., and He, Q. 2003, in press. Production and characterization of a novel chicken IgY antibody raised against C-terminal peptide from human thymidine kinase 1. *J. Immuno. Method. 277*:157–168.

Zhang, F., Li, H., Pendleton, A.R., Robison, J.G., Monson, K.O., Murray, B.K., and O'Neill K.L. 2001. Thymidine kinase 1 immunoassay: A potential maker for breast cancer. *Cancer Detect. Prev. 25*:8–15.

Zou, L., Zhang, P.G., Zou, S., Li, Y., and He, Q. 2002. The half-life of cytosolic thymidine kinase in serum by ECL dot blot: A potential marker for monitoring the response to surgery of patients with gastric cancer. *Int. J. Biolog. Marker 17*:135–140.

23

Immunohistochemical Detection of Melanoma Antigen E (MAGE) Expression in Breast Carcinoma

Antonio Juretic, Rajko Kavalar, Guilio C Spagnoli,
Bozena Sarcevic, and Luigi Terracciano

Introduction

Tumor associated antigens (TAA) of the melanoma antigen E (MAGE) family were the first described in humans (van der Bruggen *et al.*, 1991). They belong to the cancer–testis (CT) TAA subclass encoded by genes whose expression is detectable in different histologically unrelated tumor types and in a limited number of healthy tissues. Presently, over 20 related MAGE genes have been identified, a majority mapping on chromosome X (Chomez *et al.*, 2001; Zendman *et al.*, 2003).

Based on their extensive sequence homologies, four groups, MAGE-A, -B, -C, and -D have been identified. A number of immunogenic epitopes derived from these TAA and recognized by both HLA class I and class II restricted T cells have been characterized (for a review, see Renkvist *et al.*, 2001). Clinical trials suggest that specific immunization according to different procedures could lead to clinically effective antitumor immune responses (Scanlan *et al.*, 2002).

The expression of MAGE family TAA has mostly been studied in clinical materials at the gene expression level by polymerase chain reaction (PCR) (Mashino *et al.*, 2001; van der Bruggen *et al.*, 1991). This technology, however, provides only limited information, because, for instance, it does not allow the quantification of TAA positive cancer cells. Such data are of crucial relevance in clinical studies because effective immunization targeting TAA expressed in low percentages of tumor cells would likely be of modest therapeutic relevance. Thus, a major effort has been focused on the generation of serologic reagents identifying TAA determinants in clinical specimens.

A small panel of these reagents has indeed been produced and used to support PCR data and to quantify, at the cellular level, the extent of their expression (for a review, see Jungbluth *et al.*, 2000b; Juretic *et al.*, 2003). Interestingly, in specimens characterized by high infiltration of inflammatory cells, where unequivocal PCR data could not be generated, such as Hodgkin lymphoma, these reagents played a critical role in demonstrating MAGE TAA expression in neoplastic Reed-Sternberg cells (Chambost *et al.*, 2000).

Handbook of Immunohistochemistry and *in situ* Hybridization of Human
Carcinomas, Volume 1: Molecular Genetics; Lung and Breast Carcinomas

471

Copyright © 2004 by Elsevier (USA)
All rights reserved.

More recently, tissue microarray technology (TMA; Kononen *et al.*, 1998) also has been applied to the detection of cancer-testis gene expression. Availability of clinical data corresponding to individual specimens has allowed the evaluation of correlations existing between MAGE TAA expression and clinical course (Bolli *et al.*, 2002; Kocher *et al.*, 2002).

Regarding breast cancers, expression of MAGE genes has been reported by several groups (Brasseur *et al.*, 1992; Mashino *et al.*, 2001; Wascher *et al.*, 2001). In particular, MAGE-A1, -A2, -A3, -A4, -A6, and -A12 specific transcripts have been identified (Otte *et al.*, 2001). In contrast, there is a conspicuous lack of data regarding immunodetection of specific gene products. In the recent past our groups have started to produce immunohistochemical evidence of cancer/testis protein detection in breast cancers (Kavalar *et al.*, 2001).

MATERIALS

1. **Stock buffer solution (20X)**. Phosphate-buffered saline (PBS): 4 g potassium dihydrogen phosphate (KH_2PO_4), 23 g disodium hydrogen phosphate (Na_2HPO_4), 4 g potassium chloride (KCl), 160 g sodium chloride (NaCl) and 810 ml deionized glass-distilled water (pH 7.4).
Working solution: 100 ml of PBS stock solution and 1900 ml of deionized glass-distilled water.
2. **Citrate buffer stock solutions**. *Solution A:* 0.1 M citric acid (10.5 g $C_6H_8O_7 \cdot H_2O$ in 500 ml deionized glass-distilled water). *Solution B:* 0.1 M sodium citrate (14.7 g $C_6H_5O_7Na_3 \cdot 2H_2O$ in 500 ml deionized glass-distilled water).
Working solution: 4.5 ml citric acid (solution A) and 20.5 ml sodium citrate (solution B) at pH 6.0. Bring to 250 ml volume with deionized glass-distilled water.
3. **Normal rabbit (NR) serum** (Dako X0902, Dako A/S, Glostrup, Denmark) diluted 1:10 in PBS buffer.
4. **Primary antibody**: undiluted, multi-MAGE specific, 57 B hybridoma supernatant (Kocher *et al.*, 1995).
5. **Second step reagent:** Biotinylated rabbit anti-mouse Immunoglobulins (Dako E0354, Dako A/S, Glostrup, Denmark), diluted 1:300 in PBS buffer.
6. **Tris buffer:** *Tris* [hydroxymethyl] aminomethane (Sigma, St. Louis, MO, pH 7.9, Cat. No.T-4628).
7. **Tris buffered saline (TBS):** 0.18 g Tris buffer dissolved in 30 ml of deionized glass-distilled water and supplemented with 0.08 ml concentrated HCl to pH 7.6.
8. **Visualization reagents:** Avidin biotin complex (ABC), horseradish peroxidase (HRP) (Dako K 0355, Dako A/S, Glostrup, Denmark).
9. **ABC solution:** 1 ml complex A and 1 ml complex B diluted to 100 ml in TBS; ABC solution

following extensive stirring, should rest for 30 min in the dark before use.
10. **DAB:** 3,3′-diaminobenzidine (Sigma, St. Louis, MO, Cat. No. D 5637).
11. **DAB solution:** 200 ml of PBS buffer, 150 mg DAB, filtered and added to 500 ml of H_2O_2.
12. **Mayer's hematoxylin:** 1 g hematoxilin (Merck, Darmstadt, Germany, Cat. No. 1.04302), 100 ml deionized glass-distilled water, 0.2 sodium iodate-$NaIO_3$, 50 g potassium aluminium sulfate-$AlK(SO_4)_2 \cdot 12H_2O$. Mix the solution at the room temperature, then add 50 g chloralhydrate (Merck, Darmstadt, Germany, Cat. No. 1.02425) and 1 g citric acid.

METHOD

Tissue sections of 5-μm thickness are cut from paraffin-embedded tissue blocks and placed on object slides, previously treated with silane adhesive (3-aminopropyltriethoxy-silane, sigma, St. Louis, MO, Cat. No. A 3648). Slides with tissue sections are incubated overnight in a thermostat at 58°C.

Sections are then deparaffined and incubated for a total of 18 min subdivided in one 8-min and two 5-min periods in citrate buffer (pH 6.0) in a microwave oven at 800 W power. Citrate buffer with inserted tissue slides is then left to cool during the night. On the following day tissue slides are washed with PBS buffer (pH 7.2).

Endogenous peroxidase activity is blocked by a 15-min treatment at room temperature with a solution of hydrogen peroxide (H_2O_2) in methanol (3 ml of concentrated H_2O_2 and 97 ml of methanol). To minimize background, tissue sections are preincubated for 20 min with normal rabbit serum (see preceding) at room temperature. After removal of excess rabbit serum, sections are incubated for 1 hr with MAGE 57B supernatant at room temperature in a humid dark chamber.

Tissue sections are washed in PBS buffer three times for 2 min each, and then the secondary biotinylated reagent (1:300) (Biotinylated Rabbit Anti-Mouse Immunoglobulins, Dako A/S, Glostrup, Denmark, see preceding) is added dropwise onto them. Incubation with secondary biotinylated reagent lasts 30 min under the same conditions as the incubation with the primary antibody. Tissue sections are washed once more in PBS buffer for 5 min and then the ABC complex, prepared 30 min before use and kept in the dark, is added dropwise onto them. The ABC complex/HRP (avidin complexed with biotinylated peroxidase) is diluted with Tris buffer (pH 7.60) as mentioned earlier. The incubation with ABC complex lasts 30 min. Tissue sections are then washed in PBS buffer three more times for

5 min and then incubated in the dark for 15 min in DAB-solution for the visualization of MAGE-specific antibody binding. Slides are washed in distilled water three times for 5 min, stained with Mayer hematoxylin for 30 sec, washed with warm water for 5 min, dehydrated with alcohol at increasing concentrations (70%, 96%, 100%), cleared with xylene for 15 min, and mechanically covered.

The 57B hybridoma supernatant has been used in undiluted form. As positive control we use malignant melanoma sections displaying strong, diffuse, positive reactivity (Hofbauer et al., 1997). As negative control we use melanocytic nevus sections. Furthermore, we have not detected any positive staining on normal breast tissue sections.

MAGE antigen retrieval has also been attempted by using protease type XIV (Sigma P 5147) and by pressure cooking. Under these conditions, however, immunohistochemical reactions were very weak and were characterized by high background.

Immunohistochemical reactivities are semiquantitively evaluated as follows.

Negative reaction: 0
Mild reaction: + (if the reaction is positive in 20% of tumor cells)
Moderate reaction: ++ (if the reaction is positive in 20–50% of tumor cells)
Strong reaction: +++ (if the reaction is positive in more than 50% of tumor cells)

Tissue Microarrays

Tissue microarray technology (TMA) takes advantage of tissue cylinders (diameter: 0.6 mm) derived from thousands of different primary tumor blocks and subsequently brought into one empty "recipient" paraffin block (Kononen et al., 1998). Sections from such array blocks can then be used for simultaneous analysis of hundreds or thousands tumors at the deoxyribonucleic acid (DNA), ribonucleic acid (RNA), or protein level. Most importantly, specific TMA databases, including relevant clinical information related to individual specimens, have also been produced. Thus, TMAs offer the unique opportunity to combine large sets of pathological and clinical data aimed at the evaluation of the potential significance of the expression of specific markers. TMA technology is starting to be used for the evaluation of CT tumor associated antigens expression in large series of clinical specimens (Bolli et al., 2002; Kocher et al., 2002).

Materials and methods employed are similar to those described earlier for conventional sections with minor modifications.

In particular,

1. Microwave treatment in citrate buffer (pH 6) is prolonged for 30 min.
2. Incubation with the MAGE-specific reagents is prolonged overnight at 4°C.

RESULTS

We have investigated MAGE TAA expression in a group of invasive ductal breast cancers including well-differentiated (grade 1), moderately differentiated (grade 2), and poorly differentiated (grade 3) (Elston, 1987) cases. In good agreement with published classification criteria, grade 1 cancers showed smaller tumor size, lower frequency of lymphatic vessels invasion, a highly frequent intense reactivity on staining with antiestrogen receptor mAb, and lower numbers of axillary node metastases, as compared with grade 2 and 3 cases (data not shown).

Immunohistochemical detection of MAGE TAA was attempted by using 57B hybridoma supernatant. This reagent, generated using recombinant MAGE-A3 as immunogen (Kocher et al., 1995), recognizes multiple MAGE TAA in Western blot and immunocytochemistry assays (van Baren et al., 1999), but prevailingly MAGE-A4 in paraffin-embedded samples (Landry et al., 2000).

In agreement with data from other histologies (Jungbluth et al., 2000a), no staining of nontumoral cells was observed in breast tissue sections. In contrast, immunoreactivity of moderate to strong intensity, with a prevailingly cytoplasmic localization, could be observed in neoplastic cell (Figure 68).

Irrespective of percentages of stained cells, positivities were detectable with significantly higher frequency in grade 3 than in grade 1 cases (47% versus 16%) with intermediate values (25%) observed in grade 2 tumors.

Interestingly, the percentages of positively stained cells in grade 1 neoplasms never exceeded 50% (++) and were lower than 20% (+) in half the positive cases. On the other hand, in both grade 2 and 3 tumors strong positivities, involving over 50% of tumor cells (+++), were observed in a third of immunoreactive cases.

The correlation observed between 57B staining and tumor grade prompted us to investigate additional prognostic markers. Indeed, we found that MAGE TAA positivity inversely correlated with oestrogen receptor staining (Bundred, 2001). On the other hand a positive correlation with lymphatic vessel invasion and intratumoral necrosis was also observed (Gilchrist et al., 1993; Whitwell et al., 1984).

MAGE gene products encode both class I and II restricted TAA (for a review, see Renkvist et al., 2001), suggesting the possibility to detect in situ evidence of

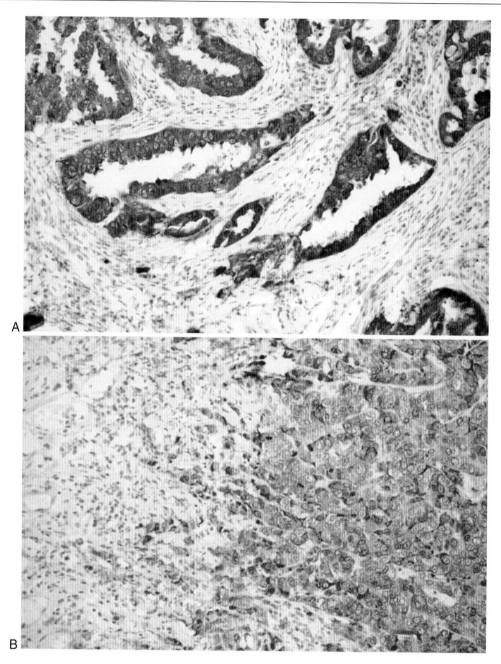

Figure 68 **A:** 57 B staining in a well-differentiated invasive ductal carcinoma. In this area nearly all neoplastic cells show a predominantly cytoplasmic strong staining. **B**: 57 B staining in a poorly differentiated invasive ductal carcinoma. In the absence of staining of normal ducts, a majority of carcinoma cells express MAGE gene products.

active immune responses. We stained a number of cases under investigation with reagents specific for CD3, CD4, CD8, and CD20. Lymphocyte infiltration to different extents was actually observed in virtually all cases but no correlation with 57B expression could be established. Furthermore, activated CD8+ cells displaying granzyme B positivity (Oberhuber *et al.*, 1998)

were found to infiltrate in low numbers breast cancer tissues with no correlation with 57B mAb staining.

DISCUSSION

The MAGE family gene products are of particular interest in tumor immunology because they encompass

HLA class I and class II restricted epitopes, recognized by specific T cells (Renkvist *et al.*, 2001; Consogno *et al.*, 2003). In particular, naturally processed peptides derived from MAGE-1, -2, -3, -4, -6, and -10 targeted by tumoricidal cytotoxic T lymphocytes (CTL) have been identified. A number of them are presently being evaluated in the context of dedicated clinical trials. On the other hand, the physiological role of MAGE gene products is still obscure. Several studies have demonstrated that DNA demethylating agents induce the expression of MAGE genes (Barker *et al.*, 2002; Zendman *et al.*, 2003). Indeed, demethylation is known to be typically associated with dedifferentiation processes, likely to include events observed in advanced stages of cancer development. Clearly, the availability of specific mAbs permits on one hand the enumeration of neoplastic cells expressing the target molecules in clonally heterogeneous clinical samples and on the other hand testing the potential role of MAGE proteins as markers of defined tumor stages.

We have documented that expression of MAGE gene products is typically detectable in poorly differentiated invasive ductal breast cancer, whereas it is rarely observed in well-differentiated cases. In keeping with this observation, MAGE immunoreactivity appears to correlate with lymphatic vessel invasion and intratumoral necrosis (Whitwell *et al.*, 1984; Gilchrist *et al.*, 1993), whereas an inverse correlation with estrogen receptor expression emerges.

No obvious correlation with the infiltration of tumors by defined lymphocyte subpopulations (CD3+, CD4+, CD8+, CD20+, or granzyme B+) suggestive of discrete specific immune responses was detected, although MAGE TAA specific CTL have been derived from TIL infiltrating breast cancers on repeated *in vitro* stimulation (Toso *et al.*, 1996).

Expression of MAGE gene products, as detectable by routine techniques on paraffin-embedded sections, could represent an important marker of dedifferentiation in breast cancers. Further studies are required to clarify the prognostic relevance, if any, of MAGE TAA immunoreactivity. Clearly, these data support the concept of specific immunotherapy procedures targeting MAGE family TAA in highly aggressive forms of breast cancer.

References

Barker, P.A., and Salehi, A. 2002. The MAGE proteins: Emerging roles in cell cycle progression, apoptosis, and neurogenetic disease. *J. Neurosci. Res. 67:*705–712.

Bolli, M., Kocher, T., Adamina, M., Guller, U., Dalquen, P., Haas, P., Mirlacher, M., Gambazzi, F., Harder, F., Heberer, M., Sauter, G., and Spagnoli, G.C. 2002. Tissue microarray evaluation of Melanoma antigen E (MAGE) tumor-associated antigen expression: Potential indications for specific immunotherapy and prognostic relevance in squamous cell lung carcinoma. *Ann. Surg. 236:*785–793; discussion 793.

Brasseur, F., Marchand, M., Vanwijck, R., Herin, M., Lethe, B., Chomez, P., and Boon, T. 1992. Human gene MAGE-1, which codes for a tumor-rejection antigen, is expressed by some breast tumors. *Int. J. Cancer 52:*839–841.

Bundred, N.J. 2001. Prognostic and predictive factors in breast cancer. *Cancer Treat. Rev. 27:*137–142.

Chambost, H., Van Baren, N., Brasseur, F., Godelaine, D., Xerri, L., Landi, S.J., Theate, I., Plumas, J., Spagnoli, G.C., Michel, G., Coulie, P.G., and Olive, D. 2000. Expression of gene MAGE-A4 in Reed-Sternberg cells. *Blood 95:*3530–3533.

Chomez, P., De Backer, O., Bertrand, M., De Plaen, E., Boon, T., and Lucas, S. 2001. An overview of the MAGE gene family with the identification of all human members of the family. *Cancer Res. 61:*5544–5551.

Consogno, G., Manici, S., Facchinetti, V., Bachi, A., Hammer, J., Conti-Fine, B., Rugarli, C., Traversari, C., and Protti, M.P. 2003. Identification of immunodominant regions among promiscuous HLA-DR-restricted CD4+ T-cell epitopes on the tumor antigen MAGE-3. *Blood 101:*1038–1044.

Elston, C.W. 1987. Grading of invasive carcinoma of the breast. In Page, D.L., and Anderson, T.J. (eds) *Diagnostic histopathology of the breast.* Edinburgh: Churchill Livingstone, 300–311.

Gilchrist, K.W., Gray, R., Fowble, B., Tormey, D.C., and Taylor, S.G. 1993. Tumor necrosis is a prognostic predictor for early recurrence and death in lymph node-positive breast cancer: A ten-year follow-up study in 723 Eastern Cooperative Oncology Group patients. *J. Clin. Oncol. 11:*1929–1935.

Hofbauer, G.F.L., Schaefer, C., Noppen, C., Böni, R., Kamarshev, J., Nestle, F.O., Spagnoli, G.C., and Dummer, R. 1997. MAGE-3 immunoreactivity in formalin-fixed, paraffin embedded primary and metastatic melanoma. Frequency and distribution. *Am. J. Pathol. 151:*1549–1553.

Jungbluth, A.A., Busam, K., Kold, D., Iversen, K., Coplan, K., Chen, Y.T., Spagnoli, G.C., and Old, L.J. 2000a. Expression of MAGE-antigens in normal tissues and cancer. *Int. J. Cancer 85:*460–465.

Jungbluth, A.A., Stockert, E., Chen, Y.T., Kolb, D., Iversen, K., Coplan, K., Williamson, B., Altorki, N., Busaam, K.J., and Old, L.J. 2000b. Monoclonal antibody MA454 reveals a heterogeneous expression pattern of MAGE-1 antigen in formalin-fixed paraffin embedded lung tumours. *Br. J. Cancer 83:*493–497.

Juretic, A., Spagnoli, G.C., Schultz-Thater, E., and Sarcevic, B. 2003. Cancer/testis tumour-associated antigens: Immunohistochemical detection with monoclonal antibodies. *Lancet Oncol. 4:*104–109.

Kavalar, R., Sarcevic, B., Spagnoli, G.C., Separovic, V., Samija, M., Terracciano, L., Heberer, M., and Juretic, A. 2001. Expression of MAGE tumour-associated antigens is inversely correlated with tumour differentiation in invasive ductal breast cancers: An immunohistochemical study. *Virchows Arch. 439:*127–131.

Kocher, T., Schultz-Thater, E., Gudat, F., Schaefer, C., Casorati, G., Juretic, A., Willimann, T., Harder, F., Heberer, M., and Spagnoli, G.C. 1995. Identification and intracellular location of MAGE-3 gene product. *Cancer Res. 55:*2236–2239.

Kocher, T., Zheng, M., Bolli, M., Simon, R., Forster, T., Schultz-Thater, E., Remmel, E., Noppen, C., Schmid, U., Ackermann, D., Mihatsch, M.J., Gasser, T., Heberer, M., Sauter, G., and Spagnoli, G.C. 2002. Prognostic relevance of MAGE-A4 tumor antigen expression in transitional cell carcinoma of the urinary bladder: A tissue microarray study. *Int. J. Cancer 100:*702–705.

Kononen, J., Bubendorf, L., Kallioniemi, A., Barlund, M., Schraml, P., Leighton, S., Torhorst, J., Mihatsch, M.J., Sauter, G., and Kallioniemi, O.P. 1998. Tissue microarrays for high-throughput molecular profiling of tumor specimens. *Nature Med.* 4:844–847.

Landry, C., Brasseur, F., Spagnoli, G.C., Marbaix, E., Boon, T., Coulie, P., and Godelaine, D. 2000. Monoclonal antibody 57B stains tumor tissues that express gene MAGE-A4. *Int. J. Cancer* 86:835–841.

Mashino, K., Sadanaga, N., Tanaka, F., Yamaguchi, H., Nagashima, H., Inoue, H., Sugimachi, K., and Mori, M. 2001. Expression of multiple cancer-testis antigen genes in gastrointestinal and breast carcinomas. *Br. J. Cancer* 85:713–720.

Oberhuber, G., Bodingbauer, M., Mosberger, I., Stolte, M., and Vogelsang, H. 1998. High proportion of granzyme B-positive (activated) intraepithelial and lamina propria lymphocytes in lymphocytic gastritis. *Am. J. Surg. Pathol.* 22:450–458.

Otte, M., Zafrakas, M., Riethdorf, L., Pichlmeier, U., Loning, T., Janicke, F., and Pantel, K. 2001. MAGE-A gene expression pattern in primary breast cancer. *Cancer Res.* 61:6682–6687.

Renkvist, N., Castelli, C., Robbins, P. F., and Parmiani, G. 2001. A listing of human tumor antigens recognized by T cells. *Cancer Immunol. Immunother. 50:*3–15.

Scanlan, M.J., Gure, A.O., Jungbluth, A.A., Old, L.J., and Chen, Y.T. 2002. Cancer/testis antigens: An expanding family of targets for cancer immunotherapy. *Immunol. Rev. 188:*22–32.

Toso, J.F., Oei, C., Oshidari, F., Tartaglia, J., Paoletti, E., Lyerly, H.K., Talib, S., and Weinhold, K.J. 1996. MAGE-1-specific precursor cytotoxic T-lymphocytes present among tumor-infiltrating lymphocytes from a patient with breast cancer: Characterization and antigen-specific activation. *Cancer Res. 56:*16–20.

van Baren, N., Brasseur, F., Godelaine, D., Hames, G., Ferrant, A., Lehmann, F., Andre, M., Ravoet, C., Doyen, C., Spagnoli, G.C., Bakkus, M., Thielemans, K., and Boon T. 1999. Genes encoding tumor-specific antigens are expressed in human myeloma cells. *Blood 94:*1156–1164.

van der Bruggen, P., Traversari, C., Chomez, P., Lurquin, C., De Plaen, E., Van den Eynde, B., Knuth, A., and Boon, T. 1991. A gene encoding an antigen recognized by cytolytic T lymphocytes on a human melanoma. *Science 254:*1643–1647.

Wascher, R.A., Bostick, P.J., Huynh, K.T., Turner, R., Qi, K., Giuliano, A.E., and Hoon, D.S. 2001. Detection of MAGE-A3 in breast cancer patients' sentinel lymph nodes. *Br. J. Cancer 85:*1340–1346.

Whitwell, H.L., Hughes, H.P., Moore, M., and Ahmed, A. 1984. Expression of major histocompatibility antigens and leukocyte infiltration in benign and malignant human breast disease. *Br. J. Cancer 49:*161–72.

Zendman, A.J., Ruiter, D.J., and Van Muijen, G.N. 2003. Cancer/testis-associated genes: Identification, expression profile, and putative function. *J. Cell. Physiol. 194:*272–288.

24

Male Breast Carcinoma: Role of Immunohistochemical Expression of Receptors in Male Breast Carcinoma

Jessica Wang-Rodriguez

Introduction

The incidence of male breast carcinoma (MBC) accounts for less than 1% of the total cases of breast cancer and 0.2% of all cancers in men (SEER, 1997), estimated at 1000–1400 new cases per year (Carmalt *et al.*, 1998; Hsing *et al.*, 1998; Ravandi-Kashani *et al.*, 1998; SEER, 1997; Williams *et al.*, 1996; Yildirim *et al.*, 1998). However, the number of reported cases in most series is small, limiting the ability to study this disease (Borgen *et al.*, 1992; Parker *et al.*, 1997). Prior studies suggest that outcomes have improved since 1985 (Donegan *et al.*, 1998). However, there is no standard therapy specifically for male breast cancer patients, and treatment selection is based on information derived from studies of female breast cancer. Although the diseases have similarities, there are notable differences in risk factors, prognosis, and survival. Risk factors that have been reported to be associated with the etiology of male breast cancer include obesity, excess estrogen, gynecomastia, infertility, ionizing radiation exposure, Jewish heritage, advanced age, liver disease, single marital status, hypoandrogenism, and Klinefelter's syndrome (Carmalt *et al.*, 1998; Donegan *et al.*, 1998; Goss *et al.*, 1999; Hsing *et al.*, 1998; Vetto *et al.*, 1999; Williams *et al.*, 1996; Yildirim *et al.*, 1998).

A number of molecular markers have been detected in female breast carcinomas recently and have contributed greatly to the understanding of the disease. However, information on molecular markers is limited in male breast carcinomas, mainly because of the rarity of the disease in any single institution. Many small, single-center studies with limited history and follow-up are reported in the literature to date. Therefore, either a large center or multicenter study combining a large number of patients is needed to have a better understanding of its phenotype. Immunophenotyping of the male breast carcinoma is best achieved through immunohistochemistry (IHC), which is relatively easy

Handbook of Immunohistochemistry and *in situ* Hybridization of Human
Carcinomas, Volume 1: Molecular Genetics; Lung and Breast Carcinomas

477

Copyright © 2004 by Elsevier (USA)
All rights reserved.

to perform, and achieve quality control and easy interpretation by surgical pathologists. However, other techniques, such as fluorescent *in situ* hybridization (FISH) or chromogenic *in situ* hybridization (CISH) (Gupta *et al.*, 2003), are becoming mainstream diagnostic tools used in surgical pathology using fresh-frozen or formalin-fixed, paraffin-embedded tissues. New molecular marker discoveries have prompted increasing use of polymerase chain reaction (PCR). More is expected in the next decade of cancer diagnostics in bringing gene expression arrays or proteonomics study to the clinical laboratory in aiding diagnosis as well as forecasting prognostic or guiding treatment.

Despite recent molecular discoveries in studying tumor pathology, none compares the widely acceptable use of IHC in routine pathology settings. Most pathologists are well versed to evaluate staining of the antigens against the familiar histologic background, thus making the distinction of certain subsets of tumor possible beyond the hematoxylin-eosin staining. However, the wide use of monoclonal antibodies on tissues also point out that certain "specific" antibodies to a cell type are not specific. Cross-reactions and nonspecific background staining also may be misinterpreted and wrong conclusions made. The possible pitfalls of IHC require the knowledge of the staining procedure, tissue histology, and expertise to troubleshoot if unexpected results occur. It is also important to validate the staining results to ensure the quality and accuracy of the interpretation.

Techniques to Study MBC

We became interested in studying MBC after finding that there is a general lack of published medical literature about this entity. We diagnosed and presented a male patient with breast carcinoma at the tumor board and concluded that there was a lack of published information of its pathophysiology and treatment. Most of the treatment is modeled after what we know in female breast carcinomas. Some of the epidemiology studies showed that there were distinct differences between male and female breast carcinomas. However, only very limited studies in the literature tried to prove the case. A recent PubMed search revealed greater than 20,000 literatures on female breast carcinomas and only 89 on MBC in the last 5 years. Knowing that it is a relatively rare disease in the male population compared to the popular cancers such as prostate, colon, and lung cancers, we took on the task of doing a large comprehensive study in the veterans population.

The Veterans Affairs (VA) Health Care System has one of the largest populations of male patients available, allowing male breast cancer to be more easily studied.

Our hypothesis based on the author's clinical experience is that male breast cancer patients present at an older age, are at a later stage of the disease, and are more likely to be symptomatic at the time of diagnosis. We have since reviewed clinical data of 241 male patients from the years 1983–2000 diagnosed with breast cancer in the VA Health Care System. Through this retrospective review, we aim to identify risk factors that would adversely affect disease-free survival, such as age, race/ethnicity, lifestyle factors (smoking and drinking), tumor characteristics (histology, size, and grade), axillary nodal status, and stage, and study the relationship of positive hormonal receptors with respect to survival and treatment.

Table 30 lists the histologic diagnosis for cases in the study. The American Joint Committee on Cancer (AJCC) has summarized stages as follows: 1 patient was stage TIS (0.4%), 40 stage 1 (16.6%), 83 stage II (34.4%), 34 stage III (14.1%), 21 stage IV (8.7%), and 62 were of unknown stage (25.7%). All the patients were diagnosed and treated in VA Health Care System facilities. Among 204 patients with therapy information, 185 (90%) received surgery as the main treatment. In addition, many patients had surgery plus radiation, 9 (4.4%); surgery plus hormonal therapy, 50 (24.5%); surgery plus chemotherapy, 17 (8.3%); or a combination of surgery plus multiple adjuvant therapies, 75 (36.7%).

Although all pathological types were reported, the majority was infiltrating ductal carcinoma and/or DCIS, which accounted for approximately 80% of the cases. Most tumors were located in the nipple, areola, and subareolar complex. Surgical excision with modified radical mastectomy was the main treatment of choice. However, no standardized adjuvant therapy was given in this cohort.

Before we address the specific protocols of the IHC, certain areas need to be discussed and familiarized.

Table 30 Histological Diagnosis

Histological diagnosis	Number of cases	Percent
Infiltrating ductal carcinoma	153	63.5
DCIS & infiltrating ductal carcinoma	33	13.7
Adenocarcinoma, not otherwise specified	17	7.1
Ductal carcinoma in situ (DCIS)	9	3.7
Invasive papillary carcinoma	4	1.7
Intracystic papillary carcinoma	3	1.2
Mucinous carcinoma	3	1.2
Other	8	3.3
Unspecified or unconfirmed	11	4.6
Total	**241**	**100%**

We have encountered several trials of methodologies of IHC and have found that the best method for MBC immunophenotyping is by the classic ABC (Avidin-Biotin Complex) method, with modifications to substitute avidin with streptavidin. Areas critical to successful IHC before the staining method include background reduction, use of positive and negative controls, and antigen retrieval method. The following paragraphs will address these issues.

Background

Background staining can vary depending on the type of tissues studied, fixation method, and the use of primary antibodies. Most of the nonspecific background staining is because of hydrophobic interactions between cell surface proteins and the reagent proteins, causing nonspecific binding and diffuse staining of the tissues. Many commercial manufacturers of IHC staining kits have attempted to eliminate the problem of hydrophobic interactions between tissue and reagent proteins by addition of detergent (i.e., Tween-20) or ethylene glycol to the diluent used to dilute the antibodies. We use the antibody diluent from Dako and achieve very good results. Additional measures of reducing background can be achieved by addition of blocking protein in a separate step immediately before the addition of the primary antibody. The blocking protein should be in serums identical to the same species from which the biotinylated secondary antibody is raised. We routinely use this step and find that this additional step reduces nonspecific binding of the secondary antibody and significantly reduces the background staining.

Controls

The reagent controls are a vital part of the IHC laboratory because the interpretation of the staining results depends on valid positive and negative controls. The positive controls validate the techniques, and the negative controls ensure that the nonspecific and background stainings are kept to minimum. All assays are conducted with both positive and negative controls. The negative control is a slide from the same patient tissue sample that is treated with a diluted serum used for protein blocking in antibody buffer solution without any primary antibody. All other steps are identical. One negative control is created per batch per antibody. The positive control is a known tissue-positive tissue sample that goes through the full protocol with the test tissue. Standard procedure is that the positive control tissue sample and the patient tissue test sample are placed on the same slide to ensure completely identical handling. Positive controls are used for all assays.

Antigen-Retrieval Method

Antigen retrieval (AR) has been a revolutionary technique in IHC and has been instrumental in unmasking low-level or formalin cross-linked antibodies (cytoplasmic or nuclear) in formalin-fixed paraffin-embedded tissues. Numerous methods are described for proteolytic digestion and heat-based antigen retrieval. We prefer using heat for antigen retrieval and have achieved much success in staining nuclear proteins such as ER, PR, Ki-67, and p53. In addition, a longer period of heat treatment is an excellent way to enhance the expression of HER-2/neu. We use an inexpensive Black and Decker Rice and Vegetable steamer for our heat epitope retrieval. Moist steam heat in our experience has resulted in better tissue histology preservation and excellent nuclear antigen immunoreactivity. In high-altitude settings, a pressure cooker may be used to keep the temperature constant. There is considerable debate in the optimal length of exposure required for heat retrieval. In general, 30–40 min are sufficient for most of the tissue types and targets. HER-2/neu stain requires 40 min of heat treatment. One important step in antigen retrieval is the cooling step, which has to take place slowly to room temperature and would require 20–30 min. The antigen-retrieval solution is generally composed of 0.01 M Tris-HCl, pH 1 or 10. However, we have had good overall results using the commercially available Target Retrieval Solution from Dako.

Staining Method

We use the popular two-step indirect method of streptavidin-biotin-horseradish peroxidase (HRP). The rational is outlined in Figure 69, where the primary antibody binds to the target antigens, and a biotinylated secondary antibody binds to the primary antibody. The streptavidin is conjugated with an enzyme-labeled HRP or alkaline phosphatase that has high affinity to biotin and binds to the secondary antibody. The color reaction is developed with the appropriate substrate–chromogen. We found that in our experience, the Dako LSAB kit using the same strategy has produced superior staining results. The kit makes it easier to quality control and to train personnel, and results are uniform and easy to interpret. The secondary antibody in the LSAB kit is already biotinylated, and the level of streptavidin/HRP has been optimized with the biotinylated secondary antibody; therefore, additional tittering of the secondary antibody with streptavidin/HRP is not required.

Autostainer Process

The Autostainer is a computerized lab tool for producing repeatable, consistent IHC assays. It is preset with staining procedures for all the IHC assays. Slides and reagents are load into designated slots, and

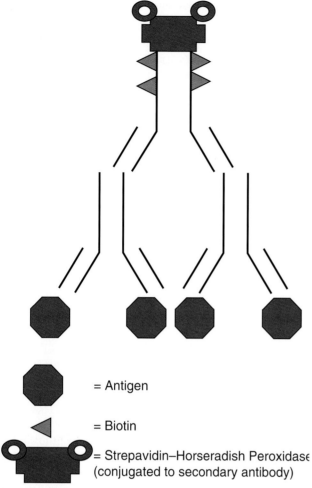

= Antigen

= Biotin

= Strepavidin–Horseradish Peroxidase
(conjugated to secondary antibody)

Figure 69 Schematic drawing of streptavidin-biotin staining method. The enzyme (horseradish peroxidase) labeled streptavidin reacts with the biotinylated secondary antibody.

the desired assay for each slide is entered into the computer controller. We use the Dako autostainer for our routine IHC stains. The autostainer can be used once each step of the IHC is optimized. The stainer uses the capillary gap principle, where reagents are filled between two microscope slides in a gap of 50-µm wide through the capillary reaction. Refilling of the capillary gaps and blotting is repeated for each reagent and wash step. Washing is achieved by the flow of buffer through the gaps. After the initial deparaffinization and antigen retrieval procedures, the basic process it goes through for all the assays in this project is as follows:

1. Buffer rinse
2. Hydrogen peroxide, to block endogenous peroxidase
3. Buffer rinse
4. Application of specific antibody
5. Buffered rinse

6. Application of labeled polymer
7. Buffered rinse × 2
8. Application of DAB color substrate
9. Distilled water (dH$_2$O) rinse

The slides are then removed from the Autostainer for final staining and mounting. This entire process takes about 90 min for a small number of slides and up to 3 hr if the machine's full 48-slide capacity is used.

MATERIALS AND METHODS

Materials (Streptavidin–Biotin Horseradish Peroxidase Method)

1. Slides: 3-aminopropyltriethoxysilane coated slides (Fisher)
2. Phosphate-buffered saline (PBS) (1 L of 10X PBS):80 g NaCl, 2 g KCl, 14.4 g Na$_2$HPO$_4$, 2.4 g KH$_2$PO$_4$, 800 ml dH$_2$O. Adjust to pH 7.4 with HCl. Add the final volume to 1 L. Autoclave before use.
3. 1X PBS: Dilute 1:10 of the 10X PBS with dH$_2$O
4. Ammonium hydroxide, 15 M diluted to 37 mM (1.2 ml 30% 15 M and 300 ml dH$_2$O)
5. Protein blockers: Dako Biotin Blocking System (Cat. No. X0590)
6. Primary antibodies: Recommended primary antibodies for MBC (See Table 31.)
7. Antibody diluent (Dako S0809) or 1% Bovine serum albumin (BSA) in 0.05 M Tris-HCl buffer (pH 7.2–7.6)
8. Blocking serum of the same species as the secondary antibody: Goat serum (Jackson Immunoreseach), diluted 1:20 in 1X PBS
9. Counterstain Gills II Hematoxylin (Fisher)
10. Coverslips
11. dH$_2$O
12. H$_2$O$_2$
13. Ethanol: 100%, 95% diluted in H$_2$O, and 70% diluted in H$_2$O

Table 31 Primary Antibodies

Antibody	Manufacturer	Clone	Dilution (In Antibody Diluent)	Species
Estrogen receptor	Dako	1D5	1:50	Mouse
Progesterone receptor	Dako	PgR636	1:50	Mouse
Ki-67	Innovex	MAB368P	1:100	Mouse
c-erbB-2 (HER2/neu)	Dako		1:250	Rabbit
p53	Novocastra	DO7	1:100	Mouse

14. Mounting media (Dako Faramount, S3025) or Gelmount (Fisher Cat. No. BM-M01)

15. Xylene

16. Hydrogen peroxide (3%)

17. Target retrieval solution (Dako S1699) 10X stock bottle: dilute to 1X before use with dH$_2$O

18. Substrate-Chromogen: 3,3'-diaminobenzidine (DAB): Dako K0673, 20 µl (one drop) of DAB in l ml of buffered substrate (included in the kit)

19. Dako LSAB2 K0675: contains biotinylated secondary goat anti-rabbit and anti-mouse antibodies and Streptavidin conjugated to horseradish peroxidase in PBS

METHOD

Paraffin-Embedded IHC

Deparaffinization Procedure

1. Sections (5-µm thick) of paraffin-embedded tissues are placed on slides that are transferred into an oven at 60°C for 1 hr (leaving slides longer will allow the sections to stick onto the slides better). Always process one to two slides extra in case tissue falls off in slide bath!

2. Allow slides to slightly cool and then place them in xylene bath for 5 min (2×) to completely deparaffinize the slides. From here on out never allow the slides to dry out.

3. Immediately place slides in 100% alcohol for 2 min (2×), 95% alcohol 2 min (2×), and 70% alcohol bath 2 min (2×).

4. Next, place slides in dH$_2$O for 2 min (2×) to remove all alcohol from the slides.

5. Place slides in an empty plastic box with the glass slide holder inside. Fill box with enough 1X Dako Target Retrieval solution to completely cover slides.

6. Place box with cover over the top in the Black and Decker Steamer and set timer for 40 min. This steamer is probably hot enough for most slides but in some cases a pressure cooker may be used in high-altitude settings.

7. Once steamed, the slide box should be removed from the steamer and left out, uncovered, for 20–30 min to cool to room temperature. The slides should remain in the target retrieval solution until the solution comes to room temperature.

8. Rinse slides 3X with dH$_2$O (1 min) to completely remove the target retrieval solution.

9. Rinse slides 3X in 1X PBS for 5 min each.

10. Quench the endogenous peroxidase activity by applying 3% H$_2$O$_2$ for 15 min.

11. Rinse off with 1X PBS using a pipette tip.

Removal of Endogenous Biotin and Protein Blocking

12. Apply Dako Protein block-Serum Free to all slides.

13. Incubate for 1 hr at room temperature (RT). Rinse the slides very briefly with 1X PBS. Or just tap off the solution and wipe well around the section.

14. Apply goat serum (Jackson Immunoreseach) for slides being treated with Dako LSAB2 kit (1:20) diluted in 1X PBS to cover tissue. Incubate serum blocker for 1 hr.

Primary Antibody

15. For negative control, place the serum from the same species in which the secondary antibody was raised.

16. Incubate serum blocker for 1 hr.

17. Rinse off with 1X PBS using a pipette tip.

18. Apply primary antibody: Dilutions of primary antibodies: See Table 34 for dilutions and identity of each antibody.

19. Incubate at RT for 30 min.

20. Wash all slides with 1X PBS using a pipette tip. Soak slides 3 times in 1X PBS for 5 min each.

Secondary Antibody and Link Step

21. Apply secondary antibody to all slides. (Dako LSAB 2 Kit, Bottle #3).

22. Incubate the slides for 30 min–1 hr at RT. Cover the slides with the box lid to keep the area humid.

23. Rinse gently with 1X PBS and place the slides 3 times in 1X PBS wash baths for 5 min each.

24. Gently tap off excess buffer and wipe well around the section.

If LSAB2 Kit Is Not Used, BD Secondary Antibodies Can Be Applied (raised in goat)

1. Biotinylated goat anti-mouse polyclonal AB for IHC (BD Biosciences/PharMingen; Cat. No. 550337). (1:200 dilution)

2. Biotinylated goat anti-rabbit polyclonal AB for IHC (BD Biosciences/PharMingen; Cat. No. 550338) (1:200 dilution)

Both antibodies were applied at ~0.6–0.7 ng/ml final concentration.

Streptavidin HRP Step

25. The slides receive ~2 drops from Dako LSAB2 kit bottle 4. Incubate for 15 min–30 min at RT.

26. Gently rinse slides 3–4 times in 1X PBS wash baths— ~5 min each.

Substrate–Chromogen Solution Step

27. Prepare fresh solution from stock solution kept at 4°C using the mixing chart in the kit (1 drop or ~20 µl of the DAB chromogen in 1 ml of the buffered substrate.

28. Incubate for 5 min at RT. Do not overincubate.

29. Collect the substrate–chromogen solution in a waste container. Immerse all slides 2–3 times in water baths, 5 min in each bath. It is also acceptable to leave the slides in the water bath longer.

Counterstaining Step

30. Immerse slides in hematoxylin: depending on the strength of the hematoxylin: Mayer's hematoxylin (Sigma;Cat. No. MHS-32): 30 sec–2 min or Gills hematoxylin II (Fisher brand); 1–3 min.

Remarks

Do not use alcohol-containing solutions for counterstaining, because the product is soluble in organic solvents.

31. Gently rinse off hematoxylin by immersing the slides in several (3–4 times) water baths or PBS.

32. Dip the slides for 10–20 sec in 35 mM NH_4OH. Rinse 1–2 times in dH_2O. Can leave the slides in water longer.

33. Mount the slides using aqueous mounting medium ~2–3 drops per slide. (Dako Faramount aqueous mounting medium; Cat. No. S302580).

Or Gelmount (Fisher Cat. No. BM-M01, 20-ml bottle).

DAKO Autostainer Procedure

For large volumes of cases, we have optimized procedures for use in the Dako Autostainer. This process reduces the time for the manual handling of the slides and ensures uniformity of the stains. This procedure can stain 48 slides at a time.

1. The slides are placed in a retrieval buffer solution and placed in a steamer for 30–40 min at 95°C.

2. After steaming, cool the slides at room temperature for 20 min, still in the retrieval buffer.

3. The slides are rinsed 3 times with dH_2O.

4. The slides are placed in the Dako Autostainer with appropriate reagents and run through antibody staining.

5. Hematoxylin counterstain is applied to the slides for 3 min, followed by dH_2O rinse.

6. The slides are placed for 1 min in Tris/Tween buffer, which acts as a bluing agent.

7. The slides are then rinsed briefly in dH_2O, then in 100% alcohol solution, and finally in xylene.

8. Cytoseal mounting media is applied to the slide face, and a coverslip is mounted. The slide/coverslip is allowed to harden for 1 hr.

9. Patient labels are printed and affixed to each slide with the appropriate information. The slides are returned to the ordering pathologist for review.

Scoring Criteria

For estrogen receptor (ER) and progesterone receptor (PR) staining, characteristic nuclear staining of greater than 50% in tumor cells are considered positive. There are a variety of ways of scoring p53 nuclear staining, with positive staining being > 5% or higher. We noticed that some nuclei stained strongly for p53, whereas other tumor nuclei stained positive but weakly for p53. Therefore, we devised a scoring scheme as positive for strong, uniform staining of the tumor nuclei, indeterminate for weak or focal nuclear staining, and negative for no staining. We use the FDA-approved Dako HercepTest scoring criteria for evaluating HER-2/neu expression. On formalin-fixed, paraffin-embedded tissue sections, HER-2/neu staining gives strong, membranous staining in more than 10% of the tumor cells (scored as 3+). A 2+ is considered as weak to moderate complete membrane staining in more than 10% of the cells; 1+ is a faint and partial membrane staining in 10% of the cells. Negative staining (0) is when there is no staining observed in tumor cells. Ki-67 is also a nuclear antigen and is interpreted as a percentage per 1000 tumor cells counted under oil immersion at 1000× magnification.

Results

We found that out of a total of 65 MBCs and 17 control male gynecomastia cases, there was clearly overexpression of p53 (MBC 15.4% versus control 6%, $p = 0.0087$), Her2-neu (MBC 9.2% versus control 0%, $p = 0.1$), and Ki-67 (MBC mean score 29.2% versus control 0%, $p = 0.0088$) in MBC cases. The ER is highly expressed in MBC cases (95.4%) and control cases (88.2%), as compared to PR (MBC 78.5% versus 88.2%, p = not significant). The findings offered potential usefulness in applying p53, HER-2/neu, and Ki-67 as diagnostic markers when diagnosis may be difficult to reach in atypical cases. However, when compared to patient outcome and survival, Ki-67 percentage was not significant in differentiating survival. Figure 70A–C shows examples of the staining on male breast carcinomas.

P53 and HER-2/neu, on the other hand, did offer prognostic information. Stage for stage, positive p53 expression in patients had a poorer outcome than p53 negative patients. Similarly, HER-2/neu expression, though low (9%), was associated with worse survival in multivariate analysis and was associated with nodal positive diseases. Therefore, HER-2/neu may be an important prognostic marker in MBCs as seen in female breast cancers.

We found that ER and PR expressions are much higher in MBCs than in female breast carcinomas or gynecomastia. This is in agreement with those previously reported in the literature. However, there is no associated prognostic significance, and our studies and others did not find a survival benefit in ER positive patients treated with tamoxifen.

CONCLUSIONS

We and others have attempted to study clinical and pathological correlation of male breast carcinomas in a multicenter setting. To date, the data are still limited, and lack of clinical trials precludes prospective clinical data collection and uniform treatment outcome. Therefore, most of the results using immunohisto-chemistry are used to correlate with other potential outcome measures, such as tumor grade, nodal status, and stage. Previous studies agree that there were distinct differences immunophenotypically between male and female breast carcinomas, suggesting that MBC needs to be addressed as a separate disease and perhaps treated based on its expression profile.

To summarize the literature review, some of the pertinent immunohistochemical markers are identified to

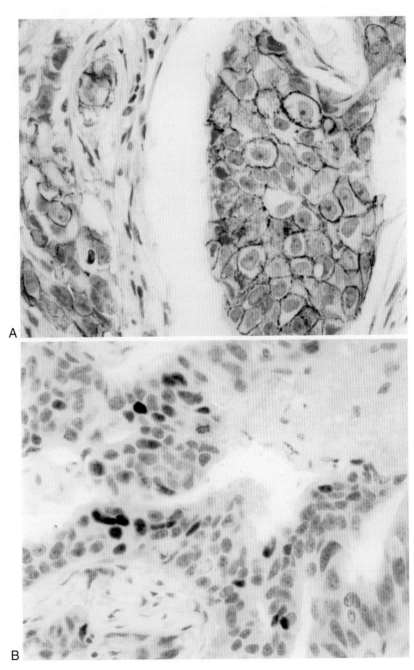

Figure 70 A: Representative HER-2/staining stain at 400X magnification. This case was scored as 3+. **B:** p53 staining in one case at 400X.

(Continued)

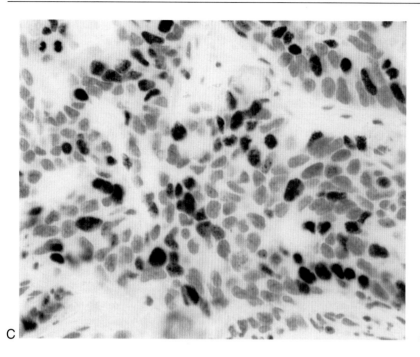

C

Figure 70—Cont'd C: Ki-67 staining at 4000X.

have potential diagnostic and prognostic importance, the following antigens are good candidates: p53, HER-2/neu, Ki-67, and Cyclin D1.

Markers found to be important to have prognostic significance in female breast carcinomas but not in MBC include cathepsin D (Youngson *et al.*, 1994), ER, PR, AR, EGFR, and bcl-2. Table 32 summarizes the percentage of a range of antigen positivity and potential prognostic significance. Note that not all immuno-histochemical staining were performed with antigen retrieval, and different methods were used in these studies.

References

Borgen, P., Wong, G., Vlamis, V., Potter, C., Hoffmann, B., Kinne, D., Osborne, M., and McKinnon, W. 1992. Current management of male breast cancer: A review of 104 cases. *Ann. Surg.* 215:451–459.

Carmalt, H., Mann, L., Kennedy, C., Fletcher, J., and Gillett, D. 1998. Carcinoma of the male breast: A review and recommendations for management. *Austral. New Zealand J. Surg.* 68:712–715.

Donegan, W., Redlich, P., Lang, P., and Gall, M. 1998. Carcinoma of the breast in males. *Cancer 83:*498–509.

Fox, S., Rogers, S., and Day, C. 1992. Oestrogen receptor and epidermal growth factor receptor expression in male breast carcinoma. *J. Pathol. 166:*13–18.

Table 32 Significance of Historical Markers

Markers	Range of positivity	Prognostic significance	References
ER	80–90%	None	(Carmalt *et al.*, 1998; Vetto *et al.*, 1999; Vinod *et al.*, 1999; Wang-Rodriguez *et al.*, 2002)
PR	70–78%	None	(Joshi, *et al.*, 1996; Margaria *et al.*, 1999; Wang-Rodriguez *et al.*, 2002; Willsher *et al.*, 1997)
HER-2/neu	17–81%	Positive gives worse prognosis	(Jimenez *et al.*, 2000; Pich *et al.*, 2000; Press *et al.*, 1990; Schnitt, 2001; Wang-Rodriguez *et al.*, 2002)
p53	2–62%	Positive gives worse prognosis	(Pich *et al.*, 2000; Shpitz *et al.*, 2000; Wang-Rodriguez *et al.*, 2002)
Cyclin D1	58%	Positive gives better prognosis	(Rayson *et al.*, 1998)
EGFR	13.8–76%	No information	(Fox *et al.*, 1992; Moore *et al.*, 1994)
Ki-67	29–38%	High percentage gives worse prognosis	(Rayson *et al.*, 1998; Wang-Rodriguez *et al.*, 2002)
bcl-2	79–94%	None	(Muir *et al.*, 2003; Pich *et al.*, 2000)

Goss, P., Reid, C., Pintilie, M., Lim, R., and Miller, N. 1999. Male breast carcinoma: A review of 229 patients who presented to the Princess Margaret Hospital during 40 years (1955–1996). *Cancer* 85:629–639.

Gupta, D., Middleton, L., Whitaker, M., and Abrams, J. 2003. Comparison of fluorescence and chromogenic in situ hybridization for detection of HER-2/neu oncogene in breast cancer. *Am. J. Clin. Pathol.* 119:381–387.

Hsing, A., McLaughlin, J., Cocco, P., Chien, H., and Fraumeni, J. 1998. Risk factors for male breast cancer (United States). *Cancer Causes and Control* 9:269–275.

Jimenez, R., Wallis, T., Tabaszka, P., and Visscher, D. 2000. Determination of Her-2/Neu status in breast carcinoma: Comparative analysis of immunohistochemistry and fluorescent in situ hybridization. *Mod. Pathol.* 13:37–45.

Joshi, M., Lee, A., Loda, M., Camus, M., Petersen, C., Heatley, G., and Hughes, K. 1996. Male breast carcinoma: An evaluation of prognostic factors contributing to a poorer outcome. *Cancer* 77:490–498.

Margaria, A., Chiusa, L., Candelaresi, G., and Canton, O. 1999. Androgen receptor expression in male breast carcinoma: Lack of clinicopathological association. *Brit. J. Cancer* 79:959–964.

Moore, J., Friedman, M., and Gramlich, T. 1994. Prognostic indicators in male breast cancer. *Lab. Invest.* 70:19A.

Muir, D., Kanthan, R., and Kanthan, S. 2003. Male versus female breast cancers: A population-based comparative immunohistochemical analysis. *Arch. Pathol. Lab. Med.* 127:36–41.

Parker, S., Tong, T., Bolden, S., and Wingo, P. 1997. Cancer Statistics. *CA Cancer J. Clin.* 47:5–27.

Pich, A., Margaria, E., and Chiusa, L. 2000. Oncogenes and male breast carcinoma: cerB-2 and p53 coexpression predicts a poor survival. *J. Clin. Oncol.* 18:2948–2965.

Press, M., Cordon-Cardo, C., and Slamon, D. 1990. Expression of the HER-2/neu proto-oncogene in normal human adult and fetal tissues. *Oncogene* 5:953–962.

Ravandi-Kashani, F., and Hayes, T. 1998. Male breast cancer: A review of the literature. *Eur. J. Cancer 34:* 1341–1347.

Rayson, D., Erlichman, C., Suman, V., Patrick, C., Roche, P., Wold, L., Ingle, J., and Donohue, J. 1998. Molecular markers in male breast carcinoma. *Cancer 83:*1947–1955.

Schnitt, S. 2001. Breast cancer in the 21st century: Neu opportunities and neu challenges. *Mod. Pathol.* 14:213–218.

SEER, Cancer incidence public-use database (CD-Rom). 1997. National Cancer Institute, version 1.1. Bethesda, MD.

Shpitz, B., Bomstein, Y., Sternberg, A., Klein, E., Liverant, S., Groisman, G., and Bernheim, J. 2000. Angiogenesis, p53, c-erbB2 immunoreactivity and clinicopathological features in male breast cancer. *J. Surg. Oncol.* 75:252–257.

Vetto, J., Jun, S., Padduch, D., Eppich, H., and Shis, R. 1999. Stages at presentation, prognostic factors, and outcome of breast cancer in males. *Am. J. Surgery* 177:379–383.

Vinod, S., and Pendlebury, S. 1999. Carcinoma of the male breast: A review of adjuvant therapy. *Austral. Radiol. 43:* 69–72.

Wang-Rodriguez, J., Cross, K., Gallagher, S., Djahanban, M., Armstrong, J., Wiedner, N., and Shapiro, D. 2002. Male Breast Carcinoma: Correlation of ER, PR, Ki-67, Her2-Neu, and p53 with treatment and survival, a study of 65 cases. *Mod. Pathol.* 15:853–861.

Williams, W., Powers, M., and Wagman, L. 1996. Cancer of the male breast: A review. *J. Nat. Med. Assoc.* 88:439–443.

Willsher, P., Leach, I., Ellis, I., Bell, J., Elston, C., Bourke, J., Blamey, R., and Robertson, J. 1997. Male breast cancer: Pathological and immunohistochemical features. *Anticancer Res.* 17:2335–2338.

Yildirim, E., and Berberoglu, U. 1998. Male breast cancer: A 22-year experience. *Eur. J. Surg. Oncol.* 24:548–552.

Youngson, B., Borgen, P., Sernie, R., et al. 1994. Prognostic markers in male breast carcinoma (MBC): An immunohistochemical study. *Lab. Invest.* 70:25A.

25

Detection of Glycoconjugates in Breast Cancer Cell Lines: Confocal Fluorescence Microscopy

Jean Guillot

Introduction

Among the disorders involved in the cancer-forming process, much work has been devoted in recent years to the study of modifications occurring in the nature of membrane glycoconjugates, glycoproteins, glycolipids, and proteoglycans in order to explain, first, specific properties of transformed cells such as loss of contact inhibition or ability for invasion and metastasis (André *et al.*, 1997; Hebert, 2000; Kannagi, 1997; Kondo *et al.*, 1998; Oates *et al.*, 1998), and, second, to improve cancer diagnosis and/or prognosis (Dallolio, 1996; Hakomori, 1998). Recently, therapeutic applications have also been envisaged through action on the tumor glycan epitopes, particularly Tn antigen, by means of naturally occurring or synthetic immunizing entities (Bay *et al.*, 1997; Cappuro *et al.*, 1998; Kudryashov *et al.*, 1998; Luo *et al.*, 1998; MacLean *et al.*, 1996; Sandmaier *et al.*, 1999) or through the use of analogues of sugars that block the activities of the glycosyltransferases or glycosidases responsible for the synthesis of potentially dangerous glycotopes involved in metastasis (Atsumi *et al.*, 1993; Goss *et al.*, 1995; Noda *et al.*, 1999; Pili *et al.*, 1995; Seftor *et al.*, 1991; Tan *et al.*,

1991; Tropea *et al.*, 1990). Besides their practical use, the expression of tumor glycan antigens raises the question of their role in directing the cell toward cancer formation.

Modifications affecting glycoconjugates include the emergence of antigens that do not normally belong to the cells of the tissues concerned or which, though initially present at the embryo stage, are repressed in the differential cells. These antigenic expressions, absent in normal cells, are generally linked to a modification in glycosyltransferase and/or glycosidase activities and often include blood group antigen modifications (Nakagoe *et al.*, 1998). Some of these antigens are especially interesting because they appear early in tumor evolution and are present in several types of cancers. This is the case for Tn antigen corresponding to the residue GalNAcα1-*O*-serine/threonine and T antigen (Thomsen-Friedenreich antigen) derived from Tn by addition of an α1-3 D-galactose. Tn antigen is the source of the biosynthesis of all glycans of mucin-type glycoproteins by addition of one or two monosaccharides bringing to core structures (Brockhausen, 1995). T antigen corresponding to core 1 is in particular the precursor of a tetrasaccharide motif present on

Handbook of Immunohistochemistry and *in situ* Hybridization of Human Carcinomas, Volume 1: Molecular Genetics; Lung and Breast Carcinomas

487

Copyright © 2004 by Elsevier (USA)
All rights reserved.

numerous cells, including erythrocytes, where it represents the saccharide part of blood group antigens M or N (Salmon *et al.*, 1991); it is also the source of a hexasaccharide specific to activated T cells (Piller *et al.*, 1988).

Expression of T antigen is related to a dysfunction of 2-3 sialyltransferase, which adds an *N*-acetylneuraminic acid onto the galactose. T antigen can temporarily appear after an infectious syndrome in which a bacterial neuraminidase separates the *N*-acetyneuraminic acid normally present from the tetrasaccharide.

Tn antigen, which is the precursor of T antigen, is a carcinoembryonic antigen present in human embryos aged 4–17 days, in particular on epithelial and mesothelial tissues (Barr *et al.*, 1989). Presence of Tn and sialosyl-Tn antigens has been shown in many cancers, including breast carcinomas (Itzkowitz *et al.*, 1989; Konska *et al.*, 1998a; Sotozono *et al.*, 1994; Springer, 1989). In malignancies, their expression is connected with the absence or inactivity of UDP-Gal:GalNAcα1-*O*-ser/thrβ1->3 galactosyltransferase (EC 2.4.1.122) or β3Gal-T (Zhuang *et al.*, 1991). In breast cancer, examination of expression and localization of Tn and T antigen differing in the progression of malignant transformation can help elucidate the mechanism underlying the formation of Tn antigen, and gain a better understanding of the relationship between Tn antigen expression and malignancy.

Dysfunction of enzyme β3GalT can herald a certain number of malignancies, but it seems to follow different mechanisms according to the case. It has thus been shown that the expression of the antigen Tn can be the result of a reversible repression under the action of an inducer such as azacytidine or sodium *n*-butyrate, as demonstrated in the Tn hematologic syndrome characterized by the polyagglutinability of a varying proportion of erythrocytes (Thurnher *et al.*, 1994), or the outcome of a mutation as demonstrated for Jurkat cells (Thurnher *et al.*, 1993). It has been shown in breast cancers that antigens T and Tn are expressed (Konska *et al.*, 1998b), but it was not known whether the tumor cells bore one or both of the antigens, the simultaneous presence of T and Tn signifying that the gene coding for β3GalT is not mutated but that the functioning of the enzyme, together with that of enzymes responsible for the lengthening of core 1, is disturbed.

With the aim of demonstrating the origin of the expression of the antigen Tn, we sought, in several cell lines derived from breast tumors (MCF-10A, HBL-100, MDA-MB 231, and MCF-7) grown *in vitro*, the simultaneous presence of antigens T and Tn by double-labeling techniques using plant lectins conjugated to two different fluorochromes, using confocal fluorescence microscopy (Konska *et al.*, 2002) (Figure 71).

MATERIALS

1. Fluorescein-isothiocyanate isomer I-Celite (FITC)
2. Rhodamine B isothiocyanate-Celite (RBITC)
3. N,N-Dimethylformamide (DMF)
4. Carbonate buffer for lectin coupling: 53 g Na_2CO_3 in 1 L of distilled water (sol. A); 42 g $NaHCO_3$ in 1 L of distilled water (sol. B): add 6 ml of A to 9 ml of B, and adjust to pH 7.2
5. Phosphate buffered saline (PBS): 8.5 g sodium chloride, 17.9 g dibasic sodium phosphate $\cdot$ 12H$_2$O, and 2.25 g monobasic sodium phosphate $\cdot$ 2H$_2$O; bring vol to 1 L with distilled water, pH 7.2
6. Lectin from *Arachis hypogaea* (PNA), lyophilized powder
7. Isolectin B$_4$ from *Vicia villosa* (VVA-B$_4$), lyophilized powder
8. Sephadex G-25 Superfine
9. DMEM/Ham's F-12 medium containing 100 ng/ml cholera endotoxine, 10 μg/ml insulin, 0.5 μg/ml hydrocortisol, 20 ng/ml EGF, 5% heat-inactivated horse serum, 1% L-glutamine, and 20 μg/ml of gentamicine
10. McCoy's 5a medium containing 10% fetal bovine serum, 2 mM L-glutamine, 2 g/L sodium pyruvate, and 20 $\mu\gamma$/ml of gentamycin
11. Leibovitch 15 medium containing 10% fetal bovine serum (FBS), 2 mM L-glutamine, and 20 μg/ml of gentamycin
12. RPMI 1640 medium supplemented with 2 g/l NaHCO$_3$, 5% heat-inactivated fetal bovine serum, 2 mM L-glutamine, 20 $\mu\gamma$/ml gentamycin, and 0.04 units/ml of insulin
13. 0.05% trypsine in distillated water
14. 0.025% versene in distillated water
15. 0.02% EDTA in distillated water
16. PBS containing 1% fetal bovine serum

METHOD

VVA-B$_4$ Coupling with FITC

1. Add 1 mg FITC-celite to 1 ml of carbonate buffer.
2. Centrifuge for 5 min at 1000, and take out the supernatant.
3. Dissolve 5 mg VVA-B$_4$ in 5 ml PBS
4. Add 500 μl of FITC supernatant to 5 ml of VVA-B$_4$ solution, and incubate for 3 hr in the dark at room temperature.
5. Add 25 g of G 25-Sephadex to 250 ml PBS; after 24 hr, make a suspension and place in a 280 × 15 mm glass column.
6. Place 5 ml of FITC-VVA-B$_4$ solution onto the column, and elute with PBS at the flow rate of 50 ml/hr. First colored fraction is VVA-B$_4$–coupled lectin.

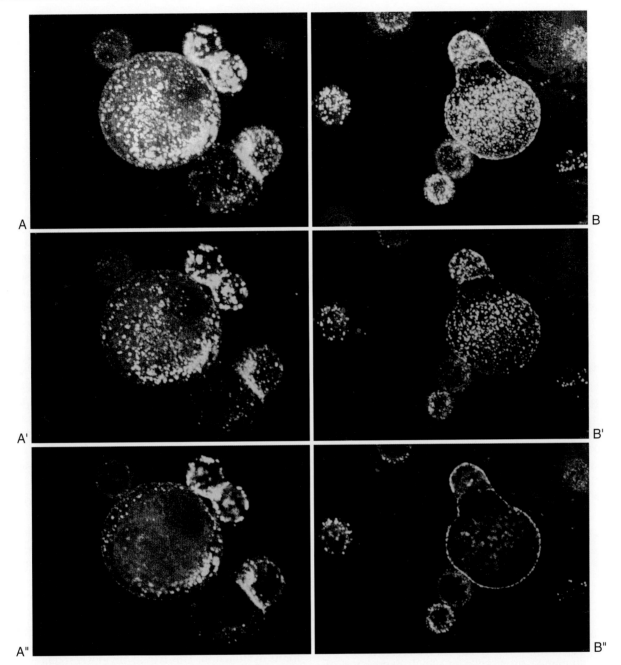

Figure 71 Localization of GalNac α1-ser/thr residues (Tn antigen) visualized by FITC-conjugated VVA-B$_4$ lectin (green spots) and Dgalβ1-3GalNAcα-*O*-ser/thr (T antigen) by RBITC-conjugated PNA lectin (red spots) in serial cross sections from the surface to the interior of MCF-7 (A-A″ and B-B″). Tn antigen residues were localized mainly at the cell membrane, whereas they were not expressed in deeper cytoplasm layers.

PNA Coupling with RBITC

7. Add 5 mg RBITC-celite to 1 ml DMF.
8. Mix RBITC-DMF solution with 4 ml carbonate buffer.
9. Dissolve as in **Step 3**.
10. Add as in **Step 4**.
11. Isolate PNA coupled lectin as in **Step 6**.

MCF-10A Cell Line Culture

12. Place the cells in 30 ml of supplemented DMEM: Ham's F-12 medium in 75-cm^2 flasks at 37°C in a humidified atmosphere of 5% CO_2.
13. At 80% confluence, treat the cells with a mixture of 1.5 ml 0.05% trypsin and 1.5 ml 0.025% versene.
14. Wash the cells 3 times with PBS.

HBL-100 Cell Line Culture

15. Place the cells in 30 ml of supplemented McCoy's 5a medium in 75-cm^2 flasks at 37°C in a humidified atmosphere of 5% CO_2.

16. At 80% confluence, treat the cells with 1.5 ml 0.05% trypsine.

17. Wash as in **Step 10**.

MDA-MB 231 Cell Line Culture

18. Place the cells in 30 ml of supplemented Leibovitch medium in 75-cm^2 flasks at 37°C in a humidified atmosphere without CO_2.

19. Treat as in **Step 16**.

20. Wash as in **Step 14**.

MCF7 Cell Line Culture

21. Place the cells in 30 ml of supplemented RPMI 1640 medium in 75-cm^2 flasks at 37°C in a humidified atmosphere of 5% CO_2.

22. Treat as in **Step 16**.

23. Wash as in **Step 10**.

Lectin Cell Labeling

24. Incubate cell suspensions (10^6 cells/ml) with VVA-B$_4$-FITC at 1:1 v/v at room temperature in absolute darkness for at least 1 hr.

25. Wash three times with PBS containing 1% fetal bovine serum.

26. Incubate with PNA-BSA as in **Step 24**.

Confocal Fluorescence Microscopy

27. Observe, immediately after **Step 26**, under a confocal microscope using appropriate filters.

RESULTS AND DISCUSSION

All cells were labelled with VVA-B$_4$ and PNA, confirmed by a light microscope. In each cell line, individual cells were shown to express Tn and T antigen simultaneously, with dissimilar levels of expression. MCF-7 contained more cells in which expression of Tn antigen exceeded expression of T antigen than MCF-l0A or HBL-100. Tn antigen was most weakly expressed in MDA-MB 231. Tn and T antigens were localized mainly on the cell membrane and in the peripheral cytoplasm in all cell lines. Expression of Tn antigen was not observed in deeper cytoplasm layers. In MCF-7 and MCF-10A, Tn antigen formed tiny clusters, irregularly scattered on the surface of cell membrane, whereas in the HBL-100, Tn antigen was concentrated over slightly larger areas.

Observation of simultaneous expression of Tn and T antigens on the same cell for all cell lines, though their

proportions ranged widely, is evidence that the changes leading to the formation of Tn antigen are not caused by irreversible genetic mutations but rather by a dysfunction of β3-galtransferase, as in the Tn syndrome. Divergent levels of expression of Tn and T antigens in different cells resulted in numerous subpopulations within each cell line, the highest heterogeneity being observed in MCF-7.

It is notworthy that the MCF-10A line, derived from fibrocystic dystrophy and commonly considered to be normal, expressed the Tn antigen. In the same way, previous studies (Konska et al., 1998a) demonstrated expression of Tn antigen on benign breast tumors such as fibroadenoma, fibrocystic dystrophy, and typical and atypical ductal hyperplasia, detected by VVA-B$_4$ and GSI-A$_4$ lectins and monoclonal antibody 83D$_4$. Positive membrane labeling with D-GalNAcα-O-ser/ thr-specific lectins and anti-Tn antibody occurred in benign cases indicating that the process of neoplastic transformation of the cell begins very early with molecular modifications of glycoconjugates, which precedes all other detectable morphological and physiological anomalies. Moreover, lectin binding indicated significant differences in the proportion of Tn and T density, which may be related to the origin of the cells. For MCF-7, derived from metastasis, expression of Tn antigen was higher than that of T antigen, in contrast to the cells obtained from fibrocystic disease (MCF-10A) or HBL-100. The increase in dysfunction of β3-galactosyltransferase thus appeared to be related to the degree of malignancy. The detection of Thomsen-Friedenreich antigen is also a sign of disturbed oligosaccharide synthesis, expression of T antigen being related to enzyme deficiencies such as α2-3 sialyltransferase inactivity. Its presence on MCF-10A and HBL-100 also showed that modifications affecting glycoconjugates in cancer occurs in subnormal cells. It would be desirable to determine whether there are genetic modifications in cells expressing antigen Tn, concerning at least one of the alleles coding for β3-galactosyltransferase (EC 2.4.1.122) that might account for its dysfunction. This would obviously require cloning of the gene and availability of probes.

References

André, S., Unverzagt, C., Kojima, S., Dong, X., Fischer, C., Kayser, K., and Gabius, H.J. 1997. Neoglycoproteins with the synthetic complex biantennary nonasaccharide or its α2.3/α2.6-sialylated derivatives: Their preparation, assessment of their ligand properties for purified lectins, for tumor cells in vitro, and in tissue sections, and their biodistribution in tumor-bearing mice. *Bioconjug. Chem.* 8:845–855.

Atsumi, S., Nosaka, C., Ochi, Y., Iinuma, H., and Umezawa, K. 1993. Inhibition of experimental metastasis by an alpha-glucosidase inhibitor, 1,6-epi-cyclophellitol. *Cancer Res.* 53:4896–4899.

Barr, N., Taylor, C.R., Young, T., and Springer, G.F. 1989. Are pancarcinoma T and Tn differentiation antigens? *Cancer* 64:834–841.

Bay, S., Lo-Man, R., Osinaga, E., Nakada, H., Leclerc, C., and Cantacuzène, D. 1997. Preparation of a multiple antigen gly-copeptide (MAG) carrying the Tn antigen. A possible approach to a synthetic carbohydrate vaccine. *J. Peptide Res.* 49:620–625.

Brockhausen, I. 1995. Biosynthesis of *O*-glycans of the *N*-acetyl-galactosamine-α-Ser/Thr linkage type. *Glycoproteins.* Elsevier Amsterdam. 201–259.

Cappuro, M., Bover, L., Portela, P., Livingston, P., and Mordoh, J. 1998. FC-2.15, a monoclonal antibody active against human breast cancer, specifically recognizes Lewis(x) hapten. *Cancer Immunol. Immunother.* 45:334–339.

Dallolio, F. 1996. Protein glycosylation in cancer biology—an overview. *J. Clin. Pathol.* 49:M126–M135.

Goss, P.E., Baker, M.A., Carver, J.P., and Dennis, J.W. 1995. Inhibitors of carbohydrate processing; a new class of anticancer agents. *Clin. Cancer Res.* 1:935–944.

Hakomori, S. 1998. Cancer-associated glycosphingolipid antigens: Their structure, organization, and function. *Acta Anat.* 161:79–90.

Hebert, E. 2000. Endogenous lectins as cell surface transducers. *Biosci. Rep.* 20:213–237.

Itzkowitz, S.H., Yuan, M., Montgomery, C.K., Kjeldsen, T., Takahashi, H.K., Bigbee, W.L., and Kim, Y.S. 1989. Expression of Tn, sialosyl-Tn, and T antigens in human colon cancer. *Cancer Res.* 49:197–204.

Kannagi, R. 1997. Carbohydrate-mediated cell adhesion involved in hematogenous metastasis of cancer. *Glycoconjugate J.* 14:577–584.

Kondo, K., Kohno, N., Yokoyama, A., and Hiwada, K. 1998. Decreased MUC1 expression induces E-cadherin-mediated cell adhesion of breast cancer lines. *Cancer Res.* 58:2014–2019.

Konska, G., Guillot, J., de Latour, M., and Fonck, Y. 1998a. Expression of Tn antigen and *N*-acetyllactosamine residues in malignant and benign human breast tumors detected by lectins and monoclonal antibody 83D4. *Int. J. Oncol.* 12:361–367.

Konska, G., Favy, D., Guillot, J., Bernard-Gallon, D., de Latour, M., and Fonck, Y. 1998b. Expression of T and Tn antigens in breast carcinomas. *C. R. Soc. Biol.* 192:733–747.

Konska, G., Vissac, C., Jagla, C., Chezet, F., Vasson, M.P., Bernard-Gallon, D., and Guillot, J. 2002. Ultrastructural local-ization of binding sites for PNA and VVA-B4 lectins in human breast cancer cell lines detected by confocal fluorescence microscopy. *Int. J. Oncol.* 21:1009–1014.

Kudryashov, V., Kim, H.M., Ragupathi, G., Danishefsky, S.J., Livingston, P.O., and Lloyd K.O. 1998. Immunogenecity of synthetic conjugates of Lewis(y) oligosaccharide with proteins in mice: Towards the design of anticancer vaccines. *Cancer Immunol. Immunother.* 45:281–286.

Luo, P., Agadjanyan, M., Qiu, J., Westerink, M. A., Steplewski, Z., and Kieber-Emmons, T. 1998. Antigenic and immunological mimicry of peptide mimotopes of Lewis carbohydrate antigens. *Mol. Immunol.* 35:865–879.

MacLean, G.D., Reddish, M.A., Koganty, R.R., and Longenecker, B.M. 1996. Antibodies against mucin-associated sialyl-Tn epitopes correlate with survival of metastatic adeno-carcinoma patients undergoing active specific immunotherapy with synthetic STn vaccine. *J. Immunother. Emphasis Tumor Immunol.* 19:59–68.

Nakagoe, T., Fukushima, K., Tuji, T., Sawai, T., Nanashima, A., Yamagushi, H., Yasutake, T., Hara, S., Ayabe, H., Matuo, T., and Kamihira, S. 1998. Immununohistochemical expression of ABH/Lewis-related antigens in primary breast carcinomas and metastatic lymph node lesions. *Cancer Detect. Prev.* 22:499–505.

Noda, I., Fujieda, S., Seki, M., Tanaka, N., Sunaga, H., Ohtsubo, T., Tsuzuki, H., Fan, G.K., and Saito, H. 1999. Inhibition of *N*-linked glycosylation by tunicamycin enhances sensitivity to cisplatin in human head-and-neck carcinoma cells. *Int. J. Cancer* 18:279–284.

Oates, A.J., Schumaker, L.M., Jenkins, S.B., Pearce, A.A., DaCosta, S.A., Arun, B., and Ellis, M.J. 1998. The mannose 6-phosphate/insuline-like growth factor2 receptor (M6P/IGF2R), a putative breast tumor suppressor gene. *Breast Cancer Res. Treat.* 47:269–281.

Pili, R., Chang, J., Partis, R.A., Mueller, R.A., Chrest, F.J., and Passaniti, A. 1995. The alpha-glucosidase I inhibitor castanosper-mine alters endothelial cell glycosylation, prevents angiogene-sis, and inhibits tumor growth. *Cancer Res.* 55:2920–2926.

Piller, F., Piller, V., Fox, I., and Fukuda, M. 1988. Human T-lymphocyte activation is associated with changes in *O*-glycan biosynthesis. *J Biol. Chem.* 263:15146–15150.

Salmon, C., Cartron, J.-P., and Rouger, P. 1991. Anomalies de la membrane du globule rouge. *Les groupes sanguins chez l'Homme.* Paris: Masson.

Sandmaier, B.M., Oparin, D.V., Holmberg, L.A., Reddish, M.A., MacLean, G.D., Longenecker, B.M. 1999. Evidence of a cellu-lar immune response against sialyl-Tn in breast and ovarian cancer. patients after high-dose chemotherapy, stem cell rescue, and immunization with Theratope STn-KLH cancer vaccine. *J. Immunother.* 22:54–66.

Seftor, R.E., Seftor, E.A., Grimes, W.J., Liotta, L.A., Stetler-Stevenson, W.G., Welch, D.R., and Hendrix, M.J. 1991. Human melanoma cell invasion is inhibited in vitro by swain-sonine and deoxymannojirimycin with a concomitant decrease in collagenase IV expression. *Melanoma Res.* 1:43–54.

Sotozono, M.A., Okada, Y., and Tsuji, T. 1994. The Thomsen-Friedenreich antigen-related carbohydrate antigens in human gastric intestinal metaplasia and cancer. *J. Histochem. Cytochem.* 42:1575–1584.

Springer, G.F. 1989. Tn epitope (*N*-acetyl-D-galactosamine alpha-*O*-serine/threonine) density in primary breast carcinoma: A functional predictor of aggressiveness. *Mol. Immunol.* 26:1–5.

Tan, A., van den Broek, L., van Boeckel, S., Ploegh, H., and Bolscher, J. 1991. Chemical modification of the glucosidase inhibitor 1-deoxynojirimycin. Structure-activity relationships. *J. Biol. Chem.* 266:14504–14510.

Thurnher, M., Rusconi, S., and Berger, E.G. 1993. Persistent repres-sion of a functional allele can be responsible for galactosyltrans-ferase deficiency in Tn syndrome. *J. Clin. Invest.* 91:2103–2110.

Thurnher, M., Fehr, J., and Berger, E.G. 1994. Differences in the regulation of specific glycosylation in the pathogenesis of paroxysmal nocturnal hemoglobinuria and the Tn syndrome. *Exp. Hematol.* 22:267–271.

Tropea, J.E., Kaushal, G.P., Pastuszak, I., Mitchell, M., Aoyagi, T., Molyneux, R.J., and Elbein, A.D. 1990. Mannostatin, a new gly-coprotein-processing inhibitor. *Biochemistry* 30:10062–10069.

Zhuang, D., Yousefi, S., and Dennis, J.W. 1991. Tn antigen and UDP-Gal: GalNAc alpha-Rbeta1-3 Galactosyltransferase expression in human breast carcinoma. *Cancer Biochem. Biophys.* 12:185–198.

26

Expression of the ETV6-NTRK3 Gene Fusion in Human Secretory Breast Carcinoma

Cristina Tognon, David Huntsman, and Poul H.B. Sorensen

INTRODUCTION

Breast Cancer and Model Systems

Familial and Sporadic Breast Cancers

Breast cancer is a complex disease that appears to be similar to the vast majority of epithelial malignancies in that it possesses numerous genetic aberrations. Familial breast cancers, or those with predisposing genes (e.g., mutations in the DNA repair–chekpoint genes *BRAC1* and *BRCA2* [Venkitaraman, 2002]), constitute only about 5–10% of total breast cancer cases (Balmain *et al.*, 2003). The other 90–95% are classified as sporadic, or caused by a combination of unknown polygenic and environmental factors. It remains unclear which of the genetic aberrations observed are causative in sporadic breast cancer tumorigenesis. A currently held model exists whereby early loss of cell cycle control, such as that perpetrated by the loss of *p53* or other cell cycle checkpoint genes, plus mitogenic signals from the surrounding tumor microenvironment, leads to unchecked cell cycling. Uncontrolled proliferation would then yield a high level of genomic instability, and continued division under these circumstances results in additional chromosomal abnormalities and genetic alterations. Increased tumor heterogeneity, in conjunction with selective pressures, results in the selection of distinct cell subpopulations within the tumor that possess progressively more aggressive phenotypes, such as those resistant to conventional therapies (Brenner *et al.*, 1997). One of the challenges facing molecular pathologists is to determine which of the many genetic abnormalities observed in breast cancer cells are causative in disease progression and to identify genetic abnormalities that can be used as diagnostic and prognostic indicators. In other words, how does one specifically identify clinically useful genetic changes from a background of mutational "noise" created by genomic instability?

Model Systems for Studying Breast Cancer Progression

To further elucidate the disease-relevant genetic changes associated with breast cancer biology and tumor progression, numerous model systems have been developed. Some of the more widely used models include

Handbook of Immunohistochemistry and *in situ* Hybridization of Human Carcinomas, Volume 1: Molecular Genetics; Lung and Breast Carcinomas

493

Copyright © 2004 by Elsevier (USA)
All rights reserved.

1) selective inbreeding of mouse strains to create ones that mimic (in frequency and in biological characteristics) spontaneous breast cancer; 2) rat breast cancer models produced by means of exposure to chemical carcinogens (reviewed by Clarke, 1996); 3) xenograft model systems created in immuno-deficient animals using breast cancer derived cell lines (reviewed by Clarke, 1996); 4) transgenic and knockout models that target genes known to be involved in familial breast cancer progression (reviewed by Hutchinson *et al.*, 2000); and 5) cell line-based model systems, such as the MCF10A cell lines series, that attempt to mimic the multistep process of breast cancer progression, which evolves from epithelial and atypical hyperplasia (AH) to ductal carcinoma *in situ* (DCIS), invasive carcinoma, and ultimately metastasis (Heppner *et al.*, 2000; Miller *et al.*, 2000; Santner *et al.*, 2001; Strickland *et al.*, 2000). These model systems, each possessing its own unique strengths and weaknesses, provide a starting point for scientists to begin posing questions about the genetics and biology involved in breast cancer progression.

Translocations: Identifying Novel Proteins Involved in Cancer Progression

Another powerful strategy used to understand how underlying genetic changes influence cancer progression is the identification of genes found at the breakpoints of recurrent translocations in specific cancers. Over the years, the study of recurrent chromosomal rearrangements has led to the recognition that genes found at the translocation breakpoint can either be disrupted, causing a loss of function (as in the case of a tumor suppressor gene), can be inappropriately expressed (through the juxtaposition of upstream promoter elements from one gene to another), or can be fused to other partners producing novel oncogenes that encode for proteins not normally observed in the cellular environment (Blume-Jensen *et al.*, 2001; Rabbitts, 1994). Bone and soft tissue sarcomas of children and young adults often possess recurrent translocations leading to expression of gene fusions thought to be causally related to oncogenesis (Ladanyi *et al.*, 2000; Sandberg, 2002; Sorensen *et al.*, 1996). In a significant proportion of these tumors, the translocation is the sole cytogenetic anomaly, indicating its probable role in the genesis of the tumor. Studying the molecular basis of the recurrent cytogenetic abnormality may provide an explicable model of cancer of general, potentially leading to the identification of common signaling pathways that are being activated in both congenital and adult tumors.

Unlike bone and soft tissue tumors, breast cancers often possess a complex array of chromosomal changes. Conventional chromosome banding techniques have documented a number of abnormalities, including isochromosomes 1q and 6p; translocations involving 1q, 8p, and 16p; deletions of 1p, 3p, 6q, 11q, and 17p; losses of chromosomes 2 and 5; and gains of 7, 18, and 20 (reviewed in Popescu *et al.*, 2002). Often, inversion of 7q and deletion of 1q and 3p were the only structural alterations and, therefore, most likely represent early events responsible for the initiation of neoplastic transformation (Mitelman *et al.*, 1997; Popescu *et al.*, 2002). Comparative genomic hybridization (CGH) data have shown the most common genomic imbalances in breast cancer to be gains of 1q, 8q, 17q, and 20q; loss of 8p, 13q, 16q, and 18q; and gain or loss of 11q (Forozan *et al.*, 2000). Cytogenetics obtained from familial breast cancer families has also identified constitutive alterations of chromosome 9p (Bergthorsson *et al.*, 1998; Savelyeva *et al.*, 2001). Until recently, the shear volume of rearrangements as well as gains and losses of chromosomal material observed cytogenetically in breast cancer cells have precluded the discovery of primary genetic lesions underlying breast cancer development (Dickson *et al.*, 2001).

A recurrent translocation breakpoint has been recently reported to be found in 4 of 34 breast cancer cell lines as well as in 2 out of 9 pancreatic cancer cell lines (Adelaide *et al.*, 2003). The translocations involve the 8p12 region and appear to target the *NRG1* gene, which encodes the growth factors of the neuregulin/heregulin-1 family of ligands that bind to the ErbB tyrosine kinase receptors. This particular study did not identify any fusion protein associated with the translocations, nor were the authors able to demonstrate differences in *NRG1* expression. However, another group previously reported a fusion protein called DOC4-NRG1 found in the MDA-MB-175 cell line. In this particular case, the promoter of *DOC4* was found to enhance the expression of the *NRG1* gene (Schaefer *et al.*, 1997).

Our group recently reported that a recurrent t(12;15) (p13;q25) translocation occurs in a rare subtype of infiltrating ductal carcinoma known as secretory breast carcinoma (SBC). This translocation results in the production of a novel gene fusion known as *ETV6-NTRK3* (described in detail following (Tognon *et al.*, 2002). Even though SBC is rare, the occurrence of this translocation has potential biological and clinical relevance and, taken together with the recurrent breakpoints in the *NRG1* gene described by Adelaide *et al.* (2003), suggests that additional recurrent translocations associated with breast cancer progression may be identified in the future. The study of these translocations may allow us to rise above the background of genetic noise typically found in sporadic breast cancers, thus providing a strategy for identifying the genes altered by such rearrangements and elucidating their potential roles in breast cancer progression. The information presented in this chapter details the methods utilized by our group in identifying and characterizing the t(12;15) translocation, the diagnostic utility of *ETV6-NTRK3* detection in SBC, and the biology of the ETV6-NTRK3 fusion protein. This serves as a useful model for other groups attempting to identify recurrent translocations in different subtypes of breast cancer.

RESULTS

ETV6-NTRK3 Expression in Pediatric Solid Tumors

Identification of the ETV6-NTRK3 Gene Fusion in Pediatric Tumors

ETV6-NTRK3 (EN) is an oncogenic fusion protein that was first identified in congenital fibrosarcoma (CFS) (Knezevich *et al.*, 1998), which is a malignant tumor of fibroblasts that occurs in patients 2 years of age or younger. It is unique among human sarcomas in that it has an excellent prognosis and a very low metastatic rate (Fisher, 1996; Pizzo *et al.*, 1997). Detection of the *ETV6-NTRK3* gene fusion in tumor specimens has enabled molecular pathologists to differentiate CFS from histologically similar childhood spindle cell tumors, including nonmalignant lesions such as infantile fibromatosis and myofibromatosis, and adult-type fibrosarcoma, which has a worse prognosis than CFS (Bourgeois *et al.*, 2000). This has allowed for the appropriate diagnosis and treatment of patients with CFS. In addition to CFS, the *ETV6-NTRK3* gene fusion has also been identified in the cellular form of an infantile spindle cell renal tumor known as congenital mesoblastic nephroma (CMN) (Knezevich *et al.*, 1998; Rubin *et al.*, 1998). The presence of *ETV6-NTRK3* in both cellular CMN and CFS has led to the recognition that these tumors are histogenetically related (Knezevich *et al.*, 1998).

The EN fusion has also been identified in one case of acute myeloid leukemia (AML) occurring in an adult patient (Eguchi *et al.*, 1999), and expression of the AML-associated variant of EN has subsequently been shown to cause a rapidly fatal myeloproliferative disease in a murine transplantation model (Liu *et al.*, 2000). However, further analysis of pediatric leukemias by two different groups has failed to demonstrate the presence of the transcripts associated with the EN fusion protein in these malignancies (Alessandri *et al.*, 2001; Eguchi *et al.*, 2001). As will be described later, we recently identified the *ETV6-NTRK3* gene fusion in SBC. This fusion gene is therefore unique in that is has been identified in malignancies derived from different cell lineages (mesenchymal, hematopoietic, and epithelial) and in both adult and pediatric populations.

ETV6-NTRK3 Oncoprotein Structure

The EN fusion protein is comprised of the sterile alpha motif (SAM) dimerization domain of the ETV6 (TEL) transcription factor linked to the protein tyrosine kinase (PTK) domain of the neurotrophin-3 receptor NTRK3 (TrkC) (Figure 72A). The ETS family transcription factor ETV6 (also known as TEL) is thought to play a major role in early hematopoieisis and angiogenesis (Edel, 1998; Wang *et al.*, 1998). The *NTRK3* gene (also known as TRKC) encodes the transmembrane surface receptor for neurotrophin-3 involved in growth, development, and cell survival in the central nervous system (reviewed in Kaplan *et al.*, 2000). The *ETV6* gene has also been identified as a fusion partner in leukemia-associated chimeric proteins, such as ETV6-PDGFR (Golub *et al.*, 1994), ETV6-AML1 (Golub *et al.*, 1995), ETV6-JAK2 (Lacronique *et al.*, 1997; Romana *et al.*, 1995), and ETV6-ARG (Iijima *et al.*, 2000), among others. The EN oncogene encompasses exons 1–5 of *ETV6* and exons 13–18 of *NTRK3*. The fusion point occurs between nucleotide 1033 of *ETV6* and nucleotide 1601 of *NTRK3* (Knezevich *et al.*, 1998). This differs from the *ETV6-NTRK3* gene fusion reported in the AML case (Eguchi *et al.*, 1999), in which *ETV6* exon 5 was not present in fusion transcripts. *ETV6-NTRK3* also does not contain *NTRK3* exon 16, which encodes a 42 base pair (bp) insert within the PTK domain (Ichaso, *et al.*, 1998). The presence of these naturally occurring inserts is known to decrease NTRK3 kinase activity (Barbacid, 1994). We hypothesize that the oncogenic properties of the EN fusion protein are mediated by mechanisms similar to those utilized by other fusion-associated protein tyrosine kinases, such as BCR-ABL and other ETV6 fusions. Namely, the fusion proteins oligomerize via the SAM dimerization motifs, causing constitutive activation of the tyrosine kinase domains leading to stimulation of downstream signal transduction cascades involved in EN oncogenesis. In fact, we have shown that the EN oncoprotein is a constitutively active protein tyrosine kinase (Wai *et al.*, 2000).

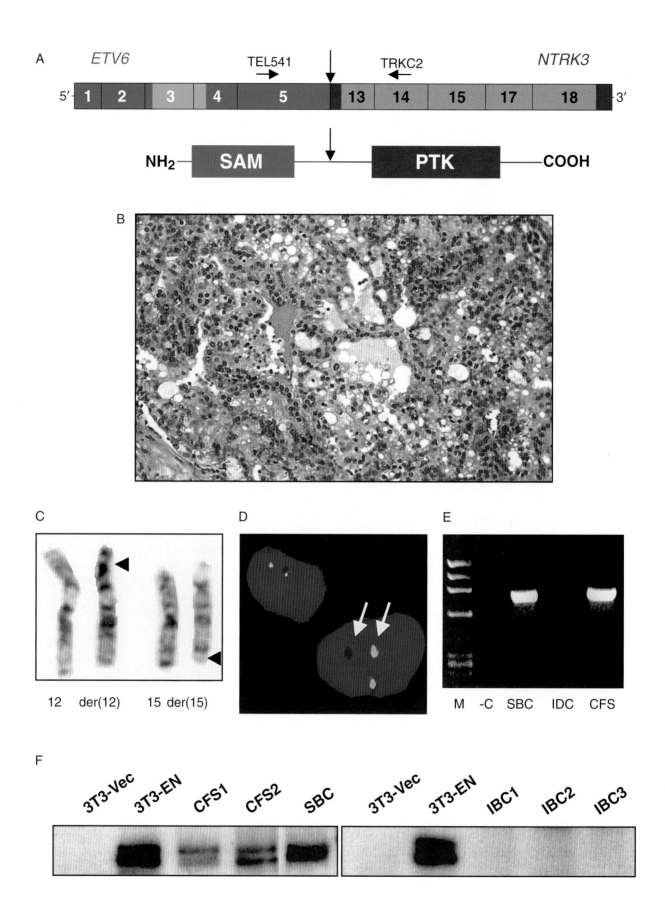

A

ETV6   TEL541   ↓   TRKC2   NTRK3

5' | 1 | 2 | 3 | 4 | 5 | | 13 | 14 | 15 | 17 | 18 | 3'

NH₂ ─ [SAM] ──↓── [PTK] ─ COOH

B

C   12   der(12)   15   der(15)

D

E   M   -C   SBC   IDC   CFS

F   3T3-Vec 3T3-EN CFS1 CFS2 SBC   3T3-Vec 3T3-EN IBC1 IBC2 IBC3

ETV6-NTRK3 Expression in Human Secretory Breast Carcinoma

Secretory Breast Carcinoma

Infiltrating ductal carcinoma (IDC), also known as invasive ductal carcinoma, represents approximately 80% of all breast cancers. Twenty percent of these cases are made up of a number of rare subtypes of IDC, one of which is known as secretory breast carcinoma (SBC). SBC was originally described in children but occurs equally in the adult population. SBC is often thought to have a favorable prognosis because of its indolent manner (Oberman, 1980; Page *et al.*, 1987). Although recurrences and nodal metastases have been observed in a number of cases in both males and females (de Bree *et al.*, 2002), distant metastases have been found to be extremely rare (Herz *et al.*, 2000). The prognosis for SBC was initially thought to be excellent compared to typical IDC, with up to a 100% 5-year survival rate (Rosen *et al.*, 1991; Tavassoli *et al.*, 1980). However, more recent studies have suggested that the favorable outcome is age related, and that in older patients the prognosis is similar to typical IDC (Maitra *et al.*, 1999).

Index Case: Identification of the ETV6-NTRK3 Oncoprotein in SBC

Our group has recently reported that SBC also expresses the EN oncoprotein (Tognon *et al.*, 2002). The index case was a 6-year-old female who presented with a unilateral breast mass. Pathology revealed an invasive breast adenocarcinoma with the histopathological features of SBC. Characteristic of SBC is an infiltrating pattern of neoplastic epithelial cells forming well-differentiated glandular lumens filled with eosinophilic, PAS-positive secretions (Rosen *et al.*, 1991) (Figure 72B). Karyotypic analysis revealed a t(12;15) (p12;q26.1) translocation as the only cytogenetic abnormality in the tumor cells (Figure 72C). This translocation was not present constitutionally. The positive karyotype prompted us to perform fluorescence *in situ* hybridization (FISH) and reverse transcriptase PCR (RT-PCR) on frozen tumor tissue from the case to screen for the ETV6-NTRK3 gene fusion. Dual-color FISH using cosmid probes from exon 1 and exon 8 of ETV6 demonstrated rearrangement of the ETV6 gene locus (Figure 72D). Moreover, RT-PCR revealed an identical 731-bp fusion transcript, which has previously been described as indicating the presence of the EN fusion in other tumor types, in the SBC case (Figure 72E; SBC lane) but not in infiltrating ductal carcinoma cases (Figure 72E; IDC lane). This was confirmed by sequencing the amplified products, which demonstrated the identical fusion point between ETV6 nucleotide 1033 and NTRK3 nucleotide 1601 as previously shown for sarcoma-associated fusions (Knezevich *et al.*, 1998).

To demonstrate that the EN protein was expressed in tumor tissue, we performed immunoprecipitations from whole-cell lysates from the primary tumor using α-NTRK3 antibodies followed by Western Blotting using α-phosphotyrosine antibodies (Tognon *et al.*, 2002). This revealed expression of the characteristic tyrosine phosphorylated 68/73 kDa doublet of EN (Figure 72F; see CFS lanes and SBC lane). The tyrosine phosphorylated doublet was not detected in three control primary invasive ductal carcinomas that were negative for the EN gene fusion by RT-PCR (Figure 72F; IBC lanes).

Figure 72 Characterization of the ETV6-NTRK3 gene fusion in the index secretory breast carcinoma (SBC) case: histology, cytogenetics, and molecular pathology. **A:** Schematic diagram showing the structure of the ETV6-NTRK3 chimeric cDNA in SBC. Exons **1–5** of ETV6 (blue boxes) are fused in frame with exons 13–15 and 17–18 of NTRK3 (red boxes). The lighter shade of blue indicates the region encoding the ETV6 sterile alpha motif (SAM) domain, whereas the lighter shade of red indicates the region encoding the NTRK3 protein tyrosine kinase (PTK) domain. The fusion point is between ETV6 nucleotide 1033 and NTRK3 nucleotide 1601 (indicated by the vertical arrow), which is identical to that observed in congenital fibrosarcoma. The positions of the TEL541 forward and the TrkC2 reverse primers used to amplify the ETV6-NTRK3 fusion transcripts are shown above the exons. A schematic of EN's protein structure is shown below the cDNA schematic. **B:** H & E of an SBC case. **C:** Partial karyogram demonstrating the t(12;15) (p13;q25) translocation in SBC occurring in a 6-year-old female. Arrowheads indicate breakpoints at derivative 12p13 and derivative 15q25. **D:** Dual-color FISH using ETV6 exon 1–containing cosmid 179A6 (green) and ETV6 exon 8–containing cosmid 148B6 (red). Arrows point out separate signals indicating disruption of the ETV6 gene. **E:** Reverse transcriptase PCR (RT-PCR) of ETV6-NTRK3 fusion transcripts using total RNA isolated from tumor tissue of the index secretory breast carcinoma case (SBC) and a congenital fibrosarcoma control (CFS), but not from an infiltrating ductal carcinoma case (IDC). M, markers; −C, negative control. **F:** Immunoprecipitation of the ETV6-NTRK3 protein using an anti-NTRK3 antibody followed by Western analysis with an antiphosphotyrosine antibody. The left panel demonstrates that characteristic tyrosine phosphorylated 68/73 kDa ETV6-NTRK3 doublet in the index SBC case, two congenital fibrosarcoma cases (CFS), and in NIH 3T3 cells expressing ETV6-NTRK3 (3T3-EN), but not in the NIH 3T3 vector control (3T3-Vec). The right panel demonstrates that the doublet is present only in the 3T3 EN control cells but not in lysates from three invasive breast carcinoma cases (IBC1-3) nor in 3T3-Vec cells.

Survey of EN Expression in a Series of SBC

To determine that EN expression represented a general finding in SBC, 12 additional breast cancer cases in which SBC was the predominant or the only histologic component were tested using an RT-PCR assay adapted to formalin-fixed, paraffin-embedded (FFPE) tumor tissue (Bourgeois *et al.*, 2000). Eleven of the 12 SBC cases were positive for the 110-bp RT-PCR fragment produced in this assay. In addition, 50 cases of typical IDC were also tested, and all were fusion negative except for one. This particular case was subsequently shown to possess several areas of well-differentiated glands containing eosinophilic secretions suggesting that it may have originated as SBC but could possibly have undergone further clonal evolution to assume a less-differentiated phenotype. To confirm these findings, FISH was performed on 9 of the 13 cases of SBC, and all were positive for the fusion protein. FISH performed on the IDC cases was negative except for the RT-PCR positive case (Tognon *et al.*, 2002). Taken altogether, these findings strongly indicate that the EN gene fusion is a nonrandom rearrangement in SBC.

Mechanisms of Action: Signaling Downstream of EN

We have previously demonstrated that EN is capable of transforming the murine NIH3T3 fibroblast cell line and have subsequently utilized this as a model system to study the signaling mediated by EN (Wai *et al.*, 2000). We hypothesized that transformation mediated by the EN fusion protein is due to activation of downstream signaling pathways normally utilized by the wild-type NTRK3 receptor. We further established that the EN oncoprotein activates two of the major downstream signaling pathways of NTRK3, the Ras/Raf/MEK proliferation pathway as well as the PI3K/AKT survival pathway (Tognon *et al.*, 2001). Furthermore, we found that blocking either pathway using chemical inhibitors or the expression of dominantly interfering proteins can block EN-mediated transformation (Tognon *et al.*, 2001). The Ras pathway impacts on the level of an important cell cycle regulatory protein known as cyclin D1, and we have hypothesized that EN-expressing cells use this mechanism to drive cells through the cell cycle. We have also determined that the EN fusion protein links to the aforementioned pathways via an adaptor molecule called IRS1 (insulin receptor substrate 1), normally utilized by the insulin and insulin-like receptors (Morrison *et al.*, 2002). In addition to this, we have determined that cells must posses the IGFIR

to be fully transformed, suggesting that, as with other dominantly acting oncoproteins, EN transformation requires a functional IGFIR signaling axis (Morrison *et al.*, 2002). Many of these pathways, identified as being activated by EN in the fibroblast system, are activated in breast epithelial cells transformed by EN (Tognon *et al.*, 2002). We are interested in studying the downstream pathways affected by EN expression to specifically block its activity as a therapeutic intervention.

EPH4 and SCG6 Murine Breast Epithelial Cells: A Model System for EN-Mediated Breast Epithelial Transformation

To demonstrate that EN expression plays a causative role in breast cancer, we created a model system by stably expressing the EN fusion protein in two different nontransformed breast epithelial cell lines using retroviral transduction. The Eph4 and Scg6 cells are immortalized nontransformed epithelial cell lines derived from normal mouse mammary epithelium (Roskelley *et al.*, 2000; Somasiri *et al.*, 2001). These two cell lines represent the opposite ends of the spectrum of mammary epithelial differentiation, with Scg6 cells possessing a stable myoepithelial or more mesenchymal features and Eph4 cells exhibiting a stable epithelial phenotype in culture. The expression of the fusion was confirmed by Western Blot analysis, and both of the EN-expressing cell lines were found to be tumorigenic in immunodeficient mice (Tognon *et al.*, 2002). Histologic and immunohistochemical analysis of sections obtained from Eph4 EN-expressing tumors demonstrated that EN expression did not affect the process of epithelial differentiation because the cells within the tumor were still capable of forming glands. Gland cells were strongly positive for cytokeratin and epithelial membrane antigen but negative for vimentin (Tognon *et al.*, 2002). These findings confirmed that EN had transformation activity in breast epithelial cells and strongly implicated EN as being causally related to oncogenesis in human SBC. The EN fusion protein uses the same signaling pathways in both fibroblasts and breast epithelial cells to mediate transformation, indicating that the model system created in the Eph4 and Scg6 breast epithelial cells may be useful for studying breast cell transformation in general. This model system can provide a useful tool to study the process of breast cancer oncogenesis in the context of SBC, thus allowing potential investigators to analyze the underlying signaling mechanisms that contribute to breast cell transformation.

MATERIALS AND METHODS

Diagnostic Molecular Pathology: Detecting the ETV6-NTRK3 Gene Fusion

Introduction to Techniques

The identification of CFS cases possessing the t(12;15) translocation were initiated from conventional cytogenetic analysis of metaphase spreads by Geimsa staining studies followed by karyotyping. However, the translocation itself can be very difficult to detect using this method and is often overlooked in standard examinations due to the positions of the breakpoints. Therefore, cytogenetically negative cases may not be negative for the ETV6-NTRK3 gene fusion at the molecular level. Cytogenetic detection of the t(12;15) can be improved by using fluorescence in situ hybridization (FISH) with probes flanking the t(12;15) breakpoints (see following). This method can be used to reconfirm the presence of the ETV6-NTRK3 gene fusion in both frozen and FFPE tumor sections (see Detailed Protocol section) (Adem et al., 2001; Knezevich et al., 1998).

The most rapid method of identifying tumors that harbor the ETV6-NTRK3 gene fusion is reverse transcriptase polymerase chain reaction (RT-PCR) (Argani et al., 2000; Bourgeois et al., 2000). Oligonucleotide primers are prepared that flank the breakpoint within ETV6-NTRK3 fusion transcripts. Using cDNA made from total RNA of tumor samples as template, a product will be amplified only from those transcripts with juxtaposed ETV6 and NTRK3 sequences binding the primers. These primers are used in an RT-PCR assay under specific conditions, and the presence of a product denotes the presence of the fusion gene in the sample. A protocol for RT-PCR from FFPE section has been created specifically for the detection of the ETV6-NTRK3 gene fusion (see Detailed Protocol section) (Bourgeois et al., 2000). The detection of ETV6-NTRK3 transcripts by RT-PCR is more specific than immunohistochemical detection of the ETV6-NTRK3 fusion protein (Bourgeois et al., 2000; Dubus et al., 2001). Although antibodies are available that recognize the EN fusion protein, such as those that bind to the C-terminal region of the NTRK3 protein (TrkC C-14; Santa Cruz Biotechnology) or those that recognize N-terminal regions of the ETV6 protein (α-HLH ETV6; Dr. P. Marynen; α-ETV6 [N-19] Ab; Santa Cruz Biotechnology), these are not useful for immunohistochemistry. Endogenous expression of ETV6 or nonspecific staining using the NTRK3 antibody, coupled with the low levels of EN protein found in tumors, preclude the use of immunohistochemical

detection of EN in paraffin tumor blocks. The aforementioned antibodies can be used to immunoprecipitate EN protein from lysates produced from frozen tumor tissue (see Detailed Protocol section) (Tognon et al., 2002). However, this technique requires large amounts of tumor tissue and is not recommended for diagnostic pathology applications. These antibodies are also used routinely to detect EN expression in cell line–based systems and used to study the signaling cascades activated downstream of EN (Morrison et al., 2002; Tognon et al., 2002; Wai et al., 2000).

A newly developed quantitative RT-PCR (qRT-PCR) assay is also being used to identify the presence of ETV6-NTRK3 transcripts. Quantitative or real-time RT-PCR is a tool that monitors the amplification of a PCR product in real time. This can be used to determine how much starting material is present in the sample at the beginning of the reaction. Quantitative RT-PCR has advantages over regular RT-PCR in that it possesses both higher specificity and sensitivity, and it is quantifiable (Joyce, 2002). The Taqman q-PCR system (Applied Biosystems) utilizes forward and reverse primers along with a specific internal probe; all three must bind to the target for the reaction to work. Internal control primers are also used, such as beta-actin (which in our hands is best for soft tissue sarcomas) or GAPDH, in order to monitor the quality of each reaction. The sensitivity of the qRT-PCR reaction removes the need to reconfirm positive results using radioactive probes, and cross contamination of samples can be avoided as the tubes remained sealed throughout the reaction. Minimal residual disease studies, to monitor the number of cells remaining in the patient after surgery or treatment, can also be performed using qRT-PCR. To date, the sensitivity of our current qRT-PCR protocol requires a minimum number of 100 fusion positive cells for detection by this assay. We have modified a qRT-PCR protocol developed by Oliver Delattre's group (Peter et al., 2001) to identify the ETV6-NTRK3 gene fusion in patients (see Detailed Protocol section).

The following sections outline different protocols (FISH, RT-PCR, Western Blot analysis, and qRT-PCR) that our laboratory has used to detect the presence of the ETV6-NTRK3 fusion gene and protein in patient samples.

Fluorescence in situ Hybridization (FISH)

Two different FISH assays were used to detect the t(12;15) EN translocation in frozen and formalin-fixed tumor samples (Knezevich et al., 1998). For frozen sections, metaphase spreads were prepared using conventional cytogenetic methods (Verma et al., 1989) and subjected to dual-color FISH as previously described (Dracopoli, 1996; Sorensen *et al.*, 1996). Probes utilized for FISH included the *ETV6* exon 1–containing cosmid 179A6 and *ETV6* exon 8–containing cosmid 148B6, which become separated in cells with the t(12;15) translocation (see Figure 72D).

For analysis of chromosome breakpoints in FFPE tumor blocks, 5-μm sections are first deparaffinized and dehydrated in 100% ethanol using standard methods. Slides are boiled in 2X SSC for 30 sec, washed, and then pretreated in 1 M sodium thiocyanate for 30 min. Sections are then digested using 5 mg/ml pepsin in 0.2 N HCl at 37°C for 20 min. This is followed by fixation in formaldehyde, dehydration with ethanol, and denaturation with 70% formamide/2X SSC (pH 7.0) at 90°C for 6 min. Dual-color FISH can be performed using the 12p13 breakpoint-spanning YAC, 817_H_1, and the 15q25 breakpoint-spanning YAC, 802_B_4 (Knezevich *et al.*, 1998). Probes can be labeled with either Spectrum Green or 0.2 Spectrum Red by nick translation following standard protocols and hybridized to denatured slides in a humid chamber at 37°C overnight. These probes become juxtaposed in cells with the t(12;15) translocation. Slides are counterstained with DAPI and analyzed using a fluorescent microscope.

More recently, new BAC probes have been used that have given better results for clinical samples. These include the following BAC clones obtained from BAC-PAC Resources Center at Children's Hospital Oakland Research Institute: RP11-434C1 and Rp11-407p10 (telomeric to *ETV6* on 12p13); RP11-52513 and RP11-267J23 (centromeric to *ETV6*); RP11-247E14 and RP11-893E1 (telomeric to *NTRK3* on 15q25); and RP11-114I9 and RP11-730G13 (centromeric to *NTRK3* on 15q25) (D. Huntsman, unpublished data).

Reverse Transcriptase–Polymerase Chain Reaction (RT-PCR)

RT-PCR can be used to screen for *ETV6-NTRK3* fusion transcripts in frozen tumor tissue as well as FFPE samples. For frozen tissue sample, total RNA (2 μg) can be isolated from primary tumor samples for cDNA synthesis as described (Sorensen *et al.*, 1996). In order to amplify the 731-bp product as shown in Figure 72E, the following PCR oligonucleotide primers are utilized: the *TEL (ETV6)* 541 forward primers (5′-CCTCCCACCATTGAACTGTT-3′) and the *TRKC (NTRK3)* 2 reverse primer (5′-CCGCA-CACTCCATAGAACTTGAC-3′) (Figure 72A), although other primers can be utilized as long as their products span the breakpoint. Cycling conditions are as follows: one cycle of 95°C × 5 min; 35 cycles of 95°C × 1 min;

60°C × 1 min; 72°C × 1 min; one cycle of 72°C × 10 min. Amplified products are visualized by electrophoresis using a 2% agarose gel stained with ethidium bromide. *ETV6-NTRK3* fusion transcripts can also be confirmed by Southern Blot analysis by blotting the PCR fragments onto Hybond-N nylon filters followed by hybridization with an oligonucleotide probe internal to the PCR product as described (Bourgeois *et al.*, 2000), potentially followed by sequencing of the PCR products.

The protocol of RT-PCR of *ETV6-NTRK3* fusion transcripts from FFPE sections is based on previously described methods (Bourgeois *et al.*, 2000). Briefly, RNA is isolated from three 30-μm paraffin sections and used directly in a one-step RT-PCR using the Biosciences Titanium One Step RT-PCR kit (Clonetech). RNA integrity can be checked by using pairs of PCR primers for the phosphoglycerate kinase (PGK) gene, as previously described (Bourgeois *et al.*, 2000). RNA is then subjected to RT-PCR using the forward primer *TEL (ETV6)* 971 (5′-ACCACATCAT GTCTCTT-GTCTCCC-3′) and the reverse primer *TRKC (NTRK3)* 1059 (5′-CAGTTCTCGCTTCAGCACGATG-3′). PCR conditions are as reported previously, with the assay detecting a 110-bp PCR product. Products can be confirmed by blotting the PCR products onto Nytran filters and hybridizing them with an oligonucleotide probe spanning the *ETV6-NTRK3* cDNA breakpoint.

Quantitative RT-PCR (qRT-PCR)

The following protocol is a modification of the qRT-PCR protocol first described by Peter *et al.* (2001). RNA can be extracted from frozen tumor tissue using standard protocols. One μg of total RNA can be reverse transcribed using random hexamers in a final volume of 20 μl using the Gene-Amp RNA PCR Kit (Applied Biosystems). qRT-PCR reactions can be carried out in a final volume of 50 μl using standard conditions with the Universal Master Mix (Applied Biosciences). The following primers and probes can be used to detect *ETV6-NTRK3* fusion transcripts: 5′ primer ETV6.1: 5′-CCCATCAACCTCTCT-CATCGG-3′; 3′ primer NTRK3.1: 5′-GGCTCCCT-CACCCAGTTCTC-3′, Taqman probe FC1 S1: CTCCCCGCCTGAAGAGCACGC. In our procedure we have not altered the Mg^{2+} concentrations, although the following cycling conditions have been slightly altered compared to those detailed by Peter *et al.* (2001). After the initial steps of the UNG reaction (2 min at 50°C) and TaqGold activation (10 min at 95°C), 40 cycles of PCR are performed according to the following conditions: 94°C × 45 sec; 60°C × 1 min; 72°C × 1 min.

Immunoprecipitations and Western Blotting Using Frozen Tumor Sections

For our studies performed on EN expression in SBC, we were able to immunoprecipitate EN from frozen tumor sections (Figure 72F; Tognon et al., 2002). Lysates were prepared from 20 × 20-μm thick scrolls of frozen tumor samples embedded in OCT using a RIPA lysis buffer consisting of 20 mM Tris (pH 7.4); 120 mM NaCl, 1% Triton X-100, 0.5% sodium deoxycholate, 0.1% SDS, 10% glycerol, 5 mM EDTA, 50 mM NaF, 0.5 mM Na_3VO_4, plus freshly added protease inhibitors (leupeptin [10 μg/ml], apoprotinin [10 μg/ml], and PMSF [250 μM]). Lysates can also be prepared from cell lines engineered to express the EN fusion protein using the following PSB lysis buffer: 50 mM HEPES, 100 mM NaF, 10 mM Na_4PO_7, 2 mM Na_3VO_4, 2 mM EDTA, 2 mM $NaMoO_4$, 0.5% NP-40, and freshly added protease inhibitors [10 μg/ml apoprotinin, 10 μg/ml leupeptin, 250 μM PMSF] (Tognon et al., 2001). In both cases, cells or tumor tissue should be solubilized in lysis buffer for 30 min at 4°C on a shaking platform. Lysates are then cleared by centrifugation at 12,000 × g for 10 min at 4°C. Protein concentration is quantified using a detergent compatible (DC) protein assay kit from Bio-Rad. Immunoprecipitation studies on total cell lysate from cell lines or tumor sections can be performed using 500 μg of total cell lysate, and 5 μl of the NTRK3 (TrkC C-14) antibody from Santa Cruz in conjunction with 20 μl of Protein A sepharose beads (Amersham Pharmacia Biotech; Buckinghamshire, United Kingdom). Washed immunoprecipitates are electrophoresed using a 10% SDS-polyacrylamide gel, transferred to Immobilon-P (Millipore) PVDF membrane, and immunoblotted using α-NTRK3 (1:500 dilution; Santa Cruz) or α-phosphotyrosine antibodies (P-Tyr 100; dilution 1:2000; NEB Cell Signaling). Proteins can be visualized using ECL (Amersham) according to the manufacturer's protocol.

CONCLUSIONS

In summary, this chapter describes our identification of the t(12;15) associated ETV6-NTRK3 gene fusion in human secretory breast carcinoma. The resulting ETV6-NTRK3 fusion protein, first discovered in pediatric spindle cell malignances, has potent transforming activity not only in fibroblasts but also in immortalized nontransformed breast epithelial cells. Injection of ETV6-NTRK3 transformed breast epithelial cells into immunocompromised mice results in the formation of tumors with the histologic features of breast carcinoma, strongly implicating ETV6-NTRK3 as playing a causative role in SBC. This is the first description of a recurrent chromosomal rearrangement and expression of a dominantly acting oncogene as a primary event in human breast carcinoma. An SBC is a rare and poorly understood subtype of IDC with, as of yet, no distinct molecular or phenotypic profile. Our results provide a strategy for pathologists to specifically identify cases of SBC by screening for the t(12;15) translocation or its resultant ETV6-NTRK3 fusion transcripts. This methodology was recently applied to a tissue microarray panel of 200 sporadic breast cancer cases. One case was positively identified as expressing an ETV6-NTRK3 gene fusion (D. Huntsman, unpublished data), and further investigation revealed that this case possessed a secretory phenotype. The ability to correctly identify cases of SBC based on the detection of the t(12;15) translocation or ETV6-NTRK3 fusion transcripts should help to prevent unwarranted chemotherapy treatment, particularly in young patients, as SBC tends to have a favorable prognosis. Such a strategy will also allow investigators to screen larger series of sporadic breast cancer cases to assess the overall incidence and prognostic implications of the ETV6-NTRK3 gene fusion in breast cancer. In addition, the identification of additional recurrent translocations associated with breast cancer is inevitable. The methods and strategies described in this chapter may therefore be useful as a model for studying recurrent translocations in other types of breast cancers.

Acknowledgments

We would like to acknowledge Chris Lannon for reviewing the manuscript and for helpful suggestions, Nikita Makretsov for the H&E photomicrograph of the SBC case and further information on new BAC probes used for ETV6-NTRK3 FISH, and Joan Mathers for information on the qRT-PCR protocol.

References

Adelaide, J., Huang, H.E., Murati, A., Alsop, A.E., Orsetti, B., Mozziconacci, M.J., Popovici, C., Ginestier, C., Letessier, A., Basset, C., Courtay-Cahen, C., Jacquemier, J., Theillet, C., Birnbaum, D., Edwards, P.A., and Chaffanet, M. 2003. A recurrent chromosome translocation breakpoint in breast and pancreatic cancer cell lines targets the neuregulin/NRG1 gene. *Genes Chromosomes Cancer 37(4):*333–345.

Adem, C., Gisselsson, D., Cin, P.D., and Nascimento, A.G. 2001. ETV6 rearrangements in patients with infantile fibrosarcomas and congenital mesoblastic nephromas by fluorescence in situ hybridization. *Mod. Pathol. 14(12):*1246–1251.

Alessandri, A.J., Knezevich, S.R., Mathers, J.A., Schultz, K.R., and Sorensen, P.H. 2001. Absence of t(12;15) associated

ETV6-NTRK3 fusion transcripts in pediatric acute leukemias. *Med. Pediatr. Oncol. 37(4):*415–416.

Argani, P., Fritsch, M., Kadkol, S.S., Schuster, A., Beckwith, J.B., and Perlman, E.J. 2000. Detection of the ETV6-NTRK3 chimeric RNA of infantile fibrosarcoma/cellular congenital mesoblastic nephroma in paraffin-embedded tissue: Application to challenging pediatric renal stromal tumors. *Mod. Pathol. 13(1):*29–36.

Balmain, A., Gray, J., and Ponder, B. 2003. The genetics and genomics of cancer. *Nat. Genet. 33 (Suppl):*238–244.

Barbacid, M. 1994. The Trk family of neurotrophin receptors. *J. Neurobiol. 25(11):*1386–1403.

Bergthorsson, J.T., Johannsdottir, J., Jonasdottir, A., Eiriksdottir, G., Egilsson, V., Ingvarsson, S., Barkardottir, R.B., and Arason, A. 1998. Chromosome imbalance at the 3p14 region in human breast tumours: High frequency in patients with inherited predisposition due to BRCA2. *Eur. J. Cancer 34(1):*142–147.

Blume-Jensen, P., and Hunter, T. 2001. Oncogenic kinase signalling. *Nature 411(6835):*355–365.

Bourgeois, J.M., Knezevich, S.R., Mathers, J.A., and Sorensen, P.H. 2000. Molecular detection of the ETV6-NTRK3 gene fusion differentiates congenital fibrosarcoma from other childhood spindle cell tumors. *Am. J. Surg. Pathol. 24(7):*937–946.

Brenner, A.J., and Aldaz, C.M. 1997. The genetics of sporadic breast cancer. *Prog. Clin. Biol. Res. 396:*63–82.

Clarke, R. 1996. Animal models of breast cancer: Their diversity and role in biomedical research. *Breast Cancer Res. Treat. 39(1):*1–6.

de Bree, E., Askoxylakis, J., Giannikaki, E., Chroniaris, N., Sanidas, E., and Tsiftsis, D.D. 2002. Secretory carcinoma of the male breast. *Ann. Surg. Oncol. 9(7):*663–667.

Dickson, R.B., and Lippiran, M. 2001. Molecular biology of breast cancer. In De vita, V.T., Jr., Hellman, S., and Rosenberg, S.A., (eds) *Cancer: Principles and Practice of Oncology*, 6th ed. Philadelphia: Lippincott Williams and Wilkins, 1633–1651.

Dracopoli, N.C. 1996. *Current Protocols in Human Genetics.* New York: John Wiley and Sons.

Dubus, P., Coindre, J.M., Groppi, A., Jouan, H., Ferrer, J., Cohen, C., Rivel, J., Copin, M.C., Leroy, J.P., de Muret, A., and Merlio, J.P. 2001. The detection of Tel-TrkC chimeric transcripts is more specific than TrkC immunoreactivity for the diagnosis of congenital fibrosarcoma. *J. Pathol. 193(1):*88–94.

Edel, M.J. 1998. The ETS-related factor TEL is regulated by angiogenic growth factor VEGF in HUVE-cells. *Anticancer Res. 18(6A):*4505–4509.

Eguchi, M., and Eguchi-Ishimae, M. 2001. Absence of t(12;15) associated ETV6-NTRK3 fusion transcripts in pediatric acute leukemias. *Med. Pediatr. Oncol. 37(4):*417.

Eguchi, M., Eguchi-Ishimae, M., Tojo, A., Morishita, K., Suzuki, K., Sato, Y., Kudoh, S., Tanaka, K., Setoyama, M., Nagamura, F., Asano, S., and Kamada, N. 1999. Fusion of ETV6 to neurotrophin-3 receptor TRKC in acute myeloid leukemia with t(12;15)(p13;q25). *Blood 93(4):*1355–1363.

Fisher, C. 1996. Fibromatosis and fibrosarcoma in infancy and childhood. *Eur. J. Cancer 32A(12):*2094–2100.

Forozan, F., Mahlamaki, E.H., Monni, O., Chen, Y., Veldman, R., Jiang, Y., Gooden, G.C., Ethier, S.P., Kallioniemi, A., and Kallioniemi, O.P. 2000. Comparative genomic hybridization analysis of 38 breast cancer cell lines: A basis for interpreting complementary DNA microarray data. *Cancer Res. 60(16):*4519–4525.

Golub, T.R., Barker G.F., Bohlander, S.K., Hiebert, S.W., Ward, D.C., Bray-Ward, P., Morgan, E., Raimondi, S.C., Rowley, J.D., and Gilliland, D.G. 1995. Fusion of the TEL gene on 12p13 to the AML1 gene on 21q22 in acute lymphoblastic leukemia. *Proc. Natl. Acad. Sci. USA 92(11):*4917–4921.

Golub, T.R., Barker, G.F., Lovett, M., and Gilliland, D.G. 1994. Fusion of PDGF receptor beta to a novel ets-like gene, tel, in chronic myelomonocytic leukemia with t(5;12) chromosomal translocation. *Cell 77(2):*307–316.

Heppner, G.H., Miller, F.R., and Shekhar P.M. 2000. Nontransgenic models of breast cancer. *Breast Cancer Res. 2(5):*331–334.

Herz, H., Cooke, B., and Goldstein, D. 2000. Metastatic secretory breast cancer. Non-responsiveness to chemotherapy: Case report and review of the literature. *Ann. Oncol. 11(10):*1343–1347.

Hutchinson, J.N., and Muller, W.J. 2000. Transgenic mouse models of human breast cancer. *Oncogene 19(53):*6130–6137.

Ichaso, N., Rodriguez, R.E., Martin-Zanca, D., and Gonzalez-Sarmiento, R. 1998. Genomic characterization of the human trkC gene. *Oncogene 17(14):*1871–1875.

Iijima, Y., Ito, T., Oikawa, T., Eguchi, M., Eguchi-Ishimae, M., Kamada, N., Kishi, K., Asano, S., Sakaki, Y., and Sato, Y. 2000. A new ETV6/TEL partner gene, ARG (ABL-related gene or ABL2), identified in an AML-M3 cell line with a t(1;12)(q25;p13) translocation. *Blood 95(6):*2126–2131.

Joyce, C. 2002. Quantitative RT-PCR: A review of current methodologies. *Methods Mol. Biol. 193:*83–92.

Kaplan, D.R., and Miller, F.D. 2000. Neurotrophin signal transduction in the nervous system. *Curr. Opin. Neurobiol. 10(3):*381–391.

Knezevich, S.R., Garnett, M.J., Pysher, T.J., Beckwith, J.B., Grundy, P.E., and Sorensen, P.H. 1998. ETV6-NTRK3 gene fusions and trisomy 11 establish a histogenetic link between mesoblastic nephroma and congenital fibrosarcoma. *Cancer Res. 58(22):*5046–5048.

Knezevich, S.R., McFadden, D.E., Tao, W., Lim, J.F., and Sorensen P.H. 1998. A novel ETV6-NTRK3 gene fusion in congenital fibrosarcoma. *Nat. Genet. 18(2):*184–187.

Lacronique, V., Boureux, A., Valle, V.D., Poirel, H., Quang, C.T., Mauchauffe, M., Berthou, C., Lessard, M., Berger, R., Ghysdael, J., and Bernard, O.A. 1997. A TEL-JAK2 fusion protein with constitutive kinase activity in human leukemia. *Science 278(5341):*1309–1312.

Ladanyi, M., and Bridge, J.A. 2000. Contribution of molecular genetic data to the classification of sarcomas. *Hum. Pathol. 31(5):*532–538.

Liu, Q., Schwaller, J., Kutok, J., Cain, D., Aster, J.C., Williams, I.R., and Gilliland, D.G. 2000. Signal transduction and transforming properties of the TEL-TRKC fusions associated with t(12;15)(p13;q25) in congenital fibrosarcoma and acute myelogenous leukemia. *EMBO. J. 19(8):*1827–1838.

Maitra, A., Tavassoli, F.A., Albores-Saavedra, J., Behrens, C., Wistuba, I.I., Bryant, D., Weinberg, A.G., Rogers, B.B., Saboorian, M.H., and Gazdar, A.F. 1999. Molecular abnormalities associated with secretory carcinomas of the breast. *Hum. Pathol. 30(12):*1435–1440.

Miller, F.R., and Heppner, G. 2000. Xenograft models of human breast cancer lines and of the MCF10AT model of human premalignant, proliferative breast disease. In IP, M. and Aseh, B., (eds) *Methods in Mammary Gland Biology and Breast Cancer Research.* New York: Kluwer Academic/Plenum Publishers, 37–50.

Mitelman, F., Mertens, F., and Johansson, B. 1997. A breakpoint map of recurrent chromosomal rearrangements in human neoplasia. *Nat. Genet. 15 Spec No:*417–474.

Morrison, K.B., Tognon, C.E., Garnett, M.J., Deal, C., and Sorensen, P.H. 2002. ETV6-NTRK3 transformation requires insulin-like growth factor 1 receptor signaling and is associated with constitutive IRS-1 tyrosine phosphorylation. *Oncogene 21(37):*5684–5695.

Oberman, H.A. 1980. Secretory carcinoma of the breast in adults. *Am. J. Surg. Pathol. 4(5):*465–470.

Page, D.L., Anderson, T.J., and Sakamoto, G. 1987. Infiltrating carcinoma: Major histological types. In Page, D.L. and Anderson, T.J., (eds) *Diagnostic Histopathology of the Breast.* Edinburgh: Churchill Livingstone.

Peter, M., Gilbert, E., and Delattre, O. 2001. A multiplex real-time pcr assay for the detection of gene fusions observed in solid tumors. *Lab. Invest. 81(6):*905–912.

Pizzo, P.A., and Poplach D.G. 1997. *Principles and Practice of Pediatric Oncology.* Philadelphia: Lippincott.

Popescu, N.C., and Zimonjic D.B. 2002. Chromosome and gene alterations in breast cancer as markers for diagnosis and prognosis as well as pathogenetic targets for therapy. *Am. J. Med. Genet. 115(3):*142–149.

Rabbitts, T.H. 1994. Chromosomal translocations in human cancer. *Nature 372(6502):*143–149.

Romana, S.P., Mauchauffe, M., Le Coniat, M., Chumakov, I., Le Paslier, D., Berger, R., and Bernard, O.A. 1995. The t(12;21) of acute lymphoblastic leukemia results in a tel-AML1 gene fusion. *Blood 85(12):*3662–3670.

Rosen, P.P., and Cranor, M.L. 1991. Secretory carcinoma of the breast. *Arch. Pathol. Lab. Med. 115(2):*141–144.

Roskelley, C.D., Wu, C., and Somasiri, A.M. 2000. Analysis of mammary gland morphogenesis. *Methods Mol. Biol. 136:*27–38.

Rubin, B.P., Chen, C.J., Morgan, T.W., Xiao, S., Grier, H.E., Kozakewich, H.P., Perez-Atayde, A.R., and Fletcher, J.A. 1998. Congenital nesoblastic mephroma t(12;15) is associated with ETV6-NTRK3 gene fusion: Cytogenetic and molecular relationship to congenital (infantile) fibrosarcoma. *Am. J. Pathol. 153(5):*1451–1458.

Sandberg, A.A. 2002. Cytogenetics and molecular genetics of bone and soft-tissue tumors. *Am. J. Med. Genet. 115(3):*189–193.

Santner, S.J., Dawson, P.J., Tait, L., Soule, H.D., Eliason, J., Mohamed, A.N., Wolman, S.R., Heppner, G.H., and Miller, F.R. 2001. Malignant MCF10CA1 cell lines derived from premalignant human breast epithelial MCF10AT cells. *Breast Cancer Res. Treat. 65(2):*101–110.

Savelyeva, L., Claas, A., Matzner, I., Schlag, P., Hofmann, W., Scherneck, S., Weber, B., and Schwab, M. 2001. Constitutional genomic instability with inversions, duplications, and amplifications in 9p23-24 in BRCA2 mutation carriers. *Cancer Res. 61(13):*5179–5185.

Schaefer, G., Fitzpatrick, V.D., and Sliwkowski, M.X. 1997. Gamma-heregulin: A novel heregulin isoform that is an autocrine growth factor for the human breast cancer cell line, MDA-MB-175. *Oncogene 15(12):*1385–1394.

Somasiri, A., Howarth, A., Goswami, D., Dedhar, S., and Roskelley, C.D. 2001. Overexpression of the integrin-linked kinase mesenchymally transforms mammary epithelial cells. *J. Cell Sci. 114(Pt 6):*1125–1136.

Sorensen, P.H., and Triche, T.J. 1996. Gene fusions encoding chimaeric transcription factors in solid tumours. *Semin. Cancer Biol. 7(1):*3–14.

Sorensen, P.H., Wu, J.K., Berean, K.W., Lim, J.F., Donn, W., Frierson, H.F., Reynolds, C.P., Lopez-Terrada, D., and Triche, T.J. 1996. Olfactory neuroblastoma is a peripheral primitive neuroectodermal tumor related to Ewing sarcoma. *Proc. Natl. Acad. Sci. USA 93(3):*1038–1043.

Strickland, L.B., Dawson, P.J., Santner, S.J., and Miller, F.R. 2000. Progression of premalignant MCF10AT generates heterogeneous malignant variants with characteristic histologic types and immunohistochemical markers. *Breast Cancer Res. Treat. 64(3):*235–240.

Tavassoli, F.A., and Norris, H.G. 1980. Secretory carcinoma of the breast. *Cancer 45:*2404–2413.

Tognon, C., Garnett, M., Kenward, E., Kay, R., Morrison, K., and Sorensen, P.H. 2001. The chimeric protein tyrosine kinase ETV6-NTRK3 requires both Ras-Erk1/2 and PI3-kinase-Akt signaling for fibroblast transformation. *Cancer Res. 61(24):*8909–8916.

Tognon, C., Knezevich, S.R., Huntsman, D., Roskelley, C.D., Melnyk, N., Mathers, J.A., Becker, L., Carneiro, F., MacPherson, N., Horsman, D., Poremba, C., and Sorensen, P.H. 2002. Expression of the ETV6-NTRK3 gene fusion as a primary event in human secretory breast carcinoma. *Cancer Cell 2(5):*367–376.

Venkitaraman, A.R. 2002. Cancer susceptibility and the functions of BRCA1 and BRCA2. *Cell 108(2):*171–182.

Verma, R.S., and Babu, H. 1989. *Human Chromosomes: Manual of Basic Techniques.* New York: Pergamon Press.

Wai, D.H., Knezevich, S.R., Lucas, T., Jansen, B., Kay, R.J., and Sorensen, P.H. 2000. The ETV6-NTRK3 gene fusion encodes a chimeric protein tyrosine kinase that transforms NIH3T3 cells. *Oncogene 19(7):*906–915.

Wang, L.C., Swat, W., Fujiwara, Y., Davidson, L., Visvader, J., Kuo, F., Alt, F.W., Gilliland, D.G., Golub, T.R., and Orkin, S.H. 1998. The TEL/ETV6 gene is required specifically for hematopoiesis in the bone marrow. *Genes Dev. 12(15):*2392–2402.

27

The Role of CA6 Protein Expression in Breast Carcinoma

Nancy Lane Smith

Introduction

DS6 is a recently developed murine monoclonal IgG1 antibody that used human serous ovarian carcinoma as the immunogen (Kearse *et al.*, 2000). In immunohistochemical studies, the DS6 antibody detects a tumor-associated glycoprotein antigen in human tissues that has been provisionally named CA6 pending full characterization. The DS6-reactive epitope on CA6 is periodic acid sensitive and neuraminidase sensitive, consistent with a sialoglycotope on the CA6 glycoprotein. Identification of the CA6 antigen is currently being investigated, and it is not currently known whether CA6 is a novel glycoprotein or represents a newly detected glycoform of a known protein.

The DS6 antibody is currently undergoing preclinical evaluation as a possible therapeutic antibody for the treatment of breast and ovarian carcinoma. It has been evaluated on a series of human carcinoma cell lines by flow cytometry and demonstrates high-affinity binding to several of the cell lines studied, including the T47D breast cancer cell line and the CaOV3 ovarian carcinoma cell line, among others (personal communication, Dr. Philip Chun, ImmunoGen, Inc., Cambridge, MA).

The DS6 antibody shows immunohistochemical reactivity in both acetone-permeablized frozen sections and in formalin-fixed, paraffin-embedded (FFPE) human tissue sections. The pattern of reactivity in cryostat sections and archival paraffin sections in normal tissues is similar, so antigen retrieval methods are not mandatory. The immunohistochemical reactivity of DS6 with normal adult human tissues and human neoplasms has been studied (Kearse *et al.*, 2000). Those studies have shown that CA6 has a limited distribution in normal human tissues. CA6 is expressed by fallopian tube epithelium (Figure 73A), type 2 pneumocytes, inner layers of urothelium, and focally in the epithelium of some pancreatic and salivary gland ducts and a few other types of epithelium. The expression in these normal epithelia is generally limited to membranes facing/lining lumens or pulmonary airspaces. CA6 is not expressed by mesothelium or mesenchymal elements. Reactivity with nonhuman species has not yet been systematically investigated.

Although not a pancarcinoma antigen, CA6 expression is not absolutely restricted to serous ovarian neoplasms. Among gynecologic neoplasms, there is strong expression in serous adenocarcinomas of ovary and endometrium, as would be expected of an antibody developed against serous ovarian adenocarcinomas.

Handbook of Immunohistochemistry and in situ Hybridization of Human Carcinomas, Volume 1: Molecular Genetics; Lung and Breast Carcinomas

Copyright © 2004 by Elsevier (USA)
All rights reserved.

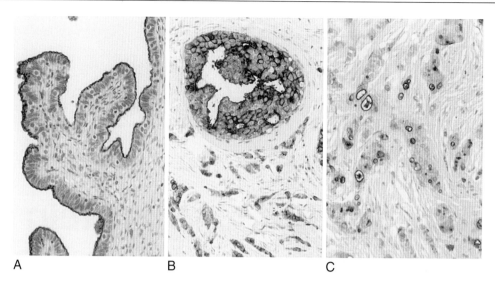

Figure 73 A: DS6 reacts strongly with the luminal surface of normal fallopian tube mucosal epithelium (DS6 immunohistochemistry with DAB. 200 × magnification). **B:** DS6 reacts strongly with infiltrating ductal carcinoma and ductal carcinoma in situ in this breast specimen. The staining is seen along peripheral cell membranes and in the cell cytoplasm (DS6 immunohistochemistry with DAB. 100 × magnification). **C:** DS6 shows reactivity in intracytoplasmic lumens and along the luminal surfaces of glands, in this case of infiltrating ductal carcinoma of the breast. This pattern of staining of intracytoplasmic lumens is also seen in some cases of infiltrating lobular carcinoma of the breast (DS6 immunohistochemistry with DAB. 200 × magnification).

CA6 is also expressed by the majority of endometrioid carcinomas of ovary and endometrial adenocarcinomas of endometrioid, mucinous, and clear cell types. However, CA6 is not detected in ovarian mucinous neoplasms of the ovary of benign, intestinal-type borderline, or malignant types or in malignant mesothelioma of the peritoneum (Smith *et al.*, 2001).

In nongynecologic neoplasms, CA6 expression is seen in the majority of carcinomas of breast, urothelium, and pancreas and is seen in some cases of renal carcinoma, pulmonary carcinoma, and some other carcinomas. CA6 expression has not been seen in pleural mesothelioma or in the majority of neoplasms of mesenchymal or hematopoietic types (Smith *et al.*, 2002).

CA6 expression as demonstrated by immunohistochemistry (IHC) can have several different patterns of distribution in carcinoma cells. Immunohistochemical staining for CA6 is often seen along the luminal cell membranes in carcinomas demonstrating a glandular or papillary growth pattern, but can also show a circumferential cell membrane staining or cytoplasmic staining in some neoplasms. The strength and pattern of DS6 immunoreactivity often varies from area to area within a tumor and between tumors of the same type.

Adenocarcinomas of the breast, and carcinomas of many other organs, have been grouped into different histologic types. Many of these different histologic types

of carcinoma have a unique behavior and prognosis, so in our previous studies of DS6 immunohistochemical reactivity, separate data were collected and analyzed for each type of breast carcinoma (Smith *et al.*, 2002). Multiple different types of neoplasm and multiple examples of each type of neoplasm were studied, which required the study of a large number of specimens. In order to make such large immunohistochemical studies more efficient, tumor sausage blocks were employed. Small sections of several different specimens were combined into a single paraffin block creating a multitumor sausage block (Battifore, 1986). Using sections of multiple sausage blocks allows the study of a very large number of specimens while conserving glass slides, technical work, and reagents. Some larger research laboratories and larger hospital pathology laboratories may be equipped with automated immunohistochemical stainers, further facilitating the study of large numbers of specimens.

MATERIALS

The DS6 monoclonal antibody is not yet commercially available but will likely be so in the near future. DS6 was prepared by our laboratory in the Department of Pathology at the Brody School of Medicine at East Carolina University. DS6 tissue culture supernatant (DMEMF12 media supplemented with 10% horse serum) was used in the immunohistochemical studies.

For manual staining of formalin-fixed, paraffin-embedded tissue sections or frozen sections:

1. DS6 murine monoclonal antibody, IgG1, as tissue culture supernatant diluted in phosphate buffered saline (PBS) to 20 µg/ml.

2. Nonimmune murine IgG1 isotype-matched negative control antibody at 20 µg/ml (Dako, Carpinteria, CA).

3. Mouse IgG peroxidase kit (PK4002 Vectastain ABC kit for mouse IgG, Vector Laboratories, Burlingame, CA).

4. 0.01 M PBS, pH 7.4 (P3818 packets, Sigma, St. Louis, MO).

5. Slide drying oven capable of maintaining temperature at 60°C or less.

6. Xylene.

7. Reagent grade 100% alcohol.

8. 95% alcohol (95 ml reagent grade alcohol and 5 ml of deionized water).

9. 70% alcohol (70 ml reagent grade alcohol and 30 ml of deionized water).

10. 30% hydrogen peroxide.

11. Choice of chromagen: Chromagens compatible with organic solvents include VIP (violet chromagen, VIP Substrate Kit SK-4600) and DAB (brown chromagen, DAB Substrate kit, 3,3′-diaminobenzidine, SK-4100). For aqueous mounting only, AEC may be used (red chromagen, AEC Substrate kit, 3-amino-9-ethylcarbazole, SK-4200). (All kits from Vector Laboratories, Burlingame, CA).

12. Gill hematoxylin (Fisher, Middletown, VA).

13. Saturated aqueous solution of lithium carbonate.

14. Acid alcohol solution (~5 ml concentrated HCl per gallon of 95% alcohol).

15. Glass coverslips.

16. Mounting media. A permanent mounting medium should be used for VIP or DAB such as Permount (Fisher, Fair Lawn, NJ). An aqueous medium such as Crystal Mount (Biomeda, Hayward, CA) should be used for AEC.

17. Plastic containers with lids to serve as humidified slide incubation chambers, plastic or glass coplin jars, pipettes.

18. One positive control slide per run. A section of human fallopian tube including the mucosal epithelium or other appropriate human tissue section containing antibody-reactive cells on the same slide as the specimen section or on a separate slide is needed for the positive control.

19. One negative control slide containing sections of the specimen tissue is needed for each specimen slide

20. Personal protective equipment to avoid contact of reagents with eyes and skin.

21. Acetone (only needed for processing of frozen sections).

For automated immunostaining on the Dako Autostainer Universal Staining System using Dako LSAB2 system (Dako, Inc.):

1. DS6 IgG1 monoclonal antibody with an approximate antibody concentration of 20 µg/mL DS6 tissue culture supernatant diluted in Dako antibody diluent (S0809) (Dako, Carpinteria, CA).

2. Mouse IgG1 isotype-matched negative control antibody (Dako, Carpinteria).

3. Xylene.

4. Reagent grade 100% alcohol.

5. 95% alcohol (95 ml reagent grade alcohol and 5 ml of deionized water).

6. 70% alcohol (70 ml reagent grade alcohol and 30 ml of deionized water).

7. Tris-buffered saline (TBS): 2.4 g tris-HCl and 8.76 g NaCl. Bring to 1 l volume with deionized water at pH 7.4. Add 0.05% Tween-20 for the Autostainer reagent container (or use Dako Tris-buffered saline S3306).

8. 3% hydrogen peroxide solution in water.

9. Gill hematoxylin (Fisher, Middletown, VA).

10. Saturated aqueous solution of lithium carbonate

11. Acid alcohol (approximately 5 ml concentrated HCl per gallon 95% alcohol).

12. Permanent mounting media such as Permount (Fisher, Fair Lawn, NJ).

13. Glass slides coated with polylysine or other coating designed for immunohistochemistry adherence.

14. Glass coverslips.

15. LSAB2 kit (includes biotinylated goat anti-mouse and goat anti-rabbit antibodies in PBS and streptavidin conjugated to horseradish peroxidase in PBS, from Dako, K0675).

16. (DAB) substrate chromogen (Dako, K3466).

17. Drying oven capable of maintaining 60°C or less.

18. Light microscope.

19. Coplin staining jars.

20. Autostainer (Dako, Carpinteria, CA).

21. One positive control slide per run. A section of human fallopian tube including the mucosal epithelium or other appropriate human tissue section containing antibody-reactive cells on the same slide as the specimen section or on a separate slide is needed for the positive control.

22. One negative control slide containing sections of the specimen tissue is needed for each specimen slide.

METHOD

Sausage Block Preparation

To facilitate the immunohistochemical study of a large number of samples, multitumor sausage blocks can be prepared (Battifore, 1986). Multitumor sausage blocks can be prepared from wet tissue fixed in formalin or from preexisting paraffin-embedded tissue blocks. To prepare a sausage block from tissue embedded in paraffin blocks, select the tissues to be studied. Review hematoxylin and eosin (H & E) stained and coverslipped sections of each block and select areas containing the tissue or neoplasm to be placed into the sausage block. A marking pen can be used to outline on the coverslip the selected portion of tissue. Later the pen mark can be wiped off the coverslip with alcohol, if needed. The portion of tissue selected should be of a size to allow embedding of multiple sections in one block, typically ~0.5–1.0 cm in greatest dimension, although larger portions could be used with fewer samples per block. Melt the paraffin blocks at a tissue-embedding station, compare the shape and appearance of the tissue with the area outlined in marker on the corresponding stained slide to select the portion of tissue to be removed, cut the tissue, remove the preselected area, and place the small samples together into the new sausage block. Record data identifying each sample and its pattern of placement in the sausage block, if needed. An asymmetric pattern of sample placement in the sausage block or use of a positive control tissue of histologically distinct type at one corner can facilitate later identification of the position of each unique sample in the block. For DS6, a section of normal fallopian tube can be placed at one corner to identify one position in the pattern of the specimens in the block and acts as an internal positive control. An H & E stained glass slide of the sausage block can be placed on a photocopier, and the copy can be labeled and used to record specimen identification and location for ease of data retrieval. Re-embed any remaining tissue in the original blocks if desired.

Manual Method, Frozen Cryostat Sections or Paraffin-Embedded Tissues

All incubations are at room temperature unless stated otherwise. Each slide should be labeled to identify each as a specimen slide, positive control slide, or negative control slide.

I. For frozen sections (mounted on glass slides):
  A. Air-dry frozen sections and fix slides in acetone for 3–5 min.
  B. Place in PBS for 5 min.
  C. Follow steps 2G–2R that follow as for paraffin method.
II. For formalin-fixed, paraffin-embedded sections (mounted on positively charges slides or poly-L-lysine coated slides):
  A. Place freshly cut slides in a 55–60°C slide dryer oven for 1 hr.
  B. Place slides in 2 changes of xylene for 10 min each.
  C. Place slides in 2 changes of 100% reagent grade alcohol for 5 min each.
  D. Place slides in 2 changes of 95% alcohol for 3 min each.
  E. Place slides in 70% alcohol for 1 min.
  F. Place slides in PBS for 5 min.
  G. Quench endogenous peroxidase activity by placing slides in 0.3% hydrogen peroxide in methanol for 15–20 min.
  H. Wash slides in three changes of PBS.
  I. Apply primary antibody (as tissue culture supernatant containing horse serum) and place slides in a covered container with moisture and incubate 30–60 min.
  J. Wash slides in three changes of PBS.
  K. Apply biotinylated horse antimouse IgG antibody at manufacturers recommended dilution (for mouse IgG peroxidase kit from Vector Laboratories, use one drop of stock blue biotinylated antibody in 10 ml PBS). Incubate in covered humidified chamber for 30–60 min.
  L. Prepare ABC reagent per manufacturers directions (for mouse IgG peroxidase kit from Vector Laboratories). Use one drop of orange label reagent A in 5 ml PBS, mix, add one drop of brown label reagent B to same container, mix and allow to stand for 30 min before use).
  M. Wash slides in three changes of PBS.
  N. Apply ABC reagent and incubate 30–60 min.
  O. Wash slides in three changes of PBS.
  P. Apply chromogen: VIP (purple chromagen) or DAB (brown chromagen) or AEC (red chromagen) per kit instructions.
  Q. Wash slides in water.
  R. Place slides in Gill's hematoxylin for 1 min and briefly wash in tap water.
  S. Briefly dip slides in acid alcohol to remove excess hematoxylin and briefly wash in tap water.
  T. Blue slides by placing in saturated aqueous lithium carbonate solution for 30 sec and wash in tap water.
  U. For slides having VIP or DAB as chromagen: dehydrate slides in 95% reagent alcohol, then absolute alcohol (2 changes each for 3–5 min each), then clear in two changes of xylene.

Place one to two drops of a permanent mounting media such as Permount on the specimen side of the slide and apply coverslip.

V. For slides having AEC as the chromagen, organic solvents must be avoided. After the final water wash, mount AEC slides in a mounting media such as Crystal Mount and apply coverslip.

Automated Immunostaining Method Using Dako LSB2

Although Dako LSB2 kits can be used by manual methods, we have used this system combined with the Dako Autostainer to allow us to stain up to 48 slides per run (all Dako products are from Dako, Carpinteria, CA). This methodology, combined with the use of sausage blocks of paraffin-embedded tissues, can result in one to a few hundred sections of tissues to be stained in one run. Although commercial autostaining systems are not in general use in most individual research laboratories, larger research core facilities and many large hospital pathology laboratories house such equipment that might be available to other users.

I. Follow method for formalin-fixed, paraffin-embedded sections under II. **steps A–E.**
II. Rinse slide in tap water and place in Tris-buffered saline while Autostainer vials/buffers are made. The quantity for each vial will depend on the number of slides to be run. The Autostainer can be programed to dispense controlled amounts with a default volume of reagent per slide of 200 µl. All reagents should be at room temperature at the initiation of the run.
  A. For quenching of endogeneous Peroxidase activity, prepare 3% hydrogen peroxide solution and place in Dako Autostainer vial.
  B. For primary antibody Autostainer vial: prepare DS6 tissue culture supernatant to a final concentration of 20 µg/ml by diluting in Dako Antibody Diluent (S0809).
  C. For negative isotype control antibody: prepare nonimmune mouse IgG1 to a final concentration of 20 µg/ml by diluting in Dako Antibody Diluent (S0809).
  D. Dako K0675 kit Biotinylated Link (biotin-labeled affinity isolated goat anti-rabbit and goat anti-mouse immunoglobulins).
  E. Dako K0675 kit Streptavidin-HRP (streptavidin conjugated to horseradish peroxidase).
  F. Reagent buffer container: fill with tris buffered saline containing 0.05% Tween-20 (Dako S3306).
  G. Dako K3466 liquid DAB chromagen (will need to be prepared per manufacturers' kit

instructions: one drop of DAB chromagen per ml of buffered substrate).

III. Follow manufacturer's protocol. User-designed programming grids can be modified to suit the user's needs. Our protocol run is set for 5 min incubation with 3% hydrogen peroxide, 30 min for primary or control antibodies, 10 min with secondary biotinylated link antibody, 10 min with streptavidin reagents, and wash steps of 3 min each.
IV. When run is complete, remove slides and place in water.
V. Place slides in Gill's hematoxylin for 1 min and briefly wash in tap water.
VI. Briefly dip slides in acid alcohol to remove excess hematoxylin and briefly wash in tap water.
VII. Blue slides by placing in saturated aqueous lithium carbonate solution for 30 sec and wash in tap water.
VIII. Dehydrate slides as in manual method/paraffin #2U and coverslip with Permount.

RESULTS AND DISCUSSION

DS6 is a newly developed antibody. The nature of the DS6-detected antigen in currently under investigation, and the role of this antigen in normal and neoplastic cells is still uncertain. DS6 reacts with an antigen that is not tumor specific, but does appear to have a limited distribution in normal and neoplastic human tissues and strong expression in selected types of human cancers. As an antibody that shows reactivity in both frozen sections and formalin-fixed sections, the antigen can be easily studied in human neoplasms using routinely processed surgical pathology specimens. If CA6 is found to have a correlation with prognosis or applications in antibody-based cancer therapy, patient tumors could be easily studied for the amount and/or pattern of CA6 expression using immunohistochemistry on routinely processed specimens.

The DS6 antibody was developed using serous adenocarcinoma of the ovary as the immunogen. Serous adenocarcinomas typically show aggressive intraperitoneal spread. The DS6 antibody is reactive with serous adenocarcinomas of the gynecological tract, and lacks reactivity with mesothelium, as studied in formalin-fixed tissue sections, so it is hoped that this antibody can have a role in intraperitoneal antibody-based therapy of ovarian carcinoma. The presence or absence of CA6 expression may also be useful in distinguishing tumor types or sites of origin. The lack of CA6 expression in normal and neoplastic mesothelial cells makes it potentially useful in distinguishing carcinomas from mesothelioma in diagnostic surgical pathology.

Correlation of the amount of CA6 expression in a tumor and the degree of tumor differentiation has not been systematically studied. However, previous studies have shown significant CA6 immunostaining in both well-differentiated and poorly differentiated carcinomas of various types, so an obvious correlation with tumor differentiation has not been observed. The preliminary data from a recent study of ovarian serous adenocarcinomas shows that the percentage of neoplastic cells immunohistochemically expressing CA6 does not correlate with tumor stage or overall survival (personal communication from Diane A. Semer, M.D., East Carolina University, Greenville, NC). The relationship of CA6 expression to tumor grade or prognosis has not been systematically evaluated in breast carcinomas, so the significance of CA6 expression in breast carcinomas is uncertain.

In evaluating immunohistochemical reactivity, the pattern and intensity of staining should be noted. In initial studies of an antibody, a grading system is typically defined to score reactivity in each sample studied. In previous studies on DS6, neoplasms were scored as nonreactive (0) if less than 1% of neoplastic cells were reactive, (1+) if 1–9% of neoplastic cells were reactive, (2+) if 10–49% of neoplastic cells were reactive, and (3+) if > 50% of neoplastic cells were reactive. For samples that showed any reactivity, a minimum number of cells was counted and evaluated to arrive at the score. Scores representing clinically significant levels of reactivity should be determined based on available data relevant to the biologic significance of the antigen. For some antibodies such as antibodies to estrogen receptors, immunoreactivity in as few as 1% of neoplastic cells in breast cancer is of potential prognostic and therapeutic significance (Harvey et al., 1999). Clinically significant levels of CA6 expression have not yet been determined. Immunohistochemical expression of CA6 in 1–9% of cells is indicative of a very low level of antigen expression. Levels of (2+) and (3+) staining are often considered significant levels of immunohistochemical reactivity for many antibodies. In previous studies of DS6, significant immunoreactivity was arbitrarily defined as (2+) and (3+) staining.

In sections of human breast tissue, CA6 expression is seen focally in benign breast epithelium of ducts and lobules. In the study by Smith et al. (2002), CA6 expression was also seen in the majority of breast carcinomas of each of the major types, with 32 of 53 cases (60% of cases) of breast carcinomas showing significant (2+ and 3+ staining) immunohistochemical reactivity. CA6 expression was seen in the majority of infiltrating ductal carcinomas, with 21 of 29 cases (72% of cases) showing significant reactivity. A small number of breast carcinomas of special types were also studied. One case of medullary carcinoma, a single case of tubular carcinoma, and a single case of invasive apocrine carcinoma were studied, and each showed CA6 expression. A single case of invasive papillary carcinoma showed only very focal CA6 reactivity. Three of 7 cases (43% of cases) of colloid carcinoma of the breast also showed significant CA6 expression. The infiltrating ductal carcinomas and ductal carcinomas of special types most commonly showed a luminal pattern of staining for DS6 in areas of gland and tubule formation, with the immunohistochemical reactivity seen along the portions of the neoplastic cell membranes lining the neoplastic glands and tubules. Cytoplasmic staining was also seen in some carcinoma cells and in some foci of ductal carcinoma in situ (Figure 73B). Staining of small intracytoplasmic round structures suggestive of intracytoplasmic lumens in the neoplastic cells was also occasionally seen in these neoplasms (Figure 73C).

In the study by Smith et al. (2002), infiltrating lobular carcinomas showed DS6 reactivity in 4 of 11 cases (36% of cases), and infiltrating pleomorphic lobular carcinomas showed reactivity in 1 of 2 cases (50% of cases). The predominant pattern of DS6 staining in these carcinomas was staining of small rounded intracytoplasmic structures suggestive of intracytoplasmic lumens (Figure 73C), with this staining pattern being extensive and quite striking in several examples lobular carcinoma.

Other monoclonal antibodies have found a role in cancer therapy for breast cancer, such as trastuzumab (Herceptin, Genentech), a humanized IgG1 monoclonal antibody directed against the Her2 receptor (Arteaga, 2003). The DS6 antibody is currently under preclinical evaluation as a potential therapeutic targeting vehicle for breast and ovarian cancer therapy, as these tumor types demonstrate substantial immunohistochemical and flow cytometric reactivity for DS6.

References

Arteaga, C.L. 2003. Trastuzumab, an appropriate first-line single agent therapy for Her2-overexpressing metastatic breast cancer. *Breast Cancer Res.* 5:96–100.

Battifore, H. 1986. The multitumor (sausage) tissue block: A novel method for immunohistochemical antibody testing. *Lab. Invest.* 55:244–248.

Harvey, J.M., Clark, G.M., Osborne, K., and Allred, D.C. 1999. Estrogen receptor status by immunohistochemistry is superior to the ligand binding assay for predicting response to adjuvant endocrine therapy in breast cancer. *J. Clin. Oncol.* 17: 1474–1481.

Kearse, K.P., Smith, N.L., Semer, D.A., Eagles, L., Finley, J.F., Kazmierczak, S., Kovacs, Rodriguez A.A., and Kellogg-Wennerberg, A.E. 2000. Monoclonal antibody DS6 detects a tumor-associated sialoglycotope expressed in human serous carcinomas. *Int. J. Cancer 88:*866–872.

Smith, N.L., Halliday, B.E., Finley, J.L., and Kellogg-Wennerberg, A.E. 2001. Immunohistochemical distribution of tumor-associated antigen CA6 in gynecologic neoplasms as detected by monoclonal antibody DS6. *Int. J. Gynecol. Pathol. 20:*260–266.

Smith, N.L., Halliday, B.E., Finley, J.L., and Kellogg-Wennerberg, A.E. 2002. The spectrum of immunohistochemical reactivity of monoclonal antibody DS6 in nongynecologic neoplasms. *Appl. Immunohistochem. Molec. Morphol. 10:*152–158.

28

Immunocytochemistry of Effusions

Adhemar Longatto Filho

Introduction

The examination of serous effusions could be part of several investigations from different lesions that induce transudate or exudate accumulation between the serous membranous space (Longatto Filho *et al.*, 2001). The regulatory turnover system could be disturbed by hydrodynamic force alterations mainly caused by cardiovascular-related diseases, renal failure, malignant primary neoplasia, or metastasis (Yarbro, 1995).

As a chronic disease, cancer is characterized by a biological behavior that could be summarized in a progressive evolution derived from cellular proliferation and differentiation processes that involve the ability of adjacent invasion of tissues and surrounded blood and lymphatic vessels and metastasis. The development of this complex and multistep phenomenon frequently assault the serosal membranes with metastasis from different organs leading to the accumulation of malignant effusion. The result of this damage is the rapid growth of serous membrane volume with signs that can be clinically observed (Nagi *et al.*, 1993).

The metastasis might appear as the first clinical manifestation of malignant neoplastic disease or primary cancer in a patient with apparently good health (Hanselaar, 2002). Hence, cytologic examination has been used as the first approach to analyze a serous effusion because of its feasibility to achieve a practical test to observe an eventual metastasis. In spite of the obvious advantages of cytology as a preferred selected method for these cases, some difficulties are encountered in this analysis (Bedrossian, 1998).

Considering the reactive mesothelial cells from non-neoplastic chronic serous effusions and the enormous variety of tumor cells in effusions as well as the similarity of neoplastic patterns of different organs, the cytologic analysis of these materials could induce errors of interpretation. The limitations of the cytologic interpretation are well known.

Many authors have described the overlapping cytologic profile of many reactive and neoplastic conditions (Bedrossian, 1998).

The summary of the major conflictive alterations involves the following diagnosis:

1. Reactive mesothelial cells versus neoplastic cells (primary or metastatic)
2. Histogenetic origin of neoplastic cells
3. Mesothelioma versus adenocarcinoma
4. Lymphoid cells versus lymphoma
5. Recognizing the primary site of malignant neoplasia

Optimizing the cytologic examination is a persistent goal of many investigators who have been working to decrease the number of uncertain cytologic reports to

Handbook of Immunohistochemistry and *in situ* Hybridization of Human Carcinomas, Volume 1: Molecular Genetics; Lung and Breast Carcinomas

513

Copyright © 2004 by Elsevier (USA)
All rights reserved.

improve the diagnosis. It is crucial to correct clinical diagnosis of patients and therapeutic decisions.

Ancillary Techniques

The importance of ancillary techniques to improve cytological diagnosis of serous effusions has been growing in the past few years. Newly developed methodologies and equipment for different laboratorial analyses have also been used to improve analysis of effusion samples. Recently, the role of these technical options involving several fields of science has been critically discussed in the context of cytology. For example, biochemistry offers hyaluronic acid assay to identify mesothelioma (Hjerpe, 1986); recent investigations have raised an old and persistent question about the general characteristics of the effusions to determine the different causes of transudates and exudates (Alexandrakis et al., 2001). Molecular markers of oncogenes and oncogene suppressors can be investigated by the polymerase chain reaction (PCR) assay. Genetic methods for mutagenesis identification using fluorescence in situ hybridization (FISH) methodology have encouraged many groups to invest time in this field because of its very specific answers related to the loss of heterozygosity (de Matos Granja et al., 2002). Transmission electronic microscopy and flow and static cytometry also have been used successfully (Hanselaar, 2002; Risberg et al., 2001). In spite of the inherent importance of these methodologies, there are several limitations in the application of these methods; these limitations include high costs, lack of specificity, and time required to carry them out.

As a practical alternative, immunocytochemistry has been regularly used as an integrate importance because of its favorable cost-effective advantage and the possibility to observe positive reactions in the cells formerly analyzed morphologically. This is a major factor to be considered because of the accessible correlation between the morphology of cellular alterations and the specific marker studied.

Immunocytochemistry (IHC) has been used for various studies involving reactive and malignant serous effusion samples (Gong et al., 2003). There are several consistent parameters that can elucidate the origin of the metastatic lesions (Jang et al., 2001), the prognosis of metastasis considering cell cycle proteins (Davidson et al., 2001), the classification of lymph proliferated lesions (Santos et al., 2000), and determination of histogenetic origin of the tumor (Longatto Filho et al., 1995). Examples of the other tumor markers for serous effusions are the role of oncogenes or suppressor oncogenes as a potential predictable factor (Simone et al., 2001), specific markers for different tumors (Zimmerman et al.,

2001), and a marker for a specific tumor (Zimmerman et al., 2001). But many of these markers lack the desirable specificity in the absence of clinical data (Longatto Filho et al., 2002).

Unfortunately, there is no marker that can differentiate a benign from a malignant condition conclusively. This is one of the most difficult problems in the examination of serous effusions (Longatto Filho et al., 2001). In spite of the benefits offered by immunocytochemistry, the real value of the markers depends on the biological meaning of the positive reactions and its direct use in the comprehension of pathological phenomena (Hayes et al., 1996).

Immunocytochemical Evaluation

First of all, immunocytochemistry should be well thought out as an ancillary technique and adjudicate its role as an optional choice between the available markers, considering the specificity and sensitivity limitations of the antibodies selected.

The choice for a specific antibody and related methodology for an optimal result depends on many circumstances (Gong et al., 2003). First, new antibodies and methods to reveal the reactions are ever improving (Abutaily et al., 2002). Ascertaining the ideal marker for a specific research is closely related to the updated information that fulfills the current necessity. Secondly, the evaluation of the positive reactions must be carefully defined in order to avoid misinterpretation.

Diest et al. (1997) have postulated seven important steps to be considered in immunoreactions.

1. Define the pattern and distribution of the immunoreactions considering the cellular biology related to the molecule to be identified (membrane, cytoplasmic, or nuclear reaction).

2. Define the area of the lesion to be evaluated.

3. Define, on this selected area, a random procedure of counting.

4. Count the positive cellular reaction.

5. Define the cells to be counted.

6. Define a cutoff point.

7. Define how to report the results.

The Impact of Immunocytochemical Panels

The reliability of immunocytochemical reactions depends on the specificity of the markers and the narrowness between positive reactions and the established goal that one wants to reach. It is not usual to use a single antibody to have a suitable result. The immunocytochemical investigation frequently involves a panel of different antibodies (Queiroz et al., 2001). This occurs because of the lack of specificity that most available antibodies present. It is very important to take into account an informative panel of antibodies to avoid misinterpretation. Regarding the serous effusions, the investigation could be even more complex because of both the several metastasis from malignant neoplasias of different organs and the special behavior of the mesothelial cells could release different products according to specific circumstances and present differences within the cytoskeletal composition (Ueda *et al.*, 2001). Despite the huge efforts of many investigators, an ideal panel for every situation has yet to be determined (Lozano *et al.*, 2001). Hence, the association of the results of immunocytochemical reactions and clinical data is crucial to spot the primary site of the malignant neoplasia found in the effusion (Longatto Filho *et al.*, 2002).

Specific Panels for Specific Settings

Looking for the Primary Sites

The immunocytochemical investigation of the cytoskeleton can be part of an important strategy to recognize primary sites of adenocarcinoma. Frequently found in the cytoplasm, cytokeratin (CK) is the component responsible for the maintenance of the cellular structure. There are several reports that discuss the positive immunoreactions of cytokeratins in effusion cytology (Abutaily *et al.*, 2002). Both mesothelial and epithelial cells and their malignant counterparts could show low and/or high CK according to the degree of cellular differentiation. Low-molecular-weight CK can be found in many adenocarcinomas, but CK7 and CK20 are believed to be more useful in a panel to discriminate adenocarcinomas (Alves *et al.*, 2001). Jang *et al.* (2001) advocate that CK7-/CK20+ immunostaining are indicative of colon adenocarcinoma. The CK7+ is more frequently found in breast and ovary adenocarcinomas, whereas CK20+ is in the colon and lungs (Longatto Filho *et al.*, 1997). Adding these cytokeratins in a panel with other markers may result in a useful adjunct proceeding to identify adenocarcinomas (Alves *et al.*, 2001), as in the following examples.

1. Breast ductal adenocarcinoma: CK7+/CK20−/ CEA+, BRST2 +/ER+ (~70%)/PR (~50%)/ Lactoferrin+/ Vimentin−
2. Breast lobular adenocarcinoma: CK7+/CK20−/ CEA+, BRST2 +/ER+ (~95%)/PR (~50%)/Lactoferrin+/ Vimentin−
3. Papillary ovary adenocarcinoma: CK7+/CK20−/ CEA−/Vimentin+/CA125+
4. Mucinous ovary adenocarcinoma: CK7+/CK20+/ CEA+/Vimentin+/−/CA125+/−
5. Colon adenocarcinoma:CK7−/CK20+/Vimentin−/ CEA+/CA19.9+

6. Pancreatic adenocarcinoma: CK7+/CK20+/−/ Vimentin +/− CEA+/CA19.9+

The primary site could be revealed by using products of secretion as specific markers of the origin of the metastatic tumors. One of the most widespread examples of that is the prostate antigen (PSA) when the suspicion obviously lies on the prostate. Thyroglobulin and calcitonin are frequently used when follicular or medullar thyroid carcinomas are suspected, respectively (Alves *et al.*, 2001). The correct identification of melanoma in serous effusion could be a very difficult issue because of the diversity of patterns that seriously impairs the interpretation of malignant cells in serous effusion. The presence of melanin pigments greatly yields the identification of the histogenetic origin of the metastasis, but there are many cases in which this element is absent. Even cytoplasmic pigments must be carefully analyzed because hemosiderin, carbon, or lipofucsin pigments could represent important biases and lead to diagnostic errors. Hence, immunocytochemical reactions are welcome as an ancillary proceeding in cases in which melanoma is one of the hypotheses. HMB45 antibody is a well-recognized antibody that distinguishes melanoma with sensitivity in about 77% of serous effusions (Longatto Filho *et al.*, 1995). However, a panel using HMB45, S-100 protein, AE1/AE3 (low- and high-molecular-weight cytokeratins pool) and MART-1 remains as another option according to Beaty *et al.* (1997), although the performance of 78% of the recognized melanoma is similar to that of HMB45 alone.

Discriminating the Cells of Mesothelial Origin

The mesothelial cells still represent a great challenge to cytologists around the world. The morphological characteristics, size, and immunocytochemical profiles of the reactive and malignant cells can vary significantly because of different causes to the differentiation degree in the case of the neoplastic cells. This fact reflects part of the problem confronted during the routine examination of serous effusions. Numerous publications have recently revealed significant immunocytochemical improvements related to the monoclonal antibodies derived from the cadherin family. Adding members of this family to a panel seems to be a very useful method to discriminate mesothelial from epithelial origin (Müller *et al.*, 2002), despite the restriction mentioned by some authors (Simsir *et al.*, 1999).

Cadherins are a family of cell-to-cell adhesion proteins whose subtypes of specific lineage are found in different tissues. A prompt use of different cadherins has been postulated to differentiate neoplastic cells of different origins with similar morphology

(Han *et al.*, 1999). The application of cadherin immuno-cytochemical reactions in serous effusions certainly has initiated valuable discussion regarding the distinction of mesothelial reactive cells, mesothelioma, and adenocarcinoma.

Positive immunoreactions for E-cadherin are very relevant to distinguish adenocarcinoma from mesothelioma. Müller *et al.* (2002) have studied a panel composed E-cadherin, E-selectin, and vascular cell adhesion molecules (VCAM). E-cadherin was positive for almost all studied pulmonary adenocarcinomas and was not reactive in mesotheliomas, whereas the two other markers had no significant immunoreaction differences in both tumors.

Davidson *et al.* (2000) have studied a large panel composed of 11 antibodies, including N-cadherin and desmin, in order to identify mesothelial reactive cells, mesothelioma, and adenocarcinoma. N-Cadherin recognized 100% of the 12 mesotheliomas, but also revealed a significant positivity in mesothelial reactive cells (86%) and marked 48% of the carcinomas as well. Considering that calretinin positivity (another important and well-recognized marker of mesothelial origin) marked more than 90% of the benign and malignant mesothelial cells and only 3% of carcinomas, desmin immunoreaction with 84% positive cases in mesothelial reactive cells, and 8% and 2% positive cases in mesothelioma and carcinoma, respectively, the N-cadherin performance was not considered consistent in this series. This is very motivating work because calretinin and desmin antibodies have played an important role in a panel of serous effusion examination. On the other hand, Thirkettle *et al.* (2000) have found 90% of N-cadherin positivity in mesothelioma against focal immunoreactivity of E-cadherin in epithelioid or mixed mesothelioma types.

Calretinin is a calcium-binding protein frequently reported as positive in mesothelioma and negative (or weakly positive) in adenocarcinoma. In the Chhieng *et al.* (2000) series, calretinin immunoreaction was positive in 87% of mesotheliomas and negative (or weakly positive) in all adenocarcinomas. The association of calretinin with E-cadherin seems to be a very powerful battery of markers in the examination of serous effusions because calretinin has proved to be very specific for mesothelial reactive cells and E-cadherin for carcinoma (Kitazume *et al.*, 2000). The use of E-cadherin is also advocated by Ordóñez (2000), who reports that its association with BG8 is very useful to discriminate epithelial mesothelioma and adenocarcinomas of different origins. Simsir *et al.* (1999) have also indicated the potential use of E-cadherin in discriminating adenocarcinoma from mesothelial reactive cells.

The Oncogenes and Oncogene Tumor-Suppressor Disorders

Malignant transformation and neoplastic progression are consequences of anomalous cellular control allied to amplification, translocation, or overexpression of oncogenes (Field *et al.*, 1996). There are four major mechanisms that involve oncogene activation and the consequent disturbed pathways: abnormal signaling by a structurally abnormal cytokine growth factor, aberrant phosphorylation of proteins by altered signal transducer kinases, abnormal transmission signals by G proteins, and finally, disturbed regulation of gene transcription by abnormal transcription factors (Piris *et al.*, 1996).

Identifying the oncogene or related proteins involved in the carcinogenesis could be used as a prognostic factor to enhance the classification from benign and malignant conditions in daily routine examination of serous effusions (Gong *et al.*, 2003; Mullick *et al.*, 1996; Thirkettle *et al.*, 2000).

There are common oncogene alterations (mutations and/or overexpression) in human tumors related to the *myc* oncogene family. Several neoplasias present amplification and overexpression of c-*myc* gene. Solid tumors associated with amplification or overexpression of c-*myc* gene could be associated with poor prognosis (Field *et al.*, 1990).

Other oncogenes are associated with the solid malignant neoplasias, lymphoma, and leukemia. Ha-*ras* is related to genitourinary, thyroid, and breast carcinoma; K-*ras*, to pancreas, colon, ovary, and lung adenocarcinomas; N-*ras*, to breast cancer and leukemia; HER-1/neu, to breast, bladder and lung carcinomas, glioma, and glioblastomas; HER-2/neu, to breast, stomach, salivary gland, kidney, and ovary carcinoma; c-*myc*, to breast, stomach, head and neck and nasopharyngeal carcinomas, and Burkitts lymphoma; N-*myc*, to neuroblastoma, retinoblastoma, and breast carcinoma; and L-*myc*, to small cell carcinoma of the lung (Field *et al.*, 1990).

Regarding the oncogene tumor suppressor, the overexpression of p53 is related to the development of neoplasias from different tissues. Significant diagnosis and prognosis of p53 do not have the same importance in all tumors, but despite these differences mutations affecting p53 can be found in bladder, breast, head and neck, ovary, colon, esophagus, lung, and uterine cervix carcinomas (Piris *et al.*, 1996).

The gene p53 encodes a 53-kDa nuclear phosphoprotein that acts as a negative regulator of the G_1/S transitional phase in the cell cycle. The activation of p53 when DNA damage occurs serves to block DNA replication and allow time for the repair. Mutagenic agents of different origins stabilize p53 protein in normal cells.

Its presence can induce delay in the G_1/S transition in cells with DNA damage, allowing the DNA repair as mentioned before or inducing apoptosis in those cells with severe and permanent damage (Piris *et al.*, 1996).

The Oncogenes and Oncogene Tumor-Suppressor Markers

Recently, de Matos Granja *et al.* (2002) have reported the importance of the PCR assay to evaluate loss of heterozygosity (LOH) in serous effusions when studying breast cancer metastasis. The authors have shown the feasibility to assess informative cases in serous effusions and the enormous potential of this research to discriminate malignant cells as well as to improve the identification of primary site in cases in which such information is not available. The catalog of tumor karyotypes and gene sequences certainly encourages the use of this information in routine examination of serous effusion. There are several possibilities to be tested in the near future, and certainly the immunocytochemical methodology seems to be a more accessible way to meet this goal.

In the last few years, oncogenes and oncogene tumor-suppressor antibodies have been constantly used to improve the cytological diagnosis. The p53 protein has been widely used with noteworthy efficacy for recognizing malignancy in serous effusion samples (Gong *et al.*, 2003). Because the wild-type p53 protein is not sufficiently accumulated in normal conditions, it cannot be easily recognized by immunocytochemical methods. Therefore, the cellular identification of this protein highly suggests malignancy (Mullick *et al.*, 1996). However, because of the low sensitivity of p53 immunocytochemical examination and the fact that p53 negative reaction does not necessarily exclude malignancy (Tiniakos *et al.*, 1995), the investigation of p53 can be associated with other tumor markers, including HER-2/neu oncogene identification (El-Habashi *et al.*, 1995) or DNA content examination that have shown very helpful results.

MATERIALS

The immunocytochemical assay of serous effusions can be performed using cytological samples obtained from different sources, such as cytospin preparations (Alves *et al.*, 2001), liquid-based system, and cell block paraffin-embedded samples (Gong *et al.*, 2003).

The effect of fixation and cell preparation can directly induce the immunostaining results (Alves *et al.*, 2001; Gong *et al.*, 2003). Retrospective analysis of previously stained samples (Papanicolaou or air-dried related methods) must be carefully thought out because of the influence of type and time of fixation, and the storage conditions of samples (Alves *et al.*, 2001).

For prospective studies, several protocols have been constantly discussed to indicate optimal conditions for immunostaining. Absolute ethanol preserves cell morphology, but there are restrictions for its use in immunocytochemical protocols. Ethanol-ether (1:1), isopropyl alcohol, methanol, polyethylene glycol, air-dried fixation, or 70% ethanol are some other possibilities to be carefully analyzed. Despite the preferred choice, the protocol also involves other variables such as the time between the preparation and immediate fixation of the sample. These parameters should be rigorously controlled to avoid false results. In retrospective analysis, however, these parameters are dramatically uncontrollable. There are a number of unknown conditions of sample preparation that could be misleading in the final result (Alves *et al.*, 2001). For this reason, antigen-retrieval strategies should be taken into consideration to improve the immunostaining (Hayat, 2002). In general, methods for epitope retrieval are used in formalin-fixed, paraffin-embedded samples, because of the well-known limitations of additive fixation for immunocytochemistry, but occasionally its use could also be applied for cytology. For instance nuclear antigens can be better recognized if microwave oven heating is used (Shi *et al.*, 1997). Another possibility for enhancing immunocytochemical reactions in retrospective studies is to incubate the primary antibody for 60 hr; however, long-standing incubation has been cautiously used to avoid excess background staining (Longatto Filho *et al.*, 1995, 1997, 2002; Santos *et al.*, 2000).

The general principles of immunocytochemical techniques used in the following protocols are derived from the Hsu *et al.* (1981) report with minor modifications. These protocols were used in studies conducted from retrospective series of serous effusion samples (pleural and peritoneal) previously stained with the Papanicolaou method (Longatto Filho *et al.*, 1997). All the slides used were prepared in both cytospin and conventional smears. The time of storage of these slides ranged from 4 to more than 20 years. All the records were reviewed and the diagnosis confirmed by histopathological correlation and/or clinical examination.

The applied primary antibodies were as follows:
1. For melanoma:
   - ▲ Primary antibody: HMB-45 monoclonal antibody (previously diluted kit) (Biogenex Laboratories, San Ramon, CA)
2. For adenocarcinoma's discrimination:
   - ▲ Anticytokeratins 7 and 20 (each diluted 1:2,000), monoclonal antibodies

▲ (DakoA/S, Glostrup, Denmark, clones OV-TL12/30 and Ks 20.8, respectively

3. For lymphoma discrimination:
  ▲ B-cell marker L-26 (diluted 1:200), monoclonal antibody (CD-20, Dako, Glostrup)
  ▲ T-cell marker UCHL-1 (diluted 1:50), monoclonal antibody (CD-45 RO, DakoA/S, Glostrup)

4. Panel for adenocarcinomas discrimination:
  ▲ Lactoferrin (diluted 1:8,000), polyclonal antibody (Dako A0186—USA)
  ▲ CA-19.9 (diluted 1:500), Clone C241:5:1:4, monoclonal antibody (Novocastra, NCI-CA, UK)
  ▲ CA125 (diluted 1:100), Clone OC 125, monoclonal antibody (Signet, 8510 USA)
  ▲ GCDPF-15 (diluted 1:10), Clone BRST2/D6, monoclonal antibody (Signet, 8510 USA)
  ▲ Mesothelial cell (diluted 1:200), Clone HBME-1, monoclonal antibody (Dako, M3505, USA)

METHODS

(Step-by-Step)

1. Remove the coverslips from the slides and rehydrate the smears that are decolorized (optional step) in alcohol–acid solution (3% HCl at 70% ethanol).

2. Thoroughly rinse with distilled water.

3. Inhibit endogeneous peroxidase activity: methanolic 3% hydrogen peroxidase solution.

4. Thoroughly rinse with distilled water.

5. TritonX-100 solution (Serva, Heidelberg, Germany) in 0.2% tamponate saline solution with phosphates 0.01 M at pH 7.4 (phosphate buffered saline [PBS]) for 20 min at 37°C.

6. Primary antibody (diluted in bovine albumin solution (BSA) at 1% in PBS:overnight or 60 hr in incubation step at 4°C.

7. Three successive 5-min-long PBS baths.

8. Secondary antibody: Biotinylated horse antimouse IgG diluted at 1:500 (Vector, Burlingame, CA; BA 2010) for 30 min at 37°C or

9. Secondary antibody: Biotinylated goat antirabbit and antimouse IgG antibody diluted at 1:200 (StrepABComplex/HRP Duet mouse/rabbit, Dako, Glostrup) for 30 min at 37°C.

10. Three successive 5-min-long PBS baths.

11. Amplification step: Avidin-biotin-peroxidase system (Vector, PK-4002) or

12. Amplification step: Streptavidin-biotin-peroxidase system (StrepABComplex/HRP Duet) diluted 1:500 for 30 min at 37°C.

13. Three successive 5-min-long PBS baths.

14. Chromogenic reagent: 60 mg% (or 50 mg%) 3,3'-diaminobenzidine tetrahydrochloride (Sigma) plus 0.1% hydrogen peroxidase solution as the chromogenic substrate for 5 min.

15. Thoroughly rinse with distilled water.

16. Counterstain with Harris or Meyer hematoxylin.

17. Three successive baths of absolute ethanol.

18. Three successive baths of xylene.

19. Coverslips mounted with Entellan (Merck 107961, Germany) or similar media.

RESULTS AND DISCUSSION

The search of primary site of metastasis in serous effusions is undoubtedly the most crucial controversial point to be considered in routine conditions. The type of cell present in the effusion, the anatomic location of the effusion as regards to the age and sex of the patient, and finally the presence and nature of possible primary site of a tumor are important factors to be investigated (Bedrossian, 1998). Immunocytochemistry is so far the most accessible ancillary technique to be used to resolve this question. Cytologists comfortably accept this technique in their routine studies because of the possibility of observing cellular details including nuclear features that indicate malignancy.

In this scenario, monoclonal cytokeratin antibodies have emerged as a powerful tool to be applied in immunocytochemical panels. As for the adenocarcinomas, low molecular weight cytokeratins can be very helpful to indicate the primary site, in spite of a certain lack of specificity. In order to achieve significant accuracy with immunocytochemical panels (with cytokeratins or not), the results must be analyzed in association with the basic information such as sex of the patient and type of effusion (pleural, peritoneal). To our understanding, the age of the patient is not a significant issue to be considered if adenocarcinoma is the histogenetic type to be studied (Longatto Filho et al., 2002). For instance, CK7 is very efficient to demonstrate ovary and breast origins. In a previous study (Longatto Filho et al., 1997), 67.6% of breast adenocarcinoma and 63.5% of ovarian adenocarcinoma were positive for CK7 antibody. The CK7 has a significant value as a variable to be kept in a prediction equation in a panel including CK20 antibody and a wide range of morphological parameters. The CK20 antibody, however, did not show the same utility. Its use, for example, was not consistent enough to support gastric origin. Besides, CK20 was positive in 32.3% of pulmonary adenocarcinomas. The differences in the immunoreaction patterns of primary and metastatic carcinomas depend on several technical and biological parameters, including the environmental conditions (Azumi *et al.*, 1987). Jang *et al.* (2001) have found that the combination of CK7–/CK20+ is very useful to identify colon adenocarcinomas. One point, though, seems to be fundamental: Cytokeratin alone, like many immunocytochemical antibodies, is not a specific marker; its value is significant only as part of a panel (Longatto Filho *et al.*, 1997).

Other markers are frequently used to identify the primary site of metastatic adenocarcinomas in the serous effusions. Lactoferrin, an iron-binding protein found in bovine and human breast glands, can be used for this

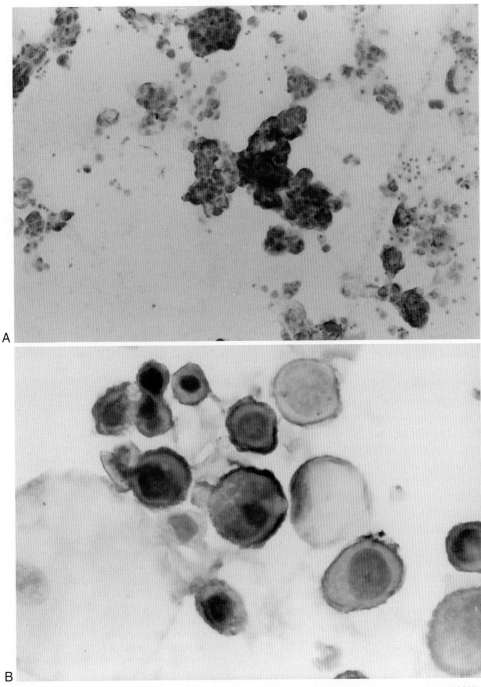

Figure 74 **A:** Shows positive immunoreaction for lactoferrin breast adenocarcinoma (400X). **B:** Shows positive immunoreaction for CA125 in ovarian adenocarcinoma (100X).

purpose. It is a polyclonal antibody very useful in the identification of breast origin if associated with BRST2 monoclonal antibody (a protein group purified in ductal cysts from the breast: gross cystic disease fluid protein). The correct identification of breast origin was possible in 83.6% of cases when immuno-reactions for both antibodies were positive and additional sample information about the patients was available (i.e., the sex female obviously) and the anatomic site of effusion (pleural in these cases) (Longatto Filho *et al.*, 2002) (Figure 74A). Regarding ovarian origin, the results are also reliable when HBME-1 (antibody originated in human mesothelioma) and CA-125[a] immunoreactions are positives (Figure 74B).

The anatomic location did not allow any additional upgrading in the discrimination of ovarian adenocarcinoma using the classification obtained by discriminant analysis because of the numerous cases of ovarian metastasis in peritoneal and/or pleural effusions (Longatto Filho *et al.*, 2002).

The CA-19.9 positive immunoreaction, at first believed to be specific to the gastrointestinal adenocarcinomas, has been recently demonstrated in breast, lung, and ovarian adenocarcinomas. Hence, it is not a useful marker for the identification of the primary site of adenocarcinoma. The CA-19.9 immunoreactivity might be considered in the clinical setting if one speculates about the gastric or intestinal origin in a peritoneal effusion; stomach is the most probable origin if lactoferrin is also positive (Longatto Filho *et al.*, 2002).

Considering the merits and limitations of immunocytochemical markers, the panels of antibodies should be judiciously selected in the clinical–laboratory context, based on carefully studied morphological features. Some promising immunocytochemical algorithmics can be used for diagnostic and prognostic assessment (Longatto Filho *et al.*, 2001), but the efficacy of the method still depends on very important technical considerations (mainly in retrospective), which is another crucial problem. But many of the numerous immunocytochemical panels studied so far encourage us to validate some of them as useful implements for the selection of the primary organ to be examined by imaging methods in patients with malignant effusion (Longatto Filho *et al.*, 2002). For targets other than, for example, the identification of specific products released by certain types of malignant neoplasia), positive immunoreaction can elucidate several situations and supply essential information for clinical management (Longatto Filho *et al.*, 1995).

---

[a] Antibody against a high-molecular-weight antigen, 200-1,000 kd, with an unknown function and found in many secretions and fluids and also in a number of adenocarcinomas of different origins.

References

Abutaily, A.S., Addis, B.J., and Roche, W.R. 2002. Immunohistochemistry in the distinction between malignant mesothelioma and pulmonary adenocarcinoma: A critical evaluation of new antibodies. *J. Clin. Pathol. 55:*662–668.

Alexandrakis, M.G., Kyriakou, D., Koutroubakis, I.E., Alexandraki, R., Vlachonikolis, I.G., and Eliopoulos, G.D. 2001. Assaying of tumor necrosis factor alpha, complement factors, alpha-1-antitrypsin in the diagnosis of malignant serous effusions. *Am. J. Clin. Oncol. 24:*562–565.

Alves, V.A.F., Kanamura, C.T., and Longatto Filho, A. 2001. Immunocytochemical in serous effusions. Bibbo, M. and Longatto Filho, A. (eds) *Clinical and Laboratorial Aspects of Serous Effusions. Therapeutics and Diagnostic and Prognostic Evaluations* (text in Portuguese). Rio de Janeiro, Brazil: Revinter, 145–171.

Azumi, N., and Battifora, H. 1987. The distribution of vimentin and keratin in epithelial and nonepithelial cells. *Am. J. Clin. Pathol. 88:*286–296.

Beaty, M.W., Fetsch, P., Wilder, A.M., Marincola, F., and Abati, A. 1997. Effusion cytology of malignant melanoma: A morphologic and immunocytochemical analysis including application on the MART-1 antibody. *Cancer (Cancer Cytopathol.) 81:*57–63.

Bedrossian, C.W.W. 1998. Diagnostic problems in serous effusions. *Diagn. Cytopathol. 19:*131–137.

Chhieng, D.C., Yee, H., Schaefer, D., Cangiarella, J.F., Jagirdar, J., Chiriboga, L.A., Jagirdar, J., Chiriboga, L.A., and Cohen, J.M. 2000. Calretinin staining pattern aids in the differentiation of mesothelioma from adenocarcinoma in serous effusions. *Cancer 90:*194–200.

Davidson, B., Nielsen, S., Christensen, J., Asschenfeldt, P., Berner, A., Risberg, B., and Johansen, P. 2001. The role of desmin and N-cadherin in effusion cytology: A comparative study using established markers of mesothelial and epithelial cells. *Am. J. Surg. Pathol. 25:*1405–1412.

Davidson, B., Risberg, B., Berner, A., Nesland, J.M., Trope, C.G., Kristensen, G.B., Bryne, M., Goscinski, M., van de Putte, G., and Florenes, V.A. 2001. Expression of cell cycle proteins in ovarian carcinoma cells in serous effusions-biological and prognostic implications. *Gynecol. Oncol. 83:*249–256.

Davidson, B., Berner. A., Nesland, J.M., Risberg, B., Berner, H.S., Trope, C.G., Kristensen, G.B., Bryne, M., and Ann Florenes, V. 2000. E-cadherin and alpha-, beta-, and gamma-catenin protein expression is up-regulated in ovarian carcinoma cells in serous effusions. *J. Pathol. 192:*460–469.

Diest, P.J., Dam, P., Henzen-Logmans, S.C., Berns, E., Burg, M.E.L., Green, J., and Vergote, I. 1997. A scoring systems for immunocytochemical staining: Consensus report to the task force for basic research of the EORTC-GCCG. *J. Clin. Pathol. 50:*801–804.

El-Habashi, A., El-Morsi, B., Freeman, S.M., El-Didi, M., and Marrogi, A.J. 1995. Tumour oncogenic expression in malignant effusions as a possible method to enhance cytologic diagnostic sensitivity. An immunocytochemical study of 87 cases. *Am. J. Clin. Pathol. 103:*206–214.

Field, J.K., and Spandidos, D.A. 1990. The role of ras and *myc* oncogenes in human solid tumors and their relevance in diagnosis and prognosis. A review. *Anticancer Res. 10:*1–22.

Field, J.K., and Spandidos, D.A. 1996. Oncogenes and tumour-suppressor genes as prognostic indicators in the development of human tumors. Pusztai, L., Lewis, C.E., and Yap, E., (eds) *Cell Proliferation in Cancer. Regulatory Mechanisms of Neoplastic Cell Growth.* Oxford: Oxford University Press, 83–104.

Gong, Y., Sun, X., Michael, C.W., Attal, S., Williamson, B.A., and Bedrossian, C.W. 2003. Immunocytochemical of serous effusion specimens: A comparison of ThinPrep® vs. cell block. *Diagn. Cytopathol.* 28:1–5.

Han, A.C., Filstein, M.R., Hunt, J.V., Soler, A.P., Knudsen, K.A., and Salazar H. 1999. N-Cadherin distinguishes pleural mesotheliomas from lung adenocarcinomas: A ThinPrep immunocytochemical study. *Cancer* 87:83–86.

Hanselaar, A.G. 2002. Additional techniques in serous effusions. *Anal. Cell. Pathol.* 24:1–4.

Hayat, M.A. 2002. *Microscopy, Immunohistochemistry and Antigen Retrieval Methods.* New York and London: Kluwer Academic/Plenum Publishers.

Hayes, D.R., Bast, R.C., Desch, C.E., Fritsche, H. Jr., Kemeny, N.E., Jessup, J.M., Locker, G.Y., MacDonald, J.S., Mennel, R.G., Norton, L., Ravdin, P., Taubes, S., and Winn, R.J. 1996. Tumor marker utility grading system: A framework to evaluate clinical utility to tumor markers. *J. Natl. Cancer Inst.* 88:1456–1466.

Hjerpe, A. 1986. Liquid-chromatographic determination of hyaluronic acid in pleural and peritoneal fluids. *Clin. Chem.* 32:952–956.

Hsu, C., Raine, L., and Fanger, H. 1981. The use of avidin-biotin-peroxidase complex (ABC) in the immunoperoxidase technique. A comparison between ABC and unlabeled procedures. *J. Histochem. Cytochem.* 29:577–580.

Jang, K.Y., Kang, M.J., Lee, D.G., and Chung, M.J. 2001. Utility of thyroid transcription factor-1 and cytokeratin 7 and 20 immunostaining in the identification of origin in malignant effusions. *Anal. Quant. Cytol. Histol.* 23:400–404.

Kitazume, H., Kitamura, K., Mukai, K., Inayama, Y., Kawano, N., Nakamura, N., Sano, J., Mitsui, K., Yoshida, S., and Nakatani, Y. 2000. Cytologic differential diagnosis among reactive mesothelial cells, malignant mesothelioma, and adenocarcinoma: Utility of combined E-cadherin and calretinin immunostaining. *Cancer* 90:55–60.

Longatto Filho, A., Alves, V.A.F., Kanamura, C.T., Nonogaki, S., Bortolan, J., Lombardo, V., and Bisi, H., 2002. Identification of the primary site of metastatic adenocarcinoma in serous effusions. Value of an immunocytochemical panel added to the clinical arsenal. *Acta Cytol.* 46:651–658.

Longatto Filho, A., Bisi, H., Alves, V.A.F., Kanamura, C.T., Oyafuso, M.S., Bortolan, J., and Lombardo, V. 1997. Adenocarcinoma in females detected in serous effusions: Cytomorphologic aspects and immunocytochemical reactivity to cytokeratins 7 and 20. *Acta Cytol.* 41:961–971.

Longatto Filho, A., Carvalho, L.V., Santos, G.C., Oyafuso, M.S., Lombardo, V., Bortolan, J., and Neves, J.I. 1995. Cytologic diagnosis of melanoma in serous effusions. A morphologic and immunocytochemical study. *Acta Cytol.* 39:481–484.

Longatto Filho, A., and Montironi, R. 2001. The role of quantitative methods in the cytopathological diagnosis in serous effusions. *J. Bras. Patol.* 37:205–212.

Lozano, M.D., Panizo, A., Toledo, G.R., Sola, J.J., and Pardo-Mindan, J. 2001. Immunocytochemical in the differential diagnosis of serous effusions: A comparative evaluation of eight monoclonal antibodies in Papanicolaou stained smears. *Cancer* 93:68–72.

Matos Granja, N de., Soares, R., Rocha, S., Paredes, J., Longatto Filho, A., Alves, V.A.F., Wiley, E., Schmitt, F.C., and Bedrossian, C.W.W. 2002. Evaluation of breast cancer metastasis in pleural effusions by molecular techniques. *Diagn. Cytopathol.* 27:210–213.

Müller, A.M., Weichert, A., and Müller, K.M. 2002. E-cadherin, E-selectin and vascular cell adhesion molecule: Immunohistochemical markers for differentiation between mesothelioma

and metastatic pulmonary adenocarcinoma? *Virchows Arch.* 441:41–46.

Mullick, S.S., Green, L.K., Ramzy, I., Brown, R.W., Smith, D., Gondo, M.M., and Cagle, P.T. 1996. p53 gene product in pleural effusions. Practical use in distinguishing benign from malignant cells. *Acta Cytol.* 40:885–860.

Nagi, J.A., Herzberg, K.T., Duorak, J.M., and Duoak, H.F. 1993. Pathogenesis of malignant ascites formation: Initiating events that lead to fluid accumulation. *Cancer Res.* 55:2631–2643.

Ordoñez, N.G. 2000. Value of thyroid transcription factor-1, E-cadherin, BG8, WT1, and CD44S immunostaining in distinguishing epithelial pleural mesothelioma from pulmonary and non pulmonary adenocarcinoma. *Am. J. Surg. Pathol.* 24:598–606.

Piris, M.A., Sanchez-Beato, M., Villuendas, R., and Martinez, J.C. 1996. Oncogenes and tumour-supressor genes. In Pusztai, L., Lewis, C.E., and Yap, E., (eds) *Cell Proliferation in Cancer. Regulatory Mechanisms of Neoplastic Cell Growth.* Oxford: Oxford University.

Queiroz, C., Barral-Netto, M., and Bacchi, C.E. 2001. Characterizing subpopulations of neoplastic cells in serous effusions. The role of immunocytochemical. *Acta Cytol.* 45:18–22.

Risberg, B., Davidson, B., Nielsen, S., Dong, H.P., Christensen, J., Johansen, P., Asschenfeldt, P., and Berner, A. 2001. Detection of monocyte/macrophage cell populations in effusions: A comparative study using flow cytometric immunophenotyping and immunocytochemical. *Diagn. Cytopathol.* 25:214–219.

Santos, G.C., Longatto Filho, A., Carvalho, L.V., Neves, J.I., and Alves, A.C. 2000. Immunocytochemical study of malignant lymphoma in serous effusions. *Acta Cytol.* 44:539–542.

Simone, G., Falco, G., Caponio, M.A., Campobasso, C., De Frenza, M., Petroni, S., Wiesel, S., and Leone, A. 2001. nm23 expression in malignant ascitic effusions of serous ovarian adenocarcinoma. *Int. J. Oncol.* 19:885–890.

Simsir, A., Fetsch, P., Mehta, D., Zakowski, M., and Abati, A. 1999. E-cadherin, N-cadherin, and calretinin in pleural effusions: The good, the bad, the worthless. *Diagn. Cytopathol.* 20:125–130.

Shi, S.R., Cote, R.J., and Taylor, C.R. 1997. Antigen retrieval immunohistochemistry: Past, present and future. *J. Histochem. Cytochem.* 45:327–343.

Tiniakos, D.G., Healicon, R.M., Hairt, T., Wadhera, V., Horne, C.H.W., and Angus, B. 1995. p53 immunostaining as a marker of malignancy in cytological preparations of body fluids. *Acta Cytol.* 39:171–176.

Thirkettle, I., Harvey, P., Hasleton, P.S., Ball, R.Y., and Warn, R.M. 2000. Immunoreactivity for cadherins, HGF/SF, met, and erbB-2 in pleural malignant mesotheliomas. *Histopathology* 36:522–528.

Ueda, J., Iwata, T., Takahashi, M., Hoshii, Y., and Ishihara, T. 2001. Comparative reactive immunochemical study of lectin-binding sites and cytoskeletal filaments in static and reactive mesothelium and adenocarcinoma. *Pathol. Int.* 51:431–439.

Yarbro, J.W. 1995. Effusions. In Abeloff, M.D., Armitage, J.O., Lichter, A.S., and Niederhuber, J.E. (eds) *Clinical Oncology.* New York: Churchill Livingstone, 709–725.

Zimmerman, R.L., and Fogt, F. 2001. Evaluation of the c-Met immunostain to detect malignant cells in body cavity effusions. *Oncol. Rep.* 8:1347–1350.

Zimmerman, R.L., Goonewardene, S., and Fogt, F. 2001. Glucose transporter Glut-1 is of limited value for detecting breast carcinoma in serous effusions. *Mod. Pathol.* 8:748–751.

29

Immunohistochemistry of Needle Cytopunctures of Breast Carcinomas

Marianne Briffod and Jean-Marc Guinebretiere

Introduction

Major progress has been made in recent years in immunohistochemical detection of various markers (cell proliferation markers, hormone receptors, etc.), improving knowledge and management of breast cancer. Most work reported in the literature has been done on frozen or paraffin-embedded tumor tissues. Fine-needle cytopuncture, that is useful for pretherapeutic diagnosis of breast cancer and for monitoring its progression, often provides highly cellular material representative of the lesions, which can be used for such analyses. Various studies have shown that nuclear and cytoplasmic markers can be detected individually by means of immunocytochemistry (IHC) on cytopuncture smears, whereas immunohistochemical studies of cells present in paraffin miniblocks can simultaneously identify several markers on consecutive sections. Whatever technique used, however, the risk of sample contamination by ductal carcinoma *in situ* (DCIS) is a major problem for breast tumors. This difficulty can be solved when simultaneous evaluations are feasible on a concomittant node metastasis. We present herein the technique we have used for years and our results from different studies have validated this technique.

MATERIALS AND METHODS

Materials

1. Fixative: Neutral formol (unbuffered formol saline). Do not use phosphate-buffered solutions during any processing step in order to avoid depolymerization of the cytoblock.
2. Shandon Cytoblock Kit containing 50 cytoblock cassettes, 1 cytoblock reagent 1 (clear fluid), 1 cytoblock reagent 2 containing formaldehyde (colored fluid). Components of these two reagents are withheld as a trade secret.
3. Tespa-precoated slides.
4. 1% hydrogen peroxide.
5. Citrate buffer (pH 6.0).
6. Tris-buffered saline.
7. Standard Labeled Streptavidin-biotin-peroxidase complex (LSAB Dako).

Handbook of Immunohistochemistry and *in situ* Hybridization of Human Carcinomas, Volume 1: Molecular Genetics; Lung and Breast Carcinomas

523

Copyright © 2004 by Elsevier (USA)
All rights reserved.

8. Anti-estrogen-receptor antibody ER 1D5 (Dako).
9. Anti-progesterone-receptor antibody, PGR 636 (Dako).
10. P 53 DO-7 antibody (Dako).
11. Anti-c-erbB-2 antibodies, polyclonal A0485 (Dako) and monoclonal CB 11 (Novocastra).
12. Anti Ki-67 antibody MIB1 (Dako).

Samples

Fine-needle cytopuncture (FNC) was carried out on each primary tumor or metastasis, according to our technique without aspiration, which regularly provides highly cellular material. Three FNCs of each lesion were performed in different areas. Part of each FNC specimen was smeared for cytologic diagnosis (including cytologic nuclear grading); the material that remained in the needle was ejected into the same tube for fixation in 0.5 ml of neutral formol for at least 1 hr. The cellularity of cytologic material influences the value of cytoblock.

Methods

The **cytoblock technique** was performed with the Shandon Cytoblock kit according to the manufacturer's instructions (Shandon Inc, Pittsburgh, PA). We slightly simplified the technique by using only one cytocentrifugation step. Consequently,

1. Add 6 drops of reagent 2 to the cell suspension and mix by vortexing.
2. Place 3 drops of reagent 1 into the center of the embedding cassette.
3. Assemble the cassette with a cytoclip and a cyto-funnel.
4. Place mixed suspension in the cytofunnel.
5. Cytocentrifuge the mixture (4 min at 1200 rev/min with low acceleration).
6. Remove the funnel.
7. Place one drop of reagent 2 and one drop of reagent 1 on the top of the cell button.
8. The button of cells formed, close the cytoblock cassette and place it in fixative to await routine processing.

The cytoblock was routinely processed and paraffin embedded. Multiple 2.5-μm paraffin sections were cut from each specimen. Sectioning requires a skilled technician because the cell button is thin and can be quickly used up by excessive trimming. Sections were mounted on TESPA-precoated slides and allowed to dry at 50°C overnight.

Immunohistochemistry (IHC)

Procedure

The simultaneous procedure was only applied to cytoblocks from paired tumors and metastatic nodes. The immunostaining was routinely realized in the same run with the other immunohistostaining of surgical paraffin-embedded fragments according to the standard procedure of our laboratory. This procedure is adapted and validated with internal and external quality assurance protocols (AFAQAP, 1997; GEFPICS-FNCLCC, 1999; UK. NEQAS, 1999).

A heat antigen-retrieval technique was used in all cases (Hayat, 2002).

1. Dewax the sections in xylene and hydrate through graded concentrations of alcohol.
2. Block the endogenous peroxidase activity with 1% hydrogen peroxide for 15 min.
3. Immerse the sections in thermoresistant plastic box containing 10 ml of buffer adapted to each antibody (EDTA buffer for ER, citrate buffer for PR, c-erbB2, p53), pH 6.0, and heat in a water bath (Stuart Scientific), at 96°C during 20 to 45 min.
4. Cool the sections at room temperature for 30 min and rinse in Tris-buffered saline.
5. Tip away the blocking reagent and add the primary antibodies for 25 min.
6. A standard streptavidin-biotin-peroxidase complex is used to reveal antibody-antigen reactions.

We use an automat for the latter steps (Dako Autostainer).

Five biological markers were investigated. The growth fraction was assessed using MIB-1 antibodies recognizing Ki-67 antigen (Dako SA, Trappes, France) at 1/100 dilution. Staining was detected with the avidin-biotin complex HRP method. Hormone receptor expression was assessed with the anti-estrogen–receptor antibody ER 1D5 at 1/30 dilution (Dako) and the anti-progesterone–receptor antibody PGR 636 at 1/75 dilution (Dako). P53 protein expression was detected using DO-7 antibody at 1/100 (Dako). C-*erb*B-2 oncoprotein expression was detected both with a polyclonal antibody (A0485) at 1/800 dilution (Dako) and a monoclonal antibody CB11 at 1/400 dilution (Novocastra).

An external positive control, including fragments of different tumors with various expression levels validated by other methods (fluorescence *in situ* hybridization [FISH] for Her2, biochemical technique for estrogen receptors [ER] and progesterone receptors [PRs]) was regularly added.

Assessment of Staining

At least 100 malignant cells were considered suitable for IHC on cytoblocks. The percentage of positive cells was scored for all the markers. For Ki-67, any nuclei with detectable staining was considered as positive. We used a two-group classification (high and low Ki-67 when the proportions of stained nuclei were $\geq$ 20% and < 20%, respectively).

For ER and PR and p53, tumors were scored as positive when nuclear staining was identified in at least 10% of cells and for c-*erb*B-2 when definite-complete membrane staining was identified in at least 10% of cells. Staining for these markers was classified as positive or negative (two-group classification) and also in three groups (< 10%, 10 to 49, $\geq$ 50%) to analyze strongly positive tumors ($\geq$ 50%). With these latter four markers we also assessed the intensity of staining as weak (= 1), moderate (= 2), or strong (= 3), and a score was calculated as the percentage of positive cells multiplied by the staining intensity; this cytoscore is particularly useful for comparison with results of RH biological assessment.

RESULTS AND DISCUSSION

Generalities on Techniques and Cytoblocks

Fine-needle cytopuncture of breast carcinoma often provides highly cellular material representative of the tumors. These samples can be used not only for cytologic diagnosis but also for obtaining information on the prognosis and likely response to therapy.

Recent years have seen major technological developments aimed at gleaning the maximum information from this type of sample, through immunocytohistochemistry, image analysis and flow cytometry, and now cytogenetics and molecular biology (Lavarino et al., 1998; Makris et al., 1997; Sauer et al., 1998; Solomides et al., 1999; Spyratos et al., 1997).

Two different approaches could be used to obtain multiple slides from cytopuncture:

1. A pure cytological approach, which allows work with intact cells, on direct and multiple smears (Marrazzo et al., 1995) or after directly collecting cells in a liquid, then cytocentrifuged on slides (Makris et al., 1997; Nizzoli et al., 2000) or by other techniques (Leung et al., 1999).

2. The other approach consists of using the histological techniques for cells embedded in paraffin miniblocks. This permits obtaining multiple slides from one block, available for simultaneously testing the expression of different markers. It also permits employing immunohistochemistry with exactly the same parameters (antigen retrieval, type and titration of antibodies) used for tissue fragments.

The possibility and the interest of including cells in miniblocks and thereby obtaining multiple sections has been the subject of several reports, but it was Pinder et al. (1995) who first studied the value of this method for analyzing multiple prognostic markers by means of immunohistochemistry in breast cancer. Pinder's study focused on samples obtained by fine-needle aspiration (FNA) of resected tumors, whereas we worked exclusively with samples obtained before therapy by *in vivo* cytopuncture, when cellularity of the sample was obvious. Consequently, we did not seek to determine what proportion of breast cancers would yield usable cytoblocks. We used Pinder's technique with slight modifications. To evaluate the reliability of the method and to identify possible sources of error in the assessment of the different immunomarkers on cytoblocks of the primary breast tumor, we undertook several studies. Our first published study (Briffod et al., 2000) compared our breast cytoblock findings with the corresponding tissue specimens (55 cases) and the concomitant node metastases when sampled (38 cases). More recently, our initial series of paired breast carcinoma and concomitant node metastasis was completed, and results were available on 117 paired cases (unpublished data). In a second study (Briffod et al., 2001), we assessed the reliability of hormonal receptors (HR) by IHC on cytoblocks in a group of 142 primary breast carcinoma by comparing the results to biological assessment through enzyme immunoassay (EIA) on their corresponding tissue samples.

Quality of Immunohistochemistry on Cytoblocks

In our experience, cytoblock immunostaining was of good quality and free of artifacts such as excessive background; its interpretation was fast and easy, and cells were located into a circular disc of 5 to 6 mm of diameter. However, cytoplasmic positivity for hormone receptors hindered interpretation in a few cases with ER and PR on the cytoblock of the primary tumor; similar positivity was always found on the cytoblock of the metastatic node or on the corresponding tumor tissue sample. These cases of cytoplasmic positivity appeared to be associated to the nature of the tumor tissue. Nevertheless, they did not lead to discrepancies.

The good interobserver reproducibility observed reflects the quality of the immunostaining and the simplicity of the interpretation; however, c-erbB-2 membrane staining was a little more difficult to interpret than nuclear staining, leading to less good reproducibility (87% versus 91–100%), in agreement with the results observed in tumor tissues (Dowsett et al., 2003).

Results of Our Different Studies and Discussion

Comparison of Immunohistochemistry on Cytoblocks and Corresponding Tissues

The comparison was studied on a series of 55 cases. Depending on the marker, a correlation of 76–96% was observed between the cytoblocks and the corresponding tumor tissues in the two-group classifications (negative versus positive). Makris et al. (1997) and Nizzoli et al. (2000) obtained a similar level of agreement for ER, PR, and p53 by means of immunocytochemistry on multiple slides of FNA material after cytocentrifugation. Our correlations were a little less satisfactory in the three-group classification, which individualizes strong positivity. When we used a score taking intensity into account, a very good correlation was obtained with p53 and c-erbB-2 (r = 0.86 and 0.80, respectively).

It is difficult to compare our results with previously published data, because the antibodies, cutoffs, immunostaining assessment (taking into account or not staining intensity), and the number of subgroups; also the populations studied vary widely from one report to another.

Positivity cutoffs for Ki-67 range from 10–25%, and the percentage of tumors with high Ki-67 ranges from 26–85% according to the authors (Clahsen et al., 1998; Lavarino et al., 1998; MacGrogan et al., 1997). With a cutoff of 20, we found that 38% tumors had high Ki-67 on both cytoblocks and tissue sections.

With p53, the reported cutoffs range from 1–75% and the proportion of positive tumors from 14–60%. Pinder et al., (1995), applying a cutoff of 5 to cytoblocks, obtained 46% of positive tumors. Six studies with a cutoff of 10 showed a proportion of positive tumors ranging from 16–51% (Barbareschi, 1996). Our histologic results for p53 are within this range, whereas the value obtained on cytoblock was higher (65%), despite the absence of values between 5 and 10%. However, our percentage of strongly positive tumors (20%) is comparable to that reported elsewhere (Lavarino et al., 1998; Rozan et al., 1998).

The c-erbB-2 cutoff is usually 1, 5, or 10. An extensive review of the literature (Ross et al., 1998) showed that 10–34% of breast cancers had HER-2/neu gene amplification or c-erbB-2 protein overexpression when examined immunohistochemically. Our percentage of positive tumors was higher on cytoblocks (56%) and even more so on tumor tissues (73%), possibly owing to the sensitivity of the antibodies at the dilution used or to the composition of our population (56% of node-positive tumors). However, Solomides et al. (1999) found percentages similar to ours in their study of cytology specimens. Finally, the percentage of highly positive tumors in our study (33%) is higher than that reported by Rozan et al. (1998).

Comparisons of hormone receptors are easier because they are better studied and coded, with well-established comparisons with biological assays (Ferrero-Poüs et al., 2001; Zafrani et al., 2000) and clinical features (Barnes et al., 1996; Blomqvist et al., 1997; Harvey et al., 1999). Although, once again, there is no international consensus on cutoff values, several authors, especially in Europe, recommend a value of 10% positive nuclei (Blomqvist et al., 1997; GEFPICS-FNCLCC, 1999; Nizzoli et al., 2000), without taking labeling intensity into account. The percentages of ER-positive tumors (82% on cytoblocks and tissues) and PR-positive tumors (67% on cytoblocks and 71% on tissues) are in keeping with those generally observed in the literature, if one considers the median age of our patients (60 years).

Comparison of Hormonal Receptor Status Between IHC and EIA

In this study, we compared on a series of 142 patients the hormonal receptors (HR) status obtained by IHC on cytoblocks and by EIA (cutoff 15 fMol/mgP) realized with a snap frozen fragment of the corresponding tumor. The concordance rates were 86.6% for ER and 76.8% for PR. These results are in agreement with those obtained on FNA material by immunocytochemistry (82–95% for ER and 71–91% for PR) performed on direct smears or slides after cytocentrifugation (Makris et al., 1997; Marrazzo et al., 1995; Sauer et al., 1998). As reported by some studies of both receptors (Makris et al., 1997; Sauer et al., 1998), agreement was also better for ER than for PR, which is possibly related to the type of antibodies used.

In this group, the percentages of IHC ER-positive tumors and PR-positive tumors were 81% and 63%, respectively; that is in keeping with the results previously observed. The proportion of highly positive IHC tumors (≥ 50%) was 75% for ER and 50% for PR. These highly positive cases were rarely negative by EIA (3.4% for ER and 6.7% for PR). Sauer et al. (1998) found the presence of major discrepancies in 7.7% of their cases.

The intensity of staining is usually taken into account in scores, according to different modalities. Like other studies, we observed a significant correlation for ER and PR between quantitative cytoscore values and titration of the biochemical technique. Among the variations between the quantitative values, some could be because of the heterogeneity of the nuclear staining, a frequent finding observed with carcinoma (Sauer *et al.*, 1998). This mosaicism implies a determination of a mean value of the staining, which could affect the reproducibility and the accuracy of the cytoscore.

The usefulness of the score is more pronounced for sequential studies of treatment follow-up: Variations of staining could bring additional information for the efficiency of hormonal treatment. Some studies have tried to use sequential samples to evaluate the effect of neoadjuvant hormonal therapy (Clarke *et al.*, 1993; Skoog *et al.*, 1992; Soubeyran *et al.*, 1996). For such studies, cytopunctures seem to be a less-traumatic and more representative method than repeated core biopsies to obtain information on tumor cells during the neoadjuvant hormonal treatment.

The Discrepancies

In addition to technical problems (which cannot always be overcome, especially on small samples) and more conventional problems such as intra- and interobserver variability, the discrepancies we observed were associated with weak staining and values close to the cutoff. In addition, small samples (cytoblocks and biopsies) may not be representative of heterogeneous breast tumors. The impact of tumor heterogeneity, especially when an *in situ* component is combined with the invasive component (nearly one-third of our cases), emerged clearly in our first series, as the percentage concordance and correlation coefficients were always better in the purely invasive tumor subgroup than in the subgroup of tumors with a DCIS component (Briffod *et al.*, 2000).

Most discrepancies observed in the hormonal status between IHC and EIA corresponded to IHC weak staining, values close to the respective cutoffs, and poor representativity of small biopsy samples. The other discrepancies were because the histologic type of the tumor and a possible contamination of samples for EIA by inflammatory of benign epithelial cells. On the other hand, tumor heterogeneity, and particularly the importance of the *in situ* component present in nearly 25% of these cases, did not significantly influence the concordance rate, inversely to what was previously observed in our comparison of hormonal status by IHC between cytoblock and the corresponding tumor tissue. Thus, in this study, both samples could be contaminated without knowing the exact cause due to the invasive and the *in situ* component, except for cytoblocks in some rare cases.

Comparison of Cytoblocks from Breast Tumors and Paired Node Metastases

The good correlation previously observed between cytoblock immunostaining of the breast tumor and the corresponding metastatic node (Briffod *et al.*, 2000) was reinforced by our larger unpublished series of 117 paired cases. Depending on the marker, a correlation of 87–97% was observed between the cytoblock of the breast tumor and the corresponding metastatic node using the two-group classifications (negative versus positive) (Fig. 75A and B). All nodes corresponding to ER-negative tumors were also negative, and all but four nodes (weakly labeled) corresponding to PR-negative tumors were also negative; three highly discrepant cases were observed for c-*erb*B-2 and only two for p53. A good correlation was also obtained with all the markers between the values of the cytoscore taking intensity into account ($r = 0.84$ to 0.90). This type of comparison is rare. Bhargava *et al.* (1994), studying a series of 24 invasive primary breast cancers with node metastases, observed full concordance with histologic sections between p53 immunopositivity of the primary tumor and metastatic node, and Barnes *et al.* (1998) obtained similar results for c-*erb*B-2. More recently, Tsutsui *et al.* (2002) also showed a good concordance for c-*erb*B-2, p53, and EGFR using a larger study. These data suggest that the metastatic cell population is very similar to the invasive breast tumor cell population, in keeping with studies based on flow cytometry (Feichter *et al.*, 1989; Goodson *et al.*, 1993) and molecular biology (Bonsing *et al.*, 1993; Chen *et al.*, 1992). Certain conflicting results in our series can be associated with the presence of a DCIS component (e.g., explaining why Ki-67 labeling was more often higher in the metastases than in the corresponding primary tumors, 90% versus 84%) or to genetic alterations occurring in some cases during the metastatic process, as shown by means of comparative genomic hybridization (Nishizaki *et al.*, 1997).

Correlations Among Various Biological Parameters in the Mammary Cytoblocks

Our correlations are in keeping with those frequently described by other authors. We found, like others, that Ki-67 was significantly linked to p53, and these two markers correlated negatively with ER and PR status (Makris *et al.*, 1997; Sirvent *et al.*, 1995). As in all previous studies, ER and PR were very significantly linked, 73% of tumors being ER-positive/PR-positive, a proportion close to that obtained by Sauer *et al.*, (1998)

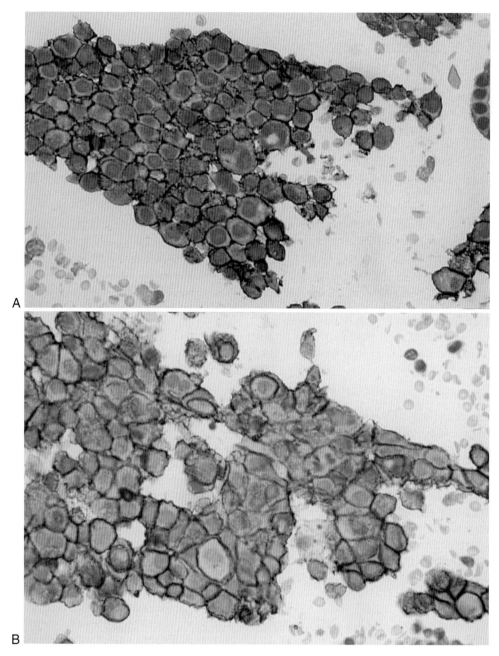

Figure 75 A concordant case, highly positive for c-*erb*B-2. **A**: On cell block from the breast tumor. **B**: On cell block from the concomitant metastatic axillary node (400X).

by immunocytochemistry on FNA slides. Finally, c-*erb*B-2 was not linked to any of the other markers, except for a negative correlation with PR, also found by other authors (Battifora *et al.*, 1991; Sirvent *et al.*, 1995). As expected, the markers of cell proliferation Ki-67 and p53 were significantly linked to our cytologic nuclear grade (Briffod *et al.*, 2000), whereas ER correlated negatively with nuclear grade.

In our study, node invasion correlated with high Ki-67 expression, PR negativity, and c-*erb*B-2 positivity. Ki-67 labeling was high in 62% of node-positive tumors and in only 21% of node-negative tumors, the latter result being close to that obtained by Dettmar *et al.* (1997). The proportion of ER-positive tumors ranged from 89–74% and that of PR-positive tumors from 79–48% according to node status, but we found that only PR was significantly linked to node status. A correlation between c-*erb*B-2 overexpression and node status was found here, but not by other authors (Battifora *et al.,* 1991; Sirvent *et al.,* 1995). This correlation was found also for highly positive tumors (≥ 50%), which were more numerous in the node-positive (54%) than in the node-negative (16%) population. Once again, the proportion of node-positive tumors that were strongly positive was far higher than that found by Thor *et al.* (1998). Although p53 did not appear to be significantly linked to node status in our study, highly p53-positive tumors were more numerous in the node-positive (29%) than in the node-negative (16%) population.

Comparison Between p53 IHC Results and Molecular Biology

Nuclear accumulation of p53 protein, revealed by means of IHC, does not always reflect a TP53 gene mutation. In a small unpublished series of 14 breast carcinomas we compared the results obtained by IHC with cytoblocks and by molecular biology (denaturing gradient gel electrophoresis [DGGE] and sequencing) in the same cytopuncture samples. Four positive cases by IHC showed a TP53 gene mutation in the corresponding samples (Fig. 76), and six negative cases by IHC did not show a gene mutation. Four cases were discrepant: two IHC positive cases without mutated gene and two mutated cases with negative IHC. These discrepancies were also observed in other studies (Geisler *et al.,* 2001). Indeed, several mutations may be missed, and p53 positive IHC may be observed without gene mutation. Moreover, not all TP53 mutations cause increased expression of a p53 protein detectable by IHC.

To conclude, these results suggest that, despite the imperfections inherent to this technique, the need to obtain highly cellular samples, and the problem of tumor heterogenity, the evaluation of immunomarkers on cytoblocks of cytopunctures is an easy, efficient, and reproducible method. Their results are correlated with those obtained on biopsies and/or surgical specimens of the corresponding tumors for each marker we tested. It can be a useful alternative for marker detection in the following circumstances.

1. Primary breast carcinoma with concurrent node metastasis, axillary, and especially supraclavicular (inoperable carcinoma) with IHC on the node cytoblock. Regarding concurrent node metastases, fine-needle cytopuncture easily allows evaluation of the metastatic population in a pretreatment diagnosis. For these cases and when neoadjuvant treatment is planned, it is not unjustified to treat according the cytological results, even if a histological proof remains preferable.

2. When microbiopsies are not conclusive, a not so rare observation especially in inflammatory carcinoma.

3. In the follow-up, for local and distant relapses, especially for deep location.

4. To follow the efficiency of neoadjuvant treatment with sequential cytopunctures.

However, a biopsy is initially required to ensure the invasiveness of the carcinoma. Moreover, the coherence of the results obtained demonstrates that useful information can be obtained on metastatic cell populations not only for diagnostic purposes (samples no longer being contaminated by a DCIS component), but also in the prognostic evaluation and in the research setting.

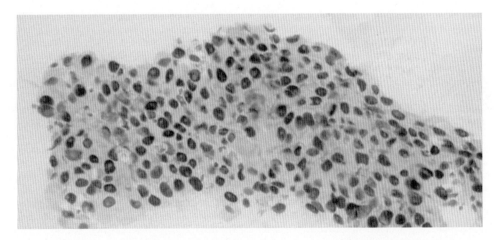

Figure 76 A highly positive breast tumor for p53 protein (400X). This case showed a TP53 gene mutation with nucleotide change [CTAG→CTAT] on intron 8/exon 9.

References

Association of Quality Assurance in Pathology and Cytology (AFAQAP-IHC). 1997. External evaluation of technical quality of immunohistochemistry. Results of a preliminary multicenter study. *Ann. Pathol. 17:*129–133.

Barbareschi, M. 1996. Prognostic value of immunohistochemical expression of p53 in breast carcinomas. A review of the literature involving over 9,000 patients. *Appl. Immunohistochem. 4:*106–116.

Barnes, D.M., Lammie, G.A., Millis, R.R., Gullick, W.L., Allen, D.S., and Altman, D.G. 1998. An immunohistochemical evaluation of c-*erb*B-2 expression in human breast carcinoma. *Br. J. Cancer 58:*448–452.

Barnes, D.M., Harris, W.H., Smith, P., Millis, R.R., and Rubens, R.D. 1996. Immunohistochemical determination of oestrogen receptor: Comparison of different methods of assessment of staining and correlation with clinical outcome of breast cancer patients. *Br. J. Cancer 74:*1445–1451.

Battifora, H., Gaffrey, M., Esteban, J., Mehta, P., Bailey, A., Faucett, C., and Niland, J. 1991. Immunohistochemical assay of neu/c-*erb*B-2 oncogene product in paraffin-embedded tissues in early breast cancer: Retrospective follow-up study of 245 stage I and II cases. *Mod. Pathol. 4:*466–474.

Bhargava, V., Thor, A., Deng, G., Ljung, B.M., Moore, D.H., Waldman, F., Benz, C., Goodson, W., Mayall, B., Chew, K., and Smith, H.S. 1994. The association of p53 immunopositivity with tumor proliferation and other prognostic indicators in breast cancer. *Mod. Pathol. 7:*361–368.

Blomqvist, C., von Boguslawski, K., Stenman, U.H., Mäenpää, H., von Smitten, K., and Nordling, S. 1997. Long-term prognostic impact of immunohistochemical estrogen receptor determinations compared with biochemical receptor determination in primary breast cancer. *Acta Oncol. 36:*530–532.

Bonsing, B.A., Devilee, P., Cleton-Jansen, A.M., Kuipers-Dijkshoorn, N., Fleuren, G.J., and Cornelisse, C.J. 1993. Evidence for limited molecular genetic heterogeneity as defined by allelotyping and clonal analysis in nine metastatic breast carcinomas. *Cancer Res. 53:*3804–3811.

Briffod, M., Hacène, K., and Le Doussal, V. 2000. Immunohistochemistry on cell blocks from fine-needle cytopunctures of primary breast carcinomas and lymph nodes metastases. *Mod. Pathol. 13:*841–850.

Briffod, M., Le Doussal, V., and Spyratos, F. 2001. Détermination des récepteurs hormonaux par immunohistochimie sur cyto-blocs de cytoponctions des cancers du sein. *Bull. Cancer 88:*1028–1035.

Chen, L., Kurisu, W., Ljung, B., Goldman, E., Moore, D., and Smith, H. 1992. Heterogeneity for allelic loss in human breast cancer. *J.N.C.I. 84:*506–510.

Clahsen, P.C., van de Velde, C.J.H., Duval, C., Pallud, C., Mandard, A.M., Delobelle-Deroide, A., van der Broek, L., Sahmoud, T.M., and van de Vijver, M.J. 1998. p53 protein accumulation and response to adjuvant chemotherapy in premenopausal women with node-negative early breast cancer. *J. Clin. Oncol. 16:*470–479.

Clarke, R.B., Laidlaw, I.J., Jones, L.J., Howell, A., and Anderson, E. 1993. Effect of tamoxifen on Ki-67 labelling index in human breast tumors and its relationship to estrogen and progesterone receptor status. *Br. J. Cancer 67:*606–611.

Dettmar, P., Harbeck, N., Thomssen, C., Pache, L., Ziffer, P., Fizi, K., Jänicke, F., Nathrath, W., Schmitt, M., Graeff, H., and Höfler, H. 1997. Prognostic impact of proliferation-associated factors MIB1 (Ki-67) and S-phase in node-negative breast cancer. *Br. J. Cancer 75:*1525–1533.

Dowsett, M., Bartlett, J., Ellis, I.O., Salter, J., Hills, M., Mallon, E., Watters, A.D., Cooke, T., Paish, C., Wencyk, P.M., and Pinder, S.E. 2003. Correlation between immunohistochemistry (HercepTest) and fluorescence in situ hybridization (FISH) for HER-2 in 426 breast carcinomas from 37 centres. *J. Pathol. 199:*411–417.

Feichter, G.E., Kaufmann, M., Muller, A., Haag, D., Eckhardt, R., and Goerttler, K. 1989. DNA index and cell cycle analysis of primary breast cancer and synchronous axillary lymph node metastases. *Breast Cancer Res. Treat. 13:*17–22.

Ferrero-Poüs, M., Trassard, M., Le Doussal, V., Hacène, K., Tubiana-Hulin, M., and Spyratos, F. 2001. Comparison of enzyme immunoassay and immunohistochemical measurements of estrogen and progesterone receptors in breast cancer patients. *Appl. Immunochistochem. Mol. Morphol. 9:*267–275.

Geisler, S., Lonning, P.E., Aas, T., Johnsen, H., Fluge, O., Haugen, D.F., Lillehaug, J.R., Akslen, L.A., and Borresen-Dale, A.L. 2001. Influence of TP53 gene alterations and c-*erb*B-2 expression on the response to treatment with doxorubicin in locally advanced breast cancer. *Cancer Res. 61:*2505–2512.

Goodson, W.H., Ljung, B.M., Moore, D.H., Mayall, B., Waldman, F.M., Chew, K., Benz, C.C., and Smith, H.S. 1993. Tumor labelling indices of primary breast cancers and their regional lymph node metastases. *Cancer 71:*3914–3919.

Group for Evaluation of Prognostic Factors using Immunohistochemistry in Breast Cancer (GEFPICS-FNCLCC). 1999. Recommendations for the immunohistochemistry of the hormonal receptors on paraffin sections in breast cancer. *Ann. Pathol. 19:*336–343.

Harvey, J.M., Clark, G.M., Osborne, C.K., and Allred, D.C. 1999. Estrogen receptor status by immunohistochemistry is superior to the ligand-binding assay for predicting response to adjuvant endocrine therapy in breast cancer. *J. Clin. Oncol. 17:*1474–1481.

Hayat, M.A. 2002. *Microscopy, Immunohistochemistry, and Antigen Retrieval Methods.* New York and London: Kluwer Academic/Plenum Publishers.

Lavarino, C., Corletto, V., Mezzelani, A., Della Torre, G., Bartoli, C., Riva, C., Pierotti, M.A., Rilke, F., and Pilotti, S. 1998. Detection of TP53 mutation, loss of heterozygosity and DNA content in fine-needle aspirates of breast carcinoma. *Br. J. Cancer 77:*125–130.

Leung, S.W., and Bédard, Y.C. 1999. Estrogen and progesterone receptor contents in ThinPrep-processed fine-needle aspirates of breast. *Am. J. Clin. Pathol. 112:*50–56.

MacGrogan, G., Jollet, I., Huet, S., Sierankowski, G., Picot, V., Bonichon, F., and Coindre, J.M. 1997. Comparison of quantitative and semiquantitative methods of assessing MIB-1 with the S-phase fraction in breast carcinoma. *Mod. Pathol. 10:*769–776.

Makris, A., Allred, D.C., Powles, T.J., Dowsett, M., Fernando, I.N., Trott, P.A., Ashley, S.E., Ormerod, M.G., Titley, J.C., and Osborne, C.K. 1997. Cytological evaluation of biological prognostic markers from primary breast carcinomas. *Breast Cancer Res. Treat. 44:*65–74.

Marrazzo, A., Taormina, P., Leonardi, P., Lupo, F., and Filosto, S. 1995. Immunocytochemical determination of estrogen and progesterone receptors on 219 fine-needle aspirates of breast cancer. A prospective study. *Anticancer Res. 15:*521–526.

Nishizaki, T., De Vries, S., Chew, K., Goodson, W.H., Ljung, B.M., Thor, A., and Waldman, F.M. 1997. Genetic alterations in primary breast cancers and their metastases: Direct comparison using modified comparative genomic hybridization. *Genes Chromosom. Cancer 19:*267–272.

Nizzoli, R., Bozzetti, C., Naldi, N., Guazzi, A., Gabrielli, M., Michiara, M., Camisa, R., Barilli, A., and Cocconi, G. 2000. Comparison of the results of immunocytochemical assays for biologic variables on preoperative fine-needle aspirates and on surgical specimens of primary breast carcinomas. *Cancer (Cancer Cytopathol.) 90:*61–66.

Pinder, S.E., Wencyk, P.M., Naylor, H.E., Bell, J.A., Elston, C.W., Robertson, J.F.R., Blamey, R.W., and Ellis, I.O. 1995. The assessment of multiple variables on breast carcinoma fine needle aspiration (FNA) cytology specimens: Method, preliminary results and prognostic associations. *Cytopathology 6:*316–324.

Ross, J.S., and Fletcher, J.A. 1998. The HER-2/neu oncogene in breast cancer. Prognostic factor, predictive factor, and target for therapy. *Oncologist 3:*237–252.

Rozan, S., Vincent-Salomon, A., Zafrani, B., Validire, P., de Cremoux, P., Bernoux, A., Nieruchalski, M., Fourquet, A., Clough, K., Dieras, V., Pouillart, P., and Sastre-Garau, X. 1998. No significant predictive value of c-erbB-2 or p53 expression regarding sensitivity to primary chemotherapy or radiotherapy in breast cancer. *Int. J. Cancer 79:*27–33.

Sauer, T., Beraki, E., Jebsen, P., Amlie, E., Harbitz, T., Karesen, R., and Naess, O. 1998. Assessing estrogen and progesterone receptor status in fine-needle aspirates from breast carcinomas. Results on six years of material and correlation with biochemical assay. *Anal. Quant. Cytol. Histol. 20:*122–126.

Sirvent, J.J., Salvado, M.T., Santafé, M., Martinez, S., Brunet, J., Alvaro, T., and Palacios, J. 1995. p53 in breast cancer. Its relation to histological grade, lymph-node status, hormone receptors, cell-proliferation fraction (Ki-67) and c-erbB-2. Immunohistochemical study of 153 cases. *Histol. Histopathol. 10:*531–539.

Skoog, L., Rutqvist, L.E., and Wilking, N. 1992. Analysis of hormone receptors and proliferative fraction in fine-needle aspirates from primary breast carcinomas during chemotherapy or tamoxifen therapy. *Acta Oncol. 31:*139–141.

Solomides, C.C., Zimmerman, R., and Bibbo, M. 1999. Semiquantitative assessment of c-erbB-2 (HER-2) status in cytology specimens and tissue sections from breast carcinoma. *Anal. Quant. Cytol. Histol. 21:*121–125.

Soubeyran, I., Quénel, N., Mauriac, L., Bonichon, F., and Coindre, J.M. 1996. Variations of hormonal receptor, pS2, C-erbB-2 and GSTπ contents in breast carcinomas under tamoxifen: A study of 74 cases. *Br. J. Cancer 73:*735–743.

Spyratos, F., and Briffod, M. 1997. DNA ploidy and S-phase fraction by image and flow cytometry in breast cancer fine-needle cytopunctures. *Mod. Pathol. 6:*556–563.

Thor, A.D., Berry, D.A., Budman, D.R., Muss, H.B., Kute, T., Henderson, I., Barcos, I.C., Cirrincione, C., Edgerton, S., Allred, C., Norton, L., and Liu, E.T. 1998. erbB-2, p53, and efficacy of adjuvant therapy in lymph node-positive breast cancer. *J.N.C.I. 90:*1346–1360.

Tsutsui, S., Ohno, S., Murakami, S., Kataoka, A., Kinoshita, J., and Hachitanda, Y. 2002. EGFR, c-erbB-2 and p53 protein in the primary lesions and paired metastatic regional lymph nodes in breast cancer. *E.J.S.O. 28:*383–387.

United Kingdom National External Quality Assessment Scheme (UK.NEQAS). 1999. Immunocytochemistry. *J. Cell. Pathol. 1:* 29–55.

Zafrani, B., Aubriot, M.H., Mouret, E., de Crémoux, P., de Rycke, Y., Nicolas, A., Boudou, E., Vincent-Salomon, A., Magdelenat, H., and Sastre-Garau, X. 2000. High sensitivity and specificity of immunohistochemistry for the detection of hormone receptors in breast carcinoma: Comparison with biochemical determination in a prospective study of 793 cases. *Histopathology 37:*536–545.

Index

W

X

Z